2008 13th International Power Electronics and Motion Control Conference

Poznan, Poland
1-3 September 2008

Pages 1033-1554

IEEE Catalog Number:	CFP0834A-PRT
ISBN 13:	978-1-4244-1741-4

**Copyright © 2008 by The Institute of Electrical and Electronics Engineers, Inc.
All Rights Reserved**

Copyright and Reprint Permissions: Abstracting is permitted with credit to the source. Libraries are permitted to photocopy beyond the limit of U.S. copyright law for private use of patrons those articles in this volume that carry a code at the bottom of the first page, provided the per-copy fee indicated in the code is paid through Copyright Clearance Center, 222 Rosewood Drive, Danvers, MA 01923.

For other copying, reprint or republications permission, write to IEEE Copyrights Manager, IEEE Operations Center, 445 Hoes Lane, Piscataway, New Jersey USA 08854. All rights reserved.

IEEE Catalog Number: CFP0834A-PRT

ISBN 13:   978-1-4244-1741-4

ISSN: 2007906910

Additional Copies of This Publication Are Available from:

IEEE Service Center
445 Hoes Lane
Piscataway, NJ 08854
Phone: (800) 678-IEEE
      (732) 981-1393
Fax: (732) 981-9667
E-mail: customer-service@ieee.org

Table of Contents

Electric Drive System for Automatic Guided Vehicles Using Contact-free Energy Transmission 1
Marcel Jufer

State-of-the-Art High Power Density and High Efficiency DC-DC Chopper Circuits for HEV and FCEV Applications 7
Atsuo Kawamura, Martin Pavlovsky, Yukinori Tsuruta

Current-Based Condition Monitoring of Electrical Machines in Safety Critical Applications 21
Thomas G. Habetler

The Essence of Three-Phase AC/AC Converter Systems 27
J. W. Kolar, T. Friedli, F. Krismer, S. D. Round

An Analysis on Turn-off Behaviour of 1.2kV NPT-CIGBT under Clamped Inductive Load Switching 43
S.T. Kong, L.Ngwendson, M. Sweet, E.M. Sankara Narayanan

Turn-off behaviour of high voltage NPT- and FS-IGBT 48
Hans-Guenter Eckel, Karl Fleisch

Exact Circuit Power Loss Design Method for High Power Density Converters Utilizing Si-IGBT/SiC-Diode Hybrid Pairs 54
Kazuto Takao, Hiromichi Ohashi

A Forward Converter with a Monolithic Cascode Device: Design and Experimental Investigation 61
F. Chimento, S. Musumeci, A. Raciti, L. Abbatelli, S. Buonomo, R. Scollo

Switching and conducting performance of SiC-JFET and ESBT against MOSFET and IGBT 69
André Knop, W. Toke Franke, Friedrich W. Fuchs

In-Service Life Consumption Estimation in Power Modules 76
Mahera Musallam, C Mark Johnson, Chunyan Yin, Hua Lu, Chris Bailey

Measurement Of Temperature Sensitive Parameter Characteristics Of Semiconductor Silicon And Silicon-Carbide Power Devices 84
Mietek Nowak, Jacek Rabkowski, Roman Barlik

Unsymmetrical Gate Voltage Drive for High Power 1200V IGBT4 Modules Based on Coreless Transformer Technology Driver 88
Piotr Luniewski, Uwe Jansen

A Novel RESURFed Double Gates IGBT with Superior Performance 97
Dongming Wu, Kaihang Li, Lingling Yang

An Empiric Approach to Establishing MOSFET Failure Rate Induced by Single-Event Burnout 102
Jeroen van Duivenbode, Bart Smet

Comparative Study on Paralleled vs. Scaled Dc-dc Converters in High Voltage Gain Applications 108
Pawel Klimczak, Stig Munk-Nielsen

A Low-Loss Dc-Dc Converter For A Renewable Energy Converter 114
David S. Thompson, Otu A. Eno

A Single Active Edge-Resonant Snubber Cell-assisted ZCS Half-Bridge DC-DC Converter with Constant Frequency Asymmetrical PWM Scheme 119
Tomokazu Mishima, Mutsuo Nakaoka, Eiji Hiraki

A New Approach to High Efficiency in Isolated Boost Converters for High-Power Low-Voltage Fuel Cell Applications 127
Morten Nymand, Michael A.E. Andersen

New Modulation Strategy with Low Switching Frequency and Minimum Baseband Distortion 132
N. E. Ruger, O. Schnick, W. Mathis, A. Mertens

Table of Contents

A Bit-Stream Based PWM Technique for Variable Frequency Sinewave Generation 139
N. D. Patel, U. K. Madawala

Control Strategies of the Quasi-Resonant DC-Link Inverter 144
Slawomir Mandrek, Piotr J. Chrzan

Consideration for Input Current-Ripple of Pulselink DC-AC Converter for Fuel Cells 148
Kentaro Fukushima, Tamotsu Ninomiya, Masahito Shoyama, Isami Norigoe, Yosuke Harada, Kenta Tsukakoshi

New Practical Approach to Input Current Shaping in AC-DC Power Converters 154
Kuno Janson, Viktor Bolgov, Lauri Kütt, Ants Kallaste, Heigo Mölder

LLCC-PWM Inverter for Driving High-Power Piezoelectric Actuators 159
Rongyuan Li, Norbert Fröhleke, Joachim Böcker

Modelling and Analysis of a Matrix-Reactance Frequency Converter Based on Buck-Boost Topology by DQ0 Transformation 165
Pawel Szczeniak, Zbigniew Fedyczak, Marius Klytta

A Modular AC/DC Rectifier Based on Cascaded H-bridge Rectifier 173
H. Iman-Eini, Sh. Farhangi, JL. Schanen

Low Loss Soft Switching Boost Converter 181
So-Ri Park, Sang-Hoon Park, Chung-Yuen Won, Yong-Chae Jung

Methods for Experimental Assessment of Component Losses to Validate the Converter Loss Model 187
Yi Wang, Sjoerd de Haan, Jan Abraham Ferreira

Modified multistage semiconductor-Fitch generator topology with magnetic compression 195
Stanislaw Kalisiak, Marcin Holub

Modeling and Measuring Results of a Shunt Current Source Active Power Filter with Series Capacitor 201
P. Parkatti, M. Salo, H. Tuusa

A Multi-Drive System Based on a Two-stage Matrix Converter 207
Dinesh Kumar, Patrick W Wheeler, Jon C Clare, Lee Empringham

Characteristics of the Single Active Bridge Converter with Voltage Doubler 213
Andreas Averberg, Axel Mertens

Analysis of Capacitor Dividers for Multilevel Inverter 221
Oleg Sivkov, Jiri Pavelka

Space Vector Modulation for a Capacitor Clamped Multi-level Matrix Converter 229
Xu Lie, Jon C. Clare, Patrick W. Wheeler, Lee Empringham

New Family of Matrix-Reactance Frequency Converters Based on Unipolar PWM AC Matrix-Reactance Choppers 236
Zbigniew Fedyczak, Pawel Szczesniak, Igor Korotyeyev

Consideration of Conduction Losses for the Series Resonant Converter by Means of a Simple Extension to the SPA Approach 244
Alexander Bucher, Thomas Duerbaum, Daniel Kuebrich, Markus Schmid

Validation and Comparison of different PWM Converter Small Signal Models 250
Alexander Bucher, Markus Schmid, Lukas Bendkowski, Thomas Duerbaum

Dynamic Behaviour of a Series - Connected Multilevel Converter with Interleaved Switching 256
C. Fahrni, A. Rufer

Simple Analysis of a Flying Capacitor Converter Voltage Balance Dynamics for DC Modulation 260
A. Ruderman, B. Reznikov, M. Margaliot

Simulation of Simplified Seven Level Multilevel Converter Circuit 268
Gerardo Ceglia, Víctor Guzmán, Carlos Sánchez, Fernando Ibáñez, Julio Walter, María Giménez

Table of Contents

SEPP High-Frequency Inverter Incorporating an Auxiliary Switch and Its Performance Evaluation......275
H.Ogiwara,Y.Fujita, R.Urabe, M.Itoi, T.Sugai, M. kuwata, M.Nakaoka

Multiphase coupled converter models dedicated to transient response and output voltage regulation studies 281
Nadia Bouhalli, Marc Cousineau, Emmanuel Sarraute, Thierry Meynard

A 13.56 MHz Current-output-type Inverter Utilizing An Immittance Conversion Element......288
Yosei Sakamoto, Keiji Wada, Toshihisa Shimizu

Voltage Fed Zero-Voltage Zero-Current Switching PWM DC-DC Converter......295
Jaroslav Dudrik, Vladimír Ru1scin

PWM Spectrum Evaluation and Over-Modulation Phenomena in a Three-Phase Inverters - Analytical Approach 301
Miro Milanovic

Experimental Study of a Matrix Converter Excited Doubly-Fed Induction Machine in Generation and Motoring 307
Ivan Shapoval, Jon Clare, Eduard Chekhet

Effect of Type and Interconnection of DG Units in the Fault Current Level of Distribution Networks 313
H.R. Baghaee, M. Mirsalim, M. J. Sanjari, G.B. Gharehpetian

An Isolated Full-Bridge DC/DC Converter..with Bidirectional Communication Capability 320
Lon-Kou Chang, Ru-Shiuan Yang

Efficiency and Power Losses in PM BLDC Motor with Variable Bridge/half-bridge Structure Electronic Commutator 326
K. Krykowski, A. Bodora

Analysis of a device for converting a unipolar input voltage into two symmetric bidirectional output voltages with a magnetically coupled coil 331
Felix. A. Himmelstoss, Wilhelm Kraeftner

Invariant Modulation Strategy for Two-stage Direct Power Converter......337
Radiy Bekbudov

Experimental Study of A Multicell ac/ac Converter Balancing Circuit......345
Robert Stala, Andrzej Mondzik

A Comparison and Optimum Design of Reluctance-Controlled Classical Load-Resonant Converters 350
Stefan V. Mollov, Michael P. Theodoridis

Capacitor Clamped Multilevel Matrix Converter Controlled with Venturini Method......357
Janina Rzasa

Reliability Consideration for a High Power Zero-Voltage-Switching Flyback Power Supply 365
Arash Rahnamaee, Jafar Milimonfared, Kaveh Malekian, Mohammad Abroushan

The Traction Drive Topology Using the Matrix Converter with Middle-Frequency Transformer 372
Martin Pittermann, Pavel Drábek, Marek Cédl

Analysis of Multipulse Rectifiers with Modulation in DC Circuit in Vector Space Approach......377
Andrzej KAPLON and Jaroslaw ROLEK

High Efficiency Soft Switching Boost Converter for Photovoltaic System......383
Gil-Ro Cha, Sang-Hoon Park, Chung-Yuen Won, Yong-Chae Jung, Sang-Hoon Song

A Power Converter For Fault Tolerant Machine Development In Aerospace Applications 388
Liliana de Lillo, Patrick Wheeler, Lee Empringham, Chris Gerada, XiaoyanHuang

Optimal Bus Capacitance Design for System Stability in On-Board Distributed Power Architecture 393
Seiya Abe, Masahiko Hirokawa, Masahito Shoyama, Tamotsu Ninomiya

Table of Contents

Steady State Analysis of Hysteretic Control Buck Converters **400**
L.K. Wong, T.K. Man

A Novel Control Method for IGBT Current Source Rectifier **405**
Longcheng Tan, Yaohua Li, Ping Wang, Congwei Liu, Zixin Li, Yonggang Chen, Wei Xu

A procedure to optimize the inductor design in boost PFC applications **409**
Florent Liffran

Electric Vehicle Drive Inverters Simulation Considering Parasitic Parameters **417**
Wen Huiqing, Liu Jun, Zhang Xuhui, Wen Xuhui

DC-DC Converters with FPGA Control for Photovoltaic System **422**
Jan Leuchter, Pavel Bauer, Vladimir Rerucha, Petr Bojda

Control of a Converter with Superconductive Energy Storage Inductor **428**
Rozanov Yurie Konstantinovich, Lepanov Michail Gennadevich, Kiselev Michail Gennadevich

FPGA-based Controllers for Switching Converters **432**
Karel Jezernik

Gamesa DAC converter: the way for REE grid code certification **437**
Itziar Martinez, Daniel Navarro

Flatness-Based Voltage-Oriented Control of Three-Phase PWM Rectifiers **444**
J. Dannehl, F.W. Fuchs

Control of a single phase H-Bridge multilevel inverter for grid-connected PV applications **451**
Elena Villanueva, Pablo Correa, Jose Rodriguez

Switching and Voltage Controls for a Flyback Switch-Mode Rectifier **456**
Yuan-Chih Chang, Chang-Ming Liaw

Method Of Designing ZVS Boost Converter **463**
Miroslaw Luft, Elzbieta Szychta, Leszek Szychta

A New DC-DC Converter with Multi Output: Topology and Control Strategies **468**
Arash A Boora, Firuz Zare, Gerard Ledwich, Arindam Ghosh

Maximum Frequency for Hysteretic Control COT Buck Converters **475**
L.K. Wong, T.K. Man

Current Control Method Based on Hysteresis Control Suitable for Single Phase Active Filter with LC Output Filter **479**
Yukinori Kobayashi, Hirohito Funato

Optimal Slope Compensation for step load in peak current controlled dc-dc Buck Converter **485**
Susovon Samanta, Pradipta Patra, Siddhartha Mukhopadhyay, Amit Patra"

Performances of a PLL Based Digital Filter for double-conversion UPS **490**
Armando Bellini, Stefano Bifaretti

10A 12V 1 chip digitally-controlled DC/DC converter IC with high resolution and high frequency DPWM **498**
Kazutoshi Nakamura*, Toshiyuki Naka*, Yuki Kamata*, Toyoki Taguchi, Takaaki Shimizu, Yoshiko Ikeda, Akio Nakagawa, Dragan Maksimovic

Modelling and Modulation of Voltage Source Converter **504**
Grzegorz Radomski

Sliding Mode Control of DC/DC Multiphase Power Converters **512**
Vadim Utkin

A New Digital Control Method for High Performance 400 Hz Ground Power Unit **515**
Zixin Li, Ping Wang, Haibin Zhu, Yaohua Li, Longcheng Tan, Yonggang Chen, Fanqiang Gao

Table of Contents

Single-phase 50-kW 16.7-Hz Four-Quadrant Line-Side Converter for Railway Traction Application 521
C. Heising, R. Bartelt, V. Staudt, A. Steimel

Technique to Improve IGBT Converter Efficiency and Transient Response 528
Robert W. Turner, Simon Walton, Richard Duke

The control of voltage converter rectifiers 536
Krzysztof Szubert

Load Voltage Regulation and Line Loss Minimization of Loop Distribution Systems Using UPFC 542
Mahmoud A. Sayed, Takaharu Takeshita

Control of Traction Single-Phase Current-Source Active Rectifier under Distorted Power Supply Voltage 550
Jan Michalík, Jan Molnár, Zdenck Peroutka

Simulation Model Of Neural Network Based Synchronous Generator Excitation Control 556
Damir Sumina, Neven Bulic, Gorislav Erceg

Predictive Current Control of a 7-level AC-DC back-to-back Converter for Universal and Flexible Power Management System 561
Stefano Bifaretti, Pericle Zanchetta, Florin Iov, Jon C. Clare

Predictive Stator Current Control For Three-Level Voltage-Source Inverters With Output LC-Filters 569
Tomasz Laczynski, Axel Mertens

Research on Dimming Control Method of Electronic Ballast for the Automotive HID Headlight 576
P. Dong, K.W.E.Cheng, S.L.Ho

Control Method for a Three-Port Interface Converter Using an Indirect Matrix Converter with an Active Snubber Circuit 581
Koji Kato, Jun-ichi Itoh

Precise Digital Control Method with Multi-rate deadbeat control for Single Phase Utility Interactive Inverter with FPGA based Hardware Controller 589
Kenta Hayashi, Tomoki Yokoyama

A Digital Current Controller for Zero-Current Transition Bidirectional Converter 595
Nobuyuki Kasa, Takahiko Iida

Control Method for a Single Phase Arbitrary Waveform-output Inverter 600
Satoshi Taniguchi, Keiji Wada, Toshihisa Shimizu

Elimination of Harmonics in Multilevel Inverters with Non-Equal DC Sources Using PSO 606
A. K. Al-Othman, Tamer H. Abdelhamid

Improved PFC Circuit Having Ladder Type Filter with Only Passive Devices 614
Kenji Ando, Keiju Matsui, Nobuhito Takeuchi, Masaru Hasegawa

Fuel Cell Current Ripple Minimization using a bi-Buck Power Interface 621
Nicu Bizon, Marian Raducu, Mihai Oproescu

Power Control Strategy of Parallel Inverter Interfaced DG Units 629
H.R. Baghaee, M.Mirsalim, M. J. Sanjari, G.B. Gharehpetian

Implementation of Nonlinear power flow controllers to control a VSC 637
Nelson L. Díaz, Fabián H. Barbosa, Cesar L. Trujillo

Harmonic Distortion Reduction Technique for Uninterruptible Power Supplies with DC Voltage Boost Technique 643
Juei Lung Shyu

Energy-based Modulation Error Control for High-Power Drives with Output LC-Filters and Synchronous Optimal Pulse Width Modulation 649
Tomasz Laczynski, Timur Werner, Axel Mertens

Table of Contents

Voltage Harmonic Control of Z-source Inverter for UPS Applications 657
Arkadiusz Kulka, Tore Undeland

A Method of Optimal Control for Switched-Mode Power Converters 663
Anatoly Bekishev, Albert Iskhakov, Leonid Klyachko, Vladimir Pospelov, Sergey Skovpen

Experiment results with modified Hybrid PWM method for three phase induction motor 669
Daniel Lewandowski, Grzegorz Lisowski

Optimized Design of a Delay line based Analog to Digital Converter for Digital Power Management Applications 674
Mukti Barai, Sabyasachi Sengupta, Jayanta Biswas

Overmodulation Region of Multi-Phase Inverters 682
S. Halasz

Optimal Control of Induction Motor Using High Performance Frequency Converter 690
Jerkovic Vedrana, Spoljaric Zeljko, Valter Zdravko

Power Electronic Converter for the Reluctance Pump Drive 695
B. J. Szymanski, K. Kompa, N. Michalke, H. Kuß, U. Schuffenhauer

A Predictive Control Scheme for Current Source Rectifiers 699
Pablo Correa, Jose Rodriguez

Analysis and Design of New Switching Table for Direct Power Control of Three-Phase PWM Rectifier 703
Abdelouahab Bouafia, Jean-Paul Gaubert, Fateh Krim

Improvement of the performance for DC-DC Converter 710
X..She, Yun She

A Drive System With High-Speed Single-Phase Supplied Three-Phase Induction Motor 714
T. Binkowski, M. Grad, M. Latka, W. Malska, D. Sobczynski

A Pulse Width Modulation Technique for a Multilevel Converter in High Voltage High Frequency Applications 718
Jafar Adabi, Hamid Soltani, Firuz Zare

Bidirectional Positive Buck-Boost Converter 723
Arash A Boora, Firuz Zare, Gerard Ledwich, Arindam Ghosh

Control system of power electronics current modulator utilized in diode rectifier with sinusoidal source current 728
Michal Gwózdz, Michal Krystkowiak

Design and control of a half-bridge converter to drive piezoelectric actuators 731
Oriol Gomis-Bellmunt, Josep Rafecas-Sabate, Daniel Montesinos-Miracle, Josep-Maria Fernandez-Mola, Joan Bergas-Jane

Online Diagnosis of PEM Fuel Cell 734
Abdellah Narjiss, Daniel Depernet, Denis Candusso, Frederic Gustin, Daniel Hissel

Application of Kalman filters to the control of independent power electronic voltage sources 740
Ryszard Porada, Lukasz Nyczkowski

Verification of the load sharing characteristics in Autonomous Decentralized UPS system using FPGA based Hardware Controller 744
Nobuaki Doi, Tsuyoshi Saito, Tomoki Yokoyama

Fault Current Reduction in Distribution Systems with Distributed Generation Units by a New Dual Functional Series Compensator 750
H.R. Baghaee, M. Mirsalim, M. J. Sanjari, G.B. Gharehpetian

Dynamic Simulation of PM Motor Drive System based on Reluctance Network Analysis 758
Kenji Nakamura, Osamu Ichinokura

Table of Contents

Performance Improvement of Direct Torque Controlled Interior Permanent Magnet Synchronous Motor Drive by Considering Magnetic Saturation 763
Behrooz Majidi, Jafar Milimonfared, Kaveh Malekian

Condition Monitoring for Mechanical Faults in Fully Integrated Servo Drive Systems 769
Jesus Arellano-Padilla, Mark Sumner, Chris Gerada

Feed-forward Compensation of Load and Parameter Variations of Electric Drive 776
Alon Kuperman, Yoram Horen, Saad Tapuchi, Uri Suissa

Thermal Effect of Short-Circuit Current in Low Power Induction Motors 782
Leo.s Beran

Generalized Model for a Class of Switched Reluctance Motors 787
Constantin Pavlitov, Yassen Gorbounov, Radoslav Rusinov, Alexandar Alexandrov, Kliment Hadjov, Dimitar Dontchev

Neural Network based Fault Detection of PMSM Stator Winding Short under Load Fluctuation 793
J. Quiroga, D.A. Cartes, C.S. Edrington, Li Liu

Review of Electrical Machine in Downhole Applications and the Advantages 799
Anyuan Chen, Ravindra. B. Ummaneni, Robert Nilssen, Arne Nysveen

Broken Rotor Bar Impact on the Closed Loop and Sensorless Control of Induction Machine 804
Piotr Kotodziejek, Elzbieta Bogalecka

Coupled Magnetic Circuit Method and Permeance Network Method Modeling of Stator Faults in Induction Machines 810
Amin Mahyob, Mohamed Y. Ould Elmoctar, Pascal Reghem, Georges Barakat

Explosion Protected Electrical Drives - Risk Assessment and Technical Diagnostics 818
Ivica Gavranic, Drago Ban, Damirarko Zarko

The effect of subharmonics on induction machine heating 826
Piotr Gnacinski, Marcin Peplinski, Mariusz Szweda

Influence of Saturation Effects in a Transverse Flux Machine 830
M. Siatkowski, B. Orlik

A Model of Semiconductor Converter-Fed Asynchronous Machines Taking into Account Energy Losses and Thermal Processes 837
M. Pronin, O. Shonin, Y. Koskin, A. Vorontsov, P. Kalatchikov

Use of an AC Self-excited Switched Reluctance Generator as a Battery Charger 845
Abelardo Martínez, Estanislao Oyarbide, Javier Vicuña, Francisco Perez, Eduardo Laloya, Bonifacio Martín-del-Brío, Tomás Pollán, Beatriz Sánchez, Juan Lladó

Direct Thrust Controlled Linear Induction Motor Including End Effect 850
Berrin Susluoglu, Vedat M. Karsli

Analysis of Short-Circuit Forces at the Top of the Low Voltage U-Type and I-Type Winding in a Power Transformer 855
Leonardo Strac, Franjo Kelemen, Damir Zarko, Josipa Mokrovica

Anisotropy Comparison of Reluctance and PM synchronous Machines for Low Speed Position Sensorless Applications 859
H.W. de Kock, M.J. Kamper, R.M. Kennel

Analysis of VSI-DTC Fed 6-phase Synchronous Machines 867
Ibrahim Abuishmais, Waqas M. Arshad, Sami Kanerva

Optimal Rotor Flux Shape for Multi-phase Permanent Magnet Synchronous Motors 874
Roberto Zanasi, Federica Grossi

Table of Contents

Modelling of Electrical Machines Using the Modelica Bond-Graph Library 880
Mieczyslaw Ronkowski

Induction Motor Parameters Identification using Genetic Algorithms for Varying Flux Levels 887
Konstantinos Kampisios, Pericle Zanchetta, Chris Gerada, Andrew Trentin, Omar Jasim

Study of the sudden symmetrical short-circuit using the mathematical models of the synchronous machine and the numerical methods 893
Petropol Serb Gabriela, Petropol Serb Ion, Campeanu Aurel, Sonia Degeratu, Anca Petrisor

Analytical Method of Calculation of the Current and Torque of a Reluctance Stepper Motor Using Fourier Complex Series 899
Pavel Zaskalicky, Maria Zaskalicka

Bearing Damage Analysis by Calculation of Capacitive Coupling between Inner and Outer Races of a Ball Bearing 903
Jafar Adabi, Firuz Zare, Gerard Ledwich, Arindam Ghosh, Robert D.Lorenz

The Model of the Squirrel Cage AC Motor including Rotor Slot Harmonics 908
Eleonora Darie, Costin Cepisca, Emanuel Darie

Identification of mathematical model induction motor's parameters with using evolutionary algorithm and multiple criteria of quality 912
Hudy Wiktor, Jaracz Kazimierz

Simulation Study on Control of Ultrahigh Speed Drives in Waste Energy Recovery Systems 916
Péter Stumpf, Miklós G. Simon, Rafael K. Járdán, István Nagy

Adaptive Back EMF Parameter Adjustment of Simplified Vector Control for Position Sensorless Permanent Magnet Synchronous Motors 924
Kiyoshi Sakamoto, Yoshitaka Iwaji, Daigo Kaneko, Toshihiro Takeuchi, Tsunehiro Endo, Atsuo Kawamura

Identification and Control of Precision XY Stages with Active Vibration Suppression System 932
Mayumi Nitta, Seiji Hashimoto

Sensitivity of the Currents Input-Output Decoupling Vector Control of the DFIM versus Current Sensors Fault 938
Meriem Abdellatif, Maria Pietrzak-David, Ilhem Slama-Belkhodja

Extended Back EMF model for PM synchronous machines with different inductances in d- and q-axis 945
Andreas Eilenberger, Manfred Schroedl

Gait generation of a two-legged robot by using adaptive network based fuzzy logic control 949
Umit Onen, Mete Kalyoncu, Mustafa Tinkir, Fatihm. Botsali

Walking robot HEXOR® II - a versatile platform for engineering education 956
M. Sajkowski, T. Stenzel, B. Grzesik

Motion Control of Steel Sheet Shears with Rocking Knife Mechanism 961
Jan Fetyko, Frantisek Durovsky, Viliam Fedak

Intelligent Adaptive Control and Monitoring Of Band Sawing 967
Ilhan Asiltürk, Ali Ünüvar

Hierarchical adaptive network based fuzzy logic controller design for a single flexible link robot manipulator 974
Mete Kalyoncu, Mustafa Tinkir

Digital Controlled High Speed Synchronous Motor 982
Zdenk Cerovský, Jaroslav Novák, Martin Novák, Marek Cambál

Analysis of combustion engine - electric Linear generator set operation 988
Jirí Pavelka

Table of Contents

Closed Loop Control of AC Drive with LC Filter..994
Jaroslaw Guzinski

Sensorless IPMSM based drive for reciprocating compressor...1002
Anton Dianov, Kim Young-Kwan, Lee Sang-Joon, Lee Sang-Taek, Yoon Tae-Ho

Controlling system of electrodynamic drive..1009
Josef Cernohorský

Expert System for Electric Drive Design...1017
Juhan Laugis, Valery Vodovozo

Improvement of Moving Characteristics of Cableless Micro-actuator and Consideration of Reversible Motion..1020
Hiroyuki Yaguchi, Kazumi Ishikawa, Toshihiro Zamma, Koichi Funayama

Sensorless Control of AC Machines using High-Frequency Excitation..........................1024
Heiko Zatocil

Adaptive PF Speed Control of SRAM Drives...1033
Laszlo Szamel

A Very Simple Fuzzy Control System for Inverter Fed Synhronous Motor.....................1040
Pawel Fabijanski, Ryszard Lagoda

Distributed control system of DC servomotors for six legged walking robot................1044
D. Belter, K. Walas, A. Kasinski

Optimization of Starting Process of the Frequency Controlled Induction Motor..........1050
I.Ya. Braslavsky, A.V. Kostylev, D.P. Stepanyuk

3-Axes Satellite Attitude Control Based on Biased Angular Momentum.......................1054
Azam Ghaedi, Mohammad Ali Nekoui

Modelling and simulation of a signal injection self-sensored drive...............................1058
Alen Poljugan, Mark Sumner, Chris Gerada, Qiang Gao

Robust PI Cascade Control for a Multi-Mass System Optimized by Evolutionary Algorithms......1064
M. Joost, K. Zielinski, B. Orlik, R. Laur

Permanent Magnet Synchronous Servo-Drive with State Position Controller...............1071
Lech M. Grzesiak, Tomasz Tarczewski, Slawomir Mandra

Closed-Loop Control of Virtual FPGA-Coded Permanent Magnet Synchronous Motor Drives using a Rapidly Prototyped Controller..1077
Christian Dufour, Vincent Lapointe, Jean Bélanger, Simon Abourida

Speed Sensorless Nonlinear Control Of Induction Motor In The Field Weakening Region......1084
MiroslawWlas, Haithem Abu-Rub, Joachim Holtz

Comparison of Dynamic Performances of Speed Control System Containing Time - Minimal Speed Controller with Control System Containing PI Speed Controller...1090
Andrzej Andrzejewski, Marian Roch Dubowski

Optimisation of Real-Time Complex Path Generation in Constrained Intelligent Motion Applications Based on IPM Motor Drives...1097
Silverio Bolognani, Roberto Petrella, Fabio Stefanutti, Piero Stocco

PMSM Sliding Mode Observer for Speed and Position Estimation Using Modified Back EMF......1105
Ilioudis Vasilios C., Margaris Nikolaos I.

Optimal Control of Electrical Drives with Induction Motors for Variable Torques.......1111
Corneliu Botan, Marcel Ratoi, Vasile Horga

Table of Contents

An Optimal Control for Saturated Interior Permanent Magnet Linear Synchronous Motors Incorporating Field Weakening 1117
Mohammad Abroshan, Jafar Milimonfared, Kaveh Malekian, Arash Rahnamaee

Improved Direct Torque Control for Induction Machine Drives using Fuzzy Logic and Particle Swarm Optimization 1123
Mohammad Mehdi Rezaei, Mojtaba Mirsalim, Kaveh Malekian

Design and Implementation of High Performance Full-Digital Spindle Drives 1128
Liu Yang, Zhao Jin

Semi hierarchical adaptive network based fuzzy logic controller design for a multi-straight-line path tracing flexible robot manipulator with rotating-prismatic joint 1132
Mete Kalyoncu, Mustafa Tinkir

Control System with the Set Point Observation 1140
Algirdas Baskys, Vitoldas Gobis, Valerijus Zlosnikas

Electropneumatic Servo System with Adaptive Force Controller 1144
Arunas Grigaitis, Vilius Antanas Gele~evicius

New fault tolerant DTC control for induction machine drives 1149
A.Ben Abdelghani Bennani, M. Ghodbane Cherif, I. Slama Belkhodja

Stability Analysis of the Natural Field Orientation Controlled Induction Machine Drive 1155
G. Mirzaeva, A. Rojas

Control of SR motor EV by instantaneous torque control using flux based commutation and phase torque distribution technique 1163
Ayumu Nishimiya, Hiroki Goto, Hai-Jiao Guo, Osamu Ichinokura

Simulation of IPM Motor by Nonlinear Magnetic Circuit Model for Comparing Direct Torque Control with Current Vector Control 1168
Hiroki Goto, Kensuke Kimura, Hai-Jiao Guo, Osamu Ichinokura

A Simplified Model for Induction Machines with Faults to Aid the Development of Fault Tolerant Drives 1173
O. Jasim, C. Gerada, M. Sumner, J. Arellano-Padilla

About the Experimental Results of an Electric Driving System Based on Asynchronous Motor and PWM Converter 1181
Petre-Marian Nicolae, Dan-Gabriel Stanescu, Ioana-Gabriela Sîrbu

Real-World Force Feedback Control for Mobile-Hapto 1187
Wataru Yamanouchi, Yuki Yokokura, Seiichiro Katsura, Kiyoshi Ohishi

The new numerical integration routine applied in sensorless drives 1193
Arkadiusz Gardecki, Krystyna Macek-Kaminska

Application of Fuzzy Logic Techniques To Robust Speed Control of PMSM 1198
Tomasz Pajchrowski, Krzysztof Zawirski

Optimal control of current commutation of high speed SRM drive 1204
Jan Deskur, Tomasz Pajchrowski, Krzysztof Zawirski

Comparison Between Direct Torque Control and Vector Control of a Permanent Magnet Synchronous Motor Drive 1209
Rafa Souad, Houcine Zeroug

Detection and self-tuning compensation of periodic disturbances by the control of DC motor 1215
Michael Ruderman, Frank Hoffmann, Johannes Krettek, Torsten Bertram

A Linear Switched Reluctance Motor Based Position Tracking System 1221
S. W. Zhao, N. C. Cheung, Y. Lu, W. C. Gan, Z. G. Sun

Table of Contents

Mobile Robot Navigation with Obstacle Avoidance Capability 1225
Anca Sorana Popa, Mircea Popa, Ioan Silea

Requirements for Power Electronics in Solid Oxide Fuel Cell System 1233
T. Riipinen, V. Väisänen, M. Kuisma, L. Seppä, P. Mustonen, P. Silventoinen

Power Supply for a IGBT-Driver with High Insulation Voltage based on a Printed Planar Transformers 1239
Günter Schmitt, Wolf Kusserow, Ralph Kennel

Variable Motor Operating Point by Integration of Power Electronic Device into Rotor 1243
Adrian Tulbure, Hans-Peter Beck, Mircea Risteiu

Magnetic Material Comparisons for High-Current Gapped and Gapless Foil Wound Inductors in High Frequency DC-DC Converters 1249
Marek S. Rylko, Brendan J. Lyons, Kevin J. Hartnett, John G. Hayes, Michael G. Egan

Feasibility Study of Half- and Full-Bridge Isolated DC/DC Converters in High-Voltage High-Power Applications 1257
Dmitri Vinnikov, Tanel Jalakas, Mikhail Egorov

Evaluation of Different Loss Calculation Methods for High-voltage IGBT-s Under Small Load Conditions 1263
T. Jalakas, D. Vinnikov, J. Laugis

Control of Power Supply Unit for Military Vehicles Based on Four-Leg Three-Phase VSI with Proportional-Resonant Controllers 1268
Tomál Glasberger, Zdenek Peroutka

Optimal Design of a Half Wave Cockroft-Walton Voltage Multiplier with Different Capacitances per Stage 1274
Ioannis C. Kobougias, Emmanuel C. Tatakis

Calculation of Leakage Inductance of Core-Type Transformers for Power Electronic Circuits 1280
Reinhard Doebbelin, Marcel Benecke, Andreas Lindemann

Enhanced Current Pulsation Smoothing Parallel Active Filter for Single Stage Grid-connected AC-PV Modules 1287
A.C. Kyritsis, N.P. Papanikolaou, E.C. Tatakis

Outline of the Design of a Cascaded H-bridge Medium Voltage STATCOM 1293
R.E. Betz, B.J. Cook, T.J. Summers, R. Fisher, A. Bastiani, S. Shao, P. Stepien, K. Willis

Investigation of High Frequency Effects on Layered Coils 1301
Georgios S. Dimitrakakis, Emmanuel C. Tatakis

Soft Switching PWM Inverter for Induction Heating Applied to Heating of Ferromagnetic Metal 1309
Sachio Kubota, Muneo Sato, Fumio Ito, Yoshihiro Shimaoka, Kunihiro Nishioka

Corona Treatment System with Resonant Inverter - Selected Proprieties 1316
Mucko Jan

Power supply unit for an electric discharge machine 1321
Wojciech Mysinski

High Power, High Voltage, High Frequency Transformer / Rectifier for HV Industrial Applications 1326
T. Filchev, D. Cook, P. Wheeler, A. Van den Bossche, J. Clare, V. Valchev

Small Power Laboratory Model and High Power Prototype of the Four-Level VSI 1332
Ryszard Michal Strzelecki, Pawel Szczepankowski, Andrzej Kasprowicz, Genady Stepanovic Zinoviev, Krzysztof Zymmer, Zbigniew Zakrzewski

AC Voltage Regulator Using PWM Technique and magnetic flux distribution 1337
A.M. Dabroom

Minimum Reactive Power Filter Design for High Power Converters 1345
Alex-Sander Amavel Luiz, Braz Jesus Cardoso Filho

Table of Contents

Injection of a carrier with higher than the PWM frequency for sensorless position detection in PM synchronous motors 1353
Roberto Leidhold, Peter Mutschler

Parallel Fixed Point FPGA Implementation of Sensorless Induction Motor Torque Control 1359
Jacek D. Lis, Czeslaw T. Kowalski

Design of an FPGA-Based Real-Time Simulator for Electrical System 1365
I. Bahri, M-W. Naouar, E. Monmasson, I. Slama-Belkhodja, L.Charaabi

A New, Ultra-low-cost Power Quality and Energy Measurement Technology 1371
Alex McEachern, Andreas Eberhard

Rotor Time Constant Adaptation Using Radial Basis Function Network 1375
Pavel Brand1tetter, Ondfej Skuta

Application of Speed and Load Torque Observers in High Speed Train 1382
Jaroslaw Guzinski, Marc Diguet, Zbigniew Krzeminski, Arka diusz Lewicki, Haithem Abu-Rub

Position Estimator including Saturation and Iron Losses for Encoder Fault Detection of Doubly-Fed Induction Machine 1390
Kai Rothenhagen, Friedrich W. Fuchs

Wide Range Low Noise Current Sensor 1398
F. Richter, C. Sourkounis

Transducerless Speed Control with Initial Position Detection for Low Cost PMSM Drives 1402
Roman Filka, Peter Balazovic, Branislav Dobrucky

Study About the Possibility of Electrodes Motion Control in the EAF Based on Adaptive Impedance Control 1409
Manuela Panoiu, Caius Panoiu, Sorin Deaconu

Asynchronous machine stator resistance estimation using integrated PWM modulator and sampler unit as FPGA application 1416
Dag Samuelsen, Waldemar Sulkowski

Development of Monitoring System for Series HEV Bus with Touch Panel 1421
Tae-Won Chun, Quang-Vinh Tran, Uk-Don Choi, Heung-Gun Kim

A Development System for Testing Integrated Circuits Used for Power and Energy Measurements 1426
Vladimir Cuk, Aleksandar Nikolic, Aleksandar Zigic

State and parameter estimation in a hydraulic system - moving horizon approach 1432
Jerzy Baranowski, Andrzej Tutaj

Technologies of Current Sensors Suitable for Hot High Density Power Electronics 1440
Filip Grecki, Grzegorz Iwanski, Wlodzimierz Koczara, Jozef Lastowiecki

Nonlinear dynamical feedback for motion control of magnetic levitation system 1446
Jerzy Baranowski, Pawel Piatek

Speed and position estimation of SRM 1454
Konrad Urbanski, Krzysztof Zawirski

Potential of Digital Gate Units in High Power Appliations 1458
Harald Kuhn, Thies Koneke, Axel Mertens

Disturbance Currents of Inverters 1465
Petr Vrana, Jiri Javurek

Improvement of the Energy Recovery of Traction Electrical Drives using Supercapacitors 1469
Diego Iannuzzi

Table of Contents

A Multi-Core PC-based Simulator for the Hardware-In-the-Loop Testing of Modern Train and Ship Traction Systems......................1475
Christian Dufour, Guillaume Dumur, Jean-Nicolas Paquin, Jean Bélanger

Energy Saving Control of Tram Motors Taking Light Signalling and City Disturbances into Account....................1481
Stanislaw Rawicki

Characterization and Improved Control of a Brushless DC Drive with In-Wheel Motor...........................1491
Manuele Bertoluzzo, Giuseppe Buja, Alessandro Pavoni

Supply of Electric Vehicles via Magnetically Coupled Air Coils.......................1497
Slawomir Judek, Krzysztof Karwowski

Sliding-Mode Approach to Control Design for Induction Motor Drive fed by a Three-Level Voltage-Source Inverter......................1505
Sergey Ryvkin, Richard Schmidt-Obermoeller, Andreas Steimel

Analysis and configuration of supercapacitor based energy storage system on-board light rail vehicles...................1512
R. Barrero, X. Tackoen, J. Van Mierlo

Design of High Power Electronic Building Block based on Parallel of IGBTs for Electric Vehicle...........................1518
Wen Huiqing, Liu Jun, Zhang Xuhui, Wen Xuhui

Stability Analysis on the DC Power Distribution System of More Electric Aircraft.......................1523
H. Zhang, C. Saudemont, B. Robyns, N. Huttin, R. Meuret

Design Considerations for Control of Traction Drive with Permanent Magnet Synchronous Machine.....................1529
Zden..k Peroutka, Karel Zeman

Control of Primary Voltage Source Active Rectifiers for Traction Converter with Medium-Frequency Transformer......................1535
Vojtech Blahník, Zdenek Peroutka, Jan Molnár, Jan Michalík

Energy management strategy for Coupling Supercapacitors and Batteries with DC-DC converters for hybrid vehicle applications......................1542
M.B. Camara, F. Gustin, H. Gualous, A. Berthon

Dual-Source Fed Multiphase Traction System with Standard and Non-Standard Control Regimes Based on Synchronized PWM......................1548
Valentin Oleschuk, Marian P. Kazmierkowski

Analysis of a H-NPC topology for an AC Traction Front-End Converter......................1555
I. Etxeberria-Otadui, A. Lopez-de-Heredia, J. San-Sebastian, H. Gaztañaga, U. Viscarret, M. Caballero

Hybrid - type system of power supply for a trolleybus with an asynchronous motor......................1562
Zygmunt Gizinski, Marcin Gasiewski, Ireneusz Mascibrodzki, Michal Zych, Krzysztof Zymmer, Marcin Zulawnik

Control of rotor flux in AC tram drive during sudden braking operation......................1568
Andrzej Debowski, Piotr Chudzik

A New Novel Power Electronic Circuit to Reduce Stray Current and Rail Potential in DC Railway......................1575
Reza Fotouhi, Siamak Farshad

Slip Control Upgrades for Light-Rail Electric Traction Drives......................1581
Madis Lehtla, Hardi Hõimoja

Practical Aspects on the Improved DC Driving System Used in Electric Urban Traction......................1585
Petre Marian Nicolae, Ioana-Gabriela Sîrbu, Ileana-Diana Nicolae, Lucian Mandache

The study of using the traction drive topology with the middle-frequency transformer......................1593
Martin Pittermann, Pavel Drábek, Marek Cédl, Jirí Foft

Control of a Linear Switched Reluctance Motor as a Propulsion System for Autonomous Railway Vehicles..........1598
L. Kolomeitsev, D. Kraynov, S. Pakhomin, F. Rednov, E. Kallenbach, V. Kireev, T. Schneider, J. Böcker

Table of Contents

Motion Copying System Based on Real-World Haptics in Variable Speed 1604
Yuki Yokokura, Seiichiro Katsura, Kiyoshi Ohishi

Adaptive Fuzzy Control of magnetically suspended Rotary Table 1610
Thomas Schallschmidt, Denis Draganov, Frank Palis

Wideband Force Sensing for Haptic Energy Transmission Utilizing FPGA 1614
Seiichiro Katsura, Masaki Kondo, Kiyoshi Ohishi

On the development of BLDC motor control run-up algorithms for aerospace application 1620
Vladimir Hubik, Martin Sveda, Vladislav Singule

Rotor Levitation by Active Magnetic Bearing Using Digital State Controller 1625
Chip Rinaldi Sabirin, Andreas Binder

Dynamical Torque-Speed-Curve Adaption To Damp Load Peaks Occuring In Drive Trains Of Shredding Plants 1633
Constantinos Sourkounis

Traction vehicle distributed control computer system architecture with auto reconfiguration features and extended DMA support 1638
Jiri Zdenek

Analysis and Position Control of a Linear Switched Reluctance Actuator Based on Sliding Mode Control 1646
António Espírito Santo, Maria R. A. Calado, Carlos M. P. Cabrita

Development and Control for a Reaction Wheel System Driven by Permanent Magnet Synchronous Motor 1652
Ming-Chang Chou, Chang-Ming Liaw, Sywe-Bin Chien, Fa-Hwa Shieh, Jih-Run Tsai, Hao-Chi Chang

Nonlinear control design for magnetic bearings via automatic differentiation 1660
Stefan Palis, Mario Stamann, Thomas Schallschmidt

Design of Energy Harvesting Generator Base on Rapid Prototyping Parts 1665
Zdenek Hadas, Jan Zouhar, Vladislav Singule, Cestmir Ondrusek

Control of Bouc-Wen hysteretic systems: Application to a piezoelectric actuator 1670
Oriol Gomis-Bellmunt, Faycal Ikhouane, Daniel Montesinos-Miracle

Electric drive for carding machine draft device 1676
Martin Diblík

Two-level and Multilevel Converters for Wind Energy Systems: A Comparative Study 1682
R. Melício, V. M. F. Mendes, J. P. S. Catalão

A Stand-alone Photovoltaic Supercapacitor Battery Hybrid Energy Storage System 1688
M.E. Glavin, Paul K.W. Chan, S. Armstrong, W.G Hurley

Integrated contactless power transmission systems with high positioning flexibility 1696
Daniel Kürschner, Christian Rathge

A Transformerless Interface Converter for a Distributed Generation System 1704
Tzung-Lin Lee, Zong-Jie Chen

A Comprehensive Analysis and Comparison Between Multilevel Space-Vector Modulation and Multilevel Carrier-Based PWM 1710
Constantinos Sourkounis, Ahmad Al-Diab

Identification of Electrical Parameters in a Power Network Using Genetic Algorithms and Transient Measurements 1716
Wei. Dong, Pericle Zanchetta, David W.P. Thomas

On Acoustic Noise Reduction Procedure for Inverter-Fed Induction Machines 1722
Weiss Helmut, Zaucher Peter, Xiao Jian

Table of Contents

Cascaded Doubly Fed Induction Generator for Mini and Micro Power Plants Connected to Grid 1729
Marek Adamowicz, Ryszard Strzelecki

Contactless power transmission with new secondary converter topology 1734
Matthias Dockhorn, Daniel Kürschner, Rudolf Mecke

Modeling Approach of a Generator with Non-linear Load in Embedded Electrical Network 1740
Nicolas Amelon, Mourad Ait-Ahmed, Mohamed-Fouad Benkhoris

Optimal Use of the 14 V Alternator in 42 V Automotive Supply Systems 1748
Vasile Comnac, Mihai Cernat, Adrian Mailat

New Dual Channel Quasi Resonant DC-DC Converter Topologies for Distributed Energy Utilization 1755
J. Hamar, I. Nagy, P. Stumpf, H. Ohsaki, E. Masada

Output Filtering of the Customer-end Inverter in a Low-Voltage DC Distribution Network 1763
Pasi Peltoniemi, Pasi Nuutinen, Pasi Salonen, Markku Niemelä, Juha Pyrhönen

Power Flow Control through a Multi-Level H-Bridge based Power Converter for Universal and Flexible Power Management in Future Electrical Grids 1771
Stefano Bifaretti, Pericle Zanchetta, Yue Fan, Florin Iov, Jon Clare

Energy Storage Systems The Flywheel Energy Storage 1779
Tomasz Siostrzonek, Stanislaw Piróg, Marcin Baszynski

Analysis of Wide Area Integration of Dispersed Wind Farms Using Multiple VSC-HVDC Links 1784
S. González-Hernández, E. Moreno-Goytia, O. Anaya-Lara

Generator Selection for Offshore Oscillating Water Column Wave Energy Converters 1790
D.L. O' Sullivan, A.W. Lewis

A Novel Approach To Photovoltaic Powered Water Pumping Design 1798
Michael James Case, Ernest Edward Denny

Direct Controls in Voltage-Source Converters - Generalizations and Deep Study 1803
Karoly Veszpremi, Istvan Schmidt

Multipolar double fed induction wind generator with a single phase secondary winding 1811
Leonids Ribickis, Guntis Dilevs, Nikolajs Levins, Vladislavs Pugachevs

The measurement on the solar cells in Liberec city 1815
Jiri Kubin

Rotor Turn-to-Turn Faults of doubly-fed Induction Generators in Wind Energy Plants - Modelling, Simulation and Detection 1819
Vincenz Dinkhauser, Friedrich W. Fuchs

Static and Dynamic Response of a Photovoltaic Characteristics Simulator 1827
Anastasios Ch. Nanakos, Emmanuel C. Tatakis

Modeling and Optimal Sizing of Hybrid Renewable Energy System 1834
Rachid Belfkira, Cristian Nichita, Pascal Reghem, Georges Barakat

Photovoltaic System MPPTracker Investigation and Implementation using DSP engine and Buck- Boost DC-DC converter 1840
Dimosthenis Peftitsis, Georgios Adamidis, Panagiotis Bakas, Anastasios Balouktsis

Multi Objective Distributed Generation Planning Using NSGA-II 1847
Muhammad Ahmadi, Ashkan Yousefi, Alireza Soroudi, Mehdi Ehsan

Testing of the Grid-connected Photovoltaic Systems Using FPGA-based Real-Time Model 1852
Robert Stala

Table of Contents

Output Maximization Using Direct Torque Control for Sensorless Variable Wind Generation System Employing IPMSG...............1859
Yukinori Inoue, Shigeo Morimoto, Masayuki Sanada

Improving Connection and Disconnection of a Small Scale Distributed Generator Using Solid-State Controller 1866
M.M.R. Ahmed

Research control of electric systems in wind generator systems...............1872
Stefan Winternheimer, Artem Kolesnikov, Evgeny Glushkin, Alexander Bukatov

Stand-alone Photovoltaic Generation System with Combined Storage using lead Battery and EDLC 1877
Hiroaki Nakayama, Eiji Hiraki, Toshihiko Tanaka, Noriaki Koda, Nobuo Takahashi, Shuji Noda

Active Filter Action of Inverter Exciting Induction Generator for Wind Power Generation 1884
Noriyuki Kimura, Tomoyuki Hamada, Katsunori Taniguchi, Toshimitsu Morizane

The Operation of Power Electronic Converters in Photovoltaic Drive Systems 1890
Marek Niechaj

Experimental results of a hybrid wind/hydro power system connected to isolated loads 1896
Mehdi Nasser, Stefan Breban, Vincent Courtecuisse, Arnaud Vergnol, Benoît Robyns, Mircea M. Radulescu

Grid Connection of Multi-Megawatt Clean Wave Energy Power Plant under Weak Grid Condition...............1904
Kai Rothenhagen, Marek Jasinski, Marian P. Kazmierkowski

Improved sizing method of storage units for hybrid wind-diesel powered system 1911
A.M. Tankari, B. Dakyo, C. Nichita

A Research Platform for a Smart-Blade Wind Generation System 1918
J. Davey, Udaya K. Madawala, R. Sharma

Soft Switching Multi-Phase Boost Converter for Photovoltaic System...............1924
Joo-Hyuk Lee, Jae-Hyung Kim, Chung-Yuen Won, Su-Jin Jang, Yong-Chae Jung

Soft Switching Boost Converter for Photovoltaic Power Generation System 1929
Doo-Yong Jung, Young-Hyok Ji, Jae-Hyung Kim, Chung-Yuen Won, Yong-Chae Jung

Optimisation Of Wind Power Pmsm To Grid Conversion System 1934
Ince Kayhan, Weiss Helmut

Analysis of Wind Farm and Multilevel Converter Interactions in Medium Voltage Networks Under Steady-State and Transient Conditions 1941
J. Sosa-Ruiz, E. Moreno-Goytia, O. Anaya-Lara

A Simple, Low Cost Design Using Current Feedback to Improve the Efficiency of a MPPT-PV System for Isolated Locations 1947
Herman Fernández, Abelardo Martínez, Víctor Guzmán, María Isabel Gímenez

A Single-Phase Active Power Filter Based in a Two Stages Grid-Connected PV System 1951
Kleber C.A. De Souza, Denizar C. Martins

Wide Bandwidth Power Flow Control Algorithm of the Grid Connected VSI under Unbalanced Grid Voltages...............1957
Zoran Ivanovic, Marko Vekic, Stevan Grabic, Evgenije Adzic, Vladimir Katic

The use of Switched Reluctance Generator in wind energy applications 1963
Eleonora Darie, Costin Cepisca, Emanuel Darie

Active Line Shaping of a Single Phase Rectifier using the Switching Function Technique 1967
Christos Marouchos

Control of Reactive Power in Double-Fed Machine Based Wind Park 1975
Elzbieta Bogalecka, Michal Kosmecki

xviii

Table of Contents

A Novel Hybrid Modulation Method for Cascaded H-bridge Active Power Filter 1981
Yonggang Chen, Ping Wang, Yaohua Li, Zixin Li, Longcheng Tan

Apparent Power Ratio of the Shunt Active Power Filter 1987
A. Kouzou, B.S Khaldi, S. Saadi, M.O. Mahmoudi, M.S. Boucherit

Shunt Active Power Filter with Improved Dynamic Performance 1995
Krzysztof Piotr Sozanski

The Research on the Active Power Filter Based on the Cascaded H-bridge Converter 2000
Yonggang Chen, Junling Chen, Ping Wang, Yaohua Li, Longcheng Tan, Zixin Li, Wei Xu

E-laboratory in the Field of Electrical Drives 2005
H.Hõimoja, A.Rosin, T.Möller, M. Müür

Laboratory Setup for Studying Ultracapacitors in Industrial Applications 2011
I. Roasto, D. Vinnikov, T. Lehtla

Synchronous machine direct axis parameters estimation module from an iterative strategy 2015
Emile Mouni, Slim Tnani, Gérard Champenois

**Determination of the Characteristic Life Time of Paper-insulated MV-Cables based on a Partial
Discharge and tan(..) Diagnosis** 2022
I. Mladenovic, Ch. Weindl

**Elimination of Increased Excitation of Common- Mode Oscillations in Electrical Drive Systems with
Active Front End and Long Motor Cables** 2028
Thomas Weidinger

Internal Short Circuit in a Tooth Wound PMSM with Stranded Conductors 2037
Damien Birolleau, Christian Chillet, Laurent Albert

Implementation of a Virtual Laboratory for Low Power Electrical Drives 2043
Gh. BALUTA, V. HORGA, C. LAZAR

**DQ-Transformation Approach for Modelling and Stability Analysis of AC-DC Power System with
Controlled PWM Rectifier and Constant Power Loads** 2049
K-N Areerak, S.V. Bozhko, G.M. Asher, D.W.P. Thomas

**Genetic Identification of Parameters the Sandwich Piezoelectric Ceramic Transducers for Ultrasonic
Systems** 2055
Pawel Fabijanski, Ryszard Lagoda

The Impact of Higher-Order Harmonics on Tripping of Residual Current Devices 2059
Stanislaw Czapp

Estimation of the Untapped Regenerative Braking Energy in Urban Electric Transportation Network 2066
Leonards Latkovskis, Linards Grigans

Performance Evaluation of Electric Power Steering with IPM Motor and Drive System 2071
Hamidreza Akhondi, Jafar Milimonfared, Kaveh Malekian

Optimal Control: Load Frequency Control of a Large Power System 2076
Sílvio José Pinto Simões Mariano, Luís António Fialho Marcelino Ferreira*

LCL-Load Modular Converter For Induction Heating 2082
Maciej A. Dzieniakowski, Jan Fabianowski, Robert Ibach

On-line PID Controller Tuning Using Genetic Algorithm and DSP PC Board 2087
Pawel Fabijanski, Ryszard Lagoda

Regulation Properties of Pumping Station Control System In The Highest Efficiency Range 2091
Szychta Leszek

xix

Table of Contents

Inner Gas Pressure Measurement Based Life-span Estimation of Electrolytic Capacitors..........................2096
A. Riz, D. Fodor, O. Klug, Z. Karaffy

Robust Control Methodologies for Optical Micro Electro Mechanical System - New approaches and Comparison..........................2102
Alireza Izadbakhsh, S.M.R. Rafiei

Modeling a Buck-Based Switching Amplifier for Sinusoid Wide Band Tracking by Using a Nonlinear Time Varying Map..........................2108
A. El Aroudi, E. Alarcón, E. Rodriguez, R. Leyva

Single Inductor Multiple Outputs Interleaved Converters Operating in CCM..........................2115
Luis Benadero, Vanessa Moreno-Font, Abdelali El Aroudi, Roberto Giral

Control of a two-cell dc/dc converter in presence of saturating duty cycle..........................2120
Moez Feki, Abdelali El Aroudi, Bruno Gerard Michel Robert, Nabil Derbel

Bifurcations and Chaotic Dynamics in a Linear Switched Reluctance Motor..........................2126
M.R. De Castro, B.G.M. Robert, C. Goeldel

Modular Architecture for Decentralized Hybrid Power Systems..........................2134
E. Ortjohann, M. Lingemann, O.Omari, A. Schmelter, N. Hmasic, A. Mohd, W. Sinsukthavorn, D. Morton

Design of a power management system for an active PV station including various storage technologies..........................2142
Di Lu, Tao Zhou, Hicham Fakham, Bruno Francois

Energy Management and Power Flow of Decoupled Generation System for Power Conditioning of Renewable Energy Sources..........................2150
Wlodzimierz Koczara, Zdzislaw Chlodnicki, Nazar Al-Khayat, Neil L.Brown

Inversion Based Control of a Diesel Fed Low Temperature Fuel Cell System..........................2156
Daniela Chrenko, Marie-Cecile Pera, Daniel Hissel

Power Management in an Autonomous Adjustable Speed Large Power Diesel Gensets..........................2164
Grzegorz Iwanski, Wlodzimierz Koczara

Cost evaluation of Generator-set with Energy Storage for 4Q-load..........................2170
Freek J.F.Baalbergen, Pavol Bauer

Integrating renewable energy sources and storage into isolated diesel generator supplied electric power systems..........................2178
Chad Abbey, Jonathan Robinson, Géza Joós

Performance comparison of different wind generator based hybrid systems..........................2184
Vincent Courtecuisse, Benoit Robyns, Marc Petit, Bruno Francois, Jacques Deuse

First Approach for a Fault Tolerant Power Converter Interface for Multi-Stack PEM Fuel Cell Generator in Transportation Systems..........................2192
Alexandre De Bernardinis, Gérard Coquery

Development of Electrical System for Hybrid Vehicles Using the Free-swinging Piston Engine and Oscillating Rotating Generator..........................2200
Sigitas Kudarauskas

Power flow control in different time scales for a wind/hydrogen/super-capacitors based active hybrid power system..........................2205
ZHOU Tao, LU Di, FAKHAM Hicham, FRANCOIS Bruno

Neuro-Fuzzy Adaptive Control of the IM Drive with Elastic Coupling..........................2211
Teresa Orlowska-Kowalska, Krzysztof Szabat, Mateusz Dybkowski

Control of Flexible Drive with PMSM employing Forced Dynamics..........................2219
Vittek Ján, Bris Peter, Makys Pavol, Stulrajter Marek, Vavrus Vladimír

Table of Contents

The problems of high dynamic drive control under circumstances of elastic transmission 2227
Jan Deskur, Roman Muszynski

Protective Predictive Control of Electrical Drives with Elastic Transmission 2235
Mario Vasak, Nedjeljko Peric

Low-Cost High-Performance Predictive Control of Drive Systems with Elastic Coupling 2241
Marcin Cychowski, Kieran Delaney, Krzysztof Szabat

Development of an Expert System for Identification, Commissioning and Monitoring of Drives 2248
Mario Pacas, Sebastian Villwock

Control of Axial Flux Permanent Magnet Motor by the PIPCRM Method at Standstill and at Low Speed 2254
Janusz Wisniewski, Wlodzimierz Koczara

Zero Speed Position Estimation of a Matrix Converter Fed AC PM Machine using PWM Excitation 2261
Q. Gao, G. M. Asher, M. Sumner

Sensorless Direct Torque and Flux Control of an IPM Synchronous Motor at Low Speed and Standstill 2269
Gilbert Foo, S. Sayeef, M.F. Rahman

Sensorless Control of PM Synchronous Motors Using a Predictive Current Controller with Integrated INFORM and EMF Evaluation 2275
Manfred Schrödl, Christian Simetzberger

Torque Sensorless Control of Induction Motor 2283
Karel Jezernik, Miran Rodic

Application of the induction motor torque - observer to the control of turbo - machines 2289
Andrzej Debowski, Daniel Lewandowski

Observer of induction motor speed based on exact disturbance model 2294
Zbigniew Krzeminski

Experimental Performance Evaluation for Low Speed and Regenerating Operation of Sensor-less Vector Control System of Induction Motor Using Observer Gain Tuning 2300
Kazuhiro Ohyama, Greg Asher, Mark Sumner

Application of the Stator Current-based MRAS Speed Estimator in the Sensorless Induction Motor Drive 2306
Mateusz Dybkowski, Teresa Orlowska-Kowalska

State and Parameter Estimation in Induction Motors using Sliding Modes 2312
Sachit Rao, Martin Buss, Vadim Utkin

Torque Transient Alleviation in Fixed Speed Wind Generators by Indirect Torque Control with STATCOM 2318
Marta Molinas, Jon Are Suul and Tore Undeland

Flicker Study on Variable Speed Wind Turbines with Permanent Magnet Synchronous Generator 2325
Weihao Hu, Zhe Chen, Yue Wang, Zhaoan Wang

Power Output Characteristics Analysis of Wind Energy Converter Control Methods 2331
Bingchang Ni, Constantinos Sourkounis

A Cooperative Control Method for Output Power Smoothing and Hydrogen Production by Using Variable Speed Wind Generator 2337
Rion Takahashi, Hirotaka Kinoshita, Toshiaki Murata, Junji Tamura Masatoshi Sugimasa, Akiyoshi Komura, Motoo Futami, Masaya Ichinose, Kazumasa Ide

A new interconnecting method for wind turbine/generators in a wind farm and basic characteristics of the integrated system 2343
Shoji Nishikata, Fujio Tatsuta

Educational aspects of mechatronic control course design for collaborative remote laboratory 2349
Andreja Rojko, Darko Hercog, Karel Jezernik

xxi

Table of Contents

PEMCWebLab - Distance and Virtual Laboratories in Electrical Engineering: Development and Trends.............2354
Pavol Bauer, Viliam Fedák, Otto Rompelman

Integrated multimedia educational program of a DC servo system for distant learning.............2360
Gabor Sziebig, Istvan Nagy, Rafael Kalman Jardan, Peter Korondi

Electromechanical Actuators WEB-lab.............2368
Dusan Maga, Jan Sitar, Juraj Dudak, Rene Hartansky, Peter Siroky, Jan Halgos, Pavol Bauer

Power Quality and Active Filters as Web-Controlled Experiment in the frame of PEMC WebLab.............2371
Volker Staudt, Andreas Steimel, Pavol Bauer, Vítezslav Hájek

Distant learning of Pulse Width Modulation Techniques for Voltage Source Converters.............2378
Bartlomiej Kamiski, Dariusz Sobczuk

Modern design optimisation exploiting field simulation.............2383
Jan K. Sykulski

Transmission-Line Modelling of Wave Propagation Effects in Machine Windings.............2385
Herbert De Gersem, Olaf Henze, Thomas Weiland, Andreas Binder

An efficient field-circuit coupling method by a dynamic lumped parameter reduction of the FE model.............2393
F. Henrotte, E. Lange, K. Hameyer

Coupled field-circuit-mechanical model of an electromagnetic actuator operating in error actuated control system.............2400
Lech Nowak

Simulation and Investigation of Magnetorheological Fluid Brake.............2406
Wieslaw Lyskawinski, Wojciech Szelag, Cezary Jedryczka

Field and Field-Circuit Description of Electrical Machines.............2412
Andrzej Demenko, Kay Hameyer

Interaction between Thermal Impedance and Parasitics in Power Sections.............2420
Stefan Forster, Andreas Lindemann

Discussion of Internal and External High Frequency Common Mode Noise Current on a Chopper Circuit.............2428
Tetsuya Mitani, Keiji Wada, Toshihisa Shimizu, Hiromichi Ohashi

A Novel Digital Control Method for DC-DC Converter.............2434
Fujio Kurokawa, Masashi Okamatsu, Yuichi Sumida, Yasuhiro Mimura, Masahiro Sasaki

A Novel Single/Three-phase Matrix Converter For High Power Integration.............2439
Makoto Saito

An Effective Design Method for High Power Density Converters.............2445
Yusuke Hayashi, Kazuto Takao, Toshihisa Shimizu, Hiromichi Ohashi

Power Devices in Polish National Silicon Carbide Program.............2452
Mariusz Sochacki, Andrzej Kubiak, Zbigniew Lisik, Jan Szmidt

SiC Power Semiconductor Devices for new Applications in Power Electronics.............2457
Dominique Planson, Dominique Tournier, Pascal Bevilacqua, Nicolas Dheilly, Herve Morel, Christophe Raynaud, Mihai Lazar, Dominique Bergogne, Bruno Allard, Jean-Pierre Chante

Silicon carbide Schottky diodes and MOSFETs: solutions to performance problems.............2464
Owen J. Guy, Michal Lodzinski, Ambroise Castaing, P. M. Igic, Amador Perez-Tomas, Michael R. Jennings, Philip A. Mawby

Characterization of the Static and Dynamic Behavior of a SiC BJT.............2472
M.M.R. Ahmed, N-A.Parker-Allotey, P.A. Mawby, Muhammed Nawaz, Carina Zaring

An active network control method using distributed energy resources in microgrids.............2478
Takayuki Tanabe, Yoshinobu Ueda, Toshihisa Funabashi, Shigeo Numata, Kimio Morino, Eisuke Shimoda

Table of Contents

Energy Management in Solar Photovoltaic Plants based on ESS 2481
M. Lafoz, L. García-Tabarés, M. Blanco

A Method of Three-Phase Balancing in Microgrid by Photovoltaic Generation Systems 2487
Masahide Hojo, Yuta Iwase, Toshihisa Funabashi, Yoshinobu Ueda

Development of HILS(Hardware In-Loop Simulation) System for MMS(Microgrid Management System) by using RTDS 2492
Jin-Hong Jeon, Jong-Yul Kim, Seul-Ki Kim, ong-Bo Ahn, JuneHo Park

Power Quality Analysis of Jeju Island Power System with Wind Farm and HVDC System 2498
Jae-Hong Kim, Eel-Hwan Kim, Se-Ho Kim, Jaeho Choi, Gil-Soo Jang, Seung-Ho Song

A New Control Method for Power Turbine Generators Using an Accurate Ship Plant System Model 2504
Nobumasa Matsui, Fujio Kurokawa, Keiichi Shiraishi

Voltage profile support in distribution networks - influence of the network R/X ratio 2510
B. Bla~ic, I. Papic

Modeling, Simulation and Analysis of Conducted Common-Mode EMI in Matrix Converters for Wind Turbine Generators 2516
S. Zhang, K.J. Tseng

Design of Frequency Shift Acceleration Contol for Anti-islanding of an Inverter-based DG 2524
Seul-Ki Kim, Jin-Hong Jeon, Heung-Kwan Choi, Jonng-Bo Ahn

Integrated Power Converter for Photovoltaic and Fuel Cell Systems in Home 2530
Yasuyuki Nishida, Shinichiro Sumiyoshi, Hideki Omori

A Comparison of Position Control Structures for Ironless Linear Synchronous Motor 2538
Martin Hrasko, Pavol Makys, Marek Franko, Jozef Kuchta

A Comparison of Sliding Mode Approaches to a Nanometre Position Control Application 2543
Paul Andreas Stadler, Stephen James Dodds

Sliding Mode Control of PMSM Drives Subject to Torsion Oscillations in the Mechanical Load 2551
Stephen J. Dodds, Jan Vittek

Sliding Mode Vector Control of PMSM Drives with Minimum Energy Position Following 2559
Stephen J. Dodds

xxiii

Adaptive PF Speed Control of SRM Drives

Laszlo Szamel

Budapest University of Technology and Economics /Department of Electric Power Engineering,
Budapest, Hungary, e-mail: *szamel@eik.bme.hu*

Abstract—This paper proposes two model reference adaptive PF speed control methods for SRM drives. Following from the structure of model reference parameter adaptive PF control it makes it easier to reach overshootless as well as fast speed changing compensation caused by jump in load. The approaching block diagram of the model reference signal adaptive control can be seen as an extended version of the PF controller, so one of the adaptation factors (which is the free parameter of the adaptive control) is given. Both model reference adaptive controls drawn up can be easily implemented because the adaptation algorithms do not need acceleration measuring.

Keywords—Adaptive control, Control of Drive, Servo-drive, Switched reluctance drive.

I. INTRODUCTION

In motion control systems there is robustness against parameter changes and disturbance rejection of main interest. The model reference adaptive control has the following features:

- It enables the compliance of the system with varying operational conditions and ensures the behavior of the controlled system according to the prescribed reference model.

- It means such a special type of adaptive systems which results in nonlinear control systems. This is the reason why the analytical analysis is completed by the Ljapunov stability criterion or by the hyper-stability principle.

- Its design and application are closely related to the using of computer methods.

- Simple implementation of the control algorithm.

In this paper applications of two model reference adaptive control methods to switched reluctance motors are presented.

II. DRIVE SYSTEM

Block scheme of the examined drive system is shown in Fig. 1.

The supply unit consists of three main blocks, namely the RECTIFIER, the FILTER and the INVERTER. The inverter is a pulsed width modulated (PWM) one, marked by QP in the figure and it contains a one-one switching transistor per phase and a brake chopper, not shown in the figure. The common point of phase windings is supplied by the PWM inverter. It is of autonomous operation and has an inner current control loop. The other ends of phase windings are connected to the phase switching transistors.

It follows from the operational principle of SRM that its phase windings are to be excited at a well determined angle of the rotor position in an appropriate order. That is

why a Rotor Position Sensor is to be mounted on the shaft of the motor. In our case the position sensor is a resolver. It can be calculated from the pole numbers that the phase switchings have to follow each other by 15 degrees. The resolver is supplied by an oscillator circuit, their signals are evaluated by a Position Decoder.

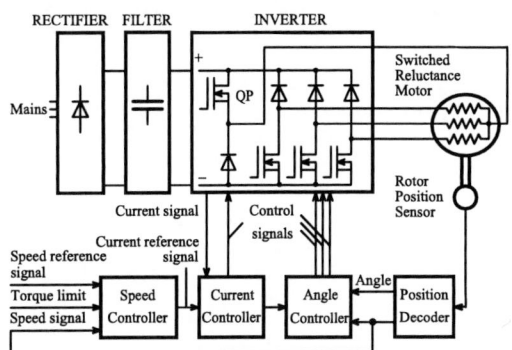

Fig. 1. Block scheme of drive system

The Position Decoder has two outputs: Angle and Speed signals. Angle Controller provides the Control signals for the phase switching transistors based on the two signals.

III. CURRENT CONTROL

The Current Controller is totally based on its hardware solution. Based on the current reference signal it controls the PWM inverter of fixed frequency by installing an analog controller. The current feedback also comes from the PWM inverter.

For the control of the sum of phase currents (Fig. 1) a simpler four-transistor inverter is suitable and a six-transistor one is not necessary as in the case of control of phase currents independently from each other. But the detriment of the previous solution is that torque pulsation can be decreased in a smaller degree by changing turn-on and turn-off angles.

Namely in case of constant current reference signal the current increase is limited by the switched-off, but the conducting phase current as regulator controls the sum of two phase currents. The increase of phase current at the beginning of the conducting state can be forced by the modification of the current reference signal:

$$i_r = u \sum_{j=1}^{3} C_j + \sum_{j=1}^{3} (1 - C_j) \cdot i_j. \tag{1}$$

where:

i_r is the current reference signal,

i_j is the current signal of phase j ,

u is the output of the speed controller,

C_j is the control signal of phase j (0 or 1).

Supplement of the first member in (1) makes the overlap of the phase conduction possible, while the effect of the second member is to increase the reference signal with the current of the switched-off but not current-free phase.

IV. SPEED CONTROL

A model reference adaptive control is used for speed control. Such an adaptive control has been successfully elaborated by using a suitable chosen Ljapunov function to compensate the gain of the speed control loop [1], [2], [6], [7].

A. Model Reference Parameter Adaptive Control

The adaptive control of servo-drives with a cascade arrangement is most effective when it is applied in the inner loop containing the effect of variable parameters directly, i.e. inertia (J_m) and/or torque factor (k_m). The speed control implemented by PI controller is of a cascade arrangement in fact as it contains an inner, proportional feedback loop (PF controller, [4], Fig. 2). A one-storage proportional element can describe this inner loop neglecting the time constant of closed current control loop. By this our adaptation algorithm is simplest.

Relation between accelerating current determining dynamic torque and angular speed can be given by the following transfer function:

$$Y_{\omega,(i-i_l)}(s) = \frac{A_i}{s} . \qquad (2)$$

where: $A_i = \dfrac{k_m}{J_m}$.

Arrangement of control circuit can be seen in the following figure:

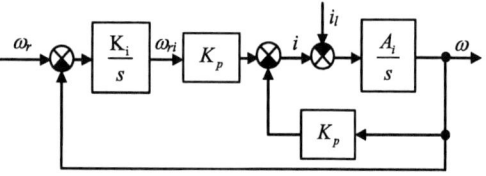

Fig. 2. Block scheme of parameter adaptive PF speed control

where

ω is the speed,

ω_r is the speed reference signal,

i is the current of the motor,

i_l is the current equivalent to the load-torque.

Section determined by the transfer function $Y_{\omega,(i-i_l)}(s)$ is fed back by a proportional member of gain K_P. The task is to change gain K_P in such a way that the product $A_i K_P$ should remain constant despite the changing of A_i parameter.

Transfer factor of the inner closed loop is given by the reciprocal ($1/K_P$) of feedback member that is not constant because of the change in torque factor. Consequently loop gain of outer speed control loop would change as well. In order to get a one-storage element with unity transfer factor we have to insert a member with gain K_P between the integrator of the PF controller and the reference signal of the inner loop. First-order reference model with time constant T_m gets sum of the input signals of above member (ω_{ri}) and signal ω_{lm} compensating the load effect for the model. So dynamics of reference model can be described by the following differential equation:

$$\dot{\omega}_m T_m + \omega_m = \omega_{ri} + \omega_{lm} . \qquad (3)$$

Dividing (3) with T_m and applying designation $q_m = 1/T_m$, we get the following equation.

$$\dot{\omega}_m + q_m \omega_m = q_m(\omega_{ri} + \omega_{lm}) . \qquad (4)$$

The differential equation of the first-order controlled plant is as follows:

$$\dot{\omega} + (K_p A_i)\,\omega = (K_p A_i)\,\omega_{ri} - A_i i_l . \qquad (5)$$

Factor K_p can be described as the sum of K_{p0} determined by mean A_i and ΔK_p accomplished by the adaptation algorithm. So:

$$K_p A_i = (K_{p0} + \Delta K_p)\,A_i = q + \Delta q , \qquad (6)$$

where K_{P0}, and q are constant.

In this case we assume that the change of A_i is slow from the viewpoint of adaptation and therefore the effect of this change can be neglected.

Substituting (6) into (5) we get:

$$\dot{\omega} + (q + \Delta q)\,\omega = (q + \Delta q)\,\omega_{ri} - A_i i_l . \qquad (7)$$

By using (4) and (7) and substituting expression of model error $\varepsilon = \omega_m - \omega$ the dynamic equation will be:

$$\dot{\varepsilon} = -q_m \varepsilon + x\,\omega - x\,\omega_{ri} + q_m \omega_{lm} + A_i i_l , \qquad (8)$$

where $x = (q + \Delta q) - q_m$.

Dynamic of model error should by asymptotically stable to follow the system with model proposed. For determination of Δq the following Ljapunov function should be composed:

$$V = \frac{1}{2}\left(\varepsilon^2 + \beta x^2\right), \qquad (9)$$

where β is a positive number.

When choosing the Ljapunov function both purposes, i.e. the termination of the model error and loop gain deviation have been taken account.

The time derivative of the Ljapunov function is:

$$\dot{V} = \varepsilon\, \dot{\varepsilon} + \beta\, x\, \dot{x}\,. \tag{10}$$

Substituting (8) into (10) the following equation is valid:

$$\dot{V} = -q_m \varepsilon^2 + \varepsilon\, x\, \omega - \varepsilon\, x\, \omega_{ri} + \\ \varepsilon\,(q_m \omega_{lm} + A_i i_l) + \beta\, x\, \dot{x}\,. \tag{11}$$

If

$$\varepsilon\, x\, \omega - \varepsilon\, x\, \omega_{ri} + \beta\, x\, \dot{x} = 0\,, \tag{12}$$

that is

$$\dot{x} = \varepsilon\,(\omega_{ri} - \omega)/\beta \tag{13}$$

and

$$\varepsilon\,(q_m \omega_{lm} + A_i i_l) < 0 \tag{14}$$

then

$$\dot{V} < -q_m \varepsilon^2\,. \tag{15}$$

The above equation is a negative definite function that shows the asymptotic stability of the error dynamic (8). By using (6), (8) and (13) the following adaptation algorithm is true:

$$\Delta \dot{K}_p = \gamma\, \varepsilon\,(\omega_{ri} - \omega)\,. \tag{16}$$

where γ may be an arbitrary positive number. The inequality (14) shows how we have to change the signal ω_{lm} representing the load of model.

If

$$\varepsilon > 0, \quad \text{then} \quad \omega_{lm} < -\frac{\left|i_l\right|_{\max} A_i}{q_m}, \tag{17}$$

respectively if

$$\varepsilon < 0, \quad \text{then} \quad \omega_{lm} > \frac{\left|i_l\right|_{\max} A_i}{q_m}\,.$$

B. Model Reference Signal Adaptive Control

The controlled loop has been approximated by an integral element. Time constant of the closed current control loop has been neglected. The control consists of a P-element with the gain K_p. Input of P-element contains not only the control error signal but an adaptation signal as well (g). Applying the signal adaptation control, a P type controller with K_p gain can ensure zero speed error as the adaptation signal can produce a current reference signal to compensate the loading current at zero speed error.

The feature of closed speed control loop has been taken into consideration by a parallel control model to be expressed by first order proportional element. The Ljapunov function has been chosen in such a way that

model error should be decreased asymptotically and gain of speed control loop and load should be compensated. In this case we have to assume that change in loop gain and in load is smaller than the speed of the adaptation.

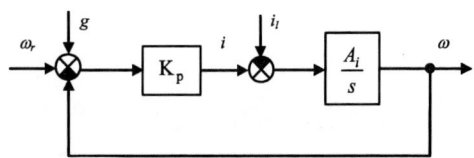

Fig. 3. Initial block scheme of signal adaptive speed control

Regarding the block diagram of control loop following differential equation is valid for the closed loop:

$$\dot{\omega} + A_i K_p \omega = A_i K_p\,(\omega_r + g) - A_i i_l\,. \tag{18}$$

The feature of the closed speed control loop has been taken into consideration by a parallel control model to be expressed by a first order proportional element. The differential equation of the first order system is:

$$\dot{\omega}_m + q_m \omega_m = q_m \omega_r\,. \tag{19}$$

where index m refers to the model and q_m is the reciprocal of model time constant.

Using (18) and (19) and introducing the expression $\varepsilon = \omega_m - \omega$ for model error, the dynamic equation for the model error is as follows:

$$\dot{\varepsilon} + q_m \varepsilon = \left(q_m - A_i K_p\right)\!\left(\omega_r - \omega\right) + A_i\!\left(i_l - K_p g\right)\!. \tag{20}$$

The adaptation signal $g(t)$ can be written in the following form:

$$g(t) = g_1(t)\left(\omega_r - \omega\right) + g_2(t)\,. \tag{21}$$

Substituting (21) for (20):

$$\dot{\varepsilon} = -q_m \varepsilon + b_1\left(\omega_r - \omega\right) + b_2\,, \tag{22}$$

where

$$b_1 = q_m - A_i K_p\left(1 + g_1(t)\right),$$

$$b_2 = A_i\!\left(i_l - K_p g_2(t)\right)\!.$$

Let us compose the following Ljapunov function to produce the signal $g_1(t)$ and $g_2(t)$:

$$V = \frac{1}{2}\varepsilon^2 + \frac{1}{2}\left(\beta_1 b_1^2 + \beta_2 b_2^2\right), \tag{23}$$

where β_1 and β_2 are positive constants.

Time-derivation of the Ljapunov function is:

$$\dot{V} = \varepsilon\dot{\varepsilon} + \beta_1 b_1 \dot{b}_1 + \beta_2 b_2 \dot{b}_2\,. \tag{24}$$

Substituting (22) for (24):

$$\dot{V} = -q_m \varepsilon^2 + (\omega_r - \omega)b_1\varepsilon + b_2\varepsilon$$
$$+ \beta_1 b_1 \dot{b}_1 + \beta_2 b_2 \dot{b}_2 . \tag{25}$$

If

$$\dot{b}_1 = -(\omega_r - \omega)\varepsilon / \beta_1 \tag{26}$$

and

$$\dot{b}_2 = -\varepsilon / \beta_2 ,$$

then

$$\dot{V} = -q_m \varepsilon^2 \tag{27}$$

and it ensures asymptotical stability of the model error. On the basis of (22), (26) and by assuming that variation of A_i can be neglected compared to the speed of adaptation the following adaptation algorithm is valid:

$$\dot{g}_1(t) = \gamma_1 \varepsilon (\omega_r - \omega) ,$$

$$\dot{g}_2(t) = \gamma_2 \varepsilon . \tag{28}$$

where γ_1 and γ_2 are positive constants, the free parameters of adaptation. Taking relations (21), (28) into consideration the following equation comes true:

$$g(t) = \gamma_1 (\omega_r - \omega) \int \varepsilon (\omega_r - \omega) dt + \gamma_2 \int \varepsilon\, dt . \tag{29}$$

A block diagram of the control circuit introducing adaptation signal $g(t)$ furthermore $g_1(t)$ = const. can be seen in Fig. 4. Taking the structure of control: it contains two parts. In the first part the reference signal is led through a first order system and a PI controller with variable gain and integration time. The second one is a differentiating filter which takes effect only on changing of reference signal. The gain and differentiation time are also changing. The adaptation gain factor γ_2 gives the reciprocal of integrating time constant of controller type PI, assuming $g_1(t) = 0$. To fulfill the constant integrating time constant it is preferable to substitute γ_2 by $\gamma_2(1 + g_1(t))$. In such a way the neglect of time constant of current control loop can be compensated.

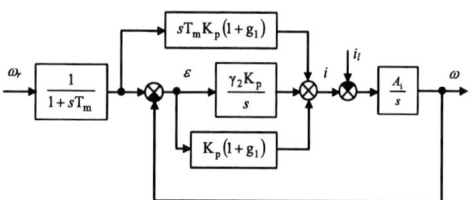

Fig. 4. Block scheme of the signal adaptive speed control with adaptation signal

The approaching block diagram of the adaptive control can be seen as an extended version of the PF (integral element with separate proportional feedback loop, [3], [6], the input of the proportional element is only the speed) controller so γ_2 one of the adaptation factors (which is

the free parameter of the adaptive control) is given. Contraction of model-filtered reference signal and PI controller can be transformed into a PF controller when integration time of PI controller equals the time constant of the model (Fig. 5).

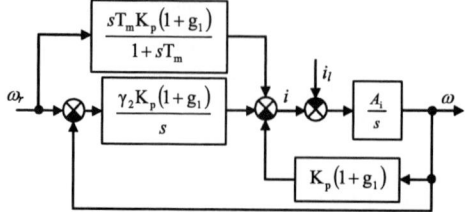

Fig. 5. Block scheme of signal adaptive PF speed control
(extended version)

The basic structure of the signal adaptive speed controller is also PF type which on the one hand provides overshootless with its structure. Moreover, it ensures fast compensation of speed variation caused by a jump in motor load.

This control has been tested by a simulation program developed in our Department. First the adaptation has been examined without current limitation and load in order to take into consideration only the nonlinearity of the motor and the adaptation. In the interest of the adaptation stability speed of change of adaptation signal g_1 has to be limited. The signal g_1 results in significant oscillations without limitations as the change of the signal is possible in discrete times.

The current limitation causes further problems. This limitation hinders the tracking of the model so the effect of above signal g_1 will be too large or it can change in the reverse direction. To eliminate the above problem signal g_1 is not to be changed in the period of the current limitation.

Digital output signal determined by software is converted by a D/A converter to an analog reference signal. Speed feedback signal is determined by calculating the difference between the actual position value and previous one of the resolver to digital converter. The measuring period ensures an accuracy of ± 5 rpm.

V. RIPPLE REDUCED CONTROL METHOD

Ripple free operation of the SRM drives can only be realized with current waveform depending on rotor angle, speed and torque [5]. Ripple reduced method proposed changes only the turn-on and the turn-off angle in function of speed and current reference. Optimum turn-on and turn-off angles of SRM drive have been determined by computer simulation based on measured results of the analyzed drive. An optimum solution has been fulfilled by four cycles embedded into each other. Two outer cycles give the current and speed reference signals, while two inner ones provide the turn-on and turn-off angles. By this one-one optimum angle pair can be determined to all operating points.

It can be considered as an interesting result that the criterion of minimal torque pulsation does not provide an optimum solution in all cases. The torque pulsation will be minimum in the speed-current plane only in those cases when torque of the motor is relatively small. For this reason a good result can be achieved if the relative, i.e. compared to the torque of motor, torque pulsation is minimized.

The angle control of the drive determines actual turn-on and turn-off angles with a two-variable interpolation from the results stored in a look-up table and calculated by the above method.

VI. SIMULATION RESULTS

According to the simulation investigations convergence of the model reference signal adaptive control at switched reluctance drives with significant torque ripples can be ensured under the following conditions:

- Input of the first-order reference model only determines its output when the filtered current reference signal is lower with a given ΔI value (in the simulation it has been set to 1 A) than its limit otherwise output is identical to the speed feedback signal. So the model works like a first-order proportional lag element only in operation without current limitations.

- Adaptation is executed when two conditions are true at the same time: filtered current reference signal is lower with a given ΔI value (in the simulation it has been set to 1 A) than current limit and absolute value of the speed error signal is higher than a given value (in the simulation it has been set to 20 revolution/min).

- The appropriate selection of the adaptation factors also has an important effect on the sufficient convergence.

Because of the restrictions described in point 2. adaptation works practically only in a relatively narrow speed error track (adaptation range) which is equal to approximately 20-100 revolution/min absolute value of speed error. The drawback of this limitation is the relatively short time for the algorithm to operate. At the same time the convergence of the algorithm is extremely fast which significantly reduces the effect of this drawback. Two more important advantages emerge when adaptation works only with small speed errors. First of all the controller at changing drive parameters adapts to parameters around the value specified by the speed reference signal which also assists to speed the adaptation. The other significant positive effect is the disappearance of the problem coming from nonlinear systems that the response of the system can even differ in its character when the value, amplitude of the reference signal is changed.

In Fig. 6 and Fig. 7 two of many executed simulations are shown. Fig. 6 shows the run-up with speed controller of PF-type (an integral element with Proportional Feedback), while Fig. 7 with model reference parameter adaptive control (16), (17) and in both cases with turn-on and turn-off angles depending on speed and current reference and with current reference compensation (1). Fig. 8 shows the run-up with model reference signal adaptive speed controller and with jump in load.

VII. RESULTS

The tests were completed by the described drive system. The test results have supported our theoretical investigations. The oscillograms in the following figures illustrate some typical starting curves and wave forms. The loading machine was a DC motor. Its inertia is about a triple of that of SRM.

Fig. 9 and Fig. 10 show the speed and current curves in the course of starting without current reference compensation.

The upper curve is the speed (1500 rpm), the lower one is the current flowing in the common point of stator windings (10 A/div). It is related to the no-load operation mode.

The experiences show that the model reference parameter adaptive and signal adaptive control suggested in this paper works without overshooting.

VIII. CONCLUSIONS

The paper proposes a simple control method for SRM drives. The proposed ripple reduced control method changes only the turn-on and the turn-off angles depending on the speed and current reference. The modification of the current reference is suggested for a simpler four-transistor inverter.

To provide constant loop gain in speed control loop with changing parameters (moment of inertia and/or torque factor), parameter and signal adaptive model reference adaptive control was developed.

Following from the structure (PF-type) of model reference parameter adaptive control was developed to provide constant loop gain in speed control loop with changing gain (moment of inertia and/or torque factor) it makes it easier to reach overshootless as well as fast speed changing compensation caused by jump in load. The algorithm even keeps its stability at fast changing, jump-like load torque.

Model reference signal adaptive control is used to provide constant loop gain in speed control loop with changing parameters (moment of inertia and/or torque factor) exposed to a significant load. The approaching block diagram of the adaptive control can be seen as an extended version of the PF controller, so one of the adaptation factors (which is the free parameter of the adaptive control) is given.

With appropriate choice of the adaptation range convergence of the adaptation can even be ensured at SRM drives with significant torque ripples. Although normally because of the nonlinear characteristic of the adaptation, the response of the system also depends on the value of the jump in the reference signal, this does not cause any problems in the proposed adaptive control as adaptation only happens in a narrow track of the reference signal called the adaptation range.

The adaptive controls suggested in this paper work without overshooting. Though these methods require a longer calculation period it is less sensitive to the variations of parameters. Both model reference adaptive controls drawn up can be easily implemented, because the adaptation algorithms do not need acceleration measuring (thanks to the first-order model). Simulation and experimental results demonstrate that the proposed methods are promising tools to control SRM drives.

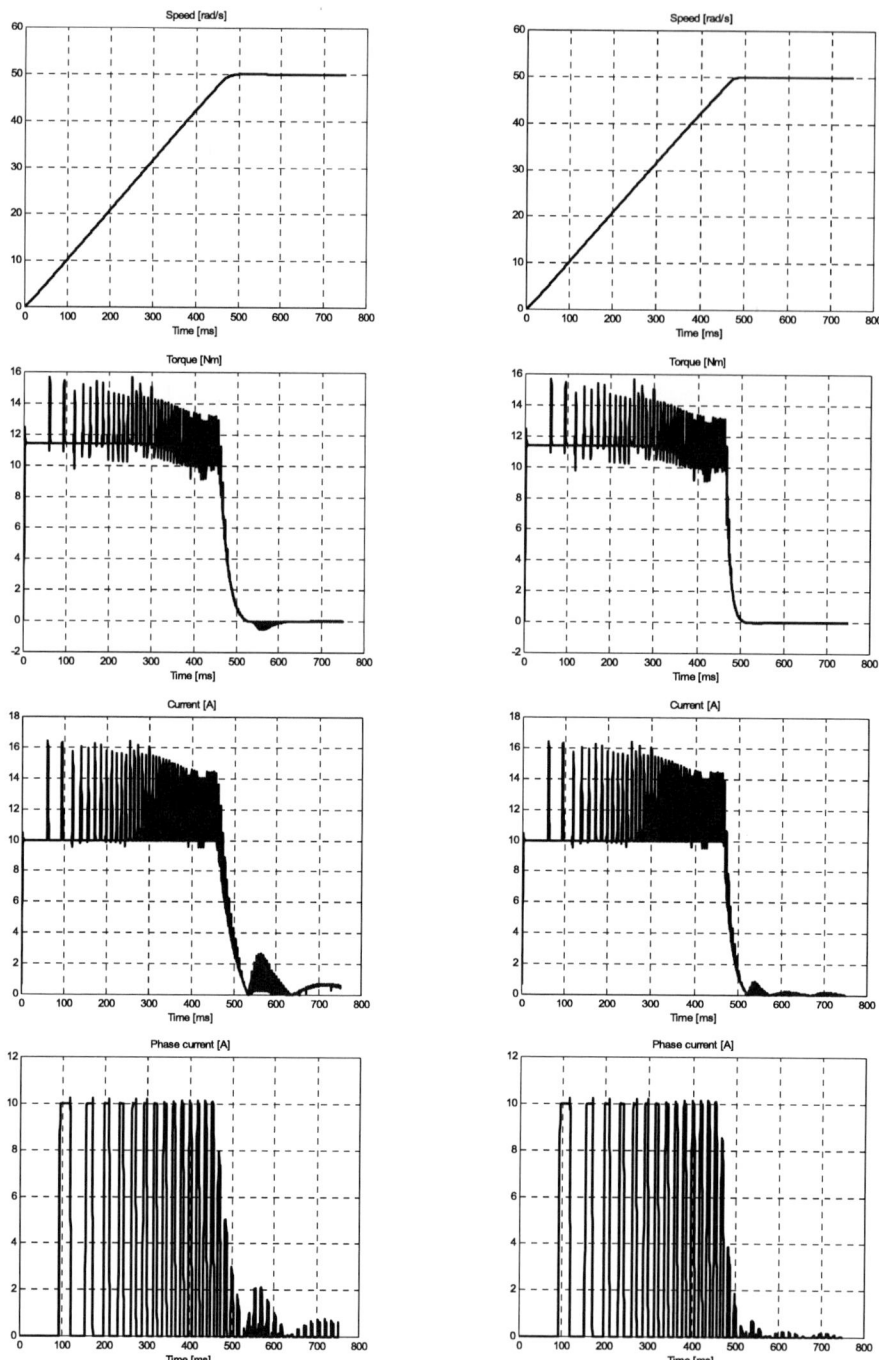

Fig. 6. Simulation results with PF-type speed control ($\Delta K_p = 0$)

Fig. 7. Simulation results with model reference parameter adaptive speed control

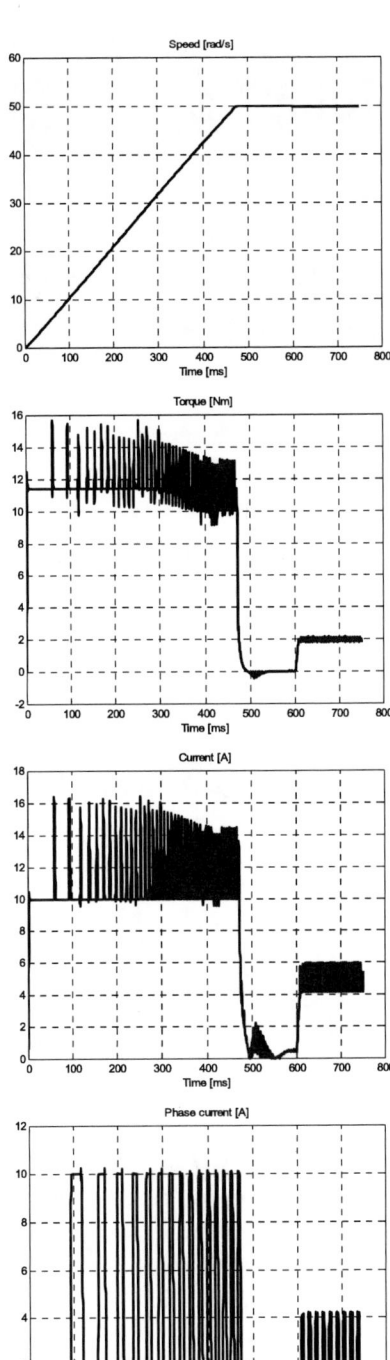

Fig. 8. Simulation results with model reference signal adaptive speed control

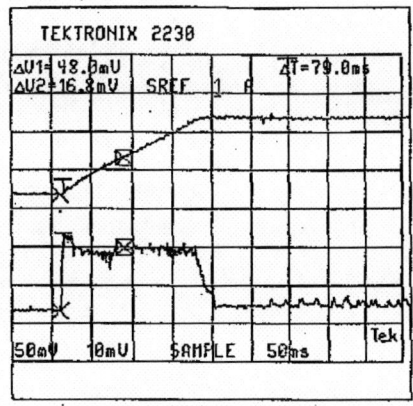

Fig. 9. Oscillogram of speed and current with model reference parameter adaptive speed control

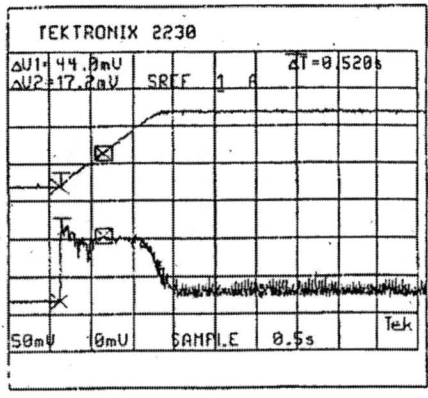

Fig. 10. Oscillogram of speed and current with model reference signal adaptive speed control

REFERENCES

[1] J. Borka, and L. Szamel, "Modern Strategy for Controlling Robot Drives," *Conference Automation'92,* Budapest, 1992, pp. 392–401.

[2] B. K. Bose, and T. G. E. Miller, "Microcomputer Control of Switched Reluctance Motor," *IEEE/IAS Annual Meeting,* 1985, pp. 542–547.

[3] N.V. Diep, and L. Szamel, "Up-to-date Control Strategy in the Regulators of Robot Drives," *PEMC'90,* Budapest, 1990, pp.811-815.

[4] G. A. Perdikaris, and K. W. VanPatten, "Computer Schemes for Modeling, Tuning and Control of DC Motor Drive Systems," *PCI Proc.,* Mar. 1982, pp. 83-96.

[5] A. Stankovic, and G. Tadmor, "On Torque Ripple Reduction in Current-Fed Switched Reluctance Motor*s," IEEE Transactions on Industrial Electronics,* Vol. 46, No. 1, February 1999. pp. 177-183

[6] L. Szamel, "Adaptive Control of SRM Drives," *EPE-PEMC 2002* , Cavtat & Dubrovnik (Croatia), 9-11 September 2002, CD-ROM: Paper No. T11-007.

[7] L. Szamel, "Investigation of Model Reference Parameter Adaptive SRM Drives," *EPE-PEMC 2004,* Riga (Latvia), 2-4 September 2004, Full paper A95117 on CD_ROM.

A Very Simple Fuzzy Control System for Inverter Fed Synhronous Motor

Paweł Fabijański*, Ryszard Łagoda†

*Institute of Control and Industrial Electronics, Warsaw University of Technology, Warsaw, Poland
e-mail :pawel@isep.pw.edu.pl
†Institute of Control and Industrial Electronics, Warsaw University of Technology, Warsaw, Poland
e-mail : lagoda@isep.pw.edu.pl

Abstract—**This paper deals with fuzzy logic control of inverter fed synchronous motor. In this paper mathematical model of the motor drive system and stability condition of fuzzy logic control algorithm are represented. The essential fuzzy logic control configuration system consists of an internal current regulation loop and an external speed regulation loop. PID controller generally controls the speed and current loop in these systems. To control a inverter fed synchronous machine a host computer and a special microprocessor system with dSPACE 1104 central unit as a micro-controller is used. Most important results of our investigation is to find a simply mathematical model of our motor drive system. Then stability conditions are determined for current and speed controller. The simulated results were compared during the laboratory test on 600 W synchronous motor.**

Keywords—**Control of drive, fuzzy control, DSP, synchronous motor.**

I. INTRODUCTION

In the last few years' fuzzy logic control was developed for some electrical drive. A fuzzy logic control and digital proportional and integral control method are proposed for electrical drive system with inverter fed synchronous motor. Our experimental drive system with fuzzy logic control of inverter fed synchronous machine consists essentially of a fully phase controlled six-pulse thyristor bridge converter operating essential as a rectifier, dc intermediate stage with smoothing inductance and a six pulse thyristor converter working as an indirect type frequency converter (Fig.1).

II. CONTROL SCHEME DESCRIPTION

To control a inverter fed synchronous machine a host computer and a special microprocessor system with dSPACE 1104 central unit as a micro-controller is used.

The host computer serves as a developing environment, it is in charge of the real-time program debugging, the execution code downloading, real-time parameter and state monitoring, and real-time gain auto-tuning. The micro-controller, on the other hand, is used to execute the real-time system control and data processing tasks.

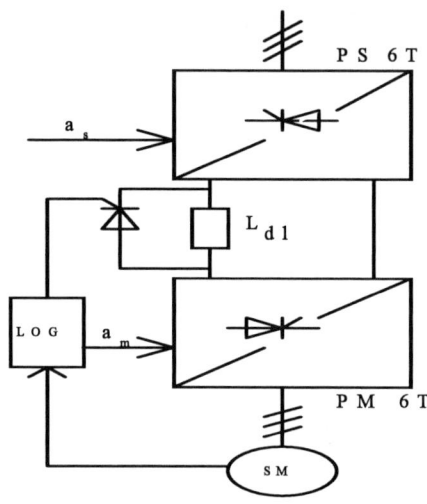

Fig. 1. General configuration of inverter fed synchronous motor

III. MATHEMATICAL MODEL OF INVERTER FED SYNCHRONOUS MOTOR

Our object consists of six-pulse thyristor converter working as an indirect type frequency converter and inverted fed synchronous motor. Main Scheme of inverter fed synchronous motor is shown on Fig.2 and simply model on Fig.3.

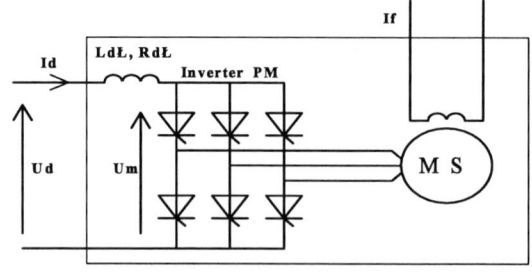

Fig.2. Main scheme of inverter fed synchronous motor

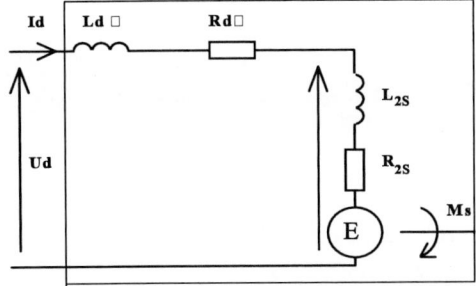

Fig.3. Simply model inverter fed synchronous motor

To find a equation describing parameter of synchronous motor we need a simply model of inverter fed synchronous motor (Fig.3) and we have :

$$\begin{cases} U_d(s) = k_e\,\omega(s) + I(s)R_z + L_z sI(s) \\ k_t\,I(s) = M_{obc}(s) + sJ\omega(s) \end{cases}$$

where $R_z = R_{dl} + R_{2s} \quad L_z = L_{dl} + L_{2s}$ (1)

$$and: T_e = \frac{L_z}{R_z}, \quad T_m = \frac{JR_z}{k_t k_e}$$

From equations (1) we have two important voltage node

$$\left[U_d(s) - k_e\omega(s) \right]\frac{1}{R_z(1+sT_e)} = I(s) \quad (2)$$

$$\left[M_e(s) - M_{obc}(s) \right]\frac{1}{Js} = \omega(s) \quad (3)$$

and well known block diagram for inverter fed synchronous machine (shown on Fig.4).

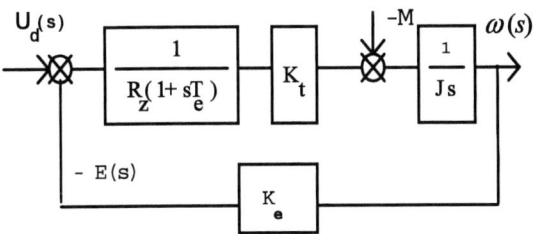

Fig. 4. Inverter fed synchronous motor- operating diagram

The transfer function of our object from Fig.2.:

$$G_{ob}(s) = \frac{k_o}{(1+sT_o)\cdot(1+sT_e)}$$

and finally digital form of transfer function $G_{ob}(z)$:

$$G_{ob}(z) = \frac{k_o T_e}{T_e - T_o}\cdot\frac{z}{z - e^{\frac{T_i}{T_e}}} - \frac{k_o T_o}{T_e - T_o}\cdot\frac{z}{z - e^{\frac{T_i}{T_o}}} \quad (4)$$

IV. CURRENT AND SPEED CONTROL

The current and speed control structure include the fuzzy logic controller. Self –tuning controller achieves the speed control. Inputs of the speed loop are the speed feedback and its reference. The output represents the references for the current loop. The inner current control loop is shown on Fig. 5. and simplified diagram of current controller is shown on Fig.6

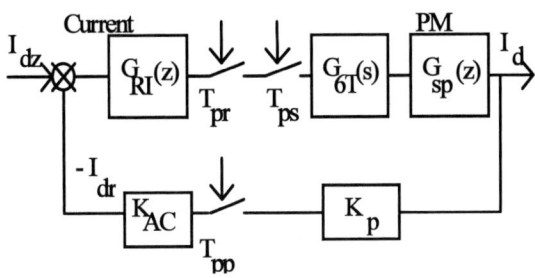

Fig. 5. Inner current control loop

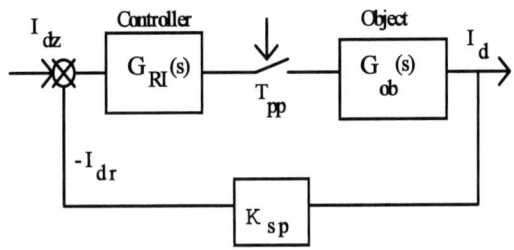

Fig. 6. Simplified diagram of current controller

The transfer function for inner current controller loop is:

$$G_1(s) = \frac{I_d(s)}{I_{dz}(s)} = \frac{G_{RI}(s)G_{ob}(s)}{1 + K_{sp}G_{RI}G_{ob}(s)}$$

where:

$$G_{ob}(s) = k\frac{1 - e^{-T_p s}}{s}\frac{1/R_z}{1 + sT_e} = \frac{k_o}{(1 + sT_0)(1 + sT_e)} \quad (5)$$

We used inverse transfer function to find time response

$$G_{oL}(s) = A\cdot\frac{1}{sT_1} + B\cdot\frac{1}{(1+sT_o)} + C\cdot\frac{1}{(1+sT_e)} \quad (6)$$

1041

where:

$$\begin{cases} A = k_o k_R \\ B = \dfrac{k_R k_o \cdot \left(T_1 T_o - T_1 T_2 - T_o^2\right)}{T_1 T_o - T_1 T_2} \\ C = \dfrac{k_R k_o \cdot \left(T_o T_2 - T_2 T_e - T_o T_e\right)}{T_1 T_o - T_1 T_2} \end{cases} \quad (7)$$

and using for Laplace transfer function table we have :

$$g(t) = A \cdot L^{-1}\left[\frac{1}{sT_1}\right] + B \cdot L^{-1}\left[\frac{1}{(1+sT_o)}\right] + C \cdot L^{-1}\left[\frac{1}{(1+sT_e)}\right] \quad (8)$$

$$g(t) = A \cdot \frac{1}{T_1} \cdot 1(t) + B \cdot e^{\frac{1}{T_o}t} + C \cdot e^{\frac{1}{T_e}t}$$

The inner and the outer loop of the control structure include the fuzzy controller. The basis of the fuzzy controller, the fuzzy rules, contain the relationship between the linguistic input variables and the linguistic output variable, which describe the conclusion to the state of the system Fig.7 and Fig.8.

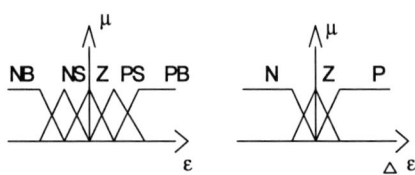

Fig.7. Membership function for control speed errors.

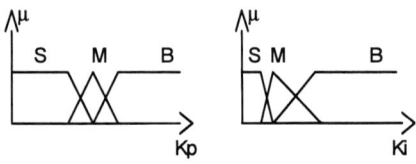

Fig.8. Membership function for parameters of speed controller

TABLE 1. CONTROL MATRIX FOR Kp AND Ki SPEED CONTROLLER PARAMETERS

$\varepsilon \backslash d\varepsilon$	N	Z	P
PB	S	S	M
PS	S	S	M
Z	S	M	B
NS	M	S	S
NB	M	S	S

$\varepsilon \backslash d\varepsilon$	N	Z	P
PB	B	S	M
PS	B	S	M
Z	B	M	S
NS	M	S	B
NB	M	S	B

V. STABILITY CONDITION FOR SPEED CONTROLLER

In the case described model of the inverted fed synchronous motor (1,2) we want to find stability condition and optimal parameter. k_{opt}, T_{opt} and ξ_{opt} for PID speed controller

$$G_{opt}(s) = \frac{k_{opt}}{T_{opt}^2 s^2 + 2 T_{opt} \xi_{opt} s + 1} \quad (9)$$

using second order model of inverted fed synchronous motor:

$$G_{ob}(s) = \frac{k_{ob}}{T_{ob}^2 s^2 + 2 T_{ob} \xi_{ob} s + 1} \quad (10)$$

Comparatively transfer function outer close loop and optimal transfer function (eq.11) we have:

$$G_{Z2} = \frac{G_R(s) G_{ob}(s)}{1 + G_R(s) G_{ob}(s)} \quad (11)$$

From this equation we can described optimal transfer function of speed PID controller by evaluate parameter controller object k_{opt}, T_{opt} and ξ_{opt}; From equations (11) we have

$$G_R(s) = \frac{k_{opt} T_{ob}^2 s^2 + 2 k_{opt} T_{ob} \xi_{ob} s + k_{opt}}{k_{ob} T_{opt}^2 s^2 + 2 k_{ob} T_{opt} \xi_{opt} s + k_{ob} - k_{ob} k_{opt}} \quad (12)$$

and after mathematical transformation we have parameter of PID speed controller

$$\begin{cases} k_P = \dfrac{T_{ob} \xi_{opt}}{k_{ob} T_{opt} \xi_{opt}} - \dfrac{1}{4 k_{ob} \xi_{opt}^2} \\ T_I = 2 k_{ob} T_{opt} \xi_{opt} \\ T_2 = \dfrac{T_{ob}^2}{2 k_{ob} T_{opt} \xi_{opt}} + \\ \quad + \dfrac{T_{opt}}{8 k_{ob} \xi_{opt}^3} - \dfrac{T_{ob} \xi_{opt}}{2 k_{ob} \xi_{op}^2} \end{cases} \quad (13)$$

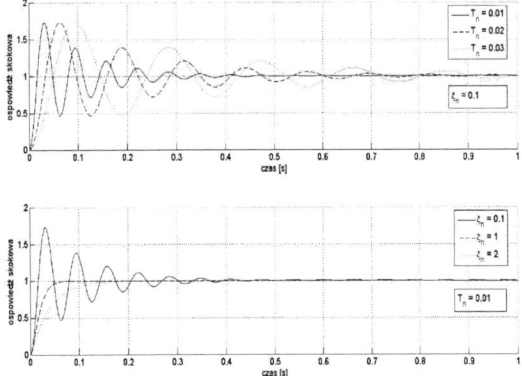

Fig.9. Step response from T_{ob}, ξ_{ob} parameter dependent

Step response our control system is stability when $T_{ob}\xi_{ob} > 0$. We have oscillation step response in case when $|\xi_{ob}| < 1$. In the case when $|\xi_{ob}| \geq 1 (\pm \infty)$, step response is aperiodic and we have only pole real part Fig.9

VI. LABORATORY TEST

The proposed control structure under application of fuzzy logic was tested with a 600W inverted fed synchronous motor. The following motor parameters were taken as a basis:

Un = 220 V, In =7.5 A, Nn = 3000 rpm, Rz = 0.87 $[\Omega]$
Te = 0.051 [s], Tm = 0.873 [s], Kt = Ke = 5 [Nm/A]

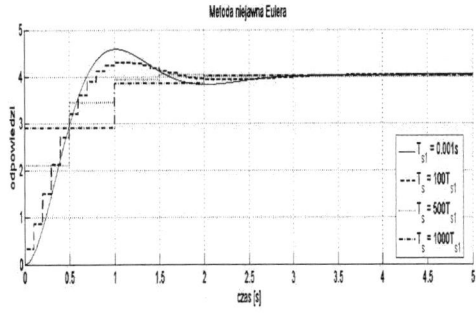

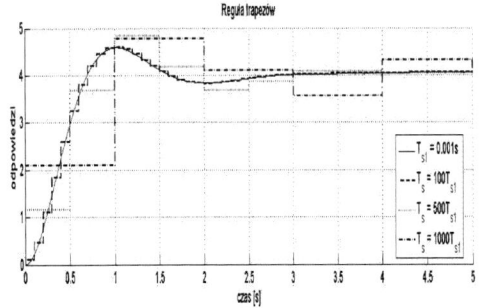

Fig. 10. Step response for some digital T_{period}

VII. CONCLUSION

The most important results of our investigation is to find a simply mathematical model of our motor drive system with inverted fed synchronous motor. Then stability condition are determined for current and speed controller. The simulated result were compared during the laboratory test on 600 W synchronous motor. The experimental results show that the proposed algorithm has the feature of simplicity. The method describing stability condition for current and speed controller has been proved to be very effective and practical. Another results of our investigation is description fuzzy logic control method of inverted fed synchronous motor.

REFERENCES

[1] J. Barrenscheen, D. Flieller , D. Kalinowski , J.P. Louis,:"A new sensorless speed and torque control for permanent magnet synchronous motors: realisation and modelling", Epe'05, Sevilla

[2] P. Brandstetter, M. Mech : " Control methods for permanent magnet synchronous motor drives with high dynamic perfomance", Proceedings of EPE' 05, Sevilla

[3] Guo Qingding, Wang Limei and Luo Ruifu . ; " Completely digital PMSM Servo System based on new self-tuning pid algorithm and dsp " Proceedings of the IEEE International Conference on industrial Technology, 1999

[4] S. Grundmann, M. Krause, V. Muller: " Application of Fuzzy control for PWM voltage

[5] source inverter fed permanent Magnet Synchronous Motor "Proceedings of the EPE 03, pp524 - 529

[6] E. Chiricozzi, F. Parasiliti, M.Tursini, D.Q. Zhang: " Implementation of a fuzzy self-tuning controller for electrical drives:Proc of the EPE 95, pp 440 - 445

[7] A. Del Pizzo, I. Marongiu, A. Perfetto; " Direct torque-control with predictive algorithm in microprocessorized permanent magnet drives" Proceedings of the EPE 03, pp 369 - 374

[8] A. Zajaczkowski " Indirect adaptive decoupling control of a permanent magnet synchronous motor", Seminar on fundamentals of electrotechnics and circuit SPETO 2002

Distributed control system of DC servomotors for six legged walking robot

D. Belter, K. Walas, A. Kasinski

Institute of Control and Information Engineering, Poznan University of Technology

Poznan, Poland, e-mail: {*Dominik.Belter, Krzysztof.Walas, Andrzej.Kasinski*}*@put.poznan.pl*

Abstract—The multi-layered drives control for walking robot is discussed. It is a six-legged robot with 18 DOF. There is one DC-Servomotor for each joint. Synchronization of drives is a matter as walking is performed in unpredictable environment, and under high measurement uncertainty. It is done through distributed control of the robot. Each leg has it's separate controller. Leg controllers are connected to robot master controller. The master controller is responsible for communication between legs and host computer. The article also addresses sensing system issues for walking robot and control loops closed through these sensors.

Index Terms—Robotics, Servo-drive, DC Machine, Control of Drive.

I. INTRODUCTION

The main challenge for autonomous mobile robots is a controlled locomotion in a rough terrain. Walking robot may be a plausible solution to that problem. Therefore a six-legged robot has been designed for that task and the reason for choosing such a type of kinematics was its static stability during a walk.

Control of walking in a six-legged robot is a complex problem. Walking involves simultaneous coordinated movements of six legs. In the reported solution each of legs has three degrees of freedom, which are actuated by DC servos. Thus the presented robot with six three-axis-legs has 18DOFs in total. As the movement of robot is performed in unstructured environment so there is a need of using specific sensors for state feedback. Nevertheless, the robot will still perform his tasks under high measurement uncertainty. However the main problem is to coordinate 18 joints while walking with particular gait. The problem was solved by designing a dedicated distributed control architecture.

II. MECHANICAL STRUCTURE OF THE ROBOT

The kinematic scheme of a single leg can be seen in figure 1. To attach coordinate system to each joint the Modified Denavit-Hartenberg Notation(MD-H) was used. The legs are biologically inspired and identical. Proportions between segments' lengths (coxa, femur and tibia) were designed in order to maximize the potential load of the robot (while using drives with given maximal torque) but without additionally restricting the maximal step length of the leg.

A DC servo with integrated metallic gear have been chosen to drive each robot joint (HS-645MG). As 18 servos were necessary the cost was also an important factor. HS-645MG has a stall torque of $0.942\,Nm$. This

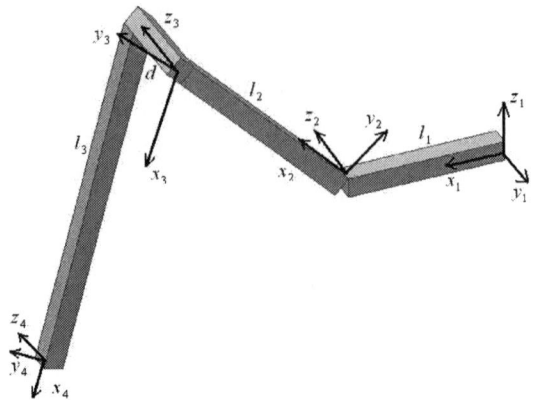

Fig. 1.   Kinematic scheme of the right leg.

servo has an integrated potentiometer to enable closing of the feedback-loop in the position control mode.

Many simulations followed by a necessary trial and error design process were required to determine the appropriate dimensions of the leg segments. The principal constant of the design was the given torque limit of a servo and its dimensions. Robot legs are based on the RRR kinematic scheme. First (hip) joint is secured from bending torques by ball bearings fixing its rotation axis. The mechanical design of the leg is shown in figure 2. Most of parts of the structure were made of duraluminium. Their shape was specially designed in order to minimize the robot proper weight and to improve its dynamics while walking. This is the reason why robot segments are thin and have large extrusion holes.

Fixture of the middle legs are not in the line which is connecting front and back legs fixtures. They are more distant from the robot center of gravity to obtain an extra advantage while tripod gait is performed. In this way every leg can be loaded equally. Such a kinestatic solution allowed to create a larger robot platform while using the same drives and to carry more sensors and devices on control board of the robot. The overall view of the robot is shown in figure 3. Each leg is equipped with ground contact sensor placed on robot foot. It is based on microswitch V4LST7 which can bear up to 2×10^5 switches and requires a small force to changeover.

978-1-4244-1741-4/08/$25.00 ©2008 IEEE

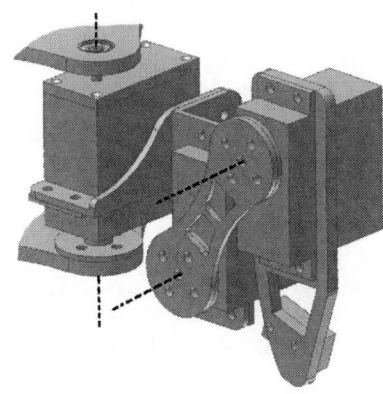

Fig. 2. Mechanical design of the leg with rotation axis shown.

Fig. 3. Robot "Ragno" with camera and bluetooth modem on board.

III. State of Art in Legged Robots

Looking at the current state of art in the design of walking robots the following classification can be made according to the number of legs:

- one - legged robot - 3D One-Leg Hoopper from Massachusetts Institute of Technology(USA). That robot was built for studying a stability problem of walking and energy efficiency issues. Single leg eliminates a problem of legs cooperation. This robot uses jumps to progress and is dynamically stable only during a move [6].
- two - legged robots - the most popular example for this group is ASIMO from HONDA laboratories in Japan. It is a humanoid robot which was built to mimic a very complicated human gait [7].
- four - legged robot - this type of robot resemble a dog or a horse. One of the examples could be BISAM (Biologically InSpired wAlking Machine) with legs that mimic horse forelegs. It was built in FZI Karlsruhe in Germany [2]. Another example could be AIBO (robotic dog) which was built in SONY laboratories in Japan.3.
- six - legged robots - there are several machines of this type all over the world. One of them is LAURON III. It was built in FZI Karlsruhe in Germany. It was designed to take part in urban rescue missions [3]. Another example is a HEXAPOD V4B which was used in a movie ("Harry Potter"). It was built by Micromagic Systems in UK. Six-legged robots roughly mimics insects locomotion.
- eight - legged robots - one of robots representing this group is SCORPION. It was built at Bremen Universitaet in Germany. Due to its 8 legs this robot can walk on various types of terrain. It can also climb and thanks to its symmetrical design it could easily continue the walk after flipping it up side down [8].

A number of control systems for walking robots have been proposed [1], [5]. The mentioned architectures were implemented and verified on the real, working robots.

IV. The Multi-leg Robot as a Control Object

A. Synchronization issues

Each of robot leg joints is driven by the servo HS-645MG. This servo has an integrated position controller. Six legged robot task control is based on positioning task of each leg. The main control task however is to move robot platform in 3D space by forcing special combination of leg positions over time. Sometimes it is necessary to move all legs at the same time. For the robot equipped with 18 drives there is a problem to control them without delays. Inaccurate simultaneous positioning of legs at the same time may cause additional forces in transverse direction and produce a robot sliding motion. Thus synchronization of drives is important issue for efficiently controlling the robot walking. Proper synchronization of drives is a case of the efficient walking control algorithm.

B. Control system architecture

Robot control architecture is shown in figure 4. Robot control system is divided into four layers:

- layer 1 - host computer (teleoperation workstation)
- layer 2 - master SPI layer (Atmega 128 on board)
- layer 3 - leg controllers (6 x Atmega 8 - SPI slaves)
- layer 4 - integrated servo controllers

The highest layer is located on a host-computer and enables the remote control of the robot. The host is dedicated to image-sequence processing tasks, trajectory planning tasks and for computing the desired reference positions for servos. Communication between the robot and the host computer is accomplished through Bluetooth channel with RS-232 protocol. At this layer data from robot sensors is read and computed reference values are sent to all servos of the robot by using Bluetooth devices. Robot control software installed on the host includes the robot kinematics simulator used to predict and reconsider every next move. It is also used to visualize robot state and in this way creates a very useful GUI in teleoperation mode.

Next two layers are embedded on the robot platform. The second layer is responsible for communication, reflex-like reactions and measurements reading from all

1045

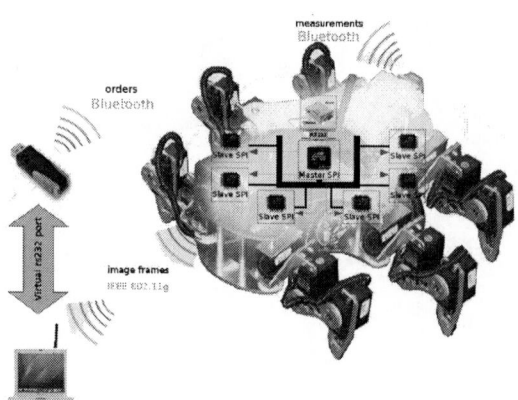

Fig. 4. Robot control architecture scheme.

robot on-board sensors. To that goal an 8-bit controller is sufficient. It receives a data frame from the host and interprets its commands. When there is a command to send new reference values to the leg controllers, it divides a received frame into appropriate parts and sends such a packet to slave processors. When the host computer requests for measurements, the master controller assembles a data frame packing measurement data in fixed order and sends the frame to the host computer.

The third layer of control consists of six controllers responsible for a single leg control. While the master controller exchanges information with the host computer and acquires data from sensors, these slave processors work independently, and in parallel at the same time. The main role of these processors is to convert the desired reference values for the joint positions into accurate motor control pulses. They also measure real state variables (drive angles) and send them to the master controller. Sending measurement data doesn't require additional time-slice because a bidirectional SPI protocol is used to communicate between layer 2 and 3. While Atmega 128 processor receives reference values, at the same time it sends measurement data from sensors in duplex mode.

The system has three communication channels. The first is the Bluetooth with RS-232 modems. It is used for sending commands to the robot and for receiving information about a current robot state. The second, implementing a SPI protocol, connects all microcontrollers on the robot controller board and it is used for the fast exchange of short and simple messages. Robot is also equipped with on-board camera. Sending image frames from the robot to the host computer requires a broadband unilateral channel. Wireless standard communication protocol is used here (IEEE 802.11).

C. On-board sensors

The robot is equipped with various types of entero- and exteroceptive sensors. Every actuator on leg joint has a feedback potentiometer which is used to measure the actual rotor angle. The output signal from this sensor is connected to the ADC converter on the appropriate

controller (Atmega 8). To allow walking on an unknown terrain robot has been provided with a contact sensor microswitch on each leg foot-tip. This gives a possibility to detect obstacles laying on the ground and to change a gait to keep robot balance and follow a desired trajectory. Moreover, the robot platform is equipped with a double-axis accelerometer (here working as robot platform inclinometer) plus a gyroscope (platform orientation sensor). The above mentioned sensors allow to calculate robot orientation in three axes. According to this data robot can modify platform orientation while climbing or walking over obstacles of some considerable height.

V. SIMULTANEOUS CONTROL OF SERVOS AND MOVEMENT COORDINATION

A. HS-645MG servo as the control object

The manufacturer of servos states the requirements for signal in a very general way:

- voltage has to be a square wave with peak-peak voltage $3 - 5\,V$.
- the length of the impulse has to be $0.9 - 2.1\,ms$ with neutral point at $1.5\,ms$
- the refresh rate is $50\,Hz$ (a pulse has to be updated every $20\,ms$)

The custom program for Atmega 8 microprocessor (joint controller) was written to meet these requirements. Two Timers of Atmega 8 were used to that goal. Timer0 was used to provide a refresh rate of $50\,Hz$, and the Timer1 for generating proper steering impulses.

Each cycle of $20\,ms$ is divided to 3 equal parts. Each of them has a length of $6.94\,ms$. The period isn't exactly $20\,ms$ but it depends on Timer settings. The longest possible steering impulse for a servo is $2.1\,ms$ so it is 3 times shorter than the steering window required. The exact method of controlling the single servo will be explained below. Each servo has an operating angle of $180°$. The impulse length ranges from $0.9 - 2.1\,ms$. Thus the range $2.1 - 0.9 = 1.2\,ms$ divided by $180°$ gives $6.7\,\mu s/°$ resolution constant. To obtain $90°$ motion of the joint the impulse of the length $0.6\,ms$ should be generated in microprocessor ($0.6\,ms = 90° \cdot 6.7\,\mu s/°$). Position control algorithm is embedded in servo controller.

Unfortunately, the estimated constants are not identical for all servos. They differ because of mechanical inaccuracy or particular feedback potentiometer. The result is that the positioning of the kinematic chain (leg) with 3DOFs doesn't meet the expected accuracy. To eliminate the resulting lack of repeatability it is necessary to calibrate every used servo separately and to determine its proper I/O characteristics(rotation angle/steering pulse width). This procedure allows to compensate for individual inaccuracies of servos and to achieve proper positioning of the leg.

While having the servos properly calibrated we are about to face the problem of simultaneous control of all servos. To achieve this the layered scheme of control system was used.

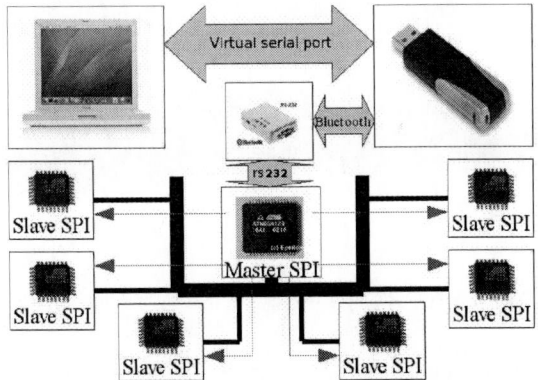

Fig. 5. Control system architecture of the robot.

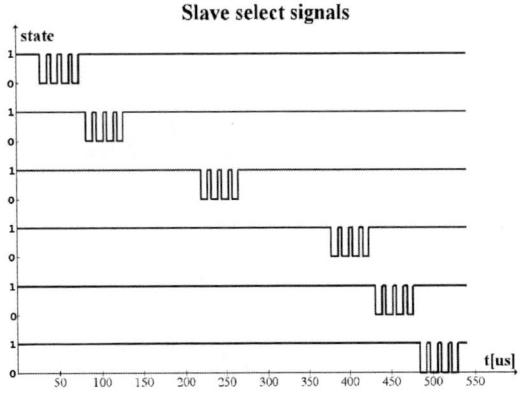

Fig. 6. Slave select signal from master SPI processor.

The general architecture scheme of the control system is in figure 5.

B. Information flow

The between-layer communication mode is bidirectional but the data frames exchanged are different for each direction of communication.

- sending reference values mode - communication starts at the host computer. Data frame is sent to the Master SPI. The frame consist of: 1 byte - a flag, 18 bytes - for holding reference values data. The Master SPI divide then the frame into 6 pieces (3 - leg - reference - value data) and sends it to the Slave SPI(leg controller).
- receiving sensors data mode - communication starts at the host computer. It sends the frame with the appropriate flag. The Master SPI collects sensors data from the Slave SPI and edits a frame to send it to the host computer. The frame consist of 18 bytes reporting joints positions, 6 bytes for microswitches statuses, 3 bytes for the gyroscope state and 2 bytes for accelerometer state.

There are three nested feedback loops in the system. The first feedback-loop is internal to the DC servo and is used to control angular position. The second feedback-loop is closed in the leg controller. It is closed with the microswitch on the leg-tip and it is responsible for a position control of a whole leg. It was implemented in a way to stop the movement of the leg while encountering an obstacle on the ground. By allocating this feedback-loop to the leg controller a fast reaction is possible. The third feedback-loop is closed within the host computer. This is also a position control loop but this time it is the one which uses accelerometer as a feedback-sensor.

Data from the accelerometers are processed by the Master SPI microcontroller. Using raw measurement results may cause inappropriate action because of relatively high random sensor noise. To remove it a moving-average filter has been implemented on robot master controller.

Output from this simple filter is defined as:

$$y_n = \sum_{i=0}^{N} b_i \cdot x_{n-1}$$

where b_i coefficient is equal:

$$b_i = \frac{1}{N+1}$$

Because filter is implemented on a 8bit controller it is recommended to use the N being a power of 2. Experimentally we have found that window of 8 samples is sufficient for a good suppression of the measurement noise. Filtered values are then sent to the host computer by using a Bluetooth channel. After computations taking into consideration the current platform orientation, reference control signals are fed back to the robot controller. These operations take approximately $0.1\,s$. Closing the loop about ten times per second is not sufficient if we want to keep the robot under control in dynamic states (eg. to stabilize robot while it is standing on two middle legs only) but it is sufficient for stabilizing robot platform in quasi-static gait while walking on a rough terrain.

C. Timing interdependencies

Once the Master SPI processor on the robot board receives a desired data, it sends it immediately to the leg controllers(Atmega 8). Sending reference values process sequence is shown in figure 6.

There are six slave select lines connecting master processor with leg controllers. Whenever main processor wants to send data to the particular leg controller it pulls a corresponding slave select line down to enable data transfer. As we can see in figure 6 the transmission could be delayed because of more important action currently taking place. For instance, after sending data to the controllers number 2 and 3, the main processor has to serve a more important task e.g. communication with the host. After completion it resumes sending of data. If we consider the worst case, which is while sending the eighteen desired reference values to the leg controllers - the delay can take about $0.85\,ms$. But when there is no

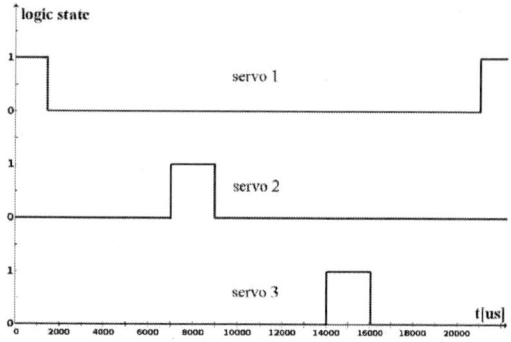

pulses send from leg controller to servo T=21042us

Fig. 7. Pulse signal send from leg controller to servo.

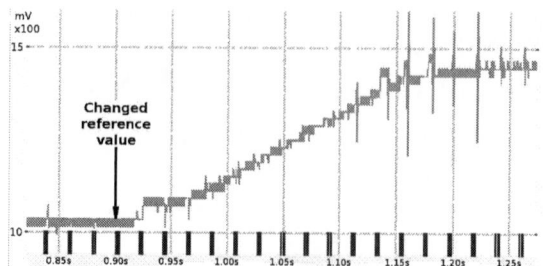

Fig. 8. Motor angle and pulse signal with reference value.

interrupt it takes only $0.3\,ms$. At the same time master processor receives data from the leg controllers including measured leg configuration and foot switch state.

Whenever leg controller receives a vector of new desired reference data for leg-joint servos, it converts it to the pulse lengths. It sends the desired reference values at frequency $47,5\,Hz$. In this time it manages to pass over three desired reference values to leg drives. Signal forms are shown on figure 7. In figure 8 we can see the angle trajectory plot after receiving the new reference value. In the background a frequency of up-dating a desired value is shown.

We consider the worst case situation, which is the delay in action between the first and the eighteenth drive equal up to $22\,ms$. It results mainly from the limited maximum achievable control frequency for servo drives. This value is acceptable for walking robot control as within such a delay time the first servo makes only a small move, being still within the limit of the gear backlash.

VI. GAITS IMPLEMENTED IN THE ROBOT "RAGNO"

The robot has been programmed to allow for three types of gaits.

- tripod gait - it is the fastest, statically stable gait for a six-legged robot. The gait is statically stable while in every moment of the movement the center of gravity lays inside of the support triangle, whose vertices are contact points of three legs currently supporting the robot. Every cycle of gait consist of two phases. First we move front and back leg on one robot side with middle leg on the other side (one triangle). Second - the other three(second triangle).

- wave gait - it is the slowest type of gait used for a six-legged robot. In each phase the robot moves only a single leg. It starts on one side from back to front of the robot, then on the other side from back to front. Normally, movement of the whole robot body is made after moving five legs. But in such a case the gait is little bit awkward. Therefor the robot uses a somehow modified type of movement. During single phase of the gait robot moves 1/6 of the road passed

for the whole gait period. This solution makes the movement of a robot more stable.

The implemented gaits allow to walk in every direction and to set any desired position and orientation of the robot platform on a flat surface. Every kind of walks have been parameterized in the control program. Due to it the robot can walk slantwise and the rhythm of the gait could be changed instantaneously.

VII. ENERGY EFFICIENCY ASPECTS

The robot uses a DC power supply of $7.4\,V$ while working on a cable in laboratory. For the outdoor tasks the robot uses five rechargeable NIMH batteries of the total capacity of $2700\,mAh$. The average current supplied is approximately $6.3\,A$. Thus theoretically the robot should work on batteries for about 25 minutes. In reality it can work only for 15 minutes, because during its work the voltage of batteries drops. When it reaches a certain threshold value the robot doesn't work properly. The drives are still able to work properly while microprocessors send inappropriate commands causing robot malfunctioning. The power supply for the controller and for the drives should be separate. A typical power consumption plot during a walk and while actively maintaining the robot in steady posture can be seen in figure 9.

VIII. CONCLUSIONS

The most challenging part of the project was to achieve the synchronization of drives. This was achieved by using the proposed hierarchical control architecture. Division of the control tasks into layers eased the simultaneous work of all 18 servomotors. Moreover the fast and reliable interconnection was established between layers in order to allow the appropriate data exchange. The properly working drive controller was connected to the robot sensing system which allowed the coordinated movement of the robot, and the rapid behavioral reactions to the environment. The results of measurements, especially those related to timing, demonstrated that a low cost robot based on popular servos while provided with the proposed distributed control architecture is able to achieve a good kinematical performance.

IX. FUTURE WORK

There are plans for a new project of the robot to overcome the limitations of the here described design. The

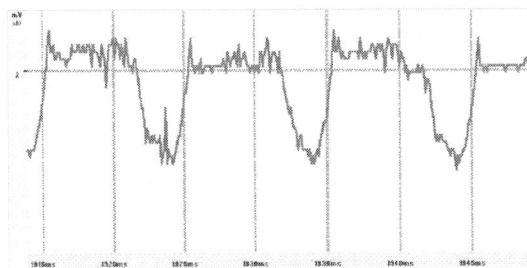

Fig. 9.   Current while walking.

[8] D. Spenneberg, K. McCullough, F. Kirchner, Stability of walking in a multilegged robot suffering leg loss Robotics and Automation, 2004. Proceedings. ICRA '04. 2004 IEEE International Conference Volume 3, Page(s):2159 - 2164 Vol.3, Apr 26-May 1, 2004

robotics servomotors with better parameters will be used. These servomotors have a built-in adjustable cascade PID controller and the direct access to the internal drive variables such as position, current, voltage. This accessability will allow a better and more robust distributed control of the whole system. Based on the experience acquired during the work with commercial drives the future plans are of building a proper servomotor in-home. It will be based on Brushless DC motor to achieve better performance than the available servomotors based on brushed DC motors. Moreover MEMS technology will be used to provide a direct access to the state variables of the new servo. Sensorless control concepts will be also tested in terms of availability (timelines) and accuracy (noise immunity) for the new servo. The problem on its own is the more involved use of exteroceptive on-board sensors such as camera, range-finder, GPS, or ground contact force sensors. Here the perspective is the augmented autonomy of the robot due to its self-localisation and obstacles detection faculties. An important issue is the gait adaptation as a function of the physical properties of the ground.

REFERENCES

[1] P. Arena, P. Di Giamberardino et al., Toward a Mobile Autonomous Robotic System for Mars Exploration, ASI International Workshop "Exploring Mars Surface and its Earth Analogues", Journal of Planetary and Space Science, Vol.52, N. 1-3, pp. 23–30, 2004

[2] K. Berns, W. Ilg, M. Deck, J. Albiez, R. Dillmann, Mechanical construction and computer architecture of the four-legged walking machine BISAM Mechatronics, IEEE/ASME Transactions on Volume 4, Issue 1, Page(s):32 - 38, March 1999

[3] E. Celaya, J. Luis Albarral, Implementation of a hierarchical walk controller for the LAURON III hexapod robot, in Proc. 6th International Conference on Climbing and Walking robots (Clawar 2003), Catania (Italy), Page(s): 409-416, 2003

[4] M. Fujita, Digital creatures for future entertainment robotics Robotics and Automation, 2000. Proceedings. ICRA '00. IEEE International Conference on Volume 1, Page(s):801 - 806 vol.1, 24-28 April 2000

[5] B. Klaassen, R. Linnemann, D. Spenneberg, F. Kirchner, Biomimetic walking robot SCORPION: Control and modeling, Proceedings of the ASME Design Engineering Technical Conference, Vol. 5, Page(s):1105-1112, 2002

[6] M. H. Raibert, H. B., Jr. Brown, M. Chepponis, Experiments in balance with a 3D one-legged hopping machine. International J. Robotics Research 3, Page(s):75-92, 1984

[7] Y. Sakagami, R. Watanabe, C. Aoyama, S. Matsunaga, N. Higaki, K. Fujimura, The intelligent ASIMO: system overview and integration, Intelligent Robots and System, 2002. IEEE/RSJ International Conference Volume 3, Page(s):2478 - 2483 vol.3, 24-28 April 2000

Optimization of Starting Process of the Frequency Controlled Induction Motor

I.Ya. Braslavsky, A.V. Kostylev, D.P. Stepanyuk

Ural State Technical University
620002, Mira str. 19, Ekaterinburg, Russia.
Tel.: +7 (343) 3754566
Fax: +7 (343) 3754566
E-Mail: braslav@ep.etf.ustu.ru

Abstract—The article deals with optimal control problem of induction motor during starting process. The method of genetic algorithm is offered as method for solution of the problem. Application of this method in order to form starting process of the induction squirrel-cage motor is examined.

Keywords—Induction motor, Optimal control.

I. INTRODUCTION

Frequency controlled induction motor drives are extremely widespread nowadays. On this reason the optimal control of such drives in wide sense is very urgent and important. As it is well known, every optimal system is designed to reach an extreme value of certain criterion which defines quality of the system. The choice of this criterion is determined by the function of electrical drive and its operating conditions. Thus all electrical drives can be divided into two very different groups. The first one is electrical drives which mainly operate under static modes with constant or slowly varying load torque. Electrical drives operating under transient modes with rapidly varying control and perturbation action belong to the second group.

Optimal control of induction motor under static modes of operation now is better investigated in comparison with transient modes [2,3] which are difficult for research and have many variants of problem definition.

II. PROBLEM DEFINITION

The problem of optimal control under transient modes (e.g. starting mode) can be presented in the following way. Technological object working with induction drive must realize technology task, which is defined by the aim of technology control. To realize this aim it is necessary to have proper control system. Optimal control problem can be illustrated by the scheme in Fig. 1.

In fig. 1 following symbols are used:

$\mathbf{X_t}=(x_{t1}(t),\dots, x_{tn}(t))$ – control aim of the technological object;

$\mathbf{X_a}=(x_{a1}(t),\dots, x_{an}(t))$ – actual states of the technological object;

$\mathbf{F}=(f_1(t),\dots, f_n(t))$ – perturbation action;

$\mathbf{U}=(\mathbf{U_1}\cup\mathbf{U_2})$ – control action;

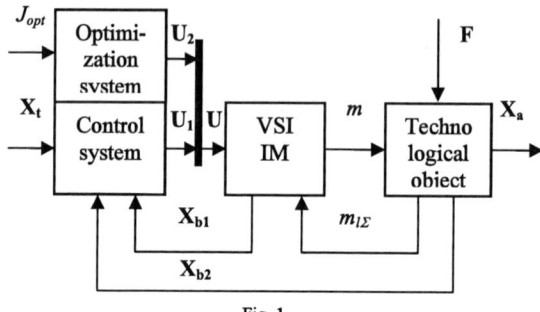

Fig. 1.

$\mathbf{X_{b1}}$, $\mathbf{X_{b2}}$ – feedbacks from IM ($\mathbf{X_{b1}}$) and technological object ($\mathbf{X_{b2}}$);

J_{opt} – optimization criteria;

M – induction motor torque;

$m_{l\Sigma}$ – total load torque.

As it is shown in Fig.1, the control system forms control action $\mathbf{U_1}$ for voltage source inverter (VSI) and induction motor (IM) system. In its turn IM forms electromagnetic torque for technological object. As it is well known, the torque of induction motor can be formed in different ways. It is possible to choose independent control action $\mathbf{U_2}$ to provide optimal control in a certain sense. The main difference in comparison with static mode is that $\mathbf{U_2}$ must depend on last and future states of induction motor. So it is necessary to form optimal trajectory of $\mathbf{U_2}$ during transient mode according to a control aim. This task is solved by optimization system.

As an independent control action we have chosen rotor flux. It is traditional way for static mode optimization. However, since the rotor flux is slow variable it is necessary to form forced stator current. This fact leads to power losses increasing. There is a problem of adequate current forcing relative to optimal power consumption. To find the optimum rotor flux trajectory during start it is possible to use the principle of extremum searching systems. Such system helps to find the optimal law for the flux and correct it in case of start conditions changing.

Taking into account high cost of electric power it is suggested to take minimum of energy losses during start

978-1-4244-1741-4/08/$25.00 ©2008 IEEE

as optimization criteria. It is obvious that considerable saving rate can be achieved in case of frequent starting of electrical drives. But in some cases optimal control during transients may be important for those drives which start not frequently. It will be shown below.

III. METHOD OF SOLUTION

As it is well known, transient modes of induction motor can be described by nonlinear differential equations and so the task of optimal control has no useful analytical solution. Therefore choice of the method of solution is very important.

The method, based on genetic algorithm [6], was proposed to solve this task. The main idea of this approach is based on the consequent processing of possible variants and the accumulation of results. Actually, it is simulation of wildlife on genetic level. The special advantage of this method is elimination of additional mathematical transformations of the object model. The number of parameters forming the optimization space is also not critical.

The utilization of this method is expected in the following way (Fig. 2). The reference flux function is divided into several nodes connected by lines (10 segments in considered case). Their coordinates (flux level and switching time) become the optimization parameters. Actually, this is the linear approximation of desirable rotor flux trajectory. Several sets of parameters are formed in a random way. In the aggregate they form the "population" (in the language of genetic algorithm).

In accordance with the method these parameters are led to Gray code. Every coded set of parameters is an "individual". After calculation process for every individual the degree of their availability is estimated. In this case the availability is considered as selected optimization criterion. The individuals corresponding the criterion in the least degree are excluded from consideration. The new individuals are generated with help of the most appropriate individuals via certain genetic operation: crossover, mutation and inversion. Every such operation is called "epoch".

The training process is rather slow. To obtain good results 8000 –10000 iterations are necessary. The calculations of transients demand long total calculation time.

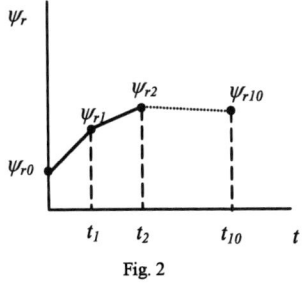

Fig. 2

Simulation have been accomplished in Matlab programme.

IV. OBTAINED RESULTS

The results of training process are shown in Fig. 3 and Fig. 4. In Fig. 3 the optimal rotor flux trajectory is shown and Fig. 4 illustrates motor starting process corresponding to this flux. The process is calculated for induction motor with rated power $P_N = 55$ kW. Time of starting process is 1 second, load torque is constant.

In Fig. 5 energy losses curves for optimal and conventional control are presented. Energy losses are calculated according to the equation:

$$\Delta W = \int \Delta P dt,$$

where: ΔW – energy losses during transients; ΔP – total power losses during transients.

Obtained energy losses decreasing is near 30%. For other motors and starting conditions energy losses decreasing is 20 – 40%. Moreover due to the optimal forcing of the rotor flux reference both magnetization time and starting time are reduced.

Obviously with other starting conditions optimal flux trajectory will be different from this one. But for practical application of the proposed method statistical data received by simulation can be used. These data are put in the optimization system. And one or another trajectory is used according to the operation conditions.

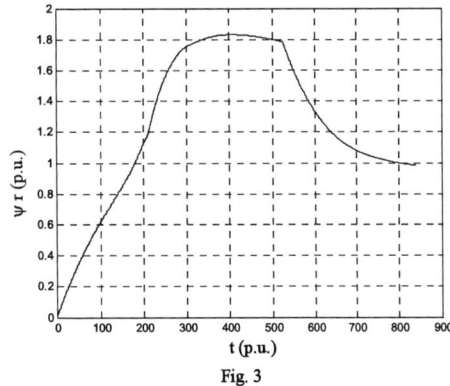

Fig. 3

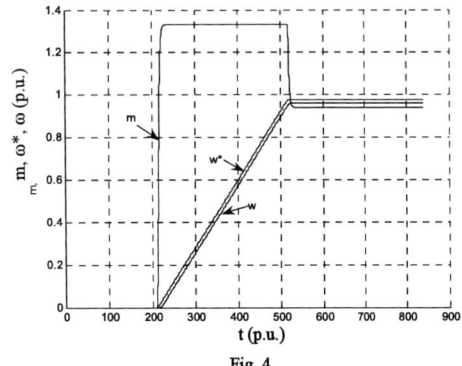

Fig. 4

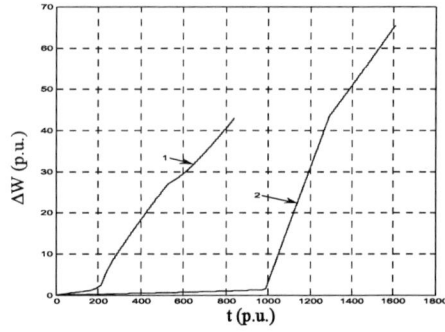

1 – energy losses curve under optimal rotor flux control with
incomplete preliminary magnetization;
2 – energy losses curve under traditional control with whole
preliminary motor magnetization.
Fig. 5

Another example of this method application is electrical drive of mill. The mill is used for pounding of building material. Pounding process takes much time (around 6-8 hours). Therefore time of starting process is short and energy optimization in it is senseless. But there is one reason to use optimal control for transient mode. The feature of this drive is great starting load torque which can reach value three times higher the rated torque. On the other hand in steady state load torque is not big and may be lower than rated torque. In this case the choice of converter is a problem. If converter is chosen in accordance with static modes of operation the converter will not be able to realize the start. The choice in accordance with transient modes leads to overstating of power rating. Therefore it is suggested to optimize starting process to reduce installed capacity of equipment. But optimization criteria will be equal $\int I^2 dt$ which define converter overheating in transient mode.

The task is to find optimal flux trajectory to provide minimum of the criteria. In contrast to the above situation the starting rate which defines dynamic torque is not adjusted and can be changed in genetic algorithm according to the optimization procedure. Intensity set-point device is linear.

Fig.6 and Fig.7 present obtained results. Starting time is near 4.5 seconds. The current is shown in Fig.8

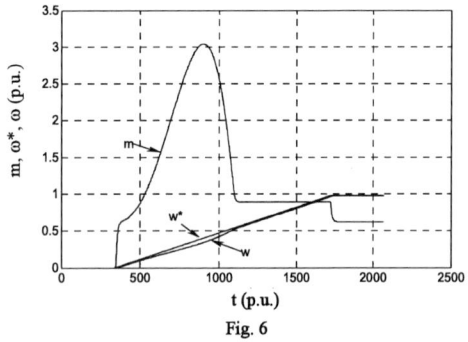

Fig. 6

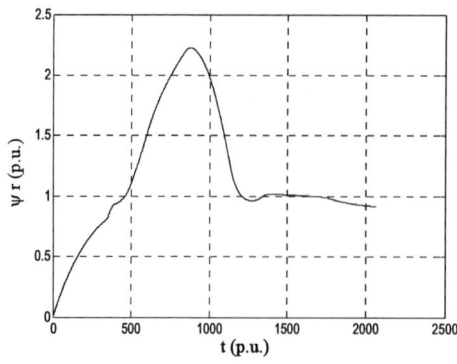

Fig. 7

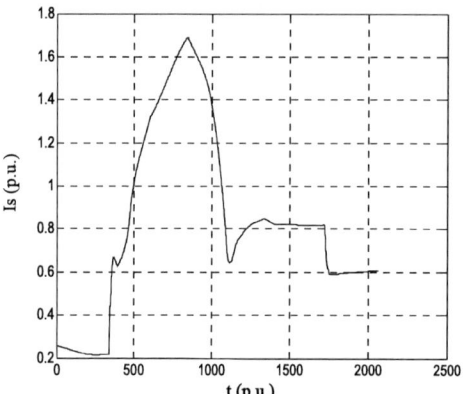

Fig. 8

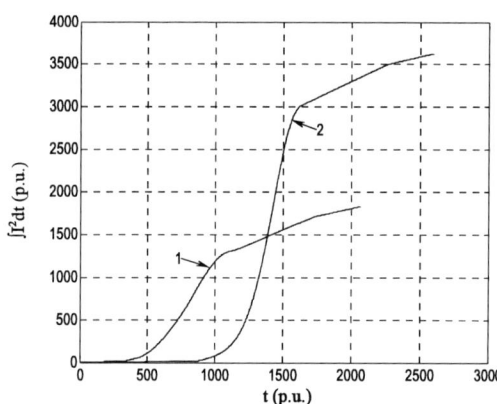

1. $\int I^2 dt$ under optimal rotor flux control with incomplete preliminary magnetization;
2. $\int I^2 dt$ under traditional control with whole preliminary motor magnetization.
Fig. 9

If starting time is also important criteria may be equal $\int I^2 t dt$. Fig. 10 - fig. 13 present obtained results. As one can see motor magnetization and start are beginning nearly simultaneously.

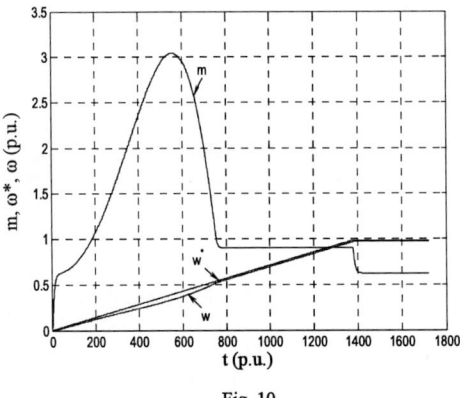

Fig. 10

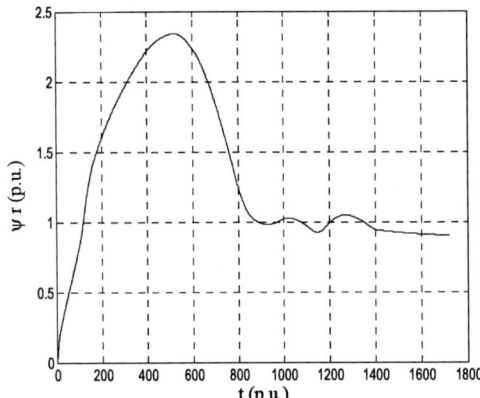

Fig. 11

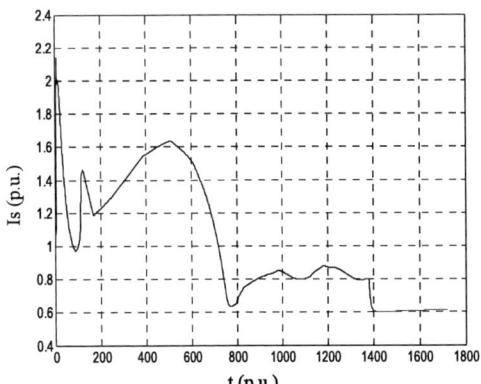

Fig. 12

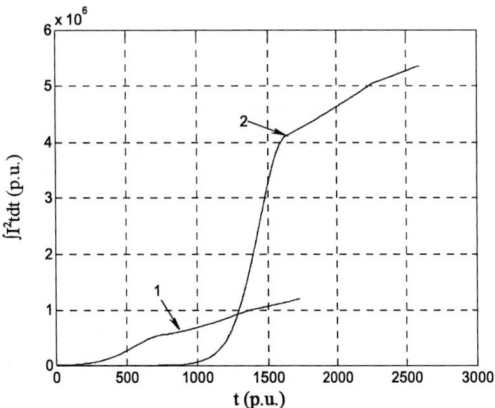

1. ∫I²tdt under optimal rotor flux control with incomplete preliminary magnetization;
2. ∫I²tdt under traditional control with whole preliminary motor magnetization.

Fig. 13

V. RESUME

- The article shows necessity of optimization under transient modes especially in starting process.
- Genetic algorithm was proposed as a solution of this task.
- Using energy losses as optimization criteria it is possible to get energy saving up to 40% under transient modes.
- Application of this method for the optimal choice is considered.

REFERENCES

[1] Braslavsky I. Ya., Kostylev A.V., Stepanuk D.P. Energy consumption optimization during of thyristor voltage convertor – induction motor system // Proceedings of symposium on power electronics, electrical drives, automation and motion. Taormina, Italy, 2006

[2] Braslavsky I. Ya. O vozmognostyah energosberegenia pri ispolsovanii reguliruemih asinhronnih electroprivodov // Electrotehnika. 1998. №8. P. 2 - 5. (rus)

[3] Polyakov V.N., Shreyner R.T. Obobshenie zadachi optimizacii ustanovivshihsya regimov electricheskih dvigateley // Electrotehnika. 2005. №9. P. 18 - 22. (rus)

[4] Braslavsky I. Ya., Kostylev A.V., Stepanuk D.P. Optimizaciya energopotreblenia pri realizacii puskovih regimov v sistemah TVC-IM // Vectnik Nacionalnogo tehnicheskogo un-ta «Chark. politech. in-t». 10, vol 1. «Problemi avtomatizirovannogo elektroprivoda». Teoria I praktika, Charkov, 2003 (rus)

[5] Kostylev A.V., Stepanuk D.P. O snigenii energopotrebleniya pri realizacii upravlyaemogo puska v sistemah TVC-IM // Electrotehnika. 2004. №9. P. 57 - 61. (rus)

[6] Geticheskye algoritmy, iskusstvennye neyronnye cety i problemy virtualnoy realnosty / G. K. Voronovsky, K. V. Machotilo, S. N. Petrashev, S. A. Sergeev. - Charkov.: OSNOVA, 1997. (rus)

3-Axes Satellite Attitude Control Based on Biased Angular Momentum

Azam Ghaedi*, Mohammad Ali Nekoui**

* Islamic Azad University , Bandarlengeh Branch, Bandarlengeh, Iran. email: azm_ghaedi@yahoo.com
**K. N. T. Univ. of Technology, Elec. Eng. Department, Tehran, Iran. email: manekoui@eetd.kntu.ac.ir.

Abstract— Satellites are used for telecommunications, telephotography and scientific experiments. In almost all of these cases, the exact orientation of satellite antennas toward the earth station is crucial. In this paper the attitude control system of the satellite will be designed in such a way that one specific point of the satellite will be always accurately oriented towards the earth station. In this way, to have the required quantity of the angular momentum, the use of momentum wheel actuator and magnetic torque generators are proposed. Three distinct modes are defined for a full positioning of the satellite to the desired direction; detumbling mode, spin-up of momentum wheel mode and the normal mode. In the first, the target is to decrease the angles ϕ and ψ and their time rate $\dot{\phi}$ and $\dot{\psi}$. This will be done by designing a PD controller in roll and yaw directions. The control torque is applied to the satellite body to achieve the desired control. In the second mode, the momentum wheel speed will be increased so high that the changes in ϕ, θ and ψ and their time rate $\dot{\phi}, \dot{\theta}$ and $\dot{\psi}$ are minimized and this gives the requested stability in the pitch direction. In the normal mode, the satellite should be oriented towards earth and hold its direction, during which ϕ, θ and ψ should be zero.

Keywords—angular momentum, attitude control, momentum wheel, Momentum biased satellite, 3-axes stabilization.

I. INTRODUCTION

Three modes are normally defined while a satellite stabilizes itself towards a specific direction; detumbling, spin-up and steady state. In the detumbling mode, ϕ, ψ and also $\dot{\phi}, \dot{\psi}$ should be decreased. By designing a PD controller in both roll and yaw direction, a control torque will be generated and applied to the satellite body, in order to achieve the desired coordinates. In spin-up mode, the speed of momentum wheel should also be high and the ϕ, θ and ψ and the time rate of changes $\dot{\phi}, \dot{\theta}$ and $\dot{\psi}$ should be at its minimum values so that the satellite will be stabled at its pitch axis. When this mode terminates, the satellite has to face the earth and be stabilized in this status. This is called the normal or steady state mode in which all of the three ϕ, θ and ψ angles should converge to zero.

The major subsystems of a satellite consist of the communication unit, energy management, architecture, heat control, electronic devices and the satellite attitude control. In the last one, the aim is to orient the satellite body to a specific reference attitude with the possibility of attitude maneuver control to go from a predetermined attitude to another desired one. Various stabilization methods are normally used, among which spin stabilization, 3-axes stabilization and gravity gradient stabilization can be named. In the first which is also the simplest one, the angular momentum vector is constant, which by its gyroscopic effect resists against rotation change.

In 3-axis stabilization method, the desired orientation for each of 3 axes is defined and the stabilization will be attained by using control components. The corresponding subsystem is much more complicated than the spin stabilization one. In this method which has been chosen in this paper, very accurate actuators and sensors are needed.

In gravity stabilization, two control booms are used in both sides of the satellite. The centralized mass at the ends of a long and light membrande will help to distribute inertial moment of the mass properly. The accuracy in this method is significantly low.

Angular momentum actuator converts only the internal angular momentum of the satellite so that the total angular momentum of the system is unchanged. In attitude control system, different actuators such as momentum wheels, thrusters, reaction wheels, control moment gyros and magnetic torque generators are used. We have chosen magnetic wheel with magnetic actuators here because this will be able to apply angular momentum almost constantly to the satellite body. If the attitude axes are parallel with the rotation axes of the momentum wheel, the maximum torque will be delivered to the satellite body. The angular momentum delivered by these wheels will be between 1 N.m sec to 300 N.m.sec.

II. ATTITUDE DYNAMIC EQUATIONS

Dynamics equations of attitude will be obtained from Euler's equation [4,5].

$$\bar{M} = \bar{\dot{h}}_I = \bar{\dot{h}}_B + \bar{\omega} * \bar{h} \qquad (1)$$

(subscript "*I*" indicates derivative in the inertial frame, while Subscript "*B*" indicates derivative in the rotating body frame.) $\bar{M}$ is the total external moment acting on the body, which is equal to the inertial momentum change of the system. A more practical form of this equation is [4,5]:

978-1-4244-1741-4/08/$25.00 ©2008 IEEE

$$\overline{T} = h_t = h + \omega * h \quad (2)$$

Where $\overline{M}$ has been replaced by $\overline{T}$ and in which h denotes differentiation of h in the body frame and is the angular momentum vector of the rigid body and $"\omega"$ is the angular velocity vector[1]. In other words,

$$\overline{h} = \overline{I}.\overline{\omega} = \begin{bmatrix} I_{xx} & -I_{xy} & -I_{xz} \\ -I_{yx} & I_{yy} & -I_{yz} \\ -I_{zx} & -I_{zy} & I_{zz} \end{bmatrix} \begin{bmatrix} \omega_x \\ \omega_y \\ \omega_z \end{bmatrix} \quad (3)$$

Where $"I"$ is the inertia matrix. The torque T can be divided into two principal parts: T_c, the control moments to be used for controlling the attitude motion of the satellite; and T_d, the moments due to the different disturbing environmental phenomena. The total torque vector is thus:

$$T = T_C + T_d. \quad (4)$$

The momentum of the entire system will be divided between the momentum of the rigid body $h_B = \begin{bmatrix} h_x & h_y & h_z \end{bmatrix}^T$ and the momentum of the moment exchange devices $h_w = \begin{bmatrix} h_{wx} & h_{wy} & h_{wz} \end{bmatrix}^T$. Finally, $\overline{h} = \overline{h}_B + \overline{h}_w$. With these definitions, the general equations of motion become[1,2,3]

$$\overline{T} = \overline{T}_c + \overline{T}_d = \left[h_x + h_{wx} + (\omega_y h_z - \omega_z h_y) + (\omega_y h_{wz} - \omega_z h_{wy}) \right] \vec{i}$$
$$+ \left[h_y + h_{wy} + (\omega_z h_x - \omega_x h_z) + (\omega_z h_{wx} - \omega_x h_{wz}) \right] \vec{j}$$
$$+ \left[h_z + h_{wz} + (\omega_x h_y - \omega_y h_x) + (\omega_x h_{wy} - \omega_y h_{wx}) \right] \vec{k} \quad (5)$$

Equation (5) summarizes the full attitude dynamics that must be implemented in the complete six degrees-of-freedom (6-DOF) simulation necessary for analyzing the attitude control system.

III. LINEARIZED EQUATIONS OF ATTITUDE DYNAMICS OF MOTION

When we work with the principal axis, the inertia cross terms will be eliminated. Moreover, the angular motion can be approximated by its infinitesimal angular motion. Then with small Euler angles and angle derivatives we will have:

$$\begin{cases} h_{wx} = I_{wx}.\omega_{wx} \\ h_{wy} = I_{wy}.\omega_{wy} \\ h_{wz} = I_{wz}.\omega_{wz} \end{cases} \quad (6) \quad , \quad \begin{cases} \omega_x = \phi - \omega_o \psi \\ \omega_y = \theta \\ \omega_z = \psi + \omega_o \phi \end{cases} \quad (7)$$

Here, we will also be concerned with the momentum biased satellite, so a constant momentum bias of h_{wyo} is applied along with the Y_B axis to give stability about

Y_B axis of the satellite. With the assumption, Eq. 5 together with Eq.7 becomes the desired linearized attitude dynamics equations of motion [1,4].

$$\begin{cases} \overline{T}_{cx} + \overline{T}_{dx} = \omega_o h_w + h_w \psi + \omega_o h_w \phi + I_{xx} \dot{\phi} + 4\omega_o^2 (I_{yy} - I_{zz}) \phi - \omega_o (I_{xx} - I_{yy} + I_{zz}) \psi \\ \overline{T}_{cy} + \overline{T}_{dy} = -h_w \quad + I_{yy} \ddot{\theta} + 3\omega_o^2 (I_{xx} - I_{zz}) \theta \\ T_{cz} + \overline{T}_{dz} = \omega_o h_w - h_w \phi + \omega_o h_w \psi + I_{zz} \ddot{\psi} + \omega_o^2 (I_{yy} - I_{xx}) \psi - \omega_o (I_{xx} - I_{yy} + I_{zz}) \phi \end{cases} \quad (8)$$

In Eq. 8, h_{wx}, h_{wy}, h_{wz} are the angular momentum components of the wheel. For a momentum biased system with a momentum wheel aligned with the satellite pitch axes, we will have[2]:

$$h_x = 0, \quad h_y = -h, \quad h_z = 0$$

It is worth to mention that the momentum wheel is in the direction of the pitch axis and the momentum vector is in opposite direction of the pitch axes.

IV. ATTITUDE CONTROL DESIGN

In 3-axes stabilization, with respect to high maneuverability and diversity of the actuators, different types of methods are in use, among which, the biased angular momentum system with a momentum wheel in the pitch axis direction has been chosen. Moreover, since the linear pitch dynamics are decoupled from the roll/yaw dynamics, their related controllers should be designed separately. Based on the simplified formerly obtained equations because of the symmetry we have; $I_{xx} = I_{zz}$ and also $T_{cy} = 0$ for uncontrolled satellite. Then the pitch loop equation controlled only by momentum biased wheel. Then, Eq. 8, simplifies to:

$$\overline{T}_{dy} = I_{yy}\ddot{\theta} - h_y \quad (9)$$

Where T_{dy} includes the external disturbance torque of the pitch axis.

Pitch error is controlled with the torque that is proportional to this error and its rate. Thus, control torque with angular momentum change of wheel is as follows:

$$h = -k_\theta \left(\tau_\theta \theta + \theta \right) \quad (10)$$

τ_θ will be found through the equation $\tau_\theta = 2\sqrt{I_{yy}/k_\theta}$ and the control parameters will be calculated through the proposed block diagram of the pitch loop, which will be as follows[2]:

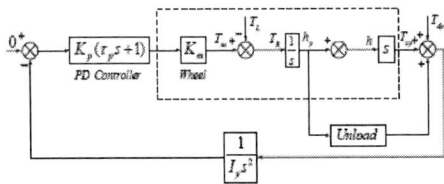

Fig1.   Block diagram of the pitch loop

The attitude control of roll/yaw loop will be done through

magnetic actuators in which the mechanical torque required for attitude control is generated by the magnetic interaction between the geomagnetic field and magnetic torque makers. Here magnetic torque makers can be used for attitude and also for wheel momentum unloading [6]. Then we have:

$I_{xx} = I_{zz}$ and $T_{cx} = T_{cz} = 0$.

The control torque from the PD controller is written as:

$$\begin{cases} T_{cx} = -k_p \left(T_{adx} \dot{\phi} + \phi \right) \\ T_{cz} = -k_p \left(T_{adz} \dot{\psi} + \psi \right) \end{cases} \quad (11)$$

The control gains T_{adx} and T_{adz} are found based on classic control methods for PD controller design.

V. SIMULATIONS

In this section, simulation results are shown in their operating modes. For this purpose satellite characteristics are simulated which are all equal in 3-modes and are given by:
Orbit attitude is 700 (km), mass is 200 (kg), size is 1.2m*0.8m*0.8m, orbit angle velocity 0.00105 (rad/sec), the moments of inertia are $I_{xx} = 80, I_{yy} = 82, I_{zz} = 40 \left(kg.m^2 \right)$ and disturbance torque is $5*10^{-5}$(N.m). It is noted that some of the initial values are changing for each mode[6].

-Simulation of PD controller in detumbling mode
Initial values in this mode and spin-up mode are the same and are given by:

$$\begin{cases} \phi(0) = 5^\circ, \theta(0) = 10^\circ, \psi(0) = 8^\circ \\ \dot{\phi}(0) = \dot{\theta}(0) = \dot{\psi}(0) = 3^\circ \end{cases}, \begin{cases} k_p = 0.001 \\ T_{adx} = 2\sqrt{I_{xx}/k_p} \\ T_{adz} = 2\sqrt{I_{zz}/k_p} \end{cases}$$

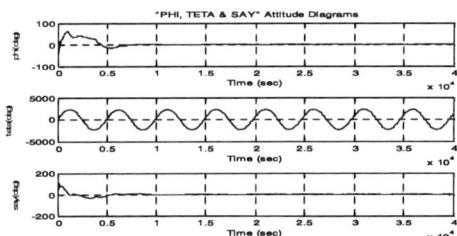

Fig 2.Diagrams of attitude

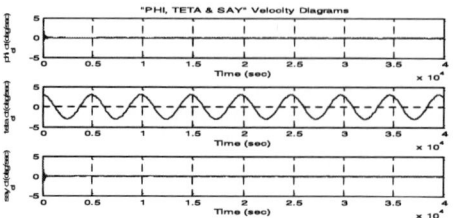

Fig 3. Diagrams of velocity

Fig 4. Diagrams of control torque makers

-Simulation of PD controller in spin up of momentum wheel mode
In this mode satellite has been stabilized on pitch axis. A switch will be used in this mode that will be activated after 1500 sec. and then the control phase will be started. As diagrams show ϕ, ψ are decreasing in start up mode and it is the same for θ angle and $\dot{\theta}$ in spin-up mode. In this mode we will have:

$k_\theta = 0.001$ and $\tau_\theta = 2\sqrt{I_{yy}/k_p}$

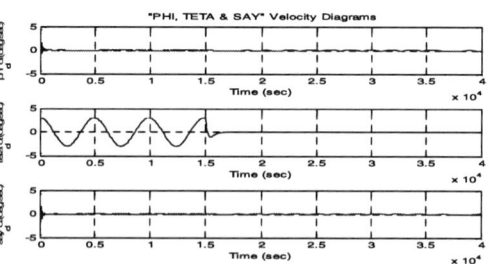

Fig5. Diagrams of attitude

Fig6. Diagrams of velocity

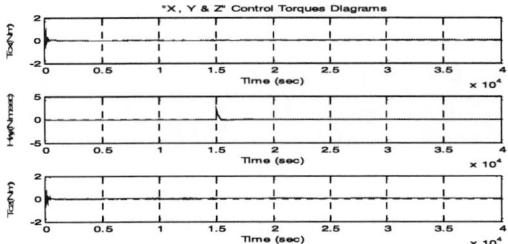

Fig7. Diagrams of control torque makers

-Simulation of PD controller in normal (steady state) mode
In this mode with using a momentum wheel on pitch axis and 3-magnetic torque makers on roll, pitch, and yaw axes must be zero ϕ, θ, ψ .
By applying controller parameters and assuming the best initial value in this mode, will have:

$$\begin{cases} \phi(0) = 2°, \theta(0) = 0.10°, \psi(0) = 3° \\ \dot{\phi}(0) = 2.4°*10^{-2}, \dot{\theta}(0) = 4°*10^{-4}, \dot{\psi}(0) = 1°*10^{-3} \end{cases} \begin{cases} k_p = 0.001 \\ k_d = 1 \end{cases}$$

T_{adx} and T_{adz} give similar to Detumbling mode.

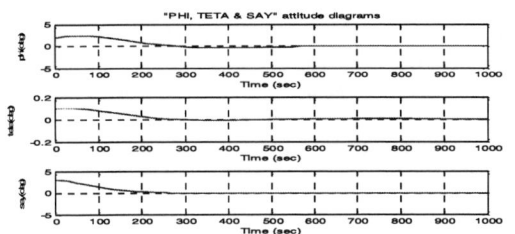

Fig 8.Diagrams of attitude

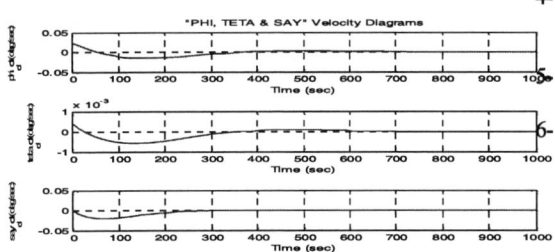

Fig 9. Diagrams of velocity

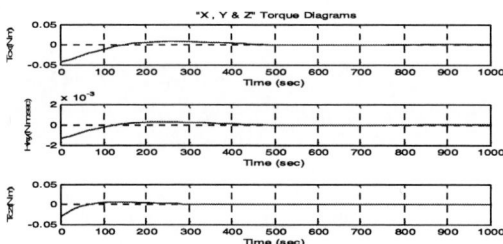

Fig 10. Diagrams of control torque makers
The same has been done for the other two modes to verify the exact function of the designed controllers.

VI. CONCLUSION

The oscillatory behavior of ϕ, θ, ψ and their rates of changes were improved, to mention better time-response, higher sustainability and decrease in system error.
Use of PD/PID control method is proposed as powerful tools to achieve better results.

REFERENCES

1- Christie schrodenger, "Guidance, Navigation and control of a spacecraft", December 1994.
2- Jianting Lv, Guangfux Ma, Dai Gao, "Bias Momentum Satellite Magnetic Attitude Control Based On GA", Department of control science and engineering, Harbin Institute of Technolgoy, Haribin 150001.
3- J. Chrone, K. Daugherty and others, "Dynamics and Control Fuctional Division Report", November 18, 2004.
4- Sidi, Marcel. J, "Space craft Dynamis and Control", A practical Engineering Approach, Combridge University Press, 1997.
5- Wertz, J.R. "Spacecraft Attitude Determination and Control"., Kluwer Academci Publishers, 1998.
6- Goal, P.S. and kudva. P, "A Delayed Pulse Roll/Yaw controller for a Momentum Biased spacecraft", IFAC, Automatic control In space, 1982.

Modelling and simulation of a signal injection self-sensored drive

Alen Poljugan, Mark Sumner, Chris Gerada, Qiang Gao

School of Electrical Engineering, University of Nottingham, Nottingham, UK, NG7 2RD
eexap5@nottingham.ac.uk, eezms@exmail.nottingham.ac.uk, eezcg@exmail.nottingham.ac.uk, qiang.gao@nottingham.ac.uk
Unska 3, 10040 Zagreb, Croatia, fax. 0038516129705, tel. 0038598694182, eexap5@nottingham.ac.uk

Abstract- **This paper presents a new induction motor model specifically developed for simulating self-sensored variable speed drives which employ signal injection based sensorless control techniques. The model is to be used to investigate improved filtering algorithms and closed loop speed controllers. The model employs a simple state space equivalent circuit based model of the induction machine, which is enhanced to include rotor slotting and main flux saturation effects. The improvement is obtained by including a variation of machine inductances variation with the rotor position and the flux position. The new model is verified against experimental results.**

Keywords- Electrical machine, Induction motor, Adjustable speed drive, Self-sensing control, Simulation

I. INTRODUCTION

In signal injection self-sensored drives, position and speed estimation is obtained by tracking the machine's response to high frequency (HF) voltages superimposed onto the normal fundamental voltage. A position dependent signal caused by the slotting of the rotor can be extracted from the HF current waveforms. However saturation effects in the machine and non-linear inverter effects such as dead time can add significant distortion to the estimated position signal [1,2,3] which can significantly impair the response of the drive.

To enhance estimation accuracy, different disturbance compensation methods can be used, such as harmonic and sideband filters [4], compensation tables [2-4], adaptive disturbance identifiers [4] etc. The main drawback of these techniques is their influence on the drive control performance, mainly observed as a reduction of the control loop bandwidth. In the commissioning process of self-sensored drives, the controller parameters need to be tuned, and this is usually achieved by trail and error. In this work the aim is to create a machine model which can be used to investigate the influence of the speed estimation algorithm (including filters etc) on the overall drive control performance with the ultimate objective being to achieve a high bandwidth speed controller.

It should be noted that the aim is not to create a physically accurate machine model. This can be achieved using finite element (FE) modelling packages, but their complexity when added to real time control algorithms means that they require very long simulation times [9]. The simplified equivalent circuit model is computationally fast but does not contain any physical detail of the machine. The Dynamic Mesh Reluctance Model (DMRM) provides a certain compromise by combining detailed machine analysis and relatively fast

computational time [7,8]. It is however, still too slow to be used for HF injection systems.

[9,10,13,14] combine a FE analysis with a phase variable model. A machine inductance variation with rotor position is obtained from FE analysis and is then incorporated as look up tables. [9,10] report a fast computation time and accurate modelling of the machine saturation, rotor slotting effect and cogging torque. A space-vector state model of an induction machine including the rotor slotting effect is described in [15]. This model considers a variation of the stator and rotor leakage inductance due to rotor and stator slotting and develops a space-vector state formulation useful for application to observers. The rotor slotting harmonic (RSH) frequency is accurately modelled but there is an amplitude mismatch when compared to experiment, and saturation effects are not included. In the winding function approach (WFA) a machine model is derived by means of winding functions where no symmetry is assumed [11,12]. All machine inductances are calculated directly from the geometry and the winding layout of the machine. The WFA shows good modelling accuracy. Its computational time is faster than FEM (2 hours vs 12 hours [11]) however it is still comparatively slow when complex realtime control is required.

The induction machine model derived here is an extension of the simple equivalent circuit model, modified to include rotor slotting and main flux saturation effects. The improvement is obtained by including a variation of machine inductance with the rotor position and the flux position. The paper is structured such that section II introduces the equations of the machine model while section III describes the key points of the modelling and implementation. In the section IV the tuning of the model parameters is presented along with a comparison of the simulation with experimental results.

II. THE INDUCTION MACHINE MODEL

The machine model proposed in this paper is derived from a general phase model of the induction machine i.e. in the a-b-c frame of reference [16]. In the following sections, the motor phase voltages and currents will be referred to as winding variables. From the equivalent circuit a general form of the equations in the a-b-c frame can be written in matrix form as follows:

978-1-4244-1741-4/08/$25.00 ©2008 IEEE

$$\begin{bmatrix} U_{abcs} \\ U_{abcr} \end{bmatrix} = \begin{bmatrix} R_s & 0 \\ 0 & R_r \end{bmatrix} \cdot \begin{bmatrix} i_{abcs} \\ i_{abcr} \end{bmatrix} + \frac{d}{dt} \begin{bmatrix} \Psi_{abcs} \\ \Psi_{abcr} \end{bmatrix} = $$
$$= \begin{bmatrix} R_s & 0 \\ 0 & R_r \end{bmatrix} \cdot \begin{bmatrix} i_{abcs} \\ i_{abcr} \end{bmatrix} + \frac{d}{dt} \left(\begin{bmatrix} L_s & L'_{sr} \\ L^T_{sr} & L'_r \end{bmatrix} \cdot \begin{bmatrix} i_{abcs} \\ i_{abcr} \end{bmatrix} \right) \quad (1)$$

where subscript s denotes stator and subscript r rotor variables.

The resistance matrices R_s and R_r as well as the inductance matrices L_s, L_r and L_{sr} in (1) can be written as follows:

$$R_s = \begin{bmatrix} R_{as} & 0 & 0 \\ 0 & R_{bs} & 0 \\ 0 & 0 & R_{cs} \end{bmatrix}, \ R_r = \begin{bmatrix} R_{ar} & 0 & 0 \\ 0 & R_{br} & 0 \\ 0 & 0 & R_{cr} \end{bmatrix},$$

$$L_s = \begin{bmatrix} L_{ls} + L_{ms} & L_{sab} & L_{sac} \\ L_{sba} & L_{ls} + L_{ms} & L_{sbc} \\ L_{sca} & L_{scb} & L_{ls} + L_{ms} \end{bmatrix} \quad (2)$$

$$L_r = \begin{bmatrix} L_{lr} + L_{mr} & L_{rab} & L_{rac} \\ L_{rba} & L_{lr} + L_{mr} & L_{rbc} \\ L_{rca} & L_{rcb} & L_{lr} + L_{mr} \end{bmatrix},$$

$$L'_{sr} = L_{ms} \begin{bmatrix} \cos\theta_r & \cos\left(\theta_r + \frac{2}{3}\pi\right) & \cos\left(\theta_r - \frac{2}{3}\pi\right) \\ \cos\left(\theta_r - \frac{2}{3}\pi\right) & \cos\theta_r & \cos\left(\theta_r + \frac{2}{3}\pi\right) \\ \cos\left(\theta_r + \frac{2}{3}\pi\right) & \cos\left(\theta_r - \frac{2}{3}\pi\right) & \cos\theta_r \end{bmatrix}$$

where R_s and R_r represent the stator and rotor resistance respectively, L_{ls} and L_{lr} are the stator and rotor leakage inductance, L_{ms} and L_{mr} are the magnetizing inductances, L_{sab}, …, L_{scb} are the stator mutual inductances and L_{rab}, …, L_{rcb} are the rotor mutual inductances. The rotor electrical angle is denoted by Θ_r. The model is implemented in the Matlab/Simulink environment.

In this basic model the rotor slotting effect is not represented and therefore it cannot be used in the simulation of a signal injection drive. In the next section this model will be extended to incorporate rotor slotting and saturation effects.

III. MODELLING OF THE MACHINE NONLINEAR EFFECTS

A. Rotor slotting and saturation effects incorporated in the stator leakage inductance

In an induction motor, a non-constant air gap permeance results from the presence of the stator and/or rotor slots, asymmetry in the stator and/or rotor, eccentricity of the rotor and/or stator and due to magnetic saturation. A general expression describing the total airgap permeance wave is quite complex and unnecessary for this work, as only speed-related terms are required. For this reason, it can be assumed that the airgap of the machine is bounded by a smooth stator and slotted rotor, and the variation of the airgap permeance is created by the rotor slot openings only. The rotor slots will produce airgap permeance waves with a spatial distribution dependent on the number used [5]. A typical permeance wave description will consist of a constant component (dc term) and a superimposed ripple (ac term). The fundamental

component of the rotor slot permeance interacts with the magnetizing component of the airgap MMF and this modulation process generates two harmonic components in the airgap flux density. The RSHs are impressed on the stator currents as a consequence of the modulation of the machine's fundamental magnetic field and the airgap permeance wave produced by the rotor slots. The magnitude of these components varies little with applied load, except in machines with the closed rotor slots. The frequencies of these harmonics components will be given by [5,6]:

$$f_{RSH} = v \cdot \frac{N_R}{pp} \cdot f_r \pm k \cdot f_e \quad (3)$$

where f_{RSH} is the rotor slot harmonic frequency, v is the RSH order, N_r is the total number of rotor slots, pp is the number of pole pairs, f_r is the mechanical frequency of the rotor rotation and f_e is the supply frequency. As can be seen from equation (3), the RSH frequencies are speed dependent and can be used to track the rotor speed [6]. In the following discussion these two components will be denoted as +1RSH and -1RSH, respectively.

To model the rotor slotting effects on the phase currents, only the stator leakage inductance in the equivalent induction motor model is considered to change in a periodic manner with rotor position. From the theoretical point of view, the rotor slotting effect will also affect other inductances in the machine [9,13,21,22] but here, for simplicity only the contribution of the stator leakage inductance was considered. The profile of the stator leakage inductance, presented in Fig. 1, is chosen following [9, 20] and also by using a DMRM simulation. The presented profile can be described with the following mathematical expression:

$$L_{lsa} = L_{ls\_DC} + k_{rs1} \sin[\frac{N_r}{pp} \cdot \Theta_r + \phi_1] + k_{rs2} \sin[(\frac{N_r}{pp}\Theta_r - pp \cdot \Theta_f) + \phi_2] \quad (4)$$

where L_{lsa} is the modified total stator leakage inductance of the phase a, L_{ls_DC} is the dc component of the leakage inductance, Θ_f is the stator flux position, k_{rs1}, k_{rs2} and Φ_1, Φ_2 are the amplitudes and phase displacements of the ac inductance components.

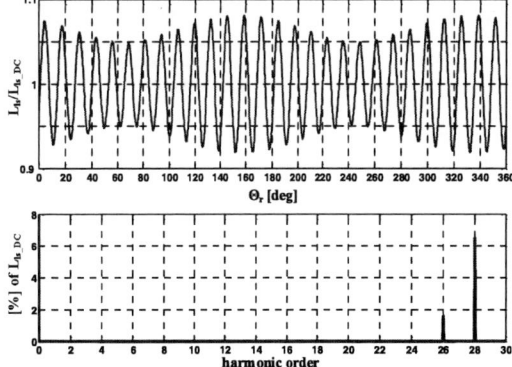

Fig. 1. Stator leakage inductance as a function of the rotor position. (top) and the frequency spectrum of the inductance profile (bottom)

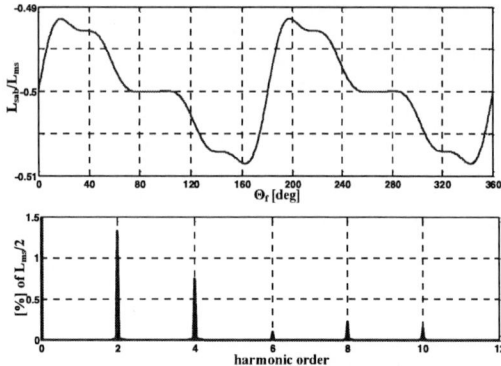

Fig. 2. Stator mutual inductance as a function of the stator flux angle (top) and the frequency spectrum of the inductance profile (bottom)

The new profile consists of a dc component superimposed with two ac components, as shown in Fig. 2. The component $28\Theta_r$ (N/pp=56/2) represents an airgap permeance wave caused by a slotted rotor, while the component $28\Theta_r$-$2\Theta_r$ represents a saturation sideband modulated by the RSH [20].

B. Saturation effect incorporated in the stator mutual inductance

Previous studies [17,18] show that saturation of the stator and rotor teeth is important as a much higher flux density exists in these locations. Since the flux density vector is rotating, the saturation regions will rotate too, modifying the tooth permeability in a cyclic manner resulting in a variation of the stator winding inductances. For simplicity, it will be assumed that only the stator mutual inductances will be modulated by the cyclic variation of tooth saturation. The authors in [17] propose a variation described with a simple mathematical equation. Although they present good results, there is only one tuning parameter available to adjust the magnitude of the saturation harmonics. For the simulation of a signal injection drive, a larger degree of fine tuning of the saturation harmonic components is desirable. Following analysis of the profiles obtained from FEM analysis

$$L_{sab} = L_{sba} = -\frac{Lms}{2}[1 + k_{s2}\sin(pp \cdot \Theta_f + \rho_1) + k_{s4}\sin(2 \cdot pp \cdot \Theta_f + \rho_2) +$$
$$+ k_{s6}\sin(3 \cdot pp \cdot \Theta_f + \rho_3) + k_{s8}\sin(4 \cdot pp \cdot \Theta_f + \rho_4) +$$
$$+ k_{s10}\sin(5 \cdot pp \cdot \Theta_f + \rho_5)] \tag{5}$$

[9,13,20,21], the following expression was chosen to incorporate the saturation effects in the stator mutual inductance:

where k_{s2}, k_{s4}, k_{s6}, k_{s8} and k_{s10} are amplitudes and ρ_1,.., ρ_5 are phase displacements of the permeance wave introduced by tooth saturation. The mutual inductance profiles $L_{sac}=L_{sca}$ and $L_{sbc}=L_{scb}$ are a similar shape. For a four-pole machine the inductance profile consists of components at 2, 4, 6, 8 and 10 Θ_r as shown in Fig. 2. The parameters of the stator leakage and mutual inductance profiles are adjusted according to experimental measurements, as presented in the next section.

IV. SIMULATION RESULTS AND EXPERIMENTAL VERIFICATION

A four-pole, 30kW delta connected machine with 56 rotor slots is modelled. The machine has non-skewed semi-closed slots, and is part of an experimental system. The basic model is extended with the stator leakage and mutual inductance

variation described in section III. The inductance variations are incorporated using (4) and (5). The simulation and experimental system also incorporated an indirect rotor field orientated control (IRFO) as well as the $\alpha\beta$ signal injection position estimation scheme described in [2-4].

A. Model parameter tuning

The parameters of the rotor slotting and saturation model are adjusted as follows. First the parameters of the rotor slotting model were tuned. The model is first simulated without the signal injection algorithm. The parameter k_{rs1} of the rotor slotting model is adjusted in such a way as to match the level of the +1RSH in the simulation to the one measured from the experimental line current. The -1RSH is present only in the phase current.

When operation with signal injection is considered, experimental investigation showed that the response of the machine model to the HF voltage does not match the experimental measurements. For the same injection voltage signal (778Hz and 37V_{peak}), it can be noticed that the current response at the HF frequency is approximately four times smaller than the experiment (1.4A compared to 5.9A). Nevertheless the relative amplitude of the RSH component modulated by injected signal is the same (approximately 3.3%). The relative amplitude is defined as the percentage of the excitation current component at the injection frequency. Although the problem with the small absolute magnitude may lead to signal processing issues, the initial results using the position estimation algorithm indicated that the absolute amplitude mismatch was not important when considering the ultimate application of the model.

Once the amplitude of the slotting harmonic has been adjusted, the saturation model can be tuned to match fundamental frequency related saturation components (i.e. 5th, 7th, etc.). However, when this method is used, the relative amplitude of the saturation components around the injection frequency will not match those observed in the experimental system. This again is due to the fact that the number of parameters adjusted in the model is kept low for simplicity, and therefore the model does not necessarily represent the true behaviour of the real machine. Non-linear frequency dependent variations neglected here include skin effect and eddy currents – to add these would over-complicate the model for the purpose intended. Therefore as this model is used specifically for investigation of signal injection techniques, the saturation model should be tuned for the response at the injection frequency.

In the position estimation process the measured line current is processed using a demodulation scheme described in [2-4] which ultimately gives the current position signals $i_{p\alpha}$ and $i_{p\beta}$, as shown Fig. 3. From the current position signals the rotor slot position is obtained using an *arctan* function or a phase locked loop. In the ideal case the current position signal spectrum will consist only of the rotor slotting component $28f_r$. However, saturation components modulated by the injected signal will appear in the current position signal and they represent a significant disturbance to the estimation process [2]. As the main aim of this research is to investigate disturbance elimination techniques and their influence on the drive control performance, the region around the injection

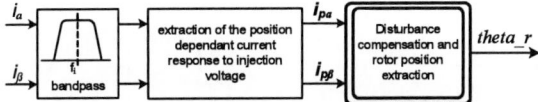

Fig. 3. Position estimation scheme in the signal injection drive

signal frequency is very important and the saturation model is adjusted according to the spectrum of the current position signal. By adjusting parameters k_{s2}, k_{s4}, k_{s6}, k_{s8} k_{s10} and k_{rs2} all the saturation harmonic components in the current position signal can be matched to experimental measurement. An adjustment of the saturation harmonic component in the current position signal will also cause a change in the harmonic spectrum around fundamental slotting harmonic +1RSH.

B. Operation around Fundamental Frequency

In the Fig. 4. and Fig. 5. results at 150 rpm and no load are presented. From the results it can be seen that parameters of the model have been adjusted such that the magnitude of the simulated +1RSH component matches the experimental results, also shown in Table I. There is slight mismatch in the amplitude of the +/-1RSH saturation sidebands because the saturation model has been tuned to match the harmonic content around the injection frequency, mentioned earlier. The biggest mismatch is in the magnitude of the 5th and 7th harmonic components. With regard to the machine model application tooth saturation is the most important saturation phenomena and core saturation in this case is neglected. By introducing a core saturation component to the model this mismatch can be reduced, at the expense of a more complicated model [17,19].

In the Fig. 6. and Fig. 7. results at 150 rpm and 30% of the nominal load are presented. From the experimental results it can be seen that loading will cause a relative amplitude decrease in all of the harmonic components considered, as shown in Table II. Although the absolute amplitude of each harmonic is increasing, their amplitude relative to the fundamental current component is decreasing. In the simulation, the magnitudes of the +1RSH component and +1RHS saturation sidebands show a tendency to decrease with load, but by a smaller amount compared to experiment. This probably results from the simplifications made in the modelling of the rotor slotting and saturation effect. The simulated fundamental saturation components (the 5th and 7th) show a tendency to increase with load whereas the opposite occurs in the experimental system as seen in Table II. Once again this can be attributed to the fact that core saturation has not been modelled.

C. Operation at signal injection frequency

The simulation model was also investigated for operation with HF voltage injection. In the drive a small magnitude (37V_{peak}) voltage signal was injected in the $\alpha\beta$ frame at the frequency of 778 Hz. The frequency spectrum of the simulated current position signal was compared with the experimental results, noting that the harmonic magnitude is expressed as a percentage of the current component at the injection frequency.

In Fig. 8. and Fig. 9. results at 150 rpm and no load are presented. The RSH component at the 28f_r and three main saturation components 2f_e, 4f_e and 8f_e match those obtained from the experiment (Table III). For the components at 10f_e and 28f_r-2f_e an amplitude mismatch can be noticed. By adjusting the parameters (k_{s10} and k_{rs2}) the amplitudes of these harmonics can be matched to experimental results but this will then create some additional harmonic components around the fundamental slotting harmonic +1RSH, namely the -1RSH-10f_e, +1RSH-10f_e and most importantly the +/-1RSH-2f_e components. As can be noticed from Table I this component is already larger than that seen in the experimental system. If for example the parameter k_{rs2} is tuned to match the 28f_r-2f_e component, the -1RSH-2f_e component will increase by up to 2% (more then half of the +1RSH), as seen in Fig. 5. A similar effect is present when the 10f_e component is tuned. Further investigation will hopefully prove that operation of the current controllers is not significantly affected by this phenomenon, and the parameters can then be tuned to match the 10f_e and 28f_r-2f_e components. In Fig. 10. and Fig. 11. results at 150 rpm and 30% load are presented. From the experimental results, the saturation components 2f_e, 4f_e and 8f_e decrease significantly with load (Table IV). In the simulation results this decrease can also be seen but it is very small. It is obvious that in the high frequency region and under the load condition the saturation model is not modelling accurately the saturation phenomenon. But from the FEM analysis of [13,14,20,21] it is clear that the magnitude of the inductance component as well as its phase displacement will vary with the load. Further investigation of the inductance variation profile can be used for additional fine tuning, either through a FE simulation or experimental testing, if required. Also it can be seen that under load two new saturation components appear at 14 and 16f_e, as shown in Fig. 9. In the simulation the magnitudes of these components are negligible. A modified saturation model was investigated which included the new harmonic components in the mutual inductance profile. The results showed that the 14 and 16f_e component can be modelled but again the tuning of these components caused additional components around fundamental RSH. As mentioned before, this only merits further investigation if a more detailed model is required.

Experimental results shown in Fig. 7 and 9, also include harmonic components at the 37, 60 and 67 Hz. The frequency of these components is not dependent on the supply frequency but the magnitudes do increase with load. The authors suspect that these components originate from asymmetry in the experimental system. The results from normal operation also showed small even harmonic components in the line current which also suggests an asymmetry problem, Fig.3 and 5.

As mentioned before, the main aim of this project was to find an induction machine model which gave a good compromise between the simulation time and modelling accuracy. The simulation model presented here is fast – 75s computation time for a 10s simulation of a DOL machine. For a 1s DMRM simulation (previously reported to be "fast"), a computation time of 946s is needed.

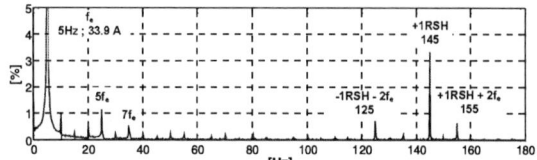

Fig. 4. Experimental results - frequency spectrum of the line current at 150 rpm and no load

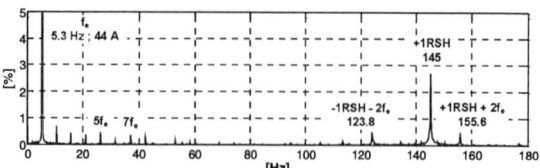

Fig. 6. Experimental results - frequency spectrum of the line current at 150 rpm and 30% of nominal load

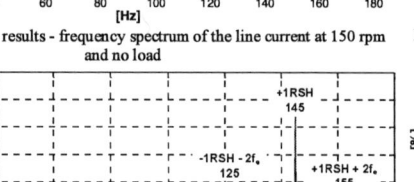

Fig. 5. Simulation results - frequency spectrum of the line current at 150 rpm and no load

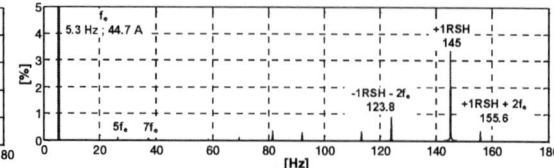

Fig. 7. Simulation results - frequency spectrum of the line current at 150 rpm and 30% of nominal load

TABLE I
HARMONIC SPECTRUM OF THE LINE CURRENT FOR NO LOAD CASE (%)

	5th	7th	+1RSH	-1RSH-2fe	+1RSH+2fe
experiment	1.13	0.48	3.35	0.72	0.63
simulation	0.06	0.07	3.37	0.98	0.39

TABLE II
HARMONIC SPECTRUM OF THE LINE CURRENT FOR 30% LOAD (%)

	5th	7th	+1RSH	-1RSH-2fe	+1RSH+2fe
experiment	0.45	0.35	2.69	0.5	0.48
simulation	0.08	0.09	3.31	0.9	0.38

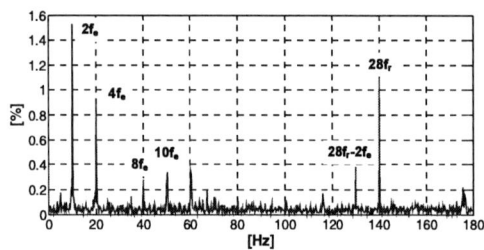

Fig. 8. Experimental results - frequency spectrum of the current position signal at 150 rpm, no load and hf injection

Fig. 10. Experimental results - frequency spectrum of the current position signal at 150 rpm, 30% of nominal load and hf injection

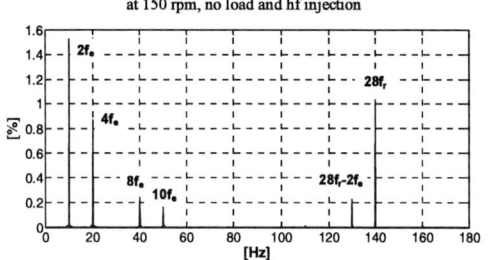

Fig. 9. Simulation results - frequency spectrum of the current position signal at 150 rpm, no load and hf injection

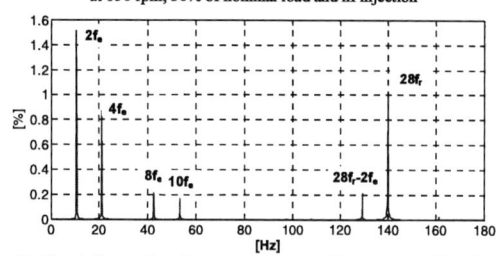

Fig. 11. Simulation results - frequency spectrum of the current position signal at 150 rpm, 30% of nominal load and hf injection

TABLE III
HARMONIC SPECTRUM OF THE LINE CURRENT FOR NO LOAD CASE AND HF INJECTION (%)

	2f$_e$	4f$_e$	8f$_e$	10f$_e$	28f$_r$-2f$_e$	28f$_r$
experiment	1.53	0.88	0.28	0.33	0.38	1.1
simulation	1.53	0.87	0.22	0.17	0.23	1.04

TABLE IV
HARMONIC SPECTRUM OF THE LINE CURRENT FOR NO LOAD CASE AND HF INJECTION (%)

	2f$_e$	4f$_e$	8f$_e$	10f$_e$	28f$_r$-2f$_e$	28f$_r$
experiment	0.84	0.22	0.32	0.42	0.3	1.12
simulation	1.51	0.86	0.21	0.16	0.21	1.02

V. CONCLUSION

This paper has presented a new induction motor model for use with developing signal injection based self-sensing induction motor control systems. A simple equivalent circuit model is improved with the rotor slotting and saturation effect.

The rotor slotting effect is incorporated as a rotor position dependant stator leakage inductance variation. The magnitudes of the rotor slotting harmonics can be adjusted to match those measured experimentally by means of tuning two parameters.

The saturation effect is incorporated as a flux position dependant component in the stator mutual inductance. The inductance profile consists of five harmonic components and the magnitude and phase displacement of each component can be adjusted. Considering the machine model application, the saturation model was tuned to match the experimental results around the injection frequency.

The results comparison shows that a relative magnitude of the rotor slotting harmonics match those obtained from the experiment for all loading conditions. The saturation model for the no load condition gives good results with a slight magnitude mismatch of two harmonic components ($10f_e$ and $28f_e\text{-}2f_e$). These components can be additionally tuned to match the experiment but this result in detuning of some harmonic components around the fundamental RSH. A further investigation will show how these components affect the current controller.

The proposed profiles of the machine inductances are not considered to be load dependant. Therefore under load conditions the saturation model shows poor results. If a more detailed model is required, variation of the inductance profiles with a load can be included (at extra computational cost).

ACKNOWLEDGMENT

The authors would like to acknowledge the support provided by the European Commission under a Marie Curie Research Training Network for realizing this work under the MEST-CT-2004-504243 Research Project.

REFERENCES

[1] J.I. Ha, S.K. Sul, "Sensorless Field-Orientation Control of an Induction Machine by High-Frequency Signal Injection", *IEEE Transactions on Industry Applications, Vol. 35, No. 1, Jan/Feb 1999*

[2] N. Teske, G. M. Asher, M. Sumner, and K. J. Bradley, "Suppression of Saturation Saliency Effects for the Sensorless Position Control of Induction Motor Drives under Loaded Conditions," *IEEE Transactions on Industrial Electronics, vol. 47, no. 5, pp. 1142 – 1150, Oct. 2000.*

[3] N. Teske, G.M. Asher, K.J. Bradley and M. Sumner, "Analysis and Suppression of Inverter Clamping Saliency in Sensorless Position Controlled Induction Machine Drives", *IEEE-IAS 2001, Proc. On CD ROM*

[4] Q. Gao, G. Asher and M. Sumner, "Sensorless Position and Speed Control of Induction Motors Using High Frequency Injection And Without off-line Pre Commissioning", *the 31st Annual Meeting of IEEE, IES 2005*

[5] A. Ferrah, P. J. Hogben-Laing, K. J. Bradley, G. M.Asher, and M. S. Woolfson, "The Effect of Rotor Design on Sensorless Speed Estimation Using Rotor Slot Harmonics Identified by Adaptive Digital Filtering Using the Maximum Likelihood Approach," *in Conf. Rec. IEEE–IAS Annual Meeting, , no. 32, pp. 128 – 135, Oct. 1997.*

[6] A. Ferrah,, K. J. Bradley, P. J. Hogben-Laing, M. S. Woolfson, G. M. Asher, M. Sumner, J. Cilia, and J. Shuli "A Speed Identifier for Induction Motor Drives Using Real-Time Adaptive Digital Filtering", , " *IEEE Transactions on Industry Applications, Vol. 34, No. 1, Jan/Feb 1998*

[7] C Gerada, K J Bradley, M Sumner and P Sewell, "Evaluation of a Vector Controlled Induction Motor Drive using the Dynamic Magnetic Circuit Model", *IEEE Transactions on Industry Applications, Vol. 43, No. 3, May/June 2007*

[8] P Sewell, KJ Bradley, JC Clare, PW Wheeler, A Ferrah, R Magill, S Sunter, "Dynamic Reluctance Mesh Modelling of Induction Motors" , *Proc ICEM'98 Istanbul, pp 1324 - 1329, Sept 1998*

[9] O. A. Mohammed, S. Liu, and Z. Liu, "Phase variable model of PM synchronous machines for integrated motor drives," *IEE Proc. Sci., Meas., Technol., vol. 151, no. 6, pp. 423–429, Nov. 2004.*

[10] O. A Mohammed, Z Liu, and S. Liu, "A novel sensorless control strategy of double-fed induction motor and its examination with the physical modeling of machines," *IEEE Transactions on Magnetics, Vol. 41, No. 5, May 2005, pp.1852-1855.*

[11] S. Nandi, "Modeling of Induction Machines Including Stator and Rotor Slot Effects" *IEEE Transactions on Industry Applications, vol. 40, no.5, Sep/Oct. 2004.*

[12] S. Nandi, "A Detailed Model of Induction Machines With Saturation Extendable for Fault Analysis" *IEEE Transactions on on Industry Applicationss, vol. 40, no.5, Sep/Oct.. 2004.*

[13] N.A. Demerdash, P. Baldassar, "A combined finite element - state modeling environment for induction motors in the abc frame: the no-load condition", *IEEE Transactions on Energy Conversion, Vol. 7, No. 4. Dec. 1992.*

[14] P. Baldassari, N.A. Demerdash, "A combined finite element - state modeling environment for induction motors in the abc frame: the blocked rotor and sinusoidally energized conditions", *IEEE Transactions on Energy Conversion, Vol. 7, No. 4. Dec 1992.*

[15] M. Cirrincione, M. Pucci, G. Cirrincione, A. Miraoui, "Space-Vector State Model of Induction Machines Including Rotor and Stator Slotting Effects", *Electric Machines & Drives Conference, 2007. IEMDC '07. IEEE International Volume 1,May 2007*

[16] S.E. Lyshevski, *Electromechanical Systems, Electric Machines and Applied Mechatronics.* CRC Press, Boca Raton, Florida, 1999.

[17] V. Donescu, A. Charette, Z. Yao, and V. Rajagopalan, "Modeling and simulation of saturated induction motors in phase quantities," *IEEE Trans. Energy Conversion, vol. 14, pp. 386–393, Sept. 1999.*

[18] J. C. Moreira and T. A. Lipo, "Modeling of saturated AC machines including air gap flux harmonic components," *IEEE Trans. Ind. Applicat., vol. 28, pp. 343–349, Mar./Apr. 1992.*

[19] J.O. Ojo, A. Consoli, T.A. Lipo, "An Improved Model of Saturated Induction Machines", *IEEE Transactions on Industry Applications, Vol. 26, No. 2, MarchlApril 1990*

[20] S. Williamson, C.I. McClay, "The effect of axial variations in saturation due to skew on induction motor equivalent-circuit parameters", *Industry Applications, IEEE Trans on , Volume: 35 Issue: 6 , Nov.-Dec. 1999*

[21] C. Gerada, K.J. Bradley, M. Sumner, G. Asher, J. Arellano-Padilla, "Permanent Magnet Synchronous machines for Saliency-based, Self-Sensored Motion Control", *IEEE IECON 2007 Proc. On CD-ROM*

Robust PI Cascade Control for a Multi-Mass System Optimized by Evolutionary Algorithms

M. Joost [*], K. Zielinski [†], B. Orlik [*] and R. Laur [†]

[*] Institute for Electrical Drives, Power Electronics and Devices, IALB, University of Bremen, Germany:
joost@ialb.uni-bremen.de
[†] Institute for Electromagnetic Theory and Microelectronics, ITEM, University of Bremen, Germany:
zielinski@item.uni-bremen.de

Abstract— Controlling a multi-mass system is a common problem in industrial automation, and cascaded PI controllers are a popular control structure for these systems. However, the parameters of a multi-mass system are often uncertain or can change. A robust H_∞-control is a powerful tool in order to control systems with varying parameters. Unfortunately, the complex structure and the mathematical theory prevent known H_∞-controllers from usage in many industrial applications. In this work two modern Evolutionary Algorithms, Differential Evolution (DE) and Particle Swarm Optimization (PSO), are used to optimize a standard PI cascade control structure with respect to the H_∞-norm, thus resulting in a robust control of simple structure.

Keywords— Robust control, Optimal control, Mechatronics

I. INTRODUCTION

Elastically coupled multi-mass systems occur in many different types of industrial applications. Such systems are found for example in robotics, machine tools or the drive train of a wind energy plant. With respect to rising performance demands, the compliant coupling of motor, gear and load must also be considered. Therefore, multi-mass models have to be used for the controller design. The principle structure of a multi-mass system is shown in Fig. 1.

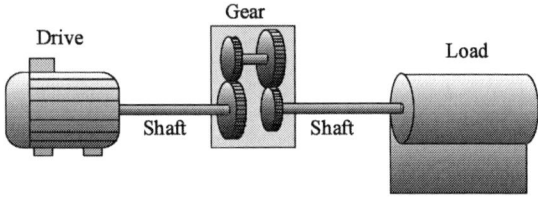

Fig. 1. Example of a multi-mass system

A simplification to a one-mass model for the controller synthesis often is not possible due to inaccuracies. An additional common problem is the variation of parameters. Parameter variations can change the resonance frequency of the system. Due to this fact a robust control is necessary.

An H_∞-control is a powerful tool in order to overcome this problem. Unfortunately, known H_∞-controllers have a complex structure and the mathematical theory is rather complicated, thus preventing their usage in many

industrial applications. On the other hand, PI controllers are frequently used because they are easy to understand and often already implemented. Consequently, in order to use PI controllers, controller parameters have to be selected to achieve a robust control. However, there is no analytical solution for an H_∞-optimal control with a fixed structure like a PI-cascade. The standard approaches to solve the H_∞-optimization problem like the gamma-iteration presented in [1] lead to complex controller structures, which consist of state controllers and observers. With the purpose of solving the optimization problem for a fixed structure a different method has to be applied.

Special search tools are needed in order to rapidly find a solution while still doing a thorough search of parameter combinations. One group of popular non-conventional search tools are Evolutionary Algorithms (EAs). Evolutionary Algorithms imitate processes from nature for the optimization of complex problems. Like various creatures have adapted to all sorts of environments by natural evolution, these algorithms use evolutionary operators to optimize different types of function. In this paper two EAs are employed: Differential Evolution (DE) and Particle Swarm Optimization (PSO).

II. MULTI-MASS SYSTEM

A. Analytical model

A multi-mass system consists of a drive motor, one or more masses in the middle, e.g. gears or couplings, and a load mass. These masses are coupled via elastic shafts. Friction is assumed to be present at each mass. For a general description of a multi-mass system, backlash has also been modeled, but it will be considered zero in this paper. Such a general structure is described by the following model (Fig. 2).

The dynamics of the frequency converter and the electric part of the drive are approximated as a PT_1-element with a time constant T_a. The friction torque m_{R1} and the shaft torque of the first shaft m_1 are subtracted from the actual drive torque m_a. The resulting torque acts on the first mass with the moment of inertia Θ_1, leading to the angular velocity ω_1. Integration of ω_1 then yields the position angle of the first mass ε_1. Subtracting from ε_1 the position angle ε_2 of the second mass leads to the difference angle $\Delta\varepsilon_{12}$, which is input for the backlash function. The resulting torsion angle is multiplied with the spring constant C_1 of the first elastic shaft resulting in the shaft torque m_1. This torque acts on the second mass with

978-1-4244-1741-4/08/$25.00 ©2008 IEEE

the moment of inertia Θ_2, leading to the angular velocity of the second mass ω_2, and so forth. At the last mass, a disturbance in the form of a load torque m_L may also appear.

The general structure of a multi-mass system can be described as being composed of separate blocks for each mass. Fig. 2 shows this model, where one such block is indicated by a dashed line. This way a general multi-mass system can be built by simply putting together an appropriate number of these blocks.

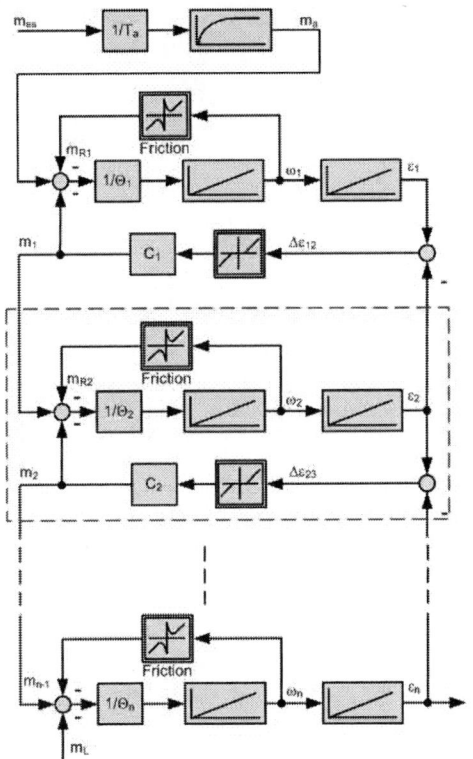

Fig. 2. Block diagram of a multi-mass system

B. Control structure

Industrial engineers often face the problem that they want to or need to use PI controllers, because they are available, easy to understand and/or already installed. For servo drive applications cascade structures of PI controllers (Fig. 3) are very common in the industry.

For the position control of a multi-mass system, the cascade usually consists of an inner control loop for the speed (K_ω), with an overlaying control loop for the position (K_ε).

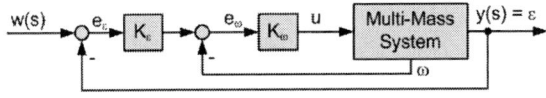

Fig. 3. Cascade control

In this work each controller is assumed to have the form

$$K_p + K_i \cdot \frac{1}{s} \qquad (1)$$

with two separate parameters: The proportional gain K_p and the integral gain K_i.

Unfortunately, the tuning of PI cascade controllers by knowledge and experience is a time consuming and difficult job when unknown parameter variations occur during operation because the multi-mass system is capable of swinging. Even with the help of experienced engineers, in most cases the result is a slow control which is far from being an optimal result regarding the robustness.

The tuning of a PI cascade control can be regarded as an optimization problem. The optimization parameters are the proportional and the integral gains of the respective controllers. The goal is to minimize some cost or objective function, describing the "fitness" of the control, by varying the controller parameters.

III. ROBUST CONTROL

In most cases the parameters of a real system will not be known exactly or can change over time. So the real system will deviate from a chosen nominal model. Hence, a robust controller is needed. A robust controller is one, which can successfully control a group of systems that deviate within a certain range from the nominal model.

A. H_∞-mixed sensitivity approach

An H_∞-mixed sensitivity approach is often used for the controller generation, because with this approach the robustness and the performance can be determined via different weighting functions by the user [1].

1) Model uncertainties

To formulate the robustness demands, the uncertainties of the nominal model have to be considered.

Let the transfer function of the nominal model be denoted as $G(s)$, and let $G^0(s)$ be the transfer function of the real system. The real transfer function can be described by the nominal transfer function together with uncertainties in additive form Δ_a and/or multiplicative form Δ_m [3].

$$G^0(s) = G(s) \cdot \left(1 + \Delta_m(s)\right) + \Delta_a(s) \qquad (2)$$

These uncertainties can be determined by making a worst case estimation: When $G^0(s)$ is chosen as the worst possible disturbed system, Δ_a and Δ_m can be estimated as

$$\Delta_a = G^0(s) - G(s) \qquad (3)$$

$$\Delta_m = \frac{G^0(s) - G(s)}{G(s)} \qquad (4)$$

2) Weighting functions

For the mixed sensitivity approach the robustness demands are usually formulated as weighting functions. An often used method is the definition of upper bounds.

It is assumed that there exist two frequency dependent bounds $W_2(s)$ and $W_3(s)$ for the model uncertainties:

$$|W_2(s)| > |\Delta_a(s)| \qquad (5)$$

$$|W_3(s)| > |\Delta_m(s)| \qquad (6)$$

Further it is assumed that the uncertainties do not change the number of instable or quasi-stable poles. These are premises for robustness criteria.

The weighting functions $W_2(s)$ and $W_3(s)$ determine how robust the control will be. It is also possible to demand a certain performance of the control, by selecting a weighting function $W_1(s)$ as a frequency dependent bound for the desired disturbance transfer function $S(s)$:

$$W_1^{-1}(s) > |S(s)| \quad \forall s = j\omega \qquad (7)$$

Accordingly, $W_1(s)$ should have an integral part in low frequencies and a unit gain in high frequencies. For the numerical controller syntheses it is sometimes useful to limit the integral part of $W_1(s)$.

Depending on the frequency, both weighting functions $W_2(s)$ and $W_3(s)$ are contradictory terms with respect to $W_1(s)$. This contradiction is mathematically an explanation for the well known practical experience, that fast controls are less robust. It is important to balance the demands and to determine which frequency response characteristic is appropriate for which weighting function. This is significant because the determination of the weighting functions is the only way to influence the control parameter syntheses.

3) Robustness criteria

The robust controller has to be able to stabilize the system with uncertainties. For the stability of a controlled system with additive uncertainties a sufficient condition is given by:

$$\left\| \Delta_a(s) \cdot \underline{K}(s) \cdot S(s) \right\|_\infty < 1 \qquad (8)$$

Where $\underline{K}(s)$ is the robust controller and $S(s)$ is the sensitivity function.

In case of multiplicative uncertainties a sufficient condition is given by:

$$\left\| \Delta_m(s) \cdot T(s) \right\|_\infty < 1 \qquad (9)$$

Where $T(s)$ is the complementary sensitivity function.

Using the weighting functions $W_2(s)$ and $W_3(s)$ in (5) and (6), (8) and (9) can be written as:

$$\left\| W_2(s) \, \underline{K}(s) \, S(s) \right\|_\infty < 1 \qquad (10)$$

$$\left\| W_3(s) \, T(s) \right\|_\infty < 1 \qquad (11)$$

B. Controller generation

The demands made by all three weighting functions can be combined to one H_∞-norm:

$$\left\| \begin{array}{c} W_1(s)\, S(s) \\ W_2(s)\, \underline{K}(s)\, S(s) \\ W_3(s)\, T(s) \end{array} \right\|_\infty < 1 \qquad \forall s = j\omega \qquad (12)$$

Equation (12) can be interpreted as the norm of a fictitious MIMO system $P(s)$, shown in Fig. 4.

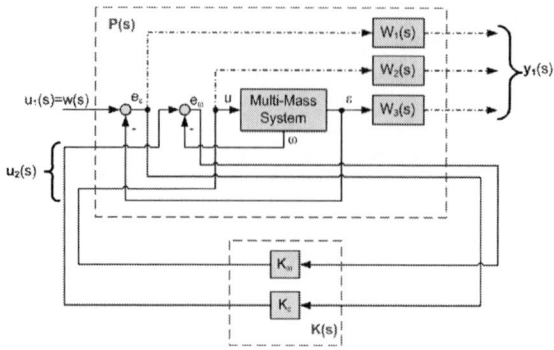

Fig. 4. Augmented system

This way (12) can be written as:

$$\left\| P_{\tilde{y}_1, u_1} \right\|_\infty < 1 \qquad \forall s = j\omega \qquad (13)$$

The task is to find a stabilizing controller $\underline{K}$ for the augmented system $P(s)$ that minimizes $\left\| P_{\tilde{y}_1, u_1} \right\|_\infty$ below one.

Known solutions for this problem, like the γ-iteration method, exist only for non-fixed controller structures $\underline{K}(s)$ and lead to complex structures that consist of state controllers and observers. In order to use simple, well understood PI-cascade controllers only the controller parameters of the PI-controllers but not the controller structure have to be optimized. The controller structure is fixed to:

$$\underline{K}(s) = \begin{bmatrix} K_{p_\omega} + K_{i_\omega} \cdot \dfrac{1}{s} & 0 \\ 0 & K_{p_\varepsilon} + K_{i_\varepsilon} \cdot \dfrac{1}{s} \end{bmatrix} \qquad (14)$$

Unfortunately, there is no analytical solution for this problem, therefore optimization algorithms have to be applied.

IV. Optimization Algorithms

Special optimization tools are needed in order to quickly find a solution while nevertheless doing a systematic and thorough search of parameter combinations. One group of popular non-conventional search tools are Evolutionary Algorithms (EAs). Evolutionary Algorithms imitate processes from nature for the optimization of complex problems. Like various creatures have adapted to all sorts of environments by natural evolution, these algorithms use evolutionary operators to optimize different types of functions. Compared to classical optimization methods, EAs have the advantage that they are able to find the global best solution even of multimodal optimization problems with a lot of local optima because of their population-based nature. Furthermore, they do not require the knowledge of any derivatives of the objective function. Therefore, they can be flexibly used for solving a large range of problems as long as there is some way to evaluate the objective and constraint functions.

One of the early EAs is the Genetic Algorithm (GA). In [4] a GA has successfully been used to optimize PI cascade controllers for a two-mass system. However, one disadvantage of the GA is that it generally uses binary representation of the parameters, so the user is required to find a suitable coding.

In contrast, newer developments like Differential Evolution (DE) use real-valued vectors, thus the need for coding is omitted, and the parameters are processed in the same form as they appear in many actual applications. Since its development in 1995, DE has shown a good performance in a large range of applications [5].

Another approach is not to rely on methods using the "survival of the fittest" principle, but to achieve optimization by cooperation of individuals. An example for such a strategy is Particle Swarm Optimization (PSO) which simulates the behavior of social groups like bird flocks or fish schools. The individuals are able to improve based on their own experience as well as knowledge from the group. Many practical optimization problems have already been successfully solved by PSO since it was first published in 1995 [6].

In the following an introduction to DE as well as to PSO is given.

A. Differential Evolution

Differential Evolution can be classified as a global population-based stochastic evolutionary optimization algorithm, meaning that

- it has the capability of finding the global optimum of a given objective function even if local optima exist,
- at each time instance not only one solution in the search space is considered but a whole so-called population that consists of several individuals which represent possible solutions of the optimization problem,
- random numbers are involved in the process,
- nature-inspired techniques are used that resemble the evolutionary process.

DE has the advantage of a high convergence speed that is achieved by using vector differences from the current population of parameter sets in the evolutionary operators, therefore step sizes are adaptively scaled. Additionally, the DE algorithm itself has only three internal variables which have to be set by the user, thus it is easy to use.

As in other evolutionary algorithms the first generation is initialized randomly in a given search space, and further generations evolve by applying specific evolutionary operators until a stopping criterion is satisfied (see Fig. 5).

The DE individuals consist of vectors with dimension D that equals the number of objective function parameters, that is the number of parameters to be varied, which is four in this case ($K_{p\varepsilon}$, $K_{i\varepsilon}$, $K_{p\omega}$, $K_{i\omega}$). The number of individuals NP has to be set by the user.

In every generation one offspring is generated for each population member x_i (with $i \in \{0,\dots,NP{-}1\}$) using the evolutionary operators mutation, recombination and selection. There are several strategies of DE that differ in the way mutation and recombination is conducted [7], [8]. They have the commonality that during mutation a mutated vector v_i is built that consists of a linear combination of several individuals, and during recombination components from the mutated vector and the target vector x_i (which is a population member) are copied to the so-called trial vector u_i which competes with x_i for a place in the next generation during selection.

The variants are specified using the notation $DE/x/y/z$ where x denotes the mutated vector, y is the number of difference vectors and z is the crossover scheme [9]. The mutated vector (also called base vector) might be a randomly chosen vector (notation: 'rand'), the best vector that was found so far (notation: 'best'), or a vector that is located on the connecting line between two solutions, e.g. between the target vector and a random vector (notation: 'current-to-rand') or between the target vector and the best vector (notation: 'current-to-best'). The number of difference vectors is normally set to one or two.

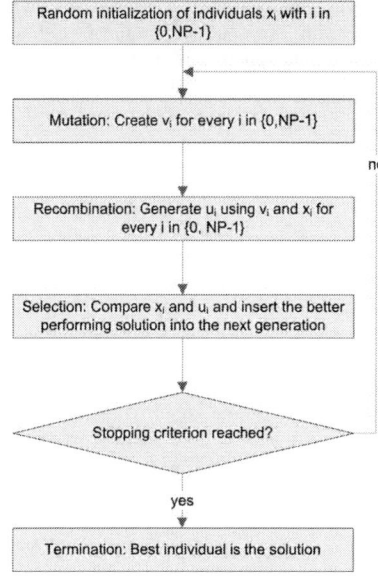

Fig. 5. General procedure of DE.

Concerning the crossover scheme, a binomial or exponential process can be used (notation: 'bin' or 'exp', respectively) which differ in the way that components are chosen for copying to the trial vector [10]. In [9] the use of binomial crossover is recommended but in [11] it is stated that there are no significant differences between the crossover methods, so in this work only the binomial process is used that will be explained later in this section.

The strategies which have been compared in former work [12] are given in Table I.

TABLE I.
STRATEGIES OF DIFFERENTIAL EVOLUTION

Notation	Equation for mutated vector
DE/rand/1/bin	$v_i = x_{r_1} + F \cdot \left(x_{r_2} - x_{r_3} \right)$
DE/rand/2/bin	$v_i = x_{r_1} + F \cdot \left(x_{r_2} - x_{r_3} \right) + F \cdot \left(x_{r_4} - x_{r_5} \right)$
DE/current-to-rand/1/bin	$v_i = x_i + K \cdot \left(x_{r_1} - x_i \right) + F \cdot \left(x_{r_2} - x_{r_3} \right)$
DE/current-to-rand/2/bin	$v_i = x_i + K \cdot \left(x_{r_1} - x_i \right) + F \cdot \left(x_{r_2} - x_{r_3} \right) + F \cdot \left(x_{r_4} - x_{r_5} \right)$
DE/best/1/bin	$v_i = x^* + F \cdot \left(x_{r_1} - x_{r_2} \right)$
DE/best/2/bin	$v_i = x^* + F \cdot \left(x_{r_1} - x_{r_2} \right) + F \cdot \left(x_{r_3} - x_{r_4} \right)$
DE/current-to-best/1/bin	$v_i = x_i + K \cdot \left(x^* - x_i \right) + F \cdot \left(x_{r_1} - x_{r_2} \right)$
DE/current-to-best/2/bin	$v_i = x_i + K \cdot \left(x^* - x_i \right) + F \cdot \left(x_{r_1} - x_{r_2} \right) + F \cdot \left(x_{r_3} - x_{r_4} \right)$

F and K in Table I are internal parameters of DE ($K = F$ is always assumed in this work as it is also often done in the literature, e.g. in [13]), the indices r_1, r_2, r_3, r_4, r_5 denote mutually different individuals which are also different from x_i, and x_i^* denotes the best individual found so far. In this work as well as in [12] binomial recombination is used for every strategy, thus trial vectors u_i (with $i \in \{0,...,NP-1\}$) are built by determining for every vector component $j \in \{0,...,D-1\}$ if the corresponding component should be copied from the target vector x_i or the mutated vector v_i.

The decision is made using a random variable $rand_j$ that is compared with the control parameter CR. However, because during selection u_i and x_i will be compared, it is ensured for every individual that at least one component of u_i is derived from v_i by a random choice of a number $k \in \{0,...,D-1\}$:

$$u_{i,j} = \begin{cases} v_{i,j} & \text{if } rand_j \leq CR \text{ or } j = k \\ x_{i,j} & \text{otherwise} \end{cases} \qquad (15)$$

The selection process is identical for all strategies. Selection is conducted by comparing the target vector x_i with the trial vector u_i. For unconstrained single-objective minimization problems, the vector that yields the smaller objective function value is chosen for the next generation. However, for constrained optimization problems the selection procedure has to be modified. In this work the feasibility rules described in [14] are applied, thus when a vector a is compared to a vector b, a is considered better if:

- Both vectors are feasible, but a yields the smaller objective function value.
- a is feasible and b is not.
- Both vectors are infeasible, but a results in the lower sum of constraint violations.

An advantage of this method is that no additional parameters are needed. As solutions with smaller constraint violations are preferred, the search is guided to feasible regions. In case of an unconstrained problem, the selection procedure is the same as for the original DE.

B. Particle Swarm Optimization

Particle Swarm Optimization imitates the behavior of social groups to solve optimization problems. Although PSO is derived from artificial life, it is usually classified as an evolutionary algorithm because its operators resemble the ones used in evolutionary computation and also because it can be argued that the self-organization present in PSO is part of the evolutionary process just as the "survival of the fittest" principle is. Adhering to this argumentation, PSO can be classified in the same way as given for DE in the beginning of Section IV.A.

Each individual (called particle) is characterized by its position x_i ($i \in \{0,...,NP-1\}$; NP is the population size), its velocity v_i, its personal best position p_i and its neighborhood best position p_g where every position and velocity is a vector with D components (D is the dimension that equals the number of objective function parameters). Several neighborhood topologies have been developed [15], and the most popular ones are the *gbest*, *lbest* and *von-Neumann* neighborhood topologies (Fig. 6).

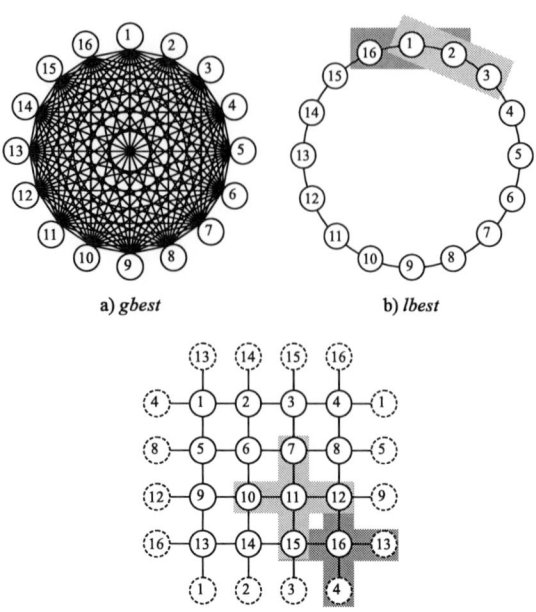

a) *gbest*   b) *lbest*

c) *von-Neumann*

Fig. 6. Neighborhood topologies (with population size $NP=16$)

In the *gbest* variant the neighborhood consists of the whole swarm (see Fig. 6 a), meaning that the particles have the information of the globally found best solution. Using other variants, the neighborhood is built by a subset of the population, thus only locally best positions are revealed to the particles. For the *lbest* neighborhood each particle generates a neighborhood consisting of itself and its two immediate neighbors where the neighbors are the particles with adjacent indices (see Fig. 6 b). Because every particle generates a neighborhood, neighborhoods are overlapping. Using the *von-Neumann* neighborhood, each particle possesses four neighbors. The topology can be visualized as a two-dimensional lattice that is wrapped on all four sides (torus), so the *von-Neumann* neighborhood of a particle consists of the particles above and below as well as the particles to the left and right (see Fig. 6 c). As for *lbest*, neighborhoods are overlapping using the *von-Neumann* neighborhood topology.

In the beginning of an optimization run, the positions of the particles are randomly initialized in the search space. The velocities are randomly chosen from $[-V_{max}, V_{max}]$ where the maximum velocity V_{max} is one half of the search space in each dimension: $V_{max,j} = (X_{max,j} - X_{min,j}) / 2$ (with $j \in \{0,...,D-1\}$). The positions and velocities of the particles are modified using the following update equations until a stopping criterion is satisfied that terminates the optimization run (see Fig. 7).

$$v_i(t+1) = w v_i(t) + c_1 r_1 [p_i(t) - x_i(t)] \\ + c_2 r_2 [p_g(t) - x_i(t)] \qquad (16)$$

$$x_i(t+1) = x_i(t) + v_i(t+1) \qquad (17)$$

The first term in (16) describes the dependence on the velocity of the previous time step that is weighted with the inertia weight w. The cognitive component is given by the second term that draws the particle towards its personal best position. The third term is the social component that causes movement towards the neighborhood best position. Besides the inertia weight w, (16) also includes the internal parameters c_1 and c_2. A stochastic effect is generated by r_1 and r_2 that are chosen randomly from [0, 1]. The position is updated by adding the new velocity to the old position in (17).

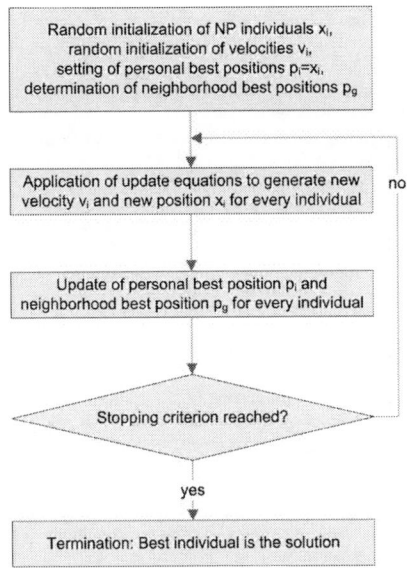

Fig. 7. General procedure of PSO.

For unconstrained single-objective minimization problems the personal and neighborhood best positions are updated by comparing the current position to the best position and keeping the one that yields the lower objective function value as best position. For constrained problems the comparison is modified in the same way as described in Section IV.A for DE, so the comparison of two solutions is based on feasibility, sum of constraint violations and objective function values. Same as for DE, for unconstrained optimization problems the original algorithm is unaltered when using this constraint-handling method.

V. EXEMPLARY THREE-MASS SYSTEM

In this paper the problem of selecting suitable parameters for an H_∞-optimal, robust PI-cascade control of an exemplary three-mass system is solved by both DE and PSO algorithms.

A. Experimental set-up of the three-mass system

For testing of the optimal PI-cascade control, an experimental set-up is used (Fig. 8). It consists of two motors and a middle mass, mechanically coupled by elastic shafts, including nonlinear friction and an adjustable backlash coupling. The elastic shafts have a spring constant of $C_{1,2} = 1383$ Nm/rev. each. The moment of inertia of the drive side Θ_1 is 39.7 Kg cm^2. The middle mass has a nominal moment of inertia Θ_2 of 251.9 Kg cm^2 and the load side has a moment of inertia Θ_3 of 63.1 Kg cm^2. Normalizing to the rated values gives the following time-constants for the three masses: $T_1 = 0.112$ s, $T_2 = 0.79$ s, $T_3 = 0.201$ s. The control is implemented on a separate microcontroller board with a Texas Instruments TMS320C40 chip. The board is connected to a frequency converter for the transmission of the control variables.

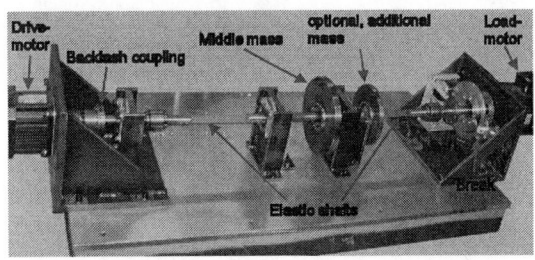

Fig. 8. Three-mass system

B. Application of the optimization algorithms

For both DE and PSO certain internal parameters of the algorithms have to be set by the user. Furthermore, for DE a strategy has to be chosen, and for PSO a neighborhood topology has to be selected. For the practitioner in the Industry who is often inexperienced in optimization, this can pose a problem, because from literature it does not become clear which settings should be used. Some standard settings can be found, but recommendations are often contradicting. However, the optimization problem at hand was carefully examined in earlier work, concerning the final objective function value as well as the number of generations for finding a robust controller (see [12], [16]).

From the applications point of view these works showed the effects of parameter settings as well as the DE strategy or the PSO neighborhood topology to be negligible.

From the optimization point of view some differences could be seen but due to their small size they do not have a practical meaning for the optimization problem.

For DE e.g. parameter settings of $F=0.9$ and $CR=0.9$ yield good results, especially when using strategy DE/best/1/bin or DE/current-to-best/1/bin. For PSO the parameter settings from the so-called constriction variant [17] can be used, which are $w=0.73$, $c_1=1.5$ and $c_2=1.5$, and the choice of the neighborhood topology is even of less importance, thus e.g. the *gbest* or the *von-Neumann* neighborhood topology can be used. The number of individuals was always set to $NP=30$ and the maximum number of generations to $G_{max}=80$.

The controller parameters obtained by DE and PSO are given in Table II. The parameters found are quite similar and also result in similar values of the objective function, showing a consistent performance of both algorithms.

TABLE II.
OPTIMIZATION RESULTS

	$K_{p\varepsilon}$	$K_{i\varepsilon}$	$K_{p\omega}$	$K_{i\omega}$	$f(x)$
DE	0.177	1.232e-4	11.446	1.073e-3	0.8188
PSO	0.177	3.418e-7	11.443	6.014e-6	0.8187

C. Measurements

In order to determine the robustness and performance of the optimized control parameters, measurements of the nominal multi-mass system and a perturbed system with a tripled load side moment of inertia are taken with the best parameters generated by DE and PSO, respectively. Step responses for an input step of one revolution are shown in Fig. 9 for control parameters generated by DE and in Fig. 10 for parameters generated by PSO. Both show good and robust performance, with nearly identical behavior as was to be expected from the very similar controller parameters.

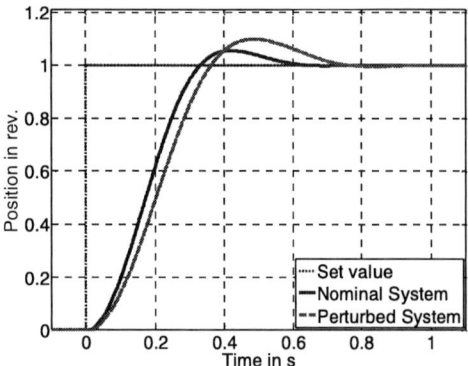

Fig. 9. Measurement of the step response with controller parameters optimized by DE

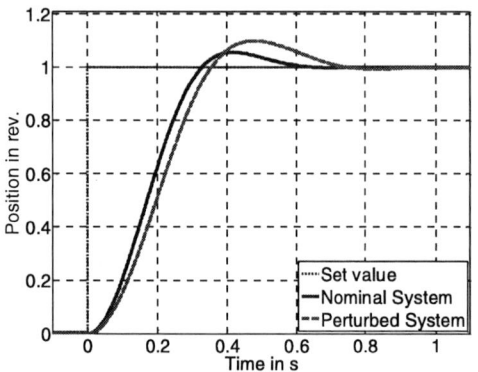

Fig. 10. Measurement of the step response with controller parameters optimized by PSO

VI. CONCLUSION

The problem was to find controller parameters for a PI-cascade controller, so that a robust control is obtained. The solution was to use the modern Evolutionary Algorithms Differential Evolution (DE) and Particle Swarm Optimization (PSO) to optimize the control.

In contrast to the Genetic Algorithms used in earlier work, DE and PSO need no binary representation of the controller parameters, making their application much easier. Both algorithms converged in all optimization runs to values of $f(x)<1$ which are needed for a robust controller and generated similar results for the controller parameters. Thus, their consistent performance makes both of them useful to field engineers without the need for special knowledge concerning optimization algorithms. The differences found in earlier work [12], [16] regarding the quality of the optimized solution and the performance of the optimization algorithms do not have a practical relevance.

The measurements show good and robust results for both algorithms, thus making both of them powerful and easy to use tools for the generation of robust controllers of a fixed, low order structure.

REFERENCES

[1] J. C. Doyle, K. Glover, P. P. Khargonekar, and B. A. Francis, "State-Space Solutions to Standard H_2 and H_∞ Control Problems," *IEEE Transactions on Automatic Control*, vol. 34, no. 8, pp. 831–847, 1989.

[2] R. Y. Chiang, M. G. Safonov *"Robust Control Toolbox User's Guide"*, The Math Works Inc., 1992

[3] S. Skogestad and I. Postlethwaite, *Multivariable Feedback Control*. John Wiley & Sons, 1996.

[4] K. Peter, K. M. Vinogradski, A. Tenhagen, M. Joost, B. Orlik, "Robust H_∞-Optimal P/PI Cascade Control for a Two-mass System." In: *Proceedings PCIM Europe 2006*. Nürnberg, May / June 2006

[5] K. Price, R. Storn, J. Lampinen, "Differential Evolution - A Practical Approach to Global Optimization." Springer-Verlag Berlin Heidelberg, 2005

[6] A. Engelbrecht, *"Fundamentals of Computational Swarm Intelligence."* John Wiley & Sons, 2006

[7] G. C. Onwubolu, "Optimizing CNC Drilling Machine Operations: Traveling Salesman Problem-Differential Evolution Approach," in *New Optimization Techniques in Engineering*, G. C. Onwubolu and B. Babu, Eds. Berlin Heidelberg: Springer-Verlag, 2004, pp. 537–566.

[8] E. Mezura-Montes, J. Velázquez-Reyes, and C. A. Coello Coello, "A Comparative Study of Differential Evolution Variants for Global Optimization," in *Proceedings of the Genetic and Evolutionary Computation Conference*, 2006.

[9] K. V. Price, "An Introduction to Differential Evolution," in *New Ideas in Optimization*, D. Corne, M. Dorigo, and F. Glover, Eds. London: McGraw-Hill, 1999, pp. 79–108.

[10] K. Zielinski, P. Weitkemper, R. Laur, and K.-D. Kammeyer, "Parameter Study for Differential Evolution Using a Power Allocation Problem Including Interference Cancellation," in *Proceedings of the IEEE Congress on Evolutionary Computation*, Vancouver, BC, Canada, 2006, pp. 6748–6755.

[11] K. Price and R. Storn, http://www.icsi.berkeley.edu/~storn/code.html, Website as in March 2008.

[12] K. Zielinski, M. Joost, R. Laur and B. Orlik, "Choosing Suitable Variants of Differential Evolution and Particle Swarm Optimization for the Optimization of a PI Cascade Control" in *Proceedings of the 11th International Conference on Optimization of Electrical and Electronic Equipment*, Brasov, Romania, 2008.

[13] R. Storn, "Designing Digital Filters with Differential Evolution," in *New Ideas in Optimization*, D. Corne, M. Dorigo, and F. Glover, Eds. London: McGraw-Hill, 1999, pp. 109–125.

[14] K. Deb, "An Efficient Constraint Handling Method for Genetic Algorithms," *Computer Methods in Applied Mechanics and Engineering*, vol. 186, no. 2-4, pp. 311–338, 2000.

[15] R. Mendes, J. Kennedy, and J. Neves, "The Fully Informed Particle Swarm: Simpler, Maybe Better," *IEEE Transactions on Evolutionary Computation*, vol. 8, no. 3, pp. 204–210, 2004.

[16] K. Zielinski, M. Joost, R. Laur and B. Orlik: "Comparison of Differential Evolution and Particle Swarm Optimization for the Optimization of a PI Cascade Control", in *Proceedings of the IEEE Congress on Evolutionary Computation 2008*, Hong Kong, 2008.

[17] M. Clerc and J. Kennedy, "The Particle Swarm - Explosion, Stability, and Convergence in a Multidimensional Complex Space," *IEEE Transactions on Evolutionary Computation*, vol. 6, no. 1, pp. 58–73, 2002.

Permanent Magnet Synchronous Servo-Drive with State Position Controller

Lech M. Grzesiak[x][xx], Tomasz Tarczewski[x], Sławomir Mandra[x]

[x]Department of Technical and Applied Physics, Toruń, Poland, e-mail: *ttarczewski@fizyka.umk.pl*
[xx]Institute of Control and Industrial Electronics, Warsaw, Poland, e-mail: *lmg@isep.pw.edu.pl*

Abstract—**This paper presents a design approach of a position controller for permanent magnet synchronous motor (PMSM) servo-drive. Firstly the well know nonlinear mathematical model of PMSM is evaluated to a linear form by introducing new variables. The state space new model presented in rotates ortogonal reference frame is decupled in means of equation in d and q axis. An internal input model have been used to achive zero steady state error for step input and as well as disturbance rejection. The proposed control method is implemented and tested in a 0,6 kW PMSM drive (TMS320F2812-DSP). The experimental test results are consistent with our computer simulation test results and validate the high performance of proposed control method.**

Index Terms—**state controller, permanent magnet motor, servo-drive system.**

I. INTRODUCTION

The permanent-magnet synchronous motor (PMS) drive has been playing an important role in motion-control applications in the wide power range. Especially servo drive with PMSM in applications such as industrial robots and machine tools have been gradually used. To achive fast four-quadrant operation the field oriented control is used in many industrial applications. However, the control performance of such PMSM drives is still influenced by the non-linearity and uncertainty of the plant which consist of power electronics converter, sensors, PMSM motor and an external load. Unpredictable plant parameter variations and stochastical disturbance cause the control to be not trivial. Therefore, many control theories [3], [4], [6], [7], [9], [10], [12], [13] such as nonlinear control, variable structure control, adaptive control, robust control have been developed for the PMSM servo-drive to deal with the uncertainties under various operating conditions. In recent years also intelligent control techniques such as fuzzy-control or neuro-control [5], [11] have been developed and applied to the position control of servo drives to obtain very good operating performance. Those advanced control techniques needed relatively complicated mathematical description. In addition an implementation in microcomputer control system is usually difficult.

From the designer's view point, linear state feedback control is an attractive method for controlling a plant which is linear and represented by a state-space model. The method has the full flexibility of shaping the dynamics of the closed-loop system to meet the desired specification. Techniques such as pole placement or LQR method have been used to achieve the designed behaviour.

Due to the coupled, non-linear, set of equations, using of the linear control theory to design precision position control system is not obvious. To solve this task we have to modified the mathematical model to form of linear one. The paper is arranged as follows: Description of the simplified linear and decoupled model of servo-drive with PMSM is presented in section II. Section III describe the proposed controller. An internal input model has been used to achieve zero steady state error for step input and as well as disturbance rejection. In section IV some simulation test results are presented. Experiments results have been presented in section V. The control method is implemented and tested in a 0,6 kW PMSM drive (TMS320F2812-DSP). Finally, in section VI some conclusions are given.

II. SERVO-DRIVE MODEL

To create a mathematical model of PMSM following assumptions are made [1]: saturation is neglected, the back emf is sinusoidal and eddy current and hysteresis losses are negligible. In orthogonal d, q coordinate system that rotates at the electrical speed of the rotor the expression of the voltage and flux equation is [1], [8]:

$$u_d = R_s i_d + \frac{d\psi_d}{dt} - p\omega_m \psi_q \tag{1}$$

$$u_q = R_s i_q + \frac{d\psi_q}{dt} + p\omega_m \psi_d \tag{2}$$

$$\psi_d = L_d i_d + \psi_f \tag{3}$$

$$\psi_q = L_q i_q \tag{4}$$

where: u_d, u_q, i_d, i_q, ψ_d, ψ_q, are voltages, currents and fluxes in d and q axis respectively, L_d, L_q, are inductance in d and q axis, R_s resistance of stator, ψ_f, is permanent magnetic flux linkage, p number of pole pairs and ω_m rotor angular speed.

The electromagnetic torque is described as follows:

$$T_e = \frac{3}{2} p[\psi_f i_q + (L_d - L_q) i_d i_q] \tag{5}$$

For a surface mounted permanent magnet machine inductance in d axis and inductance in q axis are in practical equal ($L_{sd} = L_{sq} = L_s$). In such case [10] the electromagnetic torque is proportional to the current in q axis and then the torque ca be expressed as:

$$T_e \cong \frac{3}{2} p\psi_f i_q = K_t i_q \tag{6}$$

978-1-4244-1741-4/08/$25.00 ©2008 IEEE

where: K_t is a motor torque constant.

The motor is supplied from power electronics inverter controlled by using PWM method. Because of the switching frequency, which is much higher than electrical time constant of the motor, inverter dynamic can be approximated by using proportioning element. The inverter model can be described as follows:

$$\begin{bmatrix} u_d \\ u_q \end{bmatrix} = K_p \begin{bmatrix} u_{dc} \\ u_{qc} \end{bmatrix} \tag{7}$$

where u_{dc} and u_{qc} are inverter control voltages for d and q components of stator space vector voltage.
We define new variables:

$$u_{do} = -\frac{p\omega_m \psi_q}{K_p} \tag{8}$$

$$u_{qo} = \frac{p\omega_m \psi_d}{K_p} \tag{9}$$

After defining new variables (8) and (9) and putting then into equations (1) and (2 the following expression have been obtained:

$$K_p u_{dd} = K_p (u_{dc} - u_{do}) = R_s i_d + \frac{d\psi_d}{dt} \tag{10}$$

$$K_p u_{qq} = K_p (u_{qc} - u_{qo}) = R_s i_q + \frac{d\psi_q}{dt} \tag{11}$$

$$\psi_d = L_s i_d + \psi_f \tag{12}$$

$$\psi_q = L_s i_q \tag{13}$$

with respect to magnetic symmetry in circuit:

$$L_{sd} = L_{sq} = L_s \tag{14}$$

It can be observed that cross couplings were eliminated because equation (10) consist only variables with index d and equation (11) consist only variables with index q.

To complete PMSM mathematical model, following equations of mechanical motion have been added:

$$\frac{d\omega_m}{dt} = \frac{1}{J_m}(T_e - T_l) \tag{15}$$

$$\frac{d\theta_m}{dt} = \omega_m \tag{16}$$

where: J_m, T_l, θ_m are moment of inertia, load torque and rotor angular position.

After those evaluation the model of PMSM with inverter can be described in a standard form of a linear state equation as follows:

$$\frac{dx}{dt} = Ax + Bu \tag{17}$$

The resulting mathematical model of PMSM with inverter (with assumption that external disturbance T_l is equal zero) can be written as follows:

$$A = \begin{bmatrix} -\frac{R_s}{L_s} & 0 & 0 & 0 \\ 0 & -\frac{R_s}{L_s} & 0 & 0 \\ \frac{K_t}{J_m} & 0 & 0 & 0 \\ 0 & 0 & 1 & 0 \end{bmatrix} x = \begin{bmatrix} i_q \\ i_d \\ \omega_m \\ \theta_m \end{bmatrix}$$

$$B = \begin{bmatrix} 0 & \frac{K_p}{L_s} \\ \frac{K_p}{L_s} & 0 \\ 0 & 0 \\ 0 & 0 \end{bmatrix} u = \begin{bmatrix} u_{dd} \\ u_{qq} \end{bmatrix} \tag{18}$$

The output equation can be written in following form:

$$y = Cx \tag{19}$$

where:

$$C = \begin{bmatrix} 0 & 0 & 0 & 1 \end{bmatrix} \tag{20}$$

III. STATE CONTROLLER

To control the position of the servo-drive (18), we need to build a type 1 servo system. Therefore we have introduced an internal model of the reference input in the compensator [14]. The control input for such system is:

$$u = -K_e \int_0^t e(\tau)d\tau - K_x x \tag{21}$$

The corresponding block diagram is shown in Fig. 1 and more details of controller structure is shown in Fig. 2. New state matrix (A_i) and new control matrix (B_i) of system with the integrator can be formulated as follows:

$$A_i = \begin{bmatrix} -\frac{R_s}{L_s} & 0 & 0 & 0 & 0 \\ 0 & -\frac{R_s}{L_s} & 0 & 0 & 0 \\ \frac{K_t}{J_m} & 0 & 0 & 0 & 0 \\ 0 & 0 & 1 & 0 & 0 \\ 0 & 0 & 0 & 1 & 0 \end{bmatrix} x_i = \begin{bmatrix} i_q \\ i_d \\ \omega_m \\ \theta_m \\ e_c \end{bmatrix}$$

$$B_i = \begin{bmatrix} 0 & \frac{K_p}{L_s} & 0 \\ \frac{K_p}{L_s} & 0 & 0 \\ 0 & 0 & 0 \\ 0 & 0 & 0 \\ 0 & 0 & -1 \end{bmatrix} u_i = \begin{bmatrix} u_{dd} \\ u_{qq} \\ \theta_{ref} \end{bmatrix} \tag{22}$$

where e_c is an integral of position error.

It can be observed that the controller includes an internal model (that is, an integrator) of the reference step input. The closed-loop pole locations have a direct impact on time response characteristics such as rise time, settling time, and transient oscillations. Pole placement method has been used for the controller design [2]. Assuming that the system is controllable the location of the closed-loop system poles is unrestricted. The Matlab Control System Toolbox has been used to calculate the appropriate matrix gains.

$$K_e = \begin{bmatrix} k_{e1} \\ k_{e2} \end{bmatrix} K_x = \begin{bmatrix} k_{x1} & k_{x2} & k_{x3} & k_{x4} \\ k_{x5} & k_{x6} & k_{x7} & k_{x8} \end{bmatrix} \tag{23}$$

IV. SIMULATION RESULTS

The simulation examinations were accomplished under Matlab/Simulink environment. Parameters of PMSM used in simulation are as follows: nominal power $P_n = 628\,[W]$, nominal voltage $U_n = 325\,[V]$, nominal current $I_n = 3.3\,[A]$, nominal speed $\omega_n = 418\,[rad/s]$, stator resistance $R_s = 8.5 \times 10^{-1}\,[\Omega]$, stator inductance $L_s = 2 \times 10^{-3}\,[H]$, torque constant $K_t = 3.5 \times 10^{-1}\,[Nm/A]$, moment of inertia $J_m = 1 \times 10^{-4}\,[\mathrm{kgm}^2]$, number of pole pairs $p = 3$, inverter gain $K_p = 183$.

Satisfactory control quality was achieved when arbitrary located poles of closed-loop system were: $g_1 = -8 \times 10^4$, $g_2 = -4 \times 10^4$, $g_3 = -1.5 \times 10^2$, $g_4 = -23 + 16i$, $g_5 = -23 - 16i$.

Values of state controller gain matrices $\boldsymbol{K_e}$ and $\boldsymbol{K_x}$ used in simulation are as follows: $k_{e1} = 7.5451$, $k_{e2} = 15.118$, $k_{x1} = 0.0421$, $k_{x2} = 0.8656$, $k_{x3} = 0.0092$, $k_{x4} = 0.43$, $k_{x5} = 0.4386$, $k_{x6} = 0.0413$, $k_{x7} = 0.025$, $k_{x8} = 0.9841$.

A block diagram (Simulink model) of designed servo-drive system with PMSM is depicted in Fig. 1. The Simulink model of servo drive system consist of 3 subsystems: the state controller, the decoupling unit and the plant model (permanent magnet synchronous motor and power electronics converter). More details of each subsystems are presented in next few figures.

Figure. 2 shows state-controller structure with internal model, which guarantee disturbance rejecting and zero steady state error for step inputs .

A Simulink model of proposed decoupling unit (described by equation (8),(9)) is illustrated in Fig. 3. Nonlinear model of Permanent Magnet Synchronous Motor is presented in Fig. 4.

Simulation test results are presented in figures 5 - 8. As it can be observed in Fig. 5 is no steady-state error of position for step reference signal. This features of designed state controller are consistent with our expectations. A very small position overshoot during a transient and very good dynamic control of the angular speed and the stator phase current can be observed.

V. IMPLEMENTATION RESULTS

The described in previous sections state controller (extensively tested in Matlab/Simulink environment) was implemented in TMS320F2812 DSP. Schematic diagram

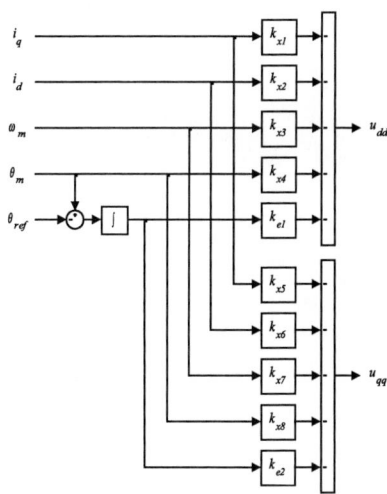

Fig. 2. Block diagram of proposed state controller.

of laboratory set-up servo-drive system is presented in Fig. 9.

Angular position and speed response of servo-drive are illustrated in Fig. 11 and in Fig. 12.

It can be seen, that the Fig. 5 and the Fig. 6 are similar to the Fig. 11 and the Fig. 12. There are no significant differences between simulink-model and real servo-drive system responses. Stator phase currents are depicted in Fig. 13 and in Fig. 14. Experiments test results are similar to simulation results presented on the Fig. 7 and the Fig. 8.

VI. CONCLUSIONS

This paper presents a design approach of a position controller for permanent magnet synchronous motor (PMSM) servo-drive. We applied well know linear control theory to simplify design process for position control of PMSM drive. Non-linear mathematical model of PMSM has been evaluated to a linear form by introducing new variables. A internal input model have been used to achive zero steady state error for step input and the disturbance

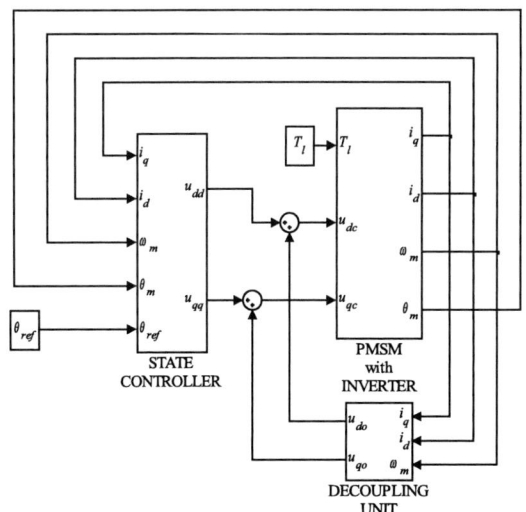

Fig. 1. Block diagram of designed servo-drive system with PMSM.

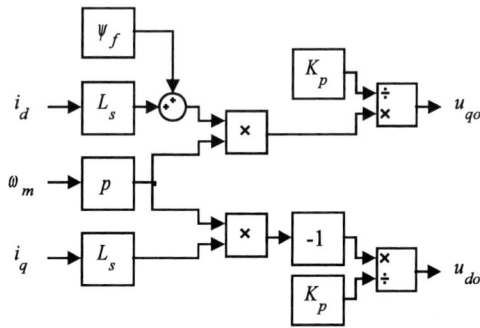

Fig. 3. Block diagram of decoupling unit.

1073

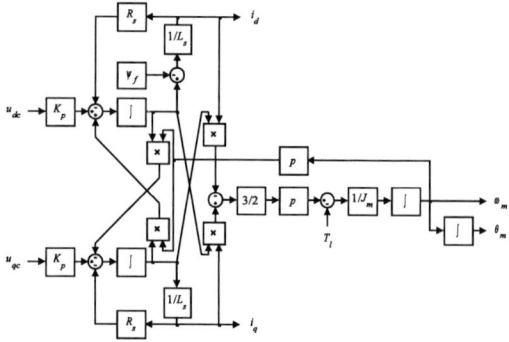

Fig. 4. Graphical representation of PMSM mathematical model.

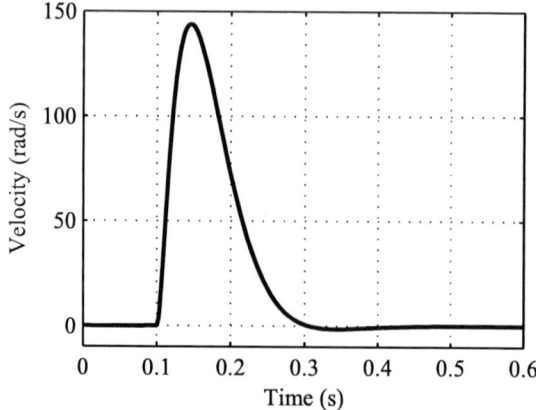

Fig. 6. Angular speed as a function of time (computer's simulation).

rejection. The simulation test results as well as experiment results validate the high performance of proposed control method.

REFERENCES

[1] P. Pillay and R. Krishnan, "Modelling of permanent magnet motor drives," *IEEE Transactions on Industrial Electronics*, vol. 35, no. 4, pp. 537–541, November 1988.

[2] J. Kautsky and N.K. Nichols, "Robust Pole Assignment in Linear State Feedback," *International Journal of Control*, vol. 41, pp. 1129–1155, Issue 5, May 1985.

[3] L. Mingji, C. Zhongqin, Ch. Ximing and O. Minggao, "Adaptive position servo control of permanent magnet synchronous motor," *Proceedings of the American Control Conference*, vol. 1, pp. 84–89, 30 June-2 July 2004.

[4] M. Sazawa, T. Yamada, K. Ohishi and S. Katsura, "Robust High Speed Positioning Servo System Considering Saturation of Current and Speed," *IEEE International Conference on Industrial Technology, ICIT 2006*, pp. 866–871, 15-17 Dec. 2006.

[5] Y.-S. Kung and P.-G. Huang, "High performance position controller for PMSM drives based on TMS320F2812 DSP," *Proceedings of the IEEE International Conference on Control Applications*, vol. 1, pp. 290–295, 2-4 Sept. 2004.

[6] K.-K. Shyu, Ch.-K. Lai, Y.-W. Tsai and D.-I. Yang, "A newly robust controller design for the position control of permanent-magnet synchronous motor," *IEEE Transactions on Industrial Electronics*, vol. 49, pp. 558–565, Issue 3, Jun 2002.

[7] F. Jiang, W. Zhang, W. Liang and X. Liu, "A MC56F8357 Based Permanent Magnet Synchronous Motor (PMSM) Servo System," *Proceedings of the Eighth International Conference on Electrical Machines and Systems, ICEMS 2005*, vol. 2, pp. 1519–1523, 27-29 Sept. 2005.

[8] Z. Dongliang, A. Xing, X. Chuanjun and Z. Chengrui, "DSP-based software AC servo systems with PM synchronous motors," *Proceedings of the Fifth International Conference on Electrical Machines and Systems, ICEMS 2001*, vol. 2, pp. 755–758, Aug 2001.

[9] S.-H. Choi, J.-S. Ko, I.-D. Kim, J.-S. Park, and S.-C. Hong, "Precise position control using a PMSM with a disturbance observer containing a system parameter compensator," *IEE Proceedings Electric Power Applications*, vol. 152, pp. 1573–1577, Issue 6, 4 Nov. 2005.

[10] F.-L. Lin and Y.-S. Lin, "A robust PM synchronous motor drive with adaptive uncertainty observer," *IEEE Transactions on Energy Conversion*, vol. 14, pp. 989–995, Issue 4, Dec 1999.

[11] F.F.M. El-Sousy, "An Intelligent Model-Following Sliding-Mode Position Controller for PMSM Servo Drives," *4th IEEE International Conference on Mechatronics, ICM2007*, pp. 1–6, 8-10 May 2007.

[12] S. Brock, J. Deskur, and K. Zawirski, "Robust speed and position control of PMSM," *Proceedings of the IEEE International Symposium on Industrial Electronics, ISIE1999*, vol. 2, pp. 667–672, 1999

[13] F.-J. Lin, K.-K. Shyu and Y.-S. Lin, "Variable structure adaptive control for PM synchronous servo motor drive," *IEE Proceedings - Electric Power Applications*, vol. 146, pp. 173–185, Issue 2, Mar

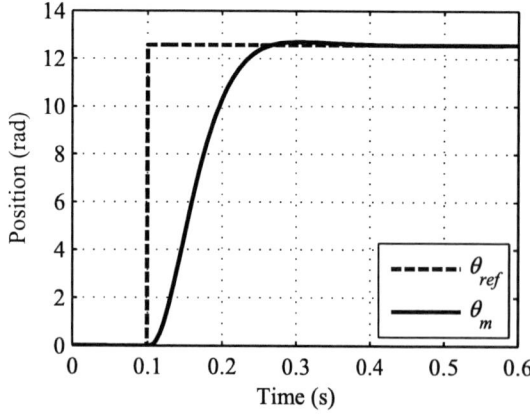

Fig. 5. Angular position as a function of time for step command (computer's simulation).

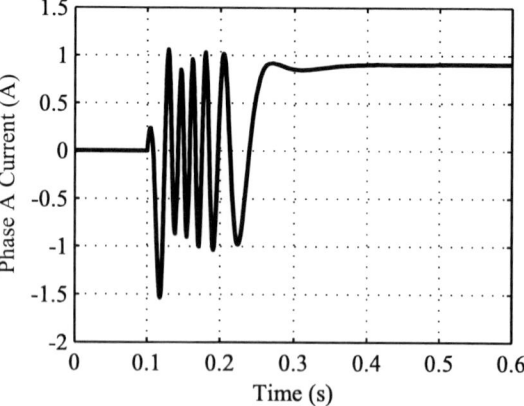

Fig. 7. Phase A current as a function of time (computer's simulation).

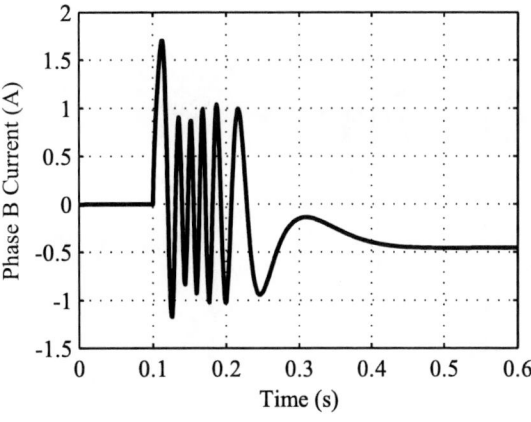

Fig. 8.   Phase B current as a function of time (computer's simulation).

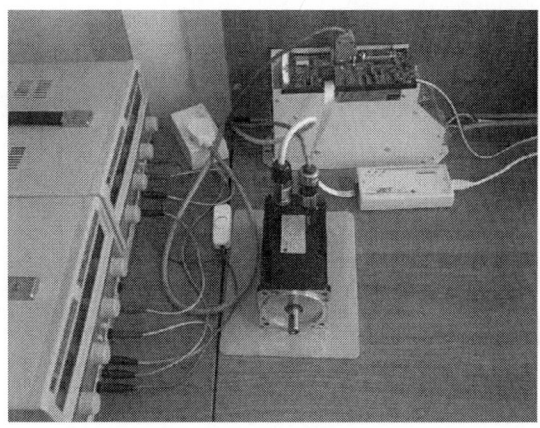

Fig. 10.   The laboratory set-up.

1999.
[14] K. Ogata, "Modern Control Engineering " *Prentice Hall International, Inc.*, 1997.

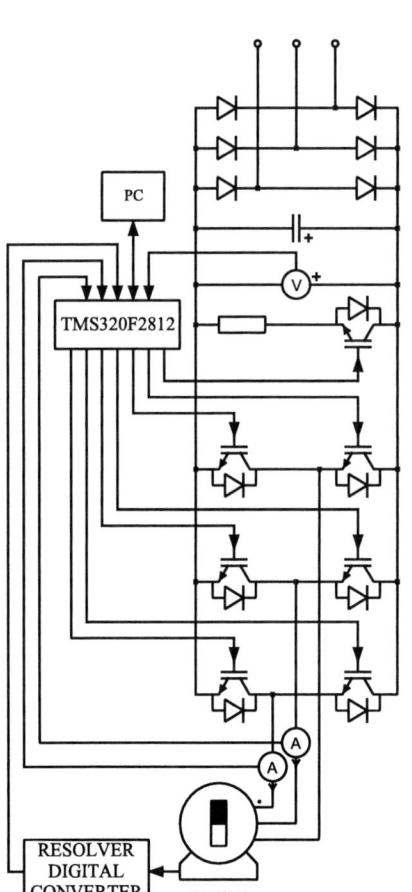

Fig. 9.   Schematic diagram of the PMSM servo-drive system.

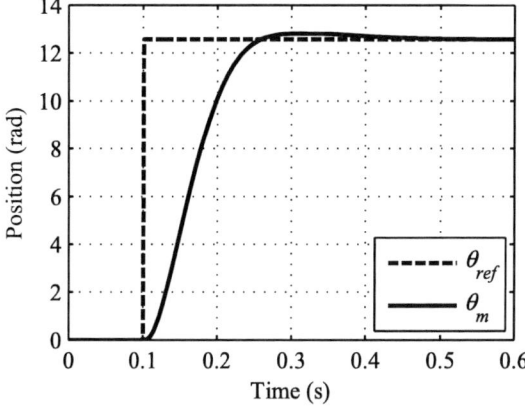

Fig. 11.   Angular position as a function of time for step command (experiment).

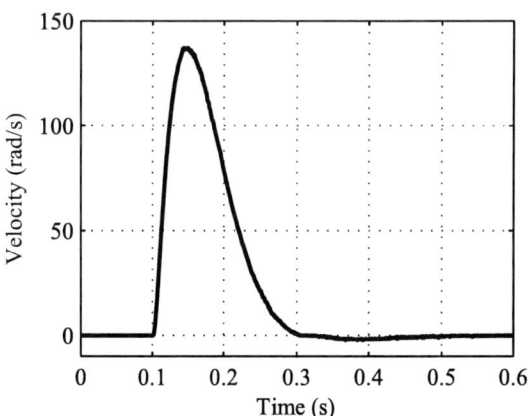

Fig. 12.   Angular speed as a function of time (experiment).

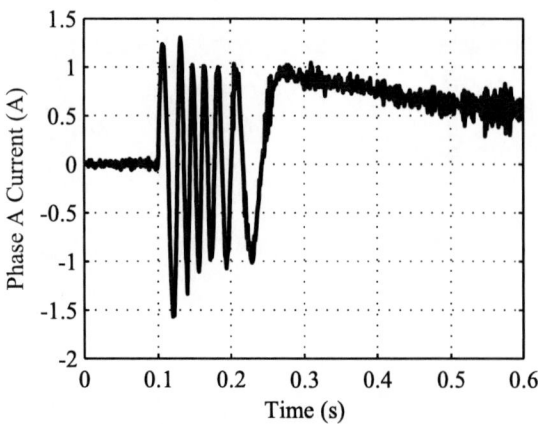

Fig. 13. Phase A current as a function of time (experiment).

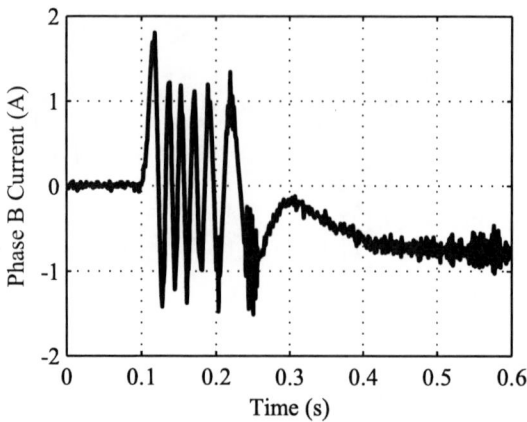

Fig. 14. Phase B current as a function of time (experiment).

Closed-Loop Control of Virtual FPGA-Coded Permanent Magnet Synchronous Motor Drives using a Rapidly Prototyped Controller

Christian Dufour   Vincent Lapointe   Jean Bélanger   Simon Abourida

Opal-RT Technologies, 1751 Richardson, suite 2525, Montréal, Québec, Canada
www.opal-rt.com
{christian.dufour, simon.abourida, jean.belanger, vincent.lapointe}@opal-rt.com

Abstract- **Presented in this paper are the results of closed-loop control experiments using a virtual permanent magnet synchronous motor (PMSM) drive implemented on a field-programmable gate array (FPGA) card connected to an external controller. The FPGA-based PMSM motor drive is implemented on an eDRIVEsim simulator, based on the RT-LAB platform. The eDRIVEsim simulator implements 2 types of motor drive models, Park (d-q) and Finite Element Analysis (FEA), on an FPGA card of the simulator.**

The FPGA-based motor model is designed with Xilinx System Generator (XSG) blockset with no HDL hand coding. Both motor models compute motor currents using a phase-domain algorithm solver that can take into account the instantaneous variation of inductance and non-sinusoidal induced voltage. The FEA-type model uses inductance and Back-EMF profiles computed with JMAG-RT. The *d-q* model uses sinusoidal induced Back-EMF voltage and phase inductance values computed from *Ld* and *Lq* using the well-known Park transformation. A 3-phase IGBT inverter implemented in the FPGA chip drives the PMSM machine.

The PWM controller is designed using Rapid Control Prototyping (RCP) methodology based on Simulink. It is implemented on a separate RT-LAB system using standard Opal-RT FPGA-based I/O cards for Analog Input capture and PWM generation.

The paper presents results from the closed-loop control of the PMSM drive in both current control and speed control modes and discusses the advantages of using such a virtual test bench for motor drives.

I. INTRODUCTION

A critical aspect in the deployment of motor drives lies in the early detection of defects in the design process. The later in the process that a problem is found, the greater the cost to fix it. Rapid prototyping of motor controllers is one methodology that enables the control engineer to quickly deploy control algorithms and find eventual problems. This is typically performed using a small real-time simulator called a Rapid Control Prototyping system (RCP) connected in closed-loop with a physical prototype of the drive to be controlled. Modern RCPs take advantage of a graphical programming language (such as Simulink) with automatic code generation support. Later in the design process, when this code has been converted and fitted into a production controller (using mass-production low-cost devices), the same engineer can verify it against the same physical motor drive, often a prototype or a pre-production unit.

This methodology implies that the real motor drive is available at the RCP stage of the design process. Furthermore, this set-up requires a 2nd drive (such as a DC motor drive) to be connected to the motor drive under test to emulate the mechanical load. This is a complex setup, however it has been proven to be very effective in detecting problems earlier in the design process.

In cases where a physical drive is not available, or where only costly prototypes are available, an HIL-simulated motor drive can be used during the RCP development stage. In such cases, the dynamometer, real IGBT converter and motor are replaced by a real-time virtual motor drive model. This approach has a number of advantages. For example, the simulated motor drive can be tested with borderline conditions that would otherwise damage a real motor. In addition, setup of the controlled-speed test bench is simplified since the virtual shaft speed is set by a single model signal, as opposed to a using real bench, where a 2nd drive would need to be used to control the shaft speed.

Other advantages of using a virtual motor drive system include the ability to easily study the impact of motor drive parameter variations on the controller itself.

Of course, the fidelity of the motor drive model is an important aspect of this process. Classical models, like the PMSM single-frame d-q model [1], are often adequate but lack some aspects of real motor drives, such as saturation and inductance variations caused by stator slots. FEA-based models handle this limitation well [2][6]. The simulator latency also adds a delay in the control loop, which may change its response.

Finally, most motor drives include high-bandwidth on-board protection that cannot be simulated on conventional RCP systems. For example, a real motor drive may include DC-link current motoring for short-circuit detection. If this condition is detected, IGBT pulses are disabled. Despite a sampling time below 10 microseconds, a high-performance classical CPU-based HIL simulation [4] is still too slow to simulate these effects. FPGA-based HIL simulation offers a solution to this challenge.

This paper presents examples of closed-loop control tests using an FPGA-based HIL motor drive controlled by an RCP system. The controller (RCP) and motor drive (HIL) are

978-1-4244-1741-4/08/$25.00 ©2008 IEEE

connected through I/O only, in the same manner as a real controller and a real drive. The motor drive model will be either a classical d-q model or a complete FEA-type model.

II. ADVANTAGES AND CHALLENGES OF AN FPGA-BASED MOTOR DRIVE SIMULATOR

Hardware-In-the-Loop simulation of motor drives is a proven test method that is used extensively in the industry and is typically conducted using CPU-based real-time simulators[3][7][8]. However, the use of an FPGA processor as the computational engine for a motor drive model still presents several advantages over classical CPU-based implementation.

- Very low input-output latency: In some applications, the delay between an IGBT gate action and the reversal of the current slope can cause the tested controller to behave incorrectly. This is notably the case when the motor current is sampled synchronously with the IGBT gating. The presented models have a computation time of 300 nanoseconds in addition to an Analog Output rate of 1 microsecond resulting in a total latency of 1.3 microseconds. Faster D/A converters are also available.

- Capacity to include on-board drive protections: Many motor drive boards include fast protection circuits which are not driven by the controller, due to their fast nature. For example, some IGBT driver boards monitor the DC-link current or stator neutral point voltage, shutting down IGBT pulses if abnormal conditions are detected. Other boards insert minimal dead-time independently of the controller. A complete test scenario for a controller should include its response to these autonomous drive board protection circuits. The presented models have been used to implement such features.

- Block diagram based coding of the FPGA: The proposed model is made with Xilinx System Generator (XSG) and therefore requires no HDL (Hardware Description Language) programming skills, since the FPGA code is automatically generated from a Simulink block diagram. This enables the engineer to quickly modify the motor drive model at will, without any dependency on an FPGA specialist.

- Smart bandwidth-related task repartition: The eDRIVEsim simulator enables users to implement multi-rate and multi-task complex models using FPGA cards and Intel/AMD compatible computers. Resource allocation is made by the user depending on the computation power and bandwidth required by individual subsystems. Separate FPGA and CPU processing allows subsystems to be optimally distributed between respectively fast and slower processors, and/or to naturally accommodate the various bandwidth requirements of a complex system (e.g., a mechanical-electric coupled model). RT-LAB allows this task repartition from within its structure(Fig. 1)

FPGA-coding of the motor drive simulator has posed many challenges in comparison to CPU-based implementation of the same models [4]. The two main causes of these challenges are 1) fixed-point calculations are used for the model and 2) FPGA firmware is not as flexible as regular software. This results in the following model design considerations:

- Per-Unit scaling: The usage of fixed-point calculation is enforced by the usage of the XSG blockset. This requires a careful design of the model. In order for the model to be adaptable to various motor drive parameters and ratings, all FPGA calculations are made on a Per-Unit basis.

- Parameter modifications: The capability for conducting on-line parameter modifications is problematic because of the firmware nature of FPGA processors. In contrast to CPU-based real-time models that allow the modification of all model variables, FPGA-based models have no such built-in features. The presented FPGA-based motor drives are designed so that important motor drive parameter modifications *do not require* the re-generation of an FPGA bitstream.

Finally, it is worth mentioning that an FPGA implementation does not automatically produce more accurate results by the virtue of the use of a small time step (10 nanoseconds). This is because of the fixed-point calculation used to compute the machine drive equations. Of course, the user can tailor the fixed-point format to meet high precision requirements by using larger number formats and FPGA resources. However, because these resources are limited, there will be some tradeoff to be made between precision and FPGA resource usage.

In the case of HIL applications, there is always an interface with real-world signals through Analog I/O channels. Since these channels have limited resolution (typically 12 to 16 bits, neglecting electrical noise), calculating at a precision higher than that of the communicated signals (i.e. the Analog I/O quantities) makes no sense. As a result, the trade-off problem is greatly simplified.

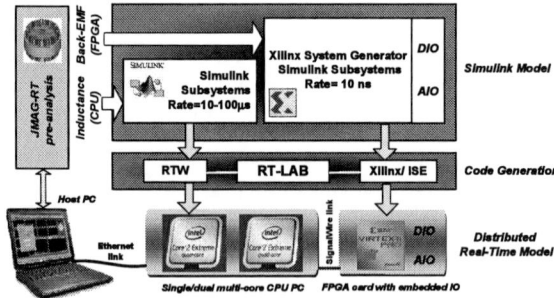

Fig. 1   Workflow structure of the RT-LAB real-time simulator from the model specification to the multi-task real-time execution

III. RT-LAB EDRIVEsim SIMULATOR MODELS

The various eDRIVEsim models used to make the HIL simulation of the PMSM motor drive are described here.

A. General PMSM machine equations

The general PMSM equation in the phase domain is:

$$[L]^{-1} \int (V_{abc} - \frac{d\psi_{abc}}{dt} - RI_{abc})dt = I_{abc} \qquad (1)$$

where

$[L]$ is the inductance matrix,
I_{abc} is the stator current inside the windings,
ψ is the magnet flux linked into the stator windings,
R is the stator resistance and
V_{abc} is the voltage across the stator windings.

Both d-q and JMAG-type models' currents are solved using equation (1). The only difference is that [L] and ψ are sinusoidal in the d-q model, while [L] and ψ are FEA-computed in the JMAG case. Also, most operations are made using 18 bit number with the notable exception of the flux integrators, which have 48 bits because of the 20 nanosecond integration step used.

B. Inverter model

The inverter model includes controlled switches with forward voltage drop, conduction losses and anti-parallel diodes. These diodes turn on only during dead time with a logic that depends on the load current (ex: if the current flows into the load, then the lower diode turns on). Since the signal resolution of the IGBT gate is 10 nanoseconds on the FPGA itself, no interpolation mechanism is used like in the CPU implementation of HIL motor drive[4]. The model does not account for switching losses[5].

The proposed inverter model can also work in fully rectifying mode with no IGBT pulses. Depending on the motor speed and the DC-link voltage, the drive can work in the mode where the motor Back-EMF voltage makes the inverter act as a diode rectifier. This is an improvement on the inverter model proposed in [1].

C. D-Q-Based PMSM model [1]

Fig. 2 describes the two-axis (d-q) PMSM and IGBT inverter models that were designed in XSG and executed with RT-LAB. With this model, all calculations are made in the FPGA itself: the model uses FPGA-stored inverse inductance and nominal-speed Back-EMF tables. During the calculations, the real Back-EMF is found by multiplication of the nominal value by the actual speed. The gate signals of the IGBT inverter can come from external I/O, from a controller model running on one CPU of the simulator, or from an internal PWM source. This internal PWM generation feature is useful for the model self-verification and to validate the model against some

reference model without the use of an externally connected controller.

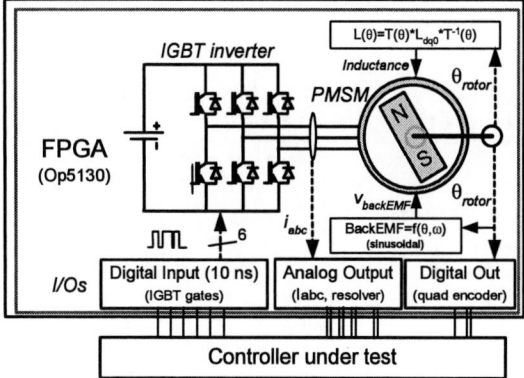

Fig. 2   D-Q-based PMSM drive implemented on the FPGA

D. Finite-Element-Analysis-Based PMSM model [2]

In the FEA-based PMSM model, the electric equations of the PMSM and its IGBT inverter are still computed on the FPGA computational engine using the same phase-domain solver as the d-q model. Notably, this includes the 1-D nominal speed Back-EMF table, derived from JMAG-RT, developed by the Japanese Research Institute. The main computational difference with the previously described d-q model is that the 2-D inductance inverse matrixes $L^{-1}(\theta, I_{abc})$ as well as the electrical torque are now computed on the CPU and transmitted on the FPGA where interpolation methods are used to up-sample the inductance at 10 nanosecond rate. This scheme is depicted in Fig. 4

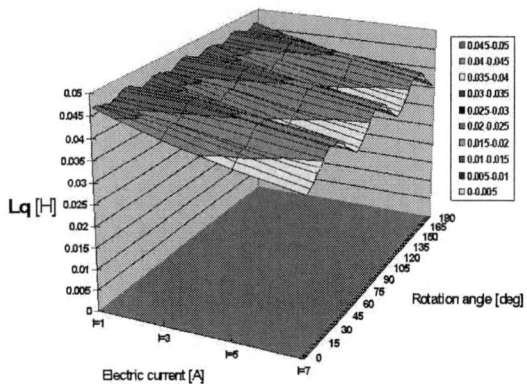

Fig. 3 L_q inductance profile computed by the JMAG model (with sqrt(3/2) factor)

The L_q profile of the motor used in this paper is shown in Fig. 3. Storing the inverse of the inductance matrix table on the FPGA itself is difficult and is under investigation. It is also normal to compute the electrical torque on the CPU because the device speed is usually computed from many different torque sources like, for example, mechanical torque produced by the engine in hybrid propulsion cars[3][8].

As for the d-q model, the storage of the nominal speed Back-EMF profile of the JMAG model on the FPGA enables the computation of real Back-EMF voltages by multiplication of the fixed table stored value by the actual speed.

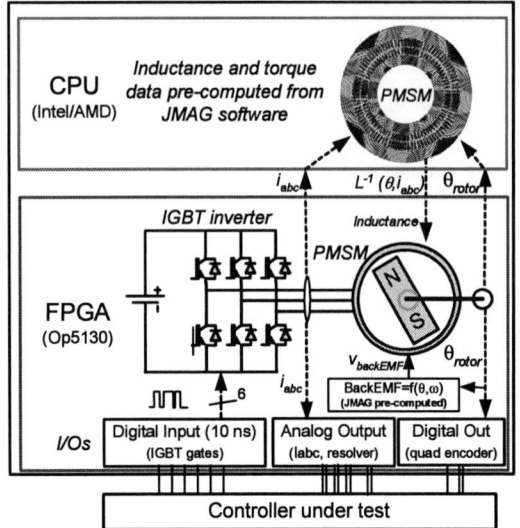

Fig. 4   Real-time simulation of FEA-based PMSM drive on an FPGA

E.   Mechanical Model

The mechanical model is computed on the main processors of the real-time simulator at 65 microseconds and can be linked to complex models implemented with Simulink or compatible tools. For example, a complete car dynamic model can be implemented using the CARSIM software package and

executed in real-time on one CPU of the simulator. This dynamic model can then be interfaced with the motor drive on the FPGA chip.

F.   DC Source and AC fed System

In the presented model, the DC-link voltage comes from one CPU of the RT-LAB simulator. If desired, a more complex feeding circuit can be implemented, for example a 6-pulse diode rectifier or a grid equivalent circuit, still computed on the CPU and interfaced with the FPGA model. This can be done using standard Simulink, with or without specialized blocksets, such as SimPowerSystems and ARTEMIS from Opal-RT Technologies.

IV. MOTOR CONTROLLER DESCRIPTION

The motor controller is a classical vector controller using the Park d-q transform to control motor currents and torque. An outer speed control loop completes the controller. This enables the torque control of the motor by acting on the Iq current component. For this experiment, no flux weakening is applied and the Id current command is therefore set to zero. PWM generation can be made in either sinus or space-vector modulation with user variable carrier frequency and dead-time. In our tests, the PWM frequency is 5 kHz and dead-time is 5 microseconds. All models are made with Simulink. The sample time of the controller is 50 microseconds.

The controller is interfaced to the virtual motor drive through a proper set of I/Os. In this case, the controller is equipped with the standard RT-LAB I/O configuration: 16 Analog Inputs, 16 Analog Outputs, 16 Digital Inputs and 16 Digital Outputs. The I/O card is based on FPGA and has a 10 nanosecond resolution for the PWM generation. The controller model and interface to the HIL motor drive is shown in more detail in Fig. 5.

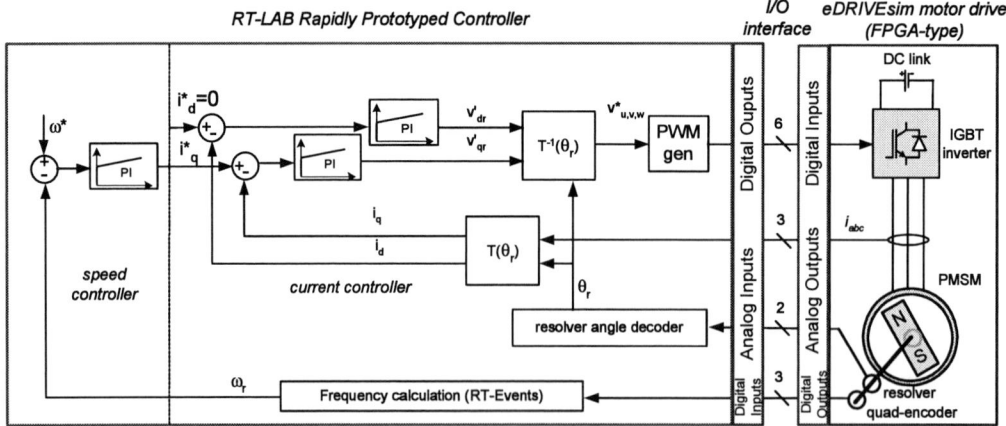

Fig. 5   Motor controller structure and interconnection with the eDRIVEsim PMSM drive

V. EXPERIMENTAL RESULTS

In this section, we describe the experiments that where conducted using the FPGA-based PMSM drive (d-q and JMAG) in closed-loop with a motor controller running on a 2nd independent simulator as a Rapidly Prototyped Controller. Using the set-up and models previously described, current and speed control tests were performed.

A. Real-Time Simulator Hardware set-up

As previously mentioned, the controller and the motor drive are implemented on two independent RT-LAB simulators, as depicted in Fig. 6. The controller is running on an MX Station-type RT-LAB simulator (middle box) with I/O connections on the back. The MX Station is a 2.3 GHz Core2 Duo-based PC. The MX Station I/Os are based on 2 different Opal-RT FPGA cards: the OP5110 (PCI) implements the digital inputs and outputs while the OP5130 implements the analog inputs. It is possible to use OP5110 for analog input, however, the OP5130 is user reconfigurable enabling it to be programmed with a closed-loop Resolver-To-Digital model for future studies. The Digital I/Os are isolated in both systems and the opto-coupler circuits are powered by the 12V power supply of the RCP.

The motor drive is running on the eDRIVEsim RT-LAB simulator (bottom box). The eDRIVEsim simulator is a 2.3 GHz dual quad-core (Intel Core2) PC with an onboard FPGA card (Opal-RT OP5130). The FPGA card holds the machine and inverters models for both d-q and JMAG-type models. One core of the PC computes the machine electrical torque with the mechanical equations as well as the inductance values in the case of the JMAG model.

A console PC (not shown) completes the simulator set-up and enables the control and monitoring of the real-time systems themselves. All of these are connected together by an Ethernet switch (top box of Fig. 6)

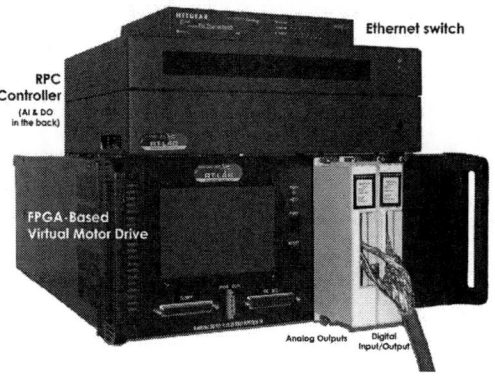

Fig. 6   Virtual motor drive closed-loop control hardware set-up

B. Machine Drive Parameters

The machine under test is a 4 poles (2 pairs) interior magnet machine (IPM) with the parameters described in Table 1. The 800V DC-link voltage is rather high and allows the motor to operate at 12000 RPM without flux weakening.

TABLE 1. MOTOR PARAMETERS

Quantity	Value
Stator resistance	3.3 Ω
Park equivalent direct inductance	0.0109 H
Park equivalent quadrature inductance	0.0310 H
Magnet flux	0.1584 Wb
Number of pair of poles	2
Total number of slots	24
Inertia	$1.854*10^{-4}$ kg.m^2
Friction	$5.396*10^{-5}$ N.m.s
Applied mechanical torque	1 N.m.
DC-link voltage	800 V
FPGA sample time	10 ns
CPU sample time	65 μs
Quadrature encoder	1024 pulses/channel

C. Controller Parameters

The controller parameters used in these tests are enumerated in Table 2. The reference Id value is set to 0 in all tests as no flux weakening is made. The Vdq limits are really the entry to the dq-abc transform that produce the modulation indexes for all phases and are therefore limited to ±1. All PI controllers have anti-windup mechanisms.

TABLE 2. CONTROLLER PARAMETERS

Quantity	Value
Current controller Ki Kp/Ki gains	22.7 0.0017
Vdq limits (from current controller)	±1 pu
Speed controller Ki Kp/Ki gains	3.5 0.026 (except section E)
Iq limits (from speed controller)	±9.5 A
PWM frequency	5 kHz
Dead time	5 μs
Sample time	50 μs

The rotor angle position, θ, sent by the motor drive analog outputs as $sin(\theta)$ and $cos(\theta)$, is read by analog inputs of the RCP. The $sin(\theta)$ and $cos(\theta)$ are then properly decoded on the controller (avoiding basic $arctan\{sin(\theta)/cos(\theta)\}$, because of the analog noise floor). This type of position encoding is similar to using a resolver with DC excitation. For its part, the quadrature encoder is used to obtain the rotor speed in the controller.

D. Current control mode tests

One great advantage of a virtual motor drive test bench over a real one is ease of the hardware set-up. This becomes even clearer when one tries to test motor current control loops. In this case, with a real bench, one really needs two motor drives: the PMSM drive itself and a second motor drive to control the shaft speed. With a virtual bench, one simply has to set the shaft speed state-variable to the desired speed!

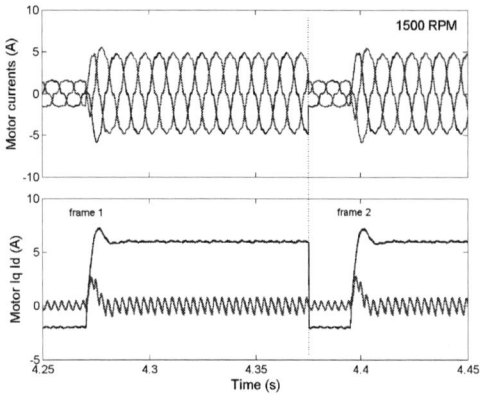

Fig. 7   Current command step (d-q model, 1500 RPM)

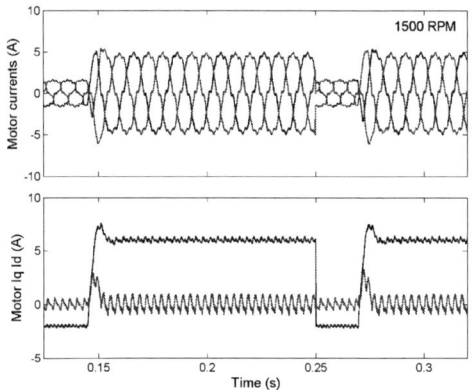

Fig. 8   Current command step (JMAG model, 1500 RPM)

Fig. 7 and Fig. 8 respectively show the d-q and JMAG model response of the current control at 1500 RPM, corresponding to a 50 Hz stator frequency and a commanded *Iq* current step from –2 to 6 A. The two responses are strikingly similar except for increased fluctuations on the *Iq* command signal of the JMAG model, most likely caused by the slot-induced current harmonic content of the JMAG model. This harmonic content is itself caused by the inductance variations (due to gap variations for example) of the JMAG model and cannot be modeled in a standard (single frame, non-saturating) d-q model. In the figures, the *Iq* component is recognizable as

the one following the current amplitude. Nevertheless, the difference between the d-q and JMAG models is quite modest. It seems that the dead-time is causing a greater impact than the effects of non-ideal Back-EMF and inductances.

The two figures actually display two acquisition frames acquired at the Windows console station of the RCP. This means that there is a time discontinuity in the figures. In Fig. 7 for example, the two frames begins at 4.25 and 4.37 sec. Since the command is applied asynchronously from the motor drive, this enables verification of the drive response relative to the motor rotor angle at the time of the command application. The results of Fig. 7 and Fig. 8 show that the controller response does not depend on the motor angle.

E. Speed control mode tests

Following the adjustment of current control parameters, closed-loop speed control tests are performed. Fig. 9 and Fig. 10 compare the speed response of the motor drive for d-q and

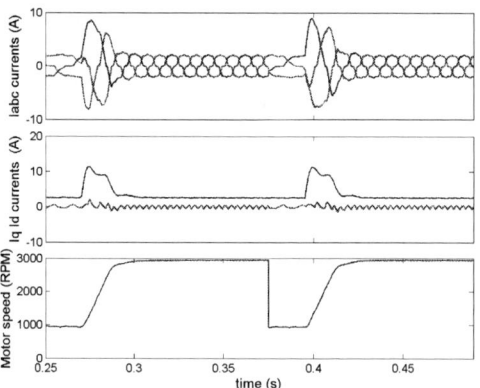

Fig. 9   Speed step (d-q model, 1000->3000 RPM)

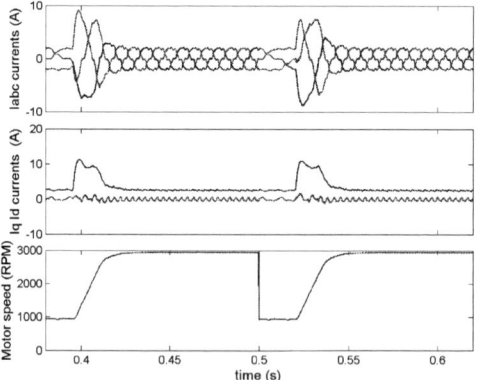

Fig. 10   Speed step (JMAG model, 1000->3000 RPM)

1082

JMAG models respectively, for a speed command step of 1000 to 3000 RPM. The responses are again very similar. Both figures show two consecutive acquisition frames. It is interesting to see that because of the asynchronous speed step application, the phase current responses differ from one frame to another mainly because of the initial conditions of the motor at each application of the speed command steps.

Again, there is not a great difference between d-q and JMAG model results. Tests on a drive with smaller inertia may reveal a greater sensibility of the drive to cogging torque.

F. HIL application: Speed Controller tuning using batch test

The controllers used in this experiment are rather simple but nevertheless have non-linearities like saturations. Because these effects are hard to account for by regular control theory, a thorough batch test procedure is often necessary to validate a controller against the various control scenarios and contingencies of the motor drive.

PI controller tuning is one good example of batch tests using a real-time simulator. In this case, a set of speed commands is applied to the motor drive for various PI control gain values. Results are analyzed for the best set of PI parameters. Fig. 11 shows that some instability results from a large speed command step (8000 to 12000 RPM). It was found that dividing the Kp gain of the speed controller by 3 stabilizes the response of the motor. One can notice that the current commands are identical in the first 20 milliseconds but diverge after. This divergence is most likely caused by a saturation or anti wind-up action in the speed control: referring to Table 2, the controller Iq limit is set to ±9.5A.

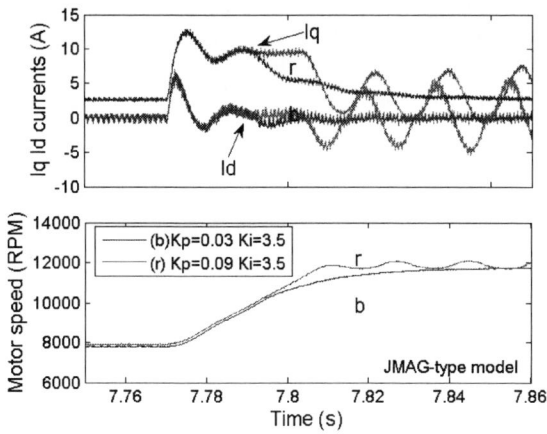

Fig. 11   Speed control PI tuning results (JMAG model)

CONCLUSION

This paper has presented experimental results of a motor controller connected to a virtual motor drive. The controller was designed in Simulink and implemented on a RT-LAB RCP system. The motor drive was also designed in Simulink, and implemented on an FPGA processor using Xilinx System Generator blockset. Both PMSM standard d-q and FEA-computed models were used for the tests.

The control was successfully tested in closed-loop for both current and speed control modes using the eDRIVEsim real-time plant simulator. The closed-loop set-up has been simply and successfully used to tune a controller for stability.

The results show certain corroboration between JMAG and d-q-type model simulation cases, as their responses are similar. This does not mean that increased precision of FEA models is not needed, but rather that the test cases presented were not chosen to highlight the benefit of high-accuracy finite-element-based models. Tests with a partially saturated machine and/or with smaller inertia should demonstrate the usefulness of FEA models.

By using a virtual motor drive, as the one presented here, early in the design process, it is expected that algorithm errors can be pinpointed earlier by the control engineer and thus produce a more efficient design methodology. Leading power electronic equipment and hybrid vehicle manufacturers now successfully use such design methodology.

REFERENCES

[1] C. Dufour, S. Abourida, J. Bélanger,V. Lapointe, "Real-Time Simulation of Permanent Magnet Motor Drive on FPGA Chip for High-Bandwidth Controller Tests and Validation", 32nd Annual Conference of the IEEE Industrial Electronics Society(IECON-06), Paris, France, November 7-10, 2006.

[2] C. Dufour, J. Bélanger, S. Abourida, V. Lapointe, "FPGA-Based Real-Time Simulation of Finite-Element Analysis Permanent Magnet Synchronous Machine Drives", Proceedings of the 2007 Power Electronics Specialists Conference (PESC-07), Orlando, Florida, June 17-21, 2007

[3] C. Dufour, T. Ishikawa, S. Abourida, J. Bélanger, "Modern Hardware-In-the-Loop Simulation Technology for Fuel Cell Hybrid Electric Vehicles", Proceeding of the 2007 IEEE Vehicle Power and Propulsion Conference (VPPC-07), Arlington, Texas, Sept. 9-12, 2007

[4] M. Harakawa, H. Yamasaki, T. Nagano, S. Abourida, C. Dufour, J. Bélanger, "Real-Time Simulation of a Complete PMSM Drive at 10 us Time Step", Proceedings of the 2005 International Power Electronics Conference (IPEC 2005) – April 4-8, 2005, Niigata, Japan.

[5] G.G. Parma, V. Dinavahi, "Real-Time Digital Hardware Simulation of Power Electronics and Drives",Transactions on Power Delivery, Vol 22, No. 2, April 2007

[6] O. A. Mohammed, Fellow IEEE, S. Liu, Member IEEE, and Z. Liu,"A Phase Variable PM Machine Model for Integrated Motor Drive Systems", 35th Annual IEEE Power Electronics Specialists Conference, Aachen, Germany, 2004, pp. 4825-4831

[7] L.-F. Pak,, V. Dinavahi, G. Chang, M. Steurer, P.F. Ribeiro, "Real-Time Digital Time-Varying Harmonic Modeling and Simulation Techniques", IEEE Trans. On Power Delivery Vol.22. No. 2, April 2007.

[8] D. Winkler, C. Gühmann, "Hardware-in-the-Loop simulation of a hybrid electric vehicle using Modelica/Dymola" Proceedings of the 22nd International Battery, Hybrid and Fuel Cell Electric Vehicle Symposium (EVS-22), Yokohama, Japan, Oct 22-26 2006, pp 1054-1063.

Speed Sensorless Nonlinear Control Of Induction Motor In The Field Weakening Region

Miroslaw Wlas [*], Haithem Abu-Rub [**], Joachim Holtz [***]

[*] Gdansk University of Technology, Gdansk, Poland, e-mail: *mlas@ely.pg.gda.pl*
[**] Texas A&M University at Qatar, Doha, Qatar, e-mail: *haitham.abu-rub@qatar.tamu.edu*
[***] Wuppertal University, Wuppertal, Germany, e-mail: *j.holtz@ieee.org*

Abstract—In the paper it is presented speed sensorless nonlinear control system using multiscalar model based MMB operating in the lower field weakening speed region. Nonlinear control methods can improve the performance of induction motor drives in transients. The maximum available output torque, which guarantees satisfactory motor dynamics in field weakening region, is calculated using the steady state dependencies of induction motor model. Full dynamic reaction to commanded changes or to disturbances that occur during operation at the voltage limit is enabled by reducing the excitation level in a fastest possible way. Experimental results are presented to demonstrate reliability of proposed controller. In the experimental implementation a 22kW induction motor is used. The whole control scheme (including multiscalar control, speed computation and space vector PWM) are implemented on a DSP and FPGA.

Keywords— Non-linear control, Sensorless control, Flux model, Voltage Source Inverters (VSI), Induction motor.

I. INTRODUCTION

Sensorless drives with induction motor, where "sensorless" means control system without speed or position sensor, became interesting for industrial applications in recent years. One of the reasons for such approach is a development of a suitable methods and algorithms for practical sensorless control of induction motor. The advanced control system for the induction motor should provide proper speed and torque control without measuring these values. The only variable that should be measured for realizing practical solutions is the stator current [12]. Increasing the effectiveness of the solution is obtained by computing the output voltage or only by using its command value while all other required variables are calculated using this voltage and the measured current.

Field oriented control (FOC) today is the most popular control method for inverter powered induction motor drives. In applications such as traction, electric vehicles and spindle drives field weakening is required to provide wide speed range operation. Stator flux oriented (SFO) control methods [1, 6, 8] are preferred for induction motor control in the field weakening region. However implementation of stator flux orientation method is more complex than rotor-flux orientation, the advantage is that at any operating condition, the stator flux level inside the machine is higher than the rotor flux. In the field-weakening region, where the lower dynamics appear after the flux command decreases, the maximum available

torque can be calculated with consideration of both the motor current limit and the inverter voltage limit using the steady state dependencies of induction motor vector model [7].

Non-linear control of induction motor has been presented first time in [2]. It has been introduced a novel model of induction motor, which was named multiscalar model based (MMB). Such model was used later by many other authors [3,4]. The resulted model consists of two completely decoupled subsystems – mechanical and electromagnetic.

Most of used nonlinear control ideas are rotor oriented. Using stator flux orientation enables faster dynamics and is less parameter sensitive. Stator flux orientation is very attractive mainly because of its insensitivity to motor parameters and fast response. Therefore our approach is to use stator dependencies for control principles. The nonlinear dependencies with stator flux were developed lastly [3]. A new multiscalar model is based on stator current and stator flux vectors. Our solution is to use the new MMB approach for sensorless drive with speed observer.

The proposed control algorithm provides limitation of possible torque in the lower field weakening region. Dynamic requirements when operating at maximum voltage are met by PI regulator of inverter output voltage [1, 7]. In this paper the stator flux based multiscalar model presented in [3] is further investigated. The maximum command output torque for the field weakening operation is calculated using the steady state dependencies of induction motor multiscalar model. Motor torque and inverter voltage limits are considered in the nonlinear controller to prevent the possibility of regulation loss and appearing of breakdown torque in the high speed operation of induction motor. The theoretical consideration will be experimentally verified. In the experimental implementation a 22kW induction motor is used. The whole control scheme (including multiscalar control, speed computation and space vector PWM) is implemented on a floating point DSP and FPGA.

II. INDUCTION MOTOR DESCRIPTION

The steady state and transient electromagnetic behaviors of an induction motor can be described by the following equations in the stationary reference frame:

$$\frac{di_{sx}}{d\tau} = -\frac{L_r R_s + L_s R_r}{w_\sigma} i_{sx} + \frac{R_r}{w_\sigma} \psi_{sx} - \omega_r i_{sy} + \omega_r \frac{L_r}{w_\sigma} \psi_{sy} + \frac{L_r}{w_\sigma} u_{sx}$$
(1)

$$\frac{di_{sy}}{d\tau} = -\frac{L_r R_s + L_s R_r}{w_\sigma} i_{sy} + \frac{R_r}{w_\sigma} \psi_{sy} + \omega_r i_{sx} - \omega_r \frac{L_r}{w_\sigma} \psi_{sx} + \frac{L_r}{w_\sigma} u_{sy}$$
(2)

$$\frac{d\psi_{sx}}{d\tau} = -R_s i_{sx} + u_{sx},$$
(3)

$$\frac{d\psi_{sy}}{d\tau} = -R_s i_{sy} + u_{sy},$$
(4)

where i_{sxy}, ψ_{sxy}, u_{sxy} denotes vectors of stator current, stator flux, and stator voltage respectively, ω_r is rotor speed. Additionally, the mechanic behavior of an induction motor can be described by the rotor speed equation:

$$\frac{d\omega_r}{d\tau} = \frac{1}{J}\left(\psi_{sx} i_{sy} - \psi_{sy} i_{sx}\right) - \frac{1}{J} m_O$$
(5)

where the parameter w_σ is defined by:

$$w_\sigma = L_s L_r - L_m^2$$
(6)

and L_s, L_r, L_m are stator, rotor and mutual inductances, R_s, R_r are stator and rotor resistances, m_O is the load torque and J is the moment of inertia. All variables and parameters are expressed in p.u. system. Stationary xy reference frame is recommended for simplifications of measurements.

III. NON-LINEAR CONTROL SYSTEMS BASED ON MULTISCALAR MODEL

A. Multiscalar Model

The variables for description of dynamics of the induction motor was defined in [2]:

$$x_{11} = \omega_r,$$
(7)

$$x_{12} = \psi_{sx} i_{sy} - \psi_{sy} i_{sx},$$
(8)

$$x_{21} = \psi_{sx}^2 + \psi_{sy}^2,$$
(9)

$$x_{22} = \psi_{sx} i_{sx} + \psi_{sy} i_{sy}.$$
(10)

These new variables (7) – (10) are called multiscalar variables because they are all scalars: rotor speed, electromagnetic torque, the square of stator flux modulus and the reactive torque respectively. They are all independent from any choice of reference frame. The differential equations for these variables form the multiscalar model of induction motor:

$$\frac{dx_{11}}{d\tau} = \frac{1}{J} \cdot x_{12} - \frac{m_0}{J},$$
(11)

$$\frac{dx_{12}}{d\tau} = -\frac{1}{T_V} \cdot x_{12} + x_{11} \cdot \left(x_{22} - \frac{L_r}{w_\sigma} \cdot x_{21}\right) + \frac{L_m}{w_\sigma} \cdot u_1,$$
(12)

$$\frac{dx_{21}}{d\tau} = -2 \cdot R_s \cdot x_{22} + 2 \cdot u_2,$$
(13)

$$\frac{dx_{22}}{d\tau} = -\frac{1}{T_V} \cdot x_{22} - x_{11} \cdot x_{12} + \frac{R_r}{w_\sigma} \cdot x_{21}$$

$$-R_s \cdot \frac{x_{12}^2 + x_{22}^2}{x_{21}} + 2 \cdot \frac{L_r}{w_\sigma} \cdot u_2 - \frac{L_m}{w_\sigma} \cdot u_1',$$
(14)

were:

$$u_1 = u_{sy} \cdot \Psi_{rx} - u_{sx} \cdot \Psi_{ry}$$
(15)

$$u_2 = u_{sx} \cdot \Psi_{sx} + u_{sy} \cdot \Psi_{sy}$$
(16)

$$u_1' = u_{sx} \cdot \Psi_{rx} + u_{sy} \cdot \Psi_{ry}$$
(17)

$$T_V = w_\delta L_r / R_r w_\delta + R_s L_r^2 + R_r L_m^2 .$$
(18)

B. Control system based on Multiscalar Model

The non-linear feedback decoupling of the form:

$$u_1 = \frac{w_\sigma}{L_m} \cdot \left[m_1 - x_{11} \cdot \left(x_{22} - \frac{L_r}{w_\sigma} \cdot x_{21} \right) \right]$$
(19)

$$u_2 = \frac{1}{2} \cdot \left(m_2 - \frac{1}{T} \cdot x_{21} \right) + R_s \cdot x_{22}.$$
(20)

transform the system (11)-(13) to the linear systems of the following form:
mechanical subsystem

$$\frac{dx_{11}}{d\tau} = \frac{1}{J} \cdot x_{12} - \frac{m_0}{J},$$
(21)

$$\frac{dx_{12}}{d\tau} = -\frac{1}{T_v} \cdot x_{12} + m_1,$$
(22)

electromechanical subsystem

$$\frac{dx_{21}}{d\tau} = -\frac{1}{T} \cdot x_{21} + m_2.$$
(23)

where m_1 and m_2 are new inputs in the linear system. In equation (23) time constant T has been accept $T = 2,5 \cdot T_v$. In this value of time constant retain fast current and slow flux variation properties. The control variables u_1 and u_2 appearing in the new multiscalar model of induction motor are transformed on stator voltage components in the following way:

$$u_{sx} = \frac{u_2 \cdot \Psi_{rx} - u_1 \cdot \Psi_{sy}}{\Psi_{sx} \cdot \Psi_{rx} + \Psi_{ry} \cdot \Psi_{sy}},$$
(24)

$$u_{sy} = \frac{u_2 \cdot \Psi_{ry} + u_1 \cdot \Psi_{sx}}{\Psi_{sx} \cdot \Psi_{rx} + \Psi_{ry} \cdot \Psi_{sy}}.$$
(25)

A scheme of the system with non-linear control based on stator flux and current is presented in Fig. 1. This system consisting of two linear subsystems may be controlled by means of cascaded controllers of mechanical subsystem but only one controllers of flux. The new state variables do not depend on the system coordinate. This is essential for the practical realization of control systems because it significantly simplifies the drive system. Sign ^ denotes variables estimated in the speed observer presented in [3, 4].

IV. THE IDEA OF FIELD WEAKENING

The maximum stator voltage is determined by the available inverter dc link voltage and PWM strategy. Control variables (18) and (19) should satisfy the following relation:

$$\frac{u_1^2 + u_2^2}{x_{21}} = U_s^2 \leq U_{smax}^2 .$$
(26)

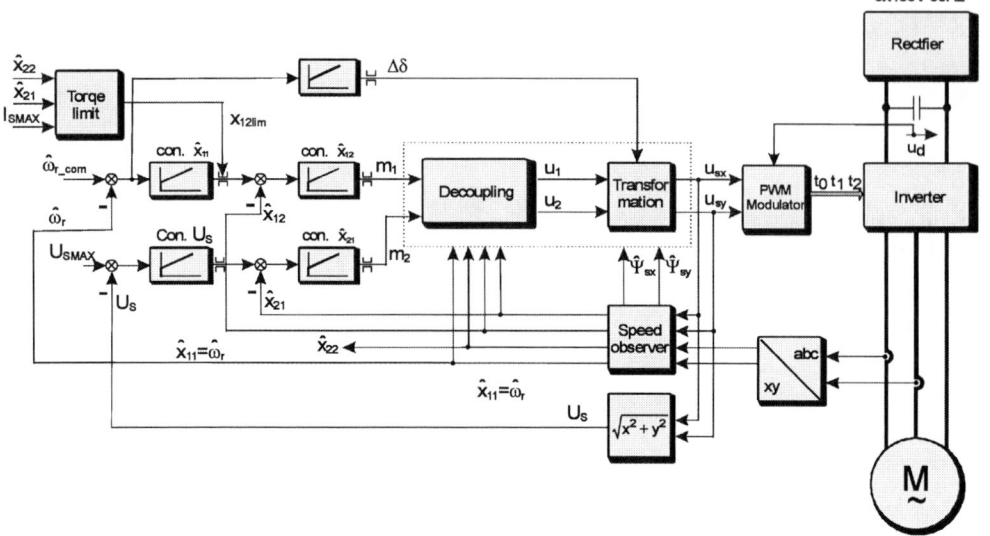

Fig. 1: Structure of the control system for the induction motor based on multiscalar model

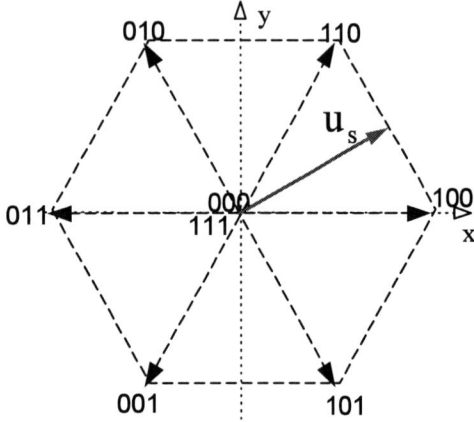

Fig. 2. Vector Voltage PWM generating without overmodulation.

The graph Fig. 2 shows how the limiting algorithm reduces an overlength reference voltage vector U_s to match the maximum output voltage of the inverter in a given operating point. The reference voltage is generally generated as the output signal of a PI controller, and this controller is normally implemented in control system (Fig. 1). Provision must be made to avoid a windup problem with this controller. The limiter of its output signal must be integrated into the controller algorithm. This requires calculation voltage U_s from stationary xy reference frame.

The maximum stator current of the motor is limited by the inverter current rating and the motor thermal rating. The relation among x_{12}, x_{21}, x_{22} and maximum stator current is given as:

$$\frac{x_{12}^2 + x_{22}^2}{x_{21}} = I_s^2 \leq I_{s\,max}^2 . \tag{27}$$

While operating in the field weakening region, the excitation level must be under all circumstances maintained as commanded by the excitation control, even at overload [1]. Overriding the excitation control would interfere with the optimal setting of the operating point. Current limitation at overload must be therefore restricted to limiting x_{12} while leaving the flux producing variable x_{21} unaffected. For limiting of state variable x_{12z} (command value of the torque) it is used form from (27) [10]:

$$x_{12\,lim} = \sqrt{I_{s\,max}^2 x_{21} - x_{22}^2} , \tag{28}$$

which establishes a priority for the commanded x22. It is only x12 must be reduced to keep the total current within the current capability of the inverter. The variable x_{21} has the priority in the control process in the field weakening range and its value is adjusted by the voltage controller.

The control structure so far described ensures operation at maximum torque throughout the field weakening range by fully utilizing the voltage and current capability of the inverter. It does not allow for a manipulative margin to enable reactions to dynamic changes. All these disturbances invariably produce a speed error $\Delta\omega$ [1]. A negative speed error indicates that the speed is going to reduce. Also the induced voltage then reduces and the voltage margin as seen from the maximum inverter voltage increases. This allows the system to react only to some limited extent, as the available voltage margin might be only small [1].

The consequence is that the stator flux controller (x_{21}) decreases the excitation level such that operation at the limit of maximum voltage can continue. However, a reaction to the dynamic requirement that had caused the speed error is not possible. The available voltage just suffices to maintain operation in a steady-state. Full dynamic reaction to commanded changes or to disturbances that occur during operation at the voltage

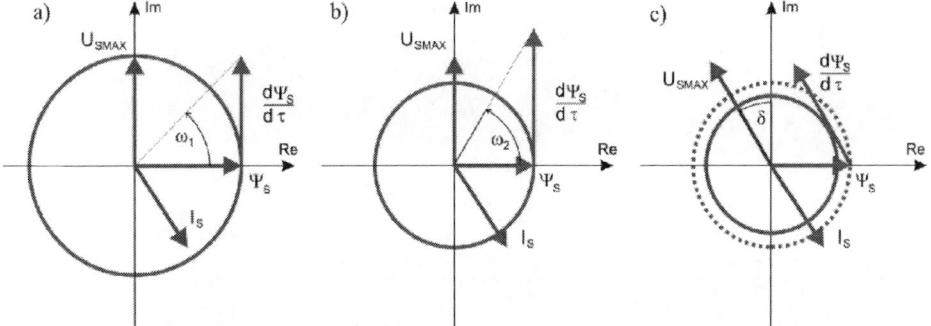

Fig. 3. Vector diagrams illustrating operation at maximum inverter voltage

limit is enabled by reducing the excitation level in a fastest possible way. A temporary voltage margin is thus created that allows the x_{12} controller to respond to the dynamic requirement [1]. The speed error $\Delta\omega$ is used as a feedforward signal for this purpose. A signal $\Delta\delta$ is then created that rotates the stator voltage vector such that the stator flux linkage vector is deviated towards a trajectory of reduced diameter. Fig. 3(a) illustrates a situation where the system operates at the voltage limit at base speed [1, 10]. As the speed increases, the magnitude ψ_s of the rotor flux linkage vector reduces since the operating point transfers to field weakening, shown in Fig. 3 (b). Also in this case is operation at the voltage limit assumed and the condition $d\psi_s/dt \approx u_{smax}$ holds. The voltage drop across the stator resistance is negligible since the speed is high. A dynamic requirement in this situation is met by photo introducing a rotation of the stator voltage vector by an angular increment $\Delta\delta$. The trajectory of ψ_s then aims towards lower flux magnitude values as seen in Fig. 3(c). The excitation of the machine is temporarily reduced which creates a voltage margin within very short time [1]. Depending on the particular conditions of operation, the generated angular increment $\Delta\delta$ may vary in a range $0 \leq \Delta\delta < 70°$.

V. ROTOR FLUX AND ROTOR SPEED OBERVER

A speed observer for estimation of rotor speed of the induction motor was proposed in [11] and developed by the first author in [4]. The speed observer consists of linear dynamical part stabilized by non-linear feedbacks and non-linear expression for calculation of the estimated rotor speed. The speed observer makes it possible to estimate the rotor speed with high accuracy in wide range of rotor speed including stopped state with the load. Differential equations of the speed observer are as follows:

$$\frac{d\hat{i}_{sx}}{d\tau} = a_1 \hat{i}_{sx} + a_2 \hat{\psi}_{rx} + a_3 \zeta_y + a_4 u_{sx} \qquad (29)$$
$$+ k_3 \left(k_1 \left(i_{sx} - \hat{i}_{sx} \right) - \hat{\omega}_r \zeta_x \right),$$

$$\frac{d\hat{i}_{sy}}{d\tau} = a_1 \hat{i}_{sy} + a_2 \hat{\psi}_{ry} - a_3 \zeta_x + a_4 u_{sy} \qquad (30)$$
$$+ k_3 \left(k_1 \left(i_{sy} - \hat{i}_{sy} \right) - \hat{\omega}_r \zeta_y \right),$$

$$\frac{d\hat{\psi}_{rx}}{d\tau} = a_5 \hat{i}_{sx} + a_6 \hat{\psi}_{rx} - \zeta_y - k_2 \left(\hat{\omega}_r \hat{\psi}_{ry} - \zeta_y \right), \quad (31)$$

$$\frac{d\hat{\psi}_{ry}}{d\tau} = a_5 \hat{i}_{sy} + a_6 \hat{\psi}_{ry} + \zeta_x + k_2 \left(\hat{\omega}_r \hat{\psi}_{rx} - \zeta_x \right), (32)$$

$$\frac{d\zeta_x}{d\tau} = k_1 \left(i_{sy} - \hat{i}_{sy} \right), \qquad (33)$$

$$\frac{d\zeta_y}{dt} = -k_1 \left(i_{sx} - \hat{i}_{sx} \right), \qquad (34)$$

$$\hat{\omega}_r = S \left(\sqrt{\frac{\zeta_x^2 + \zeta_y^2}{\hat{\psi}_{rx}^2 + \hat{\psi}_{ry}^2}} + k_4 \left(V - V_f \right) \right), \qquad (35)$$

$$\frac{dV_F}{dt} = \frac{1}{T_s} \left(V - V_F \right) \qquad (36)$$

$$V = S(\hat{\psi}_{rx}\zeta_y - \hat{\psi}_{ry}\zeta_x), \qquad (37)$$

where S is sign of the speed determined from values of the estimated rotor flux ($\hat{\psi}_{rx}, \hat{\psi}_{ry}$) and disturbances ($\zeta_y, \zeta_x$):

$$S = \begin{cases} 1 & \text{if} \quad \left(\xi_x \hat{\psi}_{rx} + \xi_y \hat{\psi}_{ry} \right) > 0 \\ -1 & \text{if} \quad \left(\xi_x \hat{\psi}_{rx} + \xi_y \hat{\psi}_{ry} \right) \leq 0 \end{cases} \qquad (38)$$

Notation of variables and coefficients are defined in [11]. The speed observer has been used in the closed control system of the induction motor.

TABLE I.
OBSERVER GAINS AND TIME CONSTANT

Name	Value
k_1	3 [-]
k_2	0.6 [-]
k_3	0.3 [-]
k_4	1 [-]
T_s	0.63 [-]

VI. EXPERIMENTAL RESULTS

A. Experimental setup

The experimental setup consists of two 22kW induction motors and two converters. The first drive system 690AC+ operates as a load while the second AMT022 is the tested drive (Photo 1). The whole control scheme (including speed computation and space vector PWM) of

Photo. 1. Switch box with power converters and 22kW electrical machines

AMT022 converter is implemented on a floating point DSP: ADSP21065L and FPGA: Flex6016. Machine operation in the field weakening region is described at Fig. 4. Operation at the voltage limit does not allow for fast dynamic reaction.

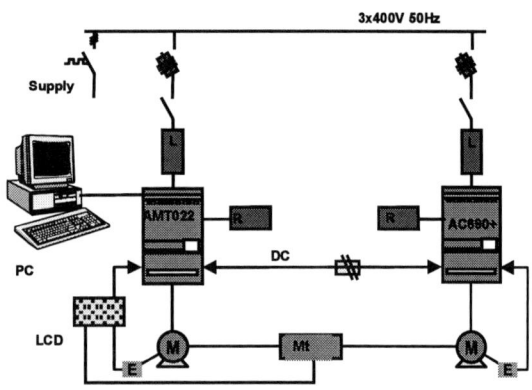

Fig. 4. Scheme of 22kW experimental setup

B. Investigation results

All system variables are presented in per unit system with base values listed in Tab. 2. Rotor speed calculated in the observer was used in speed controller and in decoupling block (19) and (20). Results obtained in system with nonlinear control system and modified speed and flux observer are presented in Fig. 5,6 and 7.

In Fig. 5 are presented transients during step changes of rotor set speed The waveforms of the estimated speed $\hat{\omega}_r$, torque limit x_{12lim}, the estimated torque $\hat{x}_{12}$, the estimated square of flux $\hat{x}_{21}$ and angular increment $\Delta\delta$ during speed change from 0.2 to 1.7 p.u. are shown in Fig. 5. When speed archive 0.8 p.u algorithm of flux-weaking starts. Square of stator flux decrease rapidly due to angular increment generation. The torque limit x_{12lim} change also what follow from equation (28). In this investigation Ismax =1.0 nad Usma x = 0.8 p.u.. Next experimental transient after change of motor load and 1.8 p.u. command for set speed in the system with speed estimated was shown in Fig. 6. The control system flux is not sensitive on disturbance – change load or set command of rotor speed. Reverse with active load are shown on fig. 7. Measured speed x11 was compared with estimated value and difference was lower then 0.5% in steady state and 1.5% in transient. Motor torque x_{12} is controlled very fast and square of rotor flux x21 is keep nearly constant except period where controllers works with field weakening region.

TABLE II.
FSLG - L4 INDUCTION MOTOR PARAMETERS AND BASE UNIT

Parameters			Value
Nominal	U_n	Voltage	400 [V]
	I_n	Current	38,8 [A]
	m_n	Torque	143,4 [Nm]
	P_n	Power	22000 [W]
	ω_n	Speed	1465 [RPM]
Base unit	U_b	Voltage	400 [V]
	I_b	Current	67,2 [A]
	m_b	Torque	171,1 [Nm]
	Ψ_b	Flux	1,27 [Wb]
	ω_b	Speed	157,07 [rad/s]

1088

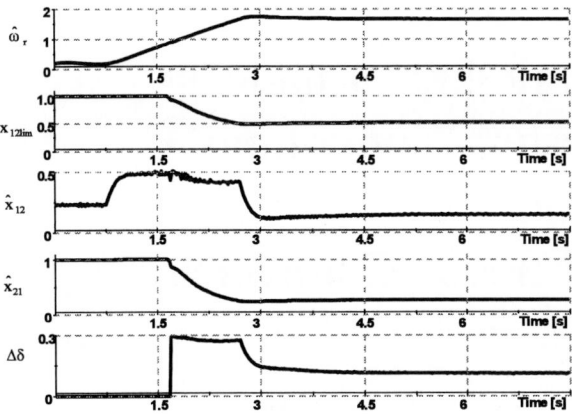

Fig. 5. Experimental transient after rotor speed change from 0.2 p.u to 1.7 p.u.
with 30% of nominal load

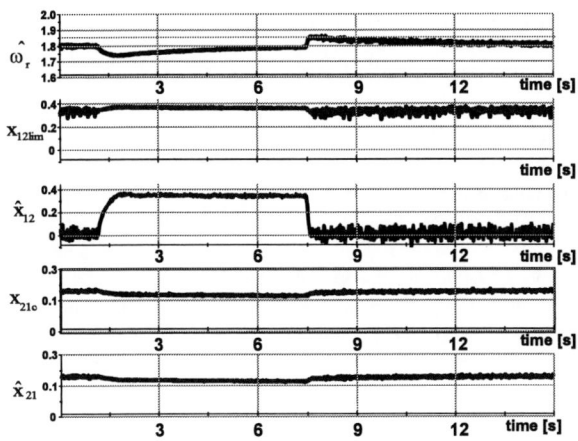

Fig. 6. Experimental transient after change of load and 1.8 p.u. command
for set speed.

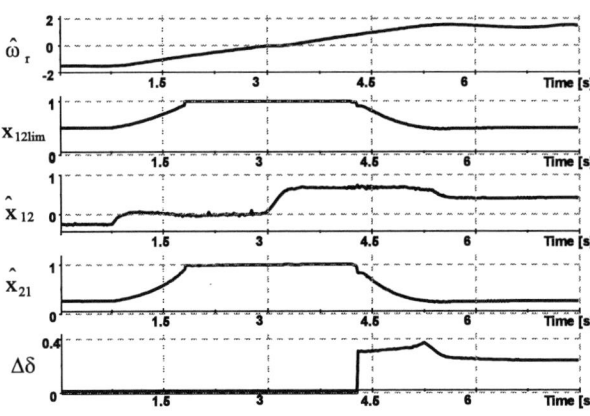

Fig. 6. Experimental transient after revers form -1.5 p.u. to 1.5 p.u.
with 40% of load

VII. SUMMARY

The paper describes nonlinear control system that utilizes the maximum voltage and current capability of the inverter for steady-state torque production in a lower speed range of field weakening operation. Maximum torque production in a field weakening is achieved by stabilizing the operating point. This is done by reducing the inverter current and adjusting the torque building variable to an optimum value. A temporary voltage margin is therefore created in a dynamic condition by deviating the trajectory of the flux linkage vector to values of lower magnitude. The nonlinear control system is fully decoupled in the base and field weakening regions.

REFERENCES

[1]. Abu-Rub H, Schmirgel H., Holz J. – „Sensorless control of induction motors for maximum steady-state torque and fast dynamics at field weakening", IEEE/IAS, 41st Annual Meeting, Tampa Florida, USA, 2006.

[2]. Krzemiński Z.: Non-linear control of induction motor, Proc. of the 10th IFAC World Congress, Munich, Germany, 1987.

[3]. Krzemiński Z., Lewicki A. Włas M: Properties of control systems based on nonlinear models of the induction motor, COMPEL The International Journal for Computation and Mathematics in Electrical and Electronic Engineering Vol. 25 No 1 2006 .

[4]. Wlas M., Lewicki A.: Dynamic Performance Of Sensorless Induction Motor Drives With Two Control Systems Based On Multiscalar Models 11th European Conference on Power Electronics EPE'2005, 10-15 09 2005, Drezdno.

[5]. Isidori A. : Nonlinear Control Systems: An Introduction. Communications and Control Engineering Series. Springer-Verlag, Berlin, 2nd edition.

[6]. J. Holtz and J. Quan,: Drift and Parameter Compensated Flux Estimator for Persistent Zero Stator Frequency Operation of Sensorless Controlled Induction Motors", IEEE Transactions on Industry Applications, Vol. 39, No. 4, July/Aug. 2003, pp. 1052-1060.

[7]. Kim S. H., Sul S.K.: Maximum torque control of an induction machine in the field-weakening region. IEEE Transactions on Industry Application, vol. 31, no. 4 1995

[8]. Kwon T.-S., M.-H. Shin and D.-S. Hyun: Speed sensorless stator flux-oriented control of induction motor in the field weakening region using Luenberger observer. IEEE Transactions on Power Electronics, vol. 20, no. 4, 2005 pp. 864 – 869.

[9]. X. Xu, D.Novotny: Selection of the flux reference for induction machine drive in the field weakening region. IEEE Transactions on industry applications, vol. 28, no 6, November/December 1992, pp. 1353.

[10]. Abu-Rub, H. and Holtz, J.: Rotor oriented nonlinear control system of induction motors operating at field weakening" 33rd Annual Conference of the IEEE Industrial Electronics Society "IECON 2007, Taipei, Taiwan

[11]. Krzemiński Z.: A new speed observer for control system of induction motor, IEEE Int. Conf. on Power Electronics and Drive Systems, PESC'99, Hong Kong, 1999

[12]. Holtz J.: Sensorless Control of Induction Motor Drives. Proceedings of IEEE, Vol. 90, No. 8, August 2002.

Comparison of Dynamic Performances of Speed Control System Containing Time - Minimal Speed Controller with Control System Containing PI Speed Controller

Andrzej Andrzejewski, Marian Roch Dubowski

Bialystok Technical University, Bialystok, Poland, e-mail: *a.andrzej@pb.bialystok.pl*
Bialystok Technical University, Bialystok, Poland, e-mail: *dubowski@pb.edu.pl*

Abstract—**In this paper the dynamic performances of two speed control systems of DC motor are compared. There are compared dynamic features of the classical speed control system containing PI speed controller and the speed control system with time - minimal speed controller. Comparison is done on the basis of simulation investigation results.**

Keywords—**Control of drive, highly dynamic drive.**

I. INTRODUCTION

The standard Proportional-Integral speed controllers are used in many electric drives [1], [2], [3], [4], [5], [6], [7], [11], [12]. This drives maintain static speed accuracy of loaded motors. This drives has short speed control time and speed overshoot.

However, the are constantly increasing expectations for electric drives especially of machine-tools and robots. The dynamic performances of speed control systems equipped with PI speed controller are not sufficient. The electric drives ought to feature with significantly shorter speed control time and with the lack of speed overshoot.

In order to fulfill this expectations, to control as fast as possible and without overshoot, the speed control system with time - minimal speed controller of DC motor was worked out [8]. The speed control system was named as time – minimal, because it features with minimal speed control time [8].

In this paper, the control time of the time – minimal speed control system is compared with control time of the standard speed control system equipped with PI speed controller. The speed overshoot of the two speed control systems responses are also compared. The Bode plots of the two speed control systems are compared as well. The speed control systems robustness on load torque change is examined too. The comparison of dynamic performances of two speed control systems is done on the basis of simulation investigations results.

The speed control systems robustness on motor parameters changes are not tested in this paper, because it is not the aim of this work.

II. CURRENT CONTROL SUBSYSTEM

The two considered speed control systems have the same type of current control subsystem and the same type of DC motor. The DC motor is supplied by four-quadrant

DC/DC converter, in Fig. 1. The sign of motor supplying voltage depends on the sign of output signal of current controller u_{ctrl}. The current controller with delta – modulation is used for two analyzed speed control systems. This current controller work as the two stable comparator with the very small delay made by sample and hold device, see in Fig.2 and in Fig.3. This current controller makes possible to obtain extremely short current response time. The current controller contains the sample and hold device in order to obtain limitation of the switching operation of the transistor converter. The sampling time of current controller is 40μs. As the result of use current controller with delta – modulation, the sign of motor supplying voltage is equal to sign of current control error after the short delay of 40μs. The value of the motor supplying voltage is equal to capacitor voltage U_{DC}.

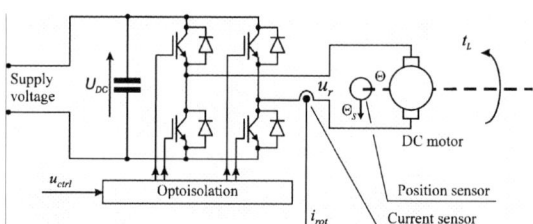

Fig. 1. DC motor and DC/DC four-quadrant converter.

III. THE STANDARD SPEED CONTROL SYSTEM WITH THE PROPORTIONAL INTEGRAL SPEED CONTROLLER

This section concerns the standard speed control system equipped with Proportional - Integral speed controller. The scheme of speed control system with PI speed controller is shown in Fig. 2. The parameters of PI speed controllers are set according to symmetric optimum [2].

There are some classical ways for improving the dynamic features of speed control system with PI speed controller. The speed overshoot is decreased from 42% to 8%, and the control time is shortened from 16,5Tσ to 13,3Tσ, where Tσ is the time constant of current control

978-1-4244-1741-4/08/$25.00 ©2008 IEEE   1090

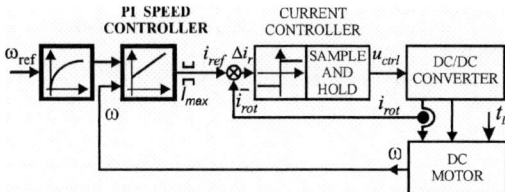

Fig. 2. Speed control system with PI speed controller.

subsystem, as the result of the use the reference speed filter [3] with time constant equal $4T\sigma$. The speed overshoot is decreased by use of *anti – windup system* [1] inside the integral element of PI speed controller. The *anti – windup system* blocks the integral element work during saturation state of speed controller.

IV. SPEED CONTROL SYSTEM WITH THE TIME – MINIMAL SPEED CONTROLLER

The second speed control system consists of the time - minimal speed controller [8] and the estimator of the electromechanical subsystem [8]. The equations of the time-minimal speed controller operation are shown in paper [8]. The estimator of the electromechanical subsystem fulfills many tasks in the time-minimal speed control system. The estimator functions as precise load torque compensator. The estimator provides the speed controller with the actual speed. The estimator adapts itself and adapts the speed controller to the actual value of moment of inertia [9]. The estimator, used as a precise load torque compensator, replaced the integral element inside the speed controller. As the result, the speed controller does not contain the integral element. The fact of elimination the integral element inside the time-minimal speed controller causes that the speed step response has not overshoot. The operation of the estimator of the electromechanical subsystem was described in details in papers [8], [9], [10]. In this paper the fourth order electromechanical subsystem estimator is used [10]. The estimator adaptation to the actual moment of inertia is omitted, because this feature is not tested is this paper.

The time-minimal speed controller does not influence the static accuracy. The speed controller influences only the transient component of the speed response. This fact was utilized to find the proper shape of the speed controller characteristics, by the use of the dynamic programming approach. As the result, the speed steep response is time - minimal [8]. The time - minimal speed

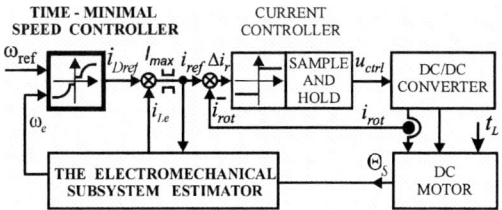

Fig. 3. Speed control system with time – minimal speed controller.

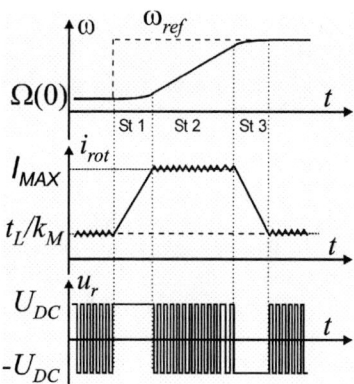

Fig. 4. The motor speed and current responses during time – minimal speed control process.

step response we can divide into three stages, see in Fig. 4. In the first stage speed St1, there is the saturation state of current and control subsystems. In the first stage, the motor current is controlled to the maximum value I_{MAX} with use of one switching process of converter supplying DC motor. This is time - minimal response, because it is not possible to control faster. To continue, the number of converter switching processes must be always integer and can not be less than one. In the second stage of speed control process, there is active state of current control subsystem and saturation state of speed control subsystem. In the second stage St2, the maximum value of current I_{MAX} is forced. This is time - minimal response, because motor current can not take higher the value I_{MAX} in order not to damage the motor or the machine tool. The value I_{MAX} can be higher or lower depending on particular type of motor or can change depending on the motor speed, in order not to damage the mechanical armature. As a matter of fact the value I_{MAX} is always maximal. In the third stage St3, there is active state of current and speed control subsystems. In the third stage, the motor current and the motor speed are simultaneously controlled with one switching process of converter. To continue, this is time - minimal speed control process. As it was mentioned before, the number of converter switching processes can not be lower than one. For this reason, the speed control process done with one converter switching process is time – minimal. If the all stages of speed control process are time – minimal, the whole speed control process is time - minimal.

In order to control speed and current with use of one switching process, the proper speed controller characteristics was found. The speed controller, for time - minimal control, has the nonlinear characteristics [8].

V. SIMULATION INVESTIGATIONS

The dynamic performance of two speed control systems are compared with means of simulation investigation. The computer program for simulation was written in a standard C++ code. The use of the C++ code was chosen because of the cheaper implementation.

1091

The maximal sample time of simulation program is 1μs.

The dynamic performances of the speed control system with time - minimal speed controller are compared with the performances of the speed control system equipped with Proportional - Integral speed controller. The simulation of the two control systems are done simultaneously. The two systems control speed of two independent DC motors. The two motors have the same parameters. The following values of DC motors parameters are assumed: R_r=4,658 Ω, L_r=0,07841 H, k_M=1,35 Nm/A; J=0,0663kg·m², I_{MAX}=5A, t_L=0,7Nm, where: J – moment of inertia, t_L – load torque.

Each speed control system contains the current control subsystem. The two current control subsystems, of two compared speed control systems, have the same type of current controller. This is the current controller with delta - modulation. The sample time of the current controllers with delta - modulation is 40μs.

The parameters of PI speed controller was set according to the to symmetric optimum [2]. In order to use the symmetric synthesis for PI speed controller parameters the current control subsystem had to be approximated with the first order lag. The current control subsystem, containing the current controller with delta - modulation, was approximated with first order lag. The gain of this lag is equal one. The time constant of this lag is equal 1,22ms. As the result, the parameters of PI speed controller are: the gain 20, whereas integrator time constant 4,9ms. For the reference speed the first order lag was used, in order to decrease the speed overshoot. The time constant of this lag is equal 4,9ms.

Firstly, speed step responses of two speed control systems are compared, see in Fig. 5. The reference speed step is so low (0,1rad/s) that it does not cause the second stage (saturation state) of speed control process. The speed response of the speed control process with PI speed controller has small overshoot equal 0,2% and relatively long speed control time 20,6ms. After the time 20,6ms speed control error is equal 2%ω_ref. Speed control time and speed overshoot, of speed control system with PI speed controller, are different from that in theory. According to the theory [2], [3] the speed control system with PI speed controller and with the filter of reference speed features with the overshoot equal to 8%. The lower value of overshoot and shorter speed control time was obtained. That is, because it is not classical speed control system with PI speed controller. According to theory the current control subsystem is replaces with the first order lag and analysis was done not for nonlinear current control subsystem, but for linear first order lag. To continue, the nonlinear current control subsystem not exactly responses like linear first order lag. The first order lag only *approximates* the current control subsystem, containing the nonlinear current controller with delta – modulation. In addition, the value of overshoot depends on the load torque. The value of speed overshoot in paper [2] was found for non-loaded motor.

In simulation investigations, the existence of load torque t_L=0,7Nm was assumed.

In contrast to speed control system with PI speed controller, the speed response of the speed control system with time – minimal speed controller has no overshot and shorter speed control time 1,96ms. After the time 1,96ms speed control error is equal to zero, as it was shown in Fig. 5.

Next, the speed step responses (1rad/s) the two speed control system are compared, in Fig. 6. In this figure, there are three stages of speed control process. The third speed control stage is only analyzed, because only in this stage the speed controllers are active.

The speed response of the speed control system with PI speed controller has significant overshoot the long speed control time in active state of speed controller, see in Fig. 6. The speed error overshoot is equal to 240 % of the speed control error $\Delta\omega_{start\_PI}$ at the beginning of the active state of speed control system. The speed control time is equal 25ms. It is worth to emphasize, that speed error is decreased to 2% of the value $\Delta\omega_{start\_PI}$ after the time 25ms. After the time 25ms the speed control error is not equal to nil, as a matter of fact. After the time 25ms, speed control error is equal to 0,02 of the reference speed. It means that after the time 25ms speed control system feature with very low accuracy 1:50. It is worth to emphasize, that speed control systems in servo-drives ought to feature with speed control accuracy 1:20000.

On the contrary, the speed response of the time - minimal speed control system has not overshoot at all and significantly shorter speed control time 1,07ms. After the time 1,07ms, motor current error is equal to load current (t_L/k_M), whereas the speed control error is equal to nil. The time T_{ctrl}, after which speed error is equal to nil, is finished and can be easily calculated from equation:

$$T_{ctrl} = L_r(I_{MAX} - I_{Le}) \big/ (U_{DC} + k_M \cdot \Omega(0) + R_r I_{MAX}) \tag{1}$$

In contrast to the time - minimal speed control system, the speed control system with linear PI speed controller

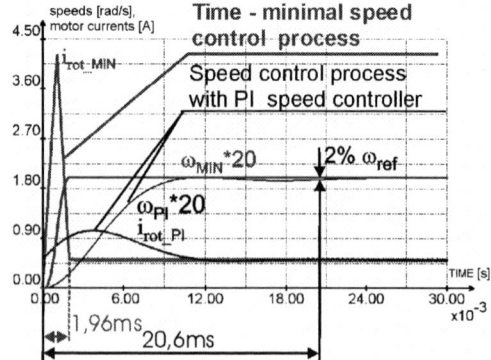

Fig. 5. The speed and current responses on small reference speed step 0,1 [rad/s].

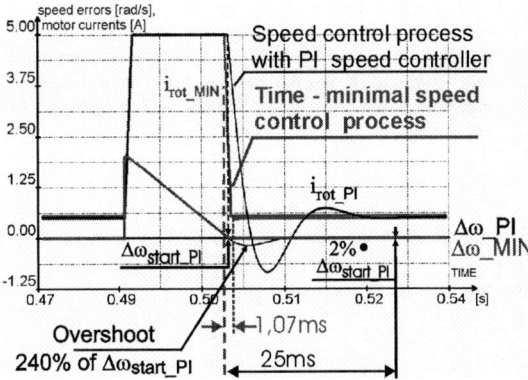

Fig. 6. The speed and current responses on reference speed step 1 [rad/s].

feature with very long time after which speed error is equal to nil. According to theory [3] for linear speed control systems with PI controller the following equation:

$$\Delta\omega_{steady} = 0 = \lim_{t \to \infty} \Delta\omega(t) \qquad (2)$$

is right. It means that speed error $\Delta\omega(t)$ is equal nil after unfinished time or speed error never reaches the value zero.

Next, Bode plots of the two systems are compared on the basis of simulation investigation results. The reference speeds of the two systems were in the shape of sinusoidal function. The amplitude was set properly low in order to ensure the active work of speed control system and to avoid the maximal value of motor current I_{MAX} in motor current response, like it was shown in Fig. 7. The reference speed amplitude was 0,1 rad/s. The reference speed had constant component 0,3rad/s in order not to cause motor reverse and to avoid motor nonlinearity. The problem is, during motor reverse the passive load torque changes its sign. The influence of passive load torque is nonlinear and causes the nonlinearity of the DC motor model. This test is usually done for linear control systems. If we do this test during motor reverse with the nonlinear influence of passive load torque, the simulation results will not be right. The Bode plot during reverse is different depending on the different values of passive load torque. The nonlinear influence of the passive load torque on Bode plots is not the aim of the test. The aim of this test is to compare the dynamic performances of the two considered speed control systems, working with linear model of loaded electric DC motor, with use of Bode plots. The motor load torque was constant, during the test.

The simulation investigation results acknowledge, that time - minimal speed control system has shorter speed control time and lower speed error than speed control system with PI speed controller. This test was done at many frequencies. Bode plots in Fig. 8 shows that the speed control error of the time minimal –speed control system is lower at every frequency than the speed error of the speed control system with PI speed controller. The phase shift versus frequency is shown in Fig. 9, too. The

phase shift of the time minimal –speed control system is lower at every frequency than the phase shift of the speed control system with PI speed controller. If the phase shift is lower, the speed control time is shorter.

On the basis of results in Fig. 8, the bandwidth of two speed control system are compared. The bandwidth of speed control system with PI speed controller is 60Hz. The bandwidth of time minimal speed control system is 215Hz for the same type DC motor.

Next, the motor speed and motor current responses on load torque change are compared of two speed control system in Fig. 10. The speed control system with time - minimal speed controller has shorter response time 1,5ms than response time 12ms of speed control system with PI

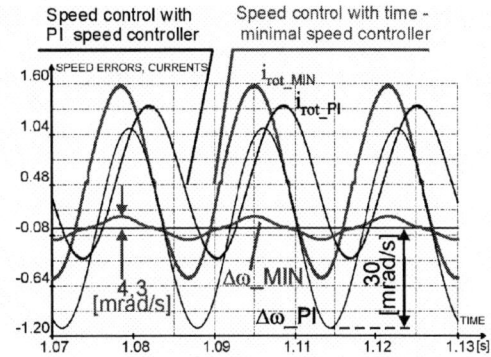

Fig. 7. The speed and current responses on the sinusoidal reference speed (frequency was 50Hz).

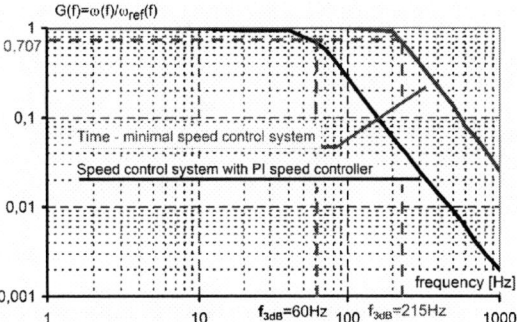

Fig. 8. Closed-loop gain versus frequency.

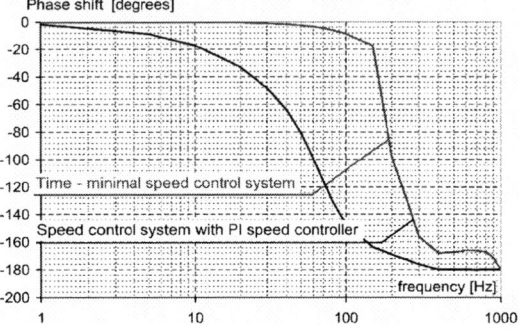

Fig. 9. Phase shift versus frequency.

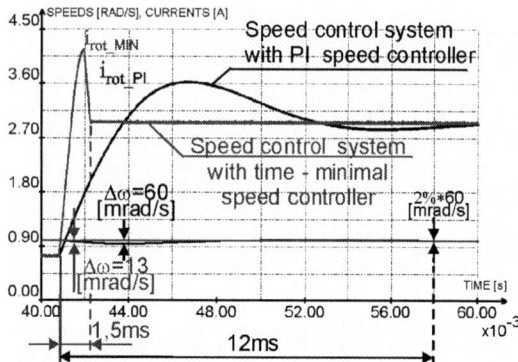

Fig. 10. The motor speed and motor current responses after load torque change form 1 to 4 Nm at reference speed 1rad/s.

speed controller. The speed control system with time - minimal speed controller has lower maximal speed error 13mrad/s than the maximal speed error 60mrad/s of speed control system with PI speed controller.

The speed error response on the sinusoidal load torque was compared for the two speed control systems. The maximal speed error of the two speed control systems changes value depending on the load torque frequency. The dependence of the maximal speed error versus the load torque frequency is shown in Fig. 11. The following simulation parameters are assumed: the reference speed 1rad/s, the load torque steady component 3,5Nm, the amplitude of the load torque sinusoidal component 2 Nm. As the result, the load torque was changed from 22% to 60% of the maximal motor torque. It is worth to mention that characteristics, shown in Fig. 11, is different for different value of the amplitude of the load torque sinusoidal component.

The simulation investigation results, shown in Fig. 11, acknowledge that the time – minimal speed control system has lower the maximal speed error than the speed control system with PI speed controller for the frequency lower than 400Hz. The maximal speed error of the speed control system with the time - minimal controller is from 10000 to 100 times lower than the maximal speed error of the speed control system with PI speed controller. Above the frequency 400 Hz is reverse. Above the frequency 400Hz, the time minimal speed control system has higher

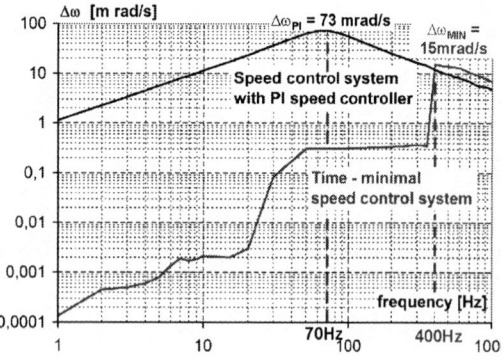

Fig. 11. Maximal speed error vs load torque frequency.

the speed control error. That is because of the high value of the current control subsystem error of the time minimal speed control system. If the current control error is high the estimator of the electromechanical subsystem is switch off automatically. Above the frequency 400Hz, the switched off electromechanical subsystem estimator does not compensate the load torque. However, the maximal speed error 15mrad/s at frequency 400Hz of the time - minimal speed control system is lower than the maximal speed error 73 mrad/s at frequency 70Hz of the speed control system with PI sped controller.

VI. SIMULATION INVESTIGATION RESULTS COMPARISON

By comparison the simulation investigations results we can come to following conclusions. The speed control system with time – minimal speed controller has about twenty times shorter speed control time (1,96ms) than the speed control system with PI speed controller (20,6ms). The speed control system with time - minimal speed controller is able to decrease speed error to nil during finished and very short time. The speed error of speed control system with PI speed controller is decreased to nil after unfinished time. The speed control system with PI speed controller has 240% speed overshoot after large reference speed step. In contrast, the speed control system with time - minimal speed controller has no overshoot, at all. The bandwidth of the speed control system with time - minimal speed controller (215Hz) is more than three times wither than the bandwidth of the speed control system with PI speed controller (60Hz). The maximal speed error after the load torque step of the speed control system with time - minimal speed controller (13mrad/s) is about four times lower than the maximal speed error (60mrad/s) of the speed control system with PI speed controller. After the load torque step, the speed control system with PI speed controller reduces the speed error to value 2%*60mrad/s in 12ms time. In contrast, the speed control system with the time – minimal speed controller reduces the speed error to nil in 1,5ms time. The speed control system with the time - minimal speed controller features also with lower maximal amplitude (73mrad/s) of the speed error response on the sinusoidal load torque than the maximal speed error amplitude (15mrad/s) of the speed control system with PI speed controller.

VII. EXPERIMENTAL RESULTS

This paper contains the experimental results in order to show the viability of proposed the time-minimal speed control method.

The experimental investigations were done on the test stand composed of two parts: the electric part and the mechanical part. The electric part of test stand is shown in Fig. 12, whereas the mechanical part of test stand is shown in Fig. 13. The laboratory test stand was built in Bialystok Technical University.

The electric part of the test stand consist of the microprocessor system based on processor DSP96002 (40MIPS), the DC/DC converter controlled by the microprocessor system, and the computer PC. The microprocessor system with DSP96002 controlled the

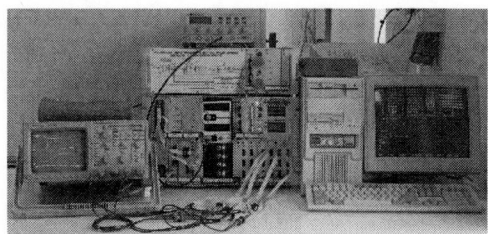

Fig. 12. The electric part of test stand.

Fig. 13. The mechanical part of test stand.

DC/DC converter. The microprocessor system contained 12-bits (1μs) Analog to Digital converters ADS-112 from company DATEL.

The DC/DC converter was used to supply the DC motor with the mechanical armature. The converter output voltage changes from +330V to – 330V and vice versa.

The maximal converter current is 5A. The main part of the converter is the transistor module IPM15CSJ600 from Mitsubishi company. The current sensors LA25NP from company LEM. The computer PC was only used to compose the program. The control program was written in the assembler language for Motorola DSP96002 (40MIPS). The sample time of the current control loop is 12μs (including interrupt subroutine service), whereas the sample time of the speed control loop is 84μs. During (84μs) one execution of the speed control loop, the current control loop was executed 7 times. By doing so, the low amplitude of the electromagnetic motor torque ripple was obtained.

The mechanical part of test stand consists of the two DC motors, the rotational encoder and the electromagnetic clutch. The DC motors have self cooling. The mounted fan on the motor shaft was increasing the moment of inertia of the rotor. The nominal parameters of the motors are: P_N=1,2kW; U_N=230V, I_N=5,2A; n_n=1450rpm. The parameters of the DC motors are: R_r=4,658 Ω, L_r=0,07841 H, k_M=1,35 Nm/A; The moment of inertia of the whole mechanical part is J=0,0663kg·m². The optical rotational encoder PFI60A2048CZ2-DAP10, from company Intron, has resolution 2048 marks per rotation. The electromagnetic clutch ESM1-5, from company Fumo-Ostrzeszów, has static torque 7Nm and dynamic torque 5Nm. The electromagnetic clutch was used to join and to disconnect two motors. In this way, the moment of inertia was changed during the rotation.

The time-minimal speed responses of motor was tested, see in Fig. 14 and Fig. 15. The experimental results shows, that the whole speed control process can be divided into three stages, like it was shown in Fig. 4. The whole speed control process is time-minimal. In order to increase the test difficulty, the DC motor was loaded. In addition test was done with the high speed and the high motor EMF. This test is heavy, because the speed control time in the active state of the speed controller strongly depends on the motor EMF. This fact could be easily noticed in the experimental results, compare Fig. 14 with Fig. 15. To continue, the speed control time (850μs) in the third stage of the speed control process in Fig. 14 is significantly shorter then the speed control time (2,2ms) in the third stage in Fig. 15. Even though the speed control times are different, the speed control processes are time-minimal. In the first stage of the speed control processes, the motor current is controlled with one converter switching process. In the second stage of the speed control processes, the maximal motor current is forced. In the third speed control processes the speed control error is decreased to nil with use of one switching process of the converter. As it was described in section IV, this is the time-minimal speed control process. It is worth to emphasize, that it is not possible to obtain the lower number switching processes than one.

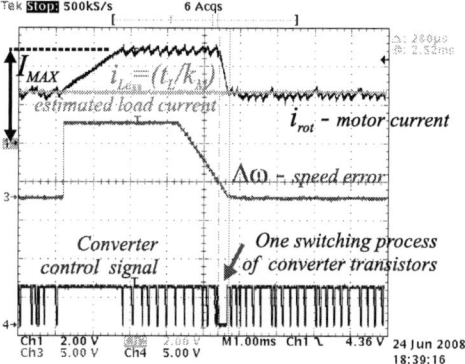

Fig. 14. The motor current and the speed error responses on the rapid decrease of the reference speed.

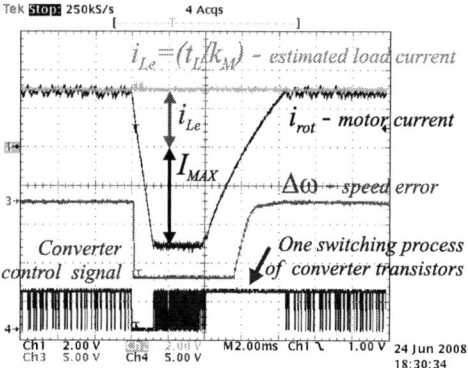

Fig. 15. The motor current and the speed error responses on the rapid decrease of the reference speed.

1095

It should be highlighted that the chattering phenomenon is not observed, even though the time-minimal speed control process was obtained. The motor current is smooth and do not cause motor torque ripples, analyze results in Fig. 14 and Fig. 15.

In addition, this speed control system is accurate even though the motor is loaded. This fact is easily noticed in Fig. 14 and Fig. 15. To continue, at the end of speed control process the motor current is equal to estimated load current i_{LE} in order to compensate motor load. At the end of speed control process the speed control error is equal to zero.

Additional explanation of the experimental results is due. In results shown in Fig. 14 and Fig. 15, the signal of speed control error is observed as being constant. In reality the speed control error was not constant. It follows from the fact that the speed control error was strongly gained for observation and next this signal was saturated in the digital to analog converter.

VIII. CONCLUSION

Dynamic performances of two speed control systems are compared on the basis of simulation investigations results. To continue, features of the time-minimal speed control system and standard speed control system with PI speed controller were compared.

The paper contains the experimental results of the time-minimal speed control system. The experimental results shows the viability of the time-minimal speed control method.

ACKNOWLEDGMENT

This scientific work is supported by funds of research project 3 T10A 008 30 in years: 2006 - 2009.

REFERENCES

[1] K. Zawirski "Control of permanent magnet synchronous motor", *Wydawnictwo Politechniki Poznańskiej*, Poznań 2005.

[2] C. Kessler, "Das symetrische optimum," *Regelungstechnik*, 1958, p. 395-400, 432-436.

[3] P. Drozdowski „Wprowadzenie do napędów elektrycznych", *Skrypt Politechniki Krakowskiej*, Kraków 1998.

[4] S. Brock, J. Deskur „Problem pomiaru i regulacji prędkości w układzie z niedokładnym przetwornikiem pomiarowym położenia", *8th National Conference on Control in Power Electronics & Electric Drives, SENE 2007*, Łódź, 21 - 23 November 2007.

[5] Taeg-Joon Kweon, Dong-Seok Hyun "High-performance speed control of electric machine using low-precision shaft encoder", *IEEE Trans. on Power Electronics*, vol. 14, no5, September 1999.

[6] Shin-ichiro Sakai, Yoichi Hori "Ultra – low speed control of servomotor using low resolution rotary encoder". *IEEE - Xplore Data Base*.

[7] W. Błasifiski, P, Makowski „Ograniczenie prądu w układzie napędowym z silnikiem prądu stałego", *8th National Conference on Control In Power Electronics & Electric Drives, SENE 2007*, Łódź, 21 - 23 November 2007.

[8] A. Andrzejewski „The time minimal speed control of dc motor without overshoot", *International Conference Eurocon 2007 on "Computer as a tool"*, Warsaw, Poland, September 9-12, 2007.

[9] A. Andrzejewski. "Speed Control System Based on Estimator of the Electromechanical Subsystem", *5th International Conference-Workshop CPE 2007*.

[10] A. Andrzejewski "Sposób na zwiększanie rzędu estymatora podsystemu elektromechanicznego", *8th National Conference on Control In Power Electronics & Electric Drives, SENE 2007*, Łódź, 21 - 23 November 2007.

[11] Z. Preitl, P. Bauer, J. Bokor "A simple control solution for traction motor used in hybrid vehicles", *IEEE Trans. on Industrial Electronics*, vol. 54, no.3. 3 June 2007.

[12] R. Pretup, S. Preitl, I. Rudas, M. Tomescu, J. Tar "Design and experiment for a class of fuzy controlled servo systems", *IEEE Trans. on Mechanics*, vol. 13, no.1, February 2008.

Optimisation of Real-Time Complex Path Generation in Constrained Intelligent Motion Applications Based on IPM Motor Drives

Silverio BOLOGNANI[†], Roberto PETRELLA[‡], Fabio STEFANUTTI[‡], Piero STOCCO[‡]

† Department of Electrical Engineering, University of Padova, Padova, Italy, e-mail: *silverio.bolognani@unipd.it*
‡ Department of Electrical, Management and Mechanical Engineering, University of Udine, Udine, Italy
e-mail: *roberto.petrella@uniud.it, Stefanutti.Fabio.1@spes.uniud.it, Stocco.Piero.1@spes.uniud.it*

Abstract—Improvements in the constrained real-time path generation for high performance motion control systems are presented in this paper. With respect to the existing solutions, all the cases generating from a set of constraints imposed on the kinematic characteristics of the movements are analysed and solved analytically. The proposed path generator has been employed inside a digital drive system based on interior PM synchronous motor and a last generation digital signal controller. Optimised torque-speed motor control is implemented and exploited to raise the accuracy and dynamical performance of the position controller. Simulation and experimental results are presented and discussed.

Keywords: intelligent drive, motion control.

I. INTRODUCTION

Motion control of industrial multi-axis machines (such as industrial robot manipulators, packing machines, etc.) consists in principle of the following tasks: optimized speed control of the axis, position control in every axis, path generation for each axis and coordination of the motion among the axes [1]. Complex paths must be employed to obtain smooth acceleration, speed and position shapes (Fig. 1), and have to be generated in real-time depending on the characteristics of the movement, such as target position and movement constraints (limitations on mechanical speed and accelerations).

Classical structure of a motion control systems allocates path generation task in an intelligent master controller and considers the drives as actuators being able to control the position, the speed and the torque in the corresponding axis [2]. Unavoidable problems arise when complex paths for multiple axes are being generated. Both the demand for real-time calculation and high communication throughputs of the master increase dramatically. Fast and accurate motion control cannot be achieved without the recourse to high performance control systems. Recent approaches transfer

path generation task into the same drive controller, as to reduce real-time communication requirements, [3], and allowing to improve control performance by means of generated motion profiles.

In this paper some improvements in the constrained real-time path generation for high performance motion control systems are presented. The problem of path generation is analytically stated for two commonly adopted shapes of acceleration (i.e. sinusoidal and raised cosine), having the movement characteristics as input and the tracking acceleration, speed and position as outputs. With respect to the existing solutions, [3][4], all the cases generated from a set of constraints imposed on the kinematic characteristics of the movements are analysed and solved analytically. The resulting movement is guaranteed to have a defined structure and to be realised always within the same phases.

The proposed path generator and drive controller have been simulated by means of a dedicated Matlab/Simulink model implementing an interior permanent magnet (IPM) synchronous motor and a last generation digital signal controller. Optimised torque-speed motor control is implemented and exploited to raise the accuracy and the dynamical performance of the position controller. Experimental results showing system performance for different movement characteristics are also presented and discussed to demonstrate the effectiveness of the proposed approach.

II. CONSTRAINED PATH GENERATION PROBLEM: GENERAL STATEMENT

A generic movement can be decomposed in a starting phase (A), a running phase (C) and a halting phase (D).

Define $\boldsymbol{\eta}(t) = [\alpha(t) \quad \omega(t) \quad \vartheta(t)]^T$ as the *vector of the mechanical paths*, where:

$$\alpha(t) = \alpha_A(t) + \alpha_C(t) + \alpha_D(t) \left[\frac{rad}{s^2}\right]$$

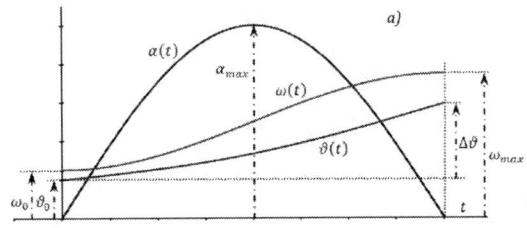 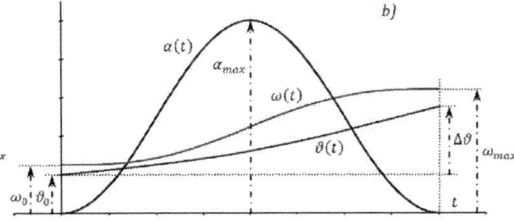

Fig. 1. typical shapes for acceleration, speed and position in a motion control systems (sinusoidal and raised cosine acceleration).

978-1-4244-1741-4/08/$25.00 ©2008 IEEE

is the *acceleration function*,

$$\omega(t) = \omega_A(t) + \omega_C(t) + \omega_D(t) \left[\frac{rad}{s}\right]$$

is the *speed function*,

$$\vartheta(t) = \vartheta_A(t) + \vartheta_C(t) + \vartheta_D(t)\ [rad]$$

is the *position function*,

$$\alpha_i(t), \omega_i(t), \vartheta_i(t)\ (i = A, C, D)$$

is a *set of continuous functions*.

By considering the vector of the initial conditions $\boldsymbol{\eta_0} = [\alpha_0 \quad \omega_0 \quad \vartheta_0]^T$, the motion model $\boldsymbol{S}$ in (1) is built.

$$S: \begin{cases} \dot{\omega}(t) = \alpha(t) \\ \dot{\vartheta}(t) = \omega(t), \\ \boldsymbol{\eta}(0) = \boldsymbol{\eta_0} \end{cases} J: \begin{cases} A\begin{bmatrix} \eta(t) \\ \dot{\alpha}(t) \end{bmatrix} \le B \\ max \begin{bmatrix} \alpha_A(t) \\ \alpha_C(t) \\ \alpha_D(t) \end{bmatrix} = C \\ \alpha(t) = \alpha^*(t, \boldsymbol{\tau}) \end{cases} \quad (1)$$

The aim of the path generator is to find out the vector of the mechanical paths $\boldsymbol{\eta}(t)$ such that the speed and position functions $\omega(t)$ and $\vartheta(t)$ are continuous, the total movement $\Delta\vartheta$ is equal to a target $\Delta\vartheta^*$ and the set of mechanical constraints $\boldsymbol{J}$ is satisfied, (1), where $\boldsymbol{A}, \boldsymbol{B}, \boldsymbol{C}$ denote generic matrices and scalar terms of constraints and

$$\alpha^*(t, \boldsymbol{\tau}) = \alpha_A^*(t, \boldsymbol{\tau}) + \alpha_C^*(t, \boldsymbol{\tau}) + \alpha_D^*(t, \boldsymbol{\tau}) \quad (2)$$

is a *piece-wise continuous function* of time having $\boldsymbol{\tau}$ as *set of parameters* and

$$\alpha_i^*(t, \boldsymbol{\tau})\ (i = A, C, D) \quad (3)$$

is a set of continuous functions of time.

As stated above, the path generation problem is equivalent to solving the set of equations on $\boldsymbol{\tau}$ generated by the motion model $\boldsymbol{S}$ once the movement constraints $\boldsymbol{J}$ and the target $\Delta\vartheta^*$ have been considered.

III. CYCLOID TRAJECTORY: STATEMENT AND SOLUTION

Let us formulate the path generation problem in a real case in which the following parametric half a period sinusoidal acceleration is used during phases A and D:

$$\alpha_i^*(t, h_i, t_{si}) = \pi \frac{h_i}{t_{si}} sin\left(\pi \frac{t}{t_{si}}\right), \\ t \in [0; t_{si}],\ i = A, D \quad (4)$$

and a null acceleration is used during phase C:

$$\alpha_C^*(t) = 0,\ t \in [0; t_C] \quad (5)$$

Smooth speed and position cycloid trajectory is obtained and different paths can be generated by choosing both the peak value and the period of the acceleration as functions of the parameters h_i and t_{si} (refer to Fig. 1a). Moreover we suppose to choose the movement constraints $\alpha_{max,A}$ (maximum acceleration), $\alpha_{max,D}$ (maximum deceleration) and $\bar{\omega}_{max}$ (maximum speed that could be reached during the movement).

The correspondent movement is shown in Fig. 2.

Once the initial conditions $\boldsymbol{\eta_0} = [0 \quad \omega_0 \quad \vartheta_0]^T$ and the constraint $\alpha(t) = \alpha^*(t, \boldsymbol{\tau})$ have been considered, system S becomes:

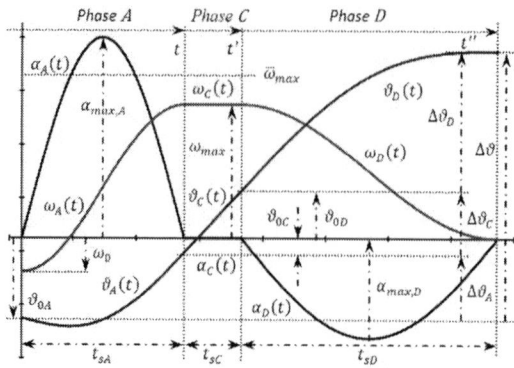

Fig. 2. Considered cycloid trajectory.

$$\begin{cases} \alpha_A(t) = \pi \frac{h_A}{t_{sA}^2} sin\left(\pi \frac{t}{t_{sA}}\right) \\ \omega_A(t) = \frac{h_A}{t_{sA}}\left[1 - cos\left(\pi \frac{t}{t_{sA}}\right)\right] + \omega_{0A} \\ \vartheta_A(t) = h_A\left[\frac{t}{t_{sA}} - \frac{1}{\pi}sin\left(\pi \frac{t}{t_{sA}}\right)\right] + \omega_{0A}t + \vartheta_{0A} \end{cases} \quad (6)$$

$$\begin{cases} \alpha_C(t') = 0 \\ \omega_C(t') = \omega_{max} \triangleq \frac{h_C}{t_{sC}} \\ \vartheta_C(t') = \omega_{max}t' + \vartheta_{0C} \end{cases} \quad (7)$$

$$\begin{cases} \alpha_D(t'') = -\pi \frac{h_D}{t_{sD}^2} sin\left(\pi \frac{t''}{t_{sD}}\right) \\ \omega_D(t'') = -\frac{h_D}{t_{sD}}\left[1 - cos\left(\pi \frac{t''}{t_{sD}}\right)\right] + \omega_{0D} \\ \vartheta_D(t'') = -h_D\left[\frac{t''}{t_{sD}} - \frac{1}{\pi}sin\left(\pi \frac{t''}{t_{sD}}\right)\right] + \omega_{0D}t'' + \vartheta_{0D} \end{cases} \quad (8)$$

where different time axes $t' = t - t_{sA}$, $t'' = t' - t_{sC}$, and

$$t \in [0; t_{sA}],\ t' \in [0; t_{sC}],\ t'' \in [0; t_{sD}] \quad (9)$$

have been considered for simplicity in each phase, additional parameters h_C and t_{sC} have been introduced for convenience as a function of the maximum (ω_{max}, constant) value of speed actually reached during running phase (C).

As stated above, the path generation problem is equivalent to solving the set of equations on

$$\boldsymbol{\tau} = [h_A \quad t_{sA} \quad h_C \quad t_{sC} \quad h_D \quad t_{sD}]^T$$

generated by the model (2), (3) and (4) once the movement constraints $\boldsymbol{J}$ defined by the following matrices

$$\boldsymbol{A} = [0 \quad 1 \quad 0 \quad 0],\ \boldsymbol{B} = \bar{\omega}_{max},\ \boldsymbol{C} = \begin{bmatrix} \alpha_{max,A} \\ 0 \\ \alpha_{max,D} \end{bmatrix} \quad (10)$$

and the target $\Delta\vartheta^*$ have been considered.
Continuity of position, speed and acceleration paths immediately leads immediately to the additional constraints:

$$\omega_{0A} = \omega_0, \omega_{0D} = \omega_{max}$$

$$\vartheta_{0A} = \vartheta_0, \vartheta_{0C} = \vartheta_A(t_{sA}), \vartheta_{0D} = \vartheta_C(t_{sC}). \quad (11)$$

Constraints on acceleration and speed therefore become:

$$\alpha_{max,A} = \left|\alpha_A\left(\frac{t_{sA}}{2}\right)\right| = \pi \frac{|h_A|}{t_{sA}^2}, \quad (12)$$

$$\omega_{max} = \omega_A(t_{sA}) = 2\frac{h_A}{t_{sA}} + \omega_0$$

$$\alpha_{max,D} = \left|\alpha_D\left(\frac{t_{sD}}{2}\right)\right| = \pi\frac{|h_D|}{t_{sD}^2},$$

$$\omega_D(t_{sD}) = 0 = -2\frac{h_D}{t_{sD}} + \omega_{max}$$

Analytically it is possible to find out an expression for the unknown parameters:

$$h_A = \frac{\pi}{4}\frac{(\omega_{max}-\omega_0)^2}{\alpha_{max,A}}sign(\omega_{max}-\omega_0)$$

$$t_{sA} = \frac{\pi}{2}\frac{|\omega_{max}-\omega_0|}{\alpha_{max,A}} = 2\frac{h_A}{\omega_{max}-\omega_0} \qquad (13)$$

$$h_C = \omega_{max}t_{sC}, t_{sC} = \frac{h_C}{\omega_{max}} \qquad (14)$$

$$h_D = \frac{\pi}{4}\frac{\omega_{max}^2}{\alpha_{max,D}}sign(\omega_{max}) \qquad (15)$$

$$t_{sD} = \frac{\pi}{2}\frac{|\omega_{max}|}{\alpha_{max,D}} = 2\frac{h_D}{\omega_{max}}$$

Moreover the position displacements that is possible to obtain within each phase are given by:

$$\Delta\vartheta_A = h_A + \underbrace{\frac{\pi}{2}\omega_0\frac{|\omega_{max}-\omega_0|}{\alpha_{max,A}}}_{h_{\omega_0}} = h_A + 2h_A\frac{\omega_0}{\omega_{max}-\omega_0} =$$

$$= h_A\frac{\omega_{max}+\omega_0}{\omega_{max}-\omega_0} \qquad (16)$$

$$\Delta\vartheta_C = h_C \qquad (17)$$

$$\Delta\vartheta_D = -h_D + 2h_D = h_D \qquad (18)$$

where the term h_{ω_0} is due to a non-zero speed initial condition. The total position displacement that is possible to obtain by the sequencing of the phases A, C and D is therefore:

$$\Delta\vartheta = \Delta\vartheta_A + \Delta\vartheta_C + \Delta\vartheta_D = h_A\frac{\omega_{max}+\omega_0}{\omega_{max}-\omega_0} + h_C + h_D \quad (19)$$

The solution of path generation problem in the considered case depends on the value of the target $\Delta\vartheta^*$ with respect to the position displacement obtained by using only phases A and D ($t_{sC} = 0$) and the maximum allowed value of the speed ($\omega_{max} = \bar{\omega}_{max}$). This displacement can be calculated by (19) as

$$\Delta\vartheta_{AD} = \Delta\vartheta_A + \Delta\vartheta_D = h_A\frac{\bar{\omega}_{max}+\omega_0}{\bar{\omega}_{max}-\omega_0} + h_D \qquad (20)$$

If the target $\Delta\vartheta^*$ is equal to $\Delta\vartheta_{AD}$, then the problem is simply solved by using the values of h_i and t_{si} ($i = A, D$) given by (13) and (15) (with $\omega_{max} = \bar{\omega}_{max}$).

The situation is more complex when $\Delta\vartheta^* > \Delta\vartheta_{AD}$ or $\Delta\vartheta^* < \Delta\vartheta_{AD}$.

In the first case the values of $\bar{\omega}_{max}$, $\alpha_{max,A}$ and $\alpha_{max,D}$ are not high enough to allow the required position displacement to be generated without introducing a constant speed phase. Hence, by assuming $\omega_{max} = \bar{\omega}_{max}$, the position displacement that must be introduced during this phase is obtained directly from (19):

$$\Delta\vartheta_C = \Delta\vartheta^* - \Delta\vartheta_{AD} = h_C =$$

$$= \Delta\vartheta^* - h_A\frac{\bar{\omega}_{max}+\omega_0}{\bar{\omega}_{max}-\omega_0} - h_D \qquad (21)$$

its application time being given by (14).

In the second case the position displacement that is possible to obtain by using only phases A and D and the maximum allowed speed ($\omega_{max} = \bar{\omega}_{max}$) is greater than the required one. Then it is required to reduce the maximum speed reached during the movement, whose value can be calculated as a function of the required $\Delta\vartheta^*$, i.e.:

$$\Delta\vartheta_{AD} = \Delta\vartheta^* = \Delta\vartheta_A + \Delta\vartheta_D = h_A\frac{\omega_{max}+\omega_0}{\omega_{max}-\omega_0} + h_D =$$

$$= \frac{\pi}{4}\frac{(\omega_{max}-\omega_0)^2}{\alpha_{max,A}}\frac{\omega_{max}+\omega_0}{\omega_{max}-\omega_0}sign(\omega_{max}-\omega_0)$$

$$+ \frac{\pi}{4}\frac{\omega_{max}^2}{\alpha_{max,D}}sign(\omega_{max}) =$$

$$= \frac{\pi}{4}\frac{\alpha_{max,D}(\omega_{max}^2-\omega_0^2)sign(\omega_{max}-\omega_0)}{\alpha_{max,A}\alpha_{max,D}} \qquad (22)$$

$$+ \frac{\pi}{4}\frac{\alpha_{max,A}\omega_{max}^2 sign(\omega_{max})}{\alpha_{max,A}\alpha_{max,D}}$$

Maximum speed can be simply calculated by inverting the previous equation:

$$\omega_{max} =$$

$$\pm\sqrt{\frac{\frac{4}{\pi}\Delta\vartheta^*\alpha_{max,A}\alpha_{max,D}+\alpha_{max,D}\omega_0^2 sign(\omega_{max}-\omega_0)}{\alpha_{max,D}sign(\omega_{max}-\omega_0)+\alpha_{max,A}sign(\omega_{max})}} \qquad (23)$$

It can be noticed that this expression contains non-linear sign functions which in turn depend on the maximum speed itself, requiring a complete study of the fraction inside the square root, that must be positive, i.e. the following conditions must be satisfied:

$$\begin{cases} \frac{4}{\pi}\Delta\vartheta^*\alpha_{max,A}\alpha_{max,D} \\ +\alpha_{max,D}\omega_0^2 sign(\omega_{max}-\omega_0) \geq 0 \\ \alpha_{max,D}sign(\omega_{max}-\omega_0) \\ +\alpha_{max,A}sign(\omega_{max}) > 0 \end{cases} \qquad (24)$$

or

$$\begin{cases} \frac{4}{\pi}\Delta\vartheta^*\alpha_{max,A}\alpha_{max,D} \\ +\alpha_{max,D}\omega_0^2 sign(\omega_{max}-\omega_0) < 0 \\ \alpha_{max,D}sign(\omega_{max}-\omega_0) \\ +\alpha_{max,A}sign(\omega_{max}) < 0 \end{cases} \qquad (25)$$

Four alternative cases can be obtained by a certain set of the input values for ω_0, $\Delta\vartheta^*$, and for the maximum acceleration and deceleration $\alpha_{max,pos}$ and $\alpha_{max,neg}$, respectively for a positive and negative acceleration. In fact parameters $\alpha_{max,A}$ and $\alpha_{max,D}$ do not take into account which movement will be generated first (A or D) as a function of the initial speed ω_0 and the target position $\Delta\vartheta^*$. So, the right acceleration must be considered if the movement is accelerating or decelerating the load and assigned to $\alpha_{max,A}$ and $\alpha_{max,D}$ for a correct profile generation.

$$if\ (\omega_0 \geq 0)\ and\ \left(\Delta\vartheta^* \geq \frac{\pi}{4}\frac{\omega_0^2}{\alpha_{max,neg}}\right)$$

$$\alpha_{max,D} = \alpha_{max,neg} \qquad (26)$$

$$if\ (\omega_0 \leq \bar{\omega}_{max})\ \alpha_{max,A} = \alpha_{max,pos}$$

$$else\ \alpha_{max,A} = \alpha_{max,neg}$$

$$\omega_{max} = \sqrt{\frac{\frac{4}{\pi}\Delta\vartheta^* \alpha_{max,A}\alpha_{max,D} + \alpha_{max,D}\omega_0^2}{\alpha_{max,D} + \alpha_{max,A}}}$$

$$if\ (\omega_0 < 0)\ and\ \left(\Delta\vartheta^* > -\frac{\pi}{4}\frac{\omega_0^2}{\alpha_{max,pos}}\right)$$

$$\alpha_{max,A} = \alpha_{max,pos},\ \alpha_{max,D} = \alpha_{max,neg} \tag{27}$$

$$\omega_{max} = \sqrt{\frac{\frac{4}{\pi}\Delta\vartheta^* \alpha_{max,A}\alpha_{max,D} + \alpha_{max,D}\omega_0^2}{\alpha_{max,D} + \alpha_{max,A}}}$$

$$if\ (\omega_0 < 0)\ and\ \left(\Delta\vartheta^* \le -\frac{\pi}{4}\frac{\omega_0^2}{\alpha_{max,pos}}\right)$$

$$\alpha_{max,D} = \alpha_{max,pos}$$

$$if\ (\omega_0 \ge -\bar{\omega}_{max})\ \alpha_{max,A} = \alpha_{max,neg} \tag{28}$$

$$else\ \alpha_{max,A} = \alpha_{max,pos}$$

$$\omega_{max} = -\sqrt{\frac{\frac{4}{\pi}\Delta\vartheta^* \alpha_{max,A}\alpha_{max,D} - \alpha_{max,D}\omega_0^2}{-\alpha_{max,D} - \alpha_{max,A}}}$$

$$if\ (\omega_0 \ge 0)\ and\ \left(\Delta\vartheta^* < \frac{\pi}{4}\frac{\omega_0^2}{\alpha_{max,neg}}\right)$$

$$\alpha_{max,A} = \alpha_{max,neg},\ \alpha_{max,D} = \alpha_{max,pos} \tag{29}$$

$$\omega_{max} = -\sqrt{\frac{\frac{4}{\pi}\Delta\vartheta^* \alpha_{max,A}\alpha_{max,D} - \alpha_{max,D}\omega_0^2}{-\alpha_{max,D} - \alpha_{max,A}}}$$

IV. "S-SHAPE TRAJECTORY: STATEMENT AND SOLUTION

Path generation problem can be stated in a similar way as compared to cycloid case, by considering half a period raised cosine acceleration during phases A and D, i.e.:

$$\alpha_i^*(t, h_i, t_{si}) = \pi\frac{h_i}{t_{si}^2}\left[\frac{1}{2} - \frac{1}{2}\cos\left(2\pi\frac{t}{t_{si}}\right)\right], \tag{30}$$
$$t \in [0; t_{si}],\ i = A, D$$

and a null acceleration during phase C:

$$\alpha_C^*(t) = 0,\ t \in [0; t_C] \tag{31}$$

The difference with respect to the previous case is the zero value of jerk at start-up and end of acceleration and deceleration phases, that can be useful for certain applications and loads (Fig. 1b). The correspondent movement is shown in Fig. 3.

Once the initial conditions $\boldsymbol{\eta}_0 = [0\ \ \omega_0\ \ \vartheta_0]^T$ and the constraint $\alpha(t) = \alpha^*(t, \boldsymbol{\tau})$ have been considered, system $\boldsymbol{S}$ becomes:

$$\begin{cases} \alpha_A(t) = \pi\frac{h_A}{t_{sA}^2}\left[\frac{1}{2} - \frac{1}{2}\cos\left(2\pi\frac{t}{t_{sA}}\right)\right] \\ \omega_A(t) = \frac{h_A}{t_{sA}}\left[\frac{\pi}{2}\frac{t}{t_{sA}} - \frac{1}{4}\sin\left(2\pi\frac{t}{t_{sA}}\right)\right] + \omega_{0A} \\ \vartheta_A(t) = h_A\left[\frac{\pi}{4}\frac{t^2}{t_{sA}^2} + \frac{1}{8\pi}\cos\left(2\pi\frac{t}{t_{sA}}\right)\right] + \omega_{0A}t \\ \qquad\qquad + \vartheta_{0A} - \frac{h_A}{8\pi} \end{cases} \tag{32}$$

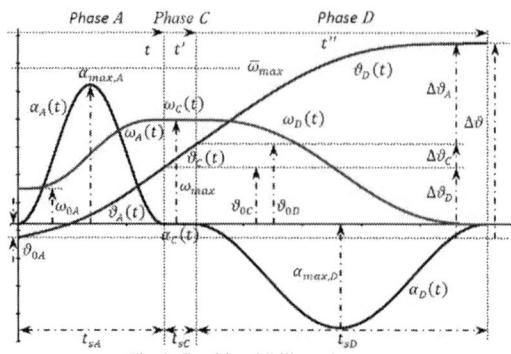

Fig. 3. Considered S-like trajectory.

$$\begin{cases} \alpha_D(t'') = -\pi\frac{h_D}{t_{sD}^2}\left[\frac{1}{2} - \frac{1}{2}\cos\left(2\pi\frac{t''}{t_{sD}}\right)\right] \\ \omega_D(t'') = -\frac{h_D}{t_{sD}}\left[\frac{\pi}{2}\frac{t''}{t_{sD}} - \frac{1}{4}\sin\left(2\pi\frac{t''}{t_{sD}}\right)\right] + \omega_{0D} \\ \vartheta_D(t'') = -h_D\left[\frac{\pi}{4}\frac{t''^2}{t_{sD}^2} + \frac{1}{8\pi}\cos\left(2\pi\frac{t''}{t_{sD}}\right)\right] + \omega_{0D}t'' \\ \qquad\qquad + \vartheta_{0D} + \frac{h_D}{8\pi} \end{cases} \tag{33}$$

where the expressions during phase C has been omitted as they are identical to (7). Constraints (10) and continuity conditions (11) still apply, leading to:

$$\alpha_{max,A} = \left|\alpha_A\left(\frac{t_{sA}}{2}\right)\right| = \pi\frac{|h_A|}{t_{sA}^2},$$

$$\omega_{max} = \omega_A(t_{sA}) = \frac{\pi}{2}\frac{h_A}{t_{sA}} + \omega_0$$

$$\alpha_{max,D} = \left|\alpha_D\left(\frac{t_{sD}}{2}\right)\right| = \pi\frac{|h_D|}{t_{sD}^2}, \tag{34}$$

$$\omega_D(t_{sD}) = 0 = -\frac{\pi}{2}\frac{h_D}{t_{sD}} + \omega_{max}$$

As before it is possible to find out an expression for the unknown parameters:

$$h_A = \frac{4}{\pi}\frac{(\omega_{max} - \omega_0)^2}{\alpha_{max,A}}sign(\omega_{max} - \omega_0)$$

$$t_{sA} = 2\frac{|\omega_{max} - \omega_0|}{\alpha_{max,A}} = \frac{\pi}{2}\frac{h_A}{\omega_{max} - \omega_0} \tag{35}$$

$$h_D = \frac{4}{\pi}\frac{\omega_{max}^2}{\alpha_{max,D}}sign(\omega_{max})$$

$$t_{sD} = 2\frac{|\omega_{max}|}{\alpha_{max,D}} = \frac{\pi}{2}\frac{h_D}{\omega_{max}}, \tag{36}$$

and position displacements that is possible to obtain within A and D phases are given by:

$$\Delta\vartheta_A = \frac{\pi}{4}h_A + \underbrace{2\omega_0\frac{|\omega_{max} - \omega_0|}{\alpha_{max,A}}}_{h_{\omega_0}} = \frac{\pi}{4}h_A +$$
$$\frac{\pi}{2}h_A\frac{\omega_0}{\omega_{max} - \omega_0} = \frac{\pi}{4}h_A\frac{\omega_{max} + \omega_0}{\omega_{max} - \omega_0} \tag{37}$$

$$\Delta\vartheta_D = -\frac{\pi}{4}h_D + \omega_{max}\frac{\pi}{2}\frac{h_D}{\omega_{max}} = \frac{\pi}{4}h_D . \tag{38}$$

The total position displacement that is possible to obtain by the sequencing of the phases A, C and D is therefore:

$$\Delta\vartheta = \Delta\vartheta_A + \Delta\vartheta_C + \Delta\vartheta_D =$$
$$\frac{\pi}{4}h_A\frac{\omega_{max} + \omega_0}{\omega_{max} - \omega_0} + h_C + \frac{\pi}{4}h_D \tag{39}$$

Same considerations as in the cycloid case apply, i.e. the solution of path generation problem depends on the value of the target $\Delta\vartheta^*$ with respect to the position displacement obtained by using only phases A and D ($t_{sC} = 0$) and the maximum allowed value of the speed ($\omega_{max} = \bar{\omega}_{max}$). This displacement can be calculated by (39) as

$$\Delta\vartheta_{AD} = \Delta\vartheta_A + \Delta\vartheta_D = \frac{\pi}{4} h_A \frac{\bar{\omega}_{max} + \omega_0}{\bar{\omega}_{max} - \omega_0} + \frac{\pi}{4} h_D \qquad (40)$$

If the target $\Delta\vartheta^*$ is equal to $\Delta\vartheta_{AD}$, then the problem is simply solved by using the values of h_i and t_{si} ($i = A, D$) given by (35) and (36).

When $\Delta\vartheta^* > \Delta\vartheta_{AD}$ the values of $\bar{\omega}_{max}$, $\alpha_{max,A}$ and $\alpha_{max,D}$ are not high enough to allow the required position displacement to be generated without introducing a constant speed phase. Hence, by assuming $\omega_{max} = \bar{\omega}_{max}$, the position displacement that must be introduced during this phase is obtained directly from (19):

$$\Delta\vartheta_C = \Delta\vartheta^* - \Delta\vartheta_{AD} = h_C =$$
$$= \Delta\vartheta^* - h_A \frac{\bar{\omega}_{max} + \omega_0}{\bar{\omega}_{max} - \omega_0} - h_D \qquad (41)$$

its application time still being given by (14).

When $\Delta\vartheta^* < \Delta\vartheta_{AD}$. the position displacement that is possible to obtain by using only phases A and D and the maximum allowed speed ($\omega_{max} = \bar{\omega}_{max}$) is greater than the required one. Then a lower value of the maximum speed ($\omega_{max} < \bar{\omega}_{max}$) has to be considered, whose value can be calculated as a function of the required $\Delta\vartheta^*$, i.e.:

$$\Delta\vartheta_{AD} = \Delta\vartheta^* = \Delta\vartheta_A + \Delta\vartheta_D =$$
$$\frac{\pi}{4} h_A \frac{\omega_{max} + \omega_0}{\omega_{max} - \omega_0} + \frac{\pi}{4} h_D =$$
$$\frac{(\omega_{max} - \omega_0)^2}{\alpha_{max,A}} \frac{\omega_{max} + \omega_0}{\omega_{max} - \omega_0} sign(\omega_{max} - \omega_0) +$$
$$\frac{\omega_{max}^2}{\alpha_{max,D}} sign(\omega_{max}) =$$
$$\frac{\alpha_{max,D}(\omega_{max}^2 - \omega_0^2) sign(\omega_{max} - \omega_0)}{\alpha_{max,A} \alpha_{max,D}} +$$
$$\frac{\alpha_{max,A} \omega_{max}^2 sign(\omega_{max})}{\alpha_{max,A} \alpha_{max,D}} \qquad (42)$$

Inversion of previous equation provides the value of the maximum speed:

$$\omega_{max} = \pm\sqrt{\frac{\Delta\vartheta^* \alpha_{max,A} \alpha_{max,D} + \alpha_{max,D} \omega_0^2 sign(\omega_{max} - \omega_0)}{\alpha_{max,D} sign(\omega_{max} - \omega_0) + \alpha_{max,A} sign(\omega_{max})}} \qquad (43)$$

Even in this case a complete study of the fraction inside the square root is required, leading to similar results as for cycloid paths. Four alternative cases can be obtained by a certain set of the input values for ω_0, $\Delta\vartheta^*$, and for the maximum acceleration and deceleration $\alpha_{max,pos}$ and $\alpha_{max,neg}$.

$$if\ (\omega_0 \geq 0)\ and\ \left(\Delta\vartheta^* \geq \frac{\omega_0^2}{\alpha_{max,neg}}\right)$$

$$\alpha_{max,D} = \alpha_{max,neg}$$
$$if\ (\omega_0 \leq \bar{\omega}_{max})\ \alpha_{max,A} = \alpha_{max,pos} \qquad (44)$$
$$else\ \alpha_{max,A} = \alpha_{max,neg}$$
$$\omega_{max} = \sqrt{\frac{\Delta\vartheta^* \alpha_{max,A} \alpha_{max,D} + \alpha_{max,D} \omega_0^2}{\alpha_{max,D} + \alpha_{max,A}}}$$

$$if\ (\omega_0 < 0)\ and\ \left(\Delta\vartheta^* \geq -\frac{\omega_0^2}{\alpha_{max,pos}}\right)$$

$$\alpha_{max,A} = \alpha_{max,pos},\ \alpha_{max,D} = \alpha_{max,neg} \qquad (45)$$
$$\omega_{max} = \sqrt{\frac{\Delta\vartheta^* \alpha_{max,A} \alpha_{max,D} + \alpha_{max,D} \omega_0^2}{\alpha_{max,D} + \alpha_{max,A}}}$$

$$if\ (\omega_0 < 0)\ and\ \left(\Delta\vartheta^* < -\frac{\omega_0^2}{\alpha_{max,pos}}\right)$$

$$\alpha_{max,D} = \alpha_{max,pos}$$
$$if\ (\omega_0 \geq -\bar{\omega}_{max})\ \alpha_{max,A} = \alpha_{max,neg} \qquad (46)$$
$$else\ \alpha_{max,A} = \alpha_{max,pos}$$
$$\omega_{max} = -\sqrt{\frac{\Delta\vartheta^* \alpha_{max,A} \alpha_{max,D} - \alpha_{max,D} \omega_0^2}{-\alpha_{max,D} - \alpha_{max,A}}}$$

$$if\ (\omega_0 \geq 0)\ and\ \left(\Delta\vartheta^* < \frac{\omega_0^2}{\alpha_{max,neg}}\right)$$

$$\alpha_{max,A} = \alpha_{max,neg},\ \alpha_{max,D} = \alpha_{max,pos} \qquad (47)$$
$$\omega_{max} = -\sqrt{\frac{\Delta\vartheta^* \alpha_{max,A} \alpha_{max,D} - \alpha_{max,D} \omega_0^2}{-\alpha_{max,D} - \alpha_{max,A}}}$$

V. EXTENDED TORQUE-SPEED CONTROL FOR IPM MOTOR

Torque-speed operation of IPM motor and its optimisation requires proper current vector control strategy. Hereafter some hints will be recalled aiming at understanding the limitations on reference trajectories that have to be introduced for a proper torque/speed operation of the motor, [5][6][7].

A. IPM operation in the i_d and i_q plane

The behavior of the IPM motor in the i_d and i_q current plane, (d axis aligned with the PM magnetic axis), can be described by considering three main loci (i.e. constant-voltage, -torque and -current) as a function of motor parameters and feeding and operating conditions.

If motor terminal voltage equations at steady-state are considered, stator resistance is neglected and by squaring and summing the two components, one obtains:

$$v^2 = v_d^2 + v_q^2 \cong \omega^2 (\psi_d^2 + \psi_q^2) \qquad (48)$$

By substitution of dq flux-linkage expressions, the constant voltage locus can be obtained, representing an ellipse in the i_d and i_q current plane:

$$\left(\frac{\bar{v}}{\omega}\right)^2 = \left(L_d i_d + \hat{\psi}_M\right)^2 + \left(\xi L_d i_q\right)^2 \qquad (49)$$

where $\xi = \frac{L_q}{L_d}$ is the saliency ratio of the machine and $\bar{v}$ is a certain (constant) value of feeding voltage.

Constant-torque and -current loci (respectively torque hyperbola and current circle) can be obtained directly from the steady-state machine model, i.e.:

$$\bar{m}_e = \hat{\psi}_M i_q + L_d(1 - \xi) i_d i_q \qquad (50)$$

$$\bar{i}^2 = i_d^2 + i_q^2 \qquad (51)$$

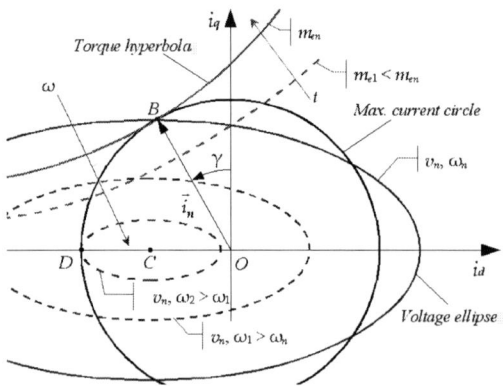

Fig. 4. Constant-torque, -voltage and -current loci in the i_d and i_q plane.

The constant loci are shown in Fig. 4, point B representing the intersection of the three curves for rated operating conditions of the motor.

According to (49), the increasing value of speed at the rated voltage condition provides a family of ellipses which collapse to their centre, i.e. point C, whose coordinates are $\left(-\frac{\psi_M}{L_d};0\right)$ and representing the theoretical operating condition of infinite speed and zero torque and also short-circuit operating point.

B. IPM optimum control

Control of the motor can realised and optimised by considering two more typical curves in the i_d and i_q current plane, [5][6][7], i.e. the *maximum-torque-per-amp* (MTPA) and the *maximum-torque-per-voltage* (MTPV) curves. The former represents the locus of the closest points to the origin of axes laying over each constant-torque hyperbola; the latter is the locus of the points producing the maximum torque for each voltage ellipse (at fixed rated voltage and increasing speed). These curves are represented in Fig. 5. Depending on the motor design the maximum-torque-per-voltage curve can be internal or external to the rated current circle: of course, only in the first case optimisation is possible with current space vector edge laying on that trajectory.

Optimum operation of the motor on both transient and steady-state conditions can be therefore obtained as represented in Fig. 5, i.e.:

a) when the speed is below the rated value (constant available torque region) there is no problem as far as the voltage limit concerns and the edge of the current space vector must lie on the MTPA locus (rated torque is achieved at point B);

b) above the rated speed (flux-weakening region), the admissible operating zone is limited by voltage limitation to its rated value, i.e. by voltage ellipses, and the rated current circle: maximum torque operation lies on the intersection of these curves (from point B to point P);

c) when the MTPV locus is attained (point P to point R), maximum torque for the given feeding voltage is possible; once the commanded speed is reached (in point R), operation moves on the rated voltage ellipse to satisfy the required torque with the minimum current (point Q).

Such a description gives an idea of the complexity of the current control for an IPM, requiring a coordinate control of both the dq components of the current space vector according to each torque/speed operating condition.

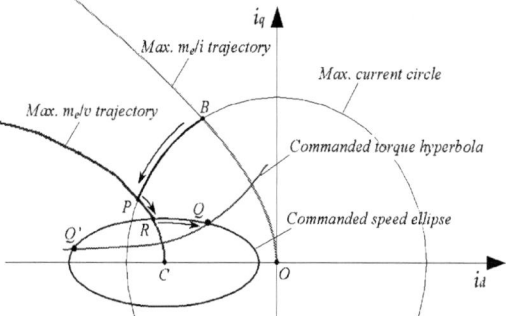

Fig. 5. Optimum trajectory and operating points in the i_d and i_q plane.

VI. POSITION CONTROLLER SCHEME

The scheme of the proposed intelligent position controller, comprising the path generator and the IPM drive, is shown in Fig. 6. IPM motor control is based on the cascaded approach and consists of three nested control loops, respectively for dq currents (in the synchronous reference frame), speed and position. The optimum control strategy of the motor described in the previous section can be implemented through three *vector-control* characteristics (F_{m_e}, F_{i_d} and F_{i_q}), according to the shown scheme. These

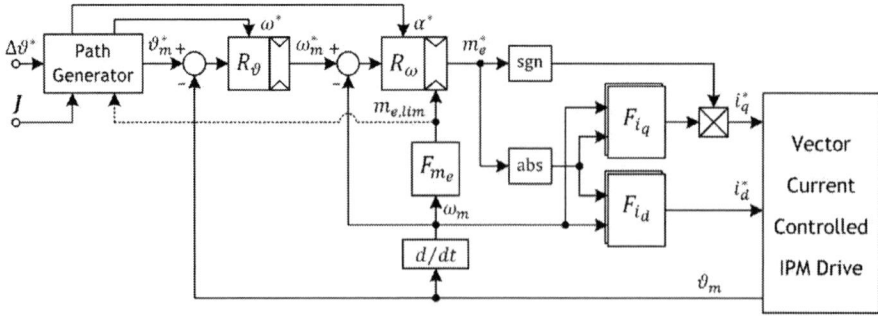

Fig. 6. Position controller scheme.

characteristics will provide the set points of the dq currents (i_d^* and i_q^*) as functions of the torque command (m_e^*) and the operating speed, i.e. $i_q^* = F_q(m_e^*, \omega_m)$ and $i_d^* = F_d(m_e^*, \omega_m)$, the value of speed affecting the control characteristics in the flux weakening region only. Moreover, in order to assure that the commanded operating point is attainable, a proper limitation of the torque reference as a function of actual motor speed is necessary, corresponding to the maximum torque vs. speed curve achieved by optimum control, $m_{e,lim} = F_{m_e}(\omega_m)$. Torque limitation can also be adopted inside path generator to verify the feasibility of the commanded trajectory, based on the knowledge of an estimate or an upper limitation of the load torque.

The inputs of the path generator are the desired displacement $\Delta\vartheta^*$ and the mechanical constraints J. Its outputs are the references for the theoretical kinematic trajectories, i.e. mechanical position ($\vartheta_m^* = \vartheta^*$), speed ($\omega^*$) and acceleration ($\alpha^*$), the first one acting as the reference for position regulator, the last two employed as feed-forward terms into position and speed regulators respectively.

VII. SIMULATION RESULTS

The proposed path generator alone has been simulated on a series of test movements, showing the feasibility of the proposed solution and the gain of performance with respect to the previous one, [4].

Stand-alone path generation is considered as well as the simulation of an actual IPM motor drive employing proposed kinematic trajectory generation.

The possibility of on-the-fly change of the speed limitation $\bar{\omega}_{max}$ and the reference position displacement $\Delta\vartheta^*$ have also been considered and analytically solved in order to assure a high degree of flexibility inside an actual industrial application.

Path generation task can be split into two main subtasks, i.e. profile calculation and generation. The first one is executed only when a new movement is requested, i.e. a new value of the target position is acquired or an admissible on-the-fly variation of speed limitation or target position is received. The second one is executed at each sampling period, providing real-time generation of position, speed and acceleration profiles within the whole duration of the requested displacement.

All the calculations within the path generator are done in per unit form, by assuming the following base system:

$$\vartheta_{base} = 50\pi \ [rad],$$
$$\omega_{base} = 200\pi \ [rad/s]$$
$$t_{base} = \frac{\vartheta_{base}}{\omega_{base}} = 0{,}25 \ [s],$$
$$\alpha_{base} = \frac{\vartheta_{base}}{t_{base}^2} = 800\pi \ [rad/s^2].$$

(52)

A first simulation investigation considers a relatively small movement ($\Delta\vartheta^* = 2\pi$) starting from null position and with the initial speed $\omega_0 = 60\pi \ [rad/s]$. The maximum speed has been set to $\bar{\omega}_{max} = 90\pi \ [rad/s]$ and positive and negative accelerations $\alpha_{max,pos} = 2400$ and $\alpha_{max,neg} = 1800 \ [rad/s^2]$ have been considered for a cycloid test case.

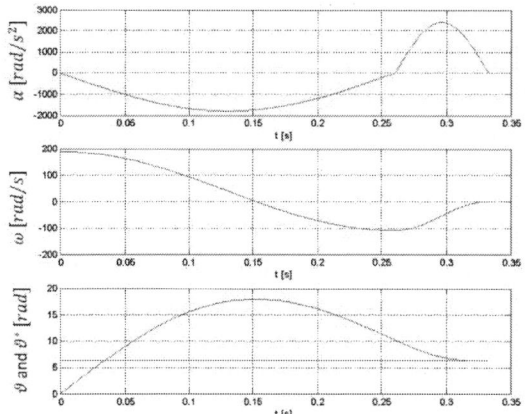

Fig. 7. Profiles generation by the proposed path generator providing only one movement.

In Fig. 7 position, speed and acceleration trajectories are shown, demonstrating that a correct movement is attained by using only deceleration and acceleration phases, i.e. without a constant speed phase.

The same operating conditions are considered in the results shown in Fig. 8, where the approach presented in [4] is considered for the path generation task. It is clearly visible how previous solution requires two different movements to reach the target position with the required constraints: the first one allows to stop to a position much higher than the required one, that is then compensated by a second movement in the opposite direction.

In Fig. 9 simulation of an actual IPM motor drive system is shown driving a non-zero passive load torque. The behavior of actual and reference direct and quadrature currents as well as speed and position shapes are shown in the case of cycloid trajectory. Accurate speed and position tracking are assured by feed-forward components coming from the path generator (refer to Fig. 6), except at motor start-up where the response is quite slow due to the latency of the speed control loop and associated speed calculation algorithm. Similar results apply when S-like acceleration is considered.

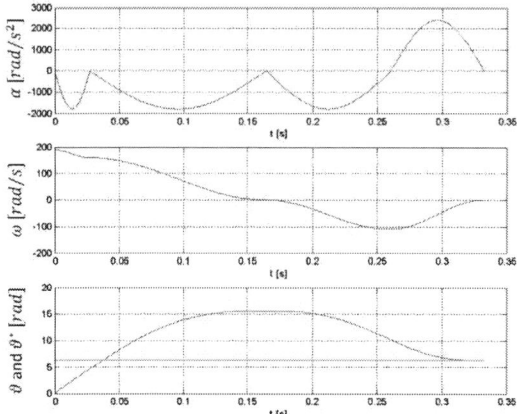

Fig. 8. Profiles generation by a standard path generator providing two consecutive movements.

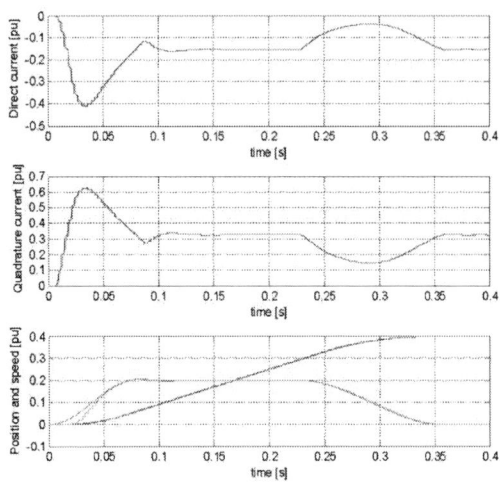

Fig. 9. Comparison between actual values and references: direct current (top), quadrature current (middle), speed and position (bottom).

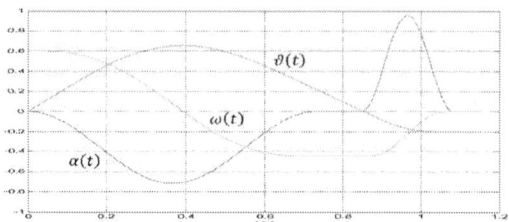

Fig. 10. Movement employing raised cosine acceleration.

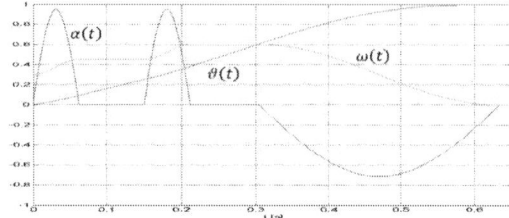

Fig. 11. Movement with on-the-fly change of speed limitation.

VIII. EXPERIMENTAL RESULTS

The proposed approach to path generation and all the features needed for an actual drive system for industrial application have been experimentally implemented by means of a 32-bit fixed-point digital signal controller (TMS320F2808). A sampling period of 50µs has been assumed, allowing the execution of both the IPM motor position controller and path generation. Per unit form and Q24 fixed-point numerical representation is adopted for all the variables. Comparison between simulated trajectories and experiments, also from a numerical point-of-view, confirm the accuracy of the implementation and the effectiveness of the proposed solutions.

Results shown hereafter refer to off-line operation of path generator as the aim of those tests was limited to the verification of path generator itself more than that of the motor drive.

In Fig. 10 S-like trajectory is considered and a simple movement with non-zero initial speed value ($\omega_0 = 120\pi$, i.e. $0.6\,p.u.$) and negative reference position displacement is commanded ($\vartheta^* = -10\pi$, i.e. $-0.2\,p.u.$). As clearly visible the operating condition requires a first phase which is indeed a deceleration (due to the positive value of initial speed and the negative value of reference position displacement) and the last one being a real acceleration. The movement needs the presence of a constant speed phase.

In Fig. 11 an on-the-fly change of speed limitation is investigated in the case of the cycloid path generator. A basic movement with positive initial speed ($\omega_0 = 60\pi$, i.e. $0.3\,p.u.$) and positive reference position displacement ($\vartheta^* = 50\pi$, i.e. $1\,p.u.$) and is initially commanded. Then, at $t = 0.15\,[s]$ a sudden variation of the speed limitation is imposed, from $\bar{\omega}_{max,1} = 90\pi$ to $\bar{\omega}_{max,2} = 120\pi$. The new profile (accounting for the new value of speed limitation and actual values of position and speed) is calculated in real-time and profile generation continues along the new profile. Similar results have been obtained if an on-the-fly variation of the target reference position is commanded.

IX. CONCLUSIONS

Constrained real-time path generation issues for high performance motion control systems have been presented in this paper. Two instances of possible shapes for the acceleration have been considered, i.e. sinusoidal and raised cosine, and all the cases generating from a certain set of constraints imposed on the kinematic characteristics of the movements have been analysed and solved analytically. The proposed path generator has been included inside a digital drive system for IPM synchronous motors based on a last generation digital signal controller. Simulation and experimental results have been presented and discussed, proving the effectiveness of the proposed approach for a number of industrial applications requiring constrained and accurate control of kinematic trajectories.

REFERENCES

[1] A. John, J.M. Pacas, "Intelligent motion: the new challenge for electrical drives," *Proc. of 7th Europen Conference on Power Electronics and Applications*, vol.3, pp.918-921, Trondheim, 1997.

[2] T. Ohmae, M. Watanabe, "Microprocessor systems for motion control," *Proc. of IFAC Workshop on Motion Control for Intelligent Automation*, pp. 253-258, Perugia, Italy, October 27-29, 1992.

[3] F. Parasiliti, R. Petrella, M. Tursini, J.M. Pacas, "Flexible High-Performance Motion Control System for Multi-Axis Applications", *Proc. of the International Conferences on Power Electronics, Intelligent Motion and Power Quality (PCIM'2002)*, Nürnberg, Germany, May 14 16, 2002.

[4] F. Parasiliti, R. Petrella, M. Tursini, "DSP Based AC Drive Controller with Real-Time Path Generation for Intelligent Motion Applications", *Proc. of the 34th IEEE Industry Application Society Annual Meeting (IAS'99)*, Phoenix, Arizona, USA, October 3-7, 1999.

[5] J. M. Kim, S. K. Sul, "Speed Control of Interior Permanent Magnet Synchronous Motor Drive for the Flux–Weakening Operation", *IEEE Trans. on Ind. Appl.*, vol. 33, pp. 43-48, Jan./Feb. 1997.

[6] S. R. MacMinn, T. M. Jahns, "Control Techniques for Improved High-speed Performance of Interior PM Synchronous Motor Drives", *IEEE Trans. on Ind. Appl.*, pp. 997-1004, 1991.

[7] M. Tursini, A. Scafati, A. Guerriero, R. Petrella , "Extended Torque-Speed Region Sensor-Less Control of Interior Permanent Magnet Synchronous Motors", *Proc. of the International Aegean Conference on Electric Machines, Power Electronics and Electromotion Joint Conference*, pp. 647 652, 10 12 September, Bodrum, Turkey, 2007 (ISBN 978-975-93410-2-2)

PMSM Sliding Mode Observer for Speed and Position Estimation Using Modified Back EMF

Ilioudis Vasilios C.*, Margaris Nikolaos I.**

* Aristotle Univ. of Thessaloniki/Dep. of ECE, Thessaloniki, Greece, e-mail: *ilioudis@otenet.gr*
**Aristotle Univ. of Thessaloniki/Dep. of ECE, Thessaloniki, Greece, e-mail: *margaris@eng.auth.gr*

Abstract—This paper introduces a novel sensorless method for speed and position estimation of permanent magnet synchronous motor (PMSM). The mathematical model of PMSM is expressed in an arbitrary rotating reference frame and no knowledge of initial rotor position is needed. The developed algorithm is based on sliding mode observer (SMO) theory and consists of two discrete steps: flux/current and speed/position estimation. Also, Lyapunov functions are suitably chosen for determining the adaptive laws for the stator resistance and the speed estimators. First stator flux/current and resistance error converges to zero and afterwards speed and position estimation is carried out using a modified back electromotive force (EMF) observer. Simulation results show the effectiveness of the proposed estimation method.

Keywords—permanent magnet synchronous motor (PMSM), sensorless control, sliding mode observer (SMO), sensorless speed and position estimation.

NOTATION

$E_\gamma = \gamma$-axis modified back EMF

$E_\delta = \delta$- axis modified back EMF

$\theta =$ angular position

$i_\gamma = \gamma$-axis current

$i_\delta = \delta$-axis current

$i_d = $ d-axis current

$i_q = $ q-axis current

$L_d = $ d-axis inductance

$L_q = $ q-axis inductance

$\lambda_\gamma = \gamma$-axis magnetic flux

$\lambda_\delta = \delta$-axis magnetic flux

$\lambda_d = $ d-axis magnetic flux

$\lambda_q = $ q-axis magnetic flux

$\lambda_m = $ permanent magnet flux

$\lambda_{m\gamma} = \gamma$-axis partial magnetic flux

$\lambda_{m\delta} = \delta$-axis partial magnetic flux

$r_s = $ stator resistance

$u_\delta = \delta$-axis voltage

$u_\gamma = \gamma$-axis voltage

$u_d = $ d-axis voltage

$u_q = $ q-axis voltage

$\omega = $ angular speed

I. INTRODUCTION

Permanent magnet synchronous motors (PMSM) are widely used in many industrial applications because of their high efficiency, reduced size and torque density. Speed control of PMSM requires accurate rotor position knowledge, which is critical information and it is obtained generally using position sensors such as optical encoders or magnetic resolvers. However sensors installed on the rotor shaft increase the machine size, cost and inserted noise. Therefore many sensorless control methods of PMSM are proposed to avoid these disadvantages.

In sensorless control, some of all machine variables are often not directly measurable, but their accurate knowledge is required for control of high-performance electrical drives. Successful implementation of sensorless vector control depends primarily on the accuracy of the rotor position estimation. PMSM rotor position is estimated using two main strategies named as saliency and signal injection and fundamental excitation method. The saliency and signal injection method estimates rotor position at low, high-speed regions even at standstill, but requires additional signals and it causes undesirable torque and noise [8], [13]. On the other hand fundamental excitation method uses back electromotive force (EMF) to estimate rotor position by means of a state observer [8]-[11]. Various approaches in this strategy are based on the estimation of the back EMF or Flux-linkage due to permanent magnet by means of a state observer or an extended Kalman filter [11]. Other simple methods are based on the voltage or current errors between the detected and calculated variables from the motor model.

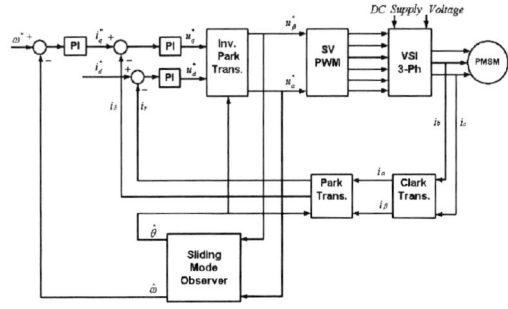

Fig. 1. Block diagram of PMSM controlled system.

978-1-4244-1741-4/08/$25.00 ©2008 IEEE

Most fundamental excitation methods are applied to a non-salient PMSM. In several sensorless control methods, some terms related to the saliency in the motor equation are ignored. However this approximation may cause estimation errors and performance decrease.

The initial rotor position estimation is one of the serious problems in sensorless PMSM speed estimation. Error in initial rotor position estimation may cause motor to rotate in wrong direction because of the torque reduction. To overcome this problem many approaches use the principle that the inductance is varied according to the rotor position.

In this paper, the proposed position and speed estimation algorithm of PMSM is based on sliding mode observers using Lyapunov stability criteria. Sliding mode observer is a promising approach because of its independence of system parameter variation, fast convergence and robustness. This algorithm is implemented by fundamental excitation method and the rotor position is detected from the back EMF. A new mathematical model of PMSM is introduced using a rotating reference frame rather than α-β stationary reference frame. Fig. 1 shows the entire control system. The sliding mode observer is divided mainly into two parts, the flux/current observer and the modified back EMF observer connected in cascade. Also a stator resistance observer is embedded into flux/current observer.

When sliding mode occurs in flux/current observer the modified back EMF and rotor speed are estimated from the observer control inputs using the equivalent control method. PMSM rotor position is extracted afterwards from the estimated back EMF without using the initial rotor position.

II. MATHEMATICAL MODEL OF PMSM

Since the rotor position cannot be detected, the d-q axis mathematical model cannot be applied directly. Most approaches are based on the estimation of the back electromotive force (EMF) in the stationary reference frame α-β. The proposed PMSM mathematical model of sliding mode observer is defined in an estimated reference frame γ-δ rotating at an angular velocity $\hat{\omega}$ and lagging behind the d-q reference frame by electrical angle $\bar{\theta}$. Fig. 2 shows schematically the relations between the synchronous (dq-axis), the estimated (γδ-axis) and the stationary (α-β) reference frames.

The sensorless approach considered uses the developed theory of reference frames. It is based on the property of d-q reference frame being transformed to an estimated γ-δ reference frame using the angle error $\bar{\theta}$ denoted as $\bar{\theta} = \theta - \hat{\theta}$. This error is the angle difference between real and estimated reference frames, which are rotating at different angular speeds ω (synchronous speed) and $\hat{\omega}$ (estimated speed) respectively. Fig. 2 shows schematically the transformation, where the vector λ_s represents the stator flux analyzed to its components in d-q and γ-δ reference frames.

The flux/current mathematical model of PMSM in d-q synchronous rotating frame is given by the system (1) and (2) in matrix form.

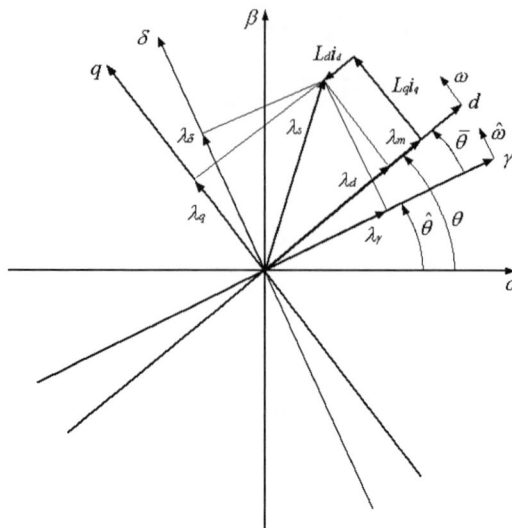

Fig. 2. The definitions of the three reference frames used in PMSM mathematical model analysis.

$$\dot{\lambda}_{dq} = \begin{bmatrix} \dot{\lambda}_d \\ \dot{\lambda}_q \end{bmatrix} = \begin{bmatrix} -r_s & +\omega L_q \\ -\omega L_d & -r_s \end{bmatrix} \begin{bmatrix} i_d \\ i_q \end{bmatrix} + \begin{bmatrix} u_d \\ u_q \end{bmatrix} + \omega \lambda_m \begin{bmatrix} 0 \\ 1 \end{bmatrix} \quad (1)$$

$$\lambda_{dq} = \begin{bmatrix} \lambda_d \\ \lambda_q \end{bmatrix} = \begin{bmatrix} L_d & 0 \\ 0 & L_q \end{bmatrix} \begin{bmatrix} i_d \\ i_q \end{bmatrix} + \lambda_m \begin{bmatrix} 1 \\ 0 \end{bmatrix} \quad (2)$$

It is obvious from Fig. 2, that

$$\lambda_{\gamma\delta} = \begin{bmatrix} \lambda_\gamma \\ \lambda_\delta \end{bmatrix} = \begin{bmatrix} \lambda_d \cos\bar{\theta} - \lambda_q \sin\bar{\theta} \\ \lambda_d \sin\bar{\theta} + \lambda_q \cos\bar{\theta} \end{bmatrix} = T(\bar{\theta}) \begin{bmatrix} \lambda_d \\ \lambda_q \end{bmatrix} \quad (3)$$

and

$$\dot{\lambda}_{\gamma\delta} = \begin{bmatrix} \dot{\lambda}_\gamma \\ \dot{\lambda}_\delta \end{bmatrix} = \dot{\bar{\theta}} J \begin{bmatrix} \lambda_\gamma \\ \lambda_\delta \end{bmatrix} + T(\bar{\theta}) \begin{bmatrix} \dot{\lambda}_d \\ \dot{\lambda}_q \end{bmatrix} \quad (4)$$

Since

$$\dot{T}(\bar{\theta}) = \dot{\bar{\theta}} \begin{bmatrix} 0 & -1 \\ 1 & 0 \end{bmatrix} \begin{bmatrix} \cos\bar{\theta} & -\sin\bar{\theta} \\ \sin\bar{\theta} & \cos\bar{\theta} \end{bmatrix} = \dot{\bar{\theta}} J T(\bar{\theta})$$

Here

$$J = \begin{bmatrix} 0 & -1 \\ 1 & 0 \end{bmatrix} \text{ and } T(\bar{\theta}) = \begin{bmatrix} \cos\bar{\theta} & -\sin\bar{\theta} \\ \sin\bar{\theta} & \cos\bar{\theta} \end{bmatrix}$$

The transformation matrix $T(\bar{\theta})$ is used to transform any vector from d-q synchronous to γ-δ estimated reference frame.

By rearranging, equation (4) could be rewritten in the following form.

$$T\left(\overline{\theta}\right)\begin{bmatrix}\dot{\lambda}_d\\\dot{\lambda}_q\end{bmatrix}=\begin{bmatrix}\dot{\lambda}_\gamma\\\dot{\lambda}_\delta\end{bmatrix}-\dot{\overline{\theta}}JT\left(\overline{\theta}\right)\begin{bmatrix}\lambda_d\\\lambda_q\end{bmatrix}\qquad(5)$$

Multiplying equation (1) by $T\left(\overline{\theta}\right)$ then results the following

$$T\left(\overline{\theta}\right)\begin{bmatrix}\dot{\lambda}_d\\\dot{\lambda}_q\end{bmatrix}=-r_sT\left(\overline{\theta}\right)\begin{bmatrix}i_d\\i_q\end{bmatrix}+T\left(\overline{\theta}\right)\begin{bmatrix}u_d\\u_q\end{bmatrix}-\omega JT\left(\overline{\theta}\right)\begin{bmatrix}\lambda_d\\\lambda_q\end{bmatrix}$$
$$(6)$$

The right-hand side parts of (5) and (6) are equals. Therefore by solving (5) with respect to $\dot{\lambda}_{\gamma\delta}$, a similar to equation (1) results.

$$\dot{\lambda}_{dq}=-r_si_{dq}+u_{dq}-\omega J\lambda_{dq}\qquad(7)$$

$$\dot{\lambda}_{\gamma\delta}=-r_si_{\gamma\delta}+u_{\gamma\delta}-\left(\omega-\dot{\overline{\theta}}\right)J\lambda_{\gamma\delta}\qquad(8)$$

Here

$$u_{dq}=\begin{bmatrix}u_d\\u_q\end{bmatrix},\ i_{dq}=\begin{bmatrix}i_d\\i_q\end{bmatrix},\ u_{\gamma\delta}=\begin{bmatrix}u_\gamma\\u_\delta\end{bmatrix}\text{ and }i_{\gamma\delta}=\begin{bmatrix}i_\gamma\\i_\delta\end{bmatrix}$$

Equations (7) and (8) represent the flux/current mathematical model of PMSM in d-q and γ-δ reference frames respectively.

It must be noted these matrix equations are very similar. It is obvious that $\left(\omega-\dot{\overline{\theta}}\right)$ is used instead of ω and the real and estimated reference frames coincide when $\dot{\overline{\theta}}=\overline{\theta}=0$. The estimation algorithm is made so as to reduce to zero the rotor angle error and its derivative.

III. ESTIMATION ARGORITHM

A. Sliding Mode Observer Analysis

By considering matrix equations (8) and (3), the estimated γ-δ axis mathematical model of PMSM is described as follows.

$$\dot{\lambda}_\gamma=-r_s\cdot i_\gamma+u_\gamma+\left(\omega-\dot{\overline{\theta}}\right)\lambda_\delta\qquad(9)$$

$$\dot{\lambda}_\delta=-r_si_\delta+u_\delta-\left(\omega-\dot{\overline{\theta}}\right)\lambda_\gamma\qquad(10)$$

$$\lambda_\gamma=L_d\cdot i_\gamma+\lambda_{m\gamma}\qquad(11)$$

$$\lambda_\delta=L_q\cdot i_\delta+\lambda_{m\delta}\qquad(12)$$

Here the partial fluxes $\lambda_{m\gamma}$ and $\lambda_{m\delta}$ are defined by

$$\lambda_{m\gamma}=L_2\left[i_\gamma\left(1-\cos2\overline{\theta}\right)-i_\delta\sin2\overline{\theta}\right]+\lambda_m\cos\overline{\theta},$$

$$\lambda_{m\delta}=L_2\left[-i_\gamma\sin2\overline{\theta}-i_\delta\left(1-\cos2\overline{\theta}\right)\right]+\lambda_m\sin\overline{\theta}\text{ and}$$

$$L_2=\left(L_q-L_d\right)/2.$$

According to previous analysis, the γ-δ stator fluxes are divided into two parts each. The first part is due to currents and impedances, while the second one is depended on saliency and rotor angle error. This property of partial fluxes is used to extract information about the rotor angle error.

In the proposed sliding mode estimation algorithm, flux/current state observer is using as sliding surfaces $s=\left[s_\gamma,s_\delta\right]^T$ where $s_\gamma=L_d\left(i_\gamma-\hat{i}_\gamma\right)=L_d\overline{i}_\gamma$ and $s_\delta=L_q\left(i_\delta-\hat{i}_\delta\right)=L_q\overline{i}_\delta$. The sliding mode observer is given by

$$\dot{\hat{\lambda}}_\gamma=-\hat{r}_si_\gamma+u_\gamma+\left(\omega-\dot{\overline{\theta}}\right)\hat{\lambda}_\delta+\left(\omega-\dot{\overline{\theta}}\right)L_q\overline{i}_\delta$$
$$+K_\gamma\operatorname{sgn}\left(L_d\overline{i}_\gamma\right)\qquad(13)$$

$$\dot{\hat{\lambda}}_\delta=-\hat{r}_si_\delta+u_\delta-\left(\omega-\dot{\overline{\theta}}\right)\hat{\lambda}_\gamma-\left(\omega-\dot{\overline{\theta}}\right)L_d\overline{i}_\gamma$$
$$+K_\delta\operatorname{sgn}\left(L_q\overline{i}_\delta\right)\qquad(14)$$

$$\hat{\lambda}_\gamma=\hat{L}_d\cdot\hat{i}_\gamma+\hat{\lambda}_{m\gamma}\qquad(15)$$

$$\hat{\lambda}_\delta=\hat{L}_q\cdot\hat{i}_\delta\qquad(16)$$

Here K_γ, $K_\delta>0$ are the observer gains.

By subtracting equations (13) from (9) and (14) from (10), the flux error dynamics due to currents is expressed by

$$\left(\dot{\overline{L_d\overline{i}_\gamma}}\right)=-\overline{r}_si_\gamma-\dot{\overline{\lambda}}_{m\gamma}+\left(\omega-\dot{\overline{\theta}}\right)\overline{\lambda}_{m\delta}-K_\gamma\operatorname{sgn}\left(L_d\overline{i}_\gamma\right)\quad(17)$$

$$\left(\dot{\overline{L_q\overline{i}_\delta}}\right)=-\overline{r}_si_\delta-\dot{\overline{\lambda}}_{m\delta}-\left(\omega-\dot{\overline{\theta}}\right)\overline{\lambda}_{m\gamma}-K_\delta\operatorname{sgn}\left(L_q\overline{i}_\delta\right)\quad(18)$$

Moreover analyzing the partial flux terms $-\dot{\overline{\lambda}}_{m\gamma}+\left(\omega-\dot{\overline{\theta}}\right)\overline{\lambda}_{m\delta}$ and $-\dot{\overline{\lambda}}_{m\delta}-\left(\omega-\dot{\overline{\theta}}\right)\overline{\lambda}_{m\gamma}$ included into (17) and (18), these could be rewritten as modified back EMF errors as follows

$$-\dot{\overline{\lambda}}_{m\gamma}+\left(\omega-\dot{\overline{\theta}}\right)\overline{\lambda}_{m\delta}=E_\gamma-\hat{E}_\gamma=\overline{E}_\gamma$$

$$-\dot{\overline{\lambda}}_{m\delta}-\left(\omega-\dot{\overline{\theta}}\right)\overline{\lambda}_{m\gamma}=E_\delta-\hat{E}_\delta=\overline{E}_\delta$$

Here

$$E_\gamma=-\omega\lambda_m\sin\overline{\theta}\text{ and }E_\delta=\omega\lambda_m\cos\overline{\theta}.$$

Finally (17) and (18) are transformed to following

$$\left(\dot{\overline{L_d\overline{i}_\gamma}}\right)=-\overline{r}_si_\gamma+\overline{E}_\gamma-K_\gamma\operatorname{sgn}\left(L_d\overline{i}_\gamma\right)\qquad(19)$$

$$\left(\dot{\overline{L_q\overline{i}_\delta}}\right)=-\overline{r}_si_\delta+\overline{E}_\delta-K_\delta\operatorname{sgn}\left(L_q\overline{i}_\delta\right)\qquad(20)$$

B. Estimated Inductances Algorithm

In the real system the stator inductances are not linearly dependent on respective stator currents. Therefore taking account of saturation in sliding mode observer described by (13) - (16), estimated stator inductances are used considering the following approximation low.

$$\hat{L}_d = \frac{\alpha_d}{1+\beta_d \cdot |i_\gamma|} \qquad (21)$$

$$\hat{L}_q = \frac{\alpha_q}{1+\beta_q \cdot |i_\delta|} \qquad (22)$$

Here $\alpha_d = 0.070H$, $\beta_d = 0.155A^{-1}$,

$\alpha_q = 0.360H$, $\beta_q = 0.155A^{-1}$

C. Sliding Mode Existence Conditions

Sliding mode existence needs $s_\gamma \dot{s}_\gamma < 0$, $s_\delta \dot{s}_\delta < 0$ and it is ensured by the following inequalities.

$$K_\gamma > |\bar{E}_\gamma| + |-\bar{r}_s i_\gamma| \qquad (23)$$

$$K_\delta > |\bar{E}_\delta| + |-\bar{r}_s i_\delta| \qquad (24)$$

Therefore inequalities (23) and (24) ensure sliding mode existence and reaching conditions. Sliding mode of the current observer is succeeded after time $t \geq t_n$, where t_n is the reaching time to the sliding manifold.

D. Stator Resistance Observer

Considering the Lyapunov function candidate V_{irs}, with $\gamma_r > 0$, it is necessary to verify that $\dot{V}_{irs}$ is definite negative.

$$V_{irs} = \left\{ \frac{1}{2}\left[\left(L_d \bar{i}_\gamma\right)^2 + \left(L_q \bar{i}_\delta\right)^2 + \frac{1}{\gamma_r}\bar{r}_s^2 \right]\right\} \qquad (25)$$

Differentiating both sides of the above equation, the V_{irs} derivative is expressed by

$$\dot{V}_{irs} = \frac{1}{\gamma_r}\bar{r}_s\left[\dot{\bar{r}}_s - \gamma_r\left(i_\gamma L_d \bar{i}_\gamma + i_\delta L_q \bar{i}_\delta\right)\right]$$
$$+ \bar{E}_\gamma L_d \bar{i}_\gamma - K_\gamma|L_d \bar{i}_\gamma| + \bar{E}_\delta L_q \bar{i}_\delta - K_\delta|L_q \bar{i}_\delta| \qquad (26)$$

In order $\dot{V}_{irs}$ to be negative, the following relations are assumed to be valid.

$$\dot{\bar{r}}_s = \gamma_r\left(i_\gamma L_d i_\gamma + i_\delta L_q \bar{i}_\delta\right) \qquad (27)$$

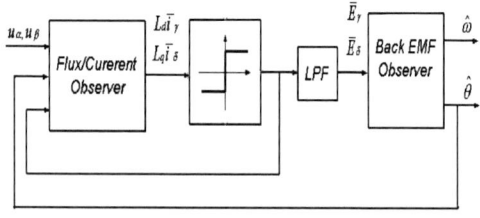

Fig. 3. Block diagram of the sliding mode observer in detailed form. Flux/current observer and modified back EMF observer are connected in cascade.

$$K_\gamma > |\bar{E}_\gamma| \qquad (28)$$

$$K_\delta > |\bar{E}_\delta| \qquad (29)$$

From (23), (24) and (28), (29) it follows that the inequalities (23) and (24) ensure asymptotic stability and sliding mode existence simultaneously.

Considering that the stator resistance r_s varies slowly, that is $\dot{r}_s \approx 0$, and adaptation law from (27), then the estimated stator resistance dynamics could be defined by

$$\dot{\hat{r}}_s = -K_{rs}\,\text{sgn}\left(i_\gamma L_d \bar{i}_\gamma + i_\delta L_q \bar{i}_\delta\right) \qquad (30)$$

In the above equation $K_{rs} > 0$ and represents the observer gain.

E. Equivalent Control Mode

When in the sliding mode, that is $s = 0$, the system trajectories satisfy the original system equation with the control replaced by its equivalent value found from the equation $\dot{s} = 0$. Therefore, once sliding mode occurs in the flux/current observer, that is $\left(\bar{i}_\gamma, \dot{\bar{i}}_\gamma\right) = 0$ and $\left(\bar{i}_\delta, \dot{\bar{i}}_\delta\right) = 0$, and stator resistance observer converges considerably fast, the error dynamics of the overall system is expressed by the following equations

$$\bar{E}_\gamma - \left[K_\gamma\,\text{sgn}\left(L_d \bar{i}_\gamma\right)\right]_{eq} = 0 \qquad (31)$$

$$\bar{E}_\delta - \left[K_\delta\,\text{sgn}\left(L_q \bar{i}_\delta\right)\right]_{eq} = 0 \qquad (32)$$

Here both second terms of right-hand side represent the filtered terms as shown by Fig. 2.

This system is used to extract information for rotor speed and rotor angle error from a modified back EMF.

F. Modified Back EMF and Speed Observer

Modified back EMF observer is based on equivalent control method applied to flux/current observer. The dynamics of modified back EMF observer is defined as follows.

$$\dot{\hat{E}}_\gamma = c_1 \bar{E}_\gamma + \hat{E}_\delta K_\theta\,\text{sgn}\,\bar{\theta} \qquad (33)$$

$$\dot{\hat{E}}_\delta = c_2 \bar{E}_\delta - \hat{E}_\gamma K_\theta\,\text{sgn}\,\bar{\theta} \qquad (34)$$

Using the Lyapunov function candidate $V_{E\omega}$, with $\gamma_\omega > 0$, determined by

$$V_{E\omega} = \frac{1}{2}\left[\bar{E}_\gamma^2 + \bar{E}_\delta^2 + \frac{1}{\gamma_\omega}\bar{\omega}^2 \right] \qquad (35)$$

then

1108

$$\dot{V}_{E\omega} = -c_1\overline{E}_\gamma{}^2 + \frac{1}{\gamma_\omega}\overline{\omega}\left[\dot{\overline{\omega}} - \gamma_\omega\left(\hat{E}_\delta\overline{E}_\gamma - \hat{E}_\gamma\overline{E}_\delta\right)\right] - c_2\overline{E}_\delta{}^2 \quad (36)$$

In order $\dot{V}_{E\omega}$ to be definite negative, the following speed adaptation law is assumed to be valid.

$$\dot{\overline{\omega}} = \gamma_\omega\left(\hat{E}_\delta\overline{E}_\gamma - \hat{E}_\gamma\overline{E}_\delta\right) \quad (37)$$

The above leads to the speed dynamics expressed by

$$\dot{\hat{\omega}} = -K_\omega\,\mathrm{sgn}\left(\hat{E}_\delta\overline{E}_\gamma - \hat{E}_\gamma\overline{E}_\delta\right) \quad (38)$$

Here, c_1, c_2 and $K_\omega > 0$ are the modified back EMF and speed observer gains respectively.

G. Rotor Angle Observer

From the previous analysis of the flux/current observer results that the rotor angle error is calculated by

$$\overline{\theta} = \tan^{-1}\left(-\frac{E_\gamma}{E_\delta}\right) \simeq \tan^{-1}\left(-\frac{\hat{E}_\gamma}{\hat{E}_\delta}\right) \quad (39)$$

Considering $\dot{\hat{\theta}} = \hat{\omega} + K_\theta\,\mathrm{sgn}\,\overline{\theta}$, then since $\dot{\theta} = \omega$, the error dynamics for $\overline{\theta}$ is as follows.

$$\dot{\overline{\theta}} = \overline{\omega} - K_\theta\,\mathrm{sgn}\,\overline{\theta} \quad (40)$$

The angle error converges to zero for $K_\theta > |\overline{\omega}|$.

IV. SIMULATION RESULTS

The parameters of the tested PMSM are listed in Table I. Experimental validation of the speed and position estimation algorithm is competently carried out by Matlab/Simulink facility.

Simulation results are presented first without external torque disturbance. Afterwards it is considered that an external torque disturbance (1.0 p.u.) is applied at time $t_1 = 1.5$s and it is removed at time $t_2 = 2.5$s. As it is shown by Fig. 4 the sliding mode phase occurs considerably fast in the flux/current observer.

Fig. 5 shows that the estimated stator resistance

TABLE I
PARAMETERS OF PERMANENT MAGNET SYNCHRONOUS MACHINE

Symbol	Quantity	Expressed in SI
λ_m	permanent magnent flux	0.213 Wb
V_{l-l}	voltage line to line	415 V
P	electric power	2.2 kW
r_s	stator resistance	3.01 Ω
L_d	d-axis inductance	0.060 H
L_q	q-axis inductance	0.340 H
J	moment of inertia	0.089 kg·s
p	magnetic pole pairs	2
ω_m	mechanical angular speed	3600 rpm

follows almost exactly any variation of the real one. In the simulation, it is assumed that the stator resistance changes between 0.8 and 1.5 of the nominal value.

Fig. 6 shows the estimated and real speed response of PMSM with external disturbance applied. Fig. 7 shows the convergence of the rotor position observer with unknown initial rotor position. Finally Fig. 8 shows γ-axis stator magnetic flux-tracking response.

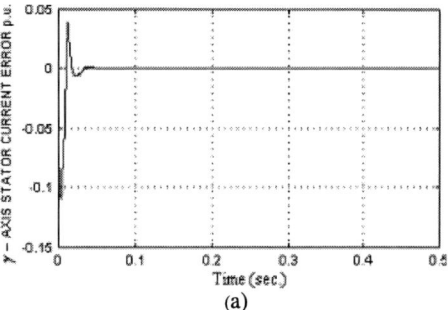

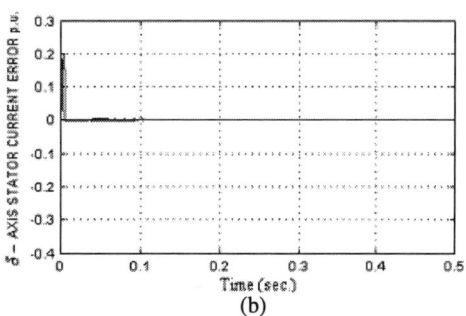

Fig. 4. γ-axis stator current error response for step change of speed reference 0-0.04pu with initial position error π/12 and (b) δ-axis stator current error response for step change of speed reference 0-0.04pu with initial position error π/12

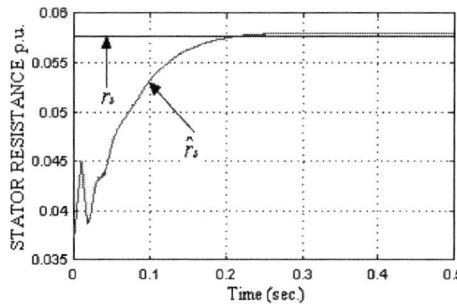

Fig. 5. Stator resistance-tracking response for step change of resistance 1.0-1.5pu with initial position error π/12 and speed reference 0.04pu

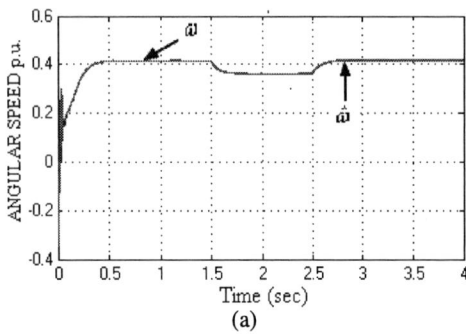

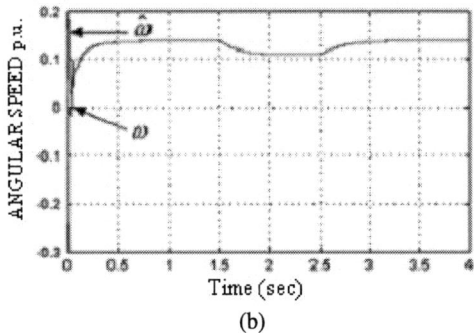

Fig. 6. Speed responses for step change of speed reference with initial position error π/12 (a) 0-0.42pu and (b) 0-0.14pu

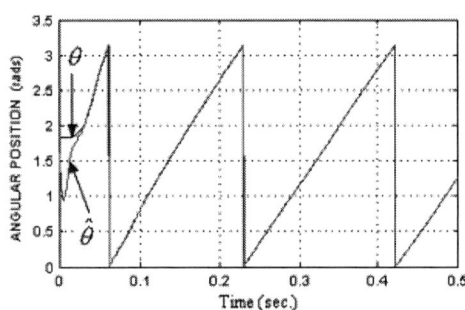

Fig. 7. Position tracking experiment for step change of speed reference with initial position error π/12 and 0-0.04pu

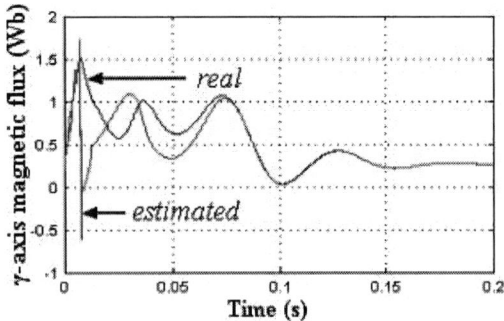

Fig. 8. γ-axis stator magnetic flux-tracking response for initial position error π/12 .

V. CONCLUSION

A new method based on rotating γ-δ reference frame is developed for speed and position estimation of a PMSM. This method uses sliding mode observers   and gives an effective approach of rotor speed at high, low and almost zeros speed. In addition the presented sliding mode observer can effectively estimate the stator resistance variation, due to temperature change.

The proposed scheme estimates the rotor position and speed from the stator voltages and currents by means of a modified back EMF observer, with initial rotor angle uncertainty. Simulation results demonstrate the efficiency and the robustness of this sliding mode estimation method.

REFERENCES

[1] P. C. Krause, "Analysis of Electric Machinery", *McGraw-Hill International Editions*, 1987

[2] J-J. E. Slotine and W. Li "Applied Non linear Control," *Prentice-Hall International, Inc.,* 1991

[3] V. I. Utkin, "Variable Structure Systems with Sliding Mode: A Survey", *IEEE Automat. Contr.*,vol   AC-22 ,no 2,pp212-222, April 1977.

[4] V. I. Utkin, "Sliding Mode Control Design Principles and Applications to Electric Drives ", *IEEE Trans.Ind. Electron.* , vol 40, no. 1, pp. 23-36, Feb. 1993.

[5] Z. Yan, C. Jin, and V. I. Utkin, "Sensorless Sliding-Mode Control of Induction Motors," *IEEE Trans. Ind. Electron.* , vol 47, no. 6, pp. 1286-1297, Dec. 2000.

[6] Zhang Yan, and Utkin V., "Sliding Mode Observers for Electric Machines – An Overview", IECON02, 28th Annual Conference of the IEEE Industrial Electronics Society, vol 3, pp. 1842-1847, 5-8 Nov. 2002.

[7] Xie Yue, D. Mahinda Vilathgumuwa, and King-Jet Tseng, "Observer-Based Robust Adaptive of PMSM With Initial Rotor Position Uncertainty", *IEEE Trans. Ind. Applic.*, vol. 39, no. 3,pp. 645-656, May/June 2003.

[8] Morimoto, S.; Kawamoto, K.; Sanada, M.; Takeda, Y., "Sensorless Control Strategy For Salient-Pole PMSM Based on Extended EMF in Rotating Reference Frame", *Industry Applications Conference, 2001*. Thirty-Sixth IAS Annual Meeting. Conference Record of the 2001 IEEE, vol. 4, pp. 2637 – 2644, 30 Sep-4 Oct 2001.

[9] Z. Chen, M. Tomita, S. Doki and. S. Okuma, "An Extended Electromotive Force Model for Sensorless Control of Interior Permanent-Magnet Synchronous Motors", *IEEE Trans. Ind. Electron.*, vol. 50, no. 2, pp. 288-295, April 2003.

[10] Z. Xu, and M. F. Rahman, "Encoder-less Operation of A Direct Torque Controlled IPM Motor Drive With A Novel Sliding Mode Observer", *Proceedings of Australasian Universities Power Engineering Conference 2004*, Brisbane, QLD, University of QLD, pp. 1131 – 1136, 26-29 Sep. 2004.

[11] Changsheng Li, and Elbuluk M., "A Sliding Mode Observer for Sensorless Control of Permanent Magnet Synchronous Motors", *Industry Applications Conference, 2001*, Thirty-Sixth IAS Annual Meeting. Conference Record of the 2001 IEEE, vol. 2, Issue, pp. 1273 – 1278, 30 Sep-4 Oct 2001.

[12] H. Kubota, K. Matsuse,and T. Nakano, "DSP-Based Speed Adaptive Flux Observer of Induction Motor ", *IEEE Trans.Ind. Applic.* , vol 29, no.2, pp. 344-348, March/April 1993.

[13] Mathew J. Corley, and Robert D. Lorenz, "Rotor Position and Velocity Estimation For a Salient-Pole Permanent Magnet Synchronous Machine at Standstill and High Speeds", *IEEE Trans. Ind. Applic.*, vol. 34, no. 4, pp. 784 – 789, July/August 1998.

[14] Shinji Ichikawa, Mutuwo Tomita, Shinji Doki, and Shinji Okuma, "Sensorless Control of Permanent-Magnet Synchronous Motors Using Online Parameter Identification Based on System Identification Teory", *IEEE Trans. Ind. Applic.*, vol. 53, no. 2, pp. 363-372, April 2006.

Optimal Control of Electrical Drives with Induction Motors for Variable Torques

Corneliu Botan, Marcel Ratoi, Vasile Horga

* Technical University /Department of Automatic Control, Iasi, Romania,
e-mail: cbotan@ac.tuiasi.ro, ratoi@tuiasi.ro, horga@tuiasi.ro

Abstract— The optimal control from the energetic point of view of the transient state of electrical drive systems with induction motors is presented. The performance criteria consider only Joule losses, since they significantly overcome other ones in the transient state. The paper refers to variable imposed electromagnetic or load torques.

Keywords—Optimal control, drive, induction motor, efficiency.

I. INTRODUCTION

The optimal control [1], [2] of the electrical drive systems represents an important way for energy saving and there are numerous studies dedicated to this problem, for different types of motors, criteria, or used methods (for instance, we mention [3], [4], [5], [6], [7], but many other papers can be indicated). Moreover, the optimization is appreciated as a main direction of the developing of the electrical drive systems in the future [8]. However, the number of applications is nowadays very small and we appreciate that a cause of the reluctance in this direction is the complexity of the algorithms. Taking into account this fact, we consider that simple implementable solutions for optimal control are necessary. The previous results obtained by authors indicate that it is possible to obtain a significant diminution of the energy losses in the motor windings in the electromechanical transient process using simple implementable optimal controllers.

It is important in many applications to achieve a control which ensure small energy losses and, at the same time, an acceptable behaviour of the drive system. However, the demands for different applications are not the same and certain differences in the adopted optimization criteria can occur. For instance, the minimization of the global losses is a goal for the steady state operation and the establishing of an adequate flux level is the most used procedure in this direction. Although a similar criterion can be also adopted for the transient period, we have to take into account that the Joule losses significantly overcame all other losses, since the start currents have very great values. Therefore, only a criterion based on Joule losses for the transient period can be adopted if this solution leads also to a good behaviour and an easy implementation can be ensured. This is a main direction of the present study.

In addition, an important approach of the paper is the considering of the stator current components in the *d-q* frame (and not voltages) as control variables. This choice ensures a significant simplification of the implementation and is justified since the motor currents are usually controlled in the vector control structures (one uses current control loops or current source inverter). In this case, a usual cascade structure with small modifications can be adopted for the optimal control.

It should be also noticed that the optimal control is useful not only for energy saving, but in many cases offers the possibility to reduce the motor rated power and therefore the weight and volume. Indeed, the motor rated power is chosen from heat consideration and the optimal control leads just to a diminished Joule losses. Moreover, the decrease of the weight of sub-ensembles leads in certain applications to the decrease of the energy consumption of all the plant.

This is the main direction of the studies performed by authors and the present paper refers to certain special applications of the electrical drive systems with induction motor. The first one considers an imposed variation of the electromagnetic torque, in order to obtain a desired variation of the speed. An example for this case is the electrical drives of the lifts. The second studied case refers to a step variation of the load torque in the course of the transient state. This situation appears, for instance, in the drive systems of the reversible rolling miles, when it is possible the increase of the load torque before the reach of the final speed. One can remark the frequent change of the speed in both mentioned examples, so that the introducing of the optimal control of the transient state is not lack of interest. Some general previous authors' results [7], [9] are used for this two main directions.

II. MAIN OPTIMALITY CONDITIONS

The mechanical equilibrium equation of the drive system is

$$\dot{\omega}(t) = \left[m_e(t) - m(t) \right] / J , \qquad (1)$$

where $\omega(t)$ is the speed, J is the inertia, $m_e(t)$ is the electromagnetic torque and $m(t)$ is the load torque (we neglect the dependence of m on the speed).

The state variable $\omega(t)$ results from (1) and, for $t = T$, one obtains

$$\omega_f = \omega(T) = \omega(0) + T(m_{em} - m_m) / J , \qquad (2)$$

where

$$m_{em} = \frac{1}{T} \int_0^T m_e(t)dt \quad \text{and} \quad m_m = \frac{1}{T} \int_0^T m(t)dt \qquad (3)$$

are the mean values of the electromagnetic and of the load torque, respectively.

978-1-4244-1741-4/08/$25.00 ©2008 IEEE

The relation (3) shows that the proposed control problem can be solved if the mean value of the load torque is beforehand known. At least the shape of the load torque must be known and its amplitude will be estimated at the beginning of the optimal control process.

The electromagnetic torque of the motor (with non-saturated iron) is

$$m_e(t) = c(i_{1q}i_{2d} - i_{1d}i_{2q}) \qquad (4)$$

where i_{1d}, i_{2d} and i_{1q}, i_{2q} are the d and q components for the stator and rotor currents (i_1 and i_2, respectively), and c is a constant of the machine. As control variables are adopted the stator current components i_{1d}, i_{1q} and not the voltages. The advantage and the justification of this choice were mentioned in the Section I.

The choice of the control currents depends on the used control technique. If the indirect rotor flux orientation is applied, $i_{2d} = 0$. The transient electromagnetic process will be neglected; the small delay of the flux has not a significant effect on the main conclusions. It this case, $i_{2q} = -\beta i_{1q}$, $\beta = L_m / L_r$ (L_m – magnetizing inductance, L_r – rotor self-inductance) and the electromagnetic torque is

$$m_e = c\beta i_{1d}i_{1q} \qquad (5)$$

The performance criterion refers to the total Joule losses

$$I = \frac{1}{2}\int_0^T \left(r_1 i_1^2 + r_2 i_2^2\right) dt, \qquad (6)$$

where r_1 and r_2 are the stator and rotor resistances, respectively.

The optimal control problem is to find the control variables i_{1q}, i_{1d} which transfer the system (1) from the initial state $\omega(0)$ (for simplification, one considers $\omega(0) = 0$) to the final state $\omega_f = \omega(T)$ so that the performance index (6) to be minimized.

The Hamiltonian [1], [2] of the problem is

$$H = \frac{1}{2}(r_1 i_1^2 + r_2 i_2^2) + \frac{\lambda}{J}\left(m_e - m\right) = $$
$$= \frac{1}{2}r_1(i_{1q}^2 + i_{2d}^2) + \frac{1}{2}r_2(i_{2q}^2 + i_{2d}^2) + \frac{\lambda}{J}(m_e - m),$$

where $\lambda \in \mathbf{R}$ is a co-state variable.

The necessary optimality conditions [1] $\partial H / \partial i_{1q} = 0$ and $\partial H / \partial i_{1d} = 0$ lead to [9]

$$i_{1d} = \pm \rho i_{1q}, \qquad (7)$$

$$\rho = \left[(r_1 + r_2\beta^2)/r_1\right]^{1/2} \qquad (8)$$

(only the sign plus will be considered in the sequel, corresponding to the motor operating mode).

One can also prove [7], [9] that in these conditions, for free final time T and constant currents,

$$m_{em} = 2m_m \qquad (9)$$

and the optimal values of currents are

$$i_{1q}^2 = 2m_m/c\beta\rho, \quad i_{1d}^2 = 2m_m\rho/c\beta \qquad (10)$$

and the performance index is

$$I^* = r_1\rho\left(J\omega_f + Tm_m\right)/c\beta. \qquad (11)$$

The last relation shows that it is indicated to choose the control variables so that T to be as smaller as possible.

The above formulae can be applied only for small load torque, in order to avoid the overcome of the limit values of the currents. Usually the limitation of the component i_{1d} firstly occurs at the maximal value i_{1dM} (in order to avoid the saturation of the machine). One proves [9] that for the optimal control with restriction, the stator current is constant and

$$m_e = 2m_m + \frac{c\beta i_{1dM}^2}{\rho^2 i_{1q}} \qquad (12)$$

$$i_{1q} = \left(m_m\rho + \sqrt{m_m^2\rho^2 + c^2\beta^2 i_{2dM}^4}\right)/c\beta i_{dM}. \qquad (13)$$

For $m_m > m_N / 2$ (m_N is the rated torque), the formula (9) is a rather good approximation for (12). In fact, the component (13) can be adopted also for small load torque, because of the negligible differences by comparison with optimal control without constraints. By this way, the implementation is simplified, because it is not necessary to switch the control law in dependence on the value of the load torque.

III. OPTIMAL CONTROL FOR IMPOSED VARIATION OF THE ELECTROMAGNETIC TORQUE

In certain applications, the diagram of the speed variation is imposed. This fact leads to an imposed variation for the electromagnetic torque. The optimal control is also possible in this case, but it is expected an increase of Joule losses, because of the supplementary constraint. Only the linear variation of the electromagnetic torque will be considered in the sequel, but the procedure can be immediately extended for other forms of variations.

We shall consider the electromagnetic torque imposed in the form

$$m_e = at + b. \qquad (14)$$

In this case

$$\dot{\omega}(t) = (at + b - m_m)/J. \qquad (15)$$

In order to avoid a shock in the initial moment, one can impose

$$\dot{\omega}(0) = 0. \qquad (16)$$

and this leads to the choice

$$b = m(0). \qquad (17)$$

For applications with $m(t)=const.$, this means $b=m_m=m$.

For small load torque, the unconstrained optimality condition (7) holds and it results

$$i_{1d} = \left(2J\omega_f \rho t \big/ c\beta T^2\right)^{1/2} \quad (18)$$

Since the component i_{1d} increases in time, it is possible to reach the maximal allowable value i_{1dM} at the moment $t_c < T$ (which results from (17) for $i_{1d} = i_{1dM}$). After this moment, the component i_{1q} will be computed with (5)

$$i_{1q} = \left[a(t-t_c)+b\right]\big/c\beta i_{1dM} . \quad (19)$$

The switch of the control law at the moment t_c is avoided if the transient time satisfies the condition

$$T \geq 2J\omega_f \rho \big/ c\beta i_{1dM}^2 ,$$

IV. OPTIMAL CONTROL FOR THE STEP VARIATION OF THE LOAD TORQUE

Let us suppose that the load torque switch from a small value m_1 to a great value m_2 at the moment $t_s \in [0,T]$. We suppose that m_1, m_2 and t_s are known.

The mean value of the load torque is

$$m_m = [m_1 t_s + m_2(T-t_s)]/T \quad (20)$$

From (3) and (12), it results

$$T = J\omega_f / (m_m + \mu), \quad (21)$$

where

$$\mu = c\beta i_{1dM}^2 / \rho i_{1q} . \quad (22)$$

Replacing T from (21) in (22), one obtains

$$m_m = (m_2 - \mu)/(\alpha + 1) , \quad (23)$$

where

$$\alpha = t_s (m_2 - m_1) / J\omega_f \quad (24)$$

has a constant value.

The above relations and (13) allow the computing of m_m and i_{1q}, but the formulae have complicated form. As an alternative, an iterative procedure can be used, with the following steps:

(a) choice an initial value for i_{1q} (for instance ρi_{1dM});

(b) compute and m_m with (22) and (23);

(c) compute the new value for i_{1q} with (13) and repeat the steps (b) and (c) until the difference between two consecutive values for i_{1q} is less than an imposed limit.

Remark: It is possible to compute separately the component i_{1q} with (12) and (13) in the intervals *[0, t_s]* and *[t_s, T]*. But this variant leads to a greater value of the energy losses. Indeed, the above presented theoretical results indicate that the optimal current has to be computed in dependence on the mean value m_m and not on

the instantaneous values of the load torque. This property is also sustained by simulation and experimental results.

V. IMPLEMENTATION OF THE OPTIMAL CONTROLLER AND EXPERIMENTAL RESULTS

The above obtained results represent an ideal optimal control. We name this control ideal since the currents cannot have an abrupt, non-inertial variation. However, if they are controlled through adequate control loops, one can obtain a very fast variation of the currents. Their small delays are negligible and the obtained control is very closely related to the optimal one. In fact, this is the main idea for an easy implementation: to impose the variation of the currents as near as possible to the ideal optimal ones. The structure is based on a usual cascade one and is presented in Fig. 1.

The cascade contains a current and a speed loops, with PI – type controllers C_1 and C_2. P_1, P_2, P_3 denote sub-ensemble of the plant (power converter, motor and driven work machine). S_1 symbolizes a proportional element with saturation; the level of saturation can be modified in concordance with the estimated $\hat{m}_r$ value of the torque, obtained from the observer OBS. The same value allows the computing of the optimal value i_{1d}^* of the reactive current; using a control closed loop CL, the desired value i_{1d} is obtained. Since S_1 is saturated, the prescribed value i_{1q}^* for the active current loop corresponds to the optimal one. Only in the last part of the transient state, when the speed error is very small, S_1 is non-saturated and the scheme runs as a cascade one, with a smooth achievement of the final imposed value ω_f.

A selection from the numerous performed simulation and experimental tests is presented below. The test refers to an induction motor with rated power 2.2 kW and speed 2855 rpm. The inertia of the drive system is seven times greater than of the motor one.

Fig. 2 and 3 present the simulation and experimental variations of the speed and of the active current (the reactive current is proportional) for optimal control, when a linear variation of the electromagnetic torque is imposed and the load torque is constant ($0.1 \, m_N$).

The energy losses are less with about 10% by comparison with the non-optimal control ($i_{1d}=const.$ and i_{1q} has a linear variation).

The next figures refer to the optimal control for the case when a step variation of the load torque appears during of the transient process. In all presented cases, the start is given at the moment $t_0 = 0.3$ s. On the interval [0, 0.3], the active

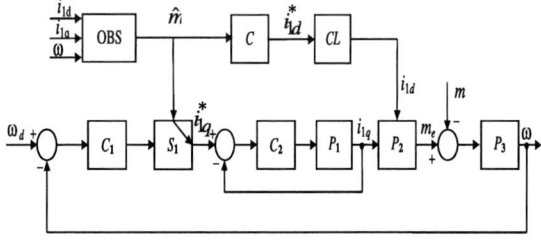

Fig. 1. Structure of the optimal controller

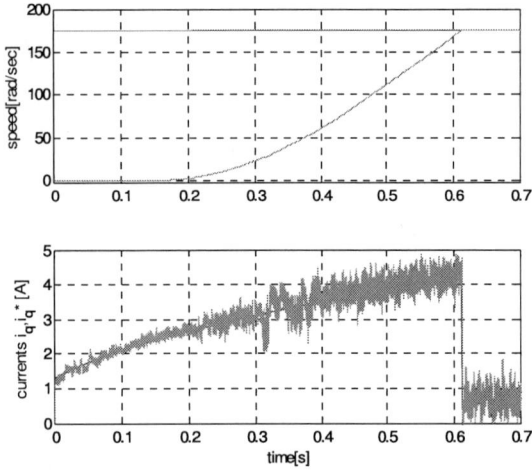

Fig. 2. Simulation results for imposed electromagnetic torque

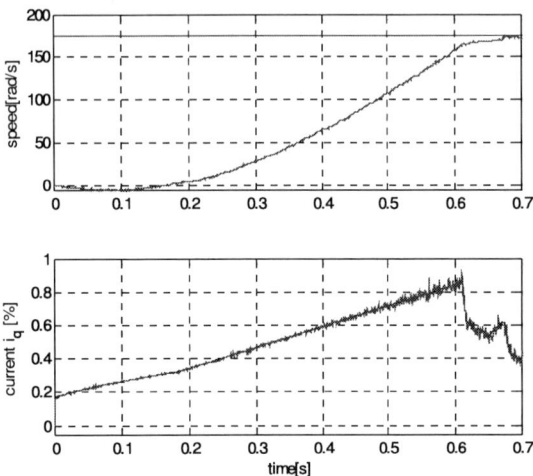

Fig. 3. Experimental results for imposed electromagnetic torque

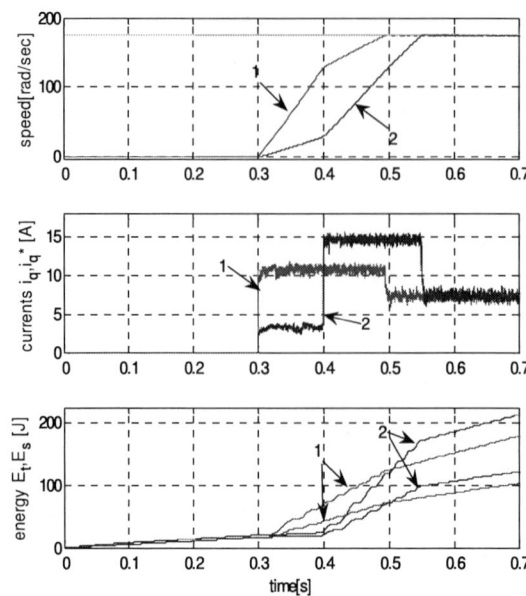

Fig. 4. Variable load torque ($m_1=0.1m_N$, $m_2=m_N$) - simulation results

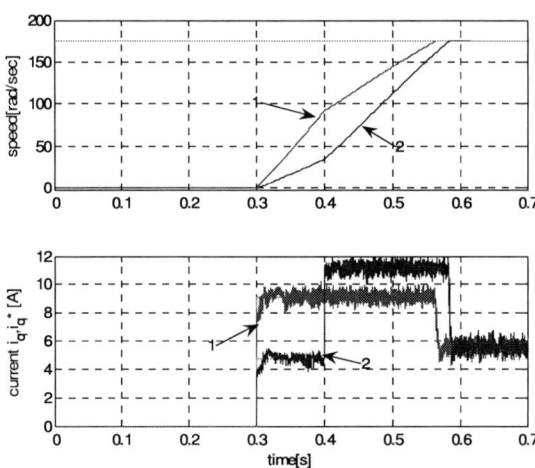

Fig. 5. Variable load torque ($m_1=0.25m_N$, $m_2=0.75m_N$) - simulation results

current is zero but $i_{1d} \neq 0$, in order to ensure the flux in the machine. Fig. 4 presents simulation results for the case $m_1= 0.1\ m_N$, $m_2= m_N$ and $t_s= 0.4$ s, being indicated the variations for the speed ω, the active current i_{1q}, and the total and stator energy losses (E_t and E_s, respectively). The prescribed values for ω and i_{1q} are indicated with green color. The optimal variations are depicted with red color (the curves 1). The blue curves (denoted with 2) correspond to the case when the active current is not calculated as a function of the mean value on the interval $[0, T]$, but is computed with the same formula (13), separately for intervals $[0, t_s]$ and $[t_s, T]$ (in fact, this is a fake optimal control).The results assert that the optimal behaviour is achieved in concordance with the mean load torque, as it was mentioned in the Remark from the Section IV.

Fig. 5 presents similar results but for $m_1= 0.25\ m_N$ and $m_2= 0.75\ m_N$. One can remark that the differences between the two cases are smaller when m_1 and m_2 are nearer (this fact can be theoretical proved).

The experimental results were performed on a set-up which uses a vector controlled PMSM as a load machine. The same motor is also used for emulation of a great value for inertia (the main components of the used experimental structure are indicated in Fig. 6). Of course, the transient behaviour of the PMSM and of the emulator affects the behaviour of the system (the obtained curves are not so smooth as in the simulation tests). But the shapes of variations and the values for the transient time, for active current and for stator energy losses (of course, the rotor losses cannot be measured) are closely related with the simulations results. These aspects are sustained by the variations presented in the Fig. 7, obtained for the same data as the simulated optimal variations from the Fig. 4 (curves 1).

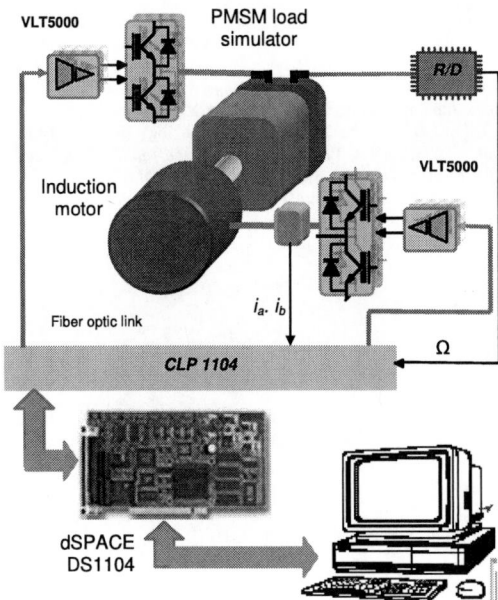

Fig. 6. The structure of the experimental set-up

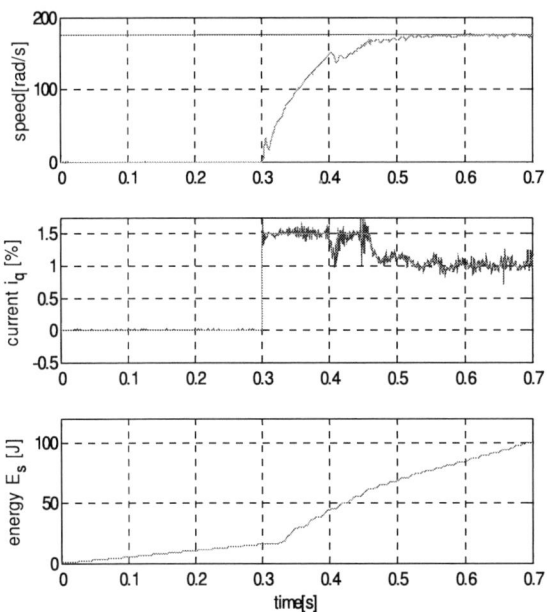

Fig. 7. Variable load torque (m_1=0.1m_N, m_2=m_N, t_S=0.4s) - experimental results

Similar closely related results with simulation ones were obtained for m_1= 0.25 m_N and m_2= 0.75 m_N. This case is not presented, but the Fig.8 shows a situation with the same load torques, but with an increased switch moment (t_s=0.5 s). By comparison with the case t_s=0.4 s, the transient time increases (since m_m and therefore the imposed i_{1q} decrease), but the stator energy losses are approximately the same).

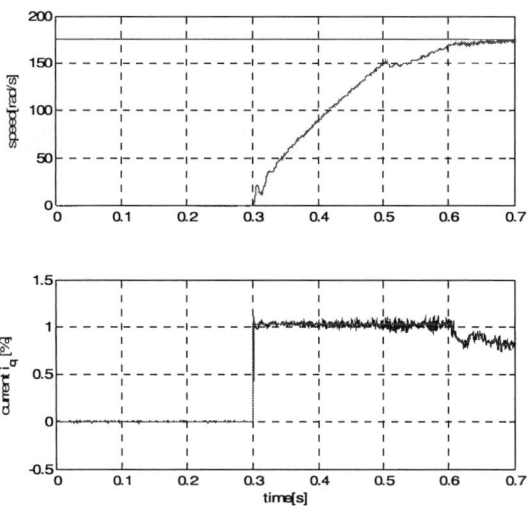

Fig. 8. Variable load torque (m_1=0.25m_N, m_2=0.75m_N, t_S=0.5s) - experimental results

The comparison of the optimal control case with the usual cascade structure indicates a near behaviour if the limit value established for the prescribed torque is chosen at the value $m_l = 2\alpha m_N$ and the load torque has the mean value $m_m = \alpha m_N$ (α is usually less than one). The explanation is the fact that, in this case, the saturation limit in the cascade structure is near with the optimal value. The greater are the differences by comparison with above described situation, the greater are the differences of the behaviour. For instance, for $\alpha > 1$, the cascade control forces too much the motor and the transient time is shorter, but the energy losses are greater. The optimal control can lead to a decrease of Joule losses up to 25%, depending on the load torque and on the established limit for the active current in the cascade structure.

VI. CONCLUSIONS

The optimal control of the electrical drives has the doubtless advantages from energetic point of view by comparison with other control strategies. In addition, a good behaviour in the transient state can be achieved.

The paper presents a simple implementable structure for optimal control of electrical drives with induction motors. This structure is based on a cascade one with certain modifications.

The proposed strategy uses stator active and reactive currents as control variables.

The performed simulation and experimental tests indicate a decrease of the energy losses and a good behaviour of the drive system.

REFERENCES

[1] B.D.O. Anderson and J.B. Moore, *Optimal Control. Linear Quadratic Methods*, Prentice Hall, 1990.
[2] M. Athans and P.L. Falb, *Optimal Control*, Mc Graw Hill, New York, 1966.
[3] C. De Capua and C. Landi, "Measurement Station Performance Optimization for Testing on High Efficiency Variable Speed Drives," *IEEE Trans. on Instrumentation and Measurement*, vol. 48, no. 6, 1999, pp. 1149-1154.

[4] B. Canudas de Wit and J. Ramirez, "Optimal torque control for current-fed induction motors," *IEEE Trans. Automat. Contr.*, vol. 44, 1084-1089, May 1999.

[5] G. Acampora, C. Landi, M. Luiso, and N. Pasquino: "Optimization of Energy Consumption in a Railway Traction System," *SPEEDAM, International Symposium on Power Electronics, Electrical Drives, Automation and Motion*, Taormina, Italy, CD-ROM (2006).

[6] I.T. Wallace, D.W.Novotny, R.D. Lorenz, and D.M. Divan: "Verification of Enhanced Dynamic Torque per Ampere Capability in Saturated Induction Machines," *IEEE Transactions on Industry*

Applications, vol. 30, no. 5, Sept./Oct., pp. 1193-1201, 1994.

[7] C. Botan, V. Horga, F. Ostafi, M. Albu, M. Ratoi, "General Aspects of the Electrical Drive Systems Optimal Control," *12th European Conf. on Power Electronics and Applications (EPE 2007)*, Aalborg, Denmark, Sept. 2007, CD-ROM

[8] R.D. Lorentz, "Future Motor Drive Technology Issues and their Evolution," *12th Int. Power Electronics and Motion Conference (EPE-PEMC)*, Portoroz, Slovenia, 2006, CD-ROM.

[9] C. Botan, V. Horga, "Optimal control of the electrical drives with induction motors," *IEEE International Symposium on Industrial Electronics (ISIE)*, Cambridge, UK, July 2008.

An Optimal Control for Saturated Interior Permanent Magnet Linear Synchronous Motors Incorporating Field Weakening

Mohammad Abroshan[†*]   Jafar Milimonfared[*]   Kaveh Malekian[*]   Arash Rahnamaee[*]

*Electrical Engineering Department, Amirkabir Uni. of Tech., Center of Excellence in Power Systems, Tehran, Iran

† Corresponding author: abroshan@aut.ac.ir
Tel: +98 21 64543550

Abstract—In this paper, the maximum thrust force per ampere (MTPA) strategy for saturated permanent magnet linear synchronous motors over the entire field weakening region is presented. In other words, influence of magnetic saturation on the MTPA and field weakening strategies as well as motor-inverter limitations is considered. Using the modified MTPA strategy leads to minimized copper losses, and so maximum capacity of motor is utilized. The proposed scheme incorporates all optimal strategies and operates the drive within the voltage and current limits of the motor/inverter. Simulation results for a prototype interior permanent magnet linear synchronous motor demonstrate that the MTPA and field weakening strategies can be successfully implemented for vector controlled saturated permanent magnet linear synchronous motor drives.

Index Terms—Field weakening, interior permanent magnet linear synchronous motor, magnetic saturation, maximum force per ampere, synchronous motor drive, velocity control.

I. INTRODUCTION

A TREND to increase use of permanent magnet linear synchronous motors (PMLSM) can be observed in widespread industrial application particularly in precision machine tools, semi-conductor manufacture equipments [1], [2]. The main benefits of the drive design based on PMLSM than indirect counter part are the high force density achievable and high positioning precision and simple mechanical structure, less loss, less friction, faster response [3]-[6]. At the same time, the end effect of PMLSM can be controlled more easily compared with the induction motor. Therefore, the PMLSM is suitable for high-performance servo applications.

Because PMLSM is not equipped with auxiliary mechanism such as gears or ball screws, its performance is affected by uncertainties in the drive system, which include force ripples, external load disturbances and unknown dynamics, which directly impose on the mover of the PMLSM. Therefore, to compensate these disturbances and achieve accurate tracking in high performance servo drive a fast response strategy is required [4], [5].

Using maximum thrust force per ampere strategy in vector control of the permanent magnet linear synchronous motor drives leads to achieve copper losses minimization and better using of motor capacity. This can be realized by control of the d-axis component of the stator current. In other words, negative d-axis current are used to minimize the sum of the core losses and the copper losses in rotary permanent magnet machines, but in linear motors because of the relatively low speeds, the core losses are mostly small compared to the copper losses, and therefore its effect is negligible [7]-[9].

When the currents are low, the motor not saturate and the force is proportional to the current, but in large acceleration or large load, the currents are so high that the magnet circuit saturates and the electrical motor parameter change so the electrical thrust force is not easily predicted, then the MTPA trajectory is different when the motor saturates [10], [11].

In this paper, a modified vector control for interior permanent magnet linear synchronous motor incorporating MTPA and field weakening strategies is presented. Implementing such optimal strategies, high performance velocity tracking and fast response system can be achieved.

II. MODEL OF SATURATED PMLSM

In interior permanent magnet synchronous motor, since the effective air-gap length on the q-axis is small and, therefore, the saturation is prominent, but the effective air-gap length on the d-axis is large and the relative magnetic permeability of the PM is close to unity, variation of the corresponding magnetizing inductance, L_{od}, due to magnetic saturation can be neglected. The L_q inductance varies with the q-axis current. In other words, the effects of saturation models as $L_q(i_q)$ which that is a nonlinear function of q-axis current component. A good approximation is that, it defines by two inductances, an unsaturated inductance L_{oq}, which applies for i_q current lower than i_{qs}, and a saturated inductance, that varies linearly with q-axis current and valuable for higher i_q current. This model can be describe as follows,

978-1-4244-1741-4/08/$25.00 ©2008 IEEE   1117

$$L_q(i_q) = \begin{cases} L_{oq} & i_q < i_{qs} \\ L_{oq} - \beta(i_q - i_{qs}) & i_q > i_{qs} \end{cases} \quad (1)$$

where i_{qs} is q-axis current component that saturates the magnetic circuit.

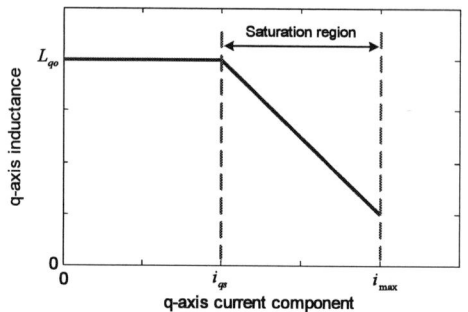

Fig.1. Variation of q-axis inductance versus i_q current.

The model of a PMLSM can be described in rotor reference frame as follows [7], [8],

$$\begin{cases} u_d = R_s i_d + \dfrac{d\psi_d}{dt} - P\dfrac{\pi}{\tau} v_m \psi_q \\ u_q = R_s i_q + \dfrac{d\psi_q}{dt} + P\dfrac{\pi}{\tau} v_m \psi_d \end{cases} \quad (2)$$

$$\begin{cases} \psi_d = L_d i_d + \psi_f \\ \psi_q = L_q i_q \end{cases}$$

$$F_T = \frac{3}{2} P \frac{\pi}{\tau} \frac{1}{L_d L_q} |\psi_S| [2\psi_f L_q \sin\delta - |\psi_S|(L_q - L_d)\sin 2\delta] \quad (3)$$

$$F_T + F_{cogging}(x) = M\frac{dv_m}{dt} + Bv_m + F_L \quad (4)$$

where,

P	number of pole pairs;
τ	pole pitch;
δ	load angle;
v_m	mover velocity;
ψ_f	permanent magnet flux linkage;
ψ_S	stator flux linkage;
R_S	phase winding resistance;
u_d, u_q	d- and q-axis voltages;
ψ_d, ψ_q	d- and q-axis linkage fluxes;
i_d, i_q	d- and q-axis current components;
L_d, L_q	d- and q-axis inductances;
M	mover mass;
B	friction factor;

F_L	load force;
F_T	electrical thrust force; and
$F_{cogging}$	cogging force.

The thrust force of the permanent magnet linear synchronous motor contains three components: main synchronous force, reluctance force, and cogging force. Cogging force is a function of position and is independent of the load angle. Due to the slotted nature of the primary core, the cogging force is periodic and repeats itself over every slot pitch [6], [4]. In this work, it is assumed that cogging force is negligible.

III. PROPOSED CURRENT VECTOR CONTROL FOR SATURATED IPMLSM

In the current vector control method, since q- and d-axis current components are directly controlled, all control strategies and motor-inverter limitations are considered in the $i_q - i_d$ plane. The configuration of the proposed vector control for an IPMLSM drive system is depicted in Fig. 2.

A. Current and Voltage Constraints

The maximum output thrust force developed by the machine is limited by allowable voltage and current rating of the machine and the inverter. In saturated IPMLSM, the q-axis inductance varies with q-axis current then voltage limit can be deduction as below,

$$U_a = \sqrt{u_d^2 + u_q^2} \le U_{am} \quad (5)$$

So,

$$(R_s i_d + \frac{d\psi_d}{dt} - P\frac{\pi}{\tau} v_m \psi_q)^2 + (R_s i_q + \frac{d(L_q(i_q).i_q)}{dt} + P\frac{\pi}{\tau} v_m \psi_d)^2 \le U_{am}^2 \quad (6)$$

The maximum voltage, U_{am}, is the maximum available output voltage of the inverter depending on the dc-link voltage.

Current limit condition is independent of saturation and give as below:

$$I_a = \sqrt{i_d^2 + i_q^2} \le I_{am} \quad (7)$$

The maximum current, I_{am}, is a continuous armature current rating in continuous operation or a maximum available current of the inverter in limited short-time operation. The critical condition of (7) (i.e., $I_a = I_{am}$) is given by the current limit circle and shown in the $i_q - i_d$ plane as in Fig. 3(a).

B. Control Strategy in the Constant Force region

The maximum available force can be produced by the MTPA strategy below the base velocity. In order to achieve the MTPA for a given force demand, the line current amplitude, or equivalently $\sqrt{i_d^2+i_q^2}$, is minimized to achieve the maximum force within the current and voltage constraints of the (5) and (7). So,

$$\left.\frac{\partial F_T}{\partial i_q}\right|_{V_m}=0 \tag{8}$$

Considering (2), (3), and (8) yields,

$$i_d=\frac{\psi_f-\sqrt{\psi_f^2+4i_q^2(L_q(i_q)-L_d)(L_q(i_q)-L_d)+i_q\dfrac{dL_q(i_q)}{di_q}}}{2(L_q(i_q)-L_d+i_q\dfrac{dL_q(i_q)}{i_q})} \tag{9}$$

If magnetic saturation is neglected, then the relation between the i_d and i_q currents for unsaturated IPMLSM is given as [1],

$$i_d=\frac{\psi_f}{2(L_q-L_d)}-\sqrt{\frac{\psi_f^2}{4(L_q-L_d)^2}+i_q^2} \tag{10}$$

This relationship is shown as the MTPA trajectory in Fig. 3(a). The maximum force, F_{max}, is produced when $I_a=\sqrt{i_d^2+i_q^2}\le I_{am}$ considering the current constraint. The current vector (i_d,i_q) pair, producing this maximum force, F_{max}, is the cross point of the MTPA trajectory and the current limit circle. This point corresponds to the operating point with the rated force and current. With rated force, the stator current is at its rated value on the MTPA trajectory.

C. Control Strategy in the Field Weakening Region

When the mover velocity is increased above the base velocity, the area within the voltage limit contour, which is allowable operation area, will be decreased. As a result, the stator flux linkage must be reduced according to (6) for FW operation. In this paper, for the sake of simplification, the

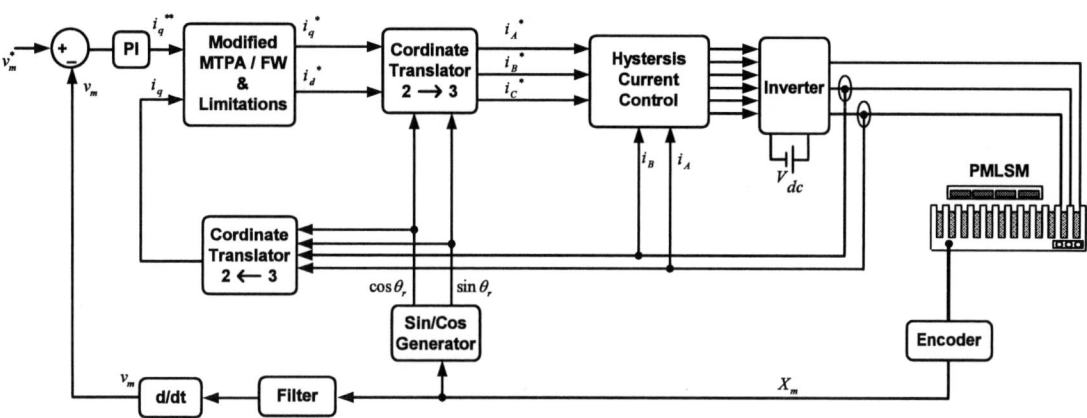

Fig. 2. Block diagram of the proposed vector control drive for an IPMLSM.

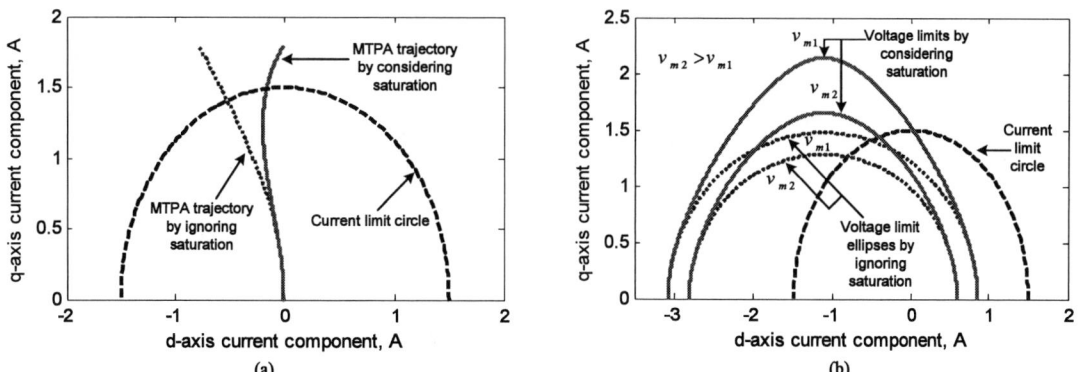

Fig. 3. Control trajectories and motor limitations: (a) MTPA by considering saturation and MTPA by ignoring saturation for a saturated PMLSM; (b) influence of saturation on voltage limitations.

voltage limits are considered in the steady state by applying (11) instead of (6),

$$(\psi_q)^2 + (\psi_d)^2 \leq \frac{\tau U_{om}^{2}}{P \pi v_m} \quad (11)$$

where,

$$U_{om} = U_{am} - I_{am} R_s \quad (12)$$

For operation above the base velocity, the amplitude of the stator flux linkage is inversely proportional (approximately) to the mover velocity, In other words, q-axis current is reduced while d-axis current becomes more negative.

When the mover velocity is below the base velocity, the intersection of the MTPA and current limit trajectories is within the voltage limit contour, and, therefore, the voltage limit is always satisfied with MTPA trajectory control.

IV. SIMULATION

To establish the effectiveness of the proposed drive system and, also, to examine the different strategies, such as the MTPA and FW strategies, for the vector controlled drive system, two "MATLAB" package simulations are designed. Parameters of the prototype IPMLSM used in simulations are shown in Table I. Also, the MTPA-based vector controlled drive system by ignoring saturation for saturated motor is simulated in order to compare the performance to those obtained from the respective proposed drive system. The following cases are tested in the simulation.

A. Starting Response

In this case, following modified and conventional conditions have been test on the MTPA-based drive systems for saturated IPMLSM in order to compare the both systems.

$$External\ Force = 150\,N \quad (13)$$

$$Velocity\ Command = \begin{cases} 0 & t < 0 \\ 1.2\ m/s & 0 < t \end{cases} \quad (14)$$

The simulation results are shown in Figs. 4(a)-(e). As shown in these figures, the modified MTPA-based drive system shows superiority over the conventional one as the actual velocity reach to command value in less time and maximum available force is greater.

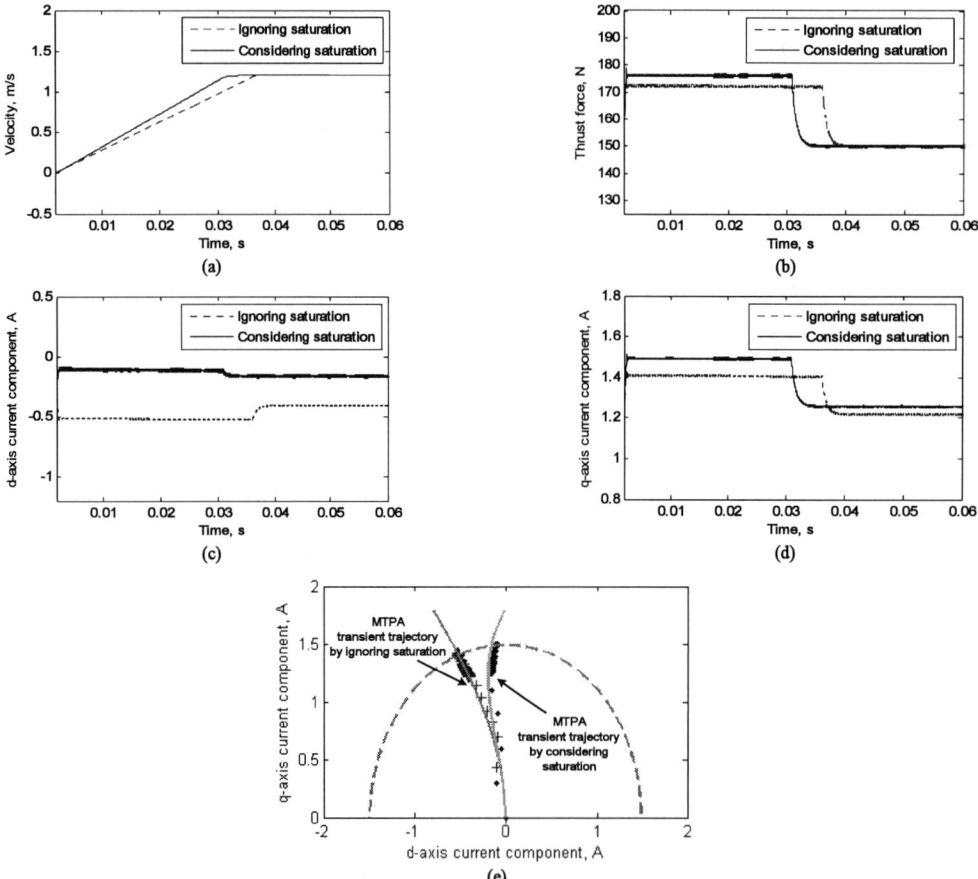

Fig. 4. Simulated starting responses of the both modified and conventional vector controlled drives for saturated IPMLSM at first case: (a) velocity responses, (b) force responses, (c) d-axis current component responses, (d) q-axis current component responses, and (e) transient trajectory in the $i_d - i_q$ plan.

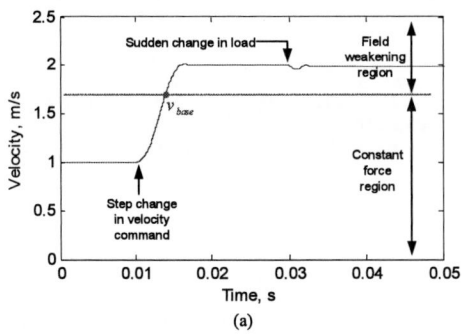 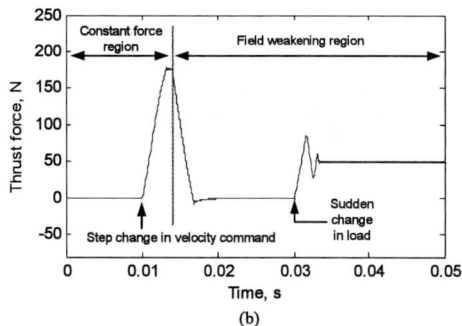

(a)   (b)

Fig. 5. Simulated responses of the modified MTPA-based vector controlled drive over wide velocity rang at second case: (a) velocity response, (b) force response.

B. Step Change in Velocity Command above the Base Velocity and External Force Disturbance

This test, which investigates the ability of the drive system incorporating MTPA strategies to reach the velocity above the base velocity, is done under following conditions,

$$External\ Force = \begin{cases} no\ load & t < 0.03s \\ 50N & 0.03s < t \end{cases} \quad (15)$$

$$Velocity\ Command = \begin{cases} 1\ m/s & t < 0.01s \\ 2\ m/s & 0.01s < t \end{cases} \quad (16)$$

Figs. 5(a)-(b), which show the velocity and force responses, indicate the proper performance of the modified MTPA-based drive system in transition from the constant force region into the field weakening region. Transition between constant force and FW operations clearly occurs at the intersections of MTPA and current limit trajectories, as indicated in Fig. 3(a).

V. CONCLUSION

In this paper, influence of the magnetic saturation on the MTPA and FW strategies as well as motor-inverter limitations has been derived and, successfully, applied in the vector controlled IPMLSM drives. The proposed drive system has found proper performance under disturbances, such as step change in the velocity command above the base velocity and external force disturbances, through the simulated tests. In addition, as demonstrated in the simulation results, in comparison with the traditional current vector control method, the maximum force capacity of IPLSM will be increased by considering saturation, and as a result, dynamic of the current vector controlled IPMLSM's responses will improve.

APPENDIX

TABLE I
DATA OF IPMLSM USED IN SIMULATIONS

Maximum current	I_{am}	1.5 A
Maximum voltage	U_{am}	240 V
Base velocity	v_{base}	1.7 m/s
Number of pole pairs	P_n	2
d-axis inductance	L_d	300 mH
q-axis inductance	L_q	is shown in Fig. 6.
Stator resistance	R	19.2 Ω
Magnetic flux-linkage	ψ_f	0.49 Wb
Mover mass	M	50 Kg
Pole pitch	τ	42 mm
Friction coefficient	B_m	0.001 N.s/m

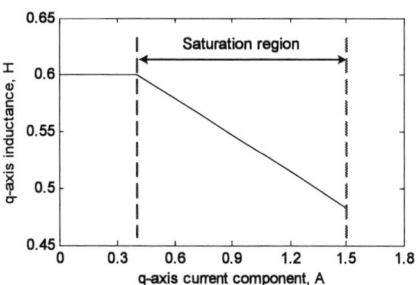

Fig. 6. Variation of q-axis inductance versus q-axis current for the prototype IPMLSM used in simulations.

REFERENCES

[1] J. Zou and S. Zhu, "High performance motion control of sensorless interior permanent magnet linear synchronous motor", *Control, Automation, Robotics and Vision Conference*, ICARCV 2004, vol. 3., pp. 1780 – 1785, 6-9 Dec. 2004.

[2] S. A. Nasar and I. Boldea, *"Linear Electric Motors: Theory, Design, and Practical Applications"*, Englewood Cliffs, NJ: Prentice-Hall, 1987.

[3] I. Boldea and S. Nasar, *"Linear Electric Actuators and Generators"*, London: Cambridge University Press, 1997.

[4] K. Yoshida, Z. Dai, "Sensorless DTC propulsion Control of PMLSM vehicle", *Power Electronics and Motion Control Conference*, Proceedings PIEMC, pp. 191-196, 2000.

[5] C. Jiefan, "Analysis of direct thrust force control for permanent magnet linear synchronous motor", *Proceedings of the 5, World Congress on Intelligent Control, and Automation*, June 15-19, 2004.

[6] C. Jiefan, "Research on Restraining Thrust Force Ripple for Permanent Magnet Linear Synchronous Motor", *IEEE IPMC*, 2007.

[7] G. Hong, X. Qiang, and J. Zhenchun, "Effects and compensation of magnetic saturation in flux-weakening controlled interior permanent magnet linear synchronous motor", *Fifth Inte. Conf. Electrical Machinesand Systems*, vol. 2, pp. 913–916, 2001.

[8] B. Chalmers, L. Musaba, and D. F. Gosden, "Variable-frequency synchronous motor drives for electric vehicles", *IEEE Trans. Ind. Applicat.*, vol. 32, pp. 896–903, July/Aug. 1996.

[9] M. Platen and G. Henneberger, "Examination of leakage and end effects in a linear synchronous motor for vertical transportation by means of finite element computation", *IEEE Trans. Magn.*, vol. 37, pp. 3640–3643, Sept. 2001.

[10] P. M. Churn , J. K. Mitchell, X. P. Xia, P. H. Mellor, and D. Howe "The control of direct drive linear actuation in point-to-point material transportation in packing applications", *Intelligent Motion*, pp. 65-74, 1997.

[11] R.S. Colby and D.W. Novotny, "Efficient operation of surface mounted PM synchronous motors", *IEEE transaction on industrial applications*, vol. 23, pp. 1048-1054, 1987.

Improved Direct Torque Control for Induction Machine Drives using Fuzzy Logic and Particle Swarm Optimization

Mohammad Mehdi Rezaei [*†]   Mojtaba Mirsalim [*‡]   Kaveh Malekian [*]

* Amirkabir University of Technology (Tehran Polytechnic),
Center of Excellence in Power Systems, No. 424, Hafez Ave., Tehran 15914, Iran.
‡ On sabbatical leave at St. Mary's University, San Antonio,TX.
† Corresponding author: mmr_soheil@aut.ac.ir

Abstract—Here, a new fuzzy direct torque control algorithm for induction motors is proposed. As in the classical direct torque control, the inverter gate control signals directly come from the optimum switching voltage vector look-up table, the best voltage space vector selection is a key factor to obtain minimum torque and flux ripples. In the proposed approach the best voltage space vector is selected using a new fuzzy method. A simulation model is built up and the torque and flux ripples of basic direct torque control and the proposed method are compared. The simulation results show that the torque and flux ripples are significantly decreased and in addition, the switching frequency can be fixed.

I. INTRODUCTION

SINCE direct torque control (DTC) of induction motors was proposed in the middle of 1980's [1]-[2], more than two decades has passed. The basic idea of DTC is slip control, which is based on the special relationship between the slip frequency and electromagnetic torque. The most common way to carry out the DTC is to use switching table and hystersis controller, as reported in [2]-[6]. Fig.1 shows the diagram for a DTC induction machine (IM) drive system. It includes flux and torque estimators, flux and torque hystersis controller and a switching table. The switching state of the inverter is updated in each sampling interval.

Although DTC is getting more popular, it also has some drawbacks, such as the torque and flux ripples [4]. In this case, none of the inverter switching vectors is able to generate the exact stator voltage required to produce the desired changes in the electromagnetic torque and stator flux linkages in most of switching instances. This undesirable ripple is of higher value when the selected state of the inverter remains unchanged for several sampling periods and especially when the stator flux linkage space vector (FLSV) is moving from one sector to the neighbouring sectors.

In this paper, a new direct torque controlled IM drive algorithm based on fuzzy logic is proposed to improve the performance of the drive in terms of ripple reduction. As an intelligent method, fuzzy control does not need an accurate mathematical model of the process to be controlled, and uses the experience of people's knowledge to form its control rule base. Fuzzy logic controllers have been used in direct torque control systems in the past few years [7]-[9]. Particle swarm optimization (PSO) is also used to tune off-line PI speed controller parameters [10].

The paper is organized as follows. Section II briefly introduces the DTC principle. The structure of the proposed fuzzy-DTC scheme is presented in section III. In section IV, a Matlab/Simulink model is built to test the algorithm. Then, steady state and dynamic responses are compared with basic DTC. The conclusions are drawn in section V.

II. DIRECT TORQUE CONTROL

A general block diagram of DTC scheme is shown in Fig.1. In this system, the instantaneous values of flux and torque are calculated from stator variables. The stator flux linkage space vector can be obtained from measured currents and voltages according to stator voltage equation of an IM in the stationary reference frame and rewritten as [4]:

$$\overline{\psi}_s = \int (\overline{u}_s - R_s \overline{i}_s)\,dt \qquad (1)$$

where $\overline{\psi}_s, \overline{u}_s, \overline{i}_s, R_s$ are the stator flux linkage vector, voltage vector, current vector, and resistance, respectively. Electromagnetic torque can be calculated from estimated flux linkages and measured stator currents as expressed in (3).

$$\psi_{sD} = \int (u_{sD} - R_s i_{sD})\,dt$$
$$\psi_{sQ} = \int (u_{sQ} - R_s i_{sQ})\,dt \qquad (2)$$

$$T_e = \frac{3}{2} p (\psi_{sD} i_{sQ} - \psi_{sQ} i_{sD}) \qquad (3)$$

where $\psi_{sD}, \psi_{sQ}, v_{sD}, v_{sQ}, i_{sD}, i_{sQ}$ are stator flux linkages, voltages, and currents in d-q coordinates, respectively.

Stator flux and torque can be controlled directly and independently by properly selecting the inverter switching configurations. If for simplicity it is assumed that the stator ohmic drop can be neglected, then $d\overline{\psi}_s/dt = \overline{u}_s$. It can be seen that the inverter voltage directly impresses the stator flux, and therefore in a short Δt time, when the voltage vector is applied, one can assume that $\Delta \overline{\psi}_s \approx \overline{u}_s \Delta t$. Thus, the stator FLSV moves by $\Delta \overline{\psi}_s$ in the direction of stator voltage space vector at a speed which is proportional to the magnitude of the stator voltage space vector. Decoupled control of the torque and stator flux is achieved by acting on the radial and tangential components of the stator FLSV in the locus. These two components are directly proportional to the

978-1-4244-1741-4/08/$25.00 ©2008 IEEE   1123

components of the stator space vector in the same direction, and thus they can be controlled by the appropriate inverter switching.

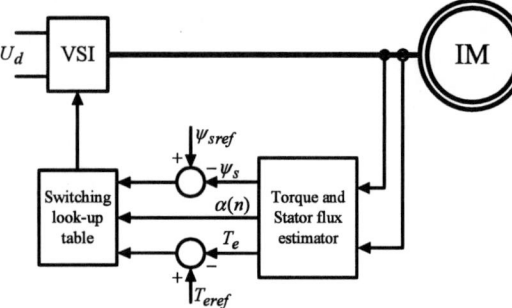

Fig. 1. Basic DTC block diagram.

In a six steps voltage source inverter (VSI), eight switching combinations can be selected, two of which determine zero voltage vectors and the remaining generate six equally spaced voltage vectors, $(u_1, u_2, ..., u_6)$ having the same magnitude, as expressed in equation (4).

$$\overline{u}_s = \overline{u}_k = \frac{2}{3} U_d \exp\left[j(k-1)\pi/3\right],$$
$$k = 1,2,...,6.$$

(4)

where, U_d is the DC link voltage of the inverter. Fig. 2 illustrates six switching voltage vectors that cover six sectors, $\alpha(1), \alpha(2), ..., \alpha(6)$. According to the principle of operation of DTC, the selection of a proper voltage vector at each sampling period is made, in order to maintain the torque and the stator flux within the limits of two hysteresis comparator bands. In particular, the selection is made on the basis of the position of the stator flux and the instantaneous errors in torque and stator flux magnitudes.

Assume that the stator flux vector is in kth sector of the d-q plane, where $k = 1,2,...,6$. The flux magnitude can be increased by selecting the voltage vectors u_k, u_{k+1} and u_{k-1} as in Fig. 2. Conversely, the decrease of $\overline{\psi}_s$ could be obtained by selecting u_{k+2}, u_{k-2} and u_{k+3}. The zero-voltage vector does not substantially affect the stator flux vector, with the exception of a small flux weakening due to the voltage drop on the stator resistance.

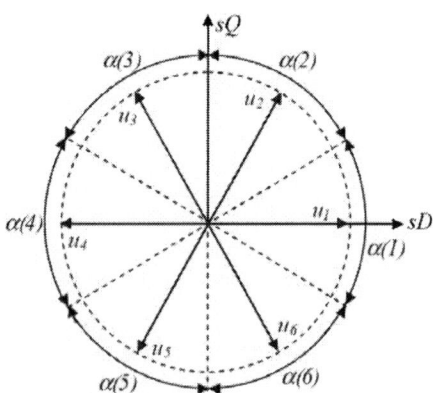

Fig. 2. Inverter output voltage space vectors.

Of course, the voltage vectors utilized to control the stator flux, also affect the torque values. An increment of torque is in general obtained by selecting the voltage vectors u_{k+1} and u_{k+2}. Conversely, a decrement of torque could be achieved by utilizing u_{k-1} and u_{k-2} or the zero voltage vectors u_0, u_7. In general, if an increase (decrease) in torque is desired, then the voltage vectors that advance (retard) the flux linkage space vector in the direction (opposite direction) of rotation are applied.

The hysteresis band comparator technique, applied to control both the stator flux and torque, leads to several possible conditions with reference to the instantaneous errors. Any voltage vector selection strategy for DTC scheme is aimed to find the most appropriate voltage vector which allows a decrease in torque and flux errors. It is then possible to drive stator flux and torque to follow any desired track curve. This allows the decoupled control of flux and torque to be achieved. Table I, shows the voltage vector selection according to stator flux and torque errors.

TABLE I
SELECTION TABLE OF VOLTAGE VECTOR

$\Delta\psi$	ΔT_e	sector					
		1	2	3	4	5	6
1	1	u_2	u_3	u_4	u_5	u_6	u_1
	0	u_0	u_7	u_0	u_7	u_0	u_7
	-1	u_6	u_1	u_2	u_3	u_4	u_5
-1	1	u_3	u_4	u_5	u_6	u_1	u_2
	0	u_7	u_0	u_7	u_0	u_7	u_0
	-1	u_5	u_6	u_1	u_2	u_3	u_4

III. FUZZY LOGIC

Fuzzy logic is able to use human reasoning not in terms of discrete symbols and numbers, but in terms of fuzzy sets. These terms are quite flexible with respect to the definition and values. From the idea of elastic sets, the concept of a fuzzy set has been proposed. Fuzzy sets are functions that map a value that might be a member of the set to a number between zero and one indicating its actual degree of membership. A degree of zero means that the value is not in the set and a degree of one means that the value is completely representative of the set. This produces a curve across the members of the set [7]-[9].

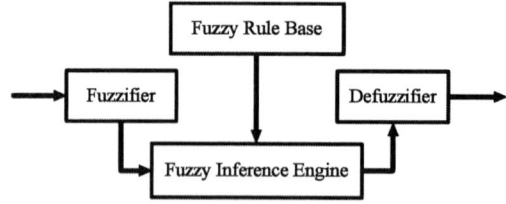

Fig. 3. Fundamental structure of fuzzy systems.

As it is depicted in Fig. 3, a fuzzy controller generally consists of four subsystems, in which two parts have the duty of transformation: fuzzifier (first transformation), fuzzy rule base, inference engine and defuzzifier (second transformation). The fuzzifier changes the input variables (crisp signals) into fuzzy values. The fuzzy rule base consists of basic data and linguistic rules. The inference engine is the brain of a fuzzy controller which has the ability to simulate the human decision based on fuzzy idea. Finally, the second transformation converts the fuzzy values into the real values.

A. Proposed Fuzzy DTC Scheme

Let's assume that the stator flux linkage space vector lies in sector 1 and rotating counter clockwise. As depicted in Fig. 4, if a torque increase with a simultaneous decrease in flux linkage is desired, then the required optimum switching voltage vector to the motor is u_3.

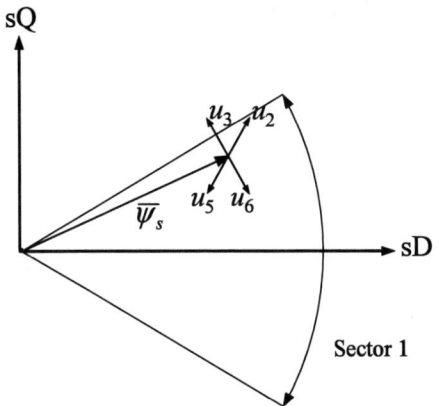

Fig. 4. Optimum switching voltage vectors for stator flux space vector in sector 1.

Now, if in this situation the stator FLSV moved from sector 1 to sector 2, then the applied vector should be changed without any effect in the desired torque and flux. For example, if according to Table I, the switching vector u_3 is applied for a desired torque increment and flux decrement in sector 1, then with the arrival of the stator FLSV in sector 2, the switching vector u_4 must be applied for the same desired conditions.

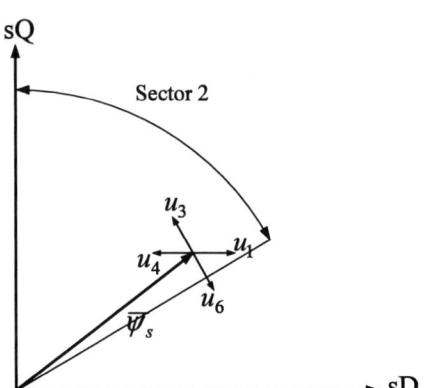

Fig. 5. Optimum switching voltage vectors for stator flux space vector in sector 2.

In general, when the stator FLSV moves from one sector to another, without any changes in $\Delta \psi$ and ΔT_e, then, the voltage vector that was applied to motor is changed. The applied voltage vectors, in the beginning of the new sector cause intense torque and stator flux

variations and create the significant ripples in the electromagnetic torque and stator flux linkages. On the other hand, the so called "best" voltage vectors in each sector defined by a lookup table cannot really be the best for various conditions of torque and flux errors.

The implementation of the above rules arise the following question: How can decease these ripples? An answer to this question can be given by using the fuzzy sets. In this new method, the membership function of the stator flux space vector in each sector is not crisp, but also is fuzzy. So, stator flux space vector's membership degree is fuzzy and between [0,1] in each sector. By implementation of this technique, the demand voltage space vector can be composed of two active voltage vectors which described in next part. Furthermore, any variation in applied voltage vector result of the changing sectors does not change discrete. This method decreases the stator flux and torque ripples significantly.

B. Fuzzification

The fuzzification is the process of mapping from estimated input to the corresponding fuzzy set in the input universe of discourse. The proposed fuzzy controller has two inputs which are defined as (5)

$$\widetilde{\psi}_{sD} = \psi_{sD} \Big/ \sqrt{(\psi_{sD}{}^2 + \psi_{sQ}{}^2)}$$
$$\widetilde{\psi}_{sQ} = \psi_{sQ} \Big/ \sqrt{(\psi_{sD}{}^2 + \psi_{sQ}{}^2)} \tag{5}$$

where $\widetilde{\psi}_{sD}$ and $\widetilde{\psi}_{sQ}$ are the normalized stator flux linkage components in the d-q coordinates [4].

The fuzzification is performed using membership function with singleton fuzzifier. With regard to the two input variables, there are two groups of membership functions as depicted in Fig. 4. Each group has a number of curves. $\widetilde{\psi}_{sQ}$ has two categories while $\widetilde{\psi}_{sD}$ has two member functions and four subsets in order to obtain a better control.

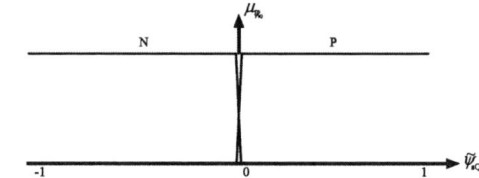

Fig. 6. Membership function of $\widetilde{\psi}_{sQ}$.

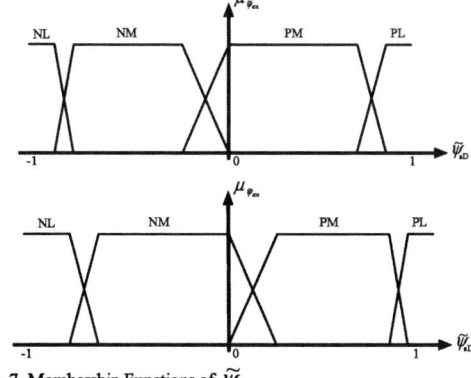

Fig. 7. Membership Functions of $\widetilde{\psi}_{sDi}$.

where $\mu_{\psi_{sQ}}, \mu_{\psi_{sDi}}$ are membership degrees of $\widetilde{\psi}_{sQ}$ and $\widetilde{\psi}_{sDi}$ respectively.

C. Fuzzy Relation and Rule Bases

The fuzzy relation and fuzzy rules are defined as sets of IF-THEN rules for fuzzy inference. Control rules are often expressed in the IF-THEN format as follows:

Rule1: IF $\widetilde{\psi}_{sD2}$ is PL and $\widetilde{\psi}_{sQ}$ is N, THEN S is S_1

Rule2: IF $\widetilde{\psi}_{sD1}$ is PL and $\widetilde{\psi}_{sQ}$ is P, THEN S is S_1

Rule3: IF $\widetilde{\psi}_{sD1}$ is PM and $\widetilde{\psi}_{sQ}$ is P, THEN S is S_2

Rule4: IF $\widetilde{\psi}_{sD1}$ is NM and $\widetilde{\psi}_{sQ}$ is P, THEN S is S_3

Rule5: IF $\widetilde{\psi}_{sD1}$ is NL and $\widetilde{\psi}_{sQ}$ is P, THEN S is S_4

Rule6: IF $\widetilde{\psi}_{sD2}$ is NL and $\widetilde{\psi}_{sQ}$ is P, THEN S is S_4

Rule7: IF $\widetilde{\psi}_{sD2}$ is NM and $\widetilde{\psi}_{sQ}$ is N, THEN S is S_5

Rule8: IF $\widetilde{\psi}_{sD2}$ is PM and $\widetilde{\psi}_{sQ}$ is N, THEN S is S_6

where $\widetilde{\psi}_{sD1}, \widetilde{\psi}_{sD2}$ are membership function of $\widetilde{\psi}_{sD}$ in $\widetilde{\psi}_{sQ} \geq 0$ and $\widetilde{\psi}_{sQ} \leq 0$ situation respectively, and S is the membership function of sector numbers.

D. Defuzzification

Generally after fuzzy reasoning, the output action is presented by the fuzzy sets, which must be converted into a non-fuzzy output in terms of certain defuzzification methods, such as Center of Area, Mean of Maximum, etc. In this system, the output of the fuzzy system is the membership degree of the stator flux space vector in each sector that is calculated by the centroid defuzzification method. The output function is given in equation (6).

$$S^* = \frac{\int S \max\{\min(\mu_{\psi_{sD}}, \mu_{\psi_{sQ}})\}}{\max\{\min(\mu_{\psi_{sD}}, \mu_{\psi_{sQ}})\}} \tag{6}$$

where S^* is the fuzzy sector number in the range of 1 to 7. The value of S^* determines which of two sectors that

the stator flux linkage space vector is located and transferred to.

$$\alpha_x = \{S^*\}$$
$$\alpha_y = \mathrm{mod}((\alpha_x + 1), 6) \tag{7}$$

$$T_x = (1 + \alpha_x - S^*) T_s$$
$$T_y = (S^* - \alpha_x) T_s \tag{8}$$

where α_x and α_y are the sector numbers that stator FLSV is transferred between them, T_x and T_y are the effective time intervals for the two voltage vectors that are applied to the motor within the sampling period T_s and $\{ \}$ denotes rounding the variable to the nearest inferior integer.

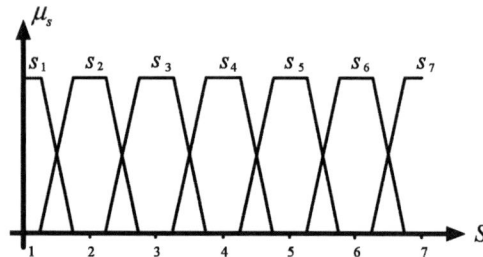

Fig. 8. Membership function of sectors.

For example, if $S^* = 1.3$, then according to (7) and (8), $\alpha_x = 1, \alpha_y = 2, T_x = 0.7 T_s$ and $T_y = 0.3 T_s$. Therefore, the stator FLSV is located at the end of $\alpha(1)$ and is transferring to $\alpha(2)$. Now, if $\Delta \psi = -1$ and $\Delta T = 1$, the voltage vector that is applied to the motor is expressed as in the following:

$$u_e = u_3 \frac{T_x}{T_s} + u_4 \frac{T_y}{T_s} \tag{9}$$

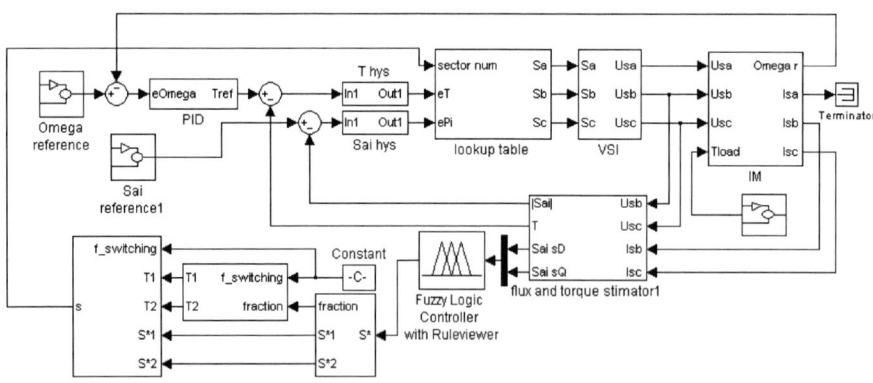

Fig. 9. Matlab/Simulink model of the modified DTC.

IV. SIMULATION RESULTS AND COMPARISONS

Two Matlab/Simulink models were developed to examine the different control algorithms; one for the basic DTC, and the other for the modified DTC. Fig. 9 shows the modified DTC Simulink model. The induction motor used in the simulations is rated at 5hp, 460v, 60Hz, 4 poles, 1475 rpm, and the band of the two hysteresis controllers are 1% of their rated values for basic DTC. In both the modified and basic DTC simulink models, sampling time is set to $20\,\mu s$. The simulation results for the two algorithms are shown in Fig. 10 for the stator flux linkage, the torque responses, and the rotor speed. One can easily deduce from the figures that the steady-state and dynamic responses of the proposed modified DTC is quite better than the basic DTC. Also, the ripples in torque and flux linkage are reduced significantly.

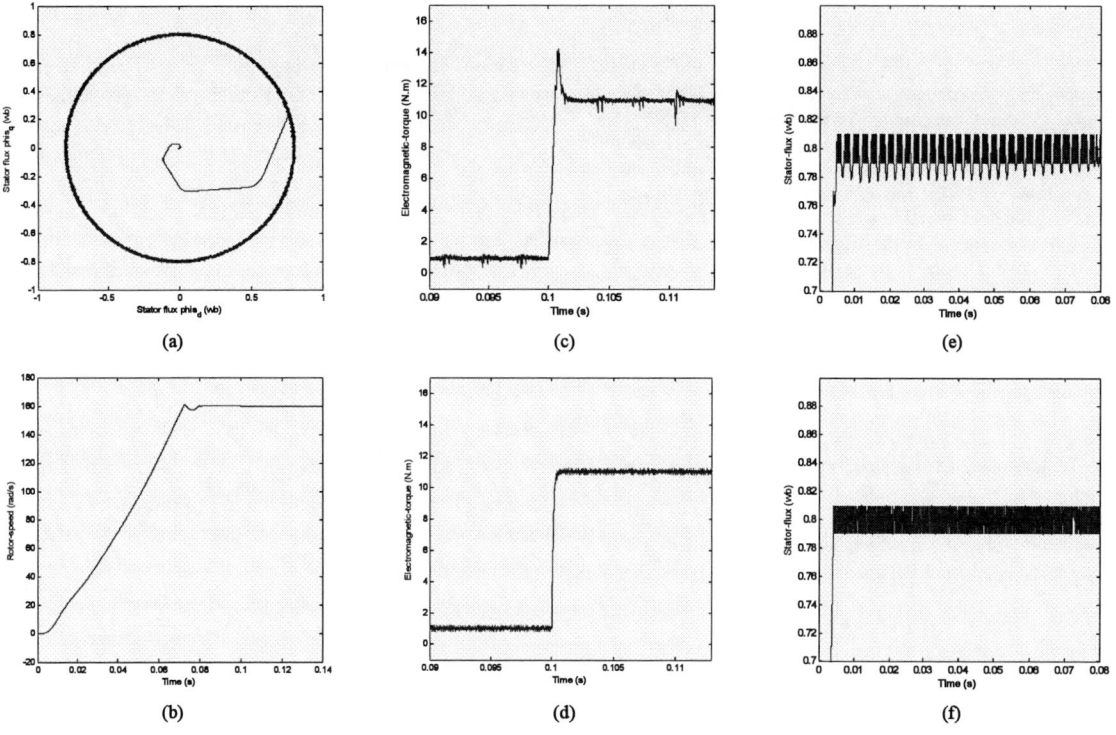

Fig. 10. The simulation results: (a) stator flux in D-Q plane, (b) rotor speed, (c) and (d) electromagnetic torque with basic and modified DTC, respectively, (e) and (f) stator flux with basic and modified DTC, respectively.

V. CONCLUSIONS

In order to improve the performance of the drive of an induction motor and to reduce the ripples in torques and flux linkages, a fuzzy logic based DTC has been proposed. The PI speed controller has been off-line tuned using PSO. As a result, both torque and flux linkage ripples are significantly reduced, and the switching frequency is kept fixed. The simulation results show the effectiveness of the new proposed control strategy.

REFERENCES

[1] M. Depenhrock, "Direct Self-control of inverter-fed machine", *IEEE Trans. Power Elec.*, vol. 3, no.4, pp.420-429, Oct. 1988.

[2] T. Naguchi and I. Takahashi, "A new quick-response and highefficiency control strategy of an induction motor", *IEEE Trans. Ind. App.*, vol. IA-22, pp. 820-827, Sept./Oct. 1986.

[3] D. Casadei, G. Serra, and A.Tani. "Implementation of A Direct Torque Control Algorithm for Induction Motors Based on Discrete Space Vector Modulation", *IEEE Trans. Power Elec.*, vol. 15, no. 4, pp. 769-777, July 2000.

[4] Peter Vas "Sensorless Vector and Direct Torque Control", *Oxford University Press*, 1998.

[5] Cristian Lascu, Ion Boldea and Frede BlaabjergI "A Modified Direct Torque Control for Induction Motor Sensorless Drive", *IEEE Transactions on Industrial Applications*, vol.36, no.1, pp. 122-130, 2000.

[6] H. Gholizad, M. Mirsalim, "A New Method for the Reduction of Flux and Torque Ripples in DTC", *ICEE 2001*, May 2001

[7] Yang Xia, Oghanna W., "Study on fuzzy control of induction machine whit direct torque control approach", *ISIE97*, vol.2, pp.625-630, July 1997.

[8] Bird I.G, Zelaya De La Parra, "Fuzzy logic torque ripple reduction for DTC based AC drives", *Electronic letters*, vol. 33, pp. 1501-1502, 14 Aug. 1997.

[9] San Dan, He Yikang , Zhi Dawei, "Direct torque control of a permanent magnet synchronous motor base on fuzzy logic", *Trans. of china Elec. Society*, vol.18, pp. 33-38, Feb. 2003.

[10] Chengzhi Cao, Bo Zhou, Min Li, Jing Du, "Digital Implementation of DTC Based on PSO for Induction Motors", *6th World Congress on Intelligent Control and Automation*, Dalian, China, June 2006.

Design and Implementation of High Performance Full-Digital Spindle Drives

Liu Yang   Zhao Jin

Dept. of Automation, Huazhong University of Science & Technology, Wuhan Hubei, China,
E-mail: yanglei_liu@163.com

Abstract—**In this paper, a sort of design scheme for full-digital spindle machine drives based on DSP is presented and a series of optimizing strategies are also proposed to improve its performance. To guarantee the rapid response and the steady operation of the spindle machine, an indirect vector control strategy is used in the system. For the sake of facilitating the implementation of the full-digital system and to increase the utilization ratio of busbar voltage, an improved space vector pulse width modulation control strategy is adopted in the system. To enhance the load capability of the motor in flux-weakening region and to spread the operating interval of the motor under constant power, an optimized flux-weakening control strategy with maximized utilization of voltage is taken. At last, the experimental results of the spindle drives show the scheme is feasible and the required performance can be achieved.**

Keywords—**DSP, AC machine, adjustable speed drives, high speed drive, optimal control**

I. INTRODUCTION

Several characteristics of spindle drives are demanded in high performance numerical control system[1]: firstly, high speed and hypervelocity operation. The maximum speed of spindle is 6000~8000r/min generally. Secondly, wide range of speed as well as constant power operation, especially in high speed for spindle drive, the smooth stable speed and maximum output power as much as possible are all necessary. In addition, the output torque in low speed is also demanded as large as possible so as to meet the demand of forceful cutting. Fast response and precise stop function are also needed. To fulfill the demands above, vector control technique is still the best control scheme. Among which indirect vector control is the easiest way for real-time realization[2]. In the paper, the full-digital indirect vector control is realized using DSP. The utilization ratio of busbar voltage is increased through optimizing overmodulation strategy of SVPWM. Meanwhile, the control method in flux- weakening region is also improved. As a result, the output torque is enhanced in the flux-weakening region and constant power range is then extended.

II. Brief principle of the full-digital indirect vector control system

The structure diagram of doubly closed-loop indirect vector control system is shown in Fig.1. In the figure, the

speed loop, current decoupling, field orientation, current loop and SVPWM are all implemented by program in DSP (pane in Fig.1), which is called full-digital design namely. Evidently, the full-digital control system can effectively reduce hardware cost and improve the reliability of system compared with the analog control mode or analog-digital mixed control mode which appeared previously. Furthermore, the control is realized by software. So various sorts of algorithms can be rebuilt and controlled with no change of hardware. Accordingly, the system has incomparable flexibility. In indirect vector control, the rotor field position($\theta_e = \theta_s + \theta_{s1}$) is calculated indirectly through the real position of rotor(θ_s) and slip angle(θ_{s1}) with d-axis of Park coordinate transformation aligned to the rotor flux vector, as shown in Fig.2 which is the sketch map of current decoupling in vector control.

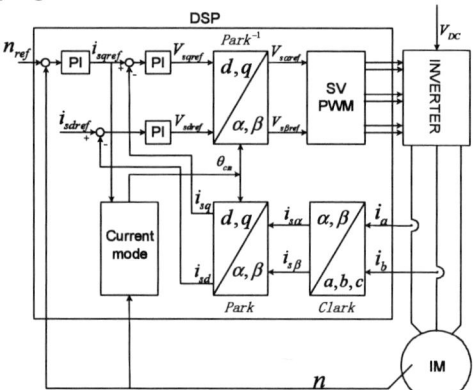

Fig. 1, Block diagram of doubly closed-loop indirect vector control

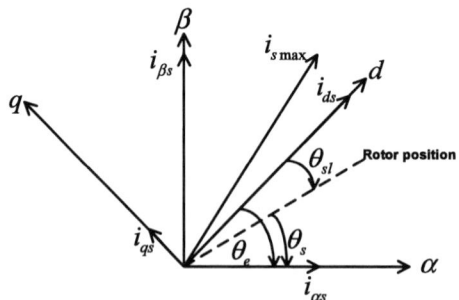

Fig. 2, Sketch map of stator current decoupling in indirect vector control

In Fig.2,

$i_{s\max}$: peak value of phase current in stator

$i_{\alpha s}, i_{\beta s}$: stator current in the stationary reference frame

i_{ds}, i_{qs} : magnetizing current and torque current in the rotate reference frame

III. The optimization of space vector pulse width modulation

The space vector pulse width modulation(SVPWM) strategy is widely used nowadays. It has two major advantages compared with the sinusoidal pulse width modulation(SPWM) strategy: higher utilization ratio of busbar voltage and easier digital implementation. The maximum modulation ratio can reach 1 in theory. Evidently, the voltage utilization ratio has great effect on the performance in high demand for overload capacity of motor or in the flux-weakening process. But the voltage utilization ratio based on the general overmodulation strategy of SVPWM cannot reach the maximum[3]. J.Holtz has proposed a novel strategy in the paper [4], which can make the voltage utilization ratio increase continuously in overmodulation range even reach the maximum 1. This method is consummated by Dong-Choon Lee et al.[5], but considering that it is derived from the open-loop control, so the improvement is developed in the paper based on the output voltage characteristic in closed-loop field-oriented control, consequently, the voltage utilization ratio reaches the maximum.

Three-phase voltage source inverter comprises of six switch components, as shown in Fig.3.

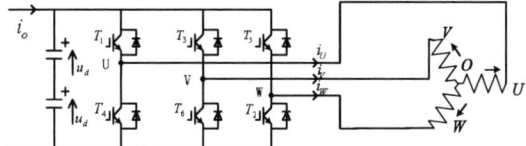

Fig. 3, Three-phase voltage source inverter

Supposing "1" stands for upper switch on and lower switch off in each leg of the bridge, thus, the inverter has eight switch states: (000), (111), (001), (010), (011), (100), (101), (110). Corresponding to these inverter switch states, there are eight switch voltage vectors shown in Fig.4, (000) as well as (111) is corresponding to zero vector, which is not shown in the figure. The six switch voltage vectors divide the plane into six sectors, and the reference voltage vector in any sector can be synthesized by the switch voltage vectors corresponding to the sector, for instance, the reference voltage vector located in sector I can be synthesized by vector $\vec{V}_6$ and $\vec{V}_2$.

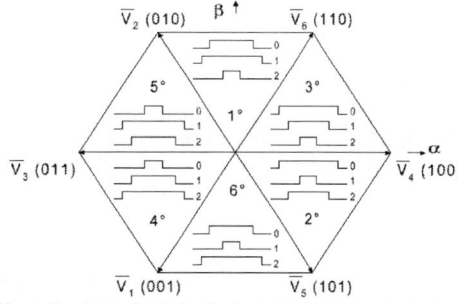

Fig. 4, Hexagon of switch voltage vector and the sector division

The principle of SVPWM digital implementation is shown in Fig.5. The time for switch corresponding

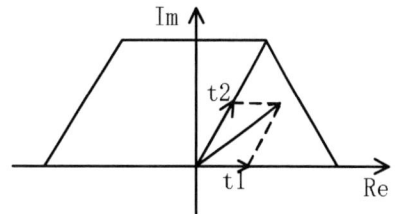

Fig. 5, Decomposition chart of SVPWM

to the voltage vector in the sector is divided into time t_1 and t_2, once $t_1 + t_2$ exceeds the PWM period, the voltage vector is then in the overmodulation region. Here, the t_1 and t_2 are generally calculated according to equations below:

$$t_1 = \frac{t_1}{t_1 + t_2} PWMPRD \qquad (1)$$

$$t_2 = \frac{t_2}{t_1 + t_2} PWMPRD \qquad (2)$$

The maximal modulation ratio is 0.9517 by adopting this method, the paper [5] proposes a novel linearized overmodulation way, in which the maximal modulation ratio can reach 1. However, the linearized method in paper [5] is given under open-loop and cannot be applied in closed-loop control. So, the linearized formula is modified below:

$if(mi > 1 \ \& \ mi < 1.0283) \ \alpha_h = 6.40 * mi - 6.40$

$if(mi >= 1.0283 \ \& \ mi < 1.0458)$

$\alpha_h = 11.75 * mi - 11.91$

$if(mi >= 1.0458 \ \& \ mi <= 1.0483)$

$\alpha_h = 48.96 * mi - 50.79$

α_h is the retention angle given in the paper [5].

Although the overmodulation scheme increases the complexity of algorithm, the voltage is enhanced obviously. It is very crucial to spindle drive system which needs flux- weakening control.

IV. The optimization of flux-weakening control

It is known that the voltage cannot be increased when motor runs over the rated speed, consequently, only by reducing the flux linkages can increase the speed, which is called flux-weakening. It is the most common flux-weakening strategy in indirect vector control that field current is inversely proportional to speed[6]. But this method doesn't consider the optimal allotment between field current and torque current, which is the problem of current utilization ratio[7,8]. Accordingly, the maximum output torque cannot be obtained, meanwhile it shortens the constant power range and lengthens the rising time of speed. Based on the current and voltage restrictions of spindle motor, the optimized scheme of field current, which is applied in the system, is derived by using the mathematics model of motor in indirect vector control system.

In indirect vector control system, the voltage equations of induction motor in steady state are:

$$V_{ds} = i_{ds} R_s - i_{qs} \sigma L_s \omega_e \qquad (3)$$

$$V_{qs} = i_{qs} R_s + i_{ds} L_s \omega_e \qquad (4)$$

where

R_s : phase resistance in stator

L_s : inductance in stator

σ : total leakage inductance coefficient

In high speed range, voltage drop of resistance is omitted generally, thus, the equations (3) and (4) can be simplified:

$$V_{ds} = -i_{qs} \sigma L_s \omega_e \qquad (5)$$

$$V_{qs} = i_{ds} L_s \omega_e \qquad (6)$$

For the sake of the restrictions of busbar voltage and SVPWM modulation strategy, the maximal phase voltage imposed on the stator has a limit value (V_{max}), thus, the V_{ds}, V_{qs} in d/q rotate reference frame must satisfy the relation below:

$$V_{ds}^2 + V_{qs}^2 \leq V_{max}^2 \qquad (7)$$

Combined the equation (5) and (6), then:

When $i_{ds} = \dfrac{V_{max}}{\sqrt{2} L_s \omega_e}$, the maximum output torque can be obtained, here

$$i_{qs} = \frac{\sqrt{V_{max}^2 - a^2 i_{ds}^2}}{b} \qquad (8)$$

where $a = L_s \omega_e$, $b = \sigma L_s \omega_e$

Using the method above to set the field current, the voltage can be fully used, furthermore, the motor capacity of lifting the load is then enhanced and the constant power range is extended, which are further proved in the experiment.

V. Experimental results

The spindle motor used in the experiment is 7.5KW and has 2 pole pairs, with rated speed 1500r/min and maximum speed 8000r/min. The motor parameters are shown in Tab.1.

TABLE I.
Motor Parameters Used In The Experiment

R_s	L_{ls}	R_r	L_{lr}	L_m
0.751Ω	3.1mH	0.547Ω	3.1mH	56.6mH

To demonstrate the effect of the optimized field weakening control, the contrast experiments of field weakening control strategies, between the one generally used(current in d-axis is inversely proportional to speed) and the one discussed in the paper, are developed. The step speed (0-8000r/min) contrast experiments are shown in Fig.6, and the contrast experiments with load are shown in Fig7.

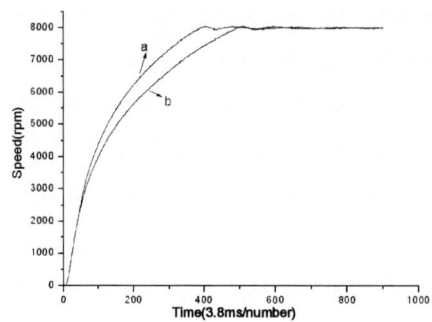

Fig. 6. the step speed(0-8000r/min) curves in contrast experiments
(a)   the step speed curve using field weakening control strategy in the paper
(b)   the step speed curve with current in d-axis inversely proportional to speed

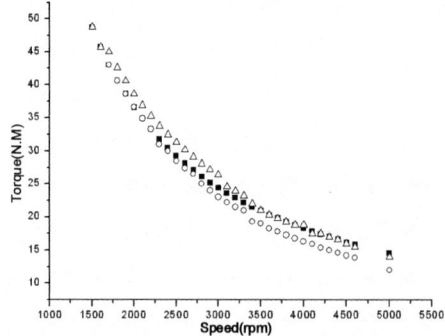

Fig. 7. the curves corresponding to two field weakening control strategies in contrast experiments with load
（■） the output torque in theory under constant power operation
（△） the output torque using field weakening control strategy in the paper
（○） the output torque with current in d-axis inversely proportional to speed

It can be seen from the figures that the field weakening control strategy discussed in the paper is superior to the one generally used, no matter in the speedability of rising speed or the capacity of lifting the load.

VI. Conclusion

The design scheme of the full-digital spindle drive is introduced in the paper. The main parts have been optimized to further improve the performance. The scheme is based on the indirect vector control system which provides a platform for the realization of various optimized algorithms. The utilization ratio of busbar voltage is further improved by optimizing overmodulation strategy of SVPWM, which is advantageous when overloaded below the rated speed as well as constant power operation in high speed range. In addition, the optimization of the flux-weakening control strategy is mainly based on the mathematics model of indirect vector control, the voltage is then fully used and the optimal allocation of the current is achieved. Finally, the physical experiment verifies the feasibility of the scheme, and the performance is improved obviously.

REFERENCES

[1] Yang Wenqiang, Jia Zhengqun, Spindle drive system of numerical control for 21st century, Marine Electric& Electric Technology, 2001,1:6-10

[2]Bose,B. K. Modern power electronics and AC drives, Beijin: China machine press, 2003,1

[3] Li Yongdong, The digital control system of induction machine, Beijin: China machine press, 2002, 4

[4] J. Holtz, W. Lotzkat, and A. Khambadkone, On Continuous Control of PWM Inverters in The Overmodulation Range Including The Six-step Mode, IEEE Trans. Power Electron,Vol 8, pp. 546-553, OCT. 1993.

[5] Dong-Choon. Lee, G-Myoung. LEE,A Novel Overmodulation Technique for Space-Vector PWM Inverters, IEEE Trans. Power Electron, Vol 13, pp. 1144-1151, NOV. 1998.

[6]R.Jotten, H.Schierling. Control of induction machine in the field weakening range. Proc. IFAC, 1983:297~304

[7]Cui Naxin, Zhang Chenghui, Du Chunshui. Advances in efficiency optimization control of inverter-fed induction motor drives. Transactions of China Electrotechnical Society, 2004, 19(5): 36~42

[8] Liu Xiaohu, Xie Shunyi, Zheng Lijie; Improvement of efficiency-optimization control of induction motor drives, Proceedings of the Csee, 2005,25(6):95-9

Semi hierarchical adaptive network based fuzzy logic controller design for a multi-straight-line path tracing flexible robot manipulator with rotating-prismatic joint

Mete KALYONCU*, Mustafa TINKIR[†]

Department of Mechanical Engineering, Faculty of Engineering and Architecture, University of Selçuk,
Alaeddin Keykubat Campus, 42079 Konya, TURKEY,
* e-mail: *mkalyoncu@selcuk.edu.tr*   † email: *mtinkir@selcuk.edu.tr*

Abstract — In this study, a semi-hierarchical adaptive network based fuzzy logic controller for a flexible robot manipulator with rotating-prismatic joint is designed. The proposed controller is aimed to get the end of the robot manipulator to the desired multi-straight-line path and terminate vibrations on arm while it moves. The controller is divided into two parts. First part is for rotation control of the arm. Second part has also two subsystems. The slow-subsystem provides desired extension and the fast-subsystem controller controls change in extension. The performance of the adaptive hierarchical control system is evaluated on the basis of the simulation results.

Keywords — Fuzzy control, neural network, adaptive control, motion control, robotics, modeling

I. INTRODUCTION

Research on the flexible manipulators has received increased attention in the last decades due to their several advantages over rigid manipulators such as light-weight, light-actuator, high speed, high-speed performance, low energy consumption, productivity, etc. Most of the investigations on the dynamics of robot manipulators with elastic arms consider manipulators with revolute joints. Whereas, dynamics and control of an elastic arm sliding in a prismatic joint is an important problem in many engineering applications, such as robot applications, telescopic members of loading vehicles, space craft antenna, magnetic tape drivers, printers, flexible transmission lines, band saws and weaving mechanisms. Most of the investigations on dynamics and control of flexible manipulators with revolute or prismatic joint considered the robot hand to follow a non-linear or non-straight-line path. Dynamics of straight-line tracing manipulators were not taken into consideration in most of the cited investigations. But, in some applications such as welding and painting, robotic arm has to follow predefined straight-line paths in the work volume. In many welding operations, the path followed by the robot hand during the process is a straight-line. In this type of operations, deviations from the predefined straight-line path may cause severe distortions in welded parts and also poor welding quality. Deviations from the predefined straight-line path can be considered as parameters to predict welding quality in such operations. Furthermore,

in cases where savings from energy and time are points of interest, the manipulator hand is expected to move through a linear trajectory between two points. As a result, dynamic analysis of flexible link manipulators with straight-line tracing hand is a significant problem for robot design and control purposes.

Existing studies on flexible robot manipulator can be divided into two groups in point of used joint: those with revolute joint, and those with prismatic joint [1-39]. Most of the investigations on the dynamics of robot manipulators with elastic arms consider manipulators with revolute joints. Some of these investigations on the control of flexible robot manipulators with revolute joint consider hierarchical control due to its several advantages over other control techniques. But, semi hierarchical adaptive network based fuzzy logic controller was not taken into consideration for single link flexible robot manipulator with prismatic joint in most of the cited investigations despite its some advantage to be indicated in this study.

The aim of this study is investigation of semi hierarchical adaptive network based fuzzy logic control of a flexible robot manipulator with rotating-prismatic joint. The tip end of the flexible robot manipulator traces a multi-straight-line path under the action of an external driving torque and an axial force. Considered robot manipulator consists of a rotating-prismatic joint and a sliding flexible arm with a tip mass. Flexible robot arm is assumed to be an Euler-Bernoulli beam. Equations of motion of the flexible manipulator are obtained by using Lagrange's equation of motion. The proposed fuzzy logic controller in the studies is aimed to get the end of the flexible robot manipulator to desired position and

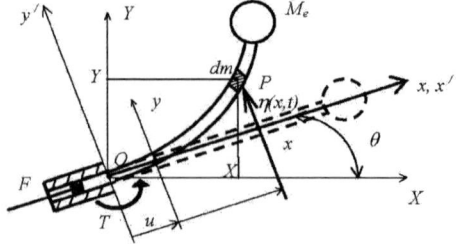

Fig. 1. Flexible robot manipulator with rotating-prismatic joint.

terminate vibrations on arm while it moves. The proposed controller is divided into two parts. First part adaptive network based fuzzy control (ANFLC) is for rotation control of the arm. Second part is a two-time-scale adaptive network based fuzzy logic controllers are applied to control extension of the system in this study. This control method has two subsystems. The slow-subsystem provides desired extension and the other fast-subsystem controller controls change in extension. Adaptive hierarchical fuzzy rules are used to reduce size of the inference engine. Adaptive network based fuzzy inference system (ANFIS) uses a hybrid learning algorithm to identify parameters of Sugeno-type fuzzy inference systems. It applies a combination of the least-squares method and the backpropagation gradient descent method for training fuzzy inference system (FIS) membership function parameters to emulate a given training data set. Change in extension affects the control of trajectory exactly. Extension control of the system is hierarchical but rotation control is not like this. So control system called as semi-hierarchical adaptive network based fuzzy logic control in this study.

II. DYNAMIC MODELLING OF A SINGLE FLEXIBLE LINK ROBOT MANIPULATOR

The physical configuration of the flexible robot manipulator considered in this study is given in Fig. 1. The flexible arm is assumed as an Euler-Bernoulli beam. The mass and flexible properties are assumed to be distributed uniformly along the flexible arm. The prismatic joint is assumed to be rigid. The flexible arm slides in the prismatic joint under the action of axial force F. The sliding motion of the flexible manipulator is assumed to be frictionless. The initial length of the beam is denoted as l_0 and the variation in the length of the manipulator is denoted as u. Torque T rotates prismatic joint about Z axis. The flexible arm experiences a combination of rotational and translational gross motion. XYZ is the global reference frame, while $x'y'z'$ is the rotating and xyz is the rotating and translating reference frame. Angle between the rotating reference frame $x'y'z'$ and the global reference frame XYZ is denoted as θ. Distance of dm to the origin in x direction is $u+x$ and the displacement from the undeformed position in y direction is denoted as η. Flexible displacements of the

manipulator can be written in the form of an infinite series:

$$\eta(x,t) = \sum_{i=1}^{\infty} \phi_i(x,t) q_i(t) = [\phi]\{q\} \qquad (1)$$

where $\{q\}$ denotes the time dependent generalized coordinate, $[\phi]$ denotes the time and space dependent assumed modes of the system [26]. Lagrange's equation of motion

$$\frac{d}{dt}\left(\frac{\partial L}{\partial \dot{q}}\right) - \frac{\partial L}{\partial q} = F \qquad (2)$$

is used in obtaining the equations of motion. Energy expressions are inserted in Lagrange's equation of motion in order to obtain the equations of motion. By using the expressions for derivatives and squares of η given below

$$\eta^2 = \{q\}^T [\phi]^T [\phi]\{q\}$$

$$\dot{\eta}^2 = \{q\}^T [\dot{\phi}]^T [\dot{\phi}]\{q\} + \{\dot{q}\}^T [\phi]^T [\phi]\{\dot{q}\} + 2\{q\}^T [\phi]^T [\dot{\phi}]\{\dot{q}\}$$

$$\dot{\eta}'^2 = \{q\}^T \left[\dot{\phi}'\right]^T \left[\dot{\phi}'\right]\{q\} + \{\dot{q}\}^T [\phi']^T [\phi']\{\dot{q}\} + 2\{q\}^T [\phi']^T \left[\dot{\phi}'\right]\{\dot{q}\}$$

$$\eta' = [\phi']\{q\}$$

$$\dot{\eta} = [\dot{\phi}]\{q\} + [\phi]\{\dot{q}\}$$

$$\eta'^2 = \{q\}^T [\phi']^T [\phi']\{q\}$$

$$\dot{\eta}' = \left[\dot{\phi}'\right]\{q\} + [\phi']\{\dot{q}\} \qquad (3)$$

a series of mathematical manipulations are performed to obtain the system equations as follows [28, 32]:

$$[M]\{\ddot{Q}\} + [H]\{\dot{Q}\} + [G]\{Q\} = [0 \quad F(t) \quad T(t)]^T \qquad (4)$$

where

$$\{Q\} = [q \quad u \quad \theta]^T$$
$$[M] = f(l_0, \rho, A, M_e, u, [I], [C_i])$$
$$[H] = f(l_0, \rho, A, M_e, u, \dot{u}, [I], [C_i])$$
$$[G] = f(l_0, \rho, A, E, M_e, u, \dot{u}, \ddot{u}, [I], [C_i], [\dot{C_i}]) \qquad (5)$$

and $[C_i]$ denote a square $\infty \times \infty$ matrix [28, 32].

III. MATHEMATICAL MODELLING OF THE STRAIGHT-LINE PATH

A flexible robot manipulator tracing a straight-line path through prescribed points $P_0, P_1, ..., P_i$ is seen in Fig. 2. Tip of the flexible manipulator experinces $P_0 P_1$, $P_1 P_2$, ..., $P_{i-1} P_i$,... straight-line paths respectively. Joint angle θ varies continously during the motion. Value of θ starts from 0 and varies depending on the input function fed through a servomotor. In order to trace a linear path, length of the flexible manipulator is changed depending on θ and so by time. This fact necessitates the determination of variation of length of the flexible manipulator. From Fig. 2, u, $\dot{u}$ and $\ddot{u}$ are expressed in [32]:

$$u = \left(u_{i-1}\sin\left(O\hat{P}_{i-1}P\right)/\sin\left(180^0 - O\hat{P}_{i-1}P_i - \theta + \theta_{i-1}\right)\right) - l_0 \qquad (6)$$

$$\dot{u} = \dot{\theta}\, u / \tan\left(180^0 - O\hat{P}_{i-1}P - \theta + \theta_{i-1}\right) \qquad (7)$$

$$\ddot{u} = \left(\ddot{\theta}\dot{u}/\dot{\theta}\right) + \left(\dot{u}^2/u\right) + \left(\dot{\theta}\,\theta u / \sin^2\left(180^0 - O\hat{P}_{i-1}P - \theta + \theta_{i-1}\right)\right) (8)$$

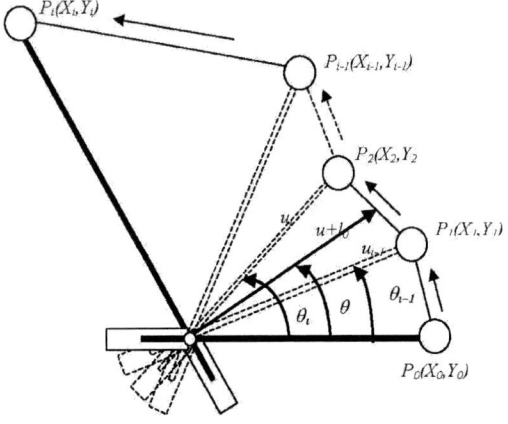

Fig. 2. Flexible robot manipulator following multi-straight-line path [32]

IV. SEMI HIERARCHICAL ADAPTIVE NETWORK BASED FUZZY LOGIC CONTROLLER (SHANFLC) DESIGN:

The field of fuzzy system and control has been making a big progress motivated by the practical success in industrial process control. Fuzzy systems can be used in as closed-loop controllers. In this case the fuzzy system measures the outputs of the process and takes control actions on the process continuously. The fuzzy controller uses a form of quantification of imprecise information (input fuzzy sets) to generate by an inference scheme, which is based on a knowledge base of control force to be applied on the system.

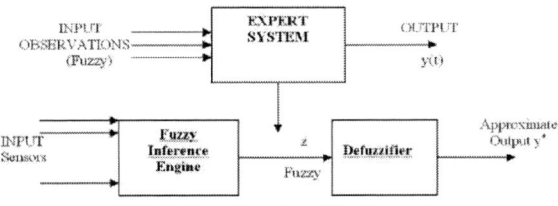

Fig. 3. The basic configuration of the fuzzy system.

The advantage of this quantification is that the fuzzy sets can be represented by a unique linguistic expression, such as small, medium, and large, etc. The linguistic representation of a fuzzy set is known as a term, and a collection of such terms defines a term-set, or library of fuzzy sets. Fuzzy control converts a linguistic control strategy usually based on expert knowledge into an automatic control strategy.

Adaptive network based fuzzy inference system (ANFIS) uses a hybrid learning algorithm to identify parameters of Sugeno-type fuzzy inference systems. It applies a combination of the least-squares method and the back propagation gradient descent method for training fuzzy inference system (FIS) membership function parameters to emulate a given training data set. Three adaptive network based fuzzy logic controllers are used in this study. Two adaptive network based fuzzy logic controllers are used in hierarchical mean. These controllers are designed by training and checking data sets that are obtained from system's dynamic model. Control system of the flexible robot manipulator with rotating-prismatic joint is shown in Fig. 4. The logical controller is made of four main components: (1) Fuzzifier; (2) Knowledge base containing fuzzy IF-THEN rules and membership functions, (3) Fuzzy reasoning; and (4) Defuzzifier interface. The basic configuration of the fuzzy system with fuzzifier and defuzzifier used in this study is shown in Fig. 3.

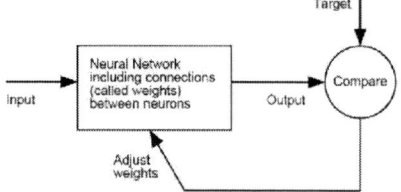

Fig. 5. Neural network structure

In this section, the notion of hierarchy is studied for fuzzy control of extension of the flexible robot manipulator with rotating-prismatic joint. Rotation control of system is only adaptive network based fuzzy logic control not hierarchical control so system control method is called as semi hierarchical adaptive network based fuzzy logic control. An attempt is made to reduce the size of the inference engine for a large-scale system. When a fuzzy controller is designed for a large-scale system often several measurable output and actuating input variables are involved. In addition, each variable is represented by a finite number l of linguistic labels which would indicate that the total number rules is equal to l^n, where n is the number of system variables. Even for a very low-order system which as few as five labels per variable, this exponential expression would become large. Consequently, we will present a hierarchical scheme for flexible robot manipulator with rotating-prismatic joint. It decomposes a large-scale system into a finite number of reduced-order subsystem, thereby eliminating the need for a large-sized inference engine.

A. Adaptive Neural Network

Neural networks are composed of simple elements operating in parallel. These elements are inspired by biological nervous systems. As in nature, the network function is determined largely by the connections between elements. We can train a neural network to perform a particular function by adjusting the values of the connections (weights) between elements. Commonly neural networks are adjusted, or trained, so that a particular input leads to a specific target output. Such a situation is shown in Fig. 5. There, the network is adjusted, based on a comparison of the output and

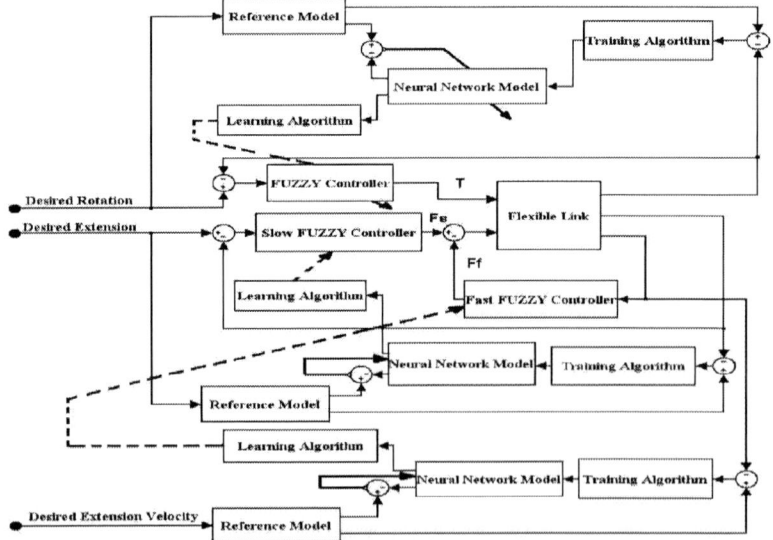

Fig. 4. Block diagram of SHANFLC for the flexible robot manipulator with rotating-prismatic joint.

the target, until the network output matches the target. Typically many such input/target pairs are needed to train a network. Neural networks have been trained to perform complex functions in various fields, including pattern recognition, identification, classification, speech, vision, and control systems.

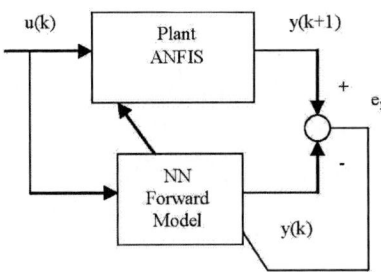

Fig. 6. Training of NN forward model

ANFIS architecture consists of five layers with the output of the nodes in each respective layer is represented by $O_{i,l}$ where i is the ith node of layer l.

Layer 1: Generate the membership grades

$$O_{i,l} = \mu_{A_i}(x), \ i = 1,2 \tag{9}$$

or

$$O_{i,l} = \mu_{B_{i-2}}(y), \ i = 3,4 \tag{10}$$

Where x (or y) is the input to the node and A_i (or B_{i-2}) is the fuzzy set associated with this node.

Layer 2: Generate the firing strengths by multiplying the incoming signals and outputs the t-norm operator result, e.g.

$$O_{2,i} = w_i = \mu_{Ai}(x) \times \mu_{Bi}(y), \ i = 1,2 \tag{11}$$

Layer 3: Normalize the firing strengths

$$O_{3,i} = \overline{w_i} = \frac{w_i}{w_1 + w_2}, \ i = 1,2 \tag{12}$$

Layer 4: Calculate rule outputs based on the consequent parameters $\{p_i, q_i, r_i\}$

$$O_{4,i} = \overline{w_i} f_i = \overline{w_i}(p_i x + q_i y + r_i) \tag{13}$$

Layer 5: Computes the overall outputs as the summation of incoming signals

$$O_{5,l} = \sum_i \overline{w_i} f_i = \frac{\sum_i w_i f}{\sum_i w_i} \tag{14}$$

We follow these steps for creating of ANFIS controller shown in below.

• 3000 training and checking data (from hierarchical PID control) are obtained for neural network of ANFIS control.

• The number and type of membership functions is determined.

• Hybrid learning algorithm and 60 epochs is chosen to train network.

B. Hybrid learning algorithm

In this study, forward hybrid learning algorithm is used for neural network part of ANFIS controllers shown in

Fig. 6. Nearly 40 epochs later error rate close to 10^{-6}. In the forward pass of the hybrid learning algorithm, node outputs go forward until layer 4 and the consequent are identified by the least-squares method. When the values of the premise parameters are fixed, the overall output can be expressed as a linear combination of the consequent parameters

$$\begin{aligned} f &= \frac{w_1}{w_1 + w_2} f_1 + \frac{w_2}{w_1 + w_2} f_2 \\ &= \overline{w_1} f_1 + \overline{w_2} f_2 \\ &= (\overline{w_1} x) p_1 + (\overline{w_1} y) q_1 + (\overline{w_1}) r_1 + (\overline{w_2} x) p_2 + (\overline{w_1} y) q_2 + (\overline{w_1}) r_2 \end{aligned} \tag{15}$$

which is linear in the consequent parameters p_1, q_1, r_1, p_2, q_2 and r_2,

$$f = XW \tag{16}$$

If X matrix is invertible then

$$W = X^{-1} f \tag{17}$$

otherwise a pseudo- inverse is used to solve for W.

$$W = (X^T X)^{-1} X^T f \tag{18}$$

Due to the adaptive capability of ANFIS, their applications to adaptive and learning control are immediate. The most common design techniques for ANFIS controllers are derived directly from neural networks counterpart methodologies. However certain design techniques apply exclusively to ANFIS.

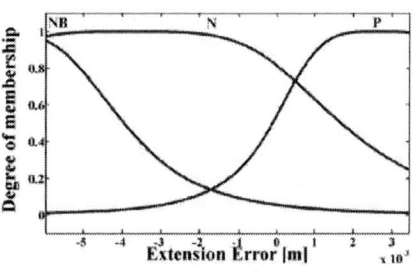

Fig. 7. Membership functions of inputs for slow subsystem fuzzy logic controller

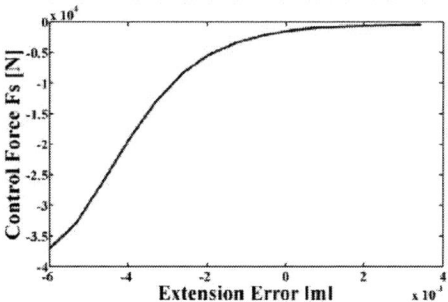

Fig. 8. Membership functions of output for slow subsystem fuzzy logic controller, (Force output, F_s [N])

C. The slow subsystem fuzzy logic controller

The slow subsystem fuzzy logic controller is designed using extension errors (e) to implement the control methodology. Hence F_s is suitable output in the slow subsystem. Extension is controlled by applied control force. In slow subsystem fuzzy logic control one input and one output occur. This control is called as SISO control system.

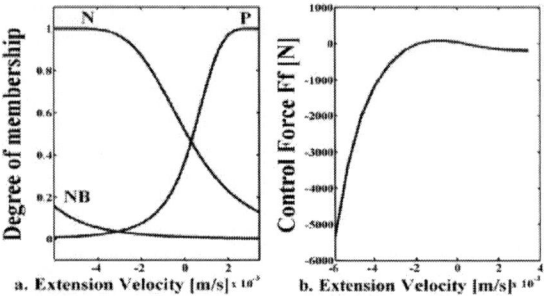

Fig. 9. Membership functions for F_f fast subsystem fuzzy logic controller

a) Extension Velocity ($\dot{u}$) input b) Force (F_f) output

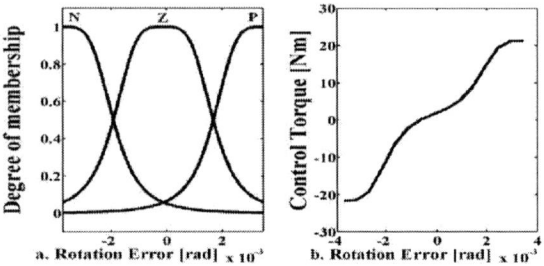

Fig. 10. Membership functions of adaptive network based fuzzy logic controller for control torque

a) Rotation error (θ_{error}) input b) Torque output

D. The fast subsystem fuzzy logic controllers

The control scheme of the system must be able to control the variations of the outputs. In this study, the fast subsystem fuzzy logic controller is designed to control extension velocity ($\dot{u}$) of the arm motion. For the approximate dynamic model, first flexible mode is chosen for the system. In the proposed fast subsystem fuzzy control structure, rules are about extension velocity and force (F_f) of the flexible robot arm. If every two variables could be combined, then for an even number of variables, the reduction is even more pronounced. For a fuzzy system with two variables ($u_{error}, \dot{u}$) and two fuzzy sets (labels), then the rules would reduce from $3^2 = 9$ to $3^1 = 3$, a 66% reduction. Clearly, depending on how many outputs can be fused and in what order, when they are put into a hierarchical structure, the size of the rule base would be reduced differently.

E. Membership Functions

Two of the difficulties with the design or any fuzzy control system are the shape of the membership functions and choice of the fuzzy rules. In fact, the decision-making logic is the way in which the controller output is generated. It uses the input fuzzy sets, and the decision is taking according to the values of the inputs. Moreover, the knowledge base comprises knowledge of application domain and the attendant control goals. It consists of a database and a fuzzy control rule base. A control system is said to be an adaptive fuzzy control system if either a set of fuzzy rules are used to modify or change an existing fuzzy controller's architecture, i.e., membership functions and/or rules.

The fuzzification uses membership functions to determine the degree of inputs. The aim of control action is to minimize the trajectory tracking error. The higher the error, the higher the control input. However, the rate of change of error also affects the value of the control input. In slow subsystem fuzzy logic controller, error and change in error rate are used in control rules as linguistic variables. In adaptive network based fuzzy inference system (ANFIS), number and type of membership functions are created by user. Then fuzzy inference system rule base is obtained automatically by ANFIS. Slow and fast adaptive network based fuzzy logic controllers have three membership functions for each inputs and outputs. Gauss type membership functions are used for fuzzy inputs in fuzzification process of slow and fast subsystem control. Sugeno type fuzzy inference system is adopted so these controllers' outputs are not fuzzy. ANFIS determine these outputs by given training and checking data sets. Figs. 7 - 10 show that membership functions and control surfaces of slow and fast subsystem adaptive network based fuzzy logic controllers.

F. Fuzzy logic controllers' rule bases

The fuzzy rule base can be illustrated as a look-up table. Slow subsystem adaptive network based fuzzy logic controller's look-up table is a vector of three rows and one column. In slow subsystem rule base is constituted three rules. Slow subsystem controller rule base is shown in Table 1. The following example demonstrates the use of the natural language modelling approach for slow subsystem fuzzy control decision-making. There are three fuzzy membership functions, that correspond to, *negative big(NB),negative(N) and positive(P)* values of extension error. The control laws of the fuzzy controller consist of a complete set of these

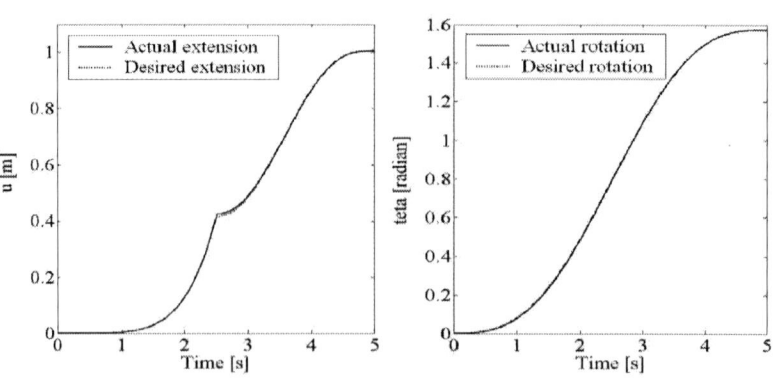

Fig. 11. Variation of length of the flexible manipulator **Fig. 12.** Angular position of the servo motor

control rules defined as:

If U_{error}=NB then F_s=Constant Force 1,

If U_{error}=P then F_s=Constant Force 3,...

Fast subsystem adaptive network based fuzzy logic controllers' has one look-up table which is made up of three rules as slow subsystem. Fast subsystem adaptive network based fuzzy logic controllers' rule base is shown in Table 2. There are three fuzzy membership functions, that correspond to, *negative big (NB), negative (N),* and *positive (P)* values of extension velocity. The control laws of the fast subsystem fuzzy controller are defined as:

If $\dot{u}$ = NB then F_f=Constant Force 1,

If $\dot{u}$ =P then F_f=Constant Force 2,...

In this study, another adaptive network based fuzzy logic controller is used as rotation control of flexible link. This controller's rule base is made up of three rules as slow and fast subsystem controllers is shown in Table 3. There are three fuzzy membership functions, that correspond to, *negative big (N), zero (Z),* and *positive (P)* values of rotation error. The control laws of the fast subsystem fuzzy controller are defined as:

The adaptive network based fuzzy logic controllers' control surfaces are shown in Fig. 8, 9b and 10b. Because of these controllers are SISO system, their control output graphics are not surface.

V. NUMERICAL EXAMPLE

In simulation studies, performance of SHANFLC is obtained and evaluated. For desired rotation motion of thirty degree, controllers' performances are applied to flexible robot arm with rotating-prismatic joint. The controllers must ensure the desired trajectory. For this desire, rotation, tip deflection and extension graphics are obtained. The controllers' responses are given in Figs. 11-13. In Fig. 13, the trajectory traced during the motion in P_0P_1 and P_1P_2 regions by the tip end of the flexible manipulator are given. The path tracking, seen in Fig. 13, is nearly perfect and sensitive.

The robot manipulator starts with its prismatic joint fully retracted at the P_0. The distance to the fixed support of the point on the path to be followed is chosen so that the flexible robot manipulator may be completely extended at the P_2, again to fully exploit the tuning capability of the hierarchical control system. As it was pointed out earlier, the goal of the hierarchical fuzzy logic control is to have the flexible robot manipulator tracking the desired multi-straight-line paths. Figs. 11-16 present the tracking behaviour of the actuator controlled by the SHANFLC designed in this study.

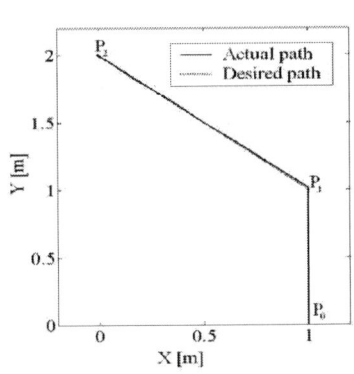

Fig. 13. End-effector path in the Cartesian space

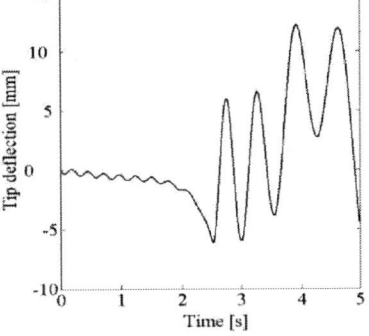

Fig. 14. Tip deflection of the flexible manipulator

The tip of the flexible manipulator is located at P_0 at the beginning of the motion and moves to P_2 along a multi-straight-line path, passing through the intermediate point P_1. Variation of $u+l_0$ versus time in order to trace the prescribed path is determined and given in Fig. 11. As seen from Fig. 11, the curve for $u+l_0$ consists of two piecewise continuous regions. Results obtained for the SHANFLC are in good agreement with the calculated reference input for the rotation and extension motion, as shown in Fig 11 and 12. Very small errors, which can be neglected, for tracking reference input signals occur in the simulations. We can see from these results that the system tracks reference position trajectory very well with hierarchical fuzzy logic control. The HFLC has been optimized to reduce the rotation and extension tracking error. The SHANFLC tracks the calculated reference trajectory almost perfectly without extreme phase difference. The rotation and extension position tracking is realized with same rate at shortening and extending motion of the manipulator with prismatic joint. The SHANFLC has good performance, especially a much smaller total motion time.

If θ_{error} = N then T=Constant Torque 1,

If θ_{error} =P then T=Constant Torque 2,...

G. *Defuzzification process*

Once the fuzzy controller is activated, rule evaluation is performed and all the rules which are true are fired. Utilizing the true output membership functions, defuzzification is then applied to determine a crisp control action. The defuzzification is to transform the control signal into an exact control output. For Sugeno-style inference, we have to choose between wtaver (weighted average) or wtsum (weighted sum) defuzzification method. In defuzzification process all adaptive network based fuzzy logic controllers, the method of weighted average (wtaver) is used:

$$u = \frac{\sum_{i=1}^{N} w_i z_i}{\sum_{i=1}^{N} w_i} \qquad (19)$$

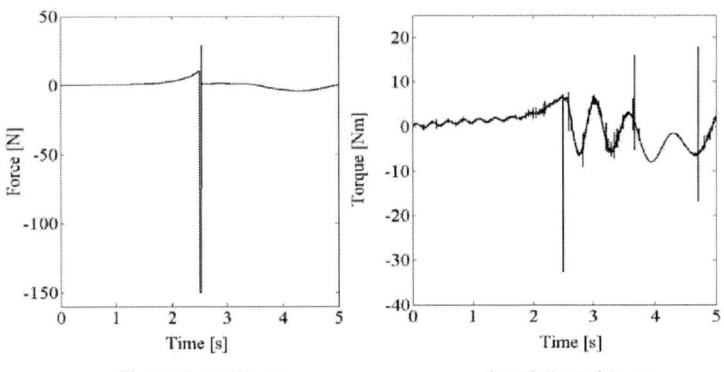

Fig. 15. Control force   **Fig. 16.** Control torque

TABLE I.
SLOW SUBSYSTEM ADAPTIVE NETWORK BASED FUZZY LOGIC
CONTROLLER RULE BASE

	Extension error (u_{error})		
	NB	**N**	**P**
Control Force (F_s)	$-3.995.10^{-4}$	1081	-160.8

TABLE II.
FAST SUBSYSTEM ADAPTIVE NETWORK BASED FUZZY LOGIC
CONTROLLER RULE BASE

	Extension Velocity ($\dot{u}$)		
	NB	**N**	**P**
Control Force (F_f)	$-4.9.10^{-4}$	990	-150.5

TABLE III.
ADAPTIVE NETWORK BASED FUZZY LOGIC CONTROLLER RULE BASE FOR
ROTATION CONTROL

	Rotation error(θ_{error})		
	N	**Z**	**P**
Control Torque (T)	-21.82	1.644	21.37

TABLE IV.
PHYSICAL AND GEOMETRIC CHARACTERISTICS OF FLEXIBLE ROBOT
MANIPULATOR

	Symbol	Value	Unit
Density	ρ	2729.5	kg/m^3
Modulus of elasticity	E	6.626×10^{10}	N/m^2
Cross sectional area	A	0.0015	m^2
Moment of inertia	I	1.095×10^{-08}	m^4

In Fig. 13, the trajectory traced during the motion in P_0P_1 and P_1P_2 regions by the tip end of the flexible manipulator for the total motion time $t_p=5$ s are given. The path tracking, seen in Fig. 13, is near perfect and sensitive. Plot of tip deflection of the flexible manipulator is given in Fig. 14. Characteristic of tip deflection of system has two different regions. Since, change of u is dissimilar for P_0P_1 and P_1P_2 paths, characteristics of the vibration of the elastic arm in P_0P_1 path and P_1P_2 path are distinct.

Vibration amplitude in P_1P_2 region is greater than the vibration amplitude in P_0P_1 region. Sudden increase in the manipulator length increases the dynamic forces thus increasing the vibration amplitude. Vibration frequency in P_1P_2 region is smaller than the vibration frequency in P_0P_1 region. In Figs. 15 and 16, depicted control force and torque respectively, similar effect is seen. Sudden increase in the manipulator length increases the control force and torque. This is the expected case since the natural frequency of the flexible manipulator decreases by increasing manipulator length. All numerical results given in Figs. 11-16 come out to be as expected physically.

VI. CONCLUSION

SHANFLC of a multi-straight-line path tracing flexible link robot manipulator with rotating-prismatic joint is investigated in this study. Numerical simulations obtained by using a developed computer program are presented and physical trend of the results are discussed. The performance of the semi-hierarchical adaptive control system is evaluated on the basis of the simulation results. Results obtained for the SHANFLC are in good agreement with the calculated reference input for the rotation and extension motion, and the desired path. Very small errors, which can be neglected, for tracking reference input signals occur in the simulations. The tip end of the flexible robot manipulator traces the desired multi-straight-line trajectory almost perfectly without extreme phase difference. The proposed controller has a good performance, especially a much smaller total motion time.

ACKNOWLEDGMENT

This work is supported by the Coordinatorship of Selçuk University's Scientific Research Projects.

REFERENCES

[1] P. C. K. Wang, J. D. Wei, "Vibration in a moving flexible robot arm", *Journal of Sound and Vibration*, vol.116, no.1, 1987, 149-160.

[2] M. Riemer, J. Wauer, "Flexible robot models with revolute and prismatic joints — handling of closed loops", *Mechanics Research Communications*, vol.15, no.6, 1988, 381-387.

[3] J. Yuh, T. Young, Y. S. Baek, "Modelling of a flexible link having a prismatic joint in robot mechanism - experimental verification", *IEEE Int. Conf. Rob. Autom.*, 1989, 722-727.

[4] S. Kalaycioglu, A. K. Misra, "Minimization of vibration of axially moving beams", *Transactions of the Canadian Society for Mechanical Engineers*, vol.13, no.4, 1990, 133-143.

[5] Y. C. Pan, R. A. Scott, A. G. Ulsoy, "Dynamic modelling and simulation of flexible robots with prismatic joints", *Journal of Mechanisms, Transmissions, and Automation in Design*, 112(3) (1990), 307-314.

[6] Y. C. Pan, A. G. Ulsoy, R. A. Scott, "Experimental model validation for a flexible robot with a prismatic joint", *Journal of Mechanisms, Transmissions, and Automation in Design*, vol.112, no.3, 1990, 315-323.

[7] J. Yuh, T. Young, "Dynamic modelling of an axially moving beam in rotation: Simulation and experiment", *Journal of Dynamic Systems, Measurement and Control*, vol.113, 1991, 34-40.

[8] K. W. Buffinton, "Dynamics of elastic manipulators with prismatic joints", *Journal of Dynamic Systems, Measurement and Control, Transactions of the ASME*, vol.114, no.1, 1992, 41-49.

[9] P. E. Gaultier, W. L. Cleghorn, "A spatially translating and rotating beam finite element for modelling flexible manipulators", *Mechanism and Machine Theory*, vol.27, no.4, 1992, 415-433.

[10] B. O. Al-Bedoor, Y. A. Khulief, "Finite element dynamic modelling of elastic beam with prismatic joint", *American Society of Mechanical Engineers, Design Engineering Division (Publication) DE*, vol.82, no.1, 1995, 609-616.

[11] R. J. Theodore, J. H. Arakeri, A. Ghosal, "Modelling of axially translating flexible beams", *Journal of Sound and Vibration*, vol.191, no.3, 1996, 363-376.

[12] B. O. Al-Bedoor, Y. A. Khulief, "Vibrational motion of an elastic beam with prismatic and revolute joints", *Journal of Sound and Vibration*, vol.190, no.2, 1996, 195-206.

[13] B. O. Al-Bedoor, Y. A. Khulief, "Finite element dynamic modelling of a translating and rotating flexible link", *Computer Methods in Applied Mechanics and Engineering*, vol.131, 1996, 173-189.

[14] B. O. Al-Bedoor, Y. A. Khulief, "General planar dynamics of a sliding flexible link", *Journal of Sound and Vibration*, 206(5) (1997), 641-660.

[15] M. Kalyoncu and F. M. Botsalı, "Dynamic modelling of a robot manipulator with a translating and rotating elastic arm (in Turkish)", *3rd International Mechatronics Design and Modelling Workshop, Ankara, Turkey*, 1997, 237-247.

[16] Ş. Yüksel, M. Gurgoze, "On the flexural vibrations of elastic manipulators with prismatic joints", *Computers and Structures*, vol.62, no.5, 1997, 897-908.

[17] Z. Yang, and J. P. Sadler, "Prediction of the dynamic response of flexible manipulators from a modal database", *Mechanism and Machine Theory*, vol.32, no.6, 1997, 679-689.

[18] R. J. Theodore, A. Ghosal, "Modelling of flexible-link manipulators with prismatic joints", *IEEE Transactions on Systems, Man, and Cybernetics, Part B: Cybernetics*, vol.27, no.2, 1997, 296-305.

[19] D. B. Marghitu, A. Guran, "Dynamics of a flexible beam with a lubricated prismatic kinematic pair", *Journal of Vibration and Acoustics, Transactions of the ASME*, vol.120, no.4, 1998, 880-885.

[20] Ş. Yuksel, "Dynamics of an elastic beam sliding through a prismatic joint)", *Journal of the Faculty of Engineering and Architecture of Gazi University*, vol.14, no.1-2, 1999, 43-55.

[21] M. Gürgöze and Ş. Yüksel, "Transverse vibrations of a flexible beam sliding through a prismatic joint", *Journal of Sound and Vibration*, vol.223, no.3, 1999, 467-482.

[22] A. M. H. Basher, "Dynamic behaviour of a translating flexible beam with a prismatic joint", *Conference Proceedings - IEEE Southeastcon*, 2000, 31-38.

[23] N. G. Chalhoub, L. Chen, "A structural flexibility transformation matrix for modelling open-kinematic chains with revolute and prismatic joints", *Journal of Sound and Vibration*, vol.218, no.1, 1998, 45-63.

[24] O. A. Bauchau, "On the modeling of prismatic joints in flexible multi-body systems", *Computer Methods in Applied Mechanics and Engineering*, vol.181, no.1-3, 2000, 87-105.

[25] M. Farid and I. Salimi, "Inverse dynamics of a planar flexible-link manipulator with revolute-prismatic joints", *American Society of Mechanical Engineers, Design Engineering Division (Publication) DE*, vol.111, 2001, 345-350.

[26] M. Kalyoncu, F. M. Botsalı, "Lateral and torsional vibration analysis of elastic robot manipulators with prismatic joint", *American Society of Mechanical Engineers, Petroleum Division (Publication) PD*, vol.2, 2002, 949-956.

[27] S. E. Khadem, A. A. Pirmohammadi, "Analytical development of dynamic equations of motion for a three-dimensional flexible link manipulator with revolute and prismatic joints", *IEEE Transactions on Systems, Man, and Cybernetics, Part B: Cybernetics*, vol.33, no.2, 2003, 237-249.

[28] M. Kalyoncu, F. M. Botsalı, "Vibration analysis of an elastic robot manipulator with prismatic joint and a time varying end mass", *The Arabian Journal For Science And Engineering*, vol.29, no.1C, 2004, 27-38.

[29] E. D. Stoenescu, D. B. Marghitu, "Effect of prismatic joint inertia on dynamics of kinematic chains", *Mechanism and Machine Theory*, vol.3, no.4, 2004, 431-443.

[30] L. Akbaba and Ş. Yüksel, "Dynamic modelling of elastic robot arm in bending and torsion", *Journal of the Faculty of Engineering and Architecture of Gazi University*, vol.21, no.2, 2006, 349-357.

[31] H. A. Basher, "Modelling and simulation of flexible robot manipulator with a prismatic joint", *Conference Proceedings - IEEE SOUTHEASTCON, 2007 IEEE Southeast Con.*, 2007, 255-260.

[32] M. Kalyoncu, "Mathematical modelling and dynamic response of a multi-straight-line path tracing flexible robot manipulator with rotating-prismatic joint", *Applied Mathematical Modelling*, vol.32, no.6, 2008, 1087-1098.

[33] Y. K. Kim, J. S. Gibson, A variable-order adaptive controller for a manipulator with a sliding flexible link, *IEEE Transactions on Robotics and Automation*, vol.7, no.6, 1991, 818-827.

[34] P. C. Müller, J. Ackermann, M. Gürgöze, "Modelling and control of an elastic robot arm with rotary and prismatic joints", *IFAC Symposia Series - Proceedings of a Triennial World Congress*, vol.5, 1991, 223-227.

[35] S. S. K. Tadikonda, H. Baruh, "Dynamics and control of a translating flexible beam with a prismatic joint", *Journal of Dynamic Systems, Measurement and Control, Transactions of the ASME*, vol.114, no.3, 1992, 422-427.

[36] I. Marom, V. J. Modi, "On the dynamics and control of the manipulator with a deployable arm", *Advances in the Astronautical Sciences*, vol.85, no.3, 1993, 2211-2227.

[37] E. Suzuki, J. Yuh, B. S. Choi, "On modelling and control of a 2-D. O. F. flexible robot having a prismatic joint", *Proceedings of the IEEE International Conference on Systems, Man and Cybernetics*, vol.3, 1993, 316-318.

[38] C. Y. Park, T. Ono, "Trajectory control of a flexible manipulator with a prismatic joint", *Bulletin of the University of Osaka Prefecture, Series A Engineering and Natural Sciences*, vol.42, no.1, 1993, 1-12.

[39] J. Lin, F. L. Lewis, "Two-time scale fuzzy logic controller of flexible link robot arm", *Fuzzy Sets and Systems*, vol.139, no.1, 2003, 125-149.

Control System with the Set Point Observation

Algirdas Baskys, Vitoldas Gobis and Valerijus Zlosnikas

Semiconductor Physics Institute/Microelectronics Laboratory, Vilnius, Lithuania, e-mail: *mel@pfi.lt*

Abstract–The closed-loop control system structure, in which the controller is provided with the information that allows the controller to know if the control system was disturbed by the set point change or not and enables it to operate with different parameters during the set point change response and the load disturbance response, has been presented. The proposed control system structure with the set point observation allows us to achieve good transient performance of the set point change response and good load disturbance rejection using the PID controller. The results of simulation, experimental investigation and industrial application of the developed control system structure have been presented.

Keywords–Control methods for electrical systems, Industrial application, Non-linear control

I. INTRODUCTION

In most control systems the primary design goal is to reject the load disturbances, which act as low frequency signals added to the control signal at the process input and drive the system away from its desired operating point. On the other hand, it is also important to have a good set point change response. However, in most cases using the PID controller it is impossible to achieve a good load disturbance rejection and a good set point change response at the same time [1–3]. Some compromise between the transient performance of the set point change response and the load disturbance response can be achieved using appropriate tuning methods of the controller [1, 2, 4, 5] but in the general case the PID controller used in the classical closed-loop control system does not allow us to achieve low overshoot (M_s) and settling time (t_s) of the set point change response and the best disturbance rejection at the same time. The modification of the PID controller with the structure of two degrees of freedom [6–8] provides a good load disturbance rejection and low M_s and t_s of the set point change response at the same time. However, the adjustment of this controller to the plant dynamics is complicated [6, 8]. On the other hand, the practicing engineers in industry prefer the controllers that can be tuned to the plant using commonly known methods developed for the PID controller.

In this work, we present the closed-loop control system structure, in which the controller is provided with the additional information. This information allows the controller to know if the control system was disturbed by the set point change or not and enables it to operate with different parameters during the set point change response and the load disturbance response. The proposed control system structure allows us to achieve good transient performance of the set point change response and good load disturbance rejection using the PID controller. The tuning technique of the controller in the suggested control system structure is the same as that used for the PID controller in the classical closed-loop control system.

II. PROBLEM FORMULATION

The equation, which relates the plant output (Y_a) with the set point (Y_d) and load disturbance (D) of the classical closed-loop control system with the unity feedback, is

$$Y_a(s) = Y_d(s) \frac{G_c(s) \, G_p(s)}{1 + G_c(s) \, G_p(s)} + $$
$$+ D(s) \frac{G_p(s)}{1 + G_c(s) \, G_p(s)} , \quad (1)$$

where $G_c(s)$ is the controller transfer function and $G_p(s)$ is the plant transfer function.

The term $G_p(s)/[(1+ G_c(s) \, G_p(s)]$ in equation (1) must be possibly small for low frequencies to guarantee a good load disturbance rejection, i.e. the controller, which provides a high $G_c(s)$ value at low frequencies, should be employed. Using the PID controller this can be done by employment of a controller with relatively high proportional (K_p), integral (K_i) and derivative (K_d) constants. However, the set point change response is determined by the $G_c(s)$ as well (1), and in many cases high values of these constants cause high M_s and long t_s of the set point change response. An example of the set point unit step response followed by the unit load disturbance of the concrete classical closed-loop control system with the unity feedback based on the PID controller with various parameters is given in Fig.1. It is seen that an increase of K_p, K_i and K_d allows us to improve the disturbance rejection, however, the set point change response is worsened significantly. The results presented in Fig.1 are

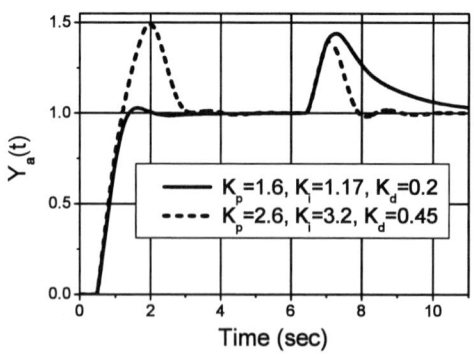

Fig. 1. The set point unit step response followed by positive unit load disturbance of the classical closed-loop control system with plant $G_{p1}(s)=e^{-0.5s}/(1+s)$ based on the PID controller with various parameters.

978-1-4244-1741-4/08/$25.00 ©2008 IEEE

obtained using the dynamic system simulation program Simulink.

The controller in the classical closed-loop control system structure works with the minimum information about the process. The only information it has is the value of error. It does not know if the control system was disturbed by the set point change or by the load disturbance. The introduction of the means of observation into the control system structure, which would provide the information about the disturbance source, could give the possibility for the controller to use different values of parameters during the set point change and the load disturbance response. This modification of the control system structure would allow us to achieve a good load disturbance rejection and good transient performance of the set point change response of the control system using the PID controller.

III. CONTROL SYSTEM WITH THE SET POINT OBSERVATION

The controller in the control system with the set point observation (Fig. 2) operates in two modes: a mode of the load disturbance rejection and a mode of the set point change response. The controller parameters are switched when the operating mode changes. We consider the situation when the load disturbance rejection is the primary purpose. Because of this, the controller operates in the load disturbance rejection mode by default. The set point observation block monitors the Y_d. It provides the signal to switch the controller parameters to the set point change response mode in case when it detects the change of Y_d. The controller operates in this mode for the given duration, which is equal to t_s, and after that the controller parameters are switched to the load disturbance rejection mode again.

The control algorithm of the PID controller in the control system with the set point observation is as follows:

$$U(t) = K_p(t)\, e(t) + \int_{t_0}^{t} K_i(\tau)\, e(\tau)\, d\tau + K_d(t) \frac{de(t)}{dt}, \quad (2)$$

$$K_p(t) = K_{ps}, \;\; K_i(\tau) = K_{is},$$

$$K_d(t) = K_{ds} \;\;\big|\, t_c < t \le t_c + t_s \;;$$

$$K_p(t) = K_{pd}, \;\; K_i(\tau) = K_{id},$$

$$K_d(t) = K_{dd} \;\;\big|\, t \le t_c \;,\; t > t_c + t_s,$$

where K_{ps}, K_{is}, K_{ds} and K_{pd}, K_{id}, K_{dd} are proportional, integral and derivative constants that function during the time when the controller operates in a mode of the set point change response and in a mode of the load disturbance rejection, respectively, t_0 is the point in time at which the algorithm starts to operate, t_c is the set point change detection time. The values of K_{ps}, K_{is} and K_{ds} should be chosen according to the requirements for the transient of the set point change response and K_{pd}, K_{id} and K_{dd} – to provide a good load disturbance rejection. The parameters of every group are tuned to the plant using methods developed for the PID controller used in the

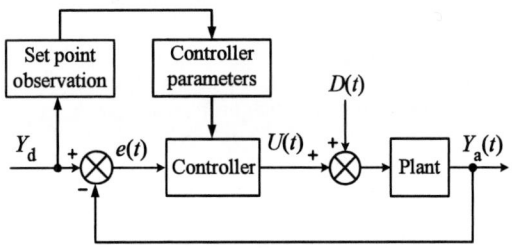

Fig.2. The block diagram of the control system with the set point observation.

classical closed-loop control system.

The set point unit step response followed by the positive and negative unit load disturbances of the classical closed-loop control system and the control system with the set point observation based on the PID controller with concrete plants was simulated. The transfer functions of plants and parameters of the PID controller, which were used for simulation, are presented in Table 1.

The parameters of the controller used in the proposed control system structure with the set point observation were tuned in a way to achieve the shortest t_s of the set point step and disturbance response under condition that M_s does not exceed 5%. The controller parameters for the classical control system structure were obtained to achieve the shortest t_s of the set point step response under condition that M_s does not exceed 5% as well, therefore parameters K_p, K_i and K_d of the controller used in a classical structure coincide with the parameters K_{ps}, K_{is}, K_{ds} of the controller employed in the proposed control system structure.

TABLE I.
PARAMETERS OF THE PID CONTROLLER

Control system structure	Plant transfer function	
	$G_{p1}(s)=e^{-0.5s}/(1+s)$	$G_{p2}(s)=1/(1+s)^4$
With the set point observation	K_{ps}=1.60, K_{is}=1.17, K_{ds}=0.20, K_{pd}=2.60, K_{id}=3.20, K_{dd}=0.45	K_{ps}=1.20, K_{is}=0.32, K_{ds}=0.90, K_{pd}=1.90, K_{id}=0.62, K_{dd}=1.10
Classical	K_p=1.60, K_i=1.17, K_d=0.20	K_p=1.20, K_i=0.32, K_d=0.90

It is seen (Fig. 3) that employment of the control system structure with the set point observation allows us to improve the load disturbance rejection of analyzed plants as compared to the case when a classical closed-loop structure is used.

The PID controller in the developed control structure has six parameters and therefore, it seems that tuning of the controller to a concrete plant can be complicated. However, the parameters K_{ps}, K_{is} and K_{ds} predominantly influence the set point response and K_{pd}, K_{id} and K_{dd} – the load disturbance response, consequently, the problem of tuning splits into two separate tuning problems.

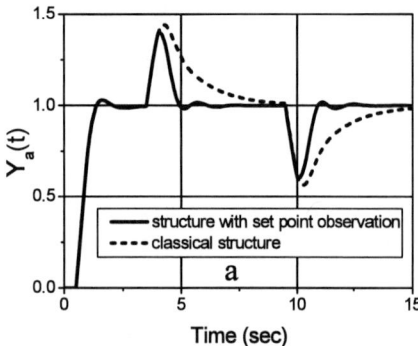

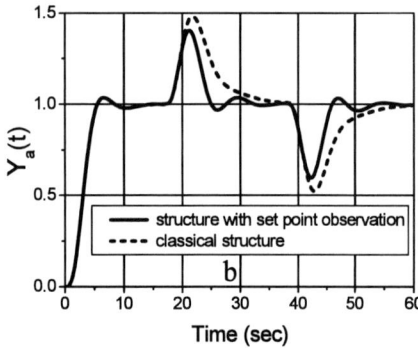

Fig. 3. The set point unit step response followed by positive and negative unit load disturbances of the control system with a classical closed-loop structure and the structure with the set point observation based on the PID controller with plants $G_{p1}(s)=e^{-0.5s}/(1+s)$ (a) and $G_{p2}(s)=1/(1+s)^4$ (b).

IV. INDUSTRIAL APPLICATION

The proposed control system structure with the set point observation (Fig. 2) with the controller, which implements the control algorithm (2), was employed in the industrial ventilation system with the AC induction motor drive for control of differential air pressure. The differential air pressure is difference between atmospheric pressure and pressure in the air inlet of the ventilation system. The AC induction motor is used as a ventilator drive. The purpose of the control system is the rotation speed control of the motor and, consequently, ventilator, with help of the frequency converter in such a way that differential air pressure would be kept close to the preset value independent of changes of ventilation system operating conditions. The step response of differential air pressure of the analyzed concrete ventilation system with the AC induction motor drive using a frequency converter is presented in Fig. 4. It is

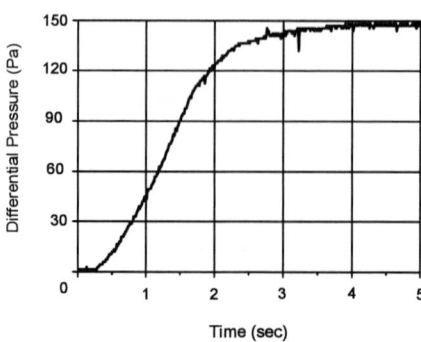

Fig. 4. The differential air pressure step response of the concrete ventilation system.

seen that the plant responses with the delay.

The experimentally obtained differential air pressure set point step response followed by positive and negative load disturbances of the analyzed concrete ventilation control system based on the PID controller with a classical closed-loop structure and the structure with the set point observation is presented in Fig. 5. The results of the investigation show that the proposed control system structure with the set point observation allows us to improve the load disturbance rejection of the ventilation system at relatively high (Fig. 5a) and low (Fig. 5b) positive and negative disturbances. However, in case when load disturbance occurs during the set point change response transient, the proposed control system structure does not have advantage in respect of the classical closed-loop structure. The results of investigation of such a situation are presented in Fig. 6. It is seen that the transient caused by the positive load disturbance for both analyzed control system structures practically coincides. The reason is that in such a situation the controller used in the structure with the set point observation operates in a mode of the set point change response, i.e. the parameters of the controller are switched to values tuned according to the requirements of the set point step response dynamics.

The ripples, which are observable on the curves presented in Figs. 4–6, are caused by the electromagnetic disturbances produced by the frequency converter.

The parameters of the PID controller, which was applied in the analyzed ventilation control system, are presented in Table 2. The parameters were obtained experimentally in process of tuning of the controller. The PID controller parameters for the classical control system structure and parameters K_{ps}, K_{is} and K_{ds} for the structure with the set point observation were obtained to achieve the shortest t_s of the set point step response under condition that M_s does not exceed 5%. The controller parameters K_{pd}, K_{id} and K_{dd} for the structure with the set point observation were tuned to obtain the shortest settling time (under condition that overshoot does not exceed 2%) of the transient caused by the load disturbances.

TABLE II.
PARAMETERS OF THE PID CONTROLLER USED IN THE VENTILATION CONTROL SYSTEM

Control system structure	Controller parameters
With the set point observation	K_{pa}=1.65, K_{ia}=1.45, K_{da}=0.40
	K_{pd}=2.20, K_{id}=3.20, K_{dd}=0.50
Classical	K_p=1.65, K_i=1.45, K_d=0.40

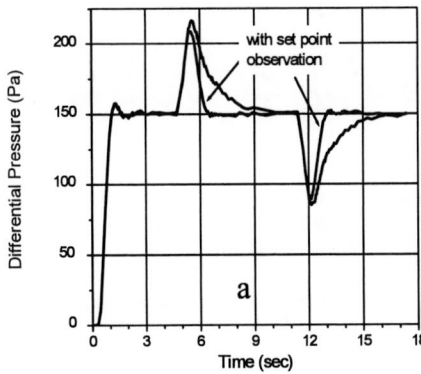

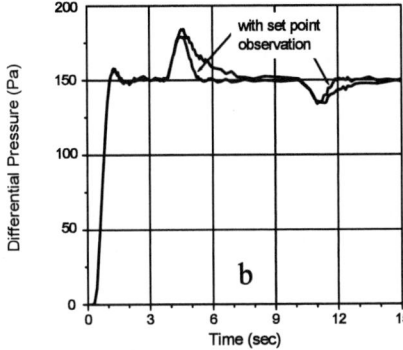

Fig. 5. The differential air pressure set point step response followed by high (a) and low (b) positive and negative load disturbances of the ventilation control system with a classical closed-loop structure and the structure with the set point observation based on the PID controller.

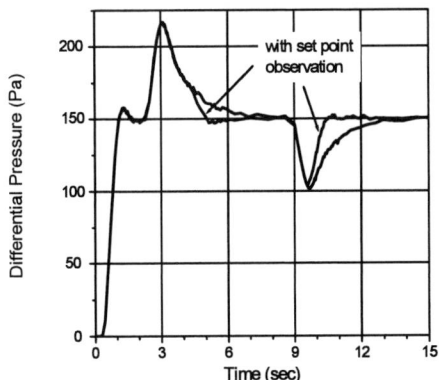

Fig. 6. The differential air pressure set point step response followed by positive and negative load disturbances of the ventilation control system with a classical closed-loop structure and the structure with the set point observation based on the PID controller (the case when positive disturbance affects during the set point change response transient).

V. CONCLUSIONS

1. The introduction of the control system structure with the set point observation as compared to the classical closed-loop structure allows us in the general case to improve the load disturbance rejection of analyzed plants using a PID controller without sacrifice of the transient performance of the set point change response.

2. In the specific case when load disturbance occurs during the set point change response transient, the proposed control system structure does not have advantage in respect of the classical closed-loop structure.

3. The control system structure with the set point observation was employed in the industrial ventilation system with the AC induction motor drive for control of differential air pressure.

ACKNOWLEDGMENT

This work was supported by the Lithuanian State Science and Studies Foundation under High–tech development program project B –13/2008.

REFERENCES

[1] D. Vrancic, J. Kocijan and S. Strmcnik, "Simplified disturbance rejection tuning method for PID controllers," in *Proceedings of the 5th Asian Control Conference*, 2004, pp.492-497.

[2] J. Shi and W. S. Lee, "Set point response and disturbance rejection tradeoff for second-order plus dead time processes," in *Proceedings of the 5th Asian Control Conference*, 2004, pp.881-887.

[3] A. Baskys and V. Zlosnikas, "Asymmetric PI controller for mechatronic systems", *Solid state phenomena*, vol.113, 2006, pp.25-28.

[4] D.H. Kim and J.H. Cho, "Robust tuning for disturbance rejection of PID controller using evolutionary algorithm", in *Proceedings of the IEEE Annual meeting NAFIPS 04*, 2004, vol.1, pp. 248-253.

[5] A. Leva, "Autotuning process controller with enhanced load disturbance rejection", in *Proceedings of the American control conference*, 2004, pp. 1400–1405.

[6] H. Panagopoulos, K.J. Astrom and T. Hagglund, "Design of PID Controllers Based on Constrained Optimisation," *IEE Proc. – Control Theory Appl.*, vol. 149, no. 1, 2002, pp. 32-40.

[7] D. Vrancic, B. Kristiansson and S. Strmcnik, "Reduced MO tuning method for PID controllers," in *Proceedings of the 5th Asian Control Conference*, 2004, pp.460-465.

[8] S. Tavakoli, I. Griffin and P.J. Fleming, "Robust PI controller for load disturbance rejection and set point regulation", in *Proceedings of the 2005 IEEE Conference on Control Applications*, 2005, pp. 1015–1020.

Electropneumatic Servo System with Adaptive Force Controller

Arūnas Grigaitis, Vilius Antanas Geleževičius
Department of Control Technology, Kaunas University of Technology,
Studentų st. 48, LT-51367 Kaunas, Lithuania, e-mail: *vilius.gelezevicius@ktu.lt*

Abstract. Due to good environmental security nature, pneumatic system attracts interest of industrial field. This requires development of modern control methods especially when pneumatic cylinders are used as force generating actuators.

Dynamics of pneumatic cylinder manifests itself by essential nonlinear features defined by initial starting position and pressure values in the working chambers of the cylinder and flow transfer characteristics of the directional valve. These nonlinearities directly influence the force generation process of the pneumatic cylinder.

To ensure required control quality of electropneumatic servo system the multiloop control system method with reference model based signal adaptive force controller is proposed and investigated in this paper.

The strategies of design of the controllers such as velocity and position are presented. The investigation results on changing initial conditions of the electropneumatic servo drive are given.

Keywords – electropneumatic, reference model, multiloop, hierarchical.

I. INTRODUCTION

Owing to their simplicity, reliability and compactness the pneumatic systems play an important role in the several industrial applications such as packaging, transferring, fixing and others. Especially pneumatic systems are effective for cylinder length defined linear motion generation do not requiring piston rod stroke regulation.

In the applications (handlers, packaging devices, robots) where steady state end position of pneumatic cylinder is to be smoothly regulated and the dynamical behaviour of positioning process state variables such as force, velocity and position is to be controlled, the multiloop hierarchical control strategies well developed for electromechanical position control systems can be applied [1,2]. The following control method requires a separate independently adjustable controller for each state variable of positioning process regulation. The good controllability of the state variables of position control system and a possibility to adjust independently the state variable controllers are the essential advantages of the method.

When developing an electropneumatic servo system with pneumatic cylinder, playing the role of actuator of the system, the essential nonlinearity of force generation process is to be considered. As it is shown in [3] the dynamical behaviour of pneumatic cylinder highly depends of initial position of piston and of starting pressures in working chambers of the cylinder.

Due to avoidance of influence of initial starting conditions on quality of force generation process of the cylinder the force control method with adaptive force controller has been proposed [2]. This method gives an excellent efficiency and ensures an invariance of force behaviour on initial control conditions defined by initial piston position and initial pressure in working chambers of pneumatic cylinder.

On the base of adaptive force control system the velocity and position of a cylinder piston rod controlling was developed [2].

Investigation results of electropneumatic servo system are given.

II. DEVELOPMENT OF PNEUMATIC SERVO SYSTEM

A. Development of adaptive force controller

The structure of multiloop electropneumatic control system with an adaptive force controller is presented in Fig. 1. There are force H_F, velocity H_S and position H_{PO} controllers. Accordingly, k_F, k_S and k_{PO} there are force, velocity and position transfer coefficients. H_R is the transfer function of the reference model, and k_R - the transfer coefficient of the reference model. H_V there is the transfer function of the proportional valve.

The force controller was adjusted under the quantitative optimum condition [2] and was expressed by transfer function:

$$H_F(p) = \frac{T_c^* p + 1}{2 k_v k_c^* k_F T_v p} \ , \qquad (1)$$

where T_c^* – the time constant defining force generation process of stopped cylinder in the middle position with equal initial pressures in the both working chambers [4], T_v - the time constant of the spool of proportional directional control valve approximately equal to 0.05s,

978-1-4244-1741-4/08/$25.00 ©2008 IEEE   1144

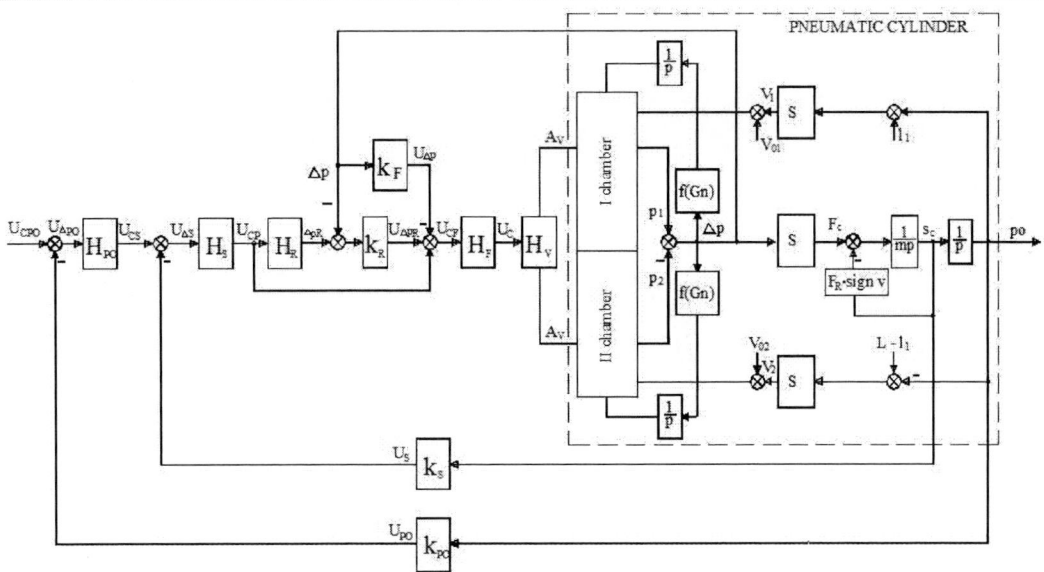

Fig. 1. The structural model of the pneumatic servo system.

$k_C^* = \dfrac{\Delta p}{A_V}$, A_V – orifice area of valve, m^2, k_V - transfer coefficients of the valve. k_F –force controller transfer coefficients.

Supposing the quality of adjusted in such way force control contour as desirable for the whole electropneumatic acting system, the transfer function of reference model for adaptive force control contour was defined as:

$$H_R(p) = \frac{1}{2T_\mu^2 p^2 + 2T_\mu p + 1} \,, \qquad (2)$$

where time constant T_μ approximately equal to T_v = 0.05s.

In order to heighten the receivable signal of the reference model was provided transfer coefficient k_R.

Design and investigation of main force controller of electropneumatic force control system are presented in the paper [3], Model investigations with adaptation contour gives proper results. Dynamic parameters of force control loop were the same as in the reference model and they were stationary Fig. 2. These results demonstrate high efficiency of proposed adaptive force regulation technique. It is distinctly seen that dynamical response curve of pneumatic cylinder developed force provoked by reference signal step mode change corresponds to quantitative optimum condition and does not depend on initial piston position and initial pressures in working chambers of the cylinder.

B. Development of velocity controller

Development of higher hierarchy level controllers of pneumatic servo system such as velocity controller of pneumatic cylinder can be carried out using well known hierarchical system design methods based on quantitative optimum condition fulfilling. Supposing the adaptive force control system being well functioning, the transfer function of the whole force control contour can be taken equal to the transfer function [3] of the reference model (2) $H_F(p) \cong H_R(p)$.

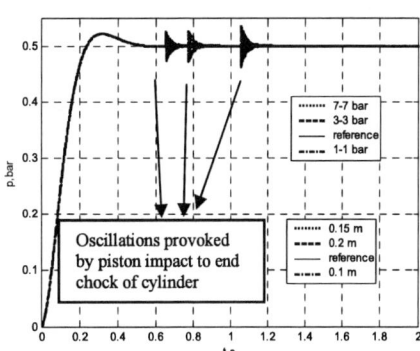

Fig. 2. Dynamical behaviours of pneumatic force control process on different initial conditions.

Approximation this function to the function of the first order, the transfer function of the force regulation system is taken as:

$$H_{FC}(p) = \frac{1}{2T_\mu p + 1}; \qquad (3)$$

Functional diagram of electropneumatic servo system with adaptive force controller and velocity controller is presented in figure 3.

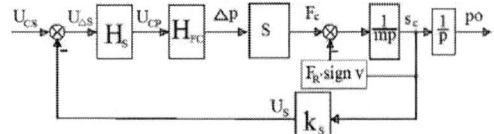

Fig. 3. Functional diagram of electropneumatic servo system with adaptive force controller and velocity controller.

According quantitative optimum condition [1] can be designed the velocity controller of pneumatic servo drive:

$$H_s(p) \cdot H_{FC}(p) \cdot S \cdot \frac{1}{mp} \cdot k_s = \frac{1}{2T_\mu p(T_\mu p + 1)} , \quad (4)$$

where: m – mass of the cylinder, kg; S– orifice area of piston, m^2, T_μ - time constant, s; k_S –velocity controller transfer coefficients.

Supposing the T_μ being equal to $2T_v = 0{,}1s$ and $H_{FC}(p) \cong \dfrac{k_R}{2T_\mu p + 1}$; simple proportional velocity controller will be :

$$H_s(p) = \frac{m}{4k_R \cdot k_s \cdot S \cdot T_\mu} . \quad (5)$$

Modelling results of velocity control system of pneumatic cylinder with adaptive force controller are presented in Fig. 4

These results demonstrate that velocity of cylinder piston do not depend on initial piston position and initial pressures in working chambers of the cylinder.

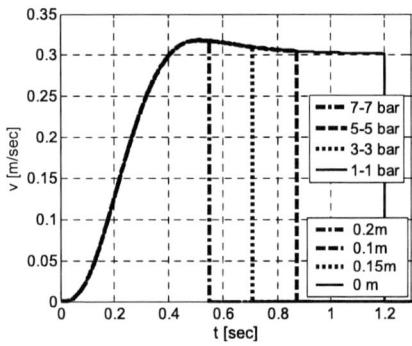

Fig. 4. Piston velocity response curves with different initial pressures in the working chambers and with different initial piston positions.

C. Development of position controller

Controller of the position was designed in same way as velocity controller:

$$H_{PO}(p) \cdot H_S(p) \cdot \frac{1}{p} \cdot k_{PO} = \frac{1}{2T_{\mu P} p(T_{\mu P} p + 1)} . \quad (6)$$

Supposing the T_μ being equal to $4T_V$ and $H_s(p) \cong \dfrac{1/k_S}{4T_R p + 1}$, position controllers will be:

$$H_{PO}(p) = \frac{k_S}{8T_R \cdot k_{PO}} . \quad (7)$$

where: k_{PO} –position controller transfer coefficients.

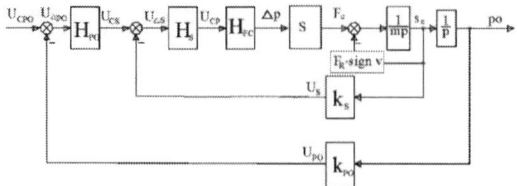

Fig. 5. Functional diagram of electropneumatic position control system with adaptive force controller, velocity controller and position controller.

II. MODELING RESULTS WITH THREE LOOP CONTROL SYSTEM

For practical investigation the electropneumatic force control system consisting of rodless pneumatic cylinder with 25 mm diameter piston and 300 mm stroke and proportional directional control valve MPYE-5-1/8 has been chosen.

Dynamic of the electropneumatic cylinder with three loop control system has been investigated on different initial conditions:

1. The investigation has been carried out when piston of the 0.3 m length cylinder was in different starting positions equal to 0 m, 0.1 m, 0.15 m and 0.2 m, and initial absolute pressure of working chambers was equal to 7 bars. Modelling results are presented in Fig. 6, 7, 8.

2. The investigation has been carried out when piston of the 0.3 m length cylinder was in middle position and initial absolute pressures of working chambers was equal to 1, 3, 5 and 7 bars. Modelling results are presented in Fig. 6, 7, 9.

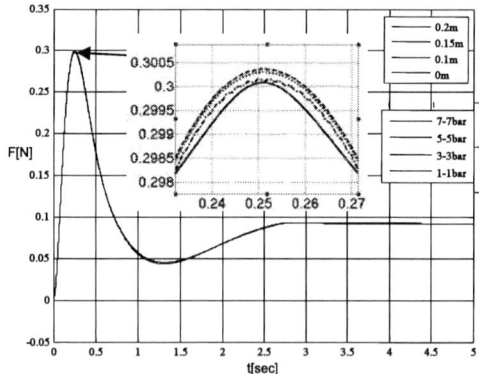

Fig. 6. Response curves of the piston generated force with different initial piston positions and with different initial pressures in working chambers of the cylinder.

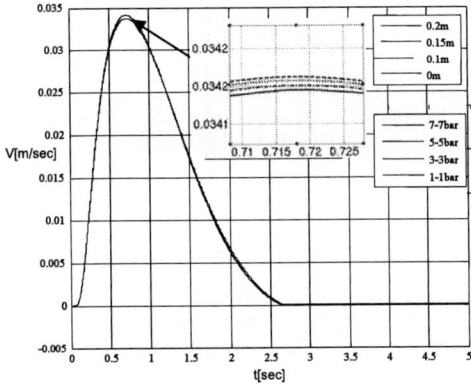

Fig. 7. Response curves of the piston velocity with different initial piston positions and with different initial pressures in working chambers of the cylinder.

Force and velocity response curves of electropneumatic position control system with adaptive force regulator confirm an efficiency of proposed method. It allows eliminate an independence of dynamical quality on either the starting position of the piston or initial pressure in working chambers of the cylinder change. This allows do not take in account of these circumstances when position controller is designed.

Investigation results given in Fig. 6, 7 and 8 confirm correctness of this assumption. The response curves of force, velocity and position of electropneumatic position control system provoked by the step mode position reference signal are identical for several starting conditions of the system.

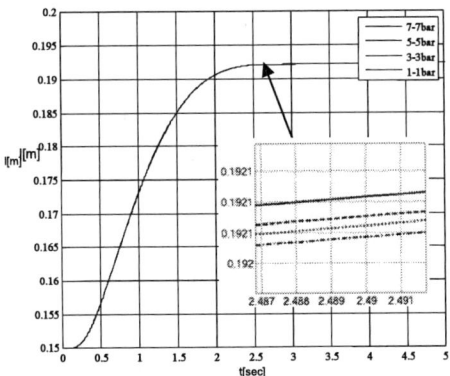

Fig. 8. Response curves of the piston position when initial pressure in cylinder chambers was 1, 3, 5 and 7bar.

The position responses presented in Fig. 9 clearly indicate uniformity of reaction of electropneumatic position control system performing preset displacement, the piston of the cylinder having the 0.3 m stroke and being in several initial positions equal to 0, 0.1, 0.15 and 0.2 m.

With this model were performed investigations with different control voltage.

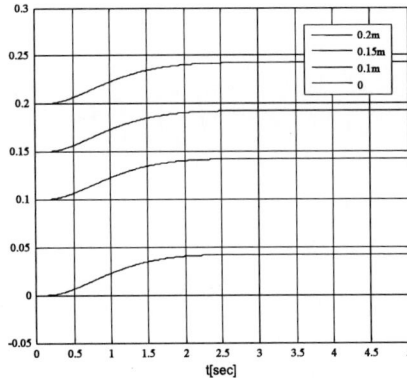

Fig. 9. Response curves of the piston position when piston is in different starting conditions.

With set different control voltage it is possible control piston velocity Fig. 10.

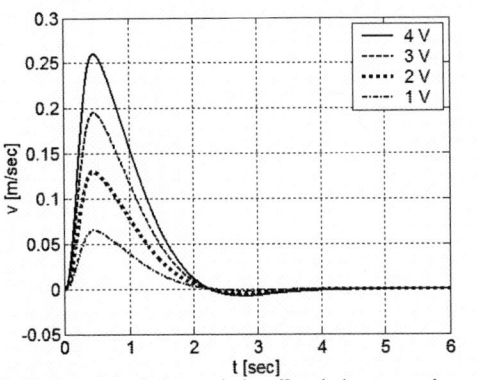

Fig.10. Response of piston velocity effected by step mode control voltage.

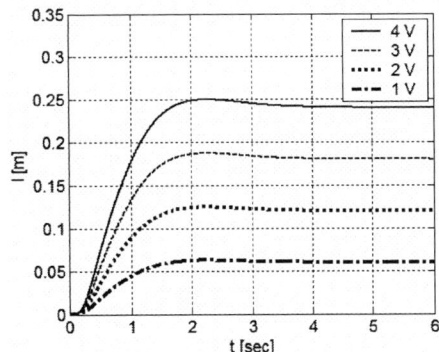

Fig.11. Piston position effected by control voltage.

Fig.11 shows position step responses. Responses for each control voltage are almost the same with that of desired model, which proves an effectiveness of proposed position control system.

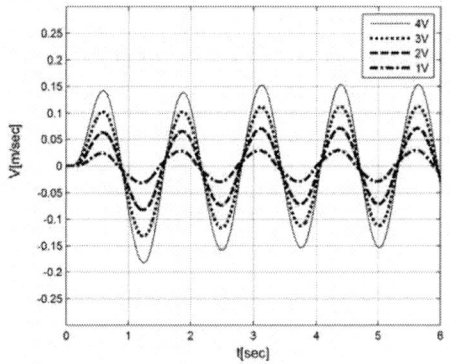

Fig.12. Velocity of the piston affected by sinusoidal control voltage.

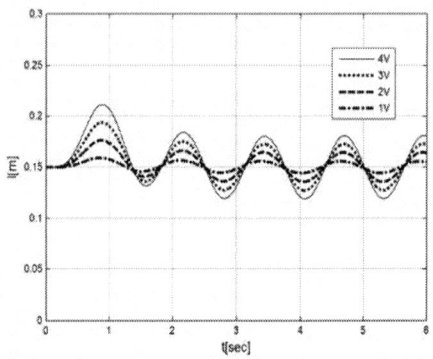

Fig.13. Responses of piston position effected by sinusoidal control voltage.

With sinusoidal control voltage Fig.12, 13 displacements are relatively smooth. With set different control voltage it is possible control piston velocity and position.

CONCLUSIONS

1. The pneumatic cylinder with tree loop (adaptive force, velocity and position) control system has been developed and investigated in this paper. Reference model based signal adaptive control method has been applied for force regulation of the position control system.
2. Force controller was designed for proportional-valve pneumatic actuators, using the detailed mathematical model developed in the first part of this work. Due to the highly nonlinear behavior and inherent uncertainties in modeling of pneumatic systems, was adjusted force controller under the quantitative optimum conditions.
3. The simulation results shown that the proposed force controller was effective to control electro-pneumatic cylinder. Furthermore, adaptive force controller could

adapt the control parameters to control the cylinder with variant starting conditions.
4. The velocity and position controllers of pneumatic servo drive can be designed using conventional methods based on respect of the quantitative optimum conditions, supposing that dynamic parameters of force control contour correspond to parameters of the reference model and are stationary.

REFERENCES

[1] Scavarda S. Les asservissements électro-pneumatiques de position, Paris, Hermès. 1989. 62 p.
[2] Geleževičius V. Elektros pavarų valdymo sistemos/ K. Kriščiūnas, V. Kubilius, Vilnius, Mokslas, 1990.360 p.
[3] A. Grigaitis, V.A. Geleževičius Development of Electropneumatic Servo System with Reference Model Based Signal Adaptive Force Controller // Journal of Vibroenginiering. – Vilnius: Vibromechanika, 2007.-No.4.-P.45-50.
[4] Geleževičius V.A, Grigaitis A. Research of Adaptive Force Control Loop of Electropneumatic Acting System // Electronics and Electrical Engineering. – Kaunas: Technologija, 2007. – No.7(79).-P.7-10.

New fault tolerant DTC control for induction machine drives

A.Ben Abdelghani Bennani[*], M. Ghodbane Cherif[†], I. Slama Belkhodja[†]

[*†]LSE, Laboratoire des Systèmes Electriques, Tunis, Tunisia.
[*]INSAT, Institut National des Sciences Appliquées et de Techonologie, Tunis, Tunisia.
[†]ENIT, Ecole Nationale d'Ingénieurs de Tunis, Tunis, Tunisia.
Afef.benabdelghani@insat.rnu.tn, Ghodbane.meriem@yahoo.fr , Ilhem.belkhodja@enit.rnu.tn

Abstract: **Fault tolerant control for induction machine drives is an interesting issue since it enhances the reliability of the drive without inducing high-cost investment. In this paper, a new induction motor drive control strategy based on DTC approach is proposed. It allows a continuous free operation of the drive even if one inverter leg is completely lost. Simulation results prove the good performances of the proposed drive, not only for the machine variables (flux and torque) but also for the converter power module ratings (voltage, current and switching frequency).**

Keywords : **Induction machine, DTC, Reconfiguration, reliability, continuous operation**

I. INTRODUCTION

Induction machine drives are the most widely used in the adjustable speed drives field. The advantages of this solution are, on one hand, their relatively low cost, simple construction and easy maintenance. On the other hand, progress made in control and algorithm implementation technologies enabled these drives to have high dynamic and static performances.They have reached a high degree of mature and become standard industrial solutions for a wide range of powers and applications (robotics, railway transportation, marine propulsion, wind power generation).

Nowadays, reliability consideration is an important issue for these systems: a fault of one power module the supply chain (rectifier, inverter) may lead the drive to shut down, causing a non-programmed maintenance, usually sensitively more expensive than the programmed ones. For some critical applications, where the system shut down is simply not acceptable, redundancy could be applied increasing the cost, volume and weight of the drive.

Improving reliability of electric motor drives can be achieved by improving the fault tolerance of converter swith modules [1-2] and the fault tolerance of the motor control algorithm. In [3] a Field Oriented fault tolerant Control has been introduced and in [4-6], PWM stategies under fault conditions are proposed. More recently, DTC fault tolerant algorithms are proposed and their performances are studied [6-8].

The work presented in this paper deals with a new DTC algorithm designed for an induction machine when fed with four switch three phase (FSTP) converter. It is used when one power module of the original SSTP six switch three phase is down: the leg of this module is isolated and the control of the converter has to be compatible with the new converter topology.

In the first section, the authors present the FSTP converter topology and its mathematical model when used to feed an induction machine. Then, the effect of one stator voltage vector, under normal operation, on the torque and flux evolutions is reminded in order to define the novel direct torque control strategy compatible with FSTP converter.Finally, simulation results are given and discussed.

II. THE FSTP TOPOLOGY AND MODELLING

Under normal operation, an induction machine drive consists normally of a rectifier and a SSTP (Six Switches Three Phases) Inverter. When a power module fault occurs, the corresponding leg is isolated and the corresponding machine phase is connected to the mid point of the DC input of the inverter (see Fig. 1). The machine is consequently fed by an FSTP (Four Switches Three Phases) inverter.

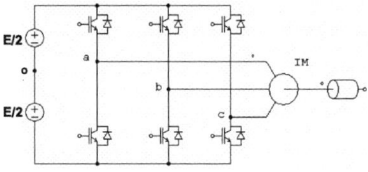

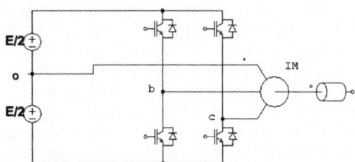

Fig. 1 Converter Topology. Up: healthy operation (SSTP). Down: After faulty leg isolation (FSTP)

Under these conditions, the expressions of the Line To Neutral voltages of the induction machine are given by equation (1).

$$\begin{pmatrix} V_{an} \\ V_{bn} \\ V_{cn} \end{pmatrix} = \frac{E}{6}\begin{pmatrix} -2 & -2 \\ 4 & -2 \\ -2 & 4 \end{pmatrix}\begin{pmatrix} C_2 \\ C_3 \end{pmatrix} + \frac{E}{6}\begin{pmatrix} 2 \\ -1 \\ -1 \end{pmatrix} \qquad (1)$$

where C2 and C3 are the control signals of phase b and phase c, respectively.

In this paper, The DC bus is modeled by two perfect voltage sources.

Within the (α,β) frame this equation leads to equation (2)

$$\begin{cases} V_{S\alpha} = \dfrac{E}{2}\left[-C_2 - C_3 + 1\right] \\ V_{S\beta} = \dfrac{\sqrt{3}E}{2}\left[C_2 - C_3\right] \end{cases} \qquad (2)$$

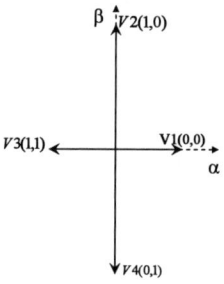

Fig. 2: FSTP converter voltage vectors

Consequently, only four active vectors are available for the FSTP converter: V1 (0,0), V2(0,1), V3(1,1)and V4(1,0). Fig. 2 shows the disposition of these vectors in the (α,β) frame.

TABLE I
FSTP CONVERTER VOLTAGE VECTORS

	C2	C3	Vs
V1	0	0	$\dfrac{E}{3}$
V2	1	0	$j\dfrac{\sqrt{3}E}{2}$
V3	1	1	$-\dfrac{E}{3}$
V4	0	1	$-j\dfrac{\sqrt{3}E}{2}$

III. EFFECT OF A STATOR VECTOR ON TORQUE ANDE STATOR FLUX

If we neglect the stotoric resisitance, stator voltage can be written as follows:

$$\underline{V}_s \approx \frac{d\underline{\phi}_s}{dt} \qquad (3)$$

So, stator flux variation between instants t1 and t2 is given by:

$$\Delta\underline{\phi}_s = \underline{\phi}_s(t_2) - \underline{\phi}_s(t_1) \approx \underline{V}_s \Delta T \qquad (4)$$

Knowing the four voltage vectors the converter can deliver, it is possible to depalce the stator flux vector in the (α,β) frame according to these four directions.

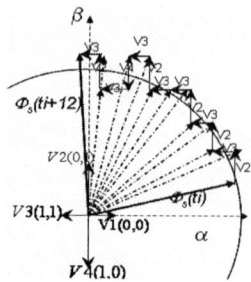

Fig. 3 Effect of a stator voltage on a the stator flux evolution

For the electromagntic torque, *Tem*, it is proportional to $\sin\delta$, where δ is the angle between the rotor flux vector, $\underline{\phi}_r$, and the stator flux vector, $\underline{\phi}_s$. One can make the approximation of neglecting rotor flux variation if compared with the stator flux one. This approximation is function of the machine speed and will be discussed within the next paragraph. So, if the stator flux angle is increased (respectively decreased), $\sin\delta$ increases (respectively decreases) and *Tem* increases (respectively dereases).

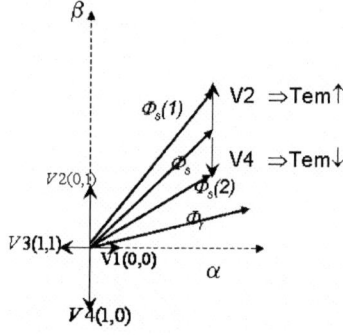

Fig. 4 Effect of a stator voltage on a the electromagnetic torque evolution

Like classic DTC approach [9], we propose, in this paper, to use a hysteresis control of both torque and stator flux. Instantaneous torque and stator flux errors are calculated and introduced to one band hysteresis regulators. 2 level (0,1) parameters, called ϕ and τ are then generated. θ identifies the angular sector of stator flux vector within (α,β) frame. A look-up table, with ϕ, τ and θ as inputs, selects the voltage vector, between the four presented above, to be applied to the machine.

IV. NEW SECTOR DEFINITION AND SWITCHING TABLE

In order to establish the switching table, Table II shows the voltage effect of each of the four voltage vectors depending on the angular sector (see Fig. 5)of stator flux.

1150

TABLE II
LOOK UP TABLE OF THE ELECTROMAGNETIC TORQUE

	θ (Sectors)			
	1	II	III	IV
V1	-	-	+	+
V2	+	-	-	+
V3	+	+	-	-
V4	-	+	+	-

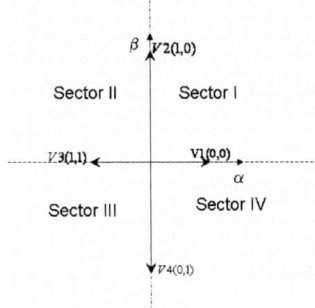

Fig. 5. (α,β) frame sectors for the torque look up table

Similarly, rigorous analysis of the four stator voltages and their effect on the stator flux evolution leads to define eight sectors in the (α,β) frame (figure 5) and the look up table shown on table III.

TABLE III
LOOK UP TABLE FOR THE STATOR FLUX

	θ (Sectors)							
	1	II	III	IV	V	VI	VII	VIII
V1	+	+	+	-	-	-	+	+
V2	+	+	+	+	+	-	-	-
V3	-	-	+	+	+	+	+	-
V4	+	-	-	-	+	+	+	+

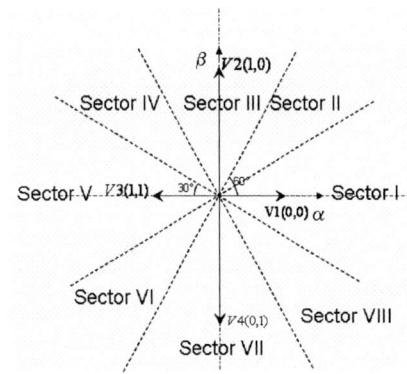

Fig. 6. (α,β) frame sectors for the stator flux look up table

Consequently, to define the DTC switching table, we have to combine between table II and table III. This involves the use of the twelve sectors shown on Fig. 7

The final switching table is shown on Table IV. Some configurations lead to impossible state (noted X): For example, when $\phi=0$, $\tau=0$ and $\theta=1$, any of the four stator voltages could imply decrease of torque and stator flux simultaneously. In this case a priority order between ϕ and τ must be defined: if the torque error is greater than the flux one, τ will be kept at 0 and ϕ will be changed to 1; otherwise, ϕ will be kept at 0 and τ will be changed to 1.

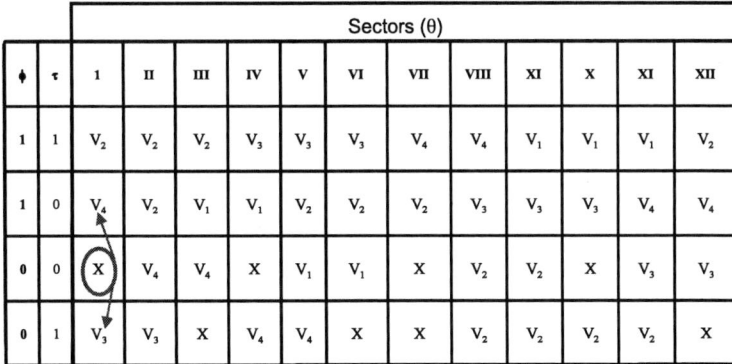

ϕ	τ	Sectors (θ)											
		1	II	III	IV	V	VI	VII	VIII	XI	X	XI	XII
1	1	V_2	V_2	V_2	V_3	V_3	V_3	V_4	V_4	V_1	V_1	V_1	V_2
1	0	V_4	V_2	V_1	V_1	V_2	V_2	V_2	V_3	V_3	V_3	V_4	V_4
0	0	X	V_4	V_4	X	V_1	V_1	X	V_2	V_2	X	V_3	V_3
0	1	V_3	V_3	X	V_4	V_4	X	X	V_2	V_2	V_2	V_2	X

TABLE IV
LOOK UP TABLE FOR THE STATOR FLUX

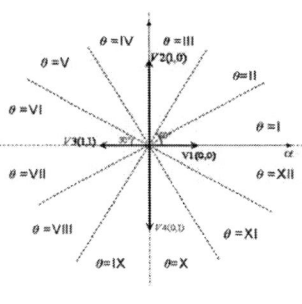

Fig. 7. (α,β) frame sectors for the proposed DTC strategy

IV. SIMULATION RESULTS

The proposed strategy is tested with a 4 pole, 1.5 kW induction machine and a 25μs sampling period. The parameters used in the simulation are: DC bus= 540V, Torque hysteresis band= 10%, Flux hysteresis band = 10%.

The simulations are carried out for a torque and a stator flux references equal to their nominal values.Fig.8 and 9 show that electromagnetic torque and the stator flux ripples are kept in their hysteresis bands before the IGBT fault. Under Fault tolerant operation ,the flux is kept within its hysteresis band. However, even the electromagnetic torque continues following its reference, it exceeds its hysteresis band. We have to notice that through these simulation tests, the global efficiency of the proposed strategy is demonstrated. Under experimental conditions, it is obvious that a faulty system wouldn't be under for nominal conditions (torque, speed and therefore power).

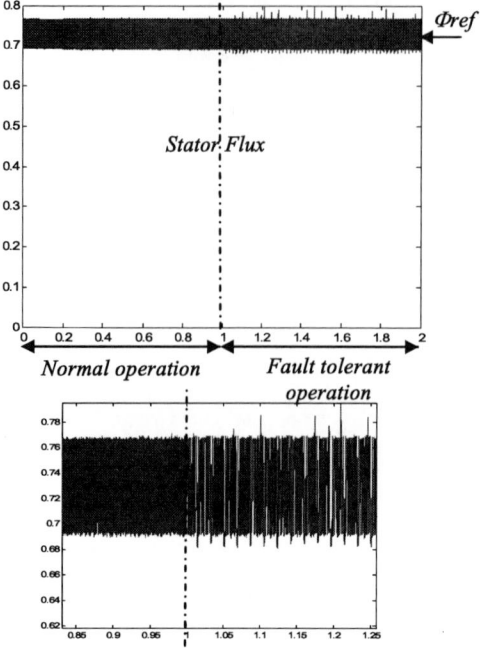

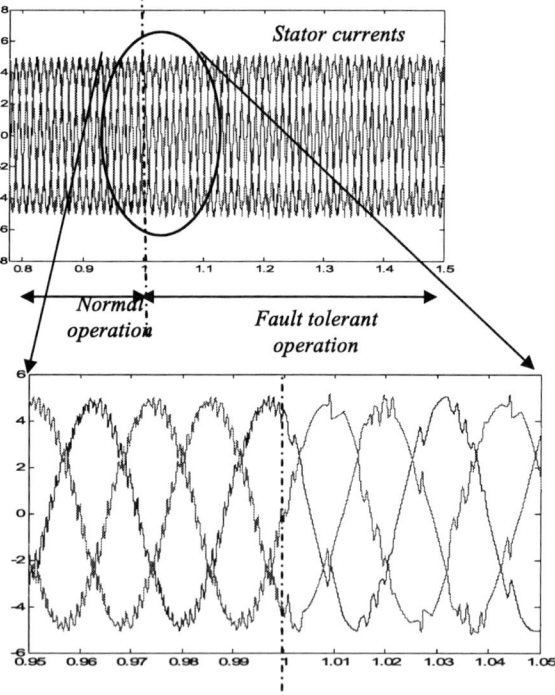

Fig. 9 Stator flux with the proposed strategy

Fig.10 gives the three stator currents. Under Fault tolerant operation, the three stator current are equibrated. We can notice that the switching frequency is affected. This is

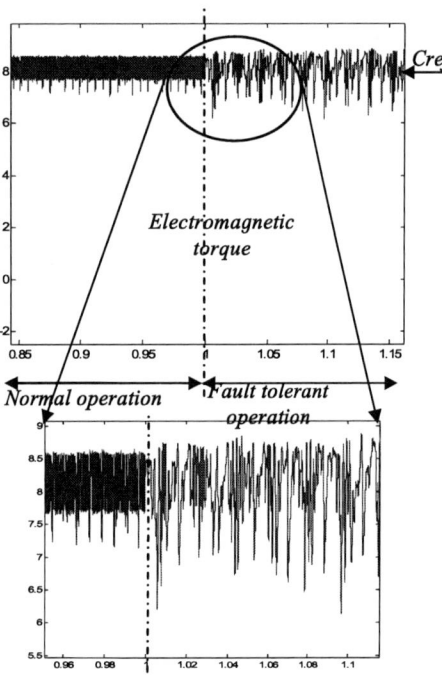

Fig. 8 Torque with the proposed strategy

Fig. 10. Stator Currents with the proposed strategy

clearer on the figure 13 a and figure 13. b.

Figure.11 shows the performances of the proposed DTC strategy when an inversion of the shaft speed occurs.

Like DTC strategies, the electromgnetic torque and the stator flux are not affected by this inversion.

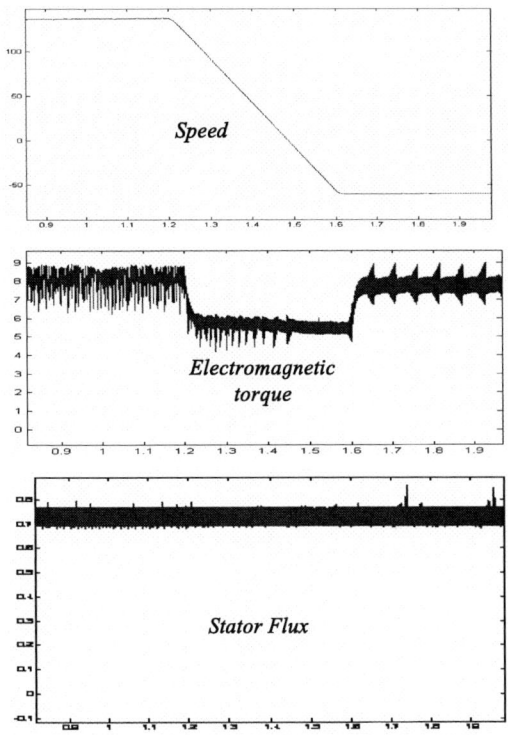

Fig. 11. Influence of the shaft speed on the electromagnetic torque and the *Stator Flux*

Figure 12 gives the torque spectra under normal operation and fault tolerant operation. We can notice this spectrum is not affected after an IGBT fault occurrence.

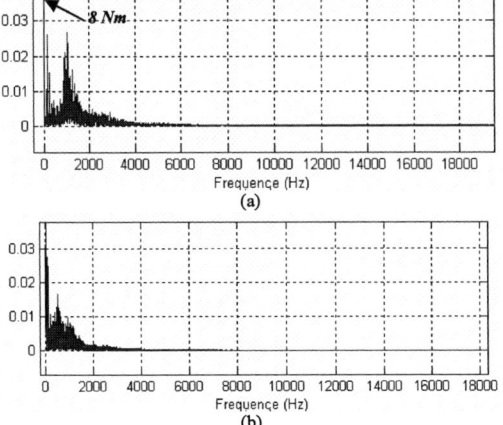

Fig. 12. , torque spectra: (a): normal operation, (b) Fault tolerant operation

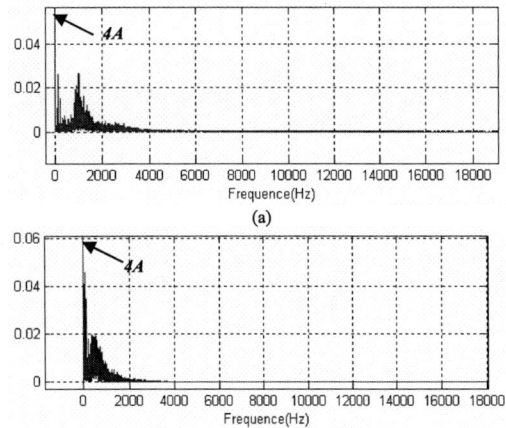

Fig. 13. , current spectra (a): normal operation, (b) Fault tolerant operation

V. CONCLUSION

In this paper, we have presented a novel DTC strategy adpted for induction machine when one leg of the the drive power converter breakes. This strategy consists on isolting the faulty leg and connecting the corresponding phae machine to the middle point of the DC bus of the inverter.

A model of this new inverter-machine association is determined to obtain the active vectors of the system in the (α, β) frame.

A new (α, β) frame sectorisation is proposed in order to optimize the use of the active vectors for the control of the machine: A new DTC strategy is then described and the simulation results prove its performances when used to improve the system availaility.

ACKNOWLEDGMENT

The work presented within this paper was partially supported by CMCU Project n°07 S 1119 "Traction Ferroviaire électrique: Continuité de service de la motorisation"

REFERENCES

[1] R.L. Ribeiro, C. B. Jacobina, E.R.C. da Silva, A.M.N. Lima "Fault-tolerant voltage-fed PWM inverter AC motor drive systems", IEEE transactions On Industrial Electronics, Vol.51, NO.2, April 2004.

[2] M. B. Corrêa, C. B. Jacobina, E. R. Cabral da SILVA, A. M. N. Lima, "An Induction Motor Drive System with Improved Fault Tolerance", IEEE Trans. On Industry Applications, Vol.37, NO.3, May/June 2001.

[3] T.-H Liu, J-R Fu, T. A. Lipo "A strategy for improving reliability of field-oriented controlled induction motor drives" IEEE Trans. On Industry Applications, Vol.29, NO.5, Sept/Oct 1993.

[4] J. Klima, "Analytical investigation of a new space-vector PWM strategy for improved induction motor

drives reliability", 10th European Conference on Power Electronics and Applications, EPE 2003, Toulouse.

[5] J. Klima, "Space-vector PWM inverter fed induction motor drive under fault conditions" Electromotion' 2001 – Pologne.

[6] A.L.Orille, M. Azab, "Implementation of direct torque control of induction motor drive using four switch three phase inverter", The 9th European Conference on Power Electronics and Applications, EPE 2001.

[7] M. Azab, A.L. Orille "Novel flux and torque control of induction motor drive using four switch three phase inverter" The 27th Annual Conference of the IEEE Industrial Electronics Society, IECON'01

[8] B. El Basdi, A. Masmoudi, "A New DTC Strategy Dedicated to the Control of four- swith three phase inverter fed induction motor drives", International Conference On ecologic Vehicles and Renewable energies, EVER'2007, Monaco.

[9] I. Takahashi, T. Noguchi "A new Quick response and high efficiency control strategy of an induction motor" IEEE transactionsOn Industry Applications, Vol I, 1986

Stability Analysis of the Natural Field Orientation Controlled Induction Machine Drive

G. Mirzaeva[†] and A. Rojas[*]

School of Electrical Engineering and Computer Science
The University of Newcastle, Callaghan, 2308 NSW, Australia
Phone: +61 (2) 49215963[†]; +61 (2) 49216023[*]
Email: Galina.Mirzaeva@newcastle.edu.au[†]; Alejandro.Rojas@newcastle.edu.au[*]

KEYWORDS

Vector control, Variable speed drive, Control of drive, Induction motor.

ABSTRACT

Natural Field Orientation (NFO) Control proposed in the 1980's is an alternative low cost control method for induction machines. The NFO Control has shown to have some stability issues in regeneration which have been partially addressed in the previous papers. In the present paper a linearised stability analysis is presented for the motor/drive system based on both the non-augmented and the augmented versions of the NFO control. Based on the results of this analysis, the limitations of the augmentation strategy are explored and conditions are identified when it does not ensure stability. Theoretical results are confirmed by simulations and experiments. Directions of further improvement of the NFO stability are outlined.

INTRODUCTION

Natural Field Orientation (NFO) is a patented control strategy proposed in 1980's ([1], [2]). It belongs to the Stator Flux Orientation (SFO) control family and is primarily distinguished by that the magnitude of the stator flux vector is not obtained by integration but is assumed to be equal to its reference value. Consider the stator voltage equation in the stator flux oriented reference frame (denoted by the index "ψ_s"):

$$\underline{e}_{s\psi_s} = \underline{u}_{s\psi_s} - R_s \underline{i}_{s\psi_s} = \frac{d|\psi_s|}{dt} + j\omega_{ms}|\psi_s| \qquad (1)$$

where $|\psi_s|$ is the magnitude of the stator flux vector $\underline{\psi}_s$; R_s is stator resistance; ω_{ms} is angular velocity of the stator flux vector $\underline{\psi}_s$. Splitting equation (1) into real and imaginary parts yields:

$$e_{sx} = u_{sx} - R_s i_{sx} = \frac{d|\psi_s|}{dt} = L_m \frac{d|i_{ms}|}{dt} \qquad (2)$$

$$e_{sy} = u_{sy} - R_s i_{sy} = \omega_{ms}|\psi_s| = \omega_{ms} L_m |i_{ms}| \qquad (3)$$

where x-axis is aligned with and y-axis is in quadrature to the vector $\underline{\psi}_s$; L_m is magnetising inductance; $|i_{ms}|$ is the magnitude of the stator magnetising current.

The magnitude of the stator flux vector can be obtained from (2) by integrating the estimated value of e_{sx}. The angular velocity of the stator flux vector can then be found from (3). The NFO control technique suggests to exclude the flux integrator and to assume that $|\psi_s|$ equals to its reference value $|\psi_s^*|$. This assumption results in elimination of the problems commonly associated with the flux integrator (see, for example, [3]), and in acquisition of some new properties as compared to the traditional SFO control (hence a separate name for the NFO control technique).

The property of "natural" frame orientation was explained in [5]. The true position of the stator flux vector and the associated (x, y) reference frame is not known to the algorithm. It uses the estimated stator flux angular position and associates with it a reference frame that we will call (d, q), or the control frame. If the control frame deviates from the true frame than an inplicit mechanism "naturally" present in the NFO algorithm would act to compensate for such misalignment.

It was also shown in [5] that for the original version of the NFO algorithm the natural self-alignment mechanism is limited to the motoring mode of operation only. Under certain conditions in regeneration, in the case of a small initial misalignment, the control frame would drift further away from the true frame and the stability of the algorithm would be lost. An augmentation to the original algorithm was proposed in [5] and [6] that overcomes this limitation (to a certain extent) and ensures stable operation (in the steady state sense only) in regeneration with limited torque.

Another desirable property of the NFO algorithm is its high tolerance of the parameter errors. The only two machine parameters that it needs to know are the stator resistance R_s and the magnetising inductance L_m. It was shown in [4] that the original and, especially, the augmented NFO versions are very little sensitive to the accurate knowledge of both parameters. The NFO algorithm compensates for the parameter errors by introducing a small angular misalignment between the control and the true reference frames, which is usually very small and does not compromise the stable operation.

The purpose of this paper is to present a more comprehensive stability analysis of the NFO-controlled induction machine including the dynamic effects. Moreover, the analysis will include not only the frame angular error but the close-loop

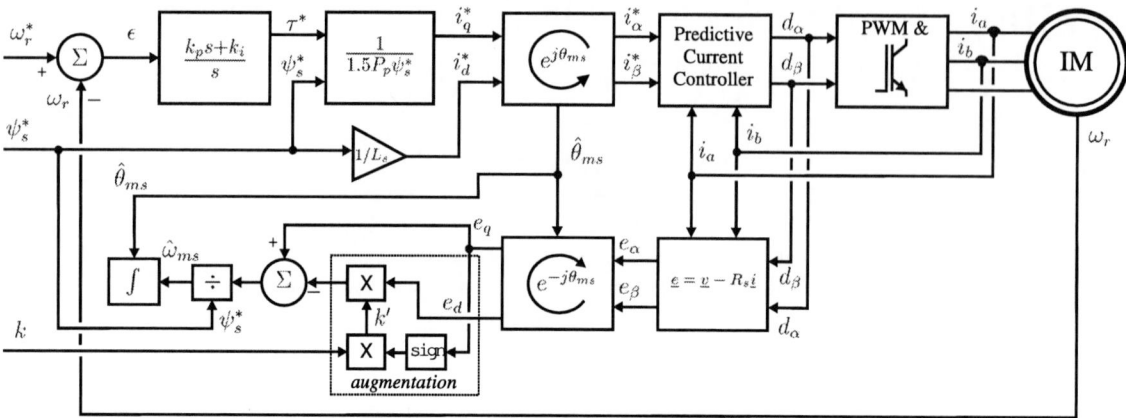

Fig. 1: Block diagram of the NFO algorithm

operation of the whole system from the speed controller to the mechanical load. The primary motivation of this detailed analysis is to explore the limits of the proposed stabilisation strategy in order to:

- conclude whether or not the NFO stability problem, in principle, can be fixed using the proposed augmentation strategy, or other solutions should be sought for;
- make sure that all the dynamic effects possibly affecting stability are included in the analysis.

The remainder of this synopsis is organised as follows. First, a model of the motor / drive system based on NFO control is developed. Small signal linearisation is then performed for the non-augmented (basic) and the augmented versions of the NFO algorithm. Following this, stability analysis based on the root locus method is presented for the linearised model. Simulation and experimental results are included to support the main theoretical conclusions. Finally, conclusions are made and the main contributions of the paper are listed together with an outline of future work.

MODEL DEVELOPMENT

The block diagram of Fig. 1 further assists in understanding of the NFO control algorithm. Indices α, β in this figure denote variables in the stationary reference frame. Indices d, q, as defined previously, refer to the rotating frame aligned with the estimated position of the stator flux vector (or the control frame). Conversion from one frame to another is performed via the estimated angular position $\hat{\theta}_{ms}$ of the stator flux vector, which is obtained by integrating the angular velocity $\hat{\omega}_{ms}$.

It was shown in [7] that current control of the inverter on both d and q axes is the preferred control option for the NFO implementation. It is only with the current-controlled inverter that the proposed augmentation strategy can be implemented. In Fig. 1 one can see the predictive current control loops around the inverter on both axes.

The optional augmentation branch enclosed in Fig. 1 in a dotted line box is based on the following idea. In the case

that the stator flux magnitude is constant or changes slowly then according to (2) the direct axis back emf e_{sx} should be close to zero. If the true and control frames are coincident, the correct stator resistance value is used and the stator voltages and currents are measured accurately, then the estimated direct axis back emf $e_d = u_d - R_s i_d$ equals to e_{sx} and is also close to zero.

In fact, the main reasons why e_d can notably differ from zero are the frame misalignment and the stator resistance error. Leaving the R_s error aside for a moment, one can use the non-zero e_d value as the indicator of the frame misalignment to adjust the position of the control frame. This can be done via the following augmentation of the frame angular velocity estimation:

$$\hat{\omega}_{ms} = \frac{e_q - k' e_d}{\psi_s^*} \qquad (4)$$

where $k' = k\, sign(e_q)$, the factor $sign(e_q)$ takes into account the direction of mechanical rotation and $k > 0$ is a tunable gain parameter. The augmentation (4) was first proposed in [5]. With relation to the R_s error, it was shown in [4] that the augmentation (4) also improves the NFO tolerance to this commonly problematic error.

Note that if $k = 0$ then one has the original (i.e. non-augmented) NFO algorithm. The further model development will be based on the augmented NFO version with the original NFO algorithm being its special case at $k = 0$. For the speed loop analysis it can be assumed that the inner current control loop is very fast and ensures that the reference currents are followed precisely: $i_\alpha = i_\alpha^*$ and $i_\beta = i_\beta^*$. Assume that angular position of the stator flux vector is estimated with an error θ_e relative to its actual position, i.e. that $\hat{\theta}_{ms} = \theta_{ms} + \theta_e$. Then it follows that the stator currents in the true reference frame (x, y) relate to those defined in the control frame (d, q) as per:

$$i_{sx} = i_d^* \cos\theta_e - i_q^* \sin\theta_e \qquad (5)$$

$$i_{sy} = i_d^* \sin\theta_e + i_q^* \cos\theta_e \qquad (6)$$

Likewise, given that the stator currents and voltages are measured accurately in the stationary frame and the correct R_s value is used, the back emfs estimated in the control and the actual back emfs defined in the true frame have the following relation:

$$e_d = e_{sx}\cos\theta_e + e_{sy}\sin\theta_e \qquad (7)$$

$$e_q = -e_{sx}\sin\theta_e + e_{sy}\cos\theta_e \qquad (8)$$

Using equation (6), electromagnetic torque can be expressed as:

$$\tau_{em} = \frac{3}{2}P_p L_m |i_{ms}| \left(i_d^*\sin\theta_e + i_q^*\cos\theta_e\right) \qquad (9)$$

To obtain a more compact torque expression, the following notations have been introduced: τ_0^* is the reference torque for the case that $i_q^* = i_d^*$; $x = \tau_{em}^*/\tau_0^* = i_q^*/i_d^*$ is the normalised reference torque or, equally, the ratio of the current reference values; $\tau_n = \tau_{em}/\tau_0^*$ is the normalised actual torque; and $y = L_m|i_{ms}|/(L_s i_d^*)$ is the normalised stator flux magnitude. Adopting these notations one obtains:

$$\tau_n = y(\sin\theta_e + x\cos\theta_e) \qquad (10)$$

Using (2) and (3), equations (7) and (8) transform into:

$$e_d = L_m\frac{d|i_{ms}|}{dt}\cos\theta_e + \omega_{ms}L_m|i_{ms}|\sin\theta_e \qquad (11)$$

$$e_q = -L_m\frac{d|i_{ms}|}{dt}\sin\theta_e + \omega_{ms}L_m|i_{ms}|\cos\theta_e \qquad (12)$$

To find the expression for $L_m|i_{ms}|$ (and consequently, for y) we will use the rotor voltage equations in the stator flux frame, which real and imaginary parts can be manipulated into (see, for example, [3]):

$$L_m\frac{d|i_{ms}|}{dt} + \frac{L_m}{T_r}|i_{ms}| = \sigma L_s\frac{di_{sx}}{dt} + \frac{L_s}{T_r}i_{sx} - \sigma L_s\omega_{sl}i_{sy} \qquad (13)$$

$$\omega_{sl}\left(L_m|i_{ms}| - \sigma L_s i_{sx}\right) = \sigma L_s\frac{di_{sy}}{dt} + \frac{L_s}{T_r}i_{sy} \qquad (14)$$

where L_m and L_s are the magnetising and the stator inductances respectively; T_r is the rotor time constant; ω_{sl} is the slip frequency of the rotor with relation to the stator flux vector; and σ is the machine leakage constant.

Substituting expressions for i_{sx}, i_{sy} and their derivatives from (5) and (6), and using the previously introduced normalised variables x and y, equations (13) and (14) can be manipulated into:

$$\frac{dy}{dt} + \frac{1}{T_r}y = \frac{1}{T_r}\left(\cos\theta_e - x\sin\theta_e\right)$$
$$- \sigma\frac{d\theta_e}{dt}\left(\sin\theta_e + x\cos\theta_e\right) - \sigma\frac{dx}{dt}\sin\theta_e \qquad (15)$$
$$- \sigma\omega_{sl}\left(\sin\theta_e + x\cos\theta_e\right)$$

$$\omega_{sl} = \qquad (16)$$
$$\frac{\frac{1}{T_r}\left(\sin\theta_e + x\cos\theta_e\right) + \sigma\frac{d\theta_e}{dt}\left(\cos\theta_e - x\sin\theta_e\right) + \sigma\frac{dx}{dt}\cos\theta_e}{y - \sigma\left(\cos\theta_e - x\sin\theta_e\right)}$$

Now, the angular misalignment between the true and the control reference frames can be expressed as follows:

$$\frac{d\theta_e}{dt} = \omega_e = \hat{\omega}_{ms} - \omega_{ms} = \hat{\omega}_{ms} - P_p\omega_r - \omega_{sl} \qquad (17)$$

where ω_e is error in the frame angular velocity estimation; ω_r is the rotor angular velocity; and P_p is the number of pole pairs. Using the augmented estimation (4) of the frame angular velocity $\hat{\omega}_{ms}$, together with expressions (11) and (12) for e_d and e_q, one obtains:

$$\hat{\omega}_{ms} = \omega_{ms}y\left(\cos\theta_e - k'\sin\theta_e\right) - \frac{dy}{dt}\left(\sin\theta_e + k'\cos\theta_e\right) \qquad (18)$$

substituting which into the angular misalignment equation (17) yields:

$$\frac{d\theta_e}{dt} = \omega_e = \left(P_p\omega_r + \omega_{sl}\right)\left[y\left(\cos\theta_e - k'\sin\theta_e\right) - 1\right]$$
$$- \frac{dy}{dt}\left(\sin\theta_e + k'\cos\theta_e\right) \qquad (19)$$

Finally, to complete the system description, the equation of mechanical rotation is added:

$$\frac{d\omega_r}{dt} = \frac{\tau_0^*}{J}\tau_n \qquad (20)$$

where J is the total mechanical intertia of the machine and the load. Other load torque components not related to the intertia are treated as a disturbance and are not included in the equation (20). This simplified load model can be replaced, if needed, by any other practical load model.

Equations (10), (15), (16), (19) and (20) fully characterise the dynamic behaviour of the NFO controlled induction machine. A separate equation (16) for ω_{sl} is kept for the sake of shorter expressions. It will be substituted into (15) and (19) when practical, and thus the variable ω_{sl} will be eliminated. Apparently, the system is non-linear. For an easier understanding of its structure, the system equations can be written in the following way:

$$\frac{J}{\tau_0^*}\frac{d\omega_r}{dt} = \tau_n = f_1\left(x, y, \theta_e\right) \qquad (21)$$

$$\frac{dy}{dt} = f_2\left(\frac{d\theta_e}{dt}, \frac{dx}{dt}, \theta_e, x, y\right) \qquad (22)$$

$$\frac{d\theta_e}{dt} = \omega_e = f_3\left(\frac{dy}{dt}, \frac{dx}{dt}, y, x, \omega_r, \theta_e\right) \qquad (23)$$

where functions f_1, f_2 and f_3 are defined, respectively, in (10), (15) and (19). A block diagram of the control system under the NFO control is shown in Fig. 2. One can see from Fig. 2 that the structure of the control system is straightforward, except for a complicated non-linear feedback present in the model of the NFO algorithm.

It would be tempting to describe the system in terms of a general state variable model, with ω_r^* as its input; ω_r as its output; and x, y, ω_r and θ_e as its state variables. However, we will refrain from using these terms as the classical definition of non-linear state equations (see, for example, [8]) requires

1157

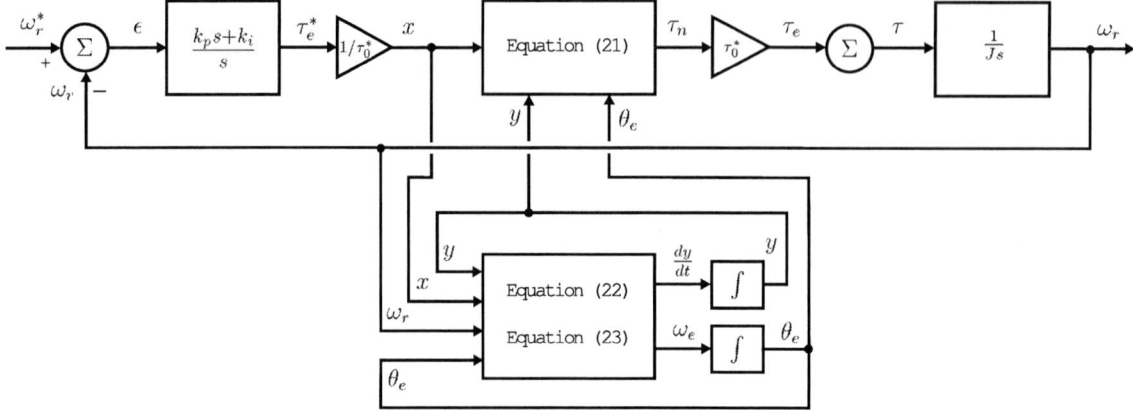

Fig. 2: Block diagram of the control system under the NFO control

that no derivatives are present in their right-hand side. We would rather say that the variables x, y, ω_r and θ_e are like the state variables and equations (21)-(23) are similar to the state equations of the system.

SMALL SIGNAL LINEARISATION AND STABILITY ANALYSIS

Based on the model developed in the previous section, we will now perform a small signal linearisation of the system equations and develop linearised transfer functions of the system. The following transfer functions will be of particular interest:

- Transfer function from the commanded torque τ_{em}^* to the actual torque τ_{em}, or, equally, between their normalised values x and τ_n. This transfer function will determine the behaviour of the NFO algorithm (plant) itself, without the PI controller. The model of the plant will include its own internal feedback from the output variable ω_r.
- Transfer function from the commanded angular velocity ω_r^* to the output angular velocity ω_r. This transfer function will include the contribution of the PI controller and will determine the close loop behaviour of the whole system under the NFO control.
- Transfer function from the angular velocity error $\epsilon = \omega_r^* - \omega_r$ to the output angular velocity ω_r. This open loop transfer function is practical to use with the Root Locus method of stability analysis.

Applying small signal linearisation, in the neighborhood of a chosen operation point, to non-linear equations (21)-(23) results in the expressions of the following form:

$$\delta\tau_n = \alpha_1\delta x + \alpha_2\delta y + \alpha_3\delta\theta_e \qquad (24)$$

$$\delta\dot{y} = \alpha_4\delta\dot{\theta}_e + \alpha_5\delta\dot{x} + \alpha_6\delta\theta_e + \alpha_7\delta x + \alpha_8\delta y \qquad (25)$$

$$\delta\dot{\theta}_e = \alpha_9\delta\dot{y} + \alpha_{10}\delta\dot{x} + \alpha_{11}\delta y + \alpha_{12}\delta x + \alpha_{13}\delta\omega_r + \alpha_{14}\delta\theta_e \qquad (26)$$

where α_1-α_{14} are linearisation coefficients dependent on the operating point. By applying Laplace transform to expressions (24)-(26) and manipulating to exclude $\delta y(s)$ and $\delta\theta_e(s)$, one obtains the following transfer function for $\delta\tau_n(s)$ with relation to $\delta x(s)$:

$$\frac{\delta\tau_n(s)}{\delta x(s)} = \frac{Js\left(a_2 s^2 + a_1 s + a_0\right)}{\tau_0^*\left(b_3 s^3 + b_2 s^2 + b_1 s + b_0\right)} \qquad (27)$$

where a_0-a_2 and b_0-b_3 are bulky expressions in terms of α_1-α_{14} omitted here for the sake of space. Now, based on the control structure of Fig. 2 the open loop transfer function of the system can be found as:

$$\frac{\delta\omega_r(s)}{\delta\epsilon(s)} = \frac{k_p}{\tau_0^*}\frac{\left(a_2 s^2 + a_1 s + a_0\right)\left(s + k_i/k_p\right)}{s\left(b_3 s^3 + b_2 s^2 + b_1 s + b_0\right)} \qquad (28)$$

where k_p and k_i are parameters of the PI controller. Finally, the close loop transfer function of the system is given by:

$$\frac{\delta\omega_r(s)}{\delta\omega_r^*(s)} = \frac{\frac{k_p}{\tau_0^*}\left(a_2 s^2 + a_1 s + a_0\right)\left(s + k_i/k_p\right)}{\frac{k_p}{\tau_0^*}\left(a_2 s^2 + a_1 s + a_0\right)\left(s + k_i/k_p\right) + \\ \overline{+ s\left(b_3 s^3 + b_2 s^2 + b_1 s + b_0\right)}} \qquad (29)$$

These transfer functions will be now examined for stability using classical tools such as Evans's root locus method and analysis of polynomial coefficients of the transfer function.

A. Non-augmented NFO

Firstly, we study the stability of the non-augmented NFO where the augmentation gain in (4) $k' = 0$. The expressions for coefficients a_0-a_2 and b_0-b_3 in equation (28) are very bulky and are not practical to be presented in this paper. It is however possible to determine, with a reasonable degree of accuracy, what parameters of the system dominate each of these coefficients and under what conditions they may turn negative. Such approximate expressions for the case when

$k' = 0$ are given below:

$$a_2 \approx y; \qquad a_1 \approx y \left(\frac{1}{T_r} + \sigma x P_p \omega_r \right) \tag{30}$$

$$a_0 \approx \frac{y}{T_r} \left(P_p \omega_r x - \frac{x^2}{y^2 T_r} - \frac{x^2}{y T_r} - \frac{1}{T_r} \right) \tag{31}$$

$$b_3 \approx \frac{J}{\tau_0^*}; \qquad b_2 \approx \frac{J}{\tau_0^*} \left(\frac{1}{T_r} + \sigma x P_p \omega_r \right) \tag{32}$$

$$b_1 \approx \frac{J}{\tau_0^* T_r} \left(P_p \omega_r x - \frac{x^2}{y^2 T_r} \right) \tag{33}$$

$$b_0 \approx \frac{P_p}{T_r} \left(y - 1 \right) \left(x^2 - y \right) \tag{34}$$

The following observations can be made from examining expressions (30)-(34):

1) Coefficients a_2 and b_3 are always positive;
2) Coefficients a_1 and b_2 are typically positive but can be negative when regenerating at high speed and torque;
3) Coefficients a_0 and b_1 can become negative when motoring at slow speed (x and ω_r are of the same sign, $|\omega_r|$ is small), and will be negative in regeneration (x and ω_r are of different signs);
4) Coefficient b_0, in both motoring and regeneration, can be positive or negative depending on flux and torque magnitudes.

Based on the above observations, the following open loop performance is expected from the non-augmented NFO algorithm. In motoring the algorithm can be stable given that torque is appropriately limited and that the relation between y and x ensures that $b_0 > 0$. The latter condition is not guaranteed but may be possible due to the natural frame alignment property, i.e. when the algorithm naturally finds a misaligned frame position corresponding to a stable operation point. In regeneration the algorithm is expected to be unstable under most conditions. As it will be shown in the next section, these expectations agree to simulation and experimental results.

Now, using the root locus method, we will explore how the gain of the PI controller affects the system stability. To vary the controller gain we will change k_p from zero to infinity while keeping the ratio k_i/k_p constant. Fig. 3a shows the root locus of the transfer function (28) in motoring mode. The operating point here is defined by $\omega_r = 100\,rad/sec$; $x = 0.8$; $y = 0.92$; and $\theta_e = -0.05\,rad$. On can see that the system is stable under all non-zero gains of the PI-controller.

Another motoring example is shown in Fig. 3b. The operating point differs from the previous case in that $\omega_r = 1\,rad/sec$. At low gain values of the controller there exist a pair of complex poles in the right half plane (RHP) and a pole at the origin. As the gain increases, the complex poles are pulled into the left half plane (LHP) but the pole at the origin moves along the real axis into the RHP. This indicates that when motoring at very slow speed the system is unstable.

Regeneration mode of the non-augmented NFO is illustrated by Fig. 3c for $\omega_r = 100\,rad/sec$; $x = -0.8$; $y = 0.92$; $\theta_e = 0.05\,rad$. For any gain of the PI-controller there exists a root in the RHP, which indicates a severe stability problem.

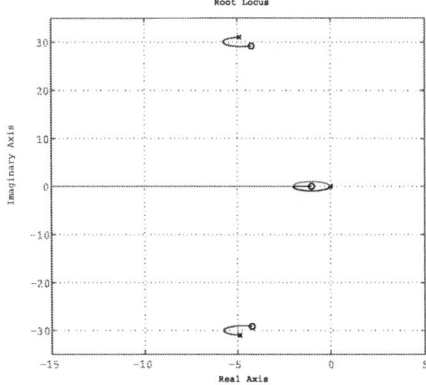

(a) Motoring, stable operation

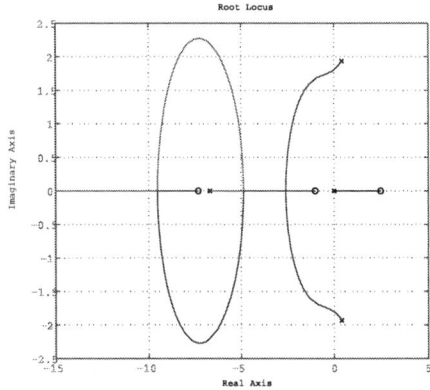

(b) Motoring, unstable operation at low speed

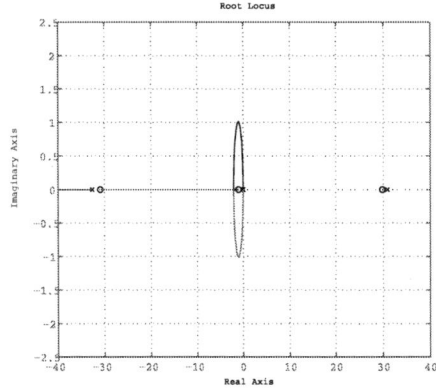

(c) Regeneration, unstable operation

Fig. 3: Root locus of the NFO controlled system, non-augmented algorithm

B. Augmented NFO

The same analysis will be now applied to the augmented version of the NFO algorithm. The approximate expressions for the polynomial coefficients for the case when $k' \neq 0$ is now given by:

$$a_2 \approx y; \qquad a_1 \approx y\left(\frac{1}{T_r} + \sigma x P_p \omega_r + k' y P_p \omega_r\right) \qquad (35)$$

$$a_0 \approx \frac{y}{T_r}\left(P_p \omega_r x - \frac{x^2}{y^2 T_r} - \frac{x^2}{y T_r} - \frac{1}{T_r} + k' y P_p \omega_r + k'\frac{x}{T_r}\right) \qquad (36)$$

$$b_3 \approx \frac{J}{\tau_0^*}; \qquad b_2 \approx \frac{J}{\tau_0^*}\left(\frac{1}{T_r} + \sigma x P_p \omega_r + k' y P_p \omega_r\right) \qquad (37)$$

$$b_1 \approx \frac{J}{\tau_0^* T_r}\left(P_p \omega_r x - \frac{x^2}{y^2 T_r} + k' y P_p \omega_r + k'\frac{x}{T_r}\right) \qquad (38)$$

$$b_0 \approx \frac{P_p}{T_r}\left(y - 1\right)\left(x^2 - y\right) \qquad (39)$$

The following can be observed from examining expressions (35)-(39):

1) Coefficients a_2 and b_3 are always positive;
2) Due to introduction of the term $k' y P_p \omega_r y$, which is always positive, coefficients a_1 and b_2 are typically positive;
3) Due to introduction of the same positive term, coefficients a_0 and b_1 can be kept positive under any conditions by using k of an appropriate magnitude;
4) The proposed augmentation had practically (within the accuracy of the approximation) no influence on the coefficient b_0 which, as in the non-augmented case, can be positive or negative depending on the flux and torque magnitudes.

From the above observations it follows that the augmentation has a positive effect on the system stability. When the speed loop is open, given that $b_0 > 0$, the NFO algorithm is expected to be stable when motoring and when regenerating with limited torque. If the gain k is allowed to change depending on the speed and torque then the region of stable regeneration can be further extended. However, the proposed augmentation has added practically no control over the sign of b_0. Therefore, with the speed loop open, there is a possibility that the NFO algorithm can be unstable regardless of the k value.

This can be illustrated with the help of the following example. The operating point of the system is chosen as: $\omega_r = 100\,rad/sec$; $x = -0.8$; $y = 0.92$; $\theta_e = -0.05\,rad$. Under these conditions, the non-augmented NFO algorithm is unstable as all the polynomial coefficients, except for a_2 and b_3, are negative. Augmentation with $k = 1$ leads to a_1 and b_2 turning positive. Increasing the gain to $k = 1.5$ makes a_0 and b_1 positive, but even then b_0 remains negative.

If the speed loop is closed then, with the appropriate gain of the PI controller, the system becomes stable. This is illustrated by the root locus of Fig. 4b plotted for the above example with $k = 1.5$. Fig. 4c shows a magnified view of the same plot near

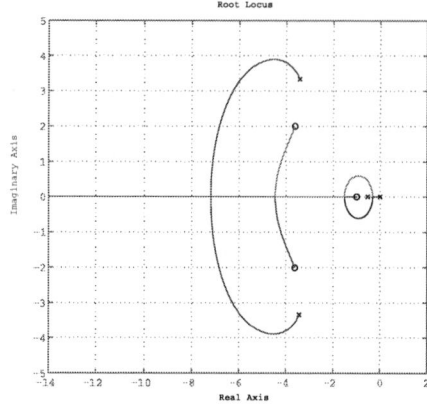

(a) Motoring, stable operation at low speed

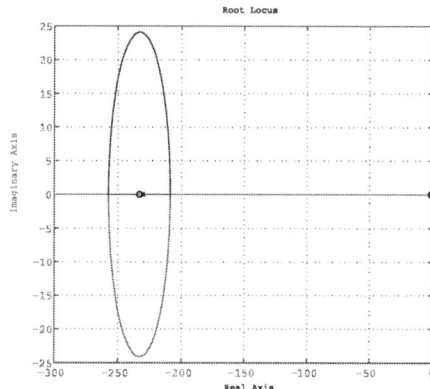

(b) Regeneration

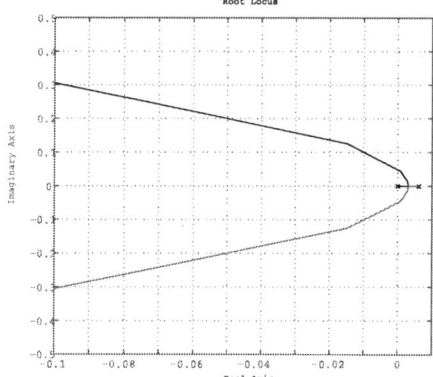

(c) Regeneration, magnified view

Fig. 4: Root locus of the NFO controlled system, augmented algorithm with $k = 1$

the origin. It can be seen how, as the k_p grows, the negative pole and the pole in the origin move towards each other along the real axis in the RHP, then break into a pair of complex poles and are pulled into the LHP. Fig. 4a illustrates stable motoring at very slow speed under the same conditions as those of Fig. 3b except that $k = 1$.

SIMULATION AND EXPERIMENTAL RESULTS

To illustrate the stability problem, performance of the NFO controlled induction machine was simulated under the following conditions: after allowing $0.5\,sec$ for fluxing, the reference speed stepped up from zero to $100\,rad/sec$ and then, after $1\,sec$ stepped down back to zero. The accurate values of the stator resistance and the magnetising inductance were used by the algorithm. Fig. 5a shows simulation results for the non-augmented NFO control and includes plots for the mechanical speed (desired vs actual), electromagnetic torque (desired vs actual) and the angular misalignment of the control frame.

It can be seen from Fig. 5a that in motoring the algorithm showed acceptable performance but immediately after entering regeneration mode NFO lost control and the system became unstable. In this simulation example, the induction motor was driving a highly inertial load, hence for a relatively long time, from $t = 0.5$ sec to $t = 1.25$ sec, the output of the PI controller was limited at its maximum positive value, i.e. the NFO algorithm was operating with the speed loop open. The same was true for the regeneration part of the plot, from $t = 1.5$ sec to $t = 2.4$ sec.

The corresponding experimental results showing the same variables appear in Fig. 6a. In general, the results are in a very good agreement with those of the simulation, except for a high initial frame misalignment observed in the experiment. This can be due to the fact that in the experiment we did not know the true frame position, and used as such the position of the stator flux vector estimated by the FOC (working in the background of the NFO).

Simulation results for the augmented algorithm with $k = 1$ are shown in Fig. 5b. As compared to the non-augmented case of Fig. 5a, with the augmentation turned on, the algorithm performance in motoring was much better, transition from motoring to regeneration was well controlled. Yet, at some point in regeneration the angular error started growing and stability was lost. The output of the PI controller at this point was limited at its maximum negative value, and the system was operating with the open speed loop.

Experimental results for the augmented algorithm with $k = 1$ are shown in Fig. 6b. The system remained stable in regeneration but only because the negative torque was limited at $-5\,Nm$ rather than $-20\,Nm$ as it was in the corresponding simulation. When a higher negative torque was allowed, the NFO algorithm was losing stability in regeneration in a manner similar to Fig. 5b. High initial frame misalignment seen in the bottom of Fig. 6b can be also attributed to the incorrect knowledge of the true frame position at start.

Finally, an attempt was made to further improve the NFO stability by using an augmentation of proportional - integral

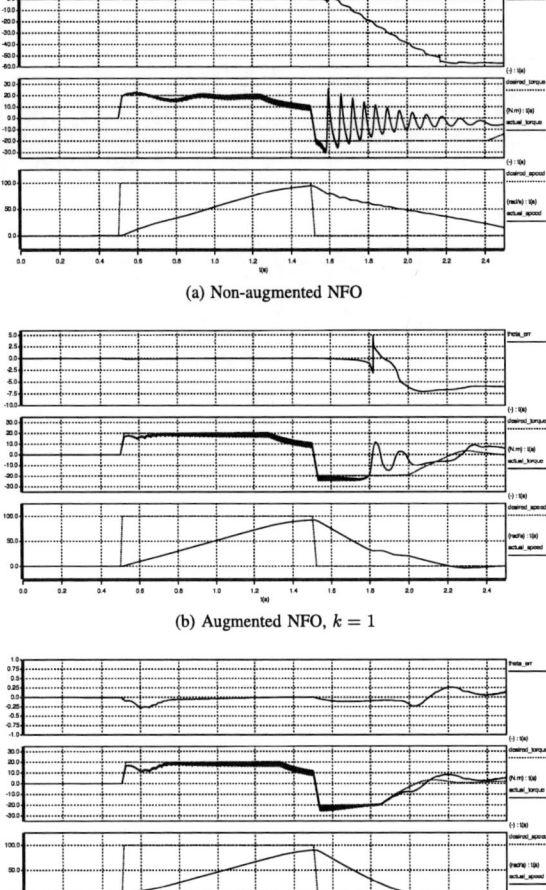

(a) Non-augmented NFO

(b) Augmented NFO, $k = 1$

(c) Augmented NFO, $k = 1$, $m = 10$

Fig. 5: Simulation plots for different strategies of the NFO control

structure, i.e. estimating the frame angular velocity according to:

$$\hat{\omega}_{ms} = \frac{e_q - k' e_d - m' \int e_d\, dt}{\psi_s^*} \qquad (40)$$

where $m' = m\, sign(e_q)$; m is the independently controlled integral gain; and the other parameters are as in equation (4). By the time the paper was submitted, the stability analysis of thus augmented system was not completed, and therefore, only simulation and experimental results are included to illustrate this case, as shown in Fig. 5c and Fig. 6c respectively. One can see that with the proportional- integral augmentation stability of the system has improved, the motor starts and stops in a controlled manner, yet some problems are evident around zero speed.

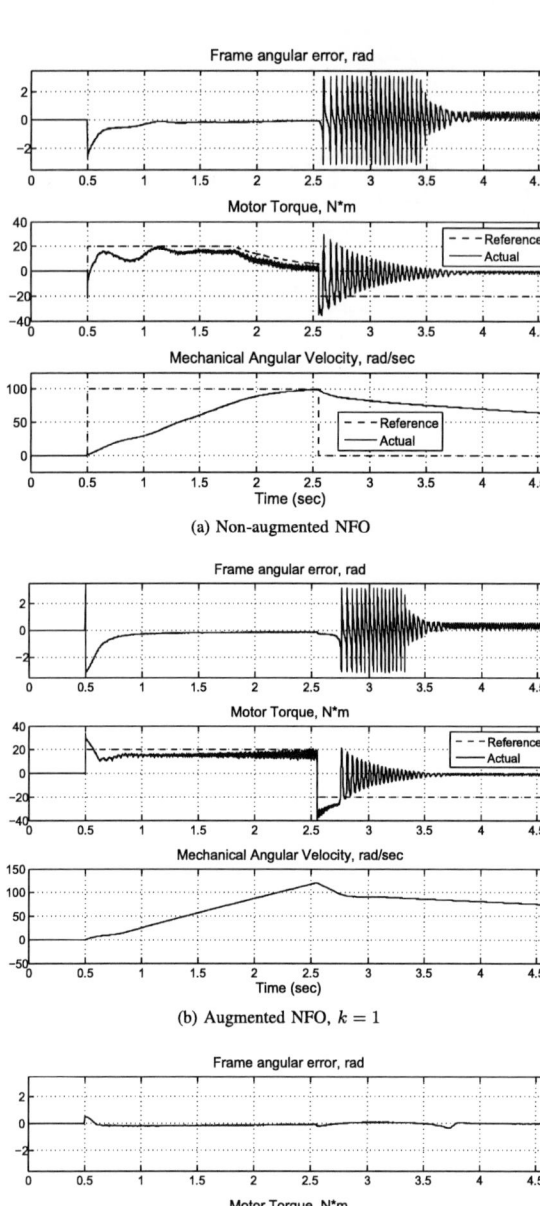

(a) Non-augmented NFO

(b) Augmented NFO, $k = 1$

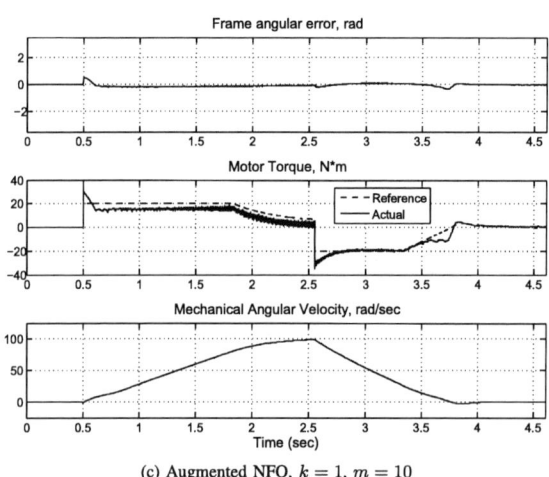

(c) Augmented NFO, $k = 1$, $m = 10$

Fig. 6: Experimental plots for different strategies of the NFO control

CONCLUSIONS AND CONTRIBUTIONS

The main conclusion of the analysis presented in this paper is that the augmentation to the NFO algorithm proposed in [5], may not guarantee the NFO stability under all conditions. This limitation comes from the fact that the algorithm offers no control over one of the essential parameters - the coefficient b_0 in the denominator of the transfer function. Particular conditions when the stability can be lost include regeneration with the open speed loop, which is likely to occur when a negative step change in speed is commanded, and the output of the PI controller is limited at its maximum negative value.

The following are the main contributions of the paper:

- The full dynamic equations describing the original and the augmented NFO algorithm have been developed.
- Linearised dynamic stability analysis of the full motor/drive system based on the NFO control algorithm has been presented. These results do not contradict but add to the previously presented in [5] static stability analysis.
- Stability problems have been identified and related to the parameters of the system.
- The limits of the augmentation strategy based on the auxiliary e_d feedback have been explored. The most critical parameter that limits the stabilisation effect of the strategy has been identified.
- Simulation and experimental results for the non-augmented and the augmented versions of the NFO algorithm have been presented. These results agree with the conclusions of the stability analysis.
- Preliminary simulation and experimental results for the proportional - integral version of the NFO augmentation have been included. These show improved performance in both motoring and regenration.

In future work the NFO-controlled system with the proportional - integral augmentation will be analysed in detail. It will be also attempted to find alternative stabilisation strategies that would provide stability of the NFO algorithm under an extended range of conditions.

REFERENCES

[1] R. Jönsson, "Method and apparatus for controlling an AC motor," United States Patent 4,458,193, July 1984.

[2] R. Jönsson, "Method and apparatus for controlling an ac induction motor by indirect measurement of the air-gap voltage," United States Patent 5,294,876, March 1994.

[3] P. Vas, *Sensorless Vector and Direct Torque Control.* Oxford University Press, 2003.

[4] G. Mirzaeva and R. E. Betz, "Parameter sensitivity issues in natural field orientation," in *Industry Applications Conference, 2007. 42nd IAS Annual Meeting. Conference Record of the 2007 IEEE*, New Orleans, LA, Sep. 23–27, 2007.

[5] R. E. Betz and G. Mirzaeva, "Frame alignment stability issues in natural field orientation," in *Industry Applications Conference, 2006. 41st IAS Annual Meeting. Conference Record of the 2006 IEEE*, vol. 5, Tampa, FL, Oct. 8–12, 2006, pp. 2411–2418.

[6] G. Mirzaeva and R. E. Betz, "An improved natural field orientation control of a current fed induction machine," in *CDROM Proceedings of IEEE-IEMDC 2007, Antalya, Turkey*, May 2007.

[7] ——, "Natural field orientation concept: a tutorial," in *CDROM Proceedings of IEEE-IEMDC 2007, Antalya, Turkey*, May 2007.

[8] G. C. Goodwin, S. F. Graebe, and M. E. Salgado, *Control System Design.* Prentice-Hall, 2001.

Control of SR motor EV by instantaneous torque control using flux based commutation and phase torque distribution technique

Ayumu Nishimiya[*], Hiroki Goto[*], Hai-Jiao Guo[†] and Osamu Ichinokura[*]

[*]Elec. and Comm. Eng. Dept., Tohoku University, Sendai, Japan, e-mail: *power18@ec.ecei.tohoku.ac.jp*
[†] Elec. and Inform. Eng. Dept., Tohoku Gakuin University, Tagajyo, Japan, e-mail: *kaku@tjcc.tohoku-gakuin.ac.jp*

Abstract—Switched reluctance motors (SR motors) have many advantages such as simple and solid construction, low-cost manufacturing, excellent reliability at high temperatures, and large torque density. However, the higher torque ripple from magnetic saliency is a serious problem preventing its applications from being expanded. To reduce torque ripple, we propose an instantaneous torque control method using flux-based commutation and phase-torque distribution. It provides efficient driving for SR motors with limited torque ripple. In this paper, we consider the application of this method to control torque of the SR motors for an electric vehicle (EV).

Keywords— Switched reluctance motor, electric vehicle, torque control, torque ripple reduction, flux control.

I. INTRODUCTION

Switched Reluctance Motors (SR Motors) are the motor that utilizes the reluctance torque originated in magnetic saliency between stator and rotor. The SR motors have neither coil nor permanent magnet on the rotor, thus it has simple structure, highly reliable at severe conditions, low-cost manufacturing, and also can be expected for high efficiency. In addition, it has high torque density because the motor can increase its torque even though stator poles are magnetically saturated. These advantages are suitable for an application of SR motors to electric vehicles (EVs). So we have also developed proto-type EVs which used SR motors [1] [2].

On the other hand, its magnetic saliency causes higher torque ripple, noise, and vibration compared to other motors. Hence SR motors have been used within limited application. Several methods have been proposed to attach those problems [3]-[5]. However, more simple and low cost method to reduce torque ripple will be useful for practice applications.

One of the major torque ripple reduction method is to control the instantaneous current by PWM [6]-[8]. This technique achieves the decrease of the torque ripple by calculating an ideal current waveform for each instruction torque beforehand from finite element method (FEM), and then request a controller to control the real current in order to trace the ideal current Therefore, the controller needs large amount of memories to storage ideal current waveforms, and request a high speed and high precision current feed-back control. Even if the ideal current profiles are obtained, near the aligned position, where the rotor poles

align the stator poles, it is not easy to control the current because of its high inductance. In addition, if the current flows at inductance decreasing area, it produces negative torque, which makes the efficient worse.

So, we have already proposed the instant torque control method using flux-based commutation and phase-torque distribution technique, which is a torque control method for switched reluctance motors without negative phase torque and ideal current profiles for constant torque [9]. This control system is simple, so the control algorithm can be implemented easily by low cost processors and minimal memories.

In this paper, an application on SR motors for an EV of the instant torque control method using flux-based commutation and phase-torque distribution are considered.

II. SWITCHED RELUCTANCE MOTORS

Fig. 1 shows the SR motor used in the paper, which is called 4-phase-16/20-SR motor has 16 stator poles and 20 rotor poles, is driven using 4-phase exciting. The rotor and the stator are made up of non-oriented silicon steel laminations.

The torque produced in SR motors can be derived from coenergy. Assuming negligible mutual coupling among phases, the phase torque produced in the kth phase is given as

$$\tau_k = \frac{\partial W_k'(\theta, i_k)}{\partial \theta}. \tag{1}$$

Where k (=A, B, C, D) represent index of phases, $W_k'(\theta, i_k)$ is the coenergy stored in the kth winding, i_k is the phase current of the kth winding, θ is the rotor position. If magnetic saturation can be neglected, then the coenergy in the kth winding is given in

$$W_k'(\theta, i_k) = \int_0^{i_k} L_k(\theta) \cdot i_k \, di = \frac{1}{2} L_k(\theta) \cdot i_k^2. \tag{2}$$

Where $L_k(\theta)$ is the linear inductance of the kth phase. Thus, the phase torque τ_k produced in the kth phase can be expressed in

$$\tau_k = \frac{1}{2} i_k^2 \frac{dL(\theta)}{d\theta} \tag{3}$$

The motor torque τ can be expressed as

978-1-4244-1741-4/08/$25.00 ©2008 IEEE

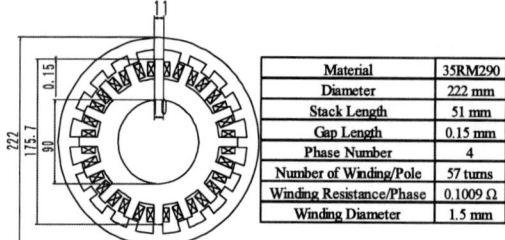

Material	35RM290
Diameter	222 mm
Stack Length	51 mm
Gap Length	0.15 mm
Phase Number	4
Number of Winding/Pole	57 turns
Winding Resistance/Phase	0.1009 Ω
Winding Diameter	1.5 mm

Fig. 1. Structure of the SR motor used in the paper.

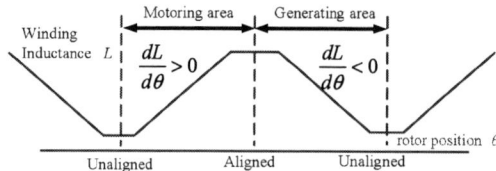

Fig. 2. Inductance profile versus rotor position of the 16/20-SR motor.

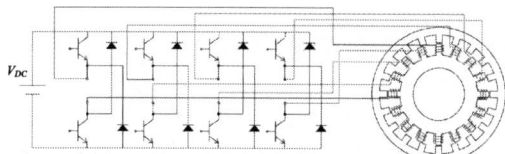

Fig. 3. Drive system of the 4-phase-16/20-SR motor using asymmetry half-bridge converter.

$$\tau = \sum_{k=A,B,C,D} \tau_k . \tag{4}$$

Fig. 2 shows the phase inductance L_k profile versus rotor position θ of the SR motor used in this paper. There is a maximum phase inductance in the aligned position ($\theta = 9.0$ deg.) where the rotor pole aligns to the stator pole, and a minimum phase inductance in the unaligned position ($\theta = 0, 18$ deg.). From (3), the positive torque is produced by excitation at the area in which the gradient of the phase inductance is positive value.

Fig. 3 shows a drive system of the SR motor using an asymmetry half bridge converter. A consecutive rotation can be obtained through switching the transistors sequentially to always produce positive torque based on the rotor position from a position sensor such as the rotary encoder.

III. PROPOSED TORQUE CONTROL METHOD

In order to reduce torque ripples, using a phase torque to calculate a phase current reference then control the phase current to track the phase current reference is a usual way. We have proposed a new method that is direct control the phase torque instead of the phase current, so complicate calculation from torque to current is not required and the implementation will be more simpler than that of the usual way.

The schematic diagram of the control system is shown as Fig. 4. The control system is mainly constructed by four components; commutation scheduler, phase torque distributor, phase torque regulator, and phase torque calculator, respectively.

A. Commutation scheduler

By this control method, the damping time of current is estimated by magnetic flux and it performs turn-off to suitable timing. When v_k is kth phase voltage, R is winding resistance and i_k is kth phase current, kth phase linkage flux ϕ_k can be expressed as (5).

$$\phi_k = \int (v_k - R \cdot i_k) dt \tag{5}$$

If the phase excitation is turn-off when time t, the DC source voltage is V_{DC} as constant value, the phase voltage v_k becomes $-V_{DC}$. Then, if $R \cdot i_k$ can be neglected compared with V_{DC}, the next step phase linkage flux ϕ_k $(t+\Delta t_k)$ can be obtained from (6).

$$\phi_k(t + \Delta t_k) \approx \phi_k(t) + \int_0^{\Delta t_k} (-V_{DC}) dt$$
$$= \phi_k(t) - V_{DC} \cdot \Delta t_k \tag{6}$$

If $\phi_k(t+\Delta t_k)$ becomes zero that means of the phase current extinction, the time for current reduction Δt_k can be estimated from (6) as follows

$$\Delta t_k = \frac{\phi_k(t)}{V_{DC}} . \tag{7}$$

Then, the position θ_{ek} where kth phase current becomes zero is shown as (8), using the rotation position $\theta(t)$ when switches are turn-off.

$$\theta_{ek} = \omega \cdot \Delta t_k + \theta(t) = \frac{\omega \cdot \phi_k(t)}{V_{DC}} + \theta(t) \tag{8}$$

Therefore, by this control method, θ_{ek} is always calculated. When the reference torque is positive, switches are turn-off at the position where θ_{ek} is aligned position.

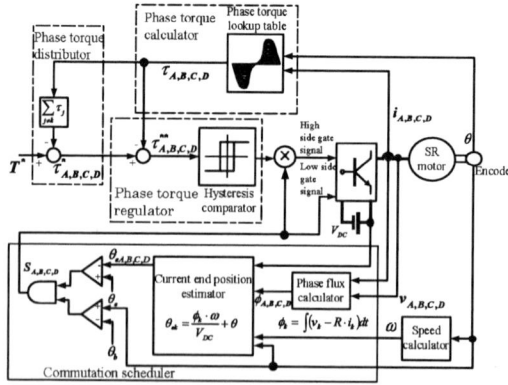

Fig. 4. Schematic diagram of proposed control system.

B. Phase torque distributor

Torque ripples are mainly generated in the domain where phase changes. Therefore, it is necessary to perform torque distribution suitably between neighboring phase torques in the domain where exciting phase changes. By the proposed technique, the value which subtracted the other actual torques from motor reference torque is given as each reference phase torque shown as Fig. 5.

C. Phase torque regulator

If the reference phase torque is obtained, the actual phase torque can be controlled with a hysteresis comparator.

If the actual phase torque is larger than a upper limit, the switch is turn-off and the actual phase torque is decreased by decreasing the phase current. Moreover, when the actual phase torque is smaller than the lower limit, the switch is turn-on and the actual phase torque is increased by increasing the phase current. The actual phase torque can be tracked by repeating the above.

D. Phase torque calculator

As mentioned above, the torque of the SR motor is expressed as (3) in linear area of the magnetic characteristic. However, strong nonlinear area is used for SR motors in actual applications. For the proposed method, each phase torque including nonlinearity is estimated using look-up tables which contain the current-position-torque characteristics of the SR motor directly obtained by FEM as shown in Fig. 6.

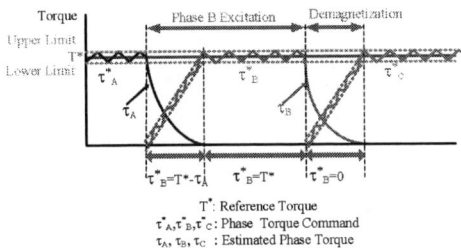

Fig. 5.   Phase-torque distribution method.

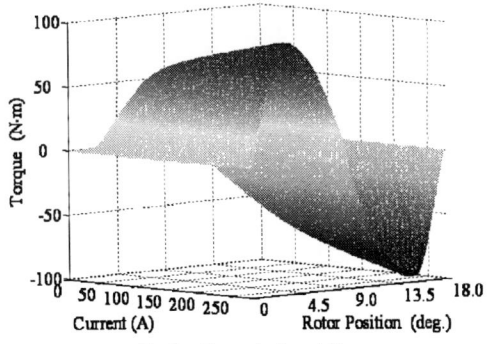

Fig. 6.   Torque look-up table.

IV.   SIMULATION

The proposed method has simulated used MATLAB/Simulink as general-purpose simulator for dynamic systems. Fig. 7. shows the flow diagram of the nonlinear simulation. We constructed a simulation model of the SR motor using results of FEM. This simulation model is composed of a converter model, a magnetic flux calculator, a look-up table for estimating the phase current, a look-up table for estimating phase torque, a speed calculator, and a position calculator respectively.

The conditions and the control parameters in the simulations are following; V_{DC} is set as 50 V and the hysteresis band for the phase torque control is ± 0.5 N·m. The hysteresis band for the current limiter is ± 6 A and the maximum current limitation I_{MAX} is 100 A.

Fig. 8 shows the waveforms of simulation results when the rotation speed $\omega = 50$ rpm and the reference motor torque $T^* = 50$ N·m. The motor torque is almost

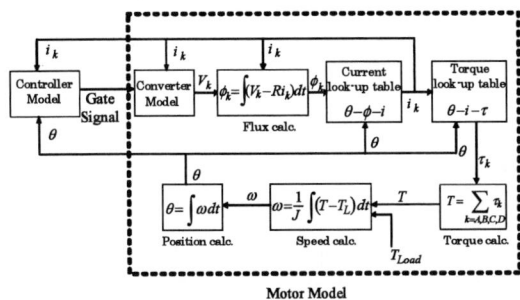

Fig.7.   Flow diagram of the nonlinear simulation.

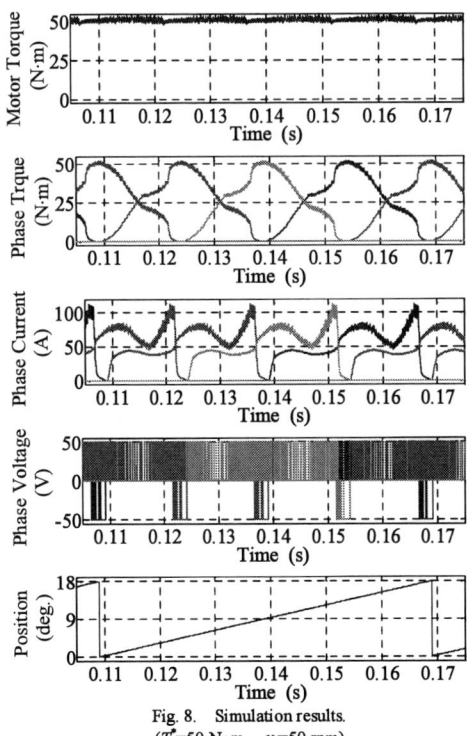

Fig. 8.   Simulation results.
(T^*=50 N·m ω=50 rpm)

controlled around 50 N·m. And neighboring phase torque compensates for the lack of each phase torque. Each phase torque has no negative part order to the commutation method based on the linkage flux.

Fig. 9 shows the waveforms of simulation results when the rotation speed $\omega = 200$ rpm and the reference motor torque $T^* = 50$ N·m. The motor torque is not controlled around 50 N·m. And torque ripple becomes large when exciting phase is changed. It is because that excited phase is forcibly turn-off before next phase current rise enough, and phase torque can not track its reference.

One of this torque ripple reduction method is by changing the current end position. If negative torque is slightly allowed, the torque producing area can be expanded and the torque drop can be improved. Fig. 10 shows the improved simulation result by changing the current end position to extend to +2.5 deg. from aligned position. The phase torque has slightly negative part, but the total torque is remarkably improved.

V. EXPERIMENT

We have actually confirmed the proposed control method through the experiments that implemented control systems to an EV. Specification of the SR motor EV is shown in Fig. 11, and Fig. 12 shows the photographs of the SR motor integrated with the wheel.

The proposed control method has been implemented using a one-chip microcomputer (SH7144F) and a FPGA (EP1K30TC144). The experimental conditions and the control parameters are following; V_{DC} is 50 V at no load supplied from Li-ion Battery. The hysteresis band for the phase torque control is ·0.78 N·m. The hysteresis band for the current limiter is ·2.2 A and the maximum current limitation I_{MAX} is 120 A. Other experimental

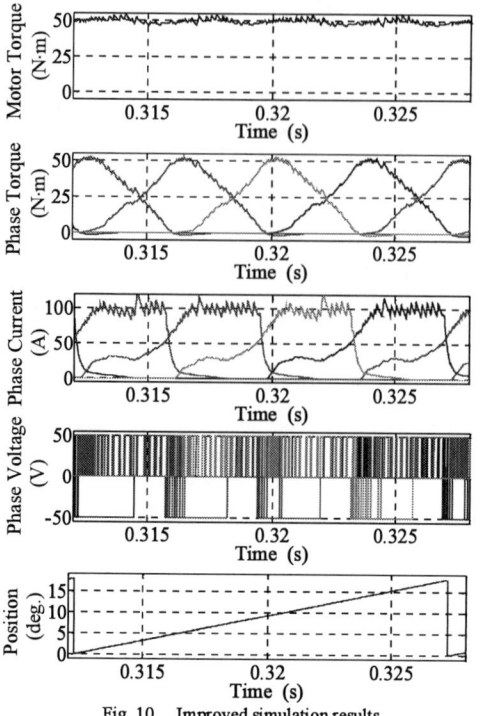

Fig. 10.   Improved simulation results.
(T^*=50 N·m ω=200 rpm θ_e =+2.5 deg.)

Weight(without ·driver)	158 kg
Body material	Carbon FRP
Battery Voltage	50 V
Current Limiter	100 A
Drive method	Direct drive
Motor type	Wheel-in Outer rotor 16/20SRM
Seating capacity	1
Wheel radius	0.2095 m

Fig. 11.   Specification of the SRM-EV to be used.

(a)The inner-stator   (b)The outer- rotor with the tyre
Fig. 12.   View of the manufactured SRM.

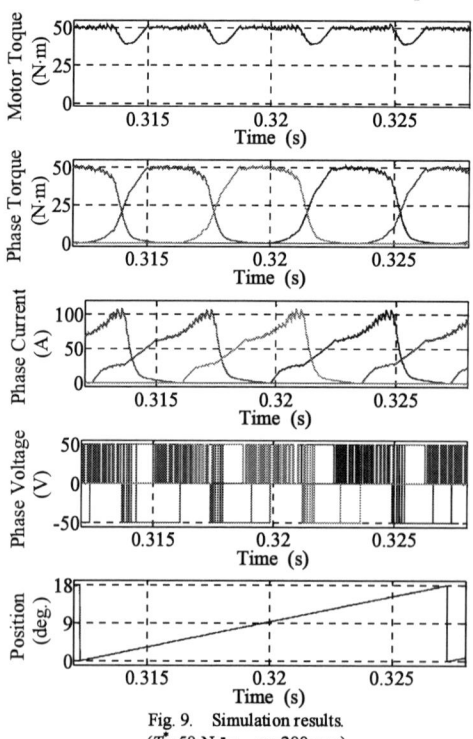

Fig. 9.   Simulation results.
(T^*=50 N·m ω=200 rpm)

conditions are same as that of the simulations.

Fig. 13 shows the waveforms of experimental results when the rotation speed ω= 50 rpm and reference motor torque T^* = 50 N·m. The motor torque is almost controlled around 50 N·m as same as the reference motor torque. The current waveforms and the phase torque waveforms are similar to the simulations.

The motor torque is slightly larger than that of the simulations. However, the proposed method can effectively reduce the torque ripple.

Fig. 14 shows the waveforms of simulation results when the rotation speed ω= 200 rpm and the reference motor torque T^* = 50 N·m. The motor torque can not

controlled around 50 N·m. And torque ripple becomes large. It is because that excited phase is forcibly turned off before next phase current rise enough. We are sure that this torque ripple is able to be improved by changing θ_e as same as simulations.

VI. CONCLUSION

In this paper, it confirmed by the simulation that an application of SR motors for an EV of the instantaneous torque control method using flux-based commutation and phase-torque distribution was possible. And it is confirmed to be able to reduce torque ripple effectively trough the simulations and the experiments.

For future work, we will consider this control method not only forward but also back and regenerating-brake and find more suitable parameters.

References

[1] K. Nakamura, Y. Suzuki, H. Goto, and O. Ichinokura, "Design of outer-rotor type multipolar SR motor for electric vehicle," *Journal of Magnetism and Magnetic Materials*, Vol. 290-291, pp. 1334-1337, 2005.

[2] H. Goto, Y. Suzuki, K. Nakamura, T. Watanabe, H. J. Guo, and O. Ichinokura, "A multipolar SR motor and its application in EV," *Journal of Magnetism and Magnetic Materials*, Vol. 290-291, pp. 1338-1342, 2005.

[3] R. C. Becerra, M. Ehsani, and T. J. E. Miller, "Commutation of SR Motors," *IEEE Trans. Power Electron.*, Vol. 8, pp. 257-263, 1993.

[4] N. Matsui and Y. Takeda, "Reluctance Motors –State-of-the-art-," *T. IEE Japan*, Vol. 118-D, No. 6, pp. 685-690, 1998 (in Japanese).

[5] C. A. Ferreira, S. R. Jones, B. T. Drager, and W. S. Heglund, "Design and Implementation of a Five-hp, Switched Reluctance, Fuel-Lube, pump Motor Drive for a Gas Turbine Engine," *IEEE Trans. Power Elect.*, Vol. 10, pp. 55-61, 1995.

[6] K. M. Rahman and S. E. Schulz, "High-Performance Fully Digital Switched Reluctance Motor Controller for Vehicle Propulsion," *IEEE Trans. Ind. Appl.*, Vol. 38, pp. 1062-1071, 2002.

[7] P. C. Kjaer, J. J. Gribble, and T. J. E. Miller, "High-Grade Control of Switched Reluctance Machines," *IEEE Trans. Ind. Appl.*, Vol. 33, pp. 1585-1593, 1999.

[8] K. Russa, I. Husain, and M. E. Elbumk, "Torque-Ripple Minimization in Switched Reluctance Motors Over a Wide Speed Range," *IEEE Trans. Ind.* Appl., Vol. 34, pp. 1105-1112, 1998.

[9] H. Goto, H. J. Guo, and O. Ichinokura, "A Novel Torque Control Method Using Direct Neighboring Phase Torque Distribution Technique for Switched Reluctance Motors," *The 2006 International Conference on Electrical Machines and Systems*, LS4C-1, 2006.

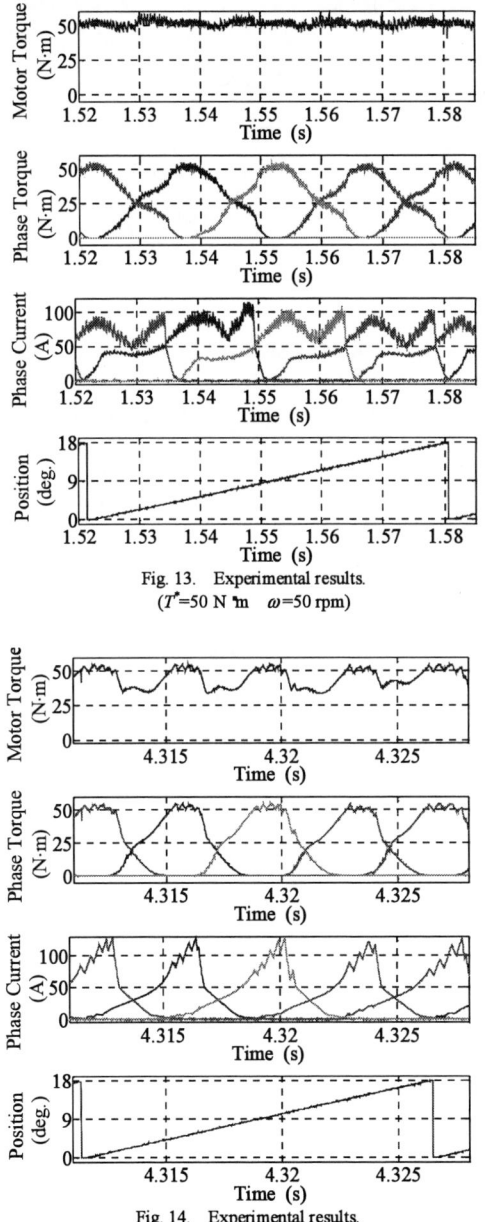

Fig. 13. Experimental results.
(T^*=50 N·m ω=50 rpm)

Fig. 14. Experimental results.
(T^*=50 N·m ω=200 rpm)

Simulation of IPM Motor by Nonlinear Magnetic Circuit Model for Comparing Direct Torque Control with Current Vector Control

Hiroki Goto[*], Kensuke Kimura[*], Hai-Jiao Guo[†] and Osamu Ichinokura[*]

[*] Dept. of Elect. & Comm. Engng., Tohoku Univ., Sendai, Japan, e-mail: *goto@ecei.tohoku.ac.jp*
[†] Dept. of Elect. & Infom. Engng., Tohoku GakuinUniv., Tagajo, Japan, e-mail: *kaku@tjcc.tohoku-gakuin.ac.jp*

Abstract— Interior Permanent Magnet (IPM) motor has high efficiency and large torque because the motor utilizes both magnet torque and reluctance torque depending on magnetic saliency. Most common methods used in control of IPM motor are current vector control and Direct Torque Control (DTC). However, it is too difficult to simulation these control methods including nonlinearity and space harmonics. In this paper, a nonlinear magnetic circuit model of the IPM motor considering space harmonics with current vector control and/or DTC is simulated and the two control methods are compared in the simulations.

Keywords—Adjustable speed drive, Direct torque and flux control, Modeling, Permanent magnet motor, Vector control.

I. INTRODUCTION

Interior Permanent Magnet (IPM) motor has high efficiency and large torque because the motor utilizes both magnet torque and reluctance torque depending on magnetic saliency. So, IPM motors are used for wide applications such as compressors, spindles, and electric vehicles [1]. On the other hands, IPM motor has large space harmonics and strong magnetic nonlinearity than those of Surface Permanent Magnet (SPM) motor. So, the space harmonics and magnetic nonlinearity of the IPM motor must be considered to control for high performance and efficient drive.

Most common methods used in control of IPM motor are current vector control and Direct Torque Control (DTC). Especially, DTC drive has possibility to apply on large space harmonics because rotational coordinate conversion is not used. However, it is too difficult to simulation these control methods including nonlinearity and space harmonics for conventional electric equivalent circuit model.

We have already proposed a nonlinear magnetic circuit model including space harmonics of IPM motor [2]. The magnetic circuit model with motion dynamics and control can be simulated by general-purpose circuit simulator.

In this paper, a nonlinear magnetic circuit model of the IPM motor considering space harmonics with current vector control and/or DTC is simulated and the two control methods are compared in the simulations.

II. MAGNETIC CIRCUIT MODEL

The schematic diagram of the IPM motor used in this study is shown in Fig. 1. The stator has three-phase

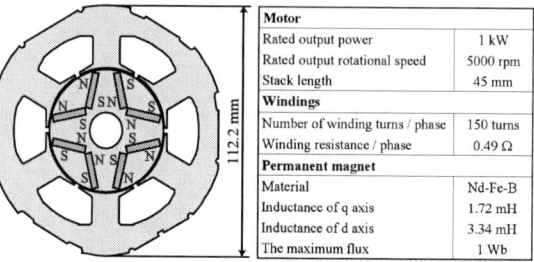

Motor	
Rated output power	1 kW
Rated output rotational speed	5000 rpm
Stack length	45 mm
Windings	
Number of winding turns / phase	150 turns
Winding resistance / phase	0.49 Ω
Permanent magnet	
Material	Nd-Fe-B
Inductance of q axis	1.72 mH
Inductance of d axis	3.34 mH
The maximum flux	1 Wb

Fig. 1. Schematic diagram of the IPM motor.

concentrated windings. The rotor has 4 poles and the rotor magnets are made of Nd-Fe-B.

The magnetic circuit of the IPM motor is shown in Fig. 2. In the figure, Magneto Motive Forces (MMFs) of each phases are named as Ni_u, Ni_v, and Ni_w, respectively. These are given from the number of winding turns N and each phase current i_u, i_v, and i_w, respectively.

The R_{sp}s, which are nonlinear reluctances on each stator poles, are given by following equation:

$$R_{sp} = \left(\frac{\beta_1 l_{sp}}{S_{sp}} + \frac{\beta_{13} l_{sp}}{S_{sp}^{13}} \phi^{12} \right). \tag{1}$$

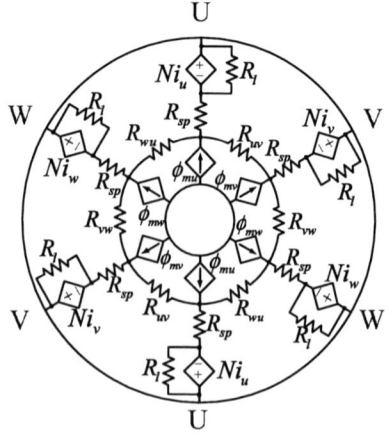

Fig. 2. Magnetic circuit model of the IPM motor.

The schematic diagram of the magnetic circuit at the stator pole is illustrated in Fig. 3. S_{sp} is cross-sectional area of the stator pole and l_{sp} is the length of the stator pole. β_1 and β_{13} are the first order and the 13th order magnetization coefficient of the material, respectively.

ϕ_{mu}, ϕ_{mv}, and ϕ_{mw} are the each phase fluxes from the permanent magnets, respectively. The reluctances between each phase are R_{uv}, R_{vw}, and R_{wu}, respectively. These reluctances and magnet fluxes change with the rotor angle. In order to obtain the reluctances and magnet fluxes, each flux passing through the stator pole are calculated using finite element method. For examples, R_{uv} is calculated from the flux ϕ_{uv} from U-phase pole at only U-phase excitation condition shown in Fig. 4 and the magnetic potential difference between A and B in Fig. 5.

In the same way, leakage reluctances R_l are calculated from the leakage flux ϕ_l and MMF N_{iu} in Fig. 5.

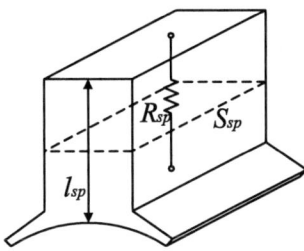

Fig. 3. Schematic diagram of the magnetic circuit at the stator pole.

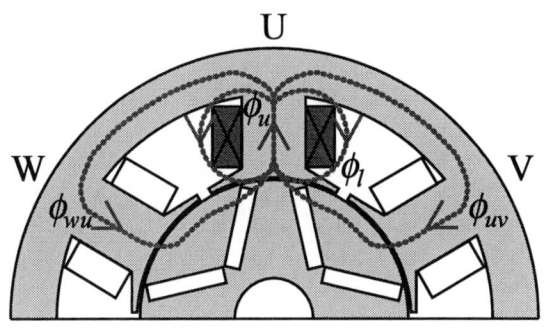

Fig. 4. Flux flow diagram when only U-phase is excited.

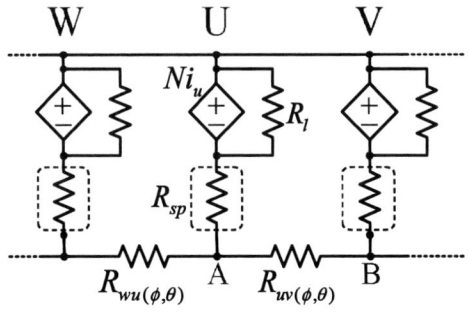

Fig. 5. Magnetic circuit model considering saturation at stator pole.

III. MODEL VERIFICATION ON SIMPLE CONTROL

The nonlinear model of IPM motor is experimental verified to check its validity. The simulations of the nonlinear model and the experiment have the same DC source voltages of 280 V. The IPM drive is worked with 120 deg. conduction drive.

The voltage waveforms and the current waveforms at load torque as 1.0 N·m are shown in Fig. 6 and Fig. 7. The torque-speed characteristics and torque-current characteristics are shown in Fig. 8 and Fig. 9.

In Fig. 8 and Fig. 9, the linear model means conventional d-q axis electrical equivalent circuit model. In the linear model, the q-axis inductance L_q and d-axis inductance L_d are 1.72 mH and 3.34 mH, respectively. The maximum stator flux from permanent magnets is 1 mWb.

Comparison of simulation results with experimental results indicates that the response of the nonlinear model is satisfactory.

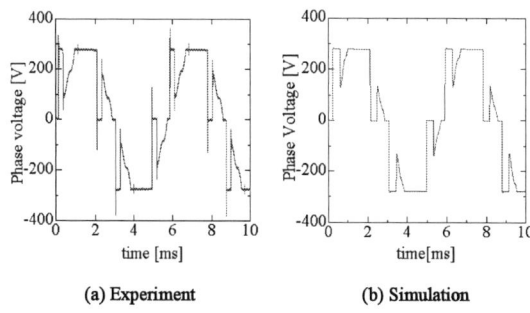

(a) Experiment (b) Simulation

Fig. 6. Phase voltage waveforms (120 deg. conduction drive).

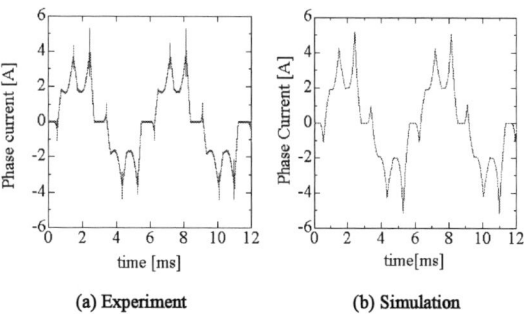

(a) Experiment (b) Simulation

Fig. 7. Phase current waveforms (120 deg. conduction drive).

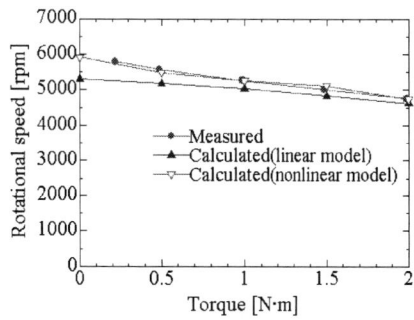

Fig. 8. Torque-speed characteristic (120 deg. conduction drive).

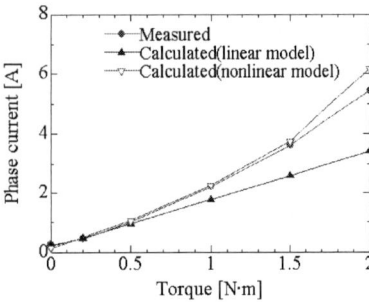

Fig. 9. Torque-current characteristic (120 deg. conduction drive).

IV. SIMULATION ON CURRENT VECTOR CONTROL

In industry applications, current vector control is most popular control method of IPM motors. The control block diagram of the current vector control is shown in Fig. 10. Measured three-phase currents are converted to d-axis current i_d and q-axis current i_q. The q-axis current reference i_q^* is given by PI control block from the speed difference between reference speed ω^* and measured actual speed ω. The d-axis current reference i_d^* is calculated from i_q. The q-axis voltage reference v_q^* and d-axis voltage reference v_d^* are given by PI control block from the errors between i_q/i_d and i_q^*/i_d^*. v_d^* and v_q^* are converted to three-phase voltage references v_u^*, v_v^*, and v_w^*. Then, each phase voltages are tracked to the references by PWM inverter.

The simulation conditions are shown as following; the DC voltage source is 280 V and the speed reference ω^* is 1500 rpm. Hence, the load torque is set as 0.2 N·m from start to 0.3 s, 0.5 N·m from 0.3 s to 0.6 s, 1.0 N·m from 0.6 s to 1.0 s, and 2.0 N·m from 1.0 s to 1.5 s, respectively.

The waveforms of the motor speed ω and torque T of the simulation results are shown in Fig. 11. The motor speed ω is controlled as almost constant even if load torque is changed. Then, current waveforms from 0.95 s to 1.0 s are shown in Fig. 12. The current waveforms are similar to three-phase sinusoidal current with low frequency harmonics.

The speed and torque waveforms of the linear model and nonlinear model to compare are shown in Fig. 13. In case of the linear model, the torque ripple is smaller than that in case of the nonlinear model. These results mean that torque ripple of the IPM motor is influenced by nonlinearity and space harmonics.

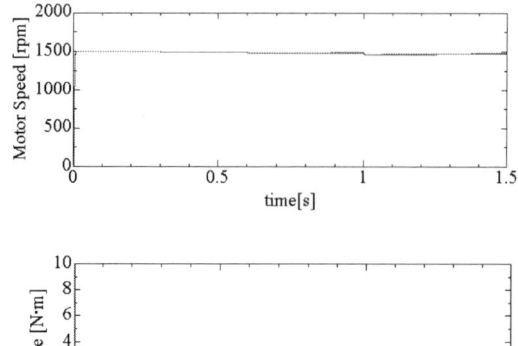

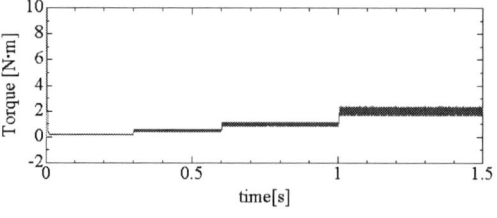

Fig. 11. Speed and torque waveforms (current vector control).

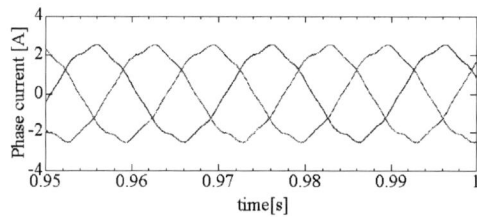

Fig. 12. Phase current waveforms (current vector control).

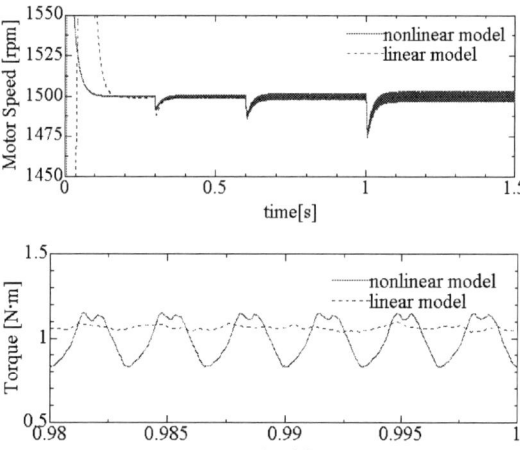

Fig. 13. Speed and torque waveforms to compare with the linear model.

V. SIMULATION ON DIRECT TORQUE CONTROL

From Fig. 13, it is too difficult to control the torque of the IPM motor including nonlinearity and space harmonics using conventional current vector control method.

Direct Torque Control (DTC) method is proposed [3][4]. The block diagram of DTC is shown in Fig. 14. The detail of the DTC algorithm is explained as following.

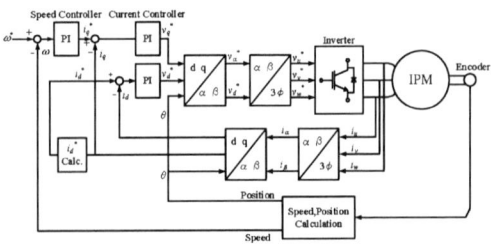

Fig. 10. Block diagram of current vector control for IPM motor.

1170

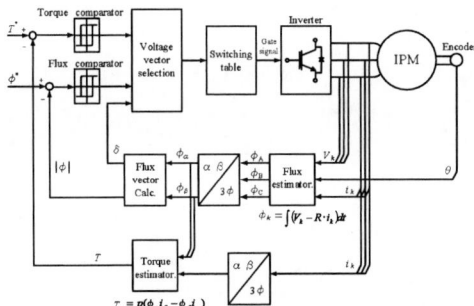

Fig. 14. Block diagram of DTC for IPM motor.

A. Flux calculation

The $k(=u,v,w)$ -phase stator flux linkage ϕ_k can be calculated as (2) from k-phase voltage v_k, k-phase current i_k, winding resistance R. And the ϕ_{k0} is initial value of ϕ_k given by the permanent magnets.

$$\phi_k = \int_0 (v_k - R \cdot i_k) dt + \phi_{k0} \qquad (2)$$

These phase flux linkages are resolved into a stator flux linkage vector $\vec{\phi}$ and expressed on α-β axis coordinates. Hence, the flux linkage magnitude $|\phi|$ and the angle δ are calculated from α-axis flux linkage ϕ_α and β-axis flux linkage ϕ_β.

B. Torque calculation

The motor torque is calculated from (3).

$$T = p(\phi_\alpha \cdot i_\beta - \phi_\beta \cdot i_\alpha) \qquad (3)$$

Where, p means number of the pole-pairs, i_α and i_β mean α- and β-axis currents, respectively.

C. Excitation pattern

The excitation patterns of the three-phase AC motors are usually selected as only six patterns which are two-phase conduction patterns. Switching state of each legs, which the high side switch is just turned on, is defined as "1". And switching state of each legs, which the low side switch is just turned on, is defined as "-1". Otherwise, switching state of each legs, which both high side and low side switches are turned off, is defined as "0". All the three-phase combinations of switching patterns and space voltage vectors are shown in TABLE I and Fig. 15. The zone number N of the flux vector angle δ are defined from $N=1$ to $N=6$ as shown in Fig. 15.

The torque reference T^* and the flux linkage reference ϕ^* are input signals, and these signals are compared with actual torque T and flux linkage ϕ using hysteresis comparators. Then, the space voltage vector is selected by Table II from these comparison signals and flux zone number N. Finally, the gate signals of the inverter are given by Table I.

As a result, the flux vector $\vec{\phi}$ tracks to circular line similar to rotational magnetic field in current vector control.

TABLE I.
CORRESPONDENCE TABLE OF SWITCHING PATTERNS AND SPACE VOLTAGE VECTORS

U,V,W	1,0,-1	0,1,-1	-1,1,0	-1,0,1	0,-1,1	1,-1,0
Space voltage vector	V_1	V_2	V_3	V_4	V_5	V_6

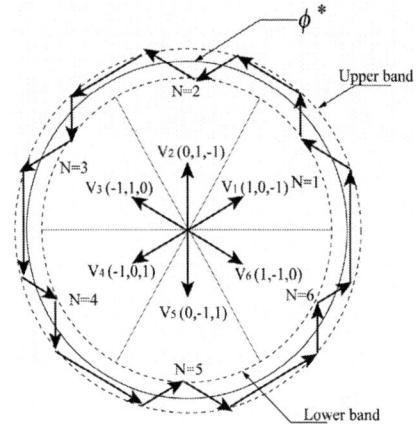

Fig. 15. Space voltage vectors and hysteresis control of the flux.

TABLE II.
SELECTION TABLE OF SPACE VOLTAGE VECTORS

Torque command	Flux command	Space voltage vector
Up	Up	V(N+1)
Up	Down	V(N+2)
Down	Up	V(N-1)
Down	Down	V(N-2)

D. Simulation Result

The simulation conditions are set as following; the DC voltage source is 280 V and the speed reference ω^* is 1500 rpm. Hence, the load torque is set as 0.2 N·m from start to 0.3 s, 0.5 N·m from 0.3 s to 0.6 s, 1.0 N·m from 0.6 s to 1.0 s, and 2.0 N·m from 1.0 s to 1.5 s, respectively. These conditions are similar to above mentioned current vector control simulation.

The motor speed and torque waveforms of the simulation results are shown in Fig. 16. The motor speed can be controlled as almost constant similar to current vector control even if load torque is changed. Then, current waveforms from 0.95 s to 1.0 s are shown in Fig. 17. However, the current waveforms are not similar to three-phase sinusoidal current with high frequency harmonics.

VI. COMPARING CURRENT VECTOR CONTROL AND DTC

In order to compare the two control methods which are current vector control and DTC, these torque waveforms,

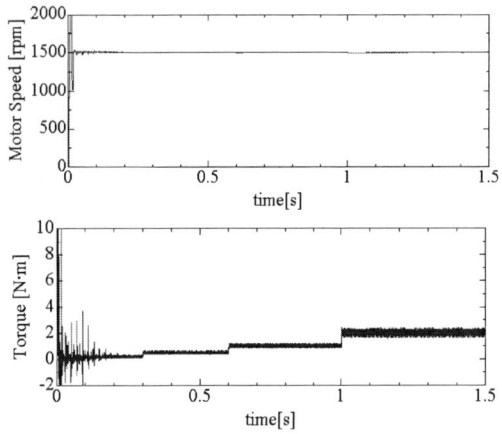

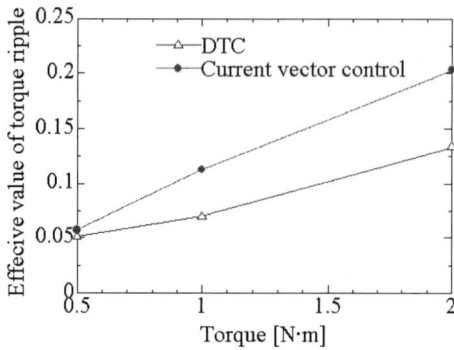

Fig. 19. Effective value of torque ripple.

Fig. 16. Speed and torque waveforms (DTC).

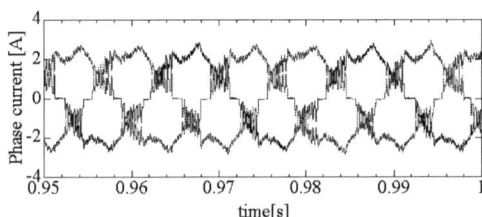

Fig. 17. Phase current waveforms (DTC).

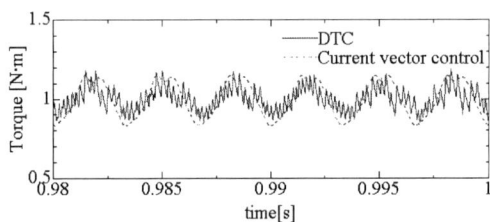

Fig. 18. Comparison of torque waveforms.

when load torque is 1.0 N·m and speed reference ω^*=1500 rpm, are shown in Fig. 18. The magnitude of the torque ripple in DTC drive is slightly smaller than that in current vector control drive. However, the torque waveform in DTC drive has larger harmonics than that of current vector control.

Effective values of the torque ripple in DTC drive is compared with those in current vector control when ω^*=1500 rpm in Fig. 19. The effective values of the torque ripple in DTC drive are smaller than those in current vector control at all load conditions. These results mean energy for vibration and noise in DTC drive is smaller than that in current vector control.

VII. CONCLUSION

In this paper, current vector control and DTC of IPM motor were simulated in order to compare their two control methods.

The simulation model was constructed based on nonlinear magnetic circuit considering space harmonics, and it could be calculated using general-purpose circuit simulator with motion dynamics and control algorithms.

The simulation model was verified with experimental results with simple control. The simulation results had higher accuracy compared with those of conventional linear equivalent circuit model.

Current vector control and DTC were compared in the simulations. The results indicated the magnitude of the torque ripple in DTC was slightly smaller than that in current vector control. And the effective values of the torque ripple in DTC drive were smaller than those in current vector control. However, the torque waveform in DTC had larger harmonics.

In the future, the two methods will be compared in experiments.

REFERENCES

[1] Y. Takeda, N. Matsui, S. Morimoto, and Y. Honda, "Design and Control of Interior Permanent Magnet Synchronous Motors," *Ohm-sha*, 2001(in Japnese).

[2] K. Nakamura, K. Saito, T. Watanabe, O. Ichinokura, "A new nonlinear magnetic circuit model for dynamic analysis of interior permanent magnet synchronous motor," *Journal of Magnetism and Magnetic Materials*, Vol. 290-291, pp. 1313-1317, 2005.

[3] Lixi Tang, "A Novel Direct Torque Controlled IPMSM Drive with Low ripple in flux and Torque and Fixed Switching Frequency," *IEEE Trans. Power Electron.*, Vol. 19, No. 2, 2004.

[4] Jawad Faiz, "A Novel Technique for Estimation and Control of Stator Flux of a Salient-Pole PMSM in DTC Method Based on MTPF," *IEEE Trans. Ind. Elect.*, Vol. 50, No. 2, 2003.

A Simplified Model for Induction Machines with Faults to Aid the Development of Fault Tolerant Drives

O. Jasim, C. Gerada, M. Sumner, J. Arellano-Padilla

School of Electrical and Electronic Engineering,
University of Nottingham, Nottingham, U.K,
e-mail: eexofj@nottingham.ac.uk, chris.gerada@nottingham.ac.uk, mark.sumner@nottingham.ac.uk,
eexja@nottingham.ac.uk

Abstract— A new mathematical model of a three-phase induction machine, suitable for the simulation of machine behaviour under fault conditions is presented. The model employs a simple state space equivalent circuit based model of the induction machine, which is enhanced to include space harmonics and main flux saturation effects. The model has been improved by including a variation of machine inductances with rotor and flux position, and creates a simulation with reasonable accuracy and fast computation time. For healthy and fault conditions, comparisons show that the simulated saturation and space harmonic frequencies and magnitudes match those obtained from experiment. This model can be used to develop and optimise control strategies for fault ride through.

Keywords—Machine modelling, Induction motor, fault tolerant drive.

I. INTRODUCTION

The design of new high performance vector control algorithms often uses simulation models for initial development. These simulations ideally use a good machine model, that takes the machine's non-linearities into account. Researchers working on condition monitoring and fault tolerant control of induction machines often need an accurate model to predict performance and to extract fault signatures [1]. Therefore, there is a real need to derive an accurate model, which can take into account the effect of saturation and field harmonics both in time and space, whilst still being embedded in a simulation, which includes real time control and a representation of power converters.

This work focuses on deriving an induction machine model, which gives a good compromise between modelling accuracy and simulation time for use with fault tolerant induction motor drives. Simplified models based on the machine equivalent circuit are computationally fast and allow integration with real time control strategies, but are not accurate because they neglect the machine geometry and nonlinear magnetization effects. At the other extreme, the finite element model (FEM) provides detailed information about most machine nonlinear effects, but its solution is time consuming especially if a control algorithm needs to be incorporated [1]. A method

providing a compromise between the speed of the conventional methods and the flexibility of finite-element analysis is the Dynamic Mesh Reluctance Modelling (DMRM) technique [2]. The analysis is not as accurate as the FEM but its computational time is significantly faster, enabling the use of DMRM models to evaluate control applications.

For simulation studies, different nonlinear models used to simulate the space harmonics have been presented based on the geometry of the squirrel cage induction motor [1-7]. These models are based on multiple coupled circuits and include the geometry and winding layout of the machine. A few restrictive assumptions are made when utilizing this technique. All inductances are derived by means of the Winding Function Approach (WFA) and are integrated using decomposition into their fourier series. These models have been extended to monitor mechanical and electrical faults such as inter-turn short circuits in the stator windings [1, 3, 4], open circuits in the stator winding [1, 3, 5], and broken rotor bars [3, 6-8].

Other work [9-12] has developed an FE-based phase-variable model, which combines the accuracy of the full FE model of the machine and fast computation time. A machine inductance variation with rotor position is obtained from finite element full cycle analysis, and then it is utilized in the phase variable model using look up tables. In this way, the effect of space harmonics can be included. [9, 10] report the use of a phase variable model, a fast computation time and accurate modelling of the machine saturation and rotor slotting effect as well as the cogging torque. By comparing the various techniques, the DMRM and the WFA showed a good modelling accuracy and computational time faster than the FEM (2 hours vs 12 hours [13]). However, it is still comparatively slow when fast response current and torque control is required.

The aim of this paper is to develop and enhance an equivalent circuit based model of the induction machine to include the space harmonics and machine saturation effects, such that the performance of motor drive systems under healthy and faulty conditions can be simulated. The model should be able to provide a useful response for the simulation of symmetrical and asymmetrical conditions, with only a short simulation time.

978-1-4244-1741-4/08/$25.00 ©2008 IEEE

II. THE INDUCTION MACHINE MODEL

In an induction motor, the three phase stator windings are designed to produce a sinusoidally distributed mmf in the space along the airgap periphery. Assuming a uniform airgap and neglecting the effects of slot harmonics, the distribution of the magnetic flux will also be sinusoidal. The machine model proposed in this work is derived from a general model of the induction machine in the a-b-c frame (i.e. phase quantities) [14]. The motor phase voltages and currents will be referred to as winding variables. From the equivalent circuit, a general form of the equations in the a-b-c frame can be written in matrix form as follows:

$$
\begin{bmatrix} U_{abcs} \\ U_{abcr} \end{bmatrix} = \begin{bmatrix} R_s & 0 \\ 0 & R_r \end{bmatrix} \cdot \begin{bmatrix} i_{abcs} \\ i_{abcr} \end{bmatrix} + \frac{d}{dt} \begin{bmatrix} \Psi_{abcs} \\ \Psi_{abcr} \end{bmatrix}
$$
$$
= \begin{bmatrix} R_s & 0 \\ 0 & R_r \end{bmatrix} \cdot \begin{bmatrix} i_{abcs} \\ i_{abcr} \end{bmatrix} + \frac{d}{dt} \left(\begin{bmatrix} L_s & L'_{sr} \\ L_{sr}^T & L'_r \end{bmatrix} \cdot \begin{bmatrix} i_{abcs} \\ i_{abcr} \end{bmatrix} \right) \tag{1}
$$

where the subscript s denotes stator and the subscript r rotor variables. The U_{abcs}, i_{abcs} and i_{abcr} are the terminal voltages, stator phase and rotor phase currents. The resistance matrices R_s and R_r as well as the inductance matrices L_s, L_r and L_{sr} from (1) can be written as follows:

$$
R_s = \begin{bmatrix} R_{as} & 0 & 0 \\ 0 & R_{bs} & 0 \\ 0 & 0 & R_{cs} \end{bmatrix}, R_r = \begin{bmatrix} R_{ar} & 0 & 0 \\ 0 & R_{br} & 0 \\ 0 & 0 & R_{cr} \end{bmatrix},
$$

$$
L_s = \begin{bmatrix} L_{ls}+L_{ms} & L_{abs} & L_{acs} \\ L_{bas} & L_{ls}+L_{ms} & L_{bcs} \\ L_{cas} & L_{cbs} & L_{ls}+L_{ms} \end{bmatrix}
$$

$$
L'_r = \frac{N_s^2}{N_r^2} L_r = \begin{bmatrix} L_{lr}+L_{ms} & L_{rab} & L_{rac} \\ L_{rba} & L_{lr}+L_{ms} & L_{rbc} \\ L_{rca} & L_{rcb} & L_{lr}+L_{ms} \end{bmatrix},
$$

$$
L'_{sr} = L_{ms} \begin{bmatrix} \cos\theta_r & \cos\left(\theta_r+\frac{2}{3}\pi\right) & \cos\left(\theta_r-\frac{2}{3}\pi\right) \\ \cos\left(\theta_r-\frac{2}{3}\pi\right) & \cos\theta_r & \cos\left(\theta_r+\frac{2}{3}\pi\right) \\ \cos\left(\theta_r+\frac{2}{3}\pi\right) & \cos\left(\theta_r-\frac{2}{3}\pi\right) & \cos\theta_r \end{bmatrix}
$$

where R_{abcs} and R_{abcr} represent the stator and rotor resistance respectively, L_{ls} and L_{lr} are the stator and rotor leakage inductance, L_{ms} is the magnetizing inductance, L_{abs}, ..., L_{cbs} are the stator mutual inductances and L_{abr}, ..., L_{cbr} are the rotor mutual inductances. The rotor electrical angle is denoted by Θ_r. Substituting (2) in (1)

$$
\frac{d}{dt}\begin{bmatrix} i_{abcs} \\ i_{abcr} \end{bmatrix} = -\begin{bmatrix} L_s & L'_{sr} \\ L_{sr}^T & L'_r \end{bmatrix}^{-1} \cdot \begin{bmatrix} R_s & 0 \\ 0 & R_r \end{bmatrix} \cdot \begin{bmatrix} i_{abcs} \\ i_{abcr} \end{bmatrix}
$$
$$
- \begin{bmatrix} L_s & L'_{sr} \\ L_{sr}^T & L'_r \end{bmatrix}^{-1} \cdot \begin{bmatrix} \frac{dL_s}{dt} & \frac{dL'_{sr}}{dt} \\ \frac{dL_{sr}^T}{dt} & \frac{dL'_r}{dt} \end{bmatrix} \cdot \begin{bmatrix} i_{abcs} \\ i_{abcr} \end{bmatrix} \tag{3}
$$
$$
+ \begin{bmatrix} L_s & L'_{sr} \\ L_{sr}^T & L'_r \end{bmatrix}^{-1} \cdot \begin{bmatrix} u_{abcs} \\ u_{abcr} \end{bmatrix}
$$

In addition to these electrical equations, the mechanical equations of the machine must be considered [14]:

$$
\frac{dw_r}{dt} = \frac{P}{2J} T_e - \frac{B_m}{J} w_r - \frac{P}{2J} T_L \tag{4}
$$

$$
\frac{d\theta_r}{dt} = w_r
$$

where, T_L is the load torque and T_e is the electromagnetic torque produced by the machine. The viscous friction is B_m and J the moment of inertia. The electromagnetic torque for a P-pole three phase induction motor is found from the magnetic coenergy W_m such that: [14]

$$
T_e = \frac{P}{2} \cdot \frac{dW_m}{d\theta_r} = \frac{P}{2} \cdot \begin{bmatrix} i_{abcs} & i_{abcr} \end{bmatrix} \frac{d\begin{bmatrix} L_s & L'_{sr} \\ L_{sr}^T & L'_r \end{bmatrix}}{d\theta_r} \cdot \begin{bmatrix} i_{abcs} \\ i_{abcr} \end{bmatrix} \tag{5}
$$

The electromagnetic torque equation simplifies to:

$$
T_e = \frac{P}{2} \cdot i_{abcs}^T \frac{dL'_{sr}}{d\theta_r} i_{abcr} \tag{6}
$$

III. MODELLING OF THE MACHINE NONLINEAR EFFECTS

A. Space Harmonic Effects Incorporated on The Stator-Rotor Mutual Inductance

An ideal sinusoidal distribution of mmf is possible only if the machine has an infinitely large number of slots and the turns of a winding are sinusoidally distributed in the slots. This is not a practical solution. As a result, when current flows through the winding, the mmf produced is not sinusoidally distributed in the air gap. The flux harmonics are a result of the combination of the spatially distributed mmf harmonics due to the discrete placement of the winding in slots and the non uniform resulting permanence seen by these fields [15]. The existence of space harmonics is well known to have a significant detrimental effect on the steady state and transient characteristics of the machine [8].

For a balanced machine the mmf distribution contains the fundamental and a family of space harmonics of order $h = 6k \pm 1$, where k is a positive integer. In a three phase machine, when sinusoidally varying currents flow through the windings, the resultant space harmonic fluxes rotate at $(1/h)$ times the speed of the fundamental wave. The space harmonic fluxes rotate in the same direction as the fundamental wave if $h = 6k + 1$ and in the opposite direction if $h = 6k - 1$ [15].

References [16, 17] indicate that any induction motor may be represented by a series of mechanically connected induction motors having different numbers of poles, whose stator windings are connected in series. This means a space harmonic flux of order (h) is equivalent to the flux created in a machine with $(h * P)$ poles. Therefore, more than one rotor circuit has been considered, each with a different numbers of poles. The equivalent circuit of such a system can be created if the classical equivalent circuit of the induction motor is extended as shown in Fig.1 [17]. In this work, the circuit is expanded to include the two principle space harmonics (5th and 7th) of a balance three-phase motor, the 3rd not being considered for healthy conditions as it cancels out for a three phase machine. It should be noted that this is only a simple approximation such that major geometrical effects can be represented in a simple model. Higher order space harmonics and space harmonics induced by the rotor have been neglected i.e. a large number of rotor bars is assumed.

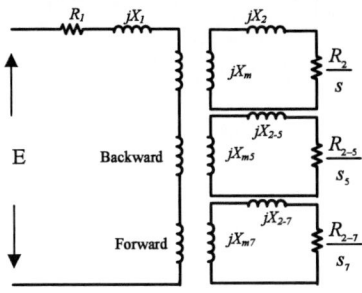

Fig.1 Equivalent Circuit of induction motor to model the space harmonics

where R is the actual resistance between terminals, X_1 is the leakage reactance of the stator winding representing the flux linking only the stator winding, except for the fundamental, 5th and 7th. X_m is the magnetizing reactance corresponding to the fundamental air gap flux wave of the winding, and the secondary reactance X_2 is the reactance of the actual rotor winding for currents with fundamental sine wave distribution. X_{m5} and X_{m7} are the magnetizing reactance corresponding to the air gap flux waves of a particular harmonic, and secondary reactances X_{2-5}, X_{2-7} are the reactance of the actual rotor winding for currents of the same number of harmonic poles. For each of these space harmonics, the corresponding rotor circuit in which currents are induced are shown. The existence of the space harmonics will impress harmonics components on the rotor currents, the frequencies of these components are calculated as:

$$ f_{rh} = \left(\frac{-m.f_e}{h} - f_r \right) h \qquad (7) $$

Where f_{rh} is the rotor current harmonic frequency, f_e is the supply frequency, f_r is the mechanical frequency of the rotor rotation, h is the order of the harmonic component, and m is a constant. (m = 1) for the positive sequence component and (m = -1) for the negative sequence component.

This work proposes a new model for the induction motor a-b-c frame (phase quantities) to model the space harmonic effect. The classical equation model will be extended such as to include these three rotor circuits instead of a single fundamental circuit. It includes space harmonic effects incorporated on the mutual inductance between stator and rotor windings as a function of rotor position. In particular, the analysis of the stator and rotor magnetically coupled system, which can be accomplished

using Fig.2, leads (1) to be modified to the following equation:

$$ \begin{bmatrix} U_{abcs} \\ U_{abcr} \\ U_{abcr5} \\ U_{abcr7} \end{bmatrix} = \begin{bmatrix} R_s & 0 & 0 & 0 \\ 0 & R_r & 0 & 0 \\ 0 & 0 & R_{r5} & 0 \\ 0 & 0 & 0 & R_{r7} \end{bmatrix} \cdot \begin{bmatrix} i_{abcs} \\ i'_{abcr} \\ i'_{abcs5} \\ i'_{abcs7} \end{bmatrix} + $$
$$ \frac{d}{dt} \left(\begin{bmatrix} L_s & L'_{sr} & 0 & 0 \\ L^T_{sr} & L'_r & 0 & 0 \\ L^T_{sr5} & 0 & L'_{r5} & 0 \\ L^T_{sr7} & 0 & 0 & L'_{r7} \end{bmatrix} \begin{bmatrix} i_{abcs} \\ i'_{abcr} \\ i'_{abcs5} \\ i'_{abcs7} \end{bmatrix} \right) \qquad (8) $$

where subscript h denotes harmonics order. $U_{abcr5,7}$, $i_{abcr5,7}$ are the rotor terminal voltages and the rotor currents for the 5th, and 7th harmonics components. L_{srh} is the new mutual inductance between the stator and rotor windings as function of the rotor position. The L_{srh} and L_{rh} matrices from (8) can be written as follows:

$$ L_{srh} = L_{msh} \begin{bmatrix} \cos\theta_{rh} & \cos\left(\theta_{rh}+\frac{2}{3}\pi\right) & \cos\left(\theta_{rh}-\frac{2}{3}\pi\right) \\ \cos\left(\theta_{rh}-\frac{2}{3}\pi\right) & \cos\theta_{rh} & \cos\left(\theta_{rh}+\frac{2}{3}\pi\right) \\ \cos\left(\theta_{rh}+\frac{2}{3}\pi\right) & \cos\left(\theta_{rh}-\frac{2}{3}\pi\right) & \cos\theta_{rh} \end{bmatrix} $$

$$ L'_{rh} = \begin{bmatrix} L'_{lr}+L_{msh} & -\frac{1}{2}L_{msh} & -\frac{1}{2}L_{msh} \\ -\frac{1}{2}L_{msh} & L'_{lr}+L_{msh} & -\frac{1}{2}L_{msh} \\ -\frac{1}{2}L_{msh} & -\frac{1}{2}L_{msh} & L'_{lr}+L_{msh} \end{bmatrix} \qquad (9) $$

where $L_{msh} = \dfrac{L_{ms}}{C}$, C is a constant and θ_{rh} is the rotor position used for the calculation of the mutual inductance. $\theta_{rh} = 7\theta_r$ for the 7th harmonic (forward) and $\theta_{rh} = -5\theta_r$ for the 5th harmonic (backward).

It can be seen that the electromagnetic torque developed in the induction motor will be modified to:

$$ T_e = \frac{P}{2} \begin{bmatrix} i^T_{abcs} \dfrac{\partial L'_{sr}(\theta_r)}{\partial \theta_r} i'_{abcr} + \frac{1}{2} i^T_{abcs} \dfrac{\partial L'_{sr5}(\theta_r)}{\partial \theta_r} i'_{abcr5} + \\ \frac{1}{2} i^T_{abcs} \dfrac{\partial L'_{sr7}(\theta_r)}{\partial \theta_r} i'_{abcr7} \end{bmatrix} \qquad (10) $$

B. Saturation Effects Incorporated in The Stator Mutual Inductance

The flux density is not uniformly distributed in the motor. For an induction motor, the motor windings generate a resultant flux density vector which will cause saturation. Since the flux density vector is rotating, the saturation regions will rotate too, modifying the permeability in a cyclic manner. This will create a cyclic variation of the stator winding inductances. For simplicity, it will be considered that only the stator mutual inductances will be modulated by the cyclic variation of the saturation.

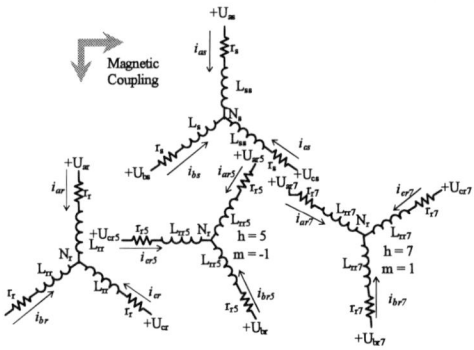

Fig.2 Three phase symmetrical induction motor including three rotor circuits to model the space harmonics

They propose a variation described for the stator mutual inductance with a simple mathematical equation for the mutual inductance. Although they present good results, effectively there is only one tuning parameter to adjust the magnitude of the saturation harmonics [18, 19]. For the simulation of the faulted drive a fine tuning of the saturation harmonic components is desirable. To achieve this fine tuning, the profiles obtained from an FEM analysis [11, 12, 20], can be used to create a mathematical expression to incorporate the saturation effects on the stator mutual inductance; as follows:

$$L_{sab} = L_{sba} = -\frac{Lms}{2}[1 + k_{s2}\sin(pp \cdot \Theta_f + \rho_1) + \qquad (11)$$
$$k_{s4}\sin(2 \cdot pp \cdot \Theta_f + \rho_2) + k_{s6}\sin(3 \cdot pp \cdot \Theta_f + \rho_3)$$

where k_{s2}, k_{s4}, and k_{s6}, are amplitudes and $\rho_1,.., \rho_3$ are phase displacements of the permanence wave introduced by the saturation effect. The mutual inductance profiles $L_{acs}=L_{cas}$ and $L_{bcs}=L_{cbs}$ are shifted by 120 and 240 degrees respectively. This profile has been included in the model proposed here, but the parameters of the stator mutual inductance profiles are adjusted according to measurements made from an experimental system, rather than FEM analysis.

IV. EXPERIMENTAL AND SIMULATION RESULTS

A four-pole, 4kW delta connected machine is modelled. The basic model is extended to incorporate the major effects of space harmonics with the stator mutual inductance variation and mutual inductance between stator and rotor variation using equations (4) and (8-11) as described in section III. All the experimental tests and simulation results were obtained with the machine incorporated in an indirect rotor field orientated control structure (IRFO). The machine electrical parameters were obtained via the conventional no-load and locked rotor tests as shown in appendix.

The saturation model (11) is adjusted according to the spectrum of the winding current signal. By adjusting parameters k_{s2}, k_{s4}, and k_{s6} and $\rho_1,.., \rho_3$ all the saturation harmonic components in the winding current can be matched to experimental measurement. The magnitude of this inductance component as well as its phase displacement will vary with load, and can be analysed using FEM analysis [20]. The proposed profiles of the machine inductances are not considered to be load dependent i.e. (11) does not include load dependency.

Therefore, for accurate results the model should be adjusted to give a good correlation with the experimental results. Tests under different loads were conducted to verify the model results. The parameters k_{s2}, k_{s4}, and k_{s6} and the phase displacements $\rho_1,.., \rho_3$ have been tuned for several values of load.

In order to calculate the electromagnetic torque developed by the motor experimentally, the estimation of the stator flux linkage is required. The flux linkage is estimated by the integration of the difference between the stator voltage and the voltage drop across the stator resistance as follows:

$$\overline{\psi}_s = \int (\overline{U}_s - R_s.\bar{i}_s)dt \qquad (12)$$

The electromagnetic torque is estimated by:

$$T_e = \frac{2}{3}\frac{P}{2}(\psi_{\alpha s}i_{\beta s} - \psi_{\beta s}i_{\alpha s}) \qquad (13)$$

where the quantities have their usual meaning.

1) Normal Operation Condition

The normal operation condition is considered first. The frequency spectrum of the simulated phase current and electromagnetic torque from the vector controlled drive is compared to those obtained from the experiment. The harmonic magnitudes of the stator current will be expressed as a percentage of its fundamental component, while the torque will be expressed as a percentage of the rated torque (27 N.m.).

The results at rated flux, 250 rad/sec, and no load are presented in Figs. 3 and 4. By adjusting parameters k_{s2}, k_{s4}, and k_{s6} and phase displacements $\rho_1,.., \rho_3$ all the magnitudes of the saturation harmonic components of the simulated phase current matched to experimental measurements. Figs. 3(b) and 4(b) show the electromagnetic torque spectra. In the simulation, the magnitude of the 6th component is slightly smaller than the experimental value. In general, current harmonics are all time harmonics. The developed torque is caused by the interaction of space harmonics and time harmonics. Ripple is created in the electromagnetic torque when the mmf distribution and flux distribution are not synchronous, but if they have the same number of poles. For example if a 5th harmonic of the flux occurs due to saturation, this interacts with the 5th harmonic of the winding distribution to create a 6th harmonic in the electromagnetic torque as shown in fig.3 (b). Because the space harmonics are not measurable quantities, the electromagnetic torque components have been used to determine the effects of the space harmonic and the time harmonics on the torque pulsation. The results at full rated flux which appear in Table 1 show that the magnitude of the simulation results match the experimental results.

Figs. 5 and 6 show test results at rated flux, 250 rad/sec, and 85% of the nominal load. The inductance profile has been modified by adjusting parameters k_{s2}, k_{s4}, and k_{s6} to match experimental measurements at this load. The results show that loading will cause a relative amplitude decrease in all of the harmonic components considered, except the 7th component (Table I) which shows a substantial increase. The authors believe that this is due to the simplification used in the model, which assumes only saturation of the mutual terms. In reality there are many other nonlinear effects in a real machine and these tend to interact with each other. They are the subject of further work.

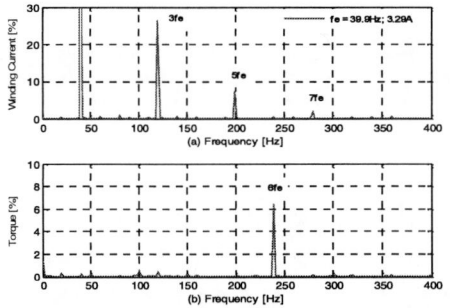

Fig.3 Experimental results – Frequency spectrum of the (a) winding current (b) electromagnetic torque at 250 rad/sec and no load

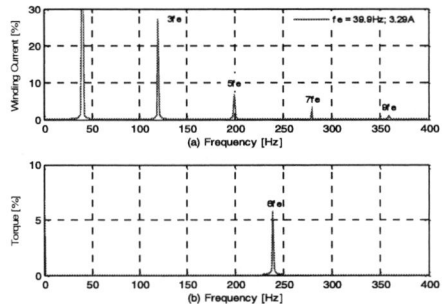

Fig.4 Simulation results – Frequency spectrum of the (a) winding current (b) electromagnetic torque at 250 rad/sec and no load

TABLE I. HARMONIC SPECTRUM OF THE WINDING CURRENT AND ELECTROMAGNETIC TORQUE FOR 100% RATED FLUX CASE

		Winding Current [%]				Electromagnetic Torque [%]	
		1st	3rd	5th	7th	DC	6th
No load	Experimental	100	26.8	7.63	2.03	1.41	6.74
	Simulation	100	27.3	7.02	2.43	6.81	5.78
Load	Experimental	100	11.5	1.68	3.42	85.95	4.56
	Simulation	100	11.6	3.01	3.3	91	4.84

The simulation model was also investigated for machine operation with 30% of the rated flux. The aim of this test is to reduce the amplitude of the winding current (time) harmonic in the experimental rig by reducing the saturation effect. Figs. 7 and 8 shows the results at 30% of the rated flux, 250 rad/sec, and no load. It is obvious that reducing the flux will cause an amplitude decrease in all of the harmonic components of the winding currents (Table II). There is a slight mismatch in the amplitude of the simulated saturation components (3rd, 5th, 7th) compared to the experimental ones. This probably results from the simplification made in the modelling of the saturation effect. The agreement between the results obtained by simulation and by experiment is fair. In fact, comparing the results presented in Fig.3 (b) with the ones of Fig.7 (b) the following conclusion can be drawn. The 6th harmonic component of the electromagnetic torque developed by the motor is much lower in the case of 30% of the full rated flux (0.74%) than the full flux (6.74%) i.e. a reduction of about 90%. This indicates the absence of saturation i.e. a much lower amplitude of the time

harmonics in the winding current, compared to full rated flux (Fig. 3 and 7).

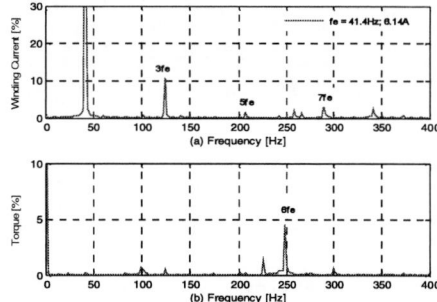

Fig.5 Experimental results – Frequency spectrum of the (a) winding current (b) electromagnetic torque at 250 rad/sec and 85% of nominal load

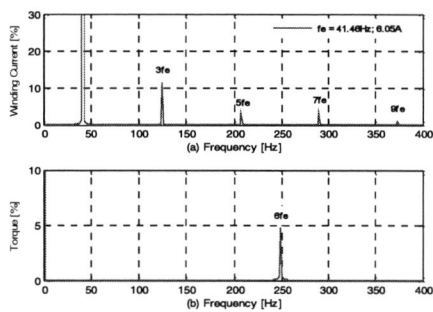

Fig.6 Simulation results – Frequency spectrum of the (a) winding current (b) electromagnetic torque at 250 rad/sec and 85% of nominal load

TABLE II. HARMONIC SPECTRUM OF THE WINDING CURRENT AND ELECTROMAGNETIC TORQUE FOR 30% RATED FLUX CASE

		Winding Current [%]				Electromagnetic Torque [%]	
		1st	3rd	5th	7th	DC	6th
No load	Experimental	100	3.77	6.18	4.31	6.04	0.74
	Simulation	100	2.6	2.86	4.98	6.78	0.43
Load	Experimental	100	7.23	2.57	2.81	28.83	0.39
	Simulation	100	8.68	2.57	2.04	27.2	0.37

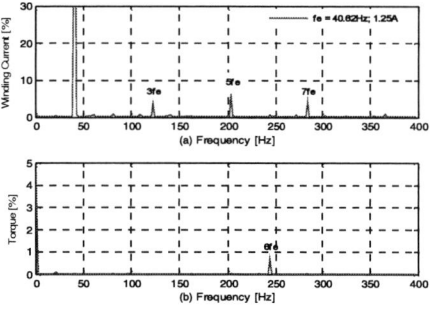

Fig.7 Experimental results – Frequency spectrum of the (a) winding current (b) electromagnetic torque at 30% of rated flux, 250 rad/sec and no load

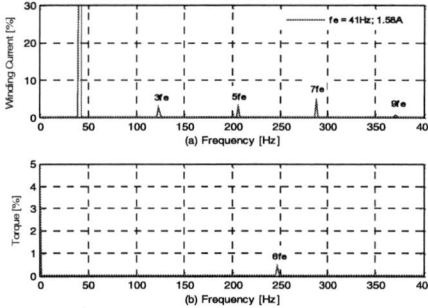

Fig.8 Simulation results – Frequency spectrum of the (a) winding current (b) electromagnetic torque at 30% of rated flux, 250 rad/sec and no load

Figs. 9 and 10 present the results at 30% of the rated flux, 250 rad/sec, and 30% of the nominal load. The loading will cause a decrease of the relative amplitude in all of the harmonic components considered, except the 3rd harmonic component and this will have no effect on the electromagnetic torque signal. The agreement between the results obtained by simulation and the experiment is reasonable as shown in Table II. Also, the 6th harmonic component of the electromagnetic torque developed by the motor is much lower than when operated at full flux. This test gave a positive indication that the 6th harmonic component of the electromagnetic torque depends on the presence of the time dependent current harmonics. As an alternative test, the frequency of the machine currents was reduced to 5Hz to ensure that any current (time) harmonics created by non-linear power converter effects are eliminated by the current controller (with bandwidth 100Hz). The flux level is set to 30% rated, as is the load as shown in Figs.11 and 12. Again, the inductance profile has been modified by adjusting parameters k_{s2}, k_{s4}, and k_{s6} to match experimental measurements. It can be observed that the 5th and 7th harmonic components of the line current reduced to less than 1% of the fundamental component. Also, the 6th harmonic components of the torque reduced to 0.14% of the rated torque as shown in Table III. Therefore, the results obtained by simulation and experiment are in reasonable agreement.

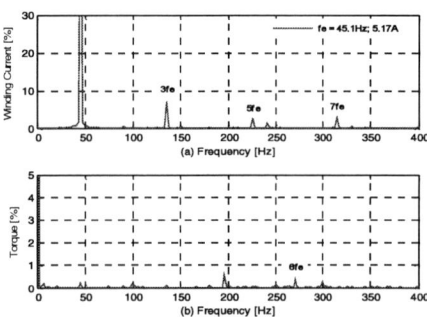

Fig.9 Experimental results – Frequency spectrum of the (a) winding current (b) electromagnetic torque at 30% of rated flux, 250 rad/sec and 30% of nominal load

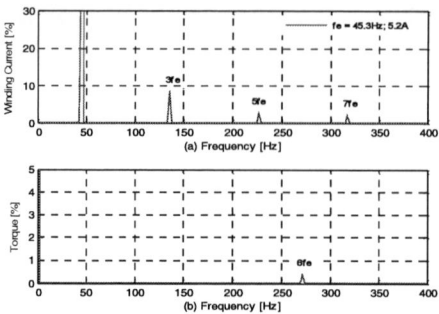

Fig.10 Simulation results – Frequency spectrum of the (a) winding current (b) electromagnetic torque at 30% of rated flux, 250 rad/sec and 30% of nominal load

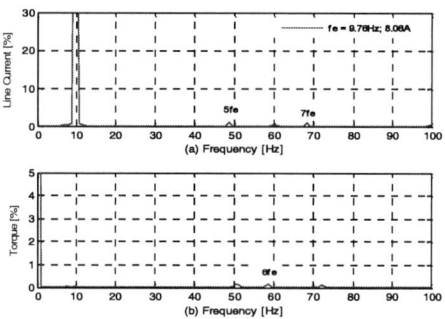

Fig.11 Experimental results – Frequency spectrum of the (a) line current (b) electromagnetic torque at 30% of rated flux, 31.4 rad/sec and 85% load

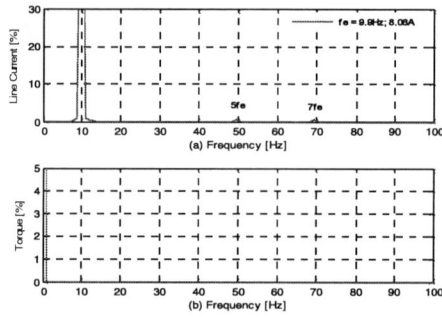

Fig.12 Simulation results – Frequency spectrum of the (a) line current (b) electromagnetic torque at 30% of rated flux, 31.4 rad/sec and 85% load

TABLE III.   HARMONIC SPECTRUM OF THE LINE CURRENT AND ELECTROMAGNETIC TORQUE FOR 30% RATED FLUX AND 31.4 RAD/SEC

		Winding Current [%]				Electromagnetic Torque [%]	
		1st	3rd	5th	7th	DC	6th
Load	Experimental	100	1.25	0.98	24.9	0.14	100
	Simulation	100	0.84	0.76	20.1	0.03	100

2) Faulty Operation Condition

Operation for an open circuit fault in one phase of the machine was investigated. Simulation results for the case of a stator open circuit fault are given for the same machine under the same experimental conditions. A phase breakdown fault has been modelled by inserting a higher resistance in to the faulted phase during the test.

The results at rated flux, 250 rad/sec, no load and phase breakdown fault are presented in Fig.13 and 14. In the presence of a stator asymmetry, there is an unbalance in the amplitudes of the three motor supply currents as a result of the existence of a negative-sequence component in these currents [21]. This negative-sequence component interacts with the stator flux thus giving rise to the appearance of an oscillatory component in the electromagnetic torque at double the supply frequency and other frequencies (4fe, 6fe, 8fe and 10fe). This is the main indicator of the presence of the fault when analysing the torque signal (Fig.13 (b)) [21]. Moreover, comparing Figs.13 (a) and 14 (a), it can be seen that the simulation results of the stator winding current match the experimental ones but with a slight mismatch in the amplitude of the -fe and -3fe components. This is probably due to the simplifications made in the modelling of the saturation effect.

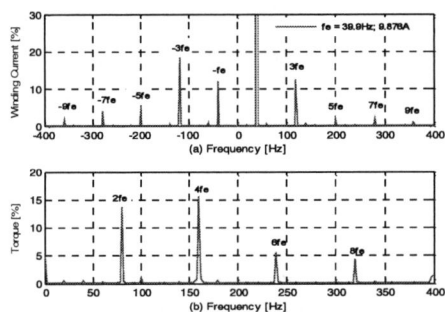

Fig.13 Experimental results – Complex frequency spectrum of the (a) winding current (b) electromagnetic torque with open circuit in phase (A) at 250 rad/sec and no load

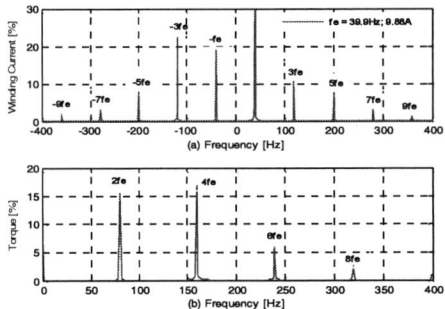

Fig.14 Simulation results – Complex frequency spectrum of the (a) winding current (b) electromagnetic torque with open circuit in phase (A) at 250 rad/sec and no load

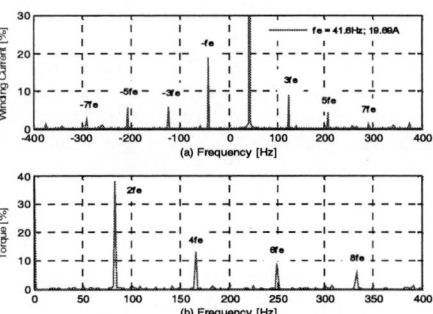

Fig.15 Experimental results – Complex frequency spectrum of the (a) winding current (b) electromagnetic torque with open circuit in phase (A) at 250 rad/sec and 85% of nominal load

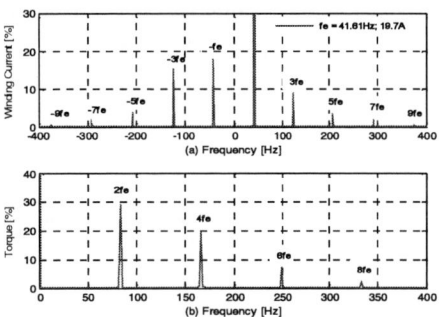

Fig.16 Simulation results – Complex frequency spectrum of the (a) winding current (b) electromagnetic torque with open circuit in phase (A) at 250 rad/sec and 85% of nominal load

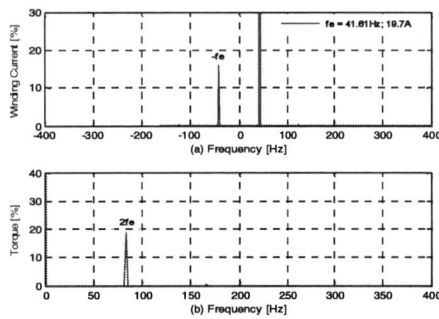

Fig.17 Simulation results – Complex frequency spectrum of the (a) winding current (b) electromagnetic torque with open circuit in phase (A) at 250 rad/sec and 85% of nominal load

Using the same condition but imposing 85% of the nominal load, Figs. 15 and 16 show the electromagnetic torque and it includes harmonic components at 2fe, 4fe, 6fe, 8fe and 10fe. The experimental and simulation results are in good agreement. Fig. 17 shows the simulation result for the same condition as in Fig. 16, but the saturation effects were eliminated to leave only the fundamental component. It can be seen that the 2nd harmonic component appears in the electromagnetic torque. This is the main indicator of the presence of the fault as mentioned early. Moreover, all other components have been eliminated by default. In summary, the main aim of this project was to find an induction machine model which gave a good compromise between the simulation time and modelling accuracy. The simulation model presented in this work is fast. It has a 150 sec computation time for a 10 sec simulation of the machine itself, (i.e. without vector control). For comparison, a 1 sec DMRM simulation (previously reported to be "fast"), requires a computation time of 946 sec.

V. Conclusion

The space harmonic effect is incorporated on the mutual inductance between stator and rotor windings as a function of rotor position. This has been done by extended the equivalent circuit of the induction motor to include the effect of the two principle (5th and 7th) space harmonics.

The saturation effect is incorporated as a flux position dependant component in the stator mutual inductance. The inductance profile consists of three harmonic components and the magnitude and phase displacement of each component can be adjusted. The saturation model was tuned to match the measured phase currents of an experimental system.

This model has been tested under healthy and faulted conditions by removing one phase in the stator circuit. The comparison of the results shows that the relative magnitude of the phase current harmonics match with those obtained from the experiment for all conditions.

The model allows the evaluation of the operation of induction motors in the presence of space harmonics and saturation effects, as well as being able to test the robustness of different field oriented controlled drives under unbalanced conditions. In addition, the proposed model can give a better evaluation of torque pulsations since it can include both time and space harmonics.

Appendix

THREE-PHASE INDUCTION MOTOR PARAMETERS

Rated Power	4 Kw
Line Voltage	415 V
Rated Torque	27 N.m
Poles, P	4
Stator resistance, R_s	5.25 Ω
Rotor resistance, R_r	3.75 Ω
Stator leakage inductance, L_{ls}	0.04 H
Rotor leakage inductance, L_{lr}	0.033 H
Stator magnetizing inductance, L_{ms}	0.356 H

References

[1] V. Devanneaux, H. Kabbaj, B. Dagues, and J. Faucher, "An accurate model of squirrel cage induction machines under rotor faults," presented at Electrical Machines and Systems, ICEMS, Proceedings of the Fifth International Conference on, 2001.

[2] C. Gerada, K. Bradley, M. Sumner, and P. Sewell, "Evaluation of a Vector Controlled Induction Motor Drive using the Dynamic Magnetic Circuit Model," in *IEEE IAS*. Salt Lake 2003.

[3] X. Luo, Y. Liao, H. A. Toliyat, A. El-Antably, and T. A. Lipo, "Multiple coupled circuit modeling of induction machines," *Industry Applications, IEEE Transactions on* vol. 31, pp. 311 - 318, 1995.

[4] G. Houdouin, G. Barakat, B. Dakyo, and E. Destobbeleer, "A Method for the Simulation of Inter-Turn Short Circuits in Squirrel Cage Induction Machines," in *EPE-PEMC*, 2002.

[5] G. Didier, H. Razik, and A. Rezzoug, "An induction motor model including the first space harmonics for broken rotor bar diagnosis," *European Transaction on Electrical Power*, vol. 15, pp. 229 - 243, 2005.

[6] M. Boucherma, M. Y. Kaikaa, and A. Khezzar, "Park model of squirrel cage induction machine including space harmonics effects," *Journal of Electrical Engineering*, vol. 57, pp. 193 - 199, 2006.

[7] G. Didier, H. Razik, A. Abed, and A. Rezzoug, "On Space Harmonics Model Of A Three Phase Squirrel Cage Induction Motor For Diagnosis Purpose," in *EPE - PEMC 2002*, 2002.

[8] M. Osama, K. Sakkoury, and T. A. Lipo, "Transient Behavior Comparison of Saturated Induction Machine Models," in *IMACS-TCI-93 Computational Aspects of Electromechanical Energy Converters and Drives*, 1993, pp. 577-580.

[9] O. A. Mohammed, S. Liu, Z. Liu, and N. Abed, "Physical phase variable models of electrical equipments and their applications in integrated drive simulation for shipboard power system," presented at Electric Ship Technologies Symposium, IEEE, 2005.

[10] O. A. Mohammed, S. Liu, and Z. Liu, "Phase-variable model of PM synchronous machines for integrated motor drives," *Science, Measurement and Technology, IEE Proceedings-* vol. 151, pp. 423 - 429, 2004.

[11] N. A. O. Demerdash and P. Baldassari, "A combined finite element-state space modeling environment for induction motors in the ABC frame of reference: the no-load condition," *Energy Conversion, IEEE Transactions on* vol. 7, pp. 698 - 709, 1992.

[12] P. Baldassari and N. A. Demerdash, "A combined finite element-state space modeling environment for induction motors in the ABC frame of reference: the blocked-rotor and sinusoidally energized load conditions " *Energy Conversion, IEEE Transactions on* vol. 7, pp. 710 - 720, 1992.

[13] S. Nandi and S. Nandi, "Modeling of induction machines including stator and rotor slot effects," *Industry Applications, IEEE Transactions on*, vol. 40, pp. 1058-1065, 2004.

[14] S. E. Lyshevski, *Electromechanical Systems, Electric Machines and Applied Mechatronics*. Florida: CRC Press, Boca Raton 1999.

[15] P. C. Sen, *Priciples of electric machines and power electronics*. USA: John Wiley & Son, 1997.

[16] P. L. Alger, *Induction machines , Their Behavior and Uses*. Australia: Gordon and Breach 1995.

[17] B. Heller and V. Hamata, *Harmonic Field effects in induction machines*. Amsterdam: Elsevier Scientific 1977.

[18] V. Donescu, A. Charette, Z. Yao, and V. Rajagopalan, "Modeling and simulation of saturated induction motors in phase quantities," *Energy Conversion, IEEE Transaction on*, vol. 14, pp. 386-393, 1999.

[19] J. C. Moreira and T. A. Lipo, "Modeling of saturated AC machines including air gap flux harmonic components," *Industry Applications, IEEE Transactions on*, vol. 28, pp. 343-349, 1992.

[20] C. Gerada, K. Bradley, M. Sumner, G. Asher, and J. Arellano-Padilla, "Permanent Magnet Synchronous machines for Saliency-based, Self-Sensored Motion Control," presented at Industrial Electronics Society, IECON, 33rd Annual Conference of the IEEE, 2007.

[21] S. M. A. Cruz and A. J. M. Cardoso, "Modelling and simulation of DTC induction motor drive for stator winding faults diagnosis," in *EPE 2003*. Toulouse, France, 2003.

About the Experimental Results of an Electric Driving System Based on Asynchronous Motor and PWM Converter

Petre-Marian Nicolae*, Dan-Gabriel Stănescu* and Ioana-Gabriela Sîrbu*

* University of Craiova/Department of Fundamental Electrotechnics, Craiova, Romania,
e-mail: *pnicolae@elth.ucv.ro; dstanescu@elth.ucv.ro; osirbu@elth.ucv.ro*

Abstract— **The paper deals with special features related to the a.c. electric driving system of a trolley-bus. One presents the driving system consisting of an asynchronous motor and inverter and some aspects regarding the PWM inverter control. The tests over the manufactured asynchronous motor were performed at a frequency equal to 50 Hz (for the rated speed) and a frequency equal to 106.6 Hz corresponding to a synchronous speed equal to 3200 rpm. The inverter was separately tested, before its inclusion in the driving system and some of the tests results presented (heating tests for the main components, the behavior at the discontinuous variation of line voltage, a.o.). Some tests concerning the behavior of the entire system at forward and backward running regimes are also presented.**

Keywords—**Electric drives, PWM inverter, Asynchronous motor, Efficiency.**

I. INTRODUCTION

The developing of public transportation systems in cities is a practice in many European countries. Its purpose is to reduce the transportation jams and intense pollution. A significant role is played by the public transportation systems that use electric energy in electro-mechanic drive systems. An example in this sense consists of the urban transportation systems by trolley-bus, where one can notice a trend in substituting the d.c. drive by a.c. drive. Traction converters for feeding asynchronous machines are for more than 20 years a dominant application for power semiconductors in the high-power range. The introduction of IGBT exhibits advantages as: control through voltage, simpler driver technology, no snubber circuit, less complexity and increased efficiency for partial load, low switching losses and therefore higher possible switching frequency.

The driving alternative that uses choppers, with performances superior to those from the classic solution (d.c. motor and velocity control through resistances), due to the d.c. motors utilization, does not entirely eliminates the potential sources of fault (collector, brushes, contactors). One can provide the trolley-buses with intelligent electronic systems by substituting the classic d.c. driving systems through their a.c. counterparts (static power converters and asynchronous motors) in order to increase the reliability, the energetic efficiency and

exploiting safeness and also to significantly reduce the maintenance and exploiting costs.

The application in traction exhibits requirements that exceed the standard industrial conditions.

II. ELECTRIC DRIVING SYSTEM CONTROL

The electric driving system control (Fig.1) is based on a structure described below.

The electronic control block is used to perform: the interface with the power electric circuits, the interface with the vehicle (running conditions), the prescription of electric quantities, the control and regulation, the determination of parameters on the vehicle, the acquisition of numerical signals and respectively the generation of signal for numerical control, diagnosis and protection.

The control equipment provides pulses to the inverter through driver circuits so as to provide continuous control for two quantities: statoric current and rotoric frequency. In Fig. 2 a block diagram of a IGBT driver is presented.

The function spectrum of these circuits mostly comprises:

- gate voltage generator;

- input for VCEsat- monitoring, sometimes also input for shunt or sense-emitter;

- monitoring of too low supply voltage;

- error memory and error feedback output ;

- adjustable dead time generation of the TOP-driver.

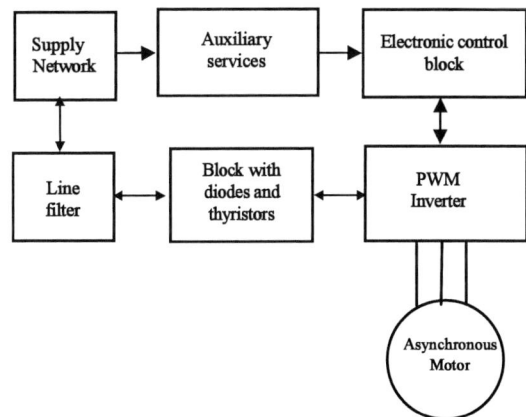

Fig. 1. Driving system structure.

This work was supported by ANCS Romania through the 4-th Program-Partnerships and under the Contract 71-145/2007

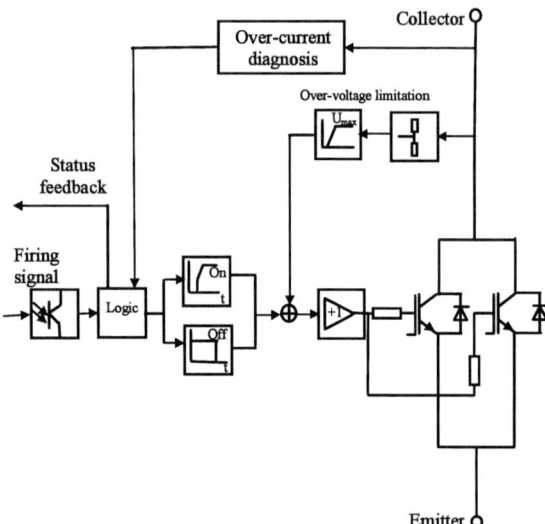

Fig. 2. Block diagram of IGBT driver.

These standard drivers do not provide a real potential isolation. For some variants, the control input may be configured for connection of opto-couplers or pulse transformers. Moreover, progress is being made in the development of fast optic couplers with power driver output which have already integrate supply under voltage- and VCEsat- or VDS (on)-monitoring. To achieve simple driver units, DC/DC-converter and few passive components merely have to be added [1].

With the growing variety of function and protection parameters in driver circuits, the assemblies necessary on the primary side also have to meet more sophisticated requirements, for example, input signal logic, short-pulse suppression, dead time generation, error memory and error evaluation and drive of the pulse transformers.

During turn-on, switching speed is limited by a ramping up of the control voltage, which allows a good compromise between soft recovery and low turn-on losses. During turn-off, a feedback-control of the collector voltage will provide a safeguard against a too high voltage [2]. A voltage divider in combination with analogue amplifier which is coupled to the gate-control voltage is used here. For short-circuit protection the collector voltage is compared to a reference voltage.

The system requires information on the angular velocity from the system so that the microcontroller should be able to perform a real time computation of the asynchronous motor control frequency as described by:

$$f_1 = f_N \pm f_2 + \Delta f_2 \qquad (1)$$

where: f1- stator supply frequency;

fN- frequency proportional to the vehicle velocity;

f2- frequency of slip;

Δf2: frequency correction;

$\pm$: +running; - braking.

After the determination of the starting period, when the motor's terminals (Fig.3.) are submitted to the entire line voltage, one performs an automated flow weakening, so that the vehicle might reach full speed.

Fig. 3. Driving asynchronous motor.

III. ELECTRIC DRIVING SYSTEM

In the brake regime, the control block controls the current through motor and the voltage in the intermediate circuit (as depicted by Fig.1). Depending on the voltages from the intermediate circuit and respectively from the supplying network, the braking can be either electrical-recovering when the braking energy is transferred in the supplying network, or electrical rheostatic when there aren't any consumers on the line and the braking energy is dissipated along the braking resistances through the braking chopper.

The driving system's design relies on the indirect vector control. The objective of the asynchronous motor's vector control is to allow it to be controlled just like a separately excited d.c. machine. This is achieved through transformations of variables to the coordinate reference frame that rotate along with the rotor flux.

An advantage of this type of control consists in the linearity of mechanical characteristics, respectively of the characteristic torque versus motor speed [3].

Further on there are given some features of the main elements from the driving systems. The inverter (Fig. 4) was designed considering the following rated data:

- supplying voltage: 750 V d.c.
- maximum apparent power: 300 kVA

Fig. 4. PWM Traction inverter.

- maximum active power: 200 kW
- maximum phase current: 410 A
- switching frequency: 2 kHz

The designed motor should meet some special requests owing to the particular features of urban electric drives:

- rated current (RMS): 228 A
- rated speed: 1479 rot/min
- rated voltage: 500 V
- power factor: 0.86
- efficiency: 91.5
- rated frequency: 50 Hz
- maximum frequency: 106.5 Hz

The following data were yielded during designing:

- stator inner diameter: 520 mm
- rotor inner diameter: 105 mm
- air gap: 2 mm
- stator phase resistance: 0.04054 Ω
- rotor phase resistance: 7.078 .10-5 Ω
- magnetization current: 40.11%

IV. EXPERIMENTAL RESULTS OF THE ELECTRIC DRIVING SYSTEM TESTING

A. Experimental Tests with Asynchronous Motor

The motor designing was realized in co-operation with S.C. Electroputere S.A. Craiova. To achieve an efficient design and manufacturing one performed numerical simulations (by means of a specialized MATHCAD software) [4]. This way the asynchronous motor characteristics were deduced. One could notice a good agreement between the designing parameters and those required for a motor used in the electric urban traction [5].

The tests over the manufactured asynchronous motor (from Fig.3) were done at S.C. Electroputere S.A. Craiova, on special stands that provide the possibility to make operational schema with energy recovering (for braking). The tests were performed at a frequency equal to 50 Hz (for the rated speed) and a frequency equal to 106.6 Hz corresponding to a synchronous speed equal to 3200 rpm. The supplying voltage was sinusoidal.

One considered the specifications of the standard EN 60034-1 (IEC 349-2). The tests results yielded are presented by Tables I and II.

B. Experimental Tests with PWM Inverter

The realized converter was submitted to a series of tests according to present standards and the results are presented below:

1) Heating tests for the main components of the inverter (according to EN 50207 par. 4.5.3.11)

The converter operated at rated parameters, the sub-ensembles temperatures being lower than the maximum values presented in Table III.

2) Tests on the behavior at the discontinuous variation of line voltage (according to EN 50207 par. 4.5.3.21.)

The verification was performed at rated load. The inverter's supplying voltage was varied in 2 steps:

- from 525 V d.c. to 1200 V d.c.;
- from 1200 V d.c. to 525 V d.c.

After tests performing one noticed that during the discontinuous variation of the supplying voltage within the range 525 V d.c. ÷ 1200 V d.c., the converter operates normally, at prescribed parameters.

3) Test of the behavior to short lasting interruption of supplying voltage (according to EN 50207 par. 4.5.3.22.)

At the interruption of converter's supplying voltage (750 V d.c.), the converter entered in the state "STOP" and the IGBT-s control pulses were blocked.

When the supplying voltage reappeared the inverter starting cycle was replayed and the inverter supplied an output voltage with the frequency given by the motor's angular velocity at the start time.

4) Tests on the behavior at start and restart moments (according to EN 50207 alin. 7.4.2.)

The tests were performed when the supplying voltage varies within the range 525 V d.c. - 1200 V d.c. The converter started directly when it got a supplying voltage within the range 525 V d.c. - 1200 V d.c. (according to EN 50207 par.7.4.2.), providing the specified voltages at the output.

TABLE I.
PARAMETERS FROM TESTS AT FUNDAMENTAL FREQUENCY

Frequency - 50 Hz; Synchronous speed - 1500 rpm									
Power (kW)	Line Voltage (V)	R1 (150^{0}C) (Ω)	Current (A)	Speed (rpm)	Mn (N.m)	Slip (%)	cos ϕ	Efficiency (%)	Mmax (N.m)
155	500	0.0413	227	1479	1000	1.4	0.86	91.5	3800

TABLE II.
PARAMETERS FROM TESTS AT A FREQUENCY OF 106.6 HZ

Frequency – 106.6 Hz; Synchronous speed - 3200 rpm									
Power (kW)	Line Voltage (V)	R1 (150^{0}C) (Ω)	Current (A)	Speed (rpm)	Mn (N.m)	Slip (%)	cos ϕ	Efficiency (%)	Mmax (N.m)
155	500	0.0413	200	3120	477	2.5	0.94	94	1097

TABLE III.
SUBENSEMBLIES TEMPERATURES OBTAINED AFTER THE HEATING TESTS

No.	Component	Imposed temperature	Registered temperature
1	radiators	<+70 °C	52 °C
2	filter's capacities	<+70 °C	50 °C
3	inductivities, transformers	<+70 °C	92 °C
4	bind bars	<+70 °C	50 °C

5) Tests on the behavior at short-circuit (according to EN 50207 par. 7.4.3.)

After these tests one could see that the equipment did not suffer any damages during external short-circuits, along the a.c. current outputs. The inverter was set in the fault state „SHORT-CIRCUIT", the IGBT control pulses were blocked and the inverter starting was accomplished through the „RESET" procedure.

6) Tests on the overload operation abilities (according to EN 50207 par. 7.4.6.)

The converter can provide on the a.c. output a current of 300 kVA for 5 sec. with no damages and no critical temperature overcoming.

The converter provides at the output a maximum current limited by 480 A a.c.+3%.

The tests revealed that the equipment supplies the overload applied on the a.c. output for 5 sec. and limits the output current to the prescribed value (480A).

C. Experimental tests of the entire driving system testing

Afterward a series of tests were made for the entire driving system according to the electric schema presented in Fig. 5. with asynchronous motor supplied through PWM inverter.

The motor speed can be estimated from the time length of a pulse received from the position sensor as Halls, using a high frequency timer to count between two consecutive transitions of a pulse from the position sensors.

The accuracy of this speed estimator depends on the maximum motor speed Ω_{max}, the number p of pole pairs of the asynchronous motor and the frequency of the count timer, f_{clk}. The above are descried by:

$$Acc_{bits} = \log 2\left(\frac{f_{clk} \cdot 2 \cdot \pi}{p \cdot \Omega_{max}}\right). \qquad (1)$$

For smaller speeds important dead times between the speed control loop and the speed estimator will arise. A possible correction of this effect can be implemented. One supposes that the speed estimate obtained from the time length of the i-th pulse is Ω_{0i}. This estimate, average speed over the previous pulse, can be quite far from the real speed Ω_{real} of the asynchronous motor, mainly during transients of motion.

Some correction terms will be used to obtain a better speed estimate Ω_{corr}, between two consecutive pulse, using the relation:

$$\Omega_{corr} = \Omega_{0i} + C_{i0} + C_{iq} \qquad (2)$$

where C_{i0} is a constant term modeling the static load torque and C_{iq} is a constant modeling the torque constant [6], [7].

The measured phase currents, i_a and i_b, are transformed into the stator reference frame components i_{ds} and i_{qs}. Then, based on the rotor field position information (computed from the rotor position and the corrective term given by the slip calculator book), these components are transformed into the rotor frame direct and quadrature components i_d and i_q.

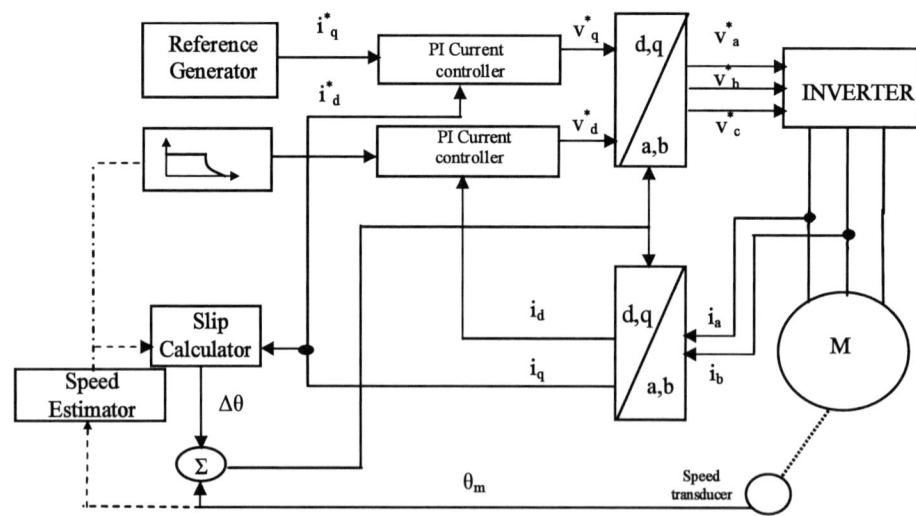

Fig. 5: Scheme of the asynchronous motor operating in torque/current mode

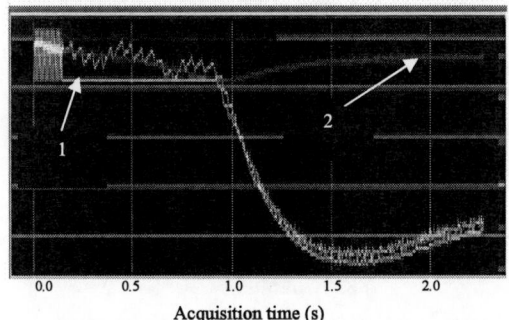

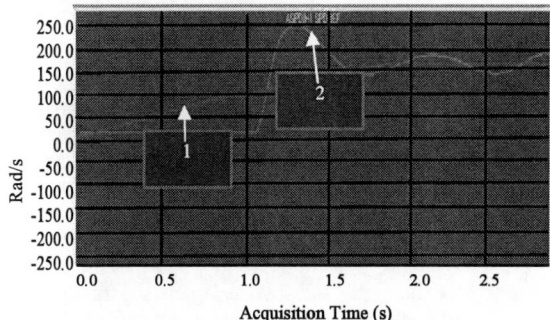

Fig. 6: The motor estimated and measured resistance

Fig. 7. Waveforms for the motor's axis reference speed ("1") and real speed ("2") for the rotation sense "forward".

The current controllers are PI discrete controllers. The inverse coordinates transformation is used for computation of the phase voltages references v^*_a, v^*_b and v^*_c applied to the inverter, starting from the values of voltage references computed in the d and q reference frame (v^*_q, v^*_d) [8]. Thus, the six full compare PWM outputs of the microcontroller are directly driven by the program, based on the reference voltages. For frequency over the rated value, the motor flux must be reduced in order to rotate the asynchronous motor at speeds greater than the rated value.

To calculate the parameters required for the implementation of flux weakening procedure one must know the magnetization current, i_m. Using i_m one can compute the parameters i_{dmax} and i_{dmin} used in the microcontroller specific language. The maximum magnetization current is calculated whereas the minimum value is imposed by user.

The following quantities are used for the magnetization current calculation: motor rated voltage U_1, rated frequency f_1, phase magnetization inductance L_m, phase stator resistance R_s and electric time-constant τ. The electromagnetic torque is calculated with the equation:

$$T_e = \frac{3}{2} \cdot p \cdot \frac{L_m^2}{L_2} \cdot |i_d| \cdot i_q \qquad (3)$$

where p is the number of pole pairs, L_m is the magnetization inductance, L_2 is the rotor equivalent self inductance, i_d is the stator reactive current and i_q is the stator active current in the rotor frame.

Fig. 6 depicts the asynchronous motor estimated and measured resistance. The signal denoted by "1" corresponds to the estimated resistance and the signal denoted by "2" corresponds to the asynchronous motor measured resistance.

The resistance from rotor reported to stator was calculated using the following equation:

$$R_{2r} = \frac{s_r \cdot k_E \cdot U_1}{I_{2r}} \qquad (4)$$

where s_r is the rated slip, k_E is a constant, I_{2r} is the rotor current reported to the stator.

During the motor idle starting at relative high acceleration one can notice the resistance diminishing up

to the moment when the motor inertial couple is cancelled when this tends to reach the minimum value orresponding to the limited maxim current supplied by the inverter.

In an ideal case both graphics must fit (the mathematic model of the control block must be very close to reality).

Fig. 7 depicts the waveform for the " forward " sense. The signal denoted by "1" stands for the motor axis reference speed whereas that denoted by "2" represents the real speed.

Motor starting was accomplished under the conditions of its blocking along axis in the first moment, followed by its sudden release.

This thing emphasizes the answer of the driving system inverter-motor in this situation.

One can notice that when the brake used to block the motor axis is released the speed tends to reach the reference speed and that after an over-regulation and a damped oscillation both values become equal.

Both the speed maximum magnitude and the oscillation period can be controlled by means of the parameters of speed regulator modeled by the inverter's control software.

Fig. 8 depicts the waveforms for the "backward" rotation sense considering that: the signal denoted by "1" stands for the motor's axis reference speed and the one denoted by "2" stands for the real speed. The characteristics speed versus time are identically shaped with those from Fig. 7, but they have reversed senses.

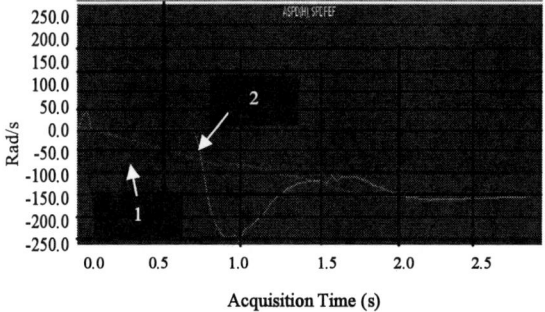

Fig. 8. Waveforms for the motor's axis reference speed ("1") and real speed ("2") for the rotation sense "backward".

V. CONCLUSIONS

The experimental determinations made on stand concerning the asynchronous motor testing revealed the existence of a power reserve that satisfy the requirements of the present application. Considering the numerical simulations results and all the factors that might influence the asynchronous motor behaviour, one can see that is possible to perform an optimised design, reducing the sizes with (8-10)% for the same characteristics of the motor [4].

The tests over the motor and inverter, made separately, were performed considering the European standards (EN 60034-1 and EN 50207). The experimental waveforms verified the validity and correctness of the driving control scheme.

The employed technology and tests were in agreement with the operational conditions of the electric driving system that is to be done.

The driving systems based on the asynchronous motor will eliminate most of the disadvantages presented by the d.c. driving systems due to the inverter control scheme based on the vectorial control.

A superior energetic efficiency will appear because:

• the electric braking might be regenerative until stop and the energy dissipation over the breaking resistances should no longer be present;

• the asynchronous motor has identical characteristics for both regimes: operating as motor and during braking;

• the asynchronous motor exhibits insignificant losses and no thermal problems; the recovered braking energy can used by auxiliary services when no other consumers are present on the line.

Considering the European trends – for the development of environmental - friendly urban transportation system in big cities, one can conclude that the developed technological solution can be improved considering both the asynchronous motor and respectively the PWM inverter (using last - generation electronic components).

ACKNOWLEDGMENT

The authors thank to their partners from S.C. INDAELTRAC S.A. and S.C. ELECTROPUTERE S.A. Craiova, who helped them to develop and implement this technological solution.

REFERENCES

[1] R. Lorentz, "Advances in electric drive control", *Proceedings of IEEE International Electric Machines and Drives Conference*, Seattle, Washington, USA, 1999, p.9-16

[2] N. Ravisekhar Raju., "Rippled dc link twelve-pulse converter with sinusoidal ac waveform", *Proceedings of IPEC 2005 - The International Power Electronics Conference*, Toki Messe, Session Paper S63, pp.16-20, Niigata, Japan, April 4-8, 2005.

[3] D. Mori and T. Ishikawa., "Force and vibration analysis of a PWM inverter-fed induction motor", *Proceedings of IPEC 2005 - The International Power Electronics Conference*, Toki Messe, Session Paper S12, pp.9-15, Niigata, Japan, April 4-8, 2005

[4] P.M. Nicolae, D.G., Stanescu and V.D.,Vitan, "Designing, realization and test of asynchronous motors used for trolley-buses electric drive", *Proceedings of Iasi Polytechnic Institue, Tomul LII(LVI), Fasc.5.A, Electrotechnics, Energetics, Electronics*, Oct. 2006, pp. 50-55

[5] M. El-Hami and S. Abu-Sharkh, "A general design model for electric motors", *Proceedings of IEEE International Electric Machines and Drives Conference*, Seattle, Washington, USA,1999, p.186-188

[6] M. Harakawa, T. Nagano and E. Hayasaka., "Improvement of vector control in consideration of voltage saturation", *Proceedings of IPEC 2005 - The International Power Electronics Conference*, Toki Messe, Session Paper S65, pp.8-14, Niigata, Japan, April 4-8, 2005

[7] U. K. Madawala and C. A. Baguley., "Transient modeling and parameter estimation of field aligned starting", *IEEE Trans. On Energy Conversion*, vol. 23, pp.15-24, March 2008

[8] B. Singh and V. Verma., "An indirect current control of hybrid power filter for varying loads", *IEEE Trans. On Power Delivery*, vol. 21, pp.178-184, January 2006

Real-World Force Feedback Control for Mobile-Hapto

Wataru Yamanouchi*, Yuki Yokokura†, Seiichiro Katsura‡, Kiyoshi Ohishi§

*Dept. of Electrical Engineering, Nagaoka University of Technology, Nagaoka, Niigata, Japan,
e-mail: *babel@stn.nagaokaut.ac.jp*
†Dept. of Electrical Engineering, Nagaoka University of Technology, Nagaoka, Niigata, Japan,
e-mail: *yuki@stn.nagaokaut.ac.jp*
‡Dept. of System Design Engineering, Keio University, Kohoku, Yokohama, Japan,
e-mail: *katsura@sd.keio.ac.jp*
§Dept. of Electrical Engineering, Nagaoka University of Technology, Nagaoka, Niigata, Japan,
e-mail: *ohishi@vos.nagaokaut.ac.jp*

Abstract—Recently, communication tools such as the portable telephone and Internet spread all over the world. These devices substitute human eye and ear. Haptic information has paid attention as the third multimedia information. This paper proposes a transmission technique of real world haptic information. This paper proposes a novel haptic transmission in different motion area, named mobile-hapto. The mobile-hapto consists of a mobile robot and a joystick. Intuitive manipulation results become possible by using mobile-hapto. Effectiveness of this proposition method is verified by driving experimental.

Keywords—Motion control, Mechatronics, Control methods for electrical system, Robotics

I. INTRODUCTION

Multimedia technologys have been widely spread as the basis for human communications. For example, telephone, television and radio broadcasting are good tools for remote communication. In other words, such tools connect a human and other humans in remote environments. Recently, Internet and the portable telephone have spread in the world. Thus, information technology and communication technology have supported growth of industry applications. Once these sensations are obtained as digital data, it is easy to save, transmit, reproduce and process. This means that human functions of eyes and ears are substituted by multimedia devices.

Haptic information has paid attention as the third multimedia information. However, since haptic information is subject to the Newton's "law of action and reaction" in the real world, it is difficult to attain a device, which acquires haptic information, Namely, haptic information is regarded as bilateral information. To acquire such a haptic information, a device should have bilateral ability in its controller.

There is much research about bilateral control for a master-slave system to manipulate remote objects [1]–[4]. In addition, researches for remote medical care and remote manipulation are presented. For example, forceps robot [5], space remote manipulation robot [6] and so on. In this way, haptic transmission of the real-world is important issue for future human assistance.

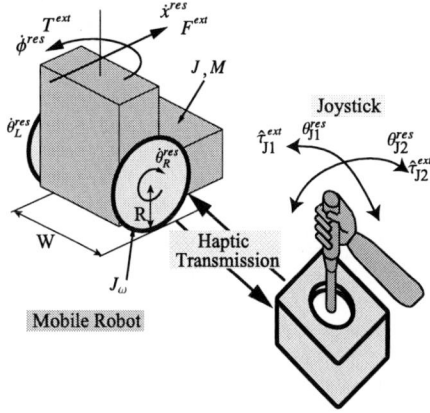

Fig. 1. Concept of mobile-hapto

Bilateral control is used for these controls mainly. Mostly, in case of bilateral control, a teleoperated master-slave system has the some structure and configurations. There are many methods for bilateral control in different configurations [7]. Macro-micro teleoperation has been proposed [8]. However, there is little research on haptic transmission between devices which have different motion areas.

For the reason, in this paper proposes sensory feedback with the movable haptic device. This concept is named "mobile-hapto". The word "hapto" means "touch" in Greece. The mobile-hapto consists of a mobile robot and joystick. A mobile robot has infinite area for its motion. On the contrary, a joystick is fixed for operation by a human. Once such a mobile haptic transmission is attained, a human can recognize remote environment without moving there. A human operates the joystick to control the mobile robot. The mobile robot recognizes the road environment and transmits it to a remote human. As a result, a human feels the remote road environment. In the paper, flat and rough terrains are used as the road environments. Real-world and real-time bilateral force feedback is attained by this frame work of mobile hapto. Usefulness of the proposed method is verified by some

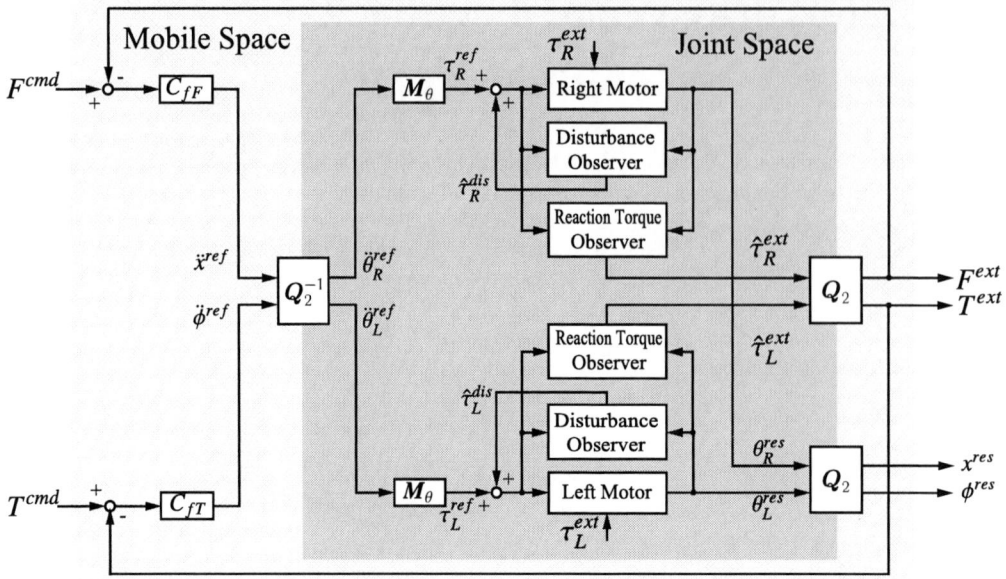

Fig. 2. Block diagram of mobile robot

experiments.

This research is organized as follows. Section II shows dynamics of a mobile robot, and model of mobile-hapto system. In Section III, real-world force feedback control is shown. The force control system is designed. Also, Section III describes bilateral control in different motion area. The experimental results are shown in Section IV. The last section summarizes this paper.

II. MODELING OF MOBILE-HAPTO SYSTEM

This paper proposes a force feedback control use of joystick and mobile robot. The concept of mobile-hapto is shown Fig. 1. The system parameters are described as follows,

θ_{J1} : joystick position for translational mode;
θ_{J2} : joystick position for rotation mode;
τ_{J1}^{ext} : reaction joystick torque of translational mode;
τ_{J2}^{ext} : reaction joystick torque of rotational mode;
$\dot{x}$: translational velocity of the mobile robot;
$\dot{\phi}$: rotational velocity of the mobile robot;
F^{ext} : reaction force of translational mode;
T^{ext} : reaction torque of rotational mode;
$\dot{\theta}_R$: velocity of right wheel;
$\dot{\theta}_L$: velocity of left wheel;
J : inertia of the mobile robot with respect to a vertical axis;
J_ω : inertia of driving motor and wheel;
R : radius of driving wheel;
W : tread of the mobile robot;
M : mass of the mobile robot.

A. Dynamic Relationship in Joint Space

This section shows an equivalent inertia matrix to construct the mobile robot control system. When the

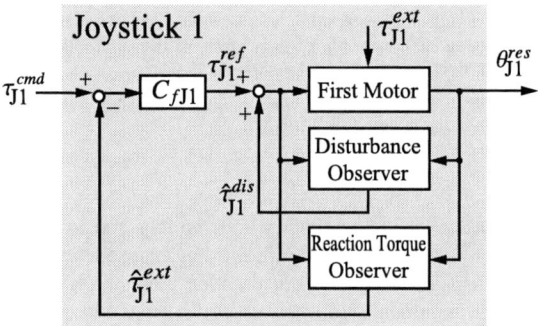

Fig. 3. Block diagram of joystick 1

parameters that are shown in Fig. 1 are fixed; movement energy function is defined by (1)

$$K = \frac{1}{2}M\left(\frac{R}{2}(\dot{\theta}_R + \dot{\theta}_L)\right)^2$$
$$+ \frac{1}{2}J\left(\frac{R}{W}(\dot{\theta}_R - \dot{\theta}_L)\right)^2$$
$$+ \frac{1}{2}J_\omega(\dot{\theta}_R^2 + \dot{\theta}_L^2). \quad (1)$$

In (1), the first term of right side is the translational movement energy, the second term is the rotational energy of the mobile robot, and third term of right side is the rotation energy of the right and left wheels. The dynamics of the mobile robot should be defined by solving the Lagrange equation from (1), yields and it represent in (2)

$$\begin{bmatrix} \tau_R \\ \tau_L \end{bmatrix} = M_\theta \begin{bmatrix} \ddot{\theta}_R \\ \ddot{\theta}_L \end{bmatrix} \quad (2)$$

where τ_R and τ_L is torque of right and left driving motor, respectively. In regard to M_θ, it is called the

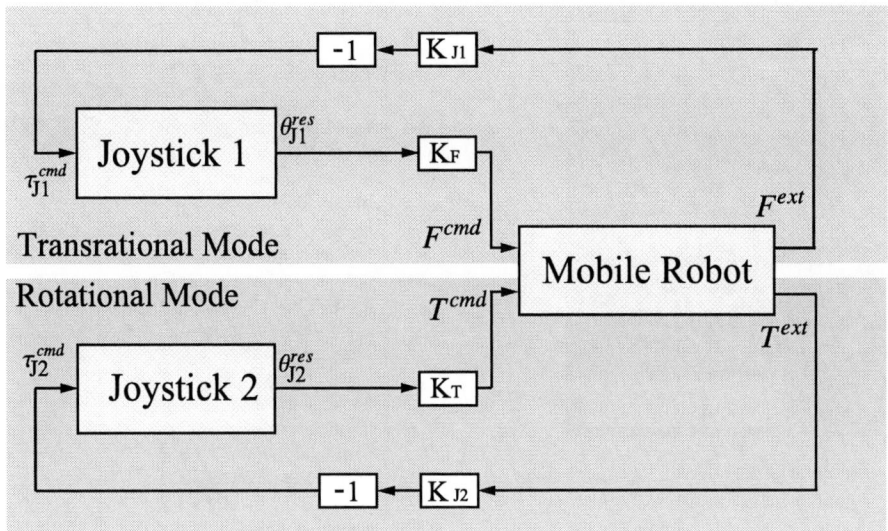

Fig. 4. Block diagram of bilateral control in different motion area

equivalent inertia matrix. The equivalent inertia matrix M_θ is expressed as (3)

$$M_\theta = R^2 \begin{bmatrix} \frac{M}{4} + \frac{J}{W^2} + \frac{J_\psi}{R^2} & \frac{M}{4} - \frac{J}{W^2} \\ \frac{M}{4} - \frac{J}{W^2} & \frac{M}{4} + \frac{J}{W^2} + \frac{J_\psi}{R^2} \end{bmatrix} \quad (3)$$

The control system of the mobile robot composes using (2) and (3).

III. REAL-WORLD FORCE FEEDBACK CONTROL

A. Force Control System Using Quarry Matrix

In order to construct force control system, robust acceleration control system is required. Robust acceleration control system is constructed by disturbance observer [9]. The mobile robot which is used in this research is controlled by force control. The right blocks in this Fig. 2 shows the block diagram of force control system for the mobile robot. The mobile robot are described as follows,

C_{fF} : Force control gain of translational mode;
C_{fT} : Force control gain of rotational mode.

The conversion from real-world joint space to virtual-world modal space is achieved by using a quarry matrix [10]. The second-order quarry matrix Q_2 is defined by (4) [11]

$$Q_2 = \frac{1}{2} \begin{bmatrix} 1 & 1 \\ 1 & -1 \end{bmatrix} \quad (4)$$

Equation (5) shows the position responses of right and left driving motor θ_R^{res}, θ_L^{res} are transformed into translational mode position response x^{res} and rotational mode position response ϕ^{res} by the second order quarry matrix Q_2

$$\begin{bmatrix} x^{res} \\ \phi^{res} \end{bmatrix} = Q_2 \begin{bmatrix} \theta_R^{res} \\ \theta_L^{res} \end{bmatrix}. \quad (5)$$

Equation (6) shows that the reaction torque responses of right and left driving motor from $\hat{\tau}_R^{ext}$, $\hat{\tau}_L^{ext}$ are transformed into translational mode reaction force response

F^{ext} and rotational mode reaction torque response T^{ext} by the second order quarry matrix Q_2

$$\begin{bmatrix} F^{ext} \\ T^{ext} \end{bmatrix} = Q_2 \begin{bmatrix} \hat{\tau}_R^{ext} \\ \hat{\tau}_L^{ext} \end{bmatrix}. \quad (6)$$

The reaction torque responses are estimated by a reaction torque observer without force sensors [12]. This system controls translational mode force F^{ext} and rotational mode torque T^{ext}, respectively. The acceleration references $\ddot{x}^{ref}$, $\ddot{\phi}^{ref}$ are denoted by (7), (8)

$$\ddot{x}^{ref} = C_{fF} \left(F^{cmd} - F^{ext} \right) \quad (7)$$

$$\ddot{\phi}^{ref} = C_{fT} \left(T^{cmd} - T^{ext} \right) \quad (8)$$

where the translational mode force response F^{ext} and rotational mode torque response T^{ext} should follow translational mode force command F^{cmd} and rotational mode torque command T^{cmd}.

On the contrary, the transformation from virtual-world modal space to real-world joint space is attained by (9)

$$\begin{bmatrix} \ddot{\theta}_R^{ref} \\ \ddot{\theta}_L^{ref} \end{bmatrix} = Q_2^{-1} \begin{bmatrix} \ddot{x}^{ref} \\ \ddot{\phi}^{ref} \end{bmatrix}$$
$$= \begin{bmatrix} 1 & 1 \\ 1 & -1 \end{bmatrix} \begin{bmatrix} \ddot{x}^{ref} \\ \ddot{\phi}^{ref} \end{bmatrix}. \quad (9)$$

where Q_2^{-1} is the inverse matrix of the quarry matrix Q_2. More specifically, the translational and rotational mode acceleration reference $\ddot{x}^{ref}$, $\ddot{\phi}^{ref}$ which are calculated by (7), (8) are converted to acceleration references of right and left driving motors $\ddot{\theta}_R^{ref}$, $\ddot{\theta}_L^{ref}$. The acceleration references are attained by robust acceleration control.

By this means, the velocity control system of the mobile robot using the quarry matrix achieves independent motion control with respect to translational mode and rotational mode.

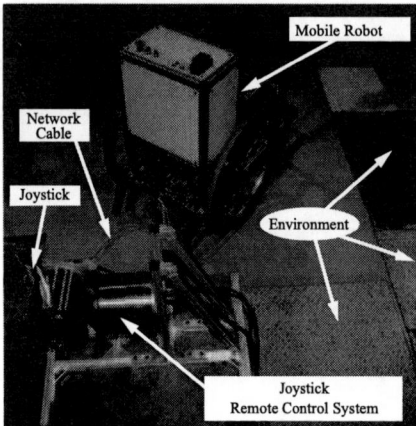

Fig. 5. Remote control system of mobile robot with haptic transmission function.

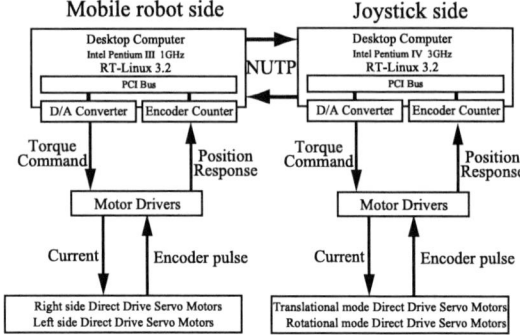

Fig. 6. Control system of bilateral control in different motion area.

B. Force Feedback Joystick

The joystick has two-degree-of-freedom. Fig. 3 shows the block diagram of the joystick 1. The joystick parameters are described as follows,

C_{fJ1} : joystick torque gain of translational mode;
C_{fJ2} : joystick torque gain of rotational mode.

The joystick is also controlled by the force control. The torque references of joystick 1 τ_{J1}^{ref} and joysthick 2 τ_{J2}^{ref} are given by (10) and (11)

$$\tau_{J1}^{ref} = C_{fJ1} \left(\tau_{J1}^{com} - \hat{\tau}_{J1}^{ext} \right) \qquad (10)$$

$$\tau_{J2}^{ref} = C_{fJ2} \left(\tau_{J2}^{com} - \hat{\tau}_{J2}^{ext} \right) \qquad (11)$$

where $\tau_{J1}^{com}, \tau_{J2}^{com}$ are torque command of joystick 1, joystick 2. Human input torque are $\hat{\tau}_{J1}^{ext}, \hat{\tau}_{J2}^{ext}$.

C. Bilateral Control in Different Motion Area

Fig. 4 shows the block diagram of the bilateral control in different motion area. Parameters are described as follows,

TABLE I
PARAMETER OF EXPERIMENT

K_{J1}	feedback force gain of translational mode	1
K_{J2}	feedback torque gain of rotational mode	1
C_{fJ1}	joystick torque gain of translational mode	15
C_{fJ2}	joystick torque gain of rotational mode	15
K_F	torque transform gain of translational mode	5
K_T	force transform gain of rotational mode	2.5
C_{fF}	joystick torque gain of translational mode	5
C_{fT}	joystick torque gain of rotational mode	5
M	Mass of the mobile robot	50[kg]
W	Tread of the mobile robot	0.580[m]
R	Radius of wheel	0.306[m]
K_t	Torque constant of the motor	4.46[Nm/A]
J_ω	Inertia of wheel and motor	0.14[kgm^2]
J_n	Inertia of body of mobile robot	5.7[kgm^2]

K_{J1} : feedback force gain of translational mode;
K_{J2} : feedback torque gain of rotational mode;
K_F : torque transform gain of translational mode;
K_T : force transform gain of rotational mode.

The reaction torque estimated in the mobile robot is feedback to the joystick. The torque commands of joystick into translational mode and rotational mode τ_{J1}^{com}, τ_{J2}^{com} are given by (12) and (13)

$$\tau_{J1}^{com} = -K_{J1} F^{ext} \qquad (12)$$

$$\tau_{J2}^{com} = -K_{J2} T^{ext}. \qquad (13)$$

Translational mode force command is given by multiplying in position θ_{J1}^{res} deviation from the initial position with the force transformation gain K_T. Rotational mode is calculated in the same way with K_T and θ_{J2}^{res}.

$$F^{cmd} = K_F \theta_{J1}^{res} \qquad (14)$$

$$T^{cmd} = K_T \theta_{J2}^{res}. \qquad (15)$$

In this way, bilateral control in different motion area is attained by bilateral force feedback.

IV. EXPERIMENT

Fig. 5 shows picture of experiments. Experimental system used mobile robot and joystick. The mobile robot and the joystick consist of two direct drive motors. The direct drive motors of mobile robot put on right and left of Wheels. The operating area of mobile robot is infinity. On the contrary, the motion area of the joystick is limited.

The control system used in this paper is shown Fig. 6. The mobile robot is controlled by a personal computer where the CPU is Pentium III. The joystick is controlled by another personal computer where the CPU is Pentium IV. These personal computers are operated by RT-Linux.

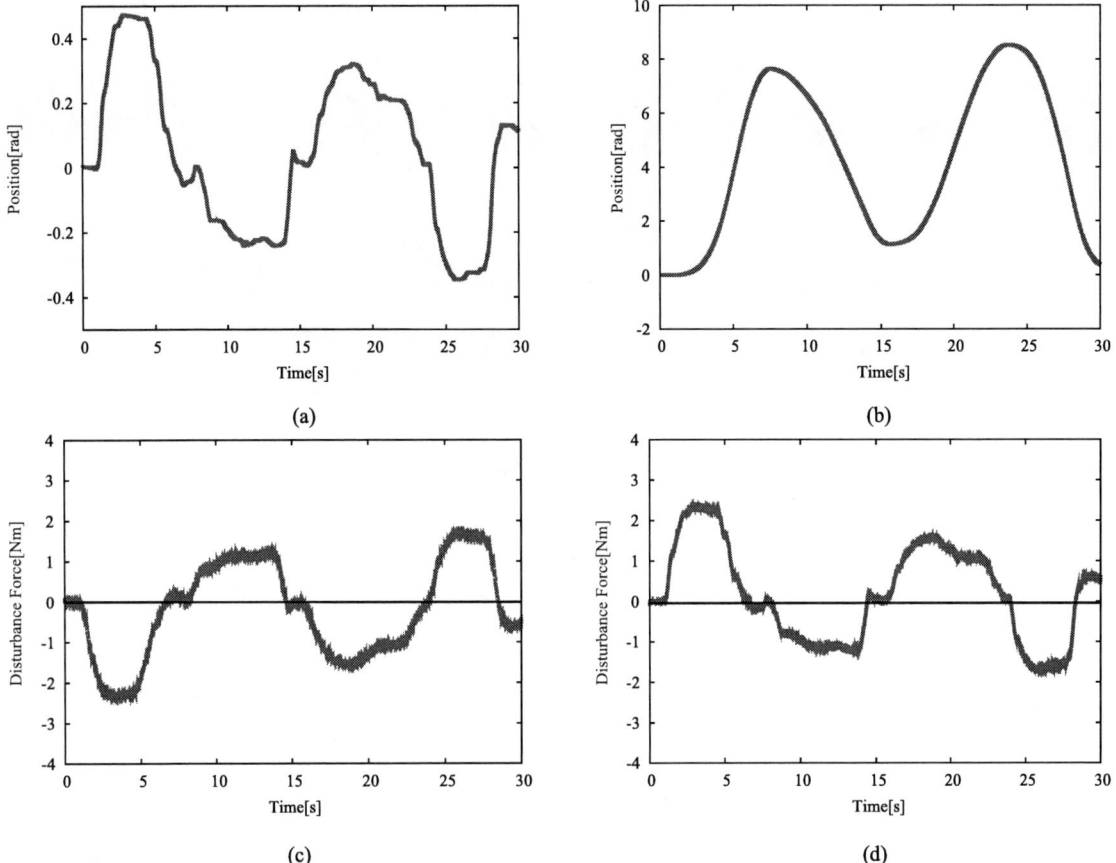

Fig. 7. Experimental results in flat area. (a) Joystick position of translational mode. (b) Mobile-robot position of translational mode. (c) Joystick force of translational mode. (d) Mobilerobot force of translational mode.

The personal computers are connected by LAN. The paper uses a new protocol in communication, named by NUTP.

A human operates the joystick to control the mobile robot. The mobile robot recognizes the road environment and transmits it to a remote human. As a result, a human feels the remote road environment.

Parameters of the experiments are shown TABLE I.

A. Experimental Results in Flat Area

First, the experiment that the mobile robot is operated in frat area is done. The joystick moves according to the force input from a human. Then, the mobile robot moves and obtains the reaction force from the road environment. Fig. 7 shows the experimental results. Fig. 7 (a) and Fig. 7 (b) show the position references of joystick and the mobile robot with respect to the translational mode It tarns out that. The mobile robot and the joystick position is different. Fig. 7 (c) and Fig. 7 (d) show the torque response of the joystick and force of the mobile robot. From this figure, transmission of bilateral haptic information is successfully attained in different area between mobile robot and joystick.

B. Experimental Results in Rough Area

Second, the experiment that the mobile robot is operated in rough area is done. Fig. 8 shows the experimental results in rough area. Fig. 8 (a) is the joystick position, and Fig. 8 (b) is the mobile robot position of translational mode. These experiment results of position analogize flat area experiment result of position. Fig. 8 (c) and Fig. 8 (d) show torque of the joystick and force of the mobile robot. It turns out that the vibration phenomenon is the force responses. They are the reaction force the road environment when the mobile robot drive the rough road

As a result, haptic information estimated by the mobile robot is transmitted to the remote joystick.

V. CONCLUSIONS

This paper proposed a new haptic transmission system, named mobile-hapto. The mobile-hapto consists of a mobile robot and joystick. The joystick has limited motion area. On the contrary the mobile robot has infinite area for operating. Thus the joystick and mobile robot have different motion area. Because, this paper proposed bilateral control in different operation area. Real-world and real-time haptic information transmitted between robots of different operation area with this control. The mobile hapto will be used effectively as new remote medical care and future remote manipulation technology.

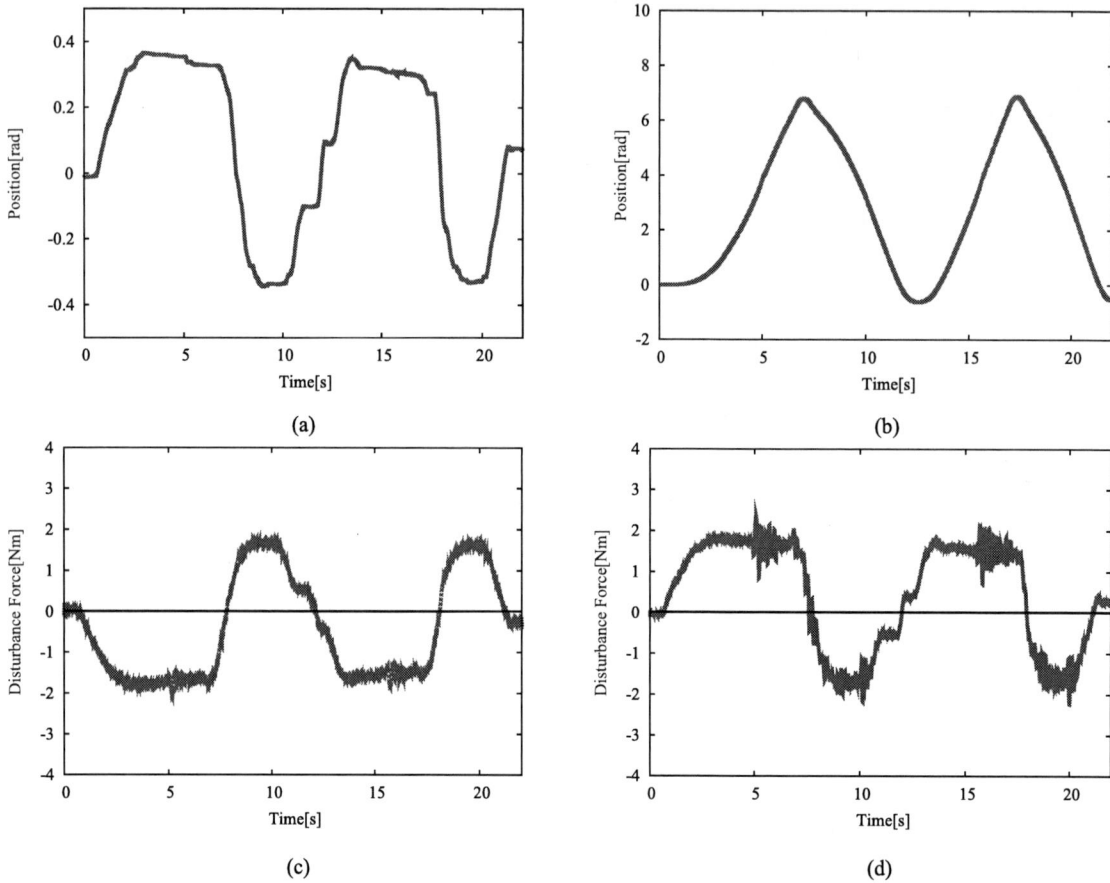

Fig. 8. Experimental results in rough area. (a) Joystick position of translational mode. (b) Mobile-robot position of translational mode. (c) Joystick force of translational mode. (d) Mobilerobot force of translational mode.

ACKNOWLEDGMENT

This work was supported in part by Industrial Technology Research Grant Program 07C46047a in 2007 from New Energy and Industrial Technology Development Organization (NEDO) of Japan.

REFERENCES

[1] R. J. Anderson and M. W. Spong: "Bilateral Control of Teleoperators with Time Delay," *IEEE Trans. on Automatic Control*, Vol. 34, No. 5, pp. 494–501 (1989).

[2] K. Takahara, T. Wakatsuki, H. Nozaki, R. Akiyama, H. Kawaguchi, Y. Ito , H. Wakamatsu: " Personal Cart System for Support of Elderly Person's Walk," *Journal of Asian Electric Vehicles*, 1, 1, pp. 417–422 2003.

[3] D. A. Lawrence: " Stability and Transparency in Bilateral Teleoperation," *IEEE Trans. on Robotics and Automation*, Vol. 9, No. 5, pp. 624–637 (1993).

[4] T. Imaida, Y. Yokokohji, T. Doi, M. Oda, T. Yoshikawa,: " Ground-Space Bilateral Teleoperation of ETS-VII Robot Arm by Direct Bilateral Coupling Under 7-s Time Delay Condition," *IEEE Transaction on Robotics and Automation*, Vol. 20, No. 3, pp. 499–511, 2004.

[5] Y. Ohba, S. Katsura, K. Ohishi: "Friction Free Bilateral Robot Based on Twin Drive Control System Considering Second Resonant Frequency," *The 31st Annual Conference of the IEEE Industrial Electronics Society IECON '05-RALEIGH, NORTH CAROLINA*, pp. 2023–2028, November 2–6, 2005.

[6] Y. Tsumaki and M. Uchiyama: "A Model-Based Space Teleoperation System with Robustness against Modeling Errors," *Proceedings of the 1997 IEEE International Conference on Robotics and Automation*,pp. 1594–1599 (1997).

[7] S. Katsura, T. Suzuyama, K. Ohishi: " Bilateral Teleoperation with Different Configurations using Interaction Mode Control," *2007 IEEE International Symposium on Industrial Electronics* ISIE '07-VIGO, pp. 3120–3125, June 4–7, 2007.

[8] K. Kaneko, H. Tokashiki, K. Tanie, K. Komoriya: "Impedance Shaping based on Force Feedback Bilateral Control in Macro-Micro Teleoperation System," *Proc. of 1997 IEEE Int. Conference on Robotics and Automation* , Vol. 1, pp. 710–717 (1997).

[9] K. Ohnishi, M. Shibata, T. Murakami : "Motion Control for Advanced Mechatronics," *IEEE/ASME Transactions on Mechatronics*, Vol. 1, No. 1, pp. 56–67, Mar. 1996.

[10] S. Katsura, K. Ohishi : "Modal System Design of Multi-Robot Systems by Interaction Mode Control," *IEEE Transactions on Industrial Electronics*, Vol. 54, No. 3, pp. 1537–1546, Jun. 2007.

[11] S. Katsura, K. Ohnishi : "Advanced Motion Control for Wheelchair in Unknown Environment," *2006 IEEE International Conference on Systems, Man, and Cybernetics, SMC '06-TAIPEI*, pp. 4926–4931, Oct. 2006.

[12] T. Murakami, F. Yu, K. Ohnishi : "Torque Sensorless Control in Multidegree-of-freedom Manipulator," *IEEE Transactions on Industrial Electronics*, Vol. 40, No. 2, pp. 259–265, Apr. 1993.

The new numerical integration routine applied in sensorless drives

Arkadiusz Gardecki[*], Krystyna Macek-Kamińska[†]

[*] University of Technology /Faculty of Electrical Engineering, Automatic Control and Computer Science, Opole,
Poland, e-mail: *a.gardecki@po.opole.pl*

[†] University of Technology /Faculty of Electrical Engineering, Automatic Control and Computer Science, Opole,
Poland, e-mail: *k.macek-kaminska@po.opole.pl*

Abstract—**The paper presents a new numerical integration routine BGKODE_DSP. This routine is based on BGKODE routine, which forms a practical application of the Krogh's algorithm. BGKODE_DSP routine realizes the predictor-corrector method and has been simplified to ensure its applicability in real time systems. The results of the comparison between the selected common numerical integration methods and the new routine is presented. The scope of this research also involves the range of accuracy of the solution resulting from the shorter computation time. Experiments have been carried out on experimental drive system with induction motor with the use of DTC method.**

Keywords—**sensorless control, electrical drive, real time processing.**

I. INTRODUCTION

Numerical methods find a common application in real-time systems, which require the solving of a system of differential equations. The most popular among them include the Runge-Kutta algorithm, representation of a function into the Taylor series [1] and the differential methods [2], which are based on polynomial approximation. This paper undertakes the application of a new BGKODE_DSP routine [3], which realizes predictor-corrector method and its comparison with the common numerical method for solving systems of differential equations. The comparison applies two routines of numerical differentiation realizing the Euler algotithm and the second-order Runge-Kutta (RK2) algorithm, as the common use of them is familiar in computations realized in real-time systems. In addition, selected comparative studies apply the procedure realizing the fourth-order Runge-Kutta (RK4) algorithm and the full version of the BGKODE routine [4, 5].

II. COMPARISON BETWEEN NUMERICAL METHODS

A. One Step Methods

Non-linear differential equations are most commonly solved with the numerical methods. Basically, the solution to the differential equation is obtained from the approximation of the time derivative with a simple difference equation

$$y_{n+1} = y_n + h\mathbf{f}(t_n, y_n) \qquad (1)$$

where local error is proportional to $O(h^2)$, and global error is proportional to $O(h)$.

The forward Euler approximation is one of the most popular methods used in real time systems [6, 7]. Besides, equation (1) constitutes a special case of the popular class of integration methods - the Runge-Kutta method. This method requires only a single evaluation of the right-hand side of the system of differential equations.

All other methods of high order, except for the forward Euler approximation, which is obtained from Taylor series, generate impediment in the practical applications during solving of the derivative function **f**.

Other popular methods used in real time systems is second-order Runge-Kutta (RK2) method. This method requires two evaluations of right-hand side of differential equations system. Runge–Kutta method, like Euler method, belong to self-starting solving techniques, where there is no requirement regarding the computation of the preceding starting points, which is the necessity in multistep methods. Used in comparison researches routine realise Euler-Cauchy algorithm:

$$\begin{aligned}
k_1 &= h\mathbf{f}(t_n, y_n) \\
k_2 &= h\mathbf{f}(t_n + 0.5h, y_n + 0.5k_1) \\
y_{n+1} &= y_n + k_2
\end{aligned} \qquad (2)$$

where the local error is proportional to $O(h^3)$, and the global error is proportional to $O(h^2)$. This method requires two evaluates of the right-hand side of the differential equations system. The Runge–Kutta methods, like the Euler method, are self-starting methods, where there is no requirement of a prior computation of the starting point, unlike multistep methods.

Some of the studies apply a procedure realizing the fourth-order Runge-Kutta algorithm (RK4) [2]:

$$\begin{aligned}
k_1 &= h\mathbf{f}(t_n, y_n) \\
k_2 &= h\mathbf{f}(t_n + 0.5h, y_n + 0.5k_1) \\
k_3 &= h\mathbf{f}(t_n + 0.5h, y_n + 0.5k_2) \\
k_4 &= h\mathbf{f}(t_n + h, y_n + k_3) \\
y_{n+1} &= y_n + (k_1 + 2k_2 + 2k_3 + k_4)/6
\end{aligned} \qquad (3)$$

where the local error is proportional to $O(h^5)$, and the global error is proportional to $O(h^4)$.

This procedure is not very commonly used in real-time systems; however, due to its greater accuracy in comparison to RK2 it offers a valuable reference.

978-1-4244-1741-4/08/$25.00 ©2008 IEEE

B. Multistep Method-BGKODE_DSP Routine

Multistep methods of computing ordinary differential equations apply information from the preceding steps. The method in consideration is based on the modified Adam's algorithm [4, 8]. Practical realisation of the multistep method applies PECE algorithm (Predict-Evaluate derivatives-Correct-Evaluate derivatives). The Adams-Bashfort modified method is used as the predictor and the Adams-Moulton modified method is the corrector. The standard procedure involves two evaluations of right-hand side of differential equations system.

The BGKODE_DSP routine has been designed for the computations in real-time systems, like drive system, based on digital signal processor (DSP). The core of BGKODE_DSP routine is based on a full set of BGKODE routines [4, 5] which were tested throughout the simulations [5]. The process of adapting this routine for the application in DSP involves the necessity of reducing the scope of the options available in the complete BGKODE routine.

The real time systems usually involve data processing over the pre-determined time span. Thus automatic step size change has been modified in order to fulfil a followed scheme of computation in real time systems. Such a modification significantly reduces code complexity; concurrently, it results in the deterioration of accuracy in comparison to various step size version. Concurrently, local error control, modification of approximation order k and coefficients relative to step size have been simplified.

Constant step size BGKODE_DSP routine algorithm is given by:

$$
\begin{aligned}
P: \ & p_{n+1} = y_n + h \sum_{i=1}^{k} g_i \Phi_i(n) \\
& \Phi_{k+1}^{e}(n+1) = 0 \\
& \Phi_i^{e}(n+1) = \Phi_{i+1}^{e}(n+1) + \Phi_i(n), \ i = k,...,1 \\
E: \ & \mathbf{f}_{n+1}^{p} = \mathbf{f}(t_{n+1}, p_{n+1}) \\
C: \ & y_{n+1} = p_{n+1} + h \, g_{k+1}(\mathbf{f}_{n+1}^{p} - \Phi_1^{e}(n+1)) \\
E: \ & \mathbf{f}_{n+1} = \mathbf{f}(t_{n+1}, y_{n+1}) \\
& \Phi_{k+1}(n+1) = \mathbf{f}_{n+1} - \Phi_1^{e}(n+1) \\
& \Phi_i(n+1) = \Phi_{i+1}^{e}(n+1) + \Phi_{k+1}(n+1), \ i = k,...,1
\end{aligned}
\tag{4}
$$

where P: denotes prediction, C: denotes correction and E: denotes evaluation of the right side of the differential equations system. The modified divided differences $\Phi_i(n+1)$ ensure the effective representation of polynomial which interpolates function $w(t)$ at the discrete points [4, 5].

The BGKODE_DSP routine involves a special procedure to be followed at the beginning of the computation. During the starting phase throughout the first two steps, the order k is increased so that it can gain the assigned value. In this way the procedure is self-starting in the sense that it does not require the use of other methods for the determination of the information demanded from the preceding steps.

In some of the research the comparison involves the full version of the BGKODE routine with a constant step [5].

III. SIMULATIONS

A. Studies of Selected Routines on the Example of an Equation System

The results of a comparative study into numerical integration routines presented below apply an example of five first-order differential equations in the form:

$$
\begin{aligned}
\frac{dy_1}{dt} &= y_2 \\
\frac{dy_2}{dt} &= -\omega_1^2 y_1 \\
\frac{dy_3}{dt} &= y_4 \\
\frac{dy_4}{dt} &= -\omega_2^2 y_3 \\
\frac{dy_5}{dt} &= y_2 y_3 + y_1 y_4
\end{aligned}
\tag{5}
$$

The solution to this system of differential equations for the following starting conditions:

$$y_1(0) = y_3(0) = y_5(0) = 0, \ y_2(0) = \omega_1 \ \text{ and } \ y_4(0) = \omega_2$$

takes the form:

$$y_5(t) = \sin(\omega_1 t) \cdot \sin(\omega_2 t). \tag{6}$$

An example of the solution curve (6) is presented in Fig. 1.

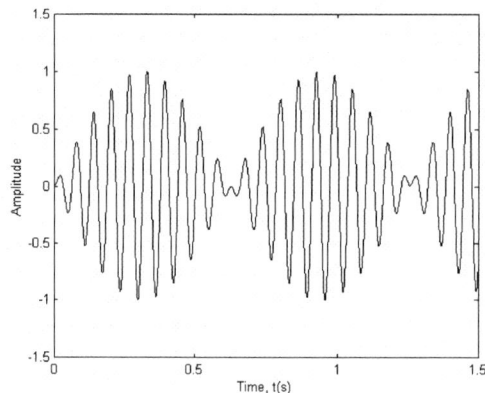

Fig. 1. Example of the solution curve (6).

The assessment of the compared procedures involves the statement of an accuracy index δ:

$$\delta = \frac{1}{m}\sqrt{\sum_{i=1}^{m} \Lambda_i^2} \tag{7}$$

where m denotes the number of steps taken during computations, and error Λ_i is defined as:

$$\Lambda_i = \frac{(W_i - Y_i)}{W_i}, \ i = 1,2,...,m \tag{8}$$

which accounts for the difference between the accurate value of the solution $W_i = W_i(t)$ and the one gained from the computations $Y_i = Y_i(t)$.

In Figs. 1. and 2. the solid line marks the results for the Euler's routine, the dashed line for the RK2 routine, the dotted line for RK4 routine, the dash-dot line for the BGKODE routine and diamond line for the BGKODE_DSP routine.

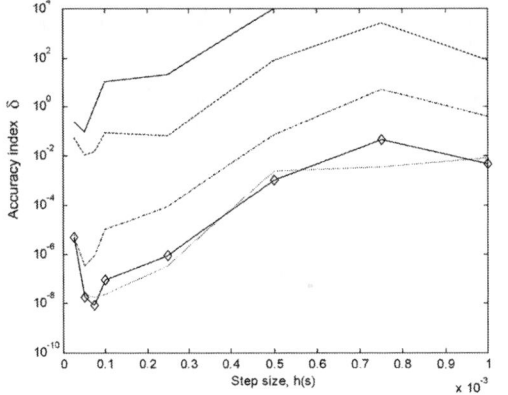

Fig. 2. Comparison of accuracy of computations for selected routines during computation of system of equations for ω_1=50 and ω_2=5.

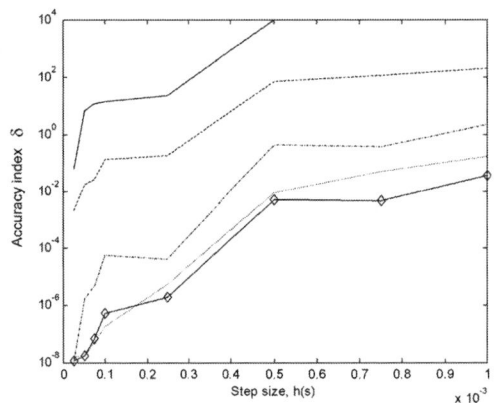

Fig. 3. Comparison of accuracy of computations for selected routines during computation of system of equations for ω_1=100 and ω_2=5.

The BGKODE_DSP routine applies the order k=2. It was experimentally selected as a compromise between the time of computations and accuracy of the result. In other applications it is possible to select the order k in the range up to 4.

The comparisons of the examined routines indicate that the accuracy rate of the solutions to the test equation are in the majority of cases similar for the BGKODE_DSP and RK4 routines. The singular advantage of the BGKODE_DSP routine over the BGKODE is easily discernible. In a wide range of the presented results the curves for the two procedures overlap. The modifications, whose effect takes the form of the BGKODE_DSP routine, render it possible to maintain or even improve the properties of the algorithm realized in the entire set of the BGKODE routines in the case when the constant step of integration is applied.

B. Testing Selected Routines on Example of Drive Control System with DTC Method

Simulations, which were conducted to verify the effectiveness of the proposed numerical integration routine, have applied drive system model [3] with a state observer [9] prepared in the Fortran programming language.

Nowadays one of the most popular induction motor drive control method is Direct Torque Control (DTC) [3, 10]. The implementation of this concept was possible due to technological developments such as the generation of the DSP controllers and the new power semiconductors. Controlled motor drives without mechanical sensors for speed display large degree of reliability and low cost. The identification of the rotor speed is generally based on the measured terminal voltages and currents. This identification system is based on the state observer [9]. The DTC method seems to be very appropriate in indicating differences between the compared numerical integration methods applied in speed observer computations.

Despite the applicability of angular velocity measurement of the rotor in the control system the users tend to prefer the sensorless systems [3, 7, 9]. The exclusion of the components, including incremental encoders in the industrial application affects the decrease of the failure rate and reduction of the exploitation cost. The observer of the angular velocity applied in the simulation and experimental studies is comprehensively described in paper [9]. The differential equations for the velocity observer have been introduced on the basis of the electromagnetic section of the model machine [6] in the stationary co-ordinate system (α-β). The system of the six differential equations, which defines the observer, is solved with the comparable numerical integration methods. The structure of the observer can be found in Fig. 4.

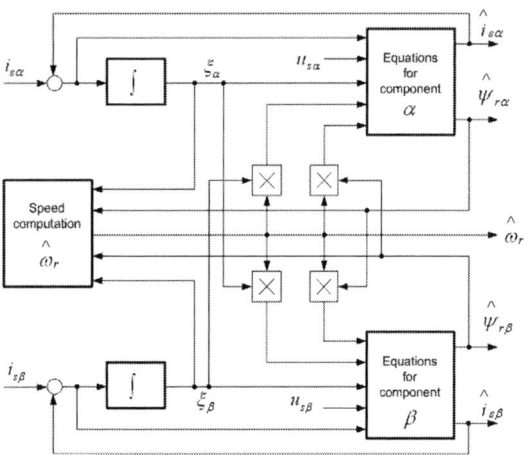

Fig. 4. Diagram of speed observer.

In order to compare the numerical integration methods, two indicators of control quality have been proposed. The quality indicators have been computed from a section of the steady-state during drive system operations. The control quality indicators include:

- estimated speed variation indicator

$$\omega_{rip} = \sqrt{\frac{\sum_{i=1}^{N}(\hat{\omega}_{ri} - \overline{\hat{\omega}}_r)^2}{N-1}} \;,\; \text{where} \quad \overline{\hat{\omega}}_r = \frac{1}{N}\sum_{i=1}^{N}\hat{\omega}_{ri}, \quad (9)$$

- estimated torque variation indicator

$$T_{rip} = \sqrt{\frac{\sum_{i=1}^{N}(T_{ei} - \overline{T}_e)^2}{N-1}} \;,\; \text{where} \quad \overline{T}_e = \frac{1}{N}\sum_{i=1}^{N}T_{ei}. \quad (10)$$

Figs. 5. and 6. present the indicators of control quality corresponding to the numerical integration method applied. The computations have been carried out for a section of the steady-state drive system operating under specific angular speed $\omega_{ref} = 0.3$ (per unit). The frequency of control system $f_{ster} = 10\,kHz$.

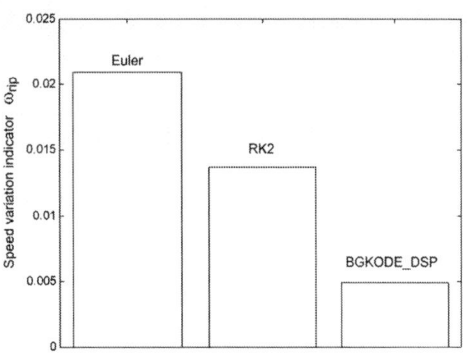

Fig. 5. Comparison of estimated speed variation indicator ω_{rip} depending on selected method.

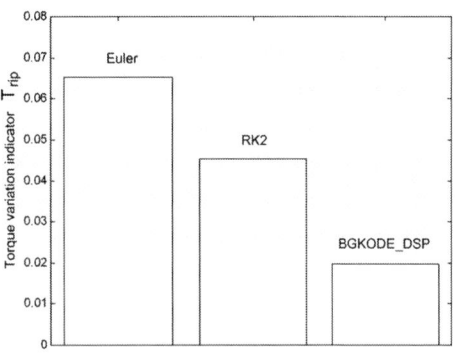

Fig. 6. Comparison of estimated torque variation indicator T_{rip} depending on selected method.

The analysis of the results presented above indicates the effect of the selection of an appropriate numerical integration method on the control quality indicators. BGKODE_DSP routine exhibits the advantage in the form of greater applicability, which is reflected in the reduction of the parameters of the estimated torque variation and estimated speed variation indicators.

IV. Experimental Results

Numerous experiments have been carried out on laboratory drive system [11], whose diagram is presented in Figure 7. The subroutines which realise DTC method and angular speed observer equations system [9] have been computed using floating-point processor MPC8240 (PowerPC 603e)/250 MHz.

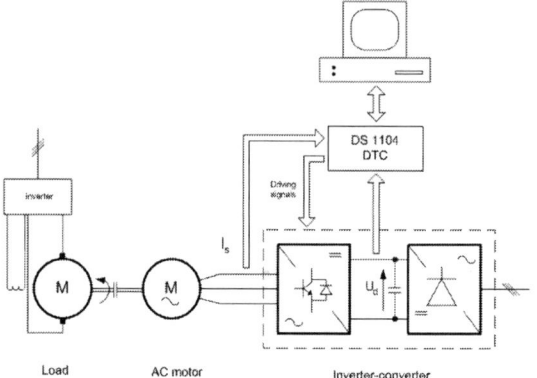

Fig. 7. Diagram of laboratory drive system.

Computations were conducted on the basis of a section of the steady-state drive system operating with a pre-set angular speed $\omega_{ref} = 0.3$ (per unit). The frequency of the control system amounted to $f_{ster} = 10$ kHz. Load torque in per unit was set at $T_{load} = 0.25$.

The experimental results presented below confirm the positive effect of an adequate selection of numerical integration method on the selected control quality indicators. The positive effect reflected in the decrease of the estimated torque variation and estimated speed variation indicators, are registered for the case when the BGKODE_DSP routine was applied. When the results of the Euler's method serve as the reference level, the decrease in the estimated speed variation is about 49% when BGKODE_DSP routine is applied and 22% when RK2 method is used. For the case of estimated torque, the decrease in the variation indicator amounts to about 25% for BGKODE_DSP routine and 16% for RK2.

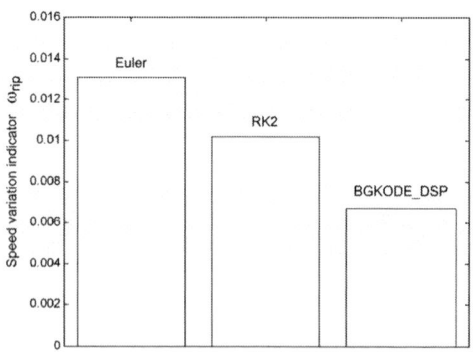

Fig. 8. Comparison of estimated speed variation indicator ω_{rip} (experimental data for $f = 10$ kHz).

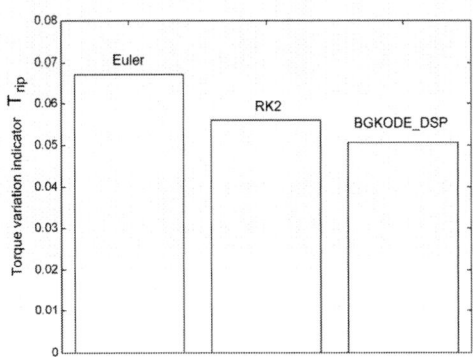

Fig. 9. Comparison of estimated torque variation indicator T_{rip}
(experimental data for f = 10 kHz).

A. Comparison of Computational Run-time for Selected Routines during Computations by Angular Velocity Observer

In the course of the operation of the control system the measurement of the actual time of computation of the angular velocity observer system was taken relative to the routines applied for the numerical solving of differential equation system. The duration of a single computation cycle was taken (Δ) resulting from the measurement of the output signal from the transducer C/A, which was activated prior to the beginning of the computations and deactivated after their termination. The difference between the minimum and the maximum time needed to execute the procedure results from the differences in the duration of the computations performed on the function which form the right-hand side of the system of differential equations. A subroutine which forms the system of the differential equations of angular velocity observer contains conditional instructions. Depending on the instantaneous values of the signal input into the system of the angular velocity observer the duration necessary for this section of the program may vary.

TABLE I.
MINIMUM AND MAXIMUM DURATION OF COMPUTATIONS FOR THE COMPARED ROUTINES

	Euler	RK2	RK4	BGKODE_DSP $k=2$
Δ_{min} (μs)	2,62	3,60	8,76	5,72
Δ_{max} (μs)	3,08	3,90	10,08	6,50

If the results gained with the Euler method are considered as the reference values, the increase in the maximum duration of the computations for RK2 routine amounts to 27%, whereas for the BGKODE_DSP $k=2$ routine the increase amounts to 110%. Definitely the highest increase in the maximum duration of the computation takes place for the case of RK4 routine, which amounts to over 227%. These values overlap with the expected duration of the computation time for the more complex numerical routines. The RK2 and BGKODE_DSP routines are associated with a necessity of twofold recall of the function, defining the right-hand

side of the differential equation system, which involves the reference to the angular velocity observer. The use of the more complex algorithm for the BGKODE_DSP routine results in the maximum increase in the duration of the computations time by nearly 67%.

V. CONCLUSIONS

The simulations and experiments involving selected numerical integration methods clearly indicate the need for an aid in the selection of the routine for real time process operations. The popular and commonly applied methods, including Euler or second-order Runge-Kutta method, in many cases ensure the required stability and accuracy. The proposed BGKODE_DSP routine, based on multistep method, has a wide area of potential application. This routine adds to the popular numerical integration methods. The most important characteristic of BGKODE_DSP routine is the improved accuracy in comparison to the popular method which is not accompanied with an increase in the number of evaluations of the right-hand side of the differential equations system. Besides, this routine offers the possibility of affecting accuracy resulting from the modification of the approximation order k [3]. The bigger effort associated with the more complicated BGKODE_DSP routine algorithm, accompanied by greater computation time in comparison to popular methods [12], is on an acceptable level.

REFERENCES

[1] L. O. Chua and P. M. Lin, *Computer Aided Analysis of Electronic Circuits: Algorithms and Computational Techniques*, Prentice-Hall, Englewood Cliff, New Jersey, 1975.

[2] A. Krupowicz, *Metody numeryczne zagadnień początkowych równań różniczkowych zwyczajnych*, PWN, Warszawa 1986.

[3] A. Gardecki and K. Macek-Kamińska, *Comparison of the influence of selected numerical integration algorithms on DTC induction motor drive*, in *Power electronics and electrical drives selected problems*, Oficyna Wydawnicza Politechniki Wrocławskiej, Wrocław 2007, pp. 411-420.

[4] F. T. Krogh, *Changing stepsize in the integration of differential equations using modified divided differences*, Lecture Notes in Math. 362, Springer-Verlag, Berlin-New York, 1974, pp. 22-71.

[5] R. Beniak and A. Gardecki, *Procedura całkująca układy równań różniczkowych zwyczajnych z wykorzystaniem zmodyfikowanego algorytmu Krogha*, Prace X Konferencji Symulacja Procesów Dynamicznych, Zakopane, 1998, pp. 23-30.

[6] M. P. Każmierkowski, R. Krishnan and F. Blaabjerg, *Control in power electronics. Selected problems*, Academic press 2001.

[7] T. Orłowska-Kowalska, *Bezczujnikowe układy napędowe z silnikami indukcyjnymi*, Oficyna Wydawnicza Politechniki Wrocławskiej 2003.

[8] L. Shampin and M. Gordon, *Computer solution of ordinary differential equations*, Freeman 1975.

[9] Z. Krzemiński, *Cyfrowe sterowanie maszynami asynchronicznymi*, Wydawnictwo PG, 2001.

[10] N. Mohan, T. Undeland and W. Robbins, *Power electronics: converters, applications and design*, NY John Wiley Sons 1989.

[11] M. Jeleń and J. Michalak, *Skompensowany przekształtnik częstotliwości – realizacja praktyczna z wykorzystaniem karty DS1104*, VII Krajowa Konferencja Naukowa "Sterowanie w Energoelektronice i Napędzie Elektrycznym", SENE 2005.

[12] A. Gardecki, M. Jeleń and J. Michalak, *Porównanie czasu obliczeń wybranych metod numerycznego rozwiązywania równań różniczkowych na przykładzie obliczeń obserwatora prędkości kątowej*, Postępy w Elektrotechnice Stosowanej PES-6, Kościelisko, 2007.

Application of Fuzzy Logic Techniques To Robust Speed Control of PMSM

Tomasz Pajchrowski, Krzysztof Zawirski

Poznan University of Technology, Institute of Control and Information Engineering,
Piotrowo 3a, 60-965 Poznań, Poland

Abstract— **The paper deals with the problem of robust speed control of electrical servodrives. A robust controller is developed using a nonlinear IP controller. The controller nonlinear characteristic is obtained due to fuzzy logic technique application. An original method of controller settings adjustment is presented. The use of this adjustment procedure ensures robust speed control against the variations of the moment of inertia and a step change of load torque. Simulations and laboratory results validate the robustness of the servodrive with Permanent Magnet Synchronous Motor.**

I. INTRODUCTION

In the last decade the use of electrical servodrives in high performance applications such as robots and machine tools has risen significantly [7]. Every control system should be stable and ensures good quality control during some perturbation and disturbances. Control performance of servodrives in such applications is very sensitive to variations of external load and system parameters. In quite many of these applications moments of inertia are variable during drive operation [5,6,8]. Moreover, steps change of load torque occurs very often. In the paper a speed controller which is robust against such moments of inertia variations and steps change of load torque is proposed. Robustness should ensure equal or similar dynamic properties of speed control – lack of overshoot and equal settling times.

In the paper a modified approach to synthesis of robust speed controller based on fuzzy logic is proposed [1,2]. Like in preceding papers [3,4,5,6] robustness is achieved by introducing a nonlinear characteristic but this time by a IP speed controller. It is assumed that the robustness of speed control guarantees robust control of position. The paper proposes a modified approach to fuzzy logic controller (FLC) synthesis, which is performed in two stages. The first stage consists in arbitrary decisions, which concerns rule base and shape of input and output variable membership functions. In this stage an expert experience is applied supported by simulation tests and trials. At the modified stage the IP controller settings are adjusted by means of simplex method (Nelder-Mead)[1] which optimises the control quality index formulated in the paper. The simple and less time consuming procedure of tuning of second stage are significant advantage of this modified method. The proposed new quality index has only one selected weight coefficient while in previous work [5,6] it was necessary to select two weight coefficients. The synthesis is performed using simulation techniques and subsequently the behaviour of a laboratory

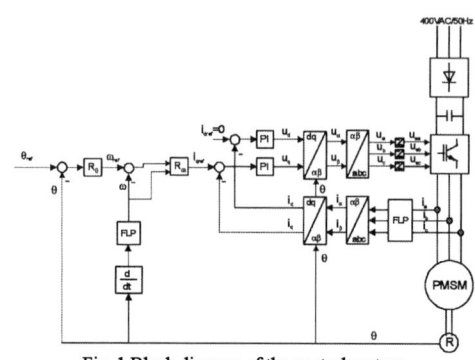

Fig. 1.Block diagram of the control system

speed control system is validated in the experimental setup. The control algorithms of the system are performed by a microprocessor floating point DSP control system. In industrial high performance applications very often are

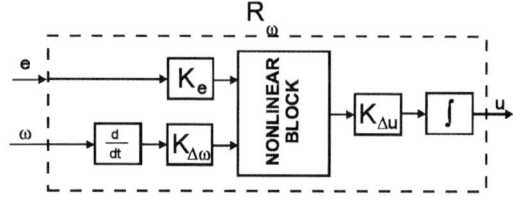

Fig.2. Structure of speed control system

used Permanent Magnet Synchronous Motors (PMSMs) due to their such favourable features as high power factor, high torque density, high efficiency and small size. For this reason such servodrive was tested in the presented work.

II. ROBUST SPEED CONTROLLER CONCEPT

The structure of the control system is shown in Fig.1. The system consists of four main parts: a motor (PMSM), a PWM inverter, current control loops and a speed control loop. The PWM inverter is modeled such as transport delay equal 100µs. Field oriented control of currents in the axes d and q is performed in the classical way by applying ordinary PI control algorithms realized by DSP with sampling period equal 100 µs. Current controller settings are adjusted according to the modulus criterion [7]. The speed controller is designed with the same sampling period equal 100µs. The low pass filter of speed measurement signal is modeled such as transport delay equal 1.5ms. The assumed concept of speed

---

978-1-4244-1741-4/08/$25.00 ©2008 IEEE   1198

controller robustness is based upon the introduction of a nonlinear module into the structure of an IP controller. The proposed structure of such a controller is shown in Figure 2. It consists of three blocks: a module for input signal conversion, a nonlinear module and a module for output signal generation. The IP dynamic characteristic is obtained by calculating the P (proportional) component of the control error and D (derivative) component of speed signal by the input module. Subsequently the output module integrates the signal, coming from the nonlinear module. Since the output signal of the FLC is a derivative of the torque command (or the current i_q), it is necessary to calculate the integral of this signal, which is done using Euler's procedure. The required robustness of the controller is achieved by proper modification of the control surface (du/dt=f(e, de/dt)) created by the nonlinear module.

III. ROBUST CONTROLLER SYNTHESIS

In the paper the process of robust controller synthesis is decomposed into two stages like in earlier works [5, 6]:

- Designing the nonlinear module by means of fuzzy logic,

- Adjusting IP controller settings K_e, $K_{\Delta\omega}$

The same rule base (Table I) and the same shape of membership functions (Fig.3 and 4) were assumed like in [6]. The assumed rule base and the membership functions create a control surface presented in Figure 5, which represents nonlinear controller characteristics.

TABLE I
RULE BASE

		Δe		
		N	Z	P
e	N	N	N	Z
	Z	N	Z	P
	P	Z	P	P

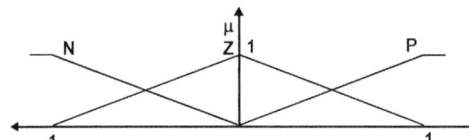

Fig.3. The shape of input membership function

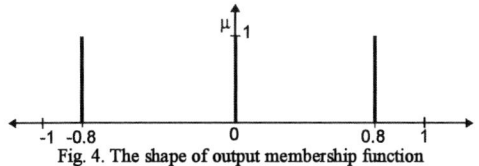

Fig. 4. The shape of output membership function

At the second stage of the controller synthesis an adjustment of two controller settings, the gain coefficient K_p and the integral time constant T_i, is performed. The controller settings are expressed by three scaling

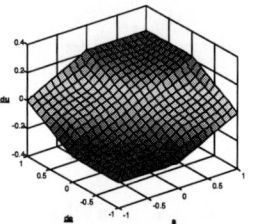

Fig. 5. The shape of control surface

coefficients K_e, $K_{\Delta\omega}$ and $K_{\Delta u}$ as follows:

$$K_P = K_{\Delta\omega} \cdot K_{\Delta u} \qquad (1)$$

$$T_i = \frac{1}{K_e \cdot K_{\Delta\omega}} \qquad (2)$$

This part of the synthesis is carried out by means of simulation techniques. Because of the nonlinear character of the controller, the less complex simplex (Nelder-Mead) procedure instead of RWC procedure is selected. The optimization criterion Q_c was modified in comparison with works [5, 6]. The criterion Q_C described by formula (3), contains the sum of three components. All three components characterize the step response of speed control, which is shown in Figure 6. The criterion Q_C is based on IAE (integral of absolute value of error). The first part consist of difference of integral of absolute value of error during the step response of speed control for both value of moment of inertia (minimum and maximum moment of inertia) in the time speed step response ($t_1 \div t_3$).The second component consist of difference of *IAE* during step change of the load torque in the time response for step change of load torque ($t_3 \div t_4$). The last part of the quality index consists of settling time.

$$Q_c = \left|\Delta IAE_{start}\right| + \left|\Delta IAE_{load}\right| + C1 \cdot \left|\Delta t_{settling}\right| \qquad (3)$$

where:

$$\Delta IAE_{start} = \int_{t1}^{t3} |e|_{J\min} dt - \int_{t1}^{t3} |e|_{J\max} dt \qquad (4)$$

$$\Delta IAE_{load} = \int_{t3}^{t4} |e|_{J\min} dt - \int_{t3}^{t4} |e|_{J\max} dt \qquad (5)$$

$$\Delta t_{settling} = t_{s\_J\min} - t_{s\_J\max} \qquad (6)$$

$e = \omega_{ref} - \omega$, (see Fig.6);

t_s – settling time (see Fig.6);

C_1 – weight coefficients.

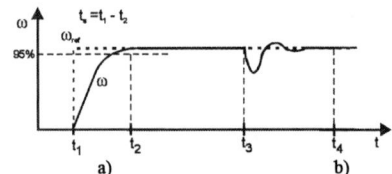

Fig. 6. Illustration of control criterion components
of speed step response

The weight coefficients C_1 were selected arbitrarily on the basis of numerous simulation tests. Some illustration of weight coefficient selection is given in figures 8–11. The simplex (method Nelder-Mead) optimisation procedure was performed by simulating the transient process of the speed control step response. During this transient process simulation the quality criterion Q_C was calculated. At each step of the optimisation procedure the simulation of the step response and the calculation of the quality criterion were repeated twice for two extreme values of the moment of inertia, small (J_{min}) and large (J_{max}). Next, the sum of both criterion values was assumed for optimisation. According to the simplex procedure, only the set, which gives improvement of the quality index (the smaller criterion value), is stored.

IV. SIMULATION RESULTS

The simulation technique was used testing the control system operation. The motor (PMSM) model is based on well known equations [7], which are valid with such simplified assumptions as lack of rotor eddy currents, constant magnetic flux and constant inductance of stator winding. The model of the PMSM and the control system was performed in MATLAB. The PWM inverter is modeled such as transport delay equal 100µs. Field oriented control of currents in the axes d and q is performed in the classical way by applying ordinary PI control algorithms realized by DSP with sampling time equal 100 µs. Current controller settings are adjusted according to the modulus criterion. The speed controller is designed with sampling time equal 100µs. The filter of speed (low past) is modeled such as transport delay equal 1.5ms. The model of the control systems takes into account the discrete character of calculations performed by the microprocessor system with proper sampling period. The motor model was calculated with a very small step of integration, which simulates its continuous character. The step value was equal 20µs. Contrary to that, the model of the control system was calculated with much higher step value (100µs), simply because it enables a better simulation of how the control system works on a signal processor with a real value of the sampling period. The control system was tested in the transient process, which was the step response to the speed reference change and the response to the step change of load torque - treated as a distortion. These tests were performed for different values of the total moment of inertia: small (J_{min}) and big (J_{max}). The simulation results are shown in figures 7–12. The figures

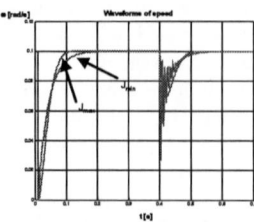

Fig. 7. Waveforms of angular speed during starting process for different values of moment of inertia with controller optimised for $|\Delta IAE_{start}| + |\Delta IAE_{load}|$ criterion. At the time 0.4s a step change of load torque occurs torque occurs.

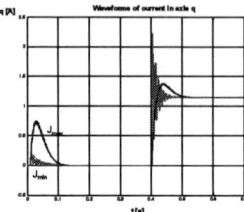

Fig. 8. Waveforms of current in axis q during starting process for different values of moment of inertia with controller optimised for $|\Delta IAE_{start}| + |\Delta IAE_{load}|$ criterion. At the time 0.4s a step change of load torque occurs torque occurs

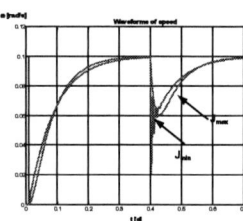

Fig. 9. Waveforms of angular speed during starting process for different values of moment of inertia with controller optimised for $|\Delta IAE_{start}| + |\Delta IAE_{load}| + |\Delta t_{settling}|$ criterion. At the time 0.4s a step change of load torque occurs.

Fig. 10. Waveforms of current in axis q during starting process for different values of moment of inertia with controller optimised for $|\Delta IAE_{start}| + |\Delta IAE_{load}| + |\Delta t_{settling}|$ criterion. At the time 0.4s a step change of load torque occurs torque occurs

TABLE II

VALUES OF SETTLING TIME FOR DIFFERENT MOMENT OF INERTIA
DURING STARTING PROCESS (ω_{ref}=0.1 rad/s)

J [kgm²]	0.75	1.65	3.3	4.95	5.83
$t_{S5\%}$ [ms]	248	243	238	234	229

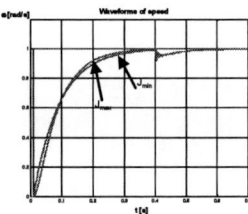

Fig. 11. Waveforms of angular speed during starting process for different values of moment of inertia with controller optimised for $|\Delta IAE_{start}| + |\Delta IAE_{load}| + |\Delta t_{settling}|$ criterion. At the time 0.4s a step change of load.

TABLE III

VALUES OF SETTLING TIME FOR DIFFERENT MOMENT OF INERTIA AFTER STEP CHANGE OF LOAD TORQUE (ω_{ref}=0.1 rad/s)

J [kgm^2]	0.75	1.65	3.3	4.95	5.83
$t_{SS5\%}$ [ms]	201	200	200	199	199

show waveforms of speed for both values (minimum and maximum) of inertia, which illustrate the influence of the optimisation process with different forms of the Q_C control criterion on controller properties. Applying only the sum of both IAE criterion (C1=0), as shown in Fig. 7

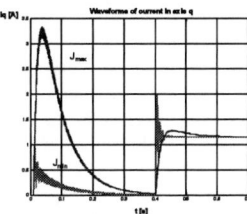

Fig. 12. Waveforms of current in axis q during starting process for different values of moment of inertia with controller optimised for $|\Delta IAE_{start}| + |\Delta IAE_{load}| + |\Delta t_{settling}|$ criterion. At the time 0.4s a step change of load torque occurs torque occurs

and 8 (waveforms of current in axis q). The response speed characterizes lack of overshoot, but the settling time for both values of inertia is different. Moreover the presented waveforms of speed for small value of moment of inertia show some oscillation after step change of load torque. Final results presented in Fig. 9 and 10, obtained for the complete form of criterion (3), show no overshoot, no oscillation and equal settling times during starting process and after step change of torque load. The presented waveforms of speed for two different values of inertia are similar so one can say that the control is robust against the variation of the moment of inertia. Results presented in Table II and III show values of settling time (5%) during starting process and during step change of torque load for different moment of inertia. This setting time values are almost equal. Waveforms of angular speed (shown in the Figure 11) and waveforms of current in axis q (shown in Figure 12) are measured during starting process to ω_{ref}=1.0 rad/s with controller optimizing a completed formula (3). Analyses of settling time values visible in figures 7, 9 and 11 points to the conclusion, that for the complete criterion settling time values obtained for different inertia are almost equal but significantly larger

than for C_I= 0. This means that the achieved robust control process leads to slightly worse dynamics. It seems to be an unavoidable side effect of achieving robustness.

V. EXPERIMENT RESULTS

The laboratory setup consists of a PMSM supplied from a PWM transistor inverter and controlled by the floating point signal processor ADSP – 21061. The control system is equipped with current and voltage sensors (LEM) and an encoder. Scheme of laboratory setup is shown in Fig. 13.

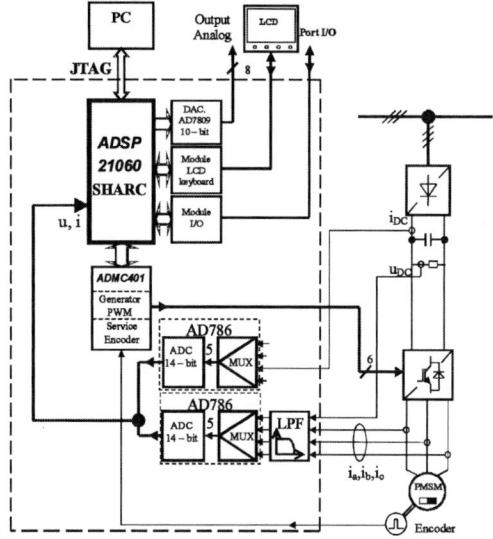

Fig. 13. Scheme of laboratory setup

The algorithms of PWM modulation, control of currents in axis d and q and fuzzy speed control are performed by DSP. The drive system was tested in the transient process involved by a step change of the speed reference signal of different values. A set of metal plates fixed to the arm mounted on the motor shaft enables to vary the moment of inertia. Some results of the laboratory tests are shown in figures 14–18. During the tests the total value of the moment of inertia of the drive was changed from small to large value. Figures 14 and 15 present waveforms of speed and reference speed, current i_q for small inertia values in the transient process involved by two different speed reference values – 0.1 rps and 0.25 rps. Figures 16 and 17 present the same waveforms in the same transient process but for the maximum value of inertia. Good behaviour of the control system is visible in the waveforms. For the small speed reference (Figs.14 and 16) when the speed controller does not reach its saturation level (torque limit), the same value of the settling time and lack of overshoot is observed for both values of inertia. In the case of the high speed reference (0.25 rps) the speed controller reaches its saturation during the transient process with large inertia so the speed change with constant electromagnetic torque has constant and limited acceleration (Fig.17). For this reason the settling time is longer than for small inertia where saturation does not

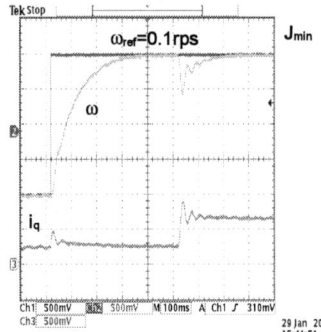

Fig. 14. Waveforms of speed and currents in axis q for speed reference change 0→0.1 rps, J_{min}. At the time 0.5s a step change of load torque occurs

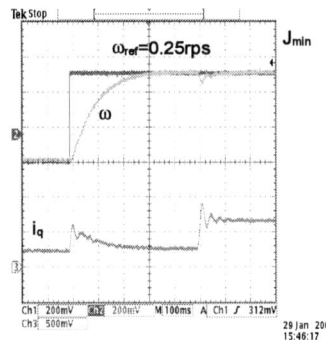

Fig. 15. Waveforms of speed and currents in axis q for speed reference change 0→0.25 rps, J_{min} At the time 0.5s a step change of load torque occurs

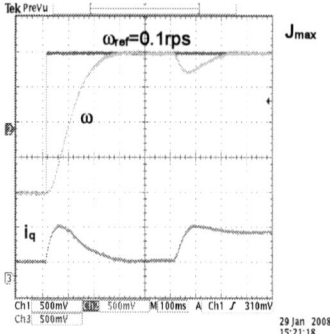

Fig. 16. Waveforms of speed and currents in axis q for speed reference change 0→0.1 rps, J_{max} . At the time 0.5s a step change of load torque occurs

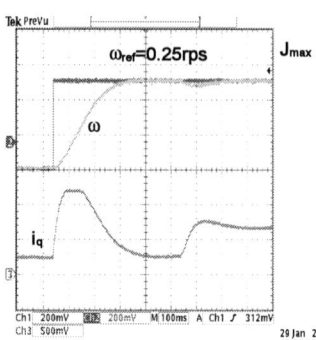

Fig. 17. Waveforms of speed and currents in axis q for speed reference change 0→0.25 rps, J_{max} At the time 0.5s a step change of load torque occurs

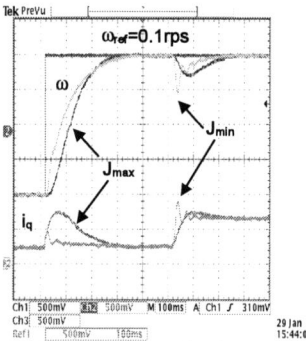

Fig. 18. Waveforms of speed and currents in axis q for speed reference change 0→0.1rps, J_{min} and J_{max} At the time 0.5s a step change of load torque occurs

occur (Fig.17). In spite of this difference, the transient processes for both values of inertia display aperiodic character. Figure 18 present waveforms of speed and currents in axis q for speed 0.1rps in the transient process for both value of moment of inertia. The present waveforms of speed are very similar.

VI. CONCLUSION

The analysis of the presented results confirms that a IP speed controller with proper but known nonlinear characteristics [5,6] allows to obtain the robustness of speed control against the variations of the moment of inertia and step change of torque load. Moreover, the IP controller settings are adjusted by means of simplex method (Nelder-Mead) which is less complex and less time consuming. The proposed new quality index in the paper has only one selected weight coefficient instead of tow selected coefficients in previous authors' methods. It is shown in the paper that the procedure of controller settings adjustment yields good results and can be recommended for such control applications.

REFERENCES

[1] Driankov D., Hollendoor H., Reinfrank M.: An Introdaction to Fuzzy Control, Springer-Verlag Berlin Heidelberg, 1993.

[2] Filev D. P, Yager R. R.:Essentials of fuzzy modelling and control, John Wiley&Sons, 1994.

[3] Pajchrowski T. Urba ski K. Zawirski K.: „Robust speed control of PMSM servodrive based on ANN application", *Electrical Power Electronics 2003. EPE 2003 2-4 September 2003, Toulouse*, paper 833 on CD-ROM

[4] Pajchrowski T., Zawirski K „Robust speed control of servodrive based on ANN", International Symposium on Industrial Electronics, ISIE 2005, June 20-23, Dubrovnik, paper A3-01 on CD-ROM.

[5] Pajchrowski T., Zawirski K.: Application of Artificial Neural Network to Robust Speed Control of Servodrive. IEEE Transaction on Industrial Electronics, Vol.54, No.1, February 2007, pp.200-207

[6] Pajchrowski T., Urbanski K., Zawirski K.,, Artifical neural network based robust speed control of permanent magnet synchronous motors" COMPEL 2006 Vol. 25 No. 1, 2006 pp. 220-234.

[7] Vas P. „Sensorless Vector and Direct Torque Control", *Oxford University Press*, 1998

[8] Vitek J., Baculak T., Dodds S. J., Perryman R. „Near-Time-Optimal control of electrical drives with permanent magnet synchronous motor", *Electrical Power Electronics 2003. EPE 2003 2-4 September 2003, Toulouse*, paper on CD-ROM.

ACKNOWLEDGMENT

This work was parity supported by grant TB 45-082/08/DS.

APPENDIX.
DATA OF INVESTIGATED DRIVE

Parameters of PMSM	Unit	Value
Moment of inertia	kgm^2	0.046
Minimum moment of inertia	kgm^2	0.75
Maximum moment of inertia	kgm^2	5.83
Torque constant	Nm/A	17.5
Voltage constant	V/1000rpm	1560
Rate load torque	Nm	50
Rated value of speed	rpm	145
Stator inductance per phase	mH	45
Stator resistance per phase	Ω	18.5
Rated current	A	1.94
Rated voltage	V	310

Optimal control of current commutation of high speed SRM drive

Jan Deskur, Tomasz Pajchrowski, Krzysztof Zawirski

Poznan University of Technology, Institute of Control and Information Engineering,
Piotrowo 3a, 60-965 Poznań, Poland

e_mail: jan.deskur@put.poznan.pl, tomasz.pajchrowski@put.poznan.pl, krzysztof.zawirski@put.poznan.pl

Abstract— The problem of optimal current commutation control, which bases on optimal off-line selection of switching angles, was investigated in the paper. Two different criteria of optimal control were taken into account: the maximum electromagnetic torque for given current and the minimum electromagnetic torque ripples. Optimization process was provided on base of computer model of SRM drive. Obtained simulation results were validated by experimental investigations.

I. INTRODUCTION

Switched reluctance motors (SRM) increase significantly their applications due to such advantages like simple rotor construction, without windings and magnets, low inertia in comparison with other motors, possibility rotate with high speed and high reliability. These advantages make the SRM an interesting alternative drive [4]. However the SRM drive has some disadvantages like significant value of torque ripples and unpleasant noise during operation. Effective control of high performance SRM drives requires precise information of instantaneous rotor position angle. This is necessary not only for speed and position control but first of all for precise control of phase-to phase current commutation. Proper values of current switching-on and switching-off angles are variable during speed and current changes [3,4,5]. The problem of optimal current commutation control, what bases on optimal off-line selection of switching angles, was investigated in the paper.

Two different criteria of optimal current commutation control were taken into account: first - the maximum electromagnetic torque for given current [5] and second - the minimum electromagnetic torque ripples. The optimal switching-on and switching-off angle values were detected as function of phase current and rotor speed. This calculated optimal values were stored in microcomputer system memory in form of two-input look-up tables.

Optimization process was carried out on the basis of SRM drive simulation model [2], and further verified experimentally. The investigated motor is high-speed SRM (see Appendix for details).

II. SRM MODEL

Usually, the following assumptions are made: stator and rotor are ideally symmetric, eddy current are neglected, electromagnetic cross interaction between motor phases are neglected. In consequence, electromagnetic torque of the SRM, T_e, can be expressed

as a sum of individual phase torques, each of which depends only on its own phase current, i, and rotor position angle, θ. The phase torque, $T(\theta,i)$, can be derived from general rules of electro-magneto-mechanic energy conversion, provided that nonlinear flux-current-angle (ψ-i-q) characteristics are know. These characteristics can be obtained from static measurements or from finite-element-analysis (FEA), in form of 2D look-up tables, e.g. $\Psi(\theta,i)$, $T(\theta,i)$. Such approach need a lot of measurements/calculations and sophisticated methods of resampling&smoothing of measured data. 2D look-up tables cannot be easily transformed (e.g. inverted, differentiated, integrated); moreover, 2D tables are inefficient in real-time implementation. To avoid difficulties mentioned above, the simplified formulas (1,2) has been proposed [2], on which 2D functions $\Psi(\theta,i)$, $T(\theta,i)$ are expressed in terms of two 1D nonlinear functions $L_\theta(\theta)$, $sat(i)$:

$$\psi(\theta,i) = L_c \cdot i + L_\theta(\theta) \cdot sat(i) \qquad (1)$$

$$T(\theta,i) = \frac{dL_\theta(\theta)}{d\theta} \cdot \int_0^i sat(\iota)d\iota \qquad (2)$$

Nonlinear functions $L_\theta(\theta)$, $sat(i)$, as well as its derivatives and integrals has been calculated from torque characteristics, measured at standstill. A fairly good approximation of characteristics has been obtained for given SRM [2] .

Simple but quite accurate model (1,2) was used for off-line calculation of optimal switching-on and switching-off angles, as well as for real-time implementation of control rules.

III. OPTIMIZATION PROCEDURE

Two criteria of SRM control were analyzed by means of computer simulation technique. The first criterion was a maximum ratio of average electromagnetic torque T_{avg} to current reference value I_{ref} and the second was a ratio of RMS value of torque ripples to average value of torque. Both criteria were analyzed as a function of speed and current values. Simulation and optimization procedure was realized with help of Matlab_Simulink program. The function *fminbnd* from toolbox OPTIMIZATION enables to solve the optimization task. In each step of optimization procedure exemplary steady state process with reference values of current and speed was provided. During this simulation period the value of selected criterion was calculated. Simulations were provided for closed loop control system for various values of reference current and

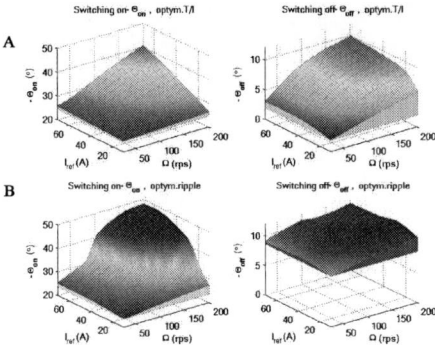

Fig. 1 Diagrams of current switching-on and switching-off angles versus angular speed and current reference value for two optimal control methods: A) maximum ratio of torque to reference current, B) minimum ripples (minimum ratio of RMS value of torque ripples to torque average value).

reference speed. Results of optimization for both used criteria can be seen in Figures 1, which present diagrams of switching-on and switching-off angles versus values of reference current and speed. Optimal switching angles were calculated for current value variable in range 10 – 70 A and speed value variable in range 25 – 200 rps. For the first criterion the switching-on angle starts from -30^0 for minimum current and speed and increases with increase of current and speed till value -42^0. For the second criterion the same changes of switching-on angle are from -25^0 till 47^0. Switching-off angle changes were obtained respectively for first criterion from value -2^0 till -9^0 and for second criterion from value -8^0 till value -13^0.

IV. SIMULATION RESULTS

Some selected simulation results illustrate SRM drive operation with different values of switching-on and switching-off angles. Two optimal control systems were compared with the control of SRM with constant switching angles, which values were equal $Q_{on} = -22.5^0$, $Q_{off} = -7.5^0$.

Figures 2, 3 and 4 show shape (waveforms) of phase current and electromagnetic torque, as a function of rotation angle θ, for variable speed (10 .. 200 rps) and constant reference current (50 A). Figure 2 presents waveforms for control with constant switching angles, figure 3 waveforms for optimal control with criterion of maximum torque and figure 4 for optimal control with criterion of minimum torque ripples. Comparison of these

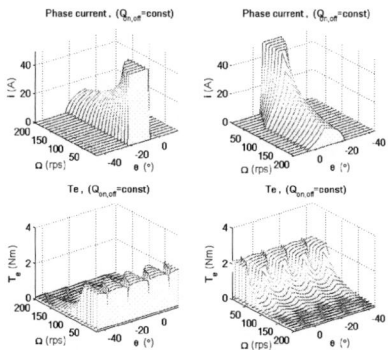

Fig.2 Waveforms of simulated phase current and electromagnetic torque component produced by phase current for constant current switching angles ($\Theta_{on} = -22.5^o$, $\Theta_{off} = -7.5^o$).

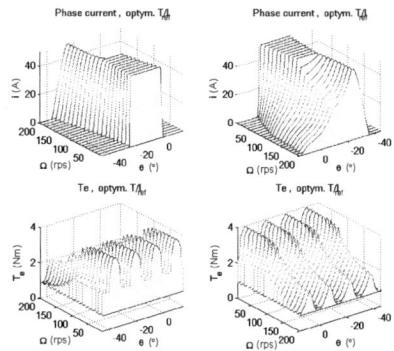

Fig.3 Waveforms of simulated phase current and electromagnetic torque component produced by phase current for current switching angles optimized for maximum ratio of torque to reference current

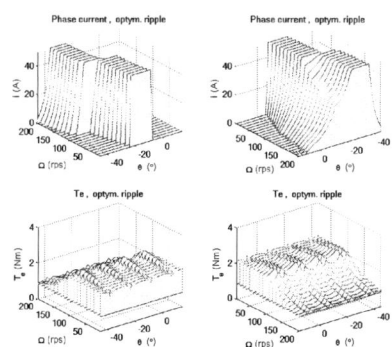

Fig.4 Waveforms of simulated phase current and electromagnetic torque component produced by phase current for current switching angles optimized for minimum ripples (minimum ratio of RMS value of torque ripples to torque average value)

three diagrams shows that proposed optimized control systems give expected results. The highest torque value was obtained for optimization with first criterion. The second criterion leads to the smallest values of torque ripples. More clear effect of optimization is visible on figures 5, where the waveforms of torque as an effect of

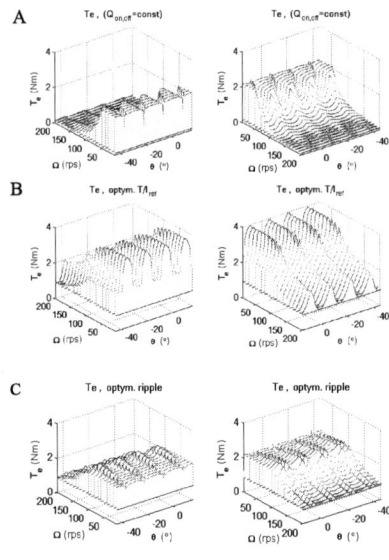

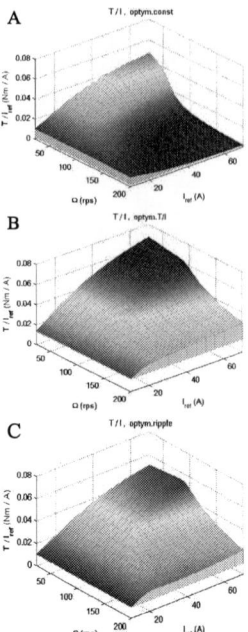

Fig. 6 Diagrams of steady state value of simulated electromagnetic torque for switching angles: A) constant (Θ_{on} = - 22.5°, Θ_{off} = - 7.5°), B) optimized for maximum ratio of torque to reference current, C) optimize for minimum ripples (minimum ratio of RMS value of torque ripples to torque average value)

summarized current of all four windings are shown. It is clearly visible that the worst result gives control with constant switching angles, for which the torque value is much smaller than for two others optimal control methods. The minimum value of torque ripples guarantees using the second criterion.

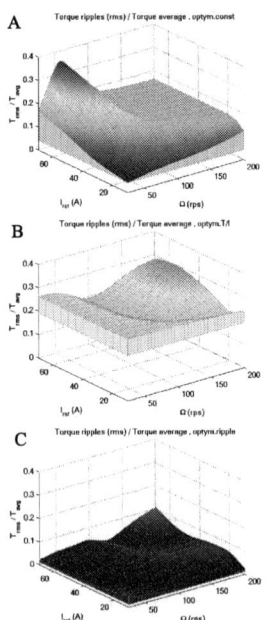

Fig. 7 Diagrams of ratio of RMS value of torque ripples to torque average value for switching angles: A) constant (Θ_{on} = - 22.5°, Θ_{off} = - 7.5°), B) optimized for maximum ratio of torque to reference current, C) optimize for minimum ripples (minimum ratio of RMS value of torque ripples to torque average value).

The same conclusions can be drown from characteristics of torque as a function of speed and reference current, shown in figures 6. The highest value of torque gives the optimization with first criterion and the lowest value gives control with constant switching angles. Calculated RMS value of torque ripples for all three analyzed control systems are clearly visible on diagrams shown in fig. 7. The optimal control with the second criterion guarantees the smallest values of torque ripples, on level about 5% of average value of torque. The torque ripples increase strongly for the criterion of maximum torque to values over 20% of average torque value. The constant switch angles give intermediate results, RMS values of torque ripples are of level between 5 – 15% of average torque.

V. EXPERIMENT VALIDATION

The laboratory setup (Fig. 8) consists of SRM with position sensor, power converter, and microprocessor control system with DSP (Analog Devices SHARC ADSP 21065L) equipped with FPGA module (Altera FLEX 6000) . This structure allows very effective control of SRM. The phase commutation scheduler and hysteresis current controller was realized in hardware, using D/A converters, comparators, and FPGA. Speed and position measurement signals are calculated in FPGA. The basic tasks realized by program implemented in the DSP are the following:

a) speed controller program, which calculate electromagnetic torque reference,

b) conversion from torque to current reference,

c) calculation of optimal switching-on and switching-off angles for each phase, depending on reference current and actual rotor speed.

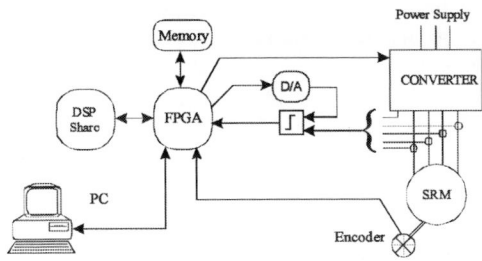

Fig. 8. Structure of the laboratory setup

Selected static characteristics measured on experimental system are compared with result obtained by simulation. These characteristics are shown in Figure 9. The measured experimental results confirm results obtained by simulation. The electromagnetic torque strongly depends on speed for constant switching angles, torque value is considerably reduced when speed increases over 100 rps. Two others optimal control methods guarantee much higher value of torque and its value do not depends on the speed increase.

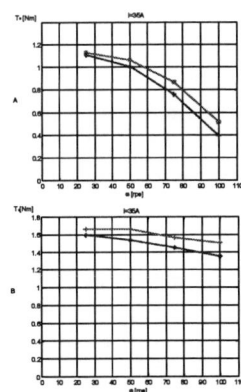

Fig. 9 Diagrams of steady state characteristics of experimentally measured and simulated values of electromagnetic torque obtained for reference current value equal 35 A versus speed for switching angles: A) constant (Θ_{on} = - 22.5°, Θ_{off} = - 7.5°), B) optimized for maximum ratio of torque to reference current

The control system operation is well illustrated on figures 10 and 11. The constant switching angles do not guarantees correct waveforms of phase currents for higher speed (100 rps), what is visible on Fig.9. In contrary to this , variable switching angles optimized for maximum torque ensure correct current waveforms (Fig. 10).

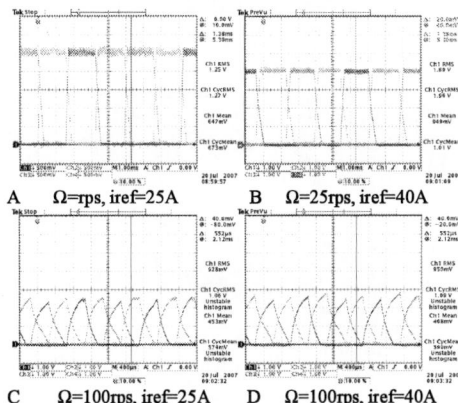

A Ω=rps, iref=25A   B Ω=25rps, iref=40A

C Ω=100rps, iref=25A   D Ω=100rps, iref=40A

Fig.10 Experimentally measured waveforms of phase currents for constant switching angles (Θ_{on} = - 22.5°, Θ_{off} = - 7.5°) and different steady state value of speed;current values: A, C) 25 A and B, D) 40 A; speed values: A, B) 25 rps and C, D) 100 rps

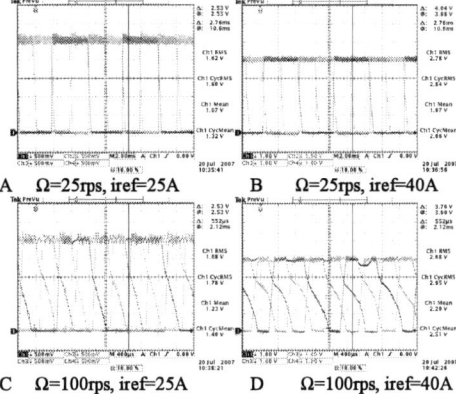

A Ω=25rps, iref=25A   B Ω=25rps, iref=40A

C Ω=100rps, iref=25A   D Ω=100rps, iref=40A

Fig.11 Experimentally measured waveforms of phase currents for switching angles optimized for maximum ratio of torque to reference current and different steady state value of speed; current values: A, C) 25 A and B, D) 40 A; speed values: A, B) 25 rps and C, D) 100 rps.

Effect of optimal control of commutation angles is well visible during step response of speed control, shown in figures 12 and 13. Waveforms of speed, current and torque for control with constant commutation angles show that the transient process for small speed (Fig. 12A) is much faster than for higher speed (Fig.12B). The optimal control of commutation angles obtained for the first criterion guarantees similar good dynamics for small and high value of speed during the same analyzed transient process (see Fig. 13).

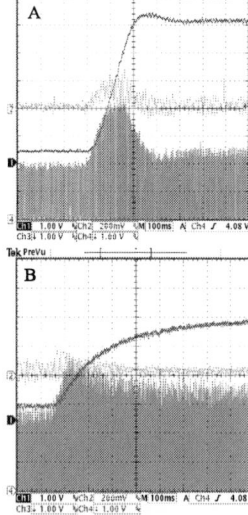

Fig.12 Experimentally measured waveforms of angular speed, phase currents and torque for constant switching angles (Θ_{on} = - 22.5°, Θ_{off} = - 7.5°) during step response of speed control: A) 25 → 50 rps, B) 75 → 100 rps.

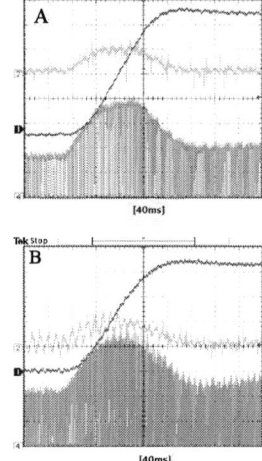

Fig.13 Experimentally measured waveforms of angular speed, phase currents and torque for current switching angles optimized for maximum ratio torque to reference current during step response of speed control: A) 25 → 50 rps, B) 75 → 100 rps.

VI. CONCLUSION

High performance of SRM drive requires precise control of current commutation angles, which should be variable during speed and current changes. Optimal selection of switching angles can be done on the base of different criteria, properly describing control requirements. In the paper two criteria were tested, criterion of maximum ratio of torque to current and criterion of minimum torque. The first criterion is

adequate for high-speed of rotation , while the second one is better for lower speed, where torque ripple can produce undesirable speed ripples. The optimization process can be carried out off-line and after that its result can be stored in look-up tables, which are used in control system.paper has only one selected weight coefficient instead of tow selected coefficients in previous authors' methods. It is shown in the paper that the procedure of controller settings adjustment yields good results and can be recommended for such control applications.

REFERENCES

[1] Deskur J., Maciejuk A.: Application of digital phase locked loop for control of SRM drive. Proceedings of 12th European Conference on Power Electronics and Applications EPE'2007, 2 - 5 September 2007, Aalborg, Denmark, CD.

[2] Deskur J., Maciejuk A.: Modelling of Switched reluctance Motor Drive" . Proceedings of XIX Symposium on Electromagnetic Phenomena in Nonlinear Circuits, EPNC'2006, Maribor, Slovenia, 28 -30 June 2006, pp.159-160.

[3] Husain I, Islam M.S.: Observers for position and speed estimation in switched reluctance motors. Decision and control, 2001. Proceedings of the 40th IEEE Conference, 4-7 Dec. 2001, vol.3, pp.2217-2222.

[4] Husain I.: Modeling, Simulation, and Control of Switched Reluctance Motor Drives. IEEE Transactions on Industrial Electronics, vol. 52, No.6, December 2005, pp.1625-1634.

[5] Krishnan R.: Switched Reluctance Motor Drives: Modelling, Simulation, Design and Applications. CRC Press, June 28, 2001.

ACKNOWLEDGMENT

This work was parity supported by grant TB 45-082/08/DS.

APPENDIX.
DATA OF INVESTIGATED DRIVE

Parameters of SRM 8/6	Value	Unit
Rated Output	1.32	kW
Maximum Output	2.27	kW
Maximum Torque	3.7	Nm
Rated voltage	48	V
Rated speed	6000	rpm
Maximum speed	15000	rpm

Comparison Between Direct Torque Control and Vector Control of a Permanent Magnet Synchronous Motor Drive

Rafa Souad * and Houcine Zeroug**

University of Sciences and Technology HB, Dpt of Electrical Engineering, Algiers, Algeria
*email: *rafasouadl@gmail.com*   **email: *zeroughoucine@hotmail.com*

Abstract—This paper presents a comprehensive comparison between two speed control techniques for permanent magnet synchronous machine using vector control (VC) and direct torque control (DTC). The main characteristics of vector control and direct torque control are studied by simulation emphasizing their advantages and disadvantages. The performances of the two control techniques are evaluated in terms of torque and current ripples, and transient responses to load toque variation. The simulation results show that the two control schemes provide in their basic configuration, comparable performances regarding the torque control, however the choice for the DTC appears advantageous in many respects, and presents a low cost solution for applications where low torque ripple is not a major issue.

Keywords— Simulation, Control of Drive, permanent magnet motor, Direct torque and flux control, vector control.

I. INTRODUCTION

Permanent magnet synchronous motor (PMSM) drives are replacing classic dc and induction machine (IM) drives in a variety of industrial applications, such as industrial robots and machine tools. Advantages of PMSM's include low inertia, high efficiency, high power density and reliability. Because of these advantages, PMSM Drives are indeed excellent for use in high-performance servo drives where a fast and accurate torque response is required [1]. The torque in PMSM is usually controlled through the armature current based on the fact that the electromagnetic torque is proportional to the armature current [2].

It is now recognized that the two high-performance control strategies for PMSM drives are vector control (VC) and direct torque control (DTC) [3,4]. These control strategies are different in their operating principles but their objectives are the same. They both aim to control effectively the motor torque and flux in order to force the motor to accurately track the speed and torque references regardless of the machine and load parameter variation [5]. In fact, there has been intensive research work to establish a fair comparison between the performances of two techniques, however, this has been mainly devoted to induction motor control [2]. For the PMSM speed control, few literature has been reported and consistent study to highlight the difference particularly in application when torque control is desirable

The purpose of this paper is to present a comprehensive comparative study by simulation between the vector control using space vector modulation technique and direct torque control technique. These two control strategies have been performed on a typical PMSMotor with 700W rating, with parameters reported in APPENDIX. In this study, a simulation model is presented. It allows a fair performance comparison in terms of torque ripple and speed response when the motor is subject to sudden load variation, as well as speed variation.

II. MODELLING OF PMSM

The electrical equations governing the behavior of the permanent magnet synchronous motor in the dq reference frame are as follows:

$$\Phi_d = L_d i_d + \Phi_f$$
$$\Phi_q = L_q i_q \tag{1}$$

$$V_d = R.I_d + \frac{d}{dt}\Phi_d - \omega.\Phi_q$$
$$V_q = R.I_q + \frac{d}{dt}\Phi_q + \omega.\Phi_d \tag{2}$$

The torque is given by :

$$T_e = \frac{3}{2}.p.\left[\left(L_d - L_q\right)I_d I_q + \Phi_f I_q\right] \tag{3}$$

Where Φ_f, Ld, and Lq, are the magnet flux constant and inductances, respectively

The mechanical equation is given by:

$$J\frac{d\Omega}{dt} = T_e - T_r - f\Omega$$
$$\Omega = \frac{\omega_r}{p} \tag{4}$$

III. PRINCIPLE OF VECTOR CONTROL

The vector control (VC) technique applied to the Permanent Magnet Synchronous Motor (PMSM) consists in maintaining the current direct (Id) null to produce a

978-1-4244-1741-4/08/$25.00 ©2008 IEEE   1209

maximum torque and to use the current (Iq) component for torque control over speed range.

The block diagram of this structure of speed control is represented in Fig. 1; here the inverter is controlled through space vector modulation (SVM) scheme. A proportional-integral regulator has been introduced in speed and current loops. In order to improve the decoupling between flux and torque a compensation scheme has been introduced.

The compensation voltage can be given by the equation (5):

$$V_d^c = -L_q \omega . i_q$$
$$V_q^c = \omega . \Phi_f + \omega . L_q . i_d \qquad (5)$$

IV. Principle of DTC

The direct torque control (DTC) method is also based on the space vector theory and has been first proposed for induction machines [5,7]. This concept can also be applied to synchronous drives [2,7,8]. The basic idea of DTC is to choose the best vector of voltage which makes the flux rotate and produce the desired torque. During this rotation, the amplitude of the flux is kept in a pre-defined band. With a three phase voltage source inverter, there are six non-zero voltage vectors and two zero voltage vectors which can be applied to the machine terminals [9].

The stator flux vector can be estimated using measured current and voltage vectors:

$$\hat{\Phi}_s = \Phi_s(0) + \int_0^t (V_s - R_s . i_s) dt \qquad (6)$$

Then torque can be calculated using the components of the estimated flux and measured currents in α, β frame:

$$\hat{T}_e = \frac{3}{2} p (\hat{\Phi}_\alpha I_\beta - \hat{\Phi}_\beta I_\alpha) \qquad (7)$$

p is the pole pair, components of the current and flux parameters in α, β frame are determined using Concordia transformation. As it can be seen, using the equations (6) and (7) to estimate flux and torque. These equations show that the voltage, current parameters measurements and the stator resistance R are involved to estimate the torque and the flux. To allow the motor to run in the four quadrants operating conditions, an adequate switching table is used according to the sector where the flux is evolving. Table I shows the switching voltage-vector lookup table. In addition, there are two hysteresis controllers for the flux and the torque. The DTC bloc diagram is shown in Fig. 2.

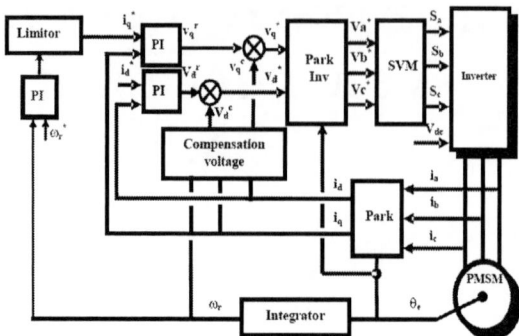

Fig. 1. Vector control bloc diagram

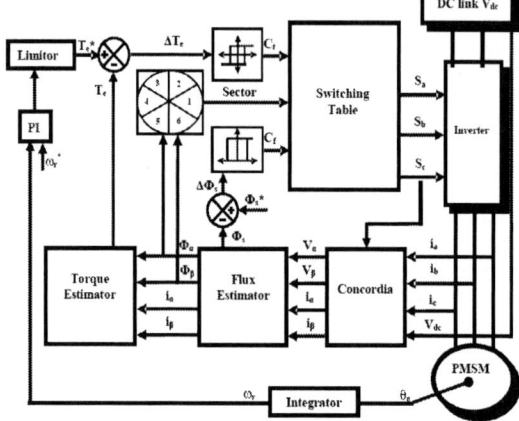

Fig. 2. Direct torque control bloc diagram

Table I.
Basic switching table. DTC control

Flux	Torque	S=1	S=2	S=3	S=4	S=5	S=6
	$C_T=1$	V_3	V_4	V_5	V_6	V_1	V_2
$C_\Phi=0$	$C_T=0$	V_0	V_7	V_0	V_7	V_0	V_7
	$C_T=-1$	V_5	V_6	V_1	V_2	V_3	V_4
	$C_T=1$	V_2	V_3	V_4	V_5	V_6	V_1
$C_\Phi=1$	$C_T=0$	V_7	V_0	V_7	V_0	V_7	V_0
	$C_T=-1$	V_6	V_1	V_2	V_3	V_4	V_5

A. Sensorless Control with DTC

Various studies have been reported, which highlight the sensorless behavior of the DTC [12, 13]. In this section, simulations have been performed in Matlab/Simulink to verify the feasibility of speed sensorless DTC of PMSM. The approach used to estimate the position and the speed are based on the estimation of the stator flux linkages in the stationary reference and the torque according to the following equations:

$$\hat{\Phi}_\alpha = \Phi_\alpha(0) + \int_0^t (V_\alpha - R_s.i_\alpha)dt$$

$$\hat{\Phi}_\beta = \Phi_\beta(0) + \int_0^t (V_\beta - R_s.i_\beta)dt \qquad (8)$$

the flux amplitude and its phase are derived respectively from:

$$\Phi_s = \sqrt{\hat{\Phi}_\alpha + \hat{\Phi}_\beta}$$

$$\theta_s = \angle\Phi_S = arctg\left(\frac{\hat{\Phi}_\beta}{\hat{\Phi}_\alpha}\right) \qquad (9)$$

The estimated electromagnetic torque can be also given by the following expression:

$$\hat{T}_e = \frac{3}{2}\frac{p}{L_d}\Phi_s\Phi_f \sin\gamma \qquad (10)$$

Φ_s and Φ_f denotes and stator and rotor flux respectively. By calculating the torque from the equation (7) above, the load angle ψ can be obtained from (10), and therefore, the estimated rotor position will be:

$$\theta_r = \theta_s - \gamma \qquad (11)$$

The rotor angular speed can be calculated as:

$$\omega = \frac{d\theta_r}{dt} \qquad (12)$$

V. SIMULATION RESULTS

A detailed comparison between the two motor control has been carried out by numerical simulations using Matlab/Simulink environment, where commutation effect are not taken into account. In this case, it is possible to perform a consistent comparison between the two control schemes under steady state and transient operating conditions. The simulation includes the model of the PMSM and a proportional integral speed regulators. The parameters of the PMSM are given in the Appendix. In order to bring the motor up to comparable performance under the two schemes, the sampling time Ts has been considered equal to 20 µs in the DTC scheme, whereas in the Vector control it has been set to 50 µs. In doing so, it was found possible to control the flux linkage adequately in the DTC technique , with a hysteresis controller set at 5% of the rated value. This is well illustrated in Fig. 3 and 4 where the locus of the flux linkage is given for two sampling times. As it is clearly shown, the stator flux follows a predetermined path, and as expected, its trajectory in the stationary reference frame is a circle.

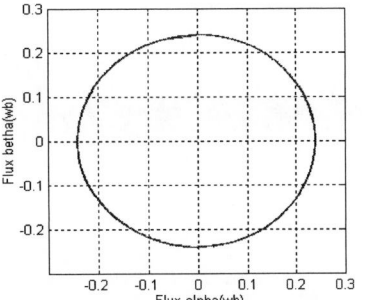

Fig. 3. Locus of stator flux in DTC scheme (T_s = 20 µs).

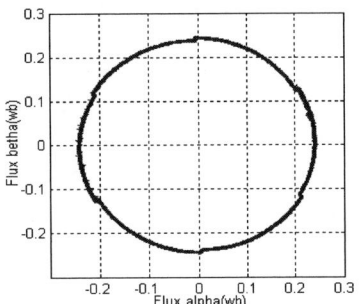

Fig. 4. Locus of stator flux in DTC scheme (T_s = 50 µs).

A. Steady-State Performance

Evaluation of the steady-state performance for a load of 0.5 N.m has been performed. In this simulation, VC only incorporates proportional-integral current regulators. The comparison between the two VC and DTC control schemes, reveals that the input current amplitude is much higher in the case of the DTC and the torque ripple is more significant in DTC scheme. The torque and the stator current waveform obtained with DTC and VC schemes are shown in Fig. 5. (a)-(d), respectively. Table II summarises the torque ripple for two operating speeds.

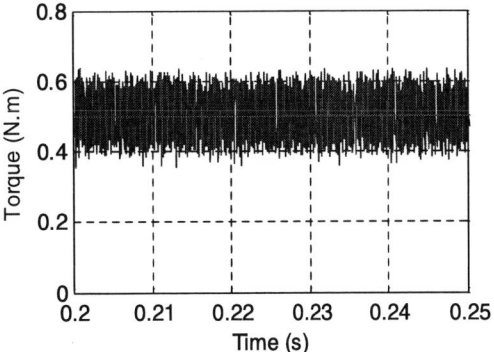

(a)

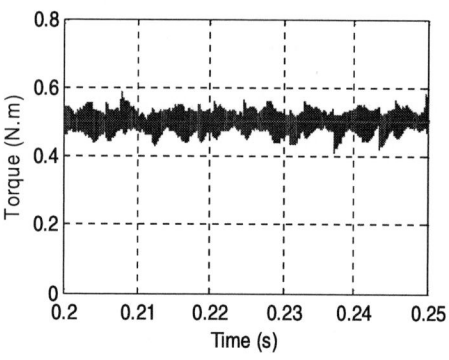

(b)

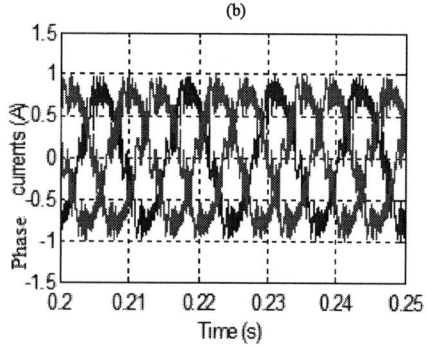

(c)

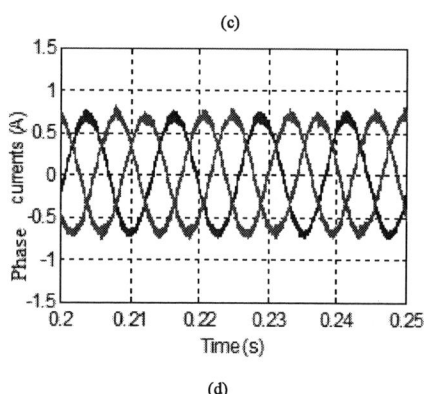

(d)

Fig. 5. (a) Torque (DTC). (b) Torque (VC).
(c) Stator currents (DTC). (d) Stator currents (VC)

Table II
Torque Ripple

Speed rd/s	125 rd/s	250 rd/s
VC	10 %	10 %
DTC	18%	20%

B. Transient Performance

The transient performance of the two schemes has been compared through the motor torque response to a variation step of the torque reference starting from 0 Nm to 0.5 N.m, Figures. 6 (a)–(d) show the torque variations

obtained for two operating speeds 250 rad/s and 125 rad/s respectively.

These results show that using the DTC scheme a better torque response can be achieved in terms of settling time. The settling times for the two cases are summarized in Table III.

Table III
Settling Time of the Torque

Speed	VC	DTC
250 rad/s	0.18s	0.1505s
125 rad/s	0.17s	0.1505s

In addition, the DTC presents a high dynamic from standstill to full speed and an effective load torque response, without major overshoot in comparison to vector control, as shown in Fig. 7.

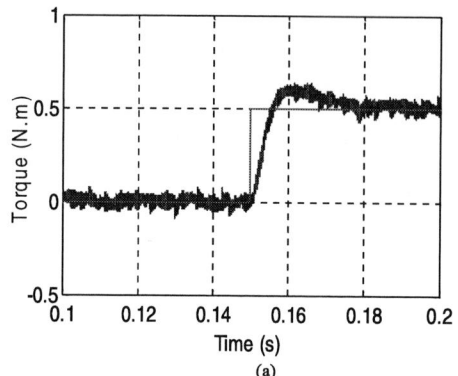

(a)

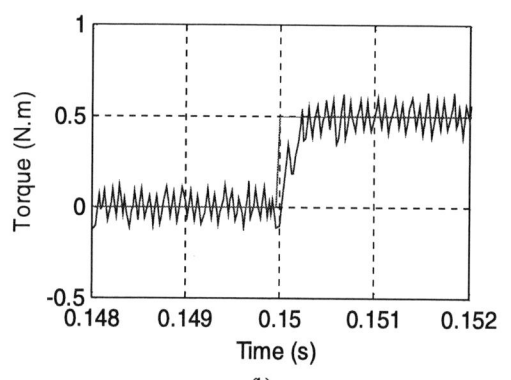

(b)

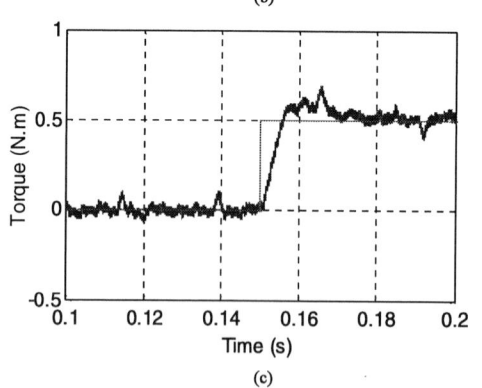

(c)

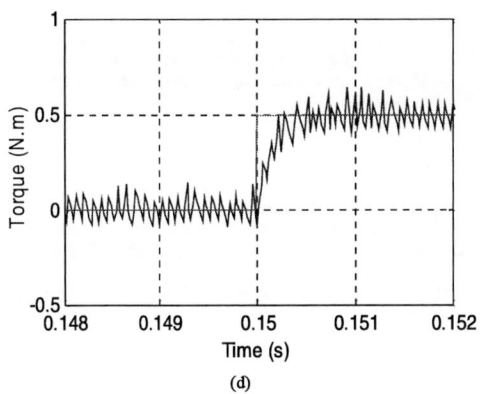

(d)

Fig. 6 : (a) Torque respone (VC), 250 rad/s. (b) Torque response (DTC), 250 rad/s. (c) Torque respone (VC), 125 rad/s. (d) Torque response (DTC), 125 rad/s.

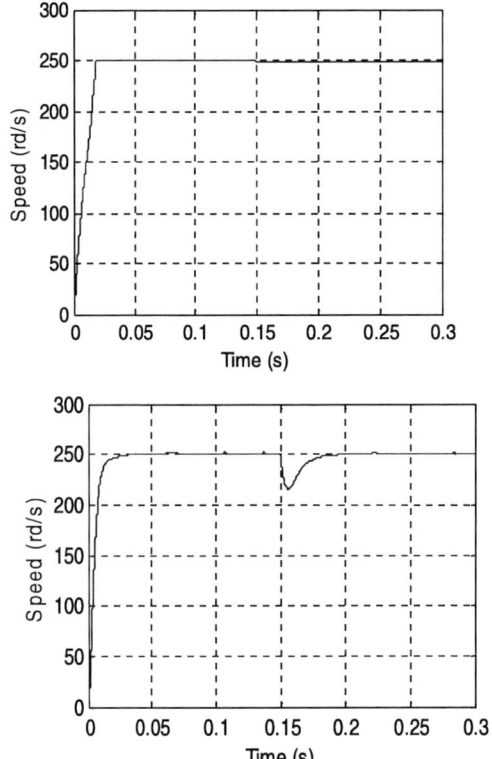

Fig. 7: (a) Speed response (DTC), 250 rad/s.
(b) Speed response (VC), 250 rad/s.

C. PMSM Sensorless control with DTC

The equations reported in section have been introduced in the bloc diagram of the DTC system shown in Fig. 2 and similar simulations were performed with and without sensor, however for two sampling frequencies 20 μs and 50μs. Fig. 8 shows the estimated and actual rotor position when the motor reaches its steady-state value in the speed of 250 rd/s. As it is clearly shown, the estimated and actual positions agree well with each other, with minor error. These errors are illustrated in Fig. 9.

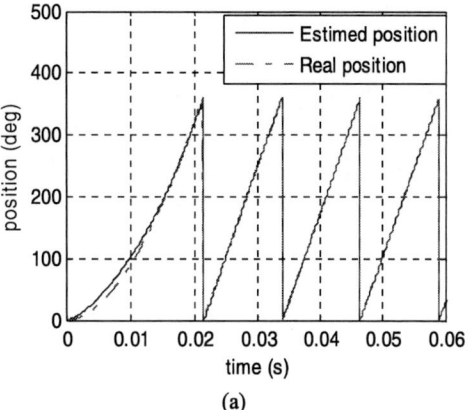

(a)

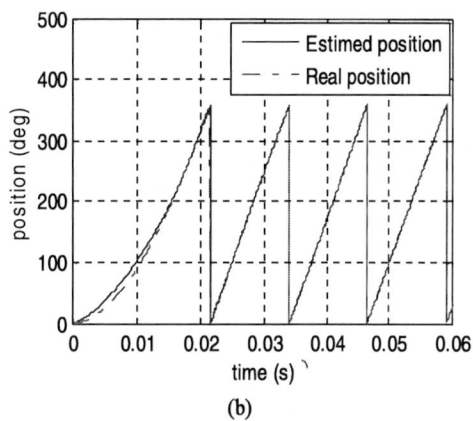

(b)

Fig. 8. Shaft estimated and actual position.
(a): Sampling time 50μs.
(b): sampling time 20 μs

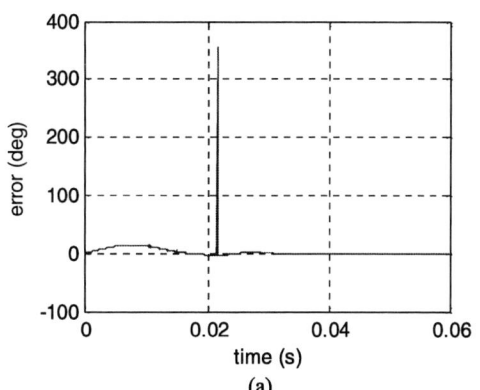

(a)

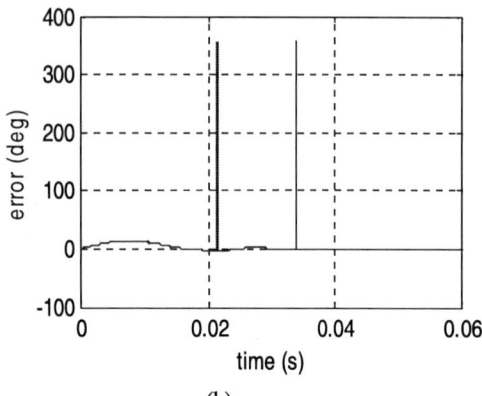

(b)

Fig. 9. Error between shaft estimated and actual position
(a): Sampling time = 50μs.
(b): Sampling time = 20 μs

CONCLUSION

The paper reports a comprehensive and fair comparison between the vector control and direct torque control when applied to PMSM speed control. The simulation results show the performances of the two schemes are comparable for the steady and dynamic operating conditions. The DTC might be preferred for high dynamic applications, but at the expense of a slight flux and torque ripple. This can be easily overcome by an additional compensation technique. In addition, this drawback is offset by the absence of d-q current regulators and position sensor, therefore leading the PMSM with DTC control to present a low cost system for many applications.

REFERENCES

[1] Thomas J. Vyncke, René K. Boel and Jan A.A. Melkebeek, " Direct Torque Control of Permanent Magnet Synchronous Motors – An Overview ", 3rd IEEE Benelux Young Researchers Symposium in Electrical Power Engineering, GHENT, BELGIUM, 27-28 April 2006.

[2] M. Vasudevan et al. "Real time implementation of viable torque and flux controllers and torque ripple minimization algorithm for induction motor drive", Energy Conversion and Management 47 (2006) pp. 1359-1371.

[3] L. Zhong, M. F. Rahman, W. Y. Hu, and K. W. Lim, « Analysis of Direct Torque Control in Permanent Magnet Synchronous Motor Drives », IEEE Tran on Power Electronics, vol. 12, no. 3, pp. 528-536, May 1997.

[4] Blaschke, " The principle of field orientation as applied to the new transvector closed loop control for rotating machines ", Siemens Review, vol. 30, no. 5, pp. 217-220, 1972.

[5] I. Takahashi and N. Noquchi, " A new quick response and high efficiency control strategy of an induction motor », IEEE Trans. on Industry Application. vol. 22, no. 5, pp. 820-827, Sept/Oct. 1986.

[6] H. Le-Huy, " Comparison of Field-Oriented Control and Direct Torque Control for Induction Motor Drives ", IEEE Industry Applications Conference, 34th IAS Annual Meeting, Vol 2, pp. 1245-1252, Oct 1999.

[7] M. Depenbrock, " Direct Self-Control (DSC) of inverter-fed induction machine ", IEEE Trans. on Power Electronics, Vol. 3, no. 4, pp. 420-429, Oct 1988.

[8] I. Boldea and S. A. Nasar, "Torque vector control (TVC)–A class of fast and robust torque speed and position digital controller for electric Drives ", in Proc. EMPS'88 Conf., vol. 15, 1988, pp. 135–148.

[9] M.R. Zolghadri, E. M. Olasagasti, and D. Roye, " Steady State Torque Correction of a Direct Torque Controlled PM Synchronous Machine ", IEEE Electrical Machines and Drives Conference Record, MC3/4.1-MC3/4.3, May 1997.

[10] C. French and P. Acarnley, "Direct torque control of permanent magnet drives ", IEEE Trans. Industry Applications, vol. 32, no. 5, pp. 1080–1088, Sept./Oct. 1996.

[11] D. Casadei, F. Profumo, G. Serra and A. Tani, " FOC and DTC: Two Viable Schemes for Induction Motors Torque Control ", IEEE Tran on power electronics, vol. 17, no 5, pp. 779-787, Sept 2002.

[12] L. Parsa, and H. A. Toliyat, " Sensorless Direct Torque Control of Five-Phase Interior Permanent-Magnet Motor Drives", IEEE Tran Industry Applications, vol. 43, no. 4, pp. 952-959, JULY/AUGUST 2007.

[13] P. Vas, Sensorless Vector and Direct Torque Control, London, UK, Oxford University Press, 1998.

APPENDIX

Rated voltage V=380 V.
Rated power = 400 W.
Rated current I_{rated}=2A.
Rated Torque T_{rated}=1.5N.m.
Self inductance L=21.5mH.
Stator Resistance=5.2Ω.
Rotor Flux Φ_f=0.24 Wb.
Rotor Inertia J=85e-6 kgm².

Detection and self-tuning compensation of periodic disturbances by the control of DC motor

Michael Ruderman, Frank Hoffmann, Johannes Krettek, Torsten Bertram

Chair for Control and Systems Engineering (RST), Technische Universität Dortmund
44221-Dortmund, Germany, e-mail: *mykhaylo.ruderman@tu-dortmund.de*

Abstract—The accuracy with which a DC motor and drive system track a reference velocity profile is limited by periodic torque disturbances. The cogging and ripples torques, and constructive imperfections in the mechanical assembly of the drive cause disturbance pulsation harmonics with a base frequency dictated by the rotational velocity. This paper describes a novel compensation technique based on disturbance observation and a self-tuning feed-forward compensation algorithm. The DC motor is modeled as a linear system augmented by the nonlinear Coulomb friction and is experimentally identified from a set of the system responses. The disturbance harmonics are detected by means of the fast Fourier transformation (FFT) and analytically described by a spatial Fourier transform with respect to the angular position of the rotor. The online algorithm tunes the parameters of the feed-forward compensator using the recursive estimation technique. The proposed self-tuning compensator is experimentally verified as part of an open control loop at different level of the rotational velocity and system load.

Index Terms—Permanent magnet motor, Control of drive, Modelling, Estimation technique, Harmonics

I. INTRODUCTION

DC motors with permanent magnet excitation are particular suitable for electrical drives that require high performance. In general, DC motors provide precise motion control at variable speed regimes with rapid dynamic responses. A persisting problem of permanent magnet machines is the nonuniformity of the developed torque which gives rise to torque pulsation. These disturbances constitute a major problem for the precise control of permanent magnet drives. Even, in cases for which the effective torque pulsation does not substantially contribute to the overall effective torque, it nevertheless may cause speed ripples and induce vibrations, particularly for small loads at low speeds [1].

A wide variety of techniques have been proposed to minimize the disturbing torque pulsation. They are basically divided into two major categories. The first described e.g. in [2],[3],[4] involves a comprehensive analysis of the fundamental mechanisms of torque pulsation generation and adjusts the motor design parameters to minimize these effects. The second category relies on feedforward compensators which augment the feedback control signal to compensate the torque pulsations with respect to their observed magnitude and phase [5], [6].

This paper proposes a simple compensation technique that combines the model based disturbance detection with a feedforward compensator to suppress torque pulsations emerging at different velocities. The method relies on a simplified dynamic model of the DC actuator system described by the authors in [7], and does not require precise knowledge about the mechanical and geometric structure of the electromechanical drive. The key feature of the proposed method is the self-tunable compensation method, which involves the gradient based recursive estimation algorithm. The method allows the application of the compensator to arbitrary rotatory drive systems with periodic disturbances for which a basic dynamic model is available.

The paper is organized as follows. Section II introduces the causes of periodic torque disturbances and illustrates their relation with the observed motor current. The dynamic model of the DC motor is described in Section III, and the identification of the unknown parameters are presented in Section IV. The DC motor is modeled as a linear second order system augmented with nonlinear Coulomb friction. The proposed online algorithm for detection and compensation of periodic torque disturbances is described in detail Section V. The experimental setup for the identification and compensation approaches is presented in Section VI. Finally, the compensation results are shown and discussed in Section VII.

II. PERIODIC DISTURBANCES

Pulsating torque is defined as the sum of cogging and ripples torque [8]. The cogging torque results from the variable airgap reluctance due to slotting (i.e. caused by the tendency of the rotor to align with the low reluctance paths). In contrast, ripple torque is generated by imperfections in the power converter and deficiencies in the geometry of permanent magnets, that causes variations of the flux density around the airgap. In addition to cogging and ripples torque, several constructive imperfections of the motor e.g. eccentricities between stator and rotor lead to a phenomenon known as mechanical beating and contribute to the periodic torque disturbances.

The pulsation torque is detected from the observation of the current in response to excitation signals that drive the actuator into steady state at different angular velocities. Assuming a linear relationship between the motor current and the torque, the pulsation torque is directly estimated from the current, and mapped into the frequency domain $T(t) \multimap T(j\omega)$ by means of a fast Fourier transformation (FFT).

Fig. 1 illustrates the power spectra of pulsation torque for different magnitudes of the average rotation frequency $\bar{f}_b = \{6.92, 25.56, 51.4\}$ Hz. The depicted circles indicate the main harmonics of the base rotation frequency. The observed spectra

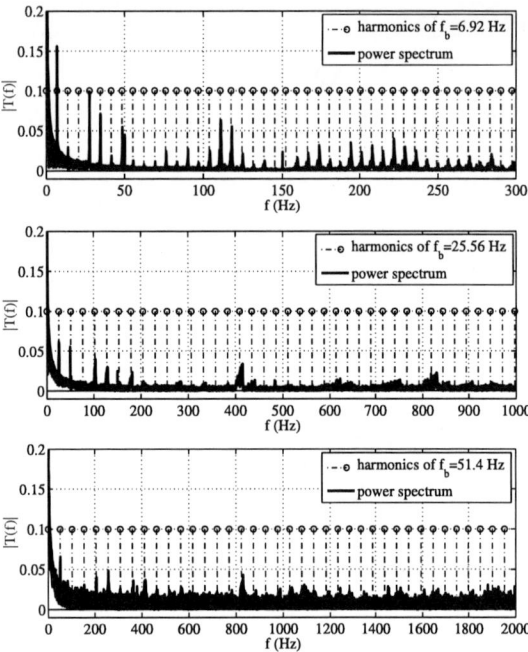

Fig. 1. Power spectra of pulsation torque at different magnitudes of the average rotation frequency $\bar{f}_b$

contain the high-frequency jitter corresponding to the white noise of the measured current. However, several peaks of the spectrum clearly coincide with harmonics of the base rotation frequency. The inspection of the peak magnitudes and the relationship between the three prototypical spectra, the most significant torque pulsations emerge at the $\{1, 4, 5, 7, 16\}$-th harmonics of the base rotation frequency. It becomes apparent that the torque pulsations do not depend on the angular velocity but are rather related to a periodic disturbance that is synchronous with the absolute rotor position.

In general there is no analytical model of the motor structure available to provide the basis for an electromechanical model of the torque pulsations. Therefore, the disturbance model is described by a general superposition of harmonic signals. The torque pulsation is represented by a Fourier series

$$T(\varphi) = T_0(u) + \sum_{n=1}^{N} T_n \sin(n\varphi + \sigma_n), \quad (1)$$

in which φ denotes the angular position of the rotor and $T_0(u)$ denotes the steady state torque as a function of the control voltage. T_n and σ_n denote the amplitude and the phase of the n-th harmonic. The number of harmonic components $N = 40$ is determined with respect to the maximal rotation frequency 50 Hz of the motor and the system sample rate of 4 kHz.

III. DC MOTOR MODEL

Under the assumption of a homogenous magnetic field, the direct current (DC) motor is modeled as a linear transducer from motor current to mechanical torque. The classical model of the DC motor, described by [9] is composed of a coupled electrical and mechanical subsystem.

The angular velocity ω is controlled by the input voltage u with a constant voltage drop attributed to the brush and rotor resistance, and a back-electromotive force (EMF) caused by the rotary armature. The motor inductance contributes proportional to the change in motor current i. The motor current couples the electrical component with its mechanical counterpart, as it generates the driving torque. This torque is antagonized by the motor inertia, structure damping, friction, and the external load. The motor dynamics are described by:

$$u(t) = L\frac{di(t)}{dt} + R_m i(t) + K_e \omega(t), \quad (2)$$

$$K_m i(t) = J\frac{d\omega(t)}{dt} + K_d \omega(t) + T_l + F, \quad (3)$$

in which K_m, K_e and K_d denote the motor torque, the back-EMF and the damping constants. J denotes the mechanical inertia including the motor armature and shaft. L and R_m represent the inductance and the total connection resistance of the motor. The external system load is denoted by T_l. The friction torque F of the motor bearing in sliding regime depends on the angular velocity only and is captured by the Coulomb and viscose friction parts.

$$F = \sigma\omega + \operatorname{sgn}(\omega)F_c, \quad (4)$$

in which F_c denotes the constant Coulomb friction and σ is the viscous friction coefficient. Note, that the viscous friction coefficient is already comprehended in the motor damping, so that the modeled friction contains the Coulomb nonlinearity only.

IV. IDENTIFICATION

The model parameters as described in Section III are identified by means of minimizing the deviation between the actual drive and the model open loop step response. The observed rotational speed ω and the motor current i are compared to the model output $\hat{\omega}$ and $\hat{i}$. The squared error between the outputs integrated over time is calculated according to:

$$\begin{aligned} E_\omega &= \int_{t=0}^{t_e} (\hat{\omega}(t) - \omega(t))^2 dt \\ E_i &= \int_{t=0}^{t_e} (\hat{i}(t) - i(t))^2 dt \end{aligned} \quad (5)$$

The identification of model parameters is done by minimizing those error functions. Both error functions are weighted and accumulated to create a scalar measure of model quality.

$$E_s = wE_\omega + (1-w)E_i. \quad (6)$$

To capture a broad range of operating conditions for parameter identification the step responses for different amplitudes of 2, 3, 4, 7 and 14 V. The total model error E is calculated as the sum of the individual step response error E_s normalized by the step amplitude.

The optimal parameters are identified with the well known Nelder-Mead simplex method [10] in which the parameter estimates are either initialized according to the vendors data sheet or are guessed from a coarse approximation.

Fig. 2 compares the model predicted step responses with the actual one in terms of velocity on the left and output current on the right. optimized step response of the system. The graphs demonstrate a good match between the model parameters and the actual system.

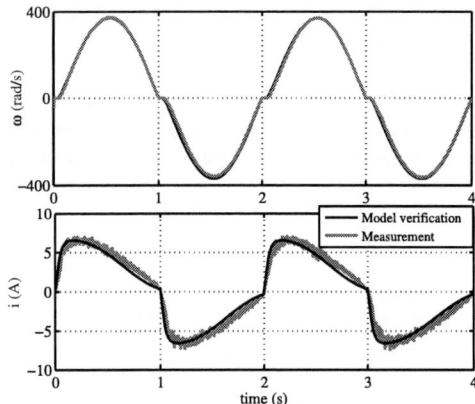

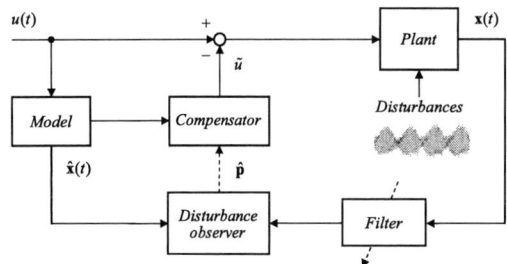

Fig. 3. Comparison of system and model responses for a sinusoidal input signal

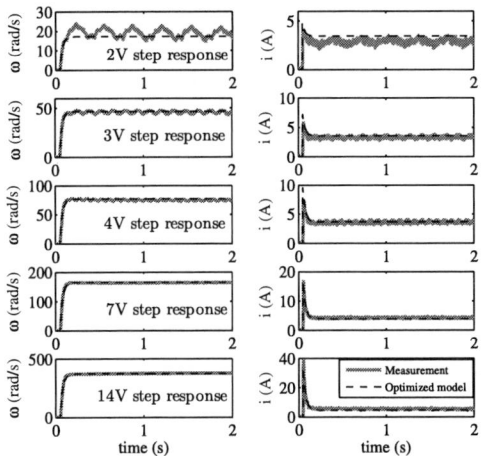

Fig. 2. System and model step responses for current and angular velocity at 5 different input step amplitudes

The model quality is verified on set of sinusoidal test signals that differ from the step responses used for identification. Figure 3 compares the outputs in terms of velocity and the current signal, which shows the consistency between model and actual system.

V. COMPENSATION ALGORITHM

The compensation scheme depicted in Fig. 4 relies on the disturbance observer and feed-forward compensation of observed periodic disturbances.

The compensator receives the predicted states $\hat{\mathbf{x}}$ and the disturbance parameters $\mathbf{p}$ as inputs. Since the modeled system is passive and the proposed compensation technique is feed-forwarding the compensation loop remains a priori stable independent of the state and parameter variations.

The self-tuning algorithm determines the parameters in an online fashion from the fast fourier transforms of the input and output signal. The feed-forward compensator aims to directly suppress the observed disturbance harmonics. The main idea of

Fig. 4. Block diagram of the detection and compensation scheme

the method is to estimate the proper magnitudes and phases of the compensation signal by means of the recursive estimation technique

$$\hat{\mathbf{p}}(t) = \hat{\mathbf{p}}(t-1) + K(t)\left(\mathbf{x}(t) - \hat{\mathbf{x}}(t)\right), \qquad (7)$$

in which $\hat{\mathbf{p}}(t)$ is a vector of parameters to be estimated, $\mathbf{x}(t)$ denotes the observation, and $\hat{\mathbf{x}}(t)$ is the model output predicted based on the current parameter estimate $\hat{\mathbf{p}}(t-1)$. The tuning gain $K(t)$ determines the step size with which the current prediction error effects the correction of the parameter estimate. The normalized gradient adaptation captures the tuning gain

$$K(t) = \Theta(t)\Psi(t), \qquad (8)$$

depending on the gradient $\Psi(t)$ of the model prediction with respect to the parameters and the function

$$\Theta(t) = \frac{\gamma}{\left|\Psi(t)\right|^2}\mathbf{I}, \qquad (9)$$

in which $\mathbf{I}$ denotes the identity matrix and γ is a gain of the recursion.

In our case, the disturbance observer estimates the compensation magnitude and phase of the torque pulsations. Since the compensation signal $\tilde{u}$ is formed by the superposition of the identified disturbance harmonics in terms of Fourier series, the individual harmonics are sequently estimated in the order from the highest frequencies to the lowest. To achieve a proper system identification an adaptive band pass filter of second order is applied to the measured current. The upper and lower cutoff frequencies are selected in the vicinity of the n-th harmonic with a bandpass for frequencies $n \cdot \hat{f}_b \pm 2$ Hz. The bandpass filter introduces no additional phase shift in the vicinity of the corresponding harmonic, and therefore reveals the correct phase of the original signal. Once the phase and magnitude estimate of the current harmonic stabilizes, the filter adaptation scheme switches to the next lower harmonic. The tuning process concludes with the estimation of all relevant harmonics. It is resumed in case of substantial variations of the load or process conditions during operation. The compensation signal

$$\tilde{u}(\varphi) = \frac{R_m}{K_m} \sum_n T_n(\varphi),\qquad(10)$$

results from the superposition of estimated harmonics $n \in \{1, 4, 5, 7, 16\}$ and added to the feedback control signal.

The convergence of the recursive estimation algorithm is first verified in an idealized system simulation with a multi-sinus excitation signal composed of multiple harmonics. Fig. 5 shows the simulation results of the recursive estimation algorithm. The upper plot depicts the parameter evolution during the five consecutive tuning steps of the individual harmonics. Both amplitudes and phases of the harmonics saturate and converge to their nominal value. The bottom part of Fig. 5 represents the simulated system response to torque ripple before and after the switch-on of the feedforward compensation signal.

VI. EXPERIMENTAL SETUP

Fig. 6 shows structure of the actuator system and the experimental setup for the identification and compensation approaches. The system is driven by the AXEM DC servo motor with a shrunk-on-disk rotor, F9M2 with rated power output of 88 W and rated speed 3000 r.p.m.

The analog and digital input and output signals are provided by the real time framework on the PC based controller with a sample rate of 4 kHz. The control signal is a voltage of ± 14 V converted by the linear current amplifier. The angular position φ of the motor shaft is measured with a single turn absolute rotary encoder with 13 bit resolution. The encoder is connected via a parallel DIO interface allowing data acquisition at the base sample rate. The motor current i is measured by the voltage drop u_s across a shunt resistance R_s connected in series with the DC motor and provided via a 16 bit analog-digital converter. The angular velocity of the motor shaft is obtained as the time derivative of measured position subsequently filtered to reduce the quantization errors.

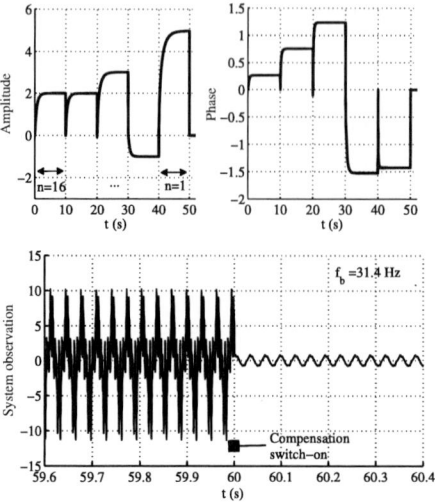

Fig. 5. Simulation of the recursive estimation algorithm: parameter evolution (top) and periodic disturbances before and after compensation switch-on (bottom)

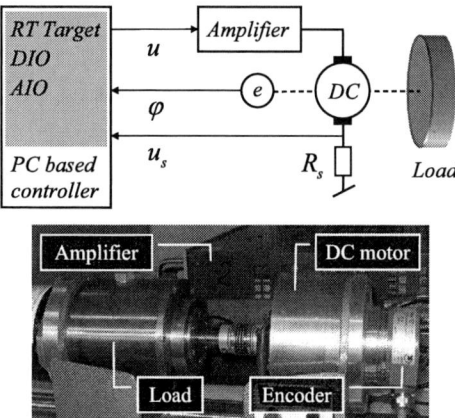

Fig. 6. DC motor actuator system (upper) and experimental setup (below)

VII. COMPENSATION RESULTS AND DISCUSSION

The performance of the proposed compensation scheme is evaluated experimentally for multiple motion trajectories at different steady states with a constant reference signal. Furthermore, the robustness of the compensation technique is verified on the drive system loaded with an additional external load, in order to impose parameter variations on the process.

Fig. 7 illustrates the evolution of the parameter estimates and the observation error for the experimental drive during the estimation and compensation phases. The upper graph shows the evolution of the estimated parameter evolution during the consecutive estimation of the harmonics. The estimation time of 30 s for each harmonic is currently limited by the controller memory required for the recursive estimation algorithm. In order to reduce the influence of unsystematic errors and noise on the parameter estimates by long term averaging a low gain of $\gamma = 0.0001$ is selected. However, the low gain slows the convergence rate the parameter estimates thus resulting in prolonged estimation periods. The magnitude estimate converges reliably for all harmonics even though a substantial oscillation in magnitude in case of the highest frequency 186 Hz ($n = 16$). The corresponding phase estimates saturate in case of $n = 1$ and $n = 5$ and tend to converge for the 4th and 7th harmonics. The phase evolution of $n = 16$ behaves erratically, due to the limited accuracy of the observed signals and the presence of non-stationary disturbances, as becomes apparent in the lower graph in Fig. 7. The graph depicts the observation error during the initial estimation phase ($t \in [0, 150]$), the noncompensated case ($t \in [150, 155]$) and the compensation phase ($t \in [155, 160]$). It becomes apparent that the high-frequency jitter for the non compensated pure feedback control is reduced after switch-on of the feedforward compensator.

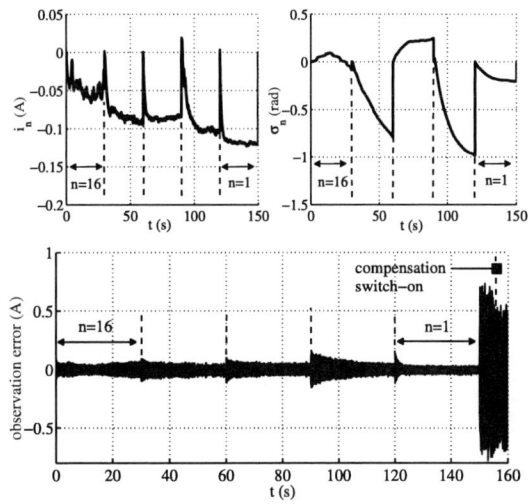

Fig. 7. Experimental results of the recursive estimation algorithm: parameter evolution (upper) and observation error during the estimation and after switch-on of the compensation signal (below)

The system behavior with and without the compensator is further analyzed in frequency domain, estimated from the current measurement during the 5 s period of compensated and non compensated control.

The power spectra of the current during pure feedback and the additional feedforward compensator are shown in Fig. 8, 9 and 10 at different stationary operating points. The average rotation frequency is $\bar{f}_b = \{11.5, 40.6\}$ Hz and the average current is $\bar{i} = \{2.17, 2.53\}$ A. Moreover the motor actuator is loaded with an additional mechanical load in an experiment with $\bar{f}_b = 24.5$ Hz and $\bar{i} = 5.56$ A.

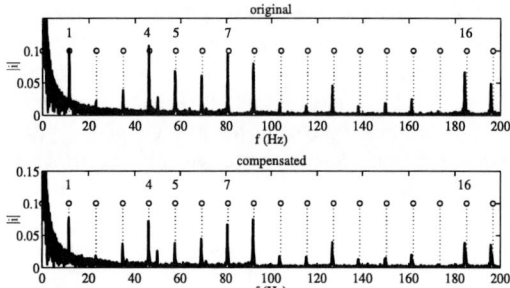

Fig. 8. Power spectrum ($\bar{f}_b = 11.5$ Hz, $\bar{i} = 2.17$ A)

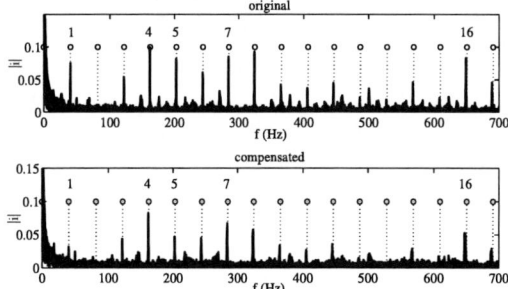

Fig. 9. Power spectrum ($\bar{f}_b = 40.6$ Hz, $\bar{i} = 2.53$ A)

All spectra show a suppression of the peaks at the compensated harmonics in the order of 20-30 %. In case of the external load the peaks are broadened and the suppression is less effective compared to the unloaded experiments which is attributed to the higher energy of the disturbances. A feasible approach to overcome this effect is an additional magnitude amplification of the compensation signal proportional to the area of the detected harmonic peaks. Notice, that the proposed compensation technique does not induce new disturbance peaks in the motor current spectra, indicating that the disturbances are truly suppressed rather than shifted to a different frequency.

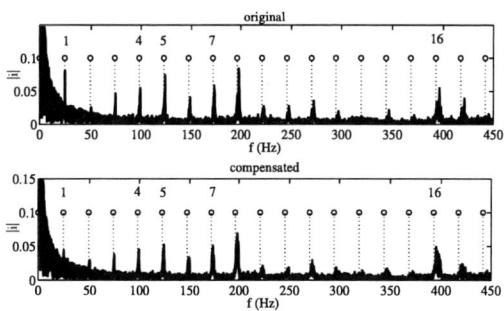

Fig. 10. Power spectrum ($\bar{f}_b = 24.5$ Hz, $\bar{i} = 5.56$ A, with load)

The proposed method is well suited for the online estimation and compensation of periodic torque disturbances. The main benefit is that the technique does not require precise knowledge or a model of the mechanical and geometric structure of the drive system and sources of the periodic disturbances. The ease of integration allows it to augment arbitrary feedback controllers with the compensation scheme and to retune the compensator parameters in case of substantial variations of the load or process conditions during operation. We assume, that an improved observation in terms of accuracy and sample rate contributes to an improved compensation. Future research is concerned with the design of an adaptive compensator in order to automatically adjust the magnitude of the compensation signal with the area and therewith the energy of the periodic disturbances.

ACKNOWLEDGMENT

This work was partially funded by the German Federal Ministry of Education and Research under BMBF-No. 02PB2197. The authors are also grateful to the Kübler GmbH company for providing the sensor.

REFERENCES

[1] Z. Q. Zhu and D. Howe, "Influence of design parameters on cogging torque in permanent magnet machines," *IEEE Transactions on energy conversion*, vol. 15, no. 4, pp. 407–412, 2000.

[2] N. Bianchi and S. Bolognani, "Design techniques for reducing the cogging torque in surface-mounted pm motors," *IEEE Transactions on industry applications*, vol. 38, no. 5, pp. 1259–1265, Sept. 2002.

[3] A. B. Proca, A. Keyhani, A. El-Antably, W. Lu, and M. Dai, "Analytical model for permanent magnet motors with surface mounted magnets," *IEEE transactions on energy conversion*, vol. 18, no. 3, pp. 386–391, Sept. 2003.

[4] K. Atallah, J. Wang, and D. Howe, "Torque-ripple minimization in modular permanent-magnet brushless machines," *IEEE transactions on industry applications*, vol. 39, no. 6, pp. 1689–1694, Nov. 2003.

[5] J. Holtz and L. Springob, "Identification and compensation of torque ripple in high-precision permanent magnet motor drives," *IEEE Transactions on Industrial Electronics*, vol. 43, no. 2, pp. 309–320, 1996.

[6] A. Seguritan and M. Rotunno, "Adaptive torque pulsation compensation for a high-torque dc brushless permanent magnet motor," in *Proc. 15th Triennial World Congress (IFAC2002)*, Barcelona, Spain, July 2002.

[7] M. Ruderman, J. Krettek, F. Hoffmann, and T. Bertram, "Optimal state space control of dc motor," in *Proc. (appears) 17th IFAC World Congress*, Seoul, South Korea, July 2008.

[8] T. Jahns and W. L. Soong, "Pulsating torque minimization techniques for permanent magnet AC motor drives a review," *IEEE Transactions on Industrial Electronics*, vol. 43, no. 2, pp. 321–330, 1996.

[9] R. Isermann, *Mechatronic Systems*, 1st ed. Berlin, Germany: Springer, 2005.

[10] J. Nelder and R. Mead, "A simplex method for function minimization," *The Computer Journal*, vol. 7, pp. 308–313, 1964.

A Linear Switched Reluctance Motor Based Position Tracking System

S. W. Zhao[*], N. C. Cheung[*], Y. Lu [*], W. C. Gan [†] and Z. G. Sun [*]

[*] Department of Electrical Engineering, Hong Kong Polytechnic University, Hong Kong, e-mail: *eencheun@polyu.edu.hk*
[†] Motion Group, ASM Assembly Automation Hong Kong Ltd., Hong Kong, e-mail: *wcgan@asmpt.com*

Abstract—**A Linear Switched Reluctance Motor (LSRM) based position tracking system and its control aspect are presented in this paper. Based on modeling analysis of the drive system, two reduced order controlled models are developed with different time scales, referred as mechanical subsystem and electrical subsystem. Two controllers are respectively designed for the two subsystems. Simulation results demonstrate the effectiveness of this control scheme.**

Keywords—**LSRM, position tracking system, winding excitation scheme.**

I. INTRODUCTION

Position tracking system is an essential element in advanced manufacturing processes. For demanding direct-drive applications, Linear Switched Reluctance Motors (LSRMs) have drawn much research attention over the past decade, due to its low cost, simple structure, ruggedness and reliability in harsh environments, and its potential for numerous industrial applications. Compared to rotary motors with mechanical transformation components for producing linear motion, LSRM has many advantages, such as quick response, high sensitivity and excellent tracking capability. Moreover, the structure of a LSRM can reduce the installation space requirement. However, the main limitation of the LSRM comes from its inherent nonlinear characteristics and force ripples problems which bring difficulties to its control.

Several control methods and schemes have been proposed in its control aspect. The most common method is to use a lookup table for the nonlinear torque/force compensation [1, 2]. Other literatures proposed nonlinear control methods for Switched Reluctance Motors (SRMs). A feedback linearization controller is designed for position tracking in [3], where two full-order nonlinear models are applied. In [4], an adaptive controller is presented to combat the nonlinear characteristics by the online estimation. In [5, 6], two passivity-based controllers are employed for SRMs and variable reluctance finger gripper, respectively. These control approaches provide possibilities to apply LSRM for building up a position tracking system. However, these control methods also require complicated control algorithms.

This paper is to propose a relatively simple control design which can be conveniently realized on a low cost embedded system for trajectory control applications.

II. SYSTEM CONSTRUCTION AND LSRM MODELING

A. Configuration of the drive system

The proposed drive system is shown in figure 1 (a). It can be seen that the mover is mounted on two linear guides, which are tightly fixed on the base. There are some ball bearings between the mover and these two linear guides. This rugged mechanical structure can effectively buffer vibration during the operations. A resolution linear optical encoder is integrated in the drive system to provide the feedback position information. The actuator of the drive system is a three-phase LSRM and its design schematic is shown in figure 1 (b). A set of three-phase coils with the same dimensions is installed on the mover.

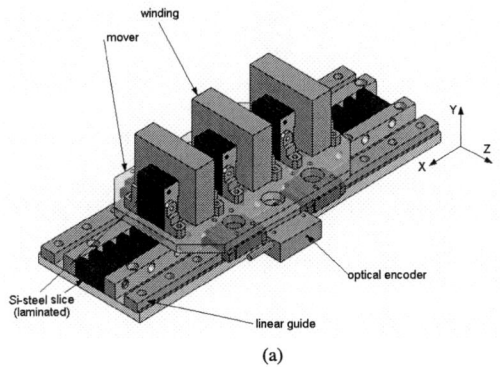

(a)

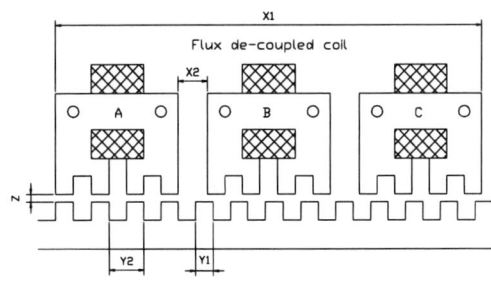

(b)

Fig. 1 Schematic of the drive system.

The body of the mover is manufactured with aluminum, so that the inertia of the mover is low and the magnetic circuits are decoupled. The stator track and the core of the windings are laminated with silicon-steel plates, by means of which the motor manufacture can be simplified and the cost is reduced.

B. Modeling of LSRM

The dynamics of the whole drive system can be described as a combination of electromagnetic behaviors by a voltage equation as (1) and the mechanical movement as (2). The voltage equation can be further expressed as (3) since the flux linkage is a function of current and position.

$$V_j = r_j i_j + \frac{d\lambda_j}{dt}, j = a,b,c. \quad (1)$$

$$M\frac{dv}{dt} = \sum f_j - Bv - f_l. \quad (2)$$

$$V_j = r_j i_j + \frac{\partial\lambda_j}{\partial i_j}\frac{di_j}{dt} + \frac{\partial\lambda_j}{\partial x}v, j = a,b,c. \quad (3)$$

V_j denotes phase voltage, i_j denotes phase current, r_j denotes the winding resistance and λ_j denotes the phase flux linkage. x denotes the position and its derivate is the velocity v. The generated electromagnetic force is the sum of each phase force f_j, f_l denotes the external load force, M and B are the mass and friction constant. In linear case, phase force produced can be expressed as (4). L_j denotes phase inductance.

$$f_j(x, i_j) = \frac{1}{2}\frac{dL_j}{dx}i_j^2, j = a,b,c. \quad (4)$$

The response times of the electromagnetic behavior and mechanical motion are quite different. This is justified for our test setup since we can achieve the current loop bandwidth up to kHz while the output mechanical bandwidth is in the order of ten Hz [2]. Depended on the fact, the two-time-scale analysis is applied to model and design the drive system. The whole drive system is divided into two subsystems with different time scales named as fast and slow subsystem. In accord with the test results, the fast subsystem describes the electromagnetic behaviors of the coils while the slow subsystem corresponds to its mechanical motion.

III. COMMUTATION AND WINDING EXCITATION SCHEME OF LSRM

As in SRMs, commutation is an important task for effectively operating LSRMs. This is derived from the fact that the direction of each phase force generated in a LSRM is dependent on its position. The desired force performance needs to be carried out by synchronous commutation with its position. However, the commutation results in the force ripples. To obtain a smooth output force, a force sharing strategy can be applied.

For any given position, there are two sets corresponding to the phases for positive force produced and the phases for negative force produced as follows,

$$\Theta^+ = \{j : \frac{\partial L_j(x)}{\partial x} \geq 0\} \text{ and } \Theta^- = \{j : \frac{\partial L_j(x)}{\partial x} < 0\}.$$

A force sharing strategy can be performed by a Force Distribution Function (FDF)

$$FDF(x, f_d) = f_d \begin{bmatrix} w_a(x) & w_b(x) & w_c(x) \end{bmatrix} \quad (5)$$

where f_d denotes the desired total force and w_j denotes the weight of force for phase j. A FDF should satisfy the principles as follows

$$\begin{cases} f_d \geq 0, w_j(x) > 0 \; \forall j \in \Theta^+ \text{ and } w_j(x) = 0 \; \forall j \in \Theta^- \\ f_d < 0, w_j(x) > 0 \; \forall j \in \Theta^- \text{ and } w_j(x) = 0 \; \forall j \in \Theta^+ \end{cases}.$$
(6)

$$\sum_{j=a}^{c} w_j(x) = 1. \quad (7)$$

The selection of weight depends on the various force sharing strategies on their design considerations. A phase inductance ratio based FDF is proposed for the phase current transition during its commutation in [8]. A simpler FDF scheme is proposed by using a linear switching in [2]. However, all of force sharing strategies should satisfy that the sum of each weight should be 1, which means that the sum of each phase force agrees with the desired total force.

As the linkage between the electrical subsystem and mechanical subsystem, a FDF and a current calculating function together are referred as the winding excitation scheme. The structure diagram of the winding excitation scheme is shown in figure 2. The FDF is used to calculate the desired phase force according to the position and the total desired force. The current calculate function is applied to calculate the desired phase current by using the desired phase force and the position.

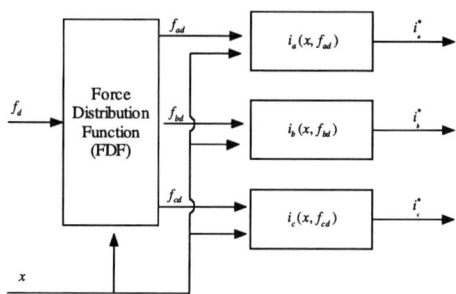

Fig. 2 The structure diagram of a winding excitation scheme.

According to force produced equation, the desired phase currents can be calculated by the output of the applied FDF and current calculating function. The precise model for flux linkage is hard to be obtained due to its inherent complexity. To calculate desired phase currents, the nonlinear flux linkage characteristics are usually approximated by a ratio function of phase inductance to displacement. Some methods have been proposed for approximating this ratio function. They can be classified into two groups: through a lookup table and through an approximation function. Lookup table only needs memory to store the characteristics data from the actual measurements; but it lacks flexibility. The approximation function needs more mathematical processing but it is more flexible for practical applications.

IV. CONTROLLER DESIGN FOR LSRM

The driving system adopts a cascaded control structure and two controllers are designed for the electromagnetic subsystem and mechanical subsystem corresponded to current control and position control, respectively. In the middle of the two control subsystems there is the applied winding excitation scheme. The block diagram of the whole driving system is shown as figure 3. In the cascaded control system, the inner loop is for current control with fast variables and the outer loop is for the position control with slow variables.

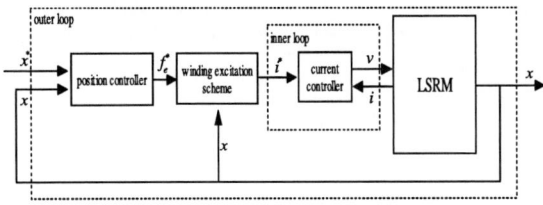

Fig. 3 Structure diagram of control scheme for position control.

For each phase coil, the relationship from terminal voltage to phase current can be approximated as a first-order differential equation. As the inner loop, it can be easily regulated by a proportional controller to guarantee both stability and quick response.

For the outer loop, the position controller can be designed from the energy dissipation viewpoint [7]. The state vector is defined as $X^T = \begin{bmatrix} x & v \end{bmatrix}$. The reduced-order model can be arranged as follows

$$D\dot{X} = [J - R]X + u + \xi \tag{8}$$

where $J = \begin{bmatrix} 0 & 1 \\ -1 & 0 \end{bmatrix}$, $R = \begin{bmatrix} 0 & 0 \\ 0 & B \end{bmatrix}$, $D = \begin{bmatrix} 1 & 0 \\ 0 & M \end{bmatrix}$,

$u = \begin{bmatrix} 0 \\ \sum f_j \end{bmatrix}$ and $\xi = \begin{bmatrix} 0 \\ x - f_l \end{bmatrix}$.

The matrix J is a skew-symmetric, satisfied $J^T = -J$. The matrix R is a semi-positive definite symmetric matrix and corresponds to the energy dissipating of the model. D is a positive definite matrix. u denotes control vector and ξ denotes disturbance.

Define the state error as $E = X - X_d$ where X_d is the reference state vector. Substituting the state error into equation (8) yields the error model as

$$D\dot{E} + [R - J]E = \Phi \tag{9}$$

where $\Phi = -D\dot{X}_d - [R - J]X_d + u + \xi$.

The energy function of state error is chosen as

$$H(E) = \frac{1}{2}E^T DE. \tag{10}$$

It is clear that the error energy function is a non-negative function and its minimum is zero. By keeping the derivative of the energy function as zero on its equilibrium and negative in other region, the error energy function would be asymptotically dissipated to its minimum and the state variables would reach their desired. The derivative of the error energy function along time can be expressed as

$$\dot{H}(E) = E^T D\dot{E}. \tag{11}$$

The derivative can be rewritten as equation (12) by substituting equation (9) into equation (11). Equation (12) can be reformatted as equation (13) in that J is a skew-symmetric matrix and $E^T JE = 0$.

$$\dot{H}(E) = -E^T[R - J]E + E^T\Phi. \tag{12}$$

$$\dot{H}(E) = -E^T RE + E^T\Phi. \tag{13}$$

The asymptotical dissipation of the error energy function can be achieved by using the control signals from $\Phi = -KE$ where $K = diag\{k_1, k_2\}$, referred as damping injection matrix, is a positive define diagonal matrix. The derivative of error energy function can be further described as equation (14).

$$\dot{H}(E) = -E^T(R + K)E \tag{14}$$

In this case, $\dot{H}(0) = 0$ and $\dot{H}(E) < 0$ when $E \neq 0$ since $R + K$ is a positive definite diagonal matrix.

V. SIMULATION RESULTS

In this section, the performance of the proposed algorithm is illustrated by simulations, which are achieved by the MATLAB software package. Figure 4 shows the position tracking results and force signals of square waveform. It is clear that the system accurately tracks the reference and actual force matches force reference well.

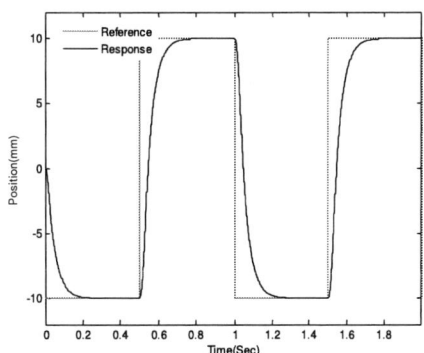

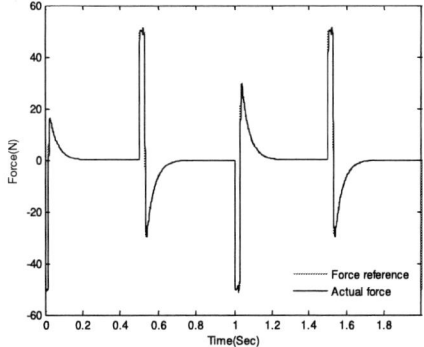

Fig. 4 Simulation: position tracking and its control signal waveforms.

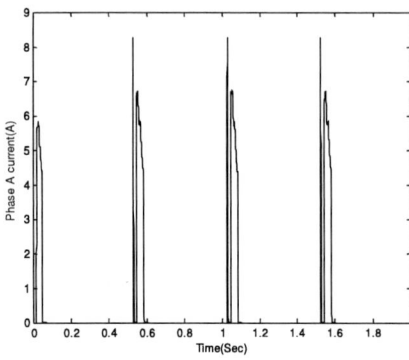

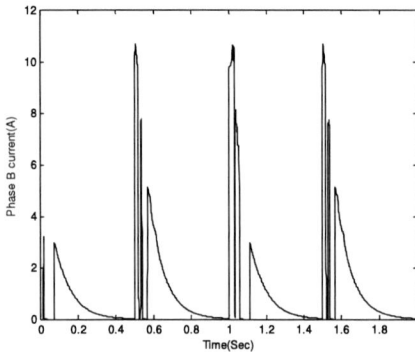

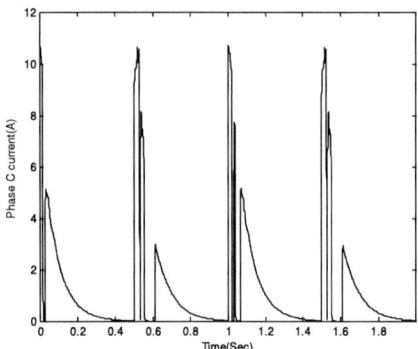

Fig. 5 Simulation: phase current waveforms.

The corresponded phase currents are shown in figure 5. Notice that all the phase currents are switched on and switched off continuously to perform the desired force.

VI. CONCLUSIONS

In this paper, a simple yet effective control design is represented for a LSRM based position tracking system is proposed. The whole drive system is decomposed into two subsystems with different time scales. Following that, the position controller and current controller are designed for the two subsystems respectively. Simulations demonstrate the effectiveness of this control scheme.

ACKNOWLEDGMENT

The authors would like to thank the University Grants Council for the funding support of this research work through project codes: PolyU 5224/04E and PolyU 5141/05E.

REFERENCES

[1] D. G. Taylor, "An experimental study on composite control of switched reluctance motors," *IEEE Contr. Sys. Magazine*, Vol. 11, Issue 2, pp. 31-36, Feb. 1991.

[2] W. C. Gan, N. C. Cheung and L. Qiu, "Position control of linear switched reluctance motors for high precision applications," *IEEE Trans. Ind. Applicat.*, Vol. 39, Issue 5, pp.1350-1362, Step. /Oct. 2003.

[3] M. Ilic'-Spong, R. Marino, S. M. Peresada and D. G. Taylor, "Feedback linearizing control of switched reluctance motors," *IEEE Trans. Automat. Contr.*, Vol. AC-32, pp. 371-379, May 1987.

[4] S. A. Bortoff, R. R. Kohan and R. Milman, "Adaptive control of variable reluctance motors: a spline function approach," *IEEE Trans. Ind. Electron.*, Vol. 45, Issue 3, pp.433-444, Jun 1998.

[5] G. Espinosa-Pérez, P. Maya-Ortiz, M. Velasco-Villa and H. Sira-Ramíez, "Passivity-based control of switched reluctance motors with nonlinear magnetic circuits," *IEEE Trans. Contr. System Tech.*, Vol. 12, Issue 3, pp. 439-448, May 2004.

[6] K. K. Chan, J. M. Yang and N. C. Cheung, "Passivity-based control for flux regulation in a variable reluctance finger gripper," *IEE Pro.-Electr. Power Appl.*, Vol. 152, Issue 3, pp. 686-694, May 2005.

[7] R. Ortega, A. J. van ser Scharf, I. Mareels and B. Maschke, "Putting energy back in control," *IEEE Contr. Sys. Magazine*, Vol. 21, Issue 2, pp. 18-33, Apr. 2001.

[8] H. K. Bae, B. S. Lee, P. Vijayraghavan and R. Krishnan, "A linear switched reluctance motor: converter and control," *IEEE Trans. Ind. Applicat.*, vol 36, pp. 1351-1359, Sep. /Oct. 2000.

Mobile Robot Navigation with Obstacle Avoidance Capability

Anca Sorana Popa*, Mircea Popa[†] and Ioan Silea[††]

* "Politehnica" University/Department of Mechatronics, Timisoara, Romania, *anca.popa@mec.upt.ro*
[†] "Politehnica" University/Department of Computers, Timisoara, Romania, *mircea.popa@ac.upt.ro*
[††] "Politehnica" University/Department of Automation, Timisoara, Romania, *ioan.silea@ac.upt.ro*

Abstract—**Mobile robots are a very dynamic part of the robotics domain. Their navigation requires solutions for: establishment of the current position, motion planning and avoidance of obstacles. The paper describes a solution for mobile robot navigation with obstacle avoidance capability based on the X80 robot from Dr. Robot. The map and the trajectory are computed on a PC which will wirelessly control the motion of the robot. The robot senses the obstacles using its sensors, sends the information to the PC and a new trajectory is computed avoiding the obstacles. The algorithm was based on the odometry method.**

Keywords—**motion control, robotics, software, sensor.**

I. INTRODUCTION

Mobile robots are a very dynamic part of the robotics domain. Unlike the static robots used in industrial environments, the mobile robots are mostly targeted to social purposes, e.g. education, medicine, assistance for elderly and with disabilities people, domestic area etc. The fundamental problem in mobile robotics is the navigation. It requires solutions for the following problems: the establishment of the current position, the motion planning and the avoidance of obstacles.

The first problem deals with the initial position of the mobile robot and with the motion orientation. Relative and absolute positioning are the two types of solutions. Relative positioning is done through odometry and inertial navigation. The odometry uses the so called positioning coders for measuring the number of degrees the wheels have turned round with or the orientation of the wheels, meaning with how many degrees the orientation changed. The main advantage is that there are not necessary external positioning equipments because all the needed modules are inside the mobile robot. The main disadvantage is that an absolute positioning method is periodically necessary for reducing the cumulated errors.

Solutions for absolute positioning are: active beacons, artificial marks, natural marks and the comparison of the models.

Two types of motion planning are described in the literature, [1]: explicit and implicit. Explicit motion planning means that the motion plan is computed before the motion is executed and is used, in general, when one wants to optimize the performance of the motion or to guarantee certain properties of the trajectory. Implicit

motion planning means that the trajectory is computed while the robot moves and is used, in general, when a desired configuration for the trajectory is needed.

Sensors are used for avoiding obstacles. The most used are video and ultrasonic.

A solution was described for mobile robot navigation with obstacle avoidance capability. It was based on the X80 robot from Dr. Robot, [2]. The map and the trajectory are computed on a PC which will control the motion of the robot through its wireless interface. The robot senses the obstacles using its sensors, sends the information to the PC and this one computes a new trajectory avoiding the obstacles. The algorithm was based on the odometry method.

The next section presents related work, the third section shortly describes the X80 robot, the fourth section gives some information on odometry and describes the navigation solution and the last one outlines the conclusions.

II. RELATED WORK

The mobile robot navigation avoiding obstacles was approached by other papers too. Reference [3] estimates the odometry error of a mobile robot during navigation. The robot is equipped with an external sensor. Two Kalman filters are used: for the systematic and for the non-systematic components of the error.

Reference [4] describes an application for tracking a moving object, by a robot with multiple sensors. A visual camera is used for sensing the movement of the desired object and a range sensor is used for detecting and avoiding obstacles .

Reference [5] presents an algorithm for robot navigation using a sensor network embedded in the environment. The robot navigates taking into account the data received from the sensors, thus obviating the need for a calculated map.

A sensors microcontroller interface for mobile robots is reported in reference [6]. In reference [7] a predictive model of sensor readings for a mobile robot is described. The model predicts sensor readings for a given time horizon based on current sensor readings and velocities of wheels assumed for this horizon.

In reference [8] a multi-agent approach for building an autonomous mobile robot capable to tackle various

---

978-1-4244-1741-4/08/$25.00 ©2008 IEEE

problems encountered during corridor navigation is described.

III. THE X80 ROBOT

X80 from Dr. Robot is an autonomous wheeled mobile robot build on the DIRAS ("Distributed Computation Robotic Architecture and System") concept. According to it the tasks needing memory space and computational power are not executed by the robot but by a host PC. They communicate through a wireless link. The robot can send video images, audio data and information from different sensors, as temperature sensors, sonar sensors, accelerometer sensors. All the collected data are available to the user through ActiveX controls which can be used together with Visual C++ or Visual Basic programming environments.

The X80 robot is made by the following modules: biprocessor CPU, local memory: 1 M x 16 bits Flash and 256 K x 16 bits SRAM, video color camera and bidirectional codec audio, WiFi 802.11b module with two communication channels and with a transfer rate up to 912.6 Kbps (the module supports the UDP and TCP/IP protocols), temperature sensors, infrared sensors, ultrasonic and presence sensors, accelerometer and bending modules and Ni – MH accumulators for independent power source.

The software of the X80 robot consists in several modules:

- the firmware of the controllers for the low level operations such as control of the motors, serial communication, local control of the CPU;
- WiRobot Gateway: ensures the communication between the low level software of the robot and the WiRobot SDK ("Software Development Kit") components running on PC;
- WiRobot SDK ActiveX Module: is an ActiveX component used for programming applications in Visual C++ or Visual Basic.

Five types of events are managed by the software:

- events generated at the reception of data from the standard sensors;
- events generated at the reception of data from the configurable sensors (data from the analog – digital converters and from the input ports);
- events generated at the reception of data from the motion sensors;
- events generated at the reception of a new video image and
- events generated at sound reception.

IV. A SOLUTION FOR MOBILE ROBOT NAVIGATION WITH OBSTACLE AVOIDANCE

A. Basics About Odometry

The odometry permits to calculate the new position of the robot with the following formulas:

$$x_i = x_{i-1} + \Delta U_i \cos \theta_i$$

$$y_i = y_{i-1} + \Delta U_i \sin \theta_i$$

x_i, y_i is the new position, x_{i-1}, y_{i-1} is the previous position, θ_i is the angle of the direction vector to the Ox axe and ΔU_i is the linear movement of the centre of the robot, computed with the following formula:

$$\Delta U_i = (\Delta U_{S,i} + \Delta U_{D,i})/2$$

$\Delta U_{S,I}$ and $\Delta U_{D,I}$ are the linear movements of the two wheels of the robot.

B. Presentation of the Solution

The first step was to develop a method for establishing the trajectory between the source and destination points situated in a planar environment. Starting from the map of the environment, the desired trajectory was indicated. A discretisation network was overlapped and the minimum way was obtained using the Lee algorithm. The obtained trajectory is a sequence of points connected by segments. All the points and the segments to close to the walls were eliminated.

The second step was to send to the robot motion commands according to the established trajectory for each point of the discretisation network. If necessary, the orientation of the wheels will be changed. Orientation errors occur because of constructive reasons and they are periodically compensated. The errors were established experimentally. The speed of the robot can be established by the user and constant or variable speeds cab be selected. In the second case the robot will accelerate on a 1 meter distance until it will reach the maximum desired speed. When the robot is close to the destination it decelerates on a 1 meter distance and finally stops.

The motion algorithm runs in an auxiliary thread. The main thread is kept for communicating with the PC and for running the obstacle detection. The sensors are continuously monitored and if an obstacle is found the robot stops, a new trajectory is computed according to the position of the obstacle and the robot and new motion commands are generated by the PC. If new trajectories can not be found the operation stops and a message is generated.

There are three types of sensors used in this application: standard, sensors and custom sensors. The standard sensors involved are: three sonar sensors for obstacle detection, one infrared sensor for obstacle detection, two sensors for detecting human presence, two sensors for detecting the motion near the robot, two bending sensors, three temperature sensors and three battery sensors.

The motor sensors which were considered are: one sensor for the turning direction of the left wheel, one sensor for the turning direction of the right wheel, the value for the left position coder, the value for the right position coder, the turning speed of the left wheel and the turning speed of the right wheel.

The custom sensors consist in seven infrared sensors

for detecting small obstacles.

C. Software of the Solution

Fig. 1. presents the PC – X80 robot communication.

The class CRobotEventReceiver is an interface that must be implemented by all the classes which want to be informed with certain events about the state of the robot or the data received by it. The robot communicates directly only with the control ActiveX CDrrobotsdkcontrolctrl1 which is part of the CRobotInterface class. When the control receives data from the robot it sends the data to all the classes wanting to be notified about that type of event. The classes that want to communicate with the robot through the CRobotInterface interface are CRobotProjectDoc and CMainRobotControl.

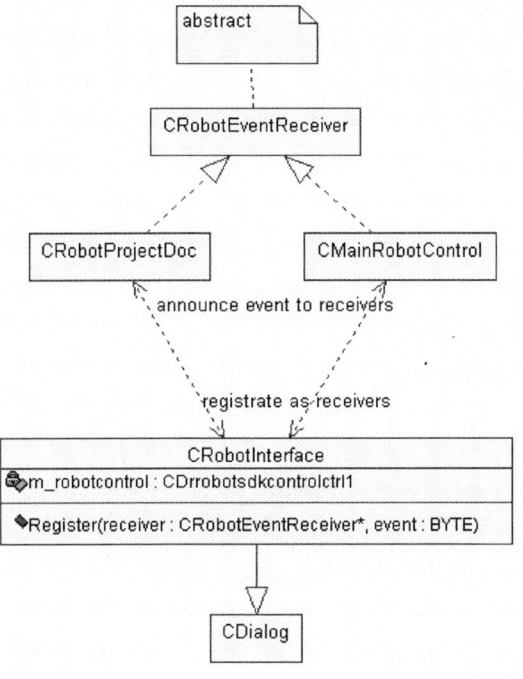

Fig. 1. Classes involved in the PC – X80 robot communication

The CRobotInterface class is a communication way between the robot and the classes collaborating with it: CMainRobotControl and CRobotProjectDoc. The CMainRobotControl class commands the indoor navigation of the robot following the minimum trajectory found between two points. It sends commands and receives information from the robot sensors. This information is necessary for correcting the motion in case some errors occur and to detect obstacles.

The CRobotProjectDoc class represents the position of the robot in the map and the information received from the ultrasonic sensors.

When the robot receives a change concerning the sensors, it notifies the DrRobot control sending a certain event, tightly connected with the type of the change. The

m_robotcontrol variable reacts when receives such an event by execution a special handler. There are five handlers corresponding to the number of events, StandardSensorEventDrrobotsdkcontrolctrl1, MotorSensorEvent Drrobotsdkcontrolctrl1, CustomSensorEvent Drrobotsdkcontrolctrl1, ImageEvent Drrobotsdkcontrolctrl1 and VoiceSegmentEvent Drrobotsdkcontrolctrl1

Inside the handlers the m_robotcontrol is interrogated about different type of information corresponding to the type of the handler. Using this information, typical structures for the three sensor types, StandardSensors, MotorSensors and CustomSensors are created and populated. After finishing this process the array with the registered receptors is verified in order to select those which have showed their interest for the event treated by the handler. For each selected receptor the method that have implemented the virtual method of the CRobotEventReceiver interface is called. The sensor structure with updated information is sent as a parameter of this method.

The robot navigates according to the commands received from the PC. When its sensors detect an obstacle it stops, a new trajectory is computed and new commands are sent to the robot. Fig. 2 presents the three structures for the sensors.

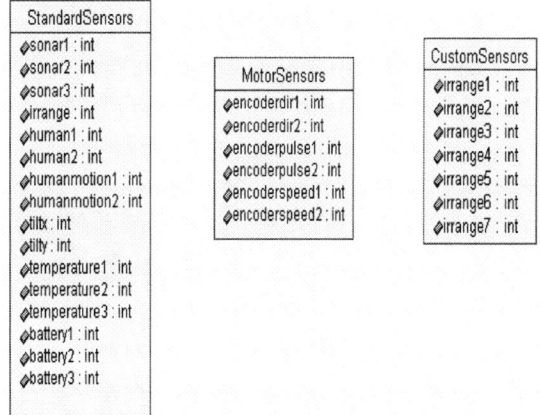

Fig. 2. Data structures for the sensors

The sonar sensors *sonar1*, *sonar2* and *sonar3* are used for detecting obstacles not found in the near vicinity of the robot. The infrared sensor *irange* is foresight for the close obstacles. There are also two sensors, *human1* and *human2* for detecting the human presence and other two sensors, *humanmotion1* and *humanmotion2* for motion detection near the robot. The bending of the robot on the two axes is sensed by *tiltx* and *tilty*.*Temperature1*, *temperature2* and *temperature3* are sensors for monitoring the environment temperature. The *battery1*, *battery2* and *battery3* are sensors for the power supply of the processor, the motors and servomechanisms.

The motor sensors are *encoderdir1* for the turning

direction of the left wheel, *encoderdir2* for the turning direction of the right wheel, *encoderpulse1* for the value for the left position coder, *encoderpulse2* for the value for the right position coder, *encoderspeed1* for the turning speed of the left wheel and *encoderspeed2* for the turning speed of the right wheel.

The custom sensors consist in infrared sensors for detecting small dimensions obstacles.

The above described solution solves the problem of receiving information from the robot sensors. The other problem consists in sending commands to the robot. For that, the simplest solution was to offer to each component, which desires, the access to the m_robotcontrol control. So the command of the robot is possible by using the functions of the control ActiveX.

The trajectory is computed on the map of environment. The following constitutive elements must be specified for building the map: the point (as an elementary unit), the wall, the door, the room, the obstacle, the robot position in the plan and the matrix points overlapped on the plan (useful for computing the minimum trajectory between two points and for generating motion commands).

Fig. 3 shows the structures involved in building the map and the relations between them. One can see the existence of an indirect relation between the structures corresponding to the wall (Wall) and door (Door) and the structure point. The reason is the fact that in the mentioned structures the instance of a point type structure is not directly memorized but the index of the corresponding point is memorized in the memory of points of the plan. This solution leads to smaller memory requirements.

The structure Point is the elementary and fundamental unit. Because the environment is planar only the coordinates on Ox and Oy axes must be specified. The structure Wall uses the extremities for specifying a wall. The extremities are not directly specified but by their positions in the m_points array of the plan.

A door is represented by the structure Door which includes the following elements:
- firstpoint: is the index of the point delimitating the first edge of the door;
- secondpoint: is the index of the point delimitating the second edge of the door;
- opened: Boolean variable indicating the status of the door: open or close;
- direction: if the door is opened specifies the direction.

The structure Room contains an array of points and an array of doors. The array of points establishes the borders of the room. Its dimension is variable according to the complexity of the room. The doors will be specified after establishing the borders of the room and according to their status (open or close) and to their orientation (if they are open). A plan may have one or more rooms and the doors are the common elements of the rooms..

In the structure Obstacle, an obstacle is identified by its positions, x and y, in the planar map of the environment.

The third parameter specifies the type of the obstacle: permanent or transient. The permanent obstacles are specified at the first generation of the map. The transient obstacles can occur anytime, are seen as mobile elements and influence the trajectory of the robot. They can be specified by directly editing the file corresponding to the plan or by the graphical interface. The obstacles can not be specified to close to the walls or outside the rooms. If they are placed close to doors the status of the door will change from open to close.

The structure MatrixPoint is an array including the trajectory of the robot. The matrix represents a network of small squares, a grid of points connected between them by adjacency relationships. The dimension of the basic unit, that is the smallest square, can be set by the user by editing the parameter 'grid distance'. The matrix will be overlapped on the map and it will be used for obtaining the minimum way between two points and for generating the motion commands, step by step from a point of the matrix to another one.

The structure RobotPosition memorizes the position of the robot, in coordinates x and y, and its orientation specified by the angle between the direction of the robot and the Ox axe.

A text type file was chosen for memorizing the data corresponding to the structure of a map and the modifications done from a session to another. The file is thought to be made only by numbers and has the following structure:
- the first line of the file contains the number of the points specified in the map; let *NrPoints* this number;
- the following *NrPoints* lines of the file contains the X points, each having a separate line; a point is specified by its two coordinates in the plan, separated by one or more spaces; these points will be used to specify the doors, the borders of the rooms, the walls and the obstacles;
- the line with number *NrPoints+1* contains the number of the walls specified in the map; let *NrWalls* this number;
- the following *NrWalls* lines of the file are used for specifying the features of the walls; a wall is delimitated by two point type structures, through its end points; a line which describes a wall will not contain the coordinates of the two points but the pointer of each point in the array of points of the map; this was specified in the first section of the file;
- the line with number *NrPoints+NrWalls+1* is foresight to specify the number of doors in the map; let *NrDoors* this number;
- the following *NrDoors* lines specify the doors, each one on a separate line; the necessary information for a door are:
  - the first point delimitating the door; it is specified by a pointer in the array of

points;

- o the second point delimitating the door also specified by a pointer in the array of points;
- o the codification of the drawing of the door: its values can be 1 to 4 and refers to the state of the door (open or close) and to its orientation in the two rooms which are separated by it;
- the line with number *NrPoints+NrWalls+NrDoors+1* specify the number of rooms in the map; let *NrRooms* this number;
- the following *NrRooms*2* lines present the features of the room; for each room two lines are foresight:
  - o the first line is for the number of points delimitating the outline of the room; there are not limitations for this number, it can have any value;
  - o the second line is for the doors from the room; they are specified by their pointers in the array of doors of the map;
- the line with number *NrPoints+NrWalls+NrDoors+NrRooms*2* specify the number of obstacles; let *NrObstacles* this number;
- the following line specify the obstacles through their pointers in the array of points of the map.

Next, a simplified version of the structure of the text type file follows:

[1. Points specification:]
Line 1: NrPoints
Line 2: x_point_1 y_point_1
Line 3: x_point_2 y_point_2
....
Line NrPoints+1: x_point_NrPoints y_points_NrPoints
[2. Walls specification:]
Line NrPoints+2: NrWalls
Line NrPoints+3: pointer_extremity1_wall1 pointer_extremity2_wall1
Line NrPoints+4: pointer_extremity1_wall2 pointer_extremity2_wall2
....
Line NrPoints+3: pointer_extremity1_wall1 pointer_extremity2_wall2
Line NrPoints+NrWalls+2: pointer_extremity1_wallNrWalls pointer_extremity2_wallNrWalls
[3. Doors specification:]
Line NrPoints+NrWalls+3: NrDoors
Line NrPoints+NrWalls+4: pointer_end1_door1 pointer_end2_door1 drawing_mode
....
[4. Rooms specification:]
Line NrPoints+NrWalls+NrDoors+2: NrRooms
Line NrPoints+NrWalls+NrDoors+3:
i_point_1_room_1   i_point_2_room_1   ...
i_point_n_room_1
Line NrPoints+NrWalls+NrDoors+4:
i_door_1_room_1   i_door_2_room_1   ...
i_door_n_room_1
Line NrPoints+NrWalls+NrDoors+5:
i_point_1_room_2   i_point_2_room_2   ...
i_point_n_room_2
....
[5. Obstacles specification:]
Line NrPoints+NrWalls+NrDoors+NrRooms*2+2: NrObstacles
Line NrPoints+NrWalls+NrDoors+NrRooms*2+3:
i_point_obstacle_1   i_point_obstacle_2   ...
i_point_obstacle_n

The plan is loaded using the main window of the application. Fig. 4 shows the dialog which is opened. The user must choose a file and press the "Open" button. If the plan will be successfully loaded it will be drew. The rest of the options (setting of the position and of the orientation, saving of the plan, setting of the destination point) will be also activated. Immediately after the drawing of the plan the user must specify the position and the orientation of the robot in plan. The symbol of the robot will permanently follow the position of the cursor from the mouse. The position is set by pushing the left button of the mouse. Next, the user will have to indicate the orientation of the robot. The operation will be done as the fig. 5 shows.

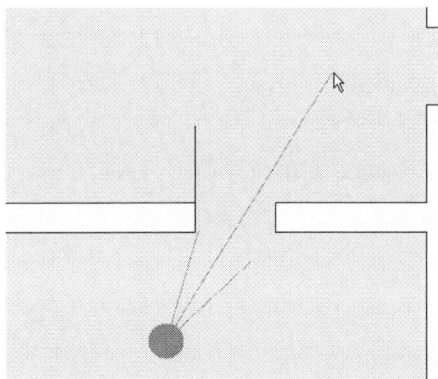

Fig. 5. Establishment of the orientation of the robot

The circle represents the robot. The movement of the mouse will appear as a line having one point in the center of the circle. The final orientation will be established when the user pushes again the mouse.

The saving of the plan is also done through the main window. A dialog window similar with the previous one will open and the user will be allowed to specify the name of the file in which the information will be saved.

D. Parameters Setting Mechanism

The application offers to the user the possibility to do its own settings concerning the robot configuration and the establishment of some parameters which will be used

later by the system. Support is offered for editing the following parameters:
- name of the robot;
- speed:
  - o maximum speed:
  - o constant or variable speed:
- way of generating the minimum way between the initial point of the robot and the destination point:
  - o side of the matrix overlapped on the map of the building: this matrix will be used for establishing the minimum way, using the Lee algorithm;
  - o optimization or not of the solution obtained with the Lee algorithm;
- for monitoring the battery:
  - o establishment of a minimum threshold of the battery; if the power supply given by the battery will be under this threshold a warning message will be generated for the user.

The dialog window for the user interface is presented in fig. 6.

The solution for establishing the speed and the level for the warning message consists in sliders. Comparing with the editing control solution, the mentioned one is easier for the user and more efficient. The minimum and maximum limits can not be outrun.

The settings necessary for the application and the needed functionalities for managing the access to them are included in the CSetting class. The associated diagram is presented in fig. 7.

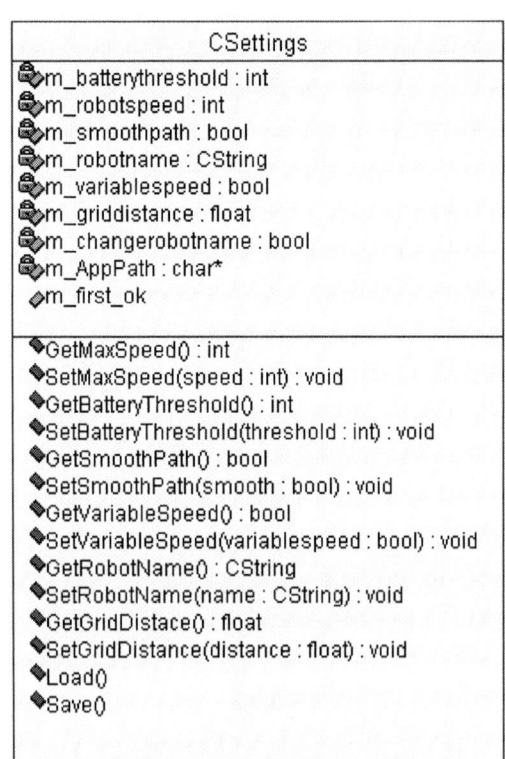

Fig. 7. Content of the CSetting class

The public variable m_first_ok has the role to indicate if a saving of the window of the user interface was done, by activating the OK button. Normally, the first saving of the settings must be done when the application is opened when the user must also specify the name of the robot. If the value of the variable is true, the window will not allow another setting for the name of the robot. If the value is false, the name can be changed.

When the application runs the necessity arise that at any moment the entities involved in monitoring operations, generating of the way or control of the robot have access to the present values of the parameters. The solution used consists in creating the instance of one CSetting class type object and to ensure the access to it at global level.

The CSetting class ensures the access of all the members of the application at the current values of the parameters and several other facilities, such as:
- the possibility to modify the settings, during a session, from an opening of a dialog window to another one; thus the modification of the settings, during a session, is allowed several times;
- the possibility to memorize the settings from a session to another one; thus the configuration of the parameters only at the beginning of a session is eliminated and it can be preserved among several sessions.

The later operation is done through a file, saved in the current directory of the application. Its name is *Settings.ini* and its format is as follows:

[SETTINGS]
RobotName=drrobot1
BateryThreshold=46
MaxRobotSpeed=780
//boolean variable 1=true
VariableSpeedControl=0
GridDistance=0.240000
//boolean variable 1=true
EnableSmoothPath=0

The modification of the settings can be achieved also before the launching of the application.

The methods for reading data from the file and saving data in the file are called *Load* and *Save*. The functions used for implementing those methods are GetPrivateProfileString, for reading from the file, and SetPrivateProfileString, for saving in the file.

V. CONCLUSIONS

A solution was described for mobile robot navigation with obstacle avoidance capability. It was based on the X80 robot from Dr. Robot. The map of the environment is built and the trajectory is computed on a PC which controls wirelessly the robot navigation. The obstacles are avoided based on the information sent by the sensors found on robot and on the odometry method.

The application can be used in monitoring purposes. Different trajectories can be established and different types of data (e.g. video, temperature) can be collected.

Further development directions can be:

- improvement of the obstacle detection mechanism by combining the web camera and the ultrasonic sensors mounted on the robot;
- increase of the positioning accuracy by combining the odometry method with the active beacons method;
- increase of the possibilities offered to the user by the graphical interface (e.g. the user might add walls, doors and rooms with the mouse).

REFERENCES

[1] J.J.A.M. Keij, "Obstacle Avoidance for Wheeled Mobile Robotic Systems", *Report No. DCT 2003.10,* Technische Universiteit Eindhoven, 2003

[2] "WiRobot X80 USER MANUAL", version 1.0.6

[3] A. Martinelli, R. Siegwart, "Estimating the Odometry Error of a Mobile Robot during Navigation", in *Proc. of European Conference on Mobile Robotics,* Warsaw, Poland, Septmber 4 – 6, 2003

[4] C, H. Chen, C. Cheng, D. Page, A. Koschan, M. Abidi, "A Moving Object Tracked by A Mobile Robot with Real-Time Obstacles Avoidance Capability", in *Proc. of IEEE International Conference on Pattern Recognition, ICPR 2006,* Hong Kong, August 2006

[5] M. A. Batalin, G. S. Sukhatme, M. Hattig, "Mobile Robot Navigation using a Sensor Network", in *Proc. of IEEE* 2004

[6] *International Conference on Robotics and Automation,* New Orleans, April 26 – May 1, 2004

[7] I. Matijevics, "Infrared Sensors Microcontroller Interface System for Mobile Robots", in *Proc. of 5th International Symposium on Intelligent Systems and Informatics SiSY 2007,* Subotica, Serbia, August 2007

[8] K. Fujarewics, "Predictive Model of Sensor Readings for a Mobile Robot", in *Proceedings of World Academy of Science, Engineering and Technology,* Volume 20, April 2007, ISSN 1307-6884

[9] Y. Ono, H. Uchiyama, W. Potter, "A MOBILE ROBOT FOR CORRIDOR NAVIGATION: A MULTI-AGENT APPROACH", in *Proc. of ACME'04,* Huntsville, Alabama, USA, April 2 – 3, 2004

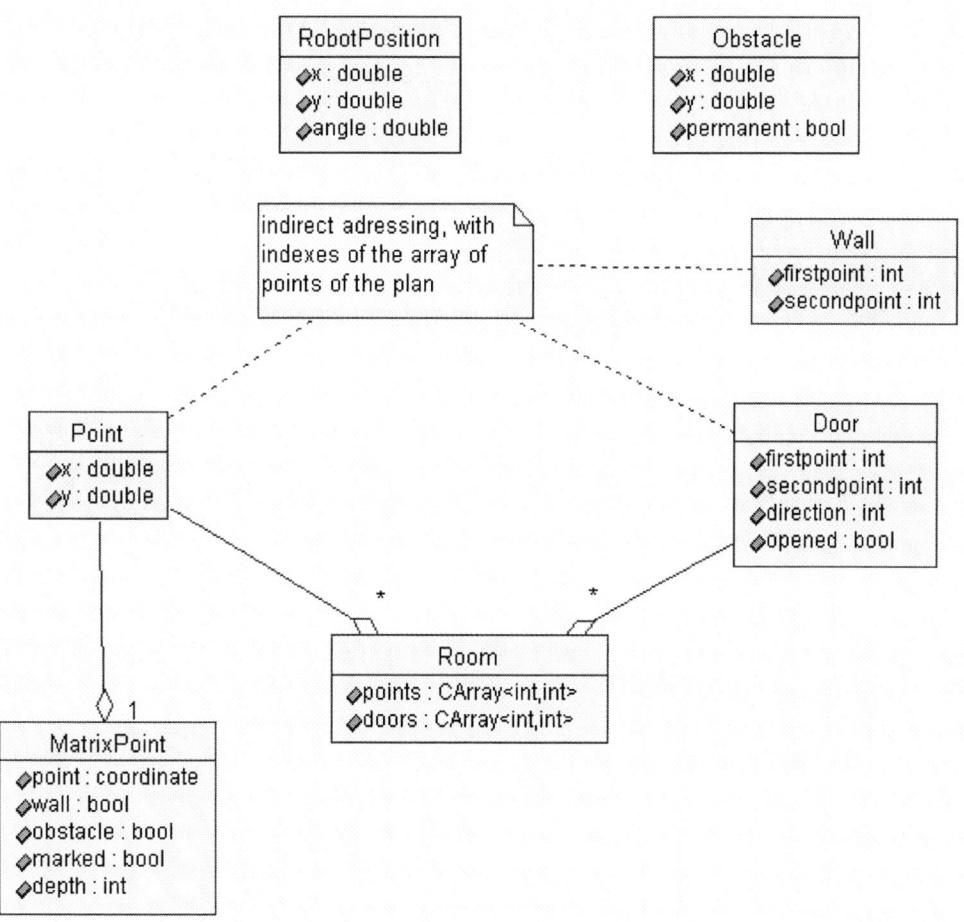

Fig. 3. Structures involved in building a map

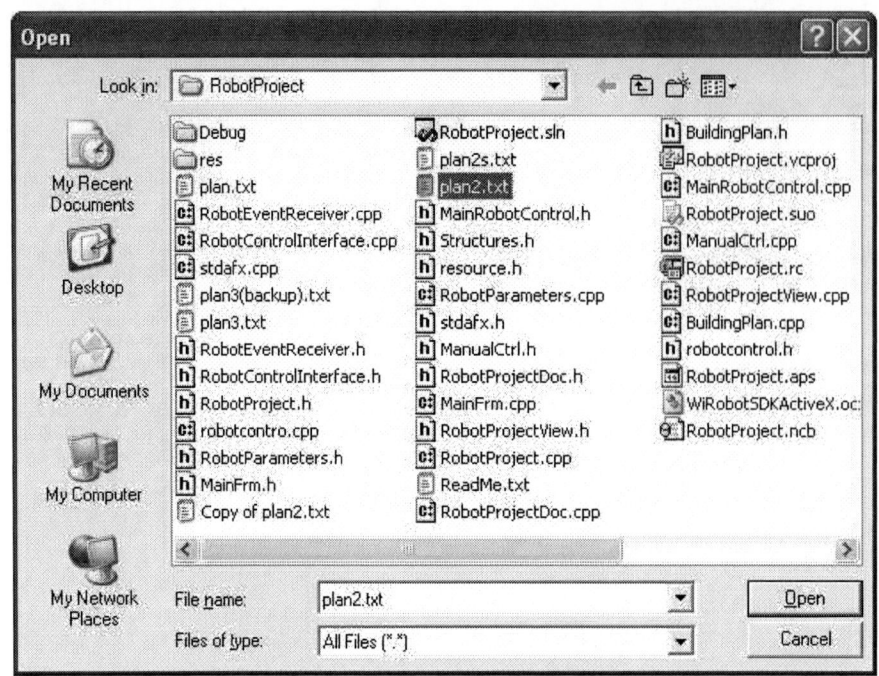

Fig. 4. Opening of a text type file

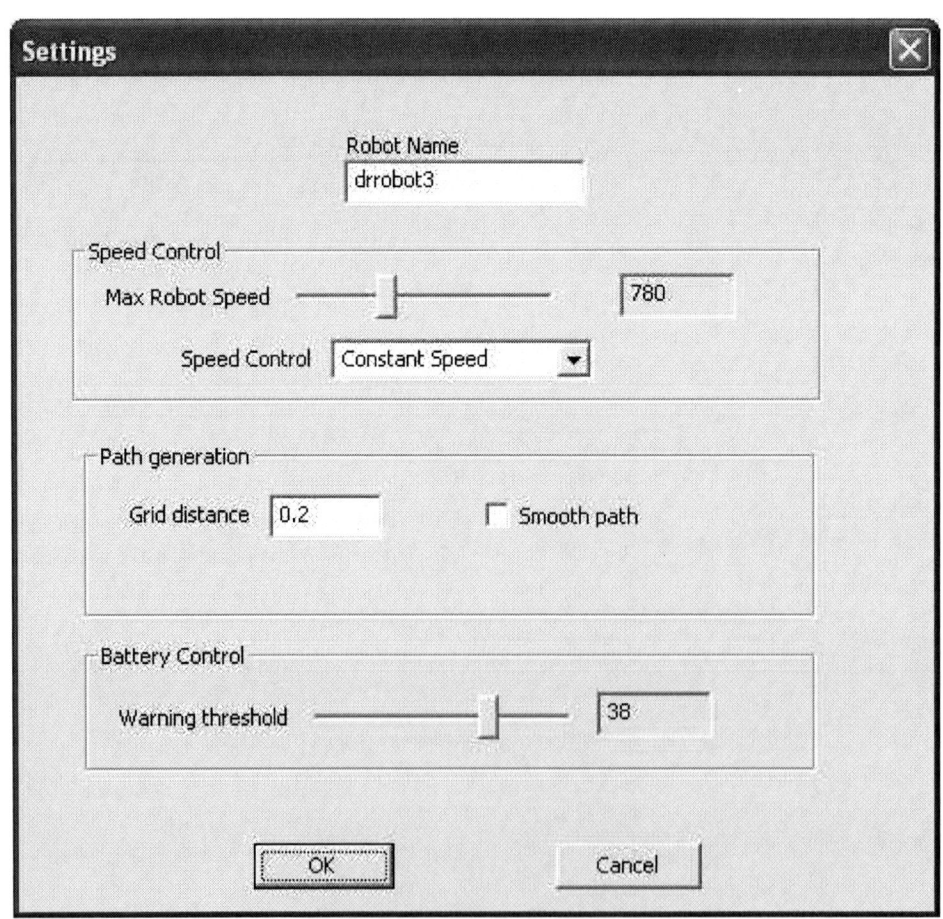

Fig. 6. Window for parameters settings

Requirements for Power Electronics in Solid Oxide Fuel Cell System

T. Riipinen, V. Väisänen, M. Kuisma, L. Seppä, P. Mustonen and P. Silventoinen

Lappeenranta University of Technology
PO Box 20, 53851 Lappeenranta, Finland

Abstract— The fuel cell has special electrical characteristics which set challenging requirements for power conditioning units. Many of these requirements origin from the operation principle and physical structure of the fuel cell and cannot be ignored in the design of power electronics. In this paper the operational requirements for power conditioning units are reviewed from the point of view of basic operation principles and characteristics of solid oxide fuel cell.

Index Terms—Fuel cell system, power conditioning, renewable energy systems.

I. INTRODUCTION

THE RESEARCH WORK in solid oxide fuel cells (SOFC) is intensive worldwide. The uses of fuel cell power systems have been predicted to dramatically increase in near future. SOFCs have several possible applications. Distributed energy generation, mobile power generation, marine electricity, stand-alone power systems and even personal electric generation are among the key issues concerning the use of fuel cell power system. Environmental aspects and green values have also accelerated the research activity during recent years. Higher efficiency is one of the major advantages when comparing fuel cell system to combustion engines or gas turbines. Electrical efficiency in SOFC is up to 60% [3] and when used with combined electrical and heat power production, overall efficiency can be up to 80% [4].

Although improvements have been made in cell structures, materials, cost and durability, the cell itself is not a perfect electrical power source. DC output voltage generated by a fuel cell is unregulated and usually too low for practical use. SOFC is not an ideal, stiff DC voltage source and the load current has significant effect on the cell output voltage. Transient response is poor and uncontrolled transients induced by power electronics can even destroy the cell. Many of these properties origin from the operation principle and the physical structure of the fuel cell.

Generally, the requirements for the power electronics in fuel cell system can be divided in three parts originating from different sources, Fig. 1. Characteristics of the fuel cell and requirements for the power electronics are discussed in this paper. Other requirements are set by the load or the network the fuel cell system is feeding. General requirements include efficiency, operating life, cost, thermal design etc.

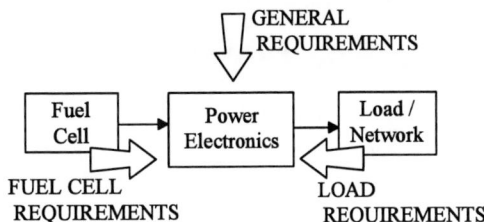

Fig 1. Different sources of requirements for power electronics in fuel cell power system.

In this paper the basic operation principle and electrical characteristics of the SOFC are introduced and power electronics is reviewed from the SOFC point of view.

II. FUEL CELL

Fuel cell converts chemical energy directly into electrical power. It uses hydrogen or hydrogen-containing fuels to produce direct current (DC), water and heat [1]. The basic principle of fuel cell was developed in the 19th century. However, it was not possible to utilize fuel cells in practical applications in early days due to corrosion and instability of materials and lack of understanding in complex electrochemical system. Achievements in material sciences, chemistry and electronics have made fuel cells an interesting resource to produce clean power for the 21st century.

There are several types of fuel cells. Classification of different types is usually made by the electrolyte of the cell: alkaline fuel cell (AFC), solid polymer fuel cell (SPFC), phosphoric acid fuel cell (PAFC), the molten-carbonate fuel cell (MCFC) and the solid-oxide fuel cell (SOFC). Another classification can be made by the fuel, i.e. reformate/air, hydrogen/air or direct oxidation fuel cells [3]. Hydrocarbon-based fuel, e.g. natural gas can be used in SOFC and MCFC, which eases the fuel management. Other cell types require the use of pure or reformed hydrogen or methanol-water solution in the case of a direct methanol fuel cell (DMFC). This paper is focused on SOFC-based fuel cell system.

SOFC has important features such as simplicity in fuel reformation and high conversion efficiency. However, high temperature (750-1000 °C), cost, corrosion and sealing problems have been problematic in system design. Increased research activity during the beginning of this century has led to important improvements in materials and fuel cell design

978-1-4244-1741-4/08/$25.00 ©2008 IEEE

[4]-[6]. Durability and reliability at reasonable costs are some of the key issues to be solved before broad commercial utilization of SOFC systems

A. Structure and operating principle

The simplified structure of single solid oxide fuel cell is presented in Fig. 2., which illustrates how electrons are generated at the anode during oxidation of the fuel and electrical current is generated to the load. The fuel and oxidant do not mix because of the isolating electrolyte, and no actual combustion occurs.

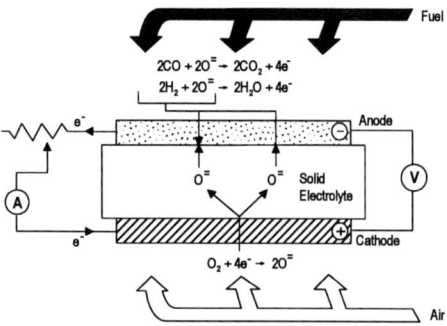

Fig 2. The simplified operation of SOFC, reproduced from [2].

The ideal standard potential, or H_2 oxidation potential, is 1.229 V. This ideal single cell voltage is reached when pure hydrogen and oxygen reacts at normal temperature and pressure [1]. In a practical case, the voltage of a single cell is usually less than 1 V, typically approximately 0.7 volts [7]. The fuel cell output voltage depends also heavily on the load.

A single fuel cell is the basic part of a fuel cell power source. Usually the fuel cells are assembled in stacks, where individual cells are electrically connected in series in order to achieve higher output voltage. Typical fuel cell stack output voltage is limited below one hundred volts. This is because of a harsh, high temperature environment, which limits the number of single cells that can be connected in series in order to prevent dielectric breakdown. This insulation-limited low voltage means high currents at higher power levels. The output current capability of the cell system can be increased by increasing the single cell area or by connecting single cells in parallel.

The fuel cell system also includes the fuel supply (pumps, valves, fuel reformation etc.) and power conditioning unit (PCU) for interfacing the load, Fig 3.

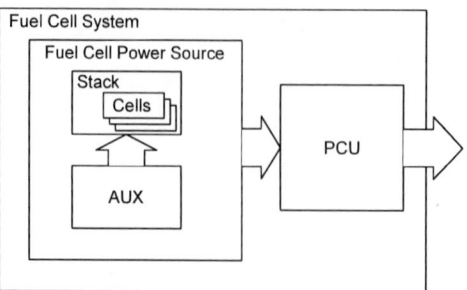

Fig 3. The diagram of fuel cell system and its subsystems. AUX contains pumps, valves, fuel reformation etc.

B. Operating areas

The voltage-current curve of a fuel cell can be divided in three different operation regions. In each region, the dominant source of power losses is different. The V-I curve of a fuel cell is presented in Fig 4.

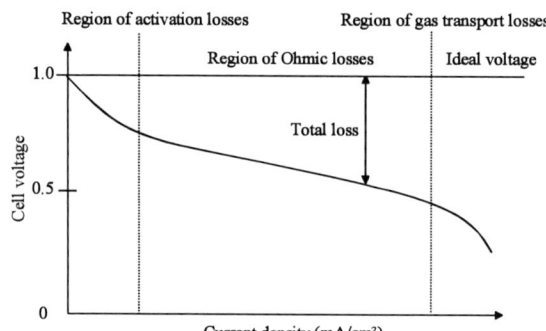

Fig. 4. Polarization curve or V-I curve

Different operating regions are [1], [8], [23]:
I.   Region of activation losses
II. Region of Ohmic losses
III. Region of gas transport losses

In region I the main source of power losses are caused by activation energy losses of the electrochemical reactions. For very small currents, the cell voltage drops rapidly as current increases. Region II is the normal operating area of a fuel cell. Here the loss mechanisms are caused by the resistance to the flow of ions in the electrolyte and resistance of the electrode, conductors and contacts. In this region, the voltage drops linearly as a function of the current. In region III, the dominant source of power losses are mass transport related losses. The losses are based on the fact that the mass transport rate is finite and it limits the supply of fresh reactants and the evacuation of products. After exceeding a particular value of current, the voltage drops rapidly. This point is also the boundary of safe operation region.

C. Loading, current ripple and fuel utilization

The output power of SOFC stack is dependent on fuel flow.

The maximum power drawn from a fuel cell stack is achieved at full rated hydrogen flow, which is often called a base flow. The fuel utilization of a cell is defined by the usage percent of the hydrogen flow and it is directly proportional to the current drawn from the cell. More power can be drawn from the fuel cell with higher hydrogen flow. However, the output voltage decreases as the output current and power increases.

In a case of load transient and change of power consumption, the flow of reactants to the fuel cell must be adjusted. The response time is dependent on the electrochemical, thermal, and mass flow of the fuel cell [12].

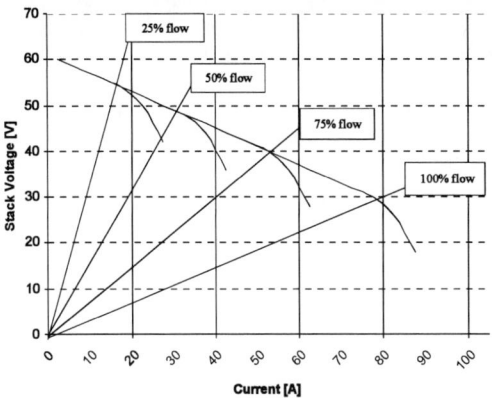

Fig. 5. Fuel cell characteristics as a function of hydrogen flow [9]. The boundaries are examined with respect to hydrogen flow. When the fuel utilization is for example 50% of the base flow, the boundary occurs at smaller output current compared to the base flow.

The hydrogen flow must be adjusted by the peak current, Fig. 5. When approaching the full utilization of hydrogen in the fuel cell, the cell output voltage collapses [8]. The boundary of voltage drop is a function of hydrogen flow, Fig. 5. In [4] it is stated that fuel utilization exceeding 80% may cause damage to the cells. The worst case for the cell is when electrical power is drawn from the cell when there is not enough fuel. This can damage the fuel cell by localized oxidization of electrolyte material [9], or even melt the cell structure.

Current ripple caused by the power conditioning unit connected to the fuel cell affects parameters like hydrogen utilization and cell temperature. Excess low frequency current ripple causes thermal stress and microcracks in the fuel cell structure thus affecting performance, efficiency and lifetime of the cell [10].

III. FUEL CELL MODELING

A fuel cell system is a complex combination of different response times. The fuel cell contains thermal, mechanical, chemical and electrical subsystems, which all have characteristic time constants [13]. Because of the complex nature of fuel cell systems, power electronics must be designed to accommodate these characteristics.

A. Fuel Cell Dynamics

The fuel cell contains thermal, mechanical, chemical and electrical subsystems, which all have characteristic time constants, [13].

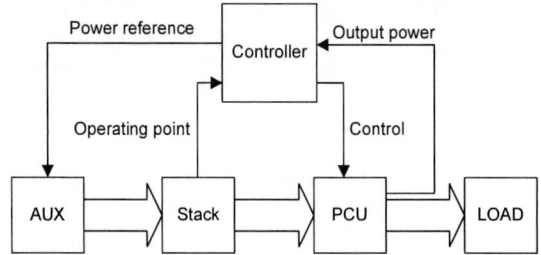

Fig. 6. The signaling and cooperation of different subsystems in fuel cell system.

The time constant of fuel supply system is many decades slower than the electrical system and it is not able to respond to fast dynamic changes. A fuel cell itself is an electrochemical device with nonlinear characteristics. Also, a fuel cell system includes many sub-systems, which interact with each other. The modeling of the fuel cell system is difficult and the model is not directly transferable to another system.

The mentioned initial startup time of SOFC to stable operation is 90 seconds and settling time in electric power change is 60 seconds [8]. In [11] the startup time for SOFC-based fuel cell power source is 20 to 30 minutes. In [13] it is mentioned that electrochemical subsystem reacts quickly to load changes, but is limited by the stack's electrical time constant, which is fractions of second. Response times for thermal, mechanical and chemical components are typically several orders longer than electrical subsystem.

In [16] it is stated based on simulation that transient time during load change is several minutes. Thermal time response is also mentioned to be several minutes. As a conclusion, several minutes are necessary to establish new steady-state conditions. For further examination, an extensive research of time scales in SOFC fuel cell systems is presented in [17].

B. Modeling

A fuel cell can be modeled in many different levels of complexity. The complex model is introduced in [19] and [29]. The detailed modeling of fuel cell's chemical, mechanical and electrical behavior is difficult. Another approach is to produce an equivalent circuit [20] [32].

When using a linearized model in the Ohmic region, the limits of the operating area are set by the hydrogen flow of the fuel cell. When leaving from the Ohmic region, the error of the model grows rapidly and the output of the model is erroneous. Consequently Thevenin's model is suitable when operating in a steady state.

Modeling of a fuel cell as a power source can be divided in individual cell modeling, fuel cell stack modeling and fuel cell system modeling. Usually, the electrical behavior of a fuel cell is modeled in two different levels:

    I.   Electrical equivalent model [20], [32], [33]
    II.   Thermal, electrochemical model [12], [13], [17],
          [19], [21], [29], [30], [34]

All models are even, in the best case only, simplifications of the real system. When designing power electronics with a fuel cell as the power source, modeling the electrical behavior of the fuel cell is crucial.

The steady state output voltage of a single fuel cell can be calculated using the Gibbs free energy principle and applying the Nerst equation as follows

$$E_{\text{cell}} = E_{0,\text{cell}} + \frac{RT}{4F}\ln\left(\frac{(p_{\text{H}_2}^{\text{ch}})^2\, p_{\text{O}_2}^{\text{ch}}}{(p_{\text{H}_2\text{O}}^{\text{ch}})^2}\right), \qquad (1)$$

where E_{cell} is cell output voltage, $E_{0,\text{cell}}$ reference output voltage in certain temperature, R gas constant, T temperature, F the Faraday constant, p_{H2} partial Hydrogen pressure, p_{O2} partial Oxygen pressure and p_{H20} partial water vapor pressure.

The fuel cell stack voltage can be represented by multiplying equation (1) with N_{cell}, which is the number of cells connected in series so that the stack voltage is acquired by applying (1) to Ohms law [34].

$$V_{\text{out}} = N_{\text{cell}}V_{\text{cell}} = E - V_{\text{act}} - V_{\text{ohm}} - V_{\text{conc}}, \qquad (2)$$

where V_{out} is stack output voltage, N_{cell} number of cells, V_{act} activation losses, V_{ohm} ohmic losses and V_{conc} concentration losses [33].

The applicability of steady state electrical fuel cell models using PEM as the source has been evaluated in [32]. It was concluded that the Thevenin equivalent model is suitable for modeling the fuel cell output voltage despite of errors in the first and the last part of the load line. In the region of Ohmic losses, a fuel cell can be modeled as a Thevenin' circuit [20], in which the system is modeled as a voltage source and series resistor.

The static model of (1) (2) does not represent the dynamic behavior of system output voltage caused by different time constants, but it models the output voltage behavior with a certain fuel, a certain fuel utilization rate and in a certain temperature.

If the fuel cell system dynamic behavior is to be modeled, the model should also include the fuel transfer, heat transfer and the electrochemical reactions [19]. This kind of model depicts the fuel cell power source in dynamic situations. In [20] the behavior of a fuel cell has been examined using two different time constants. The small time constant represents the electrochemical behavior of fuel cell and the long time constant represents the mechanical delay caused by the fuel supply.

IV. FUEL CELL POWER CONDITIONING

The fuel cell produces unregulated low voltage direct current, which is also heavily dependent on the load. Therefore the fuel cell power sources require power conditioning for interfacing the electrical load.

The usual building blocks for fuel cell PCU are a step-up DC/DC-converter for converting the low primary voltage to a higher intermediate voltage, and a voltage or a current source inverter for interfacing the three phase network, Fig. 7. DC power is acceptable in some applications, and it simplifies the converter topology and control which results in better overall efficiency because additional power conversion stages are not needed. As Siemens has demonstrated [6], 48 V or 60 V telecommunication applications are well suited to be powered by a DC fuel cell system.

Electrical isolation is difficult to achieve in high temperatures of the fuel cell stack. Many of the isolation problems can be avoided if galvanic isolation can be used in the power electronics stage. A transformer works as a galvanic isolator and it also provides the required step-up voltage conversion together with a DC/DC-converter. This requirement of galvanic isolation excludes some power electronics topologies from SOFC use and favors those with an isolating transformer. When using a transformer the converter topology sets practical limits for power output depending on how the magnetic core can be utilized. Flyback and Forward topologies are suitable for low power applications (typically below a few hundred watts). Bridge topologies are more suitable for higher power ratings [27]. At higher power levels the low DC voltage results in high currents and this increases losses in the power electronics stage.

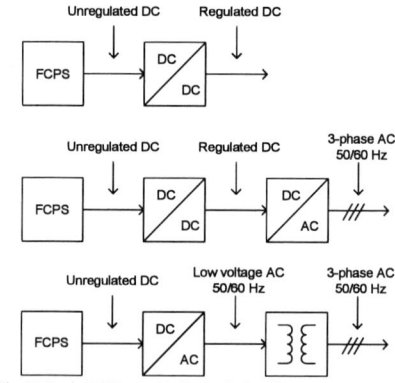

Fig. 7. Basic PCU topologies in fuel cell systems.

The control algorithms of the fuel cell system depend highly on application. If the desired output of the system is regulated DC-voltage, the control system is totally different compared to a multiple output AC-connected system.

A. Current ripple limits and limiting techniques

In [15] it is suggested that ripple current above 400 Hz has a very small or no impact at all on the operation of a fuel cell. Based on the same study, the ripple factor below 4% has no impact on the fuel cell stack's lifetime. According to [22] 6% peak-to-peak low frequency ripple should be acceptable for most SOFC power sources.

Large current ripple is not only harmful to the fuel cell, but it also results in high magnetic flux density swing in the converter transformer core. The higher the flux swing is, together with the converter switching frequency, the higher are the transformer core losses [24].

One way to reduce harmful current ripple is to use proper input filtering. The simplest approach is to use as large capacitance as possible at the input and to use current limiting in initial capacitor charging. When connecting discharged capacitor to fuel cell without any current limiting circuitry results in high inrush currents and can damage the cell.

Also to be considered in input filter design is the properties of passive components. Especially in low voltage and high current applications, the inductors can exhibit high core and winding losses at high frequencies and the high RMS ripple currents can severely degrade the lifetime of capacitors and cause significant equivalent series resistance (ESR) related losses.

Input current ripple can be limited using control techniques such as in [22] [31] and these can be combined with passive filtering. In regulated DC-usage of the fuel cell, the most simplified control system is feedback from the output voltage and feed-forward from the input current. When PI or PID controller regulates the output voltage, the hysteresis control can be used for limiting the input current of the plant. Even in this simple case, the control engineer needs to have knowledge of fuel cell as a power source. In [8] a bi-directional non-isolated DC-DC converter was used for ripple current compensation as well as transient energy compensation. The input and output filter properties will affect the control loop characteristics, so if filtering and active current shaping are used together the overall stability has to be analyzed. The interactions between input filter and current mode controlled converter have been analyzed for example in [26].

The suitability of different inverter modulation schemes in applications where there is no stiff DC intermediate circuit should also be considered especially if thinking about the fuel cell reliability in the long run.

B. Safe operating of the fuel cell

The fuel cell is sensitive for load changes and the ripple of the stacks' output current. These two characteristics need to be noticed when designing a control system for the fuel cell PCU. Because the fuel cells cannot tolerate reverse current, the fuel cell must be protected against reverse current [8]. In some topologies the fuel cell can be protected by inserting a diode in series. For power electronics and overall efficiency this means that the conduction losses of the protection diode are significant.

The fuel cell system controller should be able to operate within the following operational limits set by the fuel cell power source [19]:

1. Underused fuel. If the fuel utilization drops below a certain limit, the cell voltage would rise rapidly (boundary between region I and II in fig. 4).
2. Overused fuel. If the fuel utilization increases above a certain value, the cell output voltage drops rapidly and the cells may suffer from fuel starvation and permanent damage may occur (boundary between region II and III in fig. 4). Operating the fuel cell in region III should be avoided.
3. Undervoltage. If the fuel cell is connected to a grid and if the stack voltage drops below a certain point, the PCU will lose synchronism with the network and the whole plant will have to be disconnected.

C. Perceiving the time constants

Transient compensating energy storages, which could be used with fuel cells to provide fast transient currents and voltage stabilization, are for example batteries, flywheels, super capacitors and superconducting magnetic energy storages. They are under discussion in [18]. Capacitor is the only energy storage that could be used without additional power electronics. Other choices require additional circuitry. Even batteries need electronics to take care of charging, discharging and voltage level matching. It is preferable to use another application in power adjustment while the fuel cell works all the time with full power.

V. CONCLUSION

This paper introduced the physical phenomena of the fuel cell and the requirements for the design of power electronics to be used with the fuel cell power sources. The different time constants of the chemical, electrical and mechanical subsystems are introduced. As shown, the modeling of fuel cell plant can be done in several levels of complexity.

The ripple requirements for current drawn from the fuel cell are presented and examined from the power electronics point of view. The demand for voltage regulation is discussed and the safe operating area of the fuel cell is reviewed.

The signaling of subsystems in a fuel cell based power converter is examined and the required feedback and feed-forward information from the fuel cell to the power electronics control is introduced.

REFERENCES

[1] U.S. Department of Energy, *Fuel Cell Handbook*, seventh edition. Morgantown, West Virginia: National Energy Technology Laboratory 2004.

[2] J.T. Brown, "Solid oxide fuel cell technology," *IEEE Transactions on Energy Conversion*, vol. 3, no. 2, June 1988.

[3] M.A.Laughton, *Fuel cells*, Engineering Science and Education Journal, Feb. 2002, vol. 11, pp. 7 – 16.

[4] C. D. M. Oates, R. W. Crookes, S. H. Pyke, R. T. Leah, "Power conditioning for solid oxide fuel cells," *Power Electronics, Machines and Drives*, 16-18 April 2002, Conference Publication no. 487.

[5] B. C. LeSage, "Solid oxide fuel cell products for telecom applications," in Proc. of the 23rd IEEE International Telecommunications Energy Conference, INTELEC'01, Edinburgh, United Kingdom Oct. 14-18, 2001, pp. 667 – 670.

[6] E. Frey, "SOFC powering the telecommunication in future," Telecommunications Conference, 2005, INTELEC '05, Twenty-Seventh International, Sept. 2005, pp. 305-309, ISBN: 978-3-8007-2905-0.

[7] A. B. Cambell, J. F. Ferrall, N. Q. Minh, "Solid oxide fuel cell power system," Industrial Electronics Society, 2003. IECON '03, The 29th Annual Conference of the IEEE, Nov. 2003, vol. 2, pp. 1580 - 1584.

[8] E. Santi, D. Franzoni, A. Monti, D. Patterson, F. Ponci, N. Barry, "A fuel cell based domestic uninterruptible power supply," Applied Power Electronics Conference and Exposition, 2002. APEC 2002, Seventeenth Annual IEEE, vol. 1, pp. 605-613, ISBN: 0-7803-7404-5.

[9] S.K. Mazumder, K. Acharya, C.L. Haynes, R. Williams Jr., M.R. von Spakovsky, D.J. Nelson, D.F. Rancruel, J. Hartvigsen, R.S. Gemmen, "Solid-oxide-fuel-cell performance and durability: resolution of the effects of power-conditioning systems and application loads," IEEE Trans. on Power Electronics, vol. 19, no 5, Sept. 2004, pp. 1263 – 1278.

[10] K. Acharya, S.K. Mazumder, R.K. Burra, R. Williams, C. Haynes, "System-interaction analyses of solid-oxide fuel cell (SOFC) power-conditioning system," in Proc. Industry Applications Conference, 2003, 38th IAS Annual Meeting, Salt Lake City, Utah, USA, 12-16 Oct. 2003, vol.3, pp. 2026 – 2032.

[11] K. Rajashekara, "Power conversion and control strategies for fuel cell vehicles," Industrial Electronics Society, 2003. IECON '03. The 29th Annual Conference of the IEEE, vol. 3, pp. 2865-2870, ISBN: 0-7803-7906-3.

[12] D.J. Hall, R.G. Colclaser, "Transient modeling and simulation of a tubular solid oxide fuel cell," IEEE Transactions on Energy Conversion, vol. 14, no 3, Sept. 1999, pp. 749 – 753.

[13] M. R. von Spakovsky, D. Rancruel, D. Nelson, S. K. Mazumder, R. Burra, K. Acharya, C. Haynes, R. Williams, R. S. Gemmen, "Investigation of system and component performance and interaction issues for solid-oxide fuel cell based auxiliary power units responding to changes in application load," Industrial Electronics Society, 2003, IECON '03, The 29th Annual Conference of the IEEE, 2-6 Nov. 2003, vol 2, pp. 1574-1579, ISBN: 0-7803-7906-3.

[14] J. Lee, J. Jo, S. Choi, S-B Han, "A 10-kW SOFC low-voltage battery hybrid power conditioning system for residential use," IEEE Trans. On Energy Conversion, vol. 21, no. 2, June 2006, pp. 575-585.

[15] R. Gemmen, "Analysis for the effect of inverter ripple current on fuel cell operating conditioning," Transactions of the ASME, vol. 125, May 2003.

[16] E. Achenbach, "Response of a solid oxide fuel cell to load change," Journal of Power Sources 57, 1995, pp. 105-109.

[17] R. S. Gemmen, "Effect of load transients on SOFC operation – current reversal on loss of load," Journal of power sources 114, Issue 1, June 2005, pp. 152-164.

[18] P. F. Ribeiro, B. K. Johnsson, M. L. Crow, A. Arsoy, Y. Liu, "Energy Storage Systems for Advanced Power Applications," Proceedings of the IEEE, vol. 89, Issue 12, Dec. 2001, pp. 1744-1756, ISSN: 0018-9219.

[19] J.Padullés, G.W. Ault, J.R. McDonald, "An integrated SOFC plant dynamic model for power systems simulation," Journal of Power Sources 86, 2000, pp. 495-500.

[20] Yoon-Ho Kim, Sang-Sun Kim, "An electrical modeling and fuzzy logic control of a fuel cell generation system," IEEE Transactions on Energy Conversion, vol. 14, no. 2, June 1999, pp. 239-244.

[21] K. Sedghisigarchi, A. Feliachi, "Control of grid-connected fuel cell power plant for transient stability enhancement," Power Engineering Society Winter Meeting, 2002, IEEE, vol. 1, pp. 383-388.

[22] J.S. Lai, S.Y. Park, Seungryul Moon and Chien Liang Chen, "A high-efficiency 5-kW soft-switched power conditioning system for low-voltage solid oxide fuel cells," Power Conversion Conference, PCC '07, Nagoya, 2007, ISBN: 1- 4244-0844-X.

[23] M. E. Schenck, J. S. Lai and K. Stanton, "Fuel cell and power conditioning system interactions," Applied Power Electronics Conference and Exposition APEC 2005, 2005.. 6-10 March 2005. vol. 1, pp. 114-120, ISBN: 0-7803-8975-1.

[24] N. Mohan, W. P. Robbins, and T. M. Undeland, Power Electronics: Converters, Applications and Design, Media enhanced 3rd edition, John Wiley & Sons, 2003, ISBN 0-471-42908-2.

[25] L. Rosetto, G. Spiazzi and P. Tenti, "Control techniques for power factor correction converters," PEMC'94, 1994.

[26] T. Suntio, I. Kadoura, and K. Zenger, "Input filter fnteractions in peak-current-mode-controlled buck converter operating in CICM," IEEE Transactions on Industrial Electronics, vol. 49, no. 1, February 2002.

[27] X. Yu, M. R. Starke, L. M. Tolbert and B. Ozpineci, "Fuel cell power conditioning for electric power applications: a summary," Electric Power Applications, IET, vol 1, Issue 5, Sept. 2007, pp. 643-656.

[28] B. S. Guru, H. R. Hiziroğlu, Electric Machinery and Transformers Second Edition, New York: Oxford University Press, 1995, ISBN: 0-19-511535-X.

[29] X. Xue, J. Tang, N. Sammes, Y. Du, "Dynamic modeling of single tubular SOFC combining heat/mass transfer and electrochemical reaction effects," Journal of power sources 142, 2005, pp. 211-222.

[30] Masanori Yamaguchi, Tadayoshi Saito et al. "Analysis of Control Characteristics Using Fuel Cell Plant Simulator," IEEE Transaction on industrial electronics, Vol. 37, No. 5, October 1990, pp. 378-386.

[31] Mazumder, S.K., Burra, R.K. and Acharya, K. "A ripple-mitigating and energy-efficient fuel cell power-conditioning system". IEEE Transactions on: Power Electronics; vol. 22; issue 4; July 2007; pp. 1437-1452.

[32] Parischa S., Keppler M., Shaw. S.R, Nehrir M.H. "Comparison and Indentification of Static Electrical Terminal Fuel Cell Models". IEEE Energy Conversion Transaction, vol. 22, Sept.2007, pp. 746-754.

[33] C. Wang, M. H. Nehrir, S. R. Shaw, "Dynamic models and model validation for PEM fuel cells using electrical circuits", IEEE Energy Conversion Transaction, vol. 20, June 2005, pp. 442-451.

[34] C. Wang, M. H. Nehrir, "A Physically Based Dynamic Model for Solid Oxide Fuel Cells," IEEE Energy Conversion Transaction, vol. 22, Dec. 2007, pp. 887-897.

Power Supply for a IGBT-Driver with High Insulation Voltage based on a Printed Planar Transformers

Günter Schmitt, Wolf Kusserow and Ralph Kennel

Wuppertal University/Electrical Machines and Drives,
Wuppertal, Germany,
e-mail: *schmitt@emad.uni-wuppertal.de*

Abstract—For the operation of high-power semiconductors in medium-voltage converters, the insulation of the control signal as well as for the power-supply is indispensable. In this paper, we present a DC/DC-converter based on a printed transformer that reached an insulation voltage of 10 kV. Due to the fact that the circuit board itself is used as insulation, we avoid the use of vias or a closed magnetic core. The converter operates at a frequency of 420 kHz and provides an output power of 6 W. By minimising the amount of components and not using a standard transformer, a great cost-reduction has been achieved.

Keywords—DC power supply, IGBT, insulation, transformer, switched-mode power supply

I. INTRODUCTION

The gate drive circuits, that control the state of the high-side semiconductor in a converter, must be supplied insulated. That contributes to the fact that the reference-potential of the gate driver depends on the switching state of the semiconductor device. Insulation is necessary for the control signals and the power supply of the driver unit. Transmitting the control signals per fibre optic, even at high difference of potential, is considered state of the art [1]. Usually an insulated power-supply would be realized with an insulating transformer [2]. By increasing the operation voltage of the system, the volume and costs of such a transformer, with an adequate insulation, will also be increased.

This paper presents a DC/DC-converter based on a printed transformer [3,4]. The developed power supply purposes a high insulation voltage to supply a gate-driver

for the 6.5 kV IGBT-module. Other applications, those requiring high insulation voltage and low power are also possible. The converter operates at a frequency of 420 kHz and provides an output power of 6 W. It delivers a regulated output voltage of ±15 V. At rated output power the efficiency is 62 %.

Fig. 1 illustrates the construction of the switched-mode power supply. The primary coil is printed as a spiral on the top layer of the printed circuit board (PCB). The secondary winding is placed on the opposite position on the bottom layer. In order to increase the inductance, a ferrite foil is attached over the printed coils.

The PCB itself is used as the insulator between the different potentials of the layers. To improve the coupling between the coils of the transformer, the PCB material (RF-4) has a thickness of only 0.5 mm [5]. That ensures an electric strength of more than 18 kV.

For the electronic design only SMD-parts were used to avoid vias to hold up the insulation of the PCB. A non conductive area around the ferrite and the circuits is implemented. A leakage distance of 20 mm is realised for the DC/DC-converter that is shown in fig. 2.

The withstand-voltage of this converter has been tested at 10 kV / 50 Hz for 120 seconds without a spark-over. Higher withstand-voltages can be attained by using a thicker material for the PCB and increasing the leakage distance. This thicker material reduces the coupling of the coils and consequently the efficiency of the DC/DC-converter.

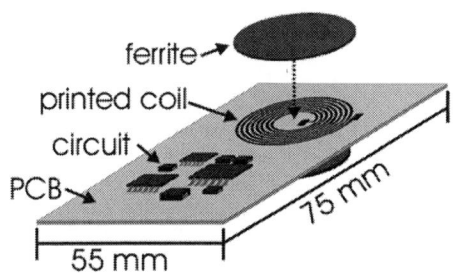

Fig. 1. Construction of the switched-mode power supply

Fig. 2. Developed power supply

II. MODELLING OF PRINTED PLANAR TRANSFORMER

In order to optimise the planar coils, a model was developed. The electrical parameters of a coil were calculated on the basis of the geometric structure. To compare the calculation methods, we built several planar transformers with different geometric structures. Some of them are shown in fig. 3. In the presented cases the primary coils have the same geometric structure, only the secondary coils are modified. N describes the number of turns, a is the with of the conductive track, b the clearance between them and d the diameter of the inner circle.

The calculated inductances L were compared with the measured values. The best result was achieved by using the following equation [6]:

$$L = 31.33 \cdot \mu_0 \cdot N^2 \cdot \frac{r^2}{8 \cdot r + 11 \cdot e} \qquad (1)$$

Here N is the number of turns, r is the average radius and e is the difference between the outer and the inner radius of the coils.

In the majority of cases the variation in the inductance value was less than 10 %. Also a graphical method to evaluate the inductance, that illustrates fig. 4, was compared to the previous investigation [7]. But the variation of the inductance is more than 15 % in most of the investigated transformer coils. Only the results of relatively large coils agreed with the measured and calculated inductances.

A ferrite foil has been added to increase the inductance of the planar coils. The difference in the inductances has been approximated by measuring several real coils with and without ferrite foil. This can be considered in the calculation of the inductance with a correction factor of 2.27 when adding the ferrite foil. This factor can be used for the primary and secondary side.

With respect to the high operating frequency of the converter the skin-effect must be considered. The skin depth δ can be calculated by:

$$\delta = \frac{1}{\sqrt{\sigma \cdot f \cdot \pi \cdot \mu_0 \cdot \mu_r}} \qquad (2)$$

Here σ is the specific conductance for copper. In our investigation we use PCB with copper thickness of 35 μm. The frequency where the skin depth is equal to the copper thickness is calculated as 3.5 MHz. Therefore the skin-effect is negligible. The resistance has been calculated by the well known equation:

$$R_{strip} = R_{copper} \cdot \frac{l}{A} \qquad (3)$$

The other parameters like the capacitance are also calculated. With this information, it was possible to build a simple PSpice simulation to get an overview of the switching behaviour of the semiconductors in combination with the tested coils.

Table I shows the parameters that for the final planar transformer. The ratio of N_2 to N_1 (1,3) guaranties a voltage above 18 V on the secondary side. Table II gives an overview of the calculated and measured inductance values. Furthermore the estimated DC-resistance is compared.

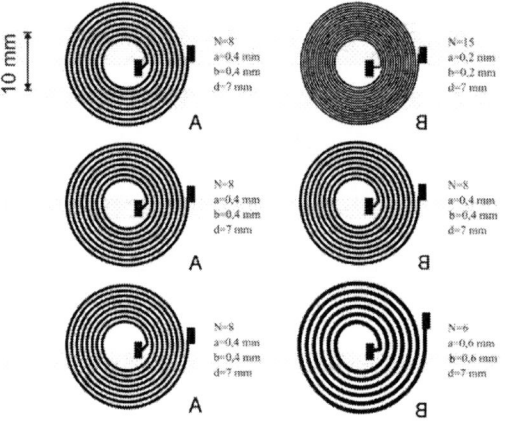

Fig. 3.   Printed test coils with different geometric structure

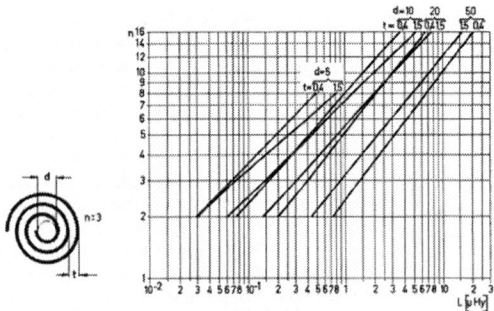

Fig. 4.   Graphical method to assign the inductance of planar coils [7]

TABLE I.

PARAMETER FOR FINAL TRANSFORMER

	prim. side	sec. side
turns, N	13	17
track with, a	0,8 mm	0,6 mm
clearance between tracks, b	0,2 mm	0,2 mm
free inner circle, d	8,3 mm	8,33 mm
outer diameter of the coil	34,0 mm	35,1 mm

TABLE II.

COMPARISON OF CALCULATED AND MEASSURED VALUES

	calculation	measurement
inductance prim. side	7,74 μH	7,41 μH
inductance sec. side	13,13 μH	12,54 μH
DC-resistance prim. side	0,631 Ω	0,67 Ω
DC-resistance sec. side	1,14 Ω	1,23 Ω

III. STRUCTURE OF THE DC/DC-CONVERTER

In order to build a low cost DC/DC-converter, linear optoelectronic couplers with high insulation voltage were avoided. Thus there is no information feedback from the secondary to the primary side of the power supply. Hence there exists no possibility to control the output voltage by

PWM on the primary side of the DC/DC-converter. The circuit on the primary side works with constant frequency. Therefore, the ratios of the windings are calculated, so that the voltage at the secondary coil is higher or lower than ±18 V respectively. Voltage regulation can be done easily by standard voltage regulators.

A. Top Layer

Figure 5 illustrates the very simple schematic of the primary side circuit. A pulse generator IC IR21531D has been chosen to generate a fixed system frequency at 420 kHz [8]. The IC has an internal zener diode for supply voltage stabilisation, a dead time controller, an adjustable frequency generator and a high-side driver. The output-driver is a standard SMD dual-MOSFET IC which can drive continuous currents up to 2 A [8]. One end of the primary inductance is connected to a capacitive voltage divider. The potential at that point is nearly half of the input voltage.

B. Bottom layer

The circuit of the secondary side is illustrated in fig. 6. The diodes D_1 and D_2 rectify the voltage of the secondary coil. The zener diode ZD_1 and ZD_2 limit the maximum voltage in front of the voltage regulators, if no load is connected.

C. Full load operation

The fig. 7 shows the function of the DC/DC converter with an input voltage of 33 V at full load operation of 6 W. In the upper part the gate-source voltage of the low-side switch u_{GS_L} and the high-side switch u_{GS_H} are illustrated. When using a switching frequency of 420 kHz the on-time is 530 ns. The interlock time between the gate signals is fixed to 660 ns by the pulse generator.

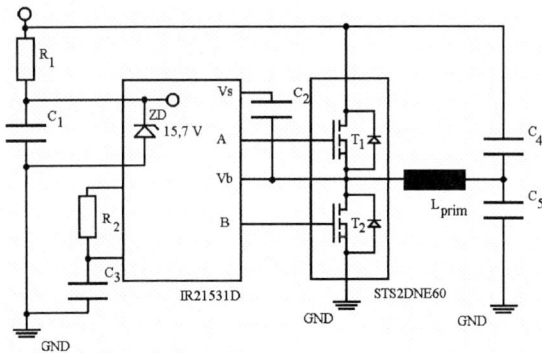

Fig. 5. circuit of the primary side

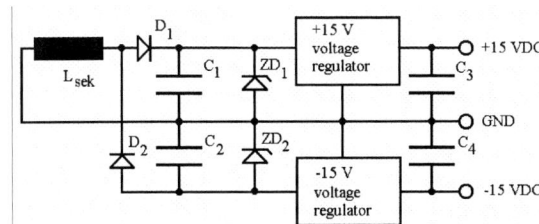

Fig. 6. circuit of the secondary side

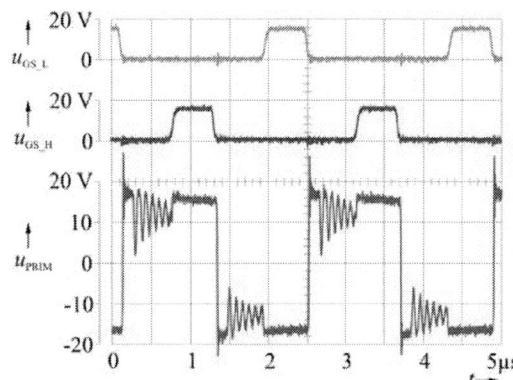

Fig. 7. Voltages on the primary side under full load operation

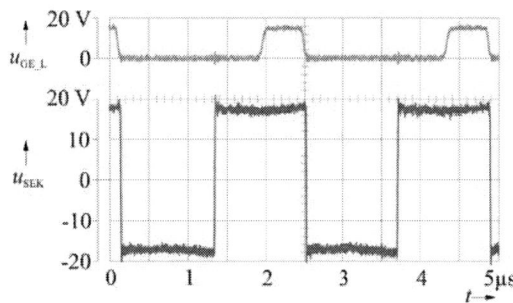

Fig. 8. Voltage at the secondary inductance in relation to the control signal of the low-side MOSFET

Furthermore, the voltage of the primary inductance L_{PRIM} is shown. With respect to the connection to the capacitive voltage divider the inductance voltage is positive and negative.

Figure 8 present the voltage at the secondary inductance L_{SEK}. The amplitude of the squared waveform is above 18 V, what guarantees a stabilised voltage of ± 15 V at the output of the voltage regulator.

D. Efficiency at different loads

The efficiency at different points of load are measured and illustrated in figure 9. In this setup the input voltage is fixed to 33 V. Even if there is no load on the secondary side, the driver on the primary side will still try to push energy to the transformer. Due to this fact lowers the efficiency of the DC/DC-converter on small loads.

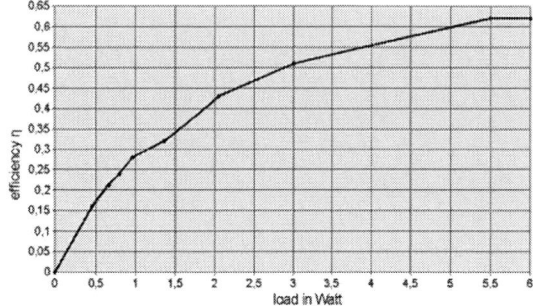

Fig. 9. Efficiency at different loads

At full load operation of 6 W the efficiency is with about 62% relative low. So the power losses of each part are estimated. Nearly 65 % of the losses are caused by the pulse generator IC, the ohmic losses of the coils and the voltage regulators.

IV. CONCLUSION

The shown DC/DC-converter uses a planar transformer and a very simple electrical structure. The PCB itself was used, to achieve a high insulation voltage of 10 kV. The converter delivers an output power of 6 W with a maximum efficiency of 62 %. The relatively low efficiency depends on the coupling of the primary and secondary coils. Tests with a closed ferrite core around the coils and the PCB have shown that the efficiency can easily be increased to more than 80 %. A closed ferrite core would however destroy the insulation. If a higher output power is needed, the dimension of the planar transformer can be increased. Then the geometric parameters can be optimised by the developed calculation method.

REFERENCES

[1] H. Rüedi, H. G. Matthes: *1.6 MW / 150 kHz Inverter for Welding Applications*, PCIM Conversion Intelligent Motion (PCIM),Nürnberg, 2000, CD-Rom.

[2] U. Nicolai, T. Reimann, J. Petzoldt, J.Lutz: *Applikationshandbuch IGBT- und MOSFET- Leistungsmodule.* http://www.semikron.de, Jan 2008

[3] Miguel J. Prieto, Juan M. Lopera, Juan A. Martin, Juan Diaz, Alberto Pernia: *Comparison of Different Technologies to Low-Power,. Low-Voltage DC/DC Converter Modules.* Power Electronics Specialist Conference (PESC), Acapulco, 2003, CD-Rom

[4] S. Y. (Ron) Hui, Henry Shu-Hung Chung, S. C. Tang: *Coreless Printed Circuit Board (PCB) Transformer for Power MOSFET/IGBT Gate Drive Circuits.* IEEE Applications Society Annual Meeting (IAS), Seattle, WA, USA, 2004, CD-Rom

[5] Fechner FS Schulte Leiterplatten GmbH. *http://www.fs-leiterplatten.de*, Nov 2007

[6] Marc T. Thompson: Inductance Calculation Techniques Part II: *Approximations and Handbook Methodes.* http://www.classictesla.com, Dec. 1999.

[7] Hans Kolbe Co: *Datasheet 10109 Gedruckte Schaltungen*, Gittelde 1976.

[8] *http://www.irf.com*, Jan 2008

Variable Motor Operating Point by Integration of Power Electronic Device into Rotor

Adrian Tulbure[1], Hans-Peter Beck[2] and Mircea Risteiu[3]

[1] University "1 Decembrie 1918" of Alba Iulia, Romania, e-mail: *aditulbure@uab.ro*
[2] Clausthal University of Technology, Germany, e-mail: *beck@iee.tu-clausthal.de*
[3] University "1 Decembrie 1918" of Alba Iulia, Romania, e-mail: *mristeiu@uab.ro*

Abstract – **The main issue of this work is to analyze and to test how the characteristic curve of the induction machine can be adjusted by the power electronic devices, when these are integrated into rotor circuits.**

In order to technically analyze the characteristic performances of the induction machine by switching between cages, the simulator software's MATLAB and NETASIM have been used. The simulated current- and torque-waveforms are compared with the experimental values from the test bench.

To practical validation of the results, a main feed induction machine with working-cage controller has been designed and tested at the Institute of Electrical Power Engineering in Clausthal, Germany.

Keywords - **induction motor, electrical drive, IGBT, power semiconductor device, industrial application, cage switching, Pulse Width Modulation (PWM), test bench, measurements, simulations.**

I. INTRODUCTION

The proposed contribution presents the behavior of the main feed induction machine with integrated power electronic devices into the rotor. These integrated devices can modify the rotor parameters as well as simultaneously the machine is directly connected to power network. Practical approach of the concept for the *Mn*-characteristic variation at an asynchronous machine is done through a rotor intervention.

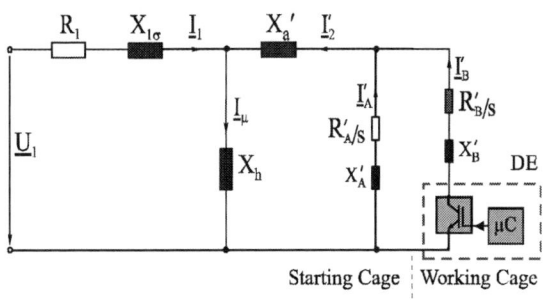

Fig.1 Equivalent electrical circuit of motor with integrated power electronics

The electronic system (DE, Figure 1) is based on two layers: power switch consisting of Schottky-diodes and IGBT systems layer, respectively microcontroller (μC)-based supervisory and control unit. For the designed machine the secondary power circuit from Fig.1 is characterized by:

- R_a, L_a - common resistance and inductance for both cages of the machine (Fig.1, X_a)
- L_A, L_B and R_A, R_B - starting cage and working cage inductance respectively reactance (Fig.1, X_A, X_B).

The performance characteristics of the machine depend on the rotor current distribution [1], [2] in the starting and working cage.

II. DESIGN OF THE PROTOTYP MACHINE

The central idea of this work has been focused on developing a structure, which allows dynamic control of the machine electric parameters, respectively the machine slipping. For practical implementation and testing of the proposed concept, a classic asynchronous machine must be implemented with an integrated switching system (DE). This module (DE from Fig.1) must be able to compute Cage-Switcher function in dynamic and stationary state, for unloaded and overloaded operation.

Figure 1 shows the concept based on the mathematic model, according to the next equations:

$$s_k = \frac{R_2'}{\Omega_1 \cdot (L_{1\sigma} + L_{2\sigma}')} \tag{1}$$

Where:

$$R_2' \approx \frac{R_A' \cdot \frac{k}{s} R_B'}{R_A' + R_B'} \tag{2}$$

Based on this approach, the main advantages of electronically-based working-cage [4] could be listed in the following:

- The electronic switcher is dimensioned only for a rate (30%) of the machine nominal power;

- Is compactly designed with advanced electronic devices, which are integrated into rotor;

- Reducing the prototyping costs by using standard devices;

- For reducing the turn-off over voltage se, by integration of electronic devices into rotor, the protection system could be easily integrated, and it is low costly;

- For designing the transmission unit from rotor to stator the only requirements are for control signals, not for power signals;

- The designed prototype has small adjusting range, constrained by the rotor and stator electric parameters. The new asynchronous machine has the ability of self stabilizing, with no propagations of damages to the power network. It means that there is no need for supplementary protection system;

The results of previous technical simulation have proved the influence among the drive system attenuation, machine slipping and the equivalent parameters of double cage induction machine. The implemented control tuning algorithm has simple loop structure suitable for real time control.

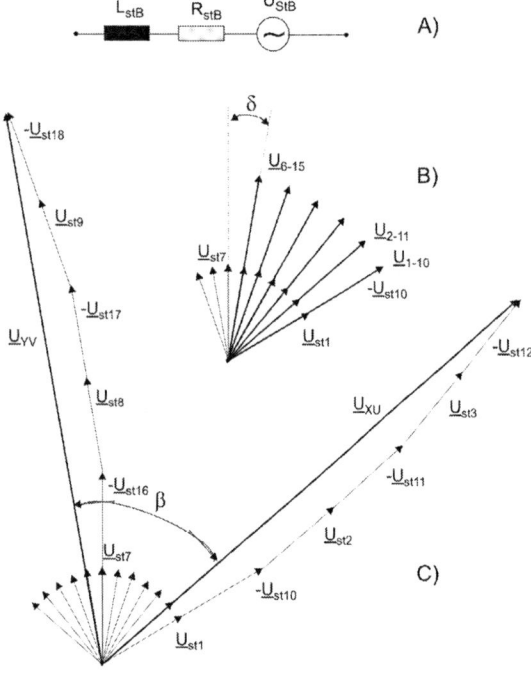

Fig.2 Vector diagram of phase voltages from working cage. $\underline{U}_{S1}$ bar voltage vector. $\underline{U}_{1-10}$ phase voltage vector for 3xB6 und $\underline{U}_{XU}$ phase voltage vector for 1xB6. (A) Equivalent circuit of working bar; (B) Vector diagram by 1xB6 version; (C) Vector diagram by 3xB6 version

When the IGBT from Fig. 1 is off, the current from working cage is cut. If the IGBT is on, the major current is driven via working cage.

In first case the rotor current is limited by starting cage impedance Z_A and for the second situation by the working cage impedance Z_B. By this setup the rotor current evolution could be controlled through an external command pattern.

The preliminary data for power electronic design are constrained by electric parameters of working cage and by maximal voltage and current induced in its circuit. The finite element from the cage metal bar is similar to an alternative power supply with own bar resistance and inductance (R_{StB}, L_{StB} from Fig. 2 A). Because of the fact that DCIM (double-cage induction machine) has 36

slots with 2 pairs of poles, in neighbored metal bars an alternative voltage delayed with 20 grades is induced.

Related voltage vectors are developing in phase and value following next formulas:

$$\beta = 6 \cdot \alpha \qquad U_{XU} = 2 \cdot U_{S1} \cdot (1 + 2\cos\alpha) \qquad (3)$$

$$\delta = \alpha \qquad U_{1-10} = 2 \cdot U_{S1} \cdot \cos\alpha \qquad (4)$$

The design of the electronic switching system has started with equivalent model of the cage bar as a voltage generator (*Fig. 2, A*), then a connecting topology has been evaluated with the vector diagrams shown in Fig. 2 B, C.

For the optimal choice of circuit topology some constrains have been taken into account [5], [10]: ripple of output voltage, total losses at nominal working point, the distribution of current harmonics, the compact construction and technical approach of the practical implementation.

III. PRACTICAL IMPLEMENTATION

The circuit topology design and semiconductor components selection for power and control, have been done into correlation with equivalent parameters of the electrical engine and technical requirements from operation conditions. The practical realization of the impedance variation by dynamic cage-switching is possible by rebuilding the double cage asynchronous machine type 225-4M, as well as by structural changes of the working-cage. At the end of copper metal bar of working cage one has replaced the typical rings with symmetric shottky diodes, as figured in Fig.3. The rectifiers' outputs are connected with IGBT's terminals through axial and radial channels, cut in main motor shaft.

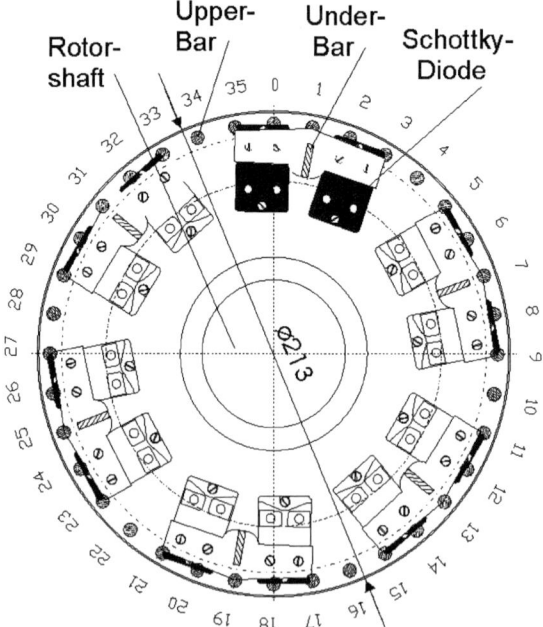

Fig.3 Symmetrical displacement of schottky-diodes (with black) in the front side of the rotor

This flexible structure with 36 metal bars (variable frequency power supply) allows a practical approach with more variants of circuit topologies. Following this idea two possible combinations have been analyzed.

The dedicated configurations with 18 concentrated power supplies [2], [8]are defined by:
- Three three-phase bridged rectifiers connected serial (Fig.2 C);
- More three-phase bridged rectifiers connected in parallel (Fig.2 B).

First concept (with serial connections) generates a pulsed waveform with 18 pulses by 900 Hz and very low ripple factor. In the same time, the input voltage is maximal, dependent on the operation point of the motor, but also enough to pass the threshold voltage level of the rectifiers' diodes.

Developing the concept of parallel connections of the rectifier bridge B6 one obtains a high ripple factor (pulse voltage only by 300Hz).

By taking into account these aspects, and in the same time by considering that one needs to design low resistance current paths, one has decided to use first described variant. So the integrated rectifier into rotor circuit will be supplied with 35 V, as in next equation:

$$U_d = \frac{3 \cdot \sqrt{6}}{\pi} U_{XU} = \frac{3 \cdot \sqrt{6}}{\pi} \cdot 2 \cdot U_S \cdot (1 + 2\cos\alpha) \quad (5)$$

According to Fig.4, in the rotor (2) are integrated the following components: IGB transistors (1), IGBT driver (3), voltage sensor (5), and schottky-diodes (4). The machine behavior has been analyzed in accordance to the same testing procedure and laboratory setup, before and after the integration of power electronics into the rotor [1], [8].

Fig.4 The rotor with integrated power electronics (1, 4) and control devices (3, 5)

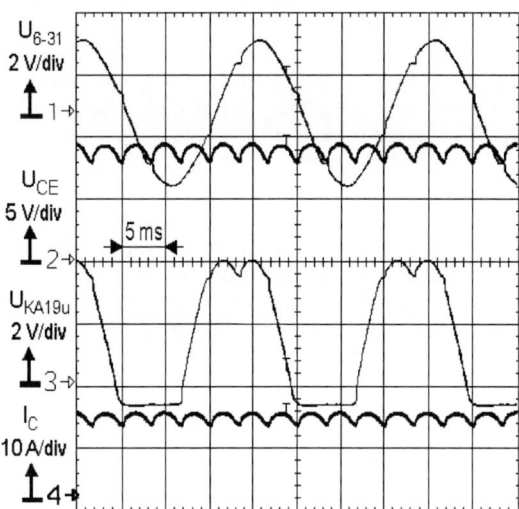

Fig.5 Voltage and current by short circuit test of the 26kW DCIM. Connection-version according to Fig. 2 C

The used sensors in rotor circuit (Fig.4) are able to measure the distribution and evolution of the current and voltage into the established ranges. The electric signals are converted into direct currents and through a specific transmission system are sent to a stationary data acquisition unit.

The output pulse current from inferior cage (working cage) is similar to the input current of the chopper module. This situation requires to select the transistors of this module for two times higher working current. For semiconductors selection one has considered that the electrical circuits are inductive, that means higher switch-off voltage [5], [7]. The main reasons of this are the pick voltages (6), which are amplified by overloaded machine.

In the next formula L_σ is the inductance of the load path and the L_{CE} is parasitic internal inductance of the semiconductor.

$$\Delta u = (L_{CE} + L_\sigma) \cdot \frac{di_c}{dt} \quad (6)$$

These pick voltages could destroy the IGBT device or upstream Shottky diodes, which are more sensitive. For them protection, one has decided to use RCD circuits consisting in MKP capacitor, metallic discharging resistor and high speed diodes type RUR.

The IGBT has threshold like characteristic [6], where the V_{CE} still depends not linearly on the current, while the MOSFET posses a resistive behavior and therefore on-state voltage increases almost linearly with the current [7], [9]. Therefore the equivalent on resistance for the IGBT (r_{CEon}) can decrease with increasing current, while the r_{DSon} for a MOSFET remains fast constant:
Therefore under circumstance is valid the eq.

$$R_{CEon}(IGBT) = \frac{1}{2...5} \cdot R_{DSon}(MOS) \quad (7)$$

1245

For actual machine under given operation conditions, one has selected Shottky diodes in SOT-case and high power IGBT in SEMITRANS-case (see Fig.4).

In order to insulate the control and measurement system against power switching perturbations, the TTL-5V level signal is increased to 15V. To the general design strategy of the control system one take into account some considerations:

- Thermal behavior by variable motor loading;
- Design of low inductivity circuits, in order to maintain defined switching frequency;
- Electromagnetic compatibility requirements;
- High quality demands by supplying of the control and measuring system.

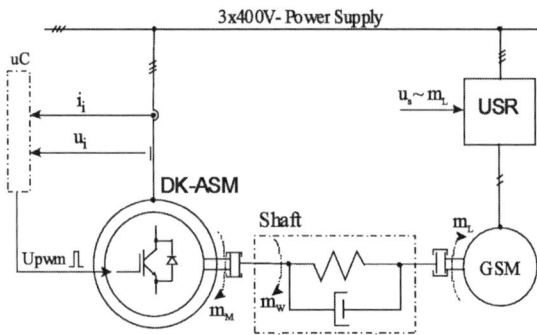

Fig.6 Preview of the measurements system

The induction machine is connected directly to the three phase (3x400V) power supply network. With the help of a metering shaft, the signal corresponding to the torque values in the shaft (m_W) is acquired and processed. The load torque (m_L) is generated with the help of a dc-machine. The switching signal on the gate of in rotor integrated transistors U_{pwm}, is delivered through the micro-controller port.

IV. MEASUREMENTS

In the first stage, the preliminary measurements of the standstill machine have been carried out. In the second stage, the induction machine has been tested with nominal voltage at different operating points from the characteristic curve. Finally, the machine has been tested with different cyclical tests (sinusoidal LIF- Load Input Function) and related control/ adjustment loops. Partial machine loading (about 30% and 70%) is ensured by a DC machine with torque/speed regulator.

The next figure shows the experimental results for a constant speed of 950 rpm and reduced stator voltage U_{12}=220V, that corresponds to the operating point P_1 from fig.8.

Through the pwm-based control signal, the IGBT-chopper forces the machine at 13,9 sec. on the starting cage for the next 400 ms. After that, the machine is leaded on the natural mechanic characteristic on the both cages.

By switching from one rotor circuit to the other, the shaft torque is more reduced than in the previous situation.

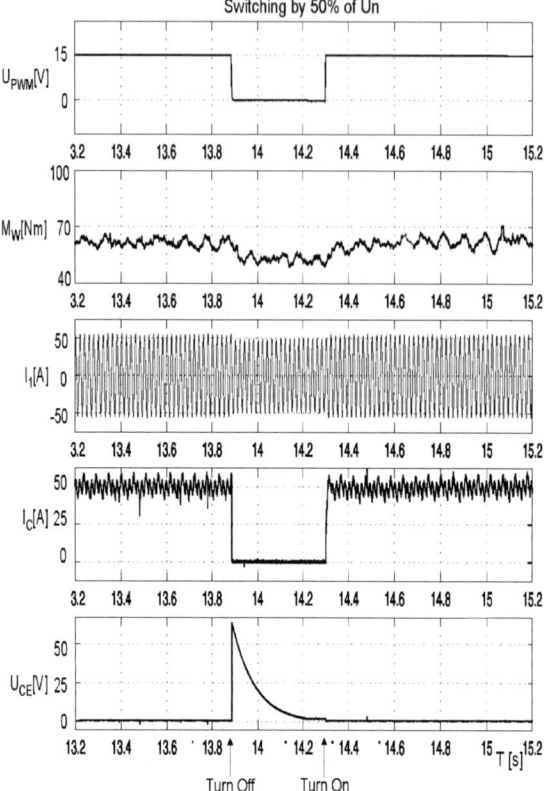

Fig.7 Measurements of control signal U_{pwm}, shaft- torque M_W, phase current I_1, collector current I_C and collector-emitter voltage U_{CE} at the 200V stator voltage

As it is known, that for each rotor winding, the voltage value and the induced frequency is proportionally with machine slipping. On the other hand the MOS controlled electronic device are very sensitive with the overstep pick voltages into the circuit network. It is the reason why the machine has been started up with low level of voltage, after that the supplying voltage has been increased up to the nominal level.

On the graph is figured that for the switching time, the output (shaft) torque (M_W) is decreased with about 10%. It can be also evaluated by the primary current (I_1). Rectified rotor current (I_c) is pulsing with 55Hz, which means that on rotor winding there are 9.55 Hz. At this point the machine works with 37% slipping.

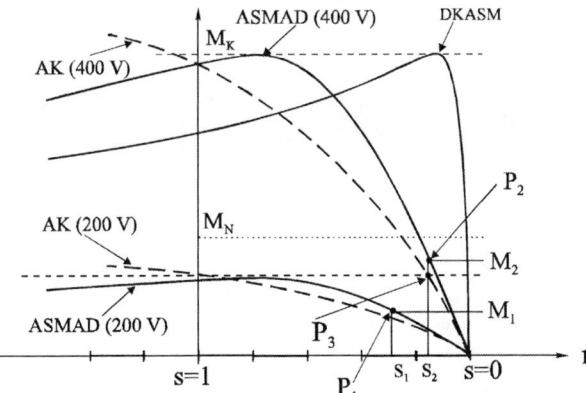

Fig.8 Adjusting the operating point P_1, P_2, P_3 of the induction machine with power electronics into the rotor by 200V supply voltage (ASMAD 200V) and by 400V supply voltage (ASMAD 400V). AK - characteristic curve according to the machine with starting cage only; DKASM - characteristic curve according to the double cage machine;

Figure 9 shows similar measurements with the nominal line voltage of 400V. For this situation, the machine is loaded at 75% (135Nm) and working between the operating points P_3 and P_2. Rectified rotor current (Ic) is 35A and is pulsing with 25Hz, that means that the voltage in rotor winding has 4,20Hz at 16,8% slipping.

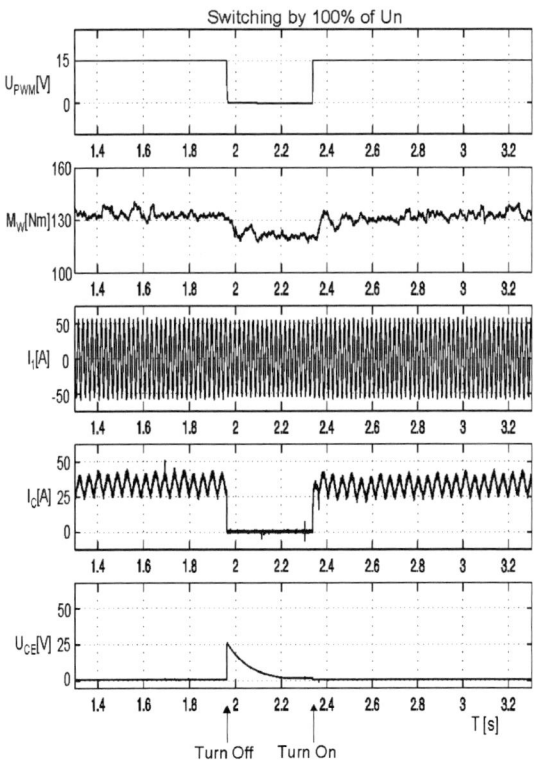

Fig.9 Measurements of control signal U_{pwm}, shaft- torque M_W, phase current I_1, collector current I_C and collector-emitter voltage U_{CE} at the 400V stator voltage

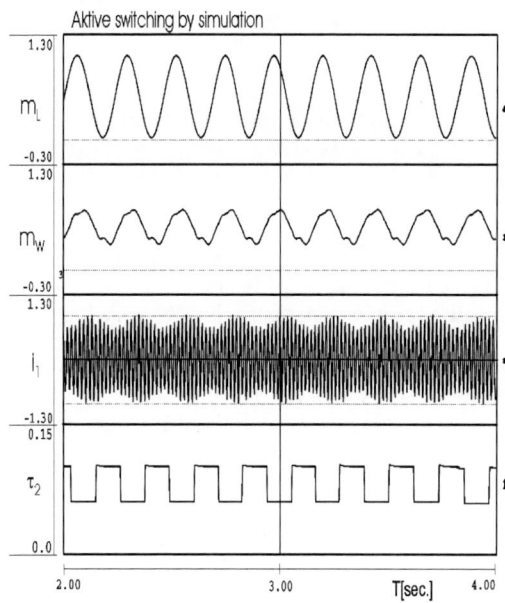

Fig.10 Simulated waveforms of load- m_L shaft- torque m_W, current i_1 and rotor time factor τ_2

For the system control and data acquisition one has built an experimental setup around dSpace-platform. This setup allowed us to acquire 8 channels simultaneously samples at 10kHz, reconfigurable, 12 bits resolution and +/- 10 V measurement range.

In Figure 10 and 11 the simulation and measurement results in time domain corresponding to the analyzed signals, load torque m_L, shaft torque m_W, phase current i_1 and control signal U_{pwm} are presented.

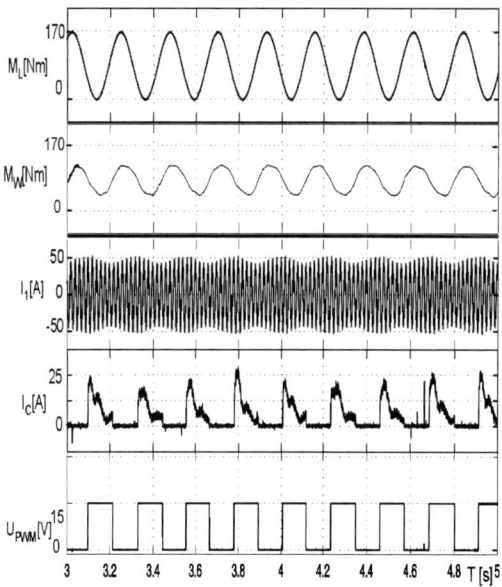

Fig.11 Measurements of load- M_L shaft- torque M_W, phase- I_1, collector current I_C and factor τ_2

V. CONCLUSIONS

In the case of simulation from Fig.10 the results are normalized to the nominal values of torque and phase current. The waveforms show a good correspondence of the simulations with the measurements from Fig.11.

Comparisons of run tests have demonstrated that the amplitude of the shaft torque oscillations could be adjusted by switching between starting and working cage of the induction machine [2], [8], [10].

Generally speaking, at nominal voltage supply, the jump of the value of shaft torque is bigger than in the situation of reduced voltage supply (Figure 7 and 9).

In order to accomplish a detailed investigation of the drive system, one needs a performing test bench. Therefore all the measurements from this paper have been carried out at the test bench from Fig. 12.

Fig.12 Preview of the test bench with the
double cage induction machine.

ACKNOWLEDGMENT

This research was possible with the support from Institute of Electrical Power Engineering in Clausthal-Zellerfeld and from the Konrad-Adenauer Foundation, Federal Republic of Germany.

REFERENCES

[1] A. Tulbure, H.-P. Beck, "Damping Torque-Oscillation at Drives with Double Cage Induction Machine". *European Power Electronic & Motion Control-Conference*, Kosice, ISBN 80-88922-22-4. Vol.5, pp.12-16.

[2] H.-P. Beck, C. Sourkounis, A.Tulbure A. Schwingungsdämpfung in Antriebssystemen mit Doppelkäfig-Asynchronmaschine. *VDI-Bericht Nr. 1606*, pp.113-126, VDI-Verlag, Duesseldorf 2001.

[3] R.Lappe & colab."Leistungselektronik." *Technik Verlag Munchen* 1994

[4] A.Tulbure "Type Test Procedure for Electronic Ground Power Units" *Proceedings of AQTR-06*, Tome II, pp176-179, Cluj Napoca 2006,

[5] D. Schroeder . "Elektrische Antriebe" *Springer Verlag Berlin* 1995

[6] R. Francis, M. Soldano "A New SMPS Non Punch Thru IGBT Replaces MOSFET in SMPS High Frequency Application" *International Rectifier Corporation*. El Segundo, CA, USA 2005

[7] N.Mohan, T. Undeland, W. Robbins "Power Electronics. Converters, applications and design." *John Wiley* New-York 1995

[8] A. Tulbure, "Netzgespeiste Asynchronmaschine mit elektronischer Käfigumschaltung". *PhD Thesis*, Clausthal University of Technology Germany 2003.

[9] S. Davis "Power Mosfet Gate Driver". *Edition OnLine, power.national.com* 2007

[10] A. Tulbure., A. Egri, R. Joldes· "Cage Switching at an induction machine using 16-bit uC". *MicroCAD-2007 International Conference* Miskolc. Hungary 2007.

Magnetic Material Comparisons for High-Current Gapped and Gapless Foil Wound Inductors in High Frequency DC-DC Converters

Marek S. Ryłko (mrylko@pei.ucc.ie), Brendan J. Lyons, Kevin J. Hartnett, John G. Hayes (john.hayes@ucc.ie), Michael G. Egan

University College Cork, Ireland

Abstract – The inductor often drives the dc-dc converter size. Thus, the inductor optimization process is required for the most effective design. The paper presents inductor analysis only. The material properties are essential for the design size. In this paper, various magnetic materials are analysed and investigated for use in a practical design. The investigation is concerned with the magnetic material selection for a dc-dc power inductor in the medium (20 kHz) to high (150kHz) frequency range. The materials under investigation are iron-based amorphous metal, silicon steel, nanocrystalline, ferrite, and gap-less powder materials. A lumped parameter algorithm is derived which includes such effects as the foil ac copper loss effects, the gap core loss, and the cooling path. The algorithm is implemented in EXCEL and generates material comparisons over a range of frequencies, ripple ratios, cooling paths. The results show that the core power loss limited inductor tends to be oversized while the minimum size is achieved for the design which is at the sweet-spot where the size is driven by the core power loss, winding power loss and core saturation limit. A 1.25 kW half-bridge dc-dc converter is built in order to proof the algorithm feasibility at the interest frequency range.

Keywords – **Magnetic device, Design**

I. INTRODUCTION

High-power-density magnetic devices are increasingly required in industrial, automotive and aerospace applications. Compact and efficient magnetic design results in converter size minimization. There are many design parameters, e.g. flux swing, material saturation level, dc-bias, operating frequency, cooling method, etc., which have an effect on final inductor dimensions. Some of these design parameters are related to the converter topology while others are mainly associated with core material properties. There is a wide variety of ferromagnetic materials available to the magnetic designer [2]-[7]. However, the optimum choice of magnetic material for a design is not straightforward. The operating current, current ripple, air-gap length, number of air-gaps and operating frequency have a strong influence on the inductor size. As shown in [1], [12] the ideal material, for the minimum size design, is with a small or distributed air-gap together with high saturation flux and generally low power loss in both the copper and core.

The analysis develops an algorithm to contrast and compare the various materials for differing electrical and thermal inputs. The investigation extends the earlier work [1] by additionally considering gapless cores and developing a simplified method for considering conduction cooling for comparison with radiation cooling. The magnetic materials of interest are presented and discussed in Section II. The underpinning design algorithm is presented in Section III. Theoretical results of the analysis are shown in Section IV. Experimental validation is presented in Section V and conclusions in Section VI.

II. MAGNETIC MATERIALS

The general specifications for the magnetic materials of interest in the medium to high frequency range are shown in Table I [2-7]. The table features three laminated cores, iron-based amorphous metal (AM) of type 26051SA1 from Metglas, silicon steel (SS) of type 10JNHF600 from JFE, and nanocrystalline of type Vitroperm 500F from Vacuumschmelze, ferrite of type 3C93 from Ferroxcube, powdered iron of type -26 from Micrometals, and a selection of pressed powdered cores of varied permeabilities, Kool Mu, MPP and High Flux from Magnetics. The core material magnetic properties depend on the basic material used and the manufacturing processes [2-8].

The iron-based amorphous metal and nanocrystalline materials are tape wound while the silicon steel is stamped and stacked. Ferrite 3C93 is a ceramic. The powdered-iron core is an iron dust bonded in epoxy. There are many different mixes of the powdered-iron cores, having different permeabilities and power losses strongly dependent on the iron content. Pressed powder cores form a significant group and have significantly lower power losses than the powdered-iron cores at 100 kHz. The permeability of the powdered cores can be adjusted during the manufacturing process as the air-gap is distributed in the material.

Silicon steel has the highest saturation flux, followed by amorphous metal, High Flux, powdered iron, nanocrystalline, Kool Mu, MPP, and ferrite.

The highest nominal power loss at 100 kHz is observed in the powdered iron core. Amorphous metal, silicon steel, Kool Mu and High Flux have similar power losses and are significantly lower than powdered iron as shown in Table I. A plot of the specific power density versus flux density for the various materials is presented in Fig. 1. The lowest power loss is reported in the nanocrystalline and ferrite materials. The MPP power loss is placed between amorphous metal and nanocrystalline.

The core maximum operating temperature is limited theoretically by the Curie temperature and practically by the lamination and coating thermal capability limits. The amorphous metal and silicon steel cores are limited by the lamination epoxy thermal capability. The iron powder and nanocrystalline cores are limited by the thermal aging effect. The pressed powder cores maximum temperature is limited by the coating thermal capability. Finally the ferrite is limited by the temperature when the power loss reaches its minimum in order to avoid the core thermal runaway [5].

The powdered iron temperature is limited by the bounding epoxy used. This epoxy is usually organic based and can undergo thermal aging at elevated temperature, making the powdered iron core temperature sensitive and limiting its reliability and lifetime [6].

The core cooling is enhanced by the thermal conductivity of the material. The powder cores have superior thermal conductivities as well as the highest maximum operating temperatures. The amorphous metal, silicon steel and nanocrystalline have anisotropic thermal conductivities, which is several times higher along the laminations than across the laminations. This thermal path anisotropy can cause difficulties in optimizing the cooling path. Ferrite and powdered iron have the lowest thermal conductivities.

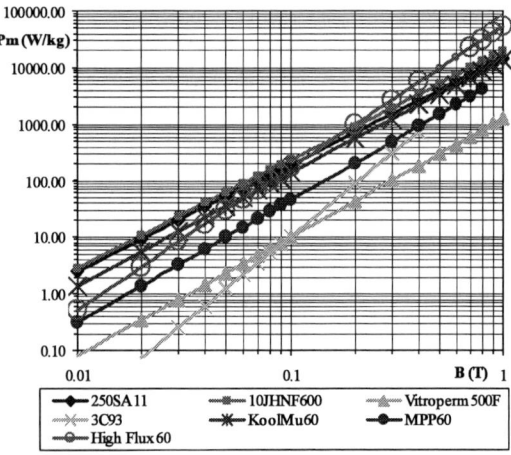

Fig. 1. Specific power density vs. flux swing at 100 kHz.

III. LUMPED PARAMETER ALGORITHM

Current inductor design flow charts involve the selection of a core size in an iterative design procedure [1, 9, 10, 11, 13, 14]. In this section, an algorithm is developed to solve for the optimum inductor design. Various electrical and magnetic parameters and restraints are input into the algorithm and the minimum sized inductor is determined based on the inputs.

Two different flow charts are used to compare the magnetic materials. The first algorithm, called A1 as

shown in Fig. 5, is based on the linear part of the magnetization curve. This algorithm is used for the high permeability materials such as amorphous metal, silicon steel, nanocrystalline and ferrite. The second algorithm, called A2, is used for the powder materials and is based on the materials dc-bias curve. The solution is obtained by seeking a minimum area product. The core's basic dimensions are the variables. The result is the core area product, AP, which is proportional to the inductor volume and mass. The algorithm relies on Excel's built in minimum seeker function. Both flowcharts allow for relatively simple use of the material parameters straight from the datasheet.

Despite the different methods of calculation, both algorithms have a similar modular structure as shown in Fig. 5 and Fig. 8, for algorithm A1 and A2, respectively. The modular structure improves the algorithm clarity as well as enabling easy modification of the design i.e. change of the core material, different input source or different winding arrangement as well as different cooling and so on.

The algorithm consists of nine main modules (A)-(I) and the minimum seeker loop with defined constraints in (J). If conditions in (J) are satisfied the minimum area product is calculated and the algorithm A2 ends. The algorithm A1 continues calculation in the next loop (K)-(M) where the additional turns are added to the design. The area product is calculated with a new number of turns and is compared with the previous iteration. If (L) is satisfied then the algorithm finishes its calculation with the final area product.

In order to set the task the following assumptions are made:

  a) boost converter;
  b) CC core set;
  c) foil winding;

The algorithm calculates such factors as:

  a) Air-gap fringing flux factor;
  b) Rounded up number of turns;
  c) Core power loss due to hysteresis and eddy currents;
  d) Core power loss due to air-gap fringing flux;
  e) Winding clearance and creepage;
  f) Distance between winding and air-gap;
  g) Winding skin and proximity effect;

Firstly the algorithm A1 is presented and then the differences in algorithm A2 are highlighted. The algorithm flowchart is described as follows:

In (A) the typical inductor electric specifications are declared: input voltage, V_{in}, output voltage, V_{out}, maximum output power capability, P_{out}, converters efficiency, η, inductors efficiency, η_i, current ripple ratio, r, and operating frequency, f. The electric specifications are generated for the design: dc input current, I_{in}, peak-to-peak inductor current, ΔI_{pp}, duty ratio, D, inductor inductance, L, inductor peak current, I_{max}, allowable inductors power loss, ΔP_i.

$$I_{in} = \frac{P_{out}}{\eta \cdot V_{in}}, \ \Delta I_{pp} = I_{in} \cdot r, \ D = 1 - \frac{V_{in}}{V_{out}}, \ L = \frac{V_{in} D}{f \cdot \Delta I_{pp}} \quad (1)$$

$$I_{max} = I_{in} + \frac{1}{2}\Delta I_{pp}, \qquad \Delta P_i = P_{out}\frac{1-\eta_i}{\eta} \qquad (2)$$

The magnetic material is defined in (B): relative permeability, μ_r, saturation flux density, B_{sat}, core fill factor, k_{cf}, core mass density, ρ_{cm}, the core power loss, P_c, is parameterised by Steimetz equation parameters: k, m, n. The saturation flux is de-rated by factor k_d in order to operate in the linear region of the materials B-H curve.

Block (C) introduces the core dimension variables: core width, a, window width, b, window length, c, and core thickness, d as shown in Fig. 2. Initial core dimensions must be set in the algorithm. The core dimension allows for further calculations as the core window ratio, r_w, core profile ratio, r_c, and the core shape ratio, r_s, as well as the core cross-sectional area, A_c, effective cross-sectional area, A_{ceff}, window area, A_w, and the area product, AP. The core mean magnetic path length, l_c, effective cross-section area, A_{ceff}, core volume, V_c, and mass, m_c, is calculated.

$$r_w = \frac{c}{b}, \quad r_c = \frac{d}{a}, \quad r_s = \frac{2a+c}{2a+b}, \quad A_c = a\cdot d, \quad A_w = b\cdot c \quad (3)$$

$$AP = A_c A_w, \quad l_c = 2(b+c)+\pi\cdot a, \quad A_{ceff} = k_{cf}\cdot A_c \quad (4)$$

$$V_c = l_c\cdot A_c, \qquad m_c = \rho_{cm}\cdot V_c \qquad (5)$$

The number of turns and air-gap is calculated in (D): The initial peak-to-peak flux swing, ΔB_{ppi}, is used to estimates initial turns, N_i. The absolute turns number equals the rounded up initial turns number and additional turns, N', from the design iteration loop. Initially N' equals zero. Then the total air-gap, l_g, is calculated. The air-gap calculation involves iteration in order to adjust the fringing field factor, F_g, which initially equals 1. Finally the adjusted ac-flux magnitude is obtained, B_{ac}, and the peak flux, B_{peak}.

$$\Delta B_{ppi} = \frac{\Delta I_{pp}}{I_{max}}B_{max}, \quad N_i = \frac{V_{in}D}{\Delta B_{ppi}A_{ceff}f}, \quad N = ceiling[N_i + N'] \quad (6)$$

$$l_g = \frac{\mu_0\cdot A_{ceff}\cdot N^2}{L}F_g - \frac{l_c}{\mu_c}, \quad F_g = 1 + \frac{l_g(j)}{\sqrt{A_{ceff}}}\ln\left(\frac{2\cdot c}{l_g(j)}\right) \quad (7)$$

$$B_{ac} = \frac{V_{in}D}{2NA_{ceff}f}, \quad B_{peak} = \frac{I_{max}}{I_{pp}}2B_{ac} \quad (8)$$

The core power loss is calculated in (E). The hysteresis and eddy current core specific power loss, P_{csp}, comes from the Steimetz equation while gap power loss is evaluated from the McLyman equation [9]. Where k_i is a core constant (i.e. 0.388 for CC core) and l_i is the lamination width.

$$P_{csp} = k\cdot f^m\cdot B_{ac}{}^n, \quad P_g = k_i\cdot l_i\cdot l_g\cdot f\cdot B_{ac}{}^2 \quad (9)$$

$$P_c = P_{csp}m_c + P_g \qquad (10)$$

Winding arrangements and geometry is defined in (F). The inputs are defined as follows: the air-gap factor, k_g, which sets the distance between the winding and the core leg, a_i, typically $k_g = 2$; the distance between the core and winding tips, c_i. The distance between the two adjacent windings on the CC core, d_w. The area for the winding, A_{wa}. The interlayer insulation thickness, t_i. The total insulation thickness, t_{itot}. The conductor fill factor, k_{con}. The conductor total cross-sectional area, A_{contot}. The single turn conductor cross-section area, A_{con}. The foil thickness, t_{con}, and foil total length, l_{con}. The conductor volume, V_w, and winding dimensions: winding thickness, t_w, winding outer width, w_w, winding outer thickness, l_w, winding inner width, w_{iw}, winding inner thickness, l_{iw}, winding height, h_w, as shown in Fig. 2.

$$a_i = 0.5\cdot k_g\cdot l_g, \quad A_{wa} = (b-2a_i-d_w)(c+0.5\cdot l_g-2c_i) \quad (11)$$

$$t_{itot} = N\cdot t_i, \quad k_{con} = 1 - \frac{t_{itot}}{b-2a_i-d_w}, \quad A_{contot} = k_{con}A_{wa} \quad (12)$$

$$A_{con} = \frac{A_{contot}}{N}, \quad t_{con} = \frac{A_{con}}{c+0.5\cdot l_g - 2\cdot h_i} \quad (13)$$

$$l_{con} = 2\{(a+d+4a_i+2t_{con})N+4(t_{con}+t_i)[1+0.5(N-2)](N-1)\} \quad (14)$$

$$V_w = l_{con}\cdot A_{con}, \quad t_w = 0.5(b-2a_i-d_w) \quad (15)$$

$$w_w = 2t_w+2a_i+a, \quad l_w = 2t_w+2a_i+d \quad (16)$$

$$w_{iw} = 2a_i+a, \quad l_{iw} = 2a_i+d, \quad h_w = c+0.5l_g-2c_i \quad (17)$$

Block (G) provides information about the conductor type and power loss in the winding. Typically the conductor of choice is copper; however, aluminium might be preferred in some applications [16]. The material input sets the material conductivity, σ_x, at the operating temperature, T_x, and the mass density, ρ_w, the conductor permeability, μ_w, and conductivity at the 20°C, σ_{20}, the temperature coefficient, k_t. Once the winding conductivity, σ_x, is known the winding resistance, R_{dc}, can be found. However, the current in the inductor contains an ac component. Therefore, the total power loss in the winding is calculated by decomposition of the triangular current waveform into the harmonics, I_{dc}, $I_{h=1}...I_{h=\infty}$ [18]. The h-th current harmonic, $I_h(h)$. The penetration depth, δ_h, for each harmonics is calculated followed by the stretch factor, η_h, and porosity factor, ξ_h. Using Dowell equations the proximity factor for each harmonic, F_h, is calculated [10][11][17]. Finally the effective proximity factor, F_{eff}, is obtained and the effective resistance, R_{eff}, calculated. Then the winding power loss, P_w, can be found.

$$m_w = \rho_w\cdot V_w, \quad \sigma_{max} = \frac{\sigma_{20}}{1+k_t(T_{max}-20)} \quad (18)$$

$$R_{dc\,max} = \frac{l_{con}}{\sigma_{max}A_{con}}, \quad I_h(h) = \Delta I_{pp}\frac{\sqrt{2[1-\cos(2\pi hD)]}}{\pi^2 h^2 D(1-D)} \quad (19)$$

$$I_{rms} = \sqrt{I_{dc}^2 + 0.5\sum_{h=1}^{\infty}I_h(h)}, \quad \delta_h(h) = \sqrt{\frac{1}{\pi\sigma_{max}\mu_0\mu_w hf}} \quad (20)$$

$$\eta = \frac{c+0.5l_g-2h_i}{0.5l_g+c}, \quad \xi_h(h) = \frac{t_{con}}{\delta_h(h)}\sqrt{\eta} \quad (21)$$

$$M = \xi_h(h)\frac{\sinh(2\xi_h(h))+\sin(2\xi_h(h))}{\cosh(2\xi_h(h))-\cos(2\xi_h(h))} \quad (22)$$

$$D = 2\xi_h(h)\frac{\sinh(\xi_h(h))-\sin(\xi_h(h))}{\cosh(\xi_h(h))+\cos(\xi_h(h))} \quad (23)$$

$$F_h(h) = M + \frac{(0.5N)^2-1}{3}D, \quad F_{eff} = \frac{I_{dc}^2+0.5\sum_{h=1}^{\infty}I_h^2(h)F_h(h)}{I_{rms}} \quad (24)$$

$$R_{eff\,max} = F_{eff} R_{dc\,max}, \quad P_w = R_{eff\,max} I_{rms}^2 \quad (25)$$

The final inductor properties are gathered in block (H). The inductor is characterized by its outer dimension: width, w_i, length, l_i, height, h_i, and followed by the inductor volume, V_i, and total mass, m_i. Finally the inductors total power loss, P_i, is calculated.

$$w_i = 2a + b + 2t_w + 2a_i, \quad l_i = 2a + c + 0.5l_g, \quad h_i = l_w \quad (26)$$

$$V_i = w_i l_i h_i, \quad m_i = m_c + m_w, \quad P_i = P_c + P_w \quad (27)$$

The last module, (I), estimates the component temperature rise. The algorithm can generate convection or conduction cooled designs. The natural convection cooling assumes that the heat removed from the object is 50 % by radiation and 50% by convection. The convective surface, SA, is calculated for the total of all surfaces exposed for heat exchange, P_i. The temperature rise, ΔT, can be approximated empirically as follows [10]:

$$SA = 2(2a + c + 0.5l_g)(2a + 3b + d) + 4(b + d)(a + b) \quad (28)$$

$$\Delta T = \left(\frac{P_i}{SA}\right)^{0.909} \quad (29)$$

The cold-plate cooling relies on conductive heat transfer as shown in Fig. 3. The heat path for the core and winding are assumed to be separate. The 2D heat transfer problem can be simplified by reduction to a 1D model. For simplicity, the core is assumed to be placed on the cold-plate such that the cold-plate makes contact with both sides of the core as shown in Fig. 4. The heat generated in the core, P_c, is removed along the core, l_c, with heat transfer surface, A_{hc}. The power loss generated due to fringing flux, P_g, is added as heat flowing through the core from the air-gap to the cold-plate. The core temperature rise is given as follows [15]:

$$\Delta T = \frac{P_c \cdot l_c}{2 \cdot k_c \cdot A_{hc}} + \frac{P_g \cdot l_c}{k_c \cdot A_{hc}} \quad (30)$$

The winding is modelled as an equivalent straight bar with a length equal to the average heat path, l_{eqCu}, from the winding hotspot to the cold-plate as shown in Fig. 6. The copper has a thermal conductivity, k_{Cu}. The total power loss generated in the copper, P_{Cu}, is transferred through the insulation layer, with thermal conductivity, k_i, between the winding and the cold-plate resulting in the following temperature rise:

$$\Delta T_{Cu} = \frac{P_{Cu} \cdot l_{eq\,Cu}}{2 \cdot k_{Cu} \cdot A_{Cu}} + \frac{P_{Cu} \cdot l_i}{k_i \cdot A_{Cu}} \quad (31)$$

Block (J) closes the first design loop in order to achieve the design constraints.

The iteration area product is stored in (K) indexed with the second loop iteration number.

The second loop, (L), iterates the turns by adding the extra turns to the design in order to decrease the flux swing. In power-limited designs local minima may occur as shown in Fig. 7. Once the absolute minimum is achieved the algorithm ends.

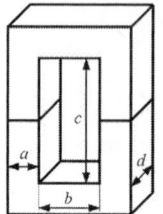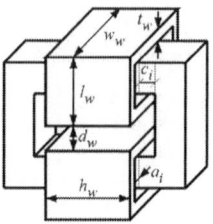

Fig. 2. Core and winding dimensions.

The second algorithm, A2, shares most of the modules with the first algorithm, A1. The electric circuit inputs in block (A) are the same. The material characteristics (B) include the magnetizing B-H curve and dc-bias μ-H curve defined as follows:

$$B(H) = \frac{H}{a_B H^2 + b_B H + c_B} \quad (32)$$

$$\mu_r(H) = \sqrt{\frac{\mu_{ir}^2 + a_\mu \mu_{ir}^3 H + b_\mu \mu_{ir}^4 H^2}{1 + c_\mu \mu_{ir} H + d_\mu \mu_{ir}^2 H^2}} \quad (33)$$

The inductors geometry (C) is defined in the same way in both algorithms. The block (D) in algorithm A2 is very different from the algorithm A1. In A1 the air-gap length is adjusted to the number of turns at μ = constant, while A2 is adjusting the number of turns to the dc-bias curve. The turns calculation is an iterative process as the cores permeability, $\mu_r(H)$, depends on the magnetizing force related to the number of turns, N, the inductors operating current, I_{in}, and the magnetic flux mean path length, l_c. The core flux bias, B_0, is calculated from the B-H curve. The peak-to- peak flux density, ΔB_{pp}, and its magnitude, B_{ac}, is found as well as the minimum and maximum flux density, B_{min} and B_{max}, respectively. The respective magnetizing force swing is found from the B-H curve.

$$N = \sqrt{\frac{L l_c}{A_{ceff} \mu_0 \mu_r}}, \quad H = \frac{I_{in} N}{l_c}, \quad \Delta B_{pp} = \frac{V_{in} D}{N A_{ceff} f} \quad (34)$$

$$B_{ac} = 0.5\Delta B_{pp}, \quad B_{min} = B_0 - B_{ac}, \quad B_{max} = B_0 + B_{ac} \quad (35)$$

$$H_{min/max} = \frac{-b_B - B_{min/max}^{-1} - \sqrt{(b_B - B_{min/max}^{-1})^2 - 4 a_B c_B}}{2 a_B} \quad (36)$$

The core power loss due to the air-gap does not apply in (E). Winding geometry (F) and power loss (G) are calculated similar as in A1. The inductor properties (H) are developed similar to A1, however, without the air-gap and core gap loss. The thermal calculations in (I) are carried out in the same way as in A1. Block (J) closes the design loop in order to achieve design constraints. Finally, in (K) the result is displayed and algorithm ends.

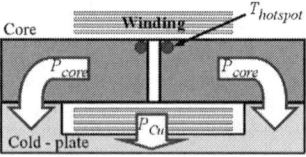

Fig. 3. Inductor mounted on the cold-plate.

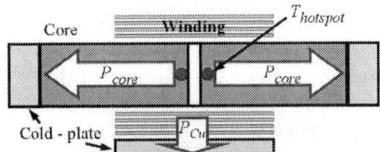

Fig. 4. Simplified core cooling model.

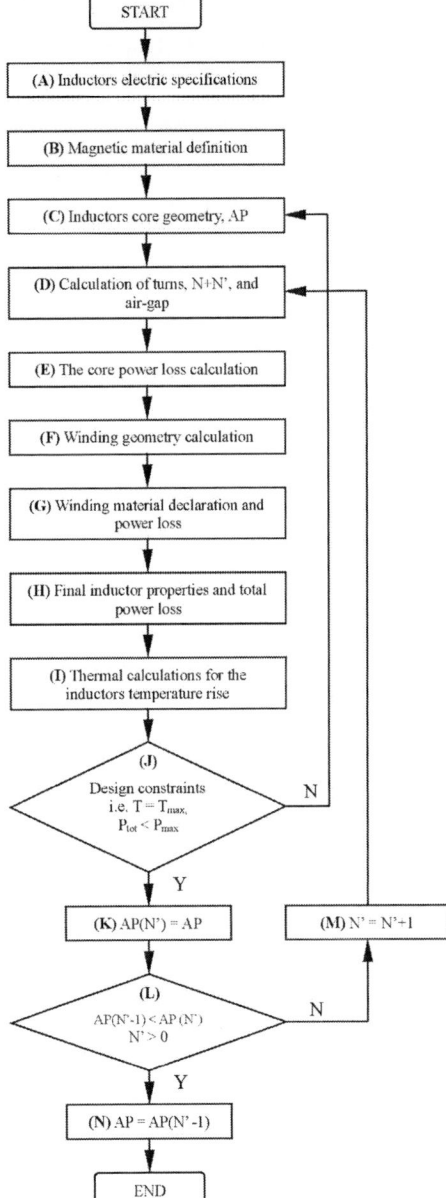

Fig. 5. Design algorithm for gapped inductors.

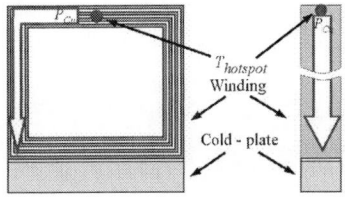

Fig. 6. Winding cooling 2-D model and its 1-D equivalent.

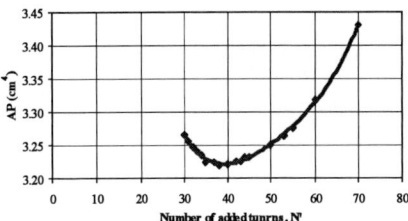

Fig. 7. Local area product minima (1 kW silicon steel at 50 kHz).

IV. CORE SIZE ANALYSIS - AREA PRODUCT

In this section the inductor size variation is investigated as a function of frequency for a constant inductance value.

The procedure presented is used to analyse an inductor for low-ripple 1 kW and 10 kW boost converters and the current ripple variation from 1 at 20 kHz to 0.13 at 150 kHz [13]. The following seven magnetic materials, from Table I are investigated: 2605SA1, 10JNHF600, Vitroperm 500F, 3C93, Kool Mu 60, MPP 60 and High Flux 60. The following are the electrical specifications for a boost converter: V_{in} = 200 V, V_{out} = 400 V, I_{in} = 50 A, L equals 1 mH and 100 µH for the 1 kW and 10 kW design, respectively. The switching frequency is increased from 20 to 150 kHz in the following steps: 20, 30, 40, 50, 75, 100, 125, 150 kHz. The ambient or cold-plate temperature is set at T_{amb} = 70 °C. The maximum temperature rise allowed for each material is dependent on the continuous operating temperature of the material shown in Table I.

The core profile ratio, r_c, is allowed to change in the range between 0.5 and 2, the core window ratio, r_w, is allowed to vary between 0.5 and 3. The air-gap is limited to 3 mm per leg with exception to the ferrite which is allowed an air-gap up to 6 mm per leg. The permeability drop of the powder materials is allowed to decrease down to 50% of its initial value [7]. The winding interlayer insulation is assumed as 0.1 mm and the distance between the core and winding tips is set to 4 mm, the while distance between windings is 2.4 mm. The algorithm has full freedom in seeking the optimum area product in the given range. This guarantees that the best performance of the each material will be developed. In some cases the significant unbalance between core and winding power loss may occur, as in the ferrite case, where almost all the power loss is generated in the winding. The results of the analyses are shown in Figs. 9-12 for convection and cold-plate conduction cooling of 1 kW and 10 kW inductor designs.

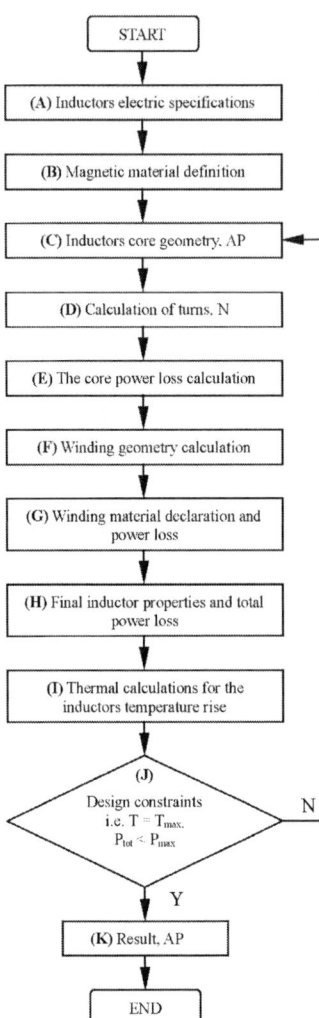

Fig. 8. Design algorithm for gapless materials.

Intuitively, the area product decreases as the frequency increases. This is due to the decrease in current ripple, ΔI_{pp}, as the frequency increases for a constant inductance.

For the natural convection cooling of the 1 kW inductors as shown in Fig. 9 almost all materials, except the silicon-steel, are saturation flux density limited. Thus, the area product is mainly driven by the material saturation flux density or the max flux density for the power materials. The silicon-steel is power limited at lower frequencies up to 60 kHz, and has the lowest area product at frequencies greater than 60 kHz. For lower frequency range, from 30 to 60 kHz, the silicon-steel is less effective than the amorphous metal and in the low range between 20 and 30 kHz the High Flux becomes the optimum solution. The ferrite, being saturation limited, results in a high area product.

A similar situation is shown in Fig. 10 for the 1 kW conduction cooled design as all materials are saturation flux limited. The cool-plate removes heat more efficiently than the convection cooling and results in decreased area product. However, the area product relation remains unchanged. The ferrite has the greatest size. Nanocrystalline core has three times lower area product, then amorphous material and the lowest is silicon-steel.

When maximum flux density limited, the MPP and High Flux materials achieve a lower area product than KoolMu. This is related to the dc-bias curve shown in Fig. 13. The permeability of the KoolMu drops at lower magnetizing force than the MPP and High Flux [7].

The powder materials are not analysed for conduction cooling. The flux lines stray near the surface of the core and will generate eddy currents in neighbouring conductive objects increasing the power loss. The low permeability powder materials are not recommended to operate in close or direct contact with the conductive surfaces, such as the metal heat sink surface [19].

The area product analysis for the 10 kW convection cooled design is shown in Fig. 11. Intuitively, the area product increases at elevated power. The amorphous metal and silicon steel are limited by specific power, which results in an area product that is significantly larger than the other considered materials. Nanocrystalline is specific power limited in the range 20 kHz to 125 kHz and then is limited by its saturation flux. The ferrite is saturation flux limited. The KoolMu is power limited in the range 20 to 40 kHz and then is limited by its maximum flux while the MPP and High Flux are power limited. The lowest area product for the convection cooled 10 kW inductor is nanocrystalline.

The conduction cooling of the 10 kW inductor is shown in Fig. 12. Due to the superior cooling of the cold-plate, the heat removed from the system ensures that all materials are flux limited. In fact in these designs the inductor's power loss is dominated by copper loss. The area product in this case is the balance between the heat capability of the winding and core window minimization. It is clear that if the design is flux limited then the inductor size will depend on the material saturation flux. As expected, the ferrite area product is the highest and then all the materials are generally in the order of the saturation flux density.

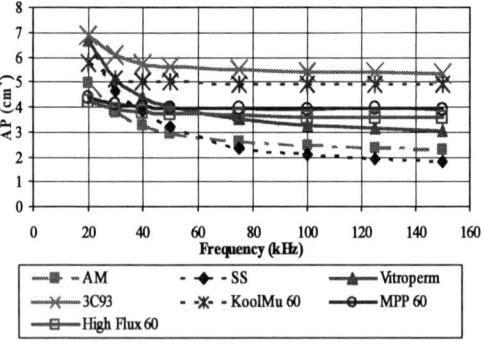

Fig. 9. Area product vs. freq. for constant inductance, natural convection cooling at 1 kW.

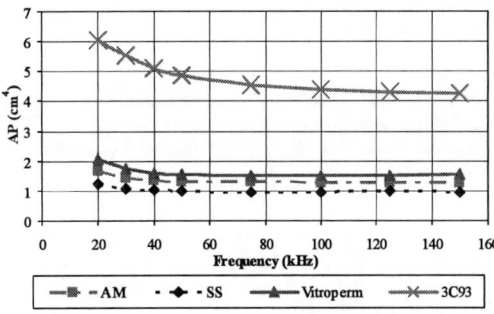

Fig. 10. Area product vs. freq. for constant inductance, cold-plate conduction cooling at 1 kW.

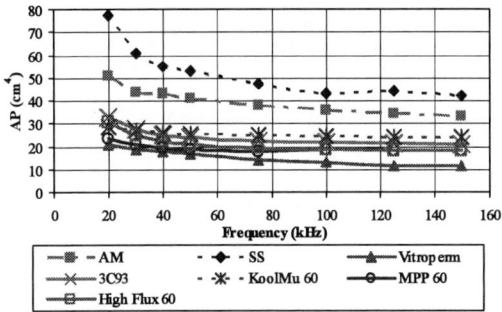

Fig. 11. Area product vs. freq. for constant inductance, natural convection cooling at 10 kW.

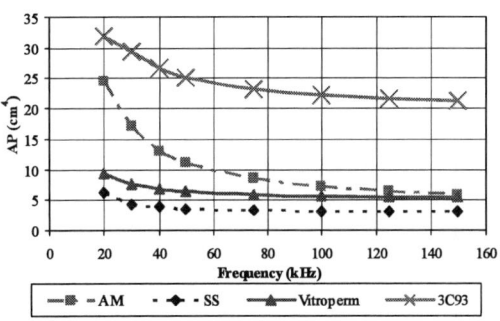

Fig. 12. Area product vs. freq. for constant inductance, cold-plate conduction cooling at 10 kW.

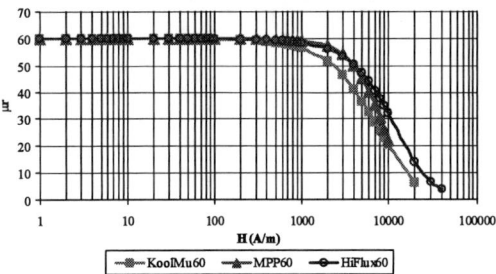

Fig. 13. Permeability vs. dc-bias curves.

V. EXPERIMENTAL VALIDATION

Three inductors are built to verify the above analysis. The amorphous and silicon steel materials are tested in a hard-switched boost converter. The single-cut amorphous core and silicon steel cores are almost identical. The multi-cut amorphous core has 33% higher profile and 35% wider window than the single-cut core. The converter specifications are $V_I = 120$ V, $V_H = 240$ V, $I_{DC} = 10$ A and $L = 80$ μH. The temperature rise is observed for three frequency points: 50 kHz, 100 kHz and 150 kHz. The single-cut amorphous and silicon steel inductors use the same foil winding and the inductor cores are almost identical. The only difference is the magnetic path length as the amorphous metal core is tape wound while the silicon steel core is composed of blocks. The multi-cut core inductor has a similar but taller foil winding. The system runs until the inductor reaches a steady state temperature. The estimated and experimental temperature rises of the cores decrease with increasing frequency. The multi-cut core has increased eddy-current losses due to surface shorts caused by cutting process and experimental results for the multi-cut amorphous metal are significantly higher than predicted. A better performance is demonstrated for the single-cut amorphous metal core.

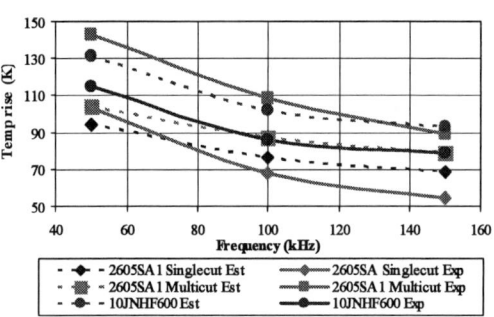

Fig. 14. Core temperature rise vs. frequency for constant inductance.

VI. CONCLUSIONS

In this paper, various magnetic materials are analysed and investigated for use in a practical design. The materials under investigation are iron-based amorphous metal, silicon steel, nanocrystalline, ferrite, and gap-less powder materials. A lumped parameter algorithm is derived which includes such effects as the foil ac copper loss, the gap core loss and the cooling path. The algorithm is implemented in EXCEL and generates material comparison over a range of frequencies, ripple ratios and cooling paths. Note that the simplified conduction cooled model may introduce a significant discrepancy in predicting the core hotspot if the design is core power loss limited.

The results show that material selection depends on the design requirement and the trade-offs between the saturation flux, specific power loss, thermal conductivity and associated anisotropy, weight, operating temperature and parameters stability under different temperature, frequency and bias condition. The core power loss limited inductor tends to be oversized while the minimum size is achieved for the design which is at the sweet-spot where the size is driven by the core power loss, winding power loss and core saturation limit. The conduction cooled designs shows that size is limited by the winding power loss and the saturation of the material. The saturation limited design can provide the most compact inductor for radiated cooling while the high-B_{sat} laminated materials are very competitive for conduction cooling as long the design is not core power loss limited

The presented results clearly show that for the given power level and ripple ratios the laminated cores are the optimum choice if conduction cooled. This situation might change at different power and ripple ratios.

REFERENCES

[1] B.J. Lyons, J.G. Hayes, M.G. Egan, "Magnetic material comparisons for high-current inductors in low-medium frequency dc-dc converters," *IEEE Applied Power Electronics Conference*, 2007, pp. 71-77.

[2] www.metglas.com

[3] www.jfe-steel.co.jp

[4] www.vacuumschmelze.de

[5] www.ferroxcube.com

[6] www.micrometals.com

[7] www.mag-inc.com

[8] J. Petzold, "Advantages of softmagnetic nanocrystaline materials for modern electronic applications," *Journal of Magnetism and Magnetic Materials*, 2002, pp. -.

[9] Colonel Wm. T. McLyman, *Transformer and Inductor Design Handbook*, 2nd ed., Marcel Dekker, Inc., 1988.

[10] Alex Van den Bossche, Vencislav Cekov Valchev, *Inductors and Transformers for Power Electronics*, CRC Press, 2005.

[11] Robert W. Erickson, *Fundamentals of Power Electronics*, Kluwer Academic Publishers, 2000.

[12] B.J. Lyons, J.G. Hayes, M.G. Egan, "Experimental Investigation of Iron-based Amorphous Metal and 6.5% Silicon Steel for high-current inductors in low-medium frequency dc-dc Converters," *IEEE Industrial Applications Conference*, 2007, pp. -.

[13] B.J. Lyons, J.G. Hayes, M.G. Egan, "Design algorithm for high-current gapped foil-wound inductors in low-to-medium frequency dc-dc converters," *IEEE Power Electronics Specialists Conference*, 2007, pp. 1760-1766.

[14] Sanjaya Maniktala, Switching Power Supply Design & Optimization, Mgraw-Hill, 2004.

[15] Michael J. Moran, Howard N. Shapiro, *Fundamentals of Engineering Thermodynamics*, VCH Wiley, 1988.

[16] Charles R. Sullivan, "Aluminium windings and other strategies for high frequency magnetics design in an era of high copper and energy costs," *IEEE Applied Power Electronics Conference*, 2007, pp. 78-84.

[17] Xi Nan, Charles R. Sullivan, "An improved calculation of proximity-effect loss in high-frequency windings of round conductors," *IEEE Power Electronics Specialist Conference*, 2003, pp. 853-860.

[18] Z. Cichowska, M. Pasko, *Wyklady z Elektrotechniki Teoretycznej*, 2nd ed., Wydawnictwo Politechniki Slaskiej, Gliwice 1998.

[19] T.E. Salem, D.P. Urciuoli, V. Lubomirsky, G.K. Overbo, "Design considerations for high power inductors in dc-dc converters," *IEEE Applied Power Electronics Conference*, 2007, pp. 1258-1263.

TABLE I

MAGNETIC MATERIAL PROPERTIES

Magnetic Material Type	Manufacturer	Material	Bsat [T]	Rel. Permeability (100C @100kHz)	Curie temp. [°C]	Cont. operating temperture [°C]	Thermal conductivity [W/(m K)]	Specific Heat Capacity [J/(K kg)]	Density [g/cm3]	Electrical Resistivity [Ω m]	Lamination Thickness [mm]	Core Fill Factor	Magnetostriction [ppm]	Core Loss @0.1T, 100kHz [W/kg]
Fe-Amorphous	Metglas	2605SA1	1.56	600	395	150*	10***	540	7.18	1.37	0.025	0.83	27	125
Silicon Steel	JFE	10JNHF600	1.88	600	700	150*	18.6***	536	7.53	0.82	0.1	0.9	0.1	233
Nanocrystalline	Vacuumschmelze	Vitroperm500F	1.2	13200	600	120*	10***	~500	7.3	1.15	0.023	0.7	0.5	10
Ferrite	Ferroxcube	3C93	0.35	1800	240	140	3.5-5	~750	4.8	5x10⁶	bulk	1	0.6	10
Powder Iron	Micrometals	Mix -26	1.38	72	-	<75**	4.2	TBD	7	TBD	bulk	1	TBD	657
Powder Core	Magnetics	KoolMu 26	1.05	26	500	200	80	TBD	6	TBD	bulk	1	TBD	263
Powder Core	Magnetics	KoolMu 60	1.05	60	500	200	80	TBD	6.8	TBD	bulk	1	TBD	136
Powder Core	Magnetics	KoolMu 125	1.05	125	500	200	80	TBD	7	TBD	bulk	1	TBD	132
Powder Core	Magnetics	MPP 26	0.75	26	460	200	80	TBD	7.5	TBD	bulk	1	TBD	80
Powder Core	Magnetics	MPP 60	0.75	60	460	200	80	TBD	8.2	TBD	bulk	1	TBD	45
Powder Core	Magnetics	MPP 125	0.75	125	460	200	80	TBD	8.7	TBD	bulk	1	TBD	57
Powder Core	Magnetics	High Flux 26	1.5	26	500	200	80	TBD	7	TBD	bulk	1	TBD	250
Powder Core	Magnetics	High Flux 60	1.5	60	500	200	80	TBD	7.7	TBD	bulk	1	TBD	169
Powder Core	Magnetics	High Flux 125	1.5	125	500	200	80	TBD	8.2	TBD	bulk	1	TBD	245

* Limited by lamination epoxy, ** Limited by thermal aging, *** Along laminations, TBD – to be determined

Feasibility Study of Half- and Full-Bridge Isolated DC/DC Converters in High-Voltage High-Power Applications

Dmitri Vinnikov[*], Tanel Jalakas[†] and Mikhail Egorov[††]

Tallinn University of Technology/Department of Electrical Drives and Power Electronics, Tallinn, Estonia
e-mail: dm.vin@mail.ee[*], tjalakas@yahoo.com[†], eg.mikh@mail.ru[††]

Abstract— This paper discusses the half- and full-bridge DC/DC converter topologies for high-power (200 kW) high-voltage (3.6 kV) applications. Focus is on the primary part of these topologies, i.e. the feasibility of replacement of two high-voltage IGBTs in the full-bridge by two high-voltage film capacitors in the half-bridge. The implementation of a half-bridge topology will lead to a sufficient simplification of a power scheme layout as well as control and protection algorithms. The full-bridge topology gives a clear advantage of the twofold reduced current of inverter switches, thus providing a possibility of implementation of smaller devices for the same transferred power as compared to a half-bridge.

Keywords— converter circuit, IGBT, high voltage power converters, voltage source converter (VSC).

I. INTRODUCTION

In half- and full-bridge DC/DC converter topologies primary inverter transistors in the off-state are subjected to a voltage stress equal to the DC input voltage and not to twice that as it is the case with the push-pull, single-ended and interleaved forward converter topologies. Since the latest improvements in the IGBT technology when high-voltage (HV) IGBTs (4.5 kV or 6.5 kV IGBTs) became available, these two topologies have gained interest even in high-power high-voltage applications. For instance, the latest auxiliary power supplies for the 3.0 kV DC trains with the direct grid connection are realized by the 2-level VSI half- or full-bridge inverter topology with HV IGBTs [1], [2], [3].

The difference between the full- and the half-bridge converter is that the latter replaces two of the inverter transistors by two capacitors and, hence, is somewhat more economical. In the case of conventional IGBTs (0.6 kV, 1.2 kV of 1.7 kV), with rapidly falling transistor prices, this advantage is not always achieved. But in converters with relatively new and costly 6.5 kV IGBTs [4], [5], [6], the issue of economic feasibility is raised again. This paper presents a feasibility study of 200 kW half-bridge and full-bridge DC/DC converters with the 6.5 kV IGBT modules. The items to be analyzed and evaluated are the volume of the primary inverter stack and component ratings, primary inverter losses, isolation transformer volume and losses, as well as the economic aspects of both topologies.

II. ANALYSIS OF HALF-AND FULL-BRIDGE TOPOLOGIES

Figure 1 shows a general arrangement of power sections for the full-bridge and the half-bridge converter topologies. As compared with the full-bridge, in the half-bridge topology switching transistors TT (Top Transistor) and TB (Bottom Transistor) form only one side of the bridge-connected circuit, the remaining half being formed by the two capacitors C1 and C2.

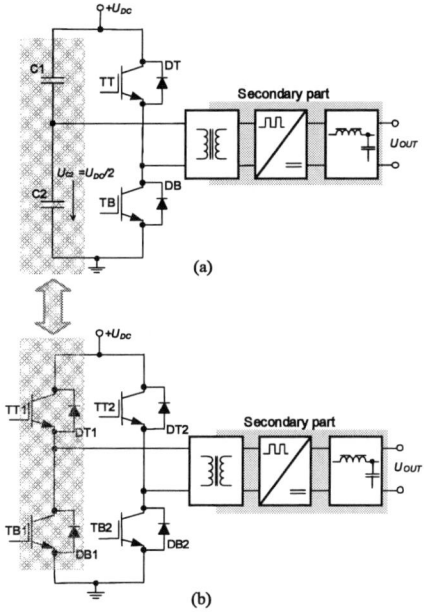

Fig. 1. Half-bridge (a) and full-bridge (b) DC/DC converter topologies.

In this paper the half-bridge and full-bridge converter topologies will be analyzed on the basis of the following assumptions:

1. the transistors used in primary inverters are to be selected from the INFINEON 6.5 kV IGBT family, the current ratings depend on the application;
2. output filter components are ideal (no power dissipation), their values as well as the output voltage values for both concurrent topologies are equal;
3. the turn ratio of the isolation transformers is 1:1;

4. nominal input DC voltage U_{DC} = 1.0 p.u. (3.6 kV, corresponds to the optimal operating voltage for the 6.5 kV IGBT);

5. maximum input DC voltage $U_{DC(max)}$ = 1.2 p.u. (permissible 20 % operating voltage deviation is considered here);

6. rated output power of converter P_{out} = 1.0 p.u. (200 kW).

The major difference between the half-bridge and full-bridge topologies is that the amplitude value of the isolation transformer supply voltage will see only half the supply voltage, and, hence the current in the primary winding and primary transistors' peak collector current $I_{Cpeak,HB}$ will be twice (1) that in the full-bridge topology $I_{Cpeak,FB}$ (2) for the same output power.

$$I_{Cpeak,HB} = \frac{2 \cdot P_{out}}{U_{DC}} = 2.0\,p.u. \qquad (1)$$

$$I_{Cpeak,FB} = \frac{P_{out}}{U_{DC}} = 1.0\,p.u. \qquad (2)$$

In the single-phase inverter, the maximum duty cycle of the switching transistors at the rated load and nominal input voltage will be 0.5 (50 %) [7]. Average current through the respective inverter switch in the half-bridge topology is

$$I_{Cav,HB} = \int_{0}^{0.5} I_{Cpeak,HB} \times dt = 1\,p.u. \qquad (3)$$

Average current through the respective inverter switch for the full-bridge topology is expressed as

$$I_{Cav,FB} = \int_{0}^{0.5} I_{Cpeak,FB} \times dt = 0.5\,p.u. \qquad (4)$$

The RMS current through the inverter switch in the half-bridge topology is

$$I_{Crms,HB} = \left[\int_{0}^{0.5} (I_{Cpeak,HB})^2 \times dt \right]^{\frac{1}{2}} = 1.414\,p.u. \qquad (5)$$

The RMS current through the inverter switch in the full-bridge topology is

$$I_{Crms,FB} = \left[\int_{0}^{0.5} (I_{Cpeak,FB})^2 \times dt \right]^{\frac{1}{2}} = 0.707\,p.u. \qquad (6)$$

The peak forward voltage through the inverter switch U_{peak} will be equal to 1.2 p.u. Thus, from the above definitions, the total inverter switch stress S_{HB} for the half-bridge topology will be

$$S_{HB} = \sum_{k=1}^{2} (I_{Crms,HB,k} \times U_{peak,k}) = 3.394\,p.u. \qquad (7)$$

The total inverter switch stress S_{FB} for the full-bridge topology will be

$$S_{FB} = \sum_{k=1}^{4} (I_{Crms,FB,k} \times U_{peak,k}) = 3.394\,p.u. \qquad (8)$$

The active switch utilization U_{HB} (converter output power obtained per unit of active switch stress) for the half-bridge topology will be

$$U_{HB} = \frac{P_{out}}{S_{HB}} = 0.295. \qquad (9)$$

The active switch utilization U_{FB} for the full-bridge topology will be

$$U_{FB} = \frac{P_{out}}{S_{FB}} = 0.295. \qquad (10)$$

Basic key points resulting from the half- and full-bridge topology analysis are submitted in Table I, while specific advantages are designated in ***Bold Italics***.

TABLE I.
SIDE-BY-SIDE COMPARISON OF HALF-AND FULL-BRIDGE
INVERTER TOPOLOGIES

	Half-bridge topology	Full-Bridge topology
Input voltage U_{DC} (p.u.)	1.0	1.0
Output power P_{out} (p.u.)	1.0	1.0
Peak collector current I_{Cpeak} (p.u.)	2.0	*1.0*
Average collector current I_{Cav} (p.u.)	1.0	*0.5*
RMS collector current I_{Crms} (p.u.)	1.414	*0.707*
Total inverter switch stress S (p.u.)	3.394	3.394
Active switch utilization U	0.295	0.295

The bottom line is that the full-bridge topology in comparison with the half-bridge exhibits an advantage of lower inverter switches ratings for the same transferred power. It directly influences the minimization of total power switch losses (e.g., minimization of steady-state losses of IGBT and free-wheeling diode) and means smaller devices to be used.

III. SIMULATION OF HALF-AND FULL-BRIDGE TOPOLOGIES

For the visual evaluation of waveforms, generalized models of the half- and full-bridge inverter topologies were developed by the help of SimPower Systems blockset of Matlab Simulink (Fig. 2). The values presented in Section II were assumed for the simulations.

If the Top and Bottom transistors in the half- and full-bridge inverter topologies are simultaneously switched on, the dead short circuit across the DC supply bus will take place and power transistors would fail. To ensure against this, the transistor on-state time t_{on}, which occurs at nominal input voltage U_{DC}, will be chosen as 90% of a half-period:

$$t_{on} = 0.9\left(\frac{T_{sw}}{2}\right), \qquad (11)$$

where T_{sw} is the switching period of the inverter.

The secondary parts of converters are represented by the equivalent load resistances, the values of which can be obtained from (12).

$$R_{ekv} = \frac{U_{TRrms}^2}{P_{out}}, \qquad (12)$$

where U_{TRrms} is the rms voltage of the isolation transformer and P_{out} is the converter output power.

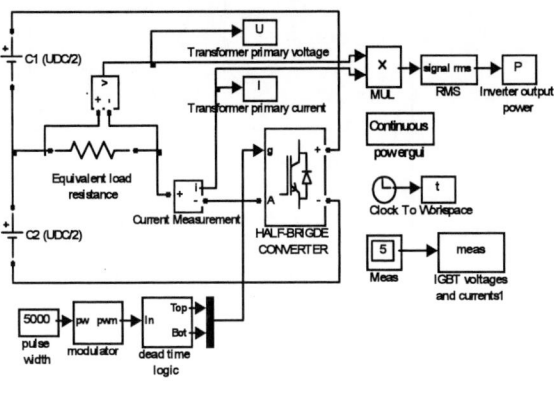

(a)

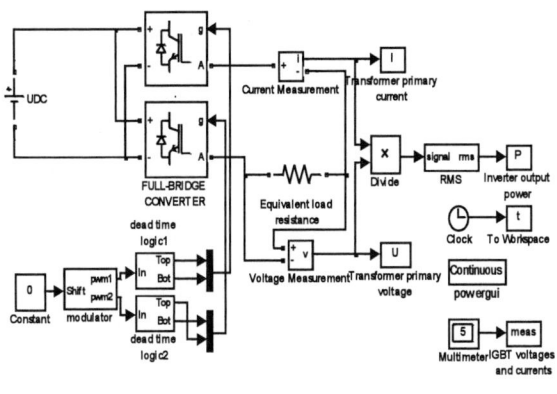

(b)

Fig. 2. Generalized half-bridge (a) and full-bridge (b) inverter models created in Matlab Simulink.

The rms voltage of the isolation transformer can be defined by (13).

$$U_{TRrms} = U_{TR} \sqrt{\frac{t_{on}}{T_{sw}/2}} = U_{TR} \sqrt{0.9}, \qquad (13)$$

where U_{TR} is the amplitude value of the isolation transformer voltage. The values of equivalent load resistances used in simulations were 58.32 Ω and 14.58 Ω for the full-bridge and half-bridge inverters, respectively.

The simulated waveforms of half- and full-bridge converters for the same operation point are presented and compared in Figs. 3 and 4.

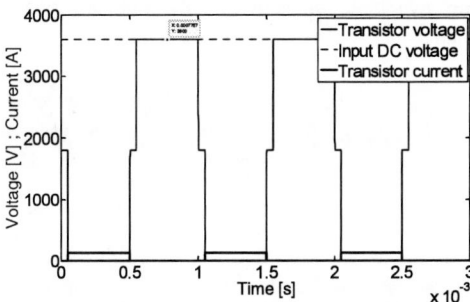

Fig. 3. General waveforms of half- and full-bridge inverters.

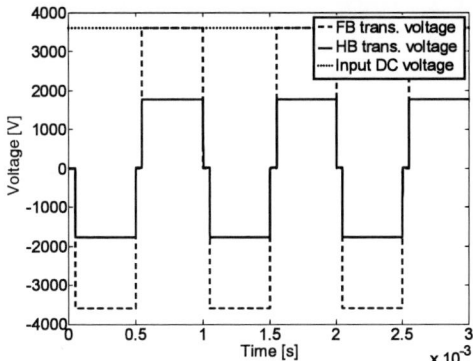

(a)

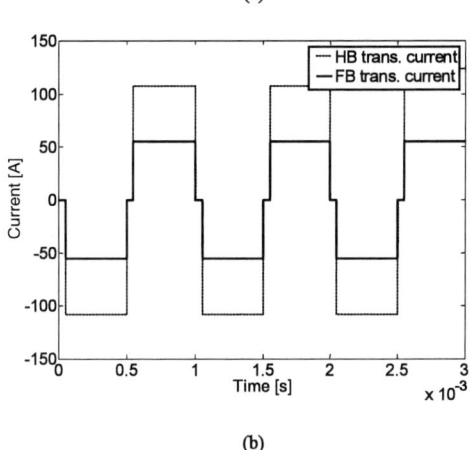

(b)

Fig. 4. Comparison of simulation results for isolation transformer supply voltages (a) and currents (b) for the half- and full-bridge converters.

IV. POWER SWITCH SELECTION

As the simulations show, the switch rms current in the full-bridge configuration is two times lower than that of the half-bridge (52 A and 104 A, respectively). An essential issue to be taken into account during the converter design with HV IGBTs is the thermal limitations. The thermal limits of IGBTs are derived by incorporating the thermal resistance Z_{thJC} and the given values for the case and maximum junction temperature (80 ^{0}C and 125 ^{0}C, respectively).

The upper switching frequency f_{up} is determined by the thermal management system of the IGBT module:

$$f_{up} = \frac{\dfrac{T_{JC}}{Z_{thJC}} - P_{cond}}{E_{on} + E_{off}}, \quad (14)$$

where P_{cond} is on-state power losses and E is the energy loss. Fig. 5 shows the maximum achievable rms current values for different switching frequencies obtained as a result of the analysis of the dynamical properties of the smallest representative from the 6.5 kV IGBT family, the 200A/6.5kV IGBT module. Operating with the different collector current values, the given IGBT module in the full-bridge configuration could be operated with more than two times higher switching frequency than in a half-bridge (1.8 kHz and 0.8 kHz, respectively).

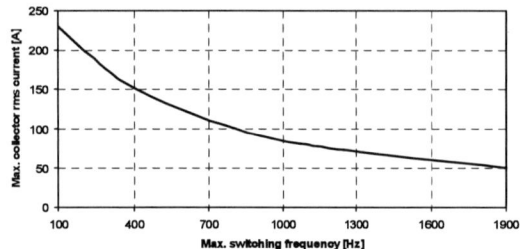

Fig. 5. Switching frequency and operation current dependence imposed by thermal limitations for 200 A 6.5 kV IGBT module.

In the case of half-bridge converter with the double operating current of primary switches, the switching frequency can be effectively increased by the help of 400A/6.5kV IGBT modules. The thermal handling capability of the 400A/6.5kV IGBTs is better than that of their counterparts with lower current ratings, thus allowing for 1.8 kHz switching frequency use in the case of the half-bridge converter (see Fig. 6).

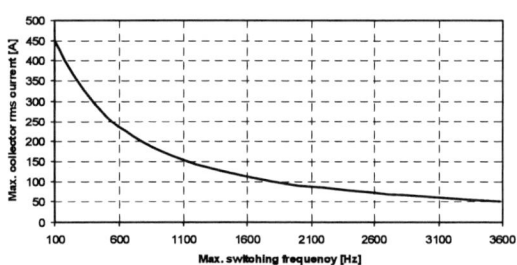

Fig. 6. Switching frequency and operation current dependence imposed by thermal limitations for 400 A 6.5 kV IGBT module.

To simplify further discussions it is assumed that 200A/6.5kV IGBT module FZ200R65KF1 and 400A/6.5kV IGBT module FZ400R65KF1 are used for the full- and half-bridge inverter topologies, respectively. Such combinations ensure the proper operation of both topologies with the switching frequency of 1.8 kHz and at the rated load conditions (200 kW). The general specifications of the discussed IGBT modules are presented in Table II.

TABLE II.
DATASHEET VALUES OF DIFFERENT IGBT MODULES USED IN THE EVALUATION

Properties of IGBT modules involved in design	Half-Bridge Inverter 2 x 400A/6.5kV IGBTs *FZ400R65KF1*	Full-Bridge Inverter 4 x 200A/6.5kV IGBTs *FZ200R65KF1*
Collector-emitter saturation voltage	5.3 V	5.3 V
Turn-on energy loss per pulse	4000 mJ	1900 mJ
Turn-off energy loss per pulse	2300 mJ	1200 mJ
Reverse recovery energy of FWD	1050 mJ	550 mJ
Therm. resistance, junction to case, IGBT	0.017 K/W	0.033 K/W
Therm. resistance, junction to case, FWD	0.032 K/W	0.063 K/W
Therm. resistance, case to heatsink	0.008 K/W	0.016 K/W
Baseplate dimensions, L x W	140 x 130 mm	140 x 73 mm

The price of the device to be designed is one of the essential aspects from the designer's point of view. IGBT is one of the main components of such devices and their prices have a sufficient impact on the final price and, hence, on the overall competitiveness of the device produced. It is obvious that the IGBT module with a lower current rating is less expensive than its more powerful counterpart (see Fig. 7). But in the inverter configurations the total price for the half-bridge topology with more powerful 400A modules will be about 20% lower than that of the 200A IGBT-based full-bridge inverter. It should be noted, that our evaluation covered only the semiconductor costs. In real design this cost difference will be further increased by the considering of costs for the gate drivers and other associated components necessary for the operation of high-voltage IGBT modules.

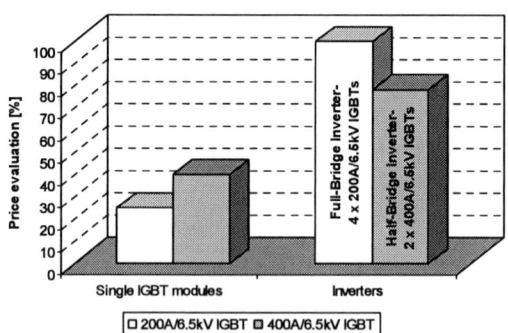

Fig. 7. Price evaluation for different inverter topologies based on different HV IGBT types.

Technical feasibility is another challenge for a designer. Fig. 8 features the requirements of the minimum installation surface area for different investigated topologies. The 200A/6.5kV IGBTs have smaller baseplate areas in comparison with 400A switches. Hence, the half-bridge configuration of 400A IGBTs requires smaller total installation surfaces than the 200A/6.5kV IGBT-based full-bridge inverter.

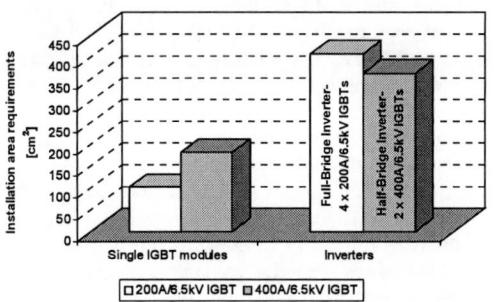

Fig. 8. Comparison of installation area requirements for different high-voltage IGBT types and inverter assemblies.

V. ANALYSIS OF INVERTER POWER DISSIPATION

The two times lower inverter switch ratings for the full-bridge topology will result in the minimization of per transistor power switch losses for the same transferred power for the half-bridge (Fig. 9). But the increased quantity of inverter switches of the full-bridge configuration (from 2 to 4), despite lower dissipation of each switch, leads to the same total inverter losses as in the half-bridge inverter topology.

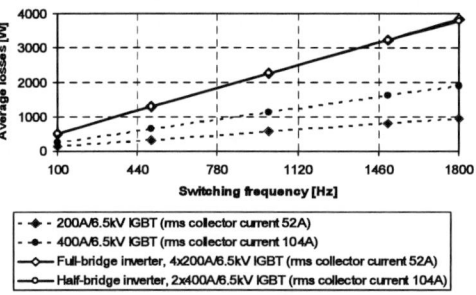

- ◆ - 200A/6.5kV IGBT (rms collector current 52A)
- ● - 400A/6.5kV IGBT (rms collector current 104A)
─◇─ Full-bridge inverter, 4x200A/6.5kV IGBT (rms collector current 52A)
─○─ Half-bridge inverter, 2x400A/6.5kV IGBT (rms collector current 104A)

Fig. 9. Power dissipation as a function of switching frequency for the 200 kW half- and full-bridge topologies.

Figure 10 shows the total losses breakdown for the 200 kW half- and full-bridge inverters and for the single switches involved in the design. The losses of both topologies were evaluated at their maximum possible switching frequency (1.8 kHz) and at the rated load (200 kW).

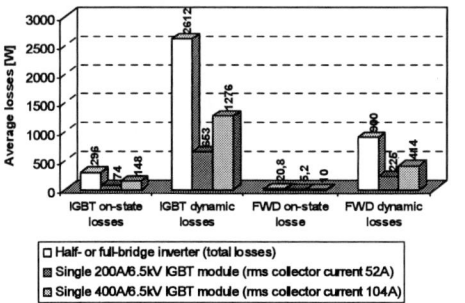

☐ Half- or full-bridge inverter (total losses)
▨ Single 200A/6.5kV IGBT module (rms collector current 52A)
▨ Single 400A/6.5kV IGBT module (rms collector current 104A)

Fig. 10. Total losses breakdown for single switches and inverter stacks of different configurations.

As it is seen from Figs. 9 and 10, the single 200A switch in the full-bridge configuration has more than 30% reduced power dissipation than the 400A switch in the half-bridge configuration. Despite the equal total power losses of both inverter configurations the full-bridge with its smaller per-switch dissipation provides an effect of "distributed power losses". By the incorporating of distributed heatsinks (for each transistor separately) it could contribute to improved converter reliability. The price for the distributed heatsinks could be also reduced, because in the case of half-bridge topology the total power loss is distributed between the two switches, thus requiring more powerful heatsinks than in full-bridge with its decentralized heat sources to ensure proper temperature mode of each switch.

VI. STUDY OF OVERALL FEASIBILITY

The half-bridge topology replaces two of the inverter switches by two capacitors and in the case of high-voltage IGBTs could provide some economic benefits. Hence, the high-voltage capacitors will occupy a larger volume than the two transistors they replace, thus reducing the half-bridge advantage.

The values of the half-bridge capacitors are calculated from the known primary current and operating frequency. The voltage change across the capacitors ($C1=C2=C$) and for the half-period ($T_{sw}/2$) is then [8]:

$$\Delta U_C = \frac{I_{in} \cdot \Delta t}{C_{tot.}} = \left[\frac{P_{out}}{(U_{DC}/2) \cdot (C1+C2)} \right] \cdot \left(\frac{1}{2 \cdot f_{sw}} \right) = \frac{P_{out}}{2 \cdot U_{DC} \cdot f_{sw} \cdot C}$$

(12)

Figure 11 gives an overview of the minimal capacitance required from $C1$ and $C2$ and the corresponding input capacitor volumes for a 200 kW half-bridge inverter estimated for different switching frequencies and for different capacitor voltage ripple factors. It must be stated here that the implementation of the half-bridge topology instead of the full-bridge leads to sufficiently increased installation volume requirements, especially at low operating frequencies (up to 500 Hz), see Fig. 12. Thus, considering the weight-volume constraints, the full-bridge topology outpaces the half-bridge in this application.

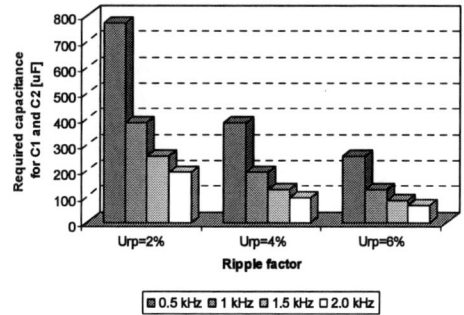

▨ 0.5 kHz ▨ 1 kHz ▨ 1.5 kHz ☐ 2.0 kHz

Fig. 11. Evaluation of half-bridge capacitor capacitances for the switching frequency range of 0.5...2.0 kHz and for different voltage ripple factors.

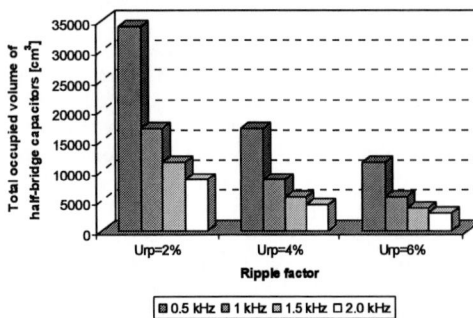

Fig. 12. Evaluation of total occupied volumes of half-bridge capacitors for the switching frequency range of 0.5...2.0 kHz and for different voltage ripple factors.

Meanwhile, the comparison of the prices for the 6.5 kV/200 A IGBTs and the replacement price for the high-voltage film capacitors shows that the half-bridge design has a real advantage for all the switching frequency ranges and capacitor voltage ripples. Fig. 13 shows that half-bridge capacitor use even for low switching frequencies and the low ripple factors provide a benefit of over 40 % as compared to the two extra transistors in the full-bridge design.

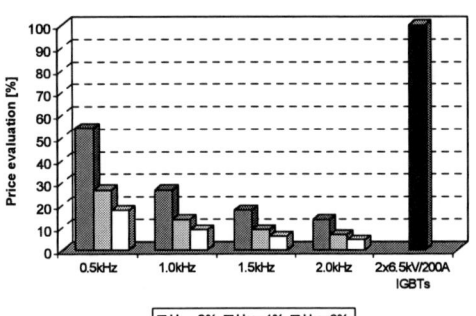

Fig. 13. Price comparison for the half-bridge capacitors and two additional IGBTs for the full-bridge design.

Moreover, the high price of the HV IGBTs (mainly due to their relative novelty and high manufacturing costs) has a substantial impact on the final converter price and, hence, on the overall competitiveness of the produced device (see Fig. 14).

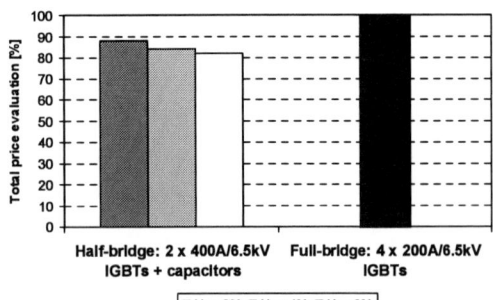

Fig. 14. Final price comparison for the half-bridge topology (with the capacitors calculated for the different voltage ripple factors) and the full-bridge topology.

VII. CONCLUSIONS

This paper has analyzed the half- and full-bridge DC/DC converter topologies for high-power (200 kW) high-voltage (3.6 kV) applications. Focus was on the primary part of these topologies, i.e. the feasibility of the replacement of two high-voltage IGBTs in the full-bridge by two high-voltage film capacitors in the half-bridge. The bottom line is that the topology must be selected in accordance with the custom design requirements and must be precisely analyzed to achieve a better price-weight-efficiency-reliability ratio. Thus, the half-bridge topology gives such benefits as simple construction, reduced complexity of control and protection circuits, and, of course, reduced converter price. The full-bridge topology provides more compact and lightweight design that would make it possible to develop a converter with a higher power density.

ACKNOWLEDGMENT

Authors thank Estonian Science Foundation (Grant No. 7425) for financial support of this study.

REFERENCES

[1] Eckel, H.-G.; Bakran, M. M.; Krafft, E. U.; Nagel, A.: *A new family of modular IGBT converters for traction applications*. Proc. of 2005 European Conference on Power Electronics and Applications. P.: 10 pp.

[2] Trivedi, M.; Shenai, K.; Joerg, P.: *Evaluation of high-voltage IGBT for series resonant DC-DC converter application*. Proc. of Sixteenth Annual IEEE Applied Power Electronics Conference and Exposition, APEC 2001. Vol. 2, pp.:1237 - 1241.

[3] Vinnikov, D.; Laugis, J.; Jalakas, T.: *Development of Auxiliary Power Supplies for the 3.0 kV DC Rolling Stock*. Proc. of IEEE International Symposium on Industrial Electronics, ISIE 2007. Pp.: 359 - 364.

[4] Kopta, A.; Rahimo, M.; Schlabach, U.; Schneider, D.; Carroll, E.; Linder, S.: *A 6.5 kV IGBT module with very high safe operating area*. Proc. of Fourtieth IAS Annual Meeting and Industry Applications Conference, 2005. Vol. 2, pp.: 794 - 798.

[5] Bauer, J. G.; Schilling, O.; Schaeffer, C.; Hille, F.: *Investigations on the ruggedness limit of 6.5 kV IGBT*. Proc. of 17th International Symposium on Power Semiconductor Devices and ICs, 2005. Pp.: 71 - 74.

[6] Mochizuki, K.; Suekawa, E.; Iura, S.; Satoh, K.: *Development of 6.5 kV class IGBT with wide safety operation area*. Proc. of Power Conversion Conference, 2002. Vol. 1, pp.: 248 - 252.

[7] Mohan, N., Undeland, T., M, Robbins, W., P.: *Power electronics: converters, applications, and design*. 3rd edition. Hoboken (N.J.): Wiley, 2003.

[8] K. H. Billings: *Switchmode Power Supply Handbook*. McGraw-Hill, 1989.

Evaluation of Different Loss Calculation Methods for High-voltage IGBT-s Under Small Load Conditions

T. Jalakas, D. Vinnikov, and J. Laugis

Tallinn University of Technology, Ehitajate tee 5, 19086 Tallinn, *Estonia*, tjalakas@yahoo.com

Abstract—This paper focuses on the evaluation of the calculated and measured values of high-voltage 6.5 kV IGBT module losses, used in mid-power applications under low load conditions. The values of the losses were calculated by use of widely known equations and **IPOSIM 6.0 software.** The experimental setup was based on a high-voltage half-bridge isolated DC/DC converter. Comparison of calculation and experimental results shows that the dynamic losses on higher switching frequencies are smaller than those calculated and that the IPOSIM software is relatively accurate. This knowledge helps developers to avoid overestimation of the losses and allows us to construct more compact power electronics applications with much higher switching frequencies.

Keywords— DC-DC power conversion, Insulated gate bipolar transistors, pulse width modulated power converters.

I. INTRODUCTION

If the IGBTs with high voltage blocking capability are used, high power transistors are required in mid-power applications, under the conditions of IGBT loading of 10-20% of the nominal. It means that in practical applications a higher switching frequency or a simpler cooling system can be used, for example, passive cooling instead of liquid- or forced air cooling. To find out the limits of the IGBTs under such operating conditions there is a need to estimate the possible loss of power. For that purpose, different methods can be used (rough estimation for general investigation, simulation for more precise results and the experimental approach for final control). Although the experimental converter uses two transistors in the half-bridge configuration, this study covers the estimation of losses of one IGBT.

The losses of a typical IGBT transistor consist of turn-on, on-state, turn-off and off-state losses. Some additional losses are created by the reverse recovery charge in the built-in freewheeling diode of the IGBT module. The main limiting factors for semiconductor devices are the conducting, switching and recovery losses, with the resulting temperature rise on the junction. The switching and recovery losses depend on the switching frequency; furthermore, the switching frequency is one of the limiting factors. At higher switching frequencies the dynamic losses can be much higher than the static ones. To produce more compact and cheaper power electronics devices with higher power density, higher switching frequency and

smaller passive components are the key elements. Experiments with high voltage IGBT transistors have shown that in low load conditions it is possible to reach switching frequencies three times higher than the maximum nominal and still be within the limits of the given temperature and loss. The switching-on, conduction, switching-off and recovery losses for an integrated freewheeling diode can be calculated using a standard set of the equations.

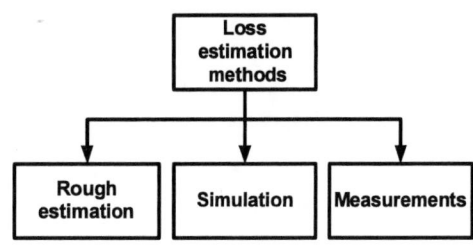

Fig. 1. Estimation of losses

It is also possible to use some specialized software packages. The software of company Infineon proposed for choosing high-voltage IGBT transistors for specific applications and for parameter estimation is IPOSIM. It allows us to compare different methods for the estimation of IGBT losses (simplified equations, simulation by IPOSIM, measured experimentally).

II. SUBJECT OF INVESTIGATION

The subjects of the study are the FZ200R65KF1 6.5kV IGBT modules produced by the company Infineon. The modules have an internal freewheeling diode. The maximum voltage for those IGBT transistors is 6.5 kV and the nominal current is 200A. The modules are intended to be used in mid-power applications with the nominal operating voltage up to 4.3 kV. The 1SC210F2 driver circuits for IGBTs are supplied by the CT-Concept Technology Ltd. The used IGBT is the punchtrough type, with smaller losses than with the non-punchtrough type IGBTS [1]. The initial values for calculations, are shown in Table I.

The experimental setup consisted of a 40 kW high-voltage isolated DC/DC converter with half-bridge input inverter topology. The tests were carried out in the switching frequency range of 1 kHz to 2.5 kHz with a 500 Hz step.

978-1-4244-1741-4/08/$25.00 ©2008 IEEE

TABLE I.
INITIAL VALUES FOR CALCULATIONS AND EXPERIMENTS

Parameter	Value
Input voltage U_{in} (V)	3000
Switching frequency f_{sw} (Hz)	1.0…2.5
Inverter power P (kW)	33
Inverter switch duty cycle D	0.65
Ambient temperature T_{amb} (^{0}C)	25

The specific feature of the half-bridge inverter topology is that the amplitude value of the isolation transformer supply voltage will be only half the supply voltage, and, hence the current in the primary winding and the peak collector current $I_{Cpeak,HB}$ of the primary transistors will be

$$I_{C(amp)} = \frac{2 \cdot P}{U_{in}} = \frac{2 \cdot 33 \cdot 10^3}{3000} = 22 \,(\text{A}). \qquad (1)$$

The average value of current through the inverter switch will be

$$I_{C(av)} = I_{C(amp)} \cdot D = 14.08 \,(\text{A}). \qquad (2)$$

The RMS value of the inverter switch current will be

$$I_{C(rms)} = I_{C(amp)} \cdot \sqrt{D} = 17.74 \,(\text{A}). \qquad (3)$$

III. ROUGH LOSS ESTIMATION

During the early stages of converter design for the initial identification of power losses and heatsink requirements designers often use a simplified or rough loss estimation method. Fig. 2 shows a survey of the possible single power dissipations during switch operation used in rough loss estimation.

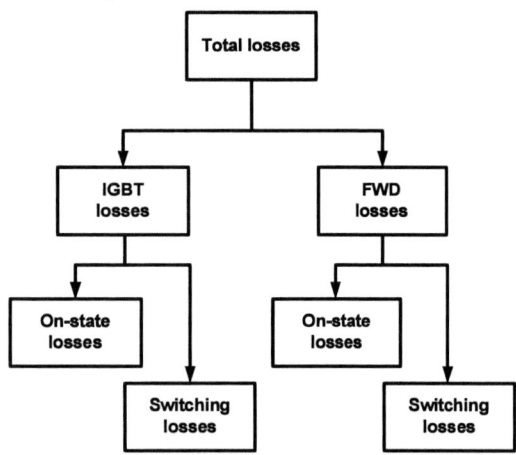

Fig. 2. Single power losses of power module switches

On-state power dissipation of IGBT depends on the load current, junction temperature, and duty cycle and can be expressed as

$$P_{on-state\_IGBT} = \frac{1}{T} \cdot \int_0^t U_{ce\_sat\_IGBT}(t) \cdot I_C(t) dt \,. \qquad (4)$$

For the given driver parameters, the turn-on and turn-off power dissipations (switching losses) are dependent on the load current, DC-link voltage, junction temperature, and switching frequency and canbe expressed as

$$P_{sw\_IGBT} = f_{sw} \cdot (E_{on\_IGBT} + E_{off\_IGBT}), \qquad (5)$$

where E_{on_IGBT} and E_{off_IGBT} are the switch-on and switch-off energies of IGBT, respectively and f_{sw} is the switching frequency.

Some additional losses are created in the integrated freewheeling diode (FWD) of the IGBT module. On-state power dissipation of the FWD depends on the load current, junction temperature, and the duty cycles and can be estimated as

$$P_{on-state\_FWD} = \frac{1}{T} \cdot \int_0^t U_{FWD}(t) \cdot I_F(t) dt \,. \qquad (6)$$

For the given driver parameters of the IGBT commutating with the diode, switching losses (turn-off power dissipation) of the FWD are dependent on the load current, DC-link voltage, junction temperature, and switching frequency is

$$P_{sw\_FWD} = f_{sw} \cdot E_{rec\_FWD}, \qquad (7)$$

where E_{rec_FWD} is the freewheeling diode reverse recovery energy [2] [3].

Typically, the values necessary for the loss estimation are provided in transistor datasheets. These data could be in the form of typical values for the nominal operation of the IGBT module (measured for the nominal collector current and collector emitter voltage, etc.) or it could be found from the detailed diagrams for any operation point of the module (Figs. 3…5).

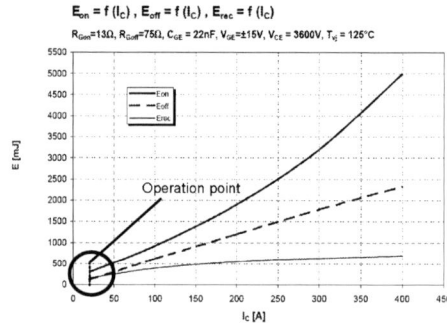

Fig. 3. Switching loss energies of the 6.5 kV 200A IGBT module Z200R65KF1 as a function of the collector RMS current [4].

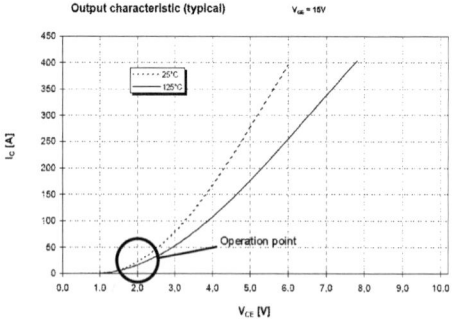

Fig. 4. Collector current of the 6.5 kV 200A IGBT module (Z200R65KF1) as a function of the collector-emitter saturation voltage [4].

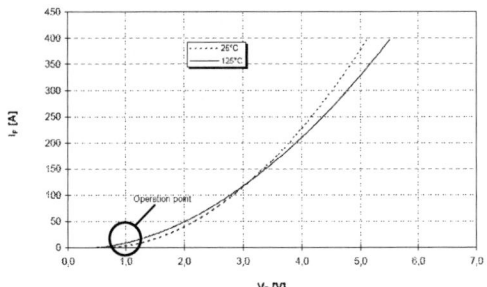

Fig. 5. Forward characteristics of the freewheeling diode [4].

Typical values and values found from the diagrams for the specific operation point of 200/6.5kV IGBT module are compared in Table II. It is remarkable that at typical values, the losses of the IGBT module could be easily overestimated.

TABLE II.
DATASHEET VALUES FOR 200 A 6.5 KV IGBT MODULE

	Transistor			FWD diode	
	E_{on} (mJ)	E_{off} (mJ)	U_{ce_sat} (V)	U_{F_FWD} (V)	E_{rec} (mJ)
Typical values U_{CE}=3.6 kV, $I_{C(rms)}$=200 A, T_{vj}=125 °C	1900	1200	5.3	3.8	550
From diagrams U_{CE}=3.0 kV, I_{C}=18 A, T_{vj}=125 °C	300	200	2.0	1.3	200

The IGBT module losses calculated by Eqs. (4 to 7) for the operation point of half-bridge converter discussed above are presented in Fig. 6.

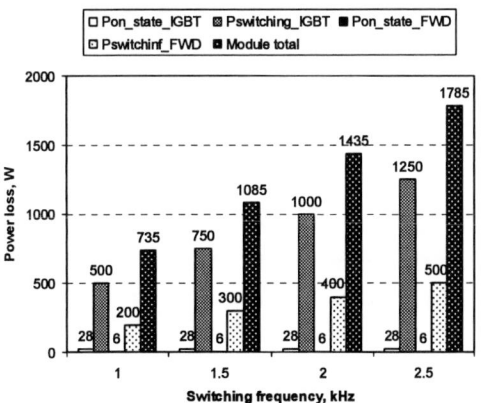

Fig. 6. IGBT and FWD losses found by the rough loss estimation method

IV. LOSS SIMULATION BY IPOSIM

To simulate the 6.5 kV IGBT module power loss, *Iposim* 6.0 software was used. *Iposim* is a Microsoft Excel based set of macros intended for quick preselection of the suited IGBT transistors for specific applications. *Iposim* was developed by Eupec Company. The average

total power dissipation of IGBT and FWD was calculated by a closed solution approach for conduction losses P_{cond} and switching losses P_{sw} during the course of one period T_0. The closed solution for P_{sw} does not require summing up several switching energies of the switched device [5]:

$$P_{loss\_mod} = P_{mod\_IGBT} + P_{mod\_FWD} = P_{on-state\_IGBT} + P_{sw\_IGBT} + \\ + P_{on-state\_FWD} + P_{sw\_FWD}, \tag{8}$$

where P_{loss_mod} is the IGBT module loss power, P_{mod_IGBT} is the IGBT total loss power, P_{mod_FWD} is the total loss power of the integrated FWD, $P_{on-state\_IGBT}$ is the IGBT conduction loss power, P_{sw_IGBT} is the IGBT switching loss power, $P_{on-state\_FWD}$ is the FWD conduction loss power, and P_{sw_FWD} is the FWD switching loss power. The conduction losses of an IGBT are expressed as follows:

$$P_{on-state\_IGBT} = \frac{1}{2}(U_{CE0} \cdot \frac{I_C}{\pi} + r \cdot \frac{I_C^2}{4}) + m \cdot \cos\varphi \times \\ \times (U_{CE0} \cdot \frac{I_C}{8} + \frac{1}{3\pi} \cdot rI_C^2), \tag{9}$$

where r is the module resistance, m is the modulation factor, U_{CE0} is the IGBT collector-emitter saturation voltage. The equation of the conduction loss energy assumes the sinusoidal modulated output current for a square wave operation , whereas the modulation factor m must be equal to $4/\pi$. The switching losses in *Iposim* are calculated by [5]:

$$P_{sw\_IGBT} = f_{sw} \cdot \frac{1}{T_0} \int_0^{T_0/2} (E_{on} + E_{off})(t, I_C)dt. \tag{10}$$

The losses in the FWD are calculated similarly to the IGBT losses. The FWD conduction losses are expressed by Eq. (11) and switching losses by Eq. (12).

$$P_{on-state\_FWD} = \frac{1}{2}(U_{T0} \cdot \frac{I}{\pi} + r_T \cdot \frac{I^2}{4}) - m \cdot \cos\varphi \times \\ \times (U_{T0} \cdot \frac{I}{8} + \frac{1}{3\pi} \cdot r_T I^2). \tag{11}$$

$$P_{sw\_FWD} = \frac{1}{\pi} \cdot f_{sw} \cdot E_{rec}(I_{nom}) \cdot (0.45\frac{I}{I_{nom}} + 0.55) \times \\ \times \frac{U_{dc}}{U_{nom}}, \tag{12}$$

where U_{T0} is the FWD forward voltage drop, f_{sw} is the switching frequency, I_{nom} is the nominal voltage of the IGBT module, U_{in} is the input voltage and U_{nom} is the nominal voltage of the IGBT module. The losses of the whole module are the sum of the IGBT and FWD losses. The parameters needed for simulation are the DC link voltage U_{dc}, maximum junction temperature T_j, maximum transistor case temperature T_c, modulation factor m and the power factor of the load $cos\varphi$. The modulated output current frequency f_0 can be set to a low value when simple square wave non-modulated output current is used. The losses of IGBT module calculated by *Iposim* software for different switching frequencies are presented in Fig. 7.

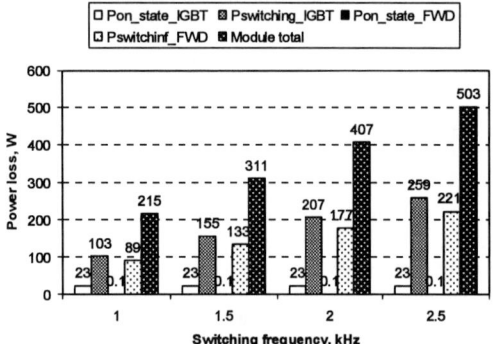

Fig. 7. IGBT and FWD losses calculated by *Iposim*.

V. MEASUREMENTS OF LOSSES

In the laboratory setup two FZ200R65KF1 6.5 kV IGBT modules in the half-bridge configuration were placed on a common forced air cooling system. During the tests of loss measurement, the heatsink temperature, ambient temperature, IGBT collector currents, and collector-emitter voltage were monitored. The temperature of the heatsink was measured in two points to calculate the heat gradient and the individual temperatures of each IGBT.

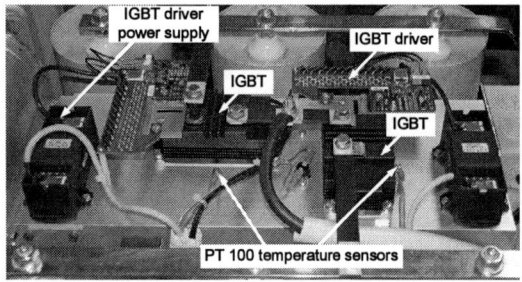

Fig. 8. Arrangements of components in the experimental converter.

Thermal measurements were conducted with FK422 PT100 temperature sensors and SEM203P measurement converters. The accuracy of the used measuring devices was 5%. The experiments were stopped 25 minutes after the thermal balance of the input inverter system was reached. The measured values were logged during the tests with switching frequencies from 1 kHz to 2.5 kHz with 500 Hz step. The total IGBT losses for one transistor were calculated by the following equation:

$$P_{loss} = \frac{T_h - T_a}{2 \cdot R_{tha}}, \qquad (13)$$

where P_{loss} is the power loss of one IGBT module, sharing common heatsink with the second transistor. The R_{tha} for the selected cooling system was 0.013 K/W. The values calculated above are considering only long-term average values. Thus thermal capacitances are not taken

into account. The simplified thermal model of the used experimental device is shown in Fig. 9 [6] [7].

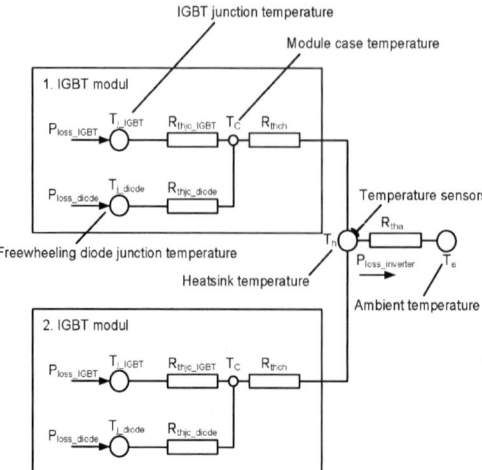

Fig. 9. Simplified thermal model of the experimental device.

The temperatures measured at different switching frequencies during the experiment are presented in Fig. 10 and the power losses calculated by Eq. (13) in Fig 11. The graph shows that the temperature stabilizes after 15 minutes of operation.

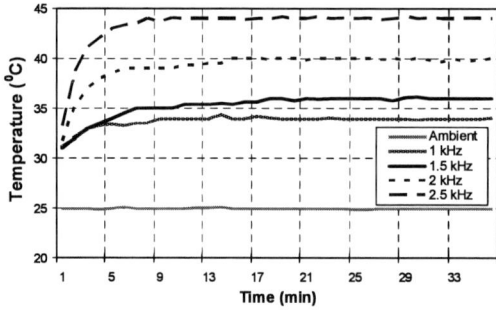

Fig. 10. Measured heatsink temperatures at various switching frequencies.

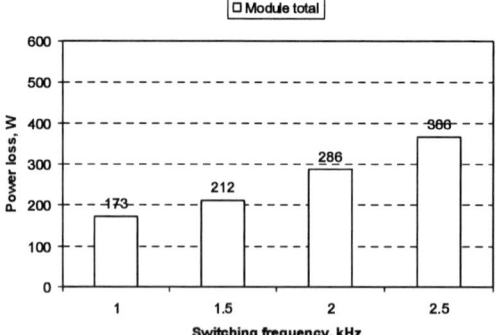

Fig. 11. Experimentally measured losses of a single IGBT

VI. COMPARISION OF RESULTS

Finally, the losses for different switching frequencies of a single switch in the half-bridge configuration calculated by different methods were compared with experimental results (Table III).

TABLE III
STEADY-STATE POWER LOSSES OF 6.5 KV 200 A IGBT MODULE

Switching frequency (kHz)	Calculated loss power (W)	Simulated loss power (W)	Measured loss power (W)
1	735	215	172
1.5	1085	311	211
2	1435	407	286
2.5	1785	503	366

As seen from Table III, the losses calculated by simplified equations (rough loss estimation method) are more than four times higher than the measured ones. This indicates that the energies of switch-on, switch-off and recovery given in manufacturer's datasheets are not accurate in the low load region of the characteristics of switching loss energies. The numbers proposed by the IPOSIM software are far more accurate, giving the difference from 20% to 37% (Fig. 12). In real applications the switching losses of an IGBT are affected by the sum of factors hard to estimate, like the capacitance and inductance of the input inverters bus-bar system, internal inductances and capacitances of the components etc. Those factors can be measured on real devices or be estimated by the geometrical dimensions of the contact wire loop [8] [9]. The hardswitching used in the application influences the overall losses of the IGBT [10]. Also, the accuracy of the measuring devices can influence the experimental results. The overall accuracy of the measurements taken was 5%.

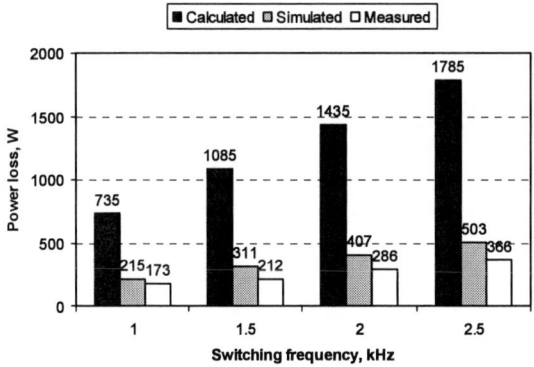

Fig. 12. Total losses estimated with different methods

In conclusion the high voltage IGBTs are performing well under low load and relatively high switching frequency conditions, although different loss estimation methods give different values. When an IGBT load is near to the nominal, the difference is reduced to the minor value.

CONCLUSIONS

Three methods of high voltage IGBT power loss estimations were compared for small load operation. Values of losses acquired experimentally were four times smaller than losses calculated by simplified equations. This was the issue due to somewhat inaccurate characteristics of switching loss energies in the small load region given in manufacturer's datasheets. However, at nominal current, this difference between the calculated and experimentally measured values is negligible. Thus, the manufacturer's datasheets could involve slightly more accurate characteristics for small currents. The IPOSIM simulation software supplied by the manufacturer was more accurate, with a difference of about 20% from experimental results. Also, the accuracy of measurements taken in the experimental part of the study is influenced by the accuracy of the used instruments (in this experiment the laboratory devices used had an accuracy of 5%). Due to the losses lower than expected in dynamic losses of IGBTs at low load conditions, higher switching frequencies can be used. In the light load operation, the FZ200R65KF1 6.5 kV IGBT modules showed good performance at the switching frequencies from 1 kHz to 2.5 kHz.

ACKNOWLEDGMENT

The authors thank Estonian Science Foundation (Grant No. 7425 "Research of dynamic performance of high-voltage IGBTs") for financial support of this study.

REFERENCES

[1] Yamashita, J; Soejima, N; Haruguchi, H: *A novel effective switching loss estimation of non-punchthrough and punchthrough IGBTs*, Proc. of Power Semiconductor Devices and IC's, ISPSD '97, 1997, pp. 109–112.

[2] Mohan, N: *Power electronics*, Second Edition, New York, Wiley, 1995.

[3] Pressman, A, I: *Switching Power Supply Design*, Second Edition, McGraw-Hill, 1998.

[4] *FZ200R65KF1 datasheet*, Infineon, 2006.

[5] *Iposim technical documentation*, Infineon, 2005.

[6] Dewei Xu; Haiwei, Lu; Lipei, Huang; Azuma, S; Kimata, M; Ushida, R: *Power Loss and Junction Temperature Analysis of Power Semiconductor Devices*, WEEE Transaction on Industry, 2002, vol. 38, pp. 1426 - 1431.

[7] Filicori, F; Guarino Lo Bianco, C; *A simplified thermal analysis approach for power transistor rating in PWM-controlled DC/AC converters*, IEEE Transactions on Volume 45, Issue 5, May 1998, pp. 557–566

[8] Winterhalter, C; Kerkman, R; Schlegel, D; Leggate, D: The effect of circuit parasitic impedance on the performance of IGBTs in voltage source inverters, APEC 2001, vol. 2., 2001, pp. 995-1001.

[9] Meade, T; O'Sullivan, D; Floley, R;Achimescu, C; Egan, M; McCloskey, P: *Parasitic inductance effect on switching losses for a high frequency Dc-Dc converter*, APEC 2008, 2008 pp. 3–9.

[10] Fujii, K.; Koellensperger, P.; De Doncker, R. W.: *Characterization and Comparison of High Blocking Voltage IGBTs and IEGTs under Hard- and Soft-Switching Conditions*, Power Electronics Specialists Conference PESC 06., 37th IEEE, 2006, pp. 1-7.

Control of Power Supply Unit for Military Vehicles Based on Four-Leg Three-Phase VSI with Proportional-Resonant Controllers

Tomáš Glasberger, Zdeněk Peroutka

University of West Bohemia/Dept. of Electromechanics and Power Electronic, Plzeň, Czech Republic,
e-mail: *tglasber@kev.zcu.cz, peroutka@ieee.org*

Abstract—This paper deals with the control of the output part of a power electronics converter, which we have designed for a diesel-electric power supply unit for military applications. The output part of the unit is composed of a four-leg three-phase voltage source inverter and a sinusoidal filter. The output part of the power supply unit with help of proposed control system allows to create three-phase symmetrical sinusoidal output voltage under arbitrary load condition (1-phase, 3-phase, linear, non-linear, symmetrical, unsymmetrical) with minimum output voltage distortion. The employed control is based on a set of proportional-resonant (PR) controllers. Behaviour of designed converter has been verified by simulation results as well as by experiments performed on designed laboratory prototype of rated power of 18kVA.

Keywords— Converter control, power supply, pulse width modulation (PWM), robust control, simulation, three-phase system, voltage source inverters (VSI).

I. INTRODUCTION

This research has been motivated by demand of our industrial partner for design of the new power electronics converter for diesel-electric power supply unit for military applications. The designed power supply unit (in this particular case of rated power of 100kW) should be used in army vehicles to supply different military devices. Moreover, the presented power supply unit is also suitable for other applications – e.g. backup power supply systems. The given military application brings demanding requirements on the power electronics converter to be designed (such as minimum dimensions and weight,

heavy overloading, wide temperature range, mechanical immunity – shock endurance, etc.).

There are several papers related to control of four-leg three-phase converters (active filters, voltage source inverters, etc). There are described different PWM strategies such as carrier-based PWM or space vector PWM in stationary coordinate system or in abc coordinate system [1] – [3]. The different types of converter control are introduced in [4] – [8]. The presented control systems are based on PI-controllers (control in the rotating frame) or on PR-controllers (control in the stationary frame); they are designed to control active filters, active rectifiers and voltage source inverters, but they are usually tested only on the lower power applications particularly. All the papers describe interesting but for industrial (robust) applications quite complicated control solutions. This paper presents an easier and robust solution for the converter control.

The proposed converter for the power supply unit is solved as an indirect frequency converter. The converter input part fed by a diesel engine driven permanent magnet synchronous generator is represented by a sensorless controlled voltage source active rectifier [9]. The output part, which is described in this paper, consists of a three-phase four-leg voltage source inverter with a sinusoidal filter at its output.

This paper presents the proposed control (including

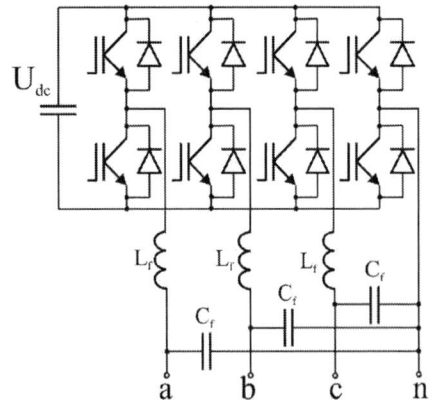

Fig. 1. Configuration of output part of designed power supply unit.

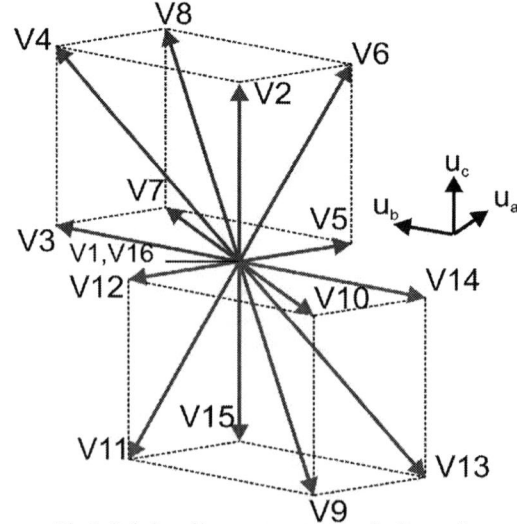

Fig. 2. Ordering of inverter output vectors in abc coordinate system.

978-1-4244-1741-4/08/$25.00 ©2008 IEEE

modulation strategy) of the three-phase four-leg voltage source inverter. The theoretical results are verified by simulation and experimental results of the laboratory prototype of rated power of 18kVA.

TABLE I.
VECTORS, SWITCHING STATES AND OUTPUT VOLTAGES

Vector	s_n	s_a	s_b	s_c	u_{an}	u_{bn}	u_{cn}
1	0	0	0	0	0	0	0
2	0	0	0	1	0	0	1
3	0	0	1	0	0	1	0
4	0	0	1	1	0	1	1
5	0	1	0	0	1	0	0
6	0	1	0	1	1	0	1
7	0	1	1	0	1	1	0
8	0	1	1	1	1	1	1
9	1	0	0	0	-1	-1	-1
10	1	0	0	1	-1	-1	0
11	1	0	1	0	-1	0	-1
12	1	0	1	1	-1	0	0
13	1	1	0	0	0	-1	-1
14	1	1	0	1	0	-1	0
15	1	1	1	0	0	0	-1
16	1	1	1	1	0	0	0

II. POWER CONVERTER TOPOLOGY

The proposed power converter topology of the power supply unit output part is shown in Fig. 1. The converter consists of four-leg three-phase voltage source inverter and the sinusoidal LC-filter, where L=6mH and C=20µF for the laboratory prototype. This LC filter is needed to obtain sinusoidal voltages at the output of the converter. The voltage source inverter employs an untypical three-dimensional space-vector PWM (3D-SVPWM) in abc system coordinates.

The most known SVPWM strategy in αβγ (dq0) coordinate system must use the transformation from phase coordinates to the given coordinate system and the switching times calculation is thereafter quite difficult to use in a real system. Further, the transformation to the rotating coordinate system or to the stationary coordinate system is not convenient for the proposed control, which uses proportional-resonant controllers. The above mentioned coordinate transformation is profitable for control systems using the control of output voltages in rotating reference frame with conventional PI controllers. This SVPWM strategy is described e.g. in [2].

The SVPWM in abc coordinates does not use any transformations; the output voltage vector is created directly from the phase voltages values. The 3-dimensional ordering of basic voltage output vectors of the four-leg inverter is shown in Fig. 2. There are 16 voltage vectors in the inverter, 14 of them are the active and two are the zero vectors. The demanded output vector is created as a combination of three selected adjacent basic vectors (and the zero vector eventually) as usually. The switching states according to the basic output vectors are described in Tab. 1. Quantities s_a, s_b, s_c, s_n present the turned-on transistor (1-top, 0-bottom transistor is switched

on), quantities u_{an}, u_{bn}, u_{cn} present normalized phase voltages. That exactly means the voltage between given phase a, b or c and neutral leg n. The exact description of this modulation strategy can be found in [3] or [8].

III. PROPORTIONAL-RESONANT CONTROLLERS

The control system of the designed power supply unit uses a set of three proportional-resonant controllers. The PR controller transfer function is described in (1):

$$F_{PR}(s) = K_p + \frac{2K_r s}{s^2 + \omega^2}, \tag{1}$$

where K_p is the proportional gain and K_r is the resonant gain, ω is the angular frequency of the output signal – the PR controller works as a filter with infinite gain for the signal of frequency ω and with zero gain for signals of other frequencies. The PR controller is ideal component for tracking sinusoidal signals. The detailed description of PR controllers theory can be found e.g. in [4].

The time domain representation of the Laplace's form (1) must be calculated in order to implement the regulator in the microcontroller:

$$u_1(t) = K_p.y(t), \tag{2a}$$

$$\frac{d u_2^2(t)}{dt^2} + u_2^2 \omega^2 = 2.Kr.\frac{dy(t)}{dt}, \tag{2b}$$

$$u(t) = u_1(t) + u_2(t), \tag{2c}$$

where $u_1(t)$ and $u_2(t)$ is the time domain function of the first and second part of (1), respectively, u(t) is the controller output, y(t) is the controller input signal (control error). The equations (2) are more suitable for implementation to a DSP instead of equations derived e.g. in [4]. The main advantage of PR controllers is that the output and input signals can be alternating. In AC systems the PR controller based system works as a PI controller based system in revolving (synchronous) coordinate system, but no transformation of controlled quantities to the rotating coordinate system is required.

IV. PROPOSED CONTROL SYSTEM

The proposed converter control is shown in Fig. 3. The main request for the designed supply unit is to create sinusoidal symmetrical output voltages under arbitrary load condition (three-phase symmetrical or unsymmetrical load, one-phase linear or nonlinear load etc.). By this reason, it is necessary to measure the output capacitor voltages in all phases (u_{ca}, u_{cb}, u_{cc}). The capacitor currents are measured in order to provide better dynamic response of the control system (i_{ca}, i_{cb}, i_{cc}). The measured voltages and demanded phase voltages u_{aw}, u_{bw}, u_{cw} are subtracted and the results represent the input quantities for the PR controllers. The control system contains a bank of three proportional-resonant controllers. Each of them controls independently the voltage on the output capacitor in involved phase. PR-controllers must usually be supported by some feedforward signal – in our case represented by signals u_{aw}, u_{bw}, u_{cw}. We have also improved the dynamic properties of the converter using the feedback from the filter capacitor currents [5].

Corrected signals u_{aref}, u_{bref}, u_{cref} are taken as PWM modulation commands into the 3D space vector PWM modulator to control the voltage source inverter. Very important advantage of the using of PR controllers in this system is that the output frequency is always constant (either f_{out}= 50 or 60 Hz).

V. SIMULATION RESULTS

Simulations have been taken to verify that the designed control system can work properly under arbitrary conditions – different kinds of the load, different transient phenomena, etc. The most important transient phenomenon is the start up of the converter with zero voltage on the output capacitors. There can appear dangerous voltage oscillations. It is the hard condition for the control system to suppress those oscillations.

The simulations have been taken under the following parameters: DC link voltage U_{dc}=600V, switching frequency f_{pwm}=7000 Hz, output frequency=50 Hz.

The first simulation result (Fig. 4) presents behaviour of the system under symmetrical load condition (a three-phase RL load; where R=150Ω, L=4mH) in the steady-state. The figure shows the phase voltages in all three phases and the current in phase a. The demanded voltage amplitude has been of 260 V.

The simulation result shown in Fig. 5 presents behaviour of the system under unsymmetrical three-phase load condition – different resistances in all phases (R_a=150 Ω, R_b=150 Ω, R_c=100 Ω) and the same inductances (L_a=L_b=L_c=4mH). There are shown all three phase voltages and currents in phases i_a and i_c, respectively.

In Fig. 6, there is analyzed behaviour of the system under no-load condition – the inverter supplies only the sinusoidal LC filter.

Fig. 7 describes the behaviour of converter feeding one-phase nonlinear load. The load is presented as a one-phase diode bridge rectifier which supplies RC load, where R_z=250 Ω and C_z= 40 µF. Configuration of this kind of load is shown in Fig. 13 (without inductor L_z).

VI. EXPERIMENTAL RESULTS

This part presents experimental results, which have been taken on the designed power supply unit laboratory prototype. The experiments have been made for several kinds of load under both steady-state and transient conditions.
The laboratory prototype is shown in Fig. 14. There can

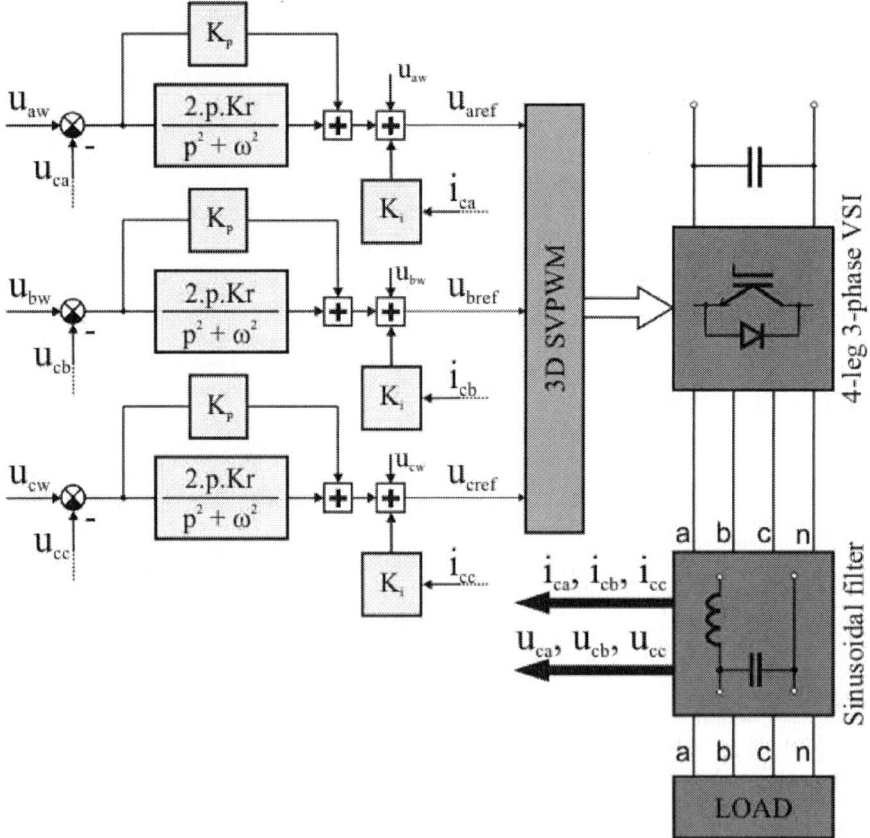

Fig. 3. Proposed control of the three-phase four-leg voltage source converter of designed power supply unit.

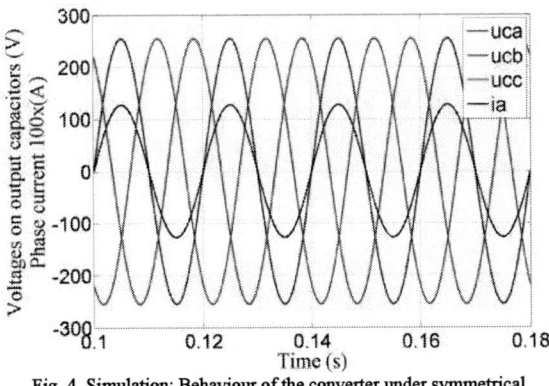

Fig. 4. Simulation: Behaviour of the converter under symmetrical three-phase load condition.

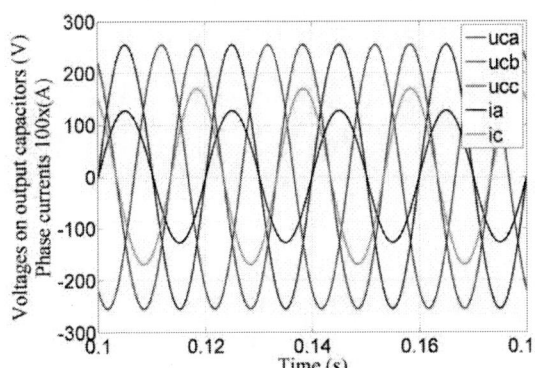

Fig. 5. Simulation: Behaviour of the converter under unsymmetrical three-phase load condition.

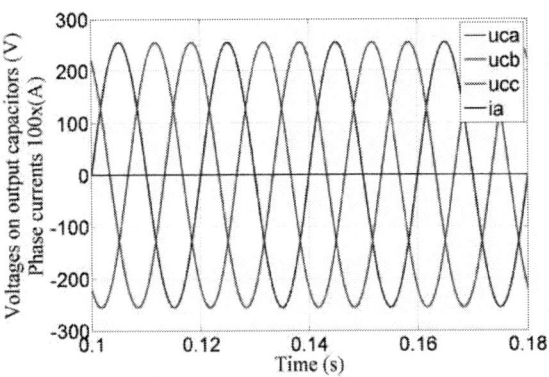

Fig. 6. Simulation: Behaviour of the converter under no-load condition.

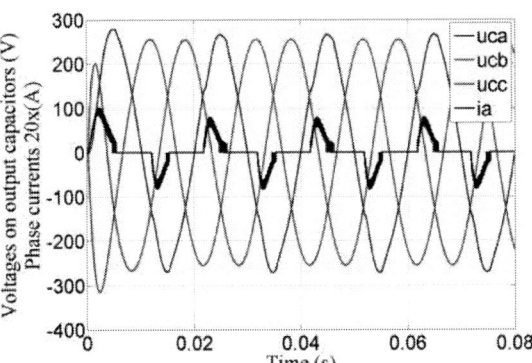

Fig.7. Simulation: Behaviour of the converter under extreme load condition – single-phase diode bridge rectifier feeding RC load.

be seen the power converter presenting the four-leg three phase voltage source inverter, the capacitors voltages and currents measurement which has been made by LEM sensors and the microprocessor board with fixed-point digital signal processor Texas Instruments TMS320LF2812. The DC link voltage has been set U_{dc}=600V, the demanded output voltage amplitude has

been u*=260 V, the switching frequency has been f_{pwm}=7000 Hz and the output frequency has been f_{out}=50Hz. The DC link voltage has been selected to be close to the rated dc-link voltage of 100kW-prototype (U_{dc}=700 V, P=100 kW).

The behaviour of designed converter with symmetrical three-phase load is analyzed in the Fig. 8. The load

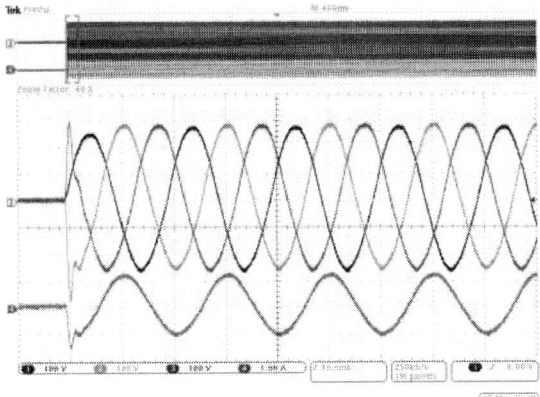

Fig. 8. Behaviour of the laboratory prototype under symmetrical three-phase load condition – start up and steady state; ch1-ch3: phase voltages, ch4: phase current.

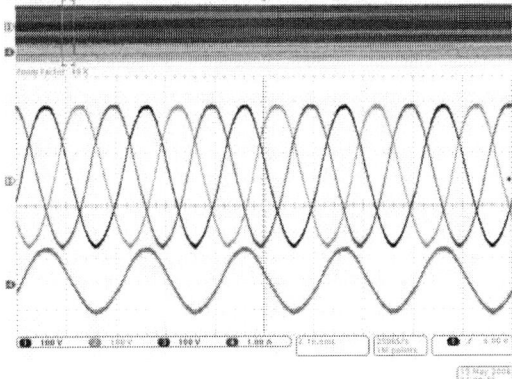

Fig. 9. Behaviour of the laboratory prototype under unsymmetrical three-phase load condition –steady state; ch1-ch3: phase voltages, ch4: phase current.

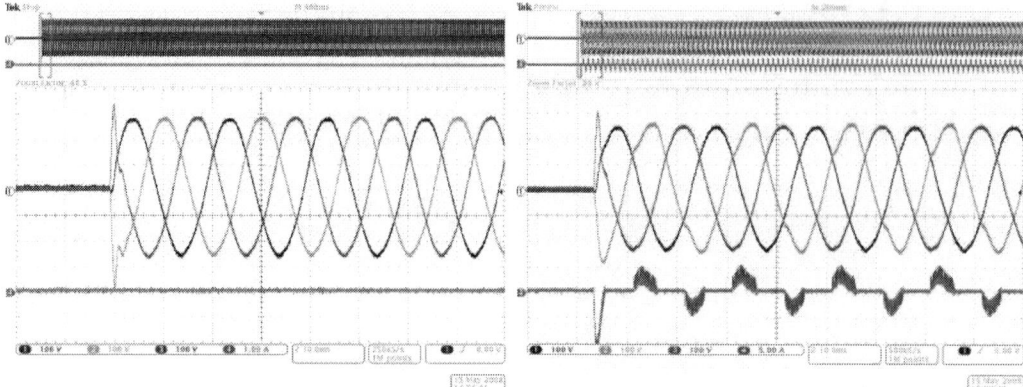

Fig. 10. Behaviour of the laboratory prototype under no-load condition
– start up and steady state;
ch1-ch3: phase voltages, ch4: phase current.

Fig. 11. Behaviour of the laboratory prototype under non-linear
one-phase load - start up and steady state;
ch1-ch3: phase voltages, ch4: phase current.

parameters have been set according to the simulation – resistors $R_a = R_b = R_c = 150\Omega$ and inductances $L_a = L_b = L_c = 4mH$. It can be seen that the converter works without problems, the output voltages are sinusoidal without any visible distortions (THDu has been demanded lower than 5%).

Fig. 9 shows converter behaviour in steady-state under unsymmetrical three-phase load condition ($R_a=R_b=150\Omega$, $R_c=100\Omega$, $L_a=L_b=L_c=4mH$). The power supply unit works without problems in this case as well. Fig. 10 presents the start-up and the steady-state behaviour of no-loaded converter. The running of the non-loaded LC filter is very hard condition for the control system, but the oscillations of the output voltage are damped very fast – in several milliseconds.

Fig. 11 illustrates behaviour of the converter supplying a one-phase diode bridge rectifier, which feeds the RC load (parameters are the same as in the simulation: $R_z=250\ \Omega$, $C_z= 40\ \mu F$). There can be seen that the rectifier input current is distorted by thin pulses. These pulses appear because the capacitor voltage (u_{ca} in Fig.13) is distorted by switching frequency and it

oscillates around the smooth sinusoidal average value (the output voltage is not fully smooth as a sinusoidal waveform) – it could be seen only in a very large detail of the voltage waveform). If the output capacitor voltage is higher than the load capacitor voltage, the current increase fast and charges the load capacitor. If the load capacitor is fully charged, the input current decreases – these two phenomena repeat dependently on the inverter output capacitor voltage distortion.

Fig. 13 presents a solution to reduce the current ripple – an additional inductor L_z is connected in the load circuit. It should be noted that the current ripple in Fig. 11 have not an influence to the load voltage (voltage on capacitor C_z).

VII. CONCLUSIONS

This paper has introduced a new design of an output part of a power supply unit, which we have developed for fighting vehicles. The converter output part has been resolved by the three-phase four-leg voltage source inverter with sinusoidal filter. We have to overcome

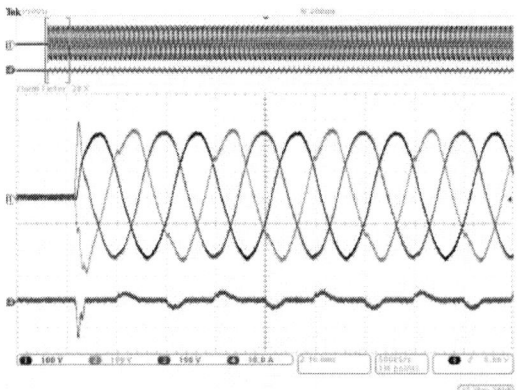

Fig. 12. Behaviour of the system under non-linear one-phase load with
additional inductor - start up and steady state;
ch1-ch3: phase voltages, ch4: phase current.

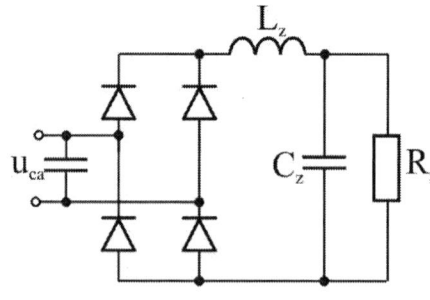

Fig. 13. Single-phase non-linear load configuration with
additional inductor L_z.

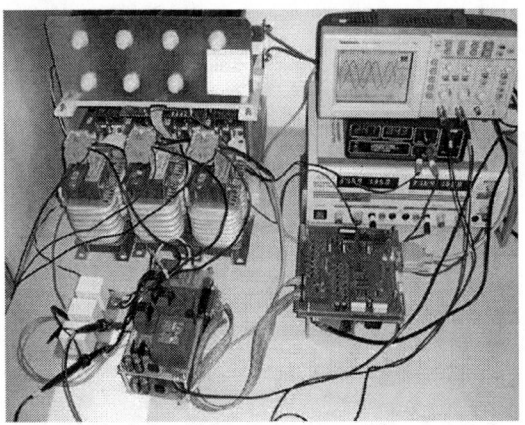

Fig. 14. The designed laboratory prototype of the power supply unit (power converter, LC filter, LEM sensors measurement, microprocessor (DSP) board).

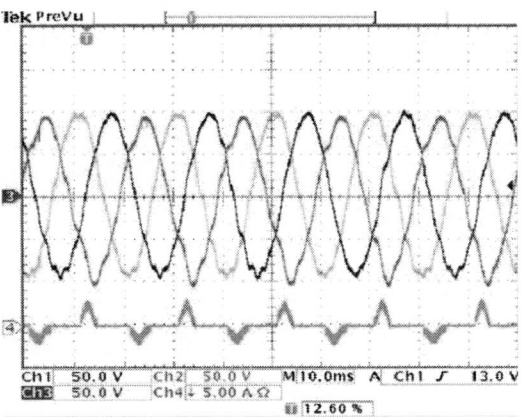

Fig. 15. Behaviour of the control system based on PI controllers under non-linear one-phase load condition - steady state; ch1-ch3: phase voltages, ch4: phase current.

several strict requirements for the design of both the converter power part and the control system: design with minimum dimensions and weight, low voltage distortion under different load condition, high temperature range, EMC, shock endurance, robustness of the proposed control, high reliability, etc.

The main emphasis has been paid in this paper to the proposed converter control. The presented control system uses the space vector PWM in abc coordinates and PR controllers to control the voltages at the output sinusoidal filter capacitors. This combination of the 3D SVPWM and the PR controllers based control system brings very robust and quite easy solution for implementation (short program code and calculation time) for the supply unit control. The converter is able to provide the symmetrical three-phase output voltage with low THDu under arbitrary load conditions (including single-phase non-linear loads). The proposed control with PR controllers reaches better results than conventional control of converter output voltage in revolving reference frame using classical PI controllers – compare the Fig. 11 and Fig. 15.

VIII. REFERENCES

[1] Kim, J.H.; Sul, S.K.: A Carrier-Based PWM Method for Three-Phase Four-Leg Voltage Source Inverters. IEEE Trans. on Power Electronics. Vol.19. No.1.pp. 66-75. January 2004.

[2] Zhang, R.; Prasad, W. H.; Boroyewich, D.: Three-Dimensional Space Vector Modulation for Four-Leg Voltage Source Converters. IEEE Trans. on Power Electronics. Vol. 17, No. 3, pp. 314-326, May 2002.

[3] Perales, M. A.; Prats, M. M.; Portillo, Ramón; Mora, José L.; León, José I.; Franquelo, Leopoldo G. : Three-Dimensional Space Vector Modulation in abc Coordinates for Four-Leg Voltage Source Converters. IEEE Power Electronics Letters. Vol.1,

No.4, pp. 104-109, December 2003.

[4] Xiaoming Y.; Merk, W.; Stemmler, H.; Allmeling, J.; Stationary-frame generalized integrators for current control of active power filters with zero steady-state error for current harmonics of concern under unbalanced and distorted operating conditions. IEEE Transactions on Industry Applications; Volume 38, Issue 2, March-April 2002; pp. 523 – 532.

[5] Demirkutlu, E.; Cetinkaya, S.; Hava, A.M.; Output Voltage Control of A Four-Leg Inverter Based Three-Phase UPS by Means of Stationary Frame Resonant Filter Banks, Electric Machines and Drives Conference 2007 IEMDC 07, Volume 1, 3-5 May 2007, pp. 880-885, Antalya, Turkey.

[6] Ryan, M.J.; De Doncker, R.W.; Lorenz, R.D: Decoupled Control of a Four-Leg Inverter Via a New 4 x 4 Transformation Matrix. IEEE Trans. on Power Electronics. Vol. 16, No. 5, September 2001, pp. 694 – 701.

[7] Zhang, R.: High Performance Power Converter Systems for Nonlinear and Unbalanced Load/Source. [Dissertation] Virginia Polytechnic Institute and State University. Blacksburg, Virginia, USA,1998.

[8] Glasberger, T.; Peroutka, Z.; Molnar, J. Comparison of 3D-SVPWM and carrier-based PWM of three-phase four-leg voltage source inverter. *In* EPE 2007. Brussels : EPE Association, 2007. s. P.1-P.9.

[9] Peroutka, Z.; Zeman, K. Sensorless control of voltage source active rectifier for PMSG based diesel-electric power supply units for military applications. *In* Electrical Drives and Power Electronics. Košice : Slovenská elektrotechnická spoločnosť, 2007. s. 1-6.

Optimal Design of a Half Wave Cockroft-Walton Voltage Multiplier with Different Capacitances per Stage

Ioannis C. Kobougias and Emmanuel C. Tatakis
University of Patras, Department of Electrical and Computer Engineering
Laboratory of Electromechanical Energy Conversion, Rion - Patras, Greece
e-mails: *ikob@ece.upatras.gr, e.c.tatakis@ece.upatras.gr*

Abstract—**Even though the Half-Wave Cockroft-Walton Voltage Multiplier (H-W C-W VM) is one of the most common AC-DC step-up topologies, most of the VM designers persist in using equal capacitances in every stage, a fact that leads to a non optimal design. The aim of this paper is to introduce a new designing method of H-W C-W VM that lays both on the choice of the adequate capacitance values to minimize the output voltage drop and ripple and the calculation of the optimal number of stages that is necessary to produce the desired output voltage with the minimum base capacitance value. In this way the voltage gain is maximized and the required capacitance value per stage is minimized. The theoretical analysis is validated by PSPICE simulations and experimental results, accomplished on laboratory prototypes.**

Keywords—**DC Power supply, Converter Circuit, High frequency power converter, High Voltage power converters.**

I. INTRODUCTION

According to the worldwide bibliography, there is an increasing interest in high voltage gain applications [1-4] such us front-end stages for battery or photovoltaic sources, DC back up systems, UPS devices, step down inverters and many more. More and more studies focus on different topologies that can offer high voltage conversion ratio, small output voltage ripple, high efficiency, simple design and of course low cost. Among these, the Half Wave Cockroft-Walton Voltage Multiplier (H-W C-W VM), shown in Fig. 1, holds one of the most prominent positions [5-9].

An extended bibliographic research revealed that almost all designed C-W VM consist of equal capacitances per stage. What is more interesting is that only one paper introduces an optimal design method [11]. However, in this analysis important parameters, such us the output voltage ripple, aren't taken into the account.

These conclusions emboldened this paper to focus on two subjects, essential for an optimal designed H-W C-W VM, namely the best choice of the capacitance values that offers the maximum voltage ratio (minimization of the output voltage drop and ripple) as well as the optimal number of stages N_{opt} that are necessary to achieve a given output voltage gain with the minimum base capacitance (capacitance of the last stage).

This work was funded by the European Social Fund (75%), the Greek Secretariat Research and Technology (25%) as well as ANCO S.A. and ENERGY SOLUTIONS S.A., within the framework of Measure 8.3 of the Operational Programme "Competitiveness" and the 3rd Community Support Programme (PENED 2003, 03EΔ400).

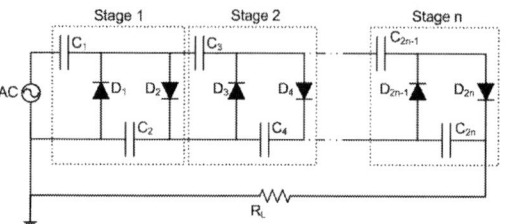

Fig. 1. The n stages Half-Wave Cockroft-Walton Voltage Multiplier with resistive load.

The theoretical analysis is validated by PSPICE simulations and experimental results, accomplished on laboratory prototypes.

II. CHOISE OF THE CAPACITANCE VALUES FOR A H-W C-W TOPOLOGY

In an unloaded H-W C-W VM consisted of n stages, the voltage ratio X of the mean value of the output voltage V_o versus the peak value E_{pk} of the sinusoidal input supply V_{in}, reaches to X_{nl}, after the successive loads of the capacitors, forced by V_{in}. The non load voltage ratio X_{nl} is given by the equation:

$$X_{nl} = \frac{V_{o,nl}}{E_{pk}} = 2 \cdot n \qquad (1)$$

However, when a load is connected creates periodical voltage fluctuations over the capacitors. As a result, V_o not only suffers by a peak to peak voltage ripple δV_o but also a voltage drop ΔV_o is a fact.

Based on the analysis presented in [9, 12 and 13] and their assumptions, namely:

- The total charge which flows in the k stage is k times smaller than the total charge that flows in the first stage.

- The duration of capacitors' charging and discharging are much smaller than the period of V_{in}.

analytical equations can be extracted that give X as a function of the values of the used capacitors.

$$X = X_{nl} - \frac{1}{E_{pk}} \cdot \left(\Delta V_o + \frac{1}{2} \cdot \delta V_o \right) \qquad (2)$$

978-1-4244-1741-4/08/$25.00 ©2008 IEEE

$$\frac{\Delta V_o}{E_{pk}} = \frac{g}{f} \cdot \left(\sum_{i=1}^{n} \frac{(n+1-i)^2}{C_{2 \cdot i-1}} + \sum_{i=1}^{n-1} \frac{(n+1-i) \cdot (n-1)}{C_{2 \cdot i}} \right) \quad (3)$$

$$\frac{\delta V_o}{E_{pk}} = \frac{g}{f} \cdot \sum_{i=1}^{n} \frac{(n+1-i)}{C_{2 \cdot i}} \quad (4)$$

where f the frequency of the voltage supply, I_L the average value of the load current and $g = I_L/E_{pk}$. It is obvious that the output voltage is directly affected by ΔV_o and δV_o and decreases when the load current increases.

Moreover, the output voltage ripple δV_o is affected only by the smoothing capacitors' (capacitors with even index) values, whereas the voltage drop ΔV_o is affected from both smoothing and transition capacitors'(capacitors with odd index) values. However, the latter ones have greater influence on ΔV_o due to their larger coefficient.

Consequently, in order to achieve a higher voltage ratio and approach an optimal H-W C-W design, the capacitance values must be chosen judiciously so as to minimize ΔV_o and δV_o. According to these remarks four different ways (Cases) to design a H-W C-W VM can be advised:

Case 1: $C_{2 \cdot i} = C_{2 \cdot i-1} = C$ (the classical case, analyzed by many authors)

Case 2: $C_1 = 2 \cdot C$ and $C_{2 \cdot i} = C_{2 \cdot i-1} = C$ for $i \neq 1$ (case often found in the bibliography)

Case 3: $C_{2 \cdot i} = C_{2 \cdot i-1} = (n+1-i) \cdot C$

Case 4: $C_{2 \cdot i} = (n+1-i) \cdot C$ and $C_{2 \cdot i-1} = (n+1-i)^2 \cdot C$

where i the number of every stage and C the capacitance of the last stage (base capacitance).

Thus, using (2)-(4) four new formulas can be extracted for the voltage gain X:

Case 1:
$$X = 2 \cdot n - \frac{g}{f \cdot C} \cdot \left(\frac{4 \cdot n^3 + 3 \cdot n^2 - n}{6} + \frac{n \cdot (n+1)}{4} \right) =$$
$$= 2 \cdot n - \frac{g}{f \cdot C} \cdot \frac{8 \cdot n^3 + 9 \cdot n^2 + n}{12} \quad (5)$$

Case 2:
$$X = 2 \cdot n - \frac{g}{f \cdot C} \cdot \left(\frac{4 \cdot n^3 - n}{6} + \frac{n \cdot (n+1)}{4} \right) =$$
$$= 2 \cdot n - \frac{g}{f \cdot C} \cdot \frac{8 \cdot n^3 + 3 \cdot n^2 + n}{12} \quad (6)$$

Case 3:
$$X = 2 \cdot n - \frac{g}{f \cdot C} \cdot \left(n^2 + \frac{n}{2} \right) =$$
$$= 2 \cdot n - \frac{g}{f \cdot C} \cdot \frac{2 \cdot n^2 + n}{2} \quad (7)$$

Case 4:
$$X = 2 \cdot n - \frac{g}{f \cdot C} \cdot \left(\frac{n \cdot (n+1)}{2} + \frac{n}{2} \right) =$$
$$= 2 \cdot n - \frac{g}{f \cdot C} \cdot \frac{n^2 + 2 \cdot n}{2} \quad (8)$$

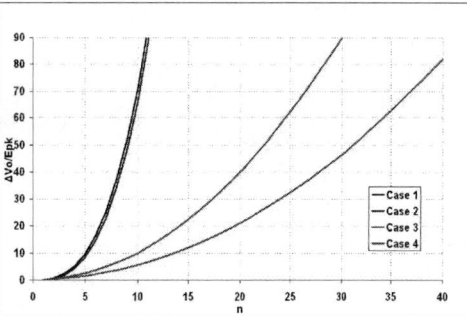

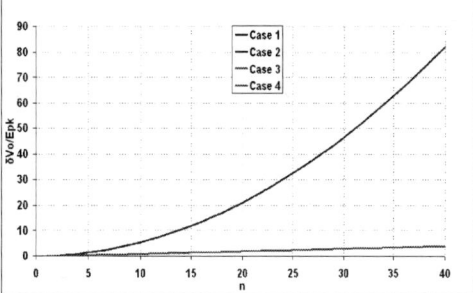

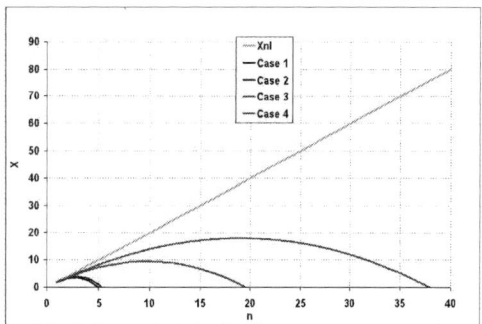

Fig. 2. $\Delta V_o/E_{pk}$, $\delta V_o/E_{pk}$ and X as a function of n for $g/f \cdot C = 0.1$.

Fig. 2 shows the per unit voltage drop $\Delta V_o/E_{pk}$, the per unit ripple $\delta V_o/E_{pk}$ and the output voltage gain X, for $g/f \cdot C = 0.1$, as a function of the number of stages.

It is obvious that Cases 3 and 4 offer lower $\Delta V_o/E_{pk}$ and $\delta V_o/E_{pk}$ as well as higher X than the common used Cases 1 and 2.

III. OPRIMUM DESIGN OF THE H-W C-W VM

Beside the electrical specifications (output voltage and current, efficiency), an optimum design of an electric device must take account of the total cost and the estimated lifetime. In the case of a C-W VM, these two parameters are strictly related to the base capacitance.

In order to determine the minimum base capacitance C that is necessary to obtain the desired X the next equation must be solved:

$$\frac{\partial Fc(n, g, X)}{\partial n} = 0 \quad (9)$$

Note that as f and C appear as a product in the equations they are initially considered as a single variable Fc. $g=I_L/E_{pk}$ and $X=V_o/E_{pk}$ are considered as constants, determined by the desired electric specifications.

Considering also that

$$\frac{\partial X}{\partial n} = \frac{\partial X}{\partial Fc} \cdot \frac{\partial Fc}{\partial n} \qquad (10)$$

and taking into the account that

$$\frac{\partial X}{\partial Fc} \neq 0 \qquad (11)$$

for any case, we conclude that:

$$\frac{\partial X}{\partial n} = 0 \qquad (12)$$

Calling N_{opt} the number of stages that corresponds to the smallest product $f\cdot C_{min}$, (12) reveals that the desired voltage gain is also the maximum one a C-W VM with N_{opt} stages can offer.

Applying (12) to the equation that corresponds to each case we obtain:

Case 1:

$$\frac{g}{Fc} \cdot \left(24 \cdot n^2 + 18 \cdot n + 1\right) = 24 \qquad (13)$$

Case 2:

$$\frac{g}{Fc} \cdot \left(24 \cdot n^2 + 6 \cdot n + 1\right) = 24 \qquad (14)$$

Case 3:

$$\frac{g}{Fc} \cdot \left(4 \cdot n + 1\right) = 4 \qquad (15)$$

Case 4:

$$\frac{g}{Fc} \cdot \left(n + 1\right) = 2 \qquad (16)$$

Combining for each case the corresponding equations (5) and (13), (6) and (14), (7) and (15), (8) and (16) an algebraic equation, having the following general form:

$$a_3 \cdot n^3 + a_2 \cdot n^2 + a_1 \cdot n + a_0 = 0 , \qquad (17)$$

is obtained. The coefficients for every single case are given in the Table 1.

TABLE I.
COEFFICIENTS OF (12), FOR EACH CASE

	Case 1	Case 2	Case3	Case 4
a_3	1	1	0	0
a_2	$\frac{3 \cdot (4 \cdot X - 3)}{16}$	$\frac{3 \cdot (4 \cdot X - 1)}{16}$	1	1
a_1	$\frac{9 \cdot X}{16}$	$\frac{3 \cdot X}{16}$	$-X$	$-X$
a_0	$\frac{X}{32}$	$\frac{X}{32}$	$\frac{X}{4}$	$-X$

Using well known formulas [14] an analytical solution can be obtained. Thus, for each case the optimum number of stages N_{opt} is given by the following equations:

Case 1:

$$N_{opt} = \frac{X}{4} \cdot \left(2 \cdot \cos\left(\frac{\theta}{3}\right) + 1\right) + \frac{3}{16} \cdot \left(2 \cdot \cos\left(\frac{\theta}{3}\right) - 1\right) \quad (18)$$

where

$$\cos(\theta) = \frac{R}{\sqrt{-Q^3}} \qquad (18a)$$

$$Q = -\left[\frac{4 \cdot X + 3}{16}\right]^2 \qquad (18b)$$

$$R = \frac{(4 \cdot X)^3 + (12 \cdot X)^2 - 44 \cdot X - 27}{16^3} \qquad (18c)$$

Case 2:

$$N_{opt} = S + T - \frac{1}{3} \cdot a_2 \qquad (19)$$

where

$$S = \sqrt[3]{R + \sqrt{D}} \qquad (19a)$$

$$T = \sqrt[3]{R - \sqrt{D}} \qquad (19b)$$

$$R = \frac{64 \cdot X^3 + 48 \cdot X^2 + 52 \cdot X - 1}{16^3} \qquad (19c)$$

$$D = Q^3 + R^2 \qquad (19d)$$

$$Q = -\left[\frac{4 \cdot X + 1}{16}\right]^2 \qquad (19e)$$

Case 3:

$$N_{opt} = \frac{X + \sqrt{X^2 + X}}{2} \qquad (20)$$

Case 4:

$$N_{opt} = \frac{X + \sqrt{X^2 + 4 \cdot X}}{2} \qquad (21)$$

It can be easily proven that the N_{opt} solution for each case can be approximated by a linear equation. So the optimal number of stages can be approximated by the following linear equations:

Case 1:

$$N_{opt} \approx \frac{3}{4} \cdot X + \frac{3}{16} \qquad (22)$$

Case 2:

$$N_{opt} \approx \frac{3}{4} \cdot X + \frac{1}{16} \tag{23}$$

Case 3:

$$N_{opt} \approx X + \frac{1}{4} \tag{24}$$

Case 4:

$$N_{opt} \approx X + \frac{371}{400} \tag{25}$$

The mathematical and graphical analysis reveal that the precision of this linear approximation is remarkable and it gives an easy tool to estimate the necessary optimal number of stages to obtain a desired voltage gain X, for given values of f and I_L.

Finally, the equations which give the minimum base capacitance C_{min} from which the values of the capacitances of each stage can be calculated, are the following:

Case 1:

$$C_{min} = \frac{g}{f} \cdot \frac{8 \cdot N_{opt}^3 + 9 \cdot N_{opt}^2 + N_{opt}}{12 \cdot (2 \cdot N_{opt} - X)} \tag{26}$$

Case 2:

$$C_{min} = \frac{g}{f} \cdot \frac{8 \cdot N_{opt}^3 + 3 \cdot N_{opt}^2 + N_{opt}}{12 \cdot (2 \cdot N_{opt} - X)} \tag{27}$$

Case 3:

$$C_{min} = \frac{g}{f} \cdot \frac{2 \cdot N_{opt}^2 + N_{opt}}{2 \cdot (2 \cdot N_{opt} - X)} \tag{28}$$

Case 4:

$$C_{min} = \frac{g}{f} \cdot \frac{N_{opt}^2 + 2 \cdot N_{opt}}{2 \cdot (2 \cdot N_{opt} - X)} \tag{29}$$

Fig. 3 depicts the quantity Fc/g as a function of n for all Cases. The points mentioned correspond to the minimum value of this quantity, for each Case. The minimum value of Fc corresponds to the optimal number of stages.

Fig. 4 gives the value of N_{opt} (rounded to the closer integer) whereas Fig. 5 gives the quantity Fc_{min}/g as a function of the voltage conversion ratio X.

Taking a closer look to the previous equations and the aforementioned figures, the following conclusions come up:

- In any Case the value of N_{opt} depends only on X and not on the output current I_L.
- According to Fig. 4, for a given value of X, Cases 1 and 2 require the fewer number of stages N_{opt}.
- On the other hand, from Fig. 5 it is obvious that Case 4 leads to the lowest value of the quantity Fc_{min}/g for a given value of X that means a

minimum base capacitance for given values of g and f.

We can conclude that the best choice for an optimized design of a Half-Wave Cockroft-Walton VM in order to obtain the minimum base capacitance C is Case 4, despite the fact it demands more stages than the classical cases. However, Case 3 can be considered as a good solution too, because the value of C_{min} is slightly greater than the one at Case 4 but the number of the necessary stages is lower.

Finally, as an example for a Case 3 H-W C-W VM, Fig. 6 represents the variation of the voltage gain X as a function of the number of stages n, for four different values of the base capacitance (C_1=5μF, C_2=4μF, C_3=3μF, C_4=2μF). For given values of f and g (f = 100kHz, g = 0.05 A/V) the desired value of X ($X_{desired}$=5.5) is obtained for C_{min} = 3μF and N_{opt} = 6. These values are also extracted by the aforementioned theoretical analysis, a fact that proves the effectiveness of the proposed method.

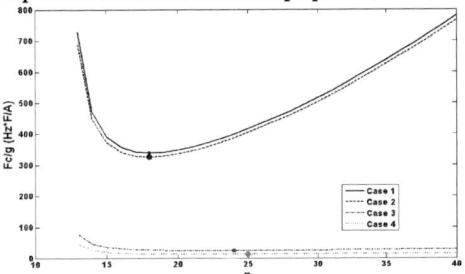

Fig. 3. Fc/g as a function of n for all Cases and X=23.8.

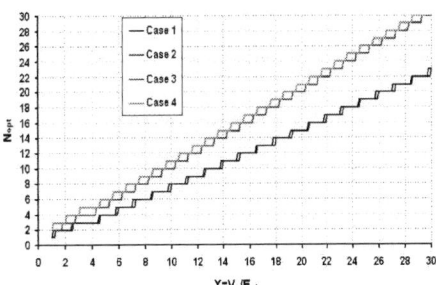

Fig. 4. N_{opt} as a function of X for all cases.

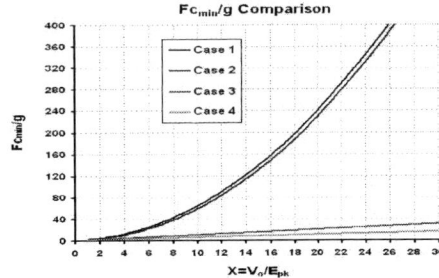

Fig. 5. Fc_{min}/g as a function of X for all cases.

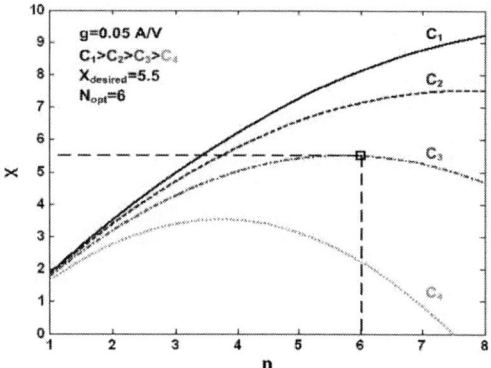

Fig. 6. C_{min} and N_{opt} of a Case3 C-W VM.

IV. SIMULATION VERIFICATION

To verify the theoretical analysis a large number of simulations is carried out using the circuit simulation software PSPICE. Four different circuits per case are simulated, consisting of 3, 5, 7 and 14 stages respectively. In order to be a comparison among the results feasible, every model uses equal capacitances per stage with base capacitances C=5μF, f=100kHz, E_{pk}=20V and I_L≈0.1A.

The results of the aforementioned simulations are depicted in Fig. 7 and Fig. 8. There are two reasons for the small divergence between the theoretical and simulation results. First of all, there is a voltage drop across the diodes, that is ignored during the theoretical analysis and secondly, the values of I_L appear an unavoidable small variation during the simulation process. Despite that, the simulation results still keep up remarkably with the theoretical ones.

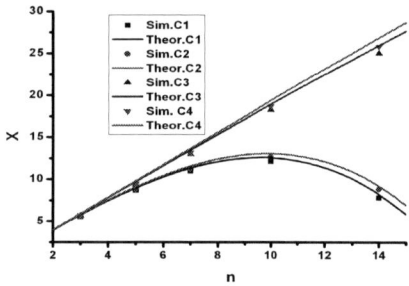

Fig. 7. X as a function of n for every case (simulation results)

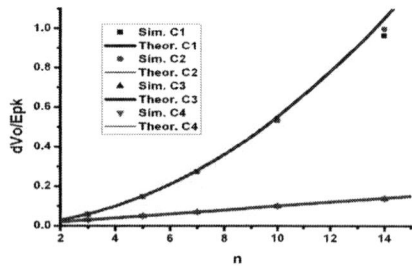

Fig. 8. $\delta V_o/E_{pk}$ as a function of n for every case (simulation results)

V. EXPERIMENTAL VERIFICATION

In an attempt to verify experimentally the theoretical analysis, five different laboratory prototypes are constructed. The first one is a 100W Full Bridge Parallel Loaded Resonant Inverter that acts as a high frequency AC power voltage supply. The values of the resonant elements are chosen at 3.7μH and 0.25μF. The inverter is driven by the dsPIC4011F Microchip microcontroller, which generates 100kHz pulses with 50% duty cycle. Fig. 9 represents a typical waveform of the produced output voltage. Because of the higher frequency harmonics, the peak value of the first harmonic was adjusted to 30V during the experimental procedure, so as to maintain an almost stable input voltage for every device.

In addition three Case 3 H-W C-W laboratory prototypes are implemented, consisting of 3, 5 and 7 stages, respectively. Every stage includes two equal MKP capacitors and two STP40NF10 diodes, with the base capacitance chosen at C=0.8μF.

Fig. 10 represent the measured voltage gain X as a function of g, of the aforementioned C-W topologies when are connected to six different resistive loads

Finally, the same resistive loads are connected to the last laboratory prototype, a 5 stages Case 1 C-W VM with base capacitance C=0.8μF. Fig. 11 verifies that a Case 3 C-W has indeed higher voltage gain than a Case 1 C-W with the same number of stages and base capacitance.

The expected divergence between the experimental and the theoretical results rely on three main reasons:

- The higher frequency harmonics of the H-W C-W VM's input voltage make difficult to choose the right E_{pk} for the theoretical analysis.
- The voltage drop across the used diodes.
- The accurancy of the measurement instruments, especially those which measure the small amplitude of I_L.

Nonetheless the experimental results fit with such a good accuracy those obtained by the theoretical analysis, confirming the aforementioned conclusions.

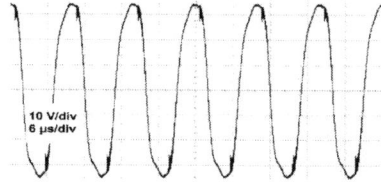

Fig. 9. Typical output voltage of the Resonant Inverter.

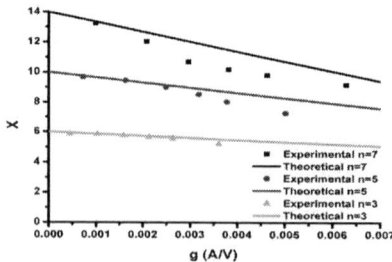

Fig. 10. X as a function of n for Case 3 an different stages

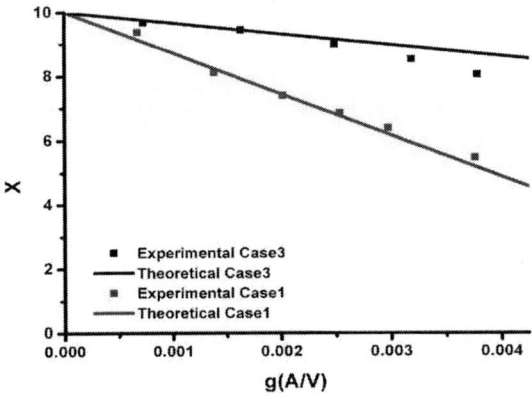

Fig. 11. X as a function of n for Cases 3 and Case 1 while n=5.

VI. CONCLUSIONS

In this paper a novel method of designing a H-W C-W VM based on using different capacitances per stage is presented. For that reason four different choises (cases) of the capacitance values of an H-W C-W VM are compared. A theoretical analysis is

held that produce new and improved equations about C-W's voltage gain. Furthemore new formulas are introduced that give the number of stages that not only provide the proper voltage gain but also minimize the base capacitance of the last stage. The analysis reveals the case which is the best choise among the four ones. These conclusions are validated by PSPICE simulations and experimental results, accomplished on laboratory prototypes.

REFERENCES

[1] D. Maksimovic, S. Cuk, "Switching Converters with Wide DC Conversion Range", *IEEE Transactions on Power Electronics*, Vol. 6, No. 1, pp. 151-157, January 1991.

[2] F. L. Luo, H. Ye, "Positive Output Super-Lift Converters", *IEEE Transactions on Power Electronics*, Vol. 18, No. 1, pp. 105-113, January 2003.

[3] K. C. Tseng, T. J. Liang, "Novel high-efficiency step-up converter", *IEE Proc.-Electr. Power Appl.*, Vol 151, No. 2, pp. 182-190, March 2004.

[4] C. Y. Inaba, Y. Konishi, M. Nakaoka, "High-frequency flyback-type soft-switching PWM DC-DC power converter with energy recovery transformer and auxiliary passive lossless snubbers", *IEE Proc.-Electr. Power Appl.*, Vol. 151, No. 1, pp. 32-37, January 2004.

[5] F. Hwang, Y. Shen, S. H. Jayaram, "Low-Ripple Compact High-Voltage DC Power Supply", *IEEE Transactions on Industry Applications*, Vol. 42, No. 5, pp. 1139-1145, Sept./Oct. 2006.

[6] A. Shenkman, Y. Berkovich, B. Axelrod, "Novel AC-DC and DC-DC Converters with a Diode-Capacitor Multiplier", *IEEE Transactions on Aerospace and Electronic Systems*, Vol. 40, No. 4, pp. 1286-1293, October 2004.

[7] K. Ogura, E. Chu, M. Ishitobi, M. Nakamura, M. Nakaoka, "Inductor Snubber-Assisted Series Resonant ZCS-PFM High Frequency Inverter Link DC-DC Converter with Voltage Multiplier", *Power Conversion Conference (PCC'2002), Osaka (Japan)*, 2-5 April, 2002, Vol. 1, pp. 110 – 114.

[8] "Experiment 19: Blaze Labs Resonant Multipliers", *Blaze Labs Research*, 12 September 2005, Internet paper, available: http://www.blazelabs.com/e-exp19.asp, accessed on 30/11/2006.

[9] M.M. Weiner, "Analysis of Cockcroft-Walton Voltage Multipliers with an Arbitrary Number of Stages", *The Review of Scientific Instruments*, Vol. 2, No. 2, pp. 330-333, February 1969.

[10] "Experiment 15: Cockroft-Walton Multiplier", *Blaze Labs Research*, 30 August 2005, Internet paper, available: http://www.blazelabs.com/e-exp15.asp, accessed on 30/11/2006.

[11] M. Julian, "Cockroft-Walton Optimum Design Guide v2.0", 10 August 2005, Internet paper, available: http://www.blazelabs.com/CWdesign.pdf, accessed on 30/11/2006.

[12] M. Khalifa, "High-Voltage Engineering, Theory and Practice", Chapter 16, *Electrical Engineering and Electronics, A series of Reference Books and Textbooks*, N°63, Marcel Decker Inc., March 1990.

[13] E. Kuffel, W.S. Zaengl, "High Voltage Engineering Fundamentals", Chapter 2, Pergamon International Library, 1984.

[14] M. R. Spiegel, "Schaum's Mathematical Handbook of Formulas and Tables", McGraw-Hill, 2nd edition, 1998.

Calculation of Leakage Inductance of Core-Type Transformers for Power Electronic Circuits

Reinhard Doebbelin*, Marcel Benecke*, Andreas Lindemann*
*Otto-von-Guericke-Universitaet Magdeburg, Chair for Power Electronics,
Magdeburg, Germany, e-mail: *reinhard.doebbelin@et.uni-magdeburg.de*

Abstract—**Leakage inductances of transformers significantly influence the behavior of power electronic circuits. This paper deals with their prediction: The application of the approximation method of Lebedev for the calculation of leakage inductance of core-type transformers as well as an extension of this method for pie winding arrangements are presented. Design aspects influencing leakage inductance are additionally investigated by means of finite element simulations.**

Keywords— **Transformer, Design, Modeling**

I. INTRODUCTION

Imperfect magnetic coupling between primary and secondary windings of transformers is represented in equivalent circuits by means of leakage inductance elements (e.g. L_{L1} and L'_{L2} in the T-type transformer equivalent circuit displayed in Fig. 1). If transformers are used in power electronic circuits, either defined leakage inductance parameters [1] or (in most cases) particularly low leakage inductance values of these transformers are required acccording to existing impacts on operation frequency or energy conversion, stress of power semiconductors etc. Often it is necessary to predict the leakage inductance of a transformer already in the design phase, especially if circuit simulation is intended [2].

II. CALCULATION METHODS

In the case of shell-type transformers with all windings arranged on the center leg, prediction of leakage inductance can be obtained using the approximation methods of Rogowski [3] and Petrov [4] which have been established in the first half of the last century. Both methods are shortly explained in [5]. However, in certain cases, the core-type transformer design has an advantage over the shell-type design of transformers. In high-current applications (e.g. inverter power sources for resistance welding processes) the implementation of the output connection of the massive secondary windings of the transformers

is an important design aspect. Contacting the secondary windings is particularly difficult in the case of shell-type transformers, if interleaving of windings is used to reduce leakage inductance and therefore secondary windings are positioned between primary winding coils. In the case of core-type transformers, resulting from the arrangement of the windings on two legs, inevitably interleaving of windings exists. Nevertheless, the secondary windings can be arranged in the external position on each leg. This way, contacting the secondary windings is simplified.

In chapter V a circuit-simulation based comparison of the operational behavior and maximum output current of a resistance welding inverter using a shell-type and a core-type transformer alternatively is presented. Therefor, the leakage inductance values of the different transformer versions have been calculated using the respectively suited approximation method.

If windings are arranged on different legs, like in the case of core-type transformers (Fig. 2a), leakage inductance can be calculated by means of an approximation method which has been established by Lebedev [6]. The basic assumptions of this method are illustrated in the different graphics of Fig. 2. The leakage field can be considered to consist of three leakage flux components (Fig. 2b). Taking the field configuration into account, the windings N_1 and N_2 are replaced by the equivalent winding N_{1-2} and a notional winding N_n is added to the transformer. Due to this model, the leakage inductance can be calculated as the sum of three parts:

$$L_L = L_{L\ \text{in}} + L_{L\ \text{y}} + L_{L\ \text{eq}} \qquad (1)$$

The parts are

- the inner leakage inductance $L_{L\ \text{in}}$ related to the leakage inductance component belonging to the section of the perimeter of the windings which is located within the core window and the leakage flux which occurs in this region. This leakage inductance component shall be calculated according to the method of Rogowski using

$$L_{L\ \text{in}} = \mu_0 \cdot w_1^2 \cdot d_c \cdot \lambda \cdot k_\sigma \qquad (2)$$

with w_1 – number of turns of the primary winding to which the leakage inductance refers, λ – relative leakage conductance and k_σ – Rogowski factor to be determined as described in [3].

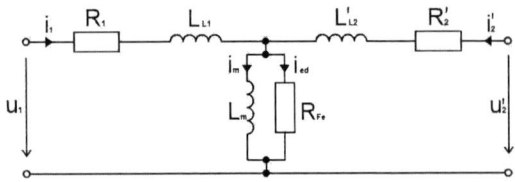

Fig. 1. T-type equivalent circuit diagram of transformers

978-1-4244-1741-4/08/$25.00 ©2008 IEEE

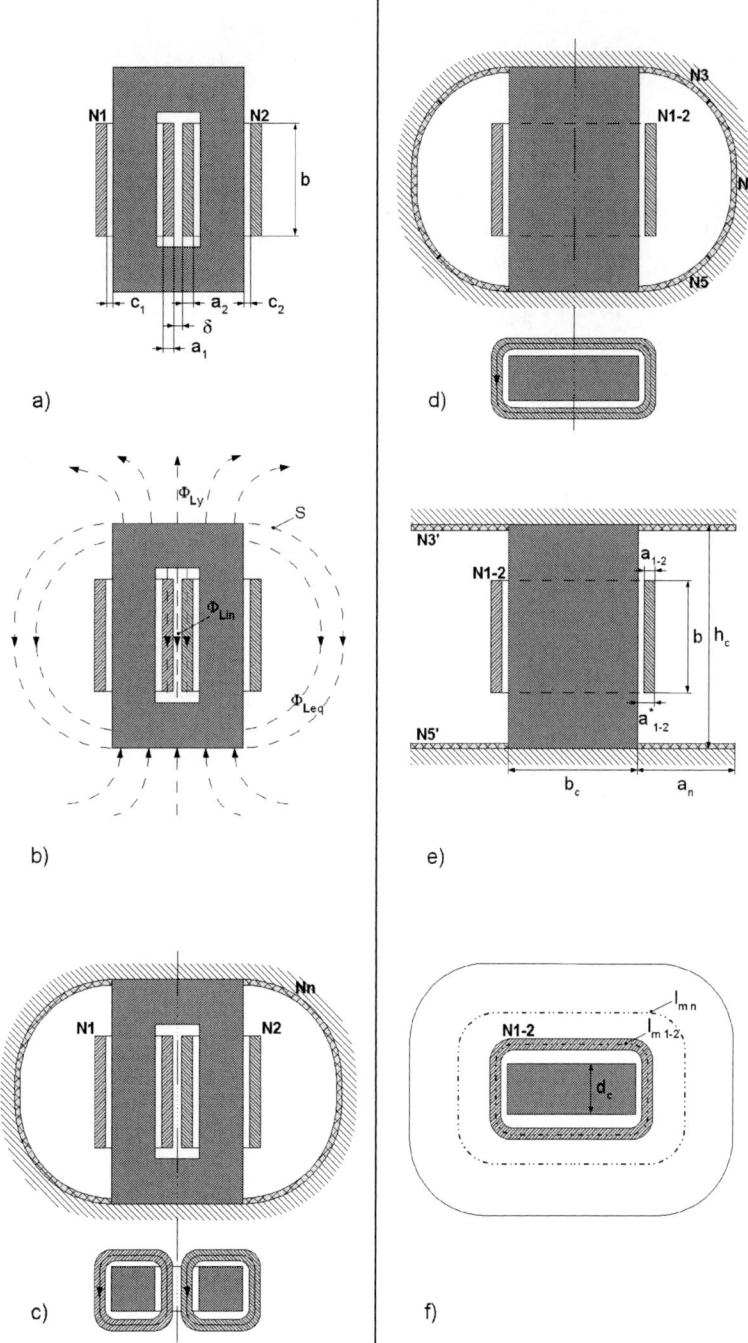

Fig. 2. Illustration of transformer modeling according to the method of Lebedev

- the leakage inductance component $L_{L\ y}$ which is related to the yoke leakage flux. This leakage inductance component can be calculated according to

$$L_{L\ y} = \mu_0 \cdot w_1^2 \cdot g_y \qquad (3)$$

The specific yoke conductivity g_y varies with the width b_c and the height h_c of the core as given in [6].

- the so-called leakage inductance of the equivalent transformer $L_{L\ eq}$. To enable the calculation of this leakage inductance component, the notional winding N_n is subdivided into the sections N_3, N_4 and N_5, and the section N_4 is divided into two equal partitions which are assigned to N_3 and N_5, respectively (Fig. 2 c-f). Based on the method of mean geometric distances [7] this leakage inductance component can be calculated according to

$$L_{L\ eq} = \frac{\mu_0}{\pi} \cdot w_1^2 \cdot l_{m\ n/1-2} \cdot \left(1,1 + \ln \frac{h_c}{b + 2a_{1-2}^*}\right) \qquad (4)$$

with $l_{m\ n/1-2} = \frac{1}{2} \cdot (l_{m\ n} + l_{m\ 1-2})$ – mean length per turn (mlt) to be calculated considering the geometry of the notional winding N_n and the equivalent winding N_{1-2} representing the average of their mlt values $l_{m\ n}$ and $l_{m\ 1-2}$.

Using the method of Lebedev as it is given in literature, its application is restricted to simple arrangements consisting of two windings (one coil on each leg of the core) which are centered on the leg in vertical direction and coincide in their dimension b. In contemporary power electronic and especially pulsed power applications, interleaving of windings is used to minimize the total leakage inductance of the applied transformers. Particularly in arrangements with the coils realized as pie windings, these windings are not centered on the leg and their dimensions are different in general. Therefore, the method has been extended to enable the leakage inductance approximation also in the case of core-type transformers with interleaving of windings and various dimensions of the coils. The idea is to combine the described calculation methods. This way the total leakage inductance values of core-type transformers have been determined and compared with the values of shell-type transformers which coincide with the core-type transformers in their cross-sectional area of the core and in the nominal data of the primary and the secondary windings. To predict the total leakage inductance of a transformer with interleaved windings, the leakage inductance values of the different couples of coils have to be calculated (by means of the methods of Rogowski or Petrov if both coils are located on the same leg, by means of the method of Lebedev if the coils are located on different legs of the transformer). Then the leakage inductance values of these couples of coils have to be merged into the total leakage inductance of the transformer using specific formulas which are related to the respective combination of the coils of the transformer. In a transformer design with two primary and two secondary coils (connected in series in each case), the total leakage inductance can be calculated according to

$$
\begin{aligned}
L_{L,\text{total}} = \frac{w_1^2}{w_{\text{ref}}^2} \cdot \Big(&\frac{w_{11} \cdot w_{21}}{w_1 \cdot w_2} \cdot L_{L\ 11/21} + \frac{w_{12} \cdot w_{21}}{w_1 \cdot w_2} \cdot L_{L\ 12/21} \\
&+ \frac{w_{11} \cdot w_{22}}{w_1 \cdot w_2} \cdot L_{L\ 11/22} + \frac{w_{12} \cdot w_{22}}{w_1 \cdot w_2} \cdot L_{L\ 12/22} \\
&- \frac{w_{11} \cdot w_{12}}{w_1^2} \cdot L_{L\ 11/12} - \frac{w_{21} \cdot w_{22}}{w_2^2} \cdot L_{L\ 21/22} \Big)
\end{aligned}
$$
(5)

with w_{xx} – number of turns of the respective coil, w_{ref} – reference number of turns to which the single leakage inductance values $L_{L\ xx/yy}$ of certain couples of coils N_{xx} and N_{yy} refer.

The calculation procedure of the method in its original and in the extended version has been implemented into a Matlab®-based computer program.

III. VERIFICATION BY EXPERIMENT

To enable an estimation of the accuracy of the leakage inductance calculation, core-type transformer versions have been realized whose coils are arranged as concentric windings and pie windings, respectively (Fig. 3).

The relatively small deviations determined from a comparison of the leakage inductance values obtained from measurements and calculations are displayed in Fig. 4, with respect to the different couples of coils which exist in the case of a transformer with four coils.

The results of calculation and measurement concerning the total leakage inductance values of the considered transformers with concentric windings are illustrated in Fig. 5. If the leakage inductance of couples of coils which are located on the same leg have been calculated according to the method of Rogowski, the coincidence with measured values is much better compared to the application of the method of Petrov.

The comparison of calculated and measured leakage inductance values of several further core-type transformers verified the applicability of the presented approximation method to these transformers.

IV. INVESTIGATION BY FEM SIMULATION

In addition to the calculated and measured values, a further result concerning the leakage inductance of the

Fig. 3. Realized core-type transformer versions

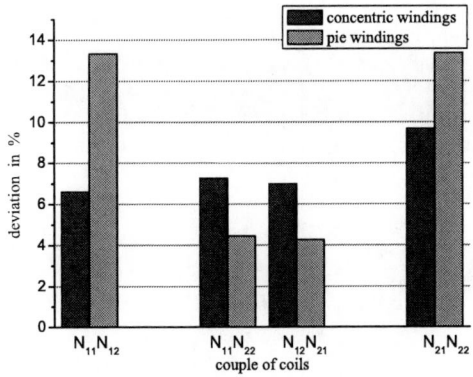

Fig. 4. Deviations between leakage inductance values obtained from measurement and calculation

the inverse connection. Considering the respective field configurations (Fig. 7) this can be explained by the number of flux lines generated by the excited coil, which do not or not completely enclose the short-circuited coil. The number of these flux lines is higher in the case of excitation of the outer coil resulting in a higher leakage inductance.

Besides the air coil arrangement, a pot-core transformer has been complementarily investigated by means of FEM simulation (Fig. 8) representing a model for the considered shell-type transformers and also for the magnetic flux within the core window of core-type transformers. In this case exciting the outer coil results in an even slightly lower value of leakage inductance. This can be explained by the pathway of the flux lines, which is slightly longer, connected with a higher magnetic resistance and the mentioned lower leakage inductance.

shell-type transformer is displayed in Fig. 5, which has been obtained from finite element (FEM) simulation of the arrangement and the configuration of its magnetic field. FEM simulations (performed by means of the program femm) have been used to clarify the background of certain results of leakage inductance determination. So, according to the measurements which have been carried out using the realized coil and transformer arrangements, the leakage inductance values of certain couples of coils are different if the position of excitation and short circuit is changed (Fig. 6). For example, in the case of a couple of concentric air coils (representing an air-core transformer), the leakage inductance value is higher if the outer coil (N_{21}) is connected with the RLC meter and the inner coil (N_{11}) is short-circuited, compared to the result of

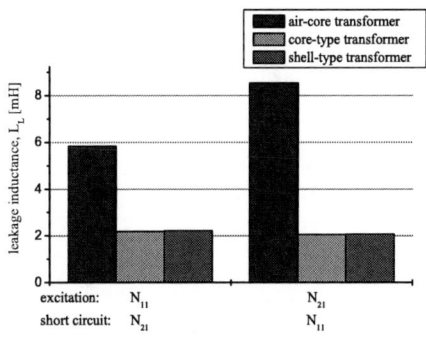

Fig. 6. Measured leakage inductance of couples of coils depending on the position of excitation and short circuit

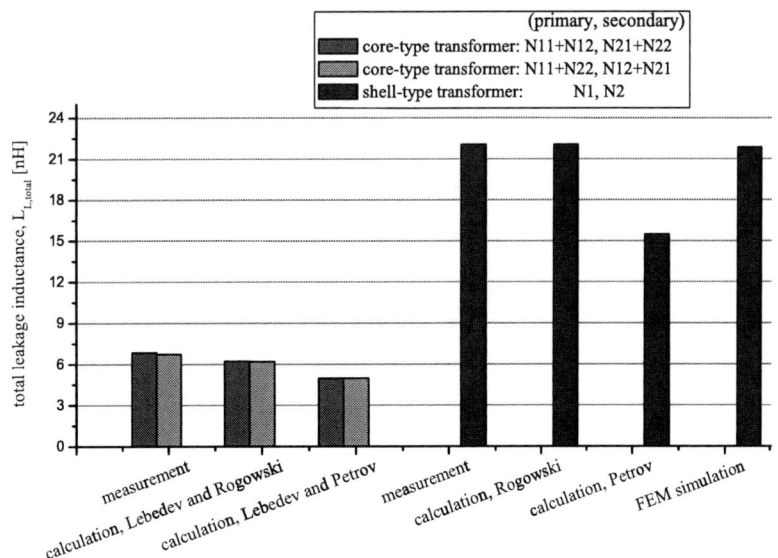

Fig. 5. Comparison of the total leakage inductance values of core-type and shell-type transformers with concentric windings (values obtained from measurement, calculation and FEM simulation, all leakage inductance values related to a reference number of turns = 1)

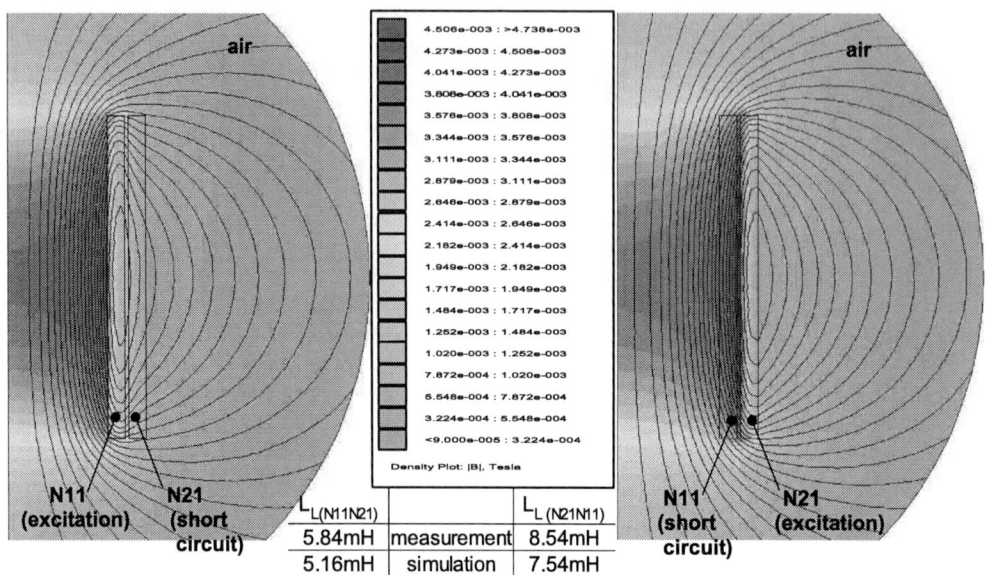

$L_{L\,(N11N21)}$		$L_{L\,(N21N11)}$
5.84mH	measurement	8.54mH
5.16mH	simulation	7.54mH

Fig. 7. Simulated magnetic field configurations of a couple of concentric air coils with changed position of excitation and short circuit (only right side of the respective rotation-symmetric arrangements displayed)

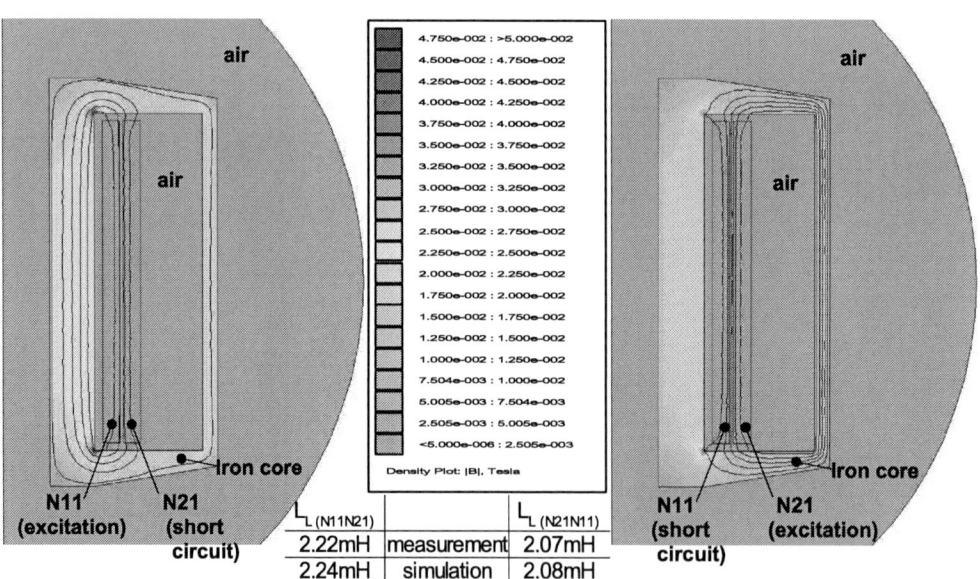

$L_{L\,(N11N21)}$		$L_{L\,(N21N11)}$
2.22mH	measurement	2.07mH
2.24mH	simulation	2.08mH

Fig. 8. Simulated magnetic field configurations of a pot-core transformer ($\mu_{\mathrm{rel\,core}} = 1200$) carrying a couple of concentric coils with changed position of excitation and short circuit (only right side of the respective rotation-symmetric arrangements displayed)

V. EXAMPLE OF A CIRCUIT-SIMULATION BASED ANALYSIS USING PREDICTED TRANSFORMER PARAMETERS

As already mentioned, in this chapter an example of the application of approximation methods for the calculation of transformer leakage inductance values, enabling a simulation based circuit analysis, shall be given.

The considered power-electronic circuit is an inverter power source for resistance welding (see the block diagram of the power section in Fig. 9).

Especially in resistance welding applications the advantage of a lightweight and small medium-frequency transformer compared to mains-frequency operated transformers is important. In the case of welding robot applications, which are typical e. g. in automotive industry, it is the precondition for an arrangement of transformer, output rectifier and welding gun as one welding unit in the "hand" of the robot.

Using this design, the complete welding unit can be moved to each welding spot and the losses in the conductors of the welding unit caused by the high-amperage output current can be minimized compared to a design with the transformer standing beside the robot and a cable connection to the welding gun in the hand of the robot. In addition to a lightweight transformer, a high maximum output current is required to offer a wide application range concerning the welding tasks to be fulfilled.

Prediction of transformer parameters is a prerequisite for the consideration of the operational behavior and the determination of maximum output current of a resistance welding inverter by means of simulation. In this investigation the shell-type and alternatively the core-type design of the medium-frequency transformer shall be taken into account. The calculation of the respective transformer parameters is based on the assumption of consistent parameters of the cross-sectional area of the core, the core window and the windings which show a concentric design.

- In the case of the shell-type transformer, the primary winding is arranged on the center leg as the inner winding. It is surrounded by the secondary winding consisting of two turns center-taped to enable the connection of the mid-point output rectifier (Fig. 9).

- In the case of the core-type transformer, the primary winding is divided into two parts (connected in series) which are arranged on both legs as the inner windings. These parts of the primary winding are surrounded by the parts of the subdivided secondary winding consisting of two center-taped turns on each leg of the transformer. The respective parts of the secondary winding on the different legs of the transformer are connected in parallel. Therefore, the cross-sectional area of one secondary winding arranged on one of the legs of the core-type transformer is 50% of the cross-sectional area of the secondary winding of the shell-type transformer.

The assumed design of these transformers shows comparable conditions concerning the output connection of the

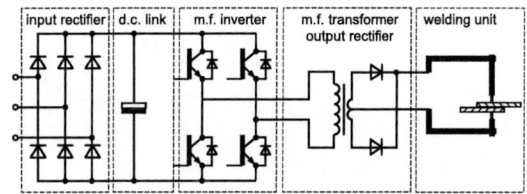

Fig. 9. Block diagram of the power section of an inverter power source for resistance welding

secondary windings.

The leakage inductance values L_L of the transformer versions have been obtained using the respective appropriate approximation method (concerning the shell-type transformer: method of Rogowski yielding $L_L = 143.3\mu H$; concerning the core-type transformer: method of Lebedev in the case of couples of windings arranged on different legs and method of Rogowski in the case of couples of windings arranged on the same leg yielding the total leakage inductance of this transformer $L_L = 33.6\mu H$). Thus, the ratio of these calculated leakage inductance values corresponds to the ratio of the leakage inductance values of the realized shell-type and core-type transformers described in chapter III (see Fig. 5). The copper resistance values of the windings have been calculated from geometry.

The simulation has been carried out approximating the power supply of the inverter unit by a constant D.C. voltage of 538V. As the switching frequency, $f = 1kHz$ has been chosen because this value is most common in resistance welding inverters. By means of the duty ratio adjustment, the operation of the inverter in continuous conduction mode is given. The consequences of the alternative application of the transformer versions can be evaluated considering the waveform of the primary currents of the transformers (Fig. 10) and the waveform of the output currents of the resistance welding inverter (Fig. 11).

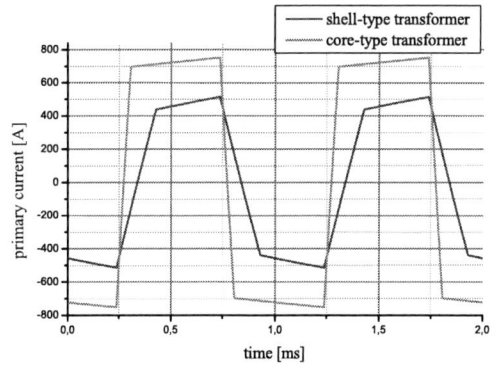

Fig. 10. Comparison of simulated primary currents for shell-type transformer and core-type transformer

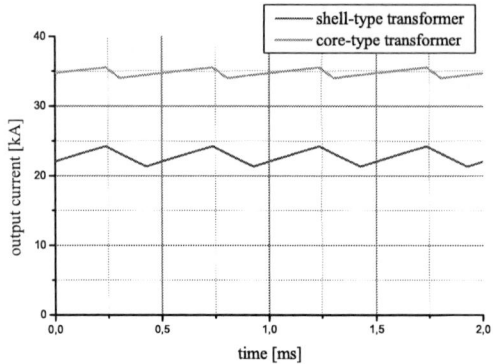

Fig. 11. Comparison of simulated output currents for shell-type transformer and core-type transformer

have been investigated using the extended calculation method as well as measurement and FEM simulation, which has further been used to illustrate the connection between magnetic field configuration and leakage inductance of transformer arrangements. These considerations are useful to predict leakage inductances of transformers already in the design phase of power electronic converters such as welding equipment or power supplies. An application example is given considering the operational behavior and the maximum reachable output current of a resistance welding inverter by means of circuit simulation if alternatively a shell-type transformer and a core-type transformer are used.

Due to the remarkable decrease of leakage inductance, the slopes of the primary current show a much higher change rate and the commutation intervals are much shorter in the case of the application of the core-type transformer compared to the shell-type transformer resulting in higher amplitudes of the primary current and a higher output current as well. The main difference between both variants can be illustrated considering the RMS value of output current which is about 22.82kA if the shell-type transformer is used and about 34.79kA if the core-type transformer is used.

VI. CONCLUSION

The approximation method of Lebedev which can be used for calculation of the leakage inductance of simple two-windings core-type transformers has been extended to enable the leakage inductance approximation also in the case of core-type transformers with interleaving of windings and various dimensions of the coils. Leakage inductance values of different transformer arrangements

REFERENCES

[1] A. Stadler, M. Albach, and S. Chromy, "The optimization of high frequency operated transformers for resonant converters," in *11th European Conference on Power Electronics and Applications (EPE 2005)*, Sept. 2005, Dresden, Germany, paper 77.
[2] H. Njiende, H. Wetzel, N. Froehleke, and W. A. Cronje, "Models of integrated magnetic components for simulation based design of SMPS with simplorer," in *10th European Conference on Power Electronics and Applications (EPE 2003)*, Sept. 2003, Toulouse, France, paper 750.
[3] W. Rogowski, *Ueber das Streufeld und den Streuinduktionskoeffizienten eines Transformators mit Scheibenwicklung und geteilten Endspulen (Dissertation)*. VdI, 1909, Mitteilung ueber Forschungsarbeiten auf dem Gebiet des Ingenieurwesens.
[4] G. N. Petrov, "Allgemeine Methode der Berechnung der Streuung von Transformatoren," *Elektrotechnik und Maschinenbau*, vol. 51, no. 25, pp. 345–350, 1933.
[5] R. Doebbelin, T. Winkler, A. Lindemann, and C. Teichert, "Design of Pulsed Power Transformers for Capacitor Discharge Resistance Welding Machines," *International Conference Power Electronics, Intelligent Motion, Power Quality (PCIM 2006)*, Nuernberg, pp. 205–210, 30 May - 01 June 2006.
[6] V. K. Lebedev, "Calculation of the short-circuit resistance of welding transformers with yoke leakage (russ.)," *Automatic Welding [Avtomaticeskaja Svarka] Kiev*, vol. 11, no. 4, pp. 37–44, 1958.
[7] J. C. Maxwell, *Lehrbuch der Elektrizitaet und des Magnetismus*. Springer-Verlag, Berlin, 1883, vol. 2.

Enhanced Current Pulsation Smoothing Parallel Active Filter for Single Stage Grid-connected AC-PV Modules

A.C. Kyritsis*, N.P. Papanikolaou[†] and E.C. Tatakis*

*University of Patras, Department of Electrical and Computer Engineering, Laboratory of Electromechanical Energy Conversion, Rion-Patras, Greece, e-mail: *Kyritsis@ece.upatras.gr, E.C.Tatakis@ece.upatras.gr*
[†] Technological Educational Institute of Lamia, Department of Electrical Engineering, Lamia, Greece,
e-mail: *Npapanikolaou@teilam.gr*

Abstract—On single stage PV Converters the output power has a large amount of power pulsation at twice line frequency, causing PV module voltage and current fluctuation. However, a PV module should operate at a sufficiently small area around the maximum power point, in order to maximize PV generation. To overcome this defect a buffering storage unit between the PV module and the inverter, which performs the well known Power Decoupling, is inevitable. This paper presents an enhanced configuration of a Current Pulsation Smoothing Parallel Active Filter (CPS-PAF) which permits the elimination of the low frequency PV current ripple with significant smaller capacitor comparatively to the classical configuration. The enhanced CPS-PAF conception, control and effectiveness are validated by PSpice simulation results as well as by experimental results accomplished on a laboratory prototype.

Keywords—Renewable energy systems, photovoltaic, active filter, design, converter control.

I. INTRODUCTION

Distributed generation (DG) has already been an important issue due to the valuable role of renewable energy resources in the battle of carbon dioxide levels stabilization. From the DG production plants that can be connected to LV network, the PV systems seem to be a rather beneficial choice since the solar power is the only renewable energy source that can be widely deployed in towns and cities. Analytically, a large number of grid-connected photovoltaic (PV) systems, mounted on residential and commercial buildings, offer the opportunity to generate considerable quantities of energy near the consumption point. These PV systems are usually interfaced to the LV distribution network by a single phase controllable power inverter. Their power production varies between 0.1 and 5kW.

The newest technology on small scale grid-connected residential PV systems is the "AC-PV Module" where the power production varies between 0.1 and 0.3kW depended on the PV module technology [1-4]. These systems due to their easily installation and incrementability (a DC wiring

This work is funded by the European Social Fund (75%), the Greek Secretariat of Research and Technology (25%) and ANCO S.A. and ENERGY SOLUTIONS S.A., within the framework of Measure 8.3 of the operational Programme "Competitiveness" and 3rd Community Support Programme (PENED03400)".

expertise is not necessary for installation and a DC safety equipment is not needed due to the very low PV output voltage), the low initial cost and their improved efficiency compared to the conventional residential PV systems (each PV module has individual M.P.P.T, while there is no power losses due to the mismatch between PV modules or due to the use of diodes), seem to be a beneficial choice for small scale grid-connected residential PV applications. In order to make AC-PV Modules more attractive, the main task is a cost reduction per inverter watt. Thus, single stage topologies seem to be a rather attractive solution, since these topologies generally characterized by high efficiency (compared to dual or multi stage topologies), low cost, simple structure and high reliability due to the reduced components count.

On the other hand, these topologies are characterized by a serious disadvantage; analytically, assuming that the fundamental inverter current flowing to the grid has sinusoidal waveform and that is in-phase with the mains voltage, the power that is transferred to the power network has a large amount of power pulsation at twice line frequency. The power pulsation, due to the single-phase power generation, causes PV module voltage and current fluctuation. This problem becomes more understandable considering that the PV module should operate at a sufficiently small area around the maximum power point (in order to exploit the maximum PV generated electricity power). Thus a buffering storage unit between the PV module and the inverter, which performs the well known Power Decoupling, is inevitable. This unit is acting as an energy buffer, whose main task is to smooth the PV module output current. Thus, the voltage at the terminals of the PV module is non-fluctuating and the PV module can be operated close enough to the maximum power point.

In contrast with indoor PV inverters, where the usual solution to reduce the PV module voltage fluctuation is a large electrolytic capacitor, in AC-PV Module systems this decoupling solution contravenes to the system lifetime, since electrolytic capacitors are very susceptible under high temperature operating conditions [5-6]. To resolve this drawback, Ripple Current Reduction (RCR) topologies in which the electrolytic capacitors are replaced with film ones have been proposed in [6-8]. These topologies limit the DC current ripple satisfyingly, but on the other hand they harm the efficiency of the whole system, since they are connected in series with the PV

Module. Furthermore, the operation of these RCR topologies depends strongly on the inverter selection and his operation mode and can be applied only for the Current Source Flyback Inverter and especially for the Discontinuous Conduction Mode. Beyond these topologies, a Current Pulsation Smoothing Parallel Active Filter (which is independent from the inverter topology and his operation mode) was presented by the authors in [9]. In the following paragraphs an enhanced configuration of the former structure of the CPS-PAF, which permits the elimination of the low frequency PV current ripple with significant smaller capacitance, is presented.

II. THE ENHANCED CPS-PAF CONFIGURATION

The enhanced Current Pulsation Smoothing Parallel Active Filter circuit is shown in fig. 1, while fig. 2 shows the former CPS-PAF configuration.

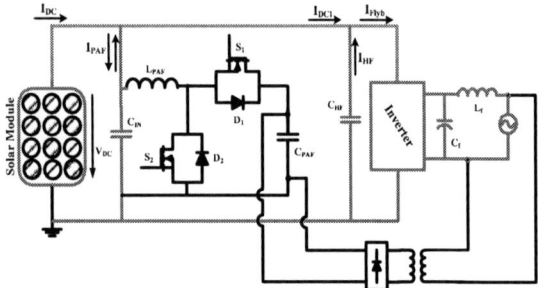

Fig. 1. Enhanced configuration of the CPS-PAF.

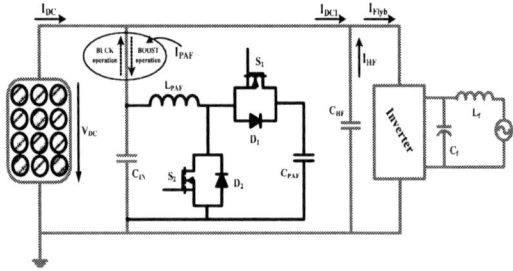

Fig. 2. Former configuration of the CPS-PAF [1].

The new configuration, as the former one, behaves as a current source and its main task is to eliminate the PV current ripple. The main difference from the former CPS-PAF configuration is that from the main side a low frequency and low power transformer feeds the C_{PAF} at the time intervals were the C_{PAF} is going to be discharged below the predefined lower capacitor voltage, V_L. The purpose of this power feedback is the substitution of the former CPS-PAF structure capacitor with significant smaller capacitance values.

In more details, this circuit equates the value of the current I_{DC} that flows from the PV with the average value of the $i_{DC1}(\omega t)$. So, in the time interval where the $i_{DC1}(\omega t)$ value is lower than I_{DC}, the PAF behaves as a boost converter (switch S_2 and diode D_1 are active) storing energy to the capacitor C_{PAF}. In the rest time intervals the PAF behaves as a buck converter (switch S_1 and diode D_2 are active) discharging the C_{PAF}. Therefore, the power pulsation due to the AC single-phase power generation is

converted into energy storage pulsation on the capacitor C_{PAF}. Fig. 3 highlights the time intervals where the PAF operates either as Boost or as Buck converter. The high frequency harmonic current content $i_{HF}(\omega t)$, due to the inverter switching operation, is eliminated with a small film capacitor C_{HF}. Last but not least, C_{IN} is also a small film capacitor which task is to eliminate the high frequency harmonic current due to the CPS-PAF switching operation (either on buck or boost operation).

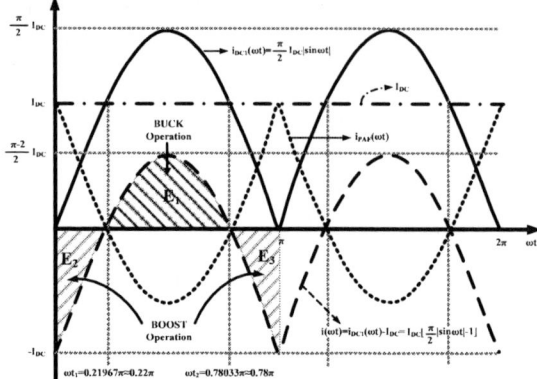

Fig. 3. Time intervals where the CPS-PAF operates either as Boost or Buck converter.

The control circuit of the Enhanced CPS-PAF is based on an hysterisis current control and is the same with the one of the former configuration. Fig. 4 shows the control scheme.

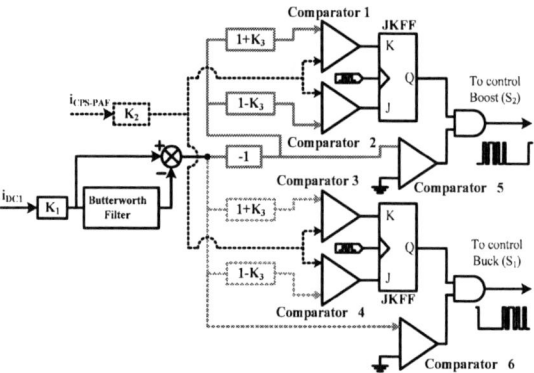

Fig.4. Block diagram of the CPS-PAF control circuit

The required CPS-PAF capacitance in order to achieve theoretical zero voltage fluctuation is given by the following equation (both for the former and the enhanced configuration) [9]:

$$C_{PAF} = \frac{0.422 \pi P_{DC}}{\omega \left(V_H^2 - V_L^2 \right)} \quad (1)$$

where V_H and V_L are the upper and lower limit of the C_{PAF} voltage ripple respectively, P_{DC} is the value of the PV generator power at the maximum power point and ω is the line angular frequency.

As it has already been presented in the same paper, the necessary electrolytic capacitor C_{in} in order to smooth the twice utility frequency current can be expressed by the following equation:

$$C_{in} = \frac{2P_{DC}}{3 \square V_{DC}\hat{u}} \qquad (2)$$

where $\hat{u}$ is the amplitude of the permissible PV voltage ripple, V_{DC} is the value of the PV generator voltage at the maximum power point.

Table 1 presents the required electrolytic capacitor C_{in} and CPS-PAF capacitor C_{PAF}, in order to achieve elimination of the low frequency PV current ripple for the case of four commercial PV Modules. The PV generators electrical characteristics P_{DC} and V_{DC} stand for 1000W/m² solar irradiation, ambient temperature 25°C and AM equal to 1.5. The electrolytic capacitor C_{in} value was calculated according to eq. (2), so as the PV voltage ripple $\Delta V\%$ will be equal to 8.5% of the PV voltage at the maximum power [3].

	C_{in} (μF)	C_{PAF} (μF) V_L=1,5V_{DC}, V_H=2V_{DC}	C_{PAF} (μF) V_L=1,5V_{DC}, V_H=3V_{DC}	C_{PAF} (μF) V_L=1,5V_{DC}, V_H=4V_{DC}
P_{DC}: 100W V_{DC}: 17V	8640	937	216	106
P_{DC}: 100W V_{DC}: 33V	2300	249	57	28
P_{DC}: 200W V_{DC}: 28V	6400	691	160	78
P_{DC}: 200W V_{DC}: 48V	2200	235	54	26

Table. 1. Comparison between the required electrolytic capacitor C_{in} and the CPS-PAF capacitor C_{PAF}, in order to achieve elimination of the low frequency PV current ripple for commercial AC-PV Modules.

Studying this table, we conclude that a high V_H value and a large divergence between V_H and V_L values lead to smaller C_{PAF} capacitance. On the other hand, a large divergence between the input and output CPS-PAF voltage values will deteriorate the whole system efficiency (due to the losses increase on the PAF circuit) and will cause ineffective PV voltage and current ripple elimination as a result of buck and boost marginal PWM operation [10, 11]. Furthermore, this marginal PWM operation encumbers the ability of the PAF configuration to absorb the whole current pulsation. Thus, a PV power fluctuation can be observed, whose value depends strongly on the system losses. Moreover, this phenomenon is enforced by the C_{PAF} capacitance deviations due to ageing and operating reasons (temperature rise, switching frequency etc). Taking into account the aforementioned comments, in the proposed CPS-PAF configuration the transformer feedback supports the current pulsation absorption, since it offers the C_{PAF} energy amount that is disposed in PAF circuit losses. Of course, zero voltage fluctuation is not possible to be achieved in practice; so, a small ripple on PV generator voltage and current is unavoidable. It is worth mentioning that in the enhanced CPS-PAF design, the rectifier bridge output voltage ought to reassure that the lower C_{PAF} discharge voltage will not be lower than the PV Module open circuit voltage. Otherwise, PAF operation in Buck Mode will come up with time intervals of almost unitary duty cycle.

Combining eq (1), (2), the required CPS-PAF capacitance C_{PAF} in order to achieve (theoretically) zero voltage fluctuation can be expressed as follows:

$$\frac{C_{PAF}}{C_{in}} \approx \frac{2\Delta V\%}{k^2} \frac{1}{\left(\frac{V_H}{V_L}+1\right)\left(\frac{V_H}{V_L}-1\right)} \qquad (3)$$

where k is the ratio V_L/V_{DC}, ($k \in R$, $k>1$), C_{in} is the required electrolytic capacitor in order to limit the PV voltage ripple below $\Delta V\%$.

Fig. 5 presents the ratio C_{PAF}/C_{in} as a function of the ratio V_H/V_L (bandwidth of the C_{PAF} voltage ripple) with k as parameter. The voltage ripple $\Delta V\%$ was selected to be 8.5% of the PV voltage at the maximum power. By studying this figure we conclude that apparently, for k values equal or higher than 2, C_{PAF} capacitance becomes small enough (smaller than 5% of the required electrolytic capacitor C_{in}) for any V_H/V_L values. Moreover, a larger divergence between the upper and lower limit of the C_{PAF} voltage ripple (graph area where V_H/V_L takes high values) leads to further capacitance decrease. On the other hand, a larger divergence between the upper and lower limit of the C_{PAF} voltage ripple leads to unacceptable duty cycle high values (both for Boost and Buck operation).

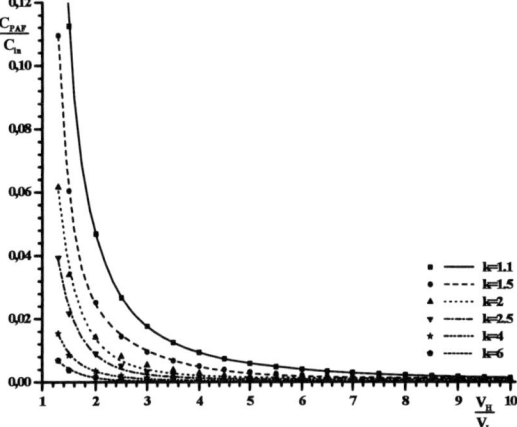

Fig. 5. C_{PAF}/C_{in} as a function of the ratio V_H/V_L (bandwidth of the C_{PAF} voltage ripple) with k as parameter.

Assuming that the PV Module voltage is constant and the power losses of the CPS-PAF switches are negligible, the C_{PAF} voltage at the time intervals where the CPS-PAF behaves as a boost converter, can be calculated by the following equation:

$$V_{PAF}(\square t)=\frac{V_{DC}}{1-d_{boost}(\square t)} \qquad (4)$$

where d_{boost} is the duty cycle value during the boost operation.

By using eq. (4) we can express the C_{PAF} maximum and minimum voltage as follows:

$$V_H=\frac{V_{DC}}{1-d_{boost,max}}, \quad d_{boost,max}=d_{boost}(\square t)\big|_{\square t=0.22\square+k\square\ k=0,1,2,.} \qquad (5)$$

and

$$V_L=\frac{V_{DC}}{1-d_{boost,min}}, \quad d_{boost,min}=d_{boost}(\square t)\big|_{\square t=0.78\square+k\square\ k=0,1,2,.} \qquad (6)$$

In order to avoid time intervals of almost unitary duty cycle in Buck mode, the lower C_{PAF} discharge voltage V_L should be greater than the PV Module open circuit voltage V_{OC}. So by using eq. (6) we can conclude that

$$d_{boost,min} \geq 1 - \frac{V_{DC}}{V_{OC}} \qquad (7)$$

Moreover, combining eq (1), (5) and (6) we can express the C_{PAF} as follows:

$$C_{PAF} = \frac{0.422 \, P_{DC}}{V_{DC}^2 \left[\frac{1}{\left(1 - d_{boost,max}\right)^2} - \frac{1}{\left(1 - d_{boost,min}\right)^2} \right]} \qquad (8)$$

For the time intervals where the CPS-PAF behaves as a buck converter, the C_{PAF} voltage can be calculated according to the following equation:

$$V_{PAF}(t) = \frac{V_{DC}}{d_{buck}(t)} \qquad (9)$$

where d_{buck} is the duty cycle value during the buck operation.

By using eq. (9) we conclude that the C_{PAF} maximum and minimum voltage can be expressed as follows:

$$V_H = \frac{V_{DC}}{d_{buck,min}}, \quad d_{buck,min} = d_{buck}(t) \big|_{t=0.22+k} \, k=0,1,2,.. \qquad (10)$$

and

$$V_L = \frac{V_{DC}}{d_{buck,max}}, \quad d_{buck,max} = d_{buck}(t) \big|_{t=0.78+k} \, k=0,1,2,.. \qquad (11)$$

Since the lower C_{PAF} discharge voltage V_L should be greater than the PV Module open circuit voltage V_{OC} we can conclude that

$$d_{buck,max} \leq \frac{V_{DC}}{V_{OC}} \qquad (12)$$

Moreover, combining eq (1), (10) and (11) we can express the C_{PAF} as follow:

$$C_{PAF} = \frac{0.422 \, P_{DC}}{V_{DC}^2 \left[\frac{1}{\left(d_{buck,min}\right)^2} - \frac{1}{\left(d_{buck,max}\right)^2} \right]} \qquad (13)$$

Last but not least, combining eq. (5) and (10) as well as (6) and (11) the two following limitations are derived:

$$d_{boost,max} + d_{buck,min} \leq 1 \qquad (14)$$

and

$$d_{boost,min} + d_{buck,max} \geq 1 \qquad (15)$$

III. SIMULATION AND EXPERIMENTAL RESULTS

The enhanced CPS-PAF operation has been investigated by PSpice simulation results, for the case of an AC-PV module consisted of a single stage Current Source Flyback Inverter and a PV module with the following electrical characteristics at 1000W/m² solar irradiation, ambient temperature 25°C and AM equal to 1.5: P_{DC}=113,6W, V_{DC}=33.235V, I_{DC}=3.418A, I_{SC}=3.6A and V_{OC}=41.3V (I_{DC} is the value of the PV generator current at the maximum power point, I_{SC} is the PV short circuit current and V_{OC} is the PV open circuit voltage). The specific single stage inverter selection was based on its high power factor regulation, very high power density and high efficiency, due to its simple structure [4, 6-8]. According to eq. (1), by selecting V_L=50V and V_H=110V, the C_{PAF} capacitance would be only 2% of the required electrolytic capacitor C_{in}. Furthermore, according to eq. (2), the required electrolytic capacitance C_{in} ought to be 2.6mF in order to achieve PV voltage ripple equal to 8.5% of the PV voltage value at the maximum power point. Thus, the C_{PAF} capacitance is calculated to be 50μF.

Fig. 6 shows the PSpice simulation waveforms of the PV Module output voltage and current, the CPS-PAF input current, the primary $I_{TR(Pri)}$ and secondary $I_{TR(Sec)}$ transformer current, the flyback inverter input current (after filtering the high frequency current content), and finally the C_{PAF} voltage and the theoretical rectifier bridge voltage (no load conditions).

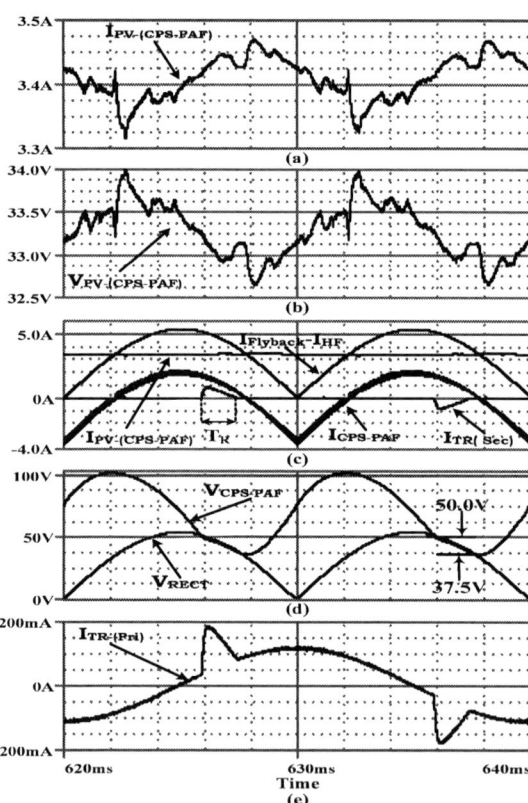

Fig. 6. Waveforms of the: a) and b) PV Module voltage and current respectively, c) CPS-PAF input current, secondary transformer current $I_{TR(Sec)}$ and flyback inverter input current (after filtering the high frequency current content), d) C_{PAF} voltage and the theoretical rectifier bridge voltage (no load conditions), e) primary transformer current $I_{TR(Pri)}$.

These simulation results prove that the enhanced CPS-PAF configuration performs satisfactory elimination of the low frequency PV current and voltage ripple. It is worth mentioning that the current ripple is only 4.39% of the I_{DC} value and the voltage ripple 4.06% of the V_{DC} value.

Fig. 7, 8, and 9 show the PSpice simulation waveforms of the PV Module output current, voltage, the CPS-PAF voltage and input current with the use of the former configuration and C_{PAF} capacitance equal to a) 50□F, b) 150□F and c) 500□F.

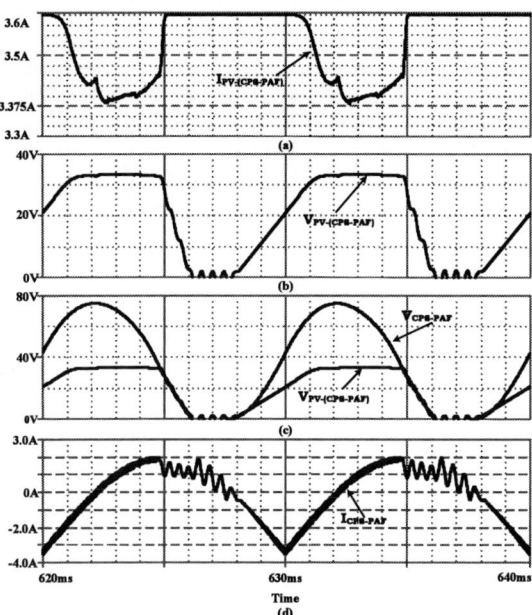

Fig. 7. Waveforms of the: a) and b) PV Module voltage and current respectively, c) C_{PAF} voltage and d) CPS-PAF input current (former CPS-PAF configuration with C_{PAF}=50□F).

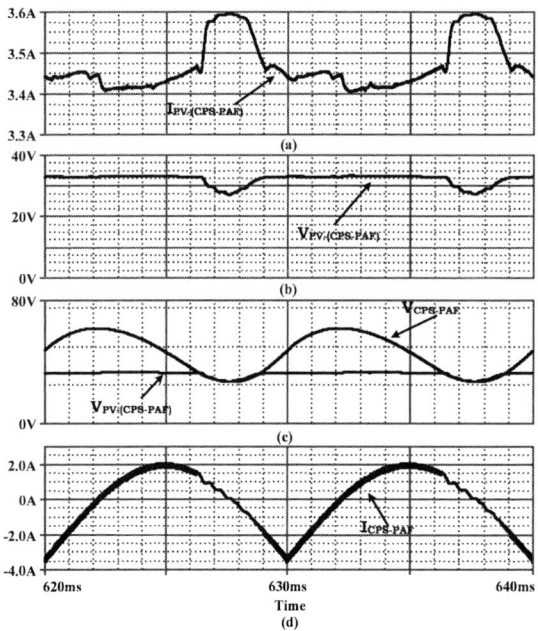

Fig. 8. Waveforms of the: a) and b) PV Module voltage and current respectively, c) C_{PAF} voltage and d) CPS-PAF input current (former CPS-PAF configuration with C_{PAF}=150□F).

By studying fig. 7 it is proven that the former CPS-PAF configuration fails to eliminate the PV voltage and current

ripple, when the C_{PAF} capacitance is small enough. In more details, although according to eq. (1) by selecting V_L and V_H values equal to 50V and 110V respectively, a 50□F C_{PAF} is the capacitance that permits theoretical zero voltage fluctuation, in practice this is impossible. The reason is the large divergence between the input and output CPS-PAF voltage values, causing a marginal boost PWM operation. Thus, the upper C_{PAF} charge level is limited at 75V, as it is shown from fig. 7.c, and so the energy that is stored to C_{PAF} during boost operation is not enough for the buck operation and the absorption of the PAF circuit losses. Consequently, the PV module operates at sort circuit region for large time interval of the buck operation (the lower C_{PAF} discharge voltage is equal to the PV Module voltage).

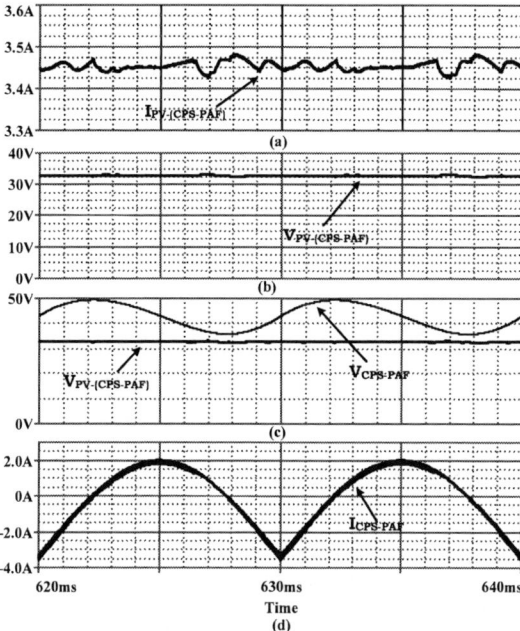

Fig. 9. Waveforms of the: a) and b) PV Module voltage and current respectively, c) C_{PAF} voltage and d) CPS-PAF input current (former CPS-PAF configuration with C_{PAF}=500□F).

On the other hand, by studying fig. 8 it is proven that a C_{PAF} capacitance of 150□F permits better PV voltage and current pulsation smoothing, since the theoretical C_{PAF} upper charge level (75V) is limited at values that can be achieved in practice with the boost operation. Of course the upper C_{PAF} charge level is smaller than the theoretical one due to the CPS-PAF circuit losses. Consequently, a considerable but not unacceptable voltage and current fluctuation is inevitable during buck operation.

Finally, by studying fig. 9 we conclude that the former CPS-PAF configuration demands a capacitance of 500□F in order to achieve approximately same PV voltage and current ripple elimination as with the use of the enhanced CPS-PAF. Taking into account the aforementioned remark, it is clear that the use of the enhanced CPS-PAF configuration instead of the former one leads to a remarkable capacitance decrease (roughly at 90%).

The enhanced CPS-PAF operation has been examined experimentally on a laboratory prototype, for the case of an AC-PV module with a BP MSX 110 PV generator and a single stage Current source flyback Inverter. The PV generator nominal electrical characteristics (1000W/m^2 solar irradiation, ambient temperature 25°C and AM equal to 1.5) were: P_{DC}=110W, V_{DC}=33.6V, I_{DC}=3.3A, I_{SC}=3.6A and V_{OC}=41.6V and the electric grid voltage was 220V$_{rms}$ – 50Hz. Fig. 10 shows, for the case of 70W PV generated electric power and the enhanced CPS-PAF utilisation, the experimental waveforms of: the PV Module output current, the enhanced CPS-PAF input current, the secondary transformer current I_{TR} and finally the flyback inverter input current (after filtering the high frequency current content).

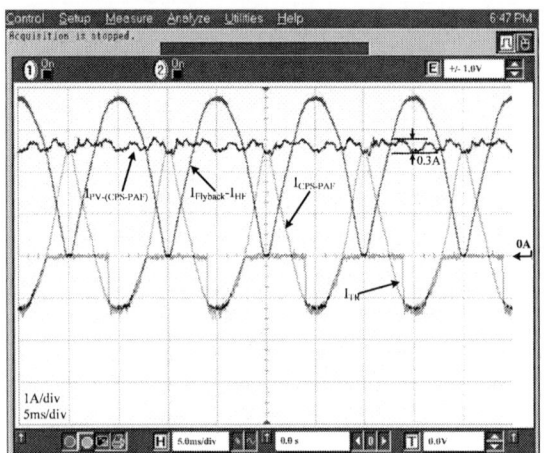

Fig. 10. Waveforms of PV Module output current, enhanced CPS-PAF input current, the secondary transformer current I_{TR} and flyback inverter input current after filtering the high frequency current content (experimental results).

Furthermore, fig. 11 illustrates the C_{PAF} voltage waveform in contrast to the enhanced CPS-PAF input current and the secondary transformer current I_{TR}, for the same PV power level (70W), ambient temperature and solar irradiation conditions. By studying these experimental results, we can conclude that the theoretical analysis is quite accurate, since the divergence among the experimental results and the corresponding theoretical and simulation values is slight. Last but not least, the whole system efficiency (from the PV dc output to the mains side) was 90.5%.

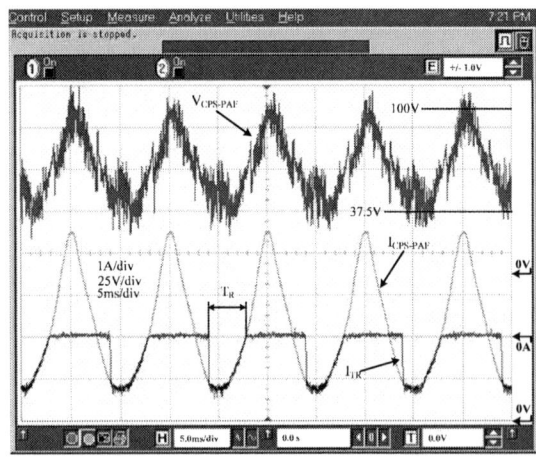

Fig. 11. Waveform of the C_{PAF} voltage in contrast to the enhanced CPS-PAF input current and the secondary transformer current I_{TR} (experimental results

IV. CONCLUSIONS

An enhanced configuration of the Current Pulsation Smoothing Parallel Active Filter (for single phase AC-PV modules), that drastically reduces the low frequency PV current ripple by using significant smaller capacitor comparatively to the basic CPS-PAF structure, has been presented. As in the basic CPS-PAF configuration, this enhanced design is independent from the inverter topology and its operation mode and hence it can be applied for various single stage topologies. Finally, experimental and simulation results have been exhibited, validating the effectiveness of the proposed circuit.

REFERENCES

[1] M. Calais, J. Myrzik, T. Spooner, and V. G. Agelidis: "Inverters for single-phase grid connected photovoltaic systems – An Overview" 33rd Annual Power Electronics Specialists Conference, PESC 02, Cairns- Queensland (Australia), 23-27 June 2002, vol. 4, pp. 1995-2000.

[2] M. Meinhrdt and G. Gramer: "Past, present and future of grid connected photovoltaic- and hybrid- power - systems", in Proc. IEEE-PES Summer Meeting, Seattle- Washington (USA) 16-20 July 2000, vol.2, pp. 1283-1288.

[3] S. B. Kjaer, J. K, Pedersen and F. Blaabjerg: "A Review of Single-Phase Grid-Connected Inverters for Photovoltaic Modules", IEEE Trans. on Industry Applications, Vol. 41, No. 5, pp. 1292-1306, September/October 2005.

[4] Kyritsis A.Ch., Tatakis E.C., Papanikolaou N.P., "Optimum Design of the Current Source Flyback Inverter for Decentralized Grid-Connected Photovoltaic Systems", *IEEE Transactions on Energy Conversion*, Vol 23, No 1, March 2008, pp. 281-293.

[5] Electrolytic Capacitors Application Guide, EVOX RIFA, ID 830G. [Online]. Available : www.evox-rifa.com

[6] T. Shimizu, K. Wada, N. Nakamura: "Flyback-type single phase utility interactive inverter with power pulsation Decoupling on the DC Input for an AC Photovoltaic Module System" IEEE Trans. On Power Electronics, Vol. 21, No. 5, pp. 1264-1272, September 2006.

[7] Soeren Baekhoej Kjaer, Frede Blaabjerg: "Design optimization of a single phase inverter for photovoltaic applications", in Proc. IEEE PESC'03, Accapulco (Mexico), 15-19 June, 2003, Vol. 3, pp. 1183-1190.

[8] G.H. Tan, J.Z. Wang and Y.C. Ji : "Soft-switching flyback inverter with enhanced power decoupling for photovoltaic applications", IET Electr. Power Appl., Vol. 1, No. 2, March 2007, pp.264-274.

[9] A.C. Kyritsis, N.P. Papanikolaou, E.C.Tatakis, "A novel Parallel Active Filter for Current Pulsation Smoothing on Single Stage Grid-connected AC-PV Modules," 11th European onference on power Electronics and Applications, EPE'2007, Aalborg (Demark), September 2007, paper on CD, Nr. 544.

[10] Mohan, Undeland, Robbins: "Power Electronics: Converters, Applications and Design", Copyright 1989, John Wiley & Sons, Inc.

[11] Muhammad H. Rashid: "Power Electronics Handbook", New York: Copyright 2001, Academic Press Inc.

Outline of the Design of a Cascaded H-bridge Medium Voltage STATCOM

R.E. Betz[†*], B.J. Cook[‡], T.J. Summers[†§], R. Fisher[‡], A. Bastiani[‡], S. Shao[‡], P. Stepien[‡], K. Willis[‡]

[†]School of Electrical Engineering and Computer Science
University of Newcastle, Australia, 2308
email:*Robert.Betz@newcastle.edu.au;
[§]Terry.Summers@newcastle.edu.au

[‡]ResTech Pty Ltd
University of Newcastle, Australia, 2308
email: [§]info@restech.net.au

Abstract—The University of Newcastle and its joint venture company, ResTech Pty Ltd, are developing a cascaded H-bridge based multilevel STATCOM. This tutorial paper outlines the salient design issues for this system. The issues covered in the paper include the choice of the converter topology, the structure of the control system hardware, the software structure and methodology, some details on the control algorithm, and the rationale behind the design decisions.

Index Terms—Static Synchronous Compensator (STATCOM), Multilevel converters, SVC Static Var Compensator, Power Quality

I. INTRODUCTION

Static Compensators (STATCOMs) are currently a topic of active research around the world [1]–[6] because they promise to be a viable and high performance replacement for the traditional thyristor based Static Var Compensator (SVCs). STATCOMs differ from SVCs in that they employ forced commutation techniques and higher switching frequencies compared to the line commutated SVC. The use of higher switching frequency and forced commutation gives the STATCOM several advantages over the SVC, namely:

- STATCOMs can respond faster to changes on the grid as they don't have to wait for 1/6th of the supply cycle as the SVC does to change their output.
- They can be used for traditional displacement power factor control, but in addition can be used as active filters to eliminate harmonics, flicker mitigation etc.
- The higher switching frequency means that their switching harmonics are easily filtered with modest sized filters between the STATCOM and the grid.
- With sophisticated control STATCOMs can rebalance unsymmetrical currents and voltages on the grid, and even provide transient dip support in the case of fault instigated grid disturbances.

In short, STATCOMs, by virtue of their higher switching frequency, are able to be controlled so that virtually any desired control objective can be achieved. SVCs are much more restricted in application because of their low effective switching frequency due to the natural commutation requirement of thyristors.

The topology and hardware of STATCOMs built to date varies depending on the voltage level that they operate at.

For example, very high voltage/high power STATCOMs are usually based on variants of GTO thyristor technology connected to the grid via transformers [7], [8]. The switching speeds of the devices is relatively low, with interleaved switching techniques and intricate transformer configurations being used to help eliminate the harmonics.

The development of power switching devices, and in particular IGBTs, has meant that IGBT based STATCOMs have been increasingly used at higher power levels. For example, a large European company uses IGBT technology for HVDC applications in the hundreds of megawatts power range. In general, IGBT systems switch faster than the GTO based topologies. The IGBT based STATCOMs are usually connected to the grid via interposing transformers.

In this paper we shall be considering an IGBT based converter topology, operating at the lowest level substation distribution voltage (in Australia this is 11kV line-to-line). Multilevel converter topologies offer the possibility of eliminating the bulky and lossy transformer from the STATCOM at such voltage levels. For higher voltage levels transformers would probably be used. Under this circumstance the advantage of using the multi-level technology is related to the above-mentioned harmonic performance, and the fact that current levels in the power electronics can be maintained at relatively lower levels.

There are three fundamental topologies of multi-level converters:

- the neutral point (or diode) clamped converter (NPC);
- the flying capacitor converter (FCC);
- and the cascaded H-bridge converter (CHC).

There are more multi-level topologies than this, but they are essentially composed of various novel combinations of these basic ones.

This paper will concentrate on the development of a STATCOM based on the use of cascaded H-bridges for the following reasons:

- The number of components in a multilevel CHC topology scales linearly with the number of output levels [5].
- High output level numbers are achievable with this topology making a medium voltage direct connection (i.e. no interposing transformer) STATCOM feasible.

978-1-4244-1741-4/08/$25.00 ©2008 IEEE

- The topology is very modular which makes construction, maintenance and provision of redundancy simple.
- The topology is very suited to applications that do not involve real power. Even though the CHC topology was conceived for medium voltage variable speed drives, the CHC is ideal for the STATCOM application because it does not require real power to be handled. This simplifies the CHC system, since a complex input transformer is not required to supply power to the individual H-bridges in the converter. VAR compensation and active filtering only mostly imaginary power to be handled, with only small amounts of real power that can be supplied from the grid.

ResTech Pty Ltd, a joint venture company between Ampcontrol Pty Ltd and the University of Newcastle, Australia, is developing a CHC direct connect STATCOM designed for 11kV applications. A detailed block diagram of the STATCOM system is shown in Fig. 1. The two main functional blocks are identified by the enclosed dashed lines. The power circuitry block consists almost entirely of the H-bridge hardware, the circuit breakers, the initial bridge charge control, and the associated transient over voltage protection circuitry. The H-bridges are connected together in a Wye configuration to minimise the number of H-bridges required for direct grid connection (although this increases their current rating).

The control hardware section of the system is quite complex, and is based on a multi-processor architecture. A Dual Processor PC is used as the main control computer, and three 32 bit microprocessors are used for the voltage control of each phase leg. This multi-processor architecture creates a high degree of decoupling between different blocks of the control algorithm. The dual processor central computer implements the highly numerical sections of the algorithm, whose output is the desired voltage to be produced by each of the phase legs over the next control cycle. The reference voltages are sent by the dual processor central computer to each of the phase control processors, whose function is to produce these voltages by selecting which H-bridges to use taking into account the voltage balancing issues for the individual H-bridge capacitors. As far as the dual processor computer is concerned the phase legs simply produce a desired voltage, and there is no need for it to be involved in the details of how this is implemented. The algorithms for H-bridge phase balancing of the three phase legs are independent of each other, therefore there is no need for any communication between the phase legs, which simplifies the communications architecture.

A design decision was made at the outset of the project to use as much off-the-shelf electronics hardware as possible so that the development time of the prototype was reduced. The choice of an industrial PC for the central control computer and Altera EPLD NIOSii development boards for the phase controllers has turned out to be an excellent decision in that these devices have successfully fulfilled the roles intended in the control system, and achieved the desired outcome of minimising the development of custom hardware.

The remainder of this paper will look in detail at each of the sub-sections in Fig. 1, and as well present an overview of the control strategies to be implemented in the system.

II. THE POWER CIRCUIT

The rationale for the choice of the CHC topology was presented in the previous section. In this particular implementation of the topology nine H-bridge modules are connected in series for each of the Wye connected phase legs. This means that each bridge supports approximately 1100 Volts DC, allowing the use of readily available 1700 Volt IGBTs as the power device, with reasonable head room still available to handle grid voltage events, the inevitable ripple on the DC link capacitor, and the voltage headroom required for control purposes. For simplicity reasons the prototype unit will use bond wire IGBT modules, but in the final system a press pack variant be considered because of the advantage that damaged press pack IGBTs tend to become a short circuit (as compared to open circuit for bond wire devices). Furthermore the final system will have redundant H-bridges modules for high reliability.

Fig. 2 shows the basic structure of a phase leg and the constituent H-bridge modules. One of the key requirements for the H-bridges were that they were completely modular, with the only connections to the units being optical fibres. All the internal power supplies are derived from their own DC bus via internal high-to-low voltage DC-DC converters. Analogue values required for the control algorithms are sampled using on-bridge module A/D converters, and the results are transmitted serially to the controlling computers.

There is a four tiered protection system for the phase legs and associated modules. The control system implements a current controlled voltage source, and the software current limits prevent large currents from being demanded. Therefore if the control system is working correctly over-currents should not occur. The second level of protection are hardware programmed current limits in the EPLD current hardware, which if exceeded trigger a software based over-current trip. Thirdly, to back-up the software over-current limit the individual modules implement hardware instigated V_{CE}^{sat} short-circuit protection, which in turn informs the control system of a failure resulting in an orderly shutdown. If the third level of protection fails, then the fourth and final layer is hardware circuit breaker protection.

III. CONTROL HARDWARE ARCHITECTURE

Fig. 3 is a detailed block diagram of the control hardware for the system. As mentioned previously it is designed as a multiprocessor architecture. This architecture was chosen because of the complexity of the control task for this system. Each phase leg of the system has to be controlled so that the capacitor voltage of each individual H-bridge is such that the total phase voltage is evenly distributed between the H-bridges, and at the same time

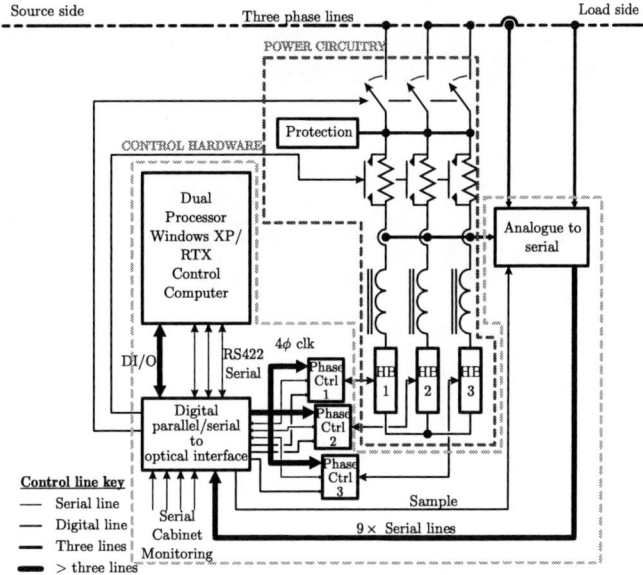

Fig. 1. Detailed block diagram of the proposed cascaded H-bridge multilevel STATCOM.

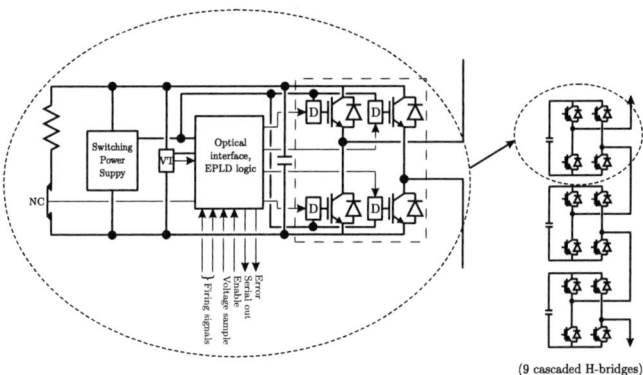

Fig. 2. Block diagram of the H-bridges used in the phase legs.

the overall desired output phase voltage is produced. The production of the desired output voltage, in this particular system, involves using PWM switching for one of the H-bridges in the H-bridge chain. Since the information to carry out these tasks only becomes available towards the end of the control cycle, considerable processing power has to be employed to achieve the this in the time allowed. This is the task of the phase controllers. As mentioned above, the calculation of the desired voltage sent to each of the phase controllers is undertaken in the central control PC, and it does not need to know the details of how this voltage is produced by the phase controller.

The distributed computing architecture requires efficient communications between the key processing units. High speed optical serial communications (3Mbits/sec) is used for sending measurements and commands between the processing units.

The phase leg processors are implemented using Altera NIOSii Development Boards which incorporate the Altera Cyclone EPLD. This approach was chosen because a full 32 bit microprocessor could be embedded into the EPLD along with the custom high speed serial communications channels required for communicating with the central PC controller and the individual H-bridges. Furthermore the use of the custom UARTs allowed non-standard length data to be efficiently sent via the serial communications (e.g. the A/D converters generate 12 bit data, and this data can be sent using a single data packet.). See Section V for information on the control algorithms implemented in the phase leg controller.

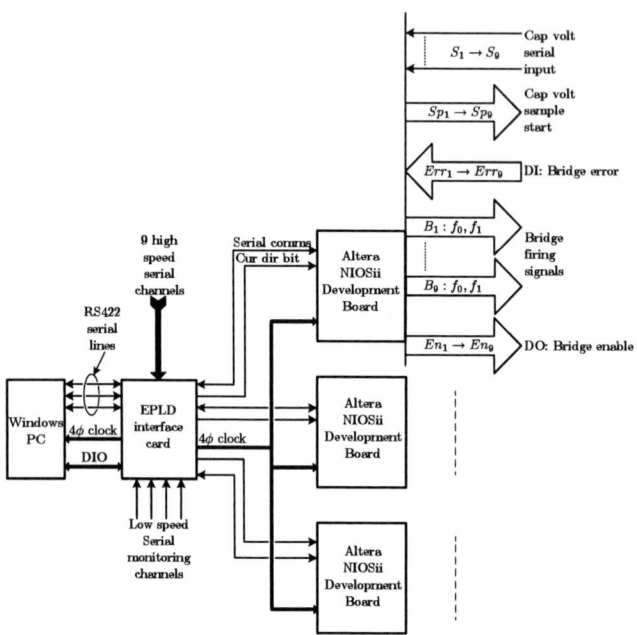

Fig. 3. Block diagram of the control hardware.

The fact that all of the communications is optical means that there is a total of 61 fibres from the phase controller, consisting of 54 fibres to the phase leg, and a further 7 fibres to the EPLD interface card to the control PC. System wide there is a total of 183 fibres involving the phase controllers. The number of fibres is this large because of a decision to use plastic fibre. This fibre has limited bandwidth, and was not fast enough to allow all of the information to be transmitted to the phase controllers and the central control PC using a single serial communications channel in the times required. Recent improvements in the bandwidth of plastic fibre my make it possible to significantly reduce the number of optical connections in the next version of the hardware.

The EPLD interface card shown in Fig. 3 has several functions. Firstly it converts the optical communications from the phase controllers to RS422 signals required for the high speed serial communications boards in the PC. It also generates the timing for the whole system in the form of a 4 phase clock that is distributed to all the control system components [1]. It also provides, via several custom programmed EPLDs, high speed serial channels to interface data from the voltage and current sensors in the system. As mentioned previously, custom serial receivers are used because the word length and data format is different than that normally supported by conventional UARTs.

The EPLD interface card can be interrogated by the central PC, which is the processing unit where the main control algorithms are executed. The central PC is an

[1]Equal length optical fibres are used so prevent any significant clock skew.

industrial PC with dual Pentium Xeon processors, a high speed RS422 serial card, and several other parallel digital I/O cards, as well as the normal networking interfaces available in a PC. The dual processor architecture was used so that the main control algorithm could run under a real-time operating system on one processor, and the other processor could attend to non-real-time tasks such as operator touch screen user interface, data logging, network interface, web server access to the system, and so on. It was felt that the Pentium processor would offer sufficient power to execute the control at the required rates. The use of a conventional processor, as compared to a DSP or embedded controller, allows the code to execute faster when written in 'C' because the more general purpose instruction set of these processor is able to produce more efficient code. In addition to this, the Pentium processor clock speeds are far higher than DSPs, and one is able to use floating point in all calculations without any degradation in speed. This makes writing the software simpler and easier to maintain.

Another significant advantage of using a standard Windows XP® PC for the control computer is that the software development tools are mature, reliable and easy to use.

IV. SOFTWARE ARCHITECTURE

The software and control hardware architectures are closely related. The phase controllers have a very specific and well specified task. Therefore the approach to the software design for this section is very simple, consisting of a small looping executive which implements a phase leg capacitor voltage balancing algorithm, as well as desired output voltage generation using PWM. The code for the phase controllers is written in 'C' to facilitate fast coding and easy maintenance.

There were several approaches available with respect to the software running on the central PC. One could again use a simple looping or interrupt driven executive, or a propriety real-time operating system, or even employ public domain operating systems such as real-time Linux. It was decided to use the proprietary Citrix/Ardence® real-time extension (RTX), which allows the execution of *true* real-time code whilst at the same time running Windows XP®. Windows XP provides powerful resources, such as networking access, a nice GUI environment, access to a disk system, and a solid and well tested software development environment.

The Ardence RTX real-time extension is not termed to be a real-time operating system, but is a Hardware Abstraction Layer/kernel extension to the standard Windows XP system that allows XP to operate as a hard real-time system. The RTX system handles all of the interrupts, and offers very fine process and thread priority control. The system comes with a set of libraries that contain routines to implement most of the normal functions of a real-time kernel – semaphores, locks, priority inheritance and disinheritance, and so on. In addition it has 'C' library emulation modules that allow real-time performance to be obtained when they are used. The RTX system is designed to integrate with the Microsoft Visual C++ .NET development environment. Code written for the system operates at Ring 0 of the processor, and therefore allows unfettered access to the hardware resources of the PC.

The Ardence RTX allows a mode of operation for multiprocessor architectures where the RTX code runs on one processor under the real-time kernel, and the non-real-time Windows code runs on the other processor under Windows XP. This approach allows a closer approximation to completely deterministic real-time performance, as compared to running the real-time software on the same processor as Windows XP. Testing of the RTX system indicated that its maximum interrupt latency is extremely low (of the order of 1μsec). Given that the control period of the STATCOM is of the order of 400μsec this is more than satisfactory.

Fig. 4 is a block diagram of the software structure for the central PC. One can see that it is designed so that all the software can operate either on a single physical dual processor machine, or if need be in client/server mode with the GUI code being executed on a separate machine connect via a local area network or via the Internet. This approach to the software architecture not only offers more options in the way the system can be physically configured, but it also decouples one section of the system from another making software development less complex.

In order to speed up the development process it was decided to write the non-real-time parts of the system in the interpretative high level language Python. This language has excellent support for multi-threaded programming, a standard GUI library based on wxWidgets, and excel-

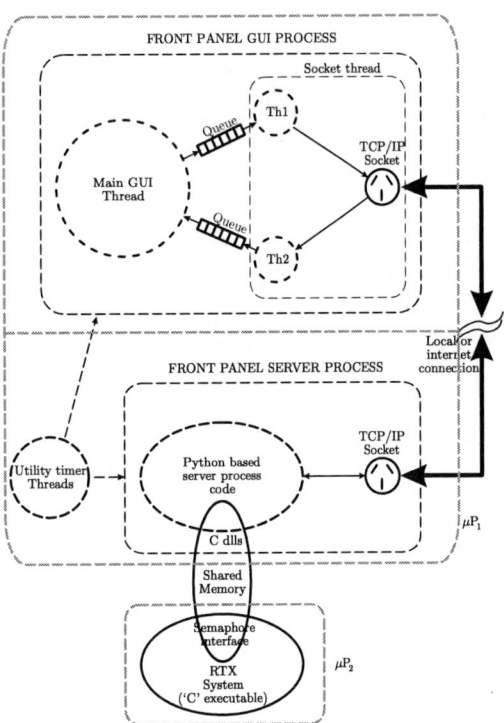

Fig. 4. Block diagram of the central computer software structure.

lent support for networking. In the applications where it is being used its interpretative nature does not cause execution speed problems. Furthermore it is reported to generate code of approximately 1/5th of the number of lines of other languages such as Java or C++ for the same functionality, and therefore allows rapid development.

The real-time code, as mentioned previously, is executed on a separate *processor*. However it also needs to communicate with the non-real-time Windows side of the system. This is implemented using a shared memory to pass data structures between the Python code and the real-time code. Particular attention was paid to ensure that under no circumstances will the real-time code be held up by this shared memory interaction. The synchronization between the two systems is implemented using RTX semaphore primitives.

Internally the Python processes have been implemented with multiple threads of execution using thread safe queue based message passing to prevent any critical section issues. There are no common shared data structures between the Python threads – the only interaction between threads is via message passing. This also means that the software threads can be developed and debugged with a large degree of independence. This approach also makes the addition of new features simple.

Fig. 5 is a conceptual diagram of the software for the front panel server process. Each circle is a separate thread in the server, and the arrowed lines indicate the

1297

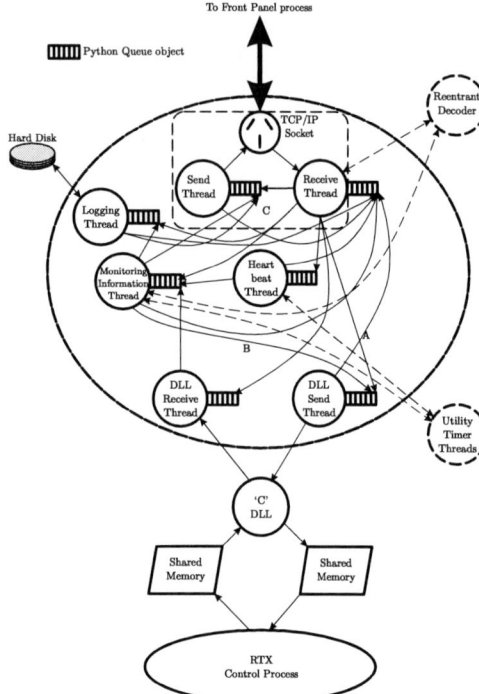

Fig. 5. Block diagram of the front panel server software.

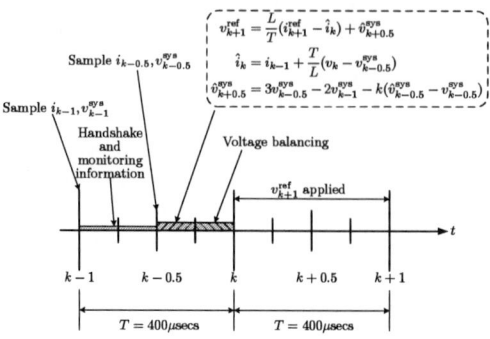

Fig. 6. Control timing for the central control computer.

various communications paths throughout the threads. As mentioned above the communications uses thread safe Queue objects which are natively supported in Python. All messaging passing has been organised using a "mailbox" and address approach, so that communications can be simply established between the various threads using the thread address that the message has to be sent to. The message format also contains a return address so that acknowledge protocols can be naturally handled. The receive thread uses a re-entrant table driven decoder to decode the address associated with a received message so that it can be sent from the front panel GUI process to the appropriate thread for action. The same decoder is also used b all the threads in the front panel server process.

The data sent via the Queue objects is serialised Python objects (encoded in ASCII). This allows simple communications of complex data structures between the server threads and between the server and front panel processes. All of the internal communications between the threads are subject to timeouts. A "heart beat" thread generates regular messages between the front panel server process and the front panel GUI process so that both processes are sure that the TCP/IP communications is working correctly and that the other process in the system is working. The user interface upon receipt of the heart beat message rotates an object on the screen so that the user has positive confirmation that the communications between the two front panel server and GUI processes is continuing to function. If a timeout occurs on the heart

beat, or if a timeout is violated between thread interactions in any of the threads of the system, the STATCOM is safely shutdown and a diagnostic message is passed via the TCP/IP channel (if available) to the front panel GUI process.

The optical serial communications channels also have error checking to make sure that data received is valid. There is a comprehensive error and warning system built into the software so that the errors are dealt with in a consistent way, and the system response is tailored to the nature of the error. There is also a comprehensive error logging system to complement this so that error data can be analysed for diagnostic purposes.

The Python front panel server process also provides an interface to the real-time code. This code is written in 'C' and physically runs on another separate process in the dual processor control computer. Windows XP cannot see the processor used for this process (this is hidden from Windows XP by the Ardence RTX), and therefore has no direct control over it. The communications between the processors is implemented using semaphore protected shared memory that is visible to both processors.

Python can interface directly to 'C' dynamic link libraries via several different techniques, and this allows data structures to be passed to and from the real-time system. It is intended in the final system that these data structures will be passed using XML so that precise alignment of the Python data structure with a 'C' data structure is not required (as would be the case if 'C' type structures are passed between the two systems). These shared memory interactions are implemented in such a way that the real-time system can never be blocked on the semaphore, ensuring that its real-time deadlines are always met.

Figure 6 shows the basic timing of the system. Samples of the currents and voltages of the system are taken precisely at the beginning and middle of the control cycles. There is a delay as the samples are serially transmitted to the control computer. The control cycle time is $T = 400\mu$sec, which is divided into four phases, each of 100μsec duration. The real-time code is written in 'C' and implemented as an event driven routine. It is suspended pending an interrupt from one of the four

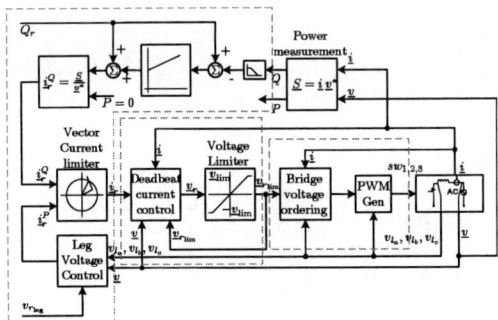

Fig. 7. Block diagram of the STATCOM control system.

phase clock edges. Upon receipt of an interrupt a port is read to check the control phase that the system is in and the appropriate code is executed based on this. The first two phases are concerned with receiving the sampled currents from the current transducers, as well as receiving monitoring information from the phase controllers. The actual current control algorithm is executed in the third phase of the control period (since it has to wait for the mid control interval voltage sample to compute the control). At the end of the third control period, the desired voltages for the phase legs are transmitted to the phase control processors. Therefore the phase control processors have a little less than $100\mu sec$ to execute the voltage balancing and PWM algorithms so that the firing times can be latched into the hardware timers that fire the power devices before the end of the control period.

V. CONTROL ALGORITHM

The control algorithm for the STATCOM is hierarchical in nature, following the structure of the hardware and software that was designed to execute it. A block diagram of the control algorithm is shown in Fig. 7. It should be noted that the entire control algorithm operates in a stationary reference frame.

The inner part of the control algorithm is associated with the control of the individual bridges and phase legs. At the lowest level this involves the control of the H-bridge capacitor voltages. These are controlled at the control rate – i.e. at 2.5kHz. In each control interval the current direction is used in conjunction with the state of charge of the H-bridge capacitors and the desired output voltage, to determine which H-bridges will be used to generate the desired output. This approach gives very accurate balancing of the capacitor voltages, and minimises the ripple current stress on the DC link capacitors. A converter model is employed to remove the effects of the diode and bulk resistance drops of the power devices. The output voltage is made more accurate by employing symmetrical PWM on one of the H-bridges being used to form the output voltage.

The next level in the control algorithm is the determination of desired voltage for the H-bridge legs. The approach taken in this STATCOM is that it operates as a

voltage controlled current source. This has the advantage that the control provides current limiting under fault conditions, and it that only instantaneous values of voltages and currents are required for the control computations, and does not require pre-computation of switching angles. The current control is a dead-beat (sometimes called predictive) current controller which is more commonly used in variable speed drive (VSD) applications [9], [10]. These algorithms have very fast transient response and are computationally very simple. In this implementation of the algorithm the "back-emf" does not have to be estimated (as is the case in the VSDs), since it is the grid system voltage and can be measured. There is however an issue that the grid voltage must be estimated one control interval ahead of where the control is evaluated. This can be seen in the control equations (see Fig. 6) where the $\hat{v}^{sys}_{k+0.5}$ value is used in the development of the reference voltage v^{ref}_{k+1}. If the $\hat{v}^{sys}_{k+0.5}$ value is not known with reasonable accuracy then the current will have significant errors. A special digital phase locked loop is used to generate these estimates, and it has the added advantage that noise on the grid supply voltage is minimised. An additional advantage of this algorithm is that precise knowledge of parameters such as the connection inductance are not required, and the control is dependent only on instantaneous samples and does not rely on precise knowledge of grid system phase. Futhermore, because of the 2.5kHz control rate the harmonics produced are small and the filtering requirements are minimal.

The outer control loops of the system look after the power control. This can be divided into the real power control, which is very important as the real power control manages the overall voltages on the individual phase legs. Because of issues such as lack of symmetry of the grid supply, and the inevitable variations in the values of the bridge components, the real power calculations are carried out on a per phase leg basis. The net result a desired two phase dq real power current that forms one component of the two phase current reference for the current controller.

The other current component is associated with the desired imaginary power that the overall control strategy desires. Using the instantaneous imaginary power expression (from PQ theory) a reference imaginary power current is generated. The combined current references are passed to a vector current limiter, which ensures that the desired current does not exceed the STATCOM current limit. This limit is carried out in such a way that the real component of the current is preserved in both magnitude and phase (if possible), but always with respect to phase. The imaginary power current component is limited before the real power current, since real power control is essential for the control of the H-bridge leg voltages. The limited value is then passed to the dead-beat current control algorithm to generate the reference output voltages.

VI. CURRENT STATUS

At the time of writing this paper a low voltage prototype of the above-mentioned system has been constructed and is currently being tested. The power hardware used for this low voltage prototype is based on MOSFETs. The unit is rated to operate from a 415VAC line-to-line

1299

Fig. 8. Photograph of the low voltage prototype STATCOM.

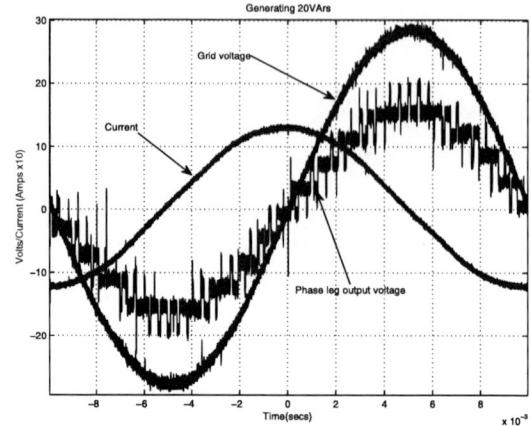

Fig. 9. Experimental result showing phase output waveforms for the low voltage prototype system.

supply. A basic VAr compensation control algorithm has been implemented on this prototype. The control system hardware is the identical hardware for the 11kV prototype system – i.e. as far as the control hardware and algorithms are concerned they are sending communications to the 11kV H-bridges. Fig. 8 is a photograph of the low voltage prototype system. The cubicle on the left contains the phase leg processors, the MOSFET based H-bridge hardware, and the central control computer. The three phase legs and their associate phase leg control boxes are clearly visible, and the box at the top of the cubicle is the control computer. The cubicle on the right contains the three phase load, a California Instruments® programmable three phase power supply, and associated contactors and circuit breakers.

Fig. 9 shows a preliminary oscilloscope trace of the output waveforms for one phase of the prototype unit. One can clearly see the multi-level phase voltage output, and the current which is 90° out of phase with this voltage. The phase leg output waveform is in phase with the system (grid) voltage, but of less magnitude. Therefore as far as the grid is concerned the STATCOM looks like an inductor in this case. At the time of writing this paper further testing is being carried out, and the control strategies are being refined as a result of this testing. More comprehensive results will be presented at the conference.

VII. Contributions

The main contributions of this paper are:

- presentation of the overall structure and design of a 19 level cascaded H-bridge STATCOM.
- discussion of the rationale behind the design decisions for the hardware, software and control algorithms.
- a more detailed description of the key functional hardware and software components of the system.
- presentation of some very preliminary output results from the prototype STATCOM.

Acknowledgment

The authors would like to acknowledge the financial support of Australian Research Council Linkage Program for this project.

References

[1] B. Blazic and I. Papic, "Improved d-statcom control for operation with unbalanced currents and voltages," *IEEE Transactions on Power Delivery*, vol. 21, no. 1, pp. 225–233, Jan. 2006.

[2] Z. Du, L. M. Tolbert, J. N. Chiasson, and B. Ozpineci, "A cascade multilevel inverter using a single DC source," in *Applied Power Electronics Conference and Exposition, 2006. APEC '06. Twenty-First Annual IEEE*, Mar. 19–23, 2006.

[3] H. Masdi, N. Mariun, S. M. Bashi, A. Mohamed, and S. Yusuf, "Design of a prototype d-statcom using DSP controller for voltage sag mitigation," in *Power Electronics and Drives Systems, 2005. PEDS 2005. International Conference on*, vol. 1, Jan. 2006, pp. 569–574.

[4] C. Hochgraf and R. H. Lasseter, "Statcom controls for operation with unbalanced voltages," *IEEE Transactions on Power Delivery*, vol. 13, no. 2, pp. 538–544, Apr. 1998.

[5] J. Rodriguez, J. S. Lai, and F. Z. Peng, "Multilevel inverters: A survey of topologies, controls, and applications," *IEEE Transactions on Industrial Electronics*, vol. 49, no. 4, pp. 724–738, August 2002.

[6] C. A. C. Cavaliere, E. H. Watanabe, and M. Aredes, "Multi-pulse STATCOM operation under unbalanced voltages," in *Power Engineering Society Winter Meeting, 2002. IEEE*, vol. 1, Jan. 2002, pp. 567–572.

[7] C. Schauder, M. Gernhardt, E. Stacey, T. Lemak, L. Gyugyi, T. W. Cease, and A. Edris, "Operation of ±100 MVAr TVA STATCON," *IEEE Trans. Power Del.*, vol. 12, no. 4, pp. 1805–1811, Oct. 1997.

[8] C. Schauder, "STATCOM for compensation of large electric arc furnace installations," in *Power Engineering Society Summer Meeting, 1999. IEEE*, vol. 2, Edmonton, Alta., Jul. 1999, pp. 1109–1112.

[9] R. Betz, B. Cook, and S. Henriksen, "A digital current controller for three phase voltage source inverters," *Conference Record of the IEEE-IAS Annual Meeting*, pp. 722–729, New Orleans, Oct. 1997.

[10] R. E. Betz, G. Mirzaeva, and D. Pulle, "Frame alignment stability issues in natural field orientation," School of Electrical Engineering and Computer Science, University of Newcastle, Australia, Tech. Rep. V1.84, May 2007. [Online]. Available: http://eecsbobb.newcastle.edu.au/rebetz/Reports/NFO_stability.pdf

Investigation of High Frequency Effects on Layered Coils

Georgios S. Dimitrakakis and Emmanuel C. Tatakis

University of Patras, Department of Electrical & Computer Engineering
Laboratory of Electromechanical Energy Conversion, 26504 Rion-Patras, GREECE
email: jimakakis@ece.upatras.gr, e.c.tatakis@ece.upatras.gr

Abstract—Copper losses in magnetic coils depend on several geometrical parameters, as well as on frequency, in a way that makes their modeling a quite difficult task. In this paper a Finite Element Analysis (FEA) software is utilized for the investigation of a series of issues critical for the accurate determination of copper losses in layered coils with round wires or foils. Some of the issues investigated are the edge effect in foil and round wire windings as well as the effect of the winding pitch on the copper losses. The results that come up from this work help to fully understand the real impact of two dimensional (2D) effects in layered windings of real rather than ideal magnetic components and they consist a tool for the accurate calculation of the encountered losses, which is necessary for an optimized magnetic component design.

Keywords—Modeling, passive component, simulation, transformer.

I. INTRODUCTION

The most extensive works on the issue of high frequency losses in copper coils, which stood as reference points for the theoretical approach of it, where those of S. Butterworth ([1], [2]), E. Bennet and S. Larson [3], P. Dowell [4] and J. Ferreira [5] as also enhanced by Bartoli *et al.* [6]. Each of these was broadly used in the design of transformers and inductors ([7], [8], [9]) for the frequency range where the capacitive currents of the coils can be neglected. For the establishment of the former analyses, several assumptions and approximations are introduced, which invalidate their predictions over numerous practical applications. These assumptions have to do with the geometry of the coil and the corresponding form of the leakage flux. More specifically, the assumption of a leakage flux through a conductor parallel to the symmetry axis of the magnetic component (one dimensional field analysis) is often violated. The actual flux through the conductors has such a form that, until now, seems very difficult to be successfully treated by any analytical method in a complete way.

Some authors ([10], [11], [12], [13]) have criticized the forementioned models and attempted to give expressions for a more accurate description of experimental or simulation results. These works added much upon the

better understanding of high frequency copper losses but they remained within the limitations of the initial models and when taking some extra 2D effects into consideration they often led to quite complicated expressions or focused only on part of the possible winding configurations. Moreover, the attempts so far of investigating layered windings seem to partially supplement the basic theoretical analysis of P. Dowell. At most of the times they offered complicated formulas and coefficient tables, without giving clear indications of when Dowell's formula is valid without modifications or what are the application limits of the new proposed expression, for the establishment of which, cases with poor practical value are often analyzed.

For layered round wire coils, copper losses calculations are carried out until now over Dowell's expression (most often used), with many designers not being aware of the possible error introduced, since literature offers only some scattered information about Dowell's model inaccuracy for coils not compactly wound ([10], [11], [12], [13]). In these works the formula of J. Ferreira is described as "inaccurate" without direct reference to the magnitude of deviation from reality or from the corresponding one of Dowell, an issue thoroughly investigated in this paper. Moreover, despite that the air gap effect seems to have been sufficiently studied ([14], [15], [16], [17]), this is not the case with edge effect ([3], [10], [12]) about which there is the generally approved impression that it increases the losses which, as will be presented, is only partially true.

Under this scope FEA is the key to investigate and quantify the effect of real rather than ideal or hypothetical winding geometries on high frequency copper losses. In this paper a full range FEA investigation of current density and magnetic flux distributions in round conductor and foil windings is carried out. As a result of this extensive work some qualitative, easily comprehensive conclusions help to fully understand the extent of impact of 2D effects on the losses encountered in layered windings. Some of the issues investigated are the edge effect in foil and round wire windings as well as the effect of the winding pitch and the interlayer distance on the overall copper losses. A quantitative comparison between the existent formulas and FEA results gives a clear picture of the most appropriate calculation method for copper losses in layered windings according to the specific values of the involved parameters. Necessary to note that, the present discussion describes classical winding configurations and does not refer to planar on-board windings.

This work was funded by the European Social Fund (75%), the Greek Secretariat Research and Technology (25%) as well as ANCO S.A. and ENERGY SOLUTIONS S.A., within the framework of Measure 8.3 of the Operational Programme "Competitiveness" and the 3rd Community Support Programme (PENED 2003, 03EΔ400).

978-1-4244-1741-4/08/$25.00 ©2008 IEEE

II. PRESENTATION OF THE BASIC THEORETICAL MODELS FOR LAYERED WINDINGS

The object of the classical theoretical analyses so far ([1], [2], [4], [5]) is the calculation of the effective resistance R_{ac} of solenoid coils with round conductors when a high frequency sinusoidal current flows through them. These coils extend between a zero and a maximum mmf point (referred to as winding portions in Dowell's work [4]). From the final expressions obtained someone can calculate the resistance factor $F_R = R_{ac}/R_{dc}$, where R_{dc} is the resistance at direct current of constant value. Basic assumption in any case is that the coil is formed by successive layers with a large number of turns, equal for all layers and with the same distance d_t between adjacent turns. For the calculation of F_R in these models, the values of four parameters are necessary: the frequency f, the radius of the round conductor r, the number of layers m and the filling factor η as defined in Dowell's work [4].

Butterworth's work ([1], [2]), compactly presented in [7], is mainly an attempt to describe with an analytical solution the available by that time experimental data on the effective resistance of solenoid coils with round conductors in the absence of magnetic core. He derived formulas for the F_R calculation of single layer as well as multilayer coils with widely spaced turns and for single layer coils with any spacing ratio. Among the goals of this paper is the comparative presentation of Butterworth's work with posterior ones aiming to the calculation of R_{ac} of coils with magnetic core. For this we will consider his results for the case where the assumed conditions ensure that the radial component of the magnetic field has minimum effect on the ohmic losses, i.e. when $l/D > 10$, where l is the length of the air cored solenoid and D its diameter. According to [1] and [2] this condition approximates what is called in literature an "infinitely long solenoid", with mmf being equal to zero right outside the solenoid and higher values of the ratio l/D do not lead to any prominent changes on R_{ac}. In the followings, s is the center-to-center distance between adjacent turns of one layer, d is the diameter of the round conductor ($d=2r$) and δ is the skin depth at given frequency. For a single layer solenoid with any spacing between conductors and $l/D > 10$, Butterworth gave the expression:

$$F_R = a(r/\delta, d/s) \cdot H(r/\delta) + \\ + K_1 \cdot \gamma(r/\delta, d/s) \cdot (d/s)^2 \cdot G(r/\delta) \quad (1)$$

For multilayer windings with widely spaced turns ($d/s < 0.6$) and for $l/D > 10$, the corresponding expression is:

$$F_R = H(r/\delta) + K_2 \cdot [p(m) \cdot m \cdot (d/s)]^2 \cdot G(r/\delta) \quad (2)$$

where the terms H and G are indicated respectively as the skin effect and the proximity effect portions on F_R and K_1, K_2 are constants. In [1] and [2] one can find the analytical expressions of all the quantities appearing in (1) and (2). Note that in (1) and (2), for the sake of uniformity with the followings, the dependence on frequency is expressed by the ratio r/δ instead of Butterworth's z ([1], [2]), since both are proportional to $\sqrt{f}$. Moreover, it must be noticed that Dowell's η parameter is proportional to the

ratio d/s used by Butterworth. The case of single layer coils with widely spaced turns ($d/s < 0.6$) is described by (2) if one substitutes $m=1$. The reason that Butterworth did not extract a formula for compactly wound multilayer coils is that main assumption at his model was the uniformity of the external proximity field acting upon each single conductor [1], which approximately holds only for sparse windings.

Dowell's analysis [4] on the high frequency copper losses offers a compact formula for the calculation of F_R for a coil with round conductors and multiple layers. The coils under study are wound on a high permeable magnetic core and the winding extends across the whole window width. These two conditions ensure that the flux is y oriented across the whole of the winding and that no edge effect is present. The several windings of it, regarding the case of a transformer, are wound successively one over the other along the x axis, perpendicular to y. For the facilitation of his calculations he considered the existence of equivalent conductors with square cross section of the same area instead of the round ones. The real part of his final equation gives F_R as:

$$F_R = Z \frac{\sinh 2Z + \sin 2Z}{\cosh 2Z - \cos 2Z} + \\ + \frac{2(m^2 - 1)}{3} \cdot \frac{\sinh Z - \sin Z}{\cosh Z + \cos Z} \quad (3)$$

where:

$$Z = (r/\delta)\sqrt{\pi\eta} \quad (4)$$

For foil windings (3) can be applied by substituting $Z=h/\delta$, with h the foil thickness. It has to be mentioned that Dowell's result for the subcase when $\eta=1$ coincides with the formula given in [3], which was a work exclusively on foil windings. Dowell's originality was just that he assumed the same flux and current density forms as in foils (y oriented flux, current density dependent only on x) when the conducting layer is only partially filled by copper, i.e. when it consists of square cross section conductors.

Finally, J. Ferreira [5], based on the orthogonality between skin and proximity effects, calculated copper losses under the same conditions as in Dowell's model (many turns per layer, high permeable core) but for the actual round shape of the conductors. Special assumption was that the magnetic flux through a conductor due to neighboring ones is parallel to the symmetry axis y of the coil. His expression for F_R did not account for the filling factor and was found to give an extremely high error [6]. In order to improve its accuracy, the authors in [6] introduced Dowell's filling factor η in the initial formula giving to it its final form:

$$F_R = \frac{q}{2}\left[r_{skin} - 2\pi\eta^2 \left(\frac{4(m^2 - 1)}{3} + 1 \right) r_{prox} \right] \quad (5)$$

where:

$$q = \frac{\sqrt{2}r}{\delta} \tag{6}$$

$$r_{skin} = \frac{ber(q)bei'(q) - bei(q)ber'(q)}{[ber'(q)]^2 + [bei'(q)]^2} \tag{7}$$

$$r_{prox} = \frac{ber_2(q)ber'(q) + bei_2(q)bei'(q)}{[ber(q)]^2 + [bei(q)]^2} \tag{8}$$

with r_{skin} and r_{prox} the skin and proximity effect terms respectively. In the next section we investigate the validity of (1), (2), (3) and (5) and comment upon previous work on this issue.

III. FEA STUDY OF THE LOSSES IN LAYERED WINDINGS – INVESTIGATION OF THE EXISTENT MODELS

While accurate experimental measurements are the most reliable way to contrast between theoretical models and practical applications, computer aided simulations are the appropriate one to validate a model created by analytical methods. Two critical factors in this process are the accuracy of the method that the software utilizes to solve a problem and the feasibility of reproducing in its environment the conditions under which a model is established. Considering FEA, the first factor is out of question since Maxwell's equations are solved, regardless of the method for solving the differential equations system or that for creating the finite element mesh. On the other hand, reproduction of the model conditions and selection of the proper boundary conditions is mainly a matter of the software user skills rather than the facilities or limitations of the software itself. For the present work the software used was Vector Fields – Opera 2D [21].

A. Coils with Square Cross Section Conductors and Foils.

As a first task, numerous finite element simulations were carried out in order to check the validity of the models about copper losses in layered windings under high frequency sinusoidal excitation. First issue of study was the leakage flux, current density and effective resistance of coils wound with square cross section conductors or conductive foils extending across the whole width of the core window. Square cross section conductors in magnetic component coils are not actually met in practice (they are used in high power – low frequency applications for the achievement of high copper filling factors) but study of this case was necessary to investigate the validity of Dowell's result. Moreover, as already mentioned, a foil winding is the limiting case of a square conductor layer when $\eta \to 1$. Maximum number of layers at the simulations was $m_{max}=4$ and at the maximum frequency it was $(h/\delta)_{max}=5$, with h the edge of the square conductor or the thickness of the foil. Windings with filling factor down to $\eta=0.5$ were studied. The following conclusions about Dowell's formula were extracted:

- For foil windings, F_R is accurately calculated by Dowell's model regardless of the number of layers or the frequency. Leakage flux is parallel to y axis and current density distribution depends only on x dimension, just as it was assumed in [4].
- For square conductors it is accurate for $\eta>0.7$. For $0.7>\eta>0.6$ it gives a good approximation with

maximum error of about +15% at the $\eta=0.6$ limit (independent of h/δ or m if $h/\delta>3$ and $m>2$). For $0.6>\eta>0.5$ the error of Dowell's formula is already between +20% and +40%.

In Fig. 1 the error E_s of Dowell's prediction compared to the FEA result is plotted versus filling factor η for $m=3$ and with the ratio h/δ as a parameter, where h in this case is the edge of the square conductor. Notice that E_s remains approximately constant for $h/\delta>3$.

It is evident from the previous simulations that the accuracy of Dowell's formula follows after the fulfillment of the condition for only a y component of the leakage flux. Current distribution, which screens the bulk of copper from this magnetic flux, has of course the same form (depends only on x). With decreasing η and increasing frequency 2D effects take place, the effective cross section increases and Dowell's model becomes inaccurate. This attribute is clearly demonstrated in Fig. 2 where a two layers winding with $\eta=0.625$ is shown, simulated at (a) $h/\delta=1$ and (b) $h/\delta=2$. Due to the relatively small η value the magnetic flux and current density distribution show a 2D form in the conductors even for $h/\delta=1$. At $h/\delta=2$ the deviation from Dowell's one-dimensional assumed forms is much more prominent.

Strictly talking, Dowell's formula gives the losses in layers of square conductors (with the limiting case of foils). It was extracted under simple electromagnetic considerations and describes accurately the current density distribution and losses for those η values where the flux is y directed. Under this scope, η has a very clear physical sense when applying Ampere's law on the whole length of a long coil with many turns per layer. This calculation indicates that proximity effect over a conductor is due to all the other conductors in the winding and a small change in the path of integration at y direction does not alter considerably the final result. What matters is whether it is the same as accurate with round conductors, since calculating η incorporates the middle logical step of considering square cross section. Thus the question is about the validity of Dowell's model on round conductors, the answer on which comes only from the result. After all, each model is an approximation of real conditions and Dowell's idea to approximate round conductors with square ones, judging from the result, proved to be an excellent idea, at least for not very sparse windings.

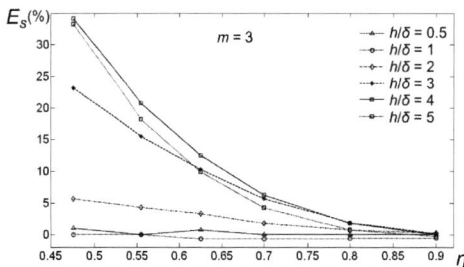

Fig. 1 – The error E_s of (3) for square conductors, versus filling factor η, with h/δ as parameter, for $m=3$.

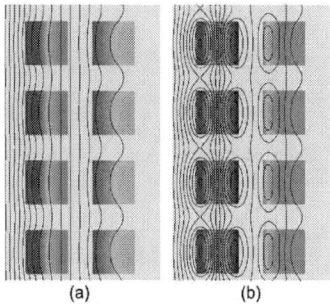

Fig. 2 – Current density and flux lines for a square conductor winding with two layers at (a) $h/\delta=1$ and (b) $h/\delta=2$. Violet color indicates high positive and blue high negative current density values.

B. Coils with Round Wire

For coils with round conductors, a large number of simulations were carried out in order to have a satisfactory sweep of the parameters involved, so as to cover all the cases referring to practical applications. More specifically, the ranges of r/δ, m and η had as following: $0<r/\delta\leq5$, $1\leq m\leq4$ and $0.47\leq\eta\leq0.865$. The layers simulated had a reasonably large number of conductors. The study excluded the first few conductors in the vicinity of the ferrite yoke since, no matter how small is the distance of the end conductors from the ferrite, there is always an imperceptible edge effect. The relative permeability of the ferrite was set to $\mu_r=2000$, while higher values or lower (down to 150) do not lead to remarkable changes on the leakage flux path or the effective resistance of the winding. The primary winding (a single sheet) was kept far from the short circuited secondary under study, despite that simulations showed only little effect on F_R for even very small distances.

At this point some remarks are necessary about the selected range for each one of the two critical parameters η and r/δ. Optimization reasons lead to SMPS designs with magnetic components operating at a fundamental frequency for which it is approximately $F_R=1.5$, a case which corresponds to frequencies close to the value $r/\delta=1$ [9]. In most of the cases the harmonic content of the current waveform gathers within the first few harmonics, although at some power electronics applications harmonics of the order of the 15th may have non-negligible magnitude. Under this point of view, study of the copper losses with $(r/\delta)_{max}=5$ is more than sufficient. Moreover, for a magnetic component designed for a switching frequency at which it is $r/\delta=1$, it is quite possible that current harmonics of order higher than 25, if present, flow as capacitive currents in through the volume of the winding window. On the other hand, in layered windings with carefully wound wires, the distance between adjacent turns is in no case much greater than twice the thickness of the wire's insulation, which is about 5-15% of its radius. If we also account for some extra distance further than the insulation and suppose a total distance between adjacent turns of $d_t=0.6r$, it will still be $\eta>0.68$. Thus the selected ranges for η and r/δ are satisfactory for studying most of the practical application designs with layered windings. If the designer consciously chooses not to pay attention on the winding method, there

will be a non-layered configuration ([8], [18]), a case not described by the forementioned models or the present analysis.

The conclusions extracted from this work can be summarized as following:

Butterworth's formula (1) for single layer infinitely long coils with any spacing:

- In the whole η range studied F_R is calculated with nearly negligible error (even for sparse windings with $\eta\cong0.5$).

Butterworth's formula (2) for multilayer, infinitely long windings with widely spaced turns in a layer:

- In the whole η range studied, F_R is calculated with an approximate error of +6% at $r/\delta=1$ but gives a considerable overestimation at higher frequencies. This high error for $r/\delta>1$ reduces to 20-40% at the low η range ($\eta<0.6$).

Dowell's formula (3):

- For $\eta>0.7$ practically coincides with the simulation results with maximum error +5% at $\eta=0.7$, $r/\delta=5$.

- For $0.7>\eta>0.6$ it is quite accurate with +5% error at $r/\delta=1$ up to +15% at $r/\delta=5$.

- For $0.6>\eta>0.5$ the error still remains at +5% at $r/\delta=1$ but can exceed +30% at higher frequencies.

Ferreira's formula (5):

- For $\eta>0.7$ overestimates F_R with an error between +15% and +140%.

- For $0.7>\eta>0.6$ shows a +8% error at $r/\delta=1$ but it reaches +70% at higher frequencies.

- For $0.6>\eta>0.5$ the error reduces to about +5% at $r/\delta=1$ but it still reaches +50% at higher frequencies.

- More or less, regardless of the specific m, η or r/δ values, Ferreira's formula (5) and Butterworth's formula (2) give approximately the same result.

It can be readily concluded that Dowell's formula is far more accurate even when low filling factors are involved. As an illustrative example of the former discussion, Fig. 3a shows a comparative graph between the result of the three models and FEA for a case when $m=3$ and $\eta=0.739$, while Figs. 3b and 3c are for $\eta=0.633$ and $\eta=0.554$ respectively. It is evident that using (5) or (2) (referred in literature also as "Bessel solutions") leads to a high error which is because of the assumption, mentioned in [5] as well as in [1], that any x component of the flux within a conductor is only due to the current in it and that proximity effect (which in any case is the dominant effect) gives only a y oriented flux. This assumption is untrue except when low frequencies and very sparse windings are involved (it is reminded though that (2) was extracted under the scope of sparse windings). This fact is clearly demonstrated in Fig. 4 where the magnetic flux lines are shown at the winding space from where a conductor has been removed. While the absence of one conductor does not considerably alter the current density distribution in the neighboring ones, the magnetic flux in the empty space is generally more than just y oriented, although this attribute is less keen for low f and η values. We can see that in Fig. 4a ($\eta=0.633$, $r/\delta=1$) the flux form is much closer to Ferreira's model assumption than in Fig 4b ($\eta=0.633$, $r/\delta=2$).

1304

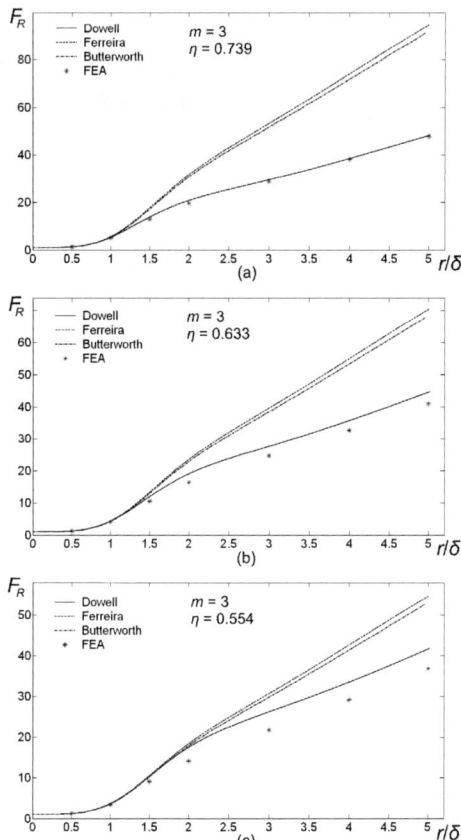

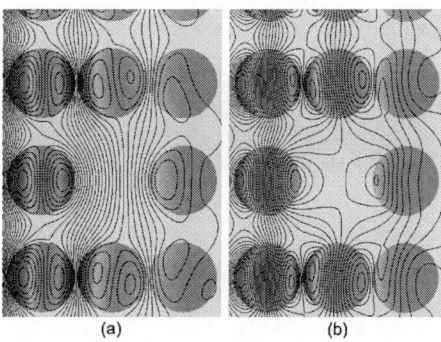

Fig. 4 – Flux lines in the space of a missing conductor in a winding with m=3 at (a) r/δ=1 and (b) r/δ=2. In the illustrated configuration it is d_f<d_l.

Fig. 3 – Comparative graph of the F_R result of the three basic theoretical models for round conductor coils for m=3 when (a) η=0.739, (b) η=0.633 and (c) η=0.554.

Necessary to remind here that, in all discussed analytical models, mmf is assumed constant between successive layers and thus F_R does not depend on the interlayer distance d_l. From the simulations it comes up that in fact it slightly affects the mmf diagram, the current distribution in the conductors and eventually the F_R value. From the case where layers are very close to each other (d_l =0.05r) up to that of very distant layers, F_R is reduced by about 10% at the most prominent case of very low η values. At high η values the effect is negligible. For the proper interpretation of all the forementioned conclusions it has to be pointed that the results of the case where d_f=d_l were used.

As an end of this section it has to be mentioned that in recent literature there have been some works in the direction of modeling high frequency losses in layered windings in the forms of analytical solutions ([13], [19]) or formulas extracted with the use of computer aided FEA ([11], [12], [20]). However, their limited application fields or the complexity in using their results considerably suppress their main advantage of improved accuracy and Dowell's formula still remains a top preference for quick calculations, especially for laboratory applications.

IV. EDGE EFFECT IN MAGNETIC COMPONENTS

The edge effect in magnetic component windings is a 2D effect, which in some cases has a negative impact on copper losses but in many others, as will be shown in the followings, it doesn't really affect them or it even improves the overall performance of the magnetic component, despite that potentially it may cause some hot spot. Even so, authors so far don't pay much attention on the real impact of edge effect, giving the general impression that it is something that has to be eliminated by all means. Unfortunately all this information is scattered in literature and most of the times the relative research suffers from limited application angle of view (as for example when focusing on single layer foils) or/and undue complexity considering the involved geometrical parameters and the form of results (expressions and tables of coefficients).

Purpose of this paper is not to repeat previously extracted results but to give some important hints in the direction of a better understanding of the edge effect. For this, several simulations were carried out, for copper foil as well as round conductor coils, where all geometrical and electromagnetic parameters were subject of change. These parameters namely were (Fig. 5): distance of the winding from the ferrite yoke d_f ($0.5 \leq d_f/r \leq 15$ for round conductors and $0.5 \leq d_f/h \leq 5$ for foils), interlayer distance d_l, ($0.05 \leq d_l/r \leq 0.8$ for round conductors and $0.1 \leq d_l/h \leq 2$ for foils), number of layers m ($1 \leq m \leq 4$), permeability of the ferrite μ_r ($150 < \mu_r < 3000$), frequency f ($0 \leq r/\delta \leq 5$ for round conductors and $0 \leq h/\delta \leq 5$ for foils). Also, for round wire windings, several η values were investigated ($0.865 > \eta > 0.633$). The windings under study are symmetrically placed within the full width of the available window, i.e. the distance d_f from the ferrite yoke is the same on top and bottom of the winding. The selected maximum values for the quantities d_f/r (or d_f/h) and d_l/r (or d_l/h) where imposed by the fact that higher values are quite unlikely to be met in practice. Magnetic component designers make an effort so that copper occupies as much as possible of the window width, within the requirements for electrical insulation from the core (e.g. use of a coil former). On the other hand a foil winding with interlayer insulation of twice the thickness of the foil shows already extremely high leakage inductance and any thicker

1305

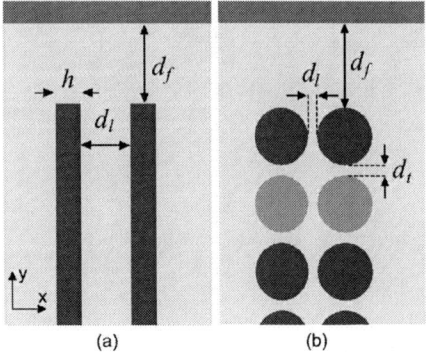

Fig. 5 – Definition of the geometrical quantities d_f, d_l, d_t and h for the better description of high frequency effects. In (b) the array with $\sigma_a=2$ is highlighted.

insulation is generally avoided. Typical interlayer insulation films are much thinner than one conductor layer.

The study of the several effects was made upon the whole of the winding, for each layer separately, for separate groups of conductors or either specific conductors. In all simulations the secondary under study was kept reasonably far from the primary. The F_R result in any of the above cases is compared to that for no edge effect present and is expressed by the quantity:

$$DF_R = \frac{F_R(\text{edge effect}) - F_R(\text{regular})}{F_R(\text{regular})} \times 100\% \quad (9)$$

A. The Edge Effect in Coils with Foil Windings

Regarding the dependence of losses on variations of the several parameters involved, the most important conclusions are summarized right below:

- Parameter d_f/h: (a) Generally, for $d_f/h<1$ edge effect can be ignored. (b) For $5>d_f/h>1$, as d_f/h increases, there is a gradual increase of DF_R towards negative values, except for low r/δ values ($h/\delta<0.8$ if $m>2$) where a slight increase of DF_R is observed. Maximum negative values of DF_R observed for $m=4$, $d_f/h=5$, $h/\delta=2$ are close to -15%.

- Parameter d_l/h: Up to the studied limit where the interlayer distance is twice the thickness of the foil ($d_l/h=2$), this parameter shows little effect on the extent of 2D effects in the vicinity of the foil edges. Only a slight increase in the negative values of DF_R as d_l/h increases is observed and respectively a slight reduction in the positive values of DF_R for low frequencies is observed.

- Parameter h/δ: As mentioned above, at low frequencies ($h/\delta<0.8$ when $m>2$) DF_R takes positive values, up to $+5\%$. For frequencies higher than this, up to $h/\delta=2$, edge effect becomes gradually more important (with negative values of DF_R) and for higher frequencies ($h/\delta>2$) no extra impact is observed or even there is a small reduction in absolute DF_R values.

- Number of layers m: For increasing m the absolute values of DF_R show a slight increase as well. However, the most important notice is that with decreasing m the frequency region for which the edge effect leads to moderately increased losses (positive DF_R) is extended to a somewhat higher limit. For $m=1$ this attribute is present up to the limit of $r/\delta=2$ and for $m=2$ the respective frequency limit is about $h/\delta=1.5$.

Illustrative of the above discussion is Fig. 6 where DF_R is plotted as a function of d_f/h with h/δ as a parameter, for $m=4$. It is also necessary to mention that the study of each single layer separately shows that all layers influence approximately in the same degree the overall DF_R. The previous conclusions give a general view of the most important aspects of the edge effect in foil windings and it is not a purpose of this work to further complicate this presentation with complex expressions about the numerical dependence of F_R upon the several parameters. Main conclusion is that at most of the practical applications the edge effect not only does not increase the losses but also improves the overall performance of the coil at the harmonic frequencies, if $h/\delta \cong 1$ is selected for the fundamental frequency. However, at $h/\delta<1$ where the greatest portion of throughput power is to be transferred, a moderate increase in losses occurs.

It is necessary to clear out that when the ratio W/h increases, with W the foil width, the edge effect becomes less important in total. For as long as W/h is not very small, the result of a change in this ratio on the value of DF_R is approximately inversely proportional. All the above numerical results refer to a typically used foil with $W/h=90$. This means that for $W/h=180$ all the DF_R values mentioned in the previous paragraphs should be halved to get the actual result. This is because 2D effects take place in a generally small depth y_e from the edges of the foils (in y direction). This depth is typically equal to a few times the copper skin depth δ at given frequency and the locally increased current densities may potentially lead to a hot spot. One way to investigate the geometrical extent of edge effect in a winding is to study the amplitude of the ratio B_x/B_y of the x and y components of the magnetic field, in y direction, along the sides of the foils. In Fig. 7 the ratio $B_{x,max}/B_{y,max}$ of the amplitudes of the x and y components of the magnetic field is plotted versus the distance y_e from the edge, expressed in δ units. The specific plot corresponds to the case of $m=4$, $d_f/h=5$, $d_l/h=2$ and it refers to the B_x and B_y components as they are in the inner side of the first foil layer (the side with the highest mmf values, the one that faces the excitation

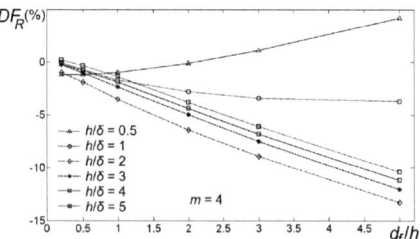

Fig. 6 – Plot of DF_R versus d_f/h, with h/δ as a parameter, for a foil winding with $m=4$.

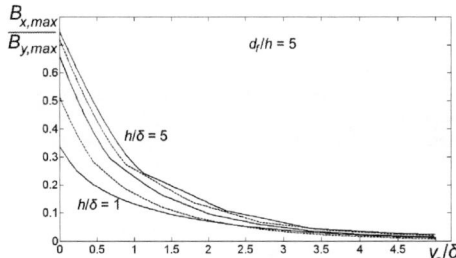

Fig. 7 – Plot of the ratio $B_{x,max}/B_{y,max}$ along the side of a foil layer versus the distance y_e from the edge, with the ratio h/δ as a parameter, for $d_l/h=5$.

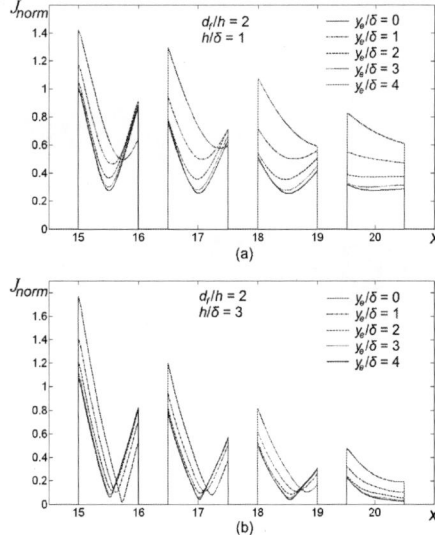

Fig. 8 – Normalized current density amplitude distribution (J_{norm}) x profile, with y_e as a parameter, at several frequencies for a winding with $m=4$, $d_l/h=2$, $d_l/h=0.5$.

primary). It is evident from these graphs that in a depth of about $y_e=4\delta$ the field component vertical to the foil becomes a negligible fraction of the parallel one. Notice that B_y and B_x are generally not in phase and since one is interested in their relative magnitudes it is more appropriate to study the ratio $B_{x,max}/B_{y,max}$ rather than the amplitude $(B_x/B_y)_{max}$ of the ratio of the two components.

Except the ratio B_x/B_y, in order to get a complete picture of the edge effect extent, it is also necessary to investigate the current density distribution J as a function of y_e. In Fig. 8 J_{norm} is the amplitude of J (J is a sinusoidal time function) normalized at the value $J(x=15, y_e\gg1)$. It corresponds to a winding with $m=4$, where the interlayer distances are half the thickness of the foils ($d_l/h=0.5$) and the inner side of the first foil (the one facing the excitation primary, region of maximum mmf) is at x=15. The J value mentioned above, used for the normalization of J, is the highest value that occurs at the first foil (and the whole of the winding too) far away from the foil edge (high y_e values), where the edge effect does not affect the current density and magnetic field. The distance y_e appears as a parameter. A general conclusion coming up from these plots is that in a depth $y_e=3\delta$ up to $y_e=4\delta$ the current density x profile takes its non-affected by the edge effect form. We can also see that as the frequency increases, except from that part of the conductor where higher J values occur, there is also another part where there is a reduction of J.

B. The Edge Effect in Coils with Round Wire

The detailed study of numerous cases of layered coils with round cross section conductors led to a couple of important conclusions:

- In all the cases studied, the result over a winding with many turns per layer is the reduction of the overall resistance of the winding.
- For $d_l/r<1$ edge effect can be ignored, while for $d_l/r>10$ it has no extra impact on copper losses.

Further than these two main conclusions some extra hints should be added that can help the designer judge about the actual impact of the edge effect on copper losses, very helpful especially when there are only few turns per layer. If we define as an "array" of conductors those ones, in all layers, with the same distance from the ferrite yoke (Fig. 5), for the sake of presentation, we need to establish two auxiliary quantities: At first, in the same way this was done in (9) for the general case of a conductor or group of conductors, we define DF_{Ra} as the

relative difference of the resistance factor F_{Ra} of a conductor array compared to the case when no edge effect is present, expressed as a percentage:

$$DF_{Ra} = \frac{F_{Ra}(\text{edge effect}) - F_{Ra}(\text{regular})}{F_{Ra}(\text{regular})} \times 100\% \qquad (10)$$

The F_{Ra} value for an array is calculated as the average of the resistance factors of each of the conductors in this array and the term F_{Ra}(regular) that appears in (10) equals the F_R of the whole winding at the absence of edge effect. Second quantity to define is σ_a, the integer index that describes the position of the array in the winding, e.g. the end array closer to the ferrite yoke has $\sigma_a=1$, the right innermost has $\sigma_a=2$ etc, as shown in Fig. 5. The following conclusions help to better understand the act of edge effect on copper losses:

- For sparse windings ($\eta<0.65$) DF_{Ra} is negative for all arrays at all frequencies. This attribute is the same at compact windings only at the relatively low frequency range, up to $r/\delta=1$.
- Focusing on each conductor separately, it is easily seen that the 2D field and current distribution effects are more prominent at the conductors belonging to the innermost layer (i.e. the layer at higher mmf values, closer to the excitation primary) and to the first few arrays. Besides that, the general concept is that DF_{Ra} follows a profile as that shown in Fig. 9, i.e. a positive value for the outermost 1-3 arrays (just one array in the case of Fig. 9, which is for $m=2$, $r/\delta=4$, $d_l/r=4$, $\eta=0.865$) and then negative. These positive values are nearly negligible for $r/\delta=1$ (typically +5% for only the first array) and may be up to +40% at $r/\delta=5$. Attention must be given to the fact that, at high frequencies, the

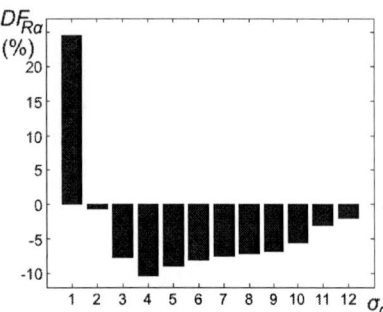

Fig. 9 – Relative difference DF_{Ra} of the F_R of a conductors array under the act of edge effect, as a function of the y position of the array in the winding (σ_a) relative to the end array for which it is $\sigma_a=1$.

first few wire turns of the innermost layer may show increased losses (compared to the no edge effect case) of up to 85%. For the conductor arrays placed deep in the winding, DF_{Ra} tends to zero.

- Keeping in mind that in a multilayer winding it is the innermost layer mainly affected, the case of a single layer winding is the one where the edge effect shows a maximum impact, which, even so, is still negligible at $r/\delta=1$.

- Copper losses even under the act of edge effect, do not depend on interlayer distance d_l. The same is true (no effect on losses) for the precise values of the core's μ_r.

Conclusively it may be said that the edge effect on windings with round conductors actually causes one single problem, that of a possible overheat of a few wire turns at the edges of the winding, especially the ones of the innermost layer. Although heat diffuses to all directions, the fact that these few turns are placed at the external parts of the winding means that effective cooling can be realized even without additional means further than just physical air flow. The most important notice though should be that, for an optimized design where it is approximately $r/\delta \cong 1$ at the fundamental frequency and the current waveform has only a moderate harmonic content, it is rather unlikely in any case to have a hot spot due to edge effect, not even on a single layer winding.

V. Conclusions

In this paper a series of issues critical for the proper calculation of high frequency copper losses in magnetic components where investigated using a FEA software. Issues such as the edge effect in foil as well as round wire windings and the effect of the winding pitch and the interlayer distance on the overall losses where thoroughly studied in a wide frequency range. The results that came out, presented in the form of a set of comprehensive conclusions, help to reliably estimate the actual impact of several 2D effects on copper losses under conditions met in real applications. A comparative presentation of the most widely used theoretical models and FEA results clearly indicated the application limits of these models and the magnitude of the error they introduce when the assumptions under which they were established are not fulfilled.

References

[1] S. Butterworth, "Eddy current losses in cylindrical conductors, with special applications to the alternating current resistances of short coils", *Phil. Trans. of the Royal Society of London*, vol. 222, pp. 57–100, 1922.

[2] S. Butterworth, "On the alternating current resistance of solenoid coils", *Proc. of the Royal Society of London*, vol. 107, No 744, pp. 693–715, Apr. 1925.

[3] E. Bennet, S. Larson, "Effective resistance to alternating currents of multilayer windings", *Trans. of American Inst. of Electr. Engineering*, vol. 59, pp. 1010–1017, 1940.

[4] P. Dowell, "Effects of eddy currents in transformer windings", *Proc. IEE*, vol. 113, No 8, pp. 1387–94, Aug. 1966.

[5] J. Ferreira, "Improved analytical modeling of conductive losses in magnetic components", *IEEE Trans. Power Electr.*, vol. 9, No 1, pp. 127–131, Jan. 1994.

[6] M. Bartoli, N. Noferi, A. Reatti, M. Kazimierczuk, "Modelling winding losses in high-frequency power inductors", *Journal of Circuits, Systems and Computers*, vol.5, No 4, pp. 607–626, 1995.

[7] F. Terman, "Circuit elements" in *Radio Engineer's Handbook*, New York, McGraw-Hill, 1943, ch. 2.

[8] E. Snelling, "Properties of windings", in *Soft Ferrites - Properties and Applications*, London, Iliffe Books Ltd, 1969, ch. 11.

[9] K. Billings, "Switcmode transformer design (general principles)" in *Handbook of Switchmode Power Supplies*, New York, McGraw Hill, 1989, Part 3, ch. 4.

[10] F. Robert, P. Mathys, "A closed form formula for 2-D ohmic losses calculation in SMPS transformer foils", *IEEE Trans. Power Electr.*, vol 16, No 3, pp. 437–444, May 2001.

[11] X. Nan, C. Sullivan, "Simplified high-accuracy calculation of eddy current loss in round-wire windings", *Proc. PESC04*, Aachen, Germany, June 2004, pp. 873–879.

[12] A. Podoltsev, I. Kucheryavaya, B Lebedev, "Analysis of effective resistance and eddy-current losses in multiturn winding of high-frequency magnetic components", *IEEE Trans. Magn.*, Vol. 39, No 1, pp. 539–548, Jan. 2003.

[13] A. Van den Bossche, V. Valchev, *Inductors and transformers for power electronics*, CRC Press - Taylor & Francis Group LLC, 2005, chapters 2, 5 and 10.

[14] W. Roshen, "Winding loss from an air-gap" *Proc. PESC04*, Aachen, Germany, June 2004, vol.2, pp. 1724–1730.

[15] M. Albach, H. Robmanith, "The influence of air gap size and winding position on the proximity losses in high frequency transformers", *Proc. PESC01*, Vancouver, Canada, June 2001, pp. 1485–1490.

[16] R. Prieto, J. Cobos, O. Garcia, R. Asensi, J. Uceda, "Optimizing the strategy of the transformer in a flyback converter", *Proc. PESC96*, Baveno, Italy, June 1996, pp. 1456–1462.

[17] A. Sinclair, J. Ferreira, "Optimum shape for AC foil conductors", *Proc. PESC95*, Atlanta, USA, June 1995, pp. 1064–1069.

[18] G. Dimitrakakis, E. Tatakis, E. Rikos, "A new model for the determination of copper losses in transformer windings with arbitrary conductor distribution under high frequency sinusoidal excitation", *12th European Conference on Power Electronics and Applications (EPE07)*, 1–4 Sept. 2007, Aalborg, Denmark, paper on CD, No 589.

[19] J. Keradec, E. Laveuve, J. Roudet, "Multipolar development of vector potential for parallel wires. Application to the study of eddy current effects in transformer windings", *IEEE Trans Magn.*, vol 27, No 5, pp. 4242–4245, Sept. 1991.

[20] F. Robert, J. Sprooten, P. Mathys, "Eddy current losses in SMPS transformers round wire windings: a semi-analytical closed-form formula", *11th European Conference on Power Electronics and Applications (EPE05)*, 11–14 Sept. 2005, Dresden, Germany, paper on CD.

[21] Opera 2D – User Guide, Vector Fields Ltd, 2003, Kidlington – Oxford.

Soft Switching PWM Inverter for Induction Heating Applied to Heating of Ferromagnetic Metal

Sachio KUBOTA, Muneo SATO, Fumio ITO, Yoshihiro SHIMAOKA and Kunihiro NISHIOKA

Toba National College of Maritime Technology/Department of Maritime Technology,
Toba, Japan, e-mail: *kubota@toba-cmt.ac.jp*

Abstract—In this paper, a soft switching PWM(Pulse Width Modulation) inverter for heating up to high temperature ferromagnetic metals is proposed as a power supply to apply to the diesel emission control system. And an optimal circuit design and an operating method about this inverter to solve several problems caused by heating up to high temperature it are described in detail. The validity of the proposed method is verified through both the simulated results and the experimental results.

Keywords—Soft switching, Induction heating, Resonant converter, Ship

I. INTRODUCTION

From viewpoints of preventing global warming and air pollution, it is necessary to reduce factors that cause these such as the toxic substances in the exhaust gas exhausted from marine diesel engines. In exhaust gas from diesel engines, nitrogen oxide (NOx), sulfur oxide (SOx) and Particulate Matters(PM) are contained as toxic substances.

Since October 2003, the bylaw which regulates discharge of NOx and PM has been enforced in the metropolitan area in Japan. In Europe, Euro-IV has been enacted as regulation of the exhaust gas for cars in 2005. Moreover, Euro-V for regulating emission of PM severely is due to be enacted in October, 2009. In the United States, US07 has been enacted as regulation of exhaust gas. And US10 which regulates it still more severely is due to be enacted in 2010. Because the regulation about cars is enacted, the necessity of regulating vessels similarly is discussed by IMO (International Maritime Organization). In the MEPC (Marine Environment Protection Committee) of IMO held in 2005, regulation of PM was added to the examination item. It is thought that the same regulation as land will be enacted at sea in near future.

In consideration of these points, we have already proposed the combustion system of PM by means of high frequency induction heating. The heating method of the proposed system is a method of heating a filter by itself without contacting the working coil by using induction heating. This system is called as diesel emission control system. The merit of an induction heating is that the response of temperature control to power control is extremely excellent.

In this paper, a method for heating up to high temperature a ferromagnetic metal is investigated and proposed as heating method of metallic filter applied to the diesel emission control system.

II. CHARACTERISTICS ON INDUCTION HEATING

A. Diesel Emission Control System

The proposed diesel emission control system is depicted in Fig. 1. The proposed system can be applied to diesel engine of land and sea. This system is premised on engines that SOx and NOx are reduced by applying the above-mentioned method, and it is a purpose to reduce an increased PM by relation with a trade-off. As for the proposed system, validity of reducing PM has been experimentally verified. [1]

In induction heating applications, it is efficiently heated when ferromagnetic materials are applied to the heating objects. Here, if a temperature of metallic filter reaches Curie temperature, the metallic characteristics are changed as are transmuted from ferromagnetism to paramagnetism, and the circuit parameters as load of induction heating are also changed rapidly. Accordingly, it is necessary to give consideration of this point to design of a power supply.

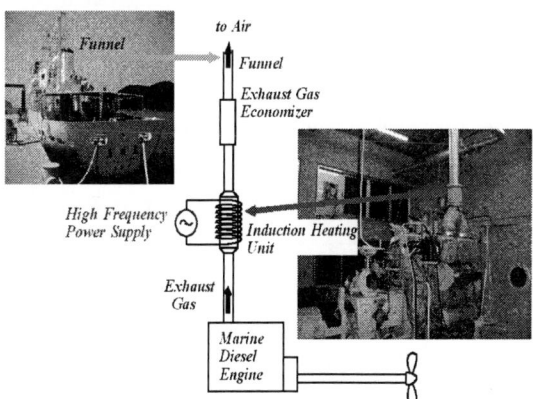

Fig. 1.   Diesel emission control system

B. Investigation of Characteristics on ferromagnetic metal

It is already known widely that ferromagnetic metals have Curie temperature and that metallic characteristics are rapidly changed at this temperature. Fig. 2 shows the

978-1-4244-1741-4/08/$25.00 ©2008 IEEE   1309

previous experimental results that investigated the tendency of the transformation, such as relation between temperature and resistance, and between temperature and inductance. [3] In this experiment, the small induction heating unit as shown in Fig. 3 is utilized. This unit is composed with the bundled pipe of stainless steel as ferromagnetic metal (SUS430), and it is applied to a heating object.

In Fig. 2, the values of resistance and inductance are rapidly decreased at approximately 700[°C], and its temperature is called Curie temperature. It is indicated that the value of an inductance is decreased to approximately 50% compared with its maximum value and that the value of resistance is decreased to approximately 10%. As these values are sharply changed at short time, an operating point is rapidly changed. At this time, a switching of each switch is converted to hard switching from soft switching, and there is a risk that power supply will be broken. Therefore a highly precise control corresponding to these changes in accordance with temperature is required to a power supply. Generally, a frequency control in accordance with temperature of heating metal is applied. However, a control circuit for accomplishing highly precise frequency control is complicated. And there is a risk that uniform heating cannot be performed in order that a depth of penetration is changed by control of frequency. To fix a depth of penetration, it is desirable that a power supply for induction heating is operating at constant frequency. From Fig. 2, it is shown that each value is stabilized if the temperature of heating becomes more than Curie temperature, and this state is called a steady state.

C. Conventional Power Supply for Induction Heating

Even if each switch is switched with soft switching, a switching stress becomes severe when a heavy current flows to switching device. Fig. 4 illustrates a conventional ZCS(Zero Current Switching) inverter, and Fig. 5 shows one example of the theoretical waveforms, in cases that this inverter is stably operated with ZCS in spite of change of temperature. Similarly, Fig. 6 illustrates a conventional ZVS(Zero Voltage Switching) inverter and Fig. 7 shows one example of the theoretical waveforms of this inverter, in cases that ZVS is always performed. In Fig. 5, it is shown that all current more than Curie temperature is decreased, compared with those at Curie temperature. In other words, although a switching stress is suppressed, output power is also reduced. From Fig. 7, it is indicated that io of output current more than Curie temperature is greatly increased, compared with it at Curie temperature, in the case of ZVS inverter. Here, not only an output current but each switching current, such as $i1$ and $i2$, is also greatly increased. Accordingly, switching stress becomes severe and there is a risk that switching devices will be broken. From these considerations, both ZCS inverter and ZVS inverter have merits and demerits, respectively.

Ideally, the power supply by which the maximum of each switch current is suppressed although an output current is increased with the rise of temperature is required. Therefore, in order to apply the advantage of both a ZCS inverter and a ZVS inverter, we have proposed the soft switching PWM inverter utilizing multi-resonance. The proposed inverter is operated at constant frequency.

In this paper, the design method and operating method for its inverter is described.

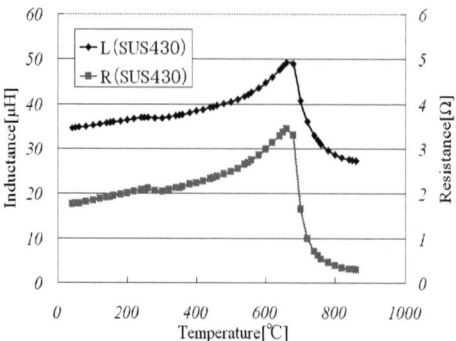

Fig. 2. Tendency of change as induction heating load in SUS430

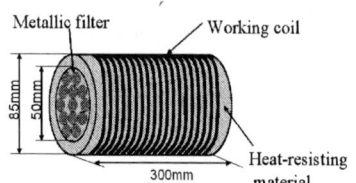

Fig. 3. Small induction heating unit for investigation

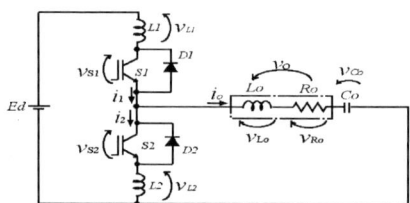

Fig. 4. Conventional ZCS inverter

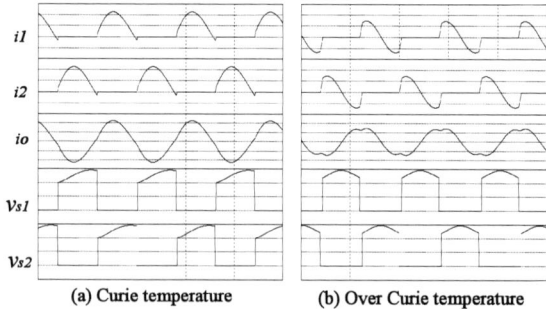

(a) Curie temperature   (b) Over Curie temperature

Fig. 5. Theoretical waveforms (ZCS inverter)

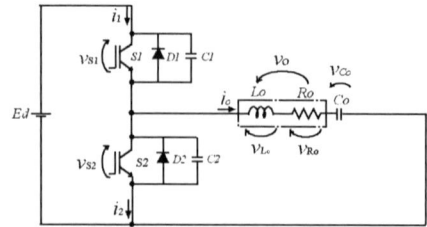

Fig. 6 Conventional ZVS inverter

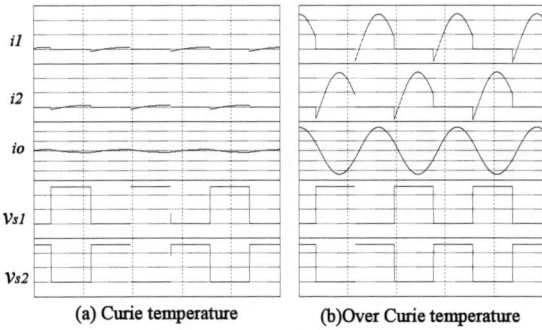

| (a) Curie temperature | (b) Over Curie temperature |

Fig. 7.   Theoretical waveforms (ZVS inverter)

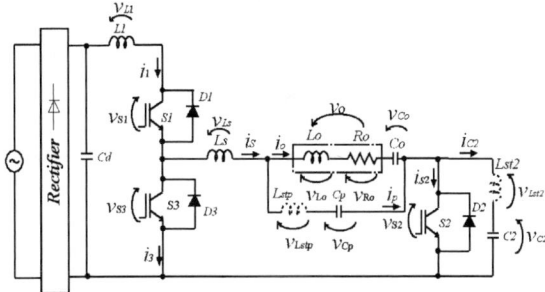

Fig. 8.   Soft switching PWM inverter for induction heating

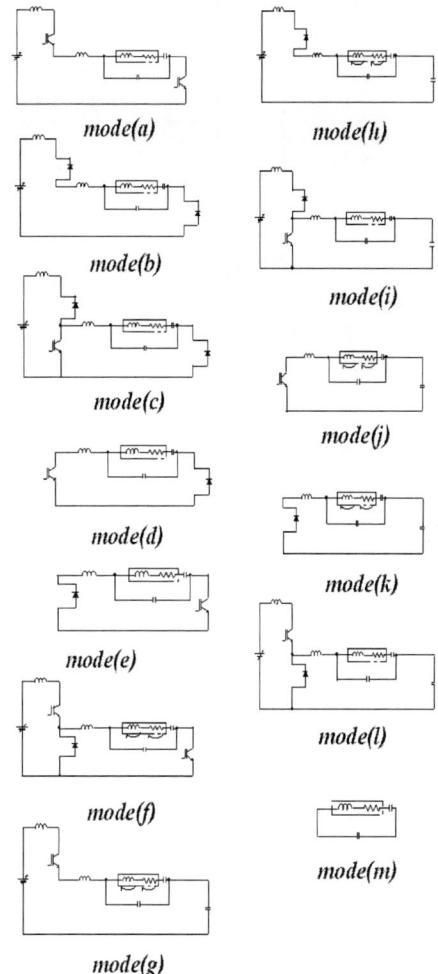

Fig. 9.   Switching operation modes

III. OPERATION PRINCIPLE

The proposed soft switching PWM (Pulse Width Modulation) inverter for induction heating is illustrated in Fig. 8. In each switching, S1 and S3 are switched with ZCS(Zero Current Switching), and S2 is switched with ZCS turn-on and is switched with ZVS(Zero Voltage Switching) turn-off. In this manner, all switches are performed with a soft switching. Moreover a switching current is suppressed although high output current is conducted to induction heating load by operating multi-resonance as tank circuit. The special feature of the proposed inverter is the point of applying PWM for suppressing the switching current that becomes high at the transient state of heating. In Fig. 8, Lstp and Lst2 mean a stray inductance. Though it is necessary to take Lst1 and Lst2 into consideration because the small induction heating unit as shown in Fig. 3 is used, these inductances can be disregarded when its unit is very large as actual unit.

The operation modes of this inverter are classified into 13 modes as below, due to the switching conditions of S1, S2, and S3 and charging condition of capacitor of C2. Each operation modes are shown in Fig. 9.

Mode (a): S1-S2 single conduction mode
Mode (b): D1-D2 single conduction mode
Mode (c):D1-D2 and S3 double conduction mode
Mode (d): D2-S3 single conduction mode
Mode (e): S2-D3 single conduction mode
Mode (f): S1-S2 and D3 double conduction mode

Mode (g): S1-C2 single conduction mode
Mode (h): D1-C2 single conduction mode
Mode (i): D1-C2 and S3 double conduction mode
Mode (j): C2-S3 single conduction mode
Mode (k): C2-D3 single conduction mode
Mode (l): S1-C2 and D3 double conduction mode
Mode (m): OFF mode

The gate pulse signal of each switch and a relation between each switching waveform and each switching mode is shown in Fig. 10. The duty factor of S2 is defined as $D=t2/T$. And the output power is controlled from minimum to maximum by modulating D from D=0 to D=0.75. The duty factors of switch of S1 and S3 are fixed on 50%, respectively. From Fig. 10, it is indicated that an output power and each switching current are controlled by conducting to C2 conduction modes, such as mode(g), mode(h), mode(i) and mode(j). In other words, these are controlled according to time conducted to capacitance of C2. An output power is controlled low if current is conducted to C2 for a long time.

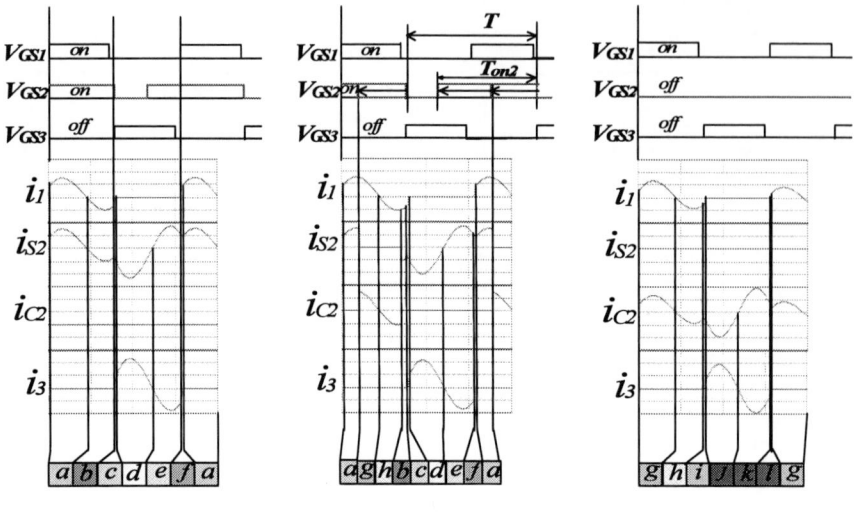

(a) Maximum output power *(b) Medium output power* *(c) Minimum output power*

Fig.10   Theoretical waveforms and each switching operation mode

TABLE I.
NORMALIZED PARAMETERS

【Normalized Value】

Frequency $f^* = 2\pi f o \sqrt{LC}$ L=Lo

Load Resistance $R^* = R/\sqrt{L/C}$ C=Co

Inductance A=Ld/L, B=Ls/L,   R=Ro

Capacitance p=Cp/C, q=C2/C

【Reference Value】

Voltage E=Ed Current $I = E/\sqrt{L/C}$

Power P=EI   Impedance $Z = \sqrt{L/C}$

Time T=1/fo   fo: output frequency

【State Variable】

Voltage $v^*(z) = v(t)/E$

Current $i^*(z) = i(t)/I$

Power Po*=Po/P   Time z=t/T=fo·t

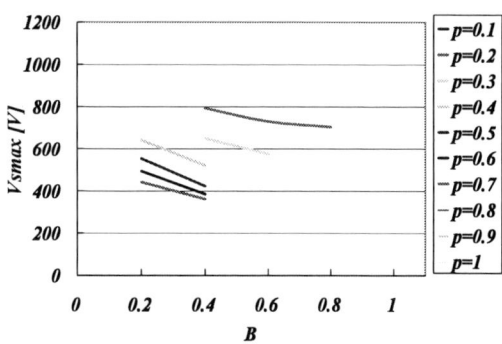

Fig. 11.   Characteristics on maximum switching voltage
(High temperature)

TABLE II.
COMBINATION OF THE NORMALIZED PARAMETERS

	p=0.1	p=0.2	p=0.3	p=0.4	p=0.5	p=0.6	p=0.7	p=0.8	p=0.9	p=1.0
B=0.1 (B=0.2)	O	O	O	O	O	O	O	O	O	O
B=0.2 (B=0.4)	O	O	O	O	O	O	O	×	×	×
B=0.3 (B=0.6)	O	O	O	×	×	×	×	×	×	×
B=0.4 (B=0.8)	O	O	×	×	×	×	×	×	×	×
B=0.5 (B=1.0)	O	×	×	×	×	×	×	×	×	×

(   ) : High temperature more than Curie temperature

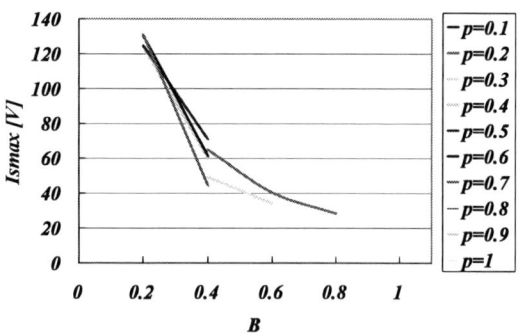

Fig. 12.   Characteristics on maximum switching current
(High temperature)

IV. DESIGN CONSIDERATION

A. Design guidelines

To design flexibly circuit, the normalized parameters are utilized in numerical analysis. Table I indicates each normalized parameters.

In the proposed inverter, it is important to analyze the circuit characteristics at both low temperature, such as Curie temperature, and high temperature, such as temperature more than Curie temperature, because the operation condition of circuit is changed according to rise of temperature, and it is necessary to suppress the switching stress. In cases where a ferromagnetic metal is heated, a load of induction heating is changed according to the rise of temperature as was previously described. For that reason, some normalized parameters are also changed according to a rise of temperature. Each value of a steady state is changed as follows as compared with each value at Curie temperature.

f^*: 70%, R^*: 15%, A: 200%, B: 200%

Table II indicates the results of investigation about the combination of an inductance ratio and a capacitance ratio by which a soft switching is always performed although a parameter changes as mentioned above. A soft switching is not always performed from low temperature to high temperature if B or p is set to high. In these combinations, it is necessary to discuss the combination by which a switching stress is most suppressed. The circuit parameters are designed at a steady state.

Fig. 11 and Fig. 12 show the influence on the switching stress, such as the maximum switching voltage as Vsmax and the maximum current as Ismax, by each combination. In these figures, the investigated parameters in Fig. 2 are used, and an output power is constant as 1.0[kW] by controlling input voltage. From Fig. 11, it is indicated that Vsmax is suppressed when high B and p are selected. And it is indicated that Ismax is hardly influenced to p and is dependent on B, as shown in Fig. 12. Ismax is enough suppressed when B is set to 0.4 or more. From the above considerations, the combination of B=0.4 and p=0.7 is selected to suppress a switching stress.

The transition of a soft switching operation region and an operating point from low temperature as Curie temperature to high temperature on the normalized R^*-f^* plane is shown in Fig. 13. From this figure, it is indicated that an operating point is always within a soft switching operation region although temperature is changed. Accordingly an operation of the proposed inverter is always stable with a soft switching from low temperature to high temperature.

B. Control scheme of maximum switching current

In the proposed inverter, each switching current is controlled by PWM. Here, an ideal method of current control is that an output power is not controlled to low power even if each maximum switching current is suppressed. The control width of a switching current is influenced to q(=C2/Co) of a capacitance ratio. An optimal q is investigated as an output power is hardly influenced in spite of suppressing the maximum switching current.

Fig. 14 and Fig. 15 show the influence on the normalized maximum switching voltage as Vsmax* by q at high temperature more than Curie temperature and at low temperature such as Curie temperature, respectively. Similarly Fig. 16 and Fig. 17 respectively show the influence on the normalized maximum switching current as Ismax* at both temperatures. When a soft switching is not performed by means of controlling D of duty factor, values of Vsmax* and Ismax* are not indicated in these figures. In Fig. 14 and Fig. 15, it is shown that Vsmax* is hardly influenced by q, even if temperature of ferromagnetic metal is changed. From Fig. 16 and Fig. 17, it is indicated that Ismax* is suppressed by setting q to a low value. However, in case that q is set up low, a soft switching is not performed if D is controlled low.

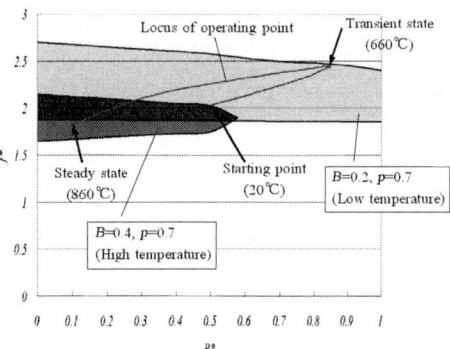

Fig. 13. Transition of soft switching operation region and operating point

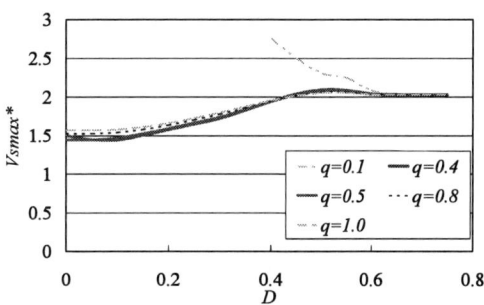

Fig. 14. Characteristics on maximum switching voltage (Low temperature)

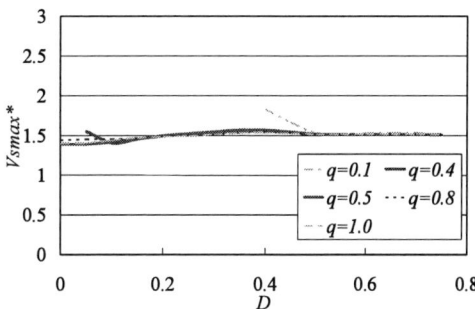

Fig. 15. Characteristics on maximum switching voltage (High temperature)

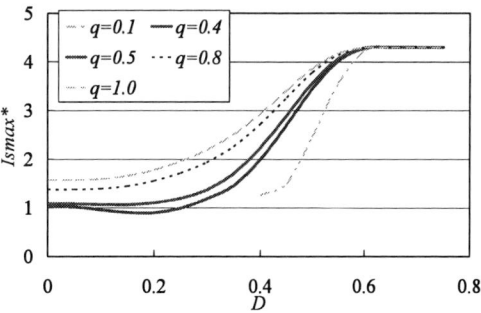

Fig. 16. Characteristics on maximum switching current (Low temperature)

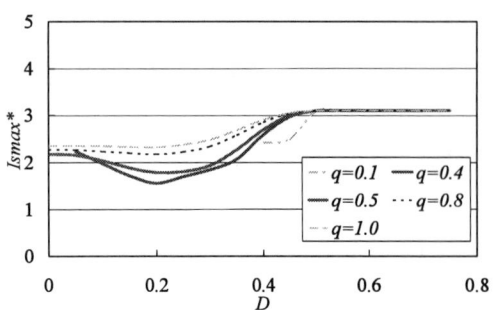

Fig. 17. Characteristics on maximum switching current
(High temperature)

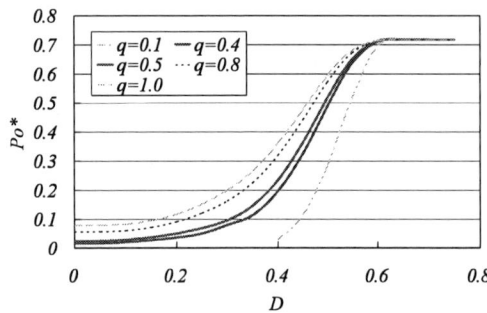

Fig. 18. Characteristics on output power
(Low temperature)

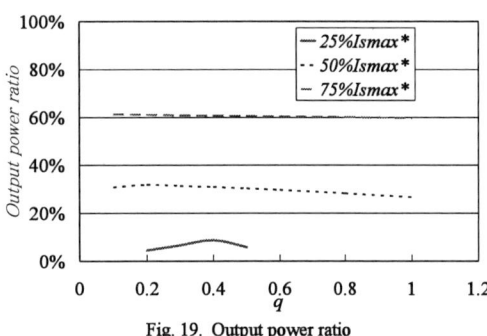

Fig. 19. Output power ratio

Consequently, it is desirable that q of capacitance ratio is set to 0.5 or more in order to always perform a soft switching from low temperature to high temperature. An influence on the normalized output power as Po* by q is shown in Fig. 18. It is indicated that Po* as well as Ismax* is suppressed by setting q low. A relationship between Ismax* and Po* at low temperature is derived from Fig. 16 and Fig. 18. Fig. 19 shows the influence on output power by q, when maximum switching current is equal to 75%, 50% and 25% of maximum value, by control of D. In this figure, an output power when maximum switching current is controlled to 75% is almost uniformly suppressed to 60% of maximum output power regardless of q. In cases that the maximum switching current is suppressed less than 50%, an output power is affected by q. In case of 25%Ismax*, the output power at q= 0.4 is the highest. However, it is already known that q has to be set to 0.5 or more, from the above reason. Therefore optimal q, which is not affected to output power even if each maximum switching current is suppressed, is set to 0.5.

TABLE III.
SPECIFICATION OF POWER SUPPLY

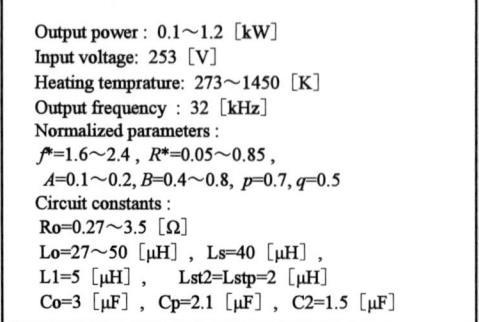

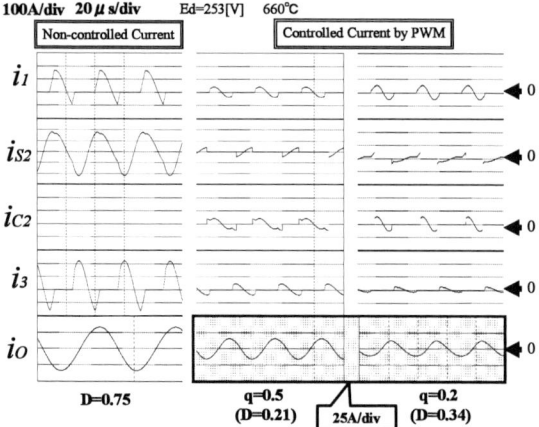

Fig. 20. Controlled theoretical current waveforms

V. SIMULATION AND EXPERIMENTAL VERIFICATION

Depending on the selected combination, a switching stress often becomes high even if a soft switching is performed, as described above. The comparisons of each switching current at D=0.75 such as maximum output power and each switching current suppressed to less than 25% of maximum switching current are shown in Fig. 20. Here, input voltage is 253[V], and a temperature of metallic filter of a ferromagnetic metal is 660[˙] at Curie temperature. And a maximum switching current is i3 at D=0.75. As shown in this figure, each switching current is suppressed to less than 25% of these at D=0.75 by PWM. The values of each switching current are different because the proposed inverter is asymmetrical SEPP inverter. At q=0.5, i3 is 25% of maximum switching current at D=0.75, and at q=0.2, i1 is 25% of it. Moreover, io of output current at q=0.5 is higher than that at q=0.2. Consequently, it is indicated that high output power is obtained by selecting optimal q if maximum switching current is suppressed to same value.

On the basis of the numerical analysis, the inverter for the induction heating is designed, and to verify the feasibility of the proposed design method and operating method, simulations and experiments of heating a ferromagnetic metal have been carried out. The specification of the designed power supply is shown in Table III.

Fig. 20 shows the theoretical comparison with the non-controlled current, such as each current at D=0.75, and the current controlled by PWM. Here, input voltage is 253[V], and a temperature of metallic filter of a ferromagnetic metal is 660[˚] at Curie temperature. As shown in this figure, all switching current controlled by PWM is suppressed to less than 25% compared with i3 that is a maximum switching current at D=0.75. Moreover, it is indicated that an output current at q=0.5 is higher than an output current at q=0.2. Consequently, high output power is obtained by selecting an optimal q, if an output current is suppressed to same value.

Fig. 21 shows the comparison of simulated waveforms and experimental waveforms at 600[˚] that is near Curie temperature using the selected parameters through analysis. The duty factor of S2 is set to D=0.2, because a switching current becomes very high in a transient state of heating. Similarly, Fig. 22 shows the comparison of simulated waveforms and experimental waveforms at 1000[˚] more than Curie temperature. As shown from these figures, the proposed inverter ensures a soft switching operation for all the switching devices and each maximum switching current at both low temperature and high temperature is controlled to almost equal by PWM. In comparison between experimental results and simulated results, there are some errors caused from the assumption that all the switches are ideal and that a stray inductance and capacitance are neglected. However each experimental waveform matches the simulated waveforms well and the tendency of both values is good agreement. Furthermore, it is shown that the output current is higher than a switching current when a ferromagnetic metal is heated up to 1000[˚]. The unit that is heated up to 1000[˚], such as more than Curie temperature, is illustrated in Fig. 23. From this figure, it is proved that a ferromagnetic metal is heated to redness at high temperature.

VI. CONCLUSION

In conclusion, the soft switching PWM inverter for induction heating is proposed as power supply for heating a ferromagnetic metal. And the characteristics of the proposed inverter have been investigated and clarified. As results, it has been proved that the proposed inverter is always a stable operation although circuit parameters are changed according to rise of temperature, and that the proposed circuit design and the operating method are valid to suppress a switching stress through the experiment.

REFERENCES

[1] S. Kubota, Y. Hatanaka, "Soft Switched High Frequency Inverter in Complex Resonance applied for an Emission Control in Diesel Engine", Proceedings of EPE-PEMC2000, 2000

[2] S. Kubota, M. Sato, F. Ito, H. Kakuage, N. Ogawa, Y. Shimaoka, "High Frequency Power Supply for Emission Control System Applied to Marine Diesel Engine," Proceedings of EPE-PEMC2004, 2004

[3] S. Kubota, M. Sato, F. Ito, N. Ogawa and Y. Shimaoka, "The Design Method of the Inverter for Heating Both a Ferromagnetic Metal and a Paramagnetic Metal", Proceedings of EPE2007, 2007

[4] J. M. Burdio, F. Monterde, J. R. Garcia, L. A. Barragan and A. Martinez, "A two-output series-resonant inverter for induction heating cooking appliances," IEEE Transactions on Power Electronics Vol. 20 No 4, pp. 815- 822

[5] N. Park, D. Lee and D. Hyun, "A Power-Control Scheme With Constant Switching Frequency in Class-D Inverter for Induction-Heating Jar Application," IEEE Transactions on Industrial Electronics Vol. 54 No 3, pp. 1252- 1260

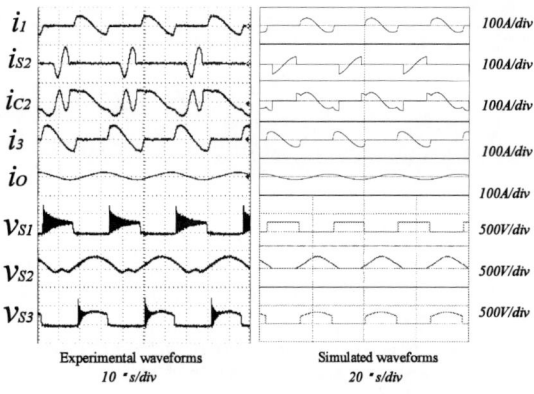

Fig. 21. Experimental results (Low temperature)

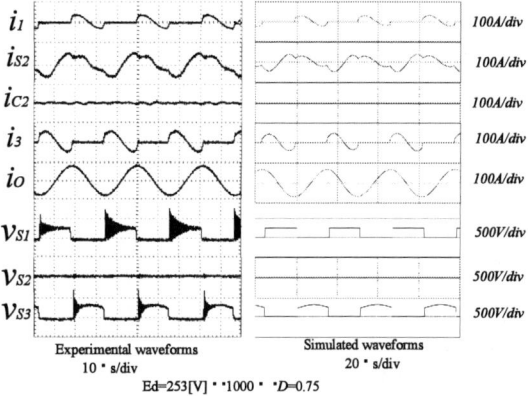

Fig. 22. Experimental results (High temperatute)

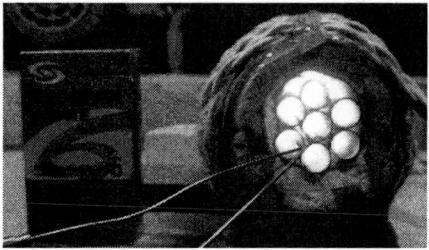

Fig. 23. Heated unit

Corona Treatment System with Resonant Inverter
- Selected Proprieties

Mućko Jan*

* University of Technology and Life Sciences, Institute of Electrical Engineering
Prof. S. Kaliskiego 7, Bydgoszcz, Poland, mucko@utp.edu.pl

Abstract—The paper describes a mathematical model and proprieties of series resonant inverters used in application for activators of polyethylene foil (corona treatment). The method of identification of corona treatment parameters and power calculation is described as well. The area of soft switching and the area with corona discharge is shown. In this paper a new control method of corona treatment based on PDM and PFM modulation is presented. The control characteristics were determined. The circuits were tested by means of simulation and experimentally by industrial applications.

Keywords— series resonant inverter, corona treatment, ZVS, ZCS, PDM, PFM..

I. INTRODUCTON

The structure of corona treatment system for polyethylene foil [1...3, 5...12] is shown on Figure 1. The main elements of corona treatment system are: voltage source inverter 1, HV transformer 2 and set of discharge electrodes. The discharge appears between roll electrode (rotating) 3 and bar (static) 4. The parameters of activators are as follows: power - 0.5...10 kVA, frequency - 5...50 kHz, voltage on the electrodes - 4...20 kV. Discharge electrodes along with two dielectric layers (silicone and air) form a set of capacitors C_1 (silicone) and C_2 (air). A third and the thinnest layer of the dielectric is a foil. A capacity related to this foil and connected in series with C_1 is greater then C_1 and can be omitted further on. Capacity of electrodes and leakage inductance of transformer constitute resonant circuit which can be used to assist switching processes in the inverter. When the main inductance of transformer is omitted, the inductance of the resonant circuit amount $L_r=L_{r1}+L_{\sigma1}+L_{\sigma2}'$ where: $L_{\sigma1}$, $L_{\sigma2}'$ – leakage inductances of transformer (seen form primary side) Additional inductor L_{r1} enables adjustment of free oscillation frequency and impedance of the resonant circuit to assumed working frequency range and assumed output power.

Figures 1b and 1c depict equivalent circuit and current-voltage characteristic of corona discharge in air. The discharge model from Figure 1b shows that ignition voltage U_z is greater than voltage U_g of suppress of discharge. In reality when electrodes are long a significant amount of ignition centers are formed. Therefore it is impossible to determine the voltage of ignition and voltage of suppression of the discharge in an unequivocal way. Intentionally simplified model (Fig. 1c) with one ignition voltage and voltage of discharge suppression is used. The value of voltage that starts corona discharge is a function of electrodes' shape and product of pressure and distance between those electrodes [4].

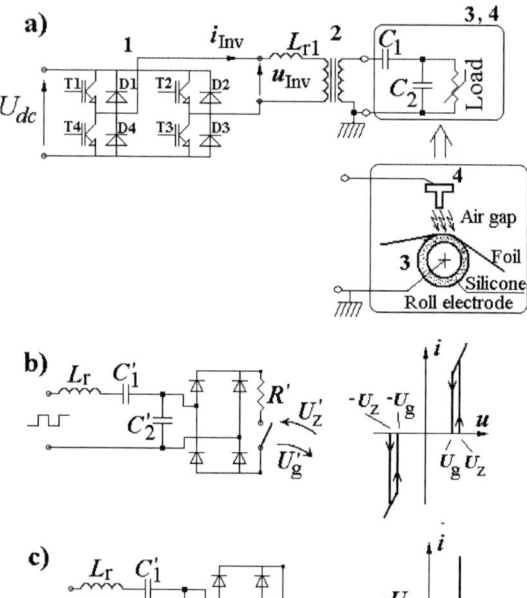

Fig.1. Series resonant inverter in corona discharge treatment system: a) inverter and load configuration, b) equivalent circuit and characteristics of corona discharge

II. THEORETICAL BACKGROUND

When the system is working, one can distinguish two operation states (Fig. 1c): state 1, without corona discharge (rectifier is off); state 2, with corona discharge (rectifier is on). In state 1, rate at which voltage $u_C(t)$ rises, describes equivalent capacity C_z of series connected capacitors C_1 and C_2 (equation 1). In turn in state 2 , rate at which voltage $u_C(t)$ rises, describes capacity of capacitor C_1 (equation 2). The operating frequency of the inverter f_{syn}, at which the voltage and output current waves are synchronized, is situated in the following range $f_{rmax} > f_{syn} > f_{rmin}$ (Eq. 3 and 4). Figure 2 and 3 depicts a wave of voltage $u_C(t)$ on the set of capacitors in function of charge $q(t)$.

978-1-4244-1741-4/08/$25.00 ©2008 IEEE

$$\frac{du_C}{dq} = \frac{1}{C_z} = \frac{C_1 + C_2}{C_1 \cdot C_2} , \quad \frac{du_C}{dq} = \frac{1}{C_1} \qquad (1), (2)$$

$$f_{r\max} = \frac{1}{2\pi\sqrt{L_r C_z}} , \quad f_{r\min} = \frac{1}{2\pi\sqrt{L_r C_1}} \qquad (3), (4)$$

For the supply voltages and frequencies at which the value of voltage amplitude on capacitor C_2 doesn't reach the value of U_p a rectifier is off (Fig. 1c). The current does not flow through the non linear load. The resonant circuit has a filtrating feature. The current flowing through this circuit as well as the voltage at the capacitors has almost a sinusoidal shape. In his case a classic circuit analysis for the fundamental harmonics can be conducted. The quotient of amplitudes at the capacitor C_2 (for fundamental harmonics) and at the inverter's output is described by the following dependence (5), where: U_{C2_1m} – the amplitude of the capacitor's C_2 voltage, U_{Inv_1m} – the amplitude of the inverter's output voltage, $\omega_s = 2\pi f_s$ – the angular frequency of inverter's voltage, f_s - the switching frequency.

The value of the fundamental harmonics' amplitude of the input voltage of the bridge inverter is described by the dependence (6). Whereas the limiting value of the amplitude of the voltage U_{C2m} at the capacitor C_2, that causes the load current to flow is described by the dependence (7).

$$\frac{U_{C2\_1m}}{U_{Inv\_1m}} = \left| \frac{1}{\omega_s^2 \cdot L \cdot C_z - 1} \cdot \frac{C_z}{C_2} \right| \qquad (5)$$

$$U_{Inv\_1m} = \frac{4}{\pi} U_{dc} , \quad U_{C2m} = U_p \approx U_{C2\_1m} \qquad (6), (7)$$

From the equations (5), (6) and 7 it is possible to determine the boundary frequencies at which corona discharge starts occurring:

$$f_{sgr1} = \frac{1}{2\pi} \sqrt{\frac{1}{L \cdot C_2} \left(\frac{C_2}{C_z} - \frac{4 \cdot U_{dc}}{\pi \cdot U_p} \right)} , \qquad (8)$$

$$f_{sgr2} = \frac{1}{2\pi} \sqrt{\frac{1}{L \cdot C_2} \left(\frac{C_2}{C_z} + \frac{4 \cdot U_{dc}}{\pi \cdot U_p} \right)} . \qquad (9)$$

Dependences shown above were confirmed by the simulations and experiments. The current flows in the non linear load (corona discharge) if the following condition is meet $f_{sgr1} < f_s < f_{sgr2}$. According to (8) and (9) the values of boundary frequencies depend on the value of the inverter's input voltage. For U_{dc} approaching to zero the boundary frequencies tend to approach to $f_{r\max}$ (Eq. 3, 8 and 9).

III. POWER IN CORONA TREATMENT PROCESS

Figure 2 shows how to determine the power of the activation process. The energy delivered to the non linear load during the one operating period is equal to the area of a rectangle described by the equation $u_{ws} = f(q_{ws})$ and amount to $4 \cdot U_p \cdot Q_0$ (Fig. 2a, u_{ws}, q_{ws} – voltage and charge of the corona discharge).

The result of the multiplication of energy and switching frequency is the power of the activation process (10). On the account of the construction of the system it is impossible to directly measure the value of Q_0 and reading the value of U_p might be difficult if not impossible (Fig. 2d). The value of Q_0 can be determined by subtracting from the registered value of Q_{max} the value of the charge stored in the capacitor C_2. The value of Q_0 can be determined from the equation (11) and according to Figure 2b. Value of U_p can be determined from the equation (12) and Q_{max} from the equation (13). The charge Q_{max} can be calculated by the numerical integration of the inverter's output current or by the voltage registered on an additional capacitor (with significant capacitance) connected in a series with the inverter's output.

$$P = 4 \cdot Q_0 \cdot U_p \cdot f_s , \quad Q_0 = Q_{max} - C_2 \cdot U_p \qquad (10), (11)$$

$$U_p = U_{max} - Q_{max} / C_1 \qquad (12)$$

$$Q_{max} = \frac{1}{2} \int_{t_0}^{T_s/2} i_{Inv} \, dt \qquad (13)$$

When taking into account the dependences (10) … (13) the power delivered to the electrodes can be determined from the equation

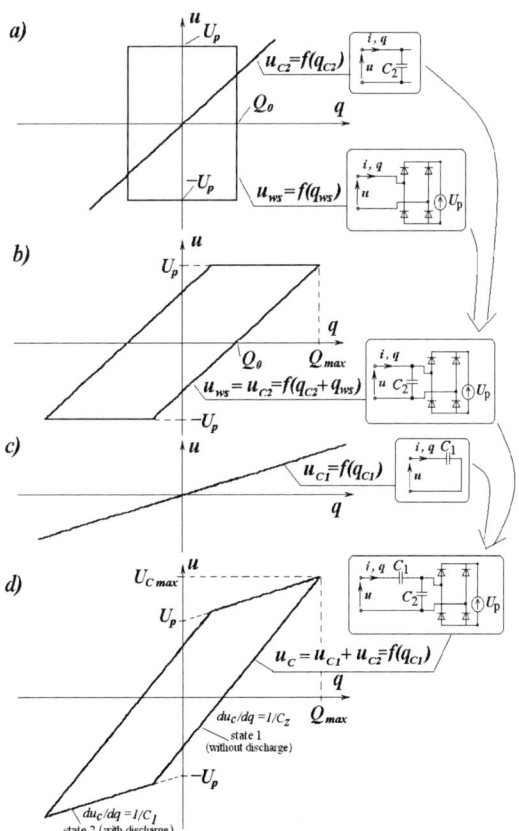

Fig. 2. Waveforms of voltage and charge on circuit's elements

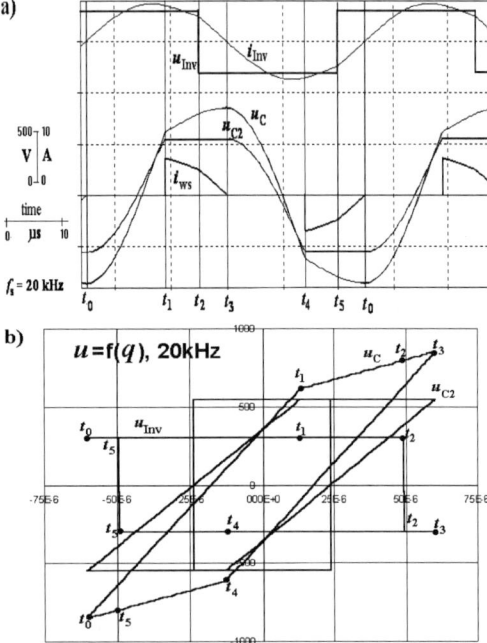

Fig. 3. Voltage and current waveforms in a corona treatment circuit obtained in simulations way: a) in function of time, b) in function of charge

$$P = 4[Q_{max} - C_2(U_{max} - Q_{max}/C_1)] \cdot$$
$$(U_{max} - Q_{max}/C_1)f_s \qquad (14)$$

To secure a proper surface treatment it is necessary to provide to it an energy flow of 0,65-1,3kJ/m2. The knowledge of the power of the process is necessary

$$P_{proc} = WD \cdot s \cdot v \cdot n \qquad (15)$$

where: WD – the energy for area unit [W·s/m²], s – width of foil [m], v – speed of technological line [m/s], n – the amount of surfaces.

The power form equation (14) should be equal to the power from (15). The inverter's output power should be greater than power from (14) and take into consideration losses in the transformer.

IV. NEW CONTROL METHOD OF CORONA TREATMENT

Control of the corona treatment process power can be run in the following ways [9]: by pulse width modulation (PWM) [5], by pulse frequency modulation (PFM), by amplitude modulation (PAM) [7], by pulse density modulation (PDM) [1, 8], by combination of the above mentioned methods [2, 3]

The control method by PWM results in losses in hard switching. The most notable are PFM, PAM, PDM methods and their combinations because of the guarantee of soft switching. Combination of PDM and PFM methods were earlier described in publications [2, 3].

Author of his paper worked out a new method [10] of control which is a combination of PDM and PFM. This method differs from the described ones in publications [2, 3] as the work of the inverter is not stopped but instead the inverter operates with much higher frequency. This mean of controlling the process couples advantages of both methods (PDM, PFM) and the control circuit is simpler then circuits described in [1, 2, 3, 8, 9].

The operating frequency of the inverter f_{syn}, at which the voltage and output current waves are synchronized, is situated in the following range $f_{rmax} > f_{syn} > f_{rmin}$ (Eq. 3 i 4). The maximal inverter output power during synchronization of inverter voltage and current waves (and PDM duty cycle D = 1) is described by equation (16).

$$P_{Inv\_max}(f_{syn}) =$$
$$\frac{1}{T_s/2}\int_0^{T_s/2} U_{dc}I_{Inv\_m}\sin(\omega_s t)\,dt \approx 0.9 U_{dc}I_{Inv} \qquad (16)$$

The inverter output power (17) can be controlled by PDM modulation and PFM modulation (Fig. 4a, 4b, 5b)

$$P_{Inv} = D \cdot P_{Inv\_max}(f_s) \qquad (17)$$

PDM method is profitable especially for low treatment power because it ensures that the discharges are stable. Other methods did not secure the stability of discharges and were not suitable for systems operating in wide range of power. Combination of PDM and PFM methods where the operations of the inverter are not stopped but

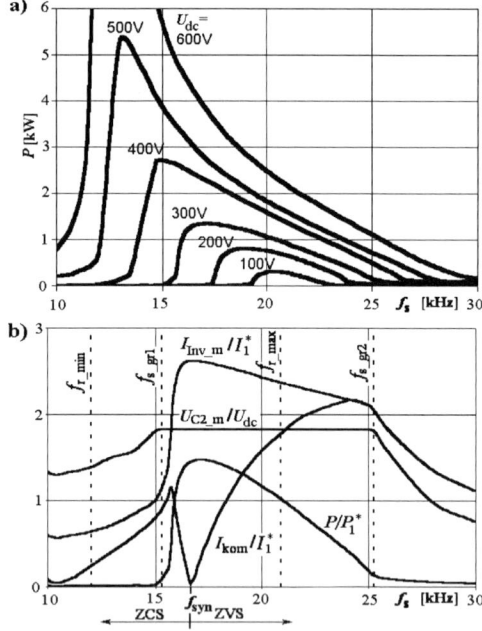

Fig. 4. Characteristics of power, voltages and currents in function of switching frequency and supply voltage obtained in simulations way (discharge chamber 1): a) discharge power, b) discharge power, transistor current during switching, amplitude of capacitor C_2 voltage and maximum inverter output current (U_{dc}=300V)

a)

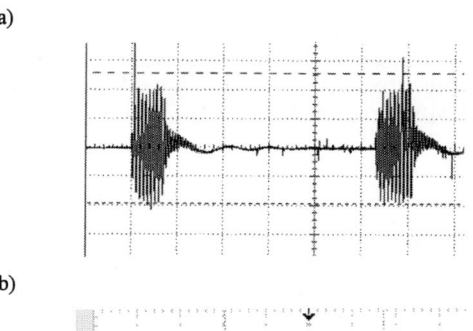

b)

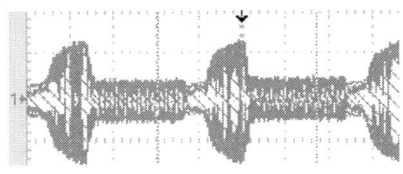

Fig.5 Output current waveforms: a) of a PDM controlled inverter [8]: i [2A/dz], f=24kHz, f_{PDM}=100Hz, b) of a PDM and PFM controlled inverter (new control strategy); i [10A/u], t [2,5ms/u]

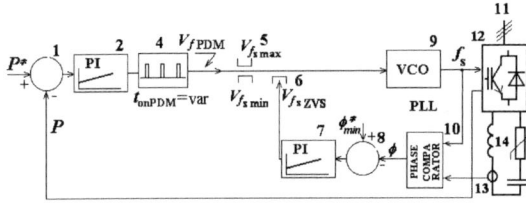

Fig. 6. Example of realization the PDM-PFM control circuit for corona discharge treatment system:
1, 8 – adders; 2, 7 – PI regulators; 4 – pulse generator with f_{PDM} frequency; 5 – frequency-limiting circuit ($f_{s min} < f_s < f_{s max}$); 6 – frequency limiting circuit to assure ZVS soft switching ($f_s > f_{s ZVS}$); 9 – voltage-controlled oscillator; 10 – phase comparator; 11 – mains; 12 – voltage source inverter, 13 – current transducer, 14 – main circuit

instead the inverter works with much higher frequency is unique (Fig.5b). This method secures stable discharges. It relies on transferring of the maximal power with controlled duty cycle but limited like in [2, 3]. In time intervals where the power has to be decreased, the inverter's operation is not suspended but instead the switching frequency of transistors is increased. The rate of changes of the frequency is limiting from the top and from the bottom the power of discharges according to characteristics depicted on figures 4a and 4b. By limiting the frequency from the top and from the bottom the power is also limited. Increasing the switching frequency above the f_{sgr2} (9) results in complete lack of discharges (Figure 4b).

Figure 6 represents a simplified block diagram of the main circuit and the control circuit, which illustrates a control strategy. In contrast to the control circuits presented in [1, 2, 3, 8] this circuit does not require any additional circuit storing the frequency from the moment before the inverter is stopped. It also does not require any circuits counting the number of half-waves of the

inverter's output voltage. The next advantage is the ease of adaptation of the existing PFM control circuit to work accordingly to the new presented method.

Signal P corresponding the power of the process is compared with the reference value P^*. Error signal P^*-P is delivered to the input of the power regulator 2. Output signal from the regulator is delivered to the PDM generator's input 4. Output signal V_{fPDM} from the generator 4 can have different shapes, especially it can be a square signal. It is limited to the value of $V_{fs min}$, $V_{fs max}$, $V_{fs ZVS}$ in blocks 5, 6 and is delivered to the VCO 9 generator's input. Signal f_s from the VCO generator controls transistors of the voltage inverter 12.

Signal V_{fPDM} from the generator 4 quickly changes the frequency f_s resulting in delivering to the load power of value oscillating with the frequency f_{PDM}. Closed loop power regulation circuit (with regulator 2) causes that the mean power of the process P is equal to the reference power P^* in spite of working generator 4.

Limits of $V_{fs min}$, $V_{fs max}$, $V_{fs ZVS}$ cause that the frequency f_s will fit the range ($f_{s min}$, $f_{s max}$) and will not be lower than frequency $f_{s ZVS}$. Blocks 6, 7, 8, 10 guarantee that the value of the phase shifting ϕ between current wave and inverter's output voltage wave does not decrease below the given value ϕ^*_{min} and frequency f_s does not decrease below the frequency $f_{s ZVS}$. That enables inverter's transistors to soft switching (ZVS type).

V. SIMULATION AND EXPERIMENTAL RESEARCH

The simulation research has been conducted with an aid of TCAD software. Data used in simulations were corresponding to the data from the experimental model (discharge chamber 1, P_N=4kW, l=3,3m, Φ=100 mm, silicone d=2 mm). Figure 3a shows current and voltage waves as a time function, that were obtained as a result of a simulation. Whereas figure 3b shows those waves as a

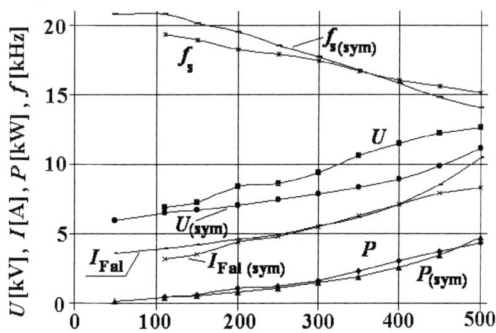

Fig.7. Characteristics of the corona treatment system and simulation model (chamber 1) during controlling input inverter voltage; $f_s > f_t$ and $f_s \approx f_t$

function of charge. Characteristic points were marked on the figures.

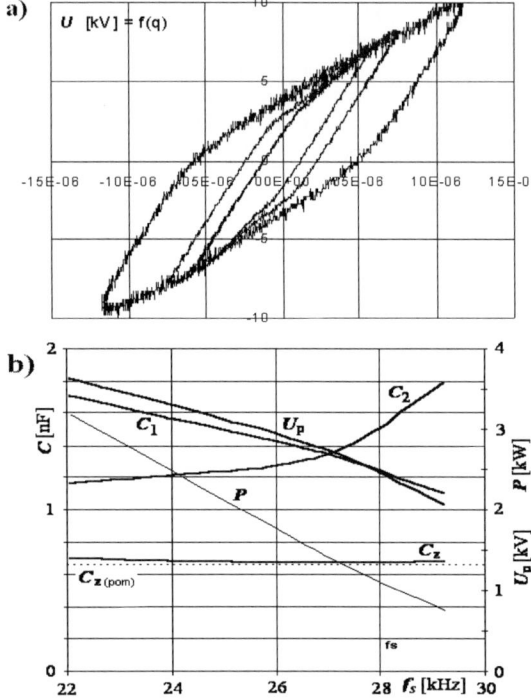

Fig. 8. Characteristics of the corona treatment system during controlling switching frequency of inverter (chamber 2): a) voltage of electrodes in function of a charge (P= 25, 50, 106% x 3kW) ; b) power, U_p voltage and capacity of electrodes in function of switching frequency f_s

Figure 4a depicts inverter's output power in a function of switching frequency for different supply voltage values. Figure 4b shows inverter's output power values, transistor's current values during the switching and voltage amplitude at the capacitor C_2 in a function of frequency. Frequencies form equation (3), (4), (8), (9) were marked. For frequency $f_s > f_{syn}$ transistors operate with ZVS soft switching.. For frequency $f_s < f_{syn}$ transistors can operate with ZCS soft switching, but this type of operations is not recommended.

On Figure 7 characteristics of prototype system (chamber 1) during controlling input inverter voltage are shown. On this figure there are also results of simulation. On Figure 8 characteristics of prototype system during controlling switching frequency of inverter are shown (chamber 2, P_N=3kW; 2 rotating electrodes 170cm, Φ100, silicone Luraflex 2mm; 2 bar toothed electrodes: 160cm x 3,6cm). Values C_1, C_2 and U_p that are depicted on Figure 8b were determined from waveforms on Figure 8a.

VI. CONCLUSIONS

Simulation and mathematical models are well suited for practical use. Analytical determination of the boundary frequencies, parameters of the discharge chamber and the power of the process basing on a trajectory of u=f(q) are important factors while designing an inverter device.

The value of the ignition voltage U_p depends on the degree of air ionization. The shorter the time for the

deionization (i.e. the higher is the switching frequency), the ignition voltage is lower. Similarly the higher the frequency, the bigger the ionization and higher capacitance of the air capacitor C_2 (Fig. 8b).

The author of this paper elaborated a new method of control which is a combination of PDM and PFM. This method enables a possibility of power regulation of the corona treatment and secure the conditions for the soft switching process (ZVS). This method was experimental investigate and secures stable discharges. This brand new method has been investigated by the author with a series of simulations and experiments. Some industrial prototype devices capable of 4kW power have been built and tested. In contrast to the control circuits presented in [1, 2, 3, 8] this circuit does not require any additional circuit storing the frequency from the moment before the inverter is stopped. It also does not require any circuits counting the number of half-waves of the inverter's output voltage. The next advantage is the ease of adaptation of the existing PFM control circuit to work accordingly to the new presented method.

REFERENCES

[1] Fujita, H., Ogasawara S., Akagi H. : An Approach to Broad Range of Power Control In Voltage-Source Series-Resonant Inverters for Corona Discharge Treatment, *IEEE Trans*, (1997), pp.1000–1006

[2] Liu Y., He X. : A Series Resonant Inverter System with PDM and PFM hybrid Control for Plastic Film Surface Treatment , *IEEE Industry Applications Conference, IAS* 2005, pp.1700–1704

[3] Liu Y. , He X.: PDM and PFM hybrid control of a series-resonant inverter for corona surface treatment, *IEE Proc.-Electr. Power Appl.*, vol.152, No. 6, November 2005, Page(s): 1445–1450

[4] Meek J., Craggs J.: *Electrical breakdown of gases*, John Wiley & Son, New York, (1978)

[5] Mućko J., Strzelecki R., Lutomirski S.,: Falownik z tranzystorami IGBT do aktywacji folii polietylenowej (IGBT Inverter for Corona Discharge Treatment of Polyethylene Foil), V Symposium PPE, Technical University of Gliwice, Proceedings, (1993) , 418-425

[6] Mućko J., Strzelecki R., Kozakiewicz J., Lutomirski S. : Resonant Inverters with Improved Output Characteristics in Application for Corona Discharge Treatment, EPE'99, Lausanne Swiitzerland, CD-ROM (1999), No 857

[7] Mućko J.: Typoszereg tranzystorowych generatorów wraz z transformatorami WN do aktywacji folii polietylenowej (A Series of Transistorized Generators for Activation of Polyethylene Foil), Report with I stage of Project KBN No 8 T10A 142 2000 C/4915, Bydgoszcz, 2000, unpublished

[8] Mućko J., Sterowanie falownika rezonansowego poprzez modulacje PDM w aplikacji aktywatora folii polietylenowej (PDM Control of a Resonant Inverter for Corona Surface Treatment of Polyethylene Foil), PES-III, Kościelisko - Poland, (2001), Proceedings

[9] Mućko J.: Aktywator folii z falownikiem rezonansowym - właściwości, metody i układy sterowania (Corona Treater for Polyethylene Foil – features, methods and control circuits), *Przegląd Elektrotechniczny*, (*Electrical Review*) no. 11/2005, pp. 42-49

[10] Mućko J.: New control methods of a series-resonant inverter for purpose of corona surface treatment. *Electrical Review* no. 4/2008, pp. 10-13

[11] Rosenthal L.A., Davis D.A. Electric Characterization of a Corona Discharge for Surface Treatment, *IEEE Trans. Industry Applik.* Vol.IA-11, , (1975), no.3, 328-335

[12] Datasheets: Corona Ahlbrandt System; Sherman Treaters; IPTS "METALCHEM

Power supply unit for an electric discharge machine

Wojciech Mysiński

Cracow University of Technology, Faculty of Electrical and Computer Engineering, Kraków, Poland,
e-mail:mysinski@usk.pk.edu.pl

Abstract: **The electrical discharge machining process provides one of the best solutions available for machining a growing number of high-strength and corrosion –and-wear resistant materials. It operates on the principle of erosion of metal surfaces by an interrupted electric spark. The power supply design and configuration requirements for producing the desired spark have always been a challenge to the researchers. This paper presents an example of a configuration of a power supply used for the EDM process. The power supply (a pulse power generator) is based on an electronic current source utilising MOSFET transistors. This article describes both simulation tests carried out on the circuit, and results of laboratory tests.**

Keywords: **Converter circuit, current source inverter (CSI), device application, power supply, simulation.**

I INTRODUCTION

The electrical discharge machining (EDM) process is one of the best solutions available for machining a growing number of high-strength and corrosion –and-wear resistance materials. It is based on the principle of erosion of metal surfaces by an interrupted electric spark.

Electronic discharge machining has become one of the most effective methods of machining hard and brittle materials. For proper machining, a high metal removal rate, and good surface finishing, it is essential to have a highly developed power supply system capable of producing the required quality and form of discharges. Controlled amount of energy during the discharge process is essential to ensure high quality and perfection in machining [1].

The spark erosion process or EDM basically consists of the following main components:

- DC power supply,
- controlled pulse generator power supply (capable of switching very fast on and off),
- control unit,
- work-piece and tool, (electrode),
- dielectric medium,
- gap control unit and servo system,
- measure system (gap voltage and gap current).

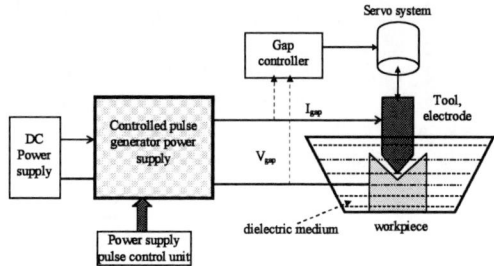

Fig.1. Schematic layout of basic EDM components

One of the key elements of an EDM is a power supply unit that feeds pulses to the electrode and the part being worked. To date, a number of power supply unit solutions have been developed, ranging from a basic resistance-capacitance RC power supply, which is still used today, to complex resonant power supply units [3]. Depending on the size of an EDM, controlled pulse generator power supply units can have power ranging from several dozen watts to several kilowatts. Electric discharge machines utilise solutions featuring a single controlled pulse generator power supply feeding pulses at output voltage of 100V to 300V; there are also EDM's incorporating two controlled pulse generator power supply units. In this case, the first power supply, the so-called ignition unit (100V-300V) feeds pulses until a spark appears, while the other unit starts operating, at output voltage of 40V- 60V, once ignition has occurred. Upon ignition, voltage at the gap drops to 20V-30V.

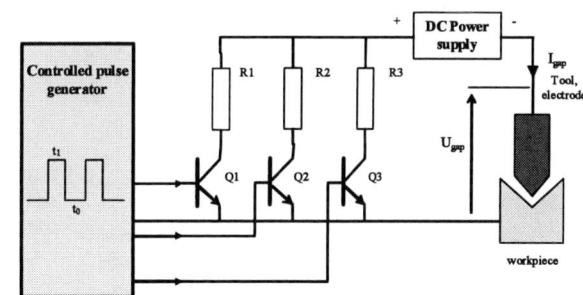

Fig.2. Simplified block diagram of a transistor power generator

978-1-4244-1741-4/08/$25.00 ©2008 IEEE

Figure 2 shows a typical lay-out of a transistor generator for an EDM machine. The generator generates impulses with adjustable duration time t_i and pause time t_0, and switches on selected transistors (Q1 to Q3). Resistors R_1 to R_3 limit transistor currents in the power circuit. If all the transistors are selected, the intensity of the duty pulse (Igap) for a given generator is at its maximum value. If a transistor is switched on during the base pulse, initial gap voltage (Vgap) equals to voltage Vps supplied by the power supply unit; if there is an electrical break-down, the voltage reaches a value of U_g=20V-30V. By adjusting pulse duration time (t1), pause duration time (t0), and gap current, the operator can ensure appropriate parameters for work-piece machining. One of the key factors in the process is gap adjustment, which, to a large extent, determines successful machining [7].

Another key issue that has to be addressed in the EDM process is maintaining a specified and stable energy level delivered to the gap. Basic RC power generators or transistor generators with a resistor operating as a current limiter are designed to ensure that gap current simply does not exceed the maximum value only. Such power generators (power supplies) can be treated just like a typical voltage source with a current limitation feature. But the process occurring inside the gap may be such that current will never reach the maximum value. There can also be voltage fluctuations at the transistor keys output, resulting from the very design of the direct current power supply.

A typical transistor power generator comprises a non-stabilized direct current power supply and a set of transistor keys that switch on appropriate resistance settings in the gap circuit. In most cases, a non-stabilized direct current power supply comprises a single-phase or a three-phase transformer reducing voltage, a bridge rectifier and a capacitance filter.

Mains voltage fluctuations (in the power grid) are transferred to the output voltage down -stream the rectifier, thus affecting output voltages of the transistor keys. There are also other difficulties involved - voltage pulsation caused by the bridge rectifier with a capacitance filter; voltage pulsation is most troublesome when high output current is applied. Therefore it is advisable to use a three-phase bridge rectifier; such a rectifier is necessary for high output power ratings – upwards of a few kilowatts.

In order to ensure stable supply voltage to be fed to the transistor keys, one can use a stabilised pulse (switching) power supply; this however, will entail higher costs involved in the construction of a such a generator.

One of the possible solutions to this problem would be to use a stabilized current source with a keying feature. The current source will feed stable current to the gap, which will in turn ensure stable energy levels provided to the gap during the actual machining process. Stabilized gap current will contribute to improved processing parameters, which will ensure adequate surface roughness of the work-piece.

Furthermore, a stabilised current source will also improve immunity to voltage fluctuations on the grid and to direct voltage pulsations downstream the rectifier.

One of the possible configurations for the current source is a circuit based on a bipolar transistor or a MOSFET transistor operating as a linear control element. This article will describe a power supply unit based on the so-called electronic current source. The prototype power supply utilizes the linear electronic current source topology for low power applications with inherent protection under short circuit conditions.

II CURRENT SOURCE LAYOUT DESIGN

Electronic current sources are widely used in measurement circuits and automatic control systems for transmission of analogue and digital signals. Electronic current sources are also to be found in practically every analogue integrated circuit, operating as one of the integrated system components.

There are quite a number of current source solutions, ranging from simple solutions based on one transistor (either bipolar or unipolar) to solutions incorporating operational amplifiers. However, these are low power systems, with maximum current reaching 1A, at supply voltage of 20V to 30V. The EDM process requires a current source that can be switched on and off for very short time periods, ranging from a few microseconds to a few milliseconds.

A simplified schematic of a current source is shown in Fig. 3

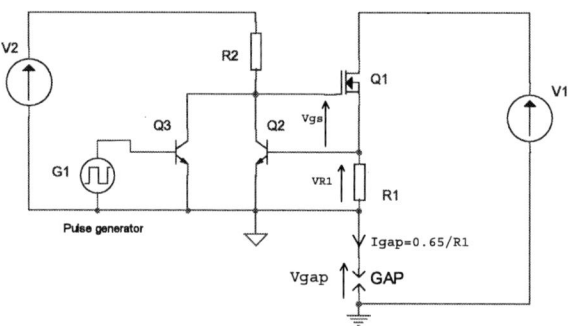

Fig. 3. Proposed current source for EDM process

Devices V1, Q1, R1 and the gap make up the main power circuit, in which transistor Q1 is designed to stabilise current. Devices V2, R2, Q2 and R1 control transistor Q1. Main current flowing through resistor R1 causes voltage drop VR1, carried to transistor Q2. If voltage across R1 exceeds 0.65V, Q2 will start conducting, and voltage across gate Q1 will decrease, thus causing the transistor to turn off. The value of current in the main circuit will equal to approx. I=0.65/R1. A signal sent by pulse generator G1, through transistor Q3, can turn on or off the control function at transistor Q2. By selecting a proper resistor R1, one can set the current value to a required value in the main circuit.

When the load is not operating like an open circuit, the gap voltage is equal to V1; during the short circuit state current is limited to value I=0.65/R1.

The EDM process does not require many output current settings. In most cases, a dozen settings or so, ranging from minimum to maximum, will suffice.

In the current source discussed, output current values can be adjusted by changing resistance R1 (e.g. with a potentiometer), but that can be difficult, as resistor R1 has a low resistance (for current above 1 A, the resistance is below 0.65 Ω). An alternative solution would be to use a resistance switch, but this would in turn increase parasitic inductance in the current measurement circuit. Another possibility would be to connect in parallel a number of current sources, and to switch them on and off electronically. This would enable the user to have a number of output current values at his disposal.

During the design of such a current source, account must be taken of power dissipated the main output transistor. The worst scenario involves a short-circuit in the gap, when the main output transistor dissipates considerable heat (Q1) (Fig.5). That is why a suitable cooling system must be implemented to counteract this shortcoming resulting from the linear (continuous) operation of the output transistor. Under normal operating conditions, the voltage drop across the gap is between 20V and 30V, and remnant of voltage V1 appears across transistor Q1. The voltage drop across measuring resistor R1 can be disregarded.

Figure 4 shows a schematic diagram of the circuit for Pspice. Figure 5 shows a DC characteristic of current source, output current versus termination resistance Rgap, which serves as a substitute for a gap. As can be seen, within the range approx. up to 20 ohms, the circuit operates as a current source with an efficiency of 2A. For the termination resistance range of 0 ohm to 18 ohms, transistor M1 dissipates a lot of heat. Figure 6 shows the result of a time analysis (TRAN) – output current waveforms under pulse control conditions. As the figure shows, the circuit can generate short current pulses.

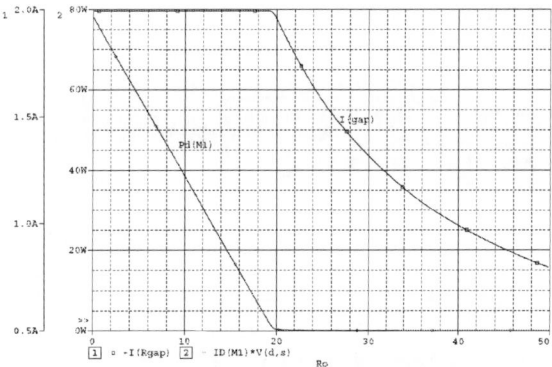

Fig. 5. DC characteristic of current source and power dissipation on M1

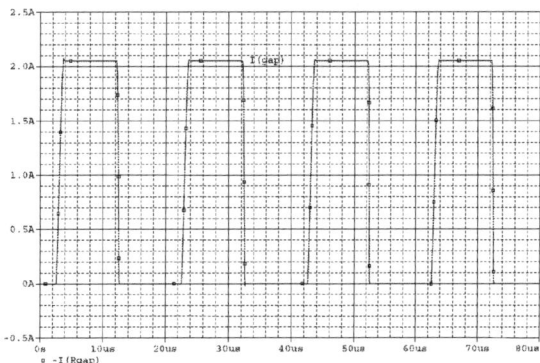

Fig.6. Output current pulses

III DESIGN OF A POWER GENERATOR BASED ON CURRENT SOURCES

The proposed power generator for a electro-discharge machine is composed of two circuits connected in parallel. The first circuit is an electronic ignition current source with a maximum output voltage of 160V/250V, and the other circuit – the second electronic operational current source with a maximum output voltage of 40V. Both circuits are to be controlled using programmable pulse generator with which the durations of state 1 and state 0 are set (Fig.7). In the electro-machining process, the duration of a pulse and the duration of a pause ranges from several microseconds to more than a dozen milliseconds.

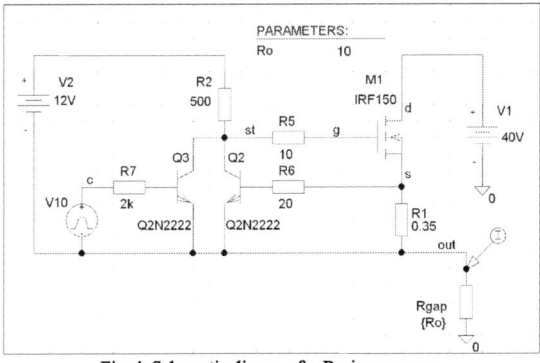

Fig. 4. Schematic diagram for Pspice program

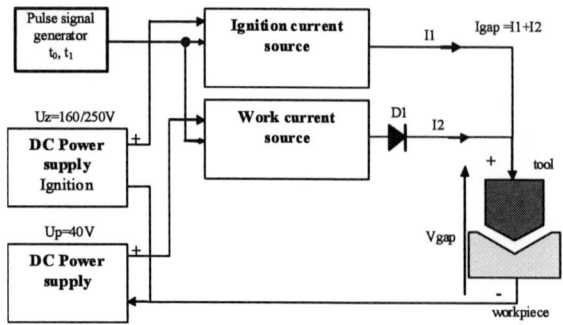

Fig.7. Block diagram of the power stage of EDM machine

When a pulse is generated, ignition voltage appears across the gap, and, if the necessary conditions are met, the so-called ignition occurs and a spark, an arc, appears. Ignition current is limited by the current source. During the duty cycle, gap current is a sum of ignition current and duty current. During the EDM process, the electrode and the work-piece are often short-circuited; when this happens, short-circuit current is limited to a sum of ignition current and duty current. A diode at the duty current output protects the circuit against excess voltage provided from the ignition current source (Fig.7).

The lay-out of the current source is not complicated; however, problems arise with parallel operation of two power supplies and transmission of current from the power supplies through cables to the gap.

Under the conditions of pulse current fluctuations, one will always have to face the problem posed by the inductance and capacitance of connecting cables. In actual EDM machines, the gap is often located several metres away from the power generator. In such cases, account has to be taken of parasitic parameters of power cables connected to the gap. The design of EDM machines provides that between the gap proper and output terminals there are various screw joints, indirect cable connections, and a gap polarization switch. These elements will reinforce parasitic parameters of the connection between the gap and power generator.

IV LABORATORY TESTS ON THE CURRENT PULSE GENERATOR

In order to test the operation of a current source, a laboratory power generator has been constructed, comprising an ignition current source and a duty current source. The maximum output current in the ignition circuit is 1A for output voltage of 160V or 250V, while the maximum output duty current is 10A, with output voltage of 40V. The two configurations are combined as shown in Fig. 7, and the output current is a sum of ignition current and duty current. In order to obtain several values of output current, the duty current module comprises five identical current sources with an efficiency of 2A, connected in parallel.

Figure 8 shows voltage and current waveforms measured at the gap which are typical in sinking EDM operations.

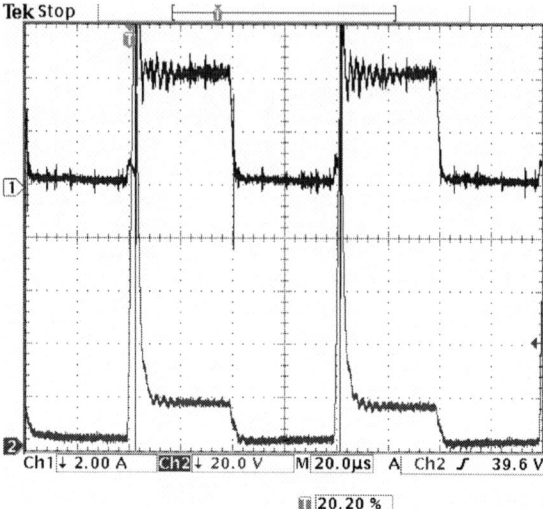

Fig. 8. Gap voltage (Ch2) and gap current (Ch1) waveforms during EDM process

The waveform in Fig.8, from channel 2, is a typical waveform representing voltage across the gap during the EDM process. The moment a pulse is generated, ignition voltage of 160 V appears across the gap. Within a few microseconds, ignition occurs and voltage drops to approx. 20V. The current flowing through the gap is stabilised, and is a sum of ignition current and duty current. After the control pulse disappears (output current being off), there is low voltage across the gap, fed from an additional low power voltage source, intended to control the gap.

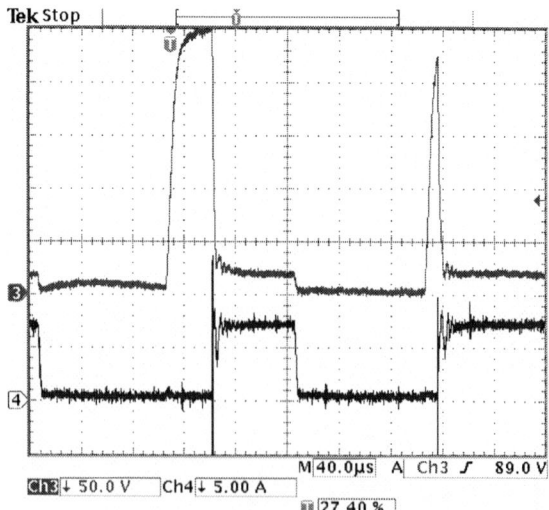

Fig. 9. Gap voltage (Ch3) and gap current (Ch4) waveforms during EDM process

The waveform in Fig. 9, from channel 3, represents gap voltage when the ignition voltage amounts to 250V. Depending on the conditions in the gap, ignition take a longer or shorter time to occur.

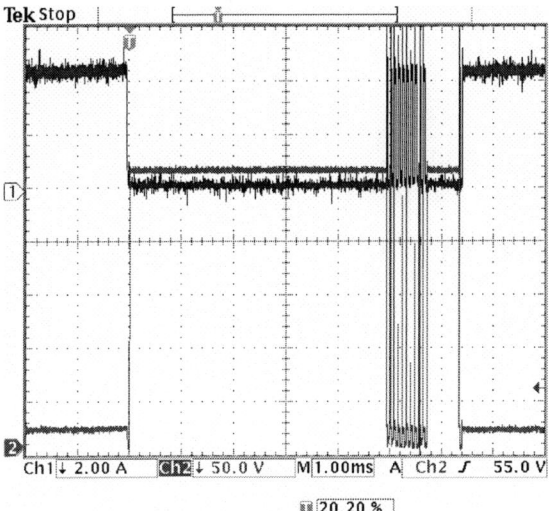

Fig. 10. Gap voltage (Ch2) and gap current (Ch1) waveforms during EDM process

While an EDM operation is in progress, there can occur short-circuit states, states without ignition and normal duty states, during which pulses appear. This is shown in Fig. 10. Initially, there is a short-circuit state the current (Ch1) is flowing but no pulses have been generated; next, high ignition voltage appears (Ch2, 250V), but without a spark; next comes an example of normal operation (current and voltage pulses), followed by a short-circuit state again at the end.

V CONCLUSION

The current source described allows for the generation of precise current pulses with duration of a few microseconds to a few milliseconds, at output current in the range of 0.1A to 10A. These power supply parameters are sufficient for some EDM's. However, the drawback of such a solution is its low efficiency, especially during the short circuit state.
Work is now underway to construct a current source with improved efficiency – up to 40A.

REFERENCES

1. B. Sen, N. Kiyawat, P. K. Singh, S. Mitra, J. H. Yeand, P. Purkait, "Developments In Electric Power Supply Configurations for Electrical Discharge Machining (EDM)."
2. M. Kunieda, B. Lauwers, K. P. Rajurkar, B. M. Schumacher, "Advancing EDM through Fundamental Insight into the Process."
3. R. Casanueva, L. A. Chiquito, F. J. Azcondo, S. Bracho, "Electrical Discharge Machining Experiences with a Resonant Power Supply."
4. R. Casanueva, F. J. Azcondo, S. Bracho, "Series–parallel resonant converter for an EDM power supply."
5. Carl Michael F. Odulio, Luis G. Sison, Ph.D., Miguel T. Escoto, Jr., "Energy-saving Flyback Converter for EDM Applications."
6. K.H. Ho, S.T. Newman, "State of the art electrical discharge machining (EDM)."
7. A. Miernikiewicz, „Doświadczalno – Teoretyczne Podstawy Obróbki Elektroerozyjnej (EDM)", Politechnika Krakowska, Kraków 2000, Monografia 274

High Power, High Voltage, High Frequency Transformer / Rectifier for HV Industrial Applications

T. Filchev [1], D. Cook [1], P. Wheeler [1], A. Van den Bossche [2], J. Clare [1], V. Valchev [3]

[1] eextf@ nottingham.ac.uk, dave.cook@nottingham.ac.uk,
Pat.Wheeler@nottingham.ac.uk, Jon.Clare@nottingham.ac.uk,
University of Nottingham, University Park, Nottingham NG7 2RD, England.
[2] Alex.VandenBossche@UGent.be, Ghent University, Department of Electrical Energy,
Systems and Automation, Sint-Pietersnieuwstraat 41, Belgium
[3] vencivalchev@hotmail.com, Technical University of Varna, Department of Electronics,
Faculty of Electronics, Studenska str., 1, Varna, 9000, Bulgaria

Abstract - This paper presents a high voltage (HV), high frequency (HF) transformer with voltage multiplier. The described systems employ a low profile HV, HF transformers and voltage bridge technology multiplier. A model of the converter of 150kW is presented and verified by Saber simulator. Design considerations and models of the HV, HF transformers are presented.

Keywords - **Transformer, Insulation, Pulsed power, High Voltage power converters**

I. INTRODUCTION

High voltage (>10kV), high power (>10kVA) supplies have a wide range of HV industrial applications including medical imaging, the aircraft industry and particulate emission control in power generations. Particulate emission control is achieved through the use of electrostatic precipitator supplies. Conventional converters for this application are based on line frequency approaches [1,2], and consequently the converters are both large and heavy. Recent advances in power electronic control techniques have allowed the development of a new type of power converter, which utilizes high switching frequencies (kHz) and high frequency transformers. Consequently, a significant reduction in the size and weight of the filters and transformers is observed. This is particularly advantageous in many applications, such as electrostatic precipitator where the size and weight of the converter is a primary concern [1, 2].

Figure 1 shows a block topology of the power supply using a high frequency (HF) transformer.

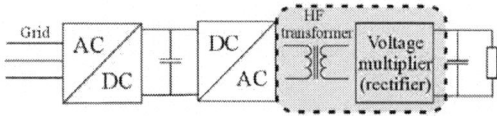

Fig.1. Block topology of the power supply for HV industrial applications

The high frequency (HF) transformer and voltage multiplier (fig.1) for high voltage applications are investigated [2].

Power supply for HV industrial applications with following electrical specifications:

V_{in} = 415V, $V_{out(DC)}$= 150 kV, $I_{out(DC)}$= 1 A, f_{sw}=20kHz has been investigated.

Two cases for transformer design will be considered and compared – HV, HF transformer with rectifier and HV, HF transformer with voltage multiplier.

The paper derives the problems involving the use of the system:

- computer simulation of the voltage multipliers;

- computer simulation and design of the high voltage, high frequency transformer.

We consider two types of voltage multipliers – full wave multiplier and multicell full wave voltage multiplier.

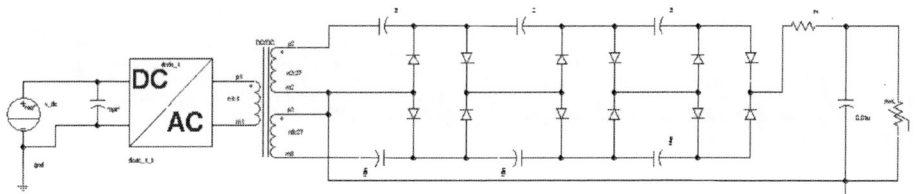

Fig.2. Cockcroft –Walton voltage multiplier - full wave

II. RECTIFIER - VOLTAGE MULTIPLIER

By introducing the voltage multiplier, the secondary turn numbers, and stray capacitance of the high-voltage high-frequency transformer as well as the rectifier diode voltage ratings may be greatly reduced. The investigated voltage multipliers are based of the well-known Cockcroft –Walton (CW) voltage multiplier [2,3,6,7]. The voltage drop and for CW voltage multiplier is given by [6,7]

$$\Delta V = \frac{I_L}{f\,C} A(n); \quad (1)$$

The output voltage ripple

$$\delta V = \frac{I_L}{f\,C} \frac{n(n+1)}{2} \quad (2)$$

Finally the output voltage is given by

$$V_{out} = 2nV_{pp} - \Delta V \ ; \ (3)$$

Where:

n is number of stages; V_{pp} is peak value of the input voltage; f is frequency.

Current dependent voltage drop (1) and the voltage ripple (2) are the main problems associated with the CW voltage multiplier [6]. To overcome the problems a high frequency (20kHz) multicell CW voltage multipliers have been presented.

A multicell circuit of CW voltage multiplier is shown on Fig.3. The supply voltage can be single phase as well as three phase voltage supply [3].

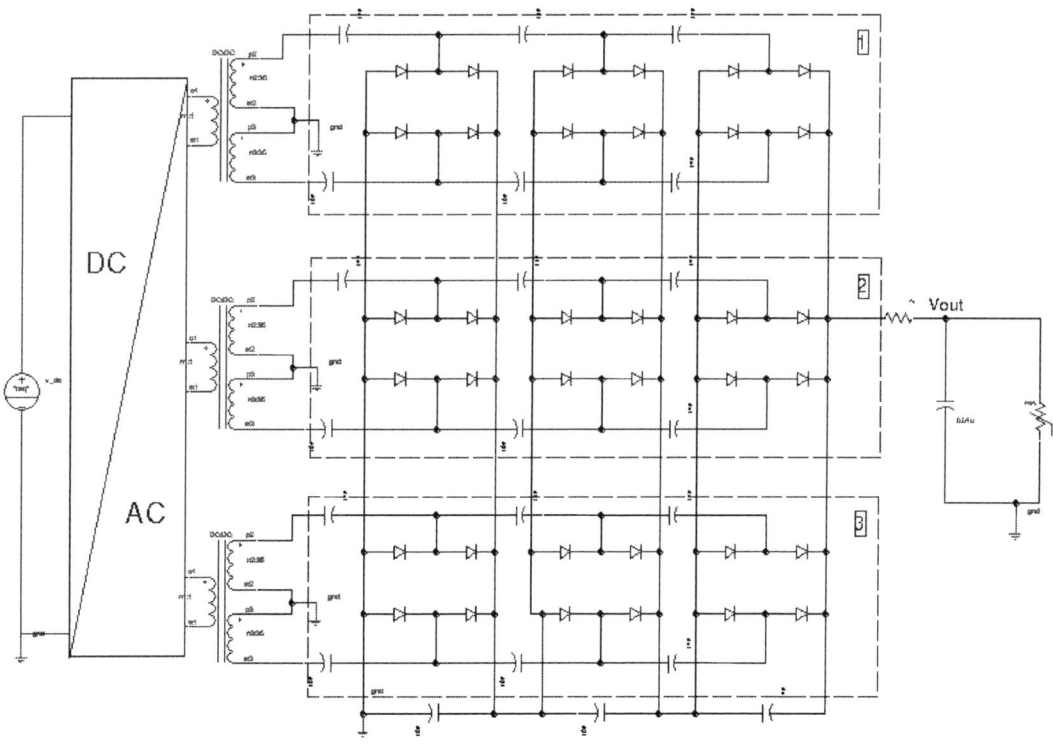

Fig. 3. Presented full wave multicell CW voltage multiplier.

1327

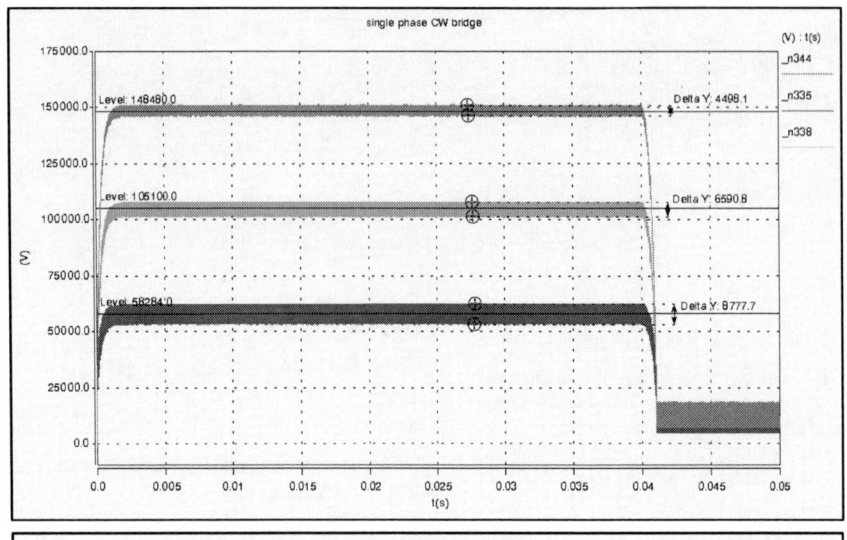

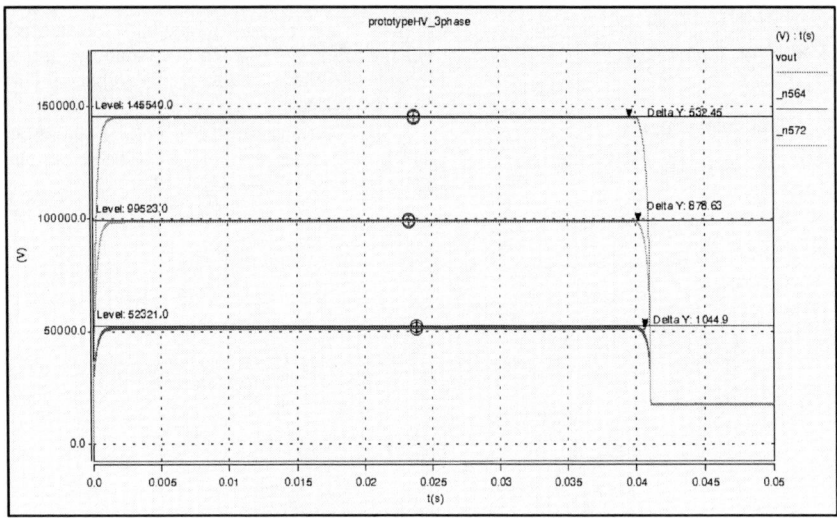

Fig. 4. Simulation results for output voltage v_{out} :

(a) single cell full wave CW multiplier

(b) three cells full wave CW multiplier

The typical transient output voltage for the multicell CW voltage multipliers (Fig.3) are shown in Fig. 4. (Simulation conditions: $V_{in\,(RMS)}$ = 415V, $V_{out\,(DC)}$= 150 kV, $I_{out\,(DC)}$= 1 A, f_{sw}=20kHz , R_{load}= 50kΩ, 50Ω,).

From the simulation results (Fig 4) we can see that there are not any big differences in the transient voltage and voltage values between both configurations of multipliers.

The advantages of the presented multicell CW voltage multiplier circuit compared to full wave multiplier are: low output voltage ripple, low transformer ratio (low stray capacitance), small winding windows and common ground (not required HV insulation). Moreover using low capacitances respectively low discharge energy in topology (Fig. 3) is a real advantage for HV application.

III. TRANSFORMER DESIGN

A high voltage HF transformer has been simulated and designed. The HV, HF transformer design requires following considerations of:

- High Voltage (HV) insulation requirements.
- Core, material, shapes and windings space.
- Parasitic capacitances and parasitic inductances.
- Windings and core losses.
- Corona effects.
- Cooling.

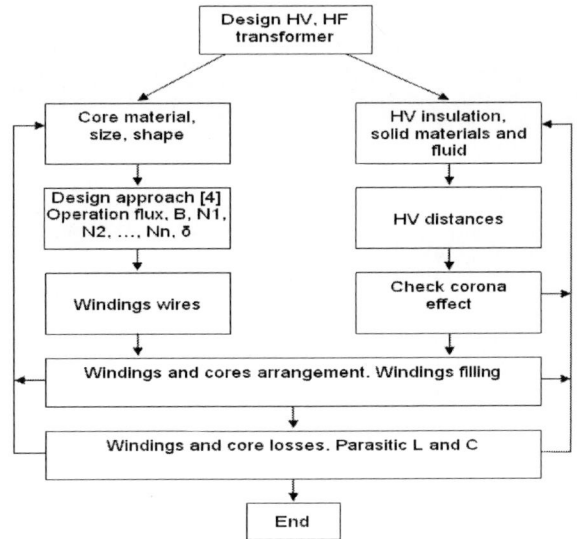

Fig. 5. Design methodology for high voltage, high frequency transformer

The design methodology with above considerations is investigated (Fig.5).

1. HV insulation requirements

It guarantees that the insulation material has sufficient dielectric strength so the electrical breakdown due to insufficient insulation thickness between primary and secondary windings and also high leakage inductance are avoided.

Practical insulation material reaches a few tens of kV/mm. Some special isolation materials have 10-40kV/mm dielectric strength. Nomex® Type 410 is an insulation paper which offers high inherent dielectric strength up to 32kV/mm (AC). This insulation material is chosen for HV windings insulation.

2. Core, material, shapes and windings space

Two types of soft magnetic materials are discussed: ferrite and nanocrystalline magnetic materials.

Ferrites are widely used for HF applications. EE, U and C core shapes of the ferrite are preferred because of the large winding room. The cores can be arranged in parallel or in series for high power VA rating and better cooling.

Nanocrystalline magnetic materials are used for toroidal cores and air gapped shapes. Some cores, U shapes are suitable for high power applications (Hitachi, Vacuumschmelze). There are difficulties in cutting and winding those (especially toroidal cores) [4].

In the present paper the transformer for high voltage industrial application, with ferrite UU core is investigated (Fig. 6).

In our transformer design the distance between parallel cores is chosen to be five millimeters [9] for better oil natural cooling .

3. Parasitic capacitances and parasitic inductances.

As we discussed above significant parasitic (leakage) inductance occurs due to HV insulation requirements between primary and secondary winding. Parasitic capacitances are also presented in the high power high frequency transformer.

4. Windings and core losses

The ferrite losses can be reduced by decreasing the induction of the transformer and choosing the right ferrite grade.

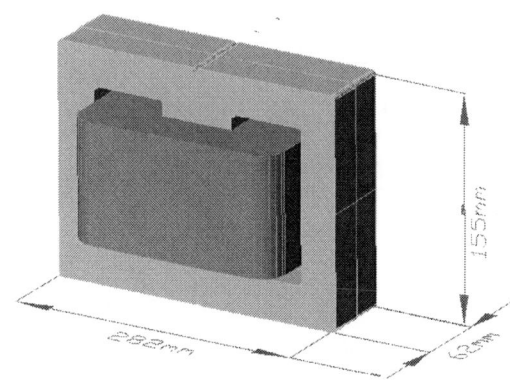

Fig. 6. Power HV, HF transformer

(UU core configuration)

Ferrite losses tend to increase with the power 2.3 ÷ 3 of the peak induction [9]. Thus decreasing the peak induction reduces the ferrite losses considerably. In practice this means an oversized core design with more turns.

The core loss is calculated using the loss curve provided by the manufacture (Epcos). The core loss is 50mW/g (f=20kHz, B=0.3T) based on the loss curve. The winding losses including copper and eddy current losses are calculated based on the method introducing in [4].

5. Corona effects.

Some insulating materials are more vulnerable to corona deterioration than others in HV applications. Improvements in electrode geometry, increasing the spacing, and reduction of applied voltage stress results in the elimination of corona.

6. Cooling.

The cooling oil in transformer is considered to avoid the corona effect and improve HV insulations. They are many fluids: mineral oil, silicon oil [General Electric], natural ester [ABB] and etc. The natural ester is chosen for being environmental friendly. Fires and explosions are also much less likely.

Following the above requirements Saber model [8] of the high voltage, high frequency transformer is developed. The model is shown in Figure 7.

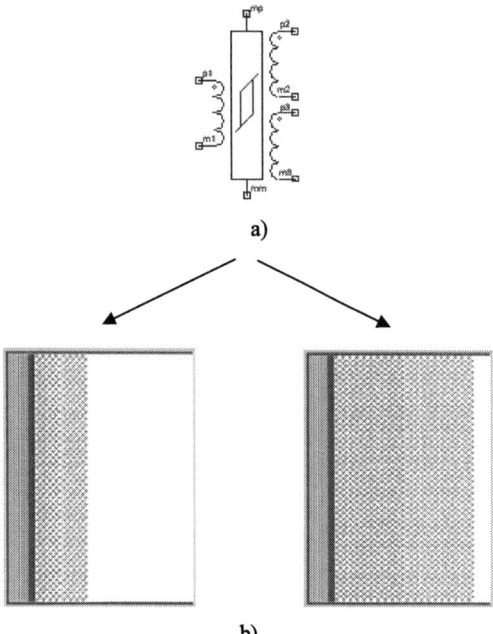

a)

b)

Fig. 7 (a) Saber model of the centre taped transformers

(b) The winding view of the models at 50 kV and 150kV

(primary – copper foil, secondary – Litz wire)

A. Design results for based HV, HF transformers

Here we present the design results for transformers without voltage multiplier (150kV) and transformer with an ideal voltage multiplier (50kV). It is necessary power transformers for these systems to have a large dimension [10] in order to have high heat capabilities. Designed transformers are constructed with the biggest commercial available ferrite U core from Epcos; 141/78/30, grade N87.

TABLE I.

COMPARISONS BETWIN THE DESIGNS

HV,HF transformer (HVHFT-1): U_{out1} = 50 kV, I_{out1} = 3 A, P_{out1}=150kVA; HV,HF transformer (HVHFT-2): U_{out2} = 150 kV, I_{out2} = 1 A, P_{out1}=150kVA

Design parameters	Transformers	
	HVHFT-1	HVHFT-2
Primary turns	6	6
Secondary turns	578	1736
Cores heat capabilities	210 W	210 W
Copper losses	157 W	113 W
Filling factor - windings	0.40	0.90
Copper weight - primary winding	234 g	234 g
Copper weight- secondary windings	150 g	436 g
Total weight - copper and cores	10 384 g	10 670 g

IV. CONCLUSION

In the present paper the trends in application of switch mode power supplies for HV application are reviewed. Two topologies of multicell CW voltage multipliers are analyzed.

Design procedure for the high voltage, high frequency transformers is suggested. The transformers are designed with copper foil for the primary winding and Litz wire for secondary winding. Thus we achieved good thermal performance.

The total weight for both transformers is almost the same because the cores weight dominates the copper one. However we can not improve significantly the weight of the transformer by using voltage multiplier techniques. The transformer with voltage multiplier has low filling factor compared to the transformer without voltage multiplier. The system with voltage multiplier can meet the high voltage insulations requirements. However the system of HV, HF transformer with voltage multiplier compared to the HV, HF transformer without voltage multiplier is preferable for HV industrial application. Saber model of the transformers for both cases is developed.

Practical results and measurements will be presented in a future paper.

ACKNOWLEDGMENT

This research is funded by STFC and dstl under the joint grant scheme (contract no.: JGS-1192) and supported by e2v Ltd.

REFERENCES

[1] John C. Fothergill, Philip W. Devine, and Paul W. Lefley "A Novel Prototype Design for a Transformer for High Voltage, High Frequency, High Power Use" Ieee Transactions on Power Delivery,vol.16 No1, Jan. (2001)

[2] Van der Broeck, H.; "Analysis of a current fed voltage multiplier bridge for high voltage applications" PESC 02. IEEE 33rd Vol. 4, 23-27 June 2002 pp :1919 - 1924 (2002)

[3] Iqbal, S. "A three-phase symmetrical multistage Voltage multiplier" Power Electronics Letters, IEEE Vol. 3, Issue 1, pp :30 – 33, March (2005)

[4] Van den Bossche A., Valchev V.C. "Transformers and Inductors for Power Electronics", CRC Press, Boca Ration, FL, USA (2005)

[5] Cook, D.J.; Clare, J.C.; Wheeler, P.W.; Pryzbyla, J.S.; Richardson, R."Design of a High Voltage Direct Resonant Converter" 12th International Power Electronics and Motion Control Conference pp :143 – 147 Aug. (2006)

[6] Sun, J.; Ding, X.; Nakaoka, M.; Takano, H. 'Series resonant ZCS-PFM DC-DC converter with multistage rectified voltage multiplier and dual-mode PFM control scheme for medical-use high-voltage X-ray power generator' ; Electric Power Applications, IEE Proceedings- Vol. 147, Issue 6, pp 527 – 534, Nov. (2000)

[7] Di Cataldo, G.; Palumbo, G.; 'Current leakage and parasitic capacitance in transient second and triple order Cockcroft-Walton voltage multiplier', Circuits and Systems, Proceedings of the 35th Midwest Symposium on 9-12 Aug. 1992 pp: 722 - 725 vol.1 (1992)

[8] T Filchev, P Wheeler, J Clare, D. Youdov, V. Valchev, A. Van den Bossche, "A LCL Resonant DC-DC Converter for Electrical Power Distribution Systems", European Power Electronics , EPE – PEMC Conference, Riga, September (2004)

[9] Van den Bossche, V. Valtchev, T.Filchev "Improved approximation for fringing permeances in gapped inductor" IAS 37th Annual Meeting Conference IEEE, Pittsburgh,USA, (2002)

[10] D. Georgiev, K. Beshinski, T. Filchev "High voltage high frequency electronic transformer" (in Bulgarian) TU-SNS. (2008)

Small Power Laboratory Model
and High Power Prototype of the Four-Level VSI

Ryszard Michał Strzelecki[*], Paweł Szczepankowski[*], Andrzej Kasprowicz[*],
Genady Stepanovic Zinoviev[†], Krzysztof Zymmer[‡], Zbigniew Zakrzewski[‡]

[*] Gdynia Maritime University/Department of Electrical Engineering, Gdynia, Poland, e-mail: _rstrzele@am.gdynia.pl_
[†] Novosibirsk Technical University/Department of Power Electronics, Novosibirsk, Russia, e-mail: _genstep@mail.ru_
[‡] Electrotechnical Institute/Department of High Power Converters, Warsaw, Poland, email: _npn@iel.waw.pl_

Abstract—In the paper, we discuss a laboratory mode of a Diode Clamped 4-level Voltage Source Inverter (VSI) of power approximately 5kVA, that is designed for initial experimental verification of solutions for "software" and "hardware" of the control components in prototype model of inverter 6kV and 0,5-1 MVA. The paper discusses tasks of control console in the PC. The article also discussed construction and development of the power part of the inverter model. In addition, it presents the voltage PWM algorithm implemented in the DSP. Experimental results are presented.

Keywords— Multilevel converters, high voltage power converters, converter control, DSP, modulation strategy.

I. INTRODUCTION

The recent popular trend in application of the power electronics systems of high power is application of a multilevel inverters [1, 2]. The aim is to achieve voltage with low higher harmonics, as well as to minimize number of power devices such as: transistors IGBT, fast switching diodes or capacitor. Along with development of new typologies, research focuses on development of efficient modulation techniques, which would minimize switches losses or decrease capacitors' capacities in DC-link circuit [3]. None the less, in the case of high and very high powers, crucial criterion for the choice of inverters is the cost of its elements and its reliability. The paper discusses laboratory model of the 4-level Diode Clamped VSI, destined for experimental verification of the functional propriety of the solutions and control algorithms implemented in DSP controller. The controller is anticipated to be implemented in prototype of the 4-level inverter of power 0,5-1 MVA and output voltage 6 kV. Sections 2 present of the information about experimental setup and power part of the inverter. The following two chapters discuss the algorithm SVPWM realized with DSP controller as well as initial experimental results.

II. LABORATORY STAND

A. Experimental Setup

Block diagram of the experimental setup, that is shown on Fig. 1, consists of six fundamental functional blocks. Block {1} consists of a PC with software Visual DSP++

This work is sponsored by the Polish Ministry of Science and Higher Education under project No. R01 002 01

(designing environment for the processor ADSP21363 by Analog Devices) and Quartus (designing environment for the system CYCLONE 2 by Altera), and operator console. From the console level the following inverter parameters are assigned: frequency carrier of the SVPWM, basic frequency of the inverter output voltage, modulation index, dead time, confirmation time of the transistors switching on, amplitude accretion speed of the basic output voltage harmonic. Besides that, the main tasks of the console are: registration and visualization of the selected processor variables, realization of the programming function and formatting of selected memories, diagnostics and event reports as text file. Transmission of the data between the PC and microprocessor controller takes place by separated USB junction with speed 1Mb/s. Blok {2} consists of the microprocessor controller, designed and constructed by „MMB Drives" in Gdansk. The main elements of the controller are: floating point 32-bits processor, working with frequency 333MHz and modern digital programmable FPGA system. The controller is installed in interface board – block {3}, with 20 two-way, optically separated transmission channels, which are used to communicate with IGBT controllers of the VSI (Fig. 2).

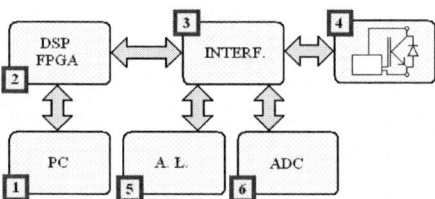

Fig. 1. Block diagram of the experimental setup.

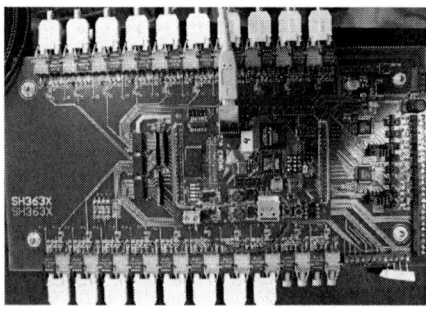

Fig. 2. Interface board.

978-1-4244-1741-4/08/$25.00 ©2008 IEEE

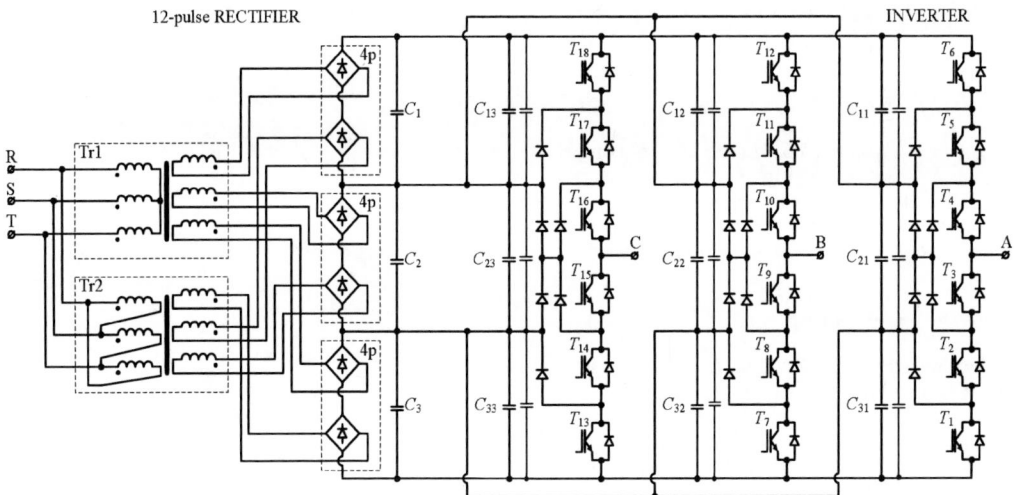

Fig. 3. The power part of the of the realized 4-level converter.

Block {4} presents a group of transistor controllers IGBT. Applied are the controllers IGD515 by CONCEPT company, which are highly integrated control system, with isolated voltage converter and complete light pipe interface. The signal FAULT, from the controller is also the signal that confirms switching on, which is very useful in system's diagnostics. Digital channels between the controller SH363 and interface board were buffered by HCT systems and LVX. One of the main tasks of the block {5} – connecting apparatus – is preliminary charge of the capacitors in linking inverter circuit. In addition elements protecting from short circuits were installed and connectors allowing quick change of the DC-link circuit configuration (for example: input chokes, Z-source, other voltage sources). Measurement block {6} consists of a part aggregating measurement signals (current and voltage sensors of type LEM and systems with optical-amplifier HCPL7800) and with analog-digital transducer. Two 14-bits, 4-channes transducers AD7865 by Analog Devices were applied, which allow quick configuration of the measurement channel range from ±5V to ±10V and a choice of number of analog channels to conversion.

B. Power Part of the VSI

The power part of the realized 4-level converter (on the base of the typical Diode Clamped 4-level VSI topology) is presented on Fig. 3. In the laboratory model, branches of the inverter were placed at separated radiators. Each of the branches consists of the three transistor modules and the three diode modules (BSM50GB120DN2 and DD46S12K) from the company EUPEC. The special three 4-pulse rectifiers ale placed on additional radiator [4]. The view of the model's design is shown on Fig. 4. It is possible to move elements on the radiator and of the radiator which facilitates change of the inverter typology. The capacitors C_{11}-C_{33} and C_1-C_3 equal 1000 μF. The capacitors that are not marked on Fig. 3, with capacity about 1 μF/1200 V, function as decoupling capacitors.

III. IMPLEMENTED MODULATION STRATEGY

To the particular states of the switches of the inverter the appropriate voltage space vector can be selected in stationary coordinates α-β [5, 6]. In the m-level voltage inverters, the area of the space vector is usually divided into 6 sectors, in which we can distinguish triangular areas among three nearest locations of the space vector. Single sector and equilateral triangles of a side a is shown on Fig. 5.

Fig. 4. Design of the laboratory model

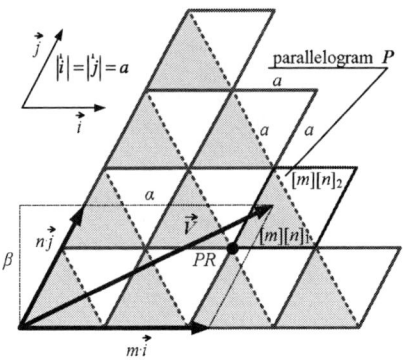

Fig. 5. Single sector for m-level VSI

Reference voltage vector V can be presented as linear combination of the vectors i and j. In accordance with Fig. 4 coordinates $[\alpha, \beta]$ of vector V are as follows:

$$\vec{V} = [\alpha, \beta] = m \cdot \vec{i} + n \cdot \vec{j} \qquad (1)$$

where

$$m = \alpha/a - \beta/\left(a\sqrt{3}\right); \; n = 2 \cdot \beta/\left(a\sqrt{3}\right) \qquad (2)$$

Integer of the m and n define coordinates of beginning of parallelogram P (point PR on Fig.5), where reference vector occurs at the time. In order to determine belonging of the reference vector V to one of the two triangles in parallelogram P, the value of the following sum D needs to be found out:

$$D = [m - int(m)] + [n - int(n)] \qquad (3)$$

If $D \leq 1$, then the reference voltage vector V occurs in the triangle with index $[m][n]_1$, else the vector V belongs to the triangle with index $[m][n]_2$. Finally, the synthesis of the reference vector V consists on addition of several states of switches, during modulator's' work, according to order defined in control strategy.

There exist two possible positions of the vector V in the parallelogram P (Fig.6). Reference vector V can be projected within area of each of the two triangles. According to the Fig.6, the reference voltage vector can be expressed as a vectors sum:

- for position of the vector V as on the Fig.6a

$$\vec{V} = \underbrace{p_1\left(\vec{V_1} - \vec{V_3}\right)}_{p_1} + \underbrace{p_2\left(\vec{V_2} - \vec{V_3}\right)}_{p_2} + \vec{V_3}$$
$$= p_1\vec{V_1} + p_2\vec{V_2} + (1 - p_1 - p_2)\vec{V_3} \qquad (4a)$$

- for position of the vector V as on the Fig.6b

$$\vec{V} = \underbrace{p_1\left(\vec{V_1} - \vec{V_4}\right)}_{p_1} + \underbrace{p_2\left(\vec{V_2} - \vec{V_4}\right)}_{p_2} + \vec{V_4}$$
$$= p_1\vec{V_1} + p_2\vec{V_2} + (1 - p_1 - p_2)\vec{V_4} \qquad (4b)$$

where p_1, p_2 is relative lengths (time duration) of the active vectors V_1 and V_2. Time duration of zero vectors V_3 and V_4 result from a difference in modulation time and duration of the active vectors. All space vectors of the 4-level VSI are presented on Fig.7.

In the selected sector 0 on the Fig.7 occurs nine, numbered regions. Specified position of the space vector is coded as follows– given number defines a point in the linking circuit connected to a load terminal of the particular phase (from left a, b, c). For example, the code „321" means, that the phase a was linked to voltage source of value $3 \cdot (U_{DC}/3)$, phase b was linked to the voltage source of value $2 \cdot (U_{DC}/3)$, and phase c was linked to the voltage source of value $1 \cdot (U_{DC}/3)$. In case of linear modulation range, value of maximum phase voltage equals $\sqrt{3} \cdot (U_{DC}/3)$, and maximum value of normalized modulation factor in this case equals $3 \cdot \sqrt{3}/2$.

Fig.8 provides exemplary positions of the space vector of four-level inverter. When the reference vector V occurs in the region 4, relative lengths of the space vectors $V_{331} = V_{220}$ and $V_{321} = V_{210}$ equal:

$$p_1 = 2 - n; \; p_2 = 1 - m \qquad (5)$$

Hence, the vector-duty factors for particular positions of the space vector in modulation period are as shown in the Table 1.

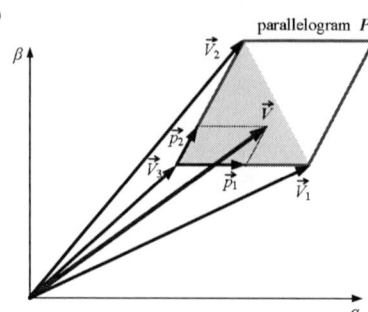

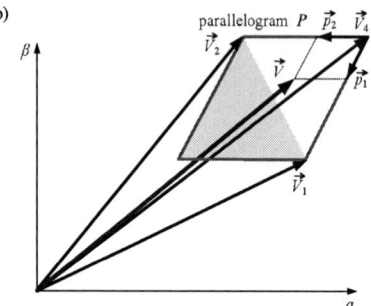

Fig. 6. Positions of the vector V in parallelogram P

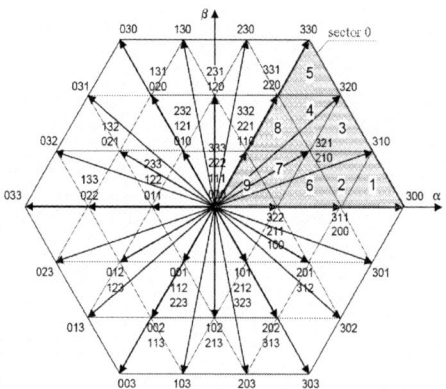

Fig. 7. Space vectors of the 4-level VSI

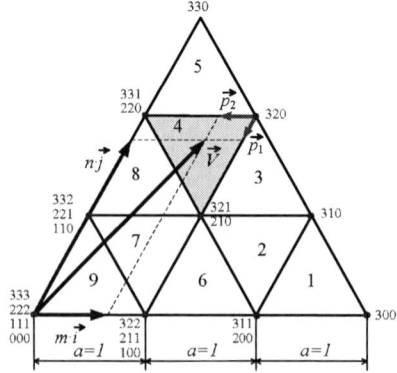

Fig. 8. Example of the position of the normalized vector V

IV. SELECTION RESULTS OF THE RESEARCH

A. Verification of the 4-Level VSI Laboratory Model

Presented oscillograme on the Fig. 9 and Fig. 10 regards preliminary phase of the experiment, where unloaded squirrel-cage motor was used. The purpose of this research was verification of the solution for the constructed laboratory model of the 4-level VSI, verification of the application of the DSP controller with own interface card, and investigation of the propriety of implemented modulation algorithms.

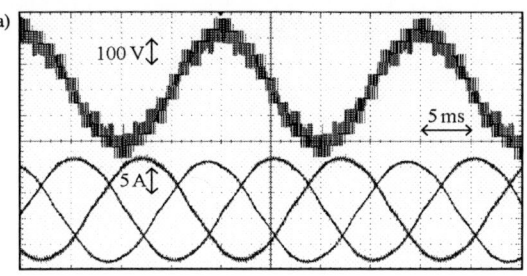

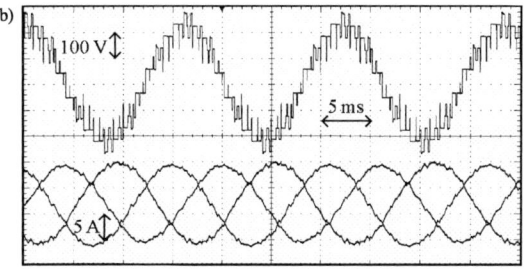

Fig. 9. Output phase voltae and currents (f_o=50 Hz) in case of relative
modulation factor m_a= 2.59 for frequency carriers
(a) f_c=4 kHz and (b) f_c=800 Hz

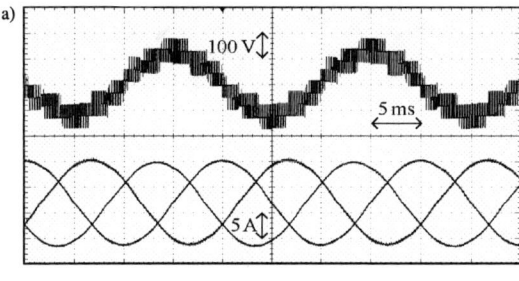

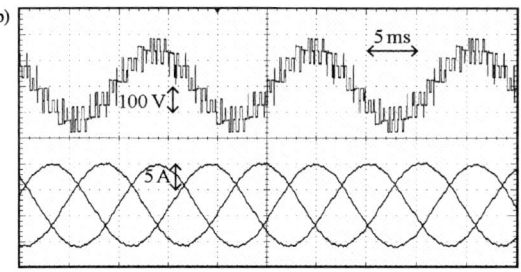

Fig. 10. Output phase voltage and currents (f_o=50 Hz) in case of relative
modulation factor m_a= 1.55 for frequency carriers
(a) f_c=4 kHz and (b) f_c=800 Hz

TABLE I.
VECTOR-DUTY FACTORS FOR REGIONS 1-9 ON FIG. 8

Region	Duty factors		
1	$d_{200/311}=3-m-n$	;	$d_{310}=n$ $d_{300}=m-2$
2	$d_{200/311}=1-n$ $d_{210/321}=2-m$	;	$d_{310}=m+n-2$
3	$d_{210/321}=3-m-n$	;	$d_{310}=m-1$ $d_{320}=n-1$
4	$d_{210/321}=2-n$ $d_{220/331}=1-m$	;	$d_{320}=m+n-2$
5	$d_{220/331}=3-m-n$	;	$d_{330}=n-2$ $d_{320}=m$
6	$d_{100/211/322}=2-m-n$	;	$d_{200/311}=m-1$ $d_{200/321}=n$
7	$d_{110/221/332}=1-m$ $d_{100/211/322}=1-n$	;	$d_{210/321}=m+n-1$
8	$d_{110/221/332}=2-m-n$	;	$d_{210/321}=m$ $d_{220/331}=n-1$
9	$d_{000/111/222/333}=1-m-n$	;	$d_{100/211/322}=m$ $d_{110/221/332}=n$

B. Verification of the 12-Pulse Rectifier

Fig. 11 - Fig. 13 presented of the selected currents of the verified rectifier [4]. This rectifier works satisfactory only under equally distributed load for every of 3 components of the 12-pulse rectifiers (see Fig. 11). It can be achieved with redundant vectors, however completely successful load distribution is only possible when relative modulation factors $m_a \leq 1,3$. One should take into account that application of the redundant vectors increases common mode voltage. Therefore, so far, supply of the linking circuit has been solved differently than it is shown on the Fig.3 (rectifier bridges).

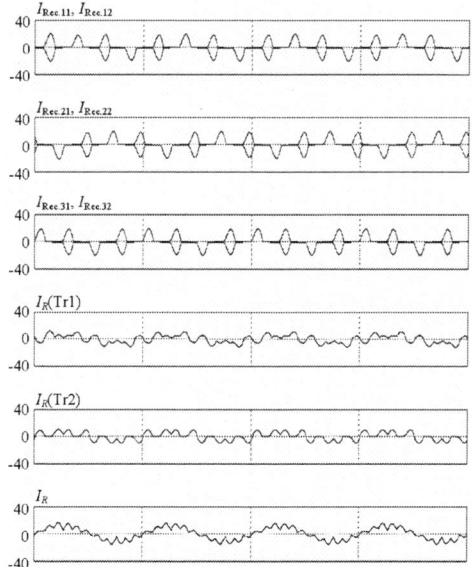

Fig. 11. Selected currents of the 12-pulse rectifier in case of relative
modulation factor m_a= 1.55 and implementation of the
SVPWM algorithm with redundant vectors.

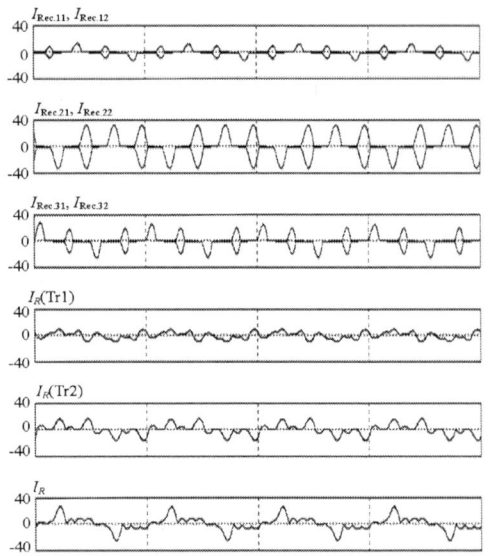

Fig. 12. Selected currents of the 12-pulse rectifier in case of relative modulation factor $m_a = 1.55$ and implementation of the SVPWM algorithm without redundant vectors

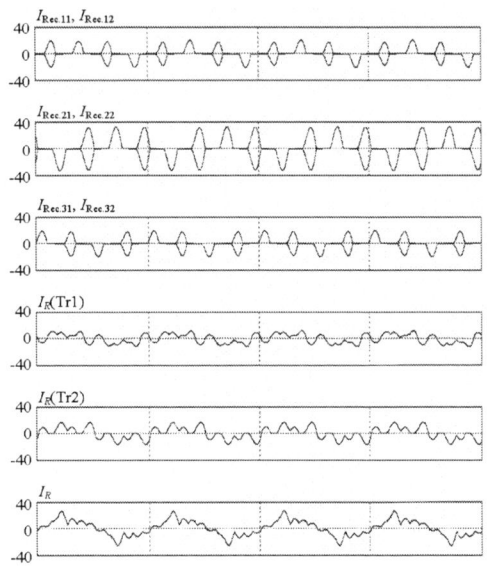

Fig. 13. Selection currents of the 12-pulse rectifier in case of relative modulation factor $m_a = 2.59$ and implementation of the any SVPWM algorithm.

V. CONCLUSIONS

Preliminary experimental results confirm propriety of suggested modulation strategy and applied solutions (excluding rectifier). The research and improvement of the model and the algorithm were continued in order to implement analog solutions in constructed prototype of the first Polish 4-level VSI 6 kV/1 MVA, based on transistors IGBT FD200R65kF1-k and FZ200R65kF1. The design of the high power, high voltage prototype presented in Fig. 14 and Fig. 15. At the moment this prototype is complete and planned for verification in industry.

a)

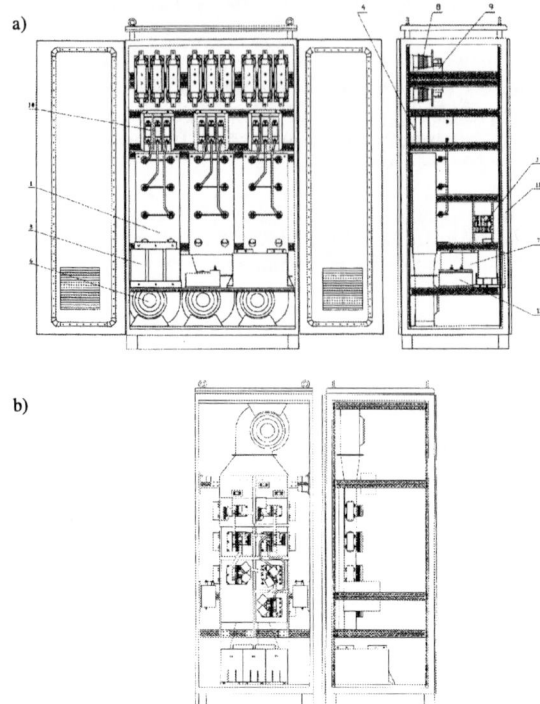

b)

Fig. 14. Design of the prototype of the AC drives converter 6kV/1MVA: 18-pulse rectifier (a), 4-level VSI branch (b)

Fig. 15. Exploded view of the constructed prototype.

REFERENCES

[1] R. Strzelecki, "Конструкция, схемотехника и применение преобразователей большой мощности и среднего напряжения с ШИМ," *Preprint IED Ukr. Academy of Science.*, Kiev, 2006.

[2] Rodriguez J., Lai J.-S., Peng F.-Z., "Multilevel Inverters: A Survey of Topologies, Controls, and Applications," *IEEE Trans. Industry Electronics*, vol.49, pp.724-738, 2002. .

[3] Perantzakis G. S, Xepapas F. H., "A Novel 4-Level VSI–Influence of Switching Strategies on the Distribution of Power Losses," *IEEE Trans Power Electronics*, vol.22, pp.149-159, 2007.

[4] Lopatkin N.N., Usachev E.P., Zinoviev G.S., Weiss H., "Three-level Rectifier Fed Four-Level Inverter for Electric Drivers," Proc. Conf. EPE-PEMC, Portoroz, pp.775-780, 2006.

[5] Pou J., Pindado R., Boroyewich D., "Voltage-Balance Limits in Four-Level Diode-Clamped Converters With Passive Front Ends," *IEEE Trans. Industry Electronics*, vol. 52, pp.190-196, 2005..

[6] Gupta A. K., Khambadkone A. M, A Space Vector PWM Scheme for Multilevel Inverters Based on Two-Level Space Vector PWM, *IEEE Trans. Industrial Electronics*, vol.53, pp.1631-1639, 2006

1336

AC Voltage Regulator Using PWM Technique and magnetic flux distribution

A.M.Dabroom
Director of Jubail Technical Institute
Industrial Jubail City – Kingdom of Saudi Arabia
Fax : 0966 03 341 87 22 - P.O.Box: 10335
E-mail: dbroommohmed@yahoo.com

Abstract:

This paper proposes an AC voltage regulator for resistive loads using the PWM technique and magnetic flux distribution. The load voltage can be controlled using one switching element(MOSFET) and three limb magnetic circuit. The proposed regulator depend on the PWM technique and control the magnetic flux distribution in the magnetic circuit sections

The suggested regulator gives smooth variation of the load voltage and nearly sinusoidal current at the supply terminals with minimum harmonics. The power factor at the supply terminals is nearly unity over the control range of the load voltage. Equivalent circuit of the regulator is proposed. The computed results are compared with those obtained experimentally using a test model, where good agreement is achieved.

Key words : AC Voltage Controllers, Power factor Correction,

1-List of symbols:

R_1, R_2, R_3, R_4 coil resistance's from 1 to 4 respectively,

L_1, L_2, L_3, L_4 coil leakage inductance's from 1 to 4 respectively ,

R_a, R_b, R_c equivalent iron losses resistance's of limb a, b, c respectively,

L_a, L_b, L_c magnetizing inductance's of limb a, b, c respectively,

i_a, i_b, i_c magnetizing current's of limb a, b, c respectively,

E_a, E_b, E_c induced E.M.F's of limb a, b, c respectively,

i_s, i_l, i_{sw}, i_5 supply, load, switch and capacitor current respectively,

R_l load resistance.

2-Introduction:

AC voltage controllers have been widely used to obtain variable ac voltage from a fixed voltage ac source. These controllers should respond to the requirements of the load rapidly and smoothly, preferably without causing noticeable distortion to the voltage waveform.

The moving- coil voltage regulator [1-4], provides smooth voltage regulation at the output terminals without any distortion. However, this regulator is cumbersome and costly. Also, a mechanical system is required for the moving coil. In reference[5], replacement of the moving coil by two multi-tapped windings such that the effect of mechanical movement can be done using triac switches between the taps. In that way, the regulator becomes static and the voltage is regulated in a stepped manner. A controlled thyristor voltage regulator with sinusoidal output is introduced in reference [6]. The principle of operation of the moving-coil regulator is exploited, but a three limb magnetic circuit without air-gap is employed. This controller gives a sinusoidal output voltage. However the control circuit is complex, the output voltage changed in stepped manner and the sequence of switches operation will affect in the output voltage.

978-1-4244-1741-4/08/$25.00 ©2008 IEEE

The phase angle control of ac voltage controllers is extensively employed in many applications such as industrial heating, lighting control and for starting and voltage control of induction motors. This technique offers the advantages of simplicity and ability of controlling large amount of power economically.

However, a delayed firing angle causes discontinuity and significant harmonics in load current, and lagging power factor. This occurs also, at the ac side even though the load is completely resistive [7]. These problems can be solved by introducing more advanced control techniques in ac chopper such as symmetrical angle control , asymmetrical angle control and modification of the power circuit with freewheeling path or using an inductance to improve the input current wave-form [8-12].

This paper introduces an ac voltage controller using PWM technique with one MOSFET and three limb magnetic circuit. The proposed system gives nearly sinusoidal output voltage, minimum distortion at the supply terminals and higher power factor over the control range. The capacitance at the load terminals is used to improve the supply quality. A prototype for the proposed system with the driving circuit of MOSFET are built and tested in the laboratory. In order to compute the system performance, the equivalent circuit is developed and examined by comparing the simulation and experimental results.

3-Description of the proposed system:

The power circuit of an ac voltage regulator with PWM technique is shown in Fig.[1]. The three limbs of the magnetic circuit are of the same cross-sectional area, the coils in the same limb are separated. Coils1 and 2 are connected in series to the supply, such that the flux in the center limb is the sum of the flux in the two outer limbs as coils 3 and 4 are opened. The four coils are designed with equal turns and with suitable core flux level as the supply voltage is applied to any coil.

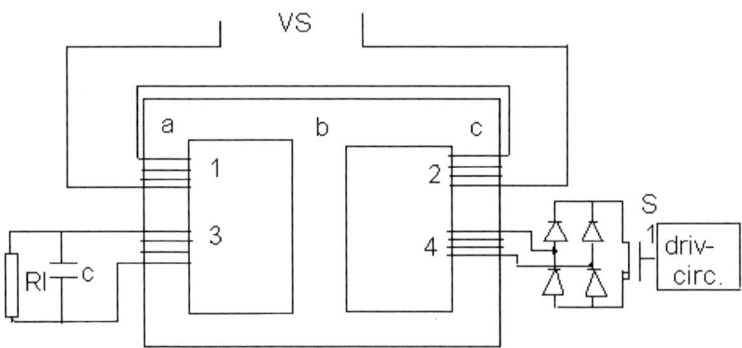

Fig.[1] The proposed system

Coil 4, can be shorted via one MOSFET controlled with trigger circuit according to PWM technique. Coil 3, is connected to the load and is shunted by a capacitor, to improve the load voltage waveform.

The mechanism of load voltage regulation is based on flux division between the two outer limbs, which is affected by the MOSFET operation modes. If the switch S_1 is off, the supply voltage applied to coils 1 and 2 will produce flux $\Phi_{\Phi b}$ will be equal to the sum of fluxes in outer limbs $\Phi_{\Phi a}$ and $\Phi_{\Phi c}$. The supply voltage is shared between the coils 3 (load) and coil 4 (switch) . Also, the voltage at the load terminals will depend on the equivalent impedance of the load and the capacitance.

When switch S_1 is on, the change in the total flux due to the supply voltage will flow in the limb a, as the flux in limb c will be nearly frozen. Hence, the load and capacitor voltages become nearly equal to the supply voltage. When the switch S_1 is turned off most of the change in the total flux produced from the supply will flow in the limb c due to the effect of the load and capacitor equivalent impedance. During this period the capacitor will be discharged into the load. The load voltage and the supply current waveforms are affected by the capacitor value.

4-Equivalent circuit and analysis:

Considering Fig.[1], and the flux distribution in the magnetic circuit, an equivalent circuit which represents the system can be proposed as shown in Fig.[2]. The two outer limbs core can be represented by two equal impedance's Z_a and Z_c , that represents the magnetizing components and iron loss components. Also, the core of the center limb can be represented by Z_b . The flux in the center limb equals the flux in the two outer limbs, so Z_a and Z_c are connected in parallel with Z_b . However, the effect of the coil resistance's, leakage inductance's, switch and load impedance can be represented as shown in Fig.[2]. The equivalent circuit parameters can be measured using the conventional short and open circuit tests. The system parameters are listed in the appendix.

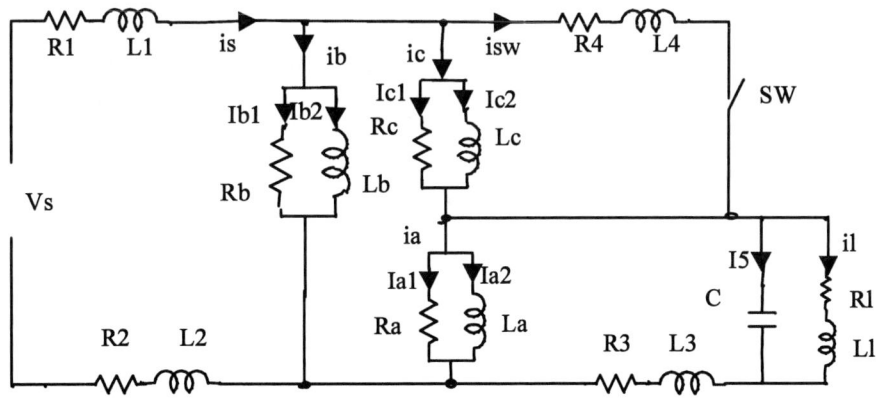

Fig.(2) The proposed system equivalent circuit

From the equivalent circuit, the system equations are :

$$V_s - E_b = i_s(R_1 + R_2) + (L_1 + L_2)\frac{di_s}{dt} \quad (1)$$

and

$$E_b = i_{b1}R_b = L_b\frac{di_{b2}}{dt} \quad (2)$$

Where:

$$i_b = i_{b1} + i_{b2} \quad (3)$$

In the outer limbs, the EMF's depend on the flux distribution. When switch S_1 is open, a higher part of the flux produced from the supply will flow in the limb c, and the minimum part will flow in the limb a, this depend on the equivalent impedance of the load and capacitance. As the equivalent impedance of the load and capacitance is increased the flux in limb a is decreased. When switch S_1 is on, the flux in limb c will be minimum, and nearly all the flux produced from the supply voltage will flow in limb a. In general case:

$$E_b = E_a + E_c \quad (4)$$

Where:

$$E_a = i_{a1}.R_a = L_a.\frac{di_{a2}}{dt} \quad (5)$$

and

$$E_c = i_{c1}.R_c = L_c.\frac{di_{c2}}{dt} \quad (6)$$

The magnetizing currents in these limb's are given as follows:

$$i_a = i_{a1} + i_{a2} \quad (7)$$
$$i_c = i_{c1} + i_{c2} \quad (8)$$

In limb a, the voltage applied to the load and capacitor can be calculated from the following equation:

$$E_a - V_l = (i_5 + i_l).R_3 + L_3.\frac{d(i_5 + i_l)}{dt} \quad (9)$$

Where i_5 is the capacitor current, the capacitor voltage will equal to the load voltage so;

$$i_5 = c.\frac{dv_c}{dt} = c\frac{dv_l}{dt} \quad (10)$$

and

$$V_l = R_l.i_l + L_l.\frac{di_l}{dt} \quad (11)$$

The supply current, load voltage and current depend on the switch S_1 operation. When switch S_1 is on , the supply current is :

$$i_s = i_b + i_c + i_{sw} \quad (12)$$

where I_{sw} is the switch current.
In this case the voltage at coil 4 terminals is given by:

$$E_c = i_{sw}.R_4 + L_4.\frac{di_{sw}}{dt} \quad (13)$$

As switch S_1 is off,

$$i_{sw} = 0.0 \quad (14)$$

and the supply current is :

$$i_s = i_b + i_a + i_5 + i_l \quad (15)$$

The differential equation describing the different modes of operation of the system are solved using simulink program with fourth order Rung-Kutta method. Calculations are carried out assuming zero initial conditions of magnetic circuit flux. Also, there is no saturation in the magnetic circuit.

5-Simulation and experimental results:

The experimental results is carried out to verify the feasibility and to investigate the validity of the computed results for the PWM technique of magnetic circuit flux control. Fig.[3-a-b] shows the supply current waveform at switching frequency of 1000 Hz. Fig.[4-a-b] shows the load current at the same conditions. Fig.[5-a-b] shows the capacitor current. It is observed that the experimental waveforms are in good agreement with the computed waveforms. Also, it is noticed that during off period of the switch, the capacitor discharge in the load to improve the load current waveforms, thus it will effect in the supply current waveform.

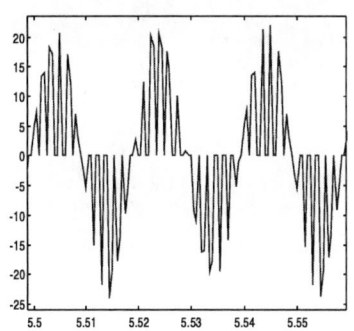

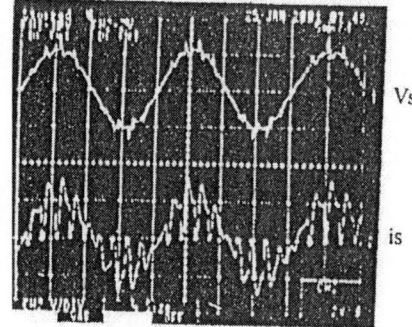

Ch2 is 2amp/div, t = 5msec/div

a- simulation waveform   b- Experimental waveform

Fig.[3] The supply current waveform at a switching frequency of 1000 Hz

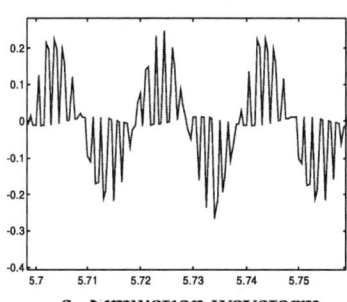

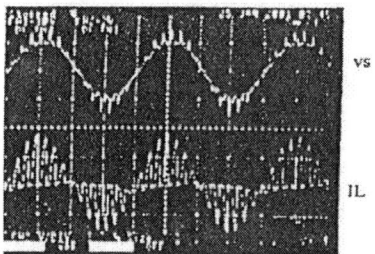

Ch2 is 0.2 amp/div, t = 5msec/div

a- Simulation waveform   b- Experimental waveform

Fig.[4] The load current waveform at a switching frequency of 1000 Hz

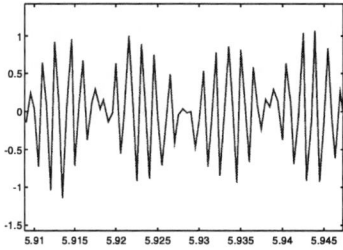

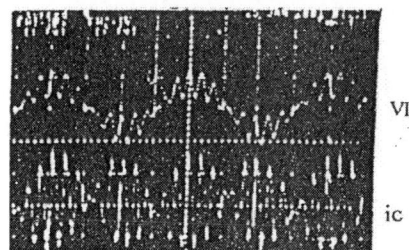

Ch2 is 1amp/div, t = 5msec/div

a- Simulation waveform   b- Experimental waveform

Fig.[5]The capacitor current waveform at a switching frequency of 1000 Hz

Figs.[6-7-8] shows the supply current, load current and capacitor current at switching frequency equals to 2500 Hz. The result shows that as the switching frequency increases the inductance of the coils will increase, thus, decreases the load voltage. Hence, the capacitor value must increases to increase the load voltage to the maximum load voltage

1341

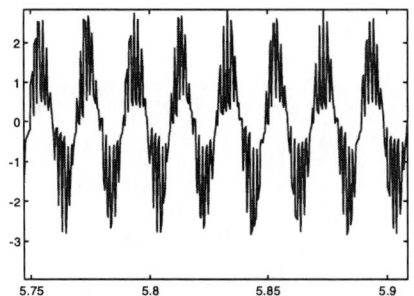 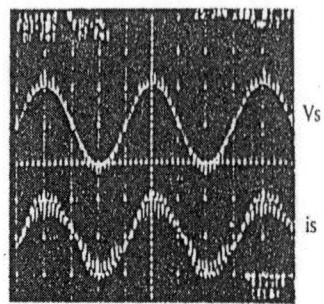

Ch2 is 2amp/div, t = 5msec/div

a- simulation waveforms   b- Experimental waveforms

Fig.[6]The supply current waveforms at switching frequency 2500 Hz ,c=50 μf

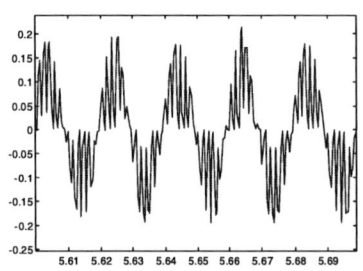 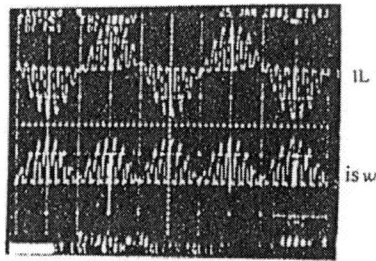

Ch1 is 0.2 amp/div, t =5msec/div

a- Simulation waveforms   b- Experimental waveforms

Fig.[7]The load current waveforms at switching frequency 2500 Hz, C=50 μf

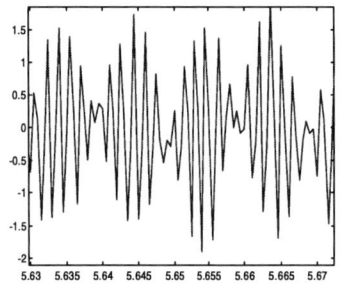 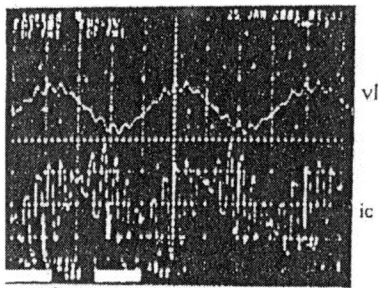

Ch2 is 1amp/div, t =5msec/div

a- theoretical waveforms   b- experimental waveforms

Fig.[8] capacitor current waveforms at switching frequency 2500 Hz, C=50 μ f

The harmonic analysis of the supply current and load current are carried out. Also, the power factor at the supply terminals is obtained. Fig.[9] shows the supply current harmonic spectrum at switching frequency 1000 Hz with capacitor value equal to 50 μ f .Fig.[10] shows the variation of the power factor at the supply terminals as function of the switching frequency at different values of capacitance(

25, 30, 50 µf). It is noticed that as the capacitor value increase the power factor at the supply will increased. Also, the power factor decreased as the switching frequency increased.

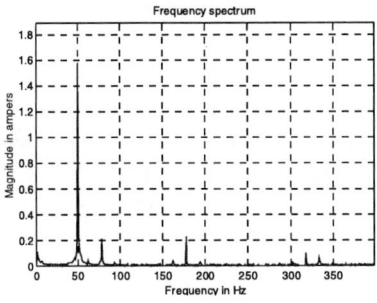 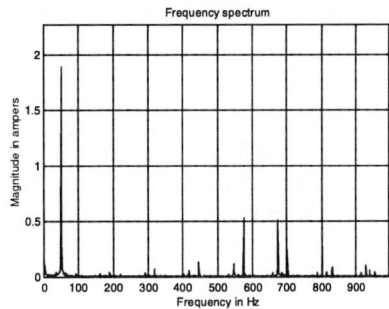

a- at switching frequency 500 Hz   b- at switching frequency 1000 Hz

Fig.[9] The supply current harmonic spectrum (c= 50 µf)

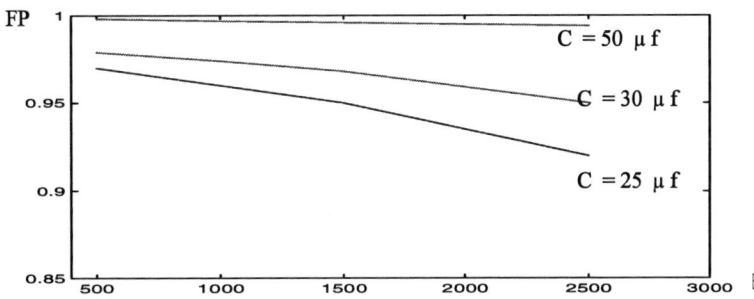

Fig.(10) Computed power factor at the supply terminals

6-Conclusion:

The ac voltage regulator for resistive loads using PWM technique proposed in this paper avoids the main demerits of moving coil voltage regulator such as the mechanical system. Also, it avoids the poor power factor at the supply terminals of the conventional thyristor voltage regulators. The proposed regulator gives smooth variation of the load voltage as the duty cycle of MOSFET operation is varied. Also, it gives sinusoidal supply current and improvement power factor at the supply terminals that in range 0.92 to 0.989 .

The suitable value of capacitance at the load terminals gives improvement in the power factor and the supply current waveforms. Equivalent circuit of the proposed regulator has been obtained and justified by comparing the computed and experimental results that are in good agreement.

7- appendix:

Parameters of the experimental set up given as follows:

V_s= 220

R_1=2.0 ohm, R_2 = 2.0 ohm,

R_3= 2.0 ohm, R_4 = 2.0 ohm,

L_1= 0.2 mH, L_2= 0.2 mH,

L_3= 0.2 mH, L_4= 0.2 mH,

R_a= 730 ohm, R_b=1085 ohm,

R_c = 730 ohm,

L_a = 13.7 mH, L_b= 28.0 mH,

L_c = 13.7 mH, R_l = 50.0 ohm

Reference:

1- J. Arrillage, B. Barrett and N.A.Vovos" Thyristor-Controlled Regulating Transformer For Variable Voltage Boosting," Proc.IEE, Vol.123, No. 10, October 1976, PP. 1005-1009.

2- G.H. Rawcliff and I.R.Smith, " The Moving-Coil Voltage Regulator: A treatmenent From first Principles," Proc. IEE. 104A, 1957, PP 68-76

3- A.S. Abdel-Karim," The Moving-Coil Regulator: A Simplified Treatment," Electric Machines and Power Systems, Vol. 8, No.4, 1983, PP 395-402.

4- Ahmet ALTINTAS, "A Comparative Study On AC Voltage Controllers In Terms Of Harmonic Effectiveness," Erciyes University For Bilimleri Enstitusu Dergisi 21(1-2), 79-86, 2005.

5-C.F.Amor," Electrical regulators," Patent Specification , No. 1353012, Patent Office, London, May 1974.

6-A.S.Abdel-Karim, A.S.Taalab and A.E.Lashine," A Thyristorized AC Voltage Controller With Sinusoidal Output," Proc. Electric Machines and Power Systems, Vol.16, 1989, PP.395-410.

7-A.S.Abdel-Karim, S.A.Mahmoud and A.El-Sabbe," A New Approach TO Capacitor Compensation of Thyristor Voltage Controller," UPEC, 1985.

8- LLya Zeltser and Sam Ben-Yaakov, " Modeling, Analysis and Simulation Of AC Inductor Based Converters, " IEEE annual meeting 2007, pp 2128-2134, 2007

9- Do-hyun Jang and Gyu-ha Choe,"improvement of input power factor in ac choppers using Asymmetrical PWM technique", IEEE. Trans. Ind. Elect. Vol.24, No. 2, april 1995, pp. 179-185.

10- G.Choe and D Jang." Asymmetrical PWM technique for ac choppers", in IEEE. IECON conf. 1991, pp 587- 594.

11- A.El-Sabbe and A. Zin El-Din, " A Novel ac voltage regulator, " IEEE. IECON conf. 1998. Germany, pp. 607-611.

12- Yim-Shu Lee, K.C.Wong and C.K.Ng," Behavior Modeling of Magnetizing Currents in Switch-Mode DC-DC Converters," IEEE. Tran. Ind. Elect. Vol. 47, No. 1, Feberuary 2000, pp. 36-44.

Minimum Reactive Power Filter Design for High Power Converters.

Alex-Sander Amavel Luiz[*] and Braz Jesus Cardoso Filho[†]

[*] CEFET-MG/Department of Electrical Engineering, Belo Horizonte-MG, Brazil, e-mail: *asal@des.cefetmg.br*
[†] UFMG/Department of Electrical Engineering, Belo Horizonte-MG, Brazil, e-mail: *cardosob@ufmg.br*

Abstract— This work proposes and evaluates an optimized filter design for high power converter with sinusoidal voltage and current. The inherent low switching frequency of this class of converters complicates the filter design and generally results in large filter components. The proposed design culminates in minimum filter reactive power, size and cost. Simulation and experimental results support those ideas.

Keywords— High Voltage power converters, Passive filter, Harmonics, Power factor correction.

I. INTRODUCTION

Many harmonic related issues have been reported recently [1] and [2]. They include excessive heating and losses in electrical machines and other electromagnetic equipments, winding over voltages, reduction of device life service and other problems. Regarding the utility grid, harmonics deteriorate the supplied voltage waveform and increase the electromagnetic emissions. On the other hand, sinusoidal voltage and current in an electrical system is highly desirable. Nevertheless, electronic apparatuses distort the grid current and voltage.

In many works, passive sinusoidal filters come out as good alternative to reduce the system harmonic content. The inherently low switching frequency of high power converters, limited from about 500 to 1000 Hz, complicates filter design, which generally results in large filter elements. Besides, due to grid resonance problems and poor filter attenuation the lower the switching frequency the more complex the filter design and worse filter performance are. Unfortunately in literature [2]-[8] only a small number of papers is devoted to multilevel structures ([2], [6] and [7]) and a number even smaller discusses the problem of filter design under very low switching frequency conditions. Still in this case design criteria result again in bulky filter components.

This paper proposes an enhanced passive filter configuration for high power converters. A complete filter design based on the optimization of the filter reactive power leads minimum cost weight and size. Besides, reasonably simple design steps culminate in unique and well-defined filter element values. The modulation strategy adopted in this paper is the Selective Harmonic Elimination-SHEPWM, which also results in simplifications on the filter design rules. Moreover, the modulator SHEPWM works under a classical voltage oriented synchronous reference frame control for a high power active rectifier. Simulation and experimental results confirm that it is possible to achieve high power factor operation with low harmonic distortion and satisfactory system performance under very low switching frequency conditions.

II. SELECTIVE HARMONIC ELIMINATION.

The power circuits in this paper consist of a Neutral Point Clamped-NPC converter working as active rectifier and a passive sinusoidal input filter. References [9]-[11] discuss the generalized method of Selective Harmonic Elimination and state that an arbitrary number M of harmonic can be controlled and/or eliminated from a standard three level output voltage waveform if it is chopped M times per half fundamental period. Expanding the waveform into Fourier series and assuming odd quarter-wave symmetry, the cosine components vanish and (1) gives n^{th} harmonic coefficient of the sine components, ($n = 1, 3, 5...$).

$$a_n = (4E/\pi n)\sum_{k=1}^{M}(-1)^{k+1} \cdot \cos(n\alpha_k) \quad (1)$$

$$0 \le \alpha_1 < \alpha_2 \cdots < \alpha_M < \pi/2$$

In order to control the fundamental component and eliminate (M-1) harmonics it is essential to solve a nonlinear and transcendental system of equations. Based on the algorithm in [9] it was developed a computer program to solve those equations intending to control the converter fundamental component and eliminate 5th and 7th harmonics.

III. SINUSOIDAL FILTER.

The sinusoidal input filter topology studied in this work takes up the classical concepts of harmonic filters from electric power systems together with a common solution for VSC active rectifiers [3]-[5]. Previous papers employed similar ideas in power electronics, but they recently reappeared under a high switching frequency modulation condition for a low power drive, [1].

The filter configuration in this work consists of a third order LCL filter plus some harmonic traps as in Fig. 1.

A. Minimum Reactive Power Optimized Filter Design.

Elementary filter theory recommends that the filter cut-off frequency f_{co} must be placed between the maximum desirable fundamental frequency f_1 and the first non-eliminated harmonic frequency f_D with separation factors between each one of them of at least 10. It results in a frequency separation equal to or greater than 100 between f_1 and f_D. This is difficult to achieve

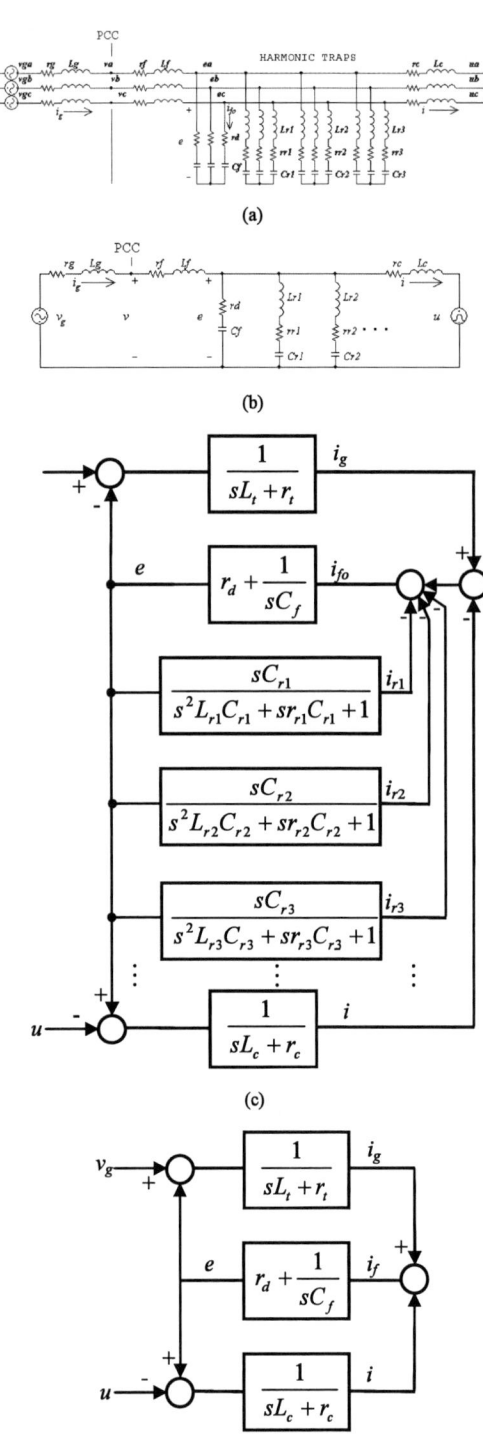

(a)

(b)

(c)

(d)

Fig. 1. Sinusoidal input filter and equivalent circuit.

with practically all modulation methods under low switching frequency conditions. For this reason, this work suggests a separation between f_1 and f_D of about 10 in the following way:
- The separation between f_1 and f_{co} must be greater than 5.
- The separation n_{scd} between f_{co} and f_D should be about 2.

Then it is possible to find out filter elements. The cut-off frequency can be generically expressed by:

$$f_{co} = 1/\sqrt{L_x C_y} \qquad (2)$$

This expression leads to an infinite set of pair of values of L_x and C_y, which simultaneously suits this equation for a given value of f_{co} It requires new constraints to solve this problem. Various guidelines for filter design are presented in literature, [5]-[7] and [12], but they are either not feasible for low switching frequency applications or not enough restrictive to come up with a specific solution to the problem.

In [7] in order to limit the converter current ripple i_{rip} the converter side inductor L_c is given by:

$$L_c \leq U_1 / 4\sqrt{6} f_c i_{rip} \qquad (3)$$

At rated voltage, commutation frequency $f_c = 180Hz$ and $i_{rip} = 10\%$ this results in $L_c < 34\%$. Yet it provides a wide set possible element values. In [12], all filter inductors should be below 10%, which is not practical for many low switching frequency applications. Reference [6] restricts all filter inductors below 10% and imposes $C_f = 5\%$, but the switching frequency is as high as 2.5 kHz. Reference [5] limits all inductors below 10% and C_f below 5%, but again the switching frequency is high, $f_{sw} = 5\text{-}8$ kHz.

The filter design optimization can be performed writing an algebraic expression for the total filter reactive power as a function of all reactive elements. This expression under the constraint of (2) represents the optimizing cost function by:

$$Q_{rt}(kVA_r) = f(L_1,...,L_i, C_1,...,C_j) \qquad (4)$$

The first derivative of Q_{rt} equated to zero leads to the cost function critical points (maximums and minimums):

$$dQ_{rt}/dL_x = f'(L_1,...,L_i, C_1,...,C_j) = 0 \qquad (5)$$

The second derivative of Q_{rt} leads to the cost function inflexions and concavity:

$$d^2Q_{rt}/dL_x^2 = f''(L_1,...,L_i, C_1,...,C_j) \qquad (6)$$

The Fig. 2a shows behavior of the reactive power Q_{rt} versus the converter side inductor L_c. The critical points achieved with the above steps represent local maximum and minimum in the case of the filter proposed here. However neglecting the negative values of inductor L_c the Q_{rt} critical point can be regarded as global minimum as in Fig. 2b for different values of the parameter β_s, (which only depends on the filter design specifications: the voltage of C_f and the current of L_c and cut-off f_{co}). β_s is in general near to "1".

With this reasoning, it is possible to derive exact and simple closed form algebraic expressions for all filter reactive elements:

$$f_{co} = f_D/n_{scd} \qquad (7)$$

$$L_c = \beta_s/f_{co} \qquad (8)$$

$$C_f = 1/(\beta_s \cdot f_{co}) \qquad (9)$$

Yet, the filter design must assure a resonant frequency f_{res} below the lower harmonic set resultant from the modulation process. Because of its superior output harmonic spectrum SHEPWM makes the choice of f_{res}

1346

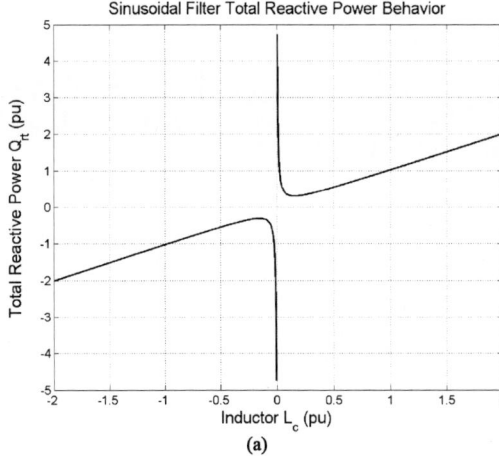

(a)

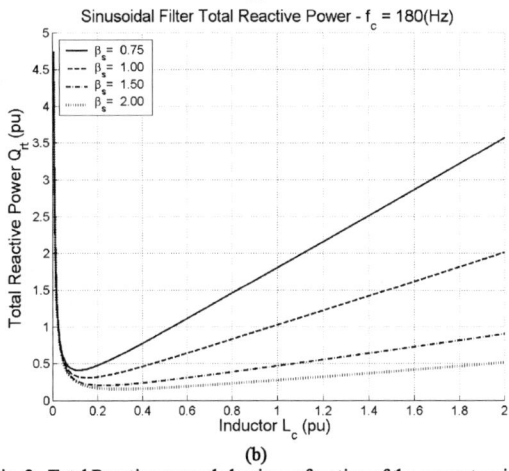

(b)

Fig. 2. Total Reactive power behavior as function of the converter side inductor and the parameter β_s.

TABLE I
MINIMUM REACTIVE POWER OPTIMIZED FILTER DESIGN.

f_c	f_D	f_{co}	f_{res}
180Hz	660Hz	390Hz	540Hz
3.0pu	11.0pu	6.5pu	9.0pu
420Hz	1380Hz	815Hz	1110Hz
7.0pu	23.0pu	13.6pu	18.5pu

TABLE II
RATINGS OF SINUSOIDAL FILTER ELEMENTS.
($U_{base} = 4160V$, $S_{base} = 5MVA$, $\beta_s = 1$).

f_c (Hz)	L_c	L_f	C_f
180Hz	1.4123mH	1.3103mH	117.92μF
	0.15385pu	0.14274pu	0.15385pu
420Hz	0.67543mH	0.56245 mH	56.398μF
	0.07358pu	0.06127pu	0.07358pu

$L_g = 0,025$ pu, $L_g = 0,22949$ mH.

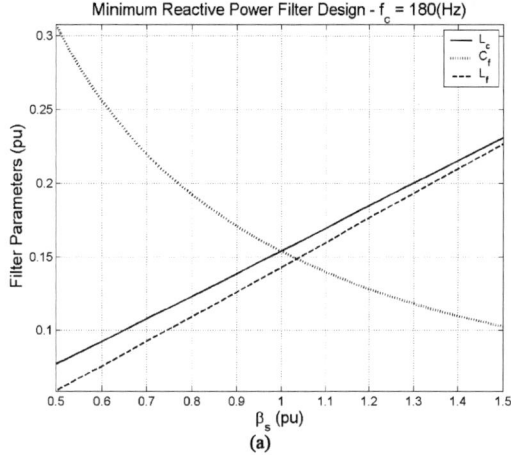

(a)

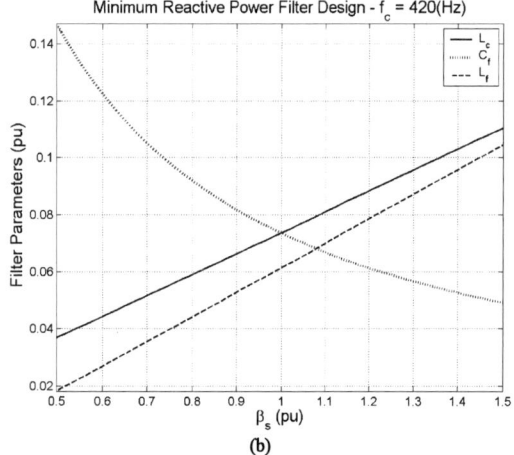

(b)

Fig. 3. Optimized filter design, a) $f_c = 180$Hz and b) $f_c = 420$Hz.

easier than other classical modulator, [8] and [11]. This work suggests the expression for f_{res}:

$$f_{res} = \left[f_D + \text{ceil}\left(f_{co}\right)\right]/2 \qquad (10)$$

which intuitively allocates f_{res} between f_{co} and f_D. Then:

$$L_{eq} = \beta_s^2 / \left(f_{res}^2 \cdot L_c\right) \qquad (11)$$

$$L_f = \left|\left(L_c \cdot L_{eq}\right)/\left(L_c - L_{eq}\right)\right| - L_g \qquad (12)$$

Fig. 3 represents a summary of the *MRP-OFD* and produces the element values presented in Table I and II. It is remarkable that the *MRP-OFD* results in all main filter elements about 15% with $f_c = 180$Hz and $f_c = 420$Hz.

Filter damping is another important designing aspect. Again, the literature presents various different recommendations, [1], [3], [5], [13]-[17]. In [13] the filter is over damped. Reference [14] designs its filter with critical damping. In contrast to them, [1] shows that under damped filters present better attenuation. Following this line [5] adopts filter quality factors $Q = 1.8$ and 3.6, and [3] chooses $Q = 7.7$. Additionally [15]-[17] discusses many alternative for resistive damping process, but according to [17] the better one consists in placing the damping resistor in series with capacitor C_f. It is recognizable that passive damping has many collateral effects over the filter. One of them is deterioration in

filter attenuation. This is reasonably compensated by the addition of series resonant harmonic traps in parallel to the shunt branch of C_f. The number of resonant traps depends on the desired system specification and they should be tuned in the first non-eliminated harmonic frequencies. Each branch has to be designed individually with the *MRP-OFD* to achieve minimum total filter reactive power. They also must be very selective, ($Q > 10$). Reduction in system efficiency is another trouble of passive damping. However if there is enough control bandwidth active damping can solve the two difficulty presented. On the contrary, this paper suggests a hybrid solution involving both passive and active damping.

B. Filter Evaluation.

After the optimized design, it is possible to evaluate the filter performance. Fig. 4 and 5 bring its frequency response for the grid current i_g and voltage v, at point common coupling PCC, in relation to converter voltage u. These figures show that damping deteriorates the filter attenuation (from expected -60dB/dec to about -40 dB/dec) though it is effective in reducing the risk of filter resonance.

The dynamic stiffness allows evaluating filter disturbance rejection capacity. Fig. 6 and 7 exhibit this feature with the *MRP-OFD* for the grid voltage with respect to the current and voltage at PCC. These figures point up that the filter dynamic stiffness is better in the high frequency region. It reflects the filter sensitivity to disturbances near the fundamental frequency, which is important to avoid. Nevertheless, it has more to do with the third order filter configuration, (whose disturbance rejection capacity is poor [17]), rather than with the proposed design procedure.

The filter installed between the utility grid and the

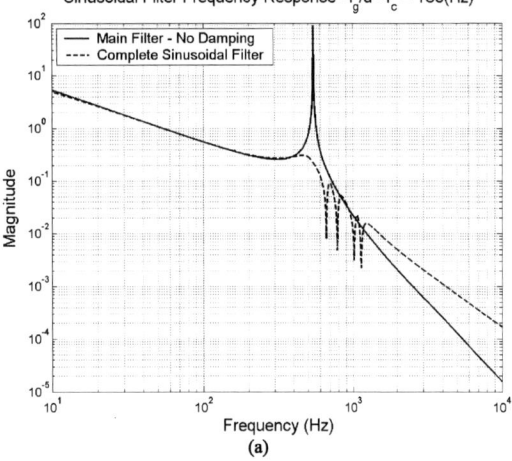

(a)

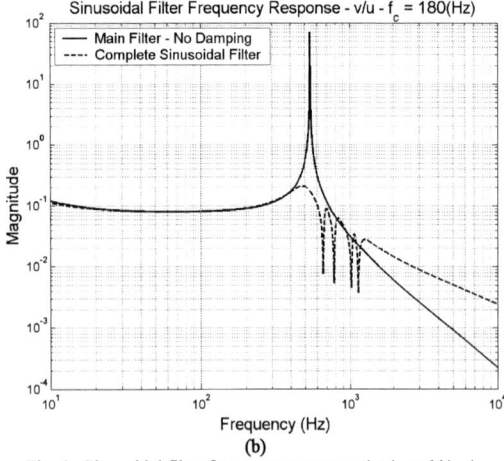

(b)

Fig. 4. Sinusoidal filter frequency response, a) i_g/u and b) v/u

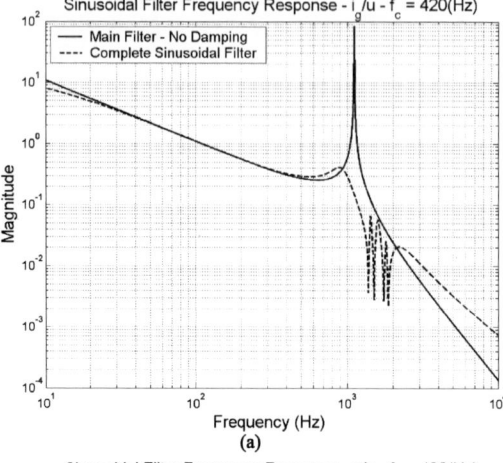

(a)

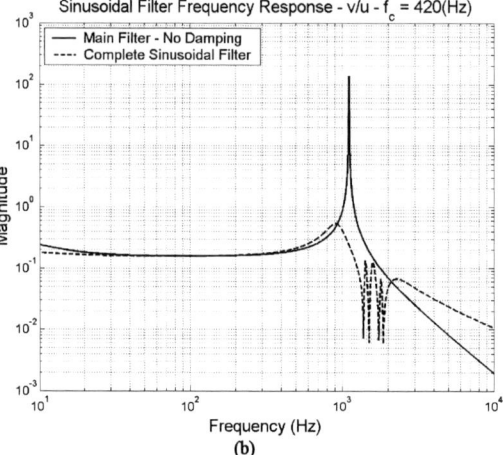

(b)

Fig. 5. Sinusoidal filter frequency response, a) i_g/u and b) v/u.

converter also increases the power demands that converter must be able to deal with. It increases the converter voltage and current ratings. It is possible to show that it depends only on the filter configuration and size of its elements. Equations (13) and (14) respectively describe the converter voltage and current at fundamental frequency of an active rectifier.

$$u = H_i(j\omega_1)v - \left[(r_c + j\omega_1 L_c) + (r_f + j\omega_1 L_f)H_i(j\omega_1)\right] \cdot i_g \quad (13)$$

$$i = -\left[\frac{j\omega_1 C_f}{(1 + j\omega_1 r_d C_f)}\right]v + \left[1 + \frac{(r_f + j\omega_1 L_f) \cdot j\omega_1 C_f}{(1 + j\omega_1 r_d C_f)}\right]i_g \quad (14)$$

$$H_i(j\omega_1) = 1 + \left[(r_c + j\omega_1 L_c) \cdot (j\omega_1 C_f)/(1 + j\omega_1 r_d C_f)\right] \quad (15)$$

There are absolute and relative increments in these two converter ratings, Δu_a, Δi_a and Δu_r, Δi_r respectively (16)-(19). The former is associated with voltage drop in the series elements and current drawn in shunt branches. The latter is an extension of the concepts of voltage and current regulation in electrical system. Table III shows those increments with the *MRP-OFD* at rated voltage and current. The table shows that the absolute increments are always positive and greater than the relative ones. The increments Δi_r are negative confirming the expectation that the filter will not require a converter over sizing. It means that converter current demand is smaller with the

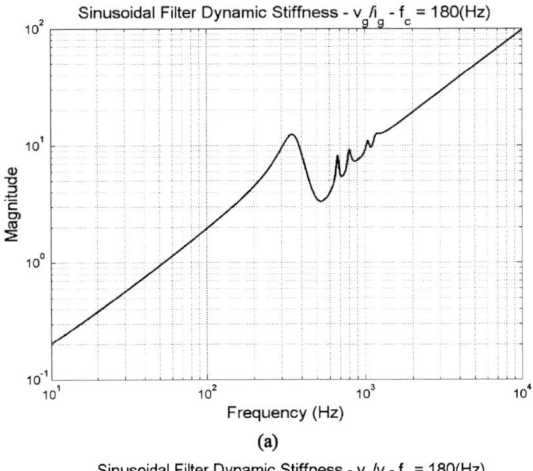

(a)

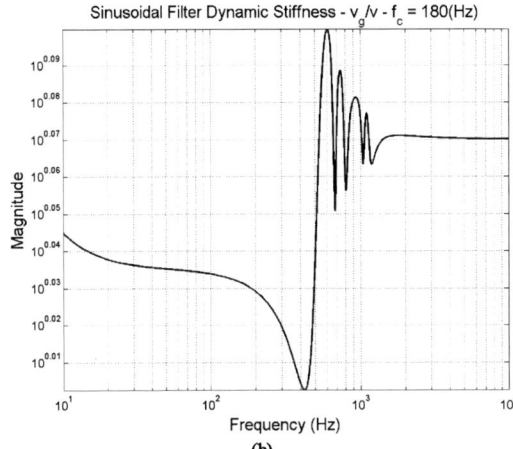

(b)

Fig. 6. Sinusoidal filter dynamic stiffness a) v_g/i_g and b) v_g/v.

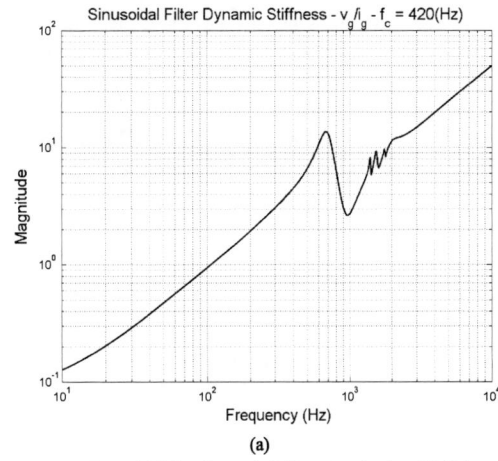

(a)

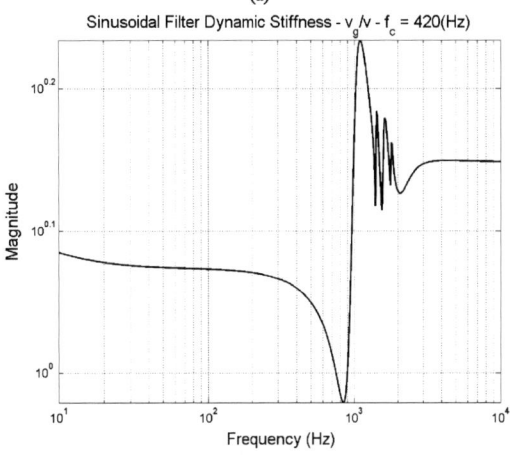

(b)

Fig. 7. Sinusoidal filter dynamic stiffness a) v_g/i_g and b) v_g/v.

filter than without it due to both the capacitive effects and the *MRP-OFD*. In addition, the converter tolerates the increment of 0.16% in its voltage after the filter installation. Here it is important to emphasize that is the relative increments in those quantities, rather than the absolute ones that must be considered for a converter design since it is more realistic, [7].

$$\Delta u_a = [H_i(j\omega_1) - 1]v - [(r_e + j\omega_1 L_e) + (r_f + j\omega_1 L_f)H_i(j\omega_1)] \cdot i_g \quad (16)$$

$$\Delta i_a = \frac{[(r_f + j\omega_1 L_f) \cdot i_g - v] \cdot j\omega_1 C_f}{(1 + j\omega_1 r_d C_f)} \quad (17)$$

$$\Delta u_r = |u| - |v| \quad (18)$$

$$\Delta i_r = |i| - |i_g| \quad (19)$$

IV. VOLTAGE ORIENTED CURRENT CONTROL.

A classical voltage oriented synchronous reference frame current loop controls a three-level NPC rectifier. In

TABLE III
VOLTAGE AND CURRENT DEMANDS AFTER FILTER INSTALLATION.

f_c (Hz)	Δu_a (%)	Δu_r (%)	Δi_a (%)	Δi_r (%)
180	29.427	0.16319	15.2670	-2.01070
420	13.642	-1.5651	7.2457	-0,41910

this case, the current components i_d and i_q controls dc bus voltage and reactive power respectively and produce the reference voltage commands u_d and u_q which are transformed from *dq* to *abc* and them applied to the modulator.

V. RESULTS.

A three-level NPC rectifier of 5MVA/4160V 60Hz with dc bus voltage v_{dc} = 3236V and switching command generated by SHEPWM at the commutation frequency f_c = 180Hz is simulated with MATLAB6.5/SIMULINK. The sinusoidal filter elements are all about 0.15pu (Table IV). The system short-circuits inductance L_g is assumed equal to 0.025 pu. The rectifier is under control of voltage oriented synchronous reference frame current loop and the converter high power factor is achieved by setting the reference current i_q =-148 A by a control action.

TABLE IV
VOLTAGE AND CURRENT DEMANDS AFTER FILTER INSTALLATION.

f_c (Hz)	L_e	L_f	C_f
180Hz	1.375mH	1.272mH	120μF
	0.1498 pu	0.1385 pu	0.1566 pu

L_g = 0,025 pu, L_g = 0,22949 mH.

In Fig. 8a, U_{ab} exhibits a typical 3-level Line voltage standard along with a grid voltage V_{ab} nearly sinusoidal. In this case the former presents a $THD = 26.61\%$ and negligible low frequency harmonics below 660Hz and the latter presents a $THD = 0.4\%$ with very low harmonic content in all frequency range, (as in Fig. 9a). In Fig. 8b V_a and i_a are basically in phase. The displacement power factor $DPF \approx 1$ and distortion factor given by $DF = (I_{s1}/I_s) = 1/[(THD)^2+1]^{1/2} \approx 1$ in [18]. It results in system operation essentially with unity power factor ($PF = DPF.DF$). Moreover i_a is highly sinusoidal with $THD = 1.37\%$ and all current harmonic amplitudes below 0.8%, (Fig. 9b). In contrast, when the filter is removed the harmonic distribution and distortion deteriorate, as one would expect. In Fig. 10, the grid voltage and current at the PCC become distorted and their THD in Fig. 11 increase to 3.77% and 11.93% respectively. The converter $THD = 26.04\%$ is practically constant.

Fig. 11 and 12 illustrate the system performance under

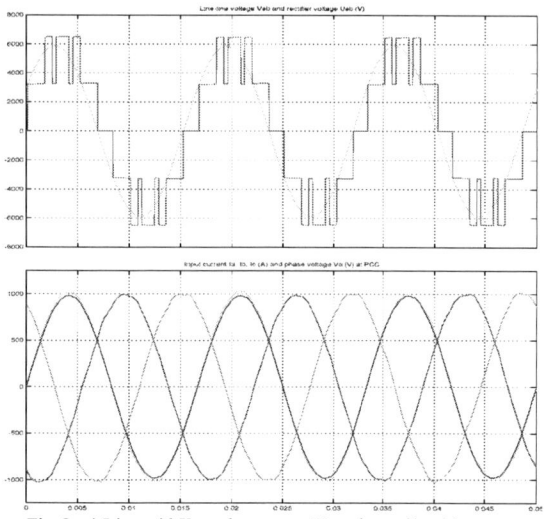

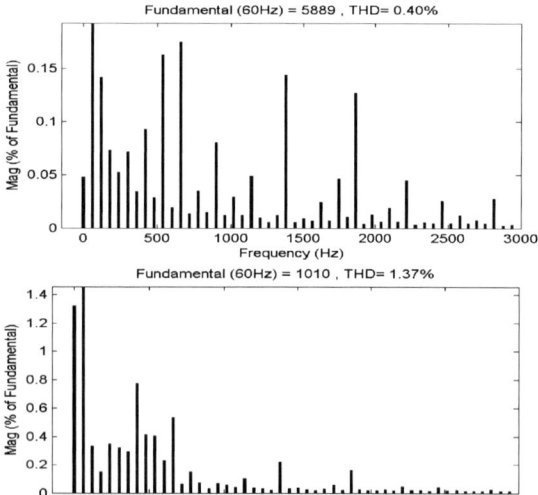

Fig. 8. a) Line grid V_{ab} and converter U_{ab} voltages b) grid currents i_a, i_b, i_c and phase voltage V_a (in red) with sinusoidal filter.

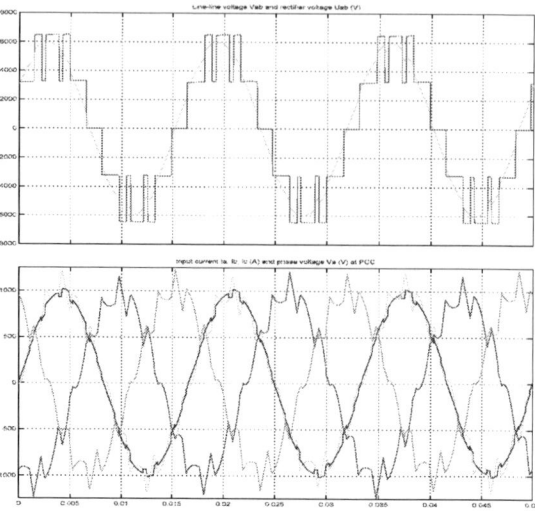

Fig. 10. a) Line grid V_{ab} and converter U_{ab} voltages b) grid currents i_a, i_b, i_c and phase voltage V_a (in red) without sinusoidal filter.

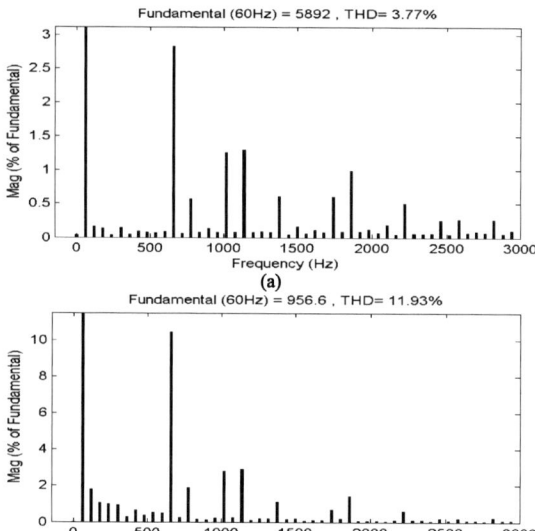

Fig. 11. a) Line grid voltage b) current spectrum without filter.

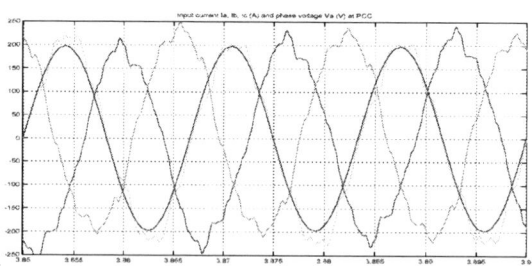

Fig. 12. Grid currents i_a, i_b, i_c and phase voltage V_a (in red) with sinusoidal filter and reduced grid current (to 0.2 pu).

Fig. 9. a) Line grid voltage b) current harmonic spectrum with filter.

light load. Following an increase in dc load to 5 pu, the grid current reduces to 0.2 pu. In this case again the V_{ab} is almost sinusoidal and its $THD = 0.42\%$ is very close to that under full-load condition. On the other hand, although the current i_a seems to be less sinusoidal their $THD = 7.79\%$ is lower than 11.93% (full-load condition

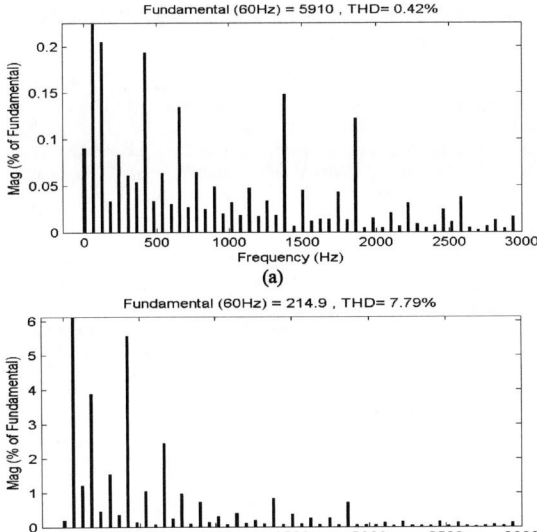

Fundamental (60Hz) = 5910 , THD= 0.42%

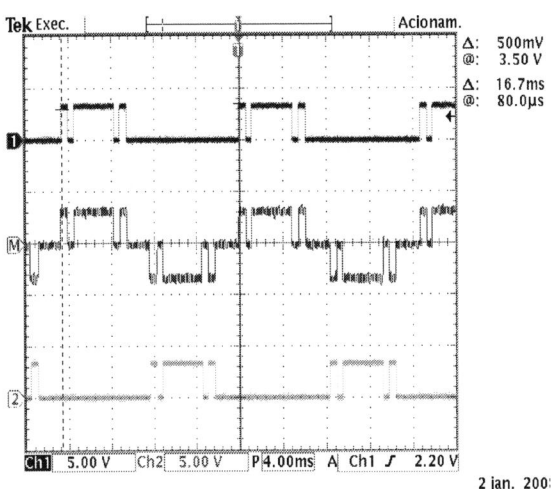

Fundamental (60Hz) = 214.9 , THD= 7.79%

Fig. 13. a) Line grid voltage V_{ab} b) current i_a, harmonic spectrum with filter and reduced load.

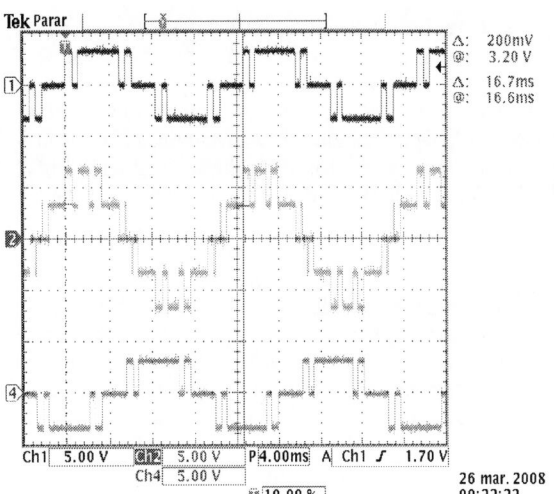

Fig. 15. SHEPWM phase and line voltages, U_{ao}, U_{bo} and U_{ab}

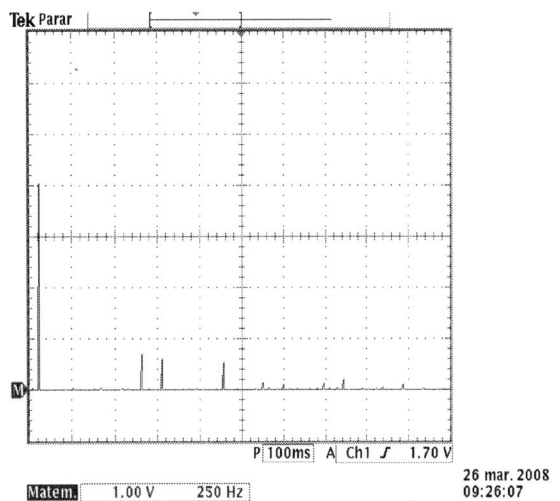

Fig. 16. SHEPWM line voltage U_{ab} harmonic spectrum.

Fig. 14. Phase converter voltage U_a and switches pulse pattern.

TABLE V
SHEPWM-LINE VOLTAGE HARMONICS.

N	Frequency f_n (Hz)	U_n (%)
1°	60	100.00000
3°	180	0.93443
5°	300	1.43095
7°	420	0.56376
9°	540	0.25467
11°	660	17.17404
13°	780	14.85169
15°	900	1.03137
17°	1020	0.63141
19°	1140	14.75186

without filter). The reduced current condition shows that system performance is also satisfactory ($DPF \approx 0.98$ and $DF \approx 1.0$) resulting in a high power factor operation even under reduced load and switching frequency conditions.

Experimental results from a lab test bench for SHEPWM with Texas DSP TMS320F2812 are in Fig. 14-16. Fig. 14 and 15 shows the converter pulse pattern a long with the phase and line voltage shapes. The line voltage spectrum in this case is in Fig. 16. The Table V shows the voltage amplitude of the first 20 odd line harmonic voltages.

VI. CONCLUSIONS

This paper presented an enhanced passive filter configuration for a high power converter. It also discussed a complete filter design, performed at a point of minimum reactive power. Besides, a reasonably simple and intuitive procedure culminates in few design steps. *MRP-OFD* permits to derive closed form expressions for the filter element, which specifies unique and well-defined element values. *MRP-OFD* states a minimum filter elements for a given design specification and since it is generically developed, it can be easily employed in adjustable speed drive filter design. After the optimized design, the paper presented an evaluation of the filter performance. The filter frequency response showed that damping deteriorates filter attenuation though it was effective in reducing the risk of filter resonance. The addition of harmonic traps in parallel to the shunt branch compensates around the fundamental frequency this loss of attenuation in high frequency region. Besides, this evaluation showed that the filter dynamic stiffness is poor. However it is influenced by the main filter (third order LCL) configuration rather than the proposed design procedure. The paper showed as well that due to capacitive effects and an adequate filter design the converter could tolerate the variations of voltage and current after the filter installation without overrating. The modulation strategy in this paper was the Selective Harmonic Elimination-SHEPWM. The modulator eliminates harmonics up to the 11th. It creates sufficient frequency separation to set cutoff and/or resonance frequencies and simplifies the filter design rules. Furthermore, the modulator SHEPWM worked in a classical voltage oriented synchronous reference frame control of a high power active rectifier. Simulation and experimental results confirmed that it is possible to achieve controllable displacement power factor with low input harmonic distortion and reduced filter kVAr rating with very low switching frequency and, in both full and reduced load conditions.

REFERENCES

[1] Skibinski, G.L.; "A series resonant sinewave output filter for PWM VSI loads" Industry Applications Conference, 2002. 37th IAS Annual Meeting. Conference Record of the Volume 1, 13-18 Oct. 2002 Page(s):247 - 256 vol.1.

[2] Steinke, J.K.; "Use of an LC filter to achieve a motor-friendly performance of the PWM voltage source inverter" IEEE Transactions on Energy Conversion, Vol. 14, No. 3, Sept. 1999, pp.: 649-654.

[3] V. Blasko, V. Kaura, "A novel control to actively damp resonance in input LC filter of a three-phase voltage source converter", IEEE Trans. on Ind. Applications, Vol. 33, No. 2, 1997, pp. 542-550.

[4] Lindgren and J. Svensson, "Control of a voltage-source converter connected to the grid through an LCL-filter-application to active filter", in Proc. of PESC 98, May 1998, vol. I, pp. 229-2315.

[5] Liserre, M.; Blaabjerg, F.; Hansen, S.; "Design and control of an LCL-filter-based three-phase active rectifier" IEEE Transactions on Industry Applications, Vol. 41, No. 5, Sept.-Oct. 2005 pp.: 1281 – 1291.

[6] Emilio José Bueno Peña "Optimization of the Behaviour of a NPC Three-Level Converter Connected to the Grid" PhD. Thesis, Universidad de Alcalá, Escuela Politécnica, Departamento de Electrónica, Spain 2005.

[7] Krug, D.; Malinowski, M.; Bernet, S.;"Design and comparison of medium voltage multi-level converters for industry applications" Industry Applications Conference, 2004. 39th IAS Annual Meeting. Conference Record of the 2004 IEEE vol.2, 2004 pp. 781-790.

[8] Rodriguez, J.; Bernet, S.; Bin Wu; Pontt, J.O.; Kouro, S., "Multilevel Voltage-Source-Converter Topologies for Industrial Medium-Voltage Drives", IEEE Trans. on Ind. Electronics, Vol. 54, No. 6, 2007, pp. 2930-2945.

[9] H. S. Patel and R. G. Hoft, "Generalized techniques of harmonic elimination and voltage control in thyristor inverters: Part I- Harmonic elimination,'' IEEE Trans. Ind. Applicat., vol. IA-9, no. 3, pp. 310-317, May/June 1973.

[10] H. S. Patel and R. G. Hoft, "Generalized techniques of harmonic elimination and voltage control in thyristor inverters: Part 11- Voltage control techniques," IEEE Trans. Ind. Applicat., vol. IA-10, no. 5, pp. 666-673, Sept./Oct. 1974.

[11] L.G. Franquelo; J. Napoles; R.C.P. Guisado; J.I. Leon; M.A. Aguirre, "A Flexible Selective Harmonic Mitigation Technique to Meet Grid Codes in Three-Level PWM Converters", IEEE Trans. on Ind. Electronics, Vol. 54, No. 6, 2007, pp. 3022-3029.

[12] S. Bernet, S. Ponnaluri, and R. Teichmann, "Design and Loss Comparison of Matrix Converters and Voltage-Source Converters for Modern AC Drives" IEEE Trans. on Industrial Electronics, vol. 49, No. 2, April 2002, pp.: 304-314.

[13] D. A. Rendusara and P. N. Enjeti "An Improved Inverter Output Filter Configuration Reduces Common and Differential Modes at the Motor Terminals in PWM Drive Systems" IEEE Transactions on Power Electronics, Vol. 13, No. 6, November 1998, pp. 1135-1143.

[14] Y. Sozer, D. A. Torrey, and S. Reva, "New Inverter Output Filter Topology for PWM Motor Drives" IEEE Trans. on Power Electronics, Vol. 15, No. 6, pp 1007–1017 Nov. 2000.

[15] P. A. Dahono, "A control method to damp oscillation in the input LC filter of AC-DC PWM converters," in Proc. PESC'02, June 2002, pp. 1630–1635.

[16] P. A. Dahono; Y. R. Bahar; Y. Sato; T. Kataoka,"Damping of transient oscillations on the output LC filter of PWM inverters by using a virtual resistor" Proceedings of 4th IEEE International Conference on Power Electronics and Drive Systems, PEDS 2001, Indonesia, pp.: 403-407

[17] L. N. Arruda, "Sistema de geração distribuída de energia fotovoltaica" Master degree dissertation, Universidade Federal de Minas Gerais, 1999.

[18] Mohan, N; Underland, T.M.; Robbins, W. P.; "Power Electronics, converters, applications and design", John Wiley & Sons, USA, 1989.

Injection of a carrier with higher than the PWM frequency for sensorless position detection in PM synchronous motors

Roberto Leidhold and Peter Mutschler

Darmstadt University of Technology / Department of Power Electronics and Control of Drives,
Darmstadt, Germany, e-mail: *leidhold@srt.tu-darmstadt.de, pmu@srt.tu-darmstadt.de*

Abstract— The injection of an alternating carrier allows detecting the rotor position in PM synchronous motors. It is usually injected through the PWM, being therefore limited to about 1/5 of the switching frequency. By increasing the carrier frequency beyond the switching one, the bandwidth of the position observer can be significantly increased and audible noise can be reduced, among other advantages. A method is proposed to inject a carrier with a higher than the PWM frequency by coupling an external source through a transformer and a filter. An analysis of the advantages and difficulties, as well as experimental results, are presented.

Keywords— Control of Drive, Permanent magnet motor, Sensorless control, Vector control.

I. INTRODUCTION

In order to drive a synchronous motor the controller must know the mover's position. The position can be determined by position sensors or by sensorless methods, which indirectly derive the position from the measured stator voltage and current [1],[2]. It is recognized that position sensors are an expensive part of the drive and that they reduce its reliability, not only because of the sensor itself, but also because of the link between the sensor and the controller.

It is expected from a sensorless drive to require only the hardware already available in standard drives. This is because the cost of additional hardware will compete with the cost of a position sensor. However, there are application where the position sensor is more expensive and frail, as for instance in linear drives [3], or applications where due to the environment it is not possible to place a position sensor. For these cases, the use of additional hardware for the sensorless position detection, as proposed in this paper, is justified. Moreover, a better dynamic performance can be reached by using the additional hardware.

Several sensorless methods have been proposed for synchronous motors. They can mainly be classified into two groups. One group is based on the fundamental field model of the motor by which the electromotive force (EMF) or the rotor flux is estimated to obtain the rotor position, e.g. [4]. These methods lose performance as speed decreases, and they do not work at standstill being therefore unsuited for position control. The other group is based on the measurement of position-dependent inductances i.e. saliencies [5]-[7]. The inductance varies with position due to salient movers, magnetic anisotropy or when the mover's flux produces a significant saturation in the stator. These methods usually require the injection of exploring signals for evaluating the inductances. They work well at low speeds and at standstill, allowing to do position control with them. As the magnetic saliencies have a periodicity two times the periodicity of one electrical turn, these methods have a 180° ambiguity (they cannot distinguish north from south pole).

Several methods have been proposed to obtain the inductance variation deducing it from the electrical variables of the motor, and by it means being able to detect the position. Most of them rely on injecting a voltage signal by the drive's inverter, and analyzing the resulting current. They can mainly be classified by the signal being injected in: revolving voltage carrier voltage [6], alternating voltage carrier [7], voltage pulse pattern [5], and without injecting a specific signal by evaluating the current ripple resulting from the standard PWM [8].

The method based on the alternating carrier [7] proved being able to track even very low saliencies [3] and will be the one analyzed in this paper. It consists on injecting an alternating voltage-carrier in the d-axis of the estimated reference frame. By demodulating the resulting q-axis a signal is obtained, that depends on the mutual inductance. This signal crosses zero when the estimated reference frame is aligned with the rotor flux, and is fed to a PLL-like observer in order to track the actual position.

If the carrier is injected in the reference voltage of the pulse width modulator, the carrier frequency is limited to about 1/5 of the switching frequency. This limitation has following consequences: First, considering the usual switching frequency range, the carrier will produce audible noise. Second, as the current controller of the drive is wanted to have the highest possible bandwidth, it would attenuate the carrier. Third, the measured current must be filtered with a band pass filter (BPF) in order to extract the carrier output signal. For a given quality factor of the BPF, as lower is the center frequency so higher is the group delay, which limits the observer's bandwidth. It can be get rid of these consequences by increasing the carrier frequency beyond the switching frequency.

In high power drives, the switching frequency is very low, being therefore impossible to inject the carrier through the inverter. The possibility of injecting a carrier with higher frequency will allow to implement sensorless position detection also in such cases.

In [9] the injection of a carrier signal of about 100 kHz is considered. However, the authors have only injected the carrier to the motor while it was disconnected from the inverter, using a signal generator. The problem of how to inject the carrier while the motor is driven by an inverter, as well as how to measure the carrier output signal, is the issue of the present paper.

It is proposed to inject the voltage carrier through a series connected transformer between the inverter and the motor. The carrier is supplied by a linear amplifier. A filter is added in order to avoid the fundamental frequency current and the ripple current flowing through the transformer and the carrier source. The current resulting from the injected carrier is sensed by an inexpensive current transformer. The carrier response is separated from the fundamental frequency current, the ripple current and its harmonics by an analog filter. The filtered current is demodulated by sampling it synchronously to the carrier, providing a signal from which the actual position can be tracked.

This paper is organized as follows: After the introduction, the alternating carrier based sensorless method is reviewed. Next, the method for injection, sensing and demodulation of a higher frequency carrier is presented. After that, the complete sensorless method is evaluated by experimental tests. Finally, some conclusions are given.

II. ALTERNATING CARRIER BASED SENSORLESS METHOD

The PM synchronous motor can be modeled in the stationary reference frame as follows:

$$\mathbf{u} = R\mathbf{i} + \frac{d}{dt}(\mathbf{L}\,\mathbf{i}) + \lambda_{PM}\omega\begin{bmatrix}-\sin\theta & \cos\theta\end{bmatrix}^T, \quad (1)$$

where $\mathbf{u} = [u_\alpha \quad u_\beta]^T$ is the voltage vector and $\mathbf{i} = [i_\alpha \quad i_\beta]^T$ is the current vector. Variables θ and ω are the rotor's position and speed, respectively. The resistance is designated by R, the inductance matrix by $\mathbf{L}$, and the flux linkage due to the permanent magnets by λ_{PM}.

In order to analyze the high frequency (HF) behavior of the motor, the voltage model of equation (1) can be reduced, by neglecting the fundamental frequency terms as follows:

$$\mathbf{u} = \mathbf{L}\frac{d\,\mathbf{i}}{dt}, \quad (2)$$

where

$$\mathbf{L} = L_0\,\mathbf{I} + L_1\begin{bmatrix}\cos 2\theta & \sin 2\theta \\ \sin 2\theta & -\cos 2\theta\end{bmatrix}. \quad (3)$$

Parameter L_0 is the mean value of the self-inductance, and L_1 is the amplitude of the inductance variation, being $L_0 \gg L_1$, and $\mathbf{I}$ is the identity matrix. The simplified model (2) is transformed to a dq reference frame oriented with an arbitrary angle γ. For low reference-frame speed, compared to the injected frequency, it yields:

$$\mathbf{u}_{dq} = \mathbf{L}_{dq}\frac{d\,\mathbf{i}_{dq}}{d\,t}, \quad (4)$$

with

$$\mathbf{L}_{dq} = L_0\,\mathbf{I} + L_1\begin{bmatrix}\cos 2(\gamma-\theta) & -\sin 2(\gamma-\theta) \\ -\sin 2(\gamma-\theta) & -\cos 2(\gamma-\theta)\end{bmatrix}. \quad (5)$$

For analyzing the current resulting from an injected HF voltage, the Laplace transform is applied to (4) and solved for the current, yielding:

$$\mathbf{i}_{dq} = \frac{1}{s}\,\mathbf{L}_{dq}^{-1}\,\mathbf{u}_{dq}, \quad (6)$$

where the inverse inductance matrix is:

$$\mathbf{L}_{dq}^{-1} = \frac{1}{L_0^2 - L_1^2}\left(L_0\,\mathbf{I} + L_1\begin{bmatrix}-\cos 2(\gamma-\theta) & \sin 2(\gamma-\theta) \\ \sin 2(\gamma-\theta) & \cos 2(\gamma-\theta)\end{bmatrix}\right). \quad (7)$$

The considered sensorless method consists in injecting a voltage carrier in the d-axis of a reference-frame oriented with an arbitrary angle γ:

$$\mathbf{u}_{cdq} = \begin{bmatrix}a_C\cos\omega_C t & 0\end{bmatrix}^T \quad (8)$$

In the usual approach, the frequency must be as high as possible to fall beyond the controller bandwidth but lower than the switching frequency in order to use the inverter to inject the signal. Frequencies between 200Hz and 2 kHz are usual. In the present paper however, the injection of frequencies up to 50 kHz will be analysed. By substituting this voltage in the HF model (6), the following current results:

$$\mathbf{i}_{dq} = \frac{a_C\sin\omega_C t}{\omega_C\,(L_0^2 - L_1^2)}\left(L_0\begin{bmatrix}1\\0\end{bmatrix} + L_1\begin{bmatrix}-\cos 2(\gamma-\theta)\\ \sin 2(\gamma-\theta)\end{bmatrix}\right) \quad (9)$$

The resulting q-axis current gives a measure of the difference between the actual position and the reference frame angle:

$$i_q = \frac{a_C\,L_1\sin\omega_C t}{\omega_C\,(L_0^2 - L_1^2)}\sin 2(\gamma-\theta). \quad (10)$$

To take advantage of this signal, it must be demodulated. The usual way for demodulation is to multiply it by $\sin(\omega_C t)$ [3]. To simplify the implementation for higher frequency, an alternative method is proposed. It consists on synchronously sampling the current at:

$$t_s = (\pi/2 + 2\pi n)/\omega_C \qquad n = 0,1,2,\cdots \quad (11)$$

By sampling (10) at t_s, it yields:

$$i_{qs} = \frac{a_C\,L_1}{\omega_C\,(L_0^2 - L_1^2)}\sin 2(\gamma_s - \theta_s), \quad (12)$$

where subindex s denotes sampled values.

The variable i_{qs} can then be used to track the unknown actual position θ by adjusting γ until i_{qs} reaches zero, in which case the reference frame angle γ will equals the angular position θ of the mover. The actual position tracking is performed by a PLL-like speed and position observer. A scheme of the sensorless position detection method is shown in Fig. 1.

III. HIGHER FREQUENCY CARRIER

As higher is the injected frequency so higher the effect of the distributed capacitance, core hysteresis and eddy currents. This would be a drawback of using higher frequencies, however this depends on the geometry, core materials and winding distribution of the motor.

In order to evaluate the demodulated carrier-response i_{qs} for different frequencies on a given motor, the following test was carried out. The test consists on injecting the carrier with a fixed reference frame angle $\gamma = 0$, i.e. the stationary reference frame $\alpha\beta$, using the

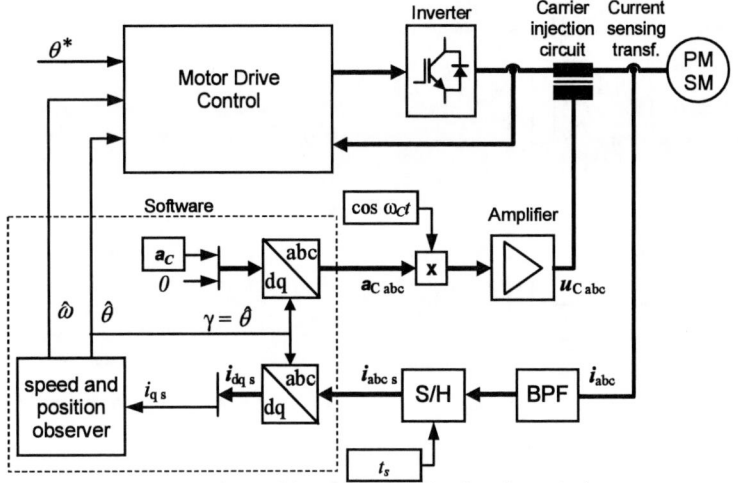

Fig. 1 Scheme of the sensorless position detection method

circuit of Fig. 2. Consequently, the carrier is injected on the α-axis and the current response is measured on the β-axis. A laboratory signal generator with low output impedance was used for the carrier voltage source u_{cd} , supplying a 15V amplitude signal. The motor under test was a synchronous one with surface PM. Its parameters are shown in Table I.

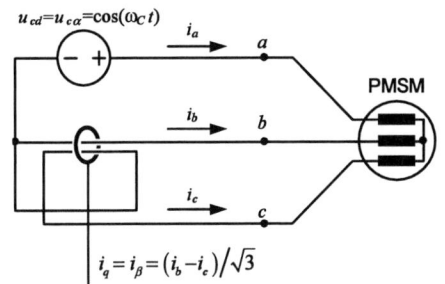

Fig. 2 Test circuit for evaluation of the position dependent variable i_{qs} for different frequencies

TABLE I

PARAMETERS OF THE PM SYNCHRONOUS MOTOR

Number pole pairs	6
Rated power	2.3 kW
Rated speed	4800 1/min
Rated current	5.5 A (rms)
Torque constant	1.12 Nm/A (rms)
Moment of inertia	4.6 Kg cm²

The demodulated carrier-response i_{qs} was obtained by sampling the current with an oscilloscope at the falling zero crossing of the supplied carrier. This variable is shown in Fig. 3 as a function of the rotor angle θ for different carrier frequencies, ranging from 10 to 100 kHz. For this motor, as can be observed in the figure, a significant signal for the rotor position detection can be obtained by using up to 100 kHz carrier frequency. It should be remarked that the motor cable had only 2 m. The consequence of using longer cables is matter of further research.

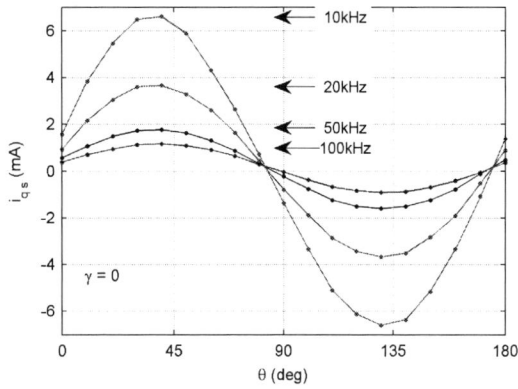

Fig. 3 Experimental test: demodulated carrier response i_{qs} as a function of the rotor angle θ

The required bandpass filter to extract the carrier output from the main current introduces a group delay in the observer loop, limiting the observation dynamics. For a filter with quality factor $Q = 10$ the group delay is $3.2\,ms$ for a center frequency $f_c = 1\,kHz$, but only $32\,\mu s$ for a center frequency $f_c = 100\,kHz$. On the one hand, a higher frequency would allow a higher observer dynamics, on the other hand as higher the carrier frequency so lower the current response of the carrier. A frequency of 49 kHz was adopted in this paper for prove of concept. Further research will allow determining an optimal frequency.

In order to inject a voltage carrier with a frequency higher than the switching frequency of the inverter, an additional hardware is required. One method is to supply the carrier with an amplifier through a transformer in series with the inverter. Even when the carrier voltage is small, the amplifier must be able to drive the fundamental current of the motor. In order to reduce the required current rating of the amplifier a filter is proposed to split the fundamental frequency current from the high frequency current, allowing to place the transformer in the high frequency branch. The proposed circuit is shown in Fig. 4.

1355

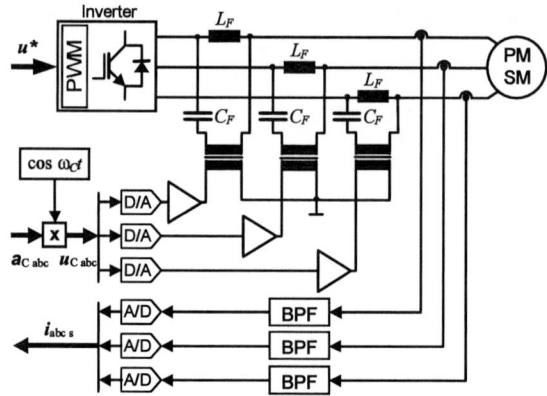

Fig. 4 Carrier injection by a transformer-coupled linear amplifier

An equivalent circuit representing one phase of the filter is shown in Fig. 5 in order to analyze the transfer characteristics. In this circuit, the inverter is represented by an ideal voltage source and the motor by an ideal current sink. The resistance R_F represents the internal impedance of the amplifier supplying the carrier. An additional resistance could be necessary if the internal resistance does not provide sufficient damping.

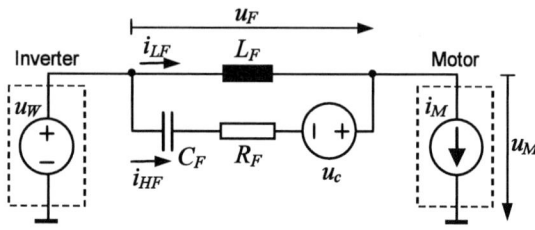

Fig. 5 One-phase equivalent circuit of the carrier injection circuit

TABLE II

PARAMETERS OF THE CARRIER INJECTION CIRCUIT

Transformer ratio	1:2
L_F	167 µH
Inductor rated current	4.3 A
C_F	0.47 µF
R_F	0.1 Ω

The transfer characteristics are analyzed using the parameters of Table II, except for the transformer that is first considered with unitary ratio. The frequency response of the voltage gain from the carrier source to the motor u_M/u_c is presented in Fig. 6, showing that for a 49 kHz carrier the gain is unitary and the phase shift almost zero. Consequently, the voltage carrier is transferred to the motor without attenuation or lag.

The current gain from the motor to the carrier source i_{HF}/i_M has the same frequency response as the voltage gain of Fig. 6. It shows how much of the motor current flows over the carrier source. For a fundamental frequency under 200 Hz, the current is attenuated more than 80 dB. For the ripple current, in this case a 5 kHz switching frequency is considered, the attenuation is 20 dB.

The series (insertion) impedance seen by the inverter u_F/i_M is presented in Fig. 7, showing 0.2 Ω for a fundamental frequency of 200Hz. In Fig. 8 the impedance

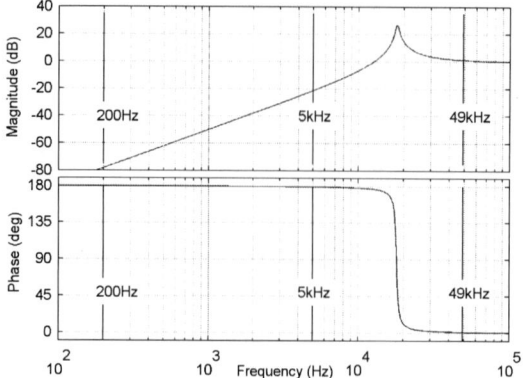

Fig. 6 Voltage gain u_M/u_c and current gain i_{HF}/i_M

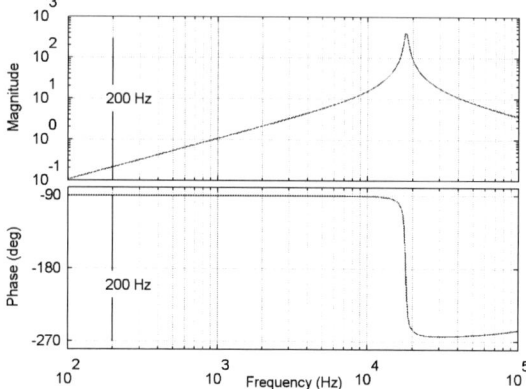

Fig. 7 Insertion impedance u_F/i_M

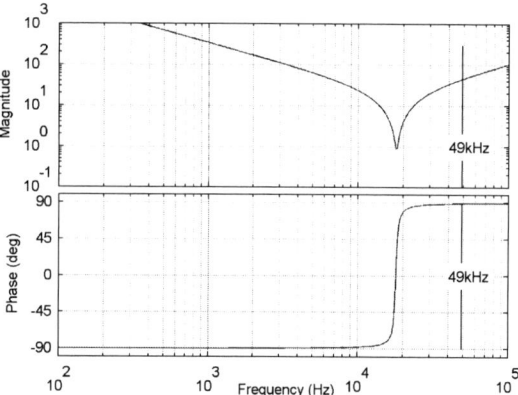

Fig. 8 Impedance seen by the carrier source u_c/i_{HF}

seen by the carrier source u_C/i_{HF} is presented, that for the carrier frequency is about 50 Ω.

A linear amplifier was considered in this paper for the carrier source in order to provide an easier implementation of the prototype. However, a small MOSFET inverter with a one-pulse width modulation would be more efficient.

In order to sense the current resulting from the injected carrier, an inexpensive current transformer with a resistor is used. First, a passive filter is used to reject the fundamental frequency component that could saturate an active filter. After the passive filter a 4th order active filter based on switched capacitors is used to isolate the carrier response.

1356

The carrier response is sampled synchronously to the carrier source according to (11). This provides a demodulation method of simple implementation for higher frequencies. The sampling interval of the controller and consequently of the position observer, is several times longer than the carrier period. From the several carrier response samples, only the most recent one is used for the observer. Alternatively, the average of the carrier response samples inside the controller's sample interval can be used to increase noise immunity.

IV. EXPERIMENTAL RESULTS

The proposed carrier injection circuit and the sensing circuit was implemented and tested in a running drive. The parameters of the motor and the carrier injection circuit are given in Table I and II, respectively. The switching frequency of the drive was 5 kHz and the sampling frequency of the controller was 10 kHz.

The first test consists in drive the motor at constant speed $\omega = 6\pi\,\text{rad/s}$, using an encoder for position and speed feedback. A 16 V, 49kHz carrier was injected in the primary of the transformer with a fixed reference frame angle $\gamma = 90°$. The resulting signals are depicted in Figures 10-13. The FFT of the A-phase voltage of the motor is shown in Fig. 10. As can be seen in the figure, the carrier frequency was chosen as to be between the harmonics of the switching frequency. In Fig. 11 and 12 the FFT of the measured current after the passive filter and after the active filter, respectively, are shown. The last one is the signal being synchronously sampled with the carrier and converted to digital to feed the position observer.

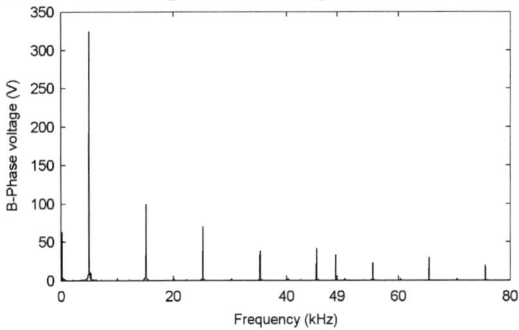

Fig. 10 FFT of the motor's B-phase voltage by carrier injection at $\gamma = 0$, with running inverter.

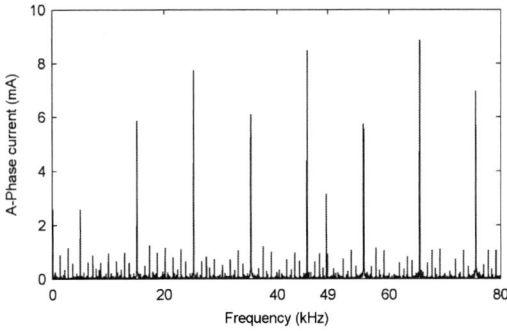

Fig. 11 FFT of the motor's A-phase current measured after the passive filter, by carrier injection at $\gamma = 0$, with running inverter.

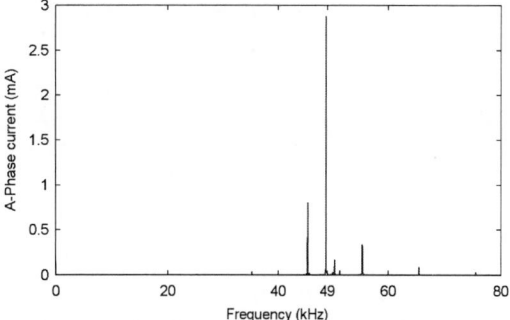

Fig. 12 FFT of the motor's A-phase current measured after the active filter, by carrier injection at $\gamma = 0$, with running inverter.

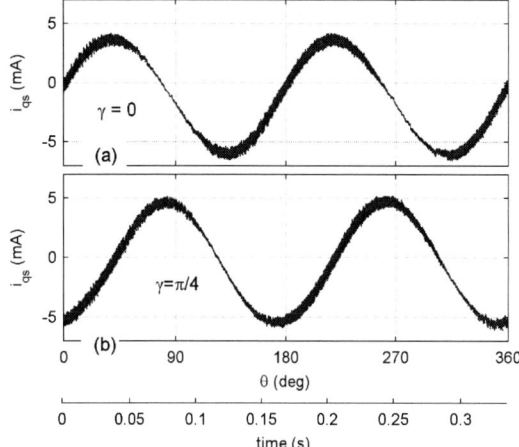

Fig. 13 Demodulated carrier response i_{qs} as a function of the rotor angle θ, by carrier injection at angle (a) $\gamma = 0$ and (b) $\gamma = \pi/4$

The sampled carrier response i_{qs} is shown in Fig. 13 as a function of the position. In Fig. 13a, the carrier was injected at a reference frame angle $\gamma = 0$ and in Fig. 13b at $\gamma = \pi/4$.

In order to track the position from the demodulated carrier-response, the following observer was implemented:

$$\frac{d\hat{\omega}}{dt} = k_2\, i_{qs}\,, \tag{13}$$

$$\frac{d\hat{\theta}}{dt} = \hat{\omega} + k_1\, i_{qs}\,, \tag{14}$$

where $\hat{\omega}$ and $\hat{\theta}$ are the estimated speed and position, respectively. The observer's dynamics is adjusted by means of the gains k_1 and k_2. As shown in Fig. 1, the estimated position is fed back to the reference frame transformation, i.e. $\gamma = \hat{\theta}$.

In a second test, the dynamics of the position estimation was evaluated. In this test, the rotor is controlled to a fixed position by using an encoder for the feedback. The observer is started from zero initial values at time $t = 10\,\text{ms}$. The measured and estimated positions are depicted in Fig. 14 as a function of the time. From this figure, it can be observed that the estimated variable settles in less than 10 ms.

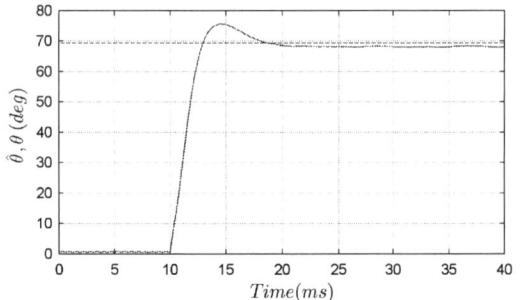

Fig. 14 Time response of the position observer. Estimated (solid) and measured (dashed) position

A third test was implemented to evaluate the position detection method used in a position control scheme. The estimated position was used as the feedback of the position controller, according to the scheme of Fig. 1. A reference position step is set from 130° to 0° and after 250 ms back to 130°. In Fig. 14a the estimated and actual positions are shown. The estimation error is shown in Fig. 14b, and the q-axis current in Fig. 14c.

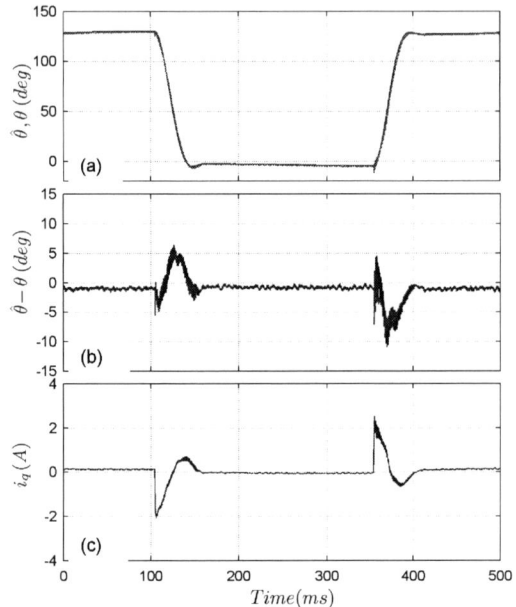

Fig. 15 Time response of the sensorless position control using higher frequency carrier, a) estimated and actual position, b) estimation error and c) q-axis current

V. CONCLUSIONS

In order to take the advantages of using a higher carrier frequency for sensorless position detection, a method for injecting a voltage carrier was proposed. This method allows injecting a carrier with higher frequency than the switching frequency of the inverter. An additional circuit is therefore required to be inserted between the inverter and the motor. The demodulation of the carrier-response is simply performed by sampling the signal synchronously with the injected carrier.

The proposed method for injection, sensing and demodulation of the carrier was analyzed and tested experimentally. It shows that a signal able to provide the

position information can be obtained independently of the inverter, PWM and control algorithm. By using the proposed method, a sensorless position control was implemented and validated experimentally.

By using carriers with a frequency higher than the switching one, the implementation of sensorless methods in high power drives with low switching frequency will be possible. The audible noise produced by the usual sensorless methods is avoided and the dynamics of the position estimation can be improved by using the proposed method. The main drawback is the additional hardware required to implement it.

Further research is necessary to optimize the carrier injection by using a small MOSFET inverter instead of a linear amplifier. A criterion for selecting the carrier frequency and the filter bandwidth of the sensing subsystem must be developed in order to optimize the estimation dynamics. Finally, the effect of using long cables between the inverter and the motor must be studied.

REFERENCES

[1] P. P. Acarnley and J. F. Watson, "Review of position-sensorless operation of brushless permanent-magnet machines," *IEEE Trans. Ind. Electron.*, vol. 53, no. 2, pp. 352–362, Apr. 2006.

[2] J. Holtz, "Perspectives of sensorless AC drive technology," *in Proc. PCIM Eur.*, 2005, pp. 80–87.

[3] R. Leidhold, P. Mutschler, "Sensorless position-control method based on magnetic saliencies for a Long-Stator Linear Synchronous-Motor". *in Proc. IECON 2006*. Paris, France, 7-10 November 2006. pp. 781-786.

[4] C. De Angelo, G. Bossio, J. Solsona, G. Garcia, M.I. Valla, "A Rotor Position and Speed Observer for Permanent Magnet Motor with Non Sinusoidal EMF Waveform," *IEEE Trans. Ind. Electron.*, vol. 52, no. 3, pp. 807-813, Jun. 2005.

[5] E. Robeischl, M. Schroedl, "Optimized INFORM measurement sequence for sensorless PM synchronous motor drives with respect to minimum current distortion," *IEEE Trans. Ind. Appl.*, vol. 40, no. 2, pp. 591-598, Mar.-Apr. 2004.

[6] Jansen, P.L.; Lorenz, R.D., "Transducerless position and velocity estimation in induction and salient AC machines," *IEEE Trans. Ind. Appl.*, vol.31, no.2pp.240-247, Mar/Apr 1995.

[7] O.C. Ferreira and R. Kennel, "Encoderless Control of Industrial Servo Drives" *in Proc. EPE-PEMC 2006*, Portoroz, Slovenia, pp. 1962-1967.

[8] Holtz, J.; Hangwen Pan, "Elimination of saturation effects in sensorless position-controlled induction motors," *IEEE Trans. Ind. Appl.*, vol.40, no.2pp. 623- 631, Mar./Apr. 2004.

[9] J. Persson, M. Markovic, Y. Perriard, "A New Standstill Position Detection Technique for Nonsalient Permanent-Magnet Synchronous Motors Using the Magnetic Anisotropy Method," *IEEE Trans. Magnetics*, vol.43, no.2, pp.554-560, Feb. 2007.

[10] J-H Jang, S-K Sul, J-I Ha, K. Ide, and M. Sawamura, "Sensorless Drive of Surface-Mounted Permanent-Magnet Motor by High-Frequency Signal Injection Based on Magnetic Saliency," *IEEE Trans. Ind. Appl.*, vol. 39, no. 4, pp. 1031-1039, Jul./Aug. 2003.

Parallel Fixed Point FPGA Implementation of Sensorless Induction Motor Torque Control

Jacek D. Lis, Czeslaw T. Kowalski

Wroclaw University of Technology, Institute of Electrical Machines, Drives and Measurements, Wroclaw, Poland

Abstract— **The paper deals with the FPGA implementation of an sensorless motor control application. The parallel processing approach is described. Few issues concerning the implementation are discussed. The method for the qualitative analysis of the considered state observer's sensitivity to the variations of the induction motor equivalent circuit parameters is presented. The experimental tests of the FPGA based implementation of the whole control structure of the sensorless DTC drive system are demonstrated.**

Index Terms—**induction motor, sensorless drive, DTC, state observer, FPGA**

I. INTRODUCTION

Modern techniques for induction motor sensorless control involve efficient computational units. Nowadays either the signal processors or microprocessor systems are used in such applications. Currently the specialized systems of 16 and 32 bit signal controllers designed for the electric drive are available on the market. The computational power they provide, is however low. Moreover the set of peripheral interfaces in case of some applications requires supplementing or even replacement.

Sufficient computational efficiency can be obtained with fast universal signal processors, but in such a case the design process is laborious and thus such applications are more expensive. An ASIC (Application Specific Integrated Circuit) can be an interesting alternative. Such solution allows arbitrary forming of the data flow scheme as well as large degree of freedom in the peripheral interfaces design. The very expensive design process can be considered the main disadvantage of the ASICs. Therefore FPGAs (Field-Programmable Gate Array) are often used in the ASIC application's prototyping stage. Due to the fact that the FPGAs are capable of completely parallel processing of many data streams, their applications are nowadays expanding. FPGAs are applied in many fields from telecommunication, intensive signal processing, to digital control. Apart from simple gate system, FPGAs contain specialized arithmetic blocks, distributed memory, clocks, advanced IO configuration blocks and many other elements.

Applying the FPGA matrices to electric drive control require a different approach to the observer and control algorithm development. In the case of the classic applications performed with the aid of the signal processors or microprocessor systems the algorithms are executed in sequence, while FPGAs are capable of parallel calculations.

In this paper the sensorless direct torque control implementation in the FPGA matrix is discussed. The paper is focused on the issues concerning both the state observer and control algorithm implementation. The problem of the state observer's sensitivity to the variations of the induction motor equivalent circuit is also addressed. The results of the experimental tests are presented.

II. FPGA APPLICATION TO THE INDUCTION MOTOR CONTROL

The speed control procedure for the induction motor drives consist of several tasks, i.e. speed control, torque or current control, state estimator or observer, inverter control algorithm, acquisition and processing of sensor signals. Usually, when a microprocessor or digital signal processor is used, the control algorithm is executed sequentially (Figure 1a). In many technical papers only chosen parts of the AC motor control structures are dedicated to FPGA implementation, like: I/O subsystems [1], space vector modulation method (SVM) for the voltage inverter [10], [16], while the main control tasks are still realised sequentially by the supervising microprocessor system (usually digital signal processor).

The FPGAs are very efficient and thus the whole control structure of the induction motor drive can be realized using parallel calculations. It forces different approach to control algorithms implementation. In order to apply the parallel processing scheme to a sensorless motor drive the control algorithm has to be decomposed to parallel tasks, as it is shown in Fig. 1b (additional internal frames specify blocks representing algorithm parts executed in parallel).

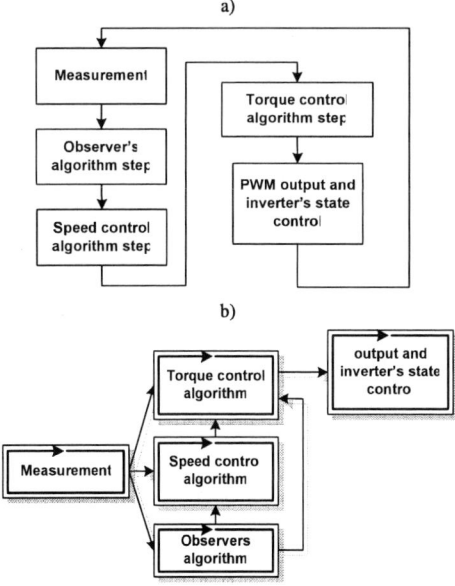

Fig. 1. Serial (a) and parallel (b) implementation of the algorithm for the induction motor sensorless control

Few stages of the algorithm's decomposition can be distinguished. In the first stage the following functional parts of the control algorithm are distinguished:
- data acquisition (IO converter's handling),
- state variables estimations (if the sensorless drive is considered),
- control signals determination,
- control of the power switches, power modules, overload protection,
- IO interfaces management.

The consecutive decomposition stage consists in distinguishing the data streams and their interaction, what enables defining the parallel processing sequences as well as the optimization of the matrix's arithmetic blocks' consumption.

To the third decomposition stage the pipeline technique for data stream decomposition is introduced [14]. Thus, the decomposed calculations can be calculated simultaneously in each algorithm's step.

It should be emphasized that the decomposed tasks are calculated in different parts of the FPGA matrix (algorithm's space positioning). The algorithm's implementation is based on the appropriate configuration of the FPGA matrix resources; hence the hardware processing sequences are developed.

The FPGA provides significantly shorter time step, comparing to the fastest signal processors on the market. High computational efficiency and short calculations steps are obtained due to changes introduced to the algorithm's architecture (parallel implementation) as well as taking advantage of the built in resources – in the case of the FPGA considered in this work, 96 arithmetic units are capable of multiplying with accumulation (equivalent of the MAC command in the case of a regular signal processor).

III. MATHEMATICAL MODEL OF THE SENSORLESS DTC INDUCTION MOTOR DRIVE

The complete FPGA implementation of the senssorless direct induction motor torque control consists of two parts: the first, which concerns the implementation of the DTC algorithm [7], and the second, which concerns the state observer implementation (Fig.2).

In the considered application the fixed point version [6] of the Kubota linear state observer with speed adaptation loop is employed [4]. The observer is derived from the induction motor's mathematical model described using the space vectors of the stator voltage, current and rotor flux.

Making the assumption that the rotational speed remains constant in the stator current sampling time, rearranging the induction motor model [11] to the form of the state equation and introducing, according to the observer's theory [11], additional feedback loop, which contains the **G** matrix, the mathematical model of the linear full order observer can be obtained. The state matrix elements are in such case dependent on the rotor speed [4]:

$$\frac{d}{dt}\begin{bmatrix} \hat{\mathbf{i}}_s \\ \hat{\boldsymbol{\Psi}}_r \end{bmatrix} = \frac{1}{T_N}\left((\mathbf{A} + \omega\mathbf{A}')\begin{bmatrix} \hat{\mathbf{i}}_s \\ \hat{\boldsymbol{\Psi}}_r \end{bmatrix} + \mathbf{B}\mathbf{u}_s \right) + \mathbf{G}(\hat{\mathbf{i}}_s - \mathbf{i}_s) \quad (1a)$$

where the appropriate matrices are as follows:

$$\mathbf{A} = \begin{bmatrix} -\left(\dfrac{r_s x_r}{w} + \dfrac{x_M^2}{w\tau_r}\right) & \dfrac{x_M}{w\tau_r} \\ \dfrac{x_M}{\tau_r} & -\dfrac{1}{\tau_r} \end{bmatrix} = \begin{bmatrix} a_{11} & a_{12} \\ a_{21} & a_{22} \end{bmatrix} \quad (1b)$$

$$\mathbf{A}' = \begin{bmatrix} 0 & j\dfrac{x_M}{w} \\ 0 & j \end{bmatrix}, \ \mathbf{B} = \begin{bmatrix} \dfrac{x_r}{w} \\ 0 \end{bmatrix} \quad (1c)$$

$$\mathbf{G} = \begin{bmatrix} (k-1)(a_{11}+a_{22})+j(k-1)\hat{\omega} \\ (k^2-1)(ca_{11}+a_{21})-c(k-1)(a_{11}+a_{22})-j(k-1)c\hat{\omega} \end{bmatrix} \quad (1d)$$

where

$\mathbf{u}_s$, $\mathbf{i}_s$, $\mathbf{i}_r$, $\boldsymbol{\Psi}_r$ – the space vectors of the rotor's voltage, current and flux, respectively,

ω – motor angular speed,

r_s, r_r, x_s, x_r, x_M – resistance and reactance of stator and rotor's windings and mutual reactance, respectively,

$T_N = 1/(2\pi f_{sN})$ – nominal integration constant,

$$w = x_s x_r - x_M^2, \ \tau_r = \frac{x_r}{r_r}, \ c = \frac{w}{x_M},$$

k – constant coefficient.

Apart from the observer gain matrix **G**, the adaptation mechanism is introduced to this observer, which provides the rotor speed estimation, taking advantage of the information on the stator current estimation error and the estimated rotor flux vector. Thus the observer state matrix is commissioned by means of the rotor speed estimate and consequently the difference in the motor and observer speed is minimized.

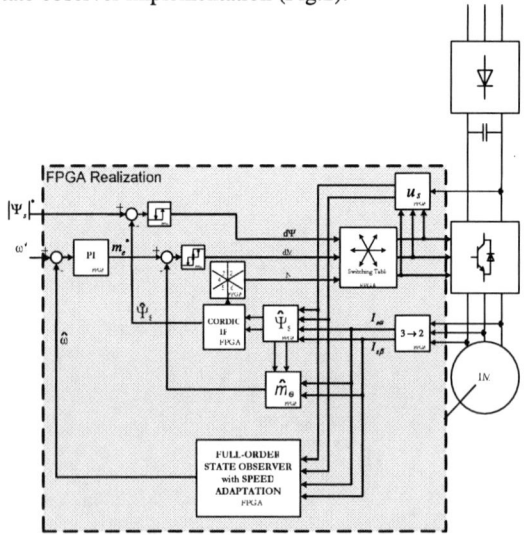

Fig. 2. The block diagram of the complete FPGA application of the sensorless direct control of the induction motor torque

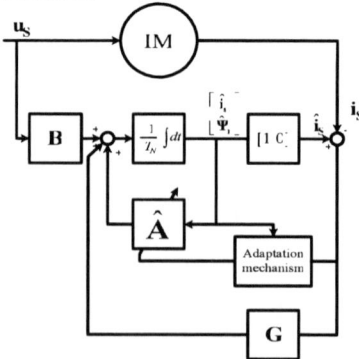

Fig. 3. Schematic diagram of the full order linear observer of the induction motor, with the speed adaptation loop

The adaptation mechanism, proposed originally in [4], is formulated as follows:

$$\hat{\omega} = K_P \, \mathrm{Im}\{e_{is}^* \, \hat{\Psi}_r\} + K_I \int_0^t \mathrm{Im}\{e_{is}^* \, \hat{\Psi}_r\} dt \qquad (2)$$

where: $\mathbf{e}_{is} = \mathbf{i}_s - \hat{\mathbf{i}}_s$

The block diagram illustrating the described solution is presented in Fig 3.

IV. SENSITIVITY OF THE OBSERVER ALGORITHM TO THE VARIANTIONS OF THE INDUCTION MOTOR EQUVALENT CIRCUIT PARAMETERS

In order that the discussed application performs properly, the induction motor equivalent circuit (1) parameters have to be known. Those parameters are usually obtained either on the basis of the motor nominal parameters or by means of the identification tests. Both approaches provide parameters determination up to some degree of the accuracy, hence the sensitivity of the concerned solution to the parameters' variations should be taken into consideration.

One of the possible approaches is conducting the sensitivity qualitative analysis. In the case of this method the sensitivity model coupled with the investigated drive system mathematical model is formulated. The primary and the coupled model are connected in such manner that the coupled one (i.e the sensitivity model) takes advantage of some transients obtained in the primary model. Both models are calculated simultaneously, thus the information on the sensitivity to particular parameter variations is obtained during the primary model computation. The results presented in this paper are obtained during the analysis conducted in the speed function.

The sensitivity models are formulated by means of calculating the appropriate derivatives of the considered model variables (1) with respect to the given motor parameter p_i:

$$\frac{\partial}{\partial p_i}\frac{d}{dt}\begin{bmatrix}\hat{\mathbf{i}}_s\\\hat{\Psi}_r\end{bmatrix} = \frac{1}{T_N}\begin{pmatrix}\left(\frac{\partial}{\partial p_i}\mathbf{A}(p_i)+\hat{\omega}\frac{\partial}{\partial p_i}\mathbf{A}'(p_i)\right)\begin{bmatrix}\hat{\mathbf{i}}_s\\\hat{\Psi}_r\end{bmatrix}+\\(\mathbf{A}(p_i)+\hat{\omega}\mathbf{A}'(p_i))\frac{\partial}{\partial p_i}\begin{bmatrix}\hat{\mathbf{i}}_s\\\hat{\Psi}_r\end{bmatrix}+\\\frac{\partial}{\partial p_i}\mathbf{B}(p_i)\mathbf{u}_s+\mathbf{B}(p_i)\frac{\partial}{\partial p_i}\mathbf{u}_s\end{pmatrix}$$
$$+\frac{\partial}{\partial p_i}\mathbf{G}(p_i)(\hat{\mathbf{i}}_s - \mathbf{i}_s)+\mathbf{G}(p_i)\frac{\partial}{\partial p_i}(\hat{\mathbf{i}}_s - \mathbf{i}_s) \qquad (3)$$

For instance, by computing the partial derivatives of the matrix from the relationship (3) with respect to r_s:

$$\frac{\partial}{\partial r_s}\mathbf{A}(p) = \frac{\partial}{\partial r_s}\begin{bmatrix}-\left(\frac{r_s x_r}{w}+\frac{x_M^2}{w\tau}\right) & \frac{x_M}{w\tau}\\\frac{x_M}{\tau} & -\frac{1}{\tau}\end{bmatrix}=\begin{bmatrix}-\frac{x_r}{w} & 0\\0 & 0\end{bmatrix} \qquad (4a)$$

$$\frac{\partial}{\partial r_s}\mathbf{A}' = \frac{\partial}{\partial r_s}\begin{bmatrix}0 & -j\frac{x_M}{w}\\0 & 0\end{bmatrix}=\begin{bmatrix}0 & 0\\0 & 0\end{bmatrix}, \qquad (4b)$$

$$\frac{\partial}{\partial r_s}\mathbf{B} = \frac{\partial}{\partial r_s}\begin{bmatrix}\frac{x_M}{w}\\0\end{bmatrix}=\begin{bmatrix}0\\0\end{bmatrix} \qquad (4c)$$

$$\frac{\partial}{\partial r_s}\mathbf{G}=\frac{\partial}{\partial r_s}\begin{bmatrix}-(k-1)\frac{r_s x_r + r_r x_s}{x_s x_r - x_M^2}+j\hat{\omega}(k-1)\\-(k^2-1)\frac{r_s x_r}{x_M}+(k-1)\frac{r_s x_r + r_r x_s}{x_M}-j\hat{\omega}\frac{x_s x_r - x_M^2}{x_M}(k-1)\end{bmatrix} \qquad (4d)$$
$$=\begin{bmatrix}-(k-1)\frac{x_r}{w}\\-(k^2-1)\frac{x_r}{x_M}+(k-1)\frac{x_r}{x_M}\end{bmatrix}$$

and introducing them to the model, according to (3), the appropriate function defining the sensitivity to the considered parameter variation can be obtained, in the function of the rotor speed, as presented in Fig. 4.

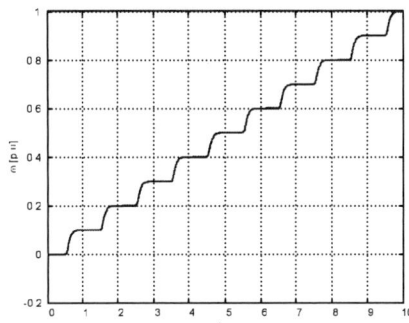

Fig. 4. The speed transient determined during the computations of the estimate's sensitivity function

The representative transients of the sensitivity function with respect to the stator resistance and the mutual reactance are presented respectively in Fig. 5 and Fig. 6.

a)

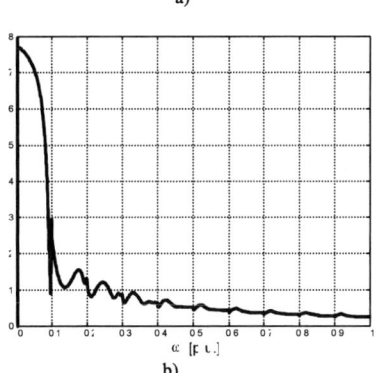

b)

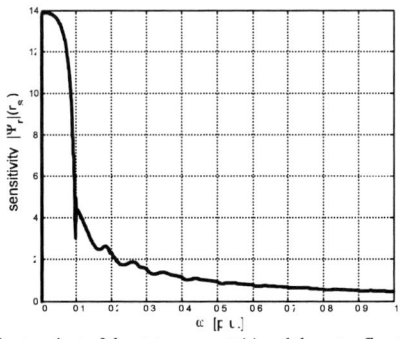

Fig. 5. The transient of the stator current (a) and the rotor flux (b) estimate sensitivity to the variations of the r_s parameter versus speed

a)

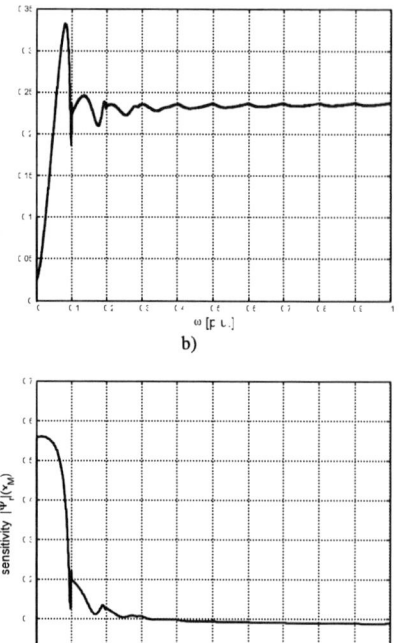

b)

Fig. 6. The transient of the stator current (a) and the rotor flux (b) estimate sensitivity to the variation of the x_M parameter versus speed

If we look at the manner in which the speed is determined (the adaptation mechanism (5)), we realize that the rotor flux and stator current estimates robustness to the motor parameter variations is crucial in the case of the considered sensorless drive. The analysis based on the sensitivity function investigations reveals that the sensitivity increases significantly along with the speed reduction. This conclusion should be taken into consideration during the algorithm's implementation.

V. FPGA APPLICATION OF THE INDUCTION MOTOR TORQUE DIRECT CONTROL

On the basis of the relationships (1), (2) and [7], the FPGA application of the complete sensorless direct control of the induction motor torque is developed. The discussed application contains the decomposed algorithms of the control system as well as the state observer's algorithm and master (superordinated) speed controller.

The DTC algorithm is decomposed to the following tasks:
- data acquisition,
- transformation from the *ABC* coordinates system to the α-β one,
- stator flux vector estimation,
- electromagnetic torque estimation,
- determination of the stator flux vector position and magnitude,
- sector determination,
- determination of the comparators' states,
- determination of the inverter switches' states

The state observer algorithm is decomposed to the following tasks:
- stator current estimation in the α axis,
- stator current estimation in the β axis,
- rotor flux estimation in the α axis,

- rotor flux estimation in the β axis,
- speed adaptation and estimation algorithms (2).

The particular tasks form the signal processing sequences in the FIFO order are computed simultaneously. The data streams are designed using the PIPELINE technique [12] and hardware arithmetic blocks (multiplication and summation with the aid of the hardware DSP block are computed in one cycle of the FPGA clock). The stator flux estimation procedure is supplemented with the additional mechanism of the numerical stabilization, which prevents the integration blocks from saturation. Otherwise the saturation would have occurred if the integrated signals had contained offsets. The stator flux vector components α and β are transformed to the polar coordinates system using the CORDIC (COordinate Rotation DIgital Computer) algorithm [1], [6], [7], [13].

The CORDIC computation method is employed in the form of the IP (Intellectual Property) kernels [3], generated in the XILINX ISE environment. The kernels are introduced to the application by means of the HDL nods, which contains appropriate interfaces programmed in the VHDL language [14] and are capable of providing the communication between the IP kernels and the rest of the application.

The stator flux vector angular position, determined using the CORDIC algorithm, enables detection of the sector. The method for the sector detection is based on the modulo division proprieties. By means of modulo dividing by $\pi/6$ (the range is from $-\pi$ to π) of the value of the stator flux vector angular position, we obtain 12 sectors, of which 6 are positive and 6 are negative numbers. The sectors can be assigned to the determined number either by using the classical DTC switching table [6], [7] or the modified DTC method with the extended sector number (12 sectors). The switching table is an array addressed with the sector number and the word defining the comparators' state. The switches' states are read from the array and written to the buffers in the FPGA's embedded RAM, where they can be read by the inverter control loop. The loop assures proper control of the inverter as well as the fault handling.

The decomposed state observer algorithm is implemented in the form of the 16 bit processing sequences (for both **A** and **A'** matrix parameters). Each sequence is supplemented with the block containing 32 bit accumulator, which converts to 1 bit output. Also the numerical stabilization loop, corresponding to the one used in the case of the stator flux observer, is introduced to each sequence. Such construction assures appropriate inner accuracy and algorithms robustness, i.e. its insensitivity to the drifts and offsets, while preserving the compatibility to the 16 bit DTC algorithm.

The communication between the observer's processing sequences is provided by the double port memory and shift registers. The data exchange between the observer and the master speed controller is performed in the FIFO order, because the observer and control system's algorithms operate in different time domains.

Apart from the elements discussed above, the FPGA application also contains some additional blocks, i.e. encoder management blocks for the speed and rotor angular position determination. Moreover, in order to enable monitoring and commissioning of some algorithm parts form the level of the master application, blocks transferring the selected variables to the d/a converter, are also employed.

VI. EXPERIMENTAL TESTS

The experimental tests are performed with the aid of the National Instruments industrial computer equipped with the RIO PXI-7833R card, which contains the FPGA Xilinx Vitex-II FPGA matrix. This FPGA contains 3 million gates. The industrial computer contains also the measurement card PXI-4472, which is used for the measurement of the signals from the PXI-7833R.

The parallel FPGA application is developed for the PXI-7833R card and subordinated to the industrial computer. The complete set of the control system – observer and controllers computations are performed, using the FPGA. The master application provides with data acquisition and supervision of FPGA application.

Appropriate hodographs of the stator and rotor flux are presented in Fig. 7a and Fig. 7b, respectively. Transients of the stator flux α–β components, stator flux magnitude and rotation angle, determined by means of CORDIC, algorithm are presented in the Fig. 7c.

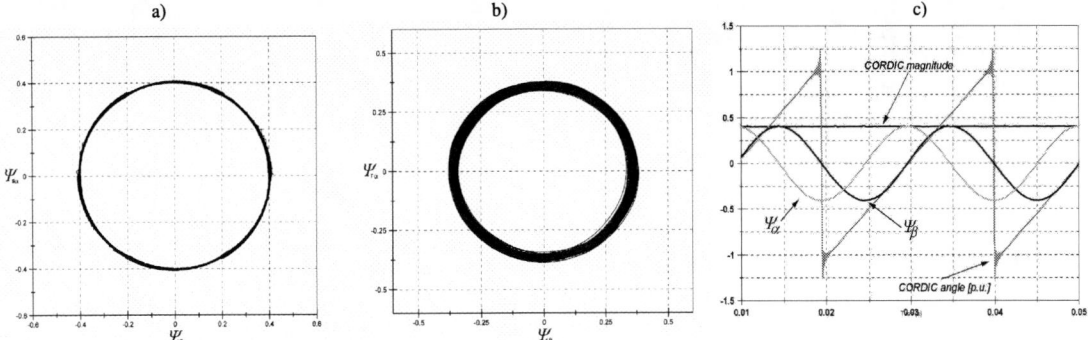

Fig. 7. The hodograph of the stator flux [p.u.] (a) and the rotor flux [p.u.] (b) vectors, stator flux α–β components [p.u.], flux magnitude [p.u.] and angle transients [p.u.], determined using the CORDIC in translation mode (c)

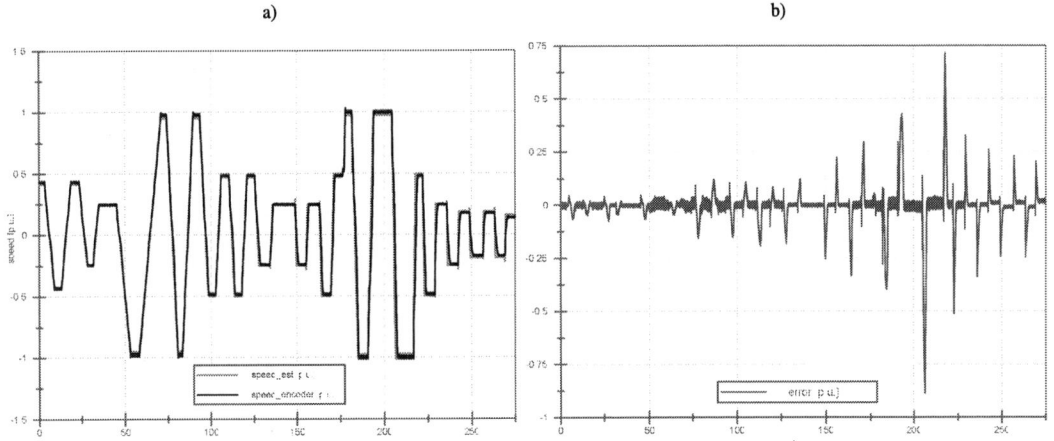

Fig. 8. The rotor speed transient for reversal operation of the DTC drive (a) and the speed estimation error (b)

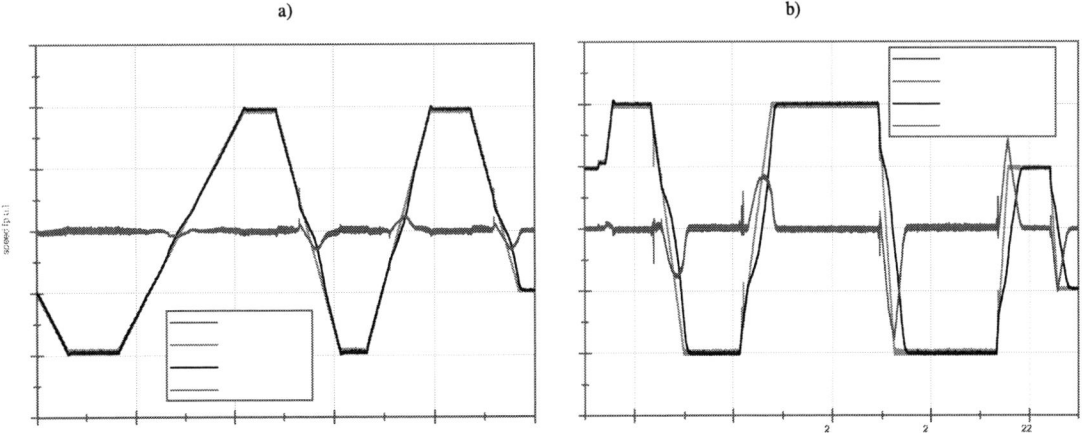

Fig 9. The rotor speed and speed estimation error for reversal operation of the drive at 50s to 100s (a) and 175s to 225 s (b) respectively

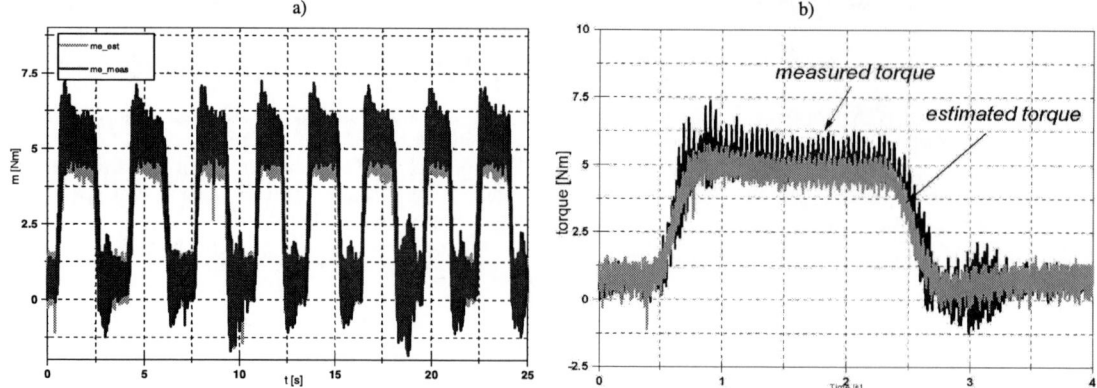

Fig. 10. Transients of the measured motor torque versus the estimated torque for reversal operation of the DTC drive under step load torque changes form 0 to 5 Nm (a), and the zoom transients (b)

The angular speed transients recorded during motor reversal operation in the DTC structure are presented in Fig. 8 and in Fig 9. The measured and estimated electromagnetic torque for the step load torque changes are presented in Fig. 10.

The presented results deal with the FPGA based DTC algorithm application, in which comparators' switching frequency is limited to 10 kHz. The time tests are also conducted and the maximal theoretical processing frequency amounts 45,031 MHz. The total consumption of the matrix resources, together with the supervision of FPGA based DTC and observer algorithms and data acquisition application reaches 99% (matrix of 3M gates).

VII. CONCLUSIONS

The conducted investigations proved that the FPGAs are suitable tool for the induction motor control applications. The parallel processing is the main advantage of the FPGA. The advanced data processing techniques (i.e. word length adjustment, PIPELINE technique) can be introduced to the FPGA matrix. The processing frequency obtained with FPGA is also significantly higher than in the case of a signal processor. The qualitative analysis of the state variables observer algorithms' sensitivity to the variations of the induction motor equivalent circuit parameters revealed that the sensitivity significantly increases with the speed reduction. This conclusion should be taken into account while designing the algorithm's implementation.

The described test implementation of the considered sensorless DTC application involves less then 3 million gates matrix. The smaller FPGA could also be used if the elements designed for the test purposes, i.e. commissioning and visualization blocks are eliminated form the structure.

ACKNOWLEDGMENT

The paper was financed by the Ministry of Sciences and Higher Education (Poland) under Grant (3T10A05230 (2006-2008)

REFERENCES

[1] Andraka R., A Survey of CORDIC Algorithms for FPGA Based Computers, *Proc. of 6th Inter. Symp. on FPGAs,* Monterey, CA, 1998, pp: 191-200.

[2] Buja G.S., Kazmierkowski M.P., Direct torque control of PWM inverter-fed AC motors - a survey, *IEEE Trans. on Industrial Electronics,* vol. 51, Aug. 2004, pp. 744-757.

[3] CORDIC-2 IP – Product Specifications, XILINX *LogiCORE, DS249* (v1.5), March 28, 2003.

[4] Kubota H., Matsuse K., Speed Sensorless Field-Oriented Control of Induction Motor with Rotor Resistance Adaptation, *IEEE Trans. on Ind. Appl.* vol. 30, no. 5, 1994, pp: 1219–1224.

[5] LabVIEW FPGA Module, *User Manual, National Instruments,* 370690B-01, March 2004.

[6] Lis J., Orłowska-Kowalska T., Kowalski Cz.T., Application of the fixed-point full-order flux observer in the sensorless SFVC induction motor drive, *Electrical Review* (Przegląd Elektrotechniczny), nr 10, 2007, pp. 1-5

[7] Orlowska-Kowalska T., Lis J., Kowalski Cz.T., FPGA Implementation of DTC Control Method for the Induction Motor Drive, *Intern. Confer. EUROCON'2007,* Warsaw, Poland, 2007.

[8] Meyer-Baese. *Digital Signal Processing with Field Programmable Gate Arrays,* Springer-Verlag, Berlin Heidelberg.

[9] Takahashi I., Noguchi T., A new quick-response and high-efficiency control strategy of an induction motor, *IEEE Trans. on Ind. Application,* vol. 22, Sept./Oct. 1986, pp: 820–827.

[10] Tonelli M., Battaiotto P., Valla M.I., FPGA implementation of an universal space vector modulator, *Proc. of 27th Annual Conf. of the IEEE Ind. Electron. Soc., IECON'01,* vol. 2, 2001 pp: 1172 - 1177.

[11] Orlowska-Kowalska T., *Sensorless Control of Induction Motor Drives,* Wroclaw Univ. Technology Press, 2003

[12] Sun W., Wirthlin M. J., Neuendorffer S., FPGA Pipeline Synthesis Design Exploration Using Module Selection and Resource Sharing, *IEEE Trans. on Computer-Aided Design of Integrated Circuits and Systems,* vol. 26, no. 2, Feb. 2007, pp: 254-265.

[13] Volder J., The CORDIC Trigonometric Computing Technique, *IEEE Trans. on Electronic Computing,* vol. EC-8, Sept. 1959, pp: 330-334.

[14] Zwolinski M. *Digital System Design with VHDL,* Prentice Hall, 2003

[15] Developing a PWM Interface using LabVIEW FPGA, *National Instruments,* http://zone.ni.com/devzone/cda/tut/p/id/3254.

[16] Zhaoyong Zhou, Tiecai Li, Takahashi T., Ho E., Design of a universal space vector PWM controller based on FPGA, *Proc. of 19th Annual Applied Power Electronics Conference and Exposition APEC '04,* vol. 3, 2004, pp: 1698 - 1700

Design of an FPGA-Based Real-Time Simulator for Electrical System

I. BAHRI[*], M-W. NAOUAR[*], E. MONMASSON[†], I. SLAMA-BELKHODJA[*] and L.CHARAABI

[*]L.S.E-ENIT BP 37-1002 Tunis le Belvédère, Tunisia,
[†]SATIE-IUP GEII, rue d'Eragny, 95031 Cergy-Pontoise, France,
email: imene.bahri@yahoo.fr, wissem_naouar@yahoo.fr, ilhem.slama@enit.rnu.tn, eric.monmasson@iupge.u-cergy.fr ,
charaabi.lotfi@enit.rnu.tn

Abstract— This paper deals with a Real-Time Simulation (RTS) able to accurately reproduce an electrical system in real-time. The RTS proposed architecture is written in VHDL and implemented in a Field Programmable Gate Array (FPGA) device. Multi-sampling approach is adopted allowing a real-time functioning with different time-steps and different operating conditions with minimized lost of accuracy. The proposed concept is illustrated by the RTS of each of a three-phase RLE load performed with a very short time step of 2.5μs, an inverter, the measurement system and the hysteresis current controllers. Comparison with experimental results shows the high performances of this RTS.

Index Terms— **Electrical system, Real-Time Simulation, multi-sampling, Field Programmable Gate Array.**

I. INTRODUCTION

During the last decade, the requirements in terms of performance, reliability, efficiency and cost reduction become particularly challenging for power electronics applications. With the rising demand of these requirements, news trends lead to the investigation of more complex algorithms such monitoring algorithms, fault-adaptive on line control, fault detection and isolation (FDI) [1], [2] and dynamic reconfiguration [3]. Thus, Real-Time Simulation (RTS) of the electrical systems allows verifying the behavior of complex systems under several operating conditions. It is also of prime importance for testing control algorithms and emulating systems [4].

Widespread availability of efficient digital technologies has made their use for Real-Time Simulation possible. Because of their fixed architecture, DSPs and microcontrollers allow a sequential process, while FPGA technology presents many attractive features to boost performances of Real-Time Simulation. These features lead from one hand to flexible dedicated parallel architectures, which significantly reduce the execution time of implemented algorithms, and from another hand, to a significant integration density and a low cost development due to the FPGA reprogrammability. Thus, FPGA technology is a valid candidate for RTS applications.

As different dynamics of different sub-parts involved in electrical systems, the most challenging task for RTS researches are modeling inaccuracies reduction and sampling period selection. The effects of the sampling period selection have been already analyzed for AC machine drives [5], [6], [7] and for power electronic converters [8], [9]. Few papers deal with the modeling inaccuracies reduction for control algorithm [10], but up

to now, no customized approach for the system modeling inaccuracies has been proposed.

The purpose of this paper is to put a special emphasis on the problematic of sampling period selection and the design chain for Real-Time Simulation of an entire electrical system. Hence, the authors present a full electrical system model written in VHDL and implemented in FPGA. In this context, the design of the entire electrical system using single time step calculation does not present the appropriate solution. To overcome this ambiguity, the adopted approach is based on the selection of the appropriate time step for each subpart of the system in order to bring the behavior of the RTS model as close as possible to real electrical system one. Additionally, a special emphasis is put on the adoption of the condition method and quantization process. So, the appropriate continuous-to-discrete method and the optimal fixed point signal formats are chosen to satisfy a balance between accuracy and minimization of the FPGA consumed resources.

The proposed concept is illustrated by a three-phase RLE load supplied by an inverter controlled by hysteresis current controllers. The chosen system is divided into three subsystems, operating at different frequencies. A multi-sampling time approach is adopted which allows a real-time functioning with different time-subsystems, The *RLE* load and the inverter are high dynamic modules; therefore they need a low time resolution to be simulated accurately. Unlike the former modules, hysteresis current controllers and measurement system are developed using a larger time resolution.

The organization of the paper is as follows: section II describes the Real-Time Simulation design. Section III discusses the time performances and FPGA consumed resources. Finally, experimental results are provided in section IV to validate the proposed simulator.

II. REAL TIME SIMULATION DESIGN

The proposed complete system is given in Fig.1. It consists of power elements (three-phase inverter, RLE load), a digital controller (hysteresis current control) and a measurement system (here only Analog Digital converters are considered). To perform an efficient RTS of the proposed system, designers must apply a methodology based on well-defined steps. The purpose of this methodology is to ensure the reusability of the design and to minimize not only the FPGA consumed resources but also the design efforts, [11]. First step is the continuous time algorithm simulation under MATLAB environment. This allows focusing on the problems of data algorithm dependencies in order to determine

978-1-4244-1741-4/08/$25.00 ©2008 IEEE

independent sub-algorithm functional blocks (functional partitioning). The next step is the sampling and quantization of the continuous time algorithm by studying the influence of sampling frequencies and the effects of fixed-point arithmetic. Finally, an optimized hardware architecture is derived from the previous step results with the help of the Adequation Algorithm Architecture (AAA) methodology [13]. The functional partitioning results here in four sub-blocks which are respectively the RLE load module, the inverter module, the ADC module and the current control module.

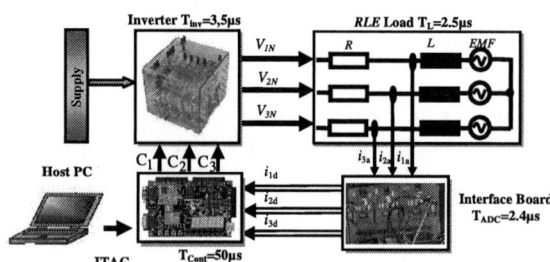

Fig.1. System components for Real-Time Simulation.

RLE Load module

A. Load Module

The general phase voltage equation for a three-phase RLE load can be written as

$$V_i(t) = L.\frac{di_i(t)}{dt} + R.i_i(t) + e_i(t) \qquad i=1, 2, 3 \quad (1)$$

Where $V_i(t)$, $i_i(t)$ and $e_i(t)$ represent respectively the ith phase voltage, current and back EMF. L and R are the inductance and resistance. For the continuous time, the load transfer function can be written as

$$\frac{i_i(t)}{V_i(t) - e_i(t)} = \frac{1/L}{s + R/L} \qquad (2)$$

The selection of the discrete method depends on the accuracy model constraints (good accuracy and simplicity). In the following, three different methods are introduced and compared

-Backaward Euler's method: $s \approx \dfrac{1 - z^{-1}}{T_s}$

-Tustin's method: $s \approx \dfrac{2}{T_s}\dfrac{1 - z^{-1}}{1 + z^{-1}}$

The performances of these methods are evaluated by comparing the load current square error obtained by using the continuous model and the developed discrete models. Although Tustin's method implies more complex equations, it gives higher precision (minimal value of square error) as shown in Fig.2.b. For a higher accuracy objective, Tustin's method is then adopted. In order to reach a high level of performance as close as possible to its analog counterpart, load model is sampled with a very small sampling period (T_s=2.5µs). Fig.2.a proves that for a reduced sampling period, the current error is minimized and the model ensures accurate results. However, as shown in Fig.2.c the reduction of the sampling period will have direct consequence on the width of the variable

fixed-point format. It can be noted that the pole location of the discrete-time becomes very close to the unit circle for a high sampling rate. This can be problematic, particularly in environments with a relatively short word length.

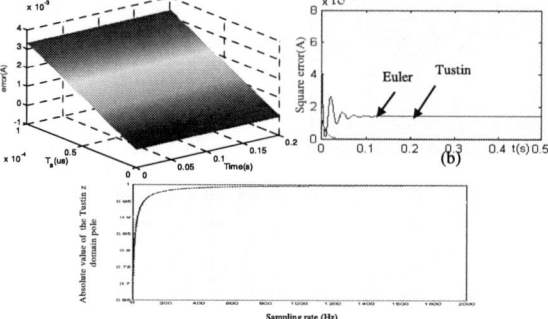

Fig.2. (a) Current error, (b) Square current error, (c) Pole location of the discrete-time.

In the following, the implemented algorithm is performed by the per-unit model. The per-unit equations are obtained by dividing all the quantities with their base values (see the appendix section). In the Tustin form, the RLE load is described by the following transfer function

$$i_i(k) = \frac{K_{pu}(1 + z^{-1})}{(1 - K_{iu}.z^{-1})}.(V_i(k) - e_i(k)) \qquad (3)$$

The quantization of the per-unit algorithm is studied under the Matlab fixed-point toolbox, which allows an algorithmic refinement, by studying the influence of the sampling frequency and the effects of fixed-point arithmetic. Actually, the fixed-point development is submitted to two main conditions: correct dynamic range and well precision evaluation [12]. The first stage of this evaluation is the estimation of the dynamic range of the data in order to determine the word-length of the integer part. This allows avoiding the overflow and it minimizes the useless bits. For a given variable x the bit number of the integer part (Fig.3 (a)), excluding the sign bit, is expressed by

$$l = \lfloor \log_2(\max(|x_{min}|, |x_{max}|)) \rfloor - 1 \qquad (4)$$

The goal of the accuracy evaluation of the fixed-point algorithms is to minimize the data word-length as long as the desired precision constraints are satisfied. For the algorithm digital, the most common criterion for evaluating signals point word length is the Signal to Quantization Noise Ratio (SQNR). The SQNR is a quantitative measurement of the signal noise level. It varies from 20 to 70 dB from a high noisy to a rather clean signal. SQNR is defined as

$$SQNR_{dB} = 10\log_{10}(\frac{\sigma^2_{signal}}{\sigma^2_{noise}}) \qquad (5)$$

Where σ_{signal} et σ_{noise} are respectively the standard deviation of the signal and of the noise. The standard deviation of the noise can be expressed by

$$\sigma^2_{noise} = \int_{-\Delta/2}^{\Delta/2} e^2.p(e).de = \frac{v^2_{max}}{12.2^m} \quad with \quad \Delta = v_{max}.2^{-m} \qquad (6)$$

Where m, e, $p(e)$ and v_{max} are respectively the bit number of the fractional part, the quantization error, the probabilistic density of the error and the maximal value of the signal. The standard deviation of the sinusoidal signal is defined as the square of the r.m.s. Therefore, the SQNR can be expressed by

$$SQNR_{dB} = 10\log_{10}\left(\frac{v^2_{max}/2}{v^2_{max}/(3.2^{2m})}\right) = (6,02.m + 1,76)dB \quad (7)$$

Using a 12 bits analog digital converter, the $SQNR_{min}$ is equal to 74dB. Taking into account that the precision constraints satisfy the following condition (8), the fractional part word length of internal signals is fixed to 16 bits.

$$\min(m_k \in Z^+) \text{ with } SQNR(m_k) \geq SQNR_{min} \quad (8)$$

Coefficients quantization must be also provided. Indeed, the pole of the RLE load z-domain transfer function is $p = 2 - (R/L)T_s/2 + (R/L)T_s = K_{iu}$. When the coefficients K_{iu} and K_{pu} are perturbed by ΔK_{iu} and ΔK_{pu} the poles of the transfer function will be also perturbed by an amount

$$\Delta p = \Delta K_{iu} \quad (9)$$

The position of the pole relative to the unit circle is critical for the performance of the discrete model of the load. Therefore, one requires

$$\frac{|\Delta p|}{\|1 - |p|\|} < \varepsilon \quad (10)$$

Where ε and $\|1 - |p|\|$ are respectively the prescribed maximum allowable percent change in pole location relative to the unit circle and the margin stability. Using (9), a condition for (10) is

$$\Delta K_{pu} < \varepsilon \|1 - |K_{iu}\| \quad (11)$$

Then, the fractional format width is given by

$$m = Log_2\{\varepsilon\|1 - |K_{iu}\|\} - 1 \quad (12)$$

Where {x} denotes the smallest integer greater than or equal to x. Setting $\varepsilon=5\%$, the condition yields m=-17 for the coefficient K_{iu}, which means that the fractional part word length is 17 bits.

A graphical representation, the Data Flow Graph (DFG), is defined in Fig.3.a, including the appropriate fixed-point format for each signals and coefficients. In order to ensure the constraints optimization, like minimization of the consumed hardware resources, the Algorithm Architecture Adequation (A^3) methodology [13] is used. The factorization process of the DFG by the A^3 methodology has been applied to the greediest operators in terms of hardware-consumed resources (multipliers). The result of the DFG factorization is the Factorized Data Flow Graph (FDFG). The consumed resources of each corresponding implemented operator for an 18 bits fixed point format are presented in Table I. Fig.3.b presents the FDFG for one phase of the RLE load.

TABLE I: Consumed hardware resources of the RLE load operator (XC3s400, Spartan 3 board)

Operations	Number of operations		consumed slices for one operation
	GFD	GFDF	
Multiplications	6	1	175 out of 5376 (3.25%)
Additions	6	2	11 out of 5376 (0.2%)
Subtraction	3	1	11 out of 5376 (0.2%)

The proposed architecture of the RLE load is deduced directly from the GFDF. It comprises two main parts, namely sequencer and data-path. The data-path is defined structurally as a network of registers, multipliers, multiplexers and buses. The sequencer is designed to control the communication inside the data-path preserving the synchronization. Fig.3.c presents the implemented load architecture.

B. Current control Algorithm

In this part, authors present the current controller module based on three identical hysteresis regulators which generate the switching states C_1, C_2 and C_3 via the comparison of the current references to the measured ones. At first, the ADCs convert the sensed currents and then the switching states resulting from hysteresis comparison are generated [6]. The development of the digital hysteresis regulator is based on

$$C_i[k] = C_i[k-1]\bar{a}_i[k] + a_i[k]b_i[k] \quad (13)$$

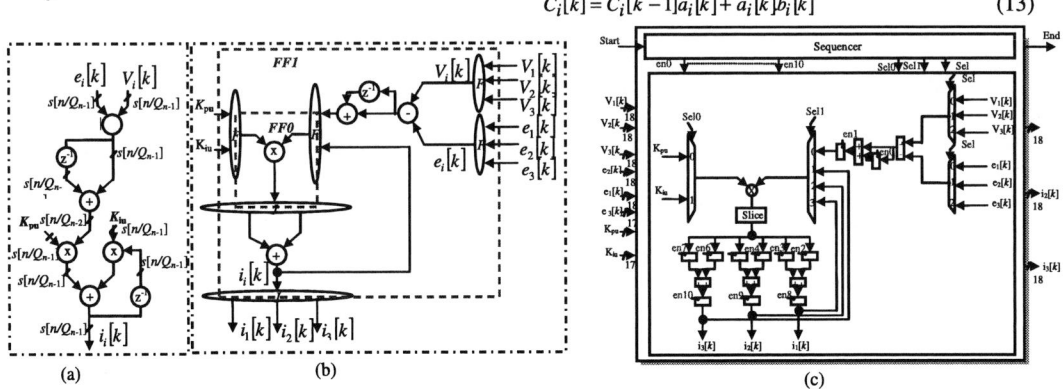

Fig.3. (a) DFG for one phase of the RLE load, (a) FDFG of the complete RLE load, (c) RLE load architecture.

Where $C_{i(i=1,2,3)}[k-1]$ and $C_{i(i=1,2,3)}[k]$ are the logical level of the switching state C_i(i=1,2,3) at the $(k-1)^{th}$ and k^{th} sampling period, respectively. The logical variables a_i and b_i are defined as follows

$$a_i[k]=0 \quad if \quad \left|i_i^*[k]-i_i[k]\right|_{(i=1,2,3))} < \frac{Bw}{2} \quad else \; a_i[k]=1$$

$$b_i[k]=0 \quad if \quad \left(i_i^*[k]-i_i[k]\right)_{(i=1,2,3))} < 0 \quad else \; b_i[k]=1$$

(14)

Where Bw is the hysteresis regulator bandwidth.

A data-path and a local sequencer are defined for the hysteresis regulators. Fig.4 shows the developed hardware architecture for one regulator. It is activated via a start signal and it generates an end signal when the computation of the switching states is achieved. The fixed-point format is set to n/Qn-1 with n equal to 13.

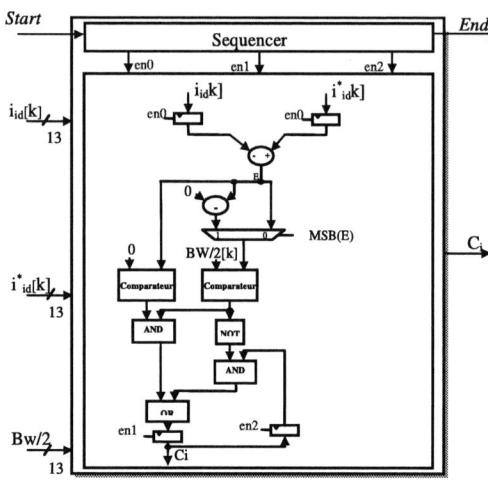

Fig.4. Hardware architecture of the hysteresis controller model.

C. ADC module

For performing the Real-Time Simulation of controlled system, it is important to take into account the influence of ADC resolution and the conversion signal delays. To reproduce faithfully the system behavior the design of the ADC model is divided in two main parts [14]:

-Truncation: Basing on the real devices used in laboratory (AD9221, 12bits), the resolution of the developed ADC is fixed to 2-12. Therefore, the modeling of the ADC requires a truncation of the current signals from 18 to 12 bits of coding (including the most significant bit, MSB).

-Conversion delays: According to the considered ADC the input signal can be converted after three periods of clock signal which is equal to 800ns. The ADC delay is modeled by using a sequencer providing a total conversion delay equal to 2.4µs. The global ADC architecture is given in Fig.5.

D. Inverter module

Depending on the desired abstraction degree and accuracy, the switching formulation of model used for representing power electronic inverters (Fig.6) can take various forms (ideal, average or switching models) [8].

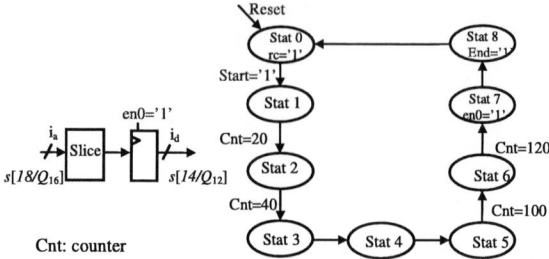

Fig.5.ADC architecture.

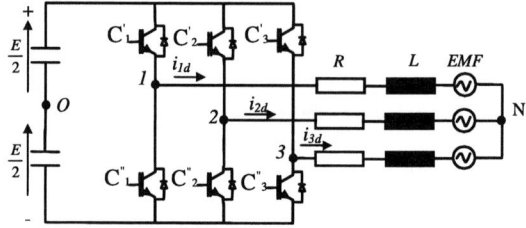

Fig.6.System components for Real-Time Simulation

In this approach, the inverter model is represented by a switching function formulation taking into account the dead time. A logical circuit allowing the generation of the dead time is then developed. A delay between the two switches of the same inverter leg is introduced in order to avoid any short-circuit. Dead time is fixed here to 3µs. In order to represent the inverter in real-time, a mathematical module with high computational efficiency is paramount. The model used in this test case can be described by using two subsets of equations as below

$$\begin{cases} V_{1N} = \frac{1}{3}\left(2.V_{1O} - \left(V_{2O} + V_{3O}\right)\right) \\ V_{2N} = \frac{1}{3}\left(2.V_{2O} - \left(V_{1O} + V_{3O}\right)\right) \\ V_{3N} = \frac{1}{3}\left(2.V_{3O} - \left(V_{1O} + V_{2O}\right)\right) \end{cases}$$

(15)

Here,

$$\begin{cases} V_{1O} = \frac{E}{2}.\left(C_1' - C_1'' + (C_1' + C_1'' - 1).sign\left(i_{1d}\right)\right) \\ V_{2O} = \frac{E}{2}.\left(C_2' - C_2'' + (C_2' + C_2'' - 1).sign\left(i_{2d}\right)\right) \\ V_{3O} = \frac{E}{2}.\left(C_3' - C_3'' + (C_3' + C_3'' - 1).sign\left(i_{3d}\right)\right) \end{cases}$$

$V_{iN}, V_{iO}, i_{id}, E, C_i'$ and C_i'' are respectively the phase voltage, the pole voltage, the phase current (i=1,2,3), DC link voltage and switching functions of IGBTs of the same inverter leg.

For hardware realization and in order to reduce design cycle a modular principle is used to split the architecture into two parts, namely sequencer and data-path. As shown in Fig.7, the top-level view of the inverter module is decomposed into sub-controllers and data-path (sub-functional modules). The inverter sampling period is chosen equal to 3.5µs (including dead time and execution time) allowing high details and precise accounting of gating signals. For performing the Real-Time Simulation of controlled system, designers develop also an ADC module taking into account the influence of the resolution and the conversion signal delays.

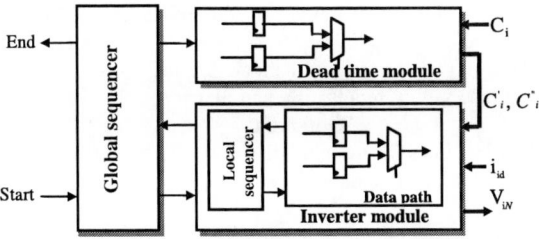

Fig.7.Modular architecture of the inverter module

Fig.8 presents the global architecture implemented on the FPGA target. This architecture results from modular partitioning, which divides the corresponding algorithm in several reusable modules. In addition to the previous models, a reference generation model is also developed to generate the current reference to be sent to the simulator.

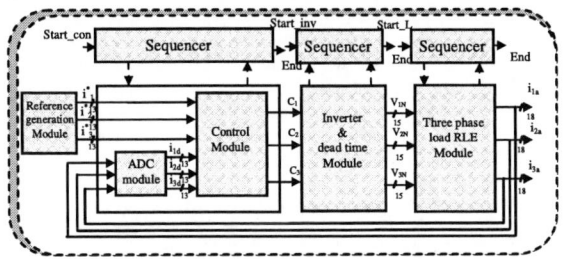

Fig.8.Modular architecture of the inverter module

III. TIME PERFORMANCE AND CONSUMED RESOURCES

Fig.9 presents the sequential timing diagrams corresponding to the Real-Time Simulation. As can be observed, multi-sampling periods are adopted. The inverter and *RLE* load modules require a high frequency (50µs) which runs at a lower to model their dynamics (respectively 3,5µs and 2,5µs), while they are directly

connected to the control module frequency. The computation cycle of different modules are activated by their appropriate start signal. For the control algorithm, this signal activates the global control unit, which activates the ADC module at first. Then, the current control module is activated and it computes the switching states. Note that all the required frequencies were derived from the common clock frequency of 50 MHz. Table II summarizes the FPGA resources used by individual components and the time performances. As it can be noted, the consumed resources rate of the Real-Time Simulator is equal to 20% permitting the implementation of other algorithms such as FDI and reconfiguration algorithms.

TABLE II: LATENCY AND CONSUMED RESOURCES OF EACH MODULE FPGA TARGET SPARTAN XC3S400 FROM XILINX, INC.

Modules	Latency	Execution Time (µs)	Fixed-point format	Consumed resources
RLE **Load**	16	0.32	18Q16	9%
Control	5	0.1	13Q12	7%
ADC	9	2.4	14Q12	1%
Dead time	0	3	17Q15	1%
Inverter	9	0.18	15Q11	2%
Global RT-Simulator				20%

IV. EXPERIMENTAL RESULTS

In order to illustrate the efficiency of the proposed approach the validation of the simulator is carried out through the comparison between the results provided by the developed RT-simulator and those provided by the actual system. The experimental set-up (Fig.10) consists of an autotransformer, a Semikron inverter, a three-phase RLE load and an Xc3s200 FPGA-target. The line currents are measured and converted with AD9221 converters.

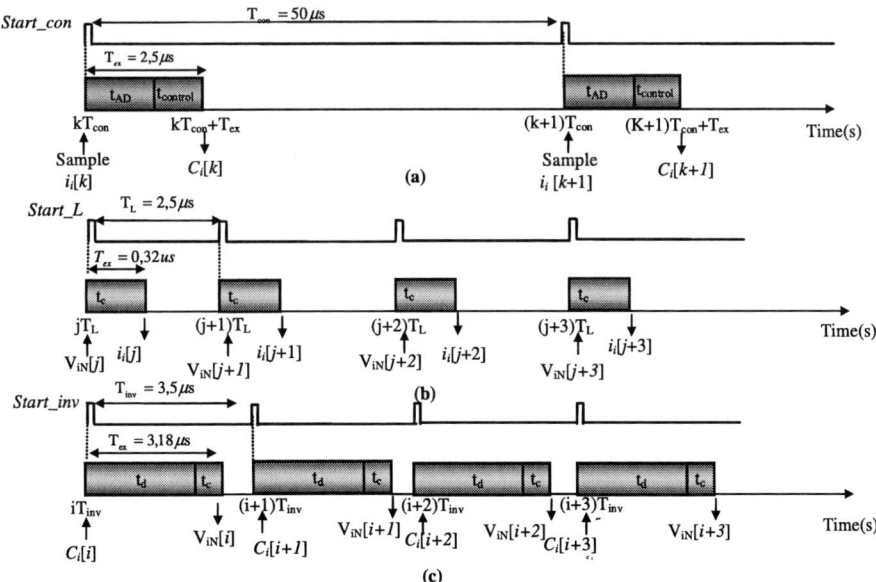

Fig.9. Timing diagrams :(a) of the control module (b) of the *RLE* load module and (c) of the three-phase inverter module

The used sampling period of the control module T_{con} is 50 s. Several simulation tests have been performed in real time on the digital simulator in order to verify the real-time model. The FPGA-based hardware architecture includes the simulator architecture and the serial RS232 interfaces. The simulation results presented in Fig.11.b and Fig.12.b show the current waveform at steady-state and during transient-state. They are obtained by using the RS232, for a DC link voltage equal to 200V and the load parameter indicated in the appendix.

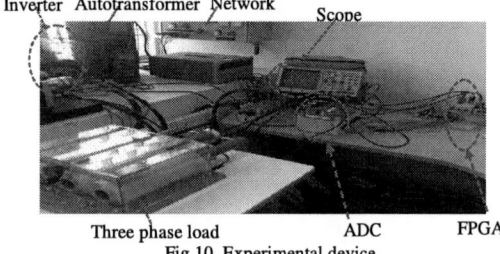

Fig.10. Experimental device

Simulation results have been validated by current response of actual system as shown in Figs. 11 and Fig.12. They clearly prove that the Real-Time Simulator is working properly. As can be observed from the current response, the controller has a fast performance during the transients. The obtained experimental results give proof of the correct functionality of the developed Real-Time Simulator and its ability to achieve efficient results under different operating conditions.

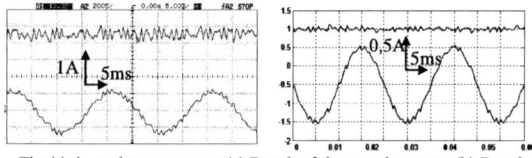

Fig.11. i_{1a} and current error (a) Result of the actual system, (b) Result of the simulator (BW= 0.1A, i*=1A, E=200V, T_{con}=50 s)

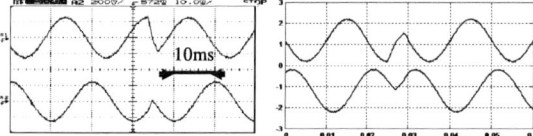

Fig.12. i_{1a} and i_{2a} for current inversion value (-1A to 1A), (a) Result of the actual system, (b) Result of the simulator (BW= 0A, i*=1A, E=200V, T_{con}=50 s).

V. CONCLUSION

The aim of this paper was to introduce the implementation of a Real-Time Simulator as an effective approach allowing design verification of the power electronic controls. The whole simulator was implemented on the FPGA target allowing very high-speed processing. Considering the requirement in terms of time accuracy a multi-sampling approach was adopted. For validation purposes, the proposed Real-Time Simulation was applied for two cases: three-phase RLE load and three-phase inverter. The proposed simulator was also performed by developing the measurement system (ADC) and by using a switching frequency

hysteresis based current controller. The close similarity between Real-Time Simulation and experimental results gave proof of the accuracy and the reliability of the proposed Real-Time Simulator. Such a Real-Time Simulator is useful in the behavioral analysis of real systems, emulation, monitoring and diagnosis.

Appendix

Following lines shows the different coefficients used for the per unit load equations:

$$K_{pu} = \frac{T_e.V_b}{(2.L+R.T_e)I_b} \ , \ K_{iu} = \frac{2.L-R.T_e}{(2.L+R.T_e)}$$

The base values are derived from the rated values by using the following expressions, where E, V are respectively the DC link voltage and the phase voltage. $V_b = E$, $I_b = 20A$ .RLE load Parameters: 220V, 1.4A, 50 Hz, 3Phases, Y connection, R= 12 Ω, L = 0.3 H

REFERENCES

[1] S.H..Kia, H.Henao, G.A.Capolino, "High Resolution Frequency Estimation Method for Three Phase Induction Machine Fault Detection", IEEE Transactions on Industrial Electronics, vol. 54, no. 4, pp.1946-1961, Aug-2007.

[2] I. Bahri, M-W. Naouar, I. Slama-Belkhodja, E. Monmasson, " FPGA-based FDI of faulty current sensor in current controlled PWM converters", IEEE-EUROCON 2007, Sept-2007, Warsaw , Poland 7, CD-ROM.

[3] E. Monmasson, B.Robyns, "Dynamic reconfiguration of control and estimation algorithm for induction motor drivers", IEEE ISIE, Conf.proc vol.1,pp.828-833,Aquila,July2002, Italy.

[4] C.Dufour, S.Abourida, J.Bélanger, "Real-Time Simulation of Permanent Magnet Motor Drive on FPGA Chip for High-Bandwidth Controller Tests and Validation", In Conf. Proc ISIE,2006 ,Montreal, Canada, CD-ROM.

[5] E.Monmasson, M.Cirstea, "FPGAs in Industrial Control Systems– a Review" IEEE Transactions on Industrial Electronics, vol. 54, no. 4, pp. 1824-1842, Aug-2007.

[6] M-W.Naouar, E.Monmasson, A.A.Naassani, I. Slama-Belkhodja,and N. Patin "FPGA-Based Current Controllers for AC Machine Drives – A Review" IEEE Transactions on Industrial Electronics, vol. 54, no. 4, pp. 1907-1925, Aug-2007.

[7] L. Charabi, E. Monmasson, I. Slama-Belkohodja, J-P. Louis, "FPGA Realization of Reconfigurable IP-Core Function for Real-Time Induction Motor Model", Proceedings of the 10th European Conference on Power Electronic and Applications, Sept-2003, Toulouse, France, CD-ROM.

[8] G.G. Parma, V.Dinavahi, "Real time Digital Hardware simulation of power electronics and Drives" IEEE Transactions on Industrial Electronics, vol.22, no.2, April 2007.

[9] P.Le-Huy, S.Guérette, L.A. Dessaint, H.Le-Huy, "Dual-Step Real-Time Simulation of Power Electronic Converters Using an FPGA" in IEEE ISIE 2006, july 9-12, Montreal, Quebec, Canada.

[10] Z. Fang, , J. E. Carletta R. and J. Veillette, " A Methodology for FPGA-Based Control Implementation" IEEE Transactions on Control systems Technology, vol.13, no.6, Nov 2005.

[11] L.Charaabi, E. Monmasson, I. Slama-Belkhodja, "Presentation of an efficient design methodology for FPGA Implementation of Control Systems. Application to the Design of an antiwindup PI controller » IECON 2002. The 28th Annual Conference of the IEEE Industrial Electronics Society, 5-8 November 5-8, 2002; Sevilla – Spain.

[12] D.Menard, O. Sentieys, "Automatic Evaluation of the Accuracy of Fixed-point Algorithms," In Proc. IEEE/ACM Conference on Design, Automation and Test in Europe, 2002, CD-ROM.

[13] T. Grandpierre, C Lavrenne and Y. Sorel, "Optimized Rapid Prototyping for Real-Time Embedded Heterogeneous Multiprocessor", CODES'99, 7th International Workshop on Hardware/ Software Co-design Conf. Rome, May 1999, CD-ROM.

[14] www.analog.com/uploadedFiles/DataDatasheet/AD9221

A New, Ultra-low-cost Power Quality and Energy Measurement Technology

Alex McEachern[1] and Andreas Eberhard[2]

[1] President
[2] Vice President of Technical Services
Power Standards Lab
1201 Marina Village Parkway #101, Alameda, California 94501 (United States)
Phone:++1-510-522-4400, e-mail: Alex@PowerStandards.com, AEberhard@PowerStandards.com

Abstract. IEC 61000-4-30 is an excellent standard that ensures that all compliant power quality instruments, regardless of manufacturer, will produce the same results when connected to the same signal. However, instruments that comply with the Class A requirements of this standard have, until now, been too expensive for common use.

Now a new set of technologies developed by an American company, in cooperation with a Japanese company, demonstrate that it is possible to manufacture three-phase power quality instruments that are fully compliant with the Class A requirements of IEC 61000-4-30 at ultra-low-cost to allow putting this monitoring devices even at entry levels of individual loads.

The development uses technologies from several fields that have not previously been related to power quality, including digital cameras, power-over-ethernet, mobile phones, and submarine sonar systems.

These new technologies have been packaged in a demonstration instrument, and may be licensed to instrument manufacturers as well. This allows gathering power quality and energy consumption information throughout manufacturing facilities or commercial buildings.

Key words

Power quality, instrument, meter, low cost

1. Introduction

Traditionally, power quality instruments have been complex and expensive – often several thousand Euro.

The cost of power quality instruments is driven by five factors:
1. The cost of developing the instruments
2. The quantity of instruments produced – the more instruments that are produced, the lower the development cost in each instrument
3. The cost of manufacturing the instruments
4. The cost of installation, especially the cost of the communication infrastructure

5. The cost of supporting the instruments, especially the cost of supporting special-purpose software, throughout the life of the instruments.

Remarkably, in the last few years, all of these costs have been driven down simultaneously.

This paper describes the technologies in a new, ultra-low-cost power quality instrument, and explains why the costs are so low.

2. Reduction in development costs

Traditional power quality instruments have been developed, from the start, as special purpose instruments. Hardware, firmware, and software have all been developed specifically for that instrument.

However, several developments in other, unrelated industries have made that approach unnecessary.

First, the wide-spread development of digital audio (mobile phones, mobile music players, digital television, etc.) has led to rapid developments in the DSP (digital signal processor) field. New DSP chips are inexpensive, use minimal power, and have built-in analog-to-digital and digital-to-analog conversion. Best of all, they are optimized for processing multiple channels of 20 Hz to 20 kHz signals. By coincidence, power quality measurements are generally made between 50 Hz and 3 kHz – right in the middle of the optimum band. The popularity of the Ipod® means cheaper, better power quality monitors.

Second, the development of relatively complex portable devices (PDA's, mobile phone that also have computer functions, digital cameras, etc.) means that extremely tiny, high-reliability electronic devices are now readily available: connectors with large numbers of pins, tiny op amps, and passive components like resistors and capacitors. Tiny means cheap, in general, if the manufacturing is completely automatic.

978-1-4244-1741-4/08/$25.00 ©2008 IEEE

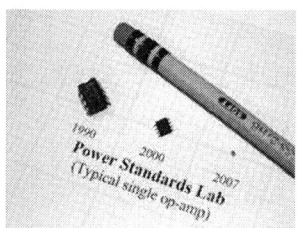

Figure 1: Reduction in component size. Smaller components reduce costs in several ways: smaller printed wiring boards, smaller plastic packages, even smaller power supply requirements. The packages shown are, from left, through-hole DIP, surface mount, and BGA or ball-grid-array.

Third, the availability of high-voltage (1kV), low current op amps, which are generally used for driving submarine sonar transducers, means that automatic test equipment for power quality instruments can be developed far more cheaply now.

Finally, software standards for file structures mean that – just like digital cameras – a power quality instrument developers no longer need to define and support their own file structures. In fact, Windows® text files and web-based graphic file formats have become virtually universal.

Of course, some extraordinarily difficult problems must still be solved by the instrument engineer. How should one deal with a 6kV lightning impulse in such a tiny package? How can one meet the creepage and clearance requirements in the safety standards? Most important of all, by definition, a power quality instrument must work when the power is bad, and other electronic devices are failing. How can an engineer design an instrument that survives? But all of these difficult challenges can be met.

3. Increase in the quantity of power quality instruments

Traditionally, each country defined its own power quality instrument requirements. This meant that an instrument optimized for France, for example, was unlikely to find acceptance in Brazil, for example. Sometimes the situation was even worse: each electric power company would define its own requirements for power quality measurements.

As a result, the quantity for each instrument design was small, and the fraction of the development cost carried in each instrument was large. A rough example: if developing an instrument costs 1.5 million Euros, and the total expected market is 2 000 instruments, each instrument must carry 750 Euros of development costs – a significant but not uncommon burden.

Recent IEC standards [1][2][3][4] have solved this problem.

IEC 61000-4-30[1][2], in particular, has defined power quality measurement methods. Class A in this standard ensures that any two instruments, when connected to the same signal, will produce the same result. Figure 2 and Figure 3 give examples.

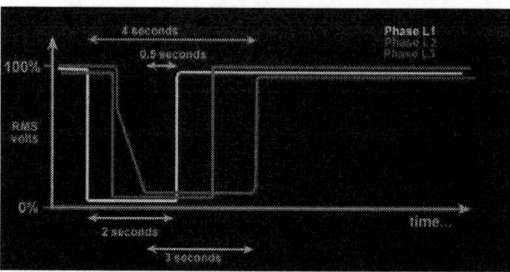

Figure 2. Example of one of the problems solved by IEC 61000-4-30. In the graph of RMS voltages above, what is the duration of the voltage dip? The answers shown range from 0,5 seconds to 4 seconds, and all of them are technically correct. Simply by designating one of these answers as the requirement (4 seconds for dip duration measurement, 0.5 seconds for interruption duration measurement), 61000-4-30 reduces cost of instrumentation.

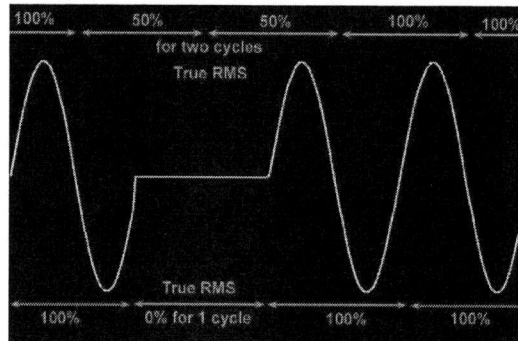

Figure 3. Example of another problem solved by IEC 61000-4-30. Is this dip 50% for two cycles, or 0% for 1 cycle? Again, both answers are technically correct. 61000-4-30 makes it clear that this is a dip to 0% for 1 cycle – thus reducing cost of instrumentation.

As a result of this IEC standardization, an instrument can be designed for world-wide acceptance, and the total quantity is much higher. This means that the development cost burden is much smaller.

4. Decrease in manufacturing costs

Globalization has driven manufacturing costs down.

It is now easy to choose the best sources for parts, world-wide. For example, in the prototype shown in Figure 4, the lower case comes from Germany, the upper case comes from the United States, the display and the memory come from Japan, and the internal electronics are automatically assembled in California using parts from the U.S., Ireland, China, Japan, and other countries.

Fully-automated manufacturing (robots for placing parts, automatic testing systems for verifying that boards are working properly, automatic calibration systems that adjust internal digital constants, etc.) means that manufacturing costs can be kept very low, even in locations with high labor costs like California, without any sacrifice in quality. Indeed, the quality is generally higher than products that are produced in regions with low labor costs, due to the highly-automated production and test procedures.

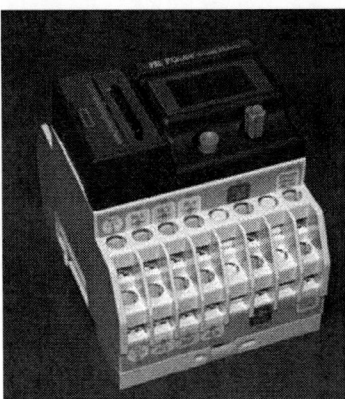

Figure 4. Prototype of the three-phase, voltage-and-current, 61000-4-30 Class A power quality monitor. The digital camera influence can be seen in the SD memory card, which holds up to 2 000 MB of data. Standard DIN-rail mounting means installation is cheap and quick.

5. Decrease in installation and communication costs

Traditional power quality instruments have been designed with their own unique packaging.

However, they are installed in locations where there are low-voltage circuit breakers. By packaging the power quality instrument in a standard 35mm DIN-rail circuit breaker package, installation is greatly simplified.

Perhaps more important, the communication cost of an installed power quality instrument, over the life of the instrument, often exceeds the cost of the instrument itself. Whether the communication is via Ethernet, or telephone modem, or short-distance radio, bringing the communication signal to the monitoring point is a significant cost.

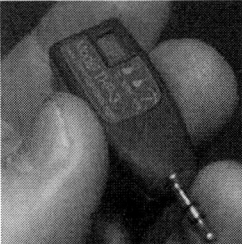

Figure 5. Probes for parameters that may be related to power quality are included: Temperature, Humidity, Barometric Pressure, etc. A GPS satellite receiver can ensure precise timing.

(We should not forget the additional hidden cost of the damage that can be caused to the communication network, during major power disturbances. In regions with strong lightning activity, for example, telephone modems have traditionally been damaged through their connection to power quality monitors.)

Furthermore, in many power quality monitoring applications, immediate communication is not necessary.

The prototype shown in Figure 4 has Ethernet connectivity (and includes a web server, an FTP server, and an e-mail generator), wireless radio connectivity, and a modem connection. However, it is optimized to function without connectivity – it can easily store a year's data on a removable SD memory card.

6. Decrease in the cost of supporting power quality instruments

Although it is often hidden from the end user, the lifetime support cost for traditional power quality instruments is significant.

This is especially true for the special-purpose software that was written for each instrument. Often, this software was written for Windows®; and the lifetime of a power quality instrument greatly exceeds the lifetime of one release of Windows®. For example, many power quality instruments are still in use that were originally issued with Windows® 3.1 software.

The lifetime costs of upgrading and supporting this software was a major cost.

However, such software is no longer necessary. By following the software model of a digital camera, the prototype power quality instrument of Figure 4 requires absolutely no software. When you connect a digital camera to your computer, you immediately see the pictures in a folder on your disk drive. The same is true for power quality data in the prototype instrument.

By eliminating the need for any software at all, we drive down the costs even further.

Again, this does not eliminate all firmware challenges. The instrument firmware must still support a wide variety of languages and character sets (Japanese, Korean, etc.), and the CSV (comma separated variable) files for spread sheets must work with European systems that use the comma symbol for other purposes. But all of these problems have been previously solved in digital cameras.

7. Conclusion

Recent changes in standards and technology have made it possible to produce a 61000-4-30 Class A power quality instrument at a very low cost.

A prototype has been constructed, and a fully-automated manufacturing system is being developed.

Acknowledgement

The first-named author (McEachern) is the Convenor of IEC TC77A Working Group 09, which is responsible for maintaining IEC 61000-4-30 Power Quality Measurement Methods. The second-named author (Eberhard) is well known in the international standard and power quality community. He is member of various power quality and EMC standard committees around the world We gratefully acknowledge the useful and knowledgeable technical contributions of members of this Working Group during discussions and exchanges over the last 10 years. Any errors that remain in this paper, are, of course, entirely the responsibility of the authors.

References

[1] IEC 61000-4-30, Ed 1, "Testing and measurement techniques – Power Quality Measurement Methods". International Electrotechnical Commission. February 2003.
[2] Corrigendum 1, IEC 61000-4-30, Ed 1. August 2006.
[3] IEC 61000-4-15, Ed 1.1, "Flickermeter – Functional and design specifications". International Electrotechnical Commission. 2003.
[4] IEC 61000-4-7 Ed 2, "General guide on harmonics and interharmonics measurements and instrumentation". International Electrotechnical Commission. 2002.

Rotor Time Constant Adaptation Using Radial Basis Function Network

Pavel Brandštetter[*], Ondřej Škuta[†]

[*] VSB – Technical University of Ostrava, Ostrava, Czech Republic, e-mail: *pavel.brandstetter@vsb.cz*
[†] VSB – Technical University of Ostrava, Ostrava, Czech Republic, e-mail: *ondrej.skuta.fei@vsb.cz*

Abstract—Our intention here was to highlight a replacement of adaptation algorithm in MRAS by the help of alternative artificial neural network (ANN) which has received great attention in recent years. The main objective was to find and design some alternative neural network within the electric drive control. After a short discussion of hardware components, an overview of Radial Basis Function (RBF) neural networks will be given. Digital signal processors TMS320F2812 are used for these electric drives control applications. The hardware accessories used within the electric control drive included: interface board for the signal processor kit--developed in our department, and an 8bit data-transfer microprocessor for data acquisition. The interface of the DSP is a general-purpose control system for power converters in the electric drives. The next section briefly outlines the estimation of the rotor time constant, which is necessary for the so-called current model. The current model is used in the vector control of the induction motor and is utilized to determine the quantities for the transformation from the stationary reference frame into the reference frame, which is oriented on the rotor flux space vector. The estimation of the rotor time constant for the adaptive model of MRAS is created with the support of a PI-controller which is then replaced with the Radial Basis Function network. The final section presents simulations results, which have been performed in the Matlab-Simulink software.

Keywords—Digital Signal Processor, Induction Motor, Electric Drives, Vector Control, Neural Networks Architecture, Radial Basis Function

I. INTRODUCTION

The demand for electrical control drives its still increasing. This is supported by the technological and economical development which has led to the decrease in expense and increase in the efficiency of electromechanical conversion. The efficiency of electromechanical conversion is also influenced by the control quality connected with the dynamics of the given system of conversion, which is ensured by modern control algorithms realized by high performance microcontrollers.

As digital signal processors become more common throughout the electric drive control, new algorithms could be used and applied. Digital Signal Processor (DSP) TMS320F2812 is one of these processors, and is a member of the TMS320C28x DSP generation. It is a highly integrated, high-performance solution for the demanding control applications. This DSP provide the high computational performance (e.g. for powerful vector control algorithms of electric drives) and has important peripherals for the control of electrical converters. DSP are very efficient C/C++ engine that enable for the users develop their own system control software in a high-level language. It allows a complex mathematical algorithm to be developed using C/C++.

Artificial intelligence has become fairly common in various fields over the last few decades. But it is relatively recent, and still to a very limited extend, that their potential in association with electric machines and drives has come to be realized and explored.

Artificial neural networks (ANN) are mainly used in these types of application, where the realization of another methods would be very difficult, expensive or even unrealizable. In these applications there is possible to take the advantage of the main features of neural networks, namely: approximation ability of different nonlinear functions, possibility to set their parameters in virtue of the experimental or learning data set, the quickness of information processing and their robustness. There is no necessary mathematical or structure description, there is possible to solve the problem just like the black box task with their inputs and outputs.

The Radial Basis Functions (RBF) emerged as a variant of artificial neural network (ANN) in late 80`s by Broomhead and Love and their work opened another ANN frontier. RBF network is a type of ANN for applications to solve problems of supervised learning regression, classification and time series prediction. The radial basis functions are powerful techniques which are built into a distance criterion with a respect to the centre. Such networks have 3 layers, the input layer, the hidden layer with the RBF non-linearity and the linear output layer. RBF networks have the advantage of non suffering from local minima in the same way as multilayer perceptrons. The most popular choice for the non-linearity is the Gaussian. The output layer is in regression problems a linear combination of hidden layer values representing mean predicted output. The RBF architecture is discussed in [1] and the training algorithms and methods are mentioned in paper [2].

In most cases, it presents higher training speed when compared with ANN based on back-propagation training methods, easier optimization of performance since the only parameter that can be used to modify its structure is the number of neurons in the hidden layer etc...

Rotor time constant adaptation methods are used in the modern control of induction drive. The value of rotor resistor changes in dependence on drive load. To improve the motor power its necessary the identification of these parameters and adjusts them.

978-1-4244-1741-4/08/$25.00 ©2008 IEEE

II. CONTROL SYSTEM

A. Characteristics of the DSP

The starter kit includes the development kit and Code Composer Studio, which is fully compatible and enables the engineers to simplify the source code. This software tool chain provides the user with a C compiler, assembler, linker, and a Windows-based debugger. The Processor TMS320F2812 is distinguished by the following characteristics:

- High-Performance static CMOS technology
- 150 MHz (6.67-ns Cycle Time)
- On-Chip memory
- Boot ROM (4K x 16)
- Three external interrupts
- Peripheral interrupt expansion (PIE) block
- Three 32-Bit CPU-timers
- Motor control peripherals
- Two event managers (EVA, EVB)
- Serial port peripherals
- Serial peripheral interface (SPI)
- Enhanced controller area network (eCAN)
- 12-Bit ADC, 16 channels
- Up to 56 General Purpose I/O (GPIO) Pins

B. Interface for a Kit with the DSP TMS320F2812

The microprocessor system with TMS320F2812 signal processor there has been used in the Department of Power Electronics and Electrical Drives recently. This electric control drive system with DSP is going to replace the previous versions. These control systems were based on microprocessors with slide-in cards. The original development kit with DSP can not be immediately used, hence there have been developed an interface for this kit (Fig. 1.) in the department. It allows the use of system in the control of a semiconductor converters and modern electric drives.

Inputs of the A/D converter were boosted and the voltage level was increased from 0 - 3V to +10V. In that point A/D converters are prepared for the connection with the current and voltage sensors utilized in power electronics. PWM outputs were boosted too and they could be directly connected to the board of IGBT drivers. This board was likewise developed in the department and is used in the majority of the currently used converters. There ware used opto-isolation elements for the Capture units of the DSP. The interface was also equipped with a serial line (SCI), which was connected to a four-channel, twelve-bit D/A converter. The D/A is important and very useful instrument for tracing internal variables, e.g. the flux in the vector control.

C. Data acquisition system board

By the reasons of computational demand and memory needs of neural network training process, there is appropriate perform the training process on personal computer. To obtain needful amount of training data, which describe all necessary drive states, there is necessary high transfer speed. The core of the transfer system is an 8-bit processor with an USB interface. This processor is especially designed for the fast communication between processor and PC. There are a few modes for the communication over the parallel bus-bar and USB port.

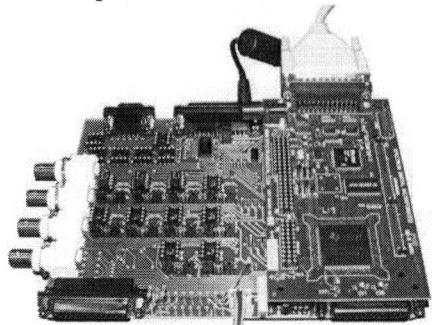

Fig. 1. Interface board for the DSP kit

In the figure 2 there is shown the simple block structure of training data acquisition. Needed data are sent toward DSP through the data acquisition board. Very important part of the communication system is an external interface of DSP for an external communication. Part of a DSP is a non-multiplexed asynchronous bus-bar. External interface communication is divided on the processor kit into seven parts. There are used two bits of address bus and 16bit of data bus. The processing data are sent step-by-step on the data bus-bar. In the personal computer there was made a program to save accepting data. Data are subsequently standardized to the necessary norm used for the artificial neural network training process.

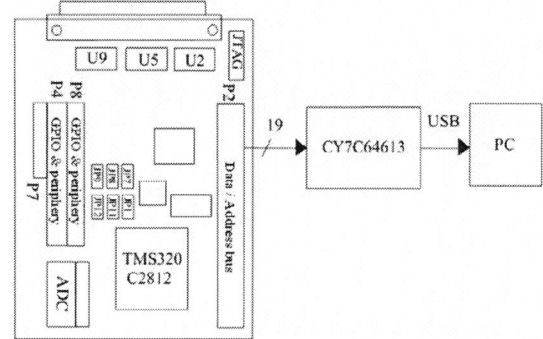

Fig. 2. Block structure of data acquisition system

III. MODERN AC DRIVE CONTROL

On the figure 4 there is shown the most common application of DSP in the control of an induction motor with voltage source inverter and modern voltage source rectifier. This system configuration of an electric drive is the recently most used.

There is indicated utilization of some DSP units, which are relevant for the control of the whole drive. Request values (e.g. the speed...) are set on PC, with user friendly interface made in LabView, through serial communication to the DSP. There are selected and traced some important values of the program to Oscilloscope via D/A converters.

$$\Phi(e) = e_\alpha \left(L_m i_{S\alpha} - \hat{\Psi}_{R\alpha} \right) + e_\beta \left(L_m i_{S\beta} - \hat{\Psi}_{R\beta} \right) \quad (1)$$

$$e_\alpha = \left(\Psi_{R\alpha} - \hat{\Psi}_{R\alpha} \right), \quad e_\beta = \left(\Psi_{R\beta} - \hat{\Psi}_{R\beta} \right) \quad (2)$$

$$\frac{1}{\hat{T}_R} = K_1 \Phi(e) + K_2 \int \Phi(e)\, dt \quad (3)$$

$$\text{, where } K_1 > 0, \quad K_2 > 0 \quad (4)$$

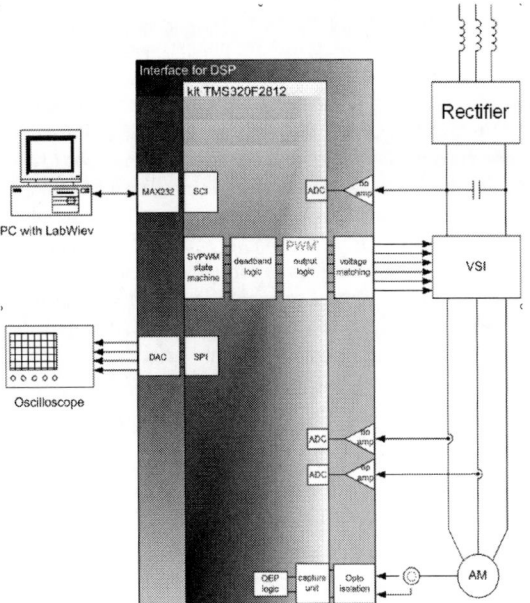

Fig. 3. Control of an induction motor

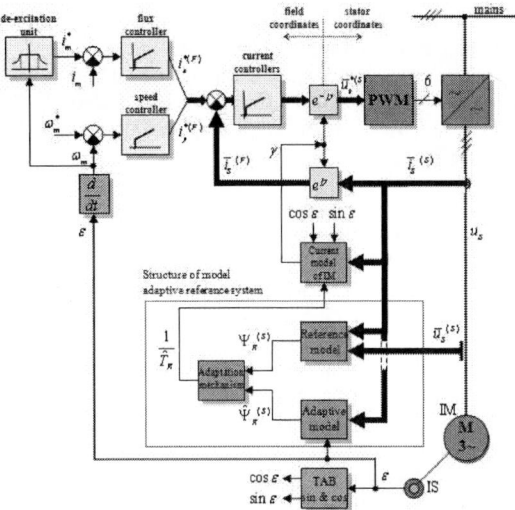

Fig. 4. Control structure with adaptive system

This DSP has a deadband logic which facilitate the control of the converters, because the preset short time period is inserted to the PWM signals automatically, so in this way it is inhibit the short of the DC circuit. The SVPWM state machine has switching combinations for voltage source inverter. Further, there is indicated the utilization of the speed sensor connection to the capture unit through the opto-isolation element. The QEP facilitate the evaluation of the pulse signals from the speed sensor.

IV. ROTOR TIME CONSTANT ADAPTATION

For determination of the value and position of the magnetizing current vector or rotor flux vector is often used so-called current model. This model contains the rotor time constant which is a changing parameter. In order to ensure optimal vector control of induction motor, the rotor time constant in the flux model has to be corrected continuously.

The adaptive reference system model is based on comparison of two estimators. The first one, so-called the reference model doesn't include rotor time constant. The other, which contains rotor time constant, is the adaptive model. The error between them is used to derive an adaptation algorithm that produces the estimated value of rotor time constant in current model, which is used in vector control of induction motor drive (Fig.4). This model contains the rotor time constant which is a changing parameter. In order to ensure the optimal vector control of induction motor, the rotor time constant has to be adjusting continuously.

To derive adaptation algorithm there were used equation of voltage model of rotor flux and the equations of current model of rotor flux. It was well described in the work [3] and the adaptation algorithm it's described by the following equations:

The adaptation mechanism consists of evaluation of adaptation signal (Eq.1) and its sequential minimalization by the help PI-controller (Eq.3).

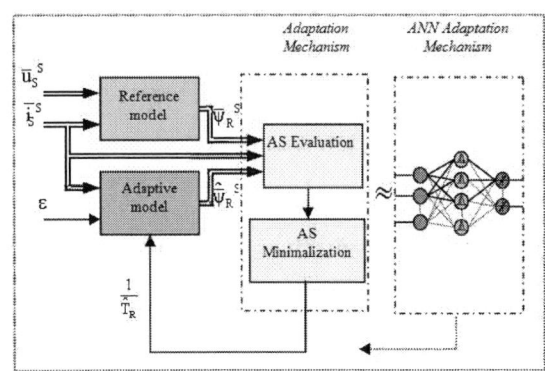

Fig. 5. Adaptation mechanism

Figure 5 shows the structure of the model reference adaptive system and also the substitution of the adaptation mechanism with the artificial neural network. Here we also can take advantage of one of the important features of ANN, which means that we don't need to know the mathematical description of the problem. We can consider Adaptation mechanism as a black box, with an input section and an output section. This is where we obtain important input/output data, which we need for the further training stage of neural networks.

V. RADIAL BASIS NEURAL NETWORK

A. Comparison of RBF Networks and Multilayer Perceptrons

Radial-basis function networks and multilayer perceptions are examples of nonlinear layered feed forward networks. They are both universal approximators. It is therefore not surprising to find that there always exists an RBF network capable of accurately mimicking a specified MLP, or vice versa. However, these two networks differ from each other in several important respects.

- An RBF network (in its most basic form) has a single hidden layer, whereas an MLP may have one or more hidden layers.
- Typically the computation nodes of an MLP, located in a hidden or an output layer, share a common neural model. On the other hand, the computation nodes in the hidden layer of an RBF networks are quite different and serve a different purpose from those in the output layer of the network.
- The hidden layer of an RBF network is nonlinear, whereas the output layer is linear. The hidden and output layers of an MLP are usually all nonlinear.
- The arguments of the activation function of each hidden unit in RBF network computes the Euclidean norm (distance) between the input vector and the center of that unit. Meanwhile, the activation function of each hidden unit in an MLP computes the inner product of the input vector and the synaptic weight vector of that unit.

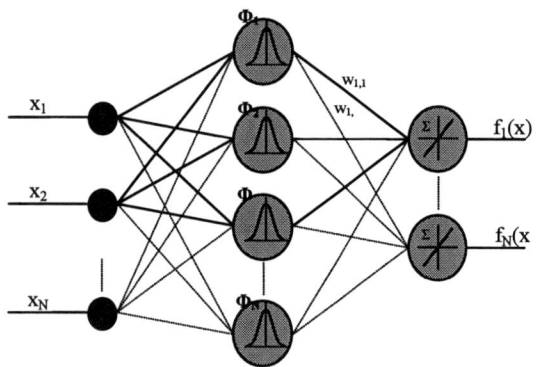

Fig. 6. Radial-basis function networks

B. Radial Basis Neural Network Training

It is interesting to compare subset selection with the standard way of optimizing neural networks. The latter involves the optimization, by gradient descent, of a non-linear sum-squared-error surface in a high-dimensional space defined by the network parameters. In RBF networks the network parameters are the centers, sizes and hidden-to-output weights. In subset selection the optimization algorithm searches in the discrete space of subsets of a set of hidden units with fixed centers and sizes and tries to find the subset with the lowest prediction error. To find the best subset is usually intractable (there are 2^M-1 subsets in a set of size M) so a heuristic method must be used to search a small but hopefully interesting fraction of the space of the entire subset. One of these heuristics is forward selection which starts with an empty subset to which is added one basis function at a time, the one which most reduces the sum-squared error, until some chosen criterion stops decreasing.

Forward criterion is also, of course, non-linear type of algorithm which has the following advantages:

- There is no need to fix the number of hidden units in advance
- The model selection criteria are tractable
- The computational requirements are relatively low

When applied to supervised learning with linear models, the least square principles lead to a particularly easy optimization problem. The choice of basis function can be based on finding the greatest decrease in sum-squared-error.

To decide when to stop adding further basis functions one of the model selection criteria must be chosen: unbiased estimate of variance (UEV), final prediction error (FPE), generalized cross-validation (GCV) or Bayesian information criterion (BIC).

Forward selection is a relatively fast algorithm but it can be sped up or improved by using some techniques, for example orthogonal least squares or combination of forward selection and ridge regression.

C. Usage of different training methods of RBF networks

Different types of training algorithms were tested and evaluated as the most fitting. Three training algorithms were used to test the main features of RBF neural networks:

- Forward subset selection
- Ridge regression
- Regression trees 1 & 2

From these training algorithms there were picked like a useful for ours purpose just the Forward subset selection algorithm. The other methods should be useful for some other problems. Figures depicts RBF networks curve, which where trained by virtue of a Forward subset selection algorithm. This algorithm was variously modified together with changes of the RBF network (e.g. activation function, radius...). Then different amount and types of training data sets were presented. In the followed figures there are always two curves. The first curve, the lighter one, shows the differences between the adaptation model output and the feed-forward ANN network. The

1378

second one shows the differences between the adaptation model output and the RBF network.

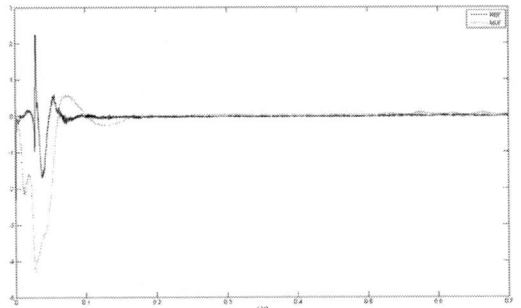

Fig. 7. Radial-basis function networks

First variation of RBF network (Fig. 7) contains 100 RBF units (neurons). This network represents better adaptation features with the variance of controlled system, but unfortunately it doesn't obtain such an accurate estimation compared to other RBF networks. Although, accuracy of estimation does achieve better results than normal feed-forward neural network. The second RBF network (Fig.8.) achieves the best curves of estimation of the rotor time constant, but the price of this accuracy is the highest number of the RBF neurons (136) in the hidden layer. It reflects higher computation power of RBF network.

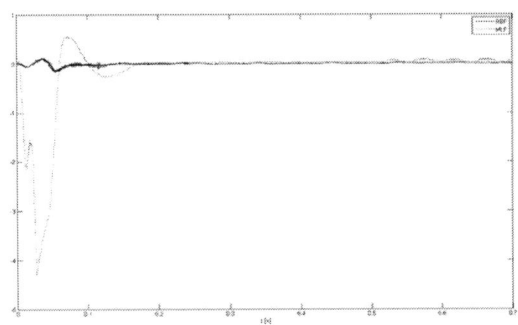

Fig. 8. Radial-basis function networks

D. Usage of different architectures of RBF networks

The aim of this work wasn't to show the all features of Radial Basis network with different architectures in the rotor time constant adaptation, but only the way of choosing a network. That's why not all obtained curves necessary to decide which architecture of proposed network is the best one are shown here, as they are in the [4] paper. RBF networks were tested, for example with one feedback (see below), two feedbacks, without feedback or with White noise added to the feedback connection.

RBF network with one feedback connection

The showed type is the most used and common RBF network with one feedback without scaling and without the white noise. The figure 9 depicts the RBF architecture

with the appropriate input variables. There are always three layers: input, hidden layer with the non-linear activation function and the output linear layer.

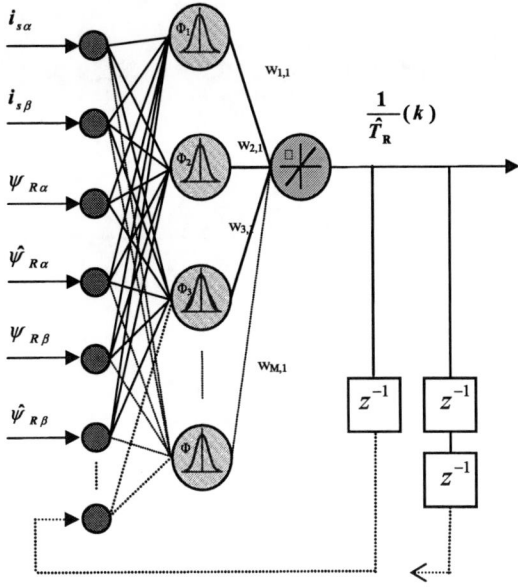

Fig. 9. Architecture of RBF neural network

There are the input data for the adaptation mechanism, which were also used like an input training data set for the neural network, in the figure 10. Output or we can say the desired output time behavior is always depicted in the figures by the red dotted line.

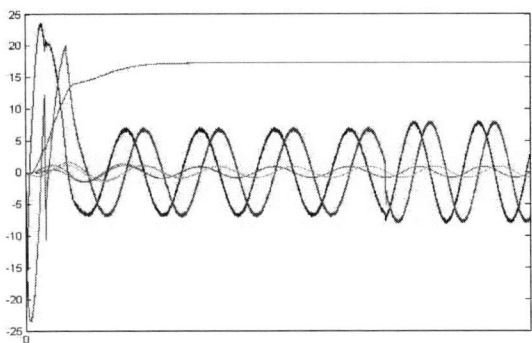

Fig. 10. Input training data set

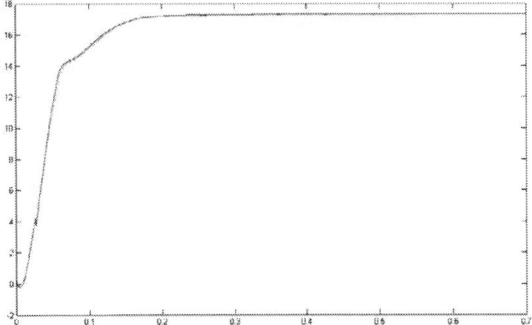

Fig. 11. Output $1/T_R$ from RBF and AM

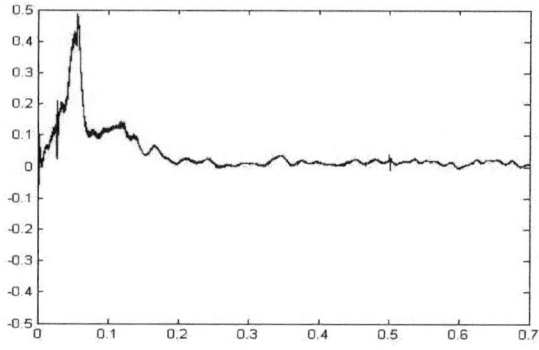

Fig. 12. Difference between RBF and AM output

In the first RBF neural network there were used 97 RBF units. The output time behavior is perfect as we can see in the figure 11, the difference between the adaptation mechanism and the RBF network is really neglect able (Fig.12). Next network was used with thinly lay-out of 33 RBF units and the response is also quite good (Fig.13).

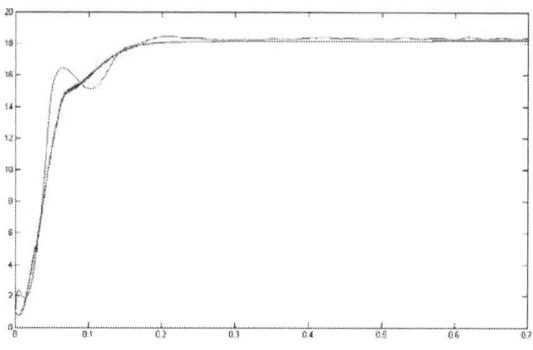

Fig. 13. Output $1/T_R$ from RBF and AM

RBF network with no feedback connection

The next architecture of the RBF network didn't include the feedback. In figure 14 it is possible to see that the output behavior is not as expected. The network contains 81 RBF units. In the next figure there is obvious improvement of the output curve, but at the cost of a high number (261) of the RBF units.

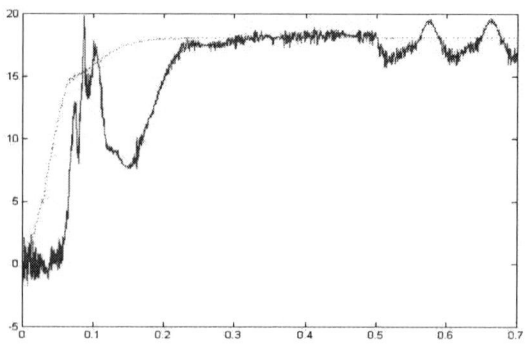

Fig. 14. Output $1/T_R$ from RBF and AM

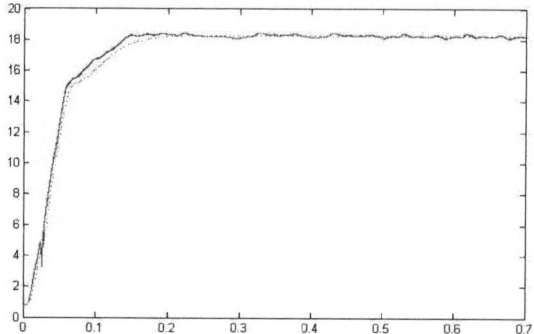

Fig. 15. Output $1/T_R$ from RBF and AM

RBF network with two feedback connection

Below, the RBF architecture with two feedbacks connections is described. As we can see the output time behavior is also very good as in the first case. There were 120 RBF units and the next was used with thin lay-out of 51 RBF units and the response is also quite good (Fig.16). Denser lay-out of RBF network disposes with 261 RBF and is the same problem as in network with one feedback.

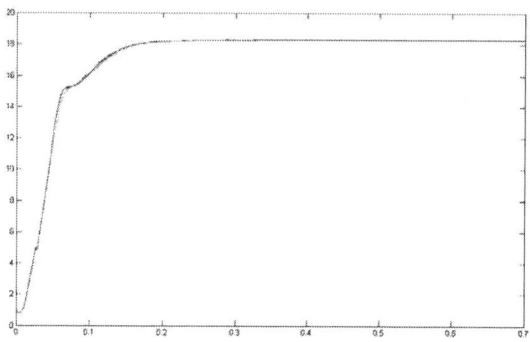

Fig. 16.: Output variables from RBF and AM

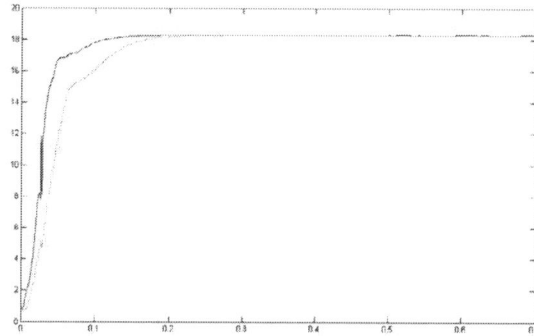

Fig. 17.: Output variables from RBF and AM

RBF network with two feedback connection

The next architecture comes from idea of feed-forward architecture, where the input values must be scaled because of their activation function. In the figure 18 the

1380

input scaled training data set for RBF neural network is depicted.

Only a dense (436 units) RBF unit lay-out provides an output curve as in the un-scaled networks. The classical one, with 116 units, can also be considered well enough (fig.19), but network with thin lay-out (32units) has a low-quality output curve.

There was one important difference in the lower values of inner network parameters such as radius, centers and weights.

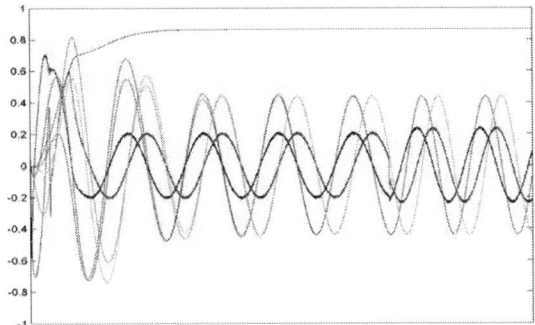

Fig. 18.: Input training data set

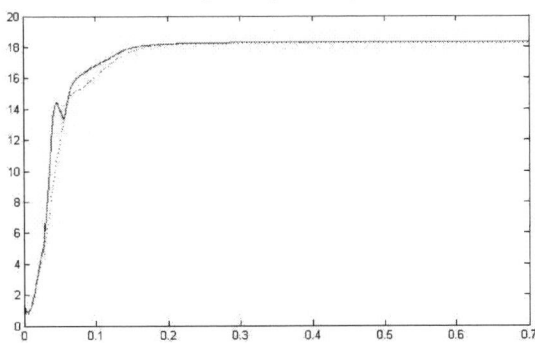

Fig. 19.: Output variables from RBF and AM

RBF network with White noise

The idea of an addition of the bounded White noise was to reduce the weight and centers importance of the feedback connection. In the feed-forward neural networks it has the less neural hidden unit foundation.

Only one type of RBF architecture with the classic lay-out of RBF units was used. The RBF neural network contains 215 activation units and so it was useless to continue with this type.

VI. CONCLUSION

This article discusses gives an overview of a digital signal processor TMS320F2812 in the control of electrical drives. This processor can be used in modern control methods of electric drives, such as vector control and direct torque control.

The question of ANN application has been thoroughly researched over the last few years, although just a few

applications regarding electric drives were designed. The above presented paper deals with an implementation method of new ANN in the control of the electrical drives. Choosing and testing procedure of alternative artificial neural networks is described. The presented implementation possibilities of radial basis function networks instead of feed-forward neural networks is advantageous but unfortunately has it's drawbacks as well. One downfall worth noting is the linear output combination of the values from the hidden nonlinear layer and their appropriate weights. A clear disadvantage seems to be a higher number of neural units.

In order to validate the work we have carried out, a more in-depth experimental investigation into research and application of RBF neural networks in the control of electrical drives is needed. In the near future a speed estimator will be created with the use of this network and important features will then be discussed.

ACKNOWLEDGMENT

The research was supported by the Grant Agency of the Czech Republic with project No. 102/08/0775.

REFERENCES

[1] O. Škuta, *Estimation of time-varying parameter of induction motor using radial basis function network*, Workshop of Faculty of Electrical Engineering and Computer Science WOFEX, Ostrava, 2006, pp. 99-104 ISBN 80-248-1152-9.

[2] O. Škuta, *Application of radial neural networks in alternate electric drives* In Conference text-book EPVE 2006, Brno (Czech Rep.): VUT Brno, ISBN 80-214-3286-1.

[3] R. Čajka, "Application of artificial intelligence in electrical drives", dissertation work, Department of Electronic, VŠB-TUO, November 2006

[4] P. Brandštetter, O. Škuta, *Variations of RBF network in rotor time constant adaptation* In 14th international symposiums on power electronics – Ee 2007

[5] P.Vas, *Artificial-intelligence-based Electrical Machines and Drives*, 1ST ED. Oxford: University Press, 1999, 630pp. ISBN 0-19-856465-1.

[6] P. Vas, *Sensorless vector and direct torque control*, 1ST ED. Oxford: University Press, 1998, 732pp. ISBN 0-19-859397 X

[7] S. Haykin, *Neural Network a Comprehensive Foundation*, 2ST ED New Jersey:Prentice-Hall, 1999, 850pp. ISBN 0-13-273350-1

[8] Orr, J.L. *Introduction to Radial Basis Function Networks*, 2ST ED. Edinburg: University of Edinburg, 1996, 68pp.

[9] Orr, J.L. *Matlab Functions for Radial Basis Function Network*, 3ST ED. Edinburg: University of Edinburg, 1996, 70pp.

[10] Anil, J., Jianchang, M. Artificial Neural Network: A Tutorial, *IEEE Computer Magazine*, 1996, Michigan(USA): Michigan State University, ISSN 0018-18-9162-96

Application of Speed and Load Torque Observers in High Speed Train*

Jaroslaw Guzinski[1], Marc Diguet[2], Zbigniew Krzemiński[1], Arkadiusz Lewicki[1], Haithem Abu-Rub[3]

[1] Gdansk University of Technology/Faculty of Electrical and Control Engineering, Gdansk, Poland, e-mail: *j.guzinski@ely.pg.gda.pl, z.krzeminski@ely.pg.gda.pl, a.lewicki@ely.pg.gda.pl,*
[2] Alstom Transport, Tarbes, France, e-mail: *marc.diguet@transport.alstom.com*
[3] Texas A&M University at Qatar, Doha, Qatar, e-mail: *haitham.abu-rub@qatar.tamu.edu*

Abstract— This paper presents a solution for observer based diagnostic system in high speed train with induction motor. The concept of the diagnostic system is dedicated to the mechanical part of the drive. The system is monitoring both motor speed sensor and torque transmission system. In case of speed sensor fault the control system could be switched into sensorless mode. Monitoring of torque transmission system allows to limit the motor torque in case of serious problem and gives an information to maintenance center for testing the suspected gear. The proposed system was verified by simulation and in test bench for 1.2 MW rail induction motor.

Keywords— rail vehicle, induction motor, sensorless control, diagnostics, mechatronics.

I. INTRODUCTION

The development of the rail transport is strongly dependent on the reliability of the rail vehicles. Nowadays, a lot of rail systems are equipped with diagnostic tools which assure safe and reliable passenger and freight transport. Most of the diagnostic systems are equipped using additional sensors. But with adding extra sensors the vehicle is less reliable and more complicated to maintain and the cost of the vehicle increases also. New modern microprocessor technique allows to eliminate additional sensors and to replace the diagnostic systems with calculation methods.

One of the most important part of the rail vehicle is the main drive and torque transmission system. Typically, the traction motors are equipped with speed sensors which are used in the control process of the drive. In case of sensor fault (e.g. disconnecting of wiring system) the drive of the train is switched off. Usually the rail vehicle is powered by few motors and few partially independent drives, while disconnecting one of the motor has an influence on the train speed and travel time. In the traction drive, it is possible to use speed calculations instead of speed measurement. A lot of speed observer systems were been presented in the literature [1-8]. Using calculated speed, it is possible to perform fully speed sensorless control system or to switch to the speed sensorless system in case of speed sensor fault.

The torque transmission system transmits the torque from motor shaft to the train wheels. It contains connection of shafts, gears and coupling. In the transmission system, the problems may appear even in

*This work was supported by the European Commission under TOK/IAP project from Marie Curie Actions.

new trains as a result of manufacturing or assembly errors e.g. misalignment of some parts of the transmission system [9-13]. After an initial period, the problems with the transmission may appear as a result of wearing and material consumption. A wear of the gears causes growing amplitudes of the frequencies related to the gear meshing. To identify problems with the transmission system ,typically an analysis of the motor or gear casing vibration, measured by the accelerometers, are done [14-22]. Instead of the vibration measurement it is possible to use the load torque observers calculation [23-29].

When the load torque is known, in case of some faults it is possible to change the limits in the traction drive control system to assure safety work of operation [30-32]. Even when load torque is measured or calculated it is difficult to unequivocally identify faults [33-34]. To define the torque transmission faults criteria it is indispensable to build a data base with a load torque waveforms for healthy and damaged mechanical system and after that to define the faults criteria.

In case of some transmission faults the train could be stopped. Such faults have also a serious influence on the safety of the train and passengers. This is a reason of the necessity to introduce the proposed diagnostic system.

The diagnostic system proposed in this paper is intended to be used to monitor the motor speed sensor and the torque transmission system. The diagnostic system is sensorless. It means that no additional sensors are used except those previously used in the rail vehicle for control purposes. The proposed system was verified in simulation and applied on test bench for a drive of 1.2MW induction motor.

II. RAIL SYSTEM DESCRIPTION

In the locomotive or in electric multiple unit few electric motors propel the train. Previously used DC motors and wired wounded induction motors are nowadays replaced by squirrel cage induction motors in mature applications or permanent magnet synchronous motors in the new rail systems. Typically one motor propels one axle in every motorized bogie. In one bogie, one, two or three motor exist. One motorized car typically has two bogies with motors.

In this paper the speed and load torque observers are applied in a high speed train (HST) drive for future diagnostic purposes. Observers systems were investigated in HST, which has two powered cars: at the beginning and at the end of multiple unit. Each motorized coach has two bogies. In each bogie each axle is propelled be one electric motor. Scheme of the HST train is presented in Fig. 1.

978-1-4244-1741-4/08/$25.00 ©2008 IEEE

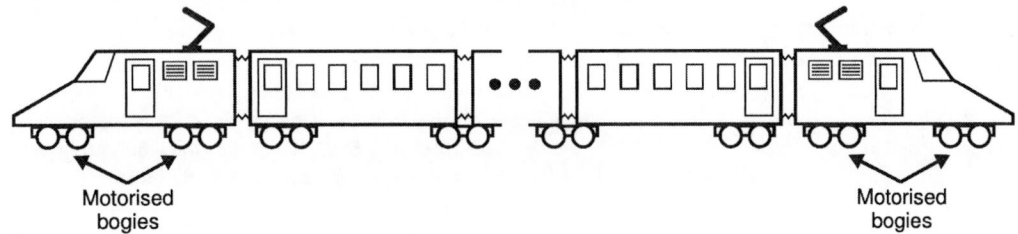

Fig. 1. Electric multiple unit of high speed train.

In the system presented in the paper 1.2 MW induction motors are used in HST. Schematic diagram of part of one from HST bogie transmission system is presented in Fig. 2.

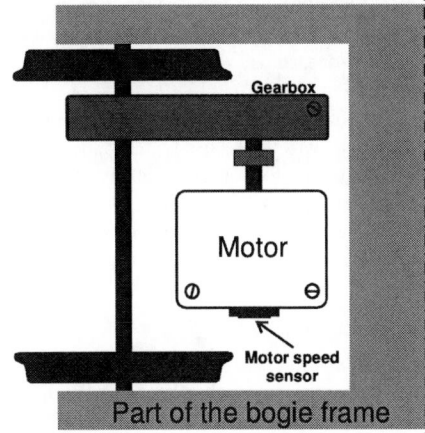

Fig. 2. Part of the HCT train bogie.

HST transmission system is complex. The motor torque is transmitted from the motor shaft to the wheels using two gears and few couplings. The motor is fixed to the car body. The position between the car body frame and the bogie frame changes during train track. To compensate such changes in the transmission system, a sliding axle and the cardans are used. On the induction motor shaft the speed sensor is mounted. It is active sensor with 240 pulses per revolution. The measured speed is used in the motor control system. Additional passive speed sensors are mounted on gears shafts and are used only in the system for anti-slip control of the wheels.

The rolling stock of the coach moves by the adhesion force between rail and driving wheel. Approximated relation between the adhesion force and the slip velocity is presented in Fig. 3 [8].

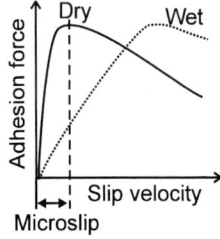

Fig. 3. Relation beteen adhesion force and slip velocity of the coach

In the slip velocity the two areas are distinguished: stable (microslip) and instable (macroslip). In the investigated HST the stable area is limited to the slip up to 2%. In the coach each driven axle has different slip phenomena mainly depending on current axle load. In the investigated HST each driving axle is propelled by one electric motor which is connected to the separate inverter. Such trend simplifies the torque control of the motor to prevent macroslip. Adhesion limit for high speed trains imposed by European Technical Specification Interoperability is 0.15 for speed up to 200km/h and decreases for higher speeds.

III. HST SIMULATION MODEL

A. Traction induction motor

In the investigated system of the HST the induction motor designed specially for the traction application was used. The power, voltage, nominal speed and maximum speed of the tested HST motor are respectively: 1.2 MW, 810 V, 280 rad/s, 397 rad/s. With used in HST gear and wheels, the motor nominal and maximum speeds are related to the train speeds of the values 225km/h and 320 km/h respectively.

The most popular mathematical model of the induction motor is the model described in the rectangular coordinates (xy). As the state variables could be choosen as: stator current and rotor flux components and motor mechanical speed noted as i_{sx}, i_{sy}, ψ_{rx}, ψ_{ry} and ω_r respectively. In this paper the parameters in per unit system are shown in Tab. I [1, 5].

TABLE I
DEFINITION OF PER UNIT VALUES

Definition	Description
$U_b = \sqrt{3} U_n$	base voltage
$I_b = \sqrt{3} I_n$	base current
$Z_b = U_b / I_b$	base impedance
$T_b = (U_b I_b p) / \omega_0$	base torque
$\Psi_b = U_b / \omega_0$	base flux
$\omega_b = \omega_0 / p$	base mechanical speed
$L_b = \Psi_b / I_b$	base inductance
$J_b = T_b / (\omega_b \omega_0)$	base inertia
$\tau = \omega_0 t$	relative time

where: t – real time, $\omega_0 = 2\pi f_n$ – grid angular frequency, f_n – motor nominal frequency.

Equations of the induction motor in rectangular frame of references xy rotating with angular speed ω_a have the form:

$$\frac{di_{sx}}{d\tau} = -\frac{R_s L_r^2 + R_r L_m^2}{L_r w_\sigma} i_{sx} + \frac{R_r L_m}{L_r w_\sigma} \psi_{rx} + \omega_a i_{sy} + \\ + \omega_r \frac{L_m}{w_\sigma} \psi_{ry} + \frac{L_r}{w_\sigma} u_{sx} \tag{1}$$

$$\frac{di_{sy}}{d\tau} = -\frac{R_s L_r^2 + R_r L_m^2}{L_r w_\sigma} i_{sy} + \frac{R_r L_m}{L_r w_\sigma} \psi_{ry} - \omega_a i_{sx} - \\ - \omega_r \frac{L_m}{w_\sigma} \psi_{rx} + \frac{L_r}{w_\sigma} u_{sy} \tag{2}$$

$$\frac{d\psi_{rx}}{d\tau} = -\frac{R_r}{L_r} \psi_{rx} + (\omega_a - \omega_r)\psi_{ry} + R_r \frac{L_m}{L_r} i_{sx}, \tag{3}$$

$$\frac{d\psi_{ry}}{d\tau} = -\frac{R_r}{L_r} \psi_{ry} - (\omega_a - \omega_r)\psi_{rx} + R_r \frac{L_m}{L_r} i_{sy}, \tag{4}$$

$$\frac{d\omega_r}{d\tau} = \frac{L_m}{L_r J}\left(\psi_{rx} i_{sy} - \psi_{ry} i_{sx}\right) - \frac{1}{J_M} T_{S1}, \tag{5}$$

where: R_r, R_s, L_r, L_s, L_m – motor equivalent circuit parameters presented in Table II, T_{S1} – load torque, J_M – motor inertia and $w_\sigma = L_r L_s - L_m^2$.

TABLE II
MOTOR EQUIVALENT CIRCUIT PARAMETERS

Parameter	Value	Description
R_s	0.01226 p.u.	stator resistance
R_r	0.01085 p.u.	rotor resistance
L_s	4.7379 p.u.	stator inductance (leakage + mutual)
L_r	4.6832 p.u.	rotor inductance (leakage + mutual)
L_m	4.5615 p.u.	mutual inductance

For simulation purposes the stationary $\alpha\beta$ coordinates were chosen allowing the motor model (1)-(5) to be rewritten using motor stator currents and rotor flux components: $i_{s\alpha}$, $i_{s\beta}$, $\psi_{r\alpha}$, $\psi_{r\beta}$.

B. Traction Power Electronics Converter

The HST induction motor is fed by a three phase voltage inverter with IGBT transistors. The supply DC voltage of the inverter depends on railway networks: 25kV AC 50Hz or 1.5kV DC. In the simulation program the voltage inverter was modeled with ideal switches controlled with a pulse width modulation method.

Regarding the high power of the drive transistor a switching frequency of the IGBT in the traction drive is variable within the range from 450 Hz for low speeds of the motor to square wave for highest motor speeds.

C. Control System

A motor control algorithm is working according to the direct field oriented control method which is very popular in industrial applications. Base scheme of the control algorithm is presented in Fig. 4.

In the traction application of the field oriented method the speed controller was not used. The operator commanded value is the motor torque which is changed using stator current component i_{sq}. The traction system of the HST is working also in the field weakening region which is used to obtain high speed of the train.

D. Torque transmission system

For simulation purposes HST transmission system was reduced to a two-mass system which is presented in Fig. 5 [23, 35]. The macroslip phenomena was omitted according to assumption that superior adhesion control is implemented in the system. The small microslip phenomena was included in the elastic model of the shaft and gear.

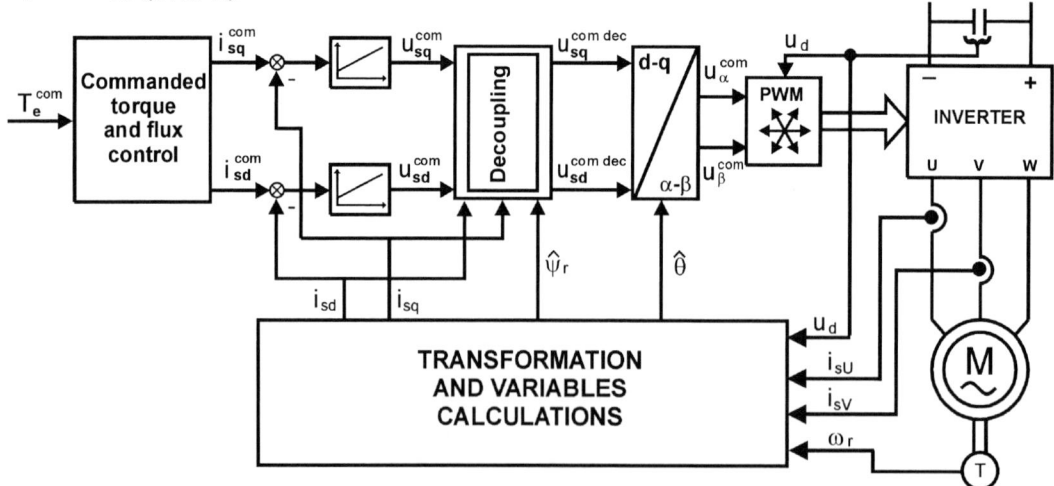

Fig. 4. Base scheme of the motor control algorithm.

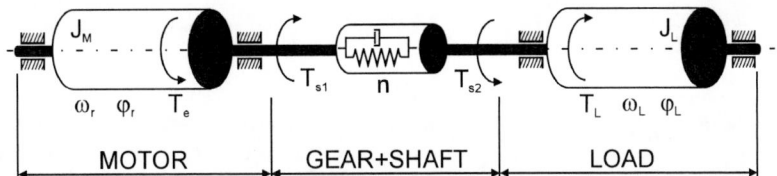

Fig. 5. Model of the torque transmission system.

The model is described by motion equations – refered to the motor shaft:

$$J_M \frac{d\omega_r}{dt} = T_e - T_{S1}, \tag{6}$$

$$J_L \frac{d\omega_L}{dt} = nT_{S2} - T_L, \tag{7}$$

$$\frac{d\varphi_r}{dt} = \omega_r, \tag{8}$$

$$\frac{d\varphi_L}{dt} = \omega_L, \tag{9}$$

$$T_{S1} = K_2(\varphi_r - n\varphi_L) + H_2(\omega_r - n\omega_L), \tag{10}$$

$$|T_{S2}| = n|T_{S1}|, \tag{11}$$

where: ω_r – rotor speed, ω_L – load speed, T_e – motor torque, T_L – load torque, T_{S1}, T_{S2} – torques transmitted trough the gear on input and output of the gear respectively, K_2 – stiffness function, H_2 – damping coefficient (170 Ns/m), J_M – motor inertia, J_L – load inertia (3000 kgm²), n- gear ratio (1.97).

Stiffness function of the gear depends on gear wheel tooth stiffness and is defined as follows [36]:

$$K_2 = K_S + K_D sin(z\varphi_r), \tag{12}$$

where: K_S – stiffness average value (3.5x10⁵ N/m), K_D – stiffness maximum value (5.7x10⁵ N/m), z – number of the gear driving wheel teeth (25 teeth).

In the (7) higher harmonics of the meshing frequency were omitted. It results from the type of wheels used inside the gear: skew teeth make the meshing amplitudes in healthy condition to be very small.

The misalignment of transmission elements was modeled as an additional load torque component which appears in the motion equation:

$$J_M \frac{d\omega_r}{dt} = T_e - T_{S1} - T_w, \tag{13}$$

where: T_w – additional load torque with sinusoidal component with frequency equal to the motor shaft rotating:

$$T_w = T_{wav} \cdot (1 + sin(\varphi_r)), \tag{14}$$

where: T_{wav} – average value of the misalignment load torque component (720 Nm).

In motion equation a viscous friction was also considered:

$$J_M \frac{d\omega_r}{dt} = T_e - T_{S1} - T_w + T_f, \tag{15}$$

where: T_f – viscous friction load torque component:

$$T_f = B_m \cdot \omega_r, \tag{16}$$

where: T_{wav} – average value of the misalignment load torque component (720 Nm).

IV. ASYNCHRONOUS MOTOR SPEED OBSERVER

An induction motor speed observer is a set of equations which are similar to the motor model equations with internal feedbacks which compare some estimated values with measured values. The speed observer needs only four input signals which are usually accessible in the control: signals of inverter output currents and signals of the PWM commanded voltages.

In the proposed system the state observer presented in [1, 37] are used. The equations of the speed observer in the stationary coordinates $\alpha\beta$ are as follows:

$$\frac{d\hat{i}_{s\alpha}}{dt} = -\frac{R_s L_r^2 + R_r L_m^2}{L_r w_\delta} \hat{i}_{s\alpha} + \frac{R_r L_m}{L_r w_\delta} \psi_{r\alpha} + \frac{L_m}{w_\delta} \xi_\beta + \frac{L_r}{w_\delta} u_{s\alpha}^{com} + k_3(i_{s\alpha} - \hat{i}_{s\alpha}) \tag{17}$$

$$\frac{d\hat{i}_{s\beta}}{dt} = -\frac{R_s L_r^2 + R_r L_m^2}{L_r w_\delta} \hat{i}_{s\beta} + \frac{R_r L_m}{L_r w_\delta} \psi_{r\beta} - \frac{L_m}{w_\delta} \xi_\alpha + \frac{L_r}{w_\delta} u_{s\beta}^{com} + k_3(i_{s\beta} - \hat{i}_{s\beta}) \tag{18}$$

$$\frac{d\psi_{r\alpha}}{dt} = -\frac{R_r}{L_r} \psi_{r\alpha} - \xi_\beta + R_r \frac{L_m}{L_r} \hat{i}_{s\alpha} - k_2 S_b \psi_{r\alpha} + S\left(\begin{matrix} k_2 k_3 \psi_{r\beta}(S_b - S_{bF}) + \\ + k_5((S_x - S_{xF})\psi_{r\alpha} - (S_b - S_{bF})\psi_{r\beta}) \end{matrix} \right) \tag{19}$$

$$\frac{d\psi_{r\beta}}{dt} = -\frac{R_r}{L_r} \psi_{r\beta} + \xi_\alpha + R_r \frac{L_m}{L_r} \hat{i}_{s\beta} - k_2 S_b \psi_{r\beta} + S\left(\begin{matrix} -k_2 k_3 \psi_{r\alpha}(S_b - S_{bF}) + \\ + k_5(-(S_x - S_{xF})\psi_{r\beta} - (S_b - S_{bF})\psi_{r\alpha}) \end{matrix} \right) \tag{20}$$

$$\frac{d\xi_\alpha}{dt} = -\hat{\omega}_{\psi r} \xi_\beta - k_1(i_{s\beta} - \hat{i}_{s\beta}) \tag{21}$$

$$\frac{d\xi_\beta}{dt} = \hat{\omega}_{\psi r} \xi_\alpha + k_1(i_{s\alpha} - \hat{i}_{s\alpha}) \tag{22}$$

$$\frac{dS_{bF}}{dt} = \frac{1}{T_{Sb}}(S_b - S_{bF}) \tag{23}$$

$$\frac{d\hat{\omega}_{rF}}{dt} = \frac{1}{T_{KT}}(\hat{\omega}_r - \hat{\omega}_{rF}) \tag{24}$$

$$\frac{dS_{xF}}{dt} = \frac{1}{T_{Sx}}(S_x - S_{xF}) \tag{25}$$

$$S = \begin{cases} 1 & if \quad \hat{\omega}_{\psi r} > 0 \\ -1 & if \quad \hat{\omega}_{\psi r} \leq 0 \end{cases} \tag{26}$$

$$S_x = \xi_\alpha \psi_{r\alpha} + \xi_\beta \psi_{r\beta} \tag{27}$$

$$S_b = \hat{\xi}_\alpha \psi_{r\beta} - \hat{\xi}_\beta \psi_{r\alpha} \qquad (28)$$

$$\hat{\omega}_{\psi r} = \omega_{rF} + R_r \frac{L_m}{L_r} \left(\frac{\psi_{r\alpha} \hat{i}_{s\beta} + \psi_{r\beta} \hat{i}_{s\alpha}}{\psi_{r\alpha}^2 + \psi_{r\beta}^2} \right) \qquad (29)$$

$$\hat{\omega}_r = \frac{\zeta_\alpha \psi_{r\alpha} + \zeta_\alpha \psi_{r\alpha}}{\psi_{r\alpha}^2 + \psi_{r\beta}^2} \qquad (30)$$

where: $\wedge$ - denotes variable calculated in the observer, ξ_α, ξ_β - components of the motor electromotive forces, $\omega_{\psi r}$ – angular speed of the motor flux vector, k_1, k_2, k_3, k_4, k_5 – observer gains (2; 1; 1.2; 1; and 0.1 respectively), S_x, S_{xF} – additional variables used to stabilize the observer work, T_{Sx} – time constant of the S_x filter (25ms).

The presented speed observer system is based on the Luenberger observer theory [2] with using of motor electromotive forces as additional state variables – Fig. 6. The mechanical equation of the system was omitted and the motor mechanical speed is treated as a variable parameter of the motor.

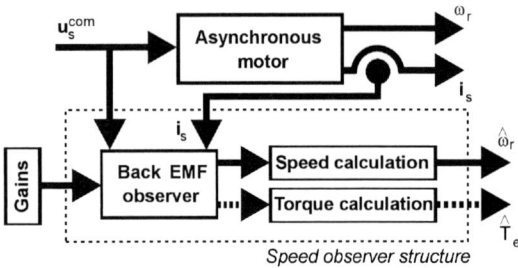

Speed observer structure

Fig. 6. Speed observer scheme.

Input/output block scheme of the speed observer is presented in Fig. 7.

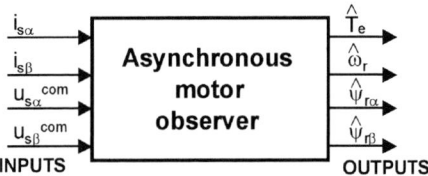

Fig. 7. Input/Output block of the induction motor observer.

The observer gains and the internal filter time constants should be properly tuned to assure fast estimation of the variables and simultaneously to guarantee stable work of the observer.

In the speed observer the motor electromagnetic torque is calculated using the observer internal variables in the following way:

$$\hat{T}_L = \frac{L_m}{L_r} \left(\psi_{r\alpha} \hat{i}_{s\beta} - \psi_{r\beta} \hat{i}_{s\alpha} \right) \qquad (31)$$

The calculated motor torque was used in the load torque observer procedure.

V. LOAD TORQUE OBSERVER

Load torque observer calculations are based on differential equations, which use the measured motor stator voltage, the motor currents and the motor speed as inputs. The output of this observer is the calculated motor load torque. The idea of the calculation of the load torque based directly on the restriction that calculated speed in the speed observer was not applicable in the presented system. This is due to the high mechanical time constant of the motor which makes the speed oscillations of the motor to be very small - less than speed estimation error.

A few methods for calculating the load torque were tested. The simplest may was to directly use the mechanical equations of the system without any correction function [5]. It is a precise and very simple method but sensitive to any inaccuracy and simplifications in the mechanical equations. The load torque observer presented in [6] was working less precisely, because important high frequencies were filtered.

The idea presented in [7] was interesting but not applicable to the use in high speed train control system. The method presented in [7] is practically applicable only to the simple control system working with U/f=constant principle. In the method from [7] the precise value of stator vector voltage angular speed should be known. Unfortunately, in the more sophisticated control systems stator voltage vector speed does not exist in the algorithm. Ex. in field oriented control methods only the angle position of stator voltage vector is accessible. In such case the angular speed of this vector has to be calculated by numerical differentiation. Such calculations are too inaccurate during transients.

Finally, the best results were obtained with an observer based on the Gopinanth's method [8]:

$$\frac{d}{dt}\begin{bmatrix} z_1 \\ z_2 \end{bmatrix} = \begin{bmatrix} 0 & -k_1 \\ 1 & -b \end{bmatrix}\begin{bmatrix} z_1 \\ z_2 \end{bmatrix} + \begin{bmatrix} k_1 b J_M \\ (k_2^2 - a) J_M \end{bmatrix}\omega_r + \begin{bmatrix} k_1 \\ k_2 \end{bmatrix}\hat{T}_e \qquad (32)$$

$$\hat{T}_L = z_2 - k_2 J_M \omega_r \qquad (33)$$

where: k_1, k_2 – observer coefficients, z_1, z_2 – internal state variables, $\hat{T}_e$ - calculated motor electromagnetic torque, $\hat{T}_L$ - calculated motor load torque.

VI. DIAGNOSTIC SYSTEM IDEA

Presented in sections IV and V speed and load torque observers are planed to be used in the HST diagnostic system. The proposed diagnostic system is going to be based on on-line methods for calculating the motor speed and the motor load torque. In a real system a speed sensor exists so that the measured motor speed could be compared with the calculated motor speed:

$$|\hat{\omega}_r - \omega_r| > E_{r\,limit} \qquad (34)$$

where $E_{r\,limit}$ is tuned level of accepted speed observer calculation error (ex. 3%).

In case of fault of the speed sensor or sensor wiring system the train control system could be switched to speed sensorless control system and may give the information to the maintenance center to fix the problem – Fig. 8.

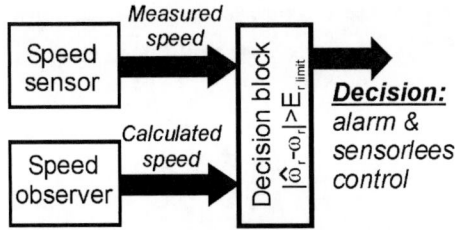

Fig. 8. Idea of the speed sensor diagnostic system.

The similar solution has been applied for the torque transmission monitoring. All the problems within the transmission system could be identified in the motor load torque. In the diagnostic system the load torque of the motor is calculated using the motor load torque observer. The analysis of the load torque is more complicated than the analysis of the calculated speed. The load torque contains a high number of frequencies whose amplitudes change depend on motor speed, motor load etc. In the proposed diagnostic system the analysis of the motor load torque is divided into two methods: on-line and off-line. In on-line method a simple comparison of the amplitudes of the characteristic frequencies is used. Using a binary criteria, it is possible to identify a serious problem with the transmission system. When a problem is identified the control system should limit the maximum torque generated by the motor to allow slower travel of the train. If harder situation occur like transmission faults, a need for more complicated and time consuming procedures, which are executed using off-line method, will be essential. The off-line method gives only the information to the maintenance center to check the suspected gear, for example by detection of particles of metallic debris in the lubricating oil – Fig. 9.

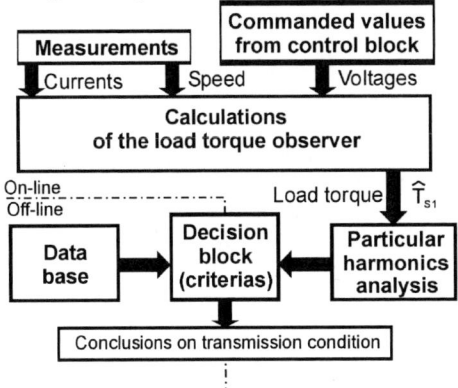

Fig. 9. Idea of the torque transmission diagnostic system.

VII. RESULTS

The presented system was verified by simulations and experiments for a 1.2 MW HST use motor drive. For simulation purposes a dedicated programs in C language

and in Matlab/Simulink were prepared. Experiments were done in the HST factory where 1.2 MW HST test bench was prepared.

In Fig. 10 are presented simulation results for the speed observer investigations.

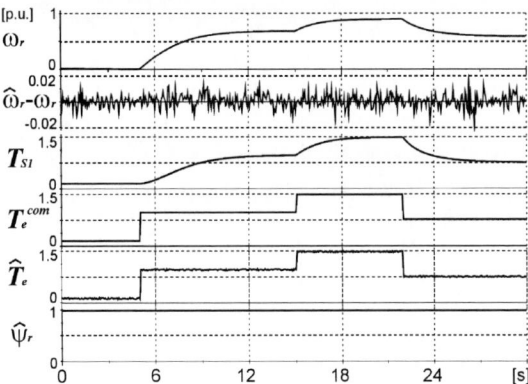

Fig. 10. Simulation results for speed observer during accelerating and decelerating of the train (30..160..210..150 km/h).

In the test presented in Fig. 7 the motor commanded load torque was changed. Adequately to the commanded torque the speed of the motor changes. The particular steady state motor speeds are related to the speed of the train: 30..160..210..150km/h. During the presented test the speed observer was working properly. The motor speed calculation error related to the nominal motor speed was less than 2% in full range of the speed. The accuracy of the calculated motor speed is better than the accuracy of the measured speed used in HST so it seem that estimated speed could be used as useful tool for diagnostic system.

In Fig. 11 are presented simulation results for the load torque observer.

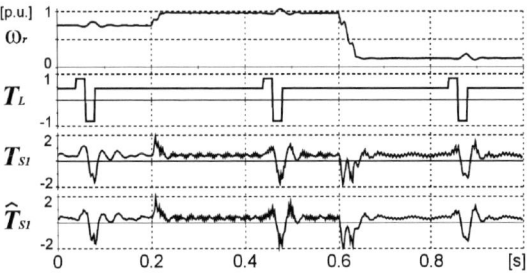

Fig. 11. Simulation results for load torque observer calculation.

During the simulation test of the load torque observer the inertia of the whole train was omitted and the main train speed controller was used. It was done in order to obtain more comprehensible waveforms of the calculated load torque presented in Fig. 11.

In the test presented in. Fig. 11 the misalignment in the system was modeled. The load torque was changed rapidly for different motor speeds. It is noticeable that estimated motor load torque follows the real one.

In Fig.12 are presented results of the motor load torque calculation in the case of healthy and faulty gear.

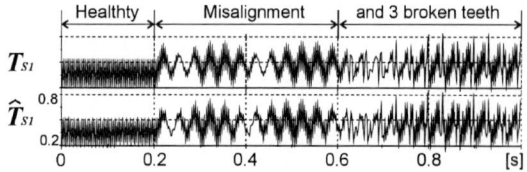

Fig. 12. Simulation results for healthy and faulty gear.

In Fig. 12 the waveforms of the real motor load torque and calculated load torque are presented. The waveforms are divided into three ranges: healthy gear, misalignment in the coupling and misalignment with three broken teeth simultaneously.

The Fourier DFT analysis results for faulty torque transmission system are presented in Fig. 13.

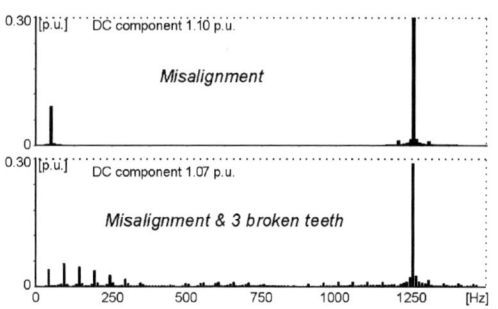

Fig. 13. Simulation results for healthy and fault gear.

In both results in Fig. 13, high amplitudes are related to motor speed (50 Hz) and misalignment frequency (1250 Hz). It is noticeable that in case of broken teeth, additional high amplitudes with frequency range up to 400Hz do appear. Also the previously meaningful amplitudes show little change.

In Fig. 14 are presented experimental results for the speed observer calculations.

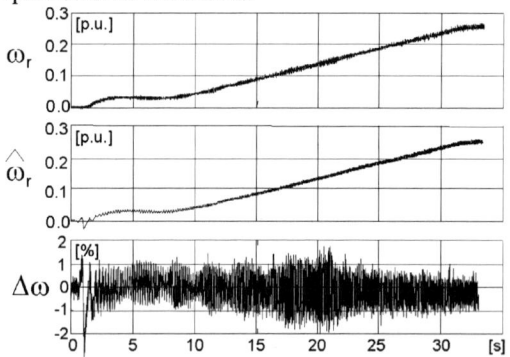

Fig. 14. Experimental results for speed observer calculation – start up of the train from 0 km/h to about 56 km/h

In Fig. 14 the speed of the train was changed from 0 km/h to about 56 km/h. Except the start of the observer procedure the speed observer is working with an error

less than 2 %. These are similar results to the simulated observer.

In 15 are presented experimental results for the load torque observer.

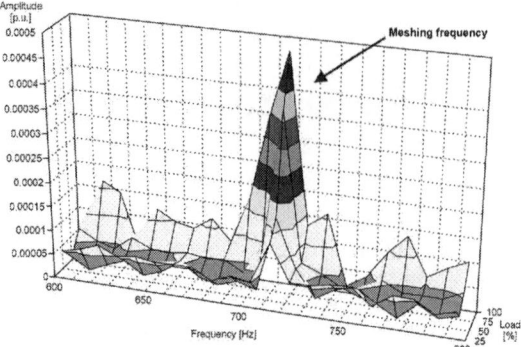

Fig. 15. Experimental results for load torque observer calculation – identification of meshing frequency – train speed about 59 km/h.

In a test whose results are presented in Fig. 15 the train was moving with constant speed 59 km/h with four different loads: 25 %, 50 %, 75 %, 100 %. In this case the calculated load torque was analyzed with discrete Fourier transformation DFT. In Fig. 15 only some interesting range of frequency is presented. This frequency range was related to the load torque meshing frequency. The train speed 59 km/h is equal to 29 Hz mechanical frequency of the motor. Regarding the number of teeth of driving wheel - 25, the meshing frequency was approximately 725 Hz. In the test bench the transmission system was healthy and the amplitude of meshing frequency was very small but was identified using the proposed calculation method. Existing load torque sensor with an accuracy of 50 Nm was also used but using the measured load torque the meshing frequency was not identified. Other results are presented in [38].

According to the limitation in the industrial setup the tests for unhealthy gear was not provided. Such tests are going to be performed in the future in the real HST systems which are repaired in the maintenance center.

VIII. CONCLUSIONS

The diagnostic system is under developing. The obtained results confirm that the presented speed and load torque observer has good property to be used in the diagnostic system. Currently, the both speed and load torque observers are under implementation as part of the HST microprocessor control system. After implementation it is our plan to apply the presented system in different trains to build a data base of the results of the calculated load torque and motor speed. The obtained data base in relation with other rail vehicles data and maintenance center information will allow defining criteria to make decision on speed sensor and torque transmission system conditions.

ACKNOWLEDGEMENTS

Authors thank Mr. Jean-Marc Descures from Alstom Transport, Semeac, France for his help with preparing the experimental tests.

REFERENCES

[1] Z. Krzemiński, "Sensorless Control of the induction motor based on new observer", in Proc. *International Conference on Power Electronics, Intelligent Motions and Power Quality, PCIM 2000*, Nuremberg, Germany, 2000.

[2] D. G. Luenberger, "An introduction to observers", *IEEE Transactions on Automatic Control*, vol. AC-16, no. 2, pp. 596-602, 1971.

[3] B. Gopinath, "On the control of linear multiple input-output systems", *The Bell Technical Journal*, vol.50, no.3, pp.1063-1081, 1971.

[4] F. Parasaliti, R. Petrella, M. Tursini, "Speed sensorless control of a PM Motor by sliding mode observer", in Proc. *IEEE International Symposium on Industrial Electronics, ISIE'97*, Guimaraes, Portugal, 1997, pp. 1106-1111.

[5] A. Lewicki, "Torque vibrations compensation on the asynchronous machine shaft", *Ph.D. dissertation*, Gdansk University of Technology, Faculty of Electrical and Control Engineering, Gdansk, Poland, 2003.

[6] J. Meditch, G. H. Hostetter, "Observers for systems with unknown and inaccessible inputs", International Journal of Control, vol. 16, no. 3, pp. 473-480, 1974.

[7] M. A. Brdyś, T. Du, "Algorithms for joint state and parameter estimation in induction motor drive systems", in Proc. *IEE International Conference on Control, Control'91*, Edinburgh, UK, 1991, pp. 915-920.

[8] K. Ohnishi, M. Shibata, T. Murakami, "Motion control for advanced mechatronics", *IEEE/ASME Transactions on Mechatronics*, vol. 1, no. 1, pp. 56-67, 1996.

[9] K. M. Al-Hussain, I. Redmond "Dynamic response of two rotors connected by rigid mechanical coupling with parallel misalignment", *Journal of Sound and Vibration*, no. 249, pp.483-498, 2002.

[10] K.M. Al-Hussain, "Dynamic stability of two rigid rotors connected by a flexible coupling with angular misalignment", *Journal of Sound and Vibration*, no. 266, pp.217–234, 2003.

[11] N. Driot, J. Perret-Liaudet, "Variability of modal behavior in terms of critical speeds of a gear pair due to manufacturing errors and shaft misalignments", *Journal of Sound and Vibration*, no. 292, pp.824–843, 2006.

[12] A. C. J. Luo, "Past, current and future on nonlinear dynamics and noise origins of non-smooth gear transmission dynamic systems", in Proc. *IEEE*, 2005, pp. 674-681.

[13] S. Prabhakar, A. S. Sekhar and A. R. Mohanty, "Crack versus coupling misalignment in a transient rotor system", *Journal of Sound and Vibration*, no. 256, pp.773-786, 2002.

[14] G. Dalpiaz, A. Rivola, R. Rubini, "Gear fault monitoring: comparison of vibration analysis techniques", pp. 623-632.

[15] D. Richards, DJ. Pines, "Passive reduction of gear mesh vibration using a periodic drive shaft", *Journal of Sound and Vibration*, no. 264, pp.317–342, 2003.

[16] A. S. Sekhar, B. S. Prabhu, "Effects of coupling misalignment on vibrations of rotating machinery", Journal of Sound and Vibration, no. 185, pp.655-671, 1995.

[17] L. Vedmar, A. Andersson, "A method to determine dynamic loads on spur gear teeth and on bearings", *Journal of Sound and Vibration*. no. 267, pp.1065–1084, 2003.

[18] B. Wu, A. Saxena, P. Sparis, "An approach to fault diagnosis of helicopter planetary gears", in Proc. *IEEE AUTOTESTCON Conference*, 2004, pp. 475-481.

[19] W. Wang, "Identification of gear mesh signals by kurtosis maximisation and its application to ch46 helicopter gearbox data", in Proc. *IEEE*, 2001, pp. 369-372.

[20] Y. Z. Li, S. X. Wang, "Error modeling and simulation analysis on CNC gear shaper", in Proc. *The 8th International Conference on Computer Supported Cooperative Work in Design Proceedings*, 2003, pp. 727-730.

[21] W. Wang, A. K. Wong, "Some new signal processing approaches for gear fault diagnosis", in Proc. *Fifth International Symposium on Signal Processing and its Applications, ISSPA'99*, Brisbane, Australia, 1999, pp. 587-590.

[22] Xiaohong Yuan, Lilong Cai, "Gearbox diagnosis using a modified Fourier series", in Proc. *International Conference on Advanced Intelligent Mechatronics, AIM'03*, 2003, pp. 193-198.

[23] M. S. Tondos, "Minimizing electromechanical oscillations in the drives with resilient couplings by means of state and disturbance observers" , in Proc. *European Power Electronics Conference, EPE1993*, Brighton, UK, 1993, pp. 360–365.

[24] P. Beccue, S. D. Pekarek, B. J. Deken, A. C. Koenig, "Compensation for asymmetries and misalignment in a Hall-effect position observer used in PMSM torque-ripple control", *IEEE Transactions on Industry Applications*, vol. 43, no. 2, pp. 560-570, 2007.

[25] N Matsui, T. Makino, H. Satoh, "Autocompensation of torque ripple of direct drive motor by torque observer", *IEEE Transactions on Industry Applications*, vol. 29, no. 1, pp. 187-194, 1993.

[26] Y. Kosaka, A. Shimada, P. Viboonchaicheep, "Vibration control without estimated disturbance feedback for robot manipulators", in Proc. *IEEE Conference*, 2003, pp. 848-853.

[27] K. Kaneko, T. Murakami, K. Ohnishi, K. Komoriya, "Torque control with nonlinear compensation for harmonic drive DC motors", in Proc. *IEEE Conference*, 1994, pp. 1022-1027.

[28] K. Peter, I. Schöling, B. Orlik, "Robust output-feedback H∞ control with a nonlinear observer for a two-mass system", *IEEE Transactions on Industry Applications*, vol. 39, no. 3, pp.637-644, 2003.

[29] A. Luca, D. Schroder, M. Thummel, "An acceleration-based state observer for robot manipulators with elastic joints", in Proc. *IEEE International Conference on Robotics and Automation*, Roma, Italy, 2007, pp. 3817-3823.

[30] M. Itoh, "Vibration suppression control of a geared mechanical system: effects of sensor-based control and installation of non-backlash gear", in Proc. *IEEE International Conference on Mechatronics and Automation*, Luoyang, China, 2006, pp. 606-611.

[31] N. Amann, J. Böcker, F. Prenner, "Active damping of drive train oscillations for an electrically driven vehicle", *IEEE/ASME Transactions on Mechatronics*, vol. 9, no. 4, pp. 697-700, 2004.

[32] S. C. Kim, Y. S. Cho, J. H. Park, S. C. Kang, K. H. Bae, "Vibration subression control for a rolling machine system using fuzzy controller", in Proc. *International Joint Conference, SICE-ICASE 2006*, Bexco, Busan, Korea, 2006, pp. 3792-3795.

[33] C. D. Begg, T. Merdes, C. Byington, K. Maynard, "Dynamics modeling for mechanical fault diagnostics and prognostics", in Proc. *Mechanical System Modeling for Failure Diagnosis and Prognosis, Maintenance and Reliability Conference, MARCON'99*, Gatlinburg, Tennessee, 1999, pp. no numbered.

[34] S. Silva, M. Dias Jr., "Statistical damage detection in stationary rotor systems through time series analysis", in Proc. *Brazilian Conference on Dynamics, Control and Their Applications, DINCON 2006*, Guaratinguetá, Brazil, 2006, pp. no numbered.

[35] P. Korondi, H. Hashimoto, V. Utkin, "Discret torsion control of flexible shaft in an observer-based discrete-time sliding mode", *IEEE Transactions on Industrial Electronics*, vol. 45, no. 2, April 1998.

[36] E. Switonski, A. Mezyk, "Dynamical analysis of optimisation of electromechanical machinery drives", in Proc. *2nd International Seminar on Vibrations and Acoustic Noise of Electrical Machinery, VANEM 2000*, Lodz, Poland 2000.

[37] M. Włas, Z. Krzeminski, J. Guzinski, H. Abu-Rub, H. A. Toliyat, "Artificial-Neural-Network-Based Sensorless Nonlinear Control of Induction Motors", *IEEE Trans on. Energy Conversion*, vol 20, no. 3, September 2005, pp. 520 - 528.

[38] J. Guzinski, M. Diguet, Z. Krzeminski, A. Lewicki, H. Abu-Rub., "Application of Speed and Load Torque Observers in High Speed Train Drive for Diagnostic Purposes", under revision in *IEEE Transaction on Industrial Electronics*.

Position Estimator including Saturation and Iron Losses for Encoder Fault Detection of Doubly-Fed Induction Machine

Kai Rothenhagen[*], Friedrich W. Fuchs[+]

[*] Institute for Power Electronics and Electrical Drives / University of Kiel, Kiel, Germany, *kro@tf.uni-kiel.de*
[+] Institute for Power Electronics and Electrical Drives / University of Kiel, Kiel, Germany, *fwf@tf.uni-kiel.de*

Abstract— Fault tolerance is gaining interest to increase reliability and availability, for example for distributed energy systems. Doubly fed induction generators need rotor position information for high performance control. As part of a multi-sensor fault tolerant generator system, fault tolerance of the position sensor is presented, using a position estimator. A known position estimator is improved and thoroughly described, including effects of machine saturation and iron losses. The fault detection and isolation is described and the reconfiguration from encoder to sensorless operation is shown. Measurements are provided and show good and fast results. Steady state properties of encoder and sensorless operation are compared.

Keywords: Doubly fed induction motor, Sensorless control, Fault handling strategy, Wind energy

I. INTRODUCTION

Energy production from renewable energy sources has developed extraordinary over the past decade, contributing to an energy mix that is less relying on carbon-dioxide emitting fuels. Wind power has gained enormous interest in the last decade and can now be considered a mature industry. One of the main types of wind generators is the Doubly Fed Induction Generator (DFIG) [1], a wound rotor induction generator that is controlled by an inverter at the rotor terminals.

In future, large offshore wind parks are expected to contribute significantly to the produced wind power. However, the remote location and high investment costs for multi-megawatt wind turbines have created an interest in more reliable, self-diagnosing and even fault tolerant wind turbines. A good introduction to diagnosis and fault detection can for example be found in [2] and [3].

One possible cause for faults is sensor failure, especially of the mechanical position sensor. Here, a position sensor fault tolerant control is presented. In case of a rotor position sensor failure, the fault is detected and isolated. A rotor position estimator using stator and rotor current measurement provides a replacement signal for the failed sensor. After fault isolation, the control is reconfigured and uses the estimated position. Using position sensorless control in the first place would also be possible. However, this work is part of a series describing a (n-1) sensor fault tolerant control, with fault tolerance for every used sensor. The position sensor is therefore needed for the fault detection and reconfiguration of the other sensors. Here, only the position sensor is treated.

Generally, fault tolerance can also be reached by implementing hardware redundancy at extra cost. The proposed method requires only extra computational power.

A thorough definition of relevant terminology has been given in [4], including:
Fault: Unpermitted deviation of at least one characteristic property or parameter of the system from acceptable / usual / standard condition.
Residual: Fault indicator, based on deviation between measurements and model-equation-based computations.
Fault Detection: Determination of faults present in a system and time of detection.
Fault Isolation: Determination of kind, location and time of detection of a fault. Follows fault detection.
This definition should be amended with the term of **Reconfiguration**: Rearranging the control structure of a system to enable continued operation in spite of a fault.

The paper is organized as follows: An introduction was given in section I. The control scheme used for the DFIG is described in section II, including a thorough description of the used position estimator. Effects of unsymmetrical machine main inductance saturation influencing the estimator are presented.

In section III, the position sensor performance is analyzed for all operating points. Steady state properties of encoder and sensorless closed loop control are compared. Section IV presents the effect of a position sensor failure and the proposed fault detection and isolation scheme. Measurements for synchronous as well as super- and subsynchronous speed are presented. A conclusion and references complete the paper.

II. POSITION SENSORLESS CONTROL OF DOUBLY FED INDUCTION GENERATOR

A. Control of the Doubly Fed Induction Generator

The considered DFIG is stator power controlled, using a voltage source inverter and rotor current control loops. The control uses a stator voltage oriented reference frame, as common for DFIG [5]. The DFIG and used sensors are shown in fig. 1. The rotor current control loops have a rise time of about 3 ms. No decoupling is used because the rotor control loop is sufficiently fast. The DC link voltage is used to calculate the duty cycle of the pulse width modulated rotor side inverter, while the stator voltage and rotor position provide the transformation angle for the used reference frame.

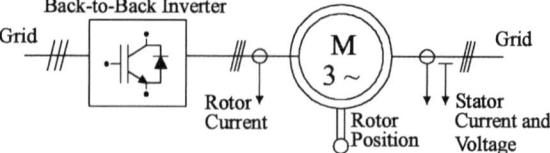

Fig. 1: Rotor controlled DFIG with sensors needed for control.

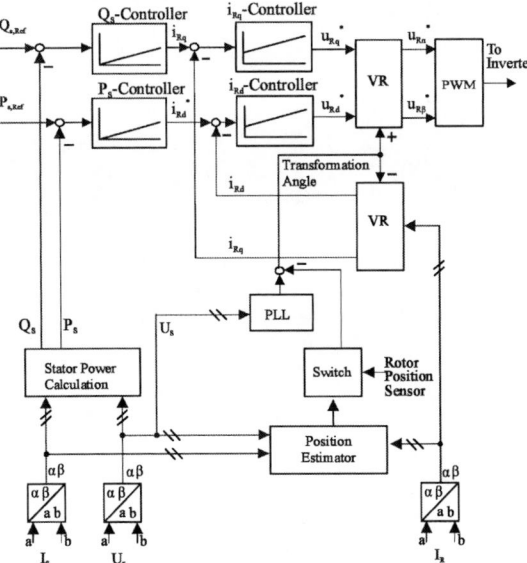

Fig. 2: Scheme used for stator power control.

$$I_{Rd}^{US} = -r_{Eff} \left.\frac{L_S}{M}\right|_d \left(I_{Sd}^{US} - I_{RFe}^{US}\right) \quad (1)$$

$$I_{Rq}^{US} = -r_{Eff} \left.\frac{L_S}{M}\right|_q \left(I_{Sq}^{US} + I_{\mu}^{US}\right) \quad (2)$$

$$\sin\left(\gamma_{EST}\right) = \sin(\varphi_2 - \varphi_1) = \dots$$
$$\dots\sin\left(\varphi_2\right)\cos\left(\varphi_1\right) - \sin\left(\varphi_1\right)\cos\left(\varphi_2\right) \quad (3)$$

$$\cos\left(\gamma_{EST}\right) = \cos(\varphi_2 - \varphi_1) = \dots$$
$$\dots\cos\left(\varphi_2\right)\cos\left(\varphi_1\right) + \sin\left(\varphi_1\right)\sin\left(\varphi_2\right) \quad (4)$$

$$\sin\left(\varphi_2\right) = \frac{I_{R\alpha}^{R}}{|I_R|}; \cos\left(\varphi_2\right) = \frac{I_{R\beta}^{R}}{|I_R|} \quad (5)$$

$$\sin\left(\varphi_1\right) = \frac{I_{R\alpha}^{S}}{|I_R|}; \cos\left(\varphi_1\right) = \frac{I_{R\beta}^{S}}{|I_R|} \quad (6)$$

They are the basis of the method used in this paper and improved by adding terms for iron losses. It is an improvement to existing methods, but still a simple approach, using equations (1) and (2).

L_S is the stator inductance, M is the mutual inductance and r_{Eff} is the effective turns ratio of stator to rotor windings, which is determined by the number of turns and type of connection, for instance YΔ. I_{RFe} models the active current due to iron losses, while the magnetizing current is represented by I_{μ}. Note that equations (1) and (2) basically consist of only a gain and an offset for the multidimensional approximation of the rotor current.

The rotor position sensor is supplying the angle γ_{Rotor} between stator fixed α axis and rotor fixed α axis. This is the angle that is to be estimated.

Stator current is inherently in stator fixed S-reference frame and rotor current is inherently in rotor fixed R-reference frame. From stator voltage measurement, a transformation angle $\gamma_{US,PLL}$ to stator voltage synchronous US-reference frame is derived using a PLL. Therefore stator voltage synchronous current I_{Sdq}^{US} is available only from electrical measurements.

Then rotor current in stator voltage synchronous reference frame is calculated from (1) and (2), as shown in figure 3. It is transformed back to stator fixed reference frame, yielding the approximated rotor current $I_{R\alpha}^{S}$, $I_{R\beta}^{S}$ in stator fixed reference frame.

This is compared to measured rotor current in rotor fixed reference frame $I_{R\alpha}^{R}$, $I_{R\beta}^{R}$. Since the rotor current is known in two reference frames, the angle γ_{EST} between these two frames can be calculated, as shown in figure 4. This leads to the sine and cosine of the estimated rotor position, out of which the angle can be derived easily. In detail, the trigonometric identities (3) and (4) are used, substituting (5) and (6).

One limitation of the used method can be derived from (5) and (6): The rotor current may not be zero, as is also later shown in fig. 12 b). This is only a minor limitation, because DFIG are usually magnetized from the rotor. Therefore, the rotor current is never zero while full control of stator active current is possible. Achievable accuracy of the position estimator is shown in fig 11.

For rotor position measurement, an encoder is used. A phased locked loop is used to generate the stator voltage angle from the voltage measurement. Stator voltage measurement is filtered with a 4th order Chebychev anti-aliasing filter. A dq-PLL as described in [7] is used to derive the stator voltage angle.

Stator power control is implemented as the outer control loop, as shown in figure 2. The stator power control loop uses stator current and voltage measurement to calculate stator power. It has a rise time of 80 ms. Integral controllers are used. The stator active power controller sets the I_{Rd} rotor current controller reference value, while the reactive power controller sets the I_{Rq} reference. Stator reactive power is controlled to zero, but other stator power factors are possible.

B. Rotor Position Estimator

There are various ways to estimate the rotor position of variable speed drives. Stator fed machines commonly rely on a flux estimator using terminal voltage and current to derive an angle for reference frame transformation, a method also called "Back-EMF method" [6]. Methods based on the back emf do have problems with low rotor speeds and are not usable at zero speed, since there is no coupling between stator and rotor. Transferring this problem to the wound rotor induction machine, the difficult point is the synchronous speed where the stator flux and the rotor rotate synchronously and the drive behaves almost like a synchronous machine. Doubly Fed Induction Generators are meant to be used in a speed region around the synchronous speed. For this reason, back emf based methods are most likely not suitable, although such a method is described in [12]. In [15], two methods for position sensorless control are proposed. For stand-alone systems, a position sensorless control scheme is presented in [14]. In [8], a current based method to estimate the rotor position is described very comprehensively and thoroughly. In [9], [10] and [11], quite similar methods are presented.

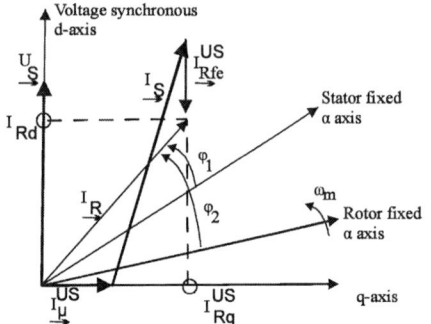

Fig. 3: Rotor current approximation from stator current measurement. For clarity, the scaling is omitted.

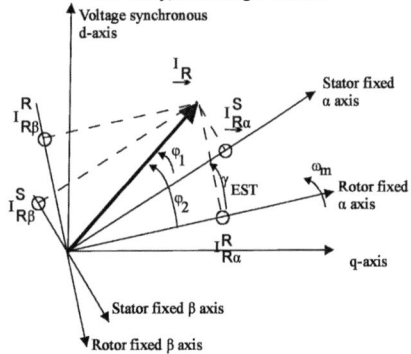

Fig. 4: Rotor current in various reference frames.

C. Rotor Speed Observer

There are two ways to derive a speed signal from the position signal, which is supplied by either the estimator or the encoder. Since the rotor angle is the integral of the angular rotor speed, differentiation of the angle will lead to the rotor speed. In discrete time systems the differential quotient (7) is used. Differentiation of the noisy estimated rotor angle may need low pass filtering, causing unwanted phase delay.

Another way is to use a speed observer to reconstruct the rotor speed, as described by (8). The internal states of the observer are observed rotor angle $\hat{\gamma}$ (not to be confused with the estimated rotor angle γ_{EST}) and observed rotor angular frequency $\hat{\omega}$. The observer error is the deviation between observed and measured γ_{Rotor} (or estimated γ_{EST}) rotor angles. It is kept within [$-\pi..+\pi$]. Once the observer is swung in, the observed rotor angular frequency is available.

The observer is tuned by the feedback vector $[L_{SO1}\ L_{SO2}]^T$, determining the eigenvalues of the observer. For a sampling time of $T_S=200\mu s$, $L_{SO1}=1$ and $L_{SO2}=20$ are used, placing both eigenvalues onto the real axis at 0.04 and 0.996 inside the unity circle. These values have been found to be a good compromise to achieve fast observer reaction time and good noise rejection.

Two identical parameterized observers are used: One uses the rotor position sensor as input, the other one uses the rotor position estimator. These speed observers provide a speed signal that might be used for speed control and also play an important part in fault isolation, as described in section VI.

$$\omega_k = \frac{\gamma_k - \gamma_{k-1}}{T_S} \tag{7}$$

$$\begin{bmatrix} \hat{\gamma}_{k+1} \\ \hat{\omega}_{k+1} \end{bmatrix} = \begin{bmatrix} 1 & T_s \\ 0 & 1 \end{bmatrix} \begin{bmatrix} \hat{\gamma}_k \\ \hat{\omega}_k \end{bmatrix} + \begin{bmatrix} L_{SO1} \\ L_{SO2} \end{bmatrix} (\hat{\gamma}_k - \gamma_k) \tag{8}$$

D. Machine Parameter Identification

The terms of equations (1) and (2) are identified by a series of measurements. Identification yields to different parameters in d- and q-axes. As a spin off, and not the focus of this work, an idea of nonlinear saturation and saliency as well as iron losses is obtained. Great care has been taken to tune in the position sensor to remove inaccuracy, and therefore coupling of the axes, due to a wrong transformation angle.

In order to test the results obtained from (1) and (2), the rotor active current I_{Rd} is varied from +50 A to -50 A continuously while changing the reactive current I_{Rq} in steps. The applied load profile is shown in figure 5, and is repeated for each speed step. The rotational speed is increased from 1000 rpm to 2000 rpm in steps of 100 rpm. Rotor and stator active and reactive currents are sampled every 200 µs for 40 s, then low pass filtered using a 20 Hz digital low pass and downsampled with 200 Hz sampling frequency to reduce data. Having measurement results for I_{Sd} and I_{Rd}, the parameters of (1) are fitted. Using I_{Sq} and I_{Rq}, the same is done for equation (2). As r_{Eff} can only be identified together with the factor L_S/M, it is treated as one parameter for each axis. The results of the fitting show some peculiar effects, such as saliency. They need to be handled the necessary care, since it is usually hard to find detailed information on the saturation of electrical machines, especially when saliency is involved. The resulting parameters are given in table I. One peculiarity is the different factor for d- and q- axes at 400 V stator voltage. The most probable reason for this is machine main inductance saturation. In order to investigate this further, the stator voltage has been varied from 66 V to 400 V. The parameters of (1) and (2) have been fitted for every voltage. Figure 9 presents the d- and q-axis factors over the stator voltage. It is clearly seen that they are similar for low voltages and that the q-axis parameter increases with higher stator voltage. Figures 6 and 7 show the deviation of (1) and (2) to the measurement of I_{Rd} and I_{Rq} after compensation. For every operational point, the same parameters of table I have been used.

The figures show that for the whole range of rotational speeds, for stator powers from +15 kW to -15 kW and for varying reactive power, the resulting deviation is close to zero in the active current, as seen in fig. 6. The reactive current balance shows deviation up to plus minus 5 ampere, as seen in fig. 7, which correlates to the active current. It is emphasized that this correlation is not due to coupling of the axes due to poor position sensor tuning, but also most likely due to saturation. This is made plausible by presenting the same graph for 200 V stator voltage in fig. 8, where only minor coupling effects are noticeable. This graph shows a much smaller deviation of up to 1 ampere.

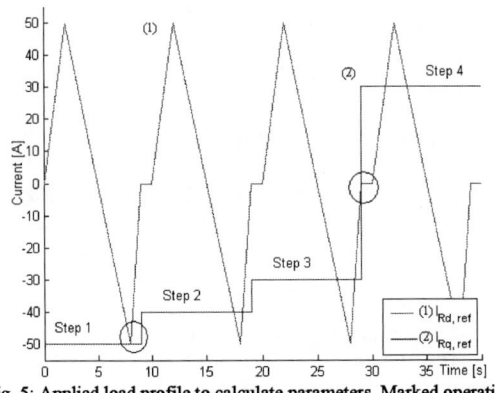

Fig. 5: Applied load profile to calculate parameters. Marked operating points are shown in fig. 12 a) and b).

TABLE I. ESTIMATED PARAMETERS

Parameter	U_S 200 V	U_S 400 V
$r_{Eff}\left.\dfrac{L_s}{M}\right|_d$	1.63	1.65
$r_{Eff}\left.\dfrac{L_s}{M}\right|_q$	1.63	1.82
I_{RFe}^S	1.02 A	2.5 A
I_μ^S	11.14 A	27.63 A
r_{Eff}	approx 1.5	approx 1.5

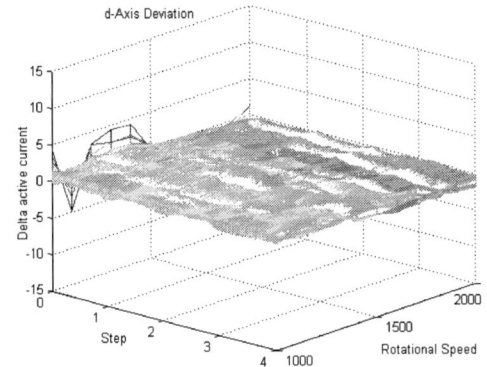

Fig. 6 Deviation of active current balance, low pass filtered at 20 Hz, at stator voltage of 400 V rms.

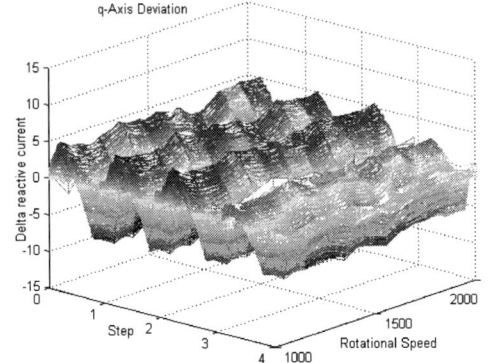

Fig. 7: Deviation of reactive current balance, low pass filtered at 20 Hz, at stator voltage of 400 V rms.

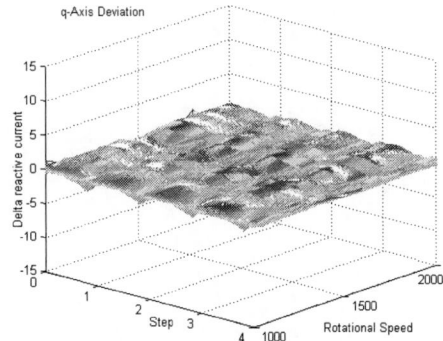

Fig. 8: Deviation of reactive current balance, low pass filtered at 20 Hz, at stator voltage of 200 V rms.

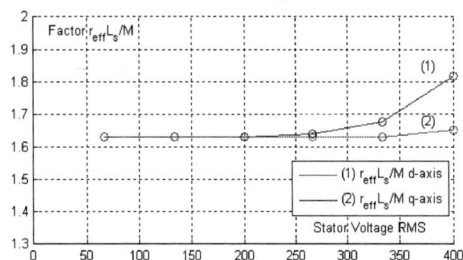

Fig. 9: Derived factors for d- and q-axis as a function of stator voltage.

III. PROPERTIES OF ESTIMATOR BASED CONTROL

A. Position Estimator Performance

Applying the load profile of fig. 5, the deviation between measured and estimated angle is presented in fig.10 without filtering, using closed loop sensorless control. Deviation is within ± 5 degrees and shows correlation with the load profile. Except for a rotor speed of 1000 rpm during step 4, the presented estimator is stable. Operation as demanded in step 4 is untypical because the generator is usually magnetized from the rotor side. The parameters of table I are used for all operating points for speeds from 1000 rpm to 2000 rpm. In the applied control structure, the power control loop corrects the remaining angle inaccuracy by demanding appropriate rotor current.

In detail shown in fig. 11, the deviation between measured and estimated angle is influenced by stator current harmonics, since stator current measurement is used to approximate the rotor current. The deviation itself therefore contains harmonics, as is shown in figure 12. The FFT of the deviation for all considered rotor speeds is calculated, showing harmonics at 100 Hz, 200 Hz and 300 Hz. They are independent of the rotor speed, which is a strong indicator that they are due to the stator current harmonics. Further research is needed to determine the exact origin. In [13], the estimated angle is filtered by a PLL to remove distortion.

The applied load profile includes the critical point of zero rotor current at 29s, when entering step 4 of the profile. As can be seen in fig. 11 b), this leads to large deviation in the estimated angle, which would most likely lead to unstable operation. Here, the critical point is only passed by during a transient. Fig. 11 a) shows the behaviour typical for all other operating points.

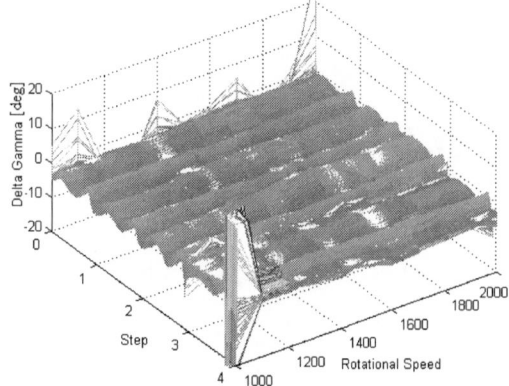

Fig 10: Unfiltered Measurement of achieved angle deviation during closed loop position sensorless operation, applying the profile of fig. 5.

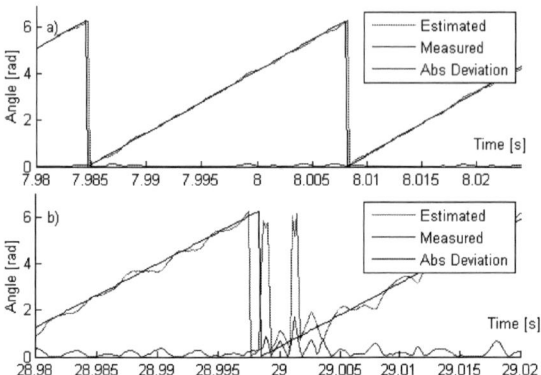

Fig 11: Typical behaviour of estimated angle deviation during closed loop position sensorless operation, at 1300 rpm and a) 15 kW stator power, b) 0 kW stator power and zero crossing of rotor current. Compare operational point with figure 5.

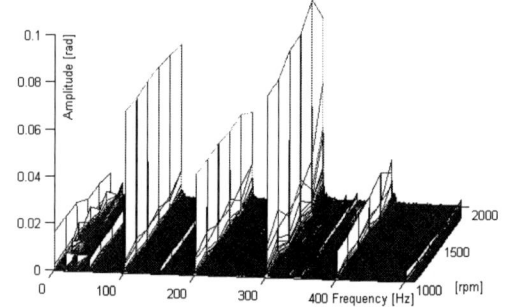

Fig. 12: Spectral analysis of deviation between measured and estimated angle for rotor speeds from 1000 rpm to 2000 rpm.

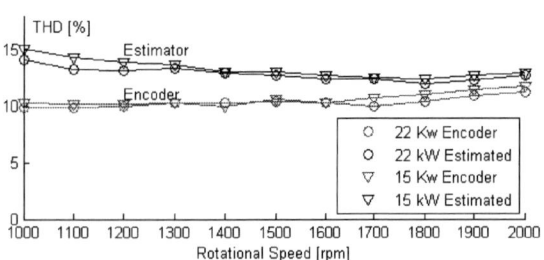

Fig. 13: Total Harmonic Distortion for encoder and estimator operation at 15 kW and 22 kW stator power over rotational speed.

B. Effect on Stator Current Harmonics

In the following, the steady state properties of the stator current for closed loop sensorless scheme are compared to the properties of encoder operation.

The grid voltage includes harmonics, such as 5th, 7th, 11th and 13th, which in return cause grid current harmonics. These are not desired, and therefore usually compensated by means of a harmonic compensator [20]. No such harmonic compensator is used, so the influence of the position estimation algorithm on grid current quality is visible. Figure 13 presents measurement of the stator current total harmonic distortion THD for encoder and estimator operation, measured at 15 and 22 kW stator power. While both THD are relatively high, the estimator operation shows persistently larger distortion than encoder operation. This is mainly due to an increased 5th harmonic, as seen in fig. 14 a) and b).

Operation using encoder operation yields to the results seen in figure 14 a) while when using the proposed position estimator, the results in figure 14 b) are obtained. The figures show terminal voltage and current of phase a. The FFT of the grid current is calculated. Power factor at machine side terminals is close to one for both methods and stator power is 15 kW at 1200 rpm. From the FFT of the stator current, it can be seen that the use of the position estimator causes increased 5th harmonics in the grid current. In the figures, the 5th harmonic has an amplitude of 1.35 A for encoder operation and 2 A for estimator operation. The other harmonics do not show significant change. The increase in 5th harmonic is possibly due to the mentioned harmonics in the angle (as shown in fig. 12), that in return, when used for closed loop current control, amplify themselves. Further research is needed to avoid this undesired effect.

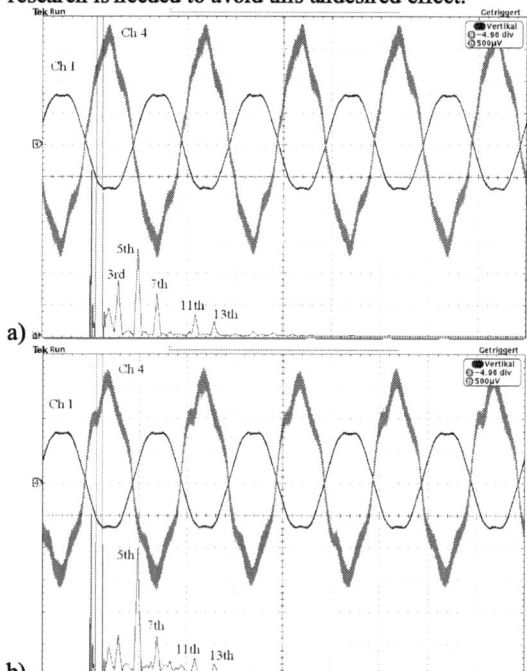

Fig 14: Operation at 1200 rpm with a) encoder feedback b) estimator feedback. Ch 1: Grid voltage phase to neutral, [200 V/div]. Ch 4: Grid current [10 A/div]. Math: FFT of Ch 4, [500 mA/ div; 250 Hz/div].

IV. ENCODER FAILURE AND RECONFIGURATION

A. Impact of Encoder Failure on Stator Power Control

A fault in any sensor used in closed loop systems is likely to cause serious malfunction. For closed loop controlled generators, sensor faults are most likely to lead to increased currents, oscillation or other unpredictable behaviour, depending on the kind and location of the fault. Conventional systems are usually turned off to for protection, once a fault is detected. Usually, conventional methods like fuses, U_{CE}-monitoring or comparison of measured current to a threshold are used. In the following, the impact of a faulty encoder on the considered control scheme is described. The rotor current control loops are affected directly. Although the stator power calculation is not directly affected, serious distortion is seen in the entire system. Without fault detection and reconfiguration, this fault leads to an inoperable generator. In figure 15, a position sensor fault at 1700 rpm and 10 kW stator power does not lead to overcurrent, but to oscillation in rotor and stator current. The drive should be switched off because of torque oscillation.

B. Proposed Fault Detection and Isolation Scheme

The proposed position sensor fault isolation is part of a scheme that covers the stator and rotor current sensors, the stator voltage sensor and the position sensor of a DFIG. A bank of observers is implemented, as shown in fig. 16. Here, only the position sensor is covered in detail, since it is one of the most fault prone sensors on a drive. Details on the whole scheme and on the other used observers can be found in [16], [17], [18] and [19].

The scheme consists of three steps: Fault detection, fault isolation and sensor reconfiguration. First step of the whole scheme is fault detection, which is realized by looking at the residuals of the electrical sensor observers. Should their residual cross a threshold, a fault is detected. As can be seen from fig. 15, a failure of the position sensor causes a change in the systems structure, leading to oscillation. This will lead to current observer residuals crossing the mentioned threshold and thereby triggering the fault detector. For the scope of this work, it can be assumed that the detection of the fault, e.g. gaining the information that there is *any* fault in the system, is realized by some external device, while the fault isolation of the position sensor, e.g. the information that the previously detected fault is due to position sensor failure, is treated here. After fault detection, the Fault Detect Counter is set to 5 ms, which is the period allowed to isolate the fault, as is explained in the next sub-section. During this period, the generator is operated in open loop without using any sensor. The rotor voltage that had been used before detecting the fault is extrapolated. If no fault is isolated within this 5ms period, it is assumed that no fault has happened. The control is switched back to normal voltage oriented control.

$$\text{abs}(\hat{\gamma}_{SO1} - \gamma_{EST}) \qquad (9)$$

$$\text{abs}(\hat{\gamma}_{SO2} - \gamma_{Rotor}) \qquad (10)$$

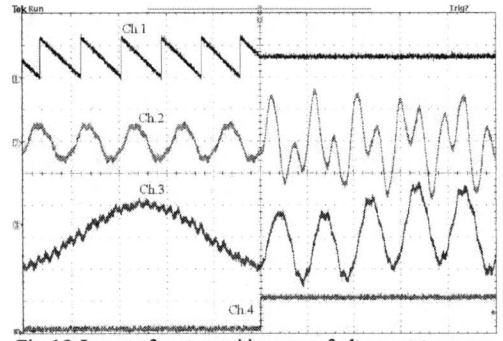

Fig. 15: Impact of a rotor position sensor fault on a stator power controlled drive at 10 kW and 1700 rpm. Ch. 1: Rotor position signal $[0..2\pi]$. Ch.2: Stator current [20 A/div] Ch.3: Rotor current [50 A/div] Ch.4: Fault On. Time [20 ms/div]

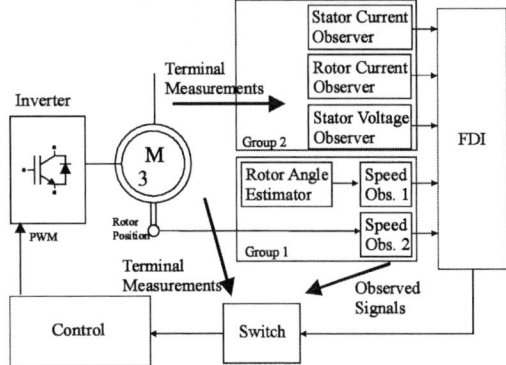

Fig. 16: Bank of observers, Fault Detection and Isolation Unit, and switching from measured to observed / estimated signals

C. Fault Isolation

In the following, the process of fault isolation is described in detail. It is started when the Fault Detect Counter is larger than zero. Real measurement data is used to illustrate how the location of the fault is isolated using the two speed observers in group 1 of fig. 16.

The measurement data is sampled using the ControlDesk environment of dSPACE, then it is plotted using MATLAB. The data is taken during the experiments shown in figure 19, so the underlying fault isolation process can be compared with the measurement results.

As mentioned two speed observers are used. The first one tracks the rotor position sensor signal γ_{Rotor}, the other one follows the estimated rotor position γ_{EST}. Faults in any of the electrical sensors lead to sudden false rotor position estimation, since their measurement is used in (1) and (2). For any fault, one observer will follow a faulty signal, either the false position sensor signal or the false estimation due to electrical fault. Thus the one observer will have a larger observer error, until it has swung in, to track the false position. Therefore, equations (9) and (10) are calculated, which are the absolute of the observer error as of eq. (8). They are used to decide whether the fault has happened in the mechanical sensor, which is the encoder, or in the electrical sensors, which are stator and rotor current as well as stator voltage sensors. No threshold is needed for fault isolation, instead eq. (9) and (10) are compared to each other. Since both observers are tuned identically, the larger one marks the faulty group, signalling either mechanical or electrical sensor fault. In order to be reliable, any result has to be steady for a defined period of time, in this case 4 ms.

```
If (Eq(10)>Eq(9))
    {
        CounterMech= CounterMech+0.0002;
        CounterElec=0;
    }
Else
    {
        CounterElec= CounterElec+0.0002;
        CounterMech=0;
    }
If ((FaultDetectCounter>0) && (CounterMech >0.004)) MechFault=1;
If ((FaultDetectCounter>0) && (CounterElec >0.004)) ElecFault=1;
```

Fig. 17: Pseudo code illustration of how mechanical faults are distinguished from electrical faults.

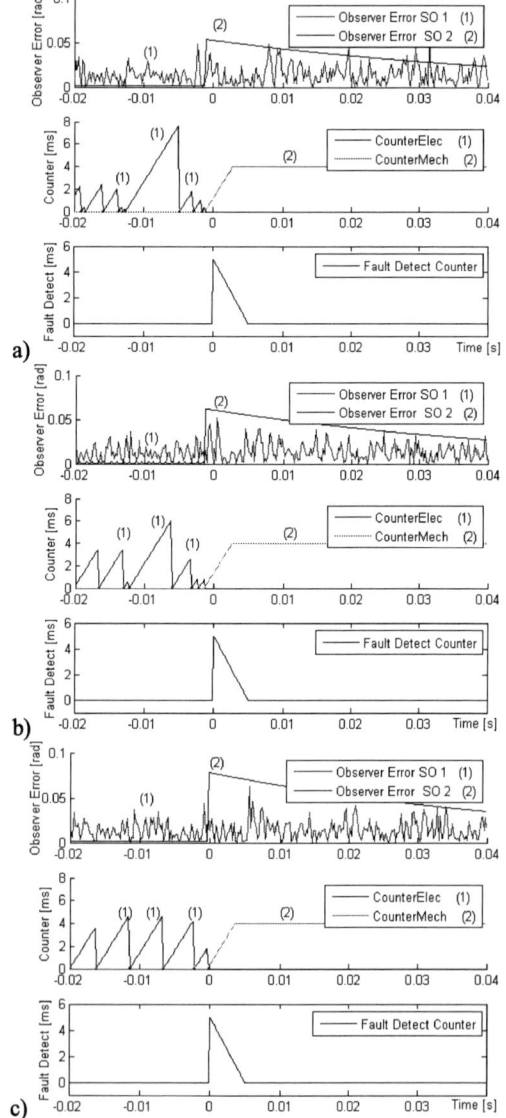

a)

b)

c)

Fig. 18 Fault Isolation of rotor position sensor at a) 1300 rpm, b) 1500 rpm, c) 1900 rpm., corresponding to fig. 17.

This is realized by counters, where each of the observers has got one counter, as is illustrated in fig. 17. For every period it is determined whether (9) or (10) is larger. One counter is increased, while the other is set to zero. Only if the same result is steadily calculated for a period of 4 ms, a fault is isolated. The fault isolation is only active after fault detection, e.g. when the Fault Detect Counter is larger than zero. If at this point the rotor position sensor is isolated as faulty, Fault Isolation is finished and Reconfiguration is started. For demonstration, the fault isolation of the rotor position sensor at 1300 rpm, 1500 rpm, and 1900 rpm is presented in fig. 17. They represent subsynchronous, synchronous and supersynchronous operation. Fig. 18 directly corresponds to fig. 19, that means both figures show different aspects of the same measurement.

The counters are only evaluated while the Fault Detect Counter is non-zero. Before that, they have random value. At all rotor speeds, the results are very comparable and isolation is clear and fast.

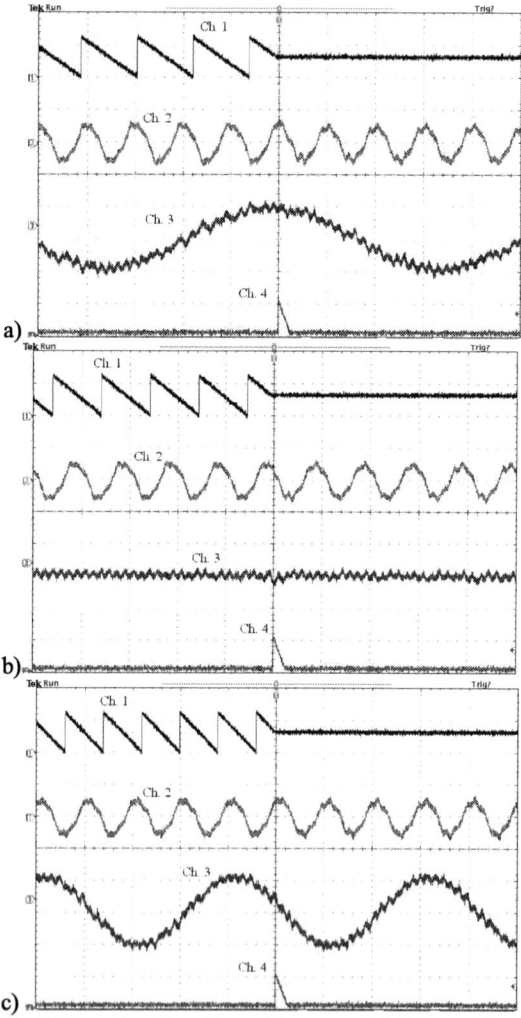

a)

b)

c)

Fig. 19: Reconfiguration of the position sensor at a) 1300 rpm, b) 1500 rpm, c) 1900 rpm. Ch. 1: Rotor position signal $[0..2\pi]$. Ch.2: Stator current [20 A/div] Ch.3: Rotor current [50 A/div] Ch.4: Fault Detection Counter [5 ms /div]. Time [20 ms/div]

For all measurements, the whole fdi scheme is active, that means including fault detection and isolation for all mentioned sensors, but only position sensor faults are considered. All measurements are taken at 400 V stator voltage, 10 kW stator power and unity stator power factor. Sample and switching frequency are 5 kHz. In figure 19 the position sensor is switched off in software. All figures show the position signal in Ch. 1, the stator current in Ch. 2, the rotor current in Ch. 3 and the Fault Detect Counter in Ch. 4. In fig. 19 b), the rotor current is almost a DC current due to operation at synchronous speed. As can be seen, the fault isolation and reconfiguration is successful at all presented rotor speeds. As can be seen in the stator current, there is a slight change in harmonic content, as has been presented in section III. Although the same fault as shown in fig. 15 has occurred, seamless continued generation is possible. It is shown that the drive is kept operational despite a position sensor fault.

V. CONCLUSION

As part of a multi-sensor fault tolerant control scheme for doubly fed induction generators, the fault detection and reconfiguration of the rotor position sensor is presented. A position estimator is presented, using current measurement on stator and rotor terminals of the generator. Iron losses and saturation have influence on the rotor position estimator.

Rotor position estimation is shown to have sufficient quality for sensorless operation, although there are characteristic harmonics in the estimated angle. Future work is necessary to derive the origin of these harmonics and remove them from the estimated position. The position estimator does not work properly for small rotor currents. Steady state sensorless operation is compared to encoder operation. The sensorless scheme shows increased stator current harmonics. Future work will have to address this issue, too.

Sensor reconfiguration is presented in detail for synchronous as well as sub- and supersynchronous speed and shows good and fast results. The machine is kept in operation in spite of position sensor faults by switching to the estimated rotor position. There are no torque spikes or other unwanted effects.

The presented scheme may help to improve the reliability and availability of doubly fed induction generator systems.

ACKNOWLEDGEMENT

This work was funded by the Deutsche Forschungsgemeinschaft (German Research Foundation).

REFERENCES

[1] A.D Hansen., F. Iov, F. Blaabjerg, L.H. Hansen – *Review of Contemporary Wind Turbine Concepts and their Market Penetration*, Journal of Wind Engineering, Vol. 28, No. 3, 2004, pp. 247-263;

[2] R. Patton, P. Frank, R. Clark: *Issues of Fault Diagnosis for Dynamic Systems*. Springer-Verlag, Berlin, 2000.

[3] R. Isermann: *Supervision, Fault-Detection and Fault Diagnosis Methods – an Introduction*, Control Engineering Practice, Vol. 5, pp. 639-652, 1997.

[4] R. Isermann, P. Balle: *Trends in the Application of Model-Based Fault Detection and Diagnosis of Technical Processes*, Control Eng. Practice, Vol. 5, No. 5, pp 709-719, 1997

[5] A. Petersson, L. Hernefors, T.Thiringer: *Evaluation of Current Control Methods for Wind Turbines Using Doubly-Fed Induction Machines*, IEEE Trans. Power Elec., Vol. 20, No. 1, pp 227-235, 2005

[6] L. Harnefors, M. Jansson, R. Ottersten, K. Pietiläinen: *Unified Sensorless Vector Control of Synchronous and Induction Motors*, Trans. Ind. Elec., Vol 50, No. 1, pp153-160,2003.

[7] A. Timbus, R. Teodorescu, F. Blaabjerg, M. Liserre: *Synchronization Methods for Three Phase Distributed Power Generation Systems. An Overview and Evaluation*, PESC07, Orlando, USA, CD-Rom Paper.

[8] R. Datta, V.T. Ranganathan: *A Simple Position Sensorless Algorithm for Rotor Side Field Oriented Control of Wound Rotor Induction Machine*, IEEE Trans. on Industrial Electronics, Band 48, No. 4, pp. 786-368, 2001.

[9] M. Abolhassani, P. Niazi, H. Toliyat, P. Enjenti: *A Sensorless Integrated Doubly-Fed Electric Alternator/Active Filter (IDEA) for Variable Speed Wind Energy System*, 38th Ind. Apl. Conference, vol. 1, pp 507-14, Salt Lake City, USA, 2003 .

[10] B. Hopfensberger, D. Atkinson, R. Lakin, *Stator-flux-oriented control of a doubly-fed induction machine with and without position sensor*, IEE Proc.-Electr. Power Apll, Vol. 147, No. 4, pp 241-250, 2000.

[11] L. Morel, H. Godfroid, A. Mirzaian, J. Kauffmann, *Double-Fed Induction machine: converter optimization and field oriented control without position sensor*, IEE Proc.-Electr. Power Apll, Vol. 145, No. 4, pp 360-368, 1998.

[12] L. Xu, W. Cheng: *Torque and Reactive Power Control of a Doubly Fed Induction Machine by Position Sensorless Scheme*, IEEE Trans. Ind. Appl. Vol. 31, No. 3, pp 630- 642, 1995

[13] D. Navarro Gevers, *Beitrag zur Regelung einer doppeltgespeisten Asynchronmaschine ohne Lagegeber für Windkraftanlagen*, (in German) PhD Thesis, University of Ilmenau, Germany, 2004.

[14] G. Iwanski, W. Koczara: *Sensorless Stand Alone Variable Speed System for Distributed Generation*, PESC'04, pp 1915-1921, Aachen, Germany 2004.

[15] E. Bogalecka, Z. Krzeminski, *Control Systems of Doubly-Fed Induction Machine Supplied by Current Controlled Voltage Source Inverter*, Int. Conf. On Electrical Machines and Drives, pp 168-172, Oxford, Uk, 1993.

[16] K. Rothenhagen, F.W. Fuchs: *Current Sensor Fault Detection by Bilinear Observer for a Doubly Fed Induction Generator*, IECON'06, Paris, France. CD-ROM Paper.

[17] K. Rothenhagen, S. Thomsen, F.W. Fuchs: *Voltage Sensor Fault Detection and Reconfiguration for a Doubly Fed Induction Generator*, SDEMPED'07, Krakow, Poland.

[18] K.Rothenhagen, F.W.Fuchs. *Current Sensor Fault Detection and Reconfiguration for a Doubly Fed Induction Machine*, PESC07, FL, USA.

[19] K. Rothenhagen, F.W: Fuchs. *Current Sensor Fault Detection, Isolation and Control Reconfiguration for Doubly Fed Induction Generators*, IECON'07, Taipei, Taiwan, 2007.

[20] R. Teodorescu, F. Blaabjerg, U. Borup, M. Liserre: *A New Control Structure for Grid-Connected LCL PV Inverters with zero Steady-State Error and Selective Harmonic Compensation*, Applied Power Electronics Conference and Exposition, Vol.1, pp. 580-586, 2004.

Wide Range Low Noise Current Sensor

F. Richter, C. Sourkounis

Ruhr-University Bochum, Research Group of Power Systems Technology and Power Mechatronics, Bochum,
Germany, e-mail: *richter@eele.rub.de*

Abstract— Compared to conventional current sensors, the proposed topology is advantageous because of its high resolution throughout its full measurement range, thus reducing offsets to a minimum. This aim is reached through active offset compensation using an analog/digital chopper/de-chopper. A further advantage is the excellent suppression of 1/f noise and a low noise density (35 nV/√Hz) down to the DC-threshold. Common current sensors may be replaced with this technology whose one output is likewise analog. The power supply requirements are also far lower than for standard topologies as only a single supply is needed.

Keywords— components for measurement, measurement, sensor

I. INTRODUCTION

Precise and robust current sensors are necessary for determination of the current state and for current control. Today current sensors based on the Hall Effect are used. The system proposed in this paper is based, in contrast to conventional systems, on an indirect measurement of the current with the combination of a shunt resistor and electronic devices (s. fig. 1).

Fig. 1. The new current sensor in a configuration for low voltage
without potential seperation.

The use of a shunt resistor features advantages over the Hall Effect based sensors in accuracy, stability, thermal coefficients, construction volume and price. Also the measurement range of the system can be adapted quite easily by changing the shunt resistor. Thereby measurement ranges from 1A to over 1500A (or more, depending on the availability of shunt resistors) are possible. However, since it has no potential separation by its function principle, this has to be done at another stage.[1]

The output signal of the shunt resistor is fed into and processed by the sensor chip's measurement electronics. Using the special sensor chip topology, the input offset and the thermal drift of the measurement system is al-most entirely eliminated. Simultaneously the 1/f-noise of the

---

The authors would like to thank the Isabellenhütte Heusler GmbH & Co. KG for supporting this research project.

CMOS-Amplifier can be suppressed to such a degree, that an extreme low noise density can be realised. Thereby currents of a few mA can be measured in a measurement range of 1500A, which has been proven by comparative research.

The amount of auxiliary power is considerably lower compared to conventional solutions. For instance, the entire measurement system only requires 10% of the energy a sensor based on the Hall Effect with its peripherals uses. Moreover, only a single power supply is needed, with few requirements regarding the power quality. For a sensor based on the Hall Effect a bipolar power supply with low noise is necessary for precise measurement.

Due to the discontinuous operation of power electronic devices, the current sensors are subject to floating potentials with gradients of up to 6 kV/µs, which result in charge displacements in the parasitic coupling capacitors, whereby the functionality may be interfered. Furthermore it is necessary to perform every current measurement using the voltage drop at the shunt resistor on the so called "high-side" because of the switching operation and the involved dynamic potential displacement. This follows from the requirement of avoiding dynamic potential displacements of the 0V-reference potential in the dimension of the energy supply system (e.g. on-board electrical system). Thereby, a potential-free measurement and data transmission to the control logic are necessary. The evidence of the isolation ability is provided within the scope of the tests.

II. STRUCTURE OF THE MEASUREMENT SYSTEM

The measurement system is based on the indirect measurement of the current over a voltage drop, which is caused by the flow of the current through a precision resistor (shunt resistor). The system consists, beside the shunt resistor, of an integrated measurement circuit, a µ-controller and a potential separation, which can be an optocoupler or a fiber optic device (s. fig. 2). Thereby an undisturbed transmission of the measurement signals is assured even over long distances.

The integrated measurement circuit represents a complete 16-bit data acquisition system in a SOIC16 package (IHM-A 1500 from Isabellenhütte Heusler GmbH & Co. KG), which is optimised for the provision of extremely low voltages in the µV-/mV-range, such as the voltage drop at a shunt resistor. With an external clock of max. 8MHz, the integrated circuit can sample the signal with a maximum conversion rate of 16 kHz (or in a special operation mode up to 64 kHz). The integrated measurement circuit is able to acquire mass related input signals with positive and negative signs even with a unipolar supply.

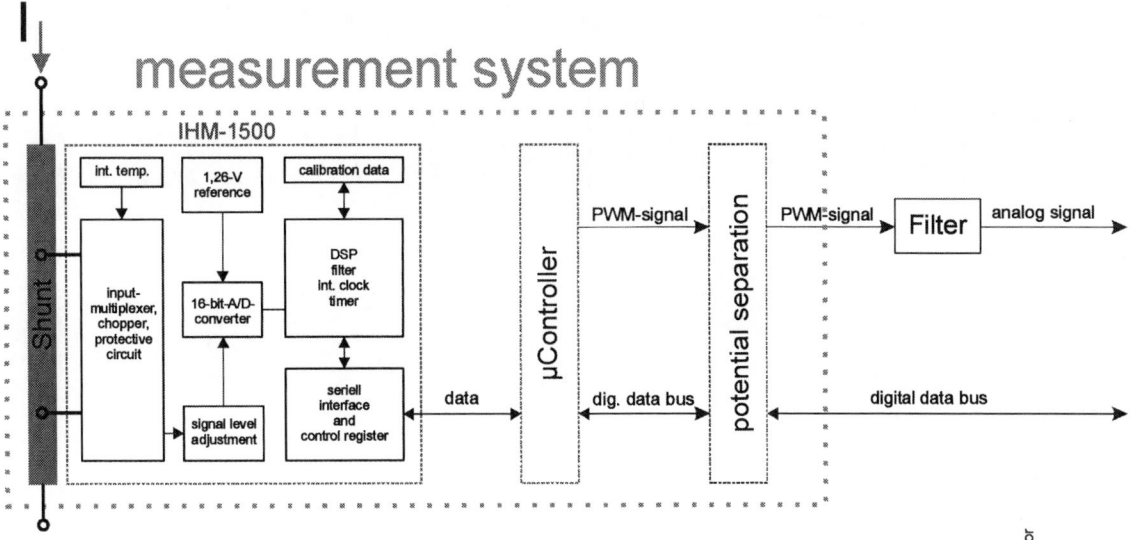

Fig. 2. Structure of the measurement system

As a special feature, the integrated measurement circuit contains an analog/digital chopper/dechopper unit for active offset compensation. Thereby the input offset and the thermal drift of the system are eliminated almost completely (offset is below 0,5μV). Through a very high chopping frequency the common 1/f noise of CMOS-amplifiers is suppressed so, that an very low noise density of only 35 nV/√Hz in the DC-range is realised.

Besides, the integrated measurement circuit contains a programmable measurement filter allowing conversion rates from 2 Hz up to 16 kHz. Thereby, the whole signal processing can be done by the integrated measurement system relieving the μ-controller of the control logic.

The integrated measurement circuit is initialised at the start-up with the internally stored calibration data and then operates autonomously as an independent A/D-converter. The result of the conversion is read out by a μ-controller and is then sent through a potential separated connection to the control logic. Thereby different types of transmission are possible: on the one hand a transmission based on PWM-signals is possible, which can be processed digitally or can be converted to an analog signal. This analog signal does not differ from the signal of a conventional current sensor based on the Hall effect, so that a care-free interchange between such systems is possible. On the other hand, a transmission via a digital data bus can take place, so that the data can be used directly in the control logic. Different bus types are possible (SPI, I²C, CAN and so on), depending on the required speed (sampling rate and data transmission).

III. RESULTS OF THE MEASUREMENTS

The setup for processing the measurements is made up of a three-phase current supply with a variable transformer, subsequent rectification and step-down chopper. The measurement systems were integrated in the design of the step-down chopper (s. fig. 3).

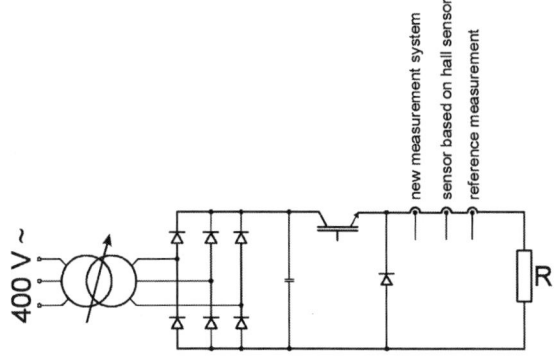

Fig. 3. Scheme of the measurement setup

First some measurements were carried out to certify the potential separation of the new measurement system. To do so, a voltage of 560V with a voltage gradient of 3700V/μs was connected to the system (s. fig. 4).

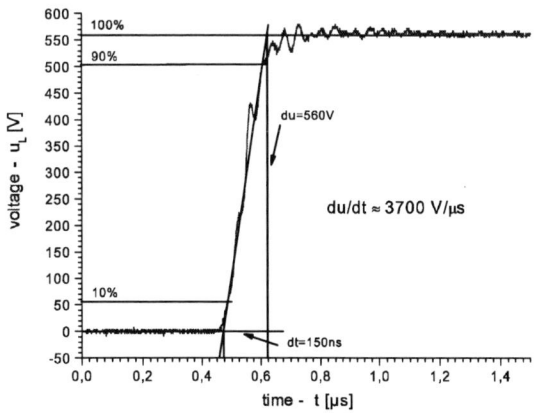

Fig. 4. Maximum voltage gradient

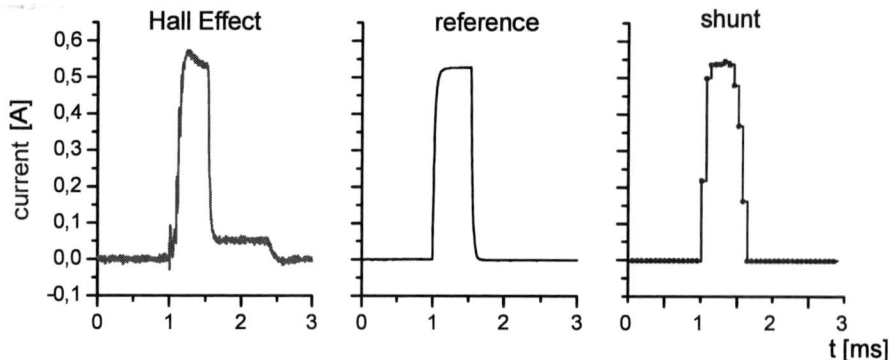

Fig. 5. Measurements at low currents

A reliable operation under this load has always been possible. The current flow through the parasitic capacitors was so low, that neither accuracy nor functionality of the measurement system were affected.

Thereafter measurements with different current amplitudes were processed in a metering range of +/- 300 A. For this measurement a "conventional" current sensor based on the Hall Effect (ABB type ES300) was also used for a comparison. A reference signal for the dynamic behavior was recorded by a current measurement system from Tektronix (A6303 with AM501).

At very low currents, the resolution of the proposed current measurement system is far better than that of the "conventional" sensor, even when the offset of the "conventional" sensor is compensated by adequate mathematic measures (s. fig. 5).

With increasing amplitude, the deviation between the "conventional" and the new sensor gets smaller and smaller until a current of 5A is reached. As it is quite visible from the current plot, the conversion rate of the new measurement system is high enough to display the current profile (s. fig. 6). If the current increases further, there are almost no deviations between the measurements (s. fig. 7).

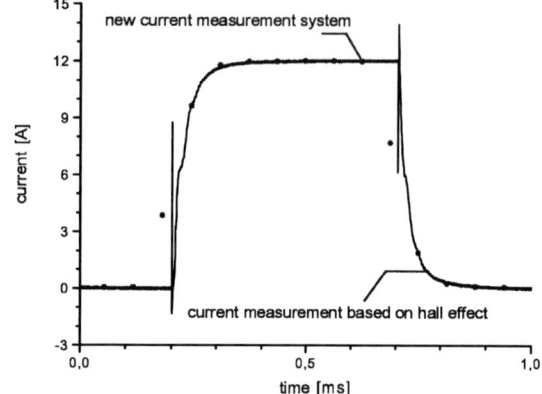

Fig. 6. Measurements at medium current

IV. CONCLUSION

The new current measurement system represents an alternative to "conventional" current sensors, which have a higher resolution and accuracy at a lower price. Besides that, not only an analog output is available, but digital data is provided from a variety of data busses. This is a considerable advantage for applications in power electronic devices, because no analog signals, which are rather susceptible to interference, have to be transmitted to the control logic. An A/D-converter in the control logic is also no longer necessary.

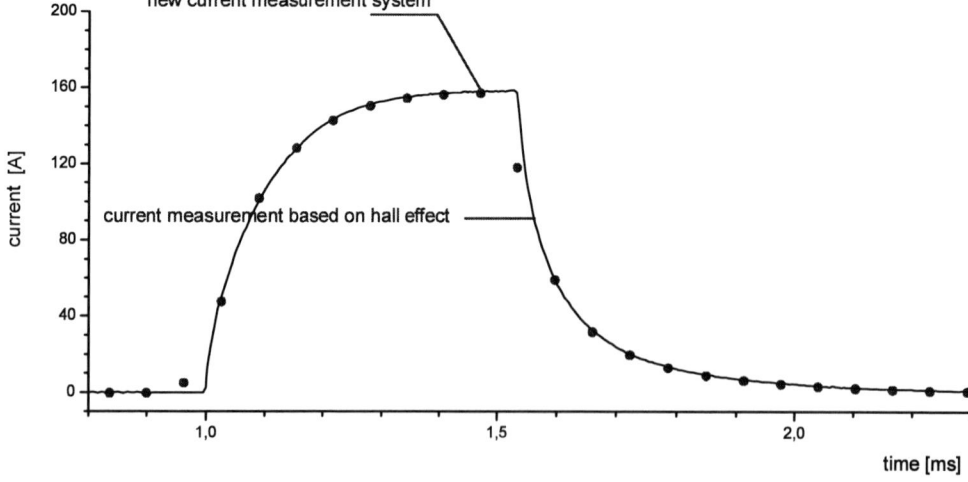

Fig. 7. Measurements at high current

REFERENCES

[1] Dr. Ullrich Hetzler, Dr. Drago Strle, Bernd Wille, "Präzisionsstrommessung von mA bis kA", Elektronik 2/2003, pages 34 – 38

[2] Wolfgang Göpel, Julian W. Gardner, Joachim Hesse (Hrsg.), "Sensors Applications",5 Volumes, Wiley-VCH, Weinheim

[3] W. Göpel, J. Hesse, N. Zemel, "Sensors A Comprehensive Survey, Vol. 1-9", VCH Verlagsgesellschaft, Weinheim

Transducerless Speed Control with Initial Position Detection for Low Cost PMSM Drives

Roman Filka*, Peter Balazovic*, Branislav Dobrucky†

*Freescale Semiconductor CZ/ Roznov pod Radhostem, CZ, e-mail: *Roman.Filka@freescale.com*
*Freescale Semiconductor CZ/ Roznov pod Radhostem, CZ, e-mail: *Peter.Balazovic@freescale.com*
†Zilinska Univerzita v Zilina/KME, Zilina, SK, e-mail: *branislav.dobrucky@fel.uniza.sk*

Abstract—This paper presents an approach for vector control of IPM synchronous motor without position transducer, applicable for low cost PMSM drives with unpredictable load and inertia. To achieve correct drive operation from zero speed, two different techniques are combined with crossover function, based on the speed reference. The rotor initial position is determined prior the control action, for the drive to deliver the full start up torque. Since back-EMF observer estimations are insufficiently accurate under limited angular speed, the zero speed startup and low speed acceleration is used. The back-EMF observer detects the generated motor voltages induced by the permanent magnets. An angle tracking observer uses the back-EMF signals to calculate the position and speed of the rotor. The experimental results based on dsp MC56F8025 show sufficient performance for washing machine, pumps, and fan applications.

Index Terms—AC machine, adjustable speed drive, control of drive, electrical drive, estimation technique, motion control, permanent magnet motor, self-sensing control, sensorless control, synchronous motor, variable speed drive, vector control.

I. INTRODUCTION

Knowing the exact position of the rotor flux, is one of the key factor enabling utilization of the Field Oriented Control (FOC) for AC motors. Because the rotor flux is mechanically linked with the rotor in case of PM synchronous motors, the position of the rotor flux can be well established by reading the position of the rotor. Rotor position mechanical sensors such as hall effect sensors, encoders or resolvers are widely used to provide information about rotor position. These sensors however, introduce additional harness and complexity into the drive system. Removing the mechanical rotor position sensor, does reduce the overall mechanical complexity of the drive and resulting cost on one side, but brings up a broad range of challenges to tackle on the other.

Even though there is a wide interest in research of sensorless control techniques, a single method enabling sensorless control across a full speed range has not yet been proposed. Therefore it has become somehow a standard practice to combine different sensorless techniques for different operational ranges of the drive. Nowadays known sensorless techniques may be broadly categorized as: motion electromotive force (EMF) based, inductance or flux linkage sensing.

The EMF estimation based sensorless drives are typically limited to operation above 1Hz. Especially at low speeds, the speed estimation of model-based approaches is poor or fail thus it is difficult to achieve correct torque and flux control. Sensorless control schemes based on fundamental frequency machine models are suitable for operation in higher speed ranges. Application requiring operation at zero and low speed cannot rely solely on these methods. This is owed to the loss of information on the rotor state when voltage generated by the revolving rotor becomes very small as the rotor speed reduces. This effect is not limited to a particular motor type, it is associated with asynchronous as well as with synchronous motors.

II. PROBLEM ANALYSIS AND DRIVE REQUIREMENTS

High frequency injection methods offer the possibility to observe the machine speed and/or position independent on the fundamental supply voltage and currents [1], [2]. These methods however, require a minimum saliency ratio to reliably track the spatial saliency position. Therefore the implementation of such method is restricted to PM motor with magnets inset or buried in the rotor and is not suitable for low saliency surface mount PM motors. Nevertheless, even in case of IPM motors, the spatial modulation of saliency rely on physical magnetic characteristics of the motor. Therefore phenomenons such as the stator core saturation with applied load, armature reaction, physical shape of the magnets, uniform distribution of the air-gap flux etc., influence the quality and shape of the high frequency response signal, generated by the *hf* excitation of the motor. Thus also the position estimation, based on analysis of such distorted high frequency motor response, is impaired, which results in poor performance of the low speed sensorless control algorithms. These facts are continually being proved by reluctance of the whole industry to fully employ such sensorless control methods.

A method, presented in this paper, is focused on sensorless control of IPM synchronous motor designed to be implemented in a standard appliance drive of washing machine. The low speed sensorless control is achieved by utilization of integration methods. This solution however, does not allow continuous control within the low speed region and is therefore limited to applications where only a "pass through" the low speed region or simple startup to higher speeds are required. Nonetheless, in order to generate full start-up torque, initial position of the rotor has to be correctly estimated. That, in case of PM synchronous motors, consists of method for initial axis alignment followed by detection of the correct magnet polarity.

The drive dynamics required for the horizontal axis washing machine applications is in general much lower than the one required by a standard servo drive [3], [4]. The typical operation profile of such washing machine can be divided into two cycles: washing and spinning. During the washing cycle the speed of the drum is kept relatively low (typically $< 100 \div 150[rpm]$). However because the motor is connected to the drum via belt, the speed of the motor during washing is in the range of 400-1000[rpm]. Moreover the wash cycle in horizontal axis washing machines does not require continuous speed reversal, instead the drum speed first drops to zero and only then the drive accelerates to the opposite direction. The spin cycle on the other hand, requires drum speed to be much higher than it was for washing. The high end washers on the market nowadays can reach as much as 1600-2000[rpm] on the drum, so given the belt ratio the speed of the rotor is well above 15000[rpm].

The torque generation capability of the horizontal axis washing machine drive is most critical during the start-up sequence. Because the soaked clothes and water are accumulated in the bottom section of the drum when not spinning, typically a maximal start-up torque is required to move the drum to the position where the clothes start to tumble. The required torque then drops as the speed increases. The typical speed-torque characteristic of horizontal axis washing machine is depicted on Fig. 1.

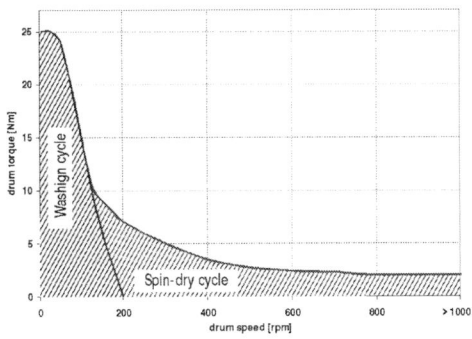

Fig. 1: Torque-speed characteristic of H-axis washer application measured on the drum.

A. Implementation of Low Cost Drive

A large number of produced units is typical sign for home appliance market. This is also valid for washing machine market. Considering the large number of washing machines produced, the overall cost of the drive is an important issue from commercial point of view. There is a number of ways to minimize the overall drive cost. Amongst these there are also removal of the position transducer and use of low end Digital Signal Controller

(DSC) for implementing the control algorithm. To further reduce the cost, the expensive hall effect sensors are substituted by shunt resistors placed in each of the inverter leg as phase current sensing [5]. Stator phase current which flows through the shunt resistor produces a voltage drop which is interfaced to the AD converter of the DSC through conditional circuitry.

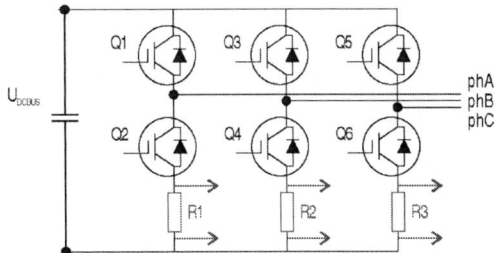

Fig. 2: Three phase voltage source inverter with shunt resistors in each inverter leg

Because the shunt resistors are placed bellow the bottom switch of the inverter leg the phase current can not be measured at an arbitrary moment. The current can only be measured when the bottom transistor of the inverter leg is switched on or when the current flows through the freewheeling diode. In order to get an actual instant of current sensing, voltage waveform analysis has to be performed. Generated duty cycles (phase A, phase B, phase C) of two different PWM periods are depicted in Fig. 3.

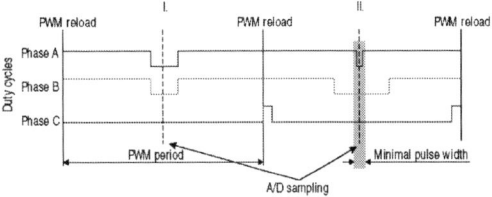

Fig. 3: Three phase voltage source inverter with shunt resistors in each inverter leg

In the case of standard motor operation where the supplied voltage is generated under space vector modulation, the sampling instant of phase current takes place in the middle of the PWM period in which all bottom transistors are switched on. If modulation index of applied SVM technique increases there is an instant when one of the bottom transistors is switched on for a very short time period. Therefore, only two currents are measured and the third current is calculated from equation:

$$i_a + i_b + i_c = 0 \qquad (1)$$

therefore, a minimum on time of the low-side switch is required for three phase current reconstruction.

Considering 3-phase VSI, six non-zero and two zero vectors can be produced. The non-zero vectors create a

1403

voltage hexagon if plotted in complex plane. Therefore the area for possible placement of the required voltage vector is divided into six sectors, each 60° wide. Which currents are measured and which is calculated, depends on PWM pulse width i.e. on required phase voltage. The simplest technique is to calculate the current of the most positive voltage phase, when the duty cycle of the given phase is high. For example, Phase A generates the most positive voltage within sector 1 (0 to 60°), Phase B within section 1 (60° to 120°), and so on. Thus the phase current can be reconstructed as follows:

- Sector 1 and 6: $i_a = -i_b - i_c$
- Sector 2 and 3: $i_b = -i_a - i_c$
- Sector 4 and 5: $i_c = -i_a - i_b$

III. TRANSDUCERLESS FIELD ORIENTED CONTROL

Low cost speed regulation of a PM synchronous motors can be achieved by employing a Field Oriented Control (FOC) method with current regulation and speed control based on estimated rotor position and speed. The FOC is an advanced control method that enables highly dynamic and efficient control of the motor. This is achieved by defining a synchronous reference frame placed such that the required current vector can be decomposed into a magnetic field generating part and a torque generating part. In order to correctly place the synchronous frame, position of the rotor flux (i.e. the rotor position in case of PM synchronous motor) must be known or estimated.

To achieve the transducerless control across the full speed range of the motor, three areas of drive operation must be addressed:

- drive initialization
- start-up
- closed loop speed control

The correct drive initialization in terms of finding the initial rotor position is essential for full torque generation during the start-up sequence. Because the speed profile of the used low cost drive does not require operation in the low speed range

A. Drive Initialization

Detection algorithm based on injection of pulsating *hf* signal, with angular frequency ω_c, in synchronous frame is used to estimate the initial rotor position [2]. Physically, this sensorless approach is based on property of d and q axes flux being decoupled. This can be mathematically explained by existence of mutual coupling between the two phases of a quadrature phase model, if the model reference frame is not aligned with the rotor dq frame. Therefore if an estimated $\hat{d}\hat{q}$ reference frame is defined and not precisely aligned with a real rotor dq reference frame, then by applying flux vector at known carrier frequency for example in the $\hat{d}$-axis, current at the injected carrier frequency can be observed in the $\hat{q}$-axis. This current is directly proportional to the misalignment angle of the estimated $\hat{d}\hat{q}$ with real rotor dq reference frame and therefore is used to drive the estimated position such that the $\hat{q}$-axis current at carrier frequency is zero.

Considering only high frequency signals, the motor model in estimated reference frame can be derived as [1]:

$$\begin{bmatrix} \hat{u}_d \\ \hat{u}_q \end{bmatrix} = j\omega_c \mathbb{L} \begin{bmatrix} \hat{i}_d \\ \hat{i}_q \end{bmatrix} \tag{2}$$

where

$$\mathbb{L} = \begin{bmatrix} L_0 + \Delta L \cos(2\theta_{err}) & -\Delta L \sin(2\theta_{err}) \\ -\Delta L \sin(2\theta_{err}) & L_0 - \Delta L \cos(2\theta_{err}) \end{bmatrix} \tag{3}$$

and

$$L_0 = \frac{L_d + L_q}{2}, \Delta L = \frac{L_d - L_q}{2}, \theta_{err} = \theta_e - \hat{\theta}_e \tag{4}$$

Applying a high frequency signal $u_{hf} = U_m \sin(\omega_c t)$ in $\hat{d}$-axis of the model (2), results in a high frequency current response:

$$\begin{bmatrix} \hat{i}_d \\ \hat{i}_q \end{bmatrix} = -\frac{\hat{U}_m}{\omega_c L_d L_q} \cos(\omega_c t) \begin{bmatrix} L_0 - \Delta L \cos(2\theta_{err}) \\ \Delta L \sin(2\theta_{err}) \end{bmatrix} \tag{5}$$

After filtering and demodulation, the current signal in $\hat{q}$-axis is described as:

$$\hat{i}_q = -\frac{U_m \Delta L}{2\omega_c L_d L_q} \sin(2\theta_{err}) \tag{6}$$

Both real rotor and estimated reference frames are aligned the current signal representing the position estimation error, as derived in (6) is of zero amplitude. Therefore a *phase locked loop* (PLL) mechanism has to be designed to drive position of the estimated reference frame $\hat{\theta}_e$ such that the amplitude of the error signal will be zero. The output of such PLL mechanism is the position of the motor saliency and therefore this PLL is called *saliency tracking observer* (STO).

As is derived in (6), the position estimation error is reflected in $\hat{i}_q$ current as a sine function with double frequency. This will produce zero position estimation error at periodic locations as follows:

$$\begin{aligned} 0 &= -\frac{\hat{U}_m \Delta L}{\omega_c L_d L_q} \cos(\omega_c t) \sin(2\theta_{err}) \\ &\Rightarrow \theta_{err} = n \cdot \frac{\pi}{2} \qquad n = 0, \pm 1, \pm 2, \pm 3 \ldots \end{aligned} \tag{7}$$

Since the position estimation error will be zero at any integer multiple of $\frac{\pi}{2}$, the closed loop tracking observer will have an infinite number of equilibriums. Each equilibrium point ($n\frac{\pi}{2}; n = 0, \pm 2, \pm 4, \pm 6 \ldots$) has an attraction region in which the STO stabilizing trajectories are forced to the equilibrium as shown in Fig. 4. However every second equilibrium is aligned with the negative direction of the permanent magnet flux vector, and therefore if the STO stabilizes in this location the estimated position will have an opposite rotation. This will result in positive feedback in the speed loop, and hence instability of the whole control system.

The PM polarity detection method is based on the stator core saturation phenomenon. The stator core magnetic operating point is placed on the knee of the hysteresis curve, making the saturation phenomenon visible even with small variations in i_d current. Therefore the *hf* carrier

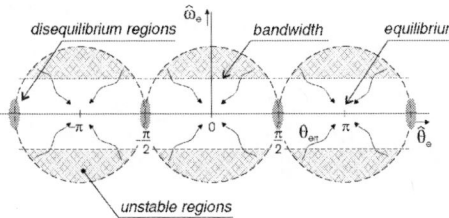

Fig. 4: STO attraction regions around each even equilibrium point.

signal voltage injection in the d-axis used for the position estimation, can also be utilized for the permanent magnet polarity detection [6]–[9].

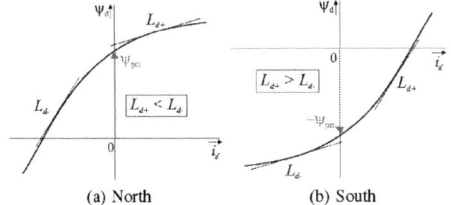

(a) North   (b) South

Fig. 5: D-axis flux and resulting inductance L_d as a function of i_d corresponding to the (a)north and (b)south pole direction.

Fig. 5a shows the theoretical relationship between d-axis flux linkage ψ_d and current i_d when the d-axis corresponds to the north pole direction and Fig. 5b when the d-axis corresponds to the south pole direction. As proposed in [7] the d-axis current i_d as a function of the d-axis flux linkage ψ_d can be approximated using a second order Taylor series expansion as follows:

$$i_d\mid_{north} \approx \frac{1}{L_d}(\psi_d - \psi_{pm}) - \frac{1}{2}\frac{d^2 i_d}{d\psi_d^2}(\psi_d - \psi_{pm})^2 \quad (8)$$

$$i_d\mid_{south} \approx \frac{1}{L_d}(\psi_d - \psi_{pm}) + \frac{1}{2}\frac{d^2 i_d}{d\psi_d^2}(\psi_d - \psi_{pm})^2 \quad (9)$$

where the second term represents the saturation effect. As can be understood from Fig. 5a and 5b, the sign of the second term is negative if the STO is aligned with the north pole (8), and positive if the STO is aligned with the south pole (9). The d-axis linkage flux is derived with considering the hf carrier signal voltage $u_d = U_m \sin(\omega_c t)$ as follows:

$$\psi_d = \int u_d dt = -\frac{U_m}{\omega_c}\cos(\omega_c t) \quad (10)$$

Substituting (10) into (9) and adopting notation of reluctance R_d instead of inductance L_d, a hf current i_d is derived as follows:

$$i_d = I_{d\_DC} + I_{d\_c}\cos(\omega_c t) + I_{d\_2c}\cos(2\omega_c t) \quad (11)$$

where I_{d_DC} is amplitude of the DC offset, I_{d_c} is amplitude at the carrier and I_{d_2c} at the second harmonic

of the carrier:

$$R_d = \frac{1}{L_d} \quad (12)$$

$$R'_d = \frac{1}{2}\frac{d^2 i_d}{d\psi_d^2} \quad (13)$$

$$I_{d\_DC} = \frac{1}{2}\frac{U_m^2 R'_d + \psi_{pm}^2 R'_d \omega_c^2 - \psi_{pm} R_d}{\omega_c^2} \quad (14)$$

$$I_{d\_c} = \frac{U_m(2R'_d\psi_{pm} - R_d)}{\omega_c} \quad (15)$$

$$I_{d\_2c} = \frac{1}{2}\frac{U_m^2 R'_d}{\omega_c^2} \quad (16)$$

The sign of the second term of the Taylor expansion in (8) and (9), which determines the rotor magnet polarity orientation, is reflected in the sign of the component at the second harmonic of the carrier I_{d_2c}. Therefore in order to determine the rotor magnet polarity the sign of I_{d_2c} has to be extracted.

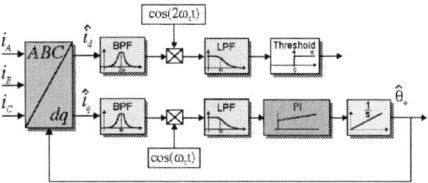

Fig. 6: Block diagram of the setup for IPM sensorless alignment with PM polarity detection to remove π ambiguity.

The block diagram of the structure for initial position estimation including both the magnet axis alignment as well as the magnet polarity detection is shown in Fig. 6. The magnet polarity information is located at $2\omega_c$ but there is also a signal at ω_c which has much greater magnitude than that at $2\omega_c$ (because $R_d \gg R'_d$). Therefore a BPF with center frequency at $2\omega_c$ has to be designed to sufficiently attenuate the carrier current $I_{d\_c}\cos(\omega_c t)$, which practically means a higher order BPF.

B. Open loop start-up

Upon completion of identifying the initial rotor position, the vector control is in open-loop mode. The vector transformations are fed by a time varying reference position derived by integrating the speed ramp reference whilst the current set-point is determined by the speed controller. The proportional integral controller of speed control loop is initialized to maximum allowable current I_{MAX}, which results in maximum torque reference current i_q^{ref}. The angular speed feedback is kept zero. This allows to deliver the necessary torque that accelerates the motor at the rate equal to the angular speed output of the back-EMF observer. The position used in vector transformations is calculated by integration of the speed ramp reference [4]. This ramp of the reference speed command is carefully chosen in order to assure safe starting with minimum oscillation up to the maximum

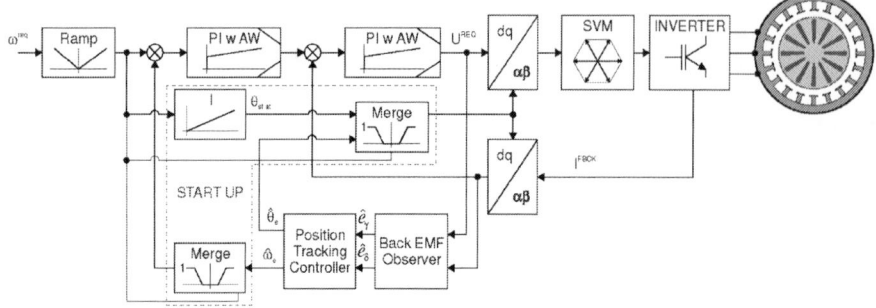

Fig. 7: Sensorless speed control with open loop startup and back-EMF observer

torque. Since the start-up torque is greater than the one required by the load, the actual rotor position advances the "forced" start-up position.

C. Closed Loop Speed Control

When the minimum operating speed is reached a measurable level of back EMF is generated by the rotor permanent magnets. The back emf observer is then gradually transition into the closed loop mode. The feedback loops are then controlled by the estimated angle and estimated speed signals from the back emf observer. The estimation

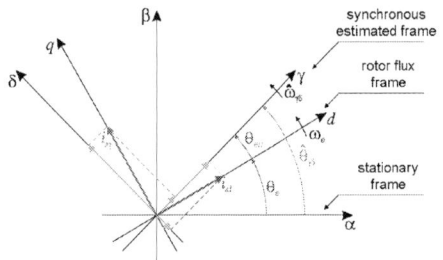

Fig. 8: Estimated $\gamma\delta$ and real rotor dq synchronous reference frames

method for the position and angular speed, considering high speeds, is based on the motor mathematical model of interior PMSM motor with an extended electro-motive force function [10], transformed into estimated quasi synchronous reference frame $\gamma\delta$, as depicted on Fig. 8. The transformed model is then derived as follows:

$$\begin{bmatrix} u_\gamma \\ u_\delta \end{bmatrix} = \begin{bmatrix} R_s + sL_d & -\hat{\omega}_{\gamma\delta}L_q \\ \hat{\omega}_{\gamma\delta}L_q & R_s + sL_d \end{bmatrix} \begin{bmatrix} i_\gamma \\ i_\delta \end{bmatrix} + E_{\text{sal}} \begin{bmatrix} -\sin(\theta_{err}) \\ \cos(\theta_{err}) \end{bmatrix} \quad (17)$$

where $\hat{\omega}_{\gamma\delta}$ is the angular speed of the estimated $\gamma\delta$ reference frame and E_{sal} the saliency based back-EMF voltage [11] also known in literature as "Extended back-EMF" (EEMF) [10] defined as:

$$E_{\text{sal}} = (L_q - L_d)si_q + (L_d - L_q)\omega_e i_d + \omega_e \psi_{pm} \quad (18)$$

This extended back-EMF model includes both position information from the conventionally defined back-EMF

and the stator inductance as well. In the steady state, the estimated reference frame rotates synchronously with the dq reference frame, i.e. $\hat{\omega}_{\gamma\delta} = \omega_e$. As can be seen from the model (17), the $2\theta_{err}$ term is not contained in the model equations anymore and the θ_{err} term only appears with saliency based back-EMF voltage. Therefore position estimation can now be performed by extracting the θ_{err} term from the model and adjusting the position of the estimated reference frame such as to achieve $\theta_{err} = 0$. Because the θ_{err} term is only included in the saliency-based EMF component of both u_γ and u_δ axis voltage equations in (17), a Luenberger based disturbance observer [12], [13] is designed to observe these voltage components.

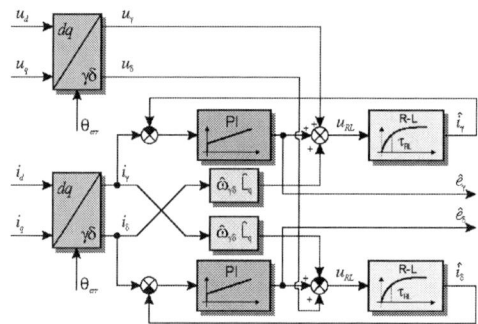

Fig. 9: Block diagram of proposed Luenberger type stator current observer acting as state filter for back-EMF.

Block diagram of the observer in the estimated reference frame is shown on Fig. 9. The observer compensator is substituted by a standard PI controller. As can be noted from Fig. 9, observer model and hence also PI controller gains in γ axis are identical to those in δ axis. The position displacement information θ_{err} is then obtained from estimated back-EMFs as follows:

$$\theta_{err} = \tan^{-1}\left(\frac{-\hat{e}_\gamma}{\hat{e}_\delta}\right) \quad (19)$$

Now, the estimated position can be obtained by driving the position of the estimated reference frame $\gamma\delta$ such as to achieve zero displacement $\theta_{err} = 0$. Therefore a

1406

phase locked loop mechanism must be adopted, where the loop compensator ensures correct tracking of the actual rotor flux position by keeping the error signal zero, $\theta_{err} = 0$. Such position tracking observer, with standard PI controller used as the loop compensator, is depicted on Fig. 10. The position tracking structure, as

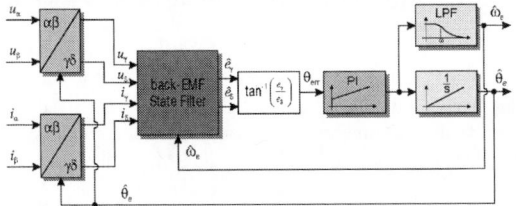

Fig. 10: Block diagram of proposed PLL scheme for position estimation at high speed region, using saliency based back-EMF state filter in rotating reference frame.

shown on Fig. 10, can be linearized around operating point $\hat{\theta}_e|_{o.p.} = \theta_e$. Hence the transfer function of the linearized position tracking observer, with a standard PI controller as a compensator, is derived as follows:

$$\frac{\hat{\theta}_{\gamma\delta}}{\theta_e} = \frac{K_p s + K_i}{s^2 + K_p s + K_i} \quad (20)$$

Neglecting the $K_p s$ term in the nominator of the transfer function (20), the PI gains K_p and K_i are calculated by comparing the characteristic polynomial of the resulting transfer function to a standard 2nd order system polynomial. Thus the PI gains are calculated as follows:

$$K_p = 4\pi\xi f_0 \quad (21)$$
$$K_i = (2\pi f_0)^2 \quad (22)$$

where ξ is the required attenuation and f_0 required bandwidth (in Hertz) of the position tracking loop.

IV. EXPERIMENTAL RESULTS

The experimental results of the STO initial alignment for two different initial rotor positions are shown in Fig. 11a and Fig. 11b. The frequency of the carrier voltage signal was set to $f_c = 500[\text{Hz}]$ and amplitude $U_m = 20[\text{V}]$. The experiments were carried out with PM axis detection loop bandwidth was set to 50[Hz]. Such bandwidth ensured the stabilization time to be less than 100[ms] even if the initial estimation error was close to $\frac{\pi}{2}$. To account for the filters initial transient, the duration of the magnet polarity detection routine was set to 300[ms]. Thus the duration of the whole sensorless alignment process was 400[ms].

In the first experiment the rotor was manually turned to 60° electrical and the observer was initialized with zero values. It can be seen from Fig. 11a, that because the initial position error was less than $\frac{\pi}{2}$ the observer stabilized into a correct equilibrium. In the second experiment Fig. 11b, the procedure was repeated but the rotor was manually positioned to unstable region $\approx 92°$ electrical. The stabilizing trajectory was attracted to the

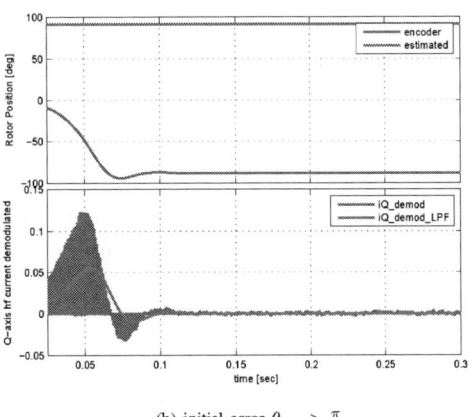

(a) initial error $\theta_{err} < \frac{\pi}{2}$

(b) initial error $\theta_{err} > \frac{\pi}{2}$

Fig. 11: Experimental results showing alignment with the (a) positive and (b) negative equilibrium.

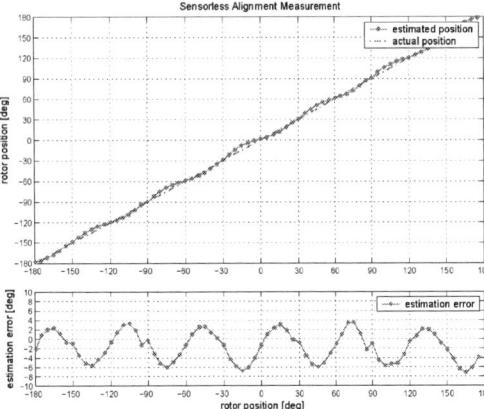

Fig. 12: Initial position alignment with PM polarity detection for full electrical revolution.

different equilibrium point then in the first experiment. In this case the position estimation error is π. Therefore the mechanism that removes such π-ambiguity is employed, where polarity of the magnet in the found equilibrium point is detected.

The experimental results of full initial position detection, which includes axis alignment followed by PM polarity detection, are shown in Fig. 12. Resulting in estimation error of $\pm 4[°\text{el.}]$ was achieved.

Fig. 13 shows experimental results of startup sequence where the time profile of the reference open loop position θ_{start} and the reference currents i_q^{ref}, i_d^{ref} are displayed. This sequence is accurately chosen in order to assure safe starting with minimum oscillation up to the maximum torque.

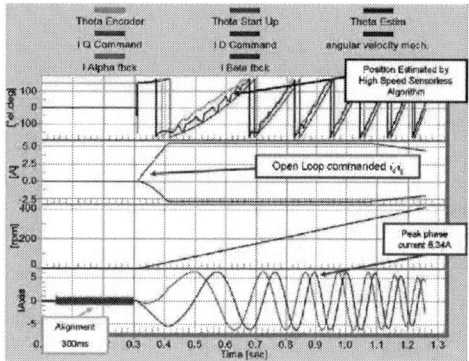

Fig. 13: Start-up sequence.

Phase currents, showing performance of the complete start-up sequence followed by transition into the closed loop speed control mode, are depicted on Fig. 14.

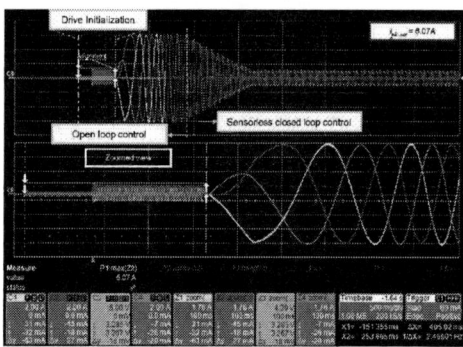

Fig. 14: Measured motor phase currents during the start-up sequence.

V. CONCLUSION

The Field Oriented Control of PM synchronous motor without position transducer has been presented in this paper. The proposed method is suitable for applications where continuous torque control in the low speed region

is not required. To achieve full start-up torque, the initial position of the rotor is determined using *hf* voltage signal injection. The start-up algorithm is based on integration method, where the rotor position is calculated by integration of the speed command. When the rotor speeds up the EEMF observer is utilized to provide estimates of the generated back-EMF, which are subsequently used for rotor position estimation. All experimental results were obtained on a standard commercial H-axis washing machine. Proposed algorithm was implemented on a low end 16-bit fixed point DSC MC56F8025. The experimental results verify this method to be successfully employed in the low cost industrial drives with PM synchronous motors.

ACKNOWLEDGEMENT

The authors wish to thank for the support to Freescale Semiconductor CZ and the Scientific Grant Agency of the Slovak Republic for project No. 1/3086/06 "Research of the new methods of modeling, control and simulation of mechatronic systems".

REFERENCES

[1] R. Filka, P. Balazovic, and B. Dobrucky, "A seamless whole speed range control of interior pm synchronous machine without position transducer," in *In Proc. of EPE-PEMC*, Portoroz, Slovenia, 2006.

[2] R. Filka and P. Balazovic, "Sensorless control of ipmsm seamless covering entire speed range without rotor position sensor," in *Power Conversion Intelligent Motion*, March-June 2006.

[3] P. Balazovic and R. Filka, "Sensorless three-phase permanent magnet synchronous motor including low and zero speed," in *International Appliance Technical Conference.* IATC, March 2005, pp. 134–143.

[4] ——, "Sensorless pmsm control for h-axis washing machine drive," in *39th IEEE Annual Power Electronics Specialists Conference.* PESC, 15-19 June 2008, pp. 4237–4242.

[5] L. Prokop and P. Grasblum, "3-phase pm synchronous motor vector control," Freescale Semiconductor Inc., Application Note, AN1931.

[6] E. Haque, L. Zhong, and F. M. Rahman, "A sensorless initial rotor position estimation scheme for a direct torque controlled interior permanent magnet synchronous motor drive," in *IEEE Transaction on Power Electronics*, vol. 18, no. 6, November 2003, pp. 1376–1383.

[7] M. C. Harke, D. Raca, , and R. D. Lorenz, "Implementation issues for fast initial position and magnet polarity identification of pm synchronous machines with near zero saliency," in *EPE Dresden*, 2005.

[8] K. Hyunbae, H. Kum-Hang, R. D. Lorenz, and T. M. Jahns, "A novel method for initial rotor position estimation for ipm synchronous machine drives," in *IEEE Transaction on*, vol. 40, no. 5, September/October 2004, pp. 1369–1378.

[9] S. Ichikawa, M. Tomita, S. Doki, and S. Okuma, "Initial position estimation and low speed sensorless control of synchronous motors in consideration of magnetic saturation based on system identification theory," in *IAS Annual Meeting*, vol. 2, October 2004, pp. 971–976.

[10] S. Morimoto, K. Kawamoto, M. Sanada, and Y. Takeda, "Sensorless control strategy for salient-pole pmsm based on extended emf in rotating reference frame," in *Trans on IA*, vol. 38, no. No. 4, July/August 2002, pp. 1054–1061.

[11] K. Hyunbae, Y. Sungmo, K. Namsu, and R. D. Lorenz, "Using low resolution position sensors in bumpless position/speed estimation methods for low cost pmsm drives," in *IEEE-IAS*, 2005.

[12] D. G. Luenberger, "An introduction to observers," in *IEEE Transactions on Automatic Control*, vol. 16, no. 6, December 1971, pp. 596–602.

[13] G. Ellis, *Observers in Control Systems - A Practical Guide.* Academic Press, 2002.

Study About the Possibility of Electrodes Motion Control in the EAF Based on Adaptive Impedance Control

Manuela Panoiu*, Caius Panoiu* and Sorin Deaconu*

* Polytechnic University of Timisoara/Electrical Engineering and Industrial Informatics Department, Hunedoara, Romania, e-mail: m.panoiu@fih.upt.ro, c.panoiu@fih.upt.ro, sorin.deaconu@fih.upt.ro

Abstract—The paper presents a study about the possibility of adaptive process control in three phased electric arc furnace. The method is based on the electrodes motion control. The control principle is depending on impedance of the electric arc. The method proposed use a data acquisition board whose input signals are taken from electric arc. These signals allow the calculation of electric arc impedance. Using a numeric computer, it can be commands the control of electrodes position independently on each one of three phases. We propose to use a static frequency converter on each phase to control the electrodes motion

Keywords— Adaptive control, motion control, power factor correction.

I. INTRODUCTION

An electrical arc furnace (EAF) changes the electrical power into thermal energy by electric arc in melting the raw materials in the furnace. During the arc furnace operation, the random property of arc melting process and the control system are the main reasons of the electrical and thermal dynamics. That will cause serious power quality problems to the supply system [1], [2], [3], [4], [5], [6], [7].

Nowadays, AC-Electric Arc Furnaces (EAF) is typically designed to melt a batch of scrap into liquid metal within 1-3 hours [8]. Therefore the installed power reaches up to 1 MW/t. Melting down the scrap bunch and superheating it is a high dynamic process. The AC arc furnace has a non-linear current-voltage characteristic. Therefore it acts as a source of disturbance in the grid from which it is supplied. It emits both harmonics and interharmonics and generates voltage unbalances, voltage dips and voltage fluctuations. Another disadvantage in the EAF is caused by the variations in the line voltage leading to flicker, which can be observed due to the luminosity fluctuation of incandescent lamps.

However, one of the most substantial disadvantages of arc furnace is caused by the reactive power due to the non-linearity of the electric arc [9], [10]. The significant values of the reactive power cause important losses of active power, therefore the efficiency are affected [1], [2], [5], [6], and [11].

The closed-up loop control is not optimized at many furnaces; the run of voltage is very dynamic. With a more optimized control closed-loop it should be possible to enhance the energy input and consequently the productivity. For improving the functioning regime by power factor correction it is possible to make an adaptive impedance control.

The proposed solution is based on some measurements made on an industrial Plant in Romania, Hunedoara where a 100t, 100 MVA UHP EAF are in function.

II. THE ELECTRICAL PARAMETERS OF THE EAF

Figure 1 shows the physical model of the electric arc furnace [8]. In this particular EAF model, there are three electrodes that are moved vertically up and down with hydraulic actuators. Each of these electrodes has a diameter of roughly 1.5 m, weighs approximately 40 tons and is 1 to 2 stories tall. The ore is melted with a huge power surge from the electrodes. The actual product is denser than the scrap and thus falls to the bottom of the furnace creating the matte. Above the matte lies the slag where the electrode tips are dipped. The tremendous heat created by these electrodes causes the ore to liquefy and separate. Thereupon more raw materials are placed in the furnace and the process repeats itself.

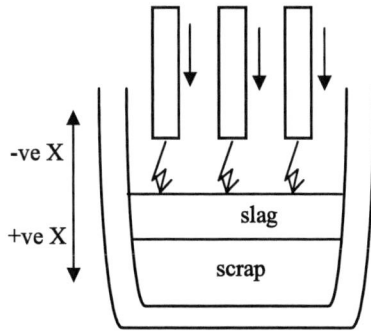

Fig. 1 The physical model of the Electric Arc Furnace

A. Arcing

Arcing is a phenomenon that occurs when the electrodes are moved above the slag. As the electrode approaches the slag, current begins to jump from the electrode to the slag, creating electric arcs. Depending on the magnitude of the input voltages of electrodes, the arcing distance can vary. Usually, arcing occurs in the region within centimeters of the slag (approximately 10 – 15 cm). Therefore, the EAF model must take into account the instances when x1, x2, x3 are negative (i.e. the electrodes are suspended above the slag), like in fig 1.

978-1-4244-1741-4/08/$25.00 ©2008 IEEE   1409

B. Characteristics of the AC electric arc

As given in [1], [9], [10], [11], [12], [13], one can consider that during the burning of the AC electric arc, the equivalent diagram of the supplying circuit can be represented as shown in figure 2.

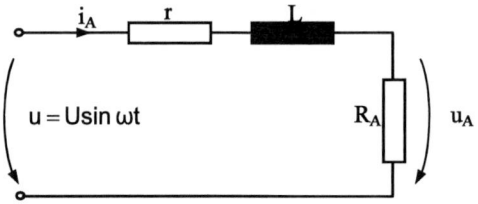

Fig. 2 The equivalent scheme of the supplying circuit of the AC electric arc

The variation curves of the electrical issues from the equivalent scheme of the supplying circuit are presented in figure 3. Analyzing the variation curves, we obtained the following conclusions: after electric arc ignition, the arc voltage u_A is practically constant and because the current is variable, the electric arc can be considered as a non-linear receiver; the arc voltage u_A and the current i_A from the circuit are in the same phase, which means that the electric arc has a resistive character; the electric current in the circuit passes through zero twice, in each period of the alternating voltage applied, which leads to the going out and re-ignition of the arc with a frequency that is double as compared to the voltage applied; the ignition voltage U_{ig} of the electric arc is higher than the work value U_A; the AC electric arc has a rectifying character [9], [10], [12]. That mean if there is a discharge between an electrode (usually made of graphite) and the metal to be heated up, due to the different thermal - physical properties of the two materials, (the temperature of the graphite electrode is higher than that of the material to be processed), the arc ignition voltage in the half-period where the metal represents the cathode is higher than the arc ignition voltage in the half-period when the cathode represents the graphite electrode, i.e. $U_{ig}^+ > |U_{ig}^-|$. Similarly, for the drop voltage in the two half-periods, relation $U_d^+ > |U_d^-|$ stands. For this reason, the amplitude of the current in the two half-periods differs, namely it is higher in the half-period when the graphite electrode is the cathode.

In figure 4 is show the dynamic characteristic of the AC electric arc, characteristic obtained according to the variation curves of $u_A(t)$ and $i_A(t)$ given in figure 3. The rectifying character of the electric arc is present because the magnitude of the ignition and drop voltage is different in the two half-periods. The burning of the electric arc can take place under the conditions of interrupted current or uninterrupted current. The burning under conditions of interrupted current leads to its unstable working and the current curve is highly distorted. For this reason it is necessary for the electric arc to burn uninterruptedly. The condition of uninterrupted burning of the arc, under the simplifying hypotheses that

$U_{ig}^+ \cong |U_{ig}^-| \cong U_{th}^+ \cong |U_{th}^-| \cong U_A$ and the feeding voltage is sinusoidal $u_s(t) = U_s \cdot \sin \omega t$ is given by

$$U_s \cdot \sin \varphi \geq U_A. \tag{1}$$

this leads to

$$\frac{U_A}{U_s} \leq \sqrt{\frac{1}{1+\frac{\pi^2}{4}}} = 0,54 \tag{2}$$

resulting that in order to have an uninterrupted current, the electrical installation must work under a natural power factor

$$\cos \varphi \leq 0,85 \tag{3}$$

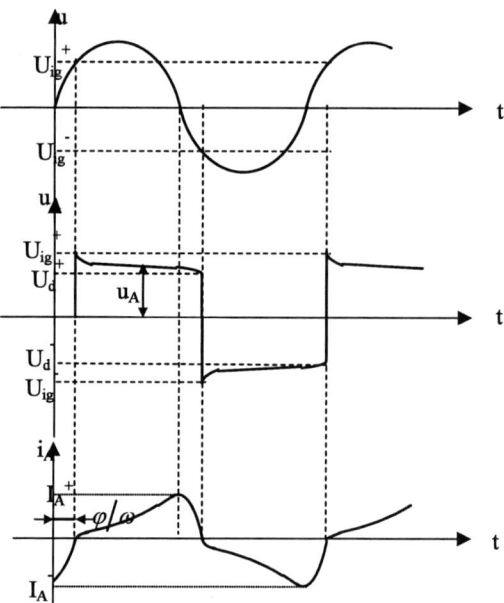

Fig. 3. The waveforms of currents and voltages

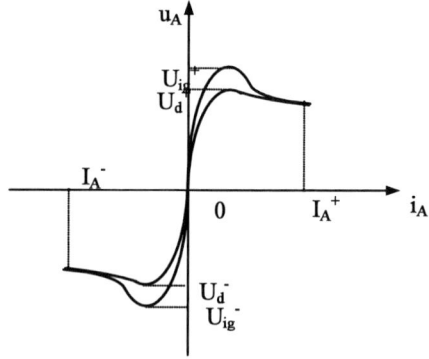

Fig. 4. The dynamic characteristic of the AC electric arc

III. MEASUREMENTS MADE ON THE EAF

The measurements were made at a 3-phase power supply installation of a 3-phase EAF of 100 t, to which were not connected the filters for the current harmonics, neither the load balancing device nor reactive power compensation. The modern methods of measuring the electric values are using numerical systems, based on data acquisition systems, and method presented in this paper are using such a system. It's been used a computer system with an ADA3100 data acquisition board. The measure scheme is show in fig. 5. The main characteristics of the ADA 3100 acquisition board are:

- 8 differential analog channels or 16 channels between ground and input
- 16 bits analog to digital conversion
- Inputs between ±5V; ± 10V
- 1, 2, 4, 8 an 16 programmable gain
- 8 ko FIFO memory
- 8 numerical inputs and 8 numerical outputs
- 2 analogical output channels with 12 bits resolution
- Outputs voltages ±5V; ± 10V, 0 – 5 V, 0 – 10 V

The acquisition board allows the simultaneous acquisition of 3 currents and 3 voltages, for the low or medium voltage lines of the transformer which supplies the furnace. The data acquisition on the 6 channels was made as follows:

- during 250 ms have been acquired simultaneously the data on the 6 channels, the selected acquisition frequency being of 5 KHz. In this way, have been acquired the signals during 12,5 periods. This fact allowed that in case the frequency of the supply voltage is different of 50 Hz, the data should contain a number of 12 full periods, selectable by program;
- the data acquisition memory the previously acquired data. In this way, results that have been acquired, on the entire duration of the heat, data in time windows of 250 ms length, the interval between two consecutive data windows being of 10 seconds.

The process was restarted, during 250 ms, at an interval of 9,75 seconds, interval during which were saved in memory the previously acquired data. In this way, results that have been acquired, on the entire duration of the heat, data in time windows of 250 ms length, the interval between two consecutive data windows being of 10 seconds. As regards to the waveforms of the currents and voltages on the low voltage supply line, presented in fig. 6, is found a strong distortion of these. Also, one can notice that because the amplitudes of the currents and voltages on the 3 phases are unequal, results that the load is also unbalanced.

The spectral characteristics of the current and voltage were achieved by using a Matlab program by processing the data acquired by using the Fourier rapid transform, and are show in fig. 7. One can observe the presence of harmonics of 3th, 5th, 7th order, but also the components of other frequencies than the harmonics' (inter-harmonics).

For the comparison of the simulation results and the performed measurement it was made simulations of the entire electric installation of the UHP electric arc furnace.

The results of these simulations are detailed in several previous papers [2], [9], [10], [11], [12], [13].

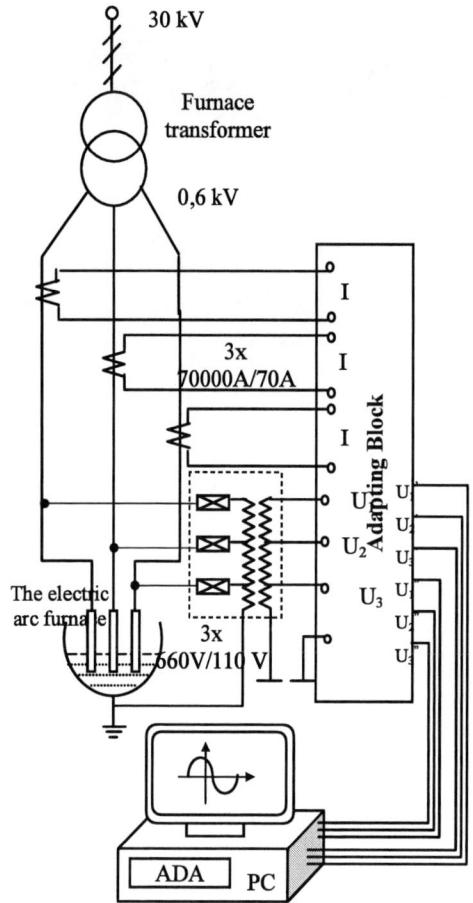

Fig. 5. The measure scheme

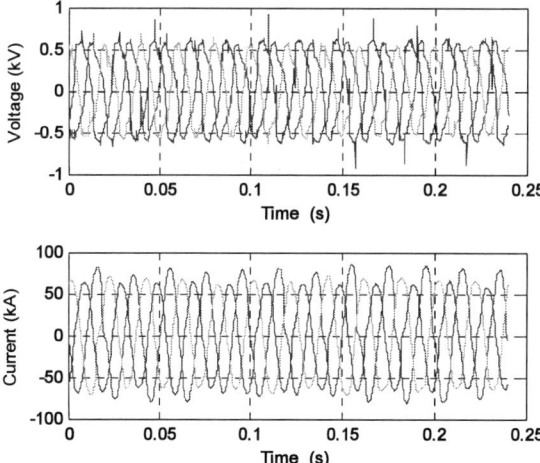

Fig. 6. The variation of measured voltages and currents for the three phases

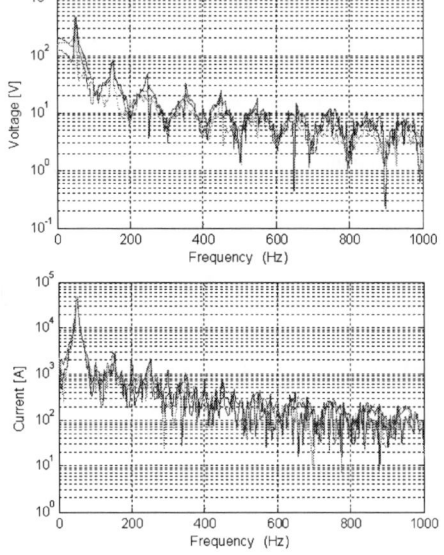

Fig. 7. The spectral characteristic of currents and voltages for measured data

It was calculate the main indicators of power quality and show in the table 1. These indicators are:

a) The active power absorbed is the mean over a period of the instantaneous power it absorbs:

$$P = \frac{1}{T}\int_0^T p \cdot dt = \frac{1}{T}\int_0^T u \cdot i \cdot dt \qquad (4)$$

$$P = U_0 \cdot I_0 + \sum_{k=1}^{\infty} U_k \cdot I_k \cdot \cos\varphi_k \qquad (5)$$

b) The reactive power absorbed by a dipole under non-sinusoidal work conditions is equal to the sum of the reactive powers corresponding to the harmonics:

$$Q = \sum_{k=1}^{\infty} U_k \cdot I_k \cdot \sin\varphi_k \qquad (6)$$

c) The total harmonic distortion, is defined taking into consideration the first *40* harmonics

$$THD = \sqrt{\sum_{k=2}^{40}\left(\frac{F_k}{F_1}\right)^2} \cdot 100\,[\%] \qquad (7)$$

d) The deforming power, specific to the non-sinusoidal, periodical work conditions, is defined by relation:

$$D^2 = S^2 - P^2 - Q^2, \qquad (8)$$

and has the expression:

$$D = \sqrt{\sum_{j>k}^{\infty}\sum_{k=1}^{\infty}\left[U_j^2 \cdot I_k^2 + U_k^2 \cdot I_j^2 - 2U_j U_k I_j I_k \cdot \cos(\varphi_j - \varphi_k)\right]} \qquad (9)$$

e) The power factor under non-sinusoidal work conditions as ratio between the active and apparent power

$$k_P = \frac{P}{S} = \frac{P}{\sqrt{P^2 + Q^2 + D^2}} \qquad (10)$$

TABLE I.

	Measure values	Simulating values
P (MW)	48,52	47,36
Q (MVAR)	52,29	55,97
D (MVAD)	11.07	9.85
THDI (%)	10,83	8,44
THDU (%)	17,3	15,17
k_p	0.72	0.75

The powers variation for 6 minutes are show in figure 8. The powers was calculated using relations (4)-(9).

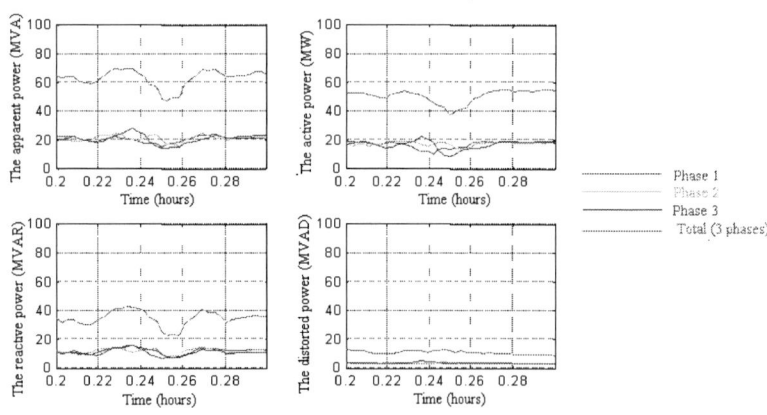

Fig. 8. Variation of powers for 6 minutes from the melting stage

IV. DEVELOPMENT A CLOSED-UP IMPEDANCE CONTROL

The reactive power has high values, so the active power has to be control in the way that the reactive power will be reduced to minimal value.

The method is based on impedance adjustment. The electrode adjustment system has to be designed so:

$$Z = \frac{U}{I} = const \qquad (11)$$

The impedance control is usual in the most of actual systems. The general control diagram is presented in figure 9.

The received dates from the electric arc furnace are transmitted to an interface block computer- frequency static converter. At the numeric output of the data acquisition board is transmitted a byte whose 6 less significant bits determine the senses of frequency static converters. The static frequency converters (SFC) are used to control the asynchrony electric drives with powers between 0.2... 280 kW to obtain maximum efficiency and without reactive power consumption. In the control scheme from fig. 9 we propose to use SFC having the main characteristics:

- the output voltage (50 Hz) 3x380 V or 3x220 V
- the output frequency: adjustable between 0.1...600 Hz
- acceleration/breaking time: 0.1... 300s
- the possibility to memorize 8 fixed speed steps
- the possibility to choice the rotation direction
- the possibility to communicate with a PC

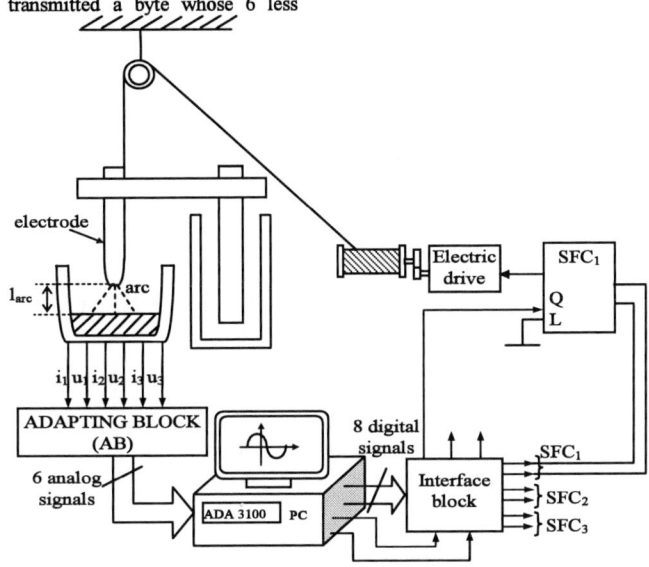

Fig. 9. The general control diagram

From interface block, every frequency static converter uses three outputs: one for control the electric drive speed and the other two for rotation command: in up and down sense of electrodes. Every frequency static converter involves an electrode through a reducer. Each frequency static converter controls an electrode using a reducer.

The proposed system functioning is based on an algorithm that is presented in figure 10.

The proposed method has the advantage allows the on-line determination of the electric arc impedance. The estimated errors are low because of the resolution of the analog-numeric converter.

$$\varepsilon_r[\%] = \frac{1}{2^{12}} \cdot 100 \approx 0.025\% \qquad (12)$$

Another advantage of the proposed method is because of using frequency static converters that allows the elimination electromagnetic coupling used at control of

electric drive – reductor movement. This electromagnetic coupling makes a heavy maintenance because of reduced reliability. Using the adequate command (through computer program) can be obtained the speed control from minimal to maximal value in 256 steps that is matching to an impose speed error of:

$$\varepsilon_r[\%] = \frac{1}{2^8} \cdot 100 \approx 0.4\% \qquad (13)$$

The reliability of the frequency static converters is eliminating the necessity of using supplementary protection diagrams at increasing the nominal values of currents and voltages.

V. SIMULATION RESULTS ON MODIFYING THE ELECTRODES POSITION

The simulations was made using PSCAD EMTDC, based on an electric arc model [9], [10], [11]. This model assumes the current – voltage characteristic of the electric arc described by the relation:

$$U_A = U_{th} + \frac{C}{D + I_A} . \tag{14}$$

$$U_{th} = A + Bl . \tag{15}$$

The electrical items variation in different functioning regimes can be done only if we consider an arc length variation between 0, corresponding to the short-circuit regime, and a maximum value. The maximum value is determined in such a way that the electric arc is burning.

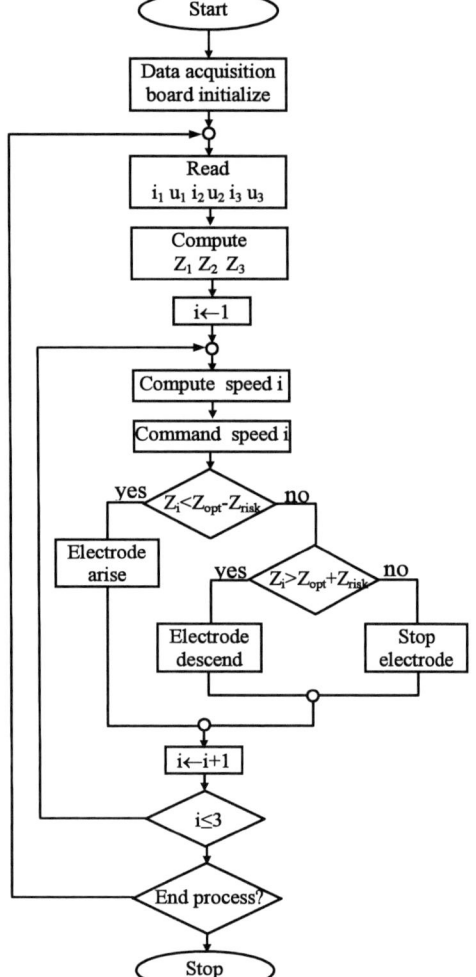

Fig. 10 The control algorithm

The electrodes position controlling is performed taking into account on the real condition existing on the considered industrial plant. The maximum motion speed of the electrodes is of *3 m/min (0.05 m/s)* and is reached in

emergency regime, its variation being achieved as in fig. 11.

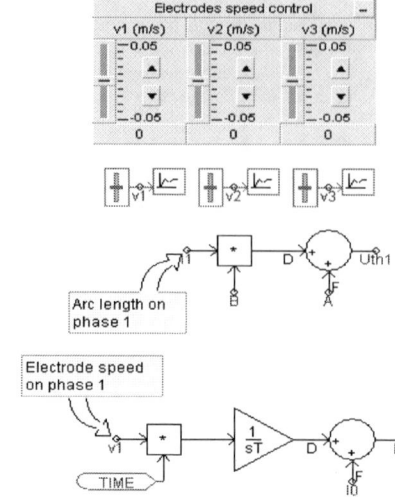

Fig. 11. The calculus of the arc length based on the electrodes speed

The electric arc's length can be modified from zero to a maximum value determined by limiting the integrator's output, fig. 11; The calculation of the drop voltage is made based on the relation (11), the implementation diagram being also in fig. 11; The control of the electrodes' position is made independently on each phase.

The simulation results are presented in fig. 12. One can observe that the highest value of the active power is obtained when the value of the threshold voltage is of *200 V;* The reactive power is positive regardless the working regime, having values between *15-100 MVAR,* being therefore necessary the utilization of the reactive power's compensation installation.

In [11] was design the reactive power compensation system and harmonics filters. This contains the 4 filters on the harmonics 5, 7, 11 and 13 and the reactive power compensation installation composed by the constant part (in Y connection) and the adjustable part in steps. Since the threshold voltage depends linearly by the electric arc's length, it results that also the active power depends on the electric arc's length. Based on these remarks, the active power's iterative control algorithm proposed by the de authors is based on the modification of the electric arc's length depending on the active power desired to be obtained. In fig. 13 are show the results of active power control simulation following the reactive power's compensation and harmonics current filter.

CONCLUSIONS

The paper presents an adaptive control method of the electrode motion control at electric arc furnace by using a numeric system that have many advantages in the field of system reliability, working speed and estimated errors. By using such a method it is possible to correct also the power factor.

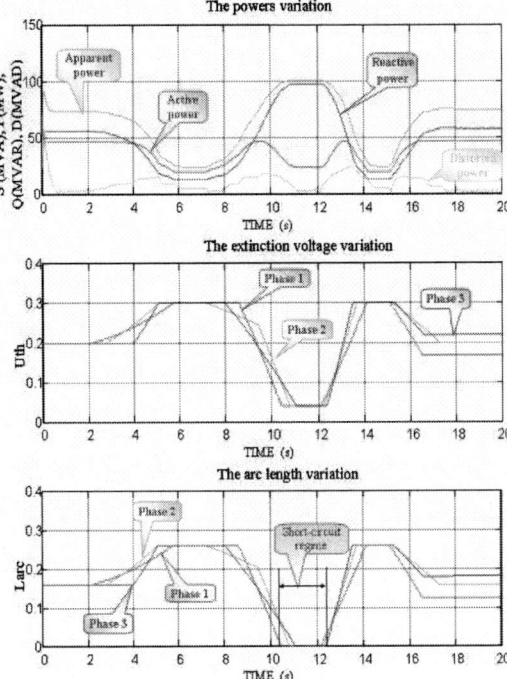

Fig. 12. The time variation of the powers, the threshold voltage and the arc lengths

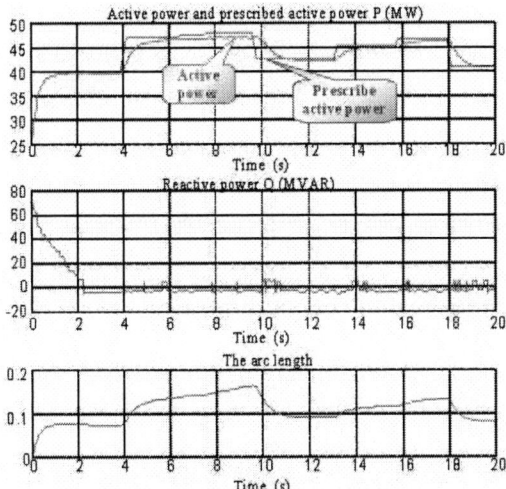

Fig. 13. The variation of active and reactive power and for the arc length obtained by simulation

REFERENCES

[1] Panoiu M., Panoiu C., Sora I., Experimental Research Concerning the Electromagnetic Pollution Generated by the 3-Phase Electric Arc Furnaces in the Electric Power Supply Networks, *Acta Electrotehnica*, nr.2, vol 47, pp 102-112, 2006.

[2] Panoiu M., Panoiu C., Osaci M, Muscalagiu I., Simulation Result about harmonics filtering using Measurements of some Electrical Items in Electrical Installation of an UHP EAF, WSEAS Trans. On Circuits and Systems, vol 7. 2008 pp 22-31

[3] Petersen, H.M., Koch, R.G., Swart, P.H., R. van Heereden, Modelling Arc Furnace Flicker and Investigating Compensation Techniques, IEEE Trans. on Power Delivery, pg. 1733-1740, 1995.

[4] Deckmann, S.M., Rabelo, G.F., A Quality Index Based on Voltage Flicker and Distortion Evaluations, IEEE Proceedings General Transmission and Distribution, vol. 2, pg. 235-241, 1997.

[5] S. Chitchian and M. Akhbari, A Simple Arc Furnace Model for Power System Harmonic Studies, Proceeding (409) Power and Energy Systems - 2003

[6] Collantes-Bellido, R. Gomez, T, Identification and modelling of a three phase arc furnace for voltage disturbance simulation, IEEE Transactions on Power Delivery ,Oct 1997 Volume: 12, Issue: 4, page(s): 1812-1817

[7] E. Emanuel, J.A: Orr "An Improved Method of Simulation of the Arc Voltage-Current Characteristic", *9th international Conference on Harmonics and Quality of Power*, Proceedings p.p. 148-150, October 1- 4, 2000, Orlando, Florida

[8] Benoit Boulet, Gino Lalli and Mark Agersch, Modeling and Control of an Electric Arc Furnace, Proc. of the American Control Conf, Denver, Colorado, June 4 –6, 2003

[9] Panoiu M., Panoiu C., Simulation Results for Modeling the AC Electric Arc as Nonlinear Element using PSCAD EMTDC, WSEAS Transaction on circuits and systems, vol 6, Jan 2007, pp 149-156

[10] Panoiu M, Panoiu C, Modeling and simulating the AC electric arc using PSCAD EMTDC, Proceedings of the 5th WSEAS Int. Conf. on System Science and Simulation in Engineering, Tenerife, Spain, Dec. 16-18, 2006

[11] Panoiu, M., Panoiu, C., Sora, I., Osaci, M., Simulations Results on the Reactive Power Compensation Process on Electric Arc Furnace Using PSCAD-EMTDC, International Journal of Modelling, Identification and Control, vol. 2, no. 3, 2007, pg. 250-257

[12] Panoiu M., Panoiu C., Sora I., Osaci M., Using a Model Based on Linearization of the Current – Voltage Characteristic for Electric Arc Simulation, Proceedings of the 16th IASTED International Conference on Applied Simulation and Modelling ~ASM 2007~ , Palma de Mallorca, Spain, August 29 – 31, 2007, pag. 99-103,

[13] Panoiu M., Panoiu C., Sora, I., Osaci M, Muscalagiu I., Modeling, Simulating And Experimental Validation of the AC Electric Arc in the Circuit of Three-Phase Electric Furnaces, EUROSIM 2007 Congress, 9-13 septembrie 2007, Ljubljana, Slovenia, 10 pg. CD Proceedings

[14] Panoiu M., Panoiu C., Sora I., Osaci M., About the possibility of power controlling in the Three-Phase Electric Arc Furnaces using PSCAD EMTDC simulation program, Advances in Electrical and Computer Engineering , vol. 7, number 1 (27), 2007

[15] Chen, F. Athreya, K.B. Sastry, V.V. Venkata, S.S., Function space valued Markov model for electric arc furnace, IEEE Transactions on Power Systems, May 2004, Volume: 19, Issue: 2, page(s): 826- 833, ISSN: 0885-8950,

[16] Harmonics Working Group IEEE PES T&D Committee,

[17] Modeling of components with nonlinear voltage current characteristics for harmonic studies, Power Engineering Society General Meeting, IEEE Publication, 6-10, June 2004, page(s): 769 – 772, Vol.1

[18] IEEE Standard 519-1992, "IEEE Recommended Practices and Requirements for Harmonic Control in Electrical Power Systems," New York, 1992.

[19] R. C. Dugan, Simulation of Arc Furnace Power Systems, IEEE Trans. on Industry Applications, IA-16(6), Nov/Dec 1980, pp.813-818.

[20] Andrews, D., Bishop, M.T., Witte, J.F. Harmonic measurements, analysis, and power factor correction in a modern steel manufacturing facility, IEEE Transactions on Industry Applications, v 32, n 3, May-June 1996, p 617-24

[21] Chi-Jui Wu1, Cheng-Ping Huang1, Tsu-Hsun Fu1, Tzu-Chih Zhao1, Hung-Shian Kuo1, Power factor definitions and effect on revenue of electric arc furnace load, 2002 International Conference on Power System Technology Proceedings , 93-7 vol.1

[22] Akdag, A.1, Cadirci, I., Nalcaci, E., Ermis, M., Tadakuma, S., Effects of main transformer replacement on the performance of an electric arc furnace system, IEEE Transactions on Industry Applications, v 36, n 2, March-April 2000, 649-58

Asynchronous machine stator resistance estimation using integrated PWM modulator and sampler unit as FPGA application

Dag Samuelsen*, Waldemar Sulkowski[†]

*Buskerud University College/ATEK, Kongsberg, Norway, e-mail: *dag.samuelsen@hibu.no*
[†]Narvik University College/IDER, Narvik, Norway, e-mail: *ws@hin.no*

Abstract—The paper demonstrates how a simple, low-cost and effective stator resistance estimation scheme for FPGA can be employed, utilizing the large degree of freedom a FPGA impose with regard to system design, while at the same time conform to the constraints the same technology infer. In a computing system, a FPGA removes the limitations of the Von Neuman-architecture. Although the (Super) Harward architecture, used by most DSP processors, relieve this limitations, this is not anywhere near the power of the parallel computing capability of FPGA. A FPGA is at the same time somewhat limited with regard to complexity of mathematical operations. Although this limitations has been removed with the introduction of FPGA with embedded CPU, there are still reasons for keeping the design simple, when overall cost should kept low. The estimator has been tested on an asynchronous machine, with satisfying results.

Keywords—Sensorless, stator resistance identification, FPGA.

I. INTRODUCTION

The use of FPGA in motor drive systems is not widely used. This is possibly due to the fact that it is more difficult to implement a complete control system in these circuits than in standard DSP. Another factor might be the problems that arise when trying to implement complex mathematical operations, demanding a huge amount of memory. This kind of operations are not very well suited for FPGA implementation.

The use of FPGA in motor drive system requires special considerations, as indicated above. In spite of these limitations, FPGA are used, and the reason for this is the degree of freedom this implementation gives. When the Von-Neuman architecture demands that all data need to be transferred over the same data bus, routed through the same ALU and fed back to the outside world through the same data bus, FPGA technology allows this to be done over several data routes, of different size, speed and, most important, simultaneously e.g. in parallel. From these data streams, data can be processed for different purposes on different special-purpose, user-specified ALU. When data are to be fed back to the outside world, new data paths can be arranged, in order not to interfere with incoming data. This is considered in [10] where different FPGA implementations of DTC control algorithm has been compared. Another example, given in [12], demonstrate the flexibility of development using IP-core functions in order to facilitate the development of real time simulation models. In [11] a FPGA based dead time compensation scheme is proposed, and show the true power of integrating several systems such as the pulswidth modulator and the dead-time compensator within the same FPGA.

The main concern in this paper is the estimation of stator resistance. A large amount of publications have been written on this subject. Some are based on MRAS theory. In [5] it is used the two-time-scale approach for estimating the stator resistance. This is a full order observer and it would involve a large amount of calculations. The authors of [8] has also employed the MRAS approach, but with a less complex algorithm. Another approach is to use signal injection. In [4] a combination of ac and dc-signals are used to reveal, via filtering, parameters of the machine. The dc-signal is used only for short periods, causing periodic disturbances in the control loop. A low frequency signal injection is implemented in [6], but the algorithm requires steady state in order to make the estimation, and the estimate is calculated as the difference between two large quantities, making the error large. As a measure to avoid disturbing the control loop, a partial connection to the center tap of the machine has been included in [9]. This gives fairly good results, but the need for center tap connection makes this approach unsuited for many applications.

Other works describe the estimator-only approach. For instance, we find a fuzzy-logic-based estimator ([1]), with the drawback that it has to be calibrated for each individual machine, using several temperature sensors inside the machine. [3] is using artificial neural network, which could be suited for parallel processing, disregarding the need for a nonlinear function in each node. While [13] and [7] requires that other parameters are kept constant during estimation, [16] is only able to correctly identify the different parameters in certain operating points of the machine.

II. ESTIMATION PRINCIPLE

Common for all successful approaches is the ability to estimate the parameters without relying on the other parameters or states of the machine, or at least be able to control for their influence. The system proposed in this paper make use of a combined pulse-width modulator

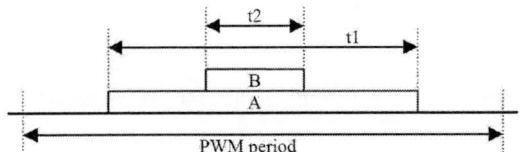

Fig. 1.   Standard model of asynchronous machine

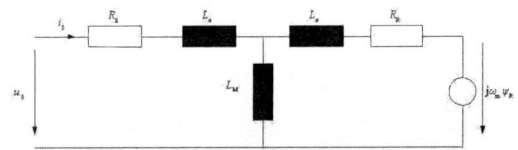

Fig. 2.   Standard model of asynchronous machine

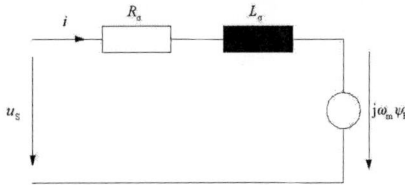

Fig. 3.   Standard model of asynchronous machine

and sampling unit [15], to retrieve certain states of an asynchronous motor within a switching interval.

By using several low-cost A/D-converters in parallel, it is possible to get several samples of stator currents and voltages within each PWM-period. This allows for more advanced analysis in each state of the inverter, where the switching pattern is shown in figure 1. The integration of the sampler unit and the PWM-unit, allows the sampler unit to access full information about the time instances when the changes in the inverter state will occur. The exact trig point of the ADC can then be shifted according to the switching pattern of the PWM-modulator, in order to get a more complete picture of the currents and voltages in the machine. This can be done in a number of ways. For instance, a sample can be done at the start and end of each state of the inverter. This will give the true derivative of the current in each state. Another idea is to trig four of the ADC within one state. This provide enough data to calculate the derivative both at the start of the state, and at the end of the state. The flexibility of the FPGA technology makes these kind of changes to the sampling pattern easy whenever needed.

An asynchronous machine can be modeled as figure 2 shows. Simplifying this model to the model shown in figure 3, the electrical circuit of the machine can be described more compact as

$$\vec{u}_s = R_\sigma \vec{i}_s + L_\sigma \frac{d\vec{i}_s}{dt} + \jmath \omega_m \vec{\psi}_R. \tag{1}$$

where $\jmath \omega_m \vec{\psi}_R$ is the back-emf produced by the rotation of the magnetic field in the rotor, as seen from stator. In the zero-sequence of the inverter, all switches of the inverter are in the same position. The stator terminals can then be regarded as short circuited, and the voltage vector $\vec{u}_s$ can be assumed constant zero during this period. As long as the rotor flux does not go to zero, the motor will produce back-emf. This will induce a current in the stator circuit, which may be identified and analyzed by the system described above. This gives the opportunity of estimating the parameters of the motor. The back-emf can be used [14] to estimate the stator resistance using the following formula:

$$\hat{R}_\sigma = \frac{L_\sigma \left(\frac{d\vec{i}_s}{dt} \big|_{t=t_1} - \frac{d\vec{i}_s}{dt} \big|_{t=t_2} \right)}{\vec{i}_s(t_2) - \vec{i}_s(t_1)} \tag{2}$$

where $\hat{L}_\sigma$ can be found in a separate operation by

$$\hat{L}_\sigma = \frac{\vec{u}_s(t_2) - \vec{u}_s(t_1)}{\left(\frac{d\vec{i}_s}{dt} \big|_{t=t_2} - \frac{d\vec{i}_s}{dt} \big|_{t=t_1} \right)} \tag{3}$$

In order to do this, the standard motor model, shown i figure 2 has been simplified to the model shown in figure 3. It is assumed that the angular velocity of the back-emf vector is equal to that of the rotor flux vector. The term $\frac{d\vec{i}_s}{dt} \big|_{t=t_1}$ denotes the derivative of the stator current at the beginning of the zero-sequence inverter state, while $\frac{d\vec{i}_s}{dt} \big|_{t=t_2}$ denotes the derivative at the end. These fractions are calculated on the basis of four current samples taken within the same zero-sequence section of one PWM-period. The trig point of the ADC are set to avoid the large disturbances seen around the switching point, while at the same time, displaced far enough from each other to get acceptable precision on the derivatives.

III. EXPERIMENTAL RESULTS

Practical tests were carried out on a 2.2kW asynchronous machine. Electrical parameters for the motor can be found in table I. In all tests the asynchronous motor was connected to a DC-motor working as a generator to simulate a load condition. In order to test if the proposed system actually can identify a change in the resistance of the asynchronous machine, a switch- and resistor network was connected as shown in figure 4. When the three switches are closed, the machine is running with nominal resistance. Opening the switch will simulate an increase in stator resistance (possibly caused by an increase in winding temperature). The nominal resistance of the stator windings were found to be 3.1Ω, and it was decided to use resistors of 2.0Ω in the network. There is not any easy way of changing the rotor resistance of a squirrel cage rotor, except for exchanging the rotor with a different type.

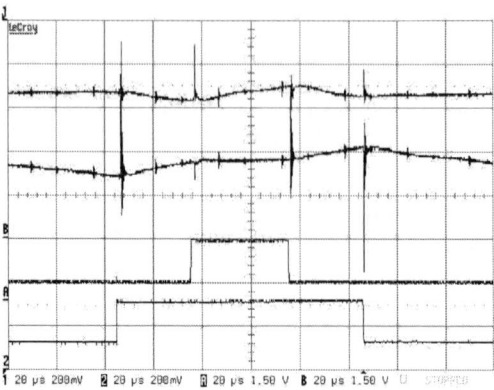

Fig. 4. Circuit showing switch for adding resistor to R_s

TABLE I
PARAMETERS FOR THE ASYNCHRONOUS MACHINE

R_S:	3.1Ω	L_S:	0.1243mH	P_N:	2.2kW
R_R:	2.44Ω	L_R:	0.1243mH	Pole pairs:	3
		L_M:	0.1098mH		

A. Control system and FPGA implementation

A direct torque control (DTC) scheme was used for controlling the asynchronous machine during the tests. See [2] for an extensive review on the subject. The control scheme has shown to be suited for implementation in a FPGA system [10], without using too much resources. However, due to the need for easy access to internal variables, parts of the control system are implemented on a dSPACE1103 system, which allows the user to fetch directly the values of any variables while the system is running. It is possible, though, to do this with only the FPGA and a computer. The DTC algorithm and the speed reference signal are implemented in the dSPACE system, while the rest of the control loop, including the integrated PWM modulator and sampling unit, as well as the estimators, are implemented in the FPGA. The FPGA prepares data for the dSPACE system in order to reduce the amount of data communication needed between the two systems.

In figure 5 the waveform of the machine currents in two legs is shown. Here it can be seen how the derivative of the currents vary during the PWM period. The two uppermost graphs show two of the three phase currents, while the two lower graphs illustrate the different states of the PWM period, as stated theoretically in figure 1. Further, the topmost graph in figure 6 shows the trigger signal for a given A/D-converter, while graph 2 and 3 show the control signals for the transistors in one leg of the power inverter. The current waveform for the actual leg of the inverter is shown in graph 4 of figure 6. This shows how the trig point of the A/D-converters can be placed arbitrary close to the switching instances of the power inverter, this being made possible through the integration of the PWM modulator and the sampling unit.

B. Resistance estimation

It has to be noted that the correct value of the resistance estimated is not equal to the actual stator resistance, as the estimation scheme is using the simplified model

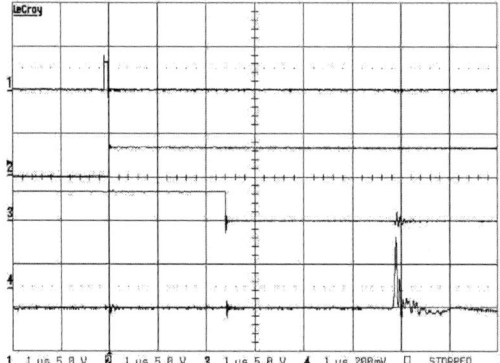

Fig. 5. Current waveforms during the PWM period

Fig. 6. Placement of the trig point on of the A/D converters

of figure 3. Here $\hat{R}_\sigma$ is estimated as a combination of rotor and stator resistance. When increasing the stator resistance alone, the increase in estimated resistance value is not equal to the increase in stator resistance, neither is the estimated resistance with no extra resistance (switch closed). To be able to know when the correct resistance value has been found, a torque transducer was mounted between the asynchronous machine and the DC-motor. For the DTC-controller to be able to estimate the correct torque, the resistance value has to be correct. The estimated torque from the DTC-controller was compared to the actual torque measured on the rotor shaft. In this way it was possible to identify the correct value of the estimated resistance.

C. Change of stator resistance

In the first test, the asynchronous motor was running at low speed, with 1/3 of rated torque. The switch was closed at first, and then opened. The result of the test is shown in figure 7, where the step-shaped line denotes the correct resistance, while the other shows $\hat{R}_\sigma$ estimated. As the figure shows, the estimate approaches the correct value after about 25 seconds, which is acceptable due to slowly temperature changes, and hence, heavy filtering can be applied.

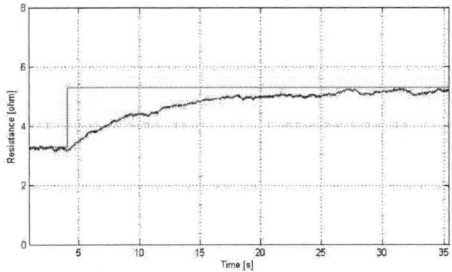

Fig. 7. Response of estimator when a 2.0Ω resistor is added to R_S

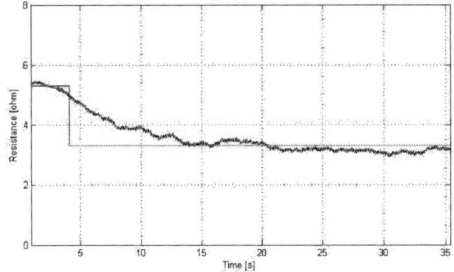

Fig. 8. Response of estimator when a 2.0Ω resistor is removed from R_S

Figure 8 shows the opposite transition. The motor were running with the extra resistors connected and the switches open. Then the resistors were short circuited, to simulate a decrease of stator resistance. The result is shown in figure 8, where the estimator is shown to approach the correct value of the resistance within 20 seconds.

D. Change of rotor speed

Another test were done to investigate how speed would affect the estimator. The estimator is using the back-emf vector during a part of the PWM-period. In (2) and (3), this back-emf vector is assumed to be constant, while in reality, it is rotating synchronously with the rotor flux vector. The initial tests show that this does have an influence on the estimator. As stated in [14], this can be overcome by a change of reference. The graphs also show that this influence behaves linearly with respect to the angular speed of the rotor flux vector, making correction easy.

In order to correct for this speed dependency, the system was modified to correct for the influence of the rotating back-emf vector, and a new test were performed. The result of a deceleration of the machine is shown in figure 9, where the topmost graph show angular velocity of the rotor flux vector and the lower graph shows the (relatively constant) estimated value. This test shows that when correcting for the influence of the rotor flux angular speed, the estimator is able to identify correctly the resistance.

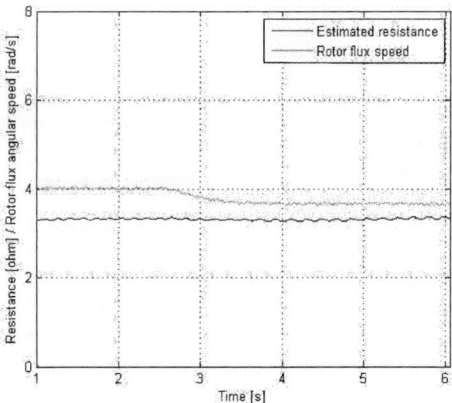

Fig. 9. Response of estimator during deceleration of rotor flux angular velocity

IV. CONCLUSION

The tests show that the system is able to estimate $\hat{R}_\sigma$ when the stator resistance changes. As the system is built using low-cost equipment, noise in the estimated resistor value requires the estimation to be filtered. This is not considered to be a problem, as the stator resistance in reality never changes rapidly. The estimator also shows dependence on the rotor flux angular speed. This is however possible to correct for, and the tests performed show that this is possible.

The experimental installation was performed on an induction machine, using the DTC-algorithm for control, and the usual "0-A-B-A-0" switching pattern for the PWM-modulator. However, the integration of the PWM-modulator and sampling unit is independent on the switching pattern of the PWM-modulator, as well as the machine type and control algorithm. This shows the versatility of the proposed system for use in other applications. The estimates are not "best in class", but gives a fairly good estimate for keeping the torque and speed within certain limits. This can be improved by building a better sampler, but this may contend the purpose of building a low-cost controller for AC-machines.

REFERENCES

[1] B.K. Bose and N.R. Patel. Quasi-fuzzy estimation of stator resistance of induction motor. *Power Electronics, IEEE Transactions on*, 13(3):401–409, May 1998.

[2] G.S. Buja and M.P. Kazmierkowski. Direct torque control of pwm inverter-fed ac motors - a survey. *Industrial Electronics, IEEE Transactions on*, 51(4):744–757, Aug. 2004.

[3] L.A. Cabrera, M.E. Elbuluk, and I. Husain. Tuning the stator resistance of induction motors using artificial neural network. *Power Electronics, IEEE Transactions on*, 12(5):779–787, Sept. 1997.

[4] L.A. de Souza Ribeiro, C.B. Jacobina, A.M.N. Lima, and A.C. Oliveira. Real-time estimation of the electric parameters of an induction machine using sinusoidal pwm voltage waveforms. *Industry Applications, IEEE Transactions on*, 36(3):743–754, May-June 2000.

[5] G. Guidi and H. Umida. A novel stator resistance estimation method for speed-sensorless induction motor drives. *Industry Applications, IEEE Transactions on*, 36(6):1619–1627, Nov.-Dec. 2000.

[6] In-Joong Ha and Sang-Hoon Lee. An online identification method for both stator-and rotor resistances of induction motors without rotational transducers. *Industrial Electronics, IEEE Transactions on*, 47(4):842–853, Aug. 2000.

[7] T.G. Habetler, F. Profumo, G. Griva, M. Pastorelli, and A. Bettini. Stator resistance tuning in a stator-flux field-oriented drive using an instantaneous hybrid flux estimator. *Power Electronics, IEEE Transactions on*, 13(1):125–133, Jan. 1998.

[8] Joachim Holtz and Juan Quan. Drift and parameter compensated flux estimator for presistent zero stator frequency operation of sensorless controlled induction motors. *IEEE Transactions on industry applications*, 39(4):1052–1060, 2003.

[9] C.B. Jacobina, J.E.C. Filho, and A.M.N. Lima. On-line estimation of the stator resistance of induction machines based on zero-sequence model. *Power Electronics, IEEE Transactions on*, 15(2):346–353, March 2000.

[10] J. Luukko K. Rauma, O. Laakkonen and O. Pyrhnen. Comparison of alternative implementations of dtc using fpga circuits. In *EPE 2003*, 2003.

[11] M. Ikonen-P. Silventoinen K. Rauma, O. Laakkonen and O. Pyrhnen. Fpga based dead-time compensation for pwm inverters. In *EPE 2005 - Dresden*, 2005.

[12] I. Slama-Belkhodja L. Charaabi, E. Monmasson and J. Louis. Fpga realization of reconfigurable ip-core function for real-time induction motor model. In *EPE 2003*, 2003.

[13] S. Koch M. Depenbrock, C. Foerth. Speed sensorless control of induction motors at very low stator frequencies. In *EPE'99*, 1999.

[14] . V.-M. Leppnen, J. Luomi. Estimating the back-emf of an induction motor. In *European conference on power electronics, EPE*, 2001.

[15] Dag Samuelsen Waldemar Sulkowski. A fpga based pwm synchronised modulator and measurement unit. In *11th International Power Electronics and Motion Control Conference EPE PEMC2004 Riga*, 2004.

[16] J.L. Zamora and A. Garcia-Cerrada. Online estimation of the stator parameters in an induction motor using only voltage and current measurements. *Industry Applications, IEEE Transactions on*, 36(3):805–816, May-June 2000.

Development of Monitoring System for Series HEV Bus with Touch Panel

Tae-Won Chun[*], Quang-Vinh Tran[**], Uk-Don Choi[***], Heung-Gun Kim[****]

[*]Department of Electrical Eng., University of Ulsan, Ulsan, Korea : twchun@mail.ulsan.ac.kr
[**]Department of Electrical Eng., University of Ulsan, Ulsan, Korea : tqvinh_vn @yahoo.com
[***]Hyundai Heavy Industrial Co. Ltd, Ulsan, Korea : udchoi@hhi.co.kr
[****]Department of Electrical Engineering, Kyungpook University, Daegu, Korea : kimhg@knu.ac.kr

Abstract— **In this paper, the monitoring strategies for a series hybrid electric vehicle (SHEV) bus are suggested for the optimal operation of powertrain and detection of abnormal conditions during the initial development of hybrid technologies. The monitoring system receives the messages from four control modules in the SHEV bus through a CAN, and then displays the data in the messages on a touch panel. In addition, all gauges and the on/off switching state of a set of telltale lamps are indicated. The monitoring system is implemented by both the 32-bit DSP and touch panel, and it is installed and tested on the real SHEV bus.**

Keywords — **DSP, monitoring, network, series hybrid electric vehicle (SHEV) bus.**

I. INTRODUCTION

The battery-power electric vehicles possess some advantages over conventional ICE vehicles, such as high energy efficiency and zero environmental pollution. However, the operation range per battery charge is far less competitive than ICE vehicles, due to the low energy content of the batteries. Hybrid electric vehicles (HEVs) which are powered by both an electric motor and an ICE, reduce emission and increase the fuel efficiency of the vehicles. The HEV can be classified into two types : series hybrid and parallel hybrid. The drawbacks of the series HEV (SHEV) over parallel types are the requirements of two electrical machines and large dimensions for the traction motor, in addition to the losses incurred during the conversion of mechanical to electrical energy. On the other hand, the engine is fully decoupled from the driven wheel, and connected to the generator for electricity production [1-3]. The series hybrid system is easier to build and does not require a transmission. In case of the heavy duty vehicles such as bus, the SHEV is much suitable for a frequent stop-and-go driving at urban routine. A growing number of companies are developing and beginning to supply commercial series hybrid-electric drive products to the bus markets.

In the exiting papers related to SHEVs, studies are mostly concentrated on controlling the engine-generator system with the consideration of the battery state-of-charge (SOC) in order to satisfy the required power from the propulsion system while keeping fuel consumption and vehicle emissions as low as possible [4-6].

The monitoring technologies are necessary for the optimal operation of powertrain in the SHEV bus and timely detection of abnormal conditions during the initial development of hybrid technologies to preclude lack of knowledge from causing program failure. This paper suggests the on-line monitoring system for the control area network (CAN) messages communicated with control modules in the SHEV bus. In addition, it can indicate all gauges and the switching state of a set of telltale lamps.

II. SHEV BUS CONTROL

A. Basic Structure of SHEV Bus

In the SHEV, many electrical modules are functionally interconnected as shown in Fig. 1. In this scheme, the ICE power controlled by an engine control unit (ECU) is converter to DC electric power with the help of an AC generator, AC-to-DC converter, and generator set unit (GENSET). The battery power is fed to an induction motor through a DC-to-AC converter and control electronics unit (CEU), which drives the wheels through a fixed gear arrangement. The vehicle's battery pack consists of 28 lead-acid batteries with a nominal voltage of 12V connected in series. The charge/discharge strategy for the battery pack is accomplished via the battery control system (BMS).

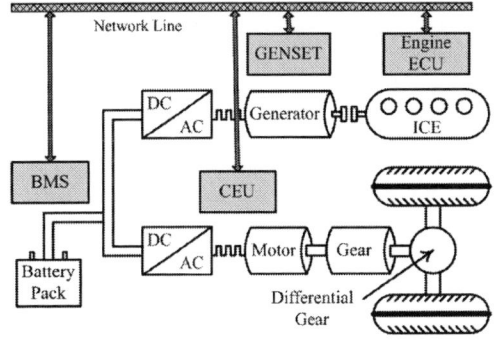

Fig. 1. Block diagram of controlling SHEV bus.

B. Data Messages in SHEV Bus

The CAN is used for real-time communication between four control modules. The CAN was originally developed to support low cost simple automotive applications, because of its performances and economic feasibility [7],[8]. J1939 is specified as the CAN protocol, designed to support real-time control functions between electronic control devices which may be distributed throughout the

978-1-4244-1741-4/08/$25.00 ©2008 IEEE

heavy-duty vehicle such as the bus and truck. Most messages defined by the J1939 standard are intended to be broadcast.

J1939 uses the 29-bit identifier defined within the CAN 2.0B protocol, as shown in Fig. 2. The identifier is used slightly different in a message with a destination address compared to a message intended for broadcast. The SOF, SRR, and IDE bits are defined by the CAN standard and will be ignored here. The RTR bit (remote request bit) is always set to zero in J1939.

bit	D28-D26	D25	D24	D23-D16	D15-D8	D7-D0
Field	Priority	R	DP	PDU Format	PDU Specific	Source Address

Fig. 2. Structure of 29-bit identifier.

The first three bits of the identifier are used for controlling a message's priority during the arbitration process. The next bit of the identifier is reserved for future use and should be set to 0 for transmitted messages. The next bit in the identifier is the data page selector. This bit expands the number of possible Parameter Groups that can be represented by the identifier. The PDU format (PF) determines whether the message can be transmitted with a destination address or the message is always transmitted as a broadcast message. The interpretation of the PDU specific (PS) field changes based on the PF value. If the PF is between 240 and 255, the message can only be broadcast and the PS field contains a Group Extension. The PF of four control modules in SHEV is determined as 255 or 254 for broadcast messages. The PS field is assigned as the number of messages and each control module has several messages.

The messages of CEU module contain the current, voltage, temperature, and status/fault conditions of the DC/AC converter-traction motor system. The BMS module transmits the voltage, current, temperature, SOC, and fault conditions of the battery pack. The GENSET module transmits the operating conditions, temperature, power, and generator speed of the GENSET system. The engine ECU has many messages, and some useful data such as the engine oil temperature, engine coolant temperature, fuel temperature, and engine oil pressure can be only received from engine ECU.

The current data rate of J1939 is 250kbps, and a typical message containing 8 data bytes and 29 bit identifier is 128 bits long. The messages of four control modules are independently transmitted on every 100msec or 200msec. The monitoring system receives the messages through a CAN, and then displays the data in the messages on the touch panel.

III. GAUGES AND TELLTALE LAMPS

The monitoring system indicates the front/rear break air pressure gauges, fuel level gauge, oil pressure gauge, and switching state for 35 telltale lamps located at the dash display in the conventional bus.

Fig. 3 shows the characteristics of a front/rear air break pressure sensor whose resistance varies as a function of the air pressure. As the air pressure may be varied from 0 to 12[Kg/cm^2], the sensor resistance is changed to 230[Ω]. So, the sensor has a positive pressure coefficient. The span of the needle angle of an air pressure gauge is 90°. The characteristics of a fuel level sensor with a negative level coefficient is provided in Fig. 4. As the diesel fuel level is higher, the sensor resistance is decreased from 80[Ω] to 10[Ω]. The span of the needle angle of a fuel level gauge is 60° . The values of both sensors are indicated by points which the needle revolves around within a circular gauge.

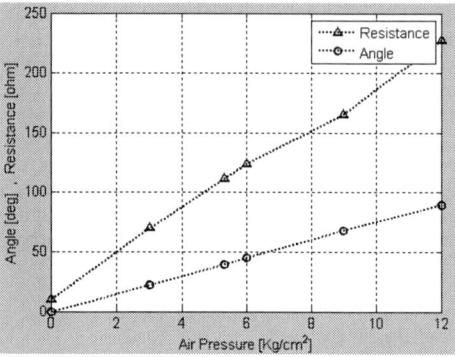

Fig. 3. Characteristics of air break pressure sensor.

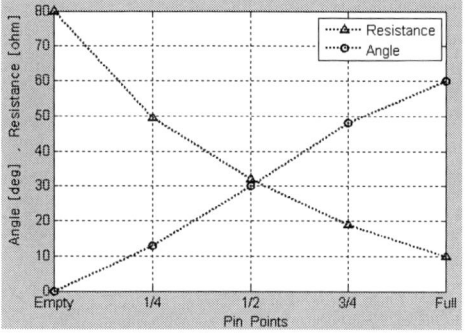

Fig. 4. Characteristics of fuel level sensor.

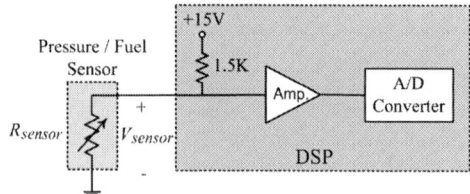

Fig. 5. Measurement circuit of sensor data.

The measurement circuit for both sensors is shown in Fig. 5. A pull-up resistor is connected in series to the sensor. After amplifying an analog measured signal taken from a midpoint lying between the resistor and the sensor, it is sent to an analog/digital converter. The resistance of sensors is converted into digital value using the measurement circuit, and the pin points of needle in the circular gauge are calculated by using a curve fitting function at the Matlab program. The oil pressure gauge is adjusted by the date received from the engine ECU through CAN.

The telltale system is used to communicate the driver's intentions to the external environment around the vehicle and generate the warning signals with the operating situations conditions of the SHEV bus. At the SHEV bus, 35 telltale signals are available, and the monitoring system can indicate on/off state for all telltale signals. The telltale signals have two types : active-high signals, active-low signals. The circuits for measuring telltale signals of both types are shown in Fig. 6. When the control signal is activated at the measurement circuit for active-high signal as shown in Fig. 6(a), the output terminal voltage becomes +24V, which is an auxiliary battery voltage. The voltage level of the signals must be reduced to 3.3V with the voltage divider, and then the on/off state of telltale lamps can be read through the I/O ports in the DSP. At the measurement circuit for active-low signal as shown in Fig. 6(b), the relay terminal is tied to +3.3V through a pull-up resistor for limiting the voltage level to 3.3V. The SHEV has 22 active-high telltale signals and 11 active-low telltale signals.

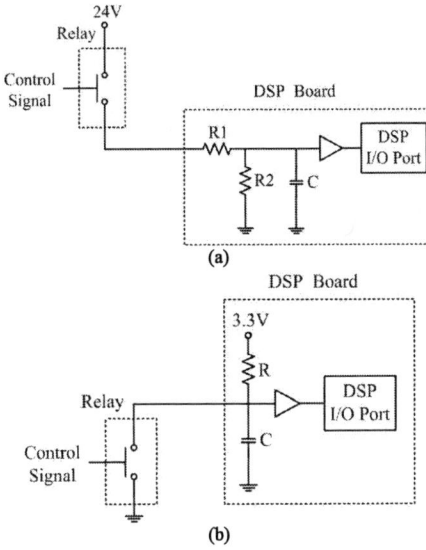

Fig. 6. Measurement circuit of telltale signals : (a) active-high signal, (b) active-low signal.

IV. SYSTEM CONFIGURATION

The monitoring system is implemented by 32-bit DSP type TMS320F2812 operating with a clock frequency of 150MHz. It provides the 16 channels 12-bit A/D converter, 56 general purpose I/O pins, CAN 2.0B controller, flesh memory, and so on. Fig. 7 shows the configuration of system hardware using the 32-bit DSP. The flowchart of the system software is shown in Fig. 8.

Whenever the CAN receiver interrupt is generated, the message can be received from control modules. The measured values of the front/rear air breaker sensors and fuel level sensor are received through A/D converter on every 100usec. The pin points of needles in the circular

gauges are calculated, where the average value of measured sensor data for 2 seconds is used for avoiding the rapid oscillation of needles. The on/off switching states of telltale lamps can be obtained from I/O ports on every 100msec.

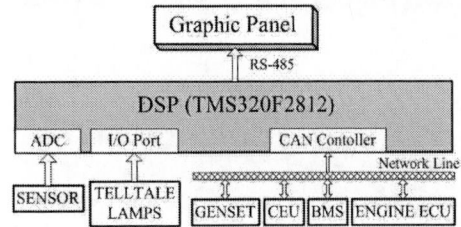

Fig. 7. Configuration of system hardware.

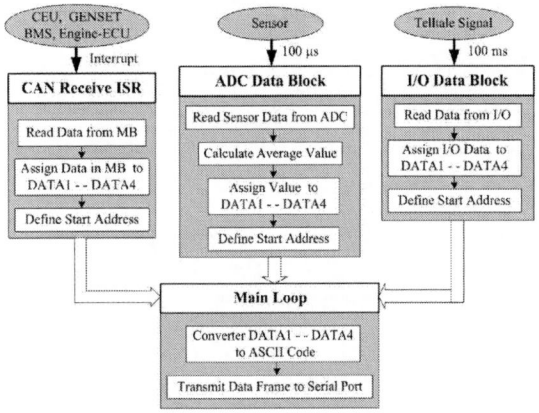

Fig. 8. Flowchart of system software.

The packages are constructed with the data frame structure required from a touch panel as shown in Fig. 9. The structure of data frame at the graphic panel is as follows.

· 'ESC' : Start of frame code
· 'W' : Write command for transmitting data
· Start address : Buffer address for storing data
· Data : Transmitting data
· 'CR' : End of frame code

ESC [1B]	W [57]	3 2 1 0 Start address	3 2 1 0 Data1	3 2 1 0 Data2	...	CR [0D]

Fig. 9. Structure of data frame.

They are sequentially transmitted to the touch panel for monitoring the operations of SHEV bus through a RS-485 serial port. The baud rate of the serial communication is 196,000bps, and each package contains 6- or 8-byte data.

V. TESTING RESULTS FOR SHEV BUS

The monitoring system developed by this paper is mounted and tested on the SHEV bus made by the Hyundai Heavy Industries Company.

Fig. 10. Main screen of the touch panel.

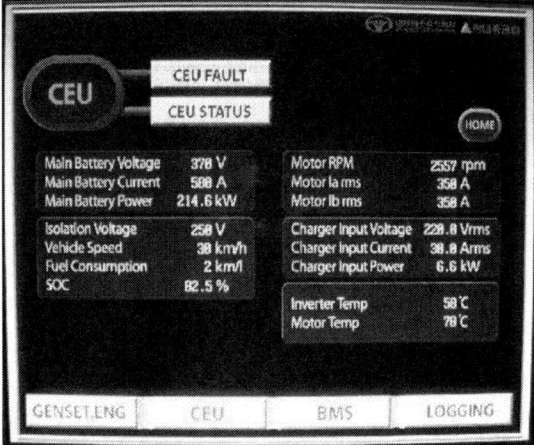

Fig. 11. Screen for CEU module.

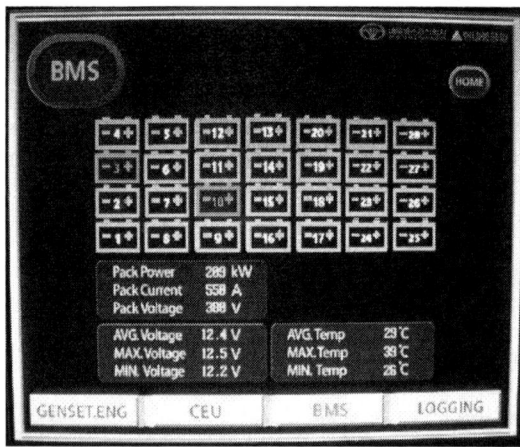

Fig. 12. Screen for BMS module.

Fig. 10 shows the main screen of the touch panel, which indicates four gauges (front/rear break air gauges, fuel level gauge, and engine oil pressure gauge), on/off states of all telltale lamps, vehicle speed, and SOC of battery pack. There are four icons at the bottom part on the main screen. Three icons such as CEU, BMS, GENSET/ECU are used for monitoring the data of messages received from four control modules.

Fig. 11 shows the CEU screen of the touch panel when the 'CEU' menu is selected. The battery voltage/current/ power, motor speed/current, charger input voltage/ current/power, temperatures of motor drive system, and fault/status conditions of CEU can be investigated.

Fig. 12 shows the BMS screen of touch panel. The battery pack power/current/voltage, average/maximum/ minimum of battery voltage, average/maximum/ minimum of battery temperature can be monitored. Also, it can be noticed which battery has the minimum voltage or maximum voltage.

Fig. 13 shows the GEMSET/ECU screen of touch panel, where the fault conditions, operating status, and various parameters in both the GENSET and Engine ECU modules can be investigated. A menu 'LOGGING' is used to record some parameters for a long time for investigation.

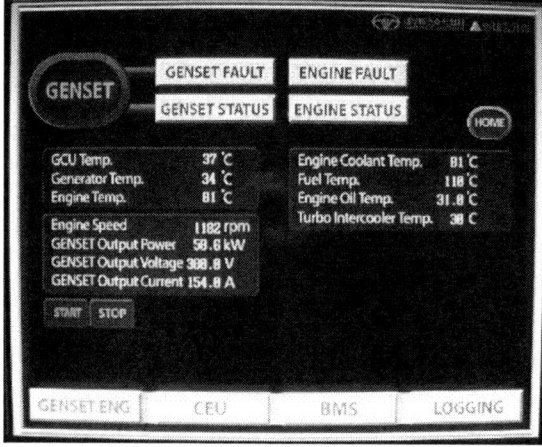

Fig. 13. Screen for GENSET and ECU modules.

Fig. 14 shows the photograph of SHEV bus including the monitoring system.

Fig. 14. Photograph of SHEV bus.

VI. CONCLUSIONS

In this paper, the on-line monitoring strategies for the SHEV bus are suggested in order to investigate an overall operation of powertrain and detect abnormal conditions. The monitoring system receives the messages from four control modules such as the GENSET, CEU, BMS, Engine ECU through the CAN, and displays the data in the messages on the touch panel. In addition, four gauges and the on/off switching states of 35 telltale lamps are indicated.

The monitoring system is implemented by both the 32-bit DSP and touch panel. Through the testing results with a real SHEV bus, it is verified that the proposed system is very useful to observe the operations of SHEV bus during the initial development of hybrid technologies.

ACKNOWLEDGMENT

This work is the outcome of a Manpower Development Program for Energy & Resources supported by the Ministry of Knowledge and Economy (MKE).

REFERENCES

[1] M.Ehsani, Y.Gao. and A.Emadi, *Modern Electric Hybrid Electric and Fuel Cell Vehicle-Fundamentals, Theory, and Design*, CRC, 2005.

[2] C.C.Chan, Y.S.Wong, "Electric Vehicle Charge Forward", Power Energy Mag., vol.2, no.6, pp.24-33, 2004.

[3] C.C.Chan and K.T.Chau, *Modern Electric Vehicle Technology*, Oxford University Press, 2001.

[4] M.Gokasan, S.Bogosyan, and D.J.Goering, "Sliding Mode Based Powertrain Control for Efficiency Improvement in Series Hybrid-Electric Vehicle", IEEE Trans. on Power Electronics, vol.21, no.3, pp.779-790, 2006.

[5] S.Barsali, M.Ceraolo, and A.Possenti, "Technique to Control the Electricity Generation in a Series Hybrid Electrical Vehicle", IEEE Trans. on Energy Conv., vol.17, no.2, pp.260-266, 2002.

[6] R.Saeks, C.J.Cox, J.Neidhoefer, P.R.Mays, and J.J.Murry, "Adaptive Control of a Hybrid Electric Vehicle", IEEE Trans. on Intell. Transport. Syst., vol.3, no.3, pp.213-234, 2002.

[7] R.V.Chacko, Z.V.Lakaparampil, and V.Chandrasekar, "CAN based Distributed Real Time Controller Implementation for Hybrid Electric Vehicle", in Proc. IEEE-IECON, pp.247-251, 2005.

[8] B.J.Ahn, B.R.Park, Y.H.Ki, H.S.Ahn, and D.H.Kim, "Development of a Controller Area Network Unit and its Application to a Fuel Cell Hybrid Electric Vehicle", in Int. Joint Conf. of SICE-ICASE, pp.5545-5549, 2006.

A Development System for Testing Integrated Circuits Used for Power and Energy Measurements

Vladimir Ćuk[*], Aleksandar Nikolić[†] and Aleksandar Žigić[*]

[*] Electrical Engineering Institute "Nikola Tesla"/Department for Electrical Measurements, Belgrade, Serbia,
e-mail: _vcuk@ieent.org_, [†] e-mail: _nba@ieee.org_, [*] e-mail: _azigic@ieent.org_

Abstract— A development system for testing integrated circuits used for power and energy measurements is presented in the paper. System is based on a DSP board, a data acquisition (DAQ) board and a computer with a real time mathematical model. The aim of the paper is to provide observing of measurement accuracy in the laboratory and to simulate different conditions with the "phantom load" technique: nonlinearity, high-order harmonics, phase unbalance, voltage sag, etc.

Keywords— Power Quality, Measurement, Real time simulation.

I. INTRODUCTION

In modern electrical networks, voltages and especially currents are often nonsinusoidal. The number of components capable of producing considerable harmonic distortion is very high and is increasing rapidly. It is very likely that the waveform distortion will continue to grow in the years to come.

Power quality problems that cause those irregularities affect power and energy measurement sometimes as well. Not all power and energy meters can measure correctly in these conditions, although they have a specified measurement precision [1], [2].

Some studies about power measurement accuracy in nonsinusoidal conditions have been done earlier [3]-[5]. The results were acceptable in most cases for commercial devices, but errors in measurement were found when the input signals had variable frequency, high crest factor or high harmonic content with quickly altering values.

That was the reason for developing the system proposed in the paper. The main requirement for this system was the ability to simulate various irregular network conditions with the "phantom load" technique: nonlinearities, high-order harmonics, phase unbalance, voltage sags, etc., which are difficult to provide in the laboratory and require very expensive equipment. The biggest advantage of the proposed system is its relatively low price. It can't be used for calibrating whole measurement systems, but it can test the accuracy of their core (the power measurement IC).

II. BASIC EXPERIMENTAL SETUP

Initial experimental setup is formed based on a proven energy measurement IC. These chips are today very cheap, but on the other side they give a lot of possibilities for developing sophisticated measurement circuits and have enough precision for billing applications. During development phase, an experiment is performed in which the Analog Devices ADE 7754 energy measurement IC is tested [6]. The same apparatus can later be used for testing similar IC's, with possible changes of communication between the DSP board and IC.

A. Real-time simulation model

One part of a desired electrical network is simulated with developed Matlab/Simulink real-time model [7], that has adjustable parameters such as RMS values of various harmonics of voltages and currents and their phase stands. In such a way nonlinearities could be simulated. All of these parameters can be time dependent, so all of the expected disturbances in a network could be simulated, too.

From the model, voltage and current signals are forwarded to the Texas Instruments F2812 DSP board that can be accessed directly from Matlab/Simulink with a corresponding toolbox [8]. This particular DSP board is used because of the existence of a targeted toolbox in Matlab for this board [7] and the necessary SPI communication that is needed to access the observed integrated circuit.

The whole experiment is schematically illustrated in Fig. 1. The code for this experiment is divided into three parts. The first program – the host side of the model, executes on the PC, and its job is to acquire data from the DSP board and the data acquisition (DAQ) board and to calculate the measurement error. The Simulink model of this program is shown in Fig. 2.

Active and reactive powers are not read from the IC directly. ADE 7754 stores these values in a 54-bit nonreadable register, so they have to be calculated from active and apparent energy. The DSP board reads values of active and apparent energy in turns from the IC and sends them to the host PC by RTDX communication. The host program accepts these values from RTDX and calculates rates of active and reactive energy flow, which represent active and reactive power measured by the IC.

On the other hand, the same program acquires instantaneous values of voltages and currents measured by the DAQ board and calculates active and reactive powers. Algorithms for power calculation found in SimPowerSystems toolbox are used for power calculation.

The error signals for active and reactive power are then calculated by subtracting reference powers (powers calculated with DAQ board measurements) from the powers measured with the IC.

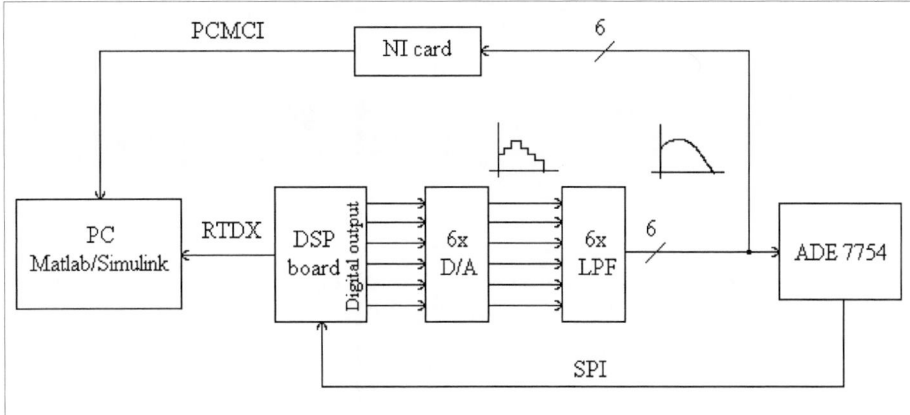

Fig.1. Block diagram of the experimental setup

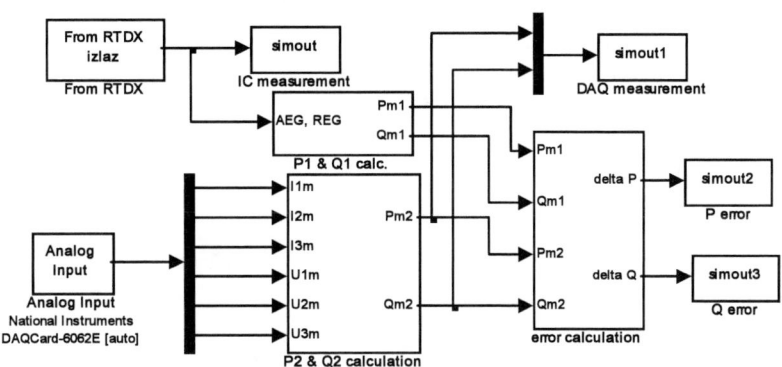

Fig.2. The host (PC) side of model

The second program – the target side of the model, executes on the DSP board. This programs job is to generate simulated signals, send them to D/A converters by general-purpose outputs (GPO's), calibrate the IC, acquire data from the IC and send it to the host by the RTDX channel. Fig. 3 shows the Simulink model of this program.

A part of this program used for signal generation works as an arbitrary waveform generator because generating 6 complicated and independent functions at once would use too much processing power. One of its modules (used for one D/A converter) is shown with more details in Fig. 4.

Pre-generated waveform signals are loaded into memory and loop continuously. They are converted into 16-bit unsigned integer numbers because the GPO uses this format. After that, they are scaled down to numbers that use only 12 bits (12 bit D/A converters are used) and sent to 12 GPO's. At the same time, one counter with 4 different states is used to determine which channel is loaded (D/A converters have 4 channels, of which 3 are used on each), the channels address, and weather latching or sending data is happening. This takes 3 additional GPO's per channel, and overall 30 GPO's are used for all 6 signals.

The SPI communication supported by the target toolbox in Simulink isn't suitable for reading ADE 7754 registers. Simulink's algorithm for SPI communication doesn't have much programmability, and on the other hand the ADE 7754 requires at least 4 μs between read and write operations with the chip select (CS) signal on the low level. This gap couldn't be filled using just Simulink. Because of that, the SPI communication part of the program had to be changed in Code Composer Studio (CCS), after the code was generated in Matlab. A part of the program that writes data in some ADE 7754 registers was also added in CCS (configuring and calibrating the IC). This part of the program isn't generated in Matlab and that is the reason it can't be seen on Fig. 3.

Configuration of the IC's measurement is done by setting the OPMODE (Operational Mode), MMODE (Measurement Mode), WATMODE (Active Power Calculation Formula) and VAMODE (Apparent Power Calculation Formula) registers, according to [6]. These registers are responsible for setting the IC in normal mode (with all of the A/D converters, LP filters and HP filters activated), choosing the line for period measurement, and choosing formulas for active and apparent power calculation (the last two settings are made corresponding to a 4 wire star (Y) meter form.

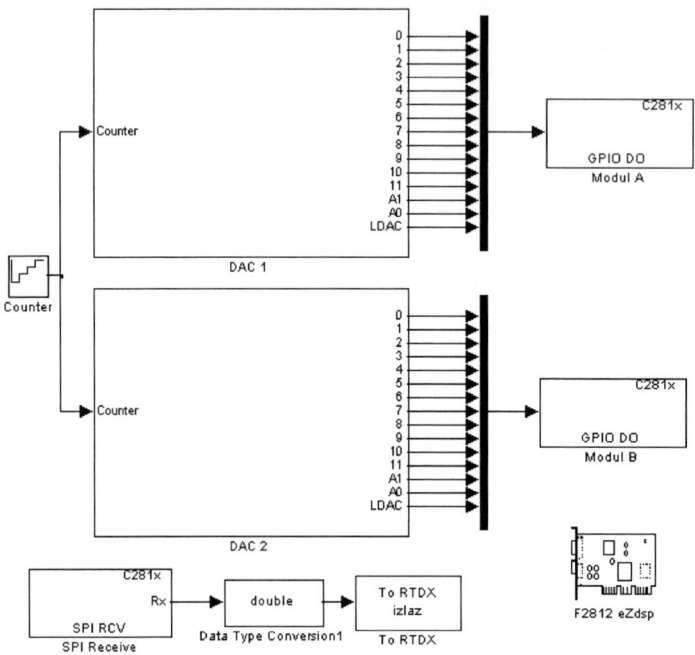

Fig.3. The target side of the model

Calibration and configuring of the chip is needed after every reset. Calibration parameters are derived from an experiment with pure sinusoidal voltage and current signals and without phase unbalance or time changes, according to [9]. Two sets of undistorted signals were used, with different power factors. In the first set, the power factor was 1, and in the second set, the power factor was 0.5.

Values of these undistorted signals applied to the IC's inputs are measured by the DAQ board as well, and readings of measurement registers are used to calculate the values of the calibration registers. The ADE 7754 has several calibration registers: GAIN (used for current and voltage gain setting), AWG, BWG and CWG (used for setting active power gain in each of the phases), AVAG, BVAG and CVAG (used for setting apparent power gain in each of the phases), APHCAL, BPHCAL and CPHCAL (used for phase calibration), AAPOS, BAPOS and CAPOS (used for power offset calibration), WDIV (sets the active energy divider) and VADIV (sets the apparent energy divider).

After the calibration, the program starts reading two registers in turns. The first of these two registers is RAENERGY. Active power is accumulated over time and stored in this 24-bit register (and some other registers as well). After a read operation, this register is reset to 0. Because of that, the average active power between two reads is equal to the contents of the RAENERGY register divided by the time between two consecutive reads. The second register being read continuously is RVAENERGY. This register is very similar to the RAENERGY. The only difference between these two registers is that the RVAENERGY

accumulates apparent power over time instead of active power. The average apparent power between two register reads is calculated the same way as the average active power.

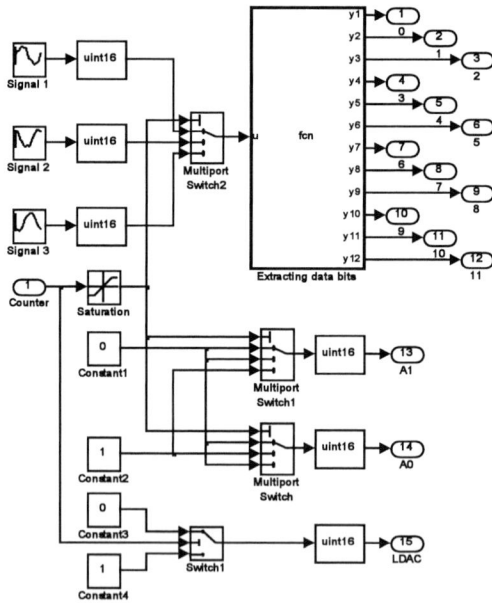

Fig. 4. Expanded model for DAC signals

The third program needed to conduct this experiment is not executed in real time. This program is a Simulink model used to generate voltage and current waveforms loaded in the arbitrary waveform generators (in the target side program). Simulink offers a wide range of simulation models. Waveforms can be generated using mathematical functions, with all of their parameters configured. This type of signal generation is presented in Fig. 5. Another way to generate signals is to make a real world model using the SimPowerSystems toolbox. This toolbox has pre built models of power system elements, power electronic devices and electrical drives. This is very convenient for simulating real world situations expected in every system. Fig. 6 shows one example of a model that can be used. Experimental results in this paper are made from the model shown in Fig. 5. Models such as the one on Fig. 6 will be used in the experiments to come.

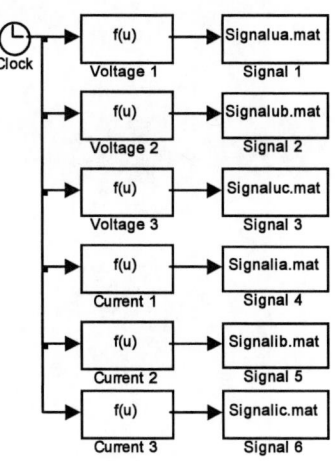

Fig. 5. Signal generator – mathematical functions

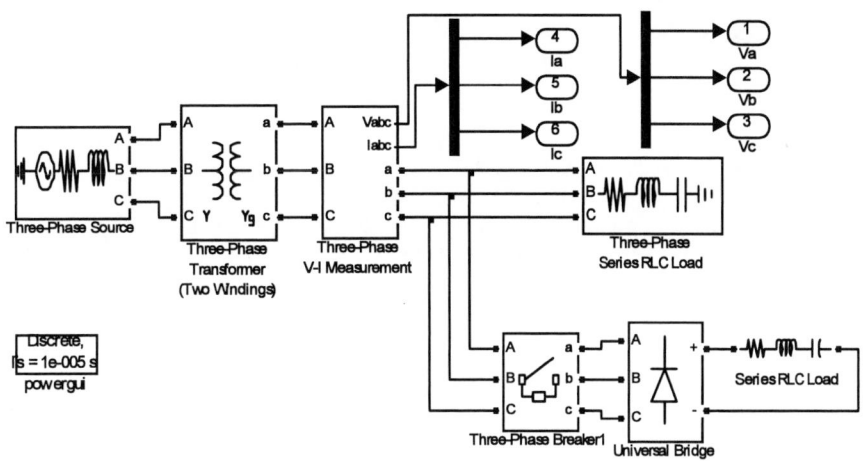

Fig. 6. Example of a signal generator – real world situation

B. Hardware used in the model

As mentioned earlier, the DSP board used for this development system is Texas Instruments TMS 320 F2812 eZdsp. Voltage and current signals are outputted from the DSP board thru digital outputs (30 GPO's), and sent to D/A converters.

Two Burr-Brown DAC 7624, 12-bit, 4 channel D/A converters are used to generate analog signals. Each of the A/D converters uses 3 of their channels. One of them is used to generate voltage signals, and the other one generates current signals.

Signals from the D/A converters are then sent to LP filters (passive, RC filters are used), and proceeded to the current and voltage inputs of the measurement IC. In this case, the ADE 7754 energy measurement IC is used, but in the future, some other power and energy measurement IC's are planed as well.

Simulated signals are not as big as the ones in real networks, but they are sent to the IC without conditioning, so they are in the same range as conditioned signals measured by a real measurement system. For this IC, input signals are in the range of – 0.5V to 0.5V.

The DAQ board used is National Instruments DAQ 6062E. This board measures voltage and current signals from the LP filters and sends them to the computer.

Fig. 7 shows the whole development system (excluding the PC).

III. RESULTS AND DISCUSSION

One set of generated voltage and current signals is presented in Fig. 8 and Fig. 9, respectively. They are generated as sums of harmonic components (by the Simulink model from Fig. 5), and recorded by the DAQ board. Harmonic distortions of current signals are bigger than voltage distortions. Also, the current signals form an unbalanced system. THD factors and phase stands of first order harmonics for all 6 signals are presented in Table I.

Fig. 10 presents active and reactive power measured by Simulink (from waveforms recorded by the DAQ board).

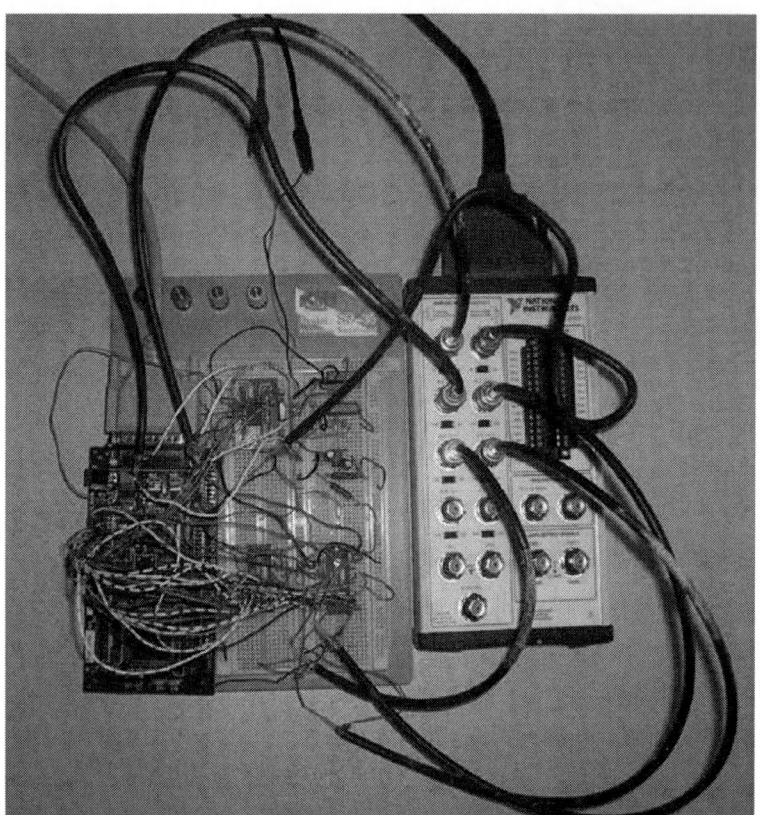

Fig.7. Hardware for the development system

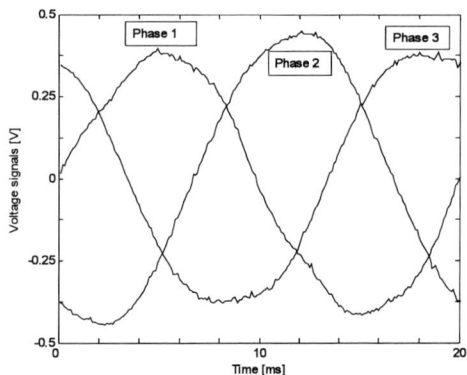

Fig. 8. Generated voltage signals

TABLE I
Signal THD's and phase stands

Signal name	THD [%]	Phase stand, first order harmonic [°]
Voltage 1	6.2	0
Voltage 2	6.4	120
Voltage 3	3.8	240
Current 1	18.1	44
Current 2	13.3	161
Current 3	15.3	285

Fig. 11 presents active and reactive power measured by the ADE 7754.

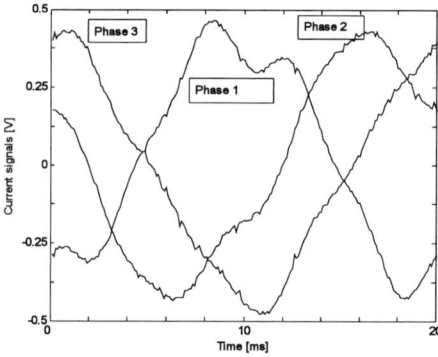

Fig. 9. Generated current signals

As mentioned earlier, those are actually powers calculated in Simulink from the active and apparent energy measured by the ADE 7754, and time.

As expected, the active power measurements had less differences then reactive power measurements. According to [6] and [7], Simulink and ADE 7754 use the same technique to calculate total active power.

On the other hand, reactive power calculation differed in these two cases. ADE 7754 is meant to be used for active and apparent power calculated, and reactive power measurement is left as an estimation option. That is way this IC can estimate reactive power only with the formula:

$$Q = sign(Q) \cdot \left(S^2 - P^2\right)^{1/2} \qquad (1)$$

Formula (1) calculates reactive power accurately only if no higher order harmonics are present in the network. In this case, the average error in reactive power measurement was 3.67%, while the average error in active power measurement was 0.2% (measurements were not conducted with full scale signals).

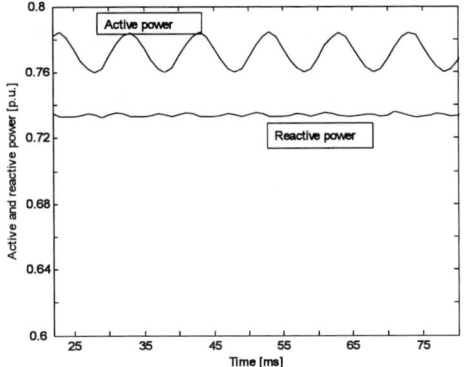

Fig. 10. Powers measured by the DAQ board

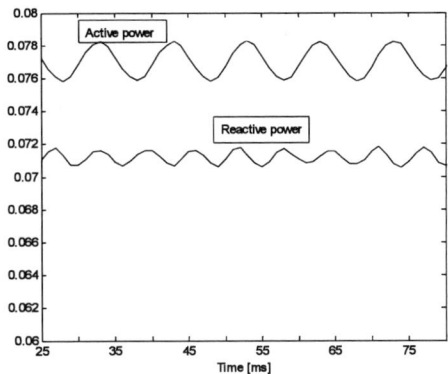

Fig. 11. Powers measured by the ADE 7754 board

IV. CONCLUSIONS

The only end conclusion in the analysis of this IC's measurement is that it measures active power correctly in the presence of high order harmonics.

Errors from reactive power measurement were large, but this IC is not meant to be used for reactive power measurement anyway, as it is clearly indicated in [6]. Its reactive power estimation is meant only as an option.

The final goal of the project presented in the paper is developing of a universal apparatus for testing power and energy measurement integrated circuits. This device should have more lines of communication (for different types of IC's). To realize complex and different situations that could be occurred in a real power network, a mathematical model with an adjustable measurement input is developed. This model is simulated using real-time simulation with data acquisition card and DSP board as a part of simulation model (Hardware-in-Loop simulation). Such approach gives opportunity to test different power measurement circuits under circumstances that could not be achieved in laboratory or could be very expensive if performed in a real world.

REFERENCES

[1] Durnte L.G., Ghosh P.K., "Active power measurement in nonsinusoidal environments", *IEEE Transactions on Power Systems*, vol. 15, August 2000.

[2] J. Driesen, T. Van Craenenbroek, D. Van Dommelen, "The registration of harmonic power by analog and digital power meters", *IEEE Trans. on Instrumentation and Measurement*, vol. 47, No. 1, pp. 195-198, February 1998.

[3] S. Santoso, W.M. Grady, E.J. Powers, J. Lamoree, S.C. Bhatt, "Characterization of distribution power quality events with Fourier and wavelet transforms", *IEEE Trans. on Power Delivery*, vol. 15, No. 1, pp. 247 – 254, January 2000.

[4] J. Driesen, R. Belmans, "Time-Frequency Analysis in Power Measurement using Complex Wavelets", *IEEE International Symposium on Circuits and Systems*, ISCAS 2002.

[5] A.A. Girgis, W.B. Chang, E.B. Makram, "A digital recursive measurement scheme for on-line tracking of power system harmonics", *IEEE Transactions on Power Delivery*, vol. 6, no. 3, July 1991.

[6] Analog Devices, *Polyphase Multifunction Energy Metering IC With Serial Port ADE 7754*, 2003.

[7] Mathworks, *Matlab/Simulink Target for TI C2000, User's guide*, March 2006.

[8] Spectrum Digital, ez*DSP F2812 Technical reference*, February 2003.

[9] Analog Devices, *Calibration of the 3-Phase Energy Measurement Board on a ADE 7754*, 2003.

[10] P.S. Filipiski, P.W. Labaj, "Evaluation of Reactive Power Meters in the Presence of High Harmonic Distorsion", *IEEE Transactions on Power Delivery*, vol. 7, No. 4, October 1992.

State and parameter estimation in a hydraulic system – moving horizon approach

Jerzy Baranowski*, Andrzej Tutaj*

* Department of Automatics, Faculty of Electrical Engineering, Automatics, Computer Science and Electronics,
AGH University of Science and Technology, Kraków, Poland, e-mail: *jb@agh.edu.pl, tutaj@ia.agh.edu.pl*

Abstract—In most control systems the best quality of control can be achieved if an exact model of a plant is known and if disturbances-free and continuous (instead of discrete) measurements of all its state space variables are available. Since it is seldom the case in real-life installations, state observers, adapting (parameters identification) algorithms, anti-noise filters and continuous signal estimators must be often employed. The paper describes an application of the moving horizon estimation (MHE) and continuous state estimation from discrete output measurements (CSE-DOM) based method to a laboratory hydraulic three-tank system. The resulting algorithm encompasses all the four above-mentioned mechanisms (state observation, parameters identification, disturbances filtration and continuous state estimation) and can be used for nonlinear systems as well, which is the case in the presented article. The paper provides a detailed explanation of how the MHE and CSEDOM methods were adapted to be applicable to the three-tank system. Results of computer simulations and practical laboratory experiments are also included. [1]

Index Terms—estimation technique, non-linear control, adaptive control, real time processing

I. INTRODUCTION

In most control systems the best quality of control can be obtained if all the following four conditions are secured:

- an exact model of a controlled process is known and parameters of the model are constant with time
- a full state of a plant is accessible and measured
- no measurement disturbances are present
- measurements are provided as a continuous time signals

In practice the above-mentioned requirements are rarely fulfilled. Usually merely a simplified process model is used and its parameters are known with a finite accuracy and may vary with time. Only partial information of the system outputs are accessible instead of a full state, there are disturbances present in measurements and only discrete samples are available instead of continuous time series (see [8]). In such situations the four following counter-measures can be adopted:

- parameters of the system may be continuously identified by an adaptive control algorithm
- unknown states can be reconstructed by an state observer
- measurement noises and disturbances may be rejected with a signal filter

[1] Work financed by state science funds for 2008-2010 as a research project. Contract no. N N514 417734

- continuous state estimates can be computed using discrete state samples

Each of the four tasks can be realized by separate algorithms. There is an alternative however where a moving horizon estimation (MHE) technique is used (see [16], [22]). The MHE algorithm can address the first three of the four listed issues. It identifies model parameters, filters disturbances and reconstructs non-measured system states. The MHE method lacks however the capability to accomplish the fourth task. An appropriate solution can be found in authors' previous works where methods for continuous state estimation from discrete output measurements (CSEDOM) has been introduced (see [2], [3], [14]) and later experimentally verified [4], [5]. In this paper the algorithm connecting MHE and CSEDOM approach is presented.

The MHE and CSEDOM algorithms can be used both for linear as well as for nonlinear systems. In the latter case this united solution presents a considerable advantage over a classical, widely used methods where a linearized system model is employed (see for example [12], [13]).

II. LABORATORY HYDRAULIC CASCADED THREE-TANK SYSTEM

A laboratory hydraulic cascaded three-tank system (see [9], [17]) considered in the paper is depicted on the photograph in the figure 1. In the same figure its schematic diagram is presented as well. The system is located in the Department of Automatics of AGH. The installation consists of three vertically arranged tanks: upper, middle and bottom. A side wall of each container have a different shape: rectangular, trapezoidal and quarter circular respectively. Their dimensions are given in the figure 2. The area of a water surface in the upper tank is constant but in two remaining tanks it varies with the water level. There is a fourth tank as well, located in the lowest position, serving as a water buffer. A sliding-vane pump driven by a electric DC permanent magnet brush motor pumps water from the buffer to the upper tank. From there the water flows through a constant valve to the middle tank and then to the lower and eventually to buffer tank in a similar fashion. These flows are driven by gravity force and governed by the Toricelli's law (see [18]). The electric motor is driven by a PWM signal from a power amplifier. Water levels in the three tanks are measured with three pressure sensors and their signals are conditioned in an appropriate electronic interface.

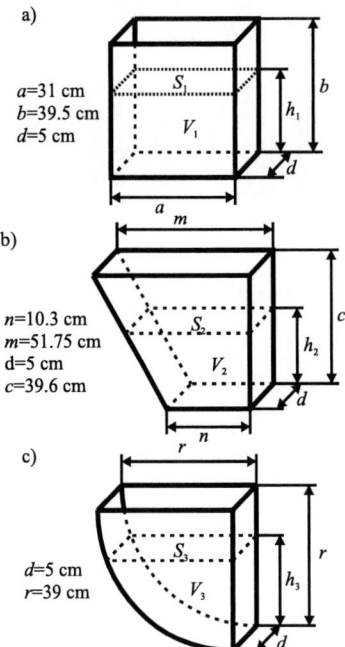

Figure 2. Shapes and dimensions of three water tanks: a) upper, b) middle, c) bottom

A. Mathematical model of the system

A mathematical model of the laboratory installation can be derived from the law of mass conservation (see [7]) which for the three cascaded tanks takes the form of the three following differential equations

$$\frac{\mathrm{d}}{\mathrm{d}t}\left(\varrho V_1\right) = \varrho q_0 - \varrho q_1 \tag{1}$$

$$\frac{\mathrm{d}}{\mathrm{d}t}\left(\varrho V_2\right) = \varrho q_1 - \varrho q_2 \tag{2}$$

$$\frac{\mathrm{d}}{\mathrm{d}t}\left(\varrho V_3\right) = \varrho q_2 - \varrho q_3 \tag{3}$$

where ϱ is a water density, V_1, V_2, V_3 are water volumes in three consecutive tanks, q_0 is a control-dependent pump voluminal flow and q_1, q_2, q_3 are level-dependent valves voluminal flows. For a partially filled open tank a derivative of a volume V with respect to time t can be expressed as follows

$$\frac{\mathrm{d}V(h(t))}{\mathrm{d}t} = \frac{\mathrm{d}V(h)}{\mathrm{d}h}\bigg|_{h=h(t)} \cdot \frac{\mathrm{d}h(t)}{\mathrm{d}t} = S(h(t))\,\dot{h}(t) \tag{4}$$

where S is an area of an open water surface in the tank and h is a water level measured from the bottom. With an additional assumption of a constant water density one can obtain from (1)–(3) and (4) the following state space equations

$$\dot{h}_1(t) = \frac{1}{S_1}\left(q_0(u) - q_1\big(h_1(t)\big)\right) \tag{5}$$

$$\dot{h}_2(t) = \frac{1}{S_2\big(h_2(t)\big)}\left(q_1\big(h_1(t)\big) - q_2\big(h_2(t)\big)\right) \tag{6}$$

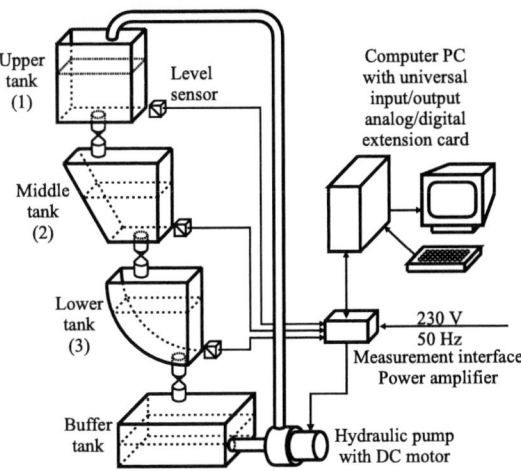

Figure 1. A photograph and a schematic diagram of the laboratory installation

There is a PC computer dedicated to the laboratory system, equipped with a universal digital-analog input-output extension card RT DAC 4 PCI which measures analog water level signals and provides digital PWM control signal for the pump DC motor. A MATLAB-Simulink (see [15], [21]) environment with *Real-Time Workshop* (RTW) and *Real-Time Windows Target* (RTWT) toolboxes is used to develop, build and test a real-time application. It runs as MS Windows XP operating system process and controls laboratory plant. The application should be coded as a Simulink graphical model and is then automatically translated into a collection of *C* language source files. Next, compilation and linkage processes are invoked giving a final executable file. *Watcom C/C++* compiler and linker are used. The whole process is conducted in a highly automatic fashion and allows a user to develop gradually his or her project in a convenient way, without necessity of textual coding (see [10]).

$$\dot{h}_3(t) = \frac{1}{S_3\big(h_3(t)\big)} \Big(q_2\big(h_2(t)\big) - q_3\big(h_3(t)\big) \Big) \quad (7)$$

where state variables $h_1(t)$, $h_2(t)$ and $h_3(t)$ are the water levels in cm in upper, middle and bottom tank respectively. Symbols S_1, $S_2(h_2)$ and $S_3(h_3)$ are the areas in cm^3 of open water surfaces given by following formulae

$$S_1 = a\,d = \text{const} \quad (8)$$

$$S_2(h_2) = d\left(n + \frac{m-n}{c}\,h_2\right) \quad (9)$$

$$S_3(h_3) = d\sqrt{2\,r\,h_3 - h_3^2} \quad (10)$$

where c, d, m, n and r are tanks physical dimensions defined on figure 2. Function $q_0(u)$ characterizes the relationship between a duty factor $u \in [0, 1]$ of a PWM signal and a voluminal water flow q_0 in cm^3/s produced by the pump. This mapping can be approximated with a polynomial of fifth degree

$$q_0(u) = w_5\,u^5 + w_4\,u^4 + w_3\,u^3 + w_2\,u^2 + w_1\,u + w_0 \quad (11)$$

Functions $q_1(h_1)$, $q_2(h_2)$ and $q_3(h_3)$ describe dependencies between water levels and water flows through valves. According to the slightly modified Toricelli's law (see [18]) these relationships can be approximated as

$$q_i(h_i) = C_i\sqrt{D_i + h_i}, \qquad i \in \{1, 2, 3\} \quad (12)$$

where C_i and D_i are constants which need to be identified and i is the tank index (1 for upper, 2 for middle and 3 for bottom one).

III. IDENTIFICATION OF MODEL PARAMETERS

One of the MHE algorithm features is the ability to continuously identify parameters of a plant model. However online identification of all the parameters (nearly 20 of them) would be very difficult if not impossible. That is why we present here the results of typical identification experiments, which are also listed below. In the following parts of the paper we will use obtained parameters along with continuously identified additional tuning values.

A. Characteristic of the pump

A static characteristic of the pump was obtained after an appropriate experiment was conducted. Several different values of control u (being the PWM duty factor) from the interval $[0, 1]$ were applied in turn to the pump. For each individual control values a corresponding pump flow was calculated with the following formula

$$q_0 = S_1\frac{\triangle h_1}{\triangle t} \quad (13)$$

where $\triangle h_1$ represents an increase in water level within the time interval of the length equal to $\triangle t$. The valve of the upper tank must be closed during this experiment to prevent water from flowing out. Several points obtained with this procedure are shown with circles on figure 3. A relationship $q_0(u)$ was approximated with a 5th degree polynomial using the least squares methods implemented in MATLAB as `polyfit` function. Resulting coefficients

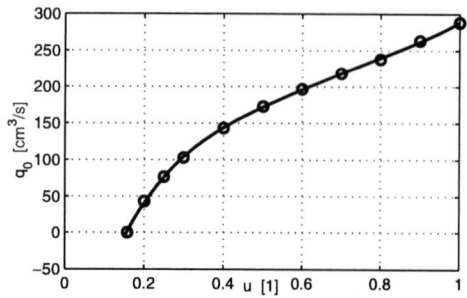

Figure 3. Static characteristic of the pump (circles – measurements, continuous line – approximating polynomial)

are given in the table I and plot of the corresponding polynomial is added to the chart in figure 3.

Table I
RESULTS OF THE IDENTIFICATION OF THE PUMP – COEFFICIENTS w_j OF THE POLYNOMIAL $q_0(u)$

w_5	1.9287
w_4	-6.5254
w_3	8.9191
w_2	-6.2047
w_1	2.4298
w_0	-0.258

B. Characteristics of valves

Values of parameters C_i and D_i in the equation (12) were obtained by applying a numerical optimization method to results of an appropriate experiment during which a piecewise constant control u was supplied and logged together with resulting water level signals h_1, h_2 and h_3. Their time series after low-pass filtering are shown in figure 4. Then a simulational model of

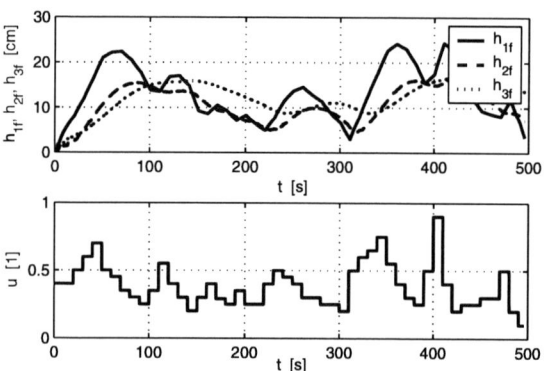

Figure 4. Time series of the control variable u and water levels h_1, h_2 and h_3 used for the identification of valves

the system was constructed in the form of C-MEX M-file for MATLAB and compiled in MS Visual C/C++ environment. It was based on the formulae (5-7), (8-10), (11) and (12). Parameters C_i and D_i were found by a numerical iterative optimization algorithm minimizing a square integral of identification error defined as the difference between the measured and approximated water level (see [19]). This procedure was applied separately to the three tanks in turn, starting from upper and ending with bottom one. Results of the identification for a preceding tank were used for the identification of a following one, hence the proper sequence must have been preserved. Initial conditions for each simulation were calculated from experimental data series as a mean value of several data points from a neighbourhood of a time point assumed as initial. In fact the time horizon [0, 500] (see figure 4) was divided into several shorter intervals and separate simulations were conducted for each of them. Initial conditions for all subintervals was being calculated independently. Consequently several indices were calculated and the eventual index in every iteration of optimization algorithm was obtained as a sum of all them. The optimization was conducted with fminsearch MATLAB function implementing the multidimensional unconstrained nonlinear Nelder-Mead minimization algorithm. Result of identification are gathered in the table II.

Table II
RESULTS OF THE IDENTIFICATION OF VALVES – PARAMETERS C_i AND D_i IN RELATIONSHIPS $h_i = C_i \sqrt{D_i + h_i}$

i	C_i	D_i
1	29.405	5.9217
2	33.260	4.0083
3	31.424	3.3205

IV. ALGORITHM OF ESTIMATION

For our problem we need to consider a continuous state estimation from discrete output measurements (CSEDOM). CSEDOM problem can be formulated as follows, let us consider a nonlinear system:

$$\dot{x} = f(x, \mu) + g(x, \mu)u \qquad (14)$$

$$y = h(x)$$

where $x(t) \in \mathbb{R}^n$ We will denote the state estimate as $\hat{x}$. $\mu \in \mathbb{R}^l$ is a vector of potentially time varying parameters, that might need identification (we will denote parameter estimate as $\hat{\mu}$). We assume that discrete measurements are taken through uniformly spaced sampling of output with sampling period of T_s. We will denote moments of sampling as $t_i = iT_s$. We want to estimate the actual state $x(t)$ of system (14), with an estimate $\hat{x}(t)$ for $t \in [t_i, t_{i+1}]$, using the output measurements for $t \leq t_i$. In general, the following iterative algorithm solves the problem:

1) PREDICTION STEP
   Solve the differential equation $\dot{\hat{x}}(t) = f(\hat{x}(t), \hat{\mu}) + g(\hat{x}(t), \hat{\mu})u(t)$ for $t \in [t_i, t_{i+1}]$ using $\hat{x}(t_i)$ as an initial condition.
2) CORRECTION STEP
   Using the measurements of $y(t)$ for $t \leq t_{i+1}$ estimate the value of $\hat{x}(t_{i+1})$ and $\hat{\mu}$.

3) Set $i = i + 1$ and go to step 1.

The concept of this algorithm is relatively simple – see figure 5, however the main problem is the choice of a estimation method used in the correction step. That is why we introduce the MHE algorithm presented below.

Remark IV.1. It should be noted, that also the prediction step requires few assumptions needed for keeping the estimation error bounded. However in case of this system we will not include specific consideration of this issue, because of asymptotic stability. This property provides a very good bound for a prediction error, because the predicted estimate becomes closer to the actual state as time follows.

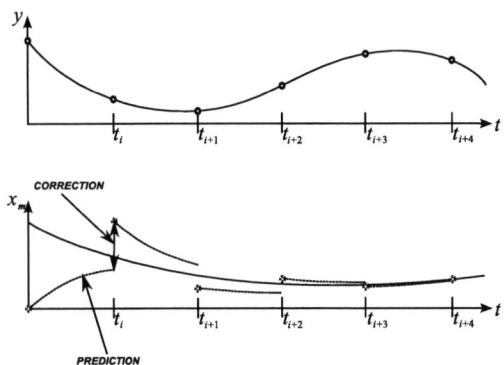

Figure 5. Illustration of CSEDOM concept – a prediction is followed by a correction

A. Moving horizon estimation (MHE)

Moving horizon estimation (see for example [22], [16]) is in a way a natural development from the linear least-squares estimation (see [6])) and can be considered a dual problem to the predictive control method (see [1], [20]). A moving window (a moving horizon) of discrete measurements is used to estimate the state at the beginning of the window and the parameters through the entire window (considering them as constant over the window). At the basis of the method lies an optimization problem solved per every window, which is formulated as follows:

$$\min_{\hat{x}_0^k, \hat{\mu}^k} J = \|\hat{x}_0^k - \hat{x}_0^{k-1}\|_{E(k)}^2 + \|\hat{\mu}^k - \hat{\mu}^{k-1}\|_{F(k)}^2$$

$$+ \sum_{j=k}^{k+m} (\hat{y}(t_j) - y(t_j))^2 \qquad (15)$$

Where

- k is a number of an iteration (a number of window)
- $\hat{x}_0^j$ is an estimate of initial state of the system at the beginning of the window using information from j-th iteration, if $j = k$ it is an optimization variable and when $j = k - 1$ it denotes a value obtained from last iteration
- $\hat{\mu}^j$ is an estimate of system parameters through the j-th window, if $j = k$ it is an optimization variable

1435

and when $j = k - 1$ it denotes parameters values obtained from last iteration

- $\|\alpha\|_A^2 = \alpha^{\mathsf{T}} A \alpha$
- $E(k)$ and $F(k)$ are time varying square $n \times n$ and $l \times l$ matrices, where n is a dimension of state space, and l is a number of parameters
- $y(t)$ is a measured output
- m is a length of the window and $m \geq n$

Optimisation problem (15) is subjected to the following constraints

$$\dot{\hat{x}}^k(t) = f(\hat{x}^k(t), \hat{\mu}^k) + g(\hat{x}^k(t), \hat{\mu}^k)u(t)$$

$$\hat{y}(t) = h(\hat{x}^k(t))$$

$$\hat{x}^k(t_k) = \hat{x}_0^k$$

$$t \in [t_k, t_{k+m}]$$

$$\hat{x}^k \in X$$

$$\mu_{min} \leq \hat{\mu} \leq \mu_{max}$$

where $X \subset \mathbb{R}^n$ denotes the state space. This constraint is usually caused by physical nature of the system (non-negative concentrations, limited tank capacity etc.).

Remark IV.2. Generally it is often advised to model the parameters μ as constant state variables. In that case above constraints are supplemented with

$$\dot{\hat{\mu}} = 0$$

This approach simplifies the gradient computation and allows easier interpretations (for parameters fixed on the entire measurement window).

Remark IV.3. Matrices $E(k)$ and $F(k)$ are used because of many reasons. One of the important interpretations are from the estimation theory.

It is obvious, that the estimation on the expanding window (from 0 till the current sample) contains full available information regarding the system state evolution. Consequently use of the entire measurement history would theoretically provide us with the best possible estimate. Unfortunately such estimation is numerically infeasible and would make the implementation of the algorithm very difficult. That is why one needs to approximate

$$J_1 = \sum_{j=0}^{k+m} (\hat{y}(t_j) - y(t_j))^2$$

with

$$J = P(\hat{x}(0), k) + \sum_{j=k}^{k+m} (\hat{y}(t_j) - y(t_j))^2$$

in a way that both problems would lead to the same solution. $P(\hat{x}(0), k)$ is called the *arrival cost* (one can find an analogy with terminal cost in nonlinear model predictive control). Optimally it should be a function $V(\xi)$, which is equal to the optimal value of the performance index on the preceding window (from $k-m$ to k). With a trajectory $\hat{x}^{k-m}$ obtained from that window it would be

$$V(\hat{x}^{k-m}(t_k)) = \min V(\xi)$$

This function is the *value function* of the problem. It is known, that for most of the cases value function cannot be directly computed, so the arrival cost has to be estimated. One of the properties of arrival cost is that

$$\bar{Q}_k \leq P(\hat{x}(0), k) \leq V(\xi)$$

where $\bar{Q}_k$ is an optimal value of the performance index at the current moment. Moreover it can be proved that

$$V(\xi) - \bar{Q}_k \leq \gamma(\|\xi - \hat{x}\|)$$

where $\gamma(\bullet)$ is a non-decreasing positive function. This implies that Q_k grows with k and is bounded, so the arrival cost also has to be bounded and non-decreasing. Two obvious choices for the arrival costs are then:

$$P(\xi, k) = \bar{Q}_k = const.$$

$$P(\xi, k) = V(\xi)$$

The first one has little beneficial properties and because it is constant it does not influence the optimisation and the latter is nearly impossible to obtain. That is why it should be locally approximated with a quadratic form

$$P(\xi, k) = \bar{Q}_k + \frac{\beta}{2}\|\xi - \hat{x}\|_{H(k)}^2$$

with $\beta \in (0, 1)$ and $H(k)$ is a positive definite matrix. Choice of the matrix is an another problem however. Most authors (see [16]) suggest, that $H(k) = \Pi^{-1}(k)$ where $\Pi(k)$ is the covariance matrix obtained from the solution of the Extended Kalman Filter. Unfortunately application of EKF requires a very good initial guess on the covariance matrix, without which appropriate Riccati equation leads to badly conditioned results. In this paper we propose $H(k)$ as a diagonal matrix with exponential entries.

$$H(k) = \begin{bmatrix} 1 - e^{(-\alpha_1(k-m))} & & \\ & \ddots & \\ & & 1 - e^{(-\alpha_N(k-m))} \end{bmatrix}$$

It can be interpreted, as a protective measure for the high impact of disturbances onto the estimate, because if last estimate was good, then big differences between it and the current one are usually caused by the measurement noise. For clearer view we have decomposed H in (15) into E and F. $\bar{Q}_k$ was omitted as it does not influence the optimisation.

V. RESULTS OF COMPUTER SIMULATIONS

A. State estimation

At the beginning we have tested the combination of CSEDOM and MHE as a tool for filtering and continuous state estimation. Results presented show the estimates of water levels in all tanks. Estimates were computed with uniformly sampled measurements of water level in the lowest tank. T_s was 2 seconds and $m = 10$. All of the estimates are presented after the first window of measurements. Optimisation was performed with Levenberg-Marquardt algorithm implemented in Matlab. Simulations were executed with Dormand-Prince 5(4) pair with a 4th order dense output (see [11]). Measurement noises were modeled as Gaussian with zero mean. All results are presented in figures 6-8.

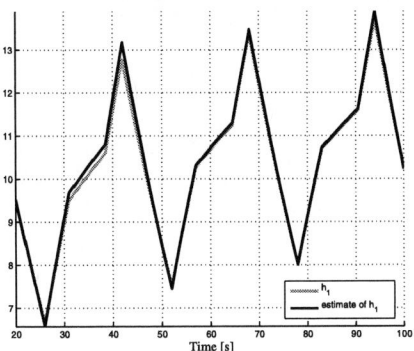

Figure 6.   State estimation - h_1 - simulation

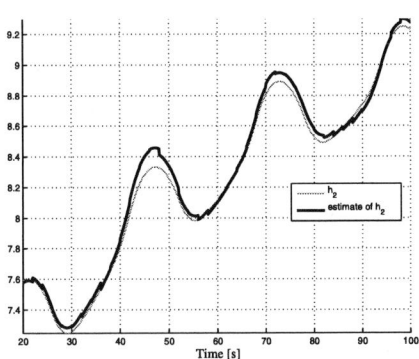

Figure 7.   State estimation - h_2 - simulation

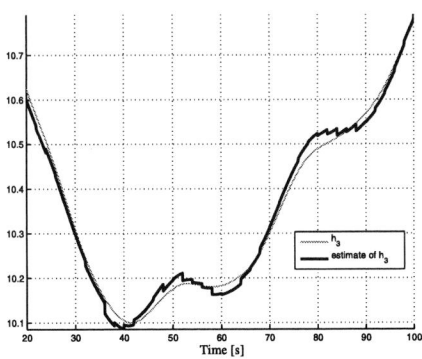

Figure 8.   State estimation - h_3 - simulation

B. State and parameter estimation

To keep the small size of the problem, we have considered on-line estimation of only parameters which can vary with time in real system. Our experience with the laboratory unit, suggested that the only element, which parameters can really change is the pump, that is why we will estimated the changes in the inlet flow modelled as

$$q(u) = aq_0(u) \qquad (16)$$

with a being an identified parameter. To check the algorithm performance we have modelled a change in parameters of the reference model. State and parameter estimates are presented in figures 9-12.   As it can be

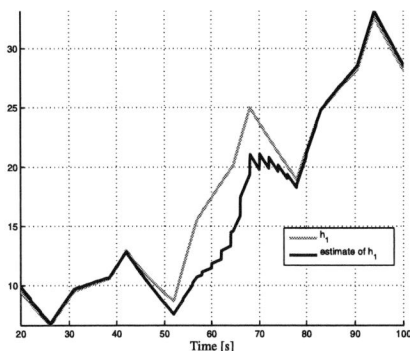

Figure 9.   State and parameter estimation - h_1 - simulation

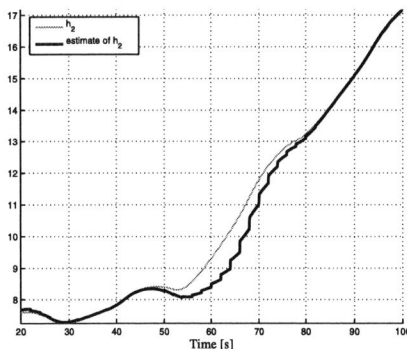

Figure 10.   State and parameter estimation - h_2 - simulation

seen on the figure 12 the change of parameter is identified, and subsequently algorithm adapts to it. Speed of adaptation is reduced by the assumption that the parameters should be constant on the window. This lag in estimation is visible especially in figure 9 where the estimate is disturbed during this transitory phase and subsequently gains convergence again.

VI. RESULTS OF LABORATORY EXPERIMENTS

Our experimental framework provided us with very noisy signals, which were introduced to the algorithm.

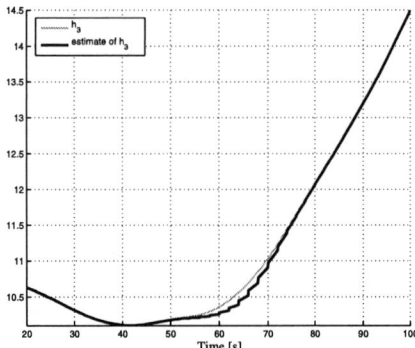

Figure 11. State and parameter estimation - h_3 - simulation

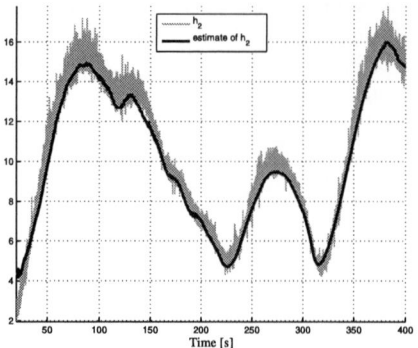

Figure 14. State estimation - h_2 - experiment

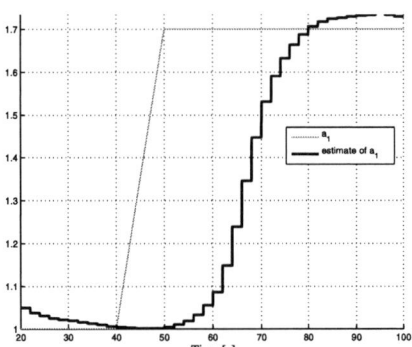

Figure 12. State and parameter estimation - a - simulation

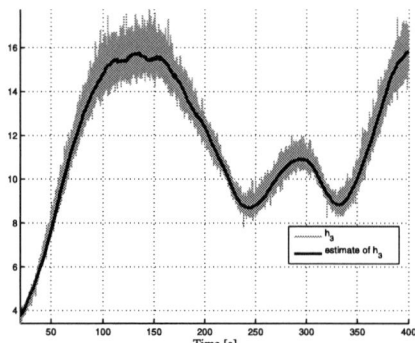

Figure 15. State estimation - h_3 - experiment

Because we did not have a reference for parameter estimation, we present here only state estimation on the longer interval. Results of this estimation are presented in figures 13-15. As it can be seen, algorithm with proposed arrival cost has very good filtering properties.

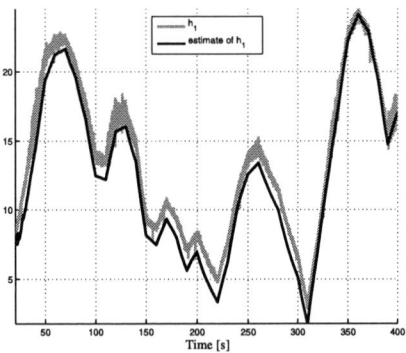

Figure 13. State estimation - h_1 - experiment

VII. SUMMARY

In the paper a developed algorithm for estimation applications was presented. It reconstructs system states, filters disturbances, estimates signals between sampling moments and has identification capabilites. The MHE and CSEDOM algorithms can be used both for linear as well as for nonlinear systems. In the latter case this approach presents a considerable advantage over classical, widely used methods where a linearized system model is employed. The usage of presented algorithm was illustrated with both computer simulations and laboratory experiments result showing its effectiveness in nonlinear system analysis.

ACKNOWLEDGMENT

Authors would like to thank Piotr Bania for his invaluable insight regarding the problems of dynamical optimisation in considered context.

REFERENCES

[1] P. Bania, *Algorytmy Sterowania Optymalnego w Nieliniowej Regulacji Predykcyjnej*, PhD thesis, Thesis supervisor - A. Korytowski. Akademia Górniczo-Hutnicza, Department of Automatics, Poland, Kraków, 2008.

[2] J. Baranowski, Continuous state estimation from discrete output measurements in a linear parabolic system, In *Materiały IX Międzynarodowych Warsztatów Doktoranckich OWD*, 20-23.10.2007, Wisła, 2007, vol 1, pp. 11-16.

[3] J. Baranowski, W. Mitkowski, New Output Approximation Method for Continuous State Estimation, In *Materiały XXXI Międzynarodowej konferencji z podstaw elektrotechniki i teorii obwodów IC-SPETO*, 2008, pp. 95-96 (extended version on CD).

[4] J. Baranowski, A. Tutaj, Continuous state estimation in water tank system, In *Proceedings of Computer Methods and Systems* 21-23.11.2007, Kraków, pp. 373-378.

[5] J. Baranowski, A. Tutaj, Comparison of continuous state estimation algorithms in a water tank system, *Recent advances in control and automation* (K. Malinowski and L. Rutkowski, eds.), Akademicka Oficyna Wydawnicza EXIT, 2008, pp. 63-72.

[6] W. Byrski, Integral description of the optimal state observers, *In Proceedings of the second European Control Conference ECC'93*, Groningen, eds. J.W. Nieuwenhuis, C. Praagman, H.L. Trentelman, 1993, vol. 4, pp. 1832-1837.

[7] D.P. Campbell, *Dynamika procesów*, Państwowe Wydawnictwo Naukowa, Warszawa 1962.

[8] J.T. Duda, *Modele matematyczne, struktury i algorytmy nadrzędnego sterowania komputerowego*, Uczelniane Wydawnictwa Naukowo-Dydaktyczne AGH, Kraków, 2003.

[9] W. Grega, *Metody i algorytmy sterowania cyfrowego w układach scentralizowanych i rozproszonych*, Uczelniane Wydawnictwa Naukowo-Dydaktyczne Akademii Górniczo-Hutniczej w Krakowie, Kraków 2004.

[10] W. Grega, K. Kołek, Simulation and real-time control: from Simulink to industrial applications, In *Proceedings of the 2002 IEEE international Conference on Control Applications and international symposium on Computer Aided Control Systems Design*, Glasgow, Scotland, U.K., 2002.

[11] E. Hairer, S.P. Nørsett, G. Wanner, *Solving Ordinary Differential Equations: I Nonstiff problems*, Springer 2000, 2nd edition.

[12] T. Kaczorek, *Teoria układów regulacji automatycznej*, WNT, Warszawa 1977.

[13] W. Mitkowski, *Stabilizacja systemów dynamicznych*, Uczelniane Wydawnictwa Naukowo-Dydaktyczne Akademii Górniczo-Hutniczej w Krakowie, Kraków 1996.

[14] W. Mitkowski, J. Baranowski, Estimation of continuous state in linear dynamical systems from discrete output measurements, In *Materiały XXX Międzynarodowej konferencji z podstaw elektrotechniki i teorii obwodów IC-SPETO*, 2007, pp. 193–194.

[15] B. Mrozek, Z. Mrozek, *MATLAB – uniwersalne środowisko do obliczeń naukowo-technicznych*, CCATIE, Kraków 1995.

[16] J.B. Rawlings, B.R. Bakshi, Particle filtering and moving horizon estimation, *Computers & Chemical Engineering*, Volume 30, Issues 10–12, 12 September 2006, Pages 1529–1541.

[17] M. Rosół, Application of nonlinear control methods for tanks system (in Polish), *Pomiary, Automatyka, Kontrola*, 2001, 7/8: 30–34.

[18] I.W. Sawieliew, *Wykłady z fizyki. Tom I. Mechanika. Fizyka cząsteczkowa*, Wydawnictwo Naukowe PWN SA, Warszawa 1998.

[19] P. Stoica, T. Söderström, *Identyfikacja systemów*, Wydawnictwo Naukowe PWN, Warszawa 1997.

[20] P. Tatjewski, *Sterowanie zaawansowane obiektów przemysłowych. Struktury i algorytmy*, Akademicka Oficyna Wydawnicza EXIT, Warszawa 2002.

[21] A. Zalewski, R. Cegieła, *MATLAB – obliczenia numeryczne i ich zastosowania*, Wydawnictwo NAKOM, Poznań 1996.

[22] V.M. Zavala, C.D. Laird, L.T. Biegler, A Fast Computational Framework for Large-Scale Moving Horizon Estimation, In *Proceedings of the 8th International Symposium on Dynamics and Control of Process Systems* 2007.

Technologies of Current Sensors Suitable for Hot High Density Power Electronics

Filip Grecki, Grzegorz Iwanski, Wlodzimierz Koczara, Jozef Lastowiecki

Warsaw University of Technology, Institute of Control and Industrial Electronics, Warsaw, Poland,

e-mail: *f.grecki@isep.pw.edu.pl, iwanskig@isep.pw.edu.pl, koczara@isep.pw.edu.pl, jlastow@isep.pw.edu.pl*

Abstract—The paper describes the review of technologies of current sensors dedicated for high temperature power electronics. Two solutions are taken into consideration as the drive system and DC/DC converter implemented in Hybrid and Electric cars requires different parameters of current measurement system - the current range and switching frequency are various in both cases. Demands for current sensors, ideas of basic system are described.

Keywords—high temperature electronics, hybrid electric vehicle (HEV), sensor, transducer, automotive application.

I. INTRODUCTION

High density power electronics demand development of more temperature resistant current sensors. It would be difficult to make a sensor appropriate for all needs, therefore universal. It is better to design a sensor for specific demands and therefore meet the needs as good as possible. This project is focusing on current sensors for new automotive power electronic devices. Within the range of automotive systems, different demands are set to different parts of the system. The demands differ in working frequency – bandwidth and amplitude of current. Different points of current measuring can be seen in Fig. 1. Current sensors are marked as CS.

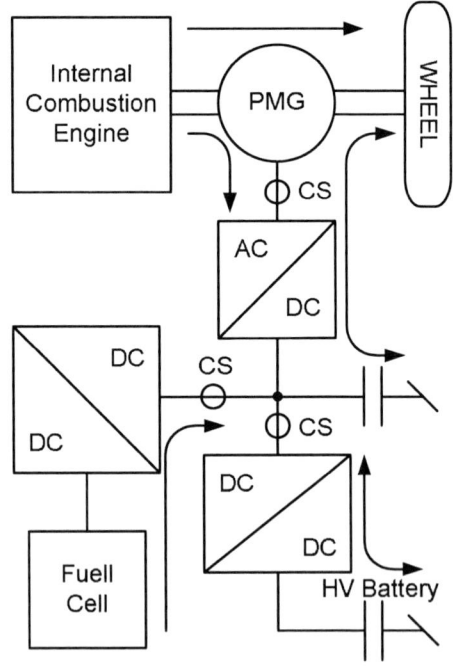

Fig. 1. Power electronics topology for hybrid drive in automotive.

The DC-DC converter is equipped with an inductive element – choke, which allows to form and manage current in the DC circuit. The size of the choke is directly dependant on the working frequency of the converter and the frequency depends on the semiconductor devices and their cooling. The higher is the working frequency, the lower inductance of coil is needed to smooth the managed current. The new SiC (Silicon Carbide) semiconductor elements have much higher working temperature, what allows smaller heat sinks, but more importantly smaller choke, because of higher working frequency. The DC-DC converter works with high frequency and the current sensor has to detect high and sudden peaks of current, in order to allow safe operation. High accuracy of the current sensor is not a strict demand, as the current control loop is mostly the inner structure with a superior controllers of DC voltage or other parameter. The superior control loop will adjust the reference current signal current, which will be adequate to the required state of the controlled object, in spite of the error provided by current measurement system. Nonetheless the current sensor has to operate at high frequency in order to sense current peaks as well as current fluctuations in DC link.

The drive has different demands to DC/DC converter. Because the drive does not operate with high switching frequency, there is no need for sensor with such a broad bandwidth. The inverter similarly to the DC/DC converter does not need high accuracy, but broad bandwidth. Table I. shows the comparison of demands for both drive inverter and DC/DC converter. Table I presents also the basic requirements for the current sensors for designed power electronics converter dedicated to the drives of hybrid vehicles.

TABLE I
COMPARISON OF THE CURRENT SENSOR DEMANDS FOR DC-DC CONVERTER AND DRIVE INVERTER.

Parameter	DC- DC Converter	Drive	Unit
Operating temperature	-40 to +150	-40 to +150	°C
Current range I	±200	±450	A
Bandwidth (-1dB)	0 to 250	0 to 20	kHz
Gain error drift	±0.05	±0.01	%/K
Gain error (% of nominal I)	±1	±1	%
Offset error	±10	±4.5	A
Supply voltage	+5 or ±15	+5 or ±15	V

Table I shows, that the difference in bandwidth is significant. There are many different physical phenomena, which allow for measurement of electric current, just to mention the most popular: Hall Effect, Faraday Effect, Magnetoresistive Effect and Gauss Law (well known current transformer). In this paper the Hall Effect and the use of current transformer will be described.

The current sensor with demand from DC (with inclusion) up to 20kHz can work in an open loop system, whereas the sensor which has to measure currents of frequency from DC up to 250kHz, has to work in an closed loop system. The closed loop system allows to use a magnetic field sensor with low bandwidth for current sensor that can operate to hundreds of kHz, thanks to combination of the field sensor and current transducer/current transformer.

II. OPEN LOOP CURRENT SENSOR - IDEA

The open loop current sensor is a simple composition of magnetic core and magnetic field sensor. The idea is very simple and well known [1]. The current conductor is encircled by a magnetic core with an air gap. The field is measured in the air gap by i.e. Hall element or any other magnetic field sensor. Depending on the level of the output signal, measured value may be further amplified if needed for example by operational amplifier. The system is presented in Fig. 2.

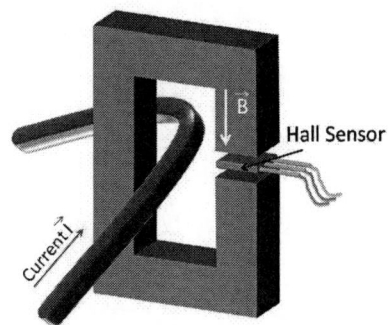

Fig. 2. Current sensor in an open loop setup.

Currently available field sensors are mostly already integrated with amplifier, what makes the task of making the current sensor easier, to some extent. The reduction of influence of magnetic interferences are reduced thanks to very short connections between Hall plate and amplifier. Long connections work like antenna and they pick up interference both form the measured current and any other outer magnetic field, what alters the correct output signal of the field sensor. In an open loop system, practically any error of the current sensor: nonlinearity, null drift or temperature error come from the integrated field sensor, which is also responsible for the bandwidth. The magnetic core is used as a concentrator and carrier of the magnetic field, but its shape also influences on the operation of the current sensor. Open loop system is simple and robust. It is also quite easy to tune in the design process. Open loop sensor is vulnerable to changes of the temperature as well as to hall sensor imperfections, nevertheless it is a sufficient solution for very many applications.

III. CLOSED LOOP CURRENT SENSOR - IDEA

The closed loop current sensor consists of the same basic elements like open loop sensor, but it also has a compensation winding and additional electronic part, which includes regulator made of operational amplifier and transistors with diodes. Closed loop sensor bases on the idea of flux compensation [2]. The idea is presented in Fig. 3.

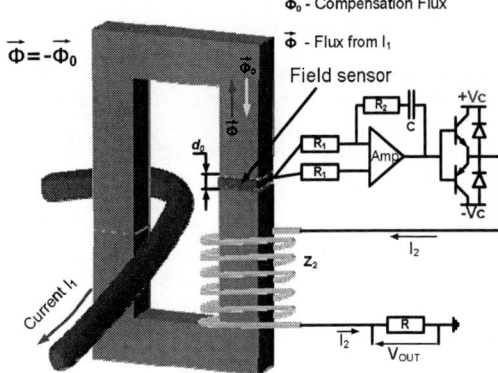

Fig. 3. Current sensor in a closed loop setup.

The hall sensor for this system should have high sensitivity and low null offset and drift. Temperature error and nonlinearity of the field measuring element are not crucial, as the compensation current does not depend on them. The value of the compensation current is controlled by the regulator.

Hall sensor is responsible for reproduction of the DC component, as well as low frequencies. Even if the Hall sensor has a low bandwidth (i.e. DC to several kHz), measurement of high frequency is still possible due to the presence of the compensation coil.

At higher frequency, (over the crossover frequency f_c) the compensation coil acts as a secondary winding of a current transducer or transformer. Proper combination of compensation coil, hall element and PI regulator results in constant amplitude of the output signal for the same amplitude of measured current in wide frequency range, what is presented in Fig. 4. Crossover frequency is marked f_c. Output signal of a closed loop sensor can be either current or voltage.

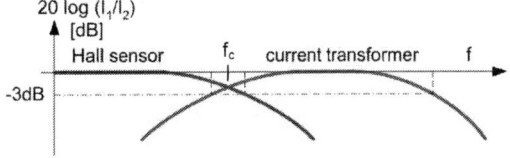

Fig. 4. Output amplitude of the closed loop current sensor consisting of Hall sensor output signal and current transformer output signal.

To protect the temperature sensitive elements of the sensor, which can be damaged by the high temperature, the sensor can be designed in a distributed form. In such setup, the delicate elements are placed on a distant and cool heat sink, thus having a better cooling and their temperature working conditions are met.

IV. LABORATORY TESTS

For the purposes of current sensor making, several issues had to be verified and the easiest and fastest way to do that, is the simulation. Influence of several factors was analyzed i.e. the size of the air gap for particular range of current, flux value for the Hall sensor, size of the magnetic core, amplitude of the current in secondary winding in the closed loop sensor.

The laboratory set up has been made of power source, high power resistors, capacitors and a switch. Capacitors were charged from the power source with constant voltage and discharged on the resistors by close of the switch. Due to the presence of some connection and cable inductance, the discharge of the capacitors has a resonant character. As a result of this discharge, current up to 600A and frequency up to 370kHz is flowing. For the measurement of DC current, separate DC power supply source is used. The scheme of the laboratory set up is presented in Fig. 5. The frequency cannot be chosen freely, as it depends on the value of capacitor and those are strictly defined (i.e. 330 nF, 470 nF).

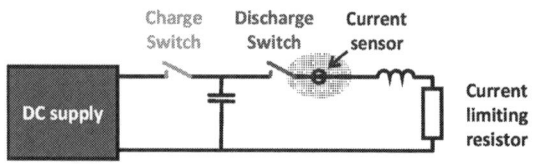

Fig. 5. Laboratory set up used in tests of current sensor and sensor elements.

High temperature tests are made with the use of high temperature air source. The temperature of the air and the air flow rate is measured and set to the desired value. It is also planned to built or use temperature/climate chamber. It will be particularly important for the tests of power elements, like transistors, which not only have to withstand given high ambient temperature, but they also heat up themselves, therefore rise the surrounding temperature.

Hall elements were tested to verify their amplitude of the output signal as well as bandwidth and the influence of high temperature on their behavior. Several sensors were tested. Results of two of them are presented in this paper.

V. OPEN LOOP SENSOR MODIFICATIONS

The Hall sensors were tested at different frequencies to see if their given bandwidth is true. As the SS495a has a lower declared working frequency and it will be used in an open loop sensor, its test result are presented in Fig. 6a and Fig. 6b. Another Hall sensor KSY44 will be used in the closed loop system.

In Fig. 6a and Fig. 6b there is a noticeable drop of output signal with the change of frequency, but it is within the (-1dB) declared by the manufacturer. At higher frequency, there is a strong disturbance from the very high change of current. The level of the disturbance depends on the position of sensor, so it is likely, that not the sensor is disturbed, but the very voltage probe and can be easily eliminated by a simple RC filter. Test results are more than satisfactory, as the declared value are true and what is more important, the Hall sensor does not have any phase delay at high frequency.

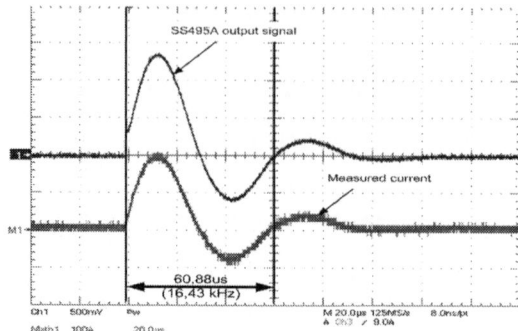

Fig. 6a. Oscillograph of Hall sensor tested at frequency of 16.43 kHz.

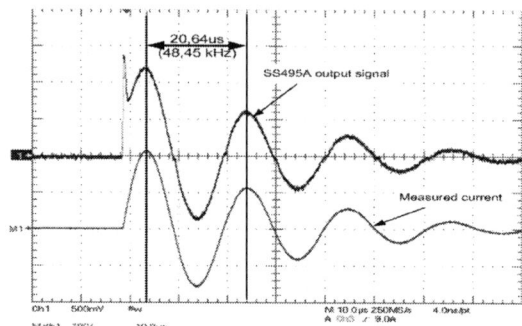

Fig. 6. Oscillogram of Hall sensor tested at frequency of 16.43kHz (a) and 48.45 kHz (b)

Tests for the drop of output amplitude were made with the current of amplitude 200A. The complete current sensor for the drive should be working with currents of ±450A and the frequency up to 20kHz. The sensor should be able to measure higher currents than 450A, that is up to 600A. As simulation showed, the amplitude of the magnetic flux, in the C shaped core with air gap of 2mm, is too high for the SS495a sensor. According to simulation calculations and lab tests, the air gap would have to be as wide as 10mm. That would bring trouble like susceptibility to external fields as well as dependence of output signal to the position of the sensor in the air gap. In order to avoid all those troubles, core modification was needed.

There are two efficient ways to weaken the magnetic field in the air gap: extend the air gap or split the magnetic flux. The second option has been chosen, as in the case of the first one, air gap would have to be too wide. The modification of the magnetic core is presented in Fig. 7

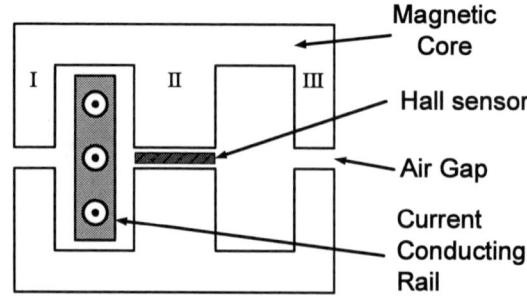

Fig. 7. Modification of magnetic core to fit the properties of the Hall sensor

In Fig. 7, columns of the core are numbered I, II and III. In the column I, there is a full flux and in columns II and III, the flux is only a precise part of the full flux. The distribution of the flux is presented in Fig. 8 as a result of simulation.

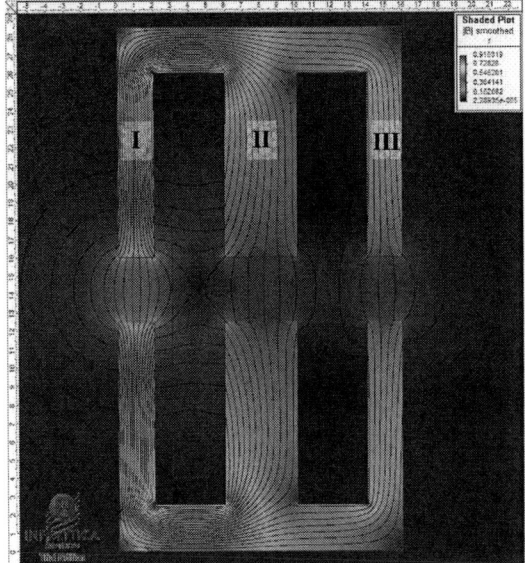

Fig. 8. Distribution of flux and magnetic field in the modified magnetic core.

Respectively, the value of the magnetic field in the air gaps is illustrated by the Fig. 9.

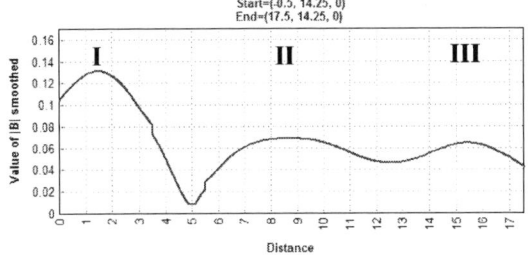

Fig. 9. Value of the magnetic field in the air gaps of the modified magnetic core

Fig. 9 shows, that the flux is divided in half in columns II and III in regard to column I. Values were obtained with the air gaps of 3.5mm width. All air gaps have to be equal in order to get even distribution in column II and III. As the middle column II is wider than two others, the field sensor will be placed there. Thicker column gives more homogenous distribution of the field (the peak of the field value is flat), what makes the output signal independent of the sensor placement. Those results were encouraging to make the test with the real sensor.

The character of the laboratory test set up does not allow for smooth change of current frequency. Only particular frequencies can be obtained, which are a result of capacitors connection combination in the resonant discharge. Three chosen frequencies were tested 7.3kHz, 10.5kHz 14.25kHz and the DC current. Test results are presented in Fig. 10 and Fig. 11.

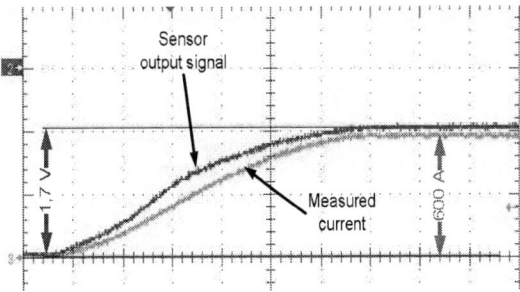

Fig. 10. Oscillogram of output signal from SS495a Hall sensor at turn on of the DC current

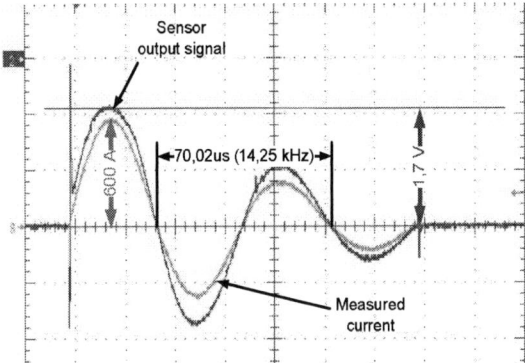

Fig. 11. Oscillogram of output signal from SS495a Hall sensor at frequency of 14.25 kHz

In both cases in Fig. 10 and Fig. 11 there are disturbances caused by "sudden" change of state, where no current is flowing, to the state where current begins to flow. That results in appearance of peak output signal and certain deviation from measured current, but in both cases application of simple RC filter for the signal will remove that distortion.

VI. CLOSED LOOP SENSOR MODIFICATIONS

For the closed loop current sensor, apart of the Hall sensor, also the current transducer had to be tested. The secondary coil is involved in current sensing above the crossover frequency, what was shown in Fig. 4, as f_c. The first tested current transducer had a secondary winding consisting of 800 turns, made of 0.2mm isolated copper wire. Fig. 12, presents laboratory test results. The y axis is scaled in a signal ratio, which represents drop of secondary current, related to the nominal output current.

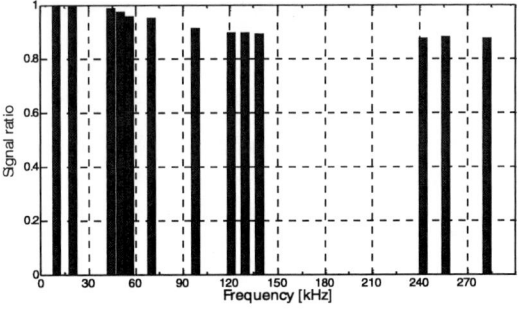

Fig. 12. Results of current transducer tests. Output signal compared to the nominal output signal presented in the function of frequency.

In case of the coil with 800 turns, the compensation current has to be 250mA in order to compensate the flux from primary current of 200A. In order to use lower compensation current, more turns is needed or a different approach – compensation of particular, proportional part of the flux. To realize that, the modified core was used. The idea is similar to the one used in the open loop system, only half of the flux is compensated by the secondary coil – the idea of differential core use has been described widely in [3][4][5]. The scheme of the set up is presented in Fig. 13.

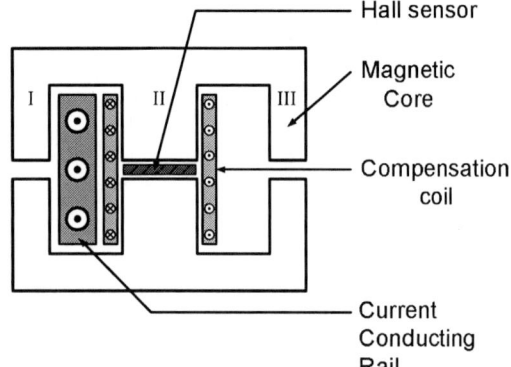

Fig. 15. Idea of closed loop system with modified magnetic core.

In the closed loop with modified core, both the sensor and the compensation coil are placed in the middle column, marked II in the Fig. 13. The whole system is similar to the one presented in Fig. 3, only fitted with a different core. The whole idea of the operation stays the same.

The modified system has been tested in the laboratory set up. Both changes have been introduced, so the tested compensation coil was placed on modified core and the number of windings was 2000 made with 80μm isolated copper wire. The bandwidth and the signal amplitude is presented in Fig. 14.

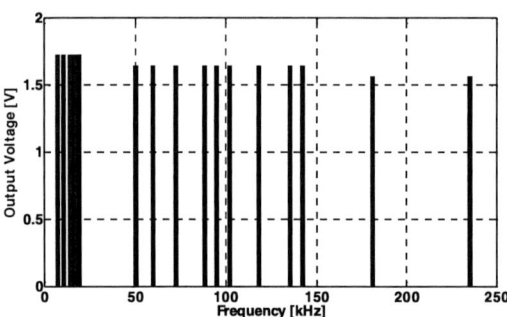

Fig. 14 Output signal change with the raise of frequency.

To show the result of placing the compensation coil on the middle column, output signal was compared to the one from the coil placed on column I, where the full flux from the primary current is flowing. Oscillograms Fig. 15 and Fig. 16 present the laboratory test results.

Fig. 15 and Fig. 16 present slight disproportion, which only comes from imperfect setting of the air gaps. It is quite difficult to set them exactly equal in three air gaps and in two spate cores.

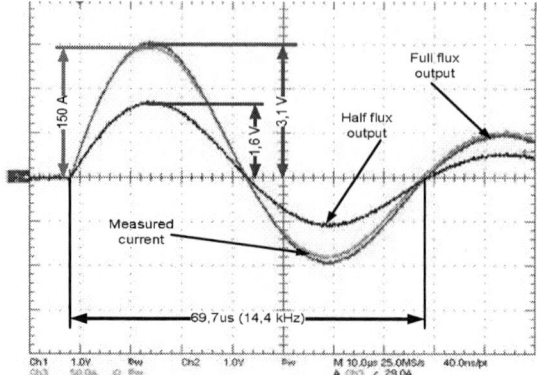

Fig. 15 Output signals from secondary coils with full flux (red) and half flux from middle column (blue) at frequency of 14,4 kHz

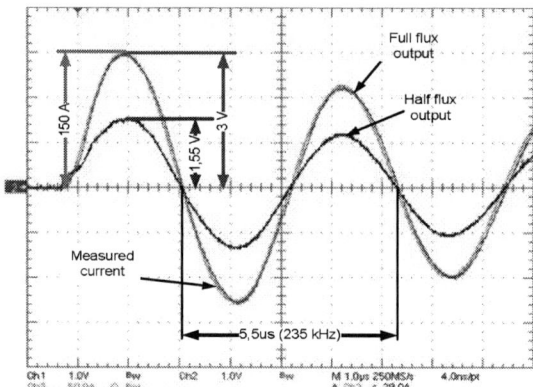

Fig. 16. Output signals from secondary coils with full flux (red) and half flux from middle column (blue) at frequency of 235 kHz

Air gaps were adapted to the needs of the system and to set them, several layers of aluminum spacers were used. Final tests will employ spacers cut out from of aluminum sheet of the same thickness. In spite that error, results are as predicted. The compensation current is 50% of that in C shaped magnetic core, so the transistors will work under lighter conditions, thus emitting less heat to ambient. The drop of output amplitude at 235kHz is still within -1dB, what is the demand for the designed sensor.

VII.   HIGH TEMPERATURE TESTS

As the current sensors are dedicated for the high temperature application, temperature tests are the part of the current sensor design process. It would be very complicated to simulate the influence of the temperature on the Hall sensor, especially that with a built-in op-amp, the temperature behavior of the device becomes rather complex. The best way to test the ready device is to make a laboratory test in the temperature.

Temperature influence on the sensor, output signal from SS459a is shown in Fig. 17 and Fig. 18. In the first oscillogram, no current was flowing and the temperature was changed from 25°C to 125°C. In the second one the temperature conditions were the same, but there was a DC current of 200A flowing. The change in the output signal was around 60mV, what is less than 2.5% of the measuring range. Considering the fact, that this sensor would be used in the open loop current sensor, this result is satisfactory.

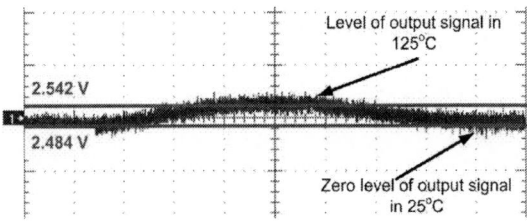

Fig. 17. Laboratory results of temperature influence on the Hall sensor.

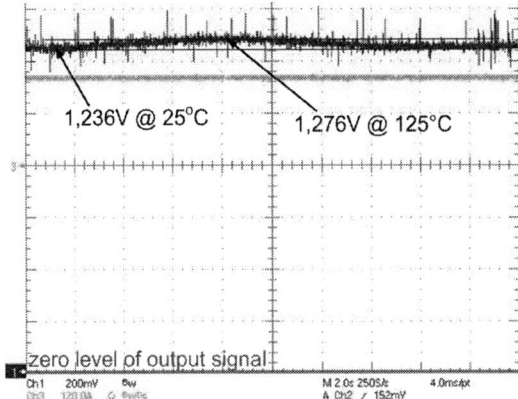

Fig. 18. Laboratory results of temperature influence on the Hall sensor at measured current of 200A.

Even though, that the field sensors are made in series, their parameters differ even within one batch. Fig. 19 illustrates changes of output signal at null current with the change of temperature. Sensors were placed in the ambient of 150°C and cooled back to room temperature.

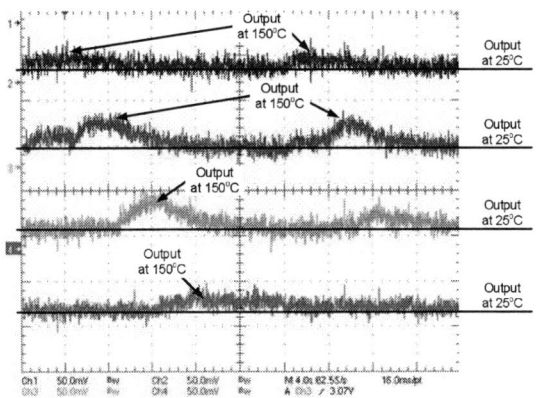

Fig. 19. Heating of SS495a sensors batch.

The change of 125°C differently influences the sensors from the same batch. The output signal differs from approximately 10mV to 30mV. That means that the current sensors made with those sensors will differently react to changes of ambient temperature. The ready sensor will be rated with certain range of output signal changes to temperature.

Fig. 20 presents a photo of tested current sensor: magnetic core with Hall sensor placed in the air gap of the middle column.

The very core used in both set ups (open and closed loop) is presented in Fig. 21. The Magnetic core is placed on an 18 mm wide current conducting copper rail.

Fig. 20. Testing of the current sensor – Hall sensor remover from the magnetic core

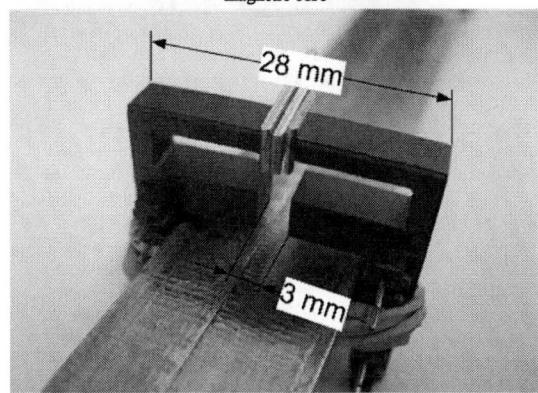

Fig. 21. Magnetic core placed on a current conducting rail

VIII. CONCLUSIONS

Two solutions of current sensor for different needs of the power electronic systems were described. For low frequency demand open loop current sensor system was suggested and for the high bandwidth demand closed loop system was recommend. In both cases new approach for the magnetic flux distribution in the magnetic core was applied. Both simulation and laboratory tests were made. High temperature test of sensors were made and results are satisfactory to meet current sensor demands.

ACKNOWLEDGEMENT

This work is financed by EC within Sixth Framework Programme project entitled "High Density Power Electronics for FC- and ICE- Hybrid Electric Vehicle Powertrains" – HOPE, Contract number FP6-019848

REFERENCES

[1] R. Dickinson, S. Milano, "Isolated Open Loop Current Sensing Using Hall Effect Technology in an Optimized Magnetic Circuit". Allegro MicroSystems, Inc., 2002

[2] J. Pankau, D. Leggate, D.W. Schlegel, R.J. Kerkman, G.L. Skibiniski, "High-frequency modeling of current sensors", *IEEE Transactions on Industry Applications*, Nov.-Dec. 1999 pp.1374 – 1382

[3] F. Grecki, G. Iwanski, W. Koczara, J. Lastowiecki, "Current Sensor Dedicated for High Temperature Automotive Power Electronics" *International Conference on Automotive Power Electronics – APE'07*, 26-27 Sept. 2007, Paris, France,.

[4] F. Grecki, G. Iwanski, W. Koczara, J. Lastowiecki, "Current Sensor Dedicated for High Temperature Integrated Power Electronics, *Proceedings of 5th International Conference on Integrated Power Electronics Systems CIPS'08*, 11-13 March 2008, Nuremberg, Germany.

[5] J. Lastowiecki, "New differential, compensative current measuring system based on amorphous magnetic circuit", *Proceedings of 2nd IEEE International Caracas Conference on Devices, Circuits and Systems*, 2-4 March, 1998, Isla de Margarita, Venezuela, pp.386 – 390

Nonlinear dynamical feedback for motion control of magnetic levitation system

Jerzy Baranowski*, Paweł Piątek*

* Department of Automatics, Faculty of Electrical Engineering, Automatics, Computer Science and Electronics,
AGH University of Science and Technology, Kraków, Poland, e-mail: *jb@agh.edu.pl, ppi@agh.edu.pl*

Abstract—**For fast, nonlinear and unstable dynamical systems, such as magnetic levitation, high quality of control is essential. Such control can be generated by different feedback structures, most of which require measurements of velocities or derivatives of the system output. In case of magnetic levitation the velocity of levitating object (a ferromagnetic sphere) is not measured. That is the reason for application of state observers which along with different controllers will create a nonlinear dynamical feedback structure. In this paper we will consider two such structures - PID controller with a nonlinear feed-forward and a cascade linearising feedback using two nonlinear state observers: High-Gain observer (HGO) and an observer with reduced, linear error dynamics (abbreviated as RO). Operation of these control algorithms will be compared with a classical PID structure, derivative part of which will be realised either as a finite difference or with an observer.** [1]

Index Terms—**Estimation technique, non-linear control, magnetic device, real time processing, motion control.**

I. MAGNETIC LEVITATION SYSTEM

We consider a laboratory magnetic levitation system presented in the figure 1. This plant consists of an electromagnet, an optic position sensor, a ferromagnetic sphere and a power interface. The power interface is used for operation of electromagnet coil, and subsequently the generation of a magnetic field that will lift the ferromagnetic sphere and allow its levitation. The measuring unit consists of the strong visible light source and the optical sensor with an appropriate electronic circuit. The sphere covers the sensor which measures the degree of light dousing. Because of the spherical shape the measurement is nonlinear and needs to be electronically or digitally linearised. This system was considered among the others in the following works [1], [5]–[7], [9], [15]–[19], [21], [22]. The ball velocity measurement is a difficult task, because the only possible options is a numerical differentiation of the position. That is why an observer design is such an attractive proposition. Among the others the observers for magnetic levitation system were considered in [3], [4], [20]. Magnetic levitation system depicted in the figure 1 can be described with the following differential equations

$$
\begin{aligned}
\dot{x}_1 &= x_2 \\
\dot{x}_2 &= \frac{1}{2m}\frac{dL(x_1)}{dx_1}x_3^2 + g \\
\dot{x}_3 &= \frac{1}{T_s}(k_s u - i_s - x_3)
\end{aligned}
\tag{1}
$$

[1]Work financed by state science funds for 2008-2011 as a research project. Contract no. N N514 414034

where:

x_1 - ball position $[m]$,
x_2 - ball velocity $[\frac{m}{s}]$,
x_3 - current $[A]$,
u - control voltage $[V]$,
m - ball mass $[kg]$,
$L(x_1)$ - electromagnet inductance $[H]$,
g - gravitational acceleration $[\frac{m}{s^2}]$,
k_s - gain of current controller $[\frac{A}{V}]$,
T_s - time constant of the linear current driver $[s]$,
i_s - zero error of current controller $[A]$,

For the purpose of this paper we use the following approximation of the electromagnet inductance (see [18]):

$$
L(x) = L_0 e^{-\frac{x_1}{b}}
\tag{2}
$$

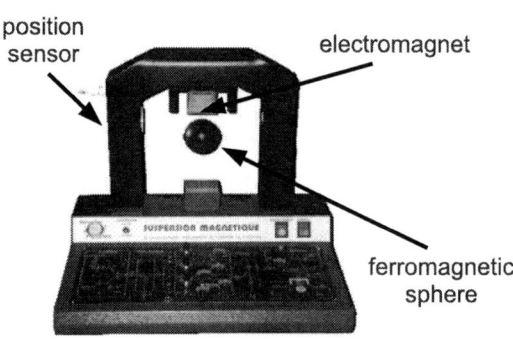

Figure 1. Laboratory magnetic levitation system

As it can be seen in the figure 1 the ball position is measured by the appropriate sensor and the coil current also is easily measured. The velocity measurement, however, is difficult to realise. That is why in this paper we consider velocity estimation with the use of only the measured position, and with the use of both measurements of position and current. However at first we will want to determine system observability.

II. OBSERVABILITY

We will verify, is the system observable (see also [3], when only the position is available (because if it is, then additional use of current measurement will be used as improvement). That is why we define the output as:

$$
y = h(x) = x_1
$$

Let us consider the dynamical system

$$\dot{x} = f(x, u) \tag{3}$$
$$y = h(x)$$

and the following matrix

$$Q(x) = \frac{\partial}{\partial x} \begin{bmatrix} h(x) \\ \mathcal{L}_f h(x) \\ \mathcal{L}_f^2 h(x) \end{bmatrix} \tag{4}$$

Where $\mathcal{L}_f h(x)$ denotes a Lie derivative of $h(x)$ in the direction of vector field f. System (3) is locally observable at x^* if $Q(x^*)$ is non-singular (see [14]).

By f we will denote the right side of (1). After the computations we can see that

$$Q(x) = \begin{bmatrix} 1 & 0 & 0 \\ 0 & 1 & 0 \\ \frac{L(x)x_3^2}{2mb^2} & 0 & -\frac{L(x)x_3}{mb} \end{bmatrix} \tag{5}$$

It is easily to verify that

$$\det(Q(x)) = -\frac{L(x)x_3}{mb} \tag{6}$$

We can see from (2) that (6) is equal to zero if x_3 is zero. Because x_3 is the coil current, it becomes zero only if the electromagnet is not operating, so the system is locally observable in all work conditions of the system.

III. HIGH-GAIN OBSERVER(HGO)

Knowing the matrix $Q(x)$ required for the determination of local observability the straightforward step is the construction of so called high-gain observer, which uses the same data. Observer is defined as (see [8], [22]):

$$\dot{\hat{x}} = f(\hat{x}, u) + Q^{-1}(\hat{x}) L_\varepsilon (y - h(\hat{x})) \tag{7}$$

where

$$f(x, u) = \begin{bmatrix} x_2 \\ -\frac{1}{2mb} L_0 e^{-\frac{x_1}{b}} x_3^2 + g \\ \frac{ku - i_s - x_3}{T} \end{bmatrix} \tag{8}$$

$$Q^{-1}(x) = \begin{bmatrix} 1 & 0 & 0 \\ 0 & 1 & 0 \\ \frac{x_3}{2b} & 0 & -\frac{mb}{L_0 x_3} e^{\frac{x_1}{b}} \end{bmatrix} \tag{9}$$

As the observer gain we define:

$$L_\varepsilon = \begin{bmatrix} \frac{l_1}{\varepsilon} \\ \frac{l_2}{\varepsilon^2} \\ \frac{l_3}{\varepsilon^3} \end{bmatrix}$$

Where

$$l_1 = 3a,$$
$$l_2 = 3a^2,$$
$$l_3 = a^3$$

and $\varepsilon \in (0, 1]$ is a tuning parameter; a should be at least 150. The gains were constructed in such way, that the observer dynamics will not be slower than the system dynamics (which is considered as time-critical (see [15])).

Verification of this type of observers is a difficult task, especially because it requires that the right side of (1) should fulfill the global Lipschitz condition. Because that is not the case, we can see that this observer is only locally stable. Moreover, error dynamics of (7) has a finite escape time (some solutions have a vertical asymptote) so once the estimation error becomes large (by the effect of measurement, or ball movement disturbances) it usually will not stabilise itself to zero. A following observer, does not have this flaw.

IV. REDUCED OBSERVER WITH LINEAR ERROR DYNAMICS(RO)

The observation task we consider requires the estimation of the ball velocity only. We have verified observability of the entire state from the position measurement only, however the coil current is also available for measurement. That is why we should consider the reduced observer which would provide only the required estimate. We propose the following construction (see [4] and a similar construction in [2], [13])

$$\dot{\hat{x}}_1 = \hat{x}_2 + k_1(y - \hat{x}_1) \tag{10}$$
$$\dot{\hat{x}}_2 = \frac{1}{2mb} L_0 e^{-\frac{x_1}{b}} x_3^2 + g + k_2(y - \hat{x}_1)$$

The error dynamics has the form

$$\dot{e} = \begin{bmatrix} -k_1 & 1 \\ -k_2 & 0 \end{bmatrix} e \tag{11}$$

Advised values of parameters are $k_1 = 2a$ and $k_2 = a^2$, where a should be at least 150 (because of the reasons stated above). As we can see the error dynamics (11) of the observer (10) is linear! So if (11) is asymptotically stable (has eigenvalues in the left open complex half-plane) the estimation error converges to zero exponentially and this property is global. It should be noted however, that while HGO used only one measurement the RO uses two of them, so it is subjected to two kinds of disturbances. That is why, great stability properties of the observer have a price of lower estimation quality.

V. CONSIDERED CONTROL STRUCTURES

In this paper we consider following control structures:

- classical PID control,
- PID control with an observer produced derivative part,
- PID control with a nonlinear static feed-forward (see Fig. 2),
- PID control with a nonlinear static feed-forward and an observer produced derivative part (see Fig.3),
- cascade linearising feedback using the velocity from an observer (see Fig. 5).

All observer based structures are considered for both observer types. All these structures, are forms of a dynamical feedback (see [10], [11]) all of which, except classical PID, are nonlinear. Dynamics of the feedback consist of integration in PID and in equations of velocity observers (7) and (10).

A. Classical and observer based PID controller

In this paper we will consider PID controller with independent gains of the following form

$$e(t) = w(t) - y(t)$$

$$u(t) = Pe(t) + I \int_0^t e(t)\mathrm{d}t + D\dot{e}(t)$$

where $w(t)$ is the reference value and $y(t)$ is the regulated output. P, I and D are real parameters. In typical implementations, the derivative $\dot{e}(t)$ is computed via finite differences. Such approach has many flaws, most notable of which are sensitivity to disturbances and numerical errors. Time series of such obtained derivative is usually non-smooth and appears noisy. Because in the magnetic levitation system the regulated value is the position x_1, its derivative (velocity x_2) can be obtained from the observer and the derivative of set point comes naturally. In that case we get:

$$e(t) = w(t) - x_1(t)$$

$$u(t) = Pe(t) + I \int_0^t e(t)\mathrm{d}t + D(\dot{w}(t) - \hat{x}_2(t))$$

B. Nonlinear feedforward

It is a known fact, that the linear controller can operate properly in the neighbourhood of a chosen steady state. Performance of classical PID can be strongly improved, if the appropriate reference control value corresponding to a reference value is added to the generated control signal. In that case, the problem of movement is reduced to problem of tracking (see [12]). Let us consider a control structure presented in figures 2 and 3. We want to find a

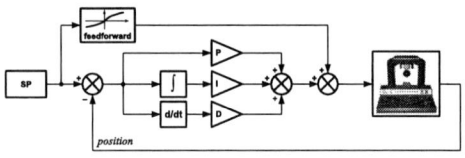

Figure 2. PID controller with a nonlinear feedforward

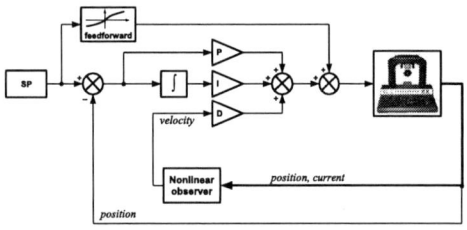

Figure 3. PID controller with a nonlinear feedforward and a nonlinear observer

function $g(x_{1r})$ (with x_{1r} is a reference steady state value of position) which fulfils

$$0 = f(x_r, g_{x_{1r}}) \qquad (12)$$

where $x_r = [x_{1r},\ 0,\ x_{3r}]^\mathsf{T}$ and $f(x, u)$ is given by (8). Such function is plotted in the figure 4.

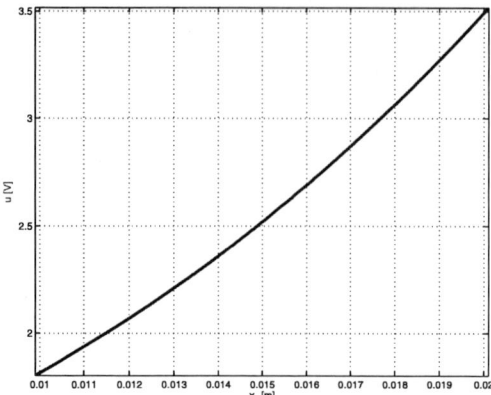

Figure 4. Characteristic of the nonlinear feedforward block

C. Cascade linearising feedback

The theory of linearising feedback is well known and different approaches lead to a result obtained from Lie derivatives (see for example [1], [9], [18], [21], [22]). Here we propose, a very intuitive (but essentially equivalent) approach of such linearising design. Let us consider first two equations of (1):

$$\dot{x}_1 = x_2$$

$$\dot{x}_2 = \frac{1}{2m}\frac{dL(x_1)}{dx_1}x_3^2 + g$$

It can be easily seen that if x_3 would be an exogenous input, system would be linearised if:

$$x_3(t) = \sqrt{2m\left(\frac{dL(x_1)}{dx_1}\right)^{-1}(-g - v(t))} \qquad (13)$$

$$v(t) = Px_1(t) + Dx_2(t) \qquad (14)$$

First two equations would take form:

$$\dot{x} = \begin{bmatrix} 0 & 1 \\ -P & -D \end{bmatrix} x \qquad (15)$$

$v(t)$ would then essentially become a PD controller, for the system linearised by feedback (14). x_3 however is not an exogenous input, but is in the first approximation a first order lag from the control voltage, and with a more precise analysis can lead to some nonlinear phenomena. The main idea of our approach is to realise (14) as a nonlinear cascade with a separate controller for tracking of coil current signal (x_3) that it would be appropriate for feedback linearisation. This control structure (supplemented with reference value and the observer is presented in the figure 5.

1448

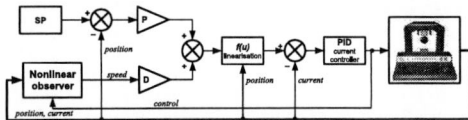

Figure 5. Cascade linearising feedback with an observer

VI. EXPERIMENTS

All of the considered control structures were validated experimentally. The laboratory system owned by the Department of Automatics of AGH, depicted on fig. 1 was used for all the experiments. It's measurements and the actuator were connected via the PCI converter card to a Windows based PC with MATLAB/Simulink environment. Experiments were performed in Simulink with RTWT. Numerical derivatives for position PID controller were realised with forward differences. Observer equations were solved with a 5th order Runge-Kutta formula implemented in Simulink. Discretisation step was $700\mu s$ for all of the experiments. Values of position on following plots should be interpreted as a distance from the electromagnet, so the lower the value the higher ball was. Wherever it reaches 0 it means that the ball was stuck to the electromagnet or bounced from it. In the experiments, controllers had a task of moving the ball with a reference value given as a rectangular wave.

A. PID control

Classical PID control is presented here for two reasons. First of all it is used as a comparison for nonlinear controllers, to show what improvements can be made. The second reason is to show how large improvements are introduced to the system through the replacement of numerical differentiation with the nonlinear observer. In the figure 6 the measurements of ball position with PID control are presented. It can be clearly seen, that introduction of the observer greatly improves the stabilisation quality resulting in drastically smaller overshoot. Also control signal quality is better (see fig. 7) - it is smooth, and because of that, better for actuator (lower energy is needed, smaller wear). Results for different observers are very similar, because considered example is located in the middle of their operation range. Differences are much more visible on the boundaries of operation ranges for different structures.

B. Estimation of velocity

Both observers provide very good velocity estimates, for the sake of completeness on figures 8 and 9 these estimates are presented and compared with numerical differentiation. It can be seen, that both provide strong improvements, and differ only in small details, caused by the additional disturbances in reduced observer from the measurement of current.

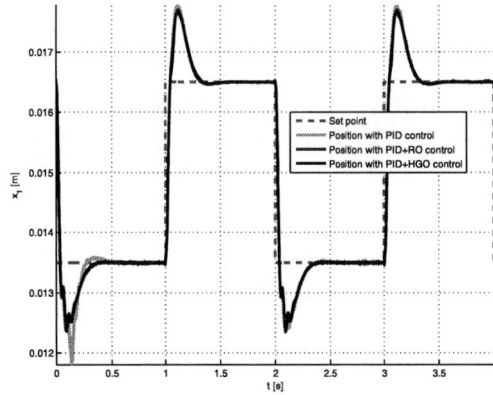

Figure 6. Comparison of PID controllers with different realisation of differentiation - ball position

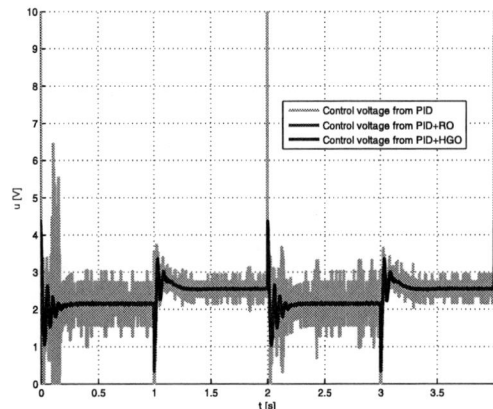

Figure 7. Comparison of PID controllers with different realisation of differentiation - controls

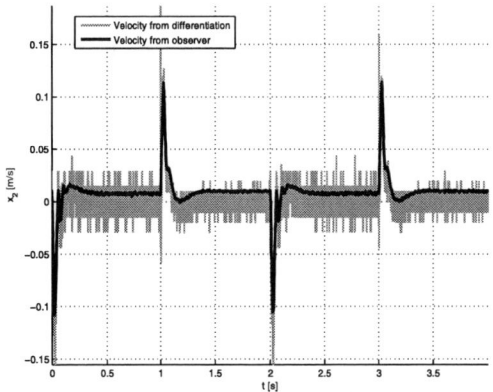

Figure 8. Velocity estimation from the reduced observer

C. PID controller and nonlinear feedforward

An addition of nonlinear feedforward to the control structure (see figures 2 and 3) leads to two direct benefits. Overshoot is smaller in the standard operating range, and

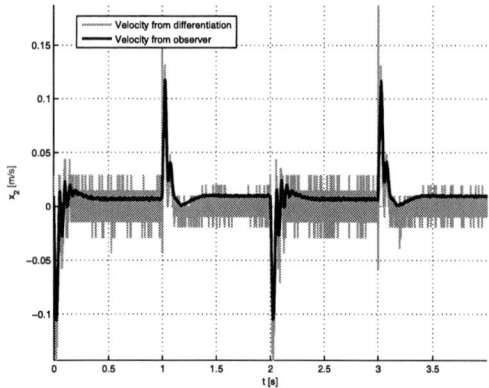

Figure 9. Velocity estimation from the High-Gain observer

D. Ranges of operation

All structures behave properly, when the control task is oriented in relatively small range. We performed a careful analysis of how the change of structure influences the operating range of the system and results of it are collected in table I. To read the table, it should be noted, that for example PID+FF+RO means a structure consisting of PID controller with a feedforward and velocity obtained from the reduced observer. Position marked N/A means, that there was no range of stable work, if the linearising feedback was using the velocity from numerical differentiation. Now we will present the

the maximal operating range is strongly increased. Joining the benefits of feedforward and velocity estimation provides the best control quality in PID based structures. We can see performance of this structure in figures 10 and 11.

Table I
RANGES OF OPERATION FOR DIFFERENT DYNAMICAL FEEDBACK STRUCTURES

Control structure	Min. distance [m]	Max distance[m]
PID	0.0125	0.0175
PID+RO	0.0112	0.0188
PID+HGO	0.0112	0.0188
PID+FF	0.0116	0.0184
PID+FF+RO	0.0103	0.0197
PID+FF+HGO	0.0099	0.0201
Linearising feedback+Diff	N/A	N/A
Linearising feedback+RO	0.0131	0.0169
Linearising feedback+HGO	0.0129	0.0171

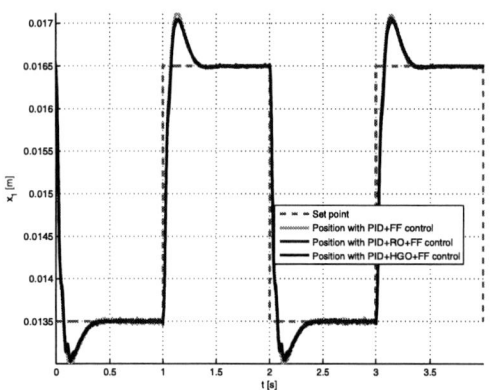

Figure 10. Comparison of PID controllers with different realisation of differentiation and a nonlinear feedforward - ball position

results of experiments comparing the structures on their boundary with a structure that provides a directly larger operation range. At the beginning we will consider PID control and PID control with the reduced observer. As it can be seen in figures 12-14 we can see that at this working point, numerical differentiation fails completely, while the observer generates smooth and proper control.

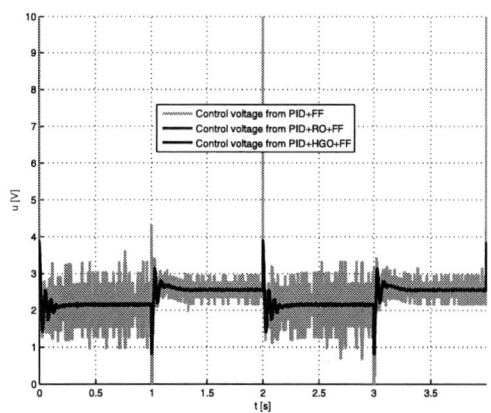

Figure 11. Comparison of PID controllers with different realisation of differentiation and a nonlinear feedforward - controls

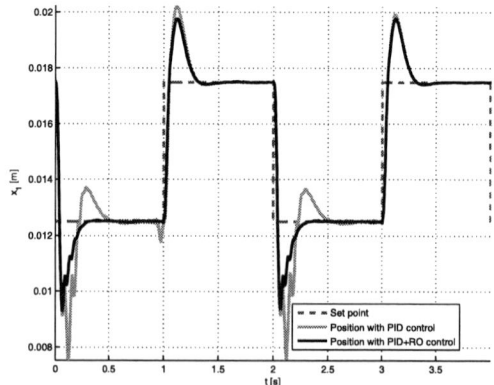

Figure 12. PID control vs PID+RO on the boundary of PID control operation range - ball position

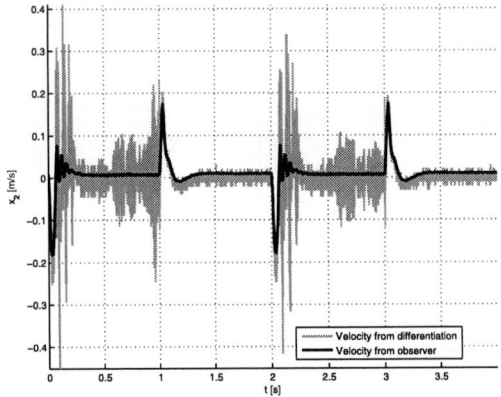

Figure 13. PID control vs PID+RO on the boundary of PID control operation range - ball velocity

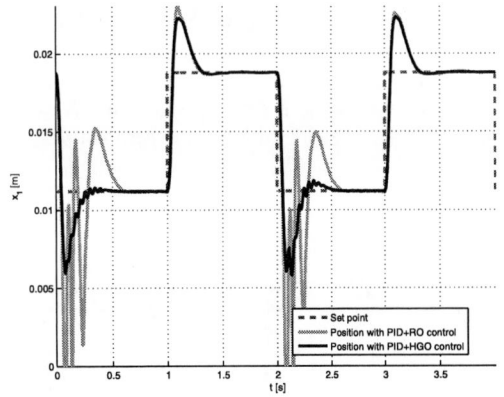

Figure 15. PID+HGO control vs PID+RO on the common boundary of their control operation range - ball position

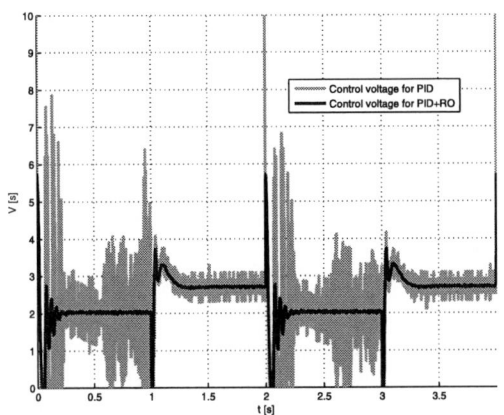

Figure 14. PID control vs PID+RO on the boundary of PID control operation range - controls

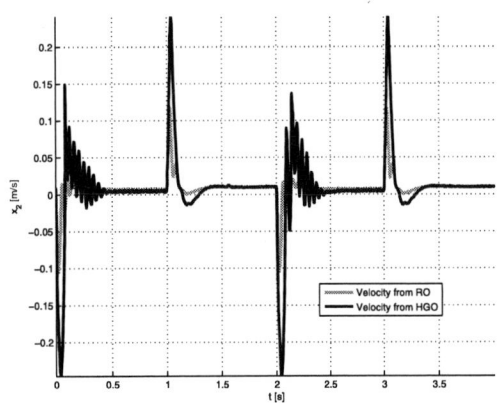

Figure 16. PID+HGO control vs PID+RO on the common boundary of their control operation range - ball velocity

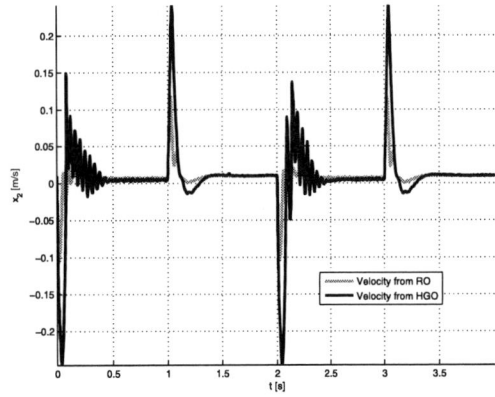

Figure 17. PID+HGO control vs PID+RO on the common boundary of their control operation range - controls

In the figures 15-17 we see both observer based structures operating on the boundary of their operation (which is equal). It can be seen that symptoms of control worsening are completely different. In case of the structure based on the reduced observer, the disturbances and general limits of linear controller are taking its toll, and system is on the verge of destabilisation. On the other hand position control of High-Gain based structure appears proper, unfortunately the problem is in the velocity estimation. As it can be seen in fig. 16 the estimated velocity is transitionally very large. This so called peaking phenomena along with observer local stability properties can lead to destabilisation, which frequently occurs on the longer time horizons.

Now we will present the comparison of PID+HGO control and a structure consisting of nonlinear feedforward, reduced observer and PID controller on the same boundary as previously. As it can be seen on figures 18-19

1451

the improvement provided by feedforward is substantial. Overshoot and oscillations are strongly reduced. This control structure has great stability properties, allowing for example to compensate for mechanical disturbances (manual touching of the ball during the experiments)

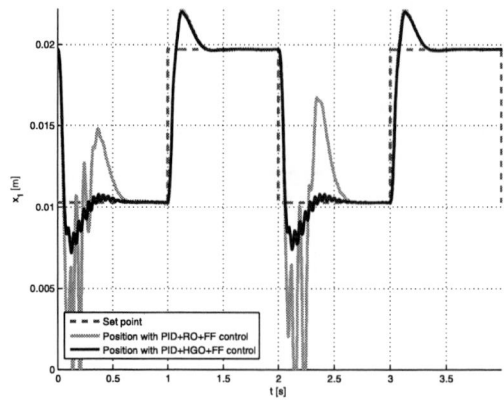

Figure 20. PID+HGO+FF control vs PID+RO+FF on the boundary of PID+RO+FF control operation range - ball position

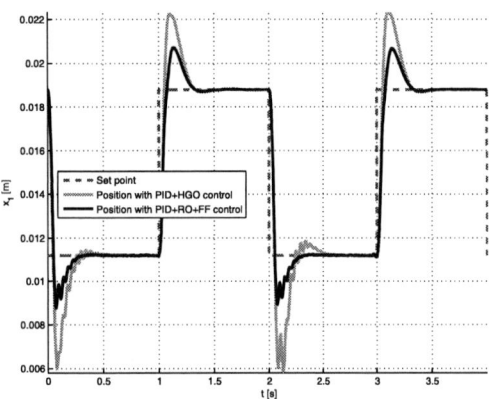

Figure 18. PID+HGO control vs PID+RO+FF on the boundary of PID+HGO control operation range - ball position

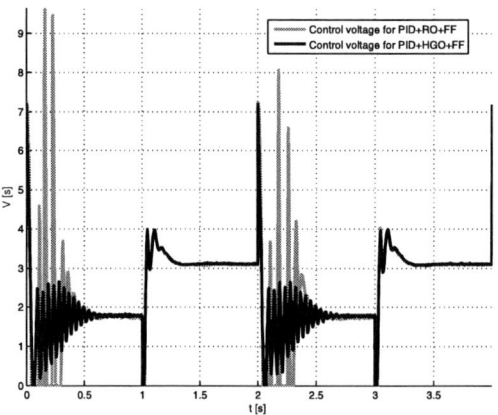

Figure 21. PID+HGO+FF control vs PID+RO+FF on the boundary of PID+RO+FF control operation range - controls

E. Cascade linearising feedback

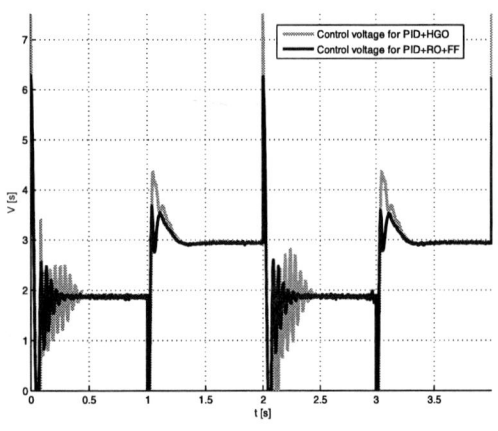

Figure 19. PID+HGO control vs PID+RO+FF on the boundary of PID+HGO control operation range - controls

Finally we present a final refinement, that can be obtained, comparing the PID+RO+FF structure with PID+HGO+FF, which "streches" the operation range a bit. This is the area close to physical capabilities of the system, because as we can see the ball is starting to bounce from the electromagnet (which is generally not the interest of magnetic levitation). As it can be seen from the table I both structures have large and close areas of operation. Comparison is illustrated in figures 20-21.

Results obtained via application of cascade linearising feedback are presented in figures . As it can be seen, there is no overshoot and the time of reaching of the set point is impressive. Unfortunately all sensitivities of this structure become apparent. First of all, control and position signals become very noisy, it is caused by strong augmentation of all present disturbances. Moreover any errors of identification are influencing the output strongly, resulting in steady state errors. Finally, the range of operation is rather small, system easily destabilises. Operation of the system is presented in figures 22 and 23. As it can be seen, HGO leads to marginally better results.

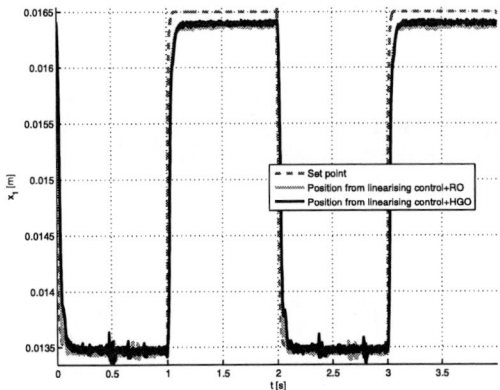

Figure 22. Cascade linearising feedback with HGO and RO observers - ball position

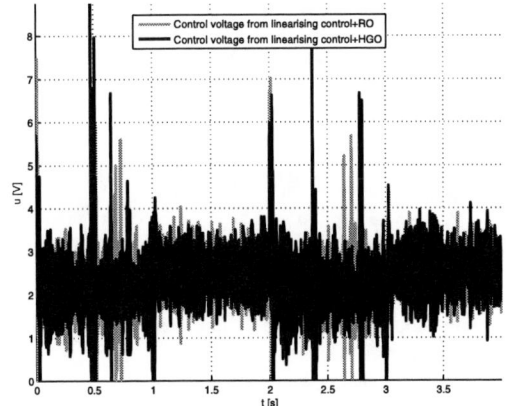

Figure 23. Cascade linearising feedback with HGO and RO observers - controls

VII. CONCLUSIONS

We have presented different structures of nonlinear dynamical feedback for the magnetic levitation system. All concepts were verified experimentally (all plots in this paper were made with real measurement data). We have shown, that introduction of the observer for velocity estimation drastically improves the regulation and the quality of control signal. Generated control signals are much smoother than those with numerical differentiation. This results in lower energy use and slower wear of actuators. Proposed application of nonlinear feedforward leads to substantial improvements in quality, and a great increase in operation range. Moreover this solution is very simple, and can be implemented as a look-up table in a programmable circuit allowing very fast implementations. Also proposed cascade structure of linearising feedback allowed easy implementation, that allowed operation regardless of modeling error. Unfortunately, it is still very sensitive as all the linearising feedback algorithms are. Further research will include attempts to implement solutions considered here in programmable circuits (like FPGA) and DSP. We think, that it would lead to further improvements in the operation of control structures.

REFERENCES

[1] P. Bania, Stabilizujące sprzężenie zwrotne dla systemu magnetycznej lewitacji, *Automatyka*, 2000, Tom 4, Zeszyt 2, pp. 117-139.

[2] J. Baranowski Projektowanie obserwatora dla silnika szeregowego prądu stałego *Automatyka*, 2006, t. 10 z. 1 pp. 33-52

[3] J. Baranowski, P. Piątek, A. Piłat , Nonlinear observer design for the magnetic levitation system, *Recent advances in control and automation* (K. Malinowski and L. Rutkowski, eds.), Akademicka Oficyna Wydawnicza EXIT, 2008, pp. 73-82.

[4] W. Barie, J. Chiasson. Linear and nonlinear state-space controllers for magnetic levitation *International Journal of Systems Science*, 1996, vol. 27, no. 11, pp. 1153-1163

[5] G. Bloch, M. Lairi, G. Millerioux, *Real Time Feedforward Neural Control of a MagLev System*, Kluwer Academic Publishers, 1999.

[6] P. Grabowski *Nonlinear systems - lecture notes*, 2008, unpublished, http://www.ia.agh.edu.pl/~pgrab/grabowski_files/nonlinear/nonlinear.xml

[7] W. Grega, A. Piłat, A Comparison of Nonlinear Controllers for Magnetic Levitation System,*5th World Multiconference on Systemics, Cybernetics and Informatics*, 22-25 July, Orlando-Florida, 2001, pp.188-193.

[8] E. H. el Yaagoubi, A. el Assoudi, H. Hammouri (2004) High gain observer : Attenuation of the peak phenomena *Proceedings of the 2004 American Control Conference ACC* (June 30-July 2, 2004, Boston Massachusetts)

[9] S.J. Joo, J.H. Seo, Design and Analysis of the Nonlinear Feedback Linearizing Control for an Electromagnetic Suspension System, *IEEE Transactions on Control Systems Technology*, 1997, Vol. 5, No. 1, pp. 136-144.

[10] T. Kaczorek *Teoria układów automatycznej regulacji*, WNT Warszawa 1977

[11] W. Mitkowski, *Stabilizacja systemów dynamicznych*. WNT, Warszawa 1991

[12] W. Mitkowski, Metody projektowania układów regulacji optymalnej (Design of optimal regulation systems). *XIV Krajowa Konferencja Automatyki, Zielona Góra*, 24-27.06.2002, Uniwersytet Zielonogórski, Inst. Sterowania i Systemów Informatycznych, Red. Z. Bubnicki i J. Korbicz, vol. 1, s. 195-204.

[13] W. Mitkowski, J. Baranowski Observer design for series DC motor - multi output approach In Proceedings of *IC - SPETO 2007 : XXX Międzynarodowa Konferencja z Podstaw Elektrotechniki i Teorii Obwodów*, Gliwice-Ustroń 23-26. 05. 2007, pp. 135-136

[14] H. Nijmejer, A.J. van der Schaft *Nonlinear Dynamical Control Systems*,Springer-Verlag 1991

[15] P. Piątek *Application of Specialized Hardware Architectures for Realization of Time-critical Control Tasks - Ph.D. Thesis (in Polish)*, Thesis supervisor - W. Grega . Akademia Górniczo-Hutnicza, Department of Automatics, Poland, Kraków 2007

[16] P. Piątek, W. Grega A speed classification method for real-time controlled dynamic systems, In Proceedings of *International Workshop on Real Time Software*, Wisła, pp. 677-684 2007

[17] P. Piątek, A. Piłat, D. Marchewka Magnetyczne zawieszenie sterowane 8-bitowym mikrokontrolerem, *V Krajowa Konferencja - Metody i systemy komputerowe w badaniach naukowych i projektowaniu inżynierskim*, Kraków, 133-135, 2005

[18] A. Piłat *Sterowanie układami magnetycznej lewitacji - Ph.D. Thesis (in Polish)*, Thesis supervisor - W. Grega Akademia Górniczo-Hutnicza, Department of Automatics, Poland, Kraków 2002

[19] A. Piłat Feedback Linearization and LQ Control for Magnetic Levitation System, In Proceedings of *Sixth International Conference on Methods and Models in Automation and Robotics, MMAR 2000*, Międzyzdroje, 2000, pp.407-412.

[20] K. Röbenack, F. Lynch Observer design using a partial nonlinear observer canonical form *Int. J. Appl. Math. Comput. Sci.*, 2006, Vol. 16, No. 3, 333-343

[21] D.L. Trumper, S.M. Olson, P.K. Subrahmanyan, Linearizing Control of Magnetic Suspension Systems, *IEEE Transactions on Control Systems Technology*, 1997, Vol. 5, No. 4, pp. 427-438.

[22] A. M. Zajączkowski, Stabilizacja układu lewitacji magnetycznej przy pomocy nieliniowego sprzężenia zwrotnego od wartości wyjściowej. In Proceedings of *IC - SPETO 2006 : XXIX Międzynarodowa Konferencja z Podstaw Elektrotechniki i Teorii Obwodów*. pp. 211-214. 2006

Speed and position estimation of SRM

Konrad Urbański, Krzysztof Zawirski

Poznan University of Technology/Institute of Control and Information Engineering, Poznań, Poland
e-mail: *konrad.urbanski@put.poznan.pl, krzysztof.zawirski@put.poznan.pl*

Abstract -The paper deals with the problem of speed and position estimation in SRM drive equipped with hysteresis band current controller. Instead of measured current the observer utilize reference current. The voltage is calculated from switching on-time. A speed observer structure which uses estimated back EMF and current of each phase is presented. The shaft position is integrated from estimated speed signal.

Keywords - drive, electrical drive, electrical machine, estimation technique, motion control, reluctance drive, sensorless control, switched reluctance drive.

I. INTRODUCTION

Switched reluctance motors (SRM) are relative simple machines. The advantages of those motors are high reliability, easy maintenance and good performance. The absence of permanent magnets and windings in rotor gives possibility to achieve very high speeds - over 10 000 rpm - and turned SRMs into perfect solution for operation in hard conditions like presence of vibrations or impacts. Such simple mechanical structure greatly reduces its price. Due those presented features, SRM drives are used more and more into aerospace, automotive and home applications. The drawback is complex algorithm to control its as a high degree of nonlinear object. SRMs must always be electronically commutated and require a shaft position sensor to operate. The other limitations are strong torque ripple and noisy effects.

II. THE SRM MODEL AND ROTOR POSITION OBSERVER

The observer is created using that motor model assumptions: eddy-current is neglected, stator and rotor are symmetric, there is no electromagnetic interaction between motor phases. Thus the electromagnetic torque can be calculated as an superposition of each phase component torque.

The well-known model utilize 2D functions as torque $T(\Theta, i)$ and flux $\Psi(\Theta, i)$. One can use a simplified 1-dimensional nonlinear equation instead of 2-dimensional [3]:

$$\Psi(\Theta, i) = L_C \cdot i + L_\Theta(\Theta) \cdot sat(i) \qquad (1)$$

and

$$T(\Theta, i) = \frac{dL_\Theta(\Theta)}{d\Theta} \cdot \int_0^i sat(\iota) \, d\iota \qquad (2)$$

where L_C is constant component of phase inductance at unaligned rotor position, L_Θ can be considered as a position-dependent component of non-saturated inductance, *sat* and *dsat* are the saturation and derivative of saturation functions.

Formula (1) and the voltage equation can be converted into (3). So, the one phase of SRM can be described by that equation:

$$\frac{di}{dt} = \frac{u - Ri - \omega \cdot \dfrac{dL_\Theta(\Theta)}{d\Theta} \cdot sat(i)}{L_c + L_\Theta(\Theta) \cdot \dfrac{dsat(i)}{di}} \qquad (3)$$

Nonlinear functions: $L_\Theta(\Theta)$, $sat(i)$ and its integrals and derivatives can be calculated from flux or torque characteristics.

An analyzed control system of SRM drive includes hysteresis band current controller. Such control method forced modification of observer's structure. Instead of measured current the observer utilize reference current (applicably modulated) and a voltage is calculated from switching on-time. Hysteresis band current controller's performance necessitate of observer algorithm change because of rapid changes of observer's feeding voltage.

Assuming description of SRM in form (3) one can create equation of observer (4-6) for stator current and speed for one stator winding (phase "n"):

equations 4, 5, 6

$$
\begin{cases}
\dfrac{d\hat{i}_n}{dt} = \dfrac{u_{n_{mean}} - R \cdot \hat{i}_n - K_e \cdot \hat{\omega} \cdot \dfrac{dL_{\Theta_n}(\hat{\Theta})}{d\hat{\Theta}} \cdot sat(\hat{i}_n)}{L_C + L_{\Theta_n}(\hat{\Theta}) \cdot \dfrac{dsat(\hat{i}_n)}{d\hat{i}_n}} + K_i \cdot \Delta i_n \\[2em]
\dfrac{d\hat{\omega}_n}{dt} = K_\omega \cdot \Delta i_n \\[1em]
\Delta i_n = \hat{i}_n - i_n
\end{cases}
$$

where $n = 1, 2, 3, 4$

The „^" symbol denotes estimated quantities, $L_{\Theta n}$ can be considered as a position-dependent component of non-saturated inductance in phase „n", u_{n_mean} is phase „n" mean value of voltage (filtered voltage). K_ω means speed correction factor, K_e means back EMF observer correction factor and K_i – current observer correction factor.

The equations of observer one can easily convert into MATLAB language (*m-file*). For example a part of equation 4, the back EMF (7):

$$EMF_n = K_e \cdot \hat{\omega} \cdot \frac{dL_{\Theta_n}(\hat{\Theta})}{d\hat{\Theta}} \cdot sat\left(\hat{i}_n\right) \qquad (7)$$

can be converted into (8):

ex(ff)=ke*wx*interp1(Q,DLq,thx(ff))*interp1(I,sat,i)

where *ex(ff)* means back EMF in *ff*-phase, *wx* means estimated speed, and *interp1* is a MATLAB interpolation function. The *interp1* function input values are x-values and adequate y-values which determine nonlinear function and actual x-value (*thx(ff)* is estimated position in phase *ff*). The "x" and "y" values are calculated from flux or torque characteristics measured at standstill.

The equations (4-6) are evaluated for each phase at every calculation step. Also speed derivative is calculated at every step - as an mean of speed derivatives of each phase. The rotor position is calculated from (9):

$$\hat{\Theta} = K_\Theta \int \hat{\omega}(t)\,dt \qquad (9)$$

where K_Θ is position correction factor.

III. Simulation Results

Simulation investigations were carried out in MATLAB-Simulink environment – as an graphic model includes text based language models. The motor and control system was modeled as an Simulink graphic model (drive parameters are given in appendix). Observer was created as an MATLAB's *m-file* to easy conversion into C language, which will be used to programming control algorithm in DSP used in a laboratory stand.

The motor model was calculated with small step of integration – $0.1 \div 1\ \mu s$ what gave its quasi-continuous character, and in opposition to this the model of control system together with observer was calculated with step about $100\ \mu s$, what simulates its microprocessor realisation.

Such preparation of control system gives opportunity to easy altering of control algorithms and fast observer program conversion into DSP system.

Some selected results of simulation are presented in the paper. The investigations consist of two parts. First – open loop mode – gives opportunity to check observer algorithm during preliminary investigations where observer not affected in drive performance but it gives possibility to control observer's calculations. The most important observer performance parameters are stable output in stady states and low estimation error. Open loop mode investigation presents prospective features of observer algorithm.

Second part– closed loop mode – gives opportunity to check observer performance where observer's output signal is used to control drive. The observer gains as a rule are should be smaller to keep stable observer (and a whole system) performance. Presented figures proofs that observer algorithm work well even in closed loop mode.

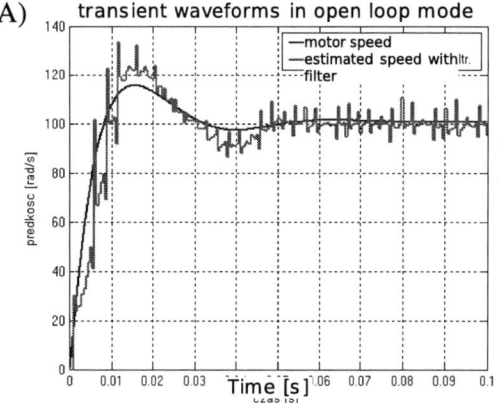

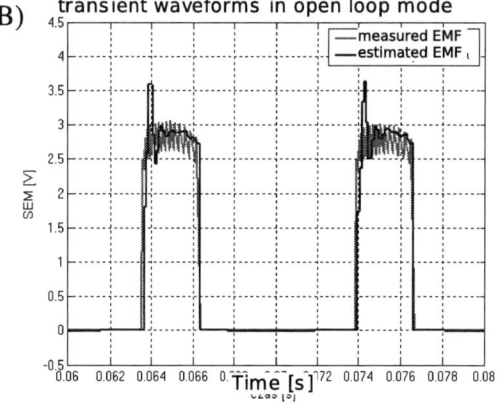

Fig. 1. Transient waveforms in open loop mode
A) motor speed and estimated speed after filtering B) one phase back EMF and estimated back EMF

Figure 1 is achieved in open loop mode - estimated speed and estimated position are not used to control the SRM drive. Figure 1A shows speed waveforms during starting a motor from zero to speed 100 rad/s without torque load and figure 1B shows waveforms of back

EMF in steady state with the same speed. That test proves the correct performance of observer estimation also in transients. Speed estimation is well performed even from zero speed (but the observer performance at zero speed is poor).

The next figures shows observer's performance in closed loop mode, where position signal and speed signal are applied to drive control. However, the utilisation of estimated speed value need a slow down of speed controller by controller settings modification. The estimated position utilisation in control loop has no restrictions and figures 2-4 presents the observer's performance.

Figure 2 presents waveforms of back EMF and current of one phase at steady state. Estimated and motor signals are presented. One can notice, the estimation is well espetially the observer response on step change of estimated value is satisfactory. The observer generates fast answer and damping is adaquate.

from zero speed and fig. 3B – at steady state. One can notice, the starting position error (fig. 3A) is noticeable however the observer decrease it into small value.

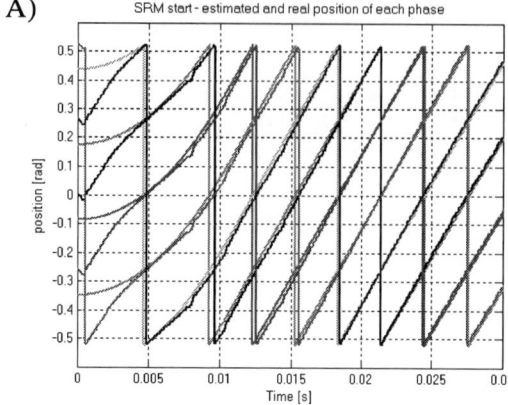

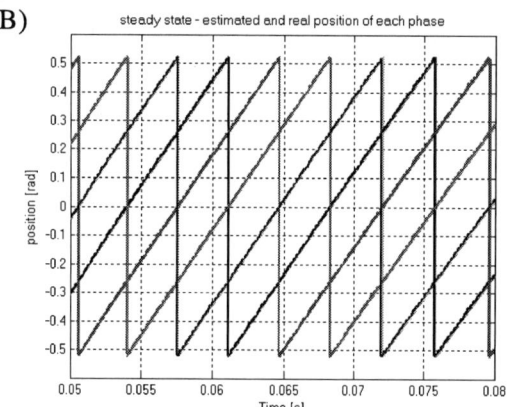

Fig. 3. Transient waveforms in closed loop mode: estimated and real position
A) during SRM start B) at steady state

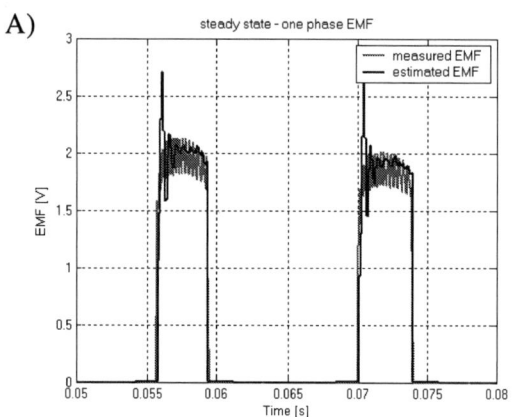

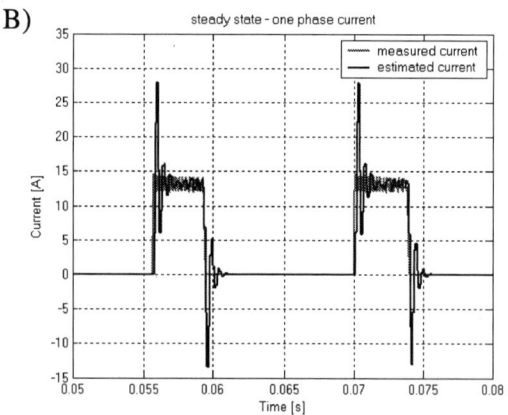

Fig. 2. Transient waveforms of one phase in closed loop mode
A) back EMF and estimated back EMF B) current and estimated current

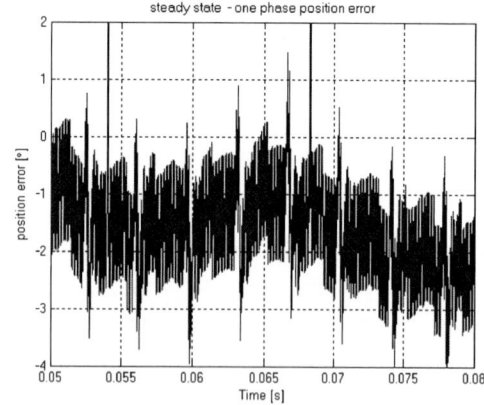

Fig. 4. Transient waveforms of one phase position error at steady state

Figure 3 presents transient waveforms of positions in each phase during SRM running. Fig. 3A presents start

Figure 4 presents waveforms of one phase position error at steady state. The error value not exceed 3.5 °.

IV. Conclusion

A concept of the observer using 1-dimensional model of SRM is presented. Also the big value of calculation step (100 μs) does not disturb the observer performance. The observer response is fast and adequate. Such observer works well in closed loop - where all required mechanical signals are estimated. Observer presents good performance even though nonlinearity of SRM model.

Appendix

Model parameters based on motor from laboratory stand:

type: SRM 8/6

model:

RA130175E from Motion System Tech, Tokyo, Japan

P_n=1.32 kW

U_n=48 V

n_n= 6000 rpm

Acknowledgements

This work was partially supported by grant TB 45082/08/DS.

References

[1] Cheok A.D., Fukuda Y.: A New Torque and Flux Control Method for Switched Reluctance Motor Drives, IEEE Trans. Power Electr., Vol.17, No.4, pp. 543-577, July 2002

[2] Krishnan R.: Switched reluctance motor drives. Modeling, Simulation, Analysis, Design, and Applications, CRC Press Boca Raton, London, New York, Washington, 2001

[3] Maciejuk A., Deskur J.: Modelling of switched reluctance motor drive, Proc. of Symposium on Electromagnetic Phenomena in Nonlinear Circuits, Maribor, Slovenia 2006.

[4] Soares F., Costa Branco P.J.: Simulation of a 6/4 switched reluctance motor based on Matlab/Simulink environment, IEEE Trans. Aerosp. Electron. Syst., vol. 37, no. 3, pp. 989–1099, Jul. 2001

[5] Urbański K., Zawirski K.: Adaptive observer of rotor speed and position for PMSM sensorless control system, COMPEL, The International Journal for Computation and Mathematics in Electrical and Electronic Engineering, Vol.23, No. 4, 2004, pp.1129 – 1145, ISSN 0332-1649, ISBN 0-84544-005-6

[6] Urbański K., Zawirski K.: Rotor position observer for sensorless control of SRM, XIX Symposium Electromagnetic Phenomena in Nonlinear Circuits, Maribor, Slovenia, 28-30 june 2006, pp.161-162

[7] Vas P.: Sensorless Vector and Direct Torque Control, Oxford University Press, 1998

[8] Zawirski K., Maciejuk A., Urbański K.: EMF observer for sensorless control of AC drives, Proc. of the 13th International Conference on Electrical Drives and Power Electronics, Dubrovnik, 26-28 September 2005, CD-ROM

Potential of Digital Gate Units in High Power Appliations

Harald Kuhn*, Thies Köneke[†], Axel Mertens[‡]
Leibniz Universität Hannover,
IAL – Institute for Drive Systems and Power Electronics, Hannover, Germany
* e-mail: *kuhn@ial.uni-hannover.de*
[†] e-mail: *koeneke@ial.uni-hannover.de*
[‡] e-mail: *mertens@ial.uni-hannover.de*

Abstract—Gate units of today's converters are in most cases voltage sources charging the input capacitance with a fixed gate resistor. Consequently, there is hardly no possibility to adapt quickly the gate unit to different semiconductor devices. During operation, there is no adaption process to adjust the switching transients to their actual optimum. The possibilities to implement new features or to adapt the units for a different device are related to time and effort. Therefore, in this paper the idea of a digital gate unit is explained and presented.

Keywords—IGBT, power semiconductor devices, signal processing.

I. INTRODUCTION

In the recent years, a tendency to digital control over power electronic applications can be observed. This trend is based on the fact that digital systems are very flexible and advantageous in further development and adaption to special implementations. The digital part in such an application cares for global tasks. It cares for the user interface and receives and sends status information etc. In power electronic systems, the digital unit sends commands to the gate unit that cares for switching on and off the power devices. By controlling the on and off states of the single switches, the whole power system is operated. Here, the basic switching operation determines a lot of parameters with respect to the quality of the total power conversion process. The basic requirements for gate units are

- turn on and off the semiconductor,
- offer different protection mechanisms such as
  - overcurrent protection (short circuit),
  - overvoltage protection (esp. for series connection in high power applications),
  - protection against undervoltage, and
- keep the switching transients in the safe operating area (SOA) within all possible voltages, currents and temperatures (often -40 °C to +120 °C).

Beside these compulsary basic elements of gate units, in literature more sophisticated methods to control the switching transients are known, [1]:

- controlling the collector current turn-on transient $\frac{di_C}{dt}$
- controlling the evolution of the collector-emitter voltage v_{CE}
- sophisticated protection mechanisms.

Nevertheless, the mentioned requirements are delivered by the most standard gate units which are nowadays implemented in an analog way. During the design process the behavior of the gate unit and the semiconductor device is determined. The fixed and hard wired parameters must be chosen in such a way, that the semiconductor works in the SOA. This may cause an ineffective operating mode in most of the working points depending on voltage, current and temperature. The variation of the turn-on process is shown in Fig. 1. The idea presented in this paper consists of implementing a digital gate unit using the benefits of digital signal processing.

II. IDEA OF A DIGITAL GATE UNIT

The heart of the digital gate unit is a field programmable gate array (FPGA), see Fig. 2. All switching-related information concerning the IGBT, e.g. collector-emitter voltage $v_{CE}(t)$, collector current $i_C(t)$, gate-emitter voltage $v_{GE}(t)$ and the junction temperature $T_{junction}$, are collected in this digital unit. This information is used to determine the optimum characteristic of the gate current to switch on or off the semiconductor device. Via the output pins of the FPGA, a current source is commanded to charge and discharge the input capacitance.

Consequently, signal conditioning, digital analog converters and a powerful current source form the periphery of the FPGA. Theses devices are the only parts that are realized analogously. All intelligence is implemented in the FPGA. Due to the fact that all parameters, control algorithms and protection mechanisms are implemented in the FPGA they can quickly be reimplemented and adopted to new applications. Using a digital gate unit may offer a lot of advantages:

- The switching process can quickly be adapted to the devices of different manufacturers.
- During the operation the switching transient can be adapted due to the varying working point (v_{CE}, i_C, $T_{junction}$).
- Different sophisticated protection mechanisms can be implemented by software.
- New monitoring procedures can be implemented.

A. Recording the Electrical Transients

To feed the FPGA during operation with the necessary information the characteristics of the voltages and currents

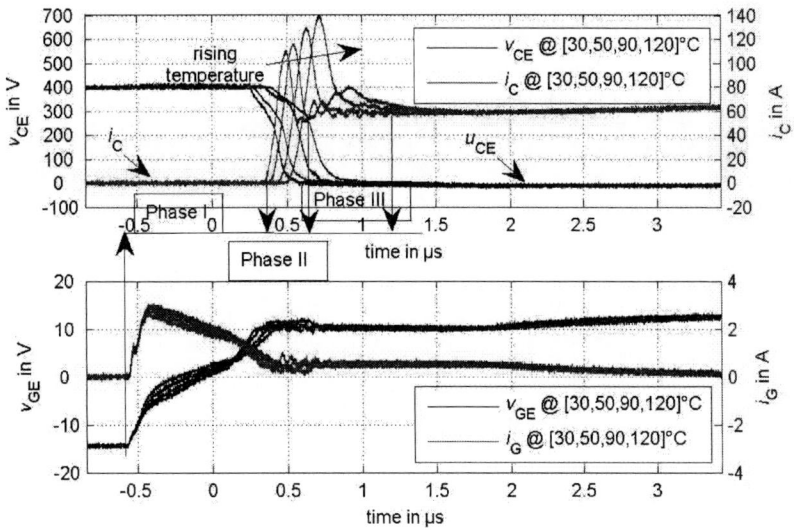

Fig. 1. Switching-on behavior of a 1.7 kV and 200 A rated IGBT at 400 V at different temperatures (30 °C, 50 °C, 90 °C and 120 °C. Device: eupec BSM200GA170DN2S.

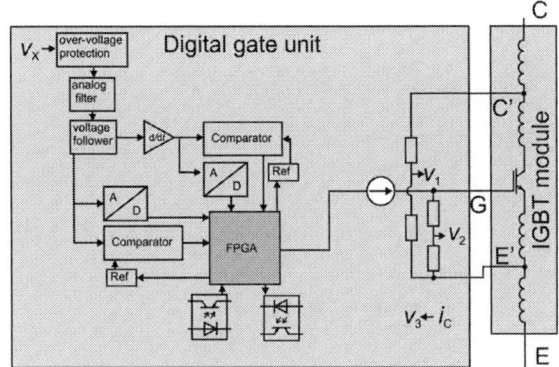

Fig. 2. Schematic of the proposed Gate Unit.

must be digitized. It is envisaged to use v_{CE}, i_{C}, v_{GE} and i_{G} as electrical values, see Fig. 2.

To measure the different voltages up to 2000 V having dynamics of 10 kV/µs, a compensated voltage divider is implemented. When using resistors as voltage divider for the mentioned dynamics, the parasitic inductance and capacitance cannot be neglected. They have to be compensated by paralleling capacitances to reach a pure resistive behavior. Analog filtering allows to inhibit noise and high frequency disturbances. A voltage follower is implemented to feed the capacitive inputs of the analog-digital converter (ADC) without loading the measuring signal. Behind the voltage follower in Fig. 2, the measuring signal is divided into two paths. On the one hand, the derivation is generated and may be completely inputted to the FPGA or optionally compared to a reference voltage with a high-speed operational amplifier (OPAmp). The reference voltage is set up with a programmable resistor. This allows changing these voltages while operating the

silicon device and therefore adjusting references according to the actual working point. On the other hand the measured signal is directly digitized and feeds the inputs of the FPGA. Here also exists the possibility to detect special levels with a high-speed OPAmp.

All these information serve the FPGA to calculate the optimum gate current and to detect the special moments in the switching transient. The method to detect special levels with an OPAmp allows to avoid the delay-afflicted analog-digital conversion. By this way, it is possible to inject the gate current as control variable exactly in time.

In a first work package, the feasibility of measuring the fast current and voltage transients was studied. Therefore, a double pulse test bench with a 1200 V and 40 A IGBT was built up. The driver in this test set-up still consisted of a bipolar voltage source. To slow down – that means adapt its switching speed to high power devices – a gate resistor of $R_{\mathrm{g}} = 56\,\Omega$ was chosen. The current was measured by inserting a shunt resistor into the power circuit. The ADC is a 90 MHz 6 bit device. The current and voltage courses of the double pulse are shown in Fig. 3. The electrical values were measured by an oscilloscope and by the digital measuring path that was implemented onto the driver board.

As can be seen, the waveforms do match very well. The signal in the current waveforms show an overcompensated behavior. Due to the layout of the driver board, it was not possible to reach better results. In Fig. 4, the collector-emitter voltage is zoomed in the switching transients. The shape of the measured signals shows a good agreement. With this figure, a delay time due to the analog digital conversion of $t_{\mathrm{delay,ADC}} = 60\,\mathrm{ns}$ can be determined. In Fig. 5, the delay time was substracted and the relative error was calculated. It is shown that even in the high dynamics part of the waveform, the error is less than 6 %.

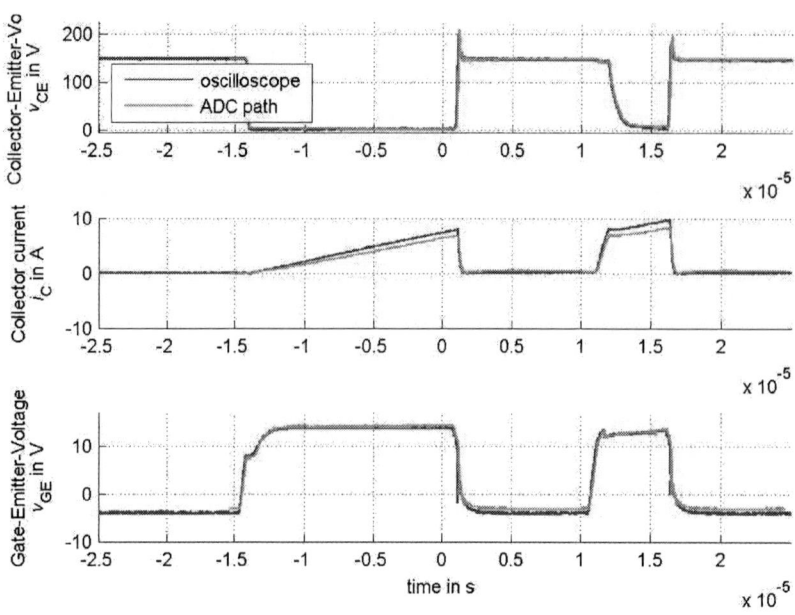

Fig. 3. Measured v_{CE}, i_C and v_{GE} in a double pulse configuration. Good results comparing the measurements with the ADC path and the waveforms recorded with an oscilloscope at the terminals of the device. Device IXSH24N60 in TO247 package.

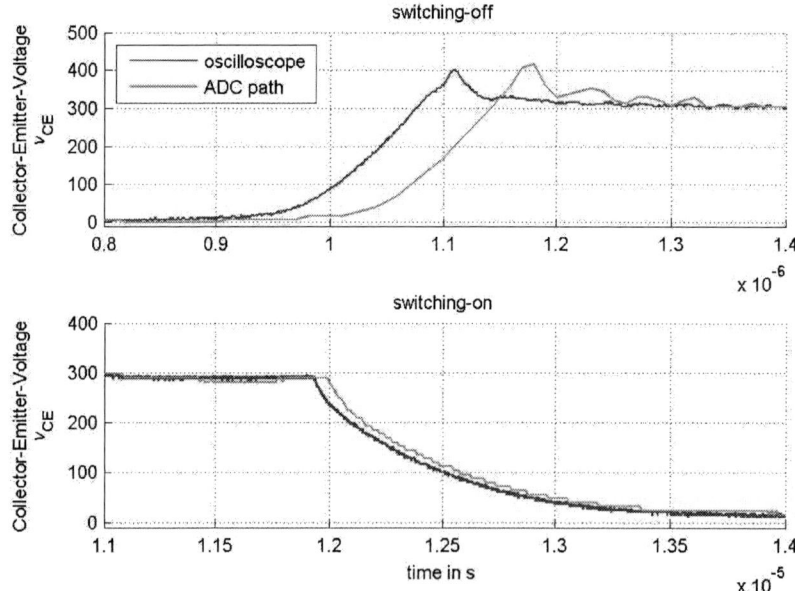

Fig. 4. Measured switching-on and switching-off characteristics. These figures allow to compare two different measurement methods. The leading course is recorded with an oscilloscope. The voltage was measured at the terminals of the IGBT device. The lagging graph shows the same voltage recorded with the digital/analog measuring path. This signal is delayed by the measuring path and is extracted out of the memory connected to the FPGA. Consequently, this signal is present in the FPGA and is the basis for future control methods.

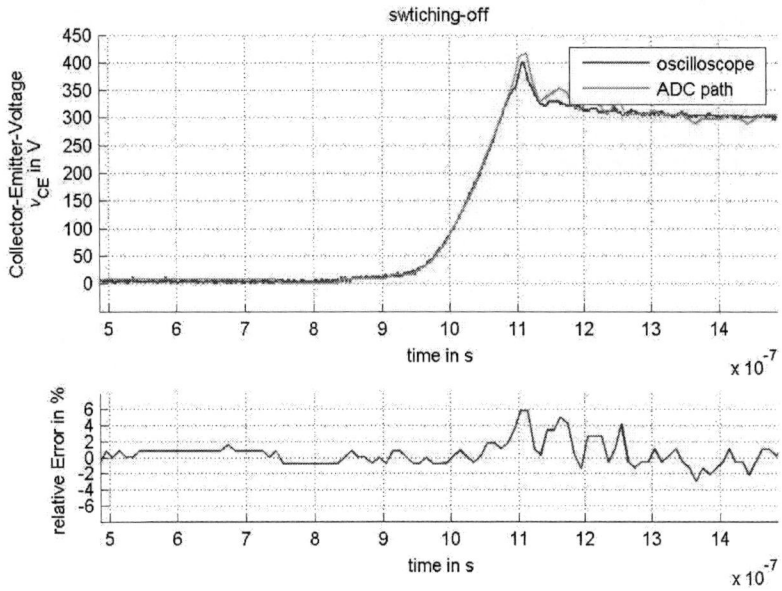

Fig. 5. Switching-on behavior with substracted delay time. The determined delay time of $t_{\text{delay,FPGA}} = 60$ ns is substracted. The measurement range was designed from -16 V to +490 V. Each digit consequently represents approx. 7.9 V. In steady state, the relative error is about 2 %, but even in the transient not inferior to 6 %.

With these experiments it was shown that it is possible to digitize the waveforms and feed them into the FPGA.

B. Reliability Aspects

In high power electronics, e.g. railway applications, offshore wind applications, minimum downtime, highest reliability and high life cycle is demanded. Thus, from the module manufacturer side a lot of effort is investigated to improve their products concerning the mentioned aspects.

Anyhow, in high power electronics the packaging of the silicon chips constitutes an important lifetime limiting factor, [2]. Due to thermal cycling and the different coefficient of thermal expansion (CTE) of the implemented materials the different layers underlay thermomechanical stress. In the sum of failure mechanisms the bond wire lift-off and the solder cracking represent two weak interconnections within the module. Lifting-off of a bond wire can be seen as the reduction of cross-sectional area of the semiconductor and thus can be detected and measured by the variation of the collector-emitter voltage v_{CE} in forward conduction mode. In [2], a significant increase due to thermal cycling tests with $\Delta T_{\text{junction}} = 130\,°C$ of the collector-emitter voltage was found: intact operation $v_{\text{CE}} = 2120$ mV; lift-off of the first bond-joint after approx. 5700 cycles leads to a risen $v_{\text{CE}} = 2140$ mV; second lift-off after 7000 cycles, $v_{\text{CE}} = 2160$ mV; third lift-off after 7150 cycles, $v_{\text{CE}} = 2220$ mV; fourth lift-off after 7000 cycles, $v_{\text{CE}} = 2250$ mV; fifth lift-off after 7000 cycles, $v_{\text{CE}} = 2350$ mV; total loss of electrical contact to the device. To detect the end of life failures of such mechanisms, modules were equipped with supplementary sense networks, [3].

The second important failure mechanism, the solder cracking, results in a shrunk thermal path to remove the dissipated power in the chip. Consequently, the degradation of the solder joints can be monitored by the variation of the thermal impedance R_{th}.

Since both of the very important failure mechanisms result from thermomechanical mechanism it may by interesting to monitor the junction temperature and to detect the level of degradation of the packaging and to predict the malfunction of the device and the system, respectively. Here it should be mentioned that the junction temperature must be measured under operating conditions, i.e. to measure the temperature under operating conditions. In [4], the correlation between different electrical parameters of a MOSFET (drain-source current $\frac{di_{\text{DS}}}{dt}$, the threshold voltage $v_{\text{GS,th}}$ and the turn on delay time $t_{\text{d,on}}$) and the temperature under operating conditions are studied. Similar analysis was done in [5] concerning the temperature-sensitive electrical parameters (TSEP) of an IGBT.

In addition to [5], supplementary measurements concerning the turn-on delay time concerning an IGBT were done with an eupec BSM200GA170DN2S device. In Fig. 6, the dependency of the turn-on delay time (time between the rising edge of the gate-emitter voltage and the startup of the rising collector current), the temperature and the collector-emitter voltage are shown. A correlation between the turn-on delay time and the collector current was not observed.

The idea is now to implement in the FPGA an algorithm that determines the junction temperature $T_{\text{junction,meas}}$, based on one of the methods presented in [4], [5]. On the other hand, the temperature can be calculated by evaluating the losses generated in the silicon chip. The product of the collector current and the collector-emitter

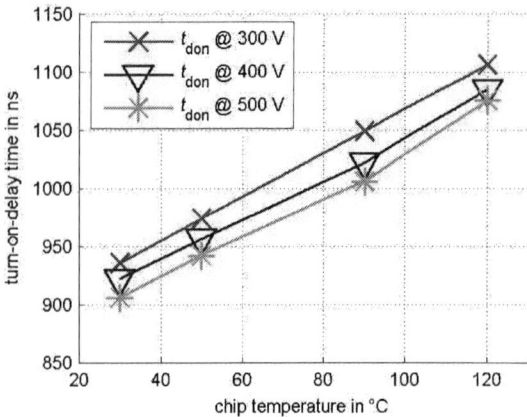

Fig. 6. Turn-on-delay time as a function of the chip temperature. The sensitivity of this method is approx. $1.862\,\mathrm{ns}/^\circ\mathrm{C}$.

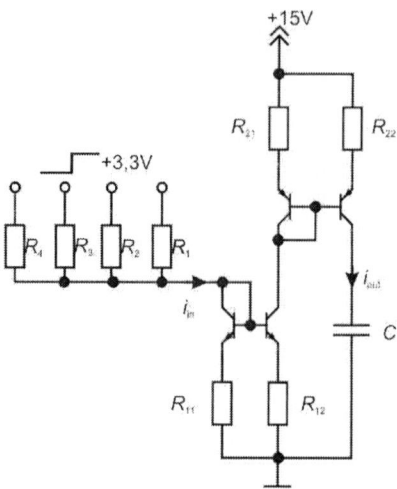

Fig. 7. unipolar two-staged current source

voltage $p = i_\mathrm{C} \cdot u_\mathrm{CE}$ can be calculated, see Fig. 3. These losses can be fed to a thermal model of the IGBT, and the theoretical junction temperature $T_\mathrm{junction,theo}$ can be calculated. Within the time due to the degradation of the solder joints the difference

$$\Delta T = T_\mathrm{junction,meas} - T_\mathrm{junction,theo} \qquad (1)$$

may rise. If this difference exceeds a certain level a highly degraded solder joint can be declared. Thus, an indicator of the degradation of the solder joints may be found.

The condition of the bond wires can - as mentioned above - be detected by the variation of the collector-emitter voltage. Using a supplementary ADC, not imperatively having a high dynamics but a higher resolution (8 bit and more), a monitoring method of the turn-on collector-emitter voltage is possible. Due to the fact that the variation is about hundreds of millivolts, the ADC measuring path for v_CE as shown in Fig. 4 cannot be used.

Implementing these algorithms the DGU will be able to monitor the degradation of the bond wire lift-offs and the solder joints. Consequently, failures can be detected by the DGU and handed on the superior control instance.

C. Current Sources

As current sources two different topologies are to be taken into account:

- Voltage-controlled current source: 8 bit of the FPGA are used to define the control variable, the output current. Consequently 127 different positive and 127 different negative output currents are possible. A fast digital-analog converter (DAC) is used to convert this information into an analog voltage. A voltage-fed bipolar current source is used to inject the current into the gate. The advantage of this realization is that very fine structured currents can be injected into the gate, whereas the DAC inserts an additional delay time $t_\mathrm{delay,DAC}$ into the system.
- Current mirrors with bipolar junction transistor (BJT): Several FPGA terminals (approx. 20 mA)

deliver an output current, which is weighted by different resistors, $R_1 \ldots R_4$, see Fig. 7. These currents are summed i_in and amplified in a double staged current mirror as shown in Fig. 7. The output current i_out is independent of the load voltage, in this case the voltage on the capacitor C. The capacitance C shown in Fig. 7 represents the input capacitance of the semiconductor device. With a complementary current source the gate can be unloaded by a negative currents. As a first approximation the gain is determined by the relation

$$i_\mathrm{out} = A \cdot i_\mathrm{in} \qquad (2)$$

$$\text{with} \quad A = \frac{R_{12}}{R_{11}} \cdot \frac{R_{22}}{R_{21}} \qquad (3)$$

In the test circuit, four FPGA terminals were used for coding the positive and four terminals for coding the negative current source. Using appropriate resistors, the current source is able to deliver 2^4 equidistant current levels. The load is a $C = 20\,\mathrm{nF}$ capacitor, simulating an IGBT gate. From (2), the gain of the current source is $A = 43$. For the measurements in Fig.8, the current sources were switched on in two different levels (positive current to load the capacitance $0000 \rightarrow 1010\ (0.5 \cdot i_\mathrm{max}) \rightarrow 1111\ (i_\mathrm{max})$ $\rightarrow 1010 \rightarrow 1111$, negative current to unload the capacitance $1010\ (-0.5 \cdot i_\mathrm{max}) \rightarrow 1111\ (-1 \cdot i_\mathrm{max})$ $\rightarrow 1010 \rightarrow 1111$). Every pulse takes 40 ns. Several measurements verify the calculated gain within the output current range of 0.7 A to 3 A. Two features were detected: For lower currents, the gain rises significantly. The dynamic performance of the current mirrors degrades with rising gain. Nevertheless, this kind of current source shows good dynamics and could be appropriate to act as a powerful current source to charge and discharge the input capacitance.

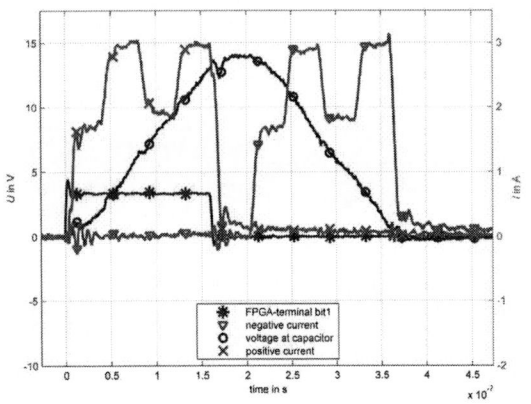

Fig. 8. Loading and unloading of a 20nF capacitor with two different current levels

D. Control Methods

To control the switching behavior of the IGBT, it is possible to section the switching behavior in three phases, [6]. Here the turning on is shortly explained, compare the curves in Fig. 1.

- Phase I switching-on is used to build up the threshold voltage and to form a channel for a current flow. In principle, this time can be described as a charging mechanism of the input capacitance of the IGBT. The higher the gate current i_G, the faster the input capacitance is charged and consequently the shorter the turn-on delay time, [7].

- At the beginning of Phase II, switching-on the channel is established and a current flow can begin. The current slope in this section can be controlled by the gate current, see (4).

$$i_G = i_{GE} = C_{GE}\frac{dv_{GE}}{dt} \quad \text{and} \quad (4a)$$

$$i_C = g_m \cdot v_{GE} \quad (4b)$$

$$\rightarrow \frac{di_C}{dt} = v_{GE} \cdot \frac{dg_m}{dv_{GE}} \cdot \frac{dv_{GE}}{dt} + g_m \cdot \frac{dv_{GE}}{dt}$$

$$= \frac{1}{C_{GE}}\left[g_m + v_{GE}\frac{dg_m}{dv_{GE}}\right] \cdot i_G \quad (4c)$$

In this phase, the complete voltage reduces by an inductive voltage drop is still blocked by the IGBT, because the freewheeling diode can not support any blocking voltage.

- The third phase is distinguished by the voltage fall at the terminals of the IGBT. The diode is now able to establish a space charge region, and consequently the voltage at the IGBT falls.

$$\frac{dv_{CE}}{dt} = \frac{1}{C_{GE}}i_{GE} + \frac{1}{C_{GC}}i_{GC} = \frac{1}{C_{GC}}i_G \quad (5)$$

The idea of the control method is to inject time-depending current amounts into the gate to achieve a control of the current and voltage slopes independently from each other, see Fig. 9, [1].

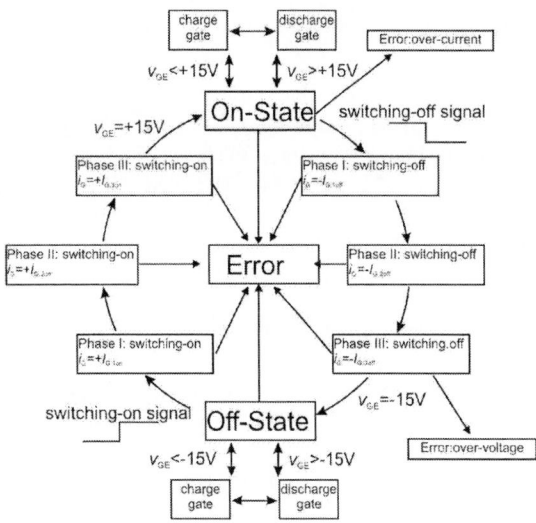

Fig. 9. State machine description of a control process.

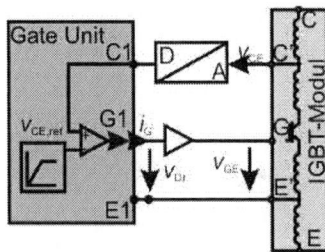

Fig. 10. State machine description of a control process.

Generally, a lot of sophisticated control mechanisms are possible to be realized. As an example, the method presented by [8], a closed-loop voltage control, can be implemented in the proposed Digital Gate Unit. A FPGA gives a reference voltage to an OPAmp. The actual curve of the collector-emitter voltage is compared with the reference, and the error is amplified as current and feeding the gate of the IGBT module. This can clearly be seen as the advantage of the digital implementation of the gate unit. Due to the fact that the FPGA can be programmed, the advantage of different control methods can be summed up in the algorithm of the FPGA. The function of the OPAmp of [8] is implemented in the FPGA by software. Furthermore, this method could be expanded to an additional closed loop current control and other features without changing the design of the gate unit board.

The reaction time and the delay time of the actuator with the delay times of the measuring path and FPGA affect the control methods that can be used to control the switching transients. The whole switching process in Fig. 1 shows a duration of 4 µs. Imagining a measuring path with a 6 bit ADC working with 100 MHz, the following considerations can be done. In the marked time intervals, the system takes the following samples:

- Phase I $\approx 2\,\mu s \rightarrow 200\,$samples
- Phase II $\approx 0.5\,\mu s \rightarrow 50\,$samples
- Phase III $\approx 1.5\,\mu s \rightarrow 150\,$samples

This shows that the number of samples is sufficient to get reliable information concerning absolute and derivational information about the current and the voltage of the device. Taking into account the following delay times:

- Analog-digital conversion $t_{\text{delay,ADC}} \approx 60\,$ns
- Calculation time of the FPGA $t_{\text{delay,FPGA}} \approx 10 \cdot 10\,\text{ns} = 100\,$ns
- current source $t_{\text{delay,CS}} \approx 50\,$ns

leads to $t_{\text{delay}} \approx 210\,$ns.

Due to the whole switching time of approximately $5\,\mu s$, it is hardly possible to react to the slopes in the current switching process. Thus, it seems to be necessary to implement a cascaded control mechanism. The digital measuring path works as digital oscilloscope scanning the courses. The FPGA calculates the characteristic of the new control variable and injects this current in the next switching transient, see Fig. 11. Doing so, the calculation time of the FPGA and the delay time of the AD path is no longer the bottleneck in this system. Nevertheless, the bandwidth of the current source must be very high, to follow quickly the new controlled gate current. At present, two different current sources are envisaged:

III. OUTLOOK

The proposed gate unit offers a wide range of different new control strategies. With the presented measuring path, it is possible to get the current and voltage waveforms with the required accuracy. These information may be the base for prospective regulation methods to get a better control over current and voltage slopes during the switching transients of high power devices. Realizing the method to detect the junction temperature would offer a good possibility to improve detecting failure mechanisms. The presented highly dynamic current source presents a good possibility to charge and discharge the input capacitance. These single components now have to be brought together to check their interaction.

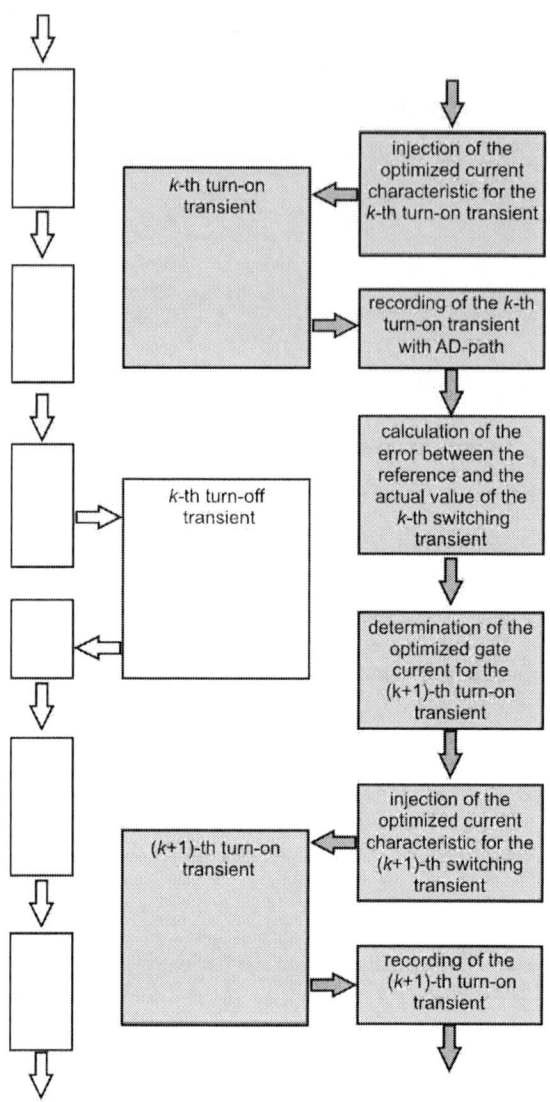

Fig. 11. The proposed control mechanism compensating the delay times in the measuring path. On the left hand side, the turning-off procedure is denoted.

REFERENCES

[1] H.Kuhn, T. Köneke, and A. Mertens, "Considerations for a Digital Gate Unit in High Power Applications," *IEEE Power Electronics Specialists Conference*, 2008.

[2] R. Amro, "Lastwechselfestigkeit von modernen Aufbau- und Verbindungstechniken bei hohen Temperaturhüben," *PhD Thesis, Chemnitz University of Technology*, 2006.

[3] J. Lehmann, M. Netzel, R. Hezer, S. Pawel, and Th. Doll "Method for Electrical Detection of end-off-life Failures in Power Semiconductors," *EPE Conference*, 2003.

[4] D. Barlini, M. Ciappa, M. Mermet-Guyennet, and W. Fichtner, "Measuerement of the transient junction temperature in MOSFET devices under operating conditions," *Microelectronics Reliability*, vol. 47, pp. 1707–1712, 2007.

[5] D. Barlini, M. Ciappa, A. Castellazi, M. Mermet-Guyennet, and W. Fichtner, "New Technique for the Measurement of the Static and of the Transient Junction Temperature in IGBT Devices under Operating Conditions," *Microelectronics Reliability*, vol. 46, pp. 1772–1777, 2007.

[6] V. John, B.-S. Suh, and T. A. Lipo, "High Performance Active Gate Drive for High Power IGBTs," *IEEE Industry and Applications Conference*, 1998.

[7] R. Chokahawala, J. Catt, and B. Pelly, "Gate Drive Considerations for IGBT Modules," *IEEE Industry and Applications Conference*, 1992.

[8] Y. Wang, P. R. Palmer, T. C. Lim, S. J. Finney, and A. T. Bryant, "Real-time Optimization of IGBT/Diode Cell Switching under Active Voltage Control," *IEEE Industry and Applications Conference*, 2006.

[9] N. Idir, R. Bausire, and J.J. Franchaud, "Active Gate Voltage Control of Turn-on di/dt and Turn-off dv/dt in Insulated Gate Tranistors," *IEEE Transactions on Power Electronics*, vol.2, no. 4, 2006.

Disturbance Currents of Inverters

Petr Vrana*, Jiri Javurek*

* Czech Technical University in Prague/Department of Electric Drives and Traction, Prague, Czech Republic, e-mail:
pvrana@polovodice.cz
* Czech Technical University in Prague/Department of Electric Drives and Traction, Prague, Czech Republic, e-mail:
jiri.javurek@skoda.cz

Abstract— This paper presents a research on a DC-AC (AC-AC) inverter used in a locomotive drive. The research is focused on creation of harmonic currents in DC-link. The content of some harmonics is restricted by standards and we call such harmonics as disturbance currents. The goal of this research is to found effects which are adherent to disturbance currents and on the basis of this to implement the changes of the power circuit, the control algorithm and the modulation.

Keywords — Converter control, Induction motor, Locomotive, Noise, Rail vehicle, Regulation, Vector control

I. Introduction

New trends in the electric drives are implemented into many branches of industry, in railways as well. The drive of a modern locomotive consists of a brushless machine supplied by an inverter. Such engine behaves as a current harmonic source. The current harmonics are switched in the same way as feeding current; it means substation, trolley wire, locomotive, rails and a substation again. Unfortunately a part of this circuit is shared with another system which performs detection of train's position, semaphore settings, automatic control of crossing-gates, switch blocking etc. The function of this railway safety appliance system may not be affected by any other equipment.

This paper discusses the introduced theme and presents the research which takes place at the Czech Technical University in Prague in co-operation with Skoda Electric a.s. The locomotive disturbance currents are crucial for the railway traffic safety and every supplier of locomotives must satisfy hard conditions given by standards. The goal of this research is finding effects which are adherent to disturbance currents and on the basis of this to implement the changes of the power circuit, the control algorithm and the modulation. There are two possible procedures how to find out the way of disturbance currents inception. The first is making a mathematical model and a simulation of generating of the disturbance currents (research on UWB Plzen). The other way (described in this paper) is measuring on the real circuits and monitoring the disturbance currents in DC-link which are produced in different states of inverter and load (CTU Prague).

II. Function of Track Circuits

Every length of track is split up into logical sections called 'track sections' or 'blocks'. Each track section has track circuit equipment (supply, receiver, transmitter, track junctions, contact converter, cabling) which recognizes whether the block is 'occupied' or 'clear' and send a signal. The signal will not allow the other train to enter the section unless it is clear. The basic principle behind the track circuit lies in the connection of two rails by the wheels and axles of locomotives and rolling stock to short out the electrical circuit. The receivers are arranged to detect the selected frequencies and to ignore DC and AC traction frequency signals. The transmitted AC signal is coded and the locomotives are equipped with inductive pickups to create a cab signaling system. "Fig. 1" illustrates the function of the track circuit. You can see the locomotive which is supplied by a two-side DC supplying system. Transmitter T2 is sending the signal into track sections Number 2. Almost the whole signal is shunted by the wheels axles of locomotive and the level of signal on receiver R2 is under the setting value. The receiver R2 is signalizing that the track section is occupied. There is a danger in case the locomotive is the source of current of the same frequency as well as the track circuit transmitter generates. It could cause wrong indication of the state of the track section.

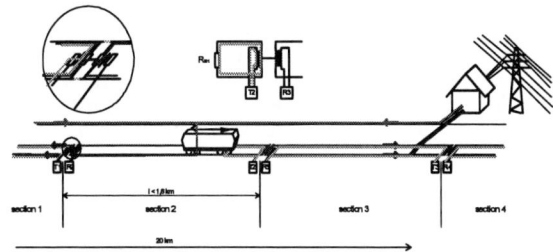

Fig. 1 Supplying and detecting the locomotive by track circuits

III. Limiting Values of Currents

The sensitivity of the receivers is given by two contradictory requirements:

- to be able to check the electrical conductivity of rails and by the help of it to check mechanical integrity of rails

- to achieve the highest shunting sensitiveness, which means to be able to detect locomotives with high shunt resistance

The mentioned requirements cause the need of high receiver sensitivity and strict limiting value of disturbance currents produced by locomotive engines. The limiting values as well as the other technical data are dependent on the type of receivers used in the track circuit. Although the new types of track circuit receivers have several time bigger noise resistance than the old types, the limiting value is generally postulated by standard ČSN 34 2613 and is 100 mA in "TABLE I". There are higher limiting values in the other countries of Europe because they do not require checking electrical conductivity of rails by the help of track circuits. They monitor the mechanical integrity in another way or they use only new types of receivers.

TABLE I
LIMITING VALUES OF HARMONIC CURRENTS INCLUDED
WITH TRACTION CURRENT

Country	Min. limiting value [A] in frequency band 0-400 Hz
Germany	2
Austria	1,5
Poland	1,2
Hungary	10
France	0,35
Switzerland	0,25
Czech	0,1

IV. THE WORKPLACE DESCRIPTION

The workplace "Fig. 2" consists of controllable source for supplying DC-link, 3mF capacitor, 5mH inductor and 3Ω discharge resistor. The resistor protects the static inverter against mismanagement effects. Inverter supplies tree-phase load. The load can be represented by tree phase resistor or motor. If we supply the DC-link with by another inverter, the measurement would be influenced because the inverter itself is the source of disturbance currents. That is why the special controllable source is used.

The purpose of this measurement is to monitor the disturbance currents (by Czech Railways 50±5, 75±6, 275±6 Hz) in DC-link which are produced in different states of inverter and load as well. Measurements are taken with current probe which makes use of Hall phenomenon or with conditioned Rogowski current probe. The Harmonic components 75Hz, 275 Hz are received by filtering. The Chebyshev analog filter was designed for this purpose. At the same time the signal is processed by sounding system which consists of a NI PXI 24 bit digitizer and PC. The digitizer communicates with PC by express card. Sampling rate is 10 kHz, each data block has 500 samples. A program written in LABView 8.5 performs each 100 ms (50ms) FFT of frame 5000 samples of measured signal and finds maxima in selected frequency band (tone detection). This way it is possible to rate an exact value of harmonic which occurs for more than 100 ms. Minimal operating period of disturbance currents 100 ms is determined by the standard and relate to a response time of track circuits receivers.

The first measurement was taken while using the regulator D8218N1 and software which is implemented in a suburban Units 471 (under operation July 2000). The old type of laboratory inverter which consists of IGBT modules FF200R12KL Eupec was used. The basic parameters of the inverter are $U_{(BR)}$= 1,2 kV, I_C=200A, $t_{(on\ delay)}$= 400 ns, $t_{(off)}$=1µs. A program Monitor performs

user interface. The modulator works with switching frequency 760 Hz. The inverter control passed over the "hand" mode in the Monitor. This mode allows setting the inverter opening (from 0 to 255 computer units [c.u.]), frequency setting (from 0 to 200 Hz output frequency with stepping 1/128 Hz) and setting of modulation switch-over position. Almost all presented measurements are taken while using "torque control" mode. At work of inverter in this mode the regulator keep a driving torque and an engine speed (output frequency of inverter) can be changed by loading engine.

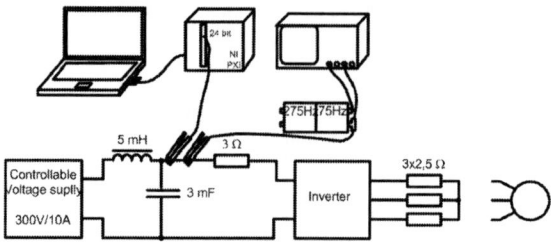

Fig. 2 The Workplace

V. RESULTS OF MEASUREMENTS

Experiments were focused on three potentialities which can influence the content of harmonics in DC-link of inverter. Firstly it is a size of dead time and minimal time of inverter. Secondly we tried to use several kinds of inverter control. Thirdly we investigated the influence of unbalance of inverter control and load.

A. Dead time and Minimal Time of Inverter

Theoretical background of influence the dead time on the content of harmonics in DC-link is stated in [1] and [2]. Both the theoretical research at University West Bohemia in Plzen (UWB) and the measurement on real drive at CTU Prague show that influence of the dead time on harmonics is minimal. The dead time bigger than 10µs approves oneself in harmonic spectrum as noisy background with minimal size in comparison with other harmonics included in spectrum.

The minimal time of inverter (time between switching on and off of one transistor in inverter) has approximately the same influence on harmonics in DC-link as dead time.

B. Vector Control and DTC

In cooperation with VŠB-Technical University of Ostrava the comparison of a vector control (regulation scheme with separated modulator) and several kinds of direct torque control was performed.

The workplace at Department of Electronics in Ostrava consists of DC link powered by an accumulator battery 340V, an inverter with 4mF condenser and an induction motor 2,7kW, 1360 nominal rpm. On Figures below you can see the harmonic analysis of current in DC link by constant torque 15 Nm and increasing motor speed by stepping 100 rpm. Same conditions were used in implementation of 5 methods (3 samples of them are on Fig. 3, Fig. 4, Fig. 5) of direct torque control (DTC) described in [3] and the vector control with a separated modulator. It is possible to see zero harmonic which corresponds to power acceptance (so to the motor speed). Each method has different demands on level of DC link voltage. Therefore some method can not reach the same level of motor speed (DC link current). The best of DTC methods seems to be either Takahashi method or PVN method. Harmonic spectrum of DC link current while using DTC is more spread. On the other hand the separate modulator which combines asynchronous and several kinds of synchronous modulations has some dominant harmonics which position is given by switching frequency and multiples of output inverter frequency. When the setting of transition of modulations is optimized (synchronous modulations do not start before the dominant harmonic is going to be higher than 275+6 Hz) it is possible to get better results than with DTC.

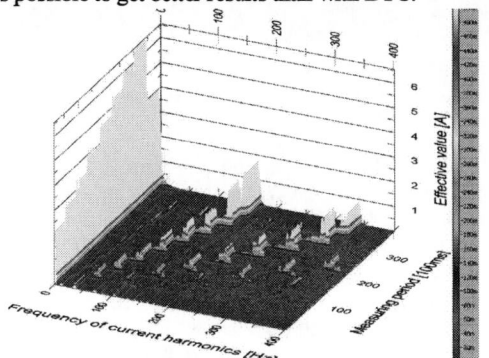

Fig. 3 Depenbrock method of DTC

"Fig. 3" Inverter output frequency 0÷30Hz, dominant 2nd, 6th, 12th and 18th harmonics

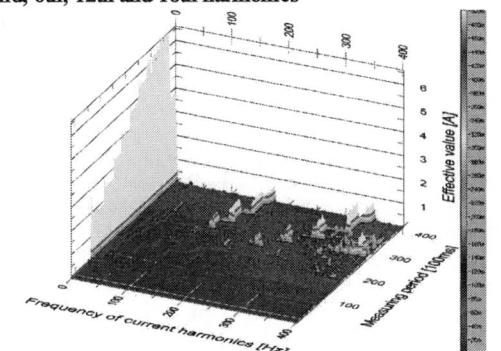

Fig. 4 Takahashi method of DTC

"Fig. 4" Inverter output frequency 0÷30Hz, dominant 2nd, 6th and 12th harmonics

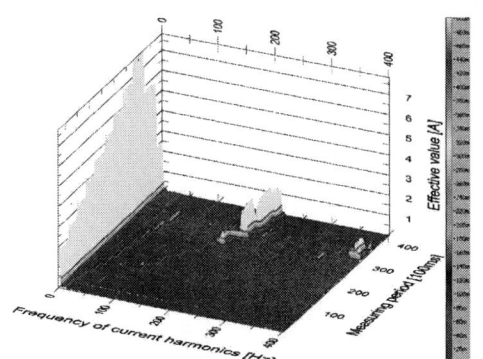

Fig. 5 PVN method of DTC

"Fig. 5" Inverter output frequency 0÷34Hz, dominant 2nd and 6th harmonics

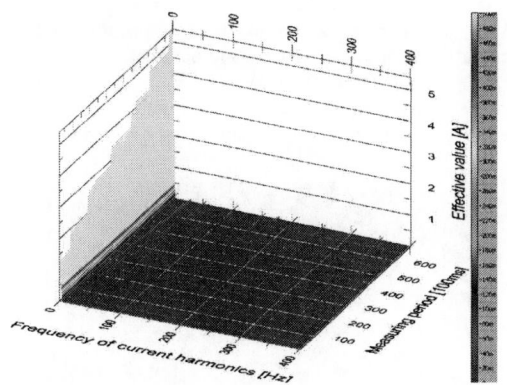

Fig. 6 The vector control

"Fig. 6" Inverter output frequency 0÷25Hz, dominant 1st, 2nd and 6th harmonics

The size of 1st, 2nd, 6th and 18th harmonics in DC-link is critical for disturbance currents of frequencies 50±5, 75±6, 275±6 Hz. The author tries to clear up the creation of mentioned harmonics in a next section.

C. Unbalance of Inverter Control and Load

On "Fig. 7" there is a scheme which was used for measuring of inverter output current unbalance and for diagnosing an influence of unbalance on harmonics in DC-link.

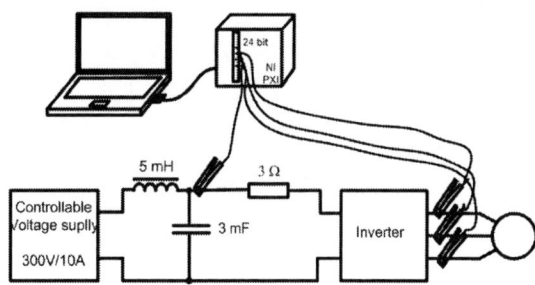

Fig. 7 Measurement and evaluation of asymmetry

Phase currents u,v,w are transformed into polar coordinates α, β. There are several kinds of asymmetry in regulation which can occur. "Fig. 8" describes three most frequently asymmetries.

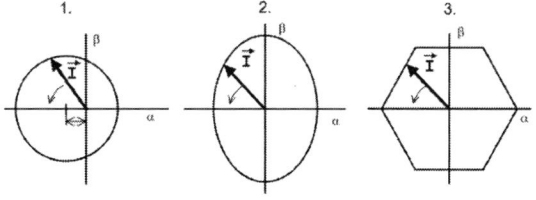

Fig. 8 Unbalanced currents

First current unbalance causes 1st harmonic (inverter output frequency) in DC-link. Amplitude of this harmonic is proportional to shift of the current space vector. Such unbalance can be invoked for example by DC component of current measuring element.

Second current unbalance causes 2nd harmonic in DC-link. This unbalance is invoked by proportional measurement error of current measuring element.

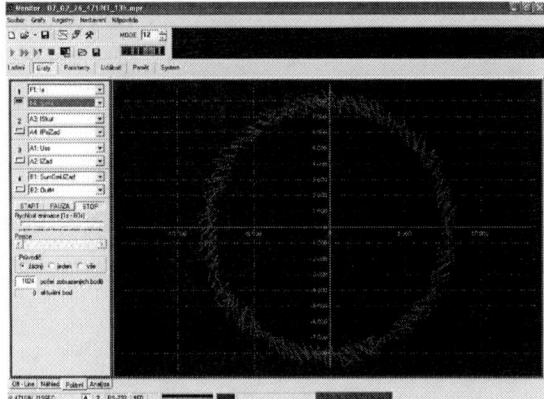

Fig. 9 Space vector of output current in the regulator

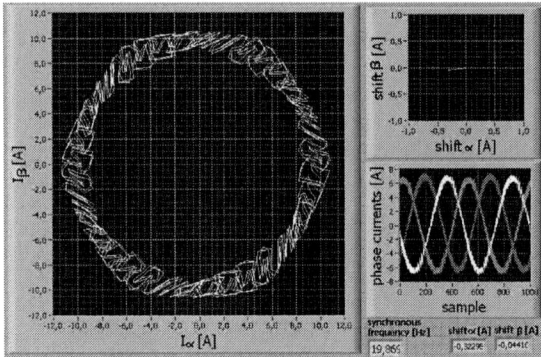

Fig. 10 Real space vector of output current

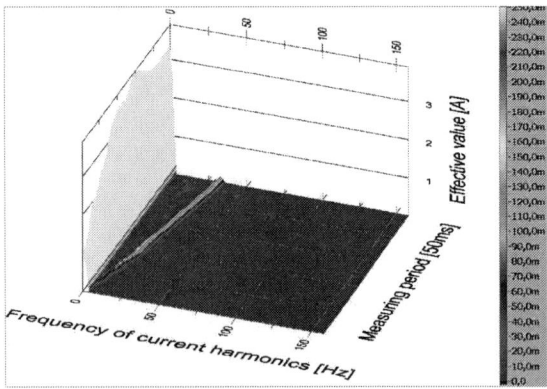

Fig. 11 Harmonic analysis of DC-link current

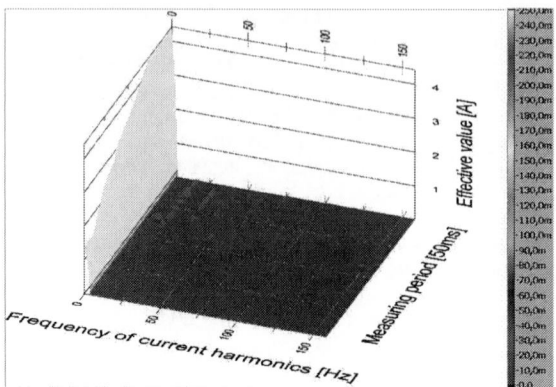

Fig. 12 Harmonic analysis of DC-link current after compensation

Third current unbalance cause 6th harmonic in DC-link. Such current space vector profile is typical for Depenbrock method of DTC or for full opening of inverter by the vector control.

On "Fig. 9" you can see a record of the current space vector how it is seen by the regulator. The regulator corrects a space vector to be symmetric but the space vector of output current is not symmetric (has current unbalance no.1 from "Fig. 8"), see "Fig. 10". Detail of spectrum of DC-current is shown on "Fig. 11". There is first harmonic which is caused by the shift of space vector along α-axis. The current unbalance is possible to be compensated by shifting the space vector over the way. "Fig. 12" shows the result of compensation.

VI. CONCLUSION

It seems that the restriction of disturbance currents in traction mains will not be successful only by using some DTC method. The comparison with the vector control shows that the spread of harmonics doesn't bring lower level of harmonics in monitored frequency bands 50±5Hz, 75±6Hz and 275±6 Hz. I tried to keep the same conditions during measurements by all tested methods. The switching frequency (one of the most important parameter) was constant only by the vector control (760 Hz). Working principle of DTC doesn't allow keeping the constant switching frequency which moved between 700 Hz and 1300 Hz. The theories based on simulations expected that harmonics are not very affected by changing of dead time setting of converter and all my measurements have validated it. The restriction of disturbance currents consists mainly in suppression of 1st, 2nd and 6th harmonics by the help of compensation of the current unbalance.

REFERENCES

[1] M. Janda, K. Zeman and Z. Peroutka, "Conductive currents in traction engine with induction motor," *30th Nation Conference about Electric Drives. Plzen*, ISBN 978-80-02-01921-3, pp. 143–148, July 2007.

[2] M. Janda, *Harmonics in DC-link of drive with induction motor*, Dissertation. Plzen, 2007

[3] L. Hrdina, "A comparison of characteristics of less common methods of direct torque control with classic Takahashi method," in *WOFEX*, ISBN 978-80-248-1571-8. VSB-Technical University of Ostrava, 2007, pp. 19-25.

Improvement of the Energy Recovery of Traction Electrical Drives using Supercapacitors

Diego Iannuzzi[*], *IEEE Member*

[*] Electrical Engineering Department, Naples, Italy, e-mail: *iandiego@unina.it*

Abstract— In the paper the possibility of improving the energy recovered during the braking of railway vehicles and reduction of power peaks during accelerating are discussed. The energy available from the regenerative braking of the motor is stored into supercapacitor sets placed on board and reused subsequently for the acceleration of the train. This auxiliary storage system allows the reduction of the losses on the line, because the power peaks are shaved by supercapacitors and, hence, the rms power supplied is reduced. In addition, also the energy consumption of the train can be reduced respect to traditional convoys, especially if the characteristic duty-cycle presents many accelerations and braking periods as the case of subways trains. The set of supercapacitors needs the use of an intermediate dc/dc converter in order to harmonize the voltage with that of the dc-link of the drive and control the power flows of the supercapacitors. The mathematical model of the whole system and the control strategy of energy management are presented.

The actual possibilities of the energy recovery are shown by means of numeric simulations and expressed in percentage respect to the energy drawn during accelerations. The control strategy has been experimentally validated on a scale system made of an asynchronous drive supplied by a dc source and a set of supercapacitors as auxiliary storage device.

Index Terms—Supercapacitors, traction drives, energy recovery

I. INTRODUCTION

The increase of the specific power demand by present-day railway traction vehicles implies to find reliable technical solution in order to reduce the energy consumption. The typical journey i.e. of subway trains, light rail vehicles (tram) is made of accelerations, coasting and braking periods. In particular, the largest part of the energy drawn by the train is ascribed to the acceleration, because of the reduced distance between two subsequent stations. Modern electrical drives for traction motors benefit from the possibility of regenerative braking [1] and the advantages related to the saving of energy has been already demonstrated attempting to inject the energy into the supplying line [2, 3].

The exchange of power flows between the vehicles and the catenary is not always allowed. It depends on if the catenary is capable of storing the amount of energy comes from more vehicles at same time. At present the main dc/dc energy substations of the catenaries are not bidirectional in the power. Indeed, the ac/dc power rectifiers have cheap diodes. As result the recovery energy can be only used if there are in the neighborhood other convoys to be supplied, otherwise line over-voltage phenomena can occurs.

At present, to avoid the over-voltage, the braking is usually rheostatic and the energy is dissipated on ballast resistors present on board.

So the recovery of the kinetic energy of the trains is subjected to the traffic conditions and dependent on phenomena not easily predictable. Such an energy management does not require any significant increase in the complexity of existing infrastructures, but the energy actually recovered is not greater than 10% of that drawn during accelerations [2].

Nevertheless different technical solutions are reliable [4,5]: use of storage devices on board of rail vehicles; use of storage devices on substation of rail vehicles; use of bidirectional electrical drives in substation.

At the present the main storage devices available are the electrochemical batteries, the electrochemical capacitors or supercapacitors, and the flywheels. The characteristics of these different devices can be compared referring to their specific energies and specific power, respectively expressed in Wh/kg and W/kg. For the use on board, only electrochemical batteries and supercapacitors seem to have practical applications, even though they present complementary characteristics. The batteries present high energy density and relatively low power density, with discharges time ranging from ten of minutes to hours. The supercapacitors have energy density (6Wh/kg) lower than that of batteries but higher power density (6kW/kg), with discharge times ranging from ten of seconds to minutes. This aspect points out that supercapacitors are suitable for supplying power peaks. In addition, their cycle life is about 106 cycles of charge discharge because the principle of the energy storage is based on electrostatic processes. So supercapacitors are, hence, the most suitable devices for the application on railway vehicles.

The first solution has been adopted by Bombardier transportation on prototype vehicle in Mannheim [7]. The prototype is a modern Light Rail Vehicle and the storage device is a stack of supercapacitors with energy content of 1kWh with a mass of 450 kg. The devices have been mounted on roof of vehicle. This solution has several advantages in terms of increase energy efficiency of the whole system, reduction of power peak demand during acceleration operations, reduction of infrastructure losses, and line voltage stabilization. The main disadvantages are the increase of train mass (approximately 2%); additional space due to energy storage container and bidirectional step/up converter, and high density supercapacitors cost.

About the second solution, the energy management of single substation has to be coordinated in real time with others on the basis of the power demand requested by vehicles traffic. So a hierarchized power control permits to reduce the power peak demand of each substation but

978-1-4244-1741-4/08/$25.00 ©2008 IEEE

without reduction of current line losses and line voltage stabilization. In latest case additional space and weight are not binding.

The last solution can be adopted just for new substations where it can be represented the cheaper solution. Nevertheless this implies an energy management on AC side of the substations. This is not always permitted by energy committing.

However the optimal solution (tradeoff between technical and economic) must be searched for on the basis of single rail services in terms of type of rail vehicles (LRV Tram, Metro, ecc...) and type of infrastructure (substation, catenary, track).

Nevertheless the considerations about the improvement of the energy recovery when there are supercapacitors on board on railway vehicles appear to be missing in the technical literature up to now published. For this reason, the author has been first presented the model of the whole system (electrical drive with induction motors, the contact line and the supercapacitors) by means of their equivalent electrical circuit and a control strategy. Numerical simulations of the electrical drive of Metropolitan Vehicles equipped on board of supercapacitors set have been presented in order to show the capability to supply power peaks and the quantification of the energy recovery referred to a standard duty cycle of the rail vehicle.

At the end experimental results on scale system are presented in order to validate the proposed control strategy.

II. MATHEMATICAL MODEL

Standard traction drives are supplied by the catenary by means of pantographs. The catenary can be supplied either in dc or in ac with different voltage levels, depending on the country where the railway is located. In case of dc lines, the catenary is directly connected to the inverters with filter capacitors, whereas in case of ac lines there is also a rectifier stage in cascade. Therefore the traction drive is always supplied by a dc line. The line is modeled considering its equivalent circuit at the pantograph terminals. For simplicity, the line has constant voltage, v_d, constant resistance, r_d, and constant inductance, l_d. The supercapacitors are modeled by their equivalent series electrical circuit with a series of their capacitance, C_{sc}, and resistance, r_{sc}. The supercapacitors set operates in dc current and is connected to the dc link of the drive. However, since the supercapacitor voltage changes with their state of charge, it is necessary an intermediate dc/dc converter capable of harmonizing voltages and regulating

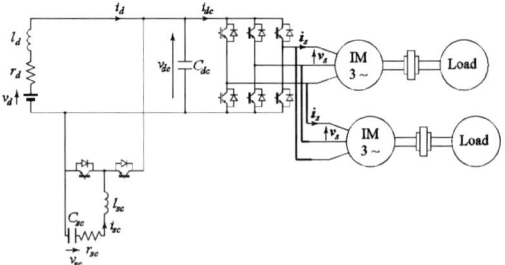

Fig. 1 – Outline of the equivalent electrical circuit of the system

power flows of supercapacitors. This converter has to boost the voltage of supercapacitors and be bidirectional

for cycling them. The system configuration is represented by the equivalent electrical circuit shown in fig.1.

From this circuit and considering the switches ideal, the mathematical model of the system can be derived as follows:

$$
\begin{cases}
v_d = r_d i_d + l_d \dfrac{di_d}{dt} + v_{dc} \; ; \\[2mm]
i_d + i_{sc}\, d(t) = i_{dc} + C_{dc} \dfrac{dv_{dc}}{dt} \; ; \\[2mm]
v_{sc} = r_{sc} i_{sc} + l_{sc} \dfrac{di_{sc}}{dt} + v_{dc} d(t) \; ; \\[2mm]
i_{sc} = -C_{sc} \dfrac{dv_{sc}}{dt} \; ; \\[2mm]
v_s = \dfrac{2}{3} m(t) v_{dc} e^{jn(t)} \; ; \\[2mm]
i_{dc} = \Re\{ m(t) i_s\, e^{-jn(t)} \} \; ; \\[2mm]
v_s = r_s\, i_s + l_s \dfrac{di_s}{dt} + \dfrac{d}{dt} L_m \big(i_s + i_r' e^{jp\theta_r} \big) \; ; \\[2mm]
0 = r_r' i_r' + l_r' \dfrac{di_r'}{dt} + \dfrac{d}{dt} L_m \big(i_s e^{-jp\theta_r} + i_r' \big) \; ; \\[2mm]
J \dfrac{d\omega_r}{dt} = \dfrac{3}{2} \Im m \{ i_s\, \breve{i}_r'\, e^{-jp\theta_r} \} - T_r \; ; \\[2mm]
\dfrac{d\theta_r}{dt} = \omega_r
\end{cases}
\tag{1}
$$

where the meaning of the parameters is given at the end of the paper, $d(t)$ is the switching function of the dc/dc converter, $m(t)$ and $n(t)$ are the switching functions of the inverter in case of SVM modulation. These functions are equal to 0 or 1 at the instant t, depending on the state of the switches and diodes.

III. CONTROL STRATEGY

The control strategy aims to minimize the power peaks of the catenary and to stabilize the line voltage. The idea is to hold a constant reference value for the line current during braking and accelerating operations in order to prevent current peaks. In addition the control unit has to regulate the level of charge of supercapacitors. This level has to be set in order that the supercapacitors can supply extra power demands and can reserve a storage capability for braking energy. Only the exceeding energy is, hence, given back into the catenary or dissipated into brake resistor.

The main difficulties of this strategy are basically due to the proper determination of the line current set-point. A current too low involves the full charge of the supercapacitors before the end of the braking process and the recovery of the remaining energy with increasing of catenary voltage and current losses; a current set-point too high implies an insufficient utilisation of the storing capacity of the supercapacitors and non-optimal average charge-discharge efficiency. The optimal current set-point has to be determined depending on the state of charge of the supercapacitors and the amount of energy to be recovered during the braking process. However, it is very difficult to estimate correctly the amount of energy because it depends strictly on the velocity profile of the vehicle.

After a load peak, the control unit has to provide to recharge partially the supercapacitor until a voltage set-point value. This value has to be a compromise between the necessities of energy enough for covering load peaks and the necessities of capacity enough for energy recovery. A simple criterion is to regulate the voltage set-point as a decreasing function of the vehicle velocity, which is directly related to the kinetic energy of the moving mass. For a low velocity, it is more logical to expect an accelerating than a braking; in addiction, only a little amount of energy is available in a braking process. Therefore, an optimised control strategy has to set a high level of charge of the supercapacitors. On the contrary, for higher velocity it is more logical to expect a braking and then it is more suitable to reserve capacity for the energy recovery. For example, if the remaining capacity is set to be proportional to the whole kinetic energy available, the voltage set-point is:

$$\frac{1}{2}C\left(V_{s,\max}^2 - V_{sp}^2\right) = \frac{1}{2}k\,m\,v^2 \Rightarrow V_{sp} = \sqrt{V_{s,\max}^2 - k\frac{m}{C}v^2} \quad (2)$$

where V_{sp}, and $V_{s,max}$ are respectively the set-point and maximum voltage of supercapacitors set, v is the vehicle velocity and k is a constant.

The second aspect of the control strategy is the regulation of the supercapacitors power flows in order to keep constant the line current. It is possible to define on the plane (i_b, v_s) the working area of the control strategy, as it is shown in fig.2. Then, the working operations are defined by the minimum and maximum threshold of the supercapacitor voltage, i.e. $V_{s,min} \leq v_s \leq V_{s,max}$. Outside this zone the supercapacitors set is disconnected from the dc-bus. The upper and lower limits of supercapacitors set voltage, $V_{s,max}$ and $V_{s,min}$, have to be chosen on the basis of the amount of energy demand during acceleration and braking operations. In particular, they define the maximum energy recovered and provided to the line during load cycle. Control automatically comes back in the working area, as soon as the motor current is lower than the reference line current and $v_s < V_{s,min}$, or the motor current is greater than the reference line current and $v_s > V_{s,max}$. In the range $V_{s,min} \leq v_s \leq V_{s,max}$, control guarantees that supercapacitors set is not charged over the upper threshold and discharged under the lower threshold. Finally, the line current limit defines the charge and discharge areas. In particular, when the load current is greater than the reference current, the supercapacitors set supports the catenary supplying the peak power demand of the motor. If the motor current is lower than reference, the catenary charges the supercapacitors. The final state of charge of supercapacitors is defined by the voltage level expressed by the relation (2) between the supercapacitors set voltage and the vehicle speed. In addition the control system verifies also other condition like: the supercapacitors set current has to be lower than the maximum current of dc-dc converter. If an over-current occurs, the control system switches to a current control of supercapacitors set.

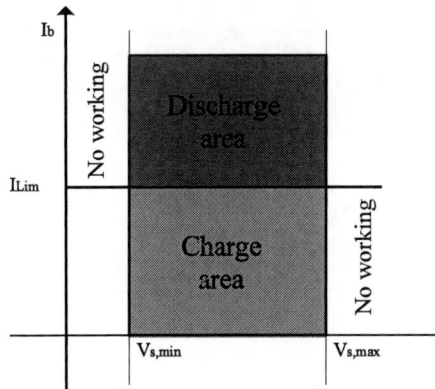

Fig.2 – working area of control strategy

IV. NUMERICAL RESULTS

The behavior of the railway vehicles equipped with supercapacitors on board has been simulated by means of the mathematical model previously presented. The induction motors are controlled by a field oriented control with feedback of the speed and indirect method of estimation of the rotor flux position. The main parameters of the motor are reported in Tab.I. The control strategy keeps constant the rotor flux up to the base speed, n_b correspondent to a train speed of 50 km/h. For speed greater than the base speed, the flux is decreased in order to keep constant the output power delivered by the motor.

TABLE I. MAIN PARAMETERS OF THE TRACTION MOTORS

P_n [kW]	V_n [kV]	n_b [rad/s]	r_s [Ω]	r'_r [Ω]	l_s [mH]	l'_r [mH]	L_m [mH]	J_m [kgm²]
300	1.5	153	0.13	0.09	1.81	2.62	72.8	3.9

The following parameters of the system have been considered for the simulation: the translating mass of the train, referred to a single wheel set of the bogie, is equal to 46000 kg; the maximum speed of the train is 70 km/h; the rated torque of each induction motor is 2000 Nm; the line voltage of the dc catenary is equal to 2.4 kV.

The supercapacitors set has been sized taking into account the energy which can be actually recovered from regenerative braking. At the speed of 70 km/h, the kinetic energy of the train, referred to a single wheel set of the boogie, is about 2.4kWh. is made of 5 modules connected in series and 2 strings in parallel. The single module is characterized by a rated operating voltage of 390 V with an energy available of 282 Wh and maximum current of 950 A, referred to the half of the operating voltage. The features of the modules are referred to a HTM390 module produced by Maxwell, see Tab.II. The size of the supercapacitors set takes into account the maximum available kinetic energy of the train and the maximum power requested by the electrical drive during the electrical braking operations. The motion resistance takes into account the friction force due to the air and the wheel rail contact.

In the simulation, the train starts from standstill, accelerates up to 70 km/h, cruise for few seconds and stops with regenerative braking. The results of the simulation are shown in fig. 3. The speed and the torque developed by each motor are shown in fig.3a. The set of supercapacitors helps the contact line in the supplying of the drive during the acceleration. The control of the dc/dc converter implies that supercapacitors discharge themselves and limit the line current (fig. 3b).

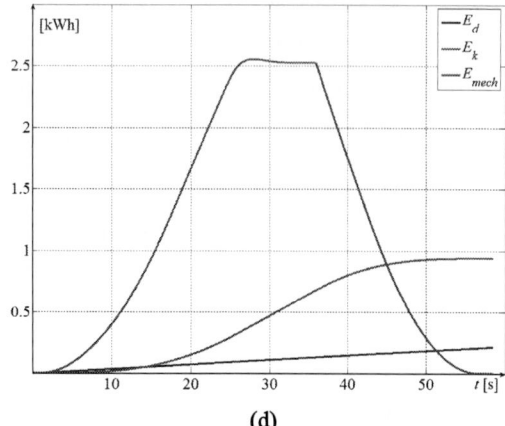

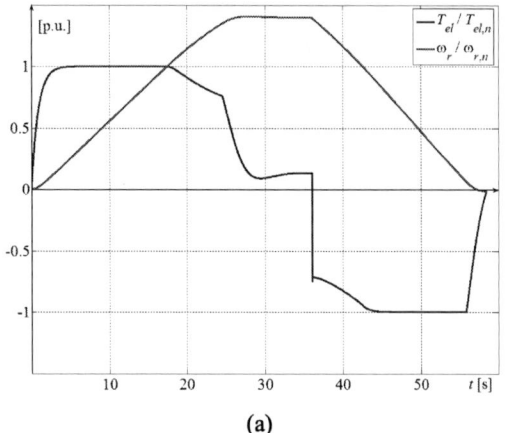

(a)

Fig.3 - numerical results for a traction cycle: motor torque and speed (a), currents supplied by the catenary, the supercapacitors and the dc-side of the inverter (b), supercapacitors set voltage (c), and diagram of energies (d).

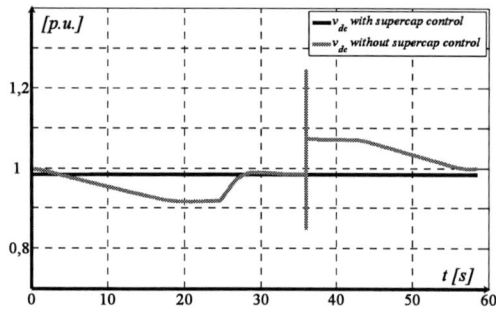

Fig.4 – dc line voltage with and in absence of supercapacitor control

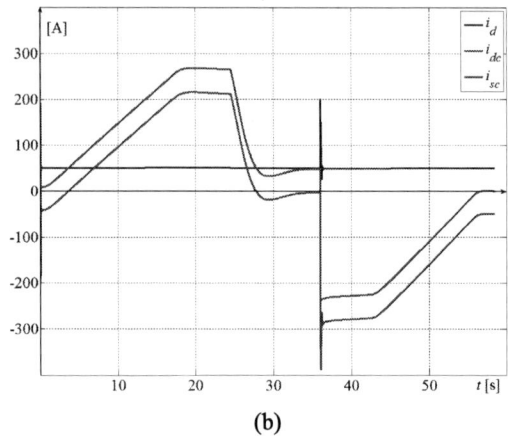

(b)

TABLE II.
FEATURES FOR HTM390 MAXWELL SUPERCAPACITOR MODULE

V_n [V]	390
C [F]	17.8
R [m□]	65
I_{max} [A]	950
E [Wh]	282
W [kg]	165
V[mm^3]	1200x629x288

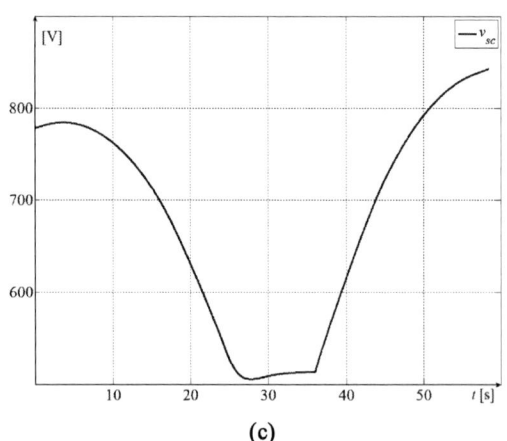

(c)

This limit current is set in order that the line supply only the average power to the drive. During the braking, the kinetic energy of the train is recovered and stored into the supercapacitors. In fact, the state of charge of supercapacitors increases because their voltage increases when the braking starts, as fig. 3c shows. In fig. 3d are finally shown the energies involved. The energy supplied by the line increases linearly, since v_d is constant and i_d is almost constant. This means that the rms power supplied by the line is considerably reduced because the peaks are supplied by supercapacitors. The reduction of the line current peak is equal to 80%, as shown in fig.3b. This implies that the droop line voltage V_{dc} is reduced compared to case of absence of the supercapacitors bank. In fact voltage droop without supercapacitors is lower than 10%, instead with the supercapacitors bank is lower

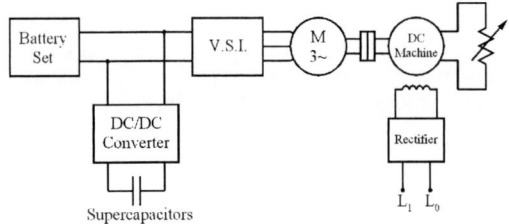

Fig. 5 – Block diagram of the test bench used for experimental tests

than 2% as shown in fig.4. The diagram of fig.4 has been carried out with line resistance and inductance equal to r_d = 0.75 Ω, and l_d =10 μH.

V. EXPERIMENTAL RESULTS

In order to verify the effectiveness of the suggested control technique a sample drive supplied by batteries and supercapacitors has been set-up. Since the impossibility to reproduce into laboratory the moving mass and the power of real rail vehicle, the electromechanical drive has been scaled in the power (1:30). The test-bench used for experimental tests is shown in fig.5. The mechanical load consists of a rotating inertia and a separately-excited dc generator of 25 kW, which simulates the behaviour of the vehicle in terms of torque required at the axis of the motor in different operating conditions. The dc generator supplies an adjustable passive dc load. Since the output power of dc machine depends on the square of the speed, it is possible to simulate the wind friction on the vehicles keeping constant the excitation voltage.

The supercapacitors set are connected in parallel with the battery by means of a three leg full bridge dc/dc converter of 20 kVA. The maximum current allowed by the converter on the supercapacitors side is 150 A. The dc bus supplies the traction drive constituted by a three-phase voltage source inverter and an induction motor of 11 kW. The supercapacitors set consists of two modules in series, each one with a rated voltage of 42 V and capacitance of a 67 F. The battery is lead-acid type for traction applications and has 33 elements of 65 Ah connected in series, reaching a rated bus voltage of 396 V. The battery set wants to simulate the dc source of the catenary.

The remote control of the inverter allows different loads by selecting the motor speed and acceleration.

The performed tests aim to evaluate the contribution of supercapacitors during accelerations of the vehicle and their recovering capability during braking operations. For this reason the supercapacitors voltages, the battery currents and the load currents have been measured by a

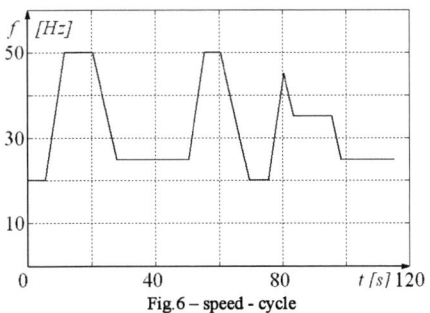

Fig.6 – speed - cycle

data acquisition board. The cycle considered has been shown in fig.6.

The first test has been performed when the rail vehicle is supplied by the battery pack and supercapacitors set. The cycle consists of three load peaks of different length and amplitude. Battery current, load current and supercapacitors set voltage have been depicted on figs.7. The initial state of charge of batteries (S.O.C.) has been set equal to 100%. The control essentially holds constant the battery current to a value equal to 10 A, also when the current load demand is two times greater than the reference value, as shown by figs.7a and 7b. These results points out of the efficiency of the control strategy when the supercapacitors voltage is inside its working area ($V_{s,min}$ = 40 $v \leq v_s \leq V_{s,max}$ = 80 v).

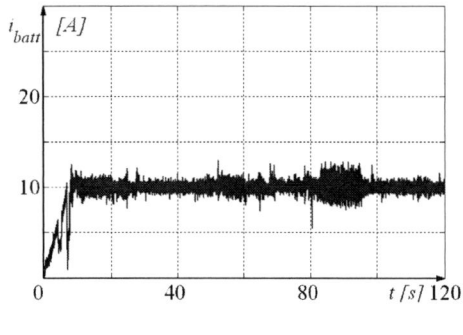

(a)

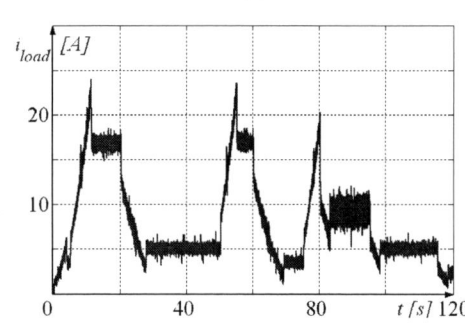

(b)

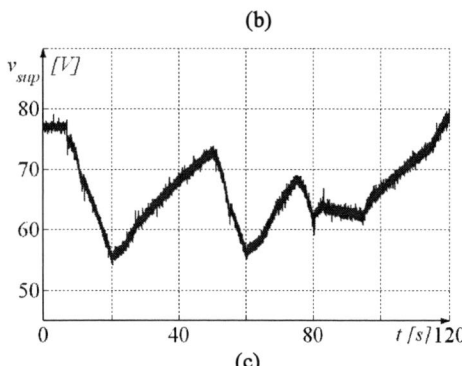

(c)

Fig.7 – Battery current (a), load current (b) and supercapacitors set voltage (c)

In addition, it is shown that during low load operations the battery-pack charges the supercapacitors with a constant current.

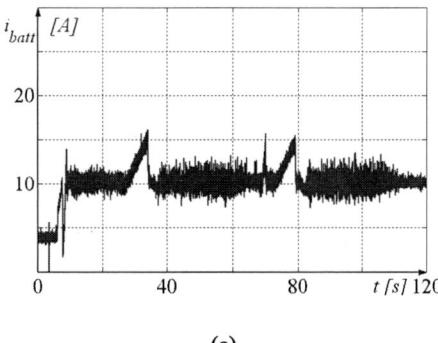

(a)

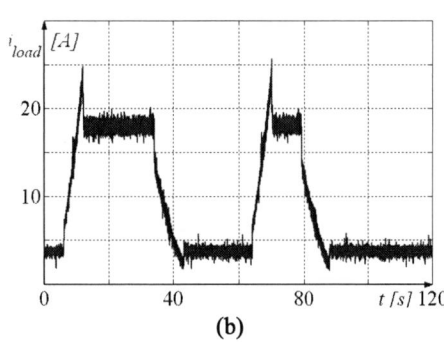

(b)

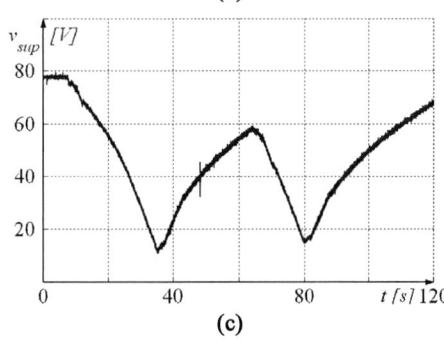

(c)

Fig.8 – Battery current (a), Load current (b) and supercapacitor voltage (c)

A second cycle has been considered with average current requested by the load higher than the battery set-point (10 A), see fig.8b. In this case the supercapacitors voltage reaches and crosses over the minimum value (V_{min} = 40 V), as shown in fig.8c, with consequence increases of battery current during load peak, as shown in fig.8a.

VI. CONCLUSION

The paper has shown why and how the use on board of supercapacitors bank represents a solution technically effective and feasible to recovery energy on board and to reduce the line power peak demands during the accelerating operations.

The use of supercapacitors on board allows to reduce the line current over the 50% (examined case 80%) which causes a reduction of the line voltage drop. It is obvious that supercapacitors storage devices on board of traction vehicles stabilize the catenary voltage. It can be confirmed by comparison of catenary voltages with and without use of supercapacitors in diagram of the fig.4.

These improvements can lead to reduction of infrastructure losses, of committing power and energy cost allowing i.e. an increasing of the distance between substations for the planned new lines; reducing of time intervals between following trains at existing lines; acceptance of longer trains on existing lines.

List of symbols used

i_d	current supplied by the line
i_{dc}	input current of the inverter
i'_r	space vector of the rotor current of the motor referred to the stator winding
i_s	space vector of the stator current of the motor
i_{sc}	supercapacitor current
l_d	equivalent input inductance of the line
l'_r	rotor leakage inductance referred to the stator winding
l_s	stator leakage inductance
l_{sc}	boost inductance of the dc/dc converter
p	pole pair number
r_d	equivalent input resistance of the line
r_{sc}	equivalent series resistance of supercapacitors
r'_r	rotor winding resistance referred to the stator winding
r_s	stator winding resistance
v_d	equivalent input voltage of the line
v_{dc}	input voltage of the inverter
v_s	space vector of the motor voltage v_{sc} supercapacitor voltage
C_{dc}	input capacitance of the inverter
C_{sc}	supercapacitor capacitance
J	rotating inertia
L_m	mutual inductance of the motor
T_r	resistant torque
θ_r	rotor angular position
ω_r	rotor angular speed

REFERENCES

[1] Bocharnikov, Y. V.; Tobias, A. M.; Roberts, C.; Hillmansen, S.; Goodman, C. J.: "Optimal driving strategy for traction energy saving on DC suburban railways", *IET Electric Power Applications (2007)*, vol. 1, issue 5, pp. 675-682;

[2] Adinolfi, A.; Lamedica, R.; Modesto, C.; Prudenzi, A.; Vimercati, S.: "Experimental assessment of energy saving due to trains regenerative braking in an electrified subway line", *IEEE Trans. Power Delivery (1998)*, vol. 13, issue 4, pp. 1536-1542;

[3] Mellitt, B.; Goodman, C. J.: "Simulation study of DC transit systems with inverting substations", *IEE Proc. (1984)*, vol. 131, issue 2, pp. 38-50;

[4] Richardson, M. B.: "Flywheel energy storage system for traction applications", *IEE Int. Conf. Power Electronics, Machines and Drives (2002)*, pp. 275-279;

[5] Lhomme, W.; Delarue, P.; Barrade, P.; Bouscayrol, A.; Rufer, A.: "Design and Control of a supercapacitor storage system for traction applications", *Conference Record of the 2005 Industry Applications Conference (2005)*;

[6] Takahara, E.; Wakasa, T.; Yamada, J.: "A study for electric double layer capacitor (EDLC) application to railway traction energy saving including change over between series and parallel modes", *Proc. of the Power Conversion*;

[7] Stanislaus Pagiela, Michael Steiner, Markus Klohr : "Energy Storage System with UltraCaps on Board of Railway Vehicles" 2007 European Conference on Power Electronics and Applications.

A Multi-Core PC-based Simulator for the Hardware-In-the-Loop Testing of Modern Train and Ship Traction Systems

Christian Dufour, Guillaume Dumur, Jean-Nicolas Paquin, Jean Bélanger

Opal-RT Technologies, 1751 Richardson, suite 2525, Montreal, Canada

Abstract— Today, the development and integration of train and ship controllers is a more difficult task than ever. Emergence of high-power switching devices has enabled the development of new solutions with improved controllability and efficiency. It has also increased the necessity for more stringent test and integration capabilities since these new topologies come with less design experience on the part of the system designers. To address this issue, a real-time simulator can be a very useful tool to test, validate and integrate the various subsystems of modern rail vehicle devices. This paper presents such a real-time simulator, based on commercial-off-the-shelf PC technology, suitable for the simulation of train and ship propulsion devices.

The requirements for rail/water vehicle test and integration reaches several levels on the control hierarchy from low-level power electronic converters used for propulsion and auxiliary systems to high-level supervisory controls. This paper places great emphasis on the real-time simulation of several high-power drives used for train and ship propulsion, including a multi-induction machine drive, a three-level GTO - PMSM drive and a high-power thyristor-based converter - synchronous machine drive. All models are designed first with the SimPowerSystems blockset and then automatically compiled and run on commercial PCs under RT-LAB. Interfaces to I/O are also made at the Simulink model level without any low-level coding required by the user. Supervisory control integration and testing can also be made using the RT-LAB real-time simulator.

The other objective of this paper is to demonstrate that HIL testing of complex drives, such as the those found on trains, can be done using commercial-off-the-shelf (COTS) software and hardware and model-based design techniques that only require high-level system models suitable for system specifications down to controller test and final system integration.

I. INTRODUCTION

The integration, test and verification of modern train and ship systems represent a serious challenge. Currently, because of the risks involved, it is not conceivable to integrate these kinds of systems with direct subsystem interconnection.

Modern design approaches mitigate these risks through the extensive use of technologies like Hardware-In-the-Loop (HIL) simulation. HIL simulation technologies enable more gradual integration, while diminishing the risk and costs of such projects. Also, more elaborate test coverage can be conducted than is possible using analog

prototypes because of the safety operational limits of real devices.

Model-based design is an approach that puts the system model at the center of the design process[7]. With this approach, the specification, controller prototype design, coding and integration tests are based on a set of reference models. At the integration stage, this approach makes extensive use of HIL simulators, with a number of objectives that are directly related to the control hierarchy of the complete train system. The control hierarchy of a train system is given in Figure 1.

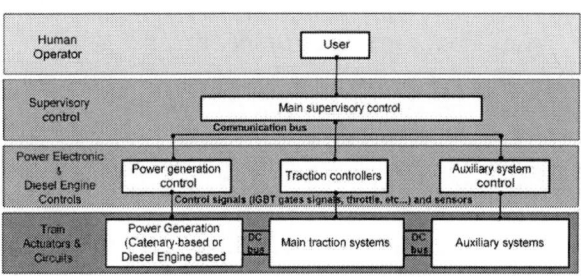

Figure 1   Train control hierarchy

A.   Individual Train Actuators and Circuits Controllers Tests:

The first stage of HIL controller test is to individually verify the controllers. At this stage, a detailed model of the subsystem is used to which the controller under test is attached, while a simplified model is made of the rest of the system. Two types of tests are then conducted:

1)   Open-loop tests: this kind of test is used to verify the functionality of the I/O of the controller by simple excitation/monitoring of the I/Os. It is also used to verify the behavior of the controller in case of short-circuit of the I/Os. In this last case, the controller should detect such conditions and output proper code to the supervisory controller.

2)   Closed-loop tests: controller behavior is tested for its control action on the power devices. The controller is connected to the HIL-simulated power devices in the exact same manner as the real device (IGBT gate signals, current sensors, etc.) For example, the acceleration/deceleration behavior of the induction traction units [4] can be tested with a simplified DC-link model.

978-1-4244-1741-4/08/$25.00 ©2008 IEEE   1475

B. Multi-subsystem Integration Tests:

The different subsystems (generation, propulsion, auxiliary) are electrically connected and may therefore interact with each other. Consequently, the next stage of test/integration is to verify the functionality of the controllers with all system interactions. Basic Supervisory control law can be tested at this stage. The scalability of the simulator is very important in this regard[8].

C. Main Supervisory Control Tests:

The main supervisory control algorithm is tested at this stage. This includes testing in normal conditions with a human operator command (start and accelerate, stopping the train) and in abnormal conditions (communication bus fault or electric fault).

D. User Control Tests:

The real-time simulator can be used to verify the overall conductibility of the train by a human operator, in normal and faulty modes, as well as for operator training. At this point, the user graphical interface becomes important because the human operator's 'I/Os' are mainly their eyes and hands.

II. CHALLENGES AND SOLUTIONS IN REAL-TIME SIMULATION OF COMPLEX ELECTRIC DRIVES

There are several challenges to achieve real-time simulation of large electric drives. The global challenge is to obtain good accuracy using fixed time step solvers and methods. Furthermore, the calculation time of *all* time steps must always be kept under a prescribed value to enable HIL interface of the simulator with external equipment or controllers. The last challenge consists of finding the right simulation platform.

Challenge 1) Keeping the accuracy of simulation with high-frequency power converter.

Given that the simulator is a sampled system, the accuracy of simulation of high-frequency PWM inverter may be compromised if the ratio of simulator sampling frequency to the PWM frequency is too low. Interpolation-capable inverter models are the solution to this problem. These inverter models are part of the RTeDRIVE[5] package from Opal-RT Technologies. This challenge notably exists in the Three-Level GTO-Inverter PMSM Drive of section IV.A.

Challenge 2) Keeping the calculation time of all time steps almost constant to achieve HIL.

The SimPowerSystems default solver takes more time to iterate whenever a switch changes position because it recalculates circuit mode on-line. The ARTEMIS plug-in for SimPowerSystems makes pre-computation of all system state-space matrices in advance to solve this problem

Challenge 3) Keeping the calculation time of large systems relatively low. Power systems are typically simulated with a time step objective typically near 25-50 μs. This objective may be difficult to reach for large networks or drives (a bigger system implies bigger set of equations to solve). The ARTEMIS[5] package provides nice solutions to this problem with distributed parameter line and stublines models that enable the decoupling of the underlying system equations.

The stubline model is particularly interesting in electric drives. The stubline is the equivalent of a distributed parameter line with an exact 1-time step propagation delay and a fully tunable inductance value[2]. It can effectively replace inductances and provide decoupling of the system equations. The stubline can be used to model transformer leakage inductance (section IV.A) or current converter chokes (section IV.B) and increases simulation speed by splitting the system equations into 2 parts.

III. HIL SIMULATION PLATFORM (RT-LAB).

RT-LAB is the real-time simulation software from Opal-RT Technologies. RT-LAB runs almost entirely on commercial-off-the-shelf hardware. The only exception, because of the extreme I/O requirements for electric drives and system applications, is the Opal-RT FPGA-based I/O card. RT-LAB supports distributed simulation through shared memory with 2/4/8/16-CPU (including multi-core CPU technology), AMD- or Intel-based systems, or through PC clusters with InfiniBand or FireWire communication links[8].

The RT-LAB real-time operating system, running on the actual simulation targets, is either QNX from QNX Software System Corp. or RedHawk-Linux from Concurrent Computer Corp.

Most commercial I/O cards are supported with RT-LAB, including cards from Acromag, DDC, Kontron, Measurement Computing, National Instruments, Quanser, RTD, Sensoray, and Softing. However, Opal-RT FPGA cards are preferred for electrical applications because such applications have unusually high switching frequencies. Opal-RT FPGA I/O cards feature 10-ns digital I/O, 1-microsecond D/A converters, and 2-microsecond A/D converters with integrated signal conditioning. XSG support enables users to fully customize I/Os for Opal-RT FPGA cards using the standard Simulink diagram editor.

IV. TRAIN AND SHIP PROPULSION DEVICES

In this section, we give examples of common train and ship propulsion drive configurations for which low-level power electronic controls are to be tested. The first model is a four-induction machine traction unit that can be driven by either an on-board synchronous generator or AC-single phase catenary system. A high-power three-level GTO-based PMSM drive is shown next. The dual-voltage DC-link of the drive is made of a 12-pulse rectifier connected to the grid by a three-phase three-winding transformer. The last drive is a very-high power current converter made of back-to-back thyristor converters (12-pulse rectification and 6-pulse inversion). The converter drives a synchronous machine. Except for the catenaries power feeds, all topologies can be studies in the context of either train or train studies. In some cases, the generator model is replaced by a simple 3-phase source. The description of the various systems is given next:

A. High-Power Three-level GTO-Based PMSM Drive Propulsion System

This type of system involves a power converter feed from a 20-kV three-phase power system which is transformed to lower voltage by 3-winding transformers. A 12-pulse thyristor rectifier is then used to control the DC-link voltages. From the bipolar DC-bus, a three-level neutral-clamped GTO inverter drives a permanent magnet synchronous machine. The machine is rated at 1 MVA with magnet flux of 2.5 Weber.

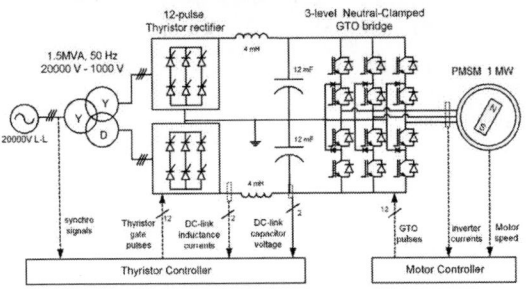

Figure 2   Three-level GTO inverter motor drive

Figure 3 shows the PMSM motor terminal voltages, currents, electric torque and speed during the drive start-up from zero speed to 4 Hz rotation frequency. Figure 4 shows the DC-link voltage and transformer secondary currents during the start-up. A system engineer might want to investigate a method to reduce the DC-link voltage oscillations during the acceleration phase of the test.

Interpolation method requirements of the RTeDRIVE package for the accurate simulation of the PMSM drive can be seen in Figure 5. For the test, the PWM frequency of the drive is 1 kHz, no dead time is applied and the sampling frequency of the model is 40 kHz (Ts=25 μs). For the purpose of the test, interpolation is disabled during the simulation. On Figure 5, one can clearly observe the increased distortion in the current and torque values when interpolation is disabled.

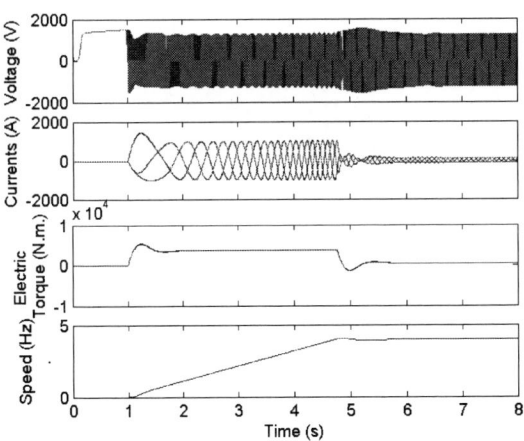

Figure 3   PMSM motor voltages, currents, electric torque and speed

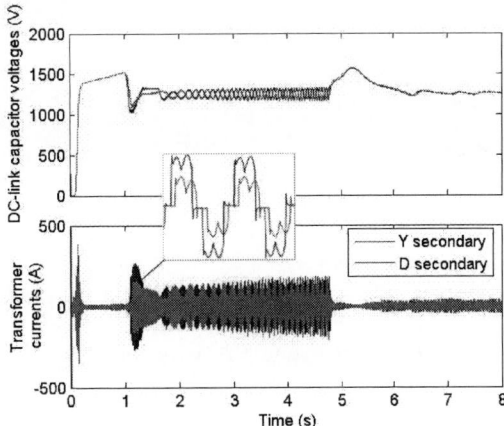

Figure 4   DC-link voltage and transformer currents.

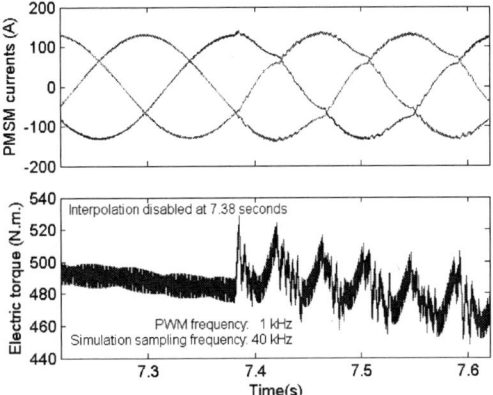

Figure 5   Effect of interpolation on the PMSM Drive accuracy

B. Very-High-Power current converter system

This type of propulsion system involves a direct AC-AC converter based on thyristor switching devices. From an AC primary feed, a step-down transformer feeds a 12-pulse thyristor rectifier. The thyristor rectifier is connected to a 6-pulse inverter through a simple smoothing reactor. The inverter drives a synchronous machine used for propulsion.

This type of drive can handle more power than its IGBT or GTO counterparts. It is however more difficult to control. For example, special techniques must be used to drive a motor at very low speed because the back-EMF is not sufficient to enable inverter thyristor commutation.

A test has been made on this model which consisted of rising the commanded DC-link current from a steady-state value of 0.5 pu to 0.8 pu. For the test, the SM machine works at a fixed speed of 50 Hz and a constant field excitation voltage is applied to the machine. Testing this device in constant speed mode is something rather difficult with real devices (requiring a test bench), but is very easy to achieve in simulation. It enables the verification of torque and current controls. The result of the test is shown in Figure 7. The test shows that the system takes less than 0.1 sec to reach the commanded current.

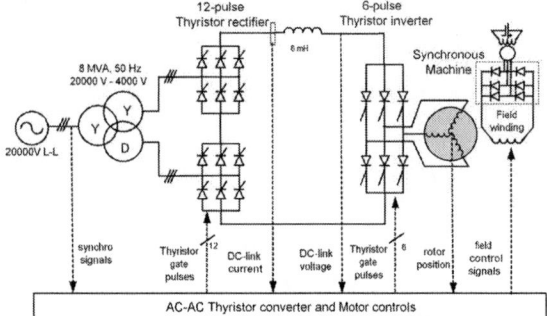

Figure 6   Thyristor-based current converter SM propulsion system

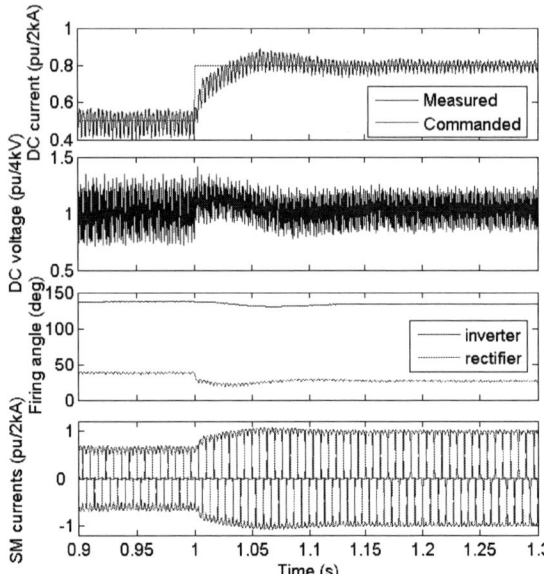

Figure 7   Current converter response to a commanded current step

C.   *Diesel-based power generation system*

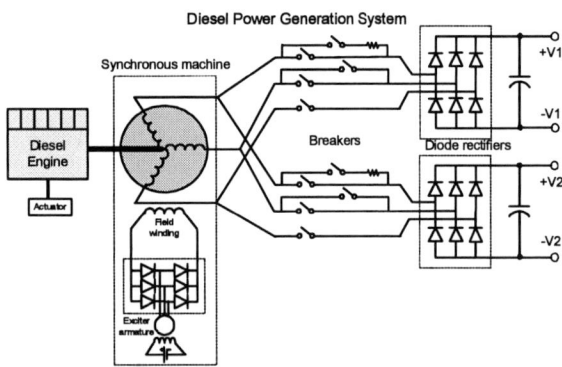

Figure 8   Diesel-based power generation systems

This system is composed of a diesel-engine-driven alternator connected to two 6-pulses diode rectifiers. The diode rectifiers produce 2 DC-link voltages. The alternator field winding is fed with the rectified voltage of an armature voltage induced by an external DC-winding. This approach avoids slip rings as the rectification circuit

is physically inside the alternator rotor. The system is protected by several breakers that control the alternator connection to the DC-links.

D.   *Catenary-based power generation systems:*

For externally powered train from an AC-catenary, this circuit uses two active-front end rectifiers to generate the 2 DC-link voltages. Breakers control the connections of the IGBTs to the catenary's transformer.

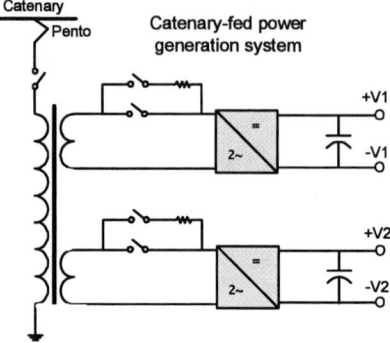

Figure 9   Caterany-based power generation system

E.   *Four Induction Motor Traction system*

This system is composed of 4 induction motors, two on each DC-link, driven by IGBT inverters. Each DC-link also has a chopper to control overvoltages and an additional inverter to feed the auxiliary systems. The challenges of conducting real-time simulation of induction motor drives are described in [2], especially with regards to the correct simulation of high-frequency PWM typically found in these applications. This type of traction system can be connected to either a Diesel-based power generation system or catenary-based power generation systems.

Figure 10   Induction machine traction system

1)   *Validation against EMTP-RV simulation*

In this sub-section, we compare the simulation results of SimPowerSystems and RTeDRIVE inverter models against a well-known reference, EMTP-RV. The model under test is a simple induction machine driven by an IGBT-inverter. The machine is driven in open loop from a DC voltage source of 700V and the inverter is modulated at 60 Hz with PWM (Carrier frequency: 4kHz, modulation index: 0.5). In particular, the use of ARTEMIS enables the induction machine equations to be simulated with the Heun 2nd order solver (ode2 Simulink solver).

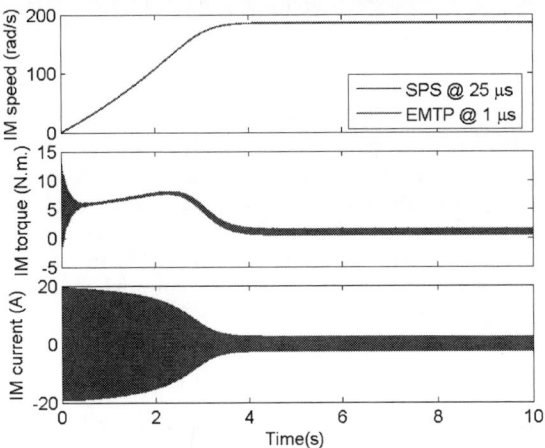

Figure 11 Comparison between EMTP and SPS of mechanical speed, electromagnetic torque and stator current (phase A)

The RTeDRIVE interpolation-capable inverter is used in the SimPowerSystems simulation. Tests have been performed at different time-steps, 1 μs and 25 μs for each package. Open-loop control of the induction machine was preferred only for the sake of simplicity[6].

Figure 11 shows results on the mechanical speed, on the electromagnetic torque and on stator currents for the test respectively. Results are exactly the same between EMTP-RV @ 1 μs and SimPowerSystems @ 25 μs.

V. SUPERVISORY SYSTEM AND HUMAN OPERATOR

The supervisory system test can be conducted by interconnecting various low-level electronic control systems together with a real-time model of the supervisory controller. The modularity of the RT-LAB simulator can be exploited in this manner. Supervisory controls can be implemented on a separate simulator and interfaced, in a asynchronous way, with the real communication protocol used on actual train like TCN[9]. The TCN is a data communication network intended to connect programmable electronic equipment on-board rail vehicles for the support of traction and vehicle control, remote diagnosis and maintenance, and other auxiliary systems. The TCN regroups two types of buses: the Multifunction Vehicle Bus (MVB), which interconnects devices within a vehicle, and the Wire Train Bus (WTB), which interconnects the vehicles in trains of variable composition. It is the equivalent of the CAN bus in automobiles.

The TCN consist of two simple serial buses at the physical level. Therefore, implementation on a real-time simulator should not be difficult.

The human-machine interface is very important when it comes to including the human factor in controllability tests. At this stage, the real-time simulation of the train system in RT-LAB can greatly benefit from the TestDrive graphical interface from Opal-RT. TestDrive is a LabVIEW-based interface to control and monitor real-time simulations, and only requires the LabVIEW runtime engine. For monitoring real-time simulations, the interface has some advantages over standard Simulink: its interface enables easy, point-and-click dynamic selection of signals to view, synchronous display of data on triggered events (like faults) and has built-in Python-based scripting capability. Figure 12 and Figure 13 show a possible human interface for the testing of train traction devices under fault conditions.

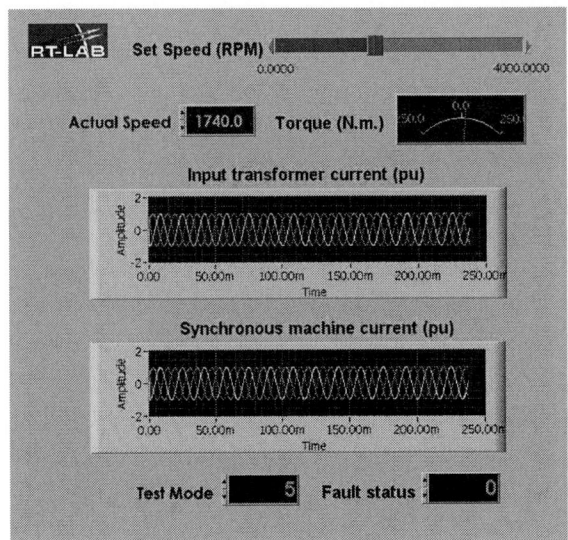

Figure 12 Main user control panel

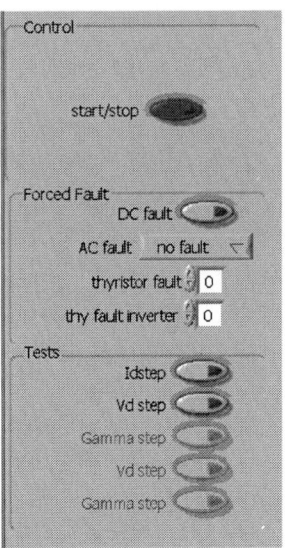

Figure 13 Fault and test control panel

VI. REAL-TIME PERFORMANCE

The various models presented in this paper where simulated on various RT-LAB real-time simulator configurations. The results are shown in TABLE 1.

TABLE I. REAL-TIME PERFORMANCE ON 2.3 GHz MULTI-CORE PCs.

Model	Sample time in μs	Configuration
Diesel-based power generation system + Four Induction Motor Traction system	55	2 PC (4 core each) w/ FireWire connection I/Os: 64 TSDI, 64 TSDO, 32 AO 6 cores used.
High Power Three-level GTO-Based PMSM Drive Propulsion System	26	2 cores used. IOs: 12 TSDI, 12TSDO, 16 AI, 16 AO
Very-High-Power current converter propulsion system	15	2 cores used. IOs: none

All the timing given represents the minimum time step achievable without overruns. An overrun means that the actual time taken to compute one (1) iteration is larger than the simulator time step.

In all cases, the model controllers are simulated with the electric apparatus. When the simulator is connected in HIL mode with an external controller, the internal controllers are no longer required and therefore the simulation speed can be increased in this case.

For testing purposes, it is very convenient to have simulated controllers. With the internal controllers running, it is possible to test the I/O by implementing loopback tests, where the IGBT gate signals (for example) are sent to Digital Output then immediately read back by Digital Inputs and fed to the model. Configurations with I/Os make use of this approach. The TSDI acronym is for 'Time Stamped Digital Input', meaning that the digital input value is measured with a high accuracy clock (100 MHz) by the FPGA-based I/O. The resulting 'Time Stamp'[3] is used by the RTeDRIVE inverter model for interpolation purposes.

CONCLUSION

In this paper, we have presented a real-time simulator capable of conducting real-time simulation of complex train and ship propulsion devices. Real-time simulation of an AC-fed three-level neutral-clamped GTO PMSM drive and a high power thyristor current converter-based synchronous machine drive were demonstrated with real-time simulation results. A third model, a diesel-power drive generation system with a four-induction motor drive was validated against EMTP validation.

The simulator is also suitable for higher-level control hierarchy. For these controllers, the simulator speed is much less critical but the human interface becomes more important. The TestDrive interface, which is based on the LabView runtime engine, provides a powerful tool for building efficient human interfaces.

REFERENCES

[1] I. Boldea, S.A. Nasar, "Electric Drives", CRC Press, ISBN: 0-8493-2521-8

[2] H.W. Dommel (Editor), EMTP Theory Book 2nd edition, MicroTran Power Analysis Corporation, May 1992.

[3] P. Terwiesch, T. Keller, E. Scheiben, "Rail Vehicle Control System Integration Testing Using Digital Hardware-in-the-Loop Simulation", IEEE Trans. On Control Systems Technology, Vol. 7, No. 3, May 1999.

[4] C. Dufour, S. Abourida, J. Bélanger, "Real-Time Simulation of Electrical Vehicle Motor Drives on a PC Cluster", Proceedings of the 10th European Conference on Power Electronics and Applications (EPE-2003), Toulouse, Sept. 2-4, 2003.

[5] C. Dufour, S. Abourida, J. Bélanger,V. Lapointe, "InfiniBand-Based Real-Time Simulation of HVDC, STATCOM, and SVC Devices with Commercial-Off-The-Shelf PCs and FPGAs", 32nd Annual Conference of the IEEE Industrial Electronics Society (IECON-06), Paris, France, November 7-10, 2006

[6] M. Ouhrouche, R. Beguenane, A., Trzynadlowski, J.S. Thongam and M., Dube-Dallaire, "PC-Cluster Based Fully Digital Real-Time Simulation of a Field-Oriented Speed Controller for an Induction Motor", International Journal of Modelling and Simulation, Vol.26, No.3, 2006, pp. 219-228.

[7] S. Abourida, C. Dufour, J. Bélanger, "Real-Time and Hardware-In-The-Loop Simulation of Electric Drives and Power Electronics: Process, problems and solutions", Proceedings of the International Power Electronics Conference (IPEC-Niigata 2005), Niigata, Japan, 2005

[8] L.-F. Pak, O. Faruque, X. Nie, V. Dinavahi, "A Versatile Cluster-Based Real-Time Digital Simulator for Power Engineering Research", IEEE Transactions on Power Systems, Vol. 21, No. 2, pp. 455-465, May 2006.

[9] Juan Carlos Moreno, Eduardo Jesús Laloya and Jesús Navarro, "Line Redundancy in MVB-TCN Devices: A Control Unit Design", IEEE MELECON 2006, May 16-19, Benalmádena (Málaga), Spain, pp. 789-794

Energy Saving Control of Tram Motors Taking Light Signalling and City Disturbances into Account

Stanisław Rawicki

Poznań University of Technology, Institute of Electrical Engineering and Electronics, Poznań, Poland
e-mail: *Stanislaw.Rawicki@put.poznan.pl*

Abstract—**The paper deals with the problems of the tram vehicle control according to the criterion of the minimum electrical energy use. The author has extended the methodics given in literature and has elaborated the algorithm of the energy saving tram traffic allowing for the light signalling and ride perturbations. During the running in the city, tram ride parameters change frequently. Mathematical optimization methods make possible the determination of the control means for the tram run with the minimum energy consumption. The influence of the following traffic disturbances was described: the light signalling, different speed limitations, stays, changes of size of the electrical energy recuperation during the braking and variations of the traction network voltage.**

Keywords—**Traction application, drive, optimal control, driver assistant systems.**

I. INTRODUCTION

The scientific work, connected with problems of the optimization strategies applied to the energy saving ride of the tram vehicle, is being realized in Poznań University of Technology (Poland). Various techniques of the tram run can cause essential differences between the values of the energy use. It generates great interest for application of better methods of the tram control (both with reference to the vehicle and to the electric drive).

Problems of the train control according to the criterion of the minimum energy use are the object of interest of many investigators [1, 4, 5, 7]. The algorithms of the energy saving traffic of tram vehicles, elaborated in literature, are connected only with the ride without the disturbances and can be referred to the separated, straight and horizontal tracks [2, 5, 8]. Except some idealized situations, dynamical changes of city traffic conditions occur [3, 6]. It causes models in literature aren't sufficient for problems of the determination of the tram traffic procedure in accordance with energy use minimization.

The author of this paper has extended the methodics given in literature and has elaborated the algorithm of the energy saving tram traffic allowing for ride perturbations. The following traffic disturbances are of great importance: influence of the light signalling, planned or unplanned speed limitations, intentional or unexpected stay, changes of a size of the energy recuperation during the braking and the variation of the network voltage. Moreover the semi-automatic control of the tram with the object of achieving the electric energy use minimization has been proposed. If the driver decides that the safety conditions are fulfilled, he can switch on the system of the automatic guidance of the tram according to the algorithm of the minimum energy consumption. After the traffic disturbance ending, the appropriate programming of the subsequent vehicle ride makes possible also the liquidation of the delay in relation to the time-table.

II. MODELLING OF TRAM VEHICLE TRAFFIC

In this chapter it's assumed that the tram vehicle contains direct current series driving motors supplied from choppers. The dynamic run of the tram vehicle can be described by the basic formula:

$$k_W m \frac{dv}{dt} = F_P - W(v) \ , \qquad (1)$$

where m is the vehicle mass, k_W is the factor of rotating masses, v denotes the vehicle speed, F_p is the tractive force, $W(v)$ describes the motion resistances. The relation between the force F_P and the useful motor torque T_U is the following:

$$F_P = \frac{n_M T_U z \eta}{r} \ , \qquad (2)$$

where n_M is the number of driving motors, z presents the transmission ratio, η is the gear efficiency, r denotes the radius of the driving wheel. In accordance with the Cooper formula, the motion resistances depend on the vehicle speed within the trinomial square:

$$W(v) = w(v)m = (p + qv + sv^2)m \ , \qquad (3)$$

where $w(v)$ – the unitary motion resistance, p, q, s – the constant factors. In comparison with the electromagnetic torque T, the useful motor torque T_U is smaller by the torque T_{mech} of the mechanical losses and the mechanical torque T_{Fe} of the iron loss:

$$T_U = T - T_{mech} - T_{Fe} \ . \qquad (4)$$

978-1-4244-1741-4/08/$25.00 ©2008 IEEE

The torque T and electromotive force E are calculated by the fundamental formulae:

$$T = k\,\phi\,I_a , \qquad E = k\,\phi\,\omega , \qquad E = U - R\,I_a , \qquad (5)$$

where: k – constant, ϕ – the magnetic flux, I_a – the armature current, U – the supply motor voltage, R – the resistance of the motor circuit. The multisegment approximation of the nonlinear magnetization curve of the DC series motor is here assumed.

The electric driving system has 3 characteristic parts within the framework of the starting stage. At supply from the converter system, during the first part of the vehicle starting the progressive increase of the motor voltage occurs; the control system maintains the determinate, constant value of the motor current I. On the basis of the approximation function, the magnetic flux Φ can be found for the given value of the field current $I_f = I$. In this stage, we calculate the motor voltage U (as a function of the speed) which is the sum of the voltage drop RI and the electromotive force depending on the magnetic flux Φ and the speed ω. During this first stage of the starting, the electromagnetic torque T is constant. While the second part of the starting at the constant motor voltage, the change of the field-weakening coefficient k_f can be realized either by jump way or fluently at the application of the modern converter system. In the second case, the network current and the electric power are constant; at the beginning we calculate the magnetic flux as a function of the velocity. In the third starting part, the motor current decreases during the velocity increase. For the known value of the voltage U, the coefficient k_f and the known magnetization characteristic, the initial problem is connected with the numerical calculation of the field current I_f and the magnetic flux ϕ.

The electrical energy En consumed by traction motors can be calculated by the following integration of the instantaneous power in relation to the time:

$$En = n_M \int_{t1}^{t2} u\, i\, dt . \qquad (6)$$

Mathematical relations describing the tram vehicle traffic contain many nonlinearities as a result of properties of motion resistances, in consequence of torques caused by iron and mechanical losses and owing to the phenomenon of the magnetic circuit saturation in the direct current series motors. Determination of the run algorithm ensuring the minimization of the electric energy consumption can be realized by application of numerical method of differential equations solving and use of optimization procedure. Calculation process must take into consideration many constraints. The important adhesion effect is connected with an existence of a limiting force acting in the wheel circumference. The following relation must be here fulfilled:

$$F_P \le \mu\, m\, g , \qquad (7)$$

where μ is the adhesion coefficient, g is the acceleration of gravity. Among others, the factor μ depends on vehicle speed. Within the adhesion phenomenon, the maximum permissible acceleration a of the tram is the following:

$$a \le \frac{\mu(v)\, g - w(v)}{k_W} . \qquad (8)$$

Additionally to ensure comfort for passengers, the tram ride manner must take into account admissible acceleration a_{com}:

$$a \le a_{com} . \qquad (9)$$

The tram speed can not exceed the permissible value V_{max}:

$$v \le V_{max} . \qquad (10)$$

For the case of the tram ride with traffic disturbances, the run between 2 stops includes ns starting phases (the starting is denoted here by s), also with the initial speed not equal zero, ncs stages of the running with the constant speed (symbol: cs), nc coastings (mark: c) and nb brakings (notation: b), sometimes at the final velocity unequal zero. Duration (Ts_i, Tcs_j, Tc_k, Tb_l) of the stages of the traffic can not have the negative value; lengthes (Ls_i, Lcs_j, Lc_k, Lb_l) of the separate segments of ride between 2 tram stops must be positive or equal zero:

$$Ts_i \ge 0, \ Ls_i \ge 0, \quad i = 1, 2, ..., ns , \qquad (11)$$

$$Tcs_j \ge 0, \ Lcs_j \ge 0, \quad j = 1, 2, ..., ncs , \qquad (12)$$

$$Tc_k \ge 0, \ Lc_k \ge 0, \quad k = 1, 2, ..., nc , \qquad (13)$$

$$Tb_l \ge 0, \ Lb_l \ge 0, \quad l = 1, 2, ..., nb . \qquad (14)$$

After disappearance of the traffic disturbance, the ride with the object of achieving the liquidation of the delay in relation to the time-table ought to be realized at the same total tram ride time T and the identical total distance L:

$$\sum_{i=1}^{ns} Ts_i + \sum_{j=1}^{ncs} Tcs_j + \sum_{k=1}^{nc} Tc_k + \sum_{l=1}^{nb} Tb_l = T , \qquad (15)$$

$$\sum_{i=1}^{ns} Ls_i + \sum_{j=1}^{ncs} Lcs_j + \sum_{k=1}^{nc} Lc_k + \sum_{l=1}^{nb} Lb_l = L . \qquad (16)$$

III. RESULTS OF CALCULATIONS AND ANALYSIS

A. Technical Data of Tram Vehicle

The computations have been done for the improved version of the tram 105N. This tram contains the choppers feeding 4 identical driving dc series motors of the total power 160 kW. The nominal data of the tram vehicle are: the traction network voltage: 600 V, total length: 13.5 m, tare mass: 16500 kg, nominal load: 8750 kg, rolling diameter of the wheel: 0.654 m, transmission ratio: 7.16, the maximum permissible speed: 72 km/h. The rated data of the driving motor are: the power: 40kW, power at one-hour duty: 41.5 kW, the voltage: 300 V, the current: 150 A, the speed: 1890 rev/min, the efficiency: 88 %, degree of excitation (with the fluent control by power electronics converter): 1.0-0.63.

The part of calculation results is here presented for the tram mass m = 22000 kg. It corresponds with the passengers number equal 80 (64% in relation to the nominal load). It was assumed that the average braking force was 23000 N for every vehicle speed. All cases of the traffic disturbances were connected with events in the ride segment I. For this segment, the distance between 2 tram stops was equal 1000 m and the primary traffic time was 100 s according to the time-table (the same average ride speed). The route between the neighbouring stops contains 1 cross-roads which beginning is 400 m from the first stop; the tram route length within the crossing is equal 60 m. The cross-roads possess the light signalling.

B. Full Privilege for Tram within Light Signalling

For results presented in Figs. 1-8, the tram ride is realized without disturbances; the vehicle is here fully privileged at the cross-roads within the light signalling (always the green light for the tram). The minimization of the energy use of the tram vehicle is possible on the ground of the suitable traffic control. Determination of the optimum duration of the starting, the runnig with the constant speed, the coasting and the braking is here necessary. Figs. 1-3 give the values of boundary speeds (diagram points) for succeeding traffic stages; moreover the recuperation coefficient kr is here the same and equal 0.5 (the factor kr determines what part of the energy is recuperated during the braking). Figs. 1-5 illustrate the influence of the coasting distance on the value of the energy consumed by the tram in spite of the identical average vehicle speed. Generally in some cases, the value of the recuperation coefficient kr can also modify the coasting length in the energy saving ride algorithm. At the recuperation factor $kr = 0.5$, the results for the extremal values of the distance sc (coasting), the distance scs (running with the constant speed) and the total energy use En are shown as a function of the terminal speed v of the vehicle starting within the following statement:

v [m/s]	11.16	14.55
sc [m]	0	758
scs [m]	869	0
En [kWh]	1.145	1.006

For different values of the recuperation factor kr, Figs. 6-8 present the total electric energy En as a function of the terminal speed v of the starting. We introduce the definition of energy increment en as difference between the maximum energy use $Enmax$ and the minimum

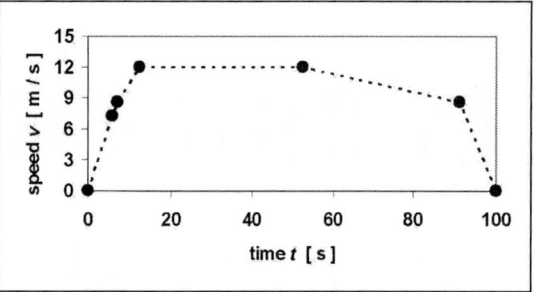

Fig. 1. The tram vehicle ride without disturbances; full privilege at the cross-roads; typical tram run with 4 stages: the starting, the running with the constant speed, the coasting and the braking; the recuperation factor $kr = 0.5$; the energy use $En = 1.023$ kWh.

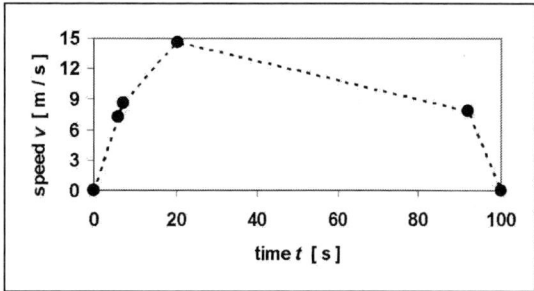

Fig. 2. The tram vehicle ride without disturbances; the vehicle is fully privileged at the cross-roads within the light signalling; the recuperation factor $kr = 0.5$; the minimum electrical energy consumption $Enmin = 1.006$ kWh.

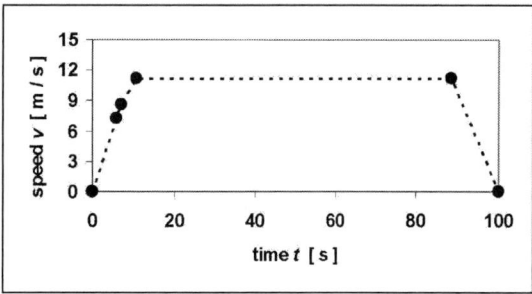

Fig. 3. The tram vehicle ride without disturbances; the vehicle is fully privileged at the cross-roads within the light signalling; the recuperation factor $kr = 0.5$; the maximum electrical energy consumption $Enmax = 1.145$ kWh.

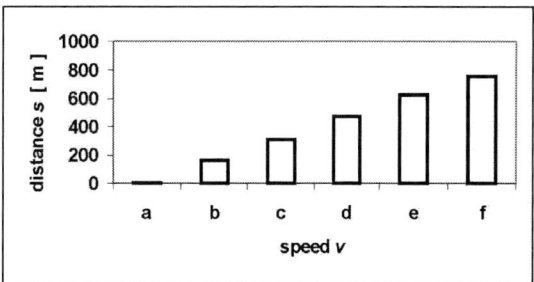

Fig. 4. The tram vehicle ride without disturbances; the recuperation factor $kr = 0.5$; the distance s of the coasting as a function of the terminal speed v of the starting: a) 11.16 m/s, b) 11.30 m/s, c) 11.70 m/s, d) 12.50 m/s, e) 13.50 m/s, f) 14.55 m/s.

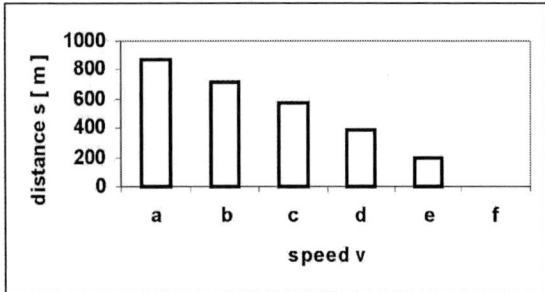

Fig. 5. The tram ride without disturbances; the recuperation factor $kr = 0.5$; the distance s of the run with the constant speed as a function of the terminal starting speed v: a) 11.16 m/s, b) 11.30 m/s, c) 11.70 m/s, d) 12.50 m/s, e) 13.50 m/s, f) the distance $s = 0$ for the speed $v = 14.55$m/s.

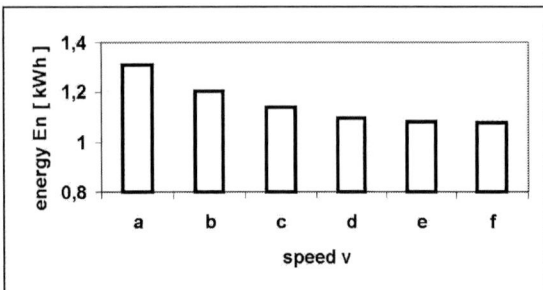

Fig. 6. The tram vehicle ride without disturbances; the recuperation factor $kr = 0$; the total electric energy En as a function of the terminal speed v of the starting: a) 11.16 m/s, b) 11.30 m/s, c) 11.70 m/s, d) 12.50 m/s, e) 13.50 m/s, e) 14.55 m/s.

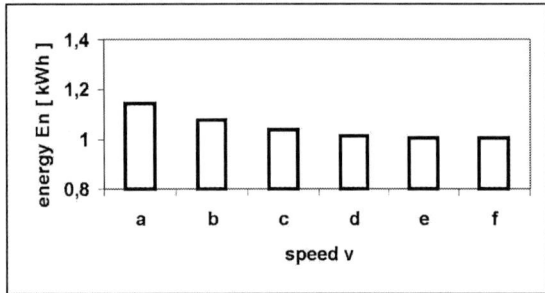

Fig. 7. The tram vehicle ride without disturbances; the recuperation factor $kr = 0.5$; the total electric energy En as a function of the terminal speed v of the starting: a) 11.16 m/s, b) 11.30 m/s, c) 11.70 m/s, d) 12.50 m/s, e) 13.50 m/s, f) 14.55 m/s.

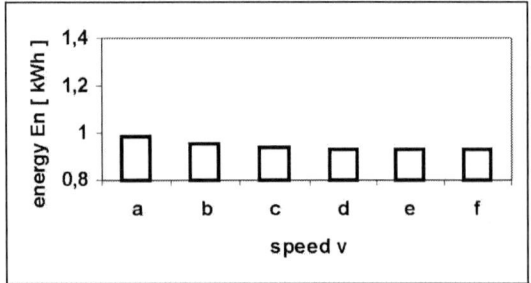

Fig. 8. The tram vehicle ride without disturbances; the recuperation factor $kr = 1$; the total electric energy En as a function of the terminal speed v of the starting: a) 11.16 m/s, b) 11.30 m/s, c) 11.70 m/s, d) 12.50 m/s, e) 13.50 m/s, f) 14.55 m/s.

energy consumption $Enmin$ with reference to $Enmin$ in percentages. At various values of the recovery coefficient kr, the maximum energy $Enmax$ and the minimum energy $Enmin$ are given in the specification:

kr [-]	0	0.5	1.0
$Enmin$ [kWh]	1.080	1.006	0.929
$Enmax$ [kWh]	1.308	1.145	0.982
en [%]	21.1	13.8	5.71 .

It was illustrated that the best control of the tram vehicle made possible to achieve the energy saving even to about 21 %.

C. Additional Tram Stop Owing to Red Signalling Light

For examples shown in this subsection, the priority for the tram ride at the green light has been removed. The tram was forced to brake in front of the cross-roads and to make the stop of the time 8 s (the vehicle speed equal zero). For the initial part of the ride in Fig. 9, the time function of the velocity is identical in comparison with the traffic without disturbances because it is assumed here that the tram ride is at the start realized according to the criterion of the minimum energy use and the necessity of the additional vehicle stop appeared as the traffic perturbation. Liquidation of the traffic delay was here planned within the sum of the roads of the segments I and II. The segment II has the length 850 m and the normal, fluent tram run was here first planned during the time 85s (the same primary, average speed in segments I and II).

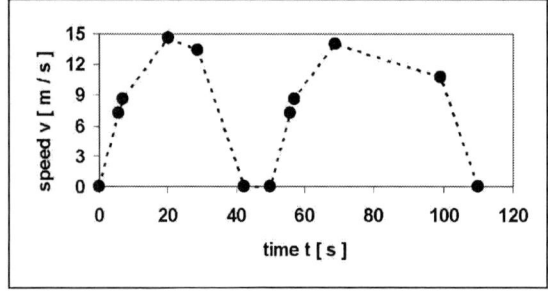

Fig. 9. The tram was forced to brake in front of the cross-roads and make the stop of the time 8 s; running with the minimum energy use at the recuperation factor $kr = 0.5$; for the second part of the traffic (after tram stop), the minimum energy consumption $Enmin = 0.824$ kWh.

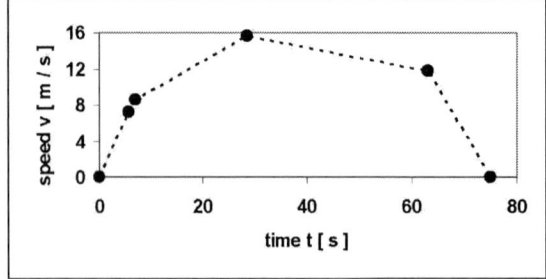

Fig. 10. The ride in the segment II with the aim of attaining liquidation of the traffic delay caused by the stop in the segment I. Running with the minimum energy use at the recuperation factor $kr = 0.5$; energy consumption $Enmin = 1.139$ kWh.

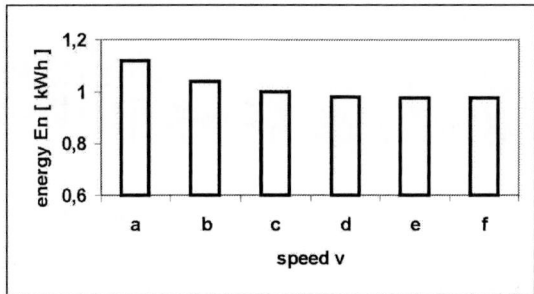

Fig. 11. The tram ride in the 2nd part (60 s) of the segment I after green light obtainment; the recuperation factor $kr = 0$; the electric energy En as a function of the terminal speed v of the renewed starting: a) 12.46 m/s, b) 12.60 m/s, c) 12.91 m/s, d) 13.34 m/s, e) 13.64 m/s, f) 13.96 m/s.

For the energy saving traffic determined on the basis of the optimization procedure, after the additional stop of 8 s the new total ride time for the segments I (Fig. 9) and II (Fig. 10) was equal suitably: 110 s and 75 s (together also 185 s – the identical sum time in comparison with the tram ride without disturbances).

Figs. 11-16 illustrate the great influence of suitable choice of the tram vehicle control on values of the electrical energy En consumption. For different values of the recuperation factor kr, these figures present the energy En as a function of the terminal speed v of the starting. For the 2nd part of the segment I and various values of recovery coefficient kr, the most characteristic results of calculations of the maximum energy $Enmax$, the minimum energy $Enmin$ and the energy increment en are given in the following list:

kr [-]	0	0.5	1.0
$Enmin$ [kWh]	0.974	0.824	0.675
$Enmax$ [kWh]	1.118	0.916	0.714
en [%]	14.8	11.2	5.78 .

It was shown that the suitably chosen control strategy for the tram vehicle enables here the energy saving even to about 15 %. Consistently in like manner, the above energies for the segment II are presented in the set:

kr [-]	0	0.5	1.0
$Enmin$ [kWh]	1.313	1.139	0.966
$Enmax$ [kWh]	1.515	1.267	1.018
en [%]	15.4	11.2	5.38 .

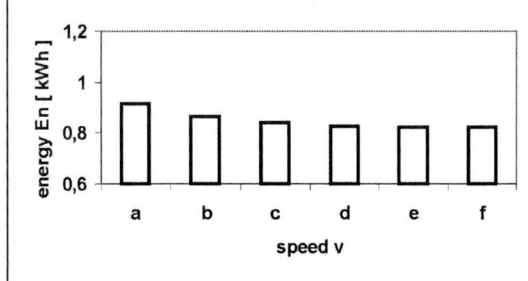

Fig. 12. The tram ride in the 2nd part (60 s) of the segment I after green light obtainment; the recuperation factor $kr = 0.5$; the electric energy En as a function of the terminal speed v of the renewed starting: a) 12.46 m/s, b) 12.60 m/s, c) 12.91 m/s, d) 13.34 m/s, e) 13.64 m/s, f) 13.96 m/s.

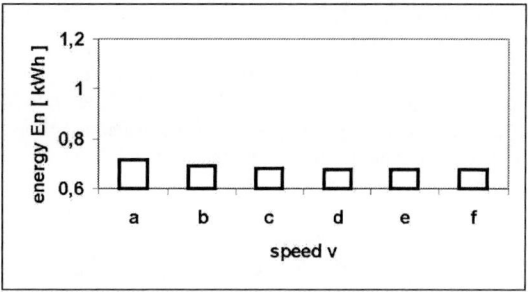

Fig. 13. The tram ride in the 2nd part (60 s) of the segment I after green light obtainment; the recuperation factor $kr = 1$; the electric energy En as a function of the terminal speed v of the renewed starting: a) 12.46 m/s, b) 12.60 m/s, c) 12.91 m/s, d) 13.34 m/s, e) 13.64 m/s, f) 13.96 m/s.

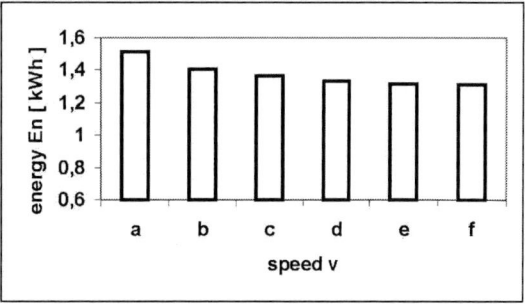

Fig. 14. The tram ride (75 s) in the segment II within the liquidation of the traffic delay; the recuperation factor $kr = 0$; the electric energy En as a function of the terminal speed v of the starting: a) 13.84 m/s, b) 14.0 m/s, c) 14.25 m/s, d) 14.60 m/s, e) 15.10 m/s, f) 15.61 m/s.

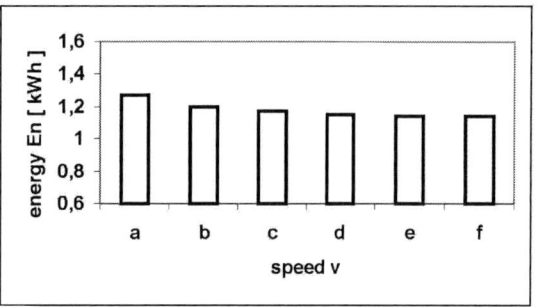

Fig. 15. The tram ride (75 s) in the segment II within the liquidation of the traffic delay; the recuperation factor $kr = 0.5$; the electric energy En as a function of the terminal speed v of the starting: a) 13.84 m/s, b) 14.0 m/s, c) 14.25 m/s, d) 14.60 m/s, e) 15.10 m/s, f) 15.61 m/s.

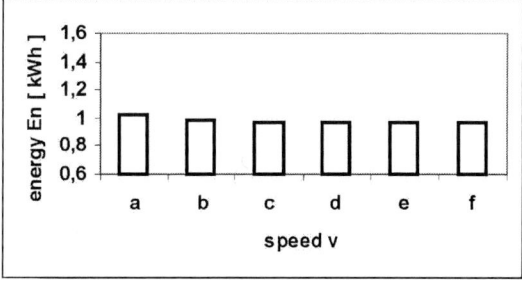

Fig. 16. The tram ride (75 s) in the segment II within the liquidation of the traffic delay; the recuperation factor $kr = 1$; the electric energy En as a function of the terminal speed v of the starting: a) 13.84 m/s, b) 14.0 m/s, c) 14.25 m/s, d) 14.60 m/s, e) 15.10 m/s, f) 15.61 m/s.

As a result of the influence of light signalling in front of the cross-roads, the tram braking, the additional starting after the stop and the necessity to quicker running in order to attain the liquidation of the time-lag generate (in comparison with the traffic without disturbances) the increase of the energy use. For the run without the recuperation of the energy during the vehicle braking (kr=0), the unplanned stop of 8 s (the segment I) causes the following rise in the energy consumption: 90.2% (in the segment I) and 29.9% (the segment II). If the energy recuperation exists and the recuperation coefficient has the average value kr equal 0.5, the increase of the energy use is smaller: 65.1% (the segment I) and 24.1% (the segment II). Together for the segments I and II, the total rise in the energy (within the global balance) is: 61% for the traffic without the energy recuperation and 45.5% if the recovery of the energy occurs at the mean value of the factor kr = 0.5. These results show clearly that the electric energy balance is more advantageous if the energy recuperation during the braking exists. When the traffic disturbance comes to an end, the computer calculates the new algorithm of the subsequent energy saving ride.

Interesting problem is connected with determination of the best, optimum period for liquidation of the traffic delay in order to attain the tram ride in accordance with the time-table. For the shorter period, the average vehicle speed must be greater and the electric energy use will be larger. The optimum choice is here conditioned by some problems. In the theory of tram system design, so-called service speed and the distance between neighbouring tram stops are of great importance. From point of view of the tram enterprise, the greater distance between stops makes possible rise in the service velocity, the better utilization of vehicles and smaller energy consumption. Of course this greater distance is uncomfortable for tram passengers and compromising solution is here necessary. It is interesting that for passengers, tram traffic regularity can be sometimes even more important than the vehicle speed. In general, ride regularity has influence not only on the traffic time. If trams run at regular intervals, the average filling of vehicles is uniform. Otherwise the trams run in groups; here the first vehicle is crowded however the next trams are almost empty. Preservation of traffic regularity is the important element for uniform vehicle load and journey comfort.

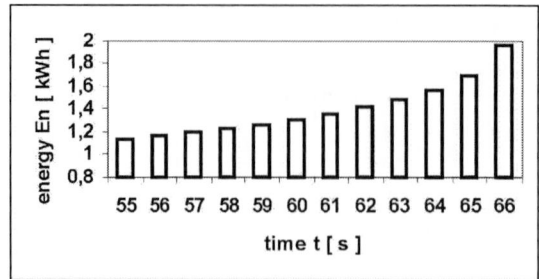

Fig. 18. The tram vehicle ride in the segment II (850 m) within the liquidation of the traffic delay; the recuperation factor kr = 0; the electric energy consumption En as a function of the traffic time in the 2nd part of the segment I.

The author has realized many computer simulations for different time, speed and distances within the framework of tram traffic delay liquidation. Part of the results of the numerical calculations is here shown in Figs. 17-24. These figures present values of the electric energy use for different traffic situations. In the case connected with Figs. 17-20, liquidation of the traffic delay (the tram vehicle stop of 8 s after additional braking) was planned within the way segments denoted by I and II. In the instance illustrated in Figs. 21-24, the time lag is eliminated during the ride in the 2nd part of the segment I and 2 identical segments (II, III) of the length equal 850 m. The normal run time (at the tram vehicle running without disturbances) is equal 85 s for each segment (II and III). In Figs. 17-19, the electrical energy is given as a function of the tram vehicle traffic time in the 2nd part of the segment I; of course, at the constant length of this part (600 m) it gives also precise information about the velocity values in this part. In Fig. 17, the maximum electric energy $Enmax$ = 1.374 kWh, the minimum energy $Enmin$ = 0.784 kWh and difference between these extreme values in percentages is equal en = 75.3%. For different values of the recuperation coefficient kr, the interesting results of calculations of the energy $Enmax$ and $Enmin$ as well as energy increment en are presented in the following statement:

kr [-]	0	0.5	1.0
$Enmin$ [kWh]	0.784	0.689	0.593
$Enmax$ [kWh]	1.374	1.099	0.825
en [%]	75.3	59.5	39.1 .

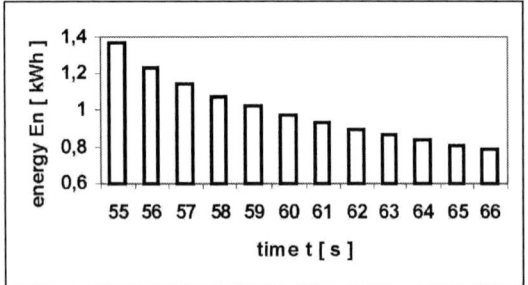

Fig. 17. The tram ride in the 2nd part of the segment I after green light obtainment; the recuperation factor kr = 0; the electric energy consumption En as a function of the traffic time in the 2nd part of the segment I of the constant length equal 600 m.

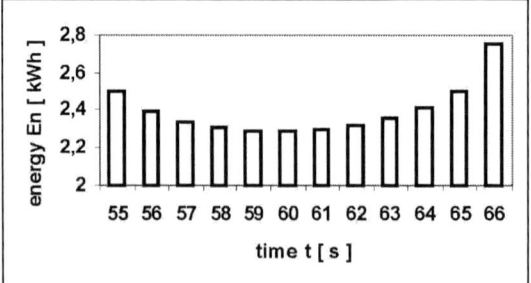

Fig. 19. The tram ride as the sum of the 2nd part of the segment I and the whole segment II within the liquidation of the traffic delay; the recuperation factor kr = 0; after summation, the electric energy use En as a function of the traffic time in the 2nd part of the segment I.

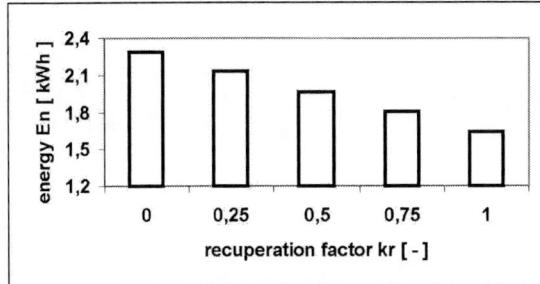

Fig. 20. The total tram ride as the sum of the 2nd part of the segment I and the whole segment II within the liquidation of the traffic delay; after summation, the minimum electric energy consumption *En* as a function of the recuperation factor *kr*.

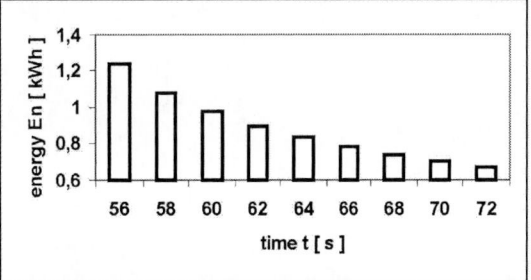

Fig. 21. The tram ride in the 2nd part of the segment I after green light obtainment; the recuperation factor *kr* = 0; the electric energy consumption *En* as a function of the traffic time in the 2nd part of the segment I (600m).

On the basis of Fig. 18 and other calculations, for the segment II the characteristic results are given in the set:

kr [-]	0	0.5	1.0
$Enmin$ [kWh]	1.131	1.007	0.883
$Enmax$ [kWh]	1.972	1.589	1.206
en [%]	74.4	57.8	36.6 .

The results in Fig. 19 dealing with the sum of the 2nd part of the segment I and the segment II are significant because they are connected with the global balance. For different factors *kr*, the specific data are shown in the list:

kr [-]	0	0.5	1.0
$Enmin$ [kWh]	2.287	1.963	1.640
$Enmax$ [kWh]	2.756	2.278	1.799
en [%]	20.5	16.1	9.70 .

Generation of the electrical energy during the recuperation process in the braking stage is advantageous for the global energy statement. For process of the liquidation of the time lag, the comparison of energy parameters, according to the various number of assumed, subsequent run segments, is the interesting problem. The comparative analysis is here possible on the ground of additional information included in Figs. 21-24. With the aim of attaining the facilitation of different variants presentation, the following denotations are introduced:

variant α – liquidation of the traffic delay both in the 2nd part of the segment I and in the segment II; the run in the segment III is executed according to the time-table;

variant β - liquidation of the delay within the ride in the 2nd part of the segment I as well as in the segments II and III (in sum: 600m + 850m + 850m = 2300m);

variant A – the ride without traffic disturbances in the whole segments: I, II and III; the total distance: 1000m + 850m + 850m = 2700m;

variant B - the ride in the whole segments: I, II and III (in sum 2700m) with the additional stop of 8 s as a consequence of influence of the light signalling; liquidation of the ride time lag is realized in the end stop of the segment II;

variant C – the traffic in the whole segments: I, II and III (in sum 2700m) with the stop of 8 s as a result of the red light in front of the cross – roads; delay elimination is executed in the end stop of the segment III.

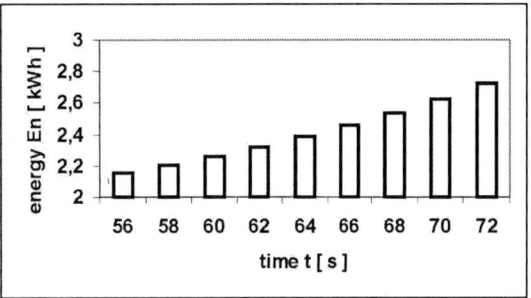

Fig. 22. The tram ride in the identical segments: II and III (each 850m) within the liquidation of the traffic delay; the recuperation factor *kr* = 0; after summation (II + III), the electric energy use *En* as a function of the traffic time in the 2nd part of the segment I.

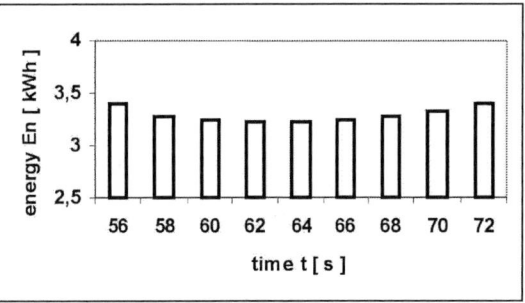

Fig. 23. The tram ride within the liquidation of the traffic delay; the recuperation factor *kr* = 0; after addition: (2nd part of I) +II + III, the electric energy consumption *En* as a function of the vehicle traffic time in the 2nd part of the segment I.

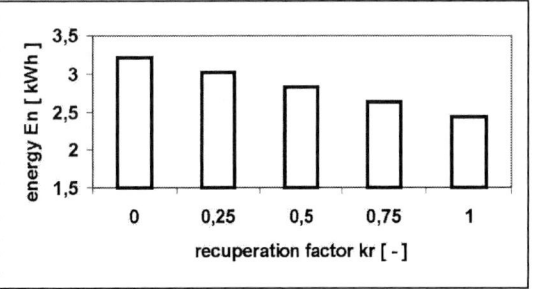

Fig. 24. The total tram ride as the sum of the 2nd part of the segment I and the segments II, III within the liquidation of the traffic delay; after addition: (2nd part of I) +II + III, the minimum electric energy consumption *En* as a function of the recuperation factor *kr*.

For the cases α and β (the common total distance 2300m) and different values of the recuperation factor *kr*, the minimum energy use *Enmin* at the best optimum control of the tram drive is given in the statement:

kr [-]	0	0.5	1.0
Enmin(α) [kWh]	3.298	2.881	2.465
Enmin(β) [kWh]	3.218	2.827	2.437 .

For the variants A, B and C, the energy consumption *Enmin* at the best control in accordance with the criterion of the minimum energy consumption is presented in the specification:

kr [-]	0	0.5
Enmin(**A**) [kWh]	3.102	2.842
Enmin(**B**) [kWh]	4.378	3.718
Enmin(**C**) [kWh]	4.226	3.664 .

Other results (not given in this paper) of calculations, executed for cases with longer tram stops (1 or even more minutes) during waiting for the green light within the light signalling, show that very long traffic delay forces the quicker tram vehicle running within some succeeding ride segments.

D. Speed Limitation as a Result of Red Light in Crossing

For the case presented in Figs. 25-28, as a result of influence of the light signalling the tram was also obliged to brake in front of the cross-roads but only the speed reduction to the value 25 km/h was necessary. For this example, in consequence of additional safety conditions the limited vehicle speed 25 km/h was obligatory for the whole crossing (60 m). The vehicle braking and the ride with smaller speed in the proximity of the cross-roads with light signalling caused the traffic delay inside the segment but this time lag wasn't great and delay removal was possible directly within the framework of the same segment. For the 2nd part (540 m) of the segment, the diagrams shown in Figs. 25, 26 are connected with the different control causing various time of the renewed starting, the running with the constant speed, the coasting and the braking. Only the optimum control ensures the minimum energy consumption (Fig. 25).

For various values of the recovery factor *kr*, Figs. 27, 28 present the energy *En* in the 2nd part of the segment as a function of the terminal speed *v* of the renewed starting. The characteristic energies are given in the following list:

kr [-]	0	0.5	1.0
Enmin [kWh]	0.565	0.443	0.321
Enmax [kWh]	0.725	0.558	0.390
en [%]	28.3	26.0	21.5 .

The best control of the tram vehicle enables the energy saving to 28% (for the 2nd part of the road).

In the whole segment and for the energy saving traffic (but at *kr* = 0), the energy use is 59% greater in comparison with the ride without disturbances (the vehicle fully privileged in the cross-roads within the light signalling). For *kr* > 0, this difference decreases.

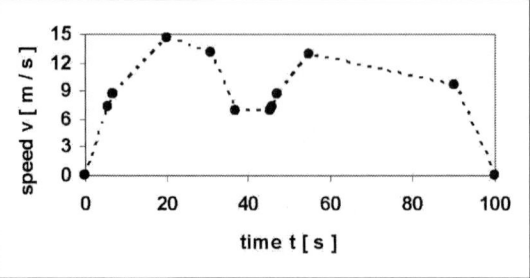

Fig. 25. Because of the light signalling, the tram was forced to brake in front of the cross-roads and reduce the speed to 25 km/h (also during the whole crossing); the minimum energy use *Enmin* = 0.443 kWh in the 2nd run part (behind the crossing); the recuperation factor *kr* = 0.5.

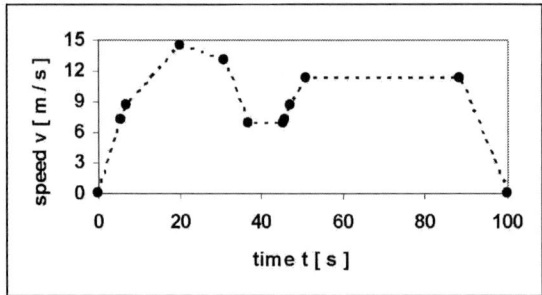

Fig. 26. Because of the light signalling, the tram was forced to brake in front of the cross-roads and reduce the speed to 25 km/h (also during the whole crossing); the maximum energy use *Enmax* = 0.558 kWh in the 2nd run part (behind the crossing); the recuperation factor *kr* = 0.5.

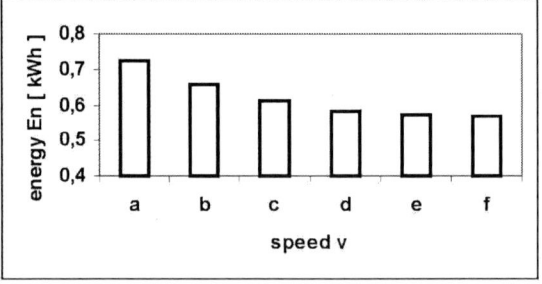

Fig. 27. The tram ride in the 2nd run part (540 m); the recuperation factor *kr* = 0; the electric energy *En* as a function of the terminal speed *v* of the renewed starting: a) 11.35 m/s, b) 11.46 m/s, c) 11.71 m/s, d) 12.13 m/s, e) 12.39 m/s, f) 12.88 m/s.

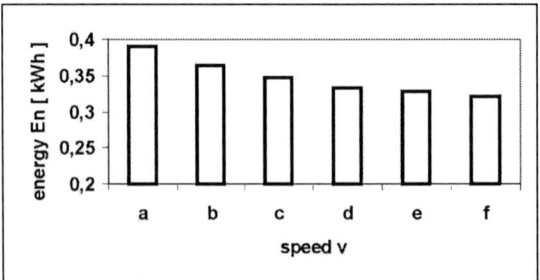

Fig. 28. The tram ride in the 2nd run part (540 m); the recuperation factor *kr* = 1; the electric energy *En* as a function of the terminal speed *v* of the renewed starting: a) 11.35 m/s, b) 11.46 m/s, c) 11.71 m/s, d) 12.13 m/s, e) 12.39 m/s, f) 12.88 m/s.

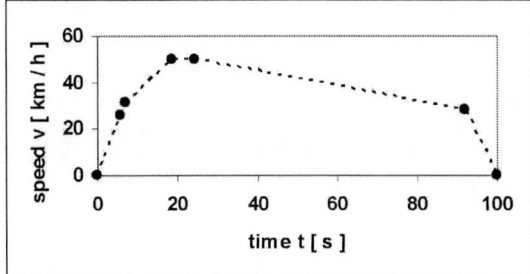

Fig. 29. The tram vehicle ride with global speed limitation to 50 km/h within whole traffic segment; the recuperation factor $kr = 0.5$; the minimum electrical energy consumption $Enmin = 1.006$ kWh.

E. Global Speed Limitation in Whole Traffic Segment

In this subsection, characteristics of the tram traffic are presented for the case of the global speed limitation between two neighbouring stops. Consistently to purpose of this work, activity in order to attain the minimum energy consumption is realized. For the given value of the permissible maximum velocity, the interesting phenomenon consisting in possibility of ride realization at the same distance and run time of the segment has been observed. The average tram speed is here the same but the lengths of the starting, running with the constant speed, coasting and braking can be different; optimization problem consists in the determination of the variant with minimum energy use. For different values of the velocity limit within the whole route (1000 m, 100 s), Figs. 29-30 present the final calculation results of boundary speed values for individual stages of the energy saving ride. For the speed limit 50 km/h and the recuperation factor $kr = 0.5$, the results for the extremal values of the coasting distance sc, the distance scs (run with the constant speed) and the energy use En are given as a function of the terminal speed v of the starting in the following set:

v [m/s]	11.16	13.89
sc [m]	0	678
scs [m]	869	124
En [kWh]	1.145	1.006

Figs. 31, 32 also confirm occurrence of dependence of the electric energy consumption on the driving system control of the tram. Here the values of the energy use are given as a function of the terminal speed v of the starting.

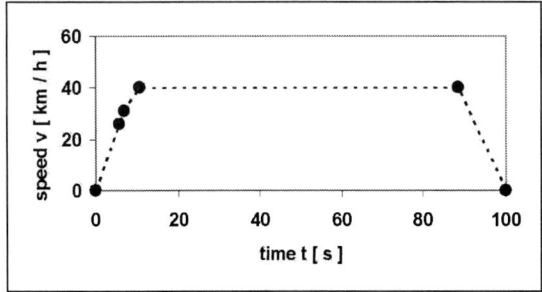

Fig. 30. The tram vehicle ride with global speed limitation to 40 km/h within whole traffic segment; the recuperation factor $kr = 0.5$; the minimum electrical energy consumption $Enmin = 1.145$ kWh.

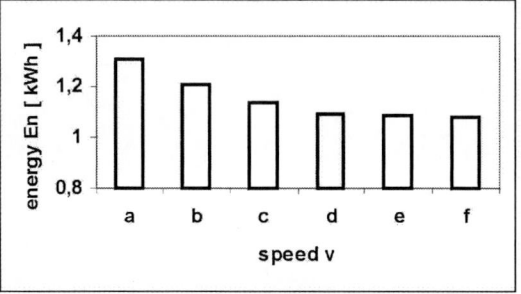

Fig. 31. The tram ride with global speed limitation to 50 km/h within whole traffic segment; the recuperation factor $kr = 0$; the total electric energy En as a function of the terminal speed v of the starting: a) 11.16 m/s, b) 11.30 m/s, c) 11.70 m/s, d) 12.50 m/s, e) 13.25 m/s, f) 13.89 m/s.

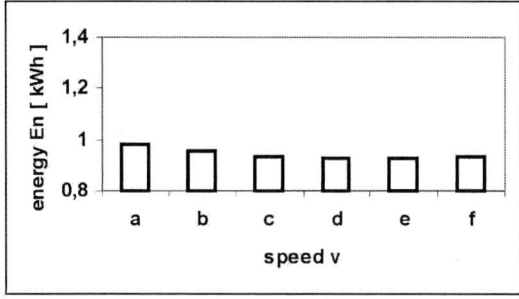

Fig. 32. The tram ride with global speed limitation to 50 km/h within whole traffic segment; the recuperation factor $kr = 1$; the total electric energy En as a function of the terminal speed v of the starting: a) 11.16 m/s, b) 11.30 m/s, c) 11.70 m/s, d) 12.50 m/s, e) 13.25 m/s, f) 13.89 m/s.

F. Influence of Change of Traction Network Voltage

It is typical for the tram starting that motor currents are great. The large voltage drop in the traction network occurs and the motor voltage can be lower compared with the rated value. To avoid great voltage drop, enlargement of the conductor cross-section area can be designed; sometimes also the supplementary wire is applied.

For the tram traffic at change of the supply voltage of the traction network, Figs. 33-35 show calculation results with the aim of attaining the minimum energy consumption. Diagrams are given for the network voltage 480 V (less by 20% in relation to the rated voltage 600 V), 540 V (less by 10% in comparison with the nominal value) and 660 V (voltage increase by 10%).

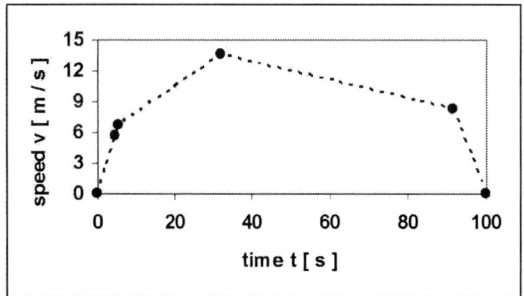

Fig. 33. The tram vehicle traffic at change of the supply voltage of the traction network to the value 480 V; terminal speed values for individual stages of the ride; results of calculations in accordance with the criterion of the minimum energy consumption.

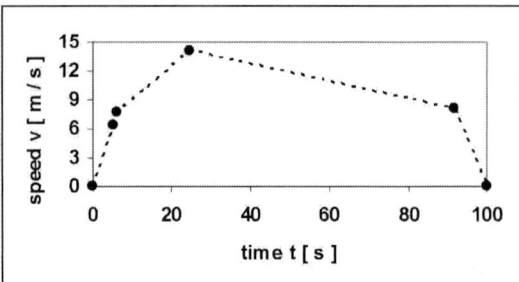

Fig. 34. The tram vehicle traffic at change of the supply voltage of the traction network to the value 540 V; terminal speed values for individual stages of the ride; results of calculations in accordance with the criterion of the minimum energy consumption.

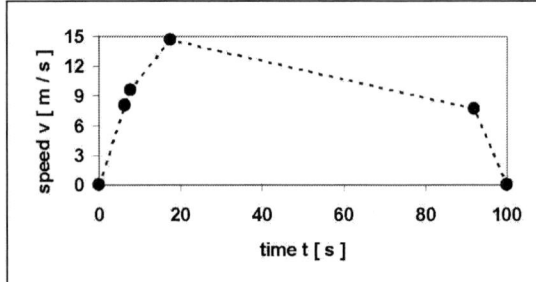

Fig. 35. The tram vehicle traffic at change of the supply voltage of the traction network to the value 660 V; terminal speed values for individual stages of the ride; results of calculations in accordance with the criterion of the minimum energy consumption.

At change of the supply voltage value, the consumed energy value modifies only on a small scale (about some per cent) but the voltage variation has great influence on the algorithm of the tram control with the aim of energy use minimization. The optimum duration of the starting, the run with constant speed, the coasting and the braking can modify considerably at the network voltage change.

For the network voltages Ut: 540V, 660V (drop or increase by 10%) and the energy saving ride, the distance $ss1$ and duration $ts1$ (the 1st part of the starting s: chopper increases the tram speed), the distance $ss2$, duration $ts2$ (the 2nd part: field weakening), the distance $ss3$ and duration $ts3$ (the 3rd starting part) are given in the list:

Ut [V]	540	660
$ss1$ [m]	16.5	26.0
$ss2$ [m]	7.6	12.1
$ss3$ [m]	218	120
$ts1$ [s]	5.10	6.41
$ts2$ [s]	1.08	1.37
$ts3$ [s]	18.2	9.72

IV. CONCLUSIONS

- Elaborated in literature algorithms of the energy saving traffic of tram vehicles are connected only with the ride without disturbances and can be referred to separated tracks. Except some situations, usually frequent and dynamical changes of city traffic conditions occur.
- The following tram traffic perturbations are of great importance for algorithm of the energy saving tram ride: influence of light signalling, changes of the network voltage, speed limitations, unexpected stays.
- New simulation models (described here scientifically) allowing for the traffic disturbances enable saving about 30% electric energy consumed by tram vehicles.
- The tram driver without additional help can not guide the vehicle in full accordance with the criterion of the minimum energy use. Particularly, it isn't possible for dynamical run of the tram during ride disturbances.
- The author proposes the new solution consisting in the semi-automatic control of the tram with the object of minimum energy use. If the driver decides that safety conditions are fulfilled, he can switch on the system of automatic tram guidance with the aim of the minimum energy consumption. The algorithm of energy saving control is often updated allowing for disturbances (computer forecasting of the subsequent ride means). At every moment the driver can switch off automatic control system and guide the tram by oneself.
- All parameters of the tram route (also the map of all segments) must be stored in vehicle computer memory.
- At change of the network voltage value, the calculation results show the consumed electric energy modifies only on a small scale. It must be here emphasized that change of the feeding voltage during the tram ride has the essential influence on the algorithm of the tram control with a view to attain the minimum energy use. It means the duration of the starting, the running with the constant speed, the coasting and the braking can considerably modify at the network voltage variation.
- For determination of the tram vehicle position, Global Positioning System (GPS) can be applied; it may be correlated with global tram ride monitoring by Traffic City Centre. Within the design process, microprocessor energy saving tram control ought to be connected with the vehicle construction optimization, parameters of power electronics systems and with traction network properties. Criteria of social costs, traffic regularity and punctuality are here also of great importance.

REFERENCES

[1] I. G. Bools and M. J. Savage, "Energy efficient driving modes and centralized train control", *Proc. ORE Coll.: Rational Use of Electrical Energy for Traction Purposes*, Vienna, 1981.

[2] J. Frąckowiak, "Optimization of electric energy use by traction vehicle driven by three-phase induction motors", *Proc. of VII Scientif. Conf. SEMTRAK'96*, Zakopane, Poland, 1996, pp. 29-34.

[3] R. S. Glia, S. Patalas, S. Rawicki and K. Rzeźnik, "Determination of Algorithms of Energy – Saving Traffic of Tram Vehicle at Unplanned Stops During Run", *Proc. of X Scientific Conf. SEMTRAK'2002*, Zakopane, Poland, 2002, pp. 289-294.

[4] P. Horn, R. Zinke, "Energieoptimale Zugsteuerung im Fernverkehr", *Scientific Journal on Traffic*, Vol. 40, No. 4, 1990.

[5] J. Kacprzak, *Automatics and control of electric traction vehicles*. WKiŁ, Warsaw, 1981.

[6] S. Rawicki, "Research of Some Particular Features of Energy Optimum Running of Tram Vehicle by Ride Simulation Method", *Proc. of International Conf. on Modelling & Simulation MS'2007*, Terni, Italy, 2007, pp. 95-100.

[7] S. Yasukawa, K. Satoh, "A study on energy saving train operation methods for the Shinkansen Express train service", QR of RTRI, Vol. 30, No. 3, 1989.

[8] M. Zielińska, "Modelling of traction vehicle traffic", *AMSE Press: Modelling, Measurement, Control*, Part B, Vol. 49, No. 2, 1993.

Characterization and Improved Control of a Brushless DC Drive with In-Wheel Motor

Manuele Bertoluzzo[*,1], Giuseppe Buja[*,2] and Alessandro Pavoni[**]

[*,1] University of Padova, Department of Electrical Engineering, Padova, Italy, e-mail: *manuele.bertoluzzo@unipd.it*
[*,2] University of Padova, Department of Electrical Engineering, Padova, Italy, e-mail: *giuseppe.buja@unipd.it*
[**] H.T.M. Company, Ronchis (UD), Italy, e-mail: *pavoni@htmworld.com*

Abstract - The paper deals with a brushless DC drive belonging to the propulsion system of an innovative light electric vehicle. The drive consists of an in-wheel motor fed by a voltage-controlled inverter. At first, the parameters of the motor and the mechanical characteristics of the drive are measured; the latter results are supported by a formulation of the torque-speed equations. Afterwards, a novel strategy for the voltage control of the drive during the commutations is presented, intended to eliminate the torque ripple and the decrease in the available torque due to the commutations. Improvement in the drive performance achieved with the novel strategy is documented by simulation.

Keywords—**Electric propulsion, brushless DC drive, in-wheel motor, drive characterization, drive control.**

I. INTRODUCTION

Light electric vehicles (LEVs) are a viable solution to cope with the traffic congestion and environment pollution in the urban areas [1]. The innovative prototype of LEV pictured in Fig.1 has been recently developed at the University of Padova [2]. The vehicle stands out because of i) the mechanical layout, which uses three wheels (one in the front module and two in the rear module) with a tilting mechanism of the front module, and ii) the propulsion system, which uses two equal brushless DC drives with in-wheel motor -one for each rear wheel- for the traction, a Lithium-ion battery for energizing the drives, and a by-wire set-up for riding the vehicle. As a matter of fact the signal coming from the handlebar is manipulated by the electronic control unit governing the vehicle before being delivered to the drives in the form of voltage reference.

This paper faces two issues, namely the characterization of the brushless DC drives and the development of a novel strategy for the voltage control of the drive during the commutations. In detail, Section II describes the devices composing the brushless DC drives (in-wheel motor, inverter and control board). Section III focuses on the measurement of the in-wheel motor parameters. Section IV provides both a measurement and a formulation of the mechanical characteristics of the drive. Section V examines the deviation of the motor current waveforms from the desired ones due to the phase inductances, and reports on the ensuing shortcomings in terms of torque ripple and of decrease in the available torque. Section VI proposes a voltage control strategy during the commutations that eliminates almost entirely these shortcomings over an extended speed range, and gives simulation results that document the effectiveness of the strategy. Section VII concludes the paper.

II. DRIVE DESCRIPTION

The brushless DC drives utilized in the LEV prototype are a commercial product of the H.T.M. Company, on-purpose manufactured for the propulsion of electric scooters. Each drive includes an in-wheel motor that motorizes the wheel, as illustrated in Figs. 2(a) and (b). The rotor with the Permanent Magnets (PMs) is fixed to the wheel rim whilst the stator is internal to the rotor and is fastened to the half shaft of the wheel. A pair of ball bearings allows rotation of the rim with respect to the half shaft. The stator is equipped with three Hall sensors that generate three $2\pi/3$-shifted, square-wave signals synchronized with the rotor. The main data of the motor are as follows: 8 pole pairs, rated voltage of 48V, rated torque of 48 N·m, rated power of 3 kW at the rated speed of 600 rpm, peak torque of 100 N·m. The outer radius of

Fig.1. LEV prototype.

(a)   (b)

Fig.2. (a) Motorized wheel and (b) in-wheel motor.

978-1-4244-1741-4/08/$25.00 ©2008 IEEE

the wheel, tire included, is of 0.2 m, resulting in a vehicle speed of 45 km/h at 600 rpm motor rotation.

The drive also includes a MOSFET inverter and a control board. The board is entered by the voltage reference for the inverter output and the Hall signals, and generates the commands for the MOSFETs. The inverter output voltage is switched at 14 kHz according to the unipolar technique and is controlled through the switching duty-cycle δ. The value of δ varies linearly with the voltage reference until the average value I_{dc} of the inverter dc-link current does not exceed the limit $I_{dc,max}$. When I_{dc} tries to go beyond $I_{dc,max}$, the duty-cycle is adjusted in order to regulate the dc-link current at $I_{dc,max}$. The control board has an input, termed torque boost, that is meant to be activated for a quick starting of the vehicle or for tackling a steep slope: when activated, the inverter dc-link current limit is increased up to twice $I_{dc,max}$ for a few seconds allowing the motor to develop the peak torque.

The brushless DC drive has the circuital representation of Fig.3, where Z is the ohmic-inductive impedance of the battery and its connections. The capacitor C across the inverter dc terminals smoothes the current i_B drawn from the battery so that i_B is nearly constant and equal to the average value of the inverter dc-link current, i.e.

$$i_B \cong I_{dc} \tag{1}$$

The voltage across the capacitor is also nearly constant and is equal to the battery voltage V_B, i.e.

$$v_{dc} \cong V_B \tag{2}$$

since the ohmic component of Z is negligible.

III. PARAMETER MEASUREMENT

The characterization of the in-wheel motor has begun with the measurement of its parameters [3]. The resistance R has been measured by executing some volt-ampere tests between two motor terminals and by dividing by 2 the resistance that best fits the experimental data.

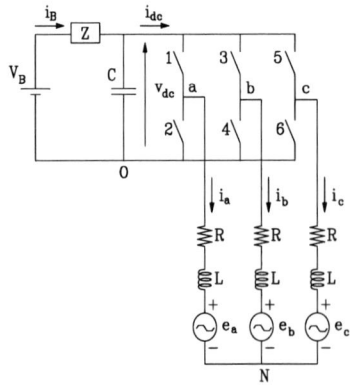

Fig.3. Circuital representation of a DC brushless drive.

The phase-to-phase inductance L_{pp} has been measured from the time response of current to a short voltage pulse applied across two motor terminals. To this aim, the motor has been blocked and the current has been detected. From L_{pp}, the phase inductance L inclusive of the mutual inductance effects has been calculated in

$$L = \frac{L_{pp}}{2} \tag{3}$$

The same experiment, executed between one motor terminal and the neutral point N, has given the phase self-inductance L_p.

The motor constant k has been measured by detecting the peak value E of the back emf of a phase, with the motor dragged at constant speed Ω, and by dividing E by Ω, i.e.

$$k = \frac{E}{\Omega} \tag{4}$$

The emf as acquired with an oscilloscope is traced in Fig. 4 and exhibits the expected trapezoidal waveform. A list of the resultant motor parameters is reported in Tab.1.

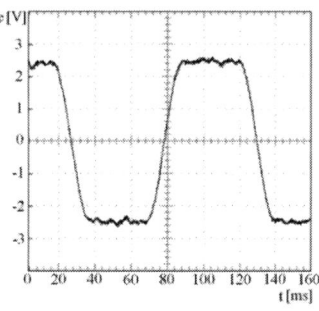

Fig.4. Back emf at 75 rpm.

Tab.I. Motor parameters

Parameter	Symbol	Value
Phase resistance	R	35 mΩ
Phase self-inductance	L_p	57 µH
Inclusive phase inductance	L	82.5 µH
Motor constant	k	0.32 Vs/rad

IV. MECHANICAL CHARACTERISTICS

A. Measurement

The mechanical characteristics of the drive have been measured by entering the control board with a constant voltage reference and by braking the motor with torques of different values; the quantity $I_{dc,max}$ has been set at its rated value (45 A). For each value of the braking torque, the speed of the in-wheel motor has been detected and the characteristics have been reported in Fig. 5, where r is the voltage reference as a percentage of its maximum value. For low torque values, the characteristics are inclined straight lines as in a voltage-controlled DC drive whilst, for high torques and for r higher than 30%, the

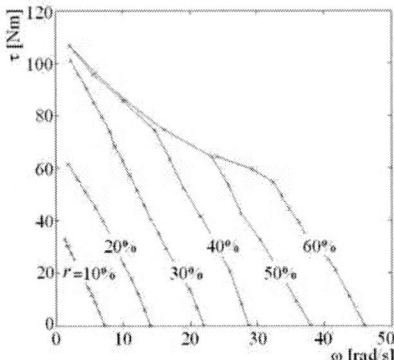

Fig. 5. Mechanical characteristics.

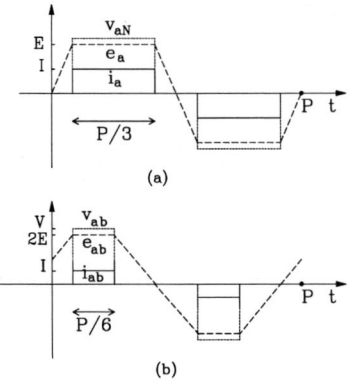

Fig.6. Plot of (a) phase variables, (b) line-to-line variables. N.B. y-scale in (b) is twice than in (a).

characteristics merge in a curve with a hyperbolic shape. The measured results are confirmed by the formulation expounded in the next Subsection.

B. Formulation

The behavior of the drive has been analyzed in steady state to formulate the mechanical characteristics. The basic assumption taken for the analysis is to neglect the motor inductances. Under this assumption, the current and back emf waveforms of a motor phase (f.i., the phase a) are shown in Fig. 6(a) by the continuous and dashed lines, respectively. The quantity I is the magnitude of i_a whilst the quantity E is the peak values of e_a. The waveforms for the phases b and c are obtained by shifting of $P/3$ those of the phase a, being P the supply period. Within P, each phase is supplied by the inverter along two intervals of length P/3; the average supply voltage v_{aN} is plotted with a dotted line. Outside the two intervals, the voltage v_{aN} is given by e_a (the relevant curve is not plotted in the figure). The corresponding line-to-line waveforms between the terminals a and b of a Y-connected motor are plotted in Fig. 6(b), where i_{ab} is the current common to the branch aN and Nb, $e_{ab}=e_a-e_b$, and V is the average magnitude of v_{ab} within the supply interval. Fig. 6(b) shows that, during the supply period, the motor phases are fed two at a time along two intervals of length P/6. Irrespectively from the interval and the supplied phase, the voltage equation of the motor is

$$V = 2E + 2RI \tag{5}$$

where

$$V = \delta V_B \tag{6}$$

$$I = \frac{I_{dc}}{\delta} \tag{7}$$

As long as $I_{dc} < I_{dc,max}$, δ is equal to r. The electric power converted into mechanical form is

$$2EI = T\Omega \tag{8}$$

where T is the motor torque. By (4) and (8), the torque can be expressed as

$$T = 2kI \tag{9}$$

Extraction of I from (5) and its substitution into (9) yield the mechanical characteristics of the drive as a function of δ

$$T = \frac{\delta k V_B}{R} - \frac{2k^2\Omega}{R} \tag{10}$$

Eq. (10) shows that the characteristics are a family of inclined straight lines with an equal slope of $-2k^2/R$.

Under I_{dc} regulation, δ is determined by (7) and (9), instead of being equal to r, and is expressed as

$$\delta = 2k\frac{I_{dc,max}}{T} \tag{11}$$

By substituting (11) into (10), it turns out that the mechanical characteristics are constituted by only one curve with the following equation:

$$T = \frac{I_{dc,max}V_B}{\Omega}\frac{2}{1+\sqrt{1+\dfrac{2RI_{dc,max}V_B}{k^2\Omega^2}}} \tag{12}$$

For $k\Omega > RI_{dc,max}$, (12) is approximated by

$$T \cong \frac{I_{dc,max}V_B}{\Omega} \tag{13}$$

showing that the toque-speed relationship is hyperbolic. Worthy to note, the numerator of the term on the right hand side of (13) represents the power delivered by the battery.

It is evident that the formulation of the mechanical characteristics, given by (10) and (13), fully supports the measured results.

V. PHASE COMMUTATION

A. Current waveforms

As outlined in Fig. 6, the phase currents of the motor should have square waveforms with instantaneous commutations. In practice, the phase inductances and the limited supply voltage alter the current waveforms as shown in Fig. 7.

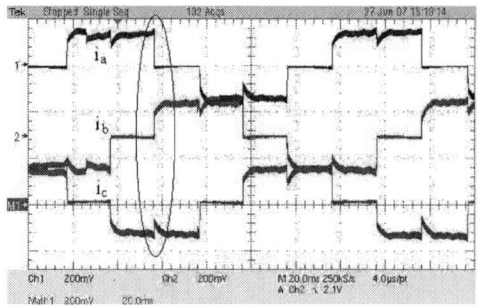

Fig.7. Phase currents at 62 rpm.

Let us consider as an example the commutation marked with an ellipse in Fig. 7, where the outgoing phase is a and the incoming phase is b. The relationships for the currents are as follows: at the beginning of the commutation, it is $i_a=I$, $i_c=-i_a$ and $i_b=0$; at the completion of the commutation, it is $i_b=I$, $i_c=-i_b$, and $i_a=0$. Before the commutation, switch 1 is closed, switch 6 is commanded with duty-cycle δ and the other switches are open. With the Conventional Voltage Control (CVC) utilized in Fig. 7, the commutation is executed by opening switch 1 and by closing switch 3, without varying the duty-cycle of switch 6. The traces of Fig. 7 point out that: i) the currents of the phases b and c during the commutation are not constant, with a built-up of the current in the incoming phase (i.e. b), and a dip of the current in the conducting phase (i.e. c), and ii) the withdrawal of the current from the outgoing phase (i.e. a) takes time; of course, this current must vanish before the occurrence of the successive commutation. Therefore the allowed time interval for completing the commutation is $\pi/(3p\Omega)$, where p is the number of pole pairs of the motor.

The deviations of the current waveforms deteriorate the drive performance [4] as they produces a torque ripple and a decrease in the available torque, i.e. in the average torque developed by the motor over a supply period, which become consistent at high speeds. The torque ripple gives rise to chassis vibrations that become annoying for the rider especially when the thrust demanded to the vehicle is notable. The torque decrease entails a lower loading capacity of the vehicle.

B. Voltage Control Strategy

To cope with the above-mentioned deterioration in the drive performance, suitable strategies have been developed to control the voltage applied to the motor during the commutations [5]-[7]. Their assessment relies on the expression of the instantaneous torque developed by the motor, given by

$$\tau = \frac{1}{\Omega}\left(i_a e_a + i_b e_b + i_c e_c\right) \tag{14}$$

During the exemplifying commutation, the back emfs e_a, e_b and e_c are given by

$$\begin{cases} e_a = k\Omega - \dfrac{6pk\Omega^2}{\pi}t \\[2mm] e_b = k\Omega \\[2mm] e_c = -k\Omega \end{cases} \tag{15}$$

provided that the origin of the time axis is set at the beginning instant of the commutation. From (14) and (15), the time rate of the torque is

$$\frac{d\tau}{dt} = \left(k - \frac{6pk\Omega}{\pi}t\right)\frac{di_a}{dt} - \frac{6pk\Omega}{\pi}i_a + k\frac{di_b}{dt} - k\frac{di_c}{dt} \tag{16}$$

For generic values of the voltages v_{aO}, v_{bO} and v_{cO} applied by the inverter between the terminals of the motor and the negative rail of the dc supply, the equations of the phase currents are

$$\begin{cases} \dfrac{di_a}{dt} = -\dfrac{R}{L}i_a + \dfrac{1}{L}\dfrac{1}{3}\left(2\dfrac{6pk\Omega^2}{\pi}t - 2k\Omega + 2v_{aO} - v_{bO} - v_{cO}\right) \\[3mm] \dfrac{di_b}{dt} = -\dfrac{R}{L}i_b + \dfrac{1}{L}\dfrac{1}{3}\left(-\dfrac{6pk\Omega^2}{\pi}t - 2k\Omega + 2v_{bO} - v_{aO} - v_{cO}\right) \\[3mm] \dfrac{di_c}{dt} = -\dfrac{R}{L}i_c + \dfrac{1}{L}\dfrac{1}{3}\left(-\dfrac{6pk\Omega^2}{\pi}t + 4k\Omega + 2v_{cO} - v_{bO} - v_{aO}\right) \end{cases} \tag{17}$$

Substituting (17) into the right hand side of (16) and equating the result to zero lead to

$$v_{aO}\left(1 - \frac{6p\Omega}{\pi}t\right) + v_{bO}\left(1 + \frac{3p\Omega}{\pi}t\right) + v_{cO}\left(-2 + \frac{3p\Omega}{\pi}t\right) =$$
$$-3Ri_c + 3Ri_a\frac{3p\Omega}{\pi}\left(\frac{L}{R} - t\right) + 4k\Omega\left(1 - \frac{3p\Omega}{\pi}t\right) + \tag{18}$$
$$-\frac{6pk\Omega^2}{\pi}t\left(1 - \frac{6p\Omega}{\pi}t\right)$$

The fulfillment of (18) assures that the torque remains constant during the commutation. This can be achieved by commanding the inverter so as to apply a suitable tern of voltages v_{aO}, v_{bO} and v_{cO}. Since the neutral point N of the motor is isolated, an arbitrary value can be added to the three voltages without affecting the currents or their derivatives. This is conveniently pursued by setting at 0 the voltage v_{cO} of the phase conducting a negative current. Then (18) is solved for v_{aO} and v_{bO}. An advantage of this approach is the reduction of the inverter power losses since switches 5 and 6 are not operated during the commutation but are kept open and closed, respectively. A disadvantage is the need of the knowledge of the phase currents to solve (18).

The solution proposed in [6] to fulfill (18) consists in setting v_{aO} at 0 and in controlling v_{bO} according to

$$v_{bO,ref} = \frac{\left(\begin{aligned} & -3Ri_c + 3Ri_a\frac{3p\Omega}{\pi}\left(\frac{L}{R} - t\right) + 4k\Omega\left(1 - \frac{3p\Omega}{2\pi}t\right) + \\ & -\frac{6pk\Omega^2}{\pi}t\left(1 - \frac{6p\Omega}{\pi}t\right) \end{aligned}\right)}{1 + \frac{3p\Omega}{\pi}t} \tag{19}$$

where $v_{bO,ref}$ is the reference of v_{bO}. Note that v_{bO} coincides with v_{ba} during the commutation, being $v_{aO}=0$. This solution is hereafter designated with One Phase Voltage Control (OPVC) strategy. With the OPVC, at low speeds the motor does not suffer from any ripple and decrease of torque, but at high speeds these shortcomings still arise because the magnitude of v_{bO}, as required by (19), would exceed V_B and then (18) can be not more fulfilled.

VI. NOVEL STRATEGY

In an attempt to reduce the torque ripple also at high speeds, a voltage control strategy during the commutations is here worked out, obtained by improving the solution in [6]. The strategy is designated with Two Phases Voltage Control (TPVC). Like the OPVC, it operates by controlling the voltages applied to the motor with the aim of fulfilling (18) on the torque time rate. Unlike the OPVC, it circumvents the limitation in the supply voltage of a phase by controlling two phases.

With reference to the exemplifying commutation, the TPVC operates as follows: it controls v_{bO} until it reaches V_B; after that, it sets v_{bO} at V_B and controls v_{aO} according to

$$v_{aO,ref} = \frac{\left(v_{bO,ref} - V_B\right)\left(1 + \frac{3p\Omega}{\pi}t\right)}{1 - \frac{6p\Omega}{\pi}t} \tag{20}$$

Similar to $v_{bO,ref}$, $v_{aO,ref}$ increases up to V_B with the speed. This value, however, can not be maintained during the whole commutation interval otherwise i_a does not vanish before the beginning of the successive commutation. To avoid this situation, the upper limit of $v_{aO,ref}$ is dynamically adjusted by imposing the fulfillment of the additional following condition that forces i_a to go down to zero within the allowed time interval:

$$\frac{di_a}{dt} \leq \frac{-i_a}{\frac{\pi}{3p\Omega} - t} \tag{21}$$

For $v_{bO}=V_B$ and $v_{cO}=0$, the expression of the upper limit of $v_{aO,ref}$, as calculated from (17) and (21), is

$$v_{aul} = \frac{V_B}{2} - \frac{6p\Omega^2 k}{\pi}t + \Omega k + \frac{3}{2}Ri_a - \frac{3}{2}L\frac{i_a}{\frac{\pi}{3p\Omega} - t} \tag{22}$$

The operation of the TPVC is illustrated by the flow-chart of Fig. 8.

The TPVC has been tested by simulation and its performance has been compared to those exhibited by the CVC and the OPVC, using the motor parameters in Tab. I and a battery of 48 V. Simulations have been carried out with the motor running at different speeds and with a current demand equal to the rated one. The torque ripple as a percentage of the motor torque and the motor torque

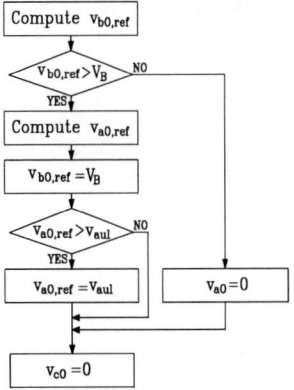

Fig.8. TPVC operation flow-chart.

are plotted in Figs. 9(a) and (b) vs. the speed. Fig. 9(a) shows that the TPCV is effective in eliminating the torque ripple up to the rated speed, whilst the OPVC does it up only to about half the rated speed. Fig. 9(b) is a further validation of the effectiveness of the TPCV since it keeps the motor torque at the rated value up to about the rated speed, whilst the OPVC produces a decrease of the available torque starting from about half the rated speed. Figs. 9 also highlight the noticeable enhancement in the drive performance (reduction of the torque ripple and increase of the available torque) achievable with suitable voltage control strategies of the drive, in comparison with the CVC.

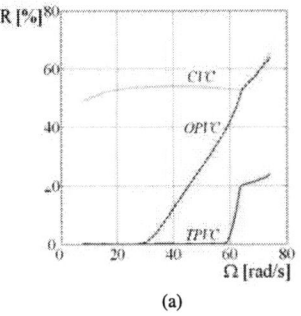

(a)

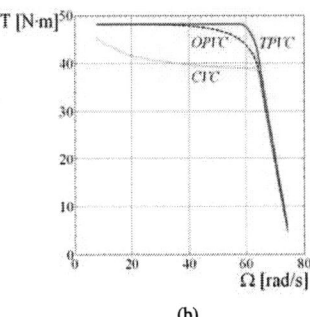

(b)

Fig. 9. (a) torque ripple and (b) motor torque vs. speed.

The results of Figs. 9 can be appreciated by looking at the time behaviors of the phase currents and of the motor torque obtained with the CVC, the OPVC and the TPVC.

The behaviors are calculated for drive operation at the rated speed and under the request of the rated current (78 A), and are plotted in Figs. 10.

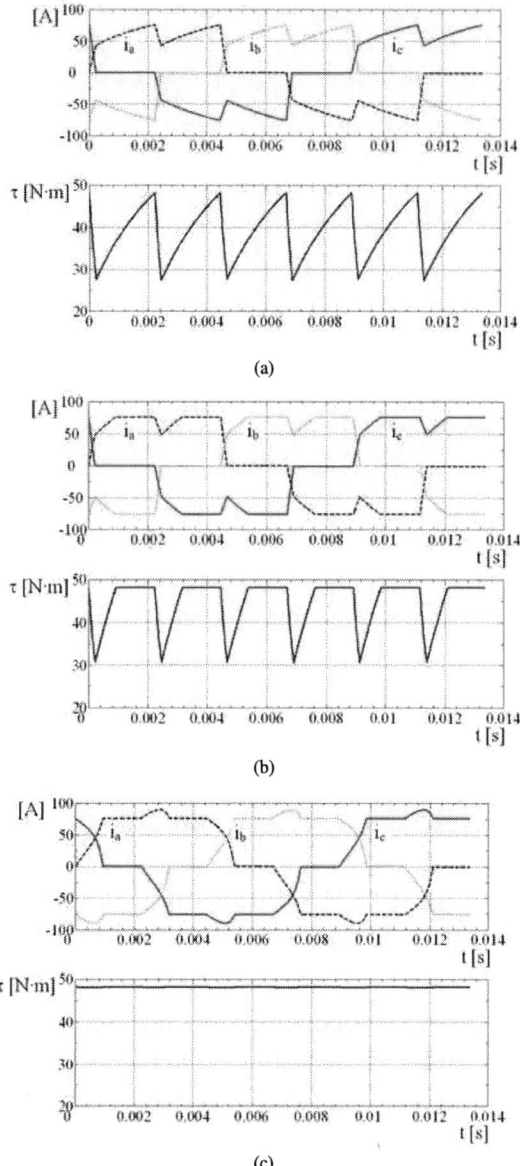

(a)

(b)

(c)

Fig. 10. Phase currents and motor torque at 560 rpm with a) the CVC, b) the OPVC, and c) the TPVC.

Fig. 10 (a) refers to the CVC. The plots show that the phase currents maintain a magnitude much lower than the requested one for most of the conducting interval. This produces a torque ripple of 20 N·m and a motor torque of 39 N·m, well below the motor capacities. Fig. 10 (b) refers to the OPVC. The plots show that the phase currents reach the requested magnitude faster than with the CVC, but still exhibit a dip in the middle of the conducting interval. This produces a torque ripple of 16

N·m and a motor torque of 44 N·m, with a progress compared to the CVC; nevertheless the motor torque is still 10% lower than the rated value. Fig. 10 (c) refers to the TPVC. The plots show the effective action played by the proposed strategy on the resultant current and torque behaviors. The magnitude of the current in the conducting phase is even increased during the commutation of the other two phases to maintain constant the motor torque. Another achievement is that the commutation takes the same time for both the outgoing and the incoming phases while, as it can be seen in Fig. 10 (a) for the CVC, i_a vanishes before i_b reaches its final value. The excellent performance of the TPVC is proved by the calculated values of the torque ripple that is of 0.1 N·m, and of the motor torque that equates the rated one.

VII. CONCLUSIONS

The paper has presented the characteristics of a brushless DC drive utilized in the propulsion system of an innovative LEV. The drive is constituted by an in-wheel motor fed by a voltage-controlled DC/AC converter. The parameters of the motor and the mechanical characteristics of the drive have been at first measured. The equations of the mechanical characteristics have been then obtained by accounting for the duty-cycle adjustment exerted by the control to limit the average current in the inverter dc-link. Afterwards, a novel strategy has been presented that controls the motor voltages during the commutations in order to eliminate the torque ripple and to develop the rated torque over a wide speed range. This is useful both to mitigate the annoying phenomenon of the chassis vibrations and to make available the rated motor torque all over the speed range. At last, the effectiveness of the novel strategy has been documented by simulation.

REFERENCES

[1] C.C.Chan, "The state of the art of electric, hybrid, and fuel cell vehicles" IEEE Proceedings, vol.95, no.4, pp. 704-718, April 2007.

[2] M.Bertoluzzo, G.Buja, V.Cossalter, A.Doria, and D.Mazzaro, "Electric tilting 3-wheel vehicle for a sustainable urban mobility", Proceedings of IEEE Workshop on Advanced Motion Control, 2008, CD, pp. 757-762.

[3] T.J.E. Miller, *Brushless Permanent-Magnet and Reluctance Motor Drives*, Clarendon Press, Oxford 1989.

[4] R.Carlson, M.Lajoie-Mazenc and J.C.D.S.Fagundes, "Analysis of torque ripple due to phase commutation in brushless dc machines", IEEE Transactions on Industry Applications, vol. 28, no. 3, pp. 632-638, May-June 1992.

[5] K.Y.Nam, W.T.Lee, C.M.Lee and J.P.Hong, "Reducing torque ripple of brushless DC motor by varying input voltage", IEEE Transactions on Magnetics, vol. 42, no. 4, pp. 1307-1310, April 2006.

[6] X.Xiao, Y.Li, M.Zhang and M.Li, "A novel control strategy for brushless DC Motor drive with low torque ripples", Proceeding of Industrial Electronics Society Annual Conference (IECON) 2005, CD, pp. 1660-1664.

[7] J.H.Song and I.Choy, "Commutation torque ripple reduction in brushless Dc motor drives using a single DC current sensor", IEEE Transactions on Power Electronics, vol. 19, no. 2l pp. 312-319, March 2004.

Supply of Electric Vehicles via Magnetically Coupled Air Coils

Slawomir Judek*, Krzysztof Karwowski[†]

*Gdansk University of Technology/Department of Electric Traction, Poland, e-mail: *s.judek@ely.pg.gda.pl*
[†]Gdansk University of Technology/Department of Electric Traction, Poland, e-mail: *k.karwowski@ely.pg.gda.pl*

Abstract—**The paper describes the contactless electrical energy transfer system (CEETS), by which electrical energy may be transmitted, without electrical connection or physical contact, through the large air gap. In this case energy is transmitted via coreless transformer. Coupling between the coils changes and depend on dimension of air gap. The efficiency of CEETS is mainly depended on the transmission frequency. Additionally large leakage inductances may be compensated also in resonance condition by adding capacitors to the coils. The article presents theoretical analyses and simulation results of transformer equivalent circuit with capacitors. Theoretical developments are compared with practical measurements from a prototype contactless system.**

Keywords—**Electric vehicle, High frequency power converter, Transmission of electrical energy.**

I. INTRODUCTION

Supply movable receivers with electrical energy is done, in vast majority of cases, via flexible connecting wires or other contact systems, for example commutators, current collectors, etc. Contactless method of supply receivers with electrical energy CEETS (Contactless Electrical Energy Transmission System), with the usage of magnetically coupled coils with wide air gap seems to be an attractive alternative to classical solutions [1, 2, 3].

Systems described in literature share a common characteristics – due to the method of their exploitation there are no significant changes in the dimensions of the air gap, and so these are the systems with weak but constant magnetic coupling. Stabilisation of the parameters of the magnetic coupling invariable in time has been achieved thanks to construction devices such as, most often, a stable and small air gap in the coupling magnetic circuit. Solutions for coupling installations, configurations of supply converter and the method of controlling energy flow in these installations have been proposed for the conditions mentioned above. The main effort of the researchers was aimed at obtaining maximum power transmission from the supply circuit to the receiver through ensuring that the installations were operating in conditions similar to full interchange of magnetic field energy and electric field energy of compensating capacitors [4, 5, 6].

As we can see, commonly known solutions do not take into consideration a system in which the value of magnetic coupling between the circuits is changeable, which results in changes of conditions regarding transmission of energy to the receiver. It has been assumed that the change of the coupling factor results from the slow movement deflecting the coils or changing the dimensions of the air gap. In particular, significant reduction of power transmitted to

the receiver may occur, with the decrease of efficiency. In order to improve the operating conditions of the CEETS, it is justified to apply a regulation system which would correct the system parameters in such a way that transmission of power at a fixed level and with high efficiency maintained throughout might be possible when changes of magnetic coupling occur.

This article presents the analysis of a contactless system transmitting electrical energy to electric vehicles, using coils with weak magnetic coupling. In particular, has been described the working conditions for the system in resonance state, including the selection of system parameters and the controlling method for transistor converter in the situation when the size of the air gap between the coils is changing.

II. CONFIGURATION OF THE SYSTEM

Air coils with weak magnetic coupling form the main part of a contactless electrical energy transmission system. This is the so-called transformer coupling. In order to analyse of such a system in details, it is necessary to replace the magnetically coupled coils with a circuital model. Fig. 1 shows an equivalent circuit of coils when supplied with AC sinusoidal voltage, where: U_p, I_p – voltage and current of primary side, U_s, I_s – voltage and current of secondary side, U_o, I_o – voltage and current of the receiver, R_p, R_s – resistance of primary and secondary winding, L_p, L_s – inductance of primary and secondary side winding, M – mutual inductance, R_o – load resistance.

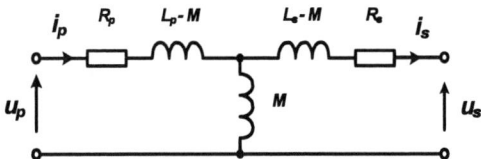

Fig. 1. Equivalent circuit diagram of CEETS system

Table 1 shows the results of exemplary power electrical analysis. In the first case, the system has been supplied with the voltage U_i = 230 V, frequency f = 50 Hz. The parameters chosen for the calculations are as follows: R_p = R_s = 0.1 Ω, L_p = L_s = 205 µH, M = 67.4 µH, R = 410 Ω Such configuration represents two coils, where the space between them is h = 50 mm (Fig. 8). For such parameters, the system functions in the state of short-circuited energy source. Therefore it does not perform the main function of transmitting energy from the primary to the secondary circuit. In such case, the efficiency is practically equal to zero.

978-1-4244-1741-4/08/$25.00 ©2008 IEEE   1497

In the second case, the system has been supplied with the voltage of frequency f = 100 kHz. This resulted in the increase of the energy transmission efficiency to 97.8 %. However, the level of transmitted power is still low. The last results presented in the table describe the condition of the system with the supply voltage of frequency f = 100 kHz, when suitably selected compensating capacitors were connected to it. One capacitor has been connected in series to the primary winding, while the second capacitor has been connected in parallel to the secondary winding (Fig. 2). The system maintains high efficiency and is capable of transmitting the power at the level of about 1 kW.

TABLE I.
RESULTS OF POWER AND EFFICIENCY ANALYSIS

Circuit model	Parameters	Efficiency
System without capacitors	f = 50 Hz U_i = 230 V, P_i = 373 kW R_o = 410 □, U_o = 4.1 V, I_o = 0.01 A, P_o = 4 W	$\eta \approx 0$ %
System without capacitors	f = 100 kHz U_i = 230 V, P_i = 13.29 W R_o = 410 □, U_o = 72.9 V, I_o = 0.178 A, P_o = 13 W	η = 97.8 %
System with capacitors connected series-parallel	f = 100 kHz U_i = 230 V, P_i = 1.188 kW R_o = 410 □, U_o = 696 V, I_o = 1.7 A, P_o = 1.182 kW	η = 99.5 %

This simple example shows that, in order to transmit energy with high efficiency, it is necessary, apart from supplying a set of coupled coils via a highfrequency voltage source, to compensate the voltage drop on leakage inductance. There are several ways of attaching compensating capacities, but only two are of interest while considering the high power systems. For that reason, a system with series-parallel compensation has been chosen for further, more detailed analysis.

Resonant circuit in series-parallel system indicates a case in which one capacitor has been connected in series to the primary side winding, while the second capacitor has been connected in parallel to the secondary side winding. The capacities of primary side capacitor C_p and secondary side capacitor C_s have been selected in such a way that voltage U_i and current I_i, which supply the system, are in phase for a certain resonant pulsation ω_o, with the assumed initial factor of magnetic coupling k_o. This condition is fulfilled when:

$$C_p = \frac{1}{\omega_o^2 L_p \left(1 - k_o^2\right)} \quad C_s = \frac{1}{\omega_o^2 L_s} \qquad (1)$$

The above relations have been obtained during the analysis of a simplified circuit, i.e. while assuming that $R_p = R_s = 0$.

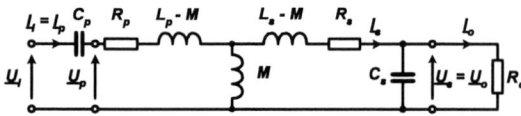

Fig. 2. Equivalent diagram of the system of air coils magnetically coupled with capacitor in series-parallel connection.

With the capacitors thus selected and while assuming that the primary and secondary coils are identical, i.e. $R_p =$

$R_s = R$, $L_p = L_s = L$, at the resonant working point $\omega = \omega_o$, $k = k_o$, a circuit like the one presented in Fig. 2 is characterised by the efficiency:

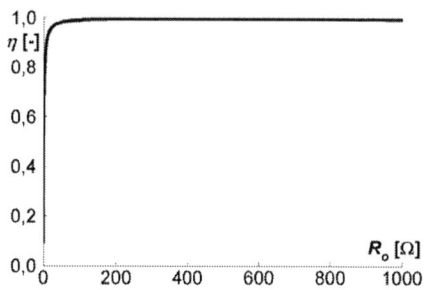

$$\eta = \frac{1}{1 + \dfrac{R}{R_o}\left(1 + \dfrac{1}{k_o} + \dfrac{1}{Q^2}\left(1 + \dfrac{1}{k_o^2}\left(\dfrac{R}{R_o}\left(2 + \dfrac{R}{R_o}\left(1 + \dfrac{1}{Q^2}\right)\right)\right)\right)\right)} \qquad (2)$$

where η – efficiency, $Q = \omega_o L / R_o$.

The efficiency defined by (2) reaches its maximum value when the load resistance equals:

$$R_o = \omega_o L \sqrt{1 + \frac{\omega_o^2 L^2}{\omega_o^2 L^2 k_o^2 + R^2}} \qquad (3)$$

Fig. 3 presents the efficiency in function of resistance R_o. The characteristic feature of this diagram is independent of load, constant value of efficiency in the wide scope of changes in resistance value R_o.

Fig. 3. Efficiency η of series-parallel system in function of load resistance R_o

For the determined working points described by (1) and (2), indicating power transmission with the highest efficiency, the power transmitted to the receiver does not reach its maximum value. In the case when $P_o = P_{omax}$ the efficiency equals 50%.

The efficiency also depends on the magnetic coupling factor, whose changeability is connected with changes in geometrical dimensions of the air gap. The changes of factor k significantly affect the parameters of the equivalent diagram. As a result, the system does not work any longer at the resonant point. Due to this, there are changes in the efficiency and the level of transmitted power. In order to maintain the working state at high efficiency, it is necessary to adjust the frequency of supply voltage in accordance with the changes of magnetic coupling factor.

Fig. 4a presents a diagram showing the changeability of efficiency in function of frequency and the coupling factor k, while the constant load resistance determined on the basis of (3) is maintained. The characteristic feature of the function drawn is a substantial range of k and f values for which the efficiency remains at a high level.

The characteristic representing the receiver power in function of f and k has also been drawn for the case analysed (Fig. 4b). The most specific feature of this diagram is the increase of the receiver power with the decrease, in a certain range of changes, of the magnetic coupling factor.

a)

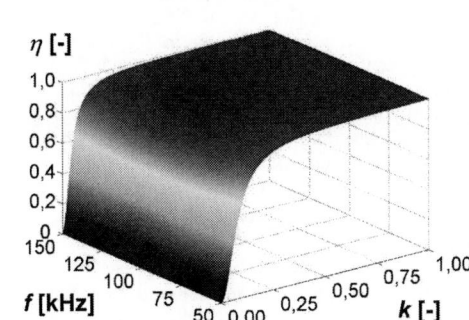

b)

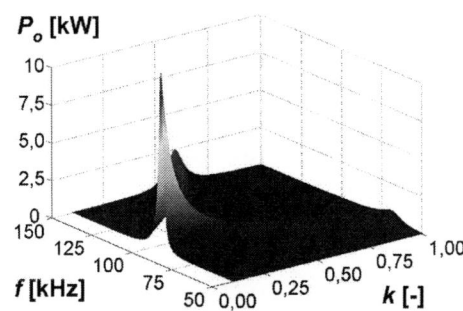

Fig. 4. Efficiency η and power of the receiver P_o of CEETS system in series-parallel configuration in function of supply voltage frequency f and coupling factor k: a) efficiency η; b) power of the receiver P_o

This denotes a situation in which the secondary circuit becomes more distant from the primary side. This results in appearance, on the terminals of the receiver, of the voltage, which is much higher than at a normal working point. This process is a disadvantageous phenomenon. For this reason, the frequency of voltage U_i must be controlled in such a way that, in case when the coils become distant from each other, the system would still be functioning with high efficiency. At the same time, the transmitted output power should be maintained at a certain, acceptable level.

This working area of the system is presented in detail by the characteristic in Fig. 5.

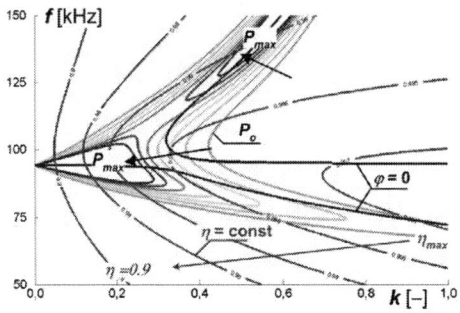

Fig. 5. The working area of CEETS system in series-parallel configuration, where ⓝefficiency, ⯑= 0 - phase shift and Po - power of the receiver with constant frequency lines marked

The characteristic representing the receiver power in function of f and k has also been drawn for the case analysed.

III. CONVERTER SYSTEM

Up till now, has been discussed functioning of the system in steady-state condition, forced by sinusoidal supply. Fig. 6 shows a circuit of CEETS system with converter supply.

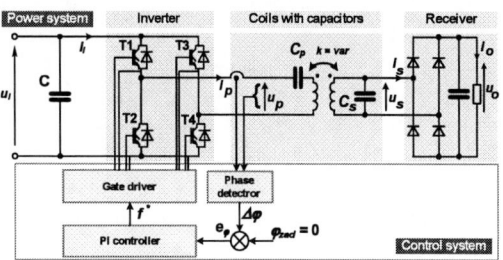

Fig. 6. Schematic diagram of CEETS with control system

Taking into consideration practical solutions used at present, the realisation of a singlephase voltage source with adjustable amplitude and frequency is possible on the basis of a voltage inverter [7, 8, 9, 10, 11]. A inverter with symmetrical voltage has been applied, which makes it necessary for the switching elements of the inverter to be arranged in either a halfbridge or a bridge configuration [12, 13, 14, 15].

A bridge configuration has been chosen for the research due to its merits in comparison with the halfbridge configuration [16, 17], which are as follows:

- the possibility of obtaining higher current values with the same supply voltage;

- filtering capacitors of lower value may be used (lack of twosection supply);

- the receiver voltage remains symmetrical even when the current flowing to the load has different average values for positive and negative semiwaves;

- there is a possibility of generating a zero vector (no energy flow).

This denotes a situation in which the circuit shown in Fig. 2 is supplied with rectangular-shaped voltage. It results in significant changes in the way the system behaves. The influence of deformed voltage on i_i current is not unimportant. The degree of this influence depends on basic frequency and the value of k factor. The analysed system may be treated as a kind of filter with changeable parameters. For example, Fig. 7a presents the amplitude frequency characteristics of the I_i current for the circuit as the one presented in Fig. 2, while assuming that $k = 0,7$. The courses of input current and voltage are shown in Fig. 7b and 7c respectively. It is worth mentioning that in the situation when the system is supplied with rectangular-shaped voltage, we cannot talk of resonant working points. Analogically, it is possible to create a situation in which the time shift between the consecutive transition of u_i and i_i through zero does not occur.

1499

a)

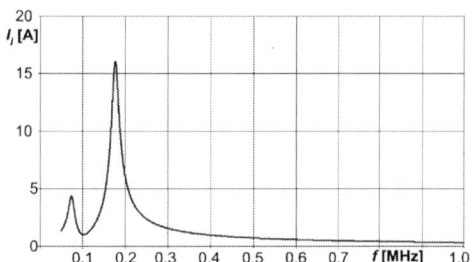

b)

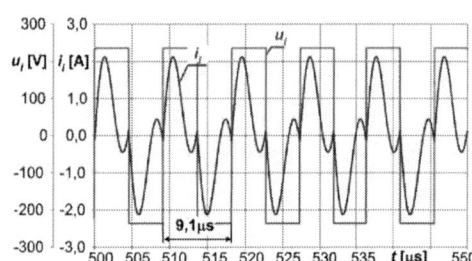

c)

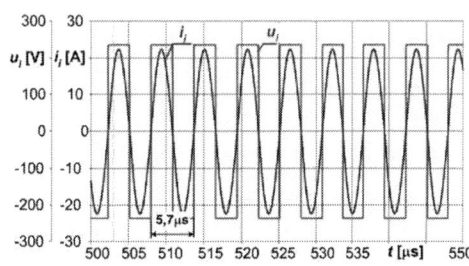

Fig. 7. Results of the analysis of the circuit with capacities in series-parallel configuration, with $k = 0,7$; a) frequency characteristics $I_i(f)$; b) courses of voltage u_i and current i_i for $f_o = 110$ kHz; c) $f_o = 175$ kHz

The effect of the system being dislocated from the working point, with no time shift between the voltage courses u_p and i_p, which accompanies the changes in air gap dimensions can be removed by the frequency regulation device [18, 19].

The contactless system of electric energy transmission with the frequency regulation system which responds to changes in time shift between the courses of current and voltage of the air transformer primary side shows the features characteristic for phase-locked loop systems PLL [20]. The function of voltage controlled frequency generator is performed in this case by a voltage inverter controlled by microprocessor. The idea of applying PLL technology in highfrequency, resonant systems is often discussed in literature. Solutions of this kind are used in constructing supply converter control systems in induction heating units [21, 22, 23]. The implementation of control via an inverter in accordance with the algorithm of software phase locked loop (SPLL) in BSPEE system is a new solution. The diagram of BSPEE model used for the analysis is presented in Fig. 10.

IV. RESEARCH OF THE CEETS MODEL

In order to verify theoretical assumptions, have been constructed a laboratory model of the CEETS system, based on the schematic diagram presented in Fig. 10. The system includes: supply source with rectifier and filter capacitor, converter with high-frequency IGBT transistors, set of magnetically coupled air-spaced coils with adjustable air gap, set of capacitors in series-parallel configuration (C_p, C_s), output rectifier with filter capacitor C and load R_o.

The test stand has been additionally equipped with PC computer with a starter-kit for DSP microcontrollers, as well as with measuring instruments.

Magnetically coupled coils are the main part of the CEETS system. It has been chosen for the research the simplest coil construction presented in Fig. 8.

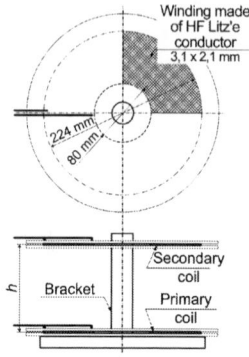

Fig. 8. Geometry of laboratory model of magnetically coupled air-spaced coil set

This laboratory model is a specific configuration of core-less transformer, whose air gap changes when the system functions normally. The coils were made of litz wire. Self inductance of each of them amounts to $L_p = L_s = L = 205$ μH, while the resistance is $R_p = R_s = R = 0.1$ Ω. Circuit also included a specially constructed set of compensating capacitors with capacities selected in accordance with (1) and (2). As a result it has been obtained $C_p = 12$ nF and $C_s = 14$ nF.

Due to specific functioning of the CEETS system where the coils are coupled with a big air gap $h = 50$ mm, has been performed calculations with regard to the distribution of magnetic field generated by the coils. The results are shown in Fig. 9.

The demonstrative case presented here corresponds to the situation when the current is $I = 4.5$ A flows through the lower winding. This current intensity is 3 times higher than that flowing through the upper coil. Both coils have the same number of turns $n = 36$. One can notice that the distribution of the magnetic field around the coils is clearly asymmetrical, where higher magnetic flux denisity values appear on that side of the winding through which flows the current of higher intensity. One of the characteristic features of the CEETS system is that the current value is higher on the primary side (the one supplied with energy) than on the secondary side. At the same time, the primary circuit is constructed as a stationary one and located in the ground, while the primary side is located in the vehicle. The phenomenon of unsymmetrical distribu-

1500

tion of the magnetic field is more advantageous as has been taken into consideration the potential influence of high frequency magnetic field on passengers. Shielding may be used as an additional treatment limiting the area of magnetic field expansion. It seems advisable to use ferrite screens.

a)

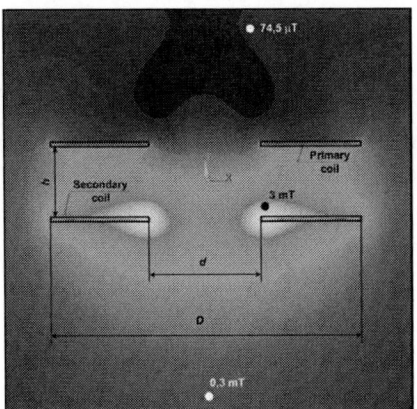

b)

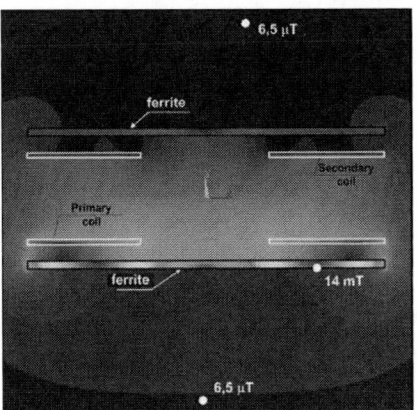

Fig. 9. Distribution of magnetic induction for: a) two magnetically coupled air-spaced coils; b) shielded coupled coils

Fig. 9b shows the distribution of the magnetic field in a situation analogous to the one shown in Fig. 9a, but making allowance for a double screen in the shape of ferrite panels. These panels, placed at the distance of 10 mm from the coils, have relative magnetic permeability $\mu_r = 3000$, thickness $b_f = 10$ mm and the diameter equal to outer diameter of the coils.

Fig. 11 presents the simulation results for functioning of CEETS control system with converter supply. The system was modelled as shown in Fig. 10. Library models, describing semiconducting power elements applied in the testing system, have been used. The simulation presents the situation where the magnetic coupling factor k changes rapidly from the level of 0.3 to 0.5 (Fig. 11a, b). This corresponds to the change of distance between the coupled circuits from 50 to 30 mm. The change in the overall dimensions of the air gap corresponds to the change in such parameters of the equivalent diagram as leakage inductances and main inductance. The resonant frequency of the circuit changes as well. The phase synchronization loop system used here increases the frequency of supply voltage. The system passes on to a new working point retaining high efficiency and increased value of the power transmitted to the receiver (Table 2).

TABLE II.
COMPILED SIMULATION RESULTS

Parameter	$k = 0.33$	$k = 0.5$
P_i	1.20 kW	1.35 kW
P_o	1.13 kW	1.27 kW
η	94.1 %	94.1 %
i_i	9.07 A	10.46 A
i_o	2.40 A	2.57 A

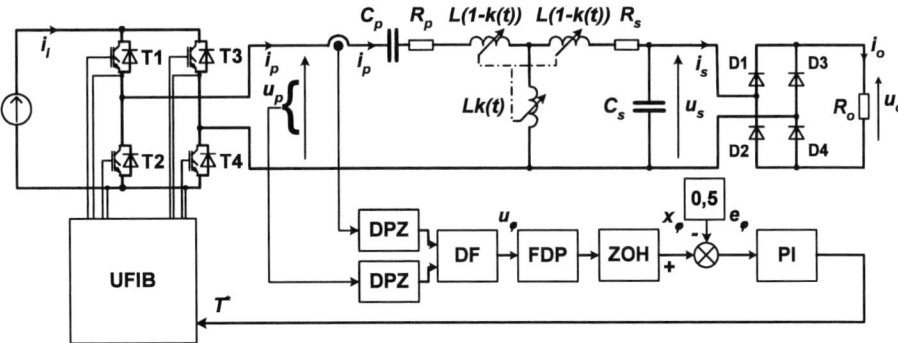

Fig. 10. Block diagram of control system, where DPZ – zero crossing detector, DF – phase detector built on the basis of XOR function, FDP – low-pass filter of the first order, ZOH – Zero -oder Hold), PI – controller of proportional-integral type, UFIB –gate-pulse formation system

1501

a)

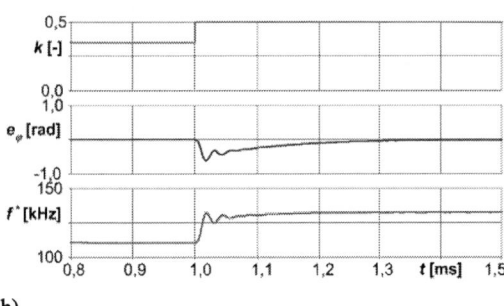

b)

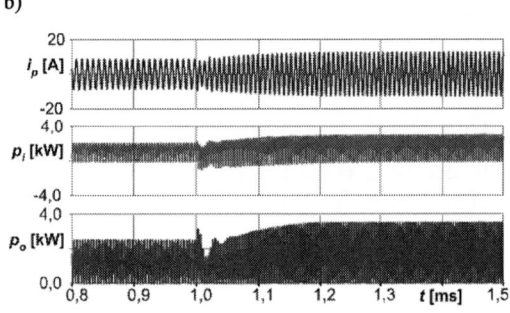

c)

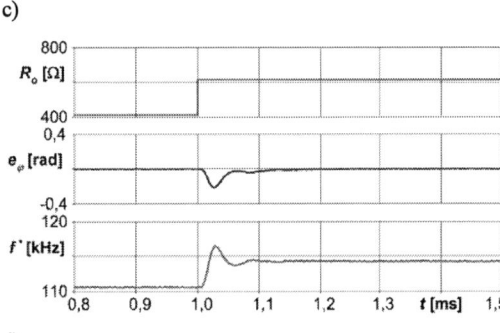

d)

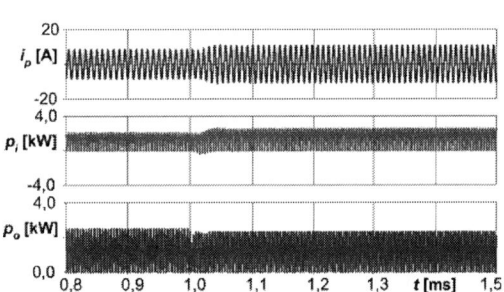

Fig. 11. Selected sequences in CEETS system: a) step change of magnetic coupling factor k, phase displacement e_f, supplied frequency f; b) primary side current I_p, power obtained from the source p_i, receiver power p_o; c) step change of output resistance R_o, phase displacement e_f, supplied frequency f; b) primary side current I_p, power obtained from the source p_i, receiver power p_o

System operation is affected not only by changes in magnetic coupling between the BSPEE windings, but also

by changes occurring on the receiver side. In order to illustrate the processes taking place in the system when the load conditions vary, calculations for sudden change of resistance R_o have been performed. The results obtained are presented in Fig. 11c, d. The courses refer to the situation when R_o resistance changes by 50%, from 410 Ω to 605 Ω. Fig. 12 presents the experimantal results of system research for $k = 0.33$, $f = 140$ kHz, with $R_o = 400$ Ω.

a)

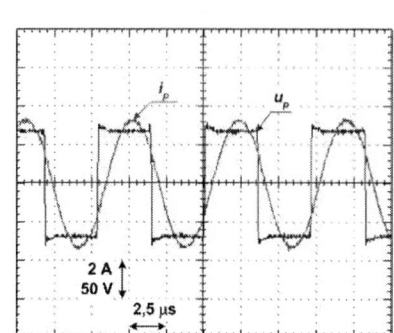

b)

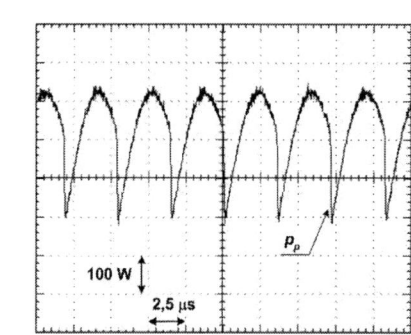

Fig. 12. Courses of selected values characteristic for BSPEE system with converter supply, where $h = 33$ mm and $f = 140$ kHz: a) voltage u_p and current i_p on the primary side; b) receiver voltage u_o; c) instantaneous power on primary side p_p.

In CEETS system working without a control unit, the change of parameters results in a disadvantageous change of working point. Exemplary courses of voltage, current and power value for increased frequency of supply voltage have been presented. These courses show that the moment of i_p current course transition through zero precedes in time the moment of u_p voltage course transition through zero. In such a situation, the active power drawn on the primary side $P_p = 132$ W and the active power of the receiver $P_o = 116$ W. Consequently, the obtained level of power transmission efficiency reached $\eta = 88$ %.

Results of measurements for the supply frequency at which there is full synchronisation between u_p voltage and i_p current are presented in Fig. 13. In such case, the active power on primary side $P_p = 840$ W, the receiver power $P_o = 764$ W and the efficiency is $\eta = 91$ %.

a)

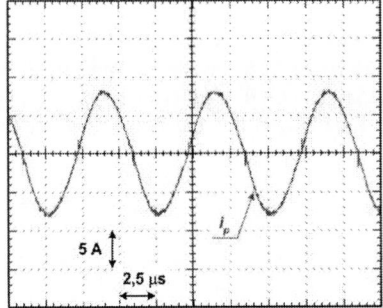

5 A

2,5 µs

b)

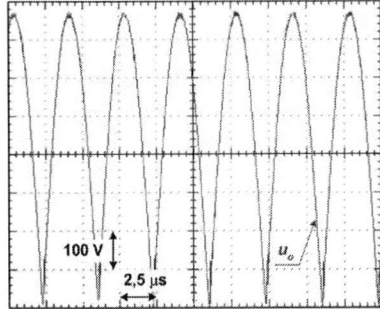

100 V

2,5 µs

u_o

Fig. 13. Courses of selected values characteristic for BSPEE system with converter supply, where $h = 33$ mm and $f = 133$ kHz: a) current i_p on the primary side; b) receiver voltage u_o (reference level below the main axis)

V. SUMMARY

Contactless transmission of electrical energy is a developing issue – particularly with reference to supplying autonomous electric vehicles. Supplementing energy during a standstill, and particularly a working standstill – at a stop, will significantly improve exploitation parameters of autonomous electric traction vehicles.

This elaboration contains theoretical, simulation and experimental analysis of contactless electrical energy transmission system (CEETS). The analysis of steadystate condition, forced by sinusoidal supply has also been presented. These calculations are based on equivalent transformer circuit of the configuration of magnetically coupled air coils. This analysis made it possible to determine the structure and the area of such functioning of the system, which is characterized by maximal efficiency.

In the operation character analysed here, it is a nonstationary system. This results from the changes of magnetic coupling factor which may be caused, for example, by slow relocation of primary winding in relation to secondary winding.

The distribution of magnetic field around the chosen geometrical configuration of air-spaced coils has been determined. It has been indicated that there is a possibility of limiting the values of spatial distribution of the magnetic field, which improves the technical parameters of CEETS.

The paper also presented the results of system simulation with converter supply, taking into consideration models of power electronics elements. The results shown refer to the functioning of the system in a quasi steady-state condition with control configuration of the phase angle between the current and the primary side voltage of magnetically coupled coil set. The results show us that the system is capable of transmitting electrical energy at high level with substantial efficiency, also in a situation when the parameters of the system, particularly the magnetic coupling factor, undergo dynamic changes.

We have determined the working area of the non-contact system of electrical energy transmission, which guarantees high value of the power transmitted coupled with high efficiency and small distortions of current from its sinusoidal course. Then we have proposed an original system controlling the frequencies of supply inverter output voltage. The only thing indispensable for the operation of the control unit is the measurement of its primary side current.

The obtained results may form the basis of further simulation and experimental research, particularly for:

- Construction and research of physical model of BSPEE unit loaded with set of (super) capacitors and/or the set of batteries;

- Simulation research of electromagnetic field distribution when it is surrounded by a different geometrical configuration of coils than the one we have suggested. The research would be focused on looking for the optimal structure, e.g. with regard to minimising the leakage magnetic flux emitted to the surroundings, forming the frequency characteristics etc.;

- Research of electromagnetic field emitted by the installation while working on a physical model. Research of this issue has not yet been undertaken, although it is extremely important for the assessment of BSPEE electromagnetic compatibility during its exploitation in the working area of passenger or industrial transport vehicles.

REFERENCES

[1] Pedder D. A. G., Brown A. D., Skinner J. A.: *A contactless electrical energy transmission system*. Industrial Electronics, IEEE Transactions on Volume 46, Issue 1, Feb. 1999 Page(s): 23 – 30

[2] Stielau O. H., Covic G. A.: *Design of loosely coupled inductive power transfer systems*. Power System Technology, 2000. Proceedings. PowerCon 2000. International Conference on Volume 1, 4-7 Dec. 2000 Page(s): 85 – 90 vol.1

[3] Mecke R., Rathge Ch.: *High frequency resonant inverter for contactless energy transmission over large air gap*. Power Electronics Specialists Conference, 2004. PESC 04. 2004 IEEE 35th Annual Volume 3, 20-25 June 2004 Page(s): 1737 - 1743 Vol.3

[4] Laouamer R., Brunello M., Ferrieux J. P., Normand O., Buchheit N.: *A multi-resonant converter for non-contact charging with electromagnetic coupling*. Industrial Electronics, Control and Instrumentation, 1997. IECON 97. 23rd International Conference on, Volume 2, 9-14 Nov. 1997 Page(s): 792 – 797 vol.2

[5] Wu Ying, Yan Luguang, Xu Shangang: *A new contactless power delivery system*. Electrical Machines and Systems, 2003. ICEMS 2003. Sixth International Conference on, Volume 1, 9-11 Nov. 2003 Page(s): 253 – 256 vol.1

[6] Wu Ying, Yan Luguang, Xu Shangang: *Modeling and Performance Analysis of the New Contactless Power Supply System*. Electrical Machines and Systems, 2005. ICEMS 2005. Proceedings of the Eighth International Conference on Volume 3, 27-29 Sept. 2005 Page(s): 1983 – 1987

[7] Harada K., Sakamoto H., Shoyama M.: *Phase-controlled DC-AC converter with high-frequency switching*. Power Electronics, IEEE Transactions on Volume 3, Issue 4, Oct. 1988 Page(s): 406 – 411

[8] Steigerwald R. L.: *A comparison of half-bridge resonant converter topologies*. Power Electronics, IEEE Transactions on Volume 3, Issue 2, April 1988 Page(s): 174 – 182

[9] Steigerwald R. L.: *Practical design methodologies for load resonant converters operating above resonance*. Telecommunications Energy Conference, 1992. INTELEC '92., 14th International 4-8 Oct. 1992 Page(s): 172 – 179

[10] Steigerwald R. L., Roshen W., Saj C. F.: *A high-density 1 kW resonant power converter with a transient boost function*. Power Electronics, IEEE Transactions on Volume 8, Issue 4, Oct. 1993 Page(s): 431 – 438

[11] Yin Y., Zane R.: *Digital phase control for resonant inverters*. Power Electronics Letters, IEEE Volume 2, Issue 2, June 2004 Page(s): 51 – 53

[12] Hayes J. G., Egan M. G., Murphy J. M. D., Schulz S. E., Hall J. T.: *Wide-load-range resonant converter supplying the SAE J-1773 electric vehicle inductive charging interface*. Industry Applications, IEEE Transactions on, Volume 35, Issue 4, July-Aug. 1999 Page(s): 884 – 895

[13] Hayes J. G., Hall J. T., Egan M. C., Murphy J. M. D.: *Full-bridge, series-resonant converter supplying the SAE J-1773 electric vehicle inductive charging interface*. Power Electronics Specialists Conference, 1996. PESC '96 Record., 27th Annual IEEE, Volume 2, 23-27 June 1996 Page(s): 1913 – 1918 vol.2

[14] Kutkut N. H., Klontz K. W.: *Design considerations for power converters supplying the SAE J-1773 electric vehicle inductive coupler*. Applied Power Electronics Conference and Exposition, 1997. APEC '97 Conference Proceedings 1997., Twelfth Annual, Volume 2, 23-27 Feb. 1997 Page(s): 841 – 847 vol.2

[15] Moisseev S., Soshin K., Sato S., Gamage L., Nakaoka M.: *Novel soft-commutation DC-DC power converter with high-frequency transformer secondary side phase-shifted PWM active rectifier*. Electric Power Applications, IEE Proceedings-Volume 151, Issue 3, 8 May 2004 Page(s): 260 – 267

[16] Citko T., Tunia H., Winiarski B.: *Układy rezonansowe w energoelektronice*. Białystok, Wydawnictwa Politechniki Białostockiej 2003

[17] Nowak M., Barlik R.: *Poradnik inżyniera energoelektronika*. Warszawa, WNT 1998

[18] Judek S, Karwowski K.: *Sterowanie falownikiem napięcia w układzie cewek sprzężonych do zasilania pojazdów elektrycznych*. Pomiary Automatyka Kontrola - Vol. 53, nr 4 (2007), s. 60-63

[19] Judek S., Karwowski K.: *Zasilanie pojazdów elektrycznych za pośrednictwem cewek powietrznych sprzężonych magnetycznie*. Eight International Conference Modern Electric Traction in Integrated XXI Century Europe, Warszawa 2007

[20] Best R. E.: *Phase-locked loops. Design, simulation, and applications*. New York, McGraw-Hill 2003

[21] Bayindir N. S., Kukrer O., Yakup M.: *DSP-based PLL-controlled 50-100 kHz 20 kW high-frequency induction heating system for surface hardening and welding applications*. Electric Power Applications, IEE Proceedings-Volume 150, Issue 3, May 2003 Page(s):365 – 371

[22] Tian J., Petzoldt J., Reimann T., Scherf M., Berger G.: *Modelling of asymmetrical pulse width modulation with frequency tracking control using phasor transformation for half-bridge series resonant induction cookers*. Power Electronics and Applications, 2005 European Conference on 11-14 Sept. 2005 Page(s):9 pp.

[23] Yu-Long Cui, Kun He, Zhi-Wei Fan, Hao-Liang Fan: *Study on DSP-based PLL-controlled superaudio induction heating power supply simulation*. Machine Learning and Cybernetics, 2005. Proceedings of 2005 International Conference on Volume 2, 18-21 Aug. 2005 Page(s):1082 - 1087 Vol. 2

Sliding-Mode Approach to Control Design for Induction Motor Drive fed by a Three-Level Voltage-Source Inverter

Sergey Ryvkin[*], Richard Schmidt-Obermoeller[†] and Andreas Steimel[†]

[*] Russian Academy of Sciences/Trapeznikov Institute of Control Sciences, Moscow, Russia, *email: rivkin@ipu.rssi.ru*
[†] Ruhr-University Bochum/Institute for Electrical Power Engineering and Power Electronics, Bochum, Germany,
emails: schmidt-o@eele.rub.de, steimel@eele.rub.de

Abstract—A novel sliding mode control for the induction motor (IM) fed by three-level voltage-source inverter (3L-VSI) is proposed in this paper. The "classical" results of sliding-mode theory were extended on the actual variable structure system with more then 2^m variable structures (m is the control space order) by using an original two-step design procedure that allows separately to use the nonlinearity of the IM and the switching character of the 3LVSI. The based on this design procedure original control algorithm that includes a choice condition for 3L-VSI input dc link voltage and a switch table for the 3L-VSI semiconductor switches is designed. The main advantages of such approach are given as follows: the control design is simpler and more graphical; the control system has such advantageous characteristics of the control plant with sliding mode control as high dynamic, low sensitivity to disturbance of both the load and the to plant parameter variations; the requirements to the quality of the input dc-link voltage and as a result to the dc-line filter capacitor are reduced; the measurement precision of the stator flux position is within $(\pi/6)$. Simulation results are presented to validate the theoretical analysis and demonstrate the performance of the proposed control.

Keywords—Control of drive, converter control, induction motor, multilevel converters, sliding mode control.

I. INTRODUCTION

Nowadays the three level voltage source inverter (3L-VSI) topology is one of the main used converter topologies in high- and medium power industrial drive applications [1]. In contrast to the classical two-level VSI such topology offers some additional benefits: a superior harmonic spectrum for a given switching frequency of the used power semiconductor switches, a lower overvoltage stress at the motor windings, a lower common-mode voltage, and substantially lower semiconductor switching losses. However for using all these advantages new advanced control techniques are needed. It connects with the complexity of a non-linear control plant having a series coupling of two non-linear objects different by nature: an induction motor (IM) and 3L-VSI equipped with power semiconductor switches as Insulated Gate-Commutated Transistors (IGCTs) or with high-voltage Insulated-Gate Bipolar Transistors (IGBTs) [2], operating in switch mode with distinctly higher switching frequency, than known from force-commutated thyristors. For the description of the IM behavior there are many different non-linear differential equations, whose elections depends on the used reference frame and the space

variables [3]. In contrast to the two-level VSI the 3L-VSI has 27 output voltage space vectors that can be used for the control design. The space vectors positions of some vectors depend on the neutral point voltage that is not constant opposite to the dc input voltage. Due the 3L-VSI aforementioned features many different control techniques for 3L-VSI were proposed that are based on the decomposition of the main high-order control task to several separated lower-order tasks, using linear control technique and heuristic solutions, e.g. Field-Oriented Control [4], Direct Self Control [5], Indirect Stator-Quantity Control [6], Direct Torque Control [7-9], various modulations techniques [10, 11].

The switching nature of the used in 3L-VSI power semiconductor switches opens the possibility to use the principal operational mode of variable structure systems – sliding mode – for solving the control task. The last decades the sliding mode control technique is an attractive one in the field of the drives fed by the converter equipped with power semiconductor switches due to the following merits. It allows reducing the system order and to simplify and decouple the control design procedure and to give the system insensitivity with respect to disturbances and plant parameter variations [12 - 14].

This paper presents a control design approach based on sliding mode control design for the IM drive fed by new class of the converters: 3L-VSI. The significantly different feature from the "classical" sliding-mode theory is that the real switched system has more than 2^m switched structure (m is the control space order). The presented solution is based on the two-step design procedure that allows separate the non-linearity of the IM and 3L-VSI.

The paper can be outlined as follows. Section II deals with a brief introduction of used sliding mode control background. A presentation und discussion of used IM and 3L-VSI models are made in Section III. Section IV presets the control task formulation and considers a two-step sliding mode control design technique. The design result is the control for the above mentioned drive system. The results of the numerical simulation examination that illustrate the properties of the suggested sliding mode control are presented in Section V followed by conclusions.

II. SLIDING MODE BACKGROUND

Sliding Mode is a special type of behavior of a control plant with variable structures:

978-1-4244-1741-4/08/$25.00 ©2008 IEEE

$$dx(t)/dt = l(x,t) + B(x,t)u(t), \qquad (1)$$

$$u_i(x,t) = \begin{cases} u_i^+(x,t) & \text{if } S_i(x,t) > 0 \\ u_i^-(x,t) & \text{if } S_i(x,t) < 0 \end{cases}, \qquad (2)$$

where $x(t)$ is a state vector, $x(t) \in R^n$; $l(x,t)$ is a system vector, $l(x,t) \in R^n$; $u(t)$ is a control vector, $u(t) \in R^m$, $n \geq m$; $u_i(t)$ is a component of the control vector $u(t)$, $i = \overline{1,m}$; $B(x,t)$ is a matrix, $B(x,t) \in R^{n \times m}$; $S_i(t)$ is a switching function.

The main feature of this mode is that none of the used general switched structures can realize such behavior. The sliding mode occurs in the intersection of all m surfaces

$$f_i = 0 \qquad (3)$$

by using a switching of the control components $u_i(x,t)$ with high frequency. The vector function $F^T = (f_1, ..., f_m)$ is a function of the system variables $x(t)$ and is usually an error function that must be led to zero by using the vector switching function $S^T = (S_1, ..., S_m)$.

Formally, the control aim is the following: the system state $x(t)$ must come to the manifold

$$F = 0 \qquad (4)$$

and "slides" on this manifold to the reference point, independently of the system dynamic.

In this case the control design procedure is decoupled to the two tasks:

- sliding mode design in the space of the vector function $F \in R^m$;

- motion design on the intersection of all m surfaces in the state space with order $(n-m)$.

The first task solution is based as it is considering in sliding mode analysis on the Lyapunov stability of the control plant in the space of the vector function $F \in R^m$ [12]. A typical sliding mode control u has the form

$$u = -U(x)\operatorname{sgn}(F), \qquad (5)$$

where; $U(x)$ is the square diagonal matrix of the control magnitude; $\operatorname{sgn}(F)$ is the vector of the signs of the error functions, $[\operatorname{sgn}(F)]^T = (\operatorname{sgn} f_1, ..., \operatorname{sgn} f_m)$. It guarantees that the system state will reach the sliding manifold in finite time from the initial condition, which has been bounded by the value of the constituent of the matrix $U(x)$, and will keep to it. This magnitude bounds the uncertainty of the system, the load value unto which the system is commonly robust.

The motion on the sliding manifold is described by using the equivalent control $u_{eq}(x)$ [12]. It is calculated from the condition that the time-derivative of the function F on the system trajectories is equal to zero

$$dF / dt = Gl + GBu = 0, \qquad (6)$$

where $G(x) = \{\partial F/\partial x\}$, $G(x) \in R^{m \times n}$ is a gradient matrix; $\det GB \neq 0$.

In this case the equivalent control $u_{eq}(x)$ that is a continuous control that would guarantee the same motion, if all needed information about the load and the system uncertainty were available is calculated

$$u_{eq} = (GB)^{-1}Gl, \qquad (7)$$

and the system motion in the sliding mode is described as

$$dx/dt = l - B(GB)^{-1}Gl \qquad (8)$$

together with (4). Using (4) the system order can be reduced to $(n-m)$, and the system description is

$$dx_1/dt = l_1[(x_1, t], \qquad (9)$$

where $x^T = (x_1, x_2)$, $x_1 \in R^{n-m}$, $x_2 \in R^m$; $l_1[(x_1, t)]$, $l_1 \in R^{n-m}$.

III. Used Mathematical Models of 3L-VSI and IM

A. 3L-VSI

At present the most extensively applied multilevel inverter topology is the 3L-VSI that is the three-phase bridge with neutral-point-clamped inverter [1] (Fig.1). Since all semiconductors are operated at a commutation voltage of half the dc-link voltage U_d, the topology offered a simple solution to extend voltage and power ranges of the existing VSI technology, which was severely limited by blocking voltage of power semiconductor with active turn-on and turn-off capabilities. Note that the switch of one leg (S1, S2, S3) has a three positions and the output terminals a,b,c of 3L-VSI are connected by using three position semiconductor switches (S1, S2, S3) either to the positive (L+) or the negative (L-) rail of the dc link or to the middle potential (M) between positive and negative potential (neutral point), accordingly the switch controls are "+", "-" and "0". Combining the state of all three phases switch controls, the 3L-VSI features 27 possible switching combinations and their corresponding output voltage space vectors.

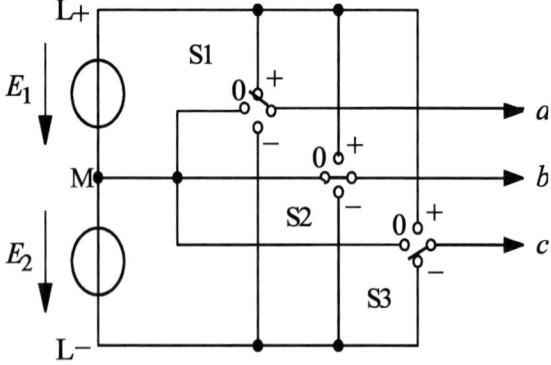

Fig. 1. . Canonical schema of neutral-point-clamped 3L-VSI

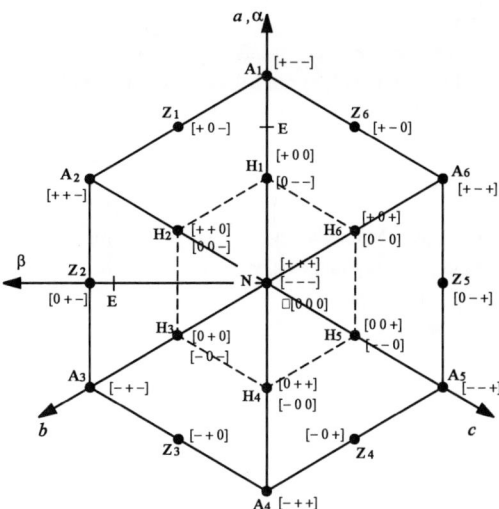

Fig. 2. Voltage space vectors generated by a 3L-VSI

If the middle potential is balanced (E1 = E2) the number of the output voltage space vectors is reduced to 19 ones. They are plotted in the (α, β) frame (Fig. 2), simply by calculating the space vector associated to each switching state. Note that some switching states have redundant space vector representation that can be used for neutral-point balance purposes. These output voltage space vectors are divided in four groups depending on their magnitudes. There are six full voltage space vectors (A) with magnitude $B_A = 2U_d/3$, where $U_d = E_1 + E_2$; six intermediate voltage space vectors (Z) with magnitude $B_Z = \sqrt{3}U_d/3$; six half-voltage space vectors with magnitude $B_H = U_d/3$, attained by two switch combinations, and the zero-voltage vector (N), with three possible switch combinations.

B. IM

One of the classic models of a symmetrical balanced squirrel-cage IM that has been written in the stationary reference frame (α, β) and with rotor quantities referred to stator [3] is used:

$$d(\omega/p)/dt = (T - T_L)/J$$
$$d\Psi_{s\alpha}/dt = -R_s i_{s\alpha} + u_{s\alpha}$$
$$d\Psi_{s\beta}/dt = -R_s i_{s\beta} + u_{s\beta}$$
$$di_{s\alpha}/dt = [R_r \Psi_{s\alpha}/(L_s L_r) + \omega \Psi_{s\beta}/L_s]/\sigma - \quad (10)$$
$$- \omega i_{s\beta} - \gamma i_{s\alpha}/L_s + u_{s\alpha}/L_s$$
$$di_{s\beta}/dt = [R_r \Psi_{s\beta}/(L_s L_r) - \omega \Psi_{s\alpha}/L_s]/\sigma +$$
$$+ \omega i_{s\alpha} - \gamma i_{s\beta}/L_s + u_{s\beta}/L_s,$$

where ω is the electrical velocity; J is the inertia and T is the electromagnetic torque

$$T = 3p(\Psi_{s\alpha} i_{s\beta} - \Psi_{s\beta} i_{s\alpha})/2, \quad (11)$$

T_L is the load torque; $\Psi_s^T = (\Psi_{s\alpha}, \Psi_{s\beta})$ is the stator flux; $i_s^T = (i_{s\alpha}, i_{s\beta})$ the stator current; $u_s^T = (u_{s\alpha}, u_{s\beta})$ the stator voltage; R and L denote resistance and self inductance, subscripts s and r stand for stator and rotor; p is the pole number; $\sigma = 1 - L_m^2/L_s L_r$, with L_m being the mutual inductance; $\gamma = L_s R_r/L_r - R_s$.

IV. SLIDING MODE CONTROL DESIGN

A. Control Task

The electrical drives must convert the input electrical form of energy in the mechanical one with the reference value of the mechanical output variable (for example rotor speed or position). One of the most effective control schemes for drives is a cascaded or nested structure with a fast inner control loop, limiting the target torque T at the motor shaft. In this case the control design is decomposed to the two more simple tasks: the torque control design and the mechanical variable one. The inner torque loop includes the converter. The inner loop controller has to deal primarily with the dynamics of the power supply and the armature, canceling the effect of the induced armature voltage; thus it provides the fastest available control action. The control space order of the inner loop (in our case of the stator voltages space) is equal to two, i.e. there are two independent controls, and it's possible to control two variables: not only the torque, but also another variable, which describes electrical or power behavior of the IM. From the viewpoint of the torque control this is the stator-flux modulus

$$|\Psi_s| = \sqrt{\Psi_{s\alpha}^2 + \Psi_{s\beta}^2} \quad (12)$$

So, the control aim can be formulated as is following: the error function F whose dimension is two and equals to the number of the independent controls

$$F = \begin{vmatrix} F_\Psi \\ F_T \end{vmatrix} = \begin{vmatrix} \|\Psi_{sz}| - |\Psi_s\| \\ T_z - T \end{vmatrix} \quad (13)$$

must be led to zero. (In all equations the reference values have subscript z).

One of the possible ways for solving such control task is the organization of the sliding mode on the intersection of the surfaces (6), (13) and the design of the control system with all sliding mode merits. However there are some problems that are connected with the using sliding-mode technique for the drives. All "classic" results have been received for the case that the number of discontinuous controls is equal to the control order. In the electrical system the situation is different: the number of discontinuous controls is bigger than the control order. They have constant directions that cannot be changed. Only the dc-link voltage value is changeable and can be used for the solving of the control task together with switch control. In the case of the IM drive fed by a 3L-VSI there are only the above-mentioned 19 different output voltage space vectors for the solving of the control task in a two-dimensional control space. The control design aims to make the synthesis of the switching law and to elect the value of the dc-line voltage that produce the sliding-mode motion on the manifold (6), (13).

1507

For the sliding mode control design a two-step design technique [13] that allows using all results of the classic sliding mode theory can be used. This technique makes it possible to decompose the control design procedure and to take separately into account the nonlinearities of IM and 3L-VSI. At the first step the control is designed for the two-phase equivalent model of the IM, in which number of discontinuous controls is equal to the order of the voltage plane, by using the standard technique. As design results the election inequalities for the magnitudes of the needed stator-voltage components and the switching law are obtained.

At second step the real discontinuous output voltage of 3L-VSI has to be taken into account. The transformation law from two two-position discontinuous controls to the three three-position controls that guaranties the above-mentioned designed sliding mode has been proposed.

B. Sliding Mode Design by Using Two-Phase Control.

The discontinuous control is designed by using the equation of the error function variation that is written subject to the IM description (10) – (13).

$$
\frac{d}{dt}\begin{vmatrix} F_\Psi \\ F_T \end{vmatrix} = \begin{vmatrix} \dfrac{d|\Psi_{sz}|}{dt} + \dfrac{R_s(\Psi_{s\alpha}i_{s\alpha} + \Psi_{s\beta}i_{s\beta})}{|\Psi_s|} \\ \dfrac{dT_z}{dt} + \dfrac{1}{\sigma}(\dfrac{\gamma}{L_s} + \dfrac{p}{L_s}\omega\Psi_s^2) - p\omega(\Psi_{s\alpha}i_{s\alpha} + \Psi_{s\beta}i_{s\beta}) \end{vmatrix} -
$$
$$
- \begin{vmatrix} 1 & 0 \\ \dfrac{T}{|\Psi_s|} & \dfrac{1}{\sigma}\dfrac{|\Psi_s|}{L_s} + \dfrac{(\Psi_{s\alpha}i_{s\alpha} + \Psi_{s\beta}i_{s\beta})}{|\Psi_s|} \end{vmatrix} \times \begin{vmatrix} u_\Psi \\ u_T \end{vmatrix},
$$
(14)

where $U_F^T = (u_\Psi, u_T)$ is the used control vector that is written in the rotating coordinate frame and connected to the phase voltages using the transformation

$$
\begin{vmatrix} u_a \\ u_b \\ u_c \end{vmatrix} = \begin{vmatrix} \cos\rho & -\sin\rho \\ \cos(\rho - 2\pi/3) & -\sin(\rho - 2\pi/3) \\ \cos(\rho + 2\pi/3) & -\sin(\rho + 2\pi/3) \end{vmatrix} \times \begin{vmatrix} u_\Psi \\ u_M \end{vmatrix}, \quad (15)
$$

where $\rho = \arctan(\psi_{s\beta}/\psi_{s\alpha})$, the stator-flux angle.

The design aim will be fulfilled, if by using a designed control law the state reaches the sliding manifold (6), (13), in finite time from the bounded initial conditions, and then remain on it, i.e. the error function is zero.

Usually a well-designed IM has a very small coefficient σ (some 5%). In this case the system (14) is the singularly perturbed systems, and for the design of the sliding motion on the manifold (6), (13) can be used a technique based on the system decomposition on the slow and fast subsystems [15]. Because of the discontinuous nature of the slow controller, the time-scale separation argument does not hold during switching. Therefore, additional technical conditions have to be satisfied to guarantee stability. In this case the problem of sliding domain is reduced to a sequential analysis of two scalar cases. The scalar sliding mode condition is

$$
\lim_{F \to +0}(dF/dt) < 0 \ \& \ \lim_{F \to -0}(dF/dt) > 0, \quad (16)
$$

i.e. the signs of the error function and its derivation must be opposite.

The control vector components u_Ψ and u_T that are satisfied the condition (16) are selected depending upon the sign of the components of the error function F :

$$
u_\Psi = U_\Psi \operatorname{sgn} F_\Psi, \quad (17)
$$

$$
u_T = U_T \operatorname{sgn} F_T, \quad (18)
$$

and their selected magnitudes U_Ψ and U_T must be satisfied the following inequalities

$$
U_\Psi \ge U_{\Psi eq} = \left\| d|\Psi_{sz}|/dt + R_s(\Psi_{s\alpha}i_{s\alpha} + \Psi_{s\beta}i_{s\beta})/|\Psi_s| \right\|,
$$
(19)

$$
U_T \ge U_{Teq} = |dT_z/dt - T(d|\Psi_{sz}|/dt)/|\Psi_{sz}| + (\gamma +
$$
$$
+ p\omega\Psi_s^2)/(\sigma L_s) - (p\omega + R_s M/\Psi_{sz}^2)(\Psi_{s\alpha}i_{s\alpha} + \Psi_{sb}i_{s\beta})|
$$
. (20)

The magnitudes U_Ψ and U_T bound the initial condition, from which the state will reach the sliding manifold in finite time, and the uncertainty of the system and the load value to which the system is robust in general.

It must be paid attention to the fact that the control (17), (18) and inequalities (19), (20) determine in the control space rotating with the stator-flux four control areas $U_1^*, U_2^*, U_3^*, U_4^*$ (Tab. I):

$$
U^* = \{U^*\} = U_1^* \cup U_2^* \cup U_3^* \cup U_4^*, \qquad U_1^* \cap U_2^* = 0,
$$
$$
U_1^* \cap U_3^* = 0, \ U_1^* \cap U_4^* = 0, \ U_2^* \cap U_3^* = 0, \ U_2^* \cap U_4^* = 0,
$$
$$
U_3^* \cap U_4^* = 0.
$$

TABLE I.
SLIDING-MODE CONTROL AREAS

	U_1^*	U_2^*	U_3^*	U_4^*
sgn F_Ψ	1	1	-1	-1
sgn F_T	1	-1	1	-1

And any combination of the four control vectors each of them belongs to own control area guarantees sliding motion on the intersection of the surfaces (6) (13). This fact will be used on the second step of the control design.

C. Design of the Sliding Mode Control for 3L-VSI

However, as it is shown in the Section II.A, in the fact the IM control is a set of 19 different output voltage space vectors with four different magnitudes, produced by discontinuous control (switching) of the 3L-VSI. And it must be designed the transformation law that connects the fictive two dimension two positions discontinuous controls and actual three position discontinuous phase controls of 3L-VSI that attains the above designed sliding motion with these output vectors. There are two ways. The first way is to use the inverse transformation (15). However, the needed three-phase voltage would be sine wave. Using a feed-forward PWM could generate it. And the sine-wave voltages would be the mean average values of the output voltages of the semiconductor 3L-VSI with high-frequency switches. The organization of such PWM needs additional information about the average value of the output voltage space vector U_{eq} during any time period, a calculation of the switching time for each switch and a discipline of the sequence of their switching. The designed control does not reach any of the basic properties of sliding mode: the realization simplification.

The alternative control design approach, i.e. transferring the two-dimensional control (17) – (20) to the 3L-VSI control, is based on the above-mentioned fact, that the selection conditions of the magnitudes of the formally entered controls in the reference frame rotating with stator-flux uses the inequalities (19) and (20). In this case there is the area of allowable controls U^* in the space of the formally entered controls u_Ψ, u_T. It is obvious, if we design the real discontinuous voltages thus that their projections on suitable axes of the stator-flux-rotating frame have their marks and sizes, which are needed by the control algorithm with the formally entered controls, the sliding mode on crossing before the chosen surfaces will take place. Of course, the sizes of the formally entered controls will change during work.

These four control areas $U_1^*, U_2^*, U_3^*, U_4^*$ are transformed to the three-phase stator-winding-fixed frame (a, b, c) by using (15). They have the same form in this frame, but move with the stator-flux velocity (Fig. 3).

The sliding mode that has been synthesized on the fist design step can be secured by using the discontinuous voltages of the 3L-VSI, if each control area has at any time at minimum one of the 3L-VSI output voltage vectors. Fulfilling this condition it is necessary to calculate the 3L-VSI input dc voltage and to design a transition law between the designed control (17), (18) and the switching control of 3L-VSI switches. In contrast to [16] there are only two design assumptions: the used 3L-VSI output voltage space vectors are full ones and intermediate ones and the value of the 3L-VSI dc-link voltage is the minimal possible. A selection condition of the 3LVSI input dc-voltage value for geometrical reasons is

$$U_d \geq U_{d\min} = \max \left\{ \begin{array}{l} \sqrt{4U_T^2 + 3U_\Psi^2 + 2\sqrt{3}U_T U_\Psi}, \\ \sqrt{3U_T^2 + 4U_\Psi^2 + 2\sqrt{3}U_T U_\Psi} \end{array} \right\}, \quad (21)$$

and the transfer strategy from the controls $\operatorname{sgn} F_T$ and $\operatorname{sgn} F_\Psi$ to control of the switches S1, S2 and S3 is a logical Tables II and III. There are 12 $(\pi/6)$-sectors, where switch control is equal to one of the controls $\operatorname{sgn} F_T$ and $\operatorname{sgn} F_\Psi$ or their opposite values. The positions of the $(\pi/6)$-sectors depend on the values of the selected magnitudes U_Ψ, U_T and U_d and is moved opposite the $(\pi/6)$-stator-flux-angle ρ zones on the angle ξ

$$\xi = \pi/6 - \arcsin(\sqrt{3}U_\Psi / U_{d\min}). \quad (22)$$

TABLE II.
SWITCHES CONTROLS

ρ	S_a	S_b	S_c
(-ξ...π/6-ξ)	sgn F$_\Psi$	sgn F$_T$	C
(π/6-ξ...π/3-ξ)	D	sgn F$_T$	-sgn F$_\Psi$
(π/3-ξ...π/2-ξ)	-sgn F$_T$	A	-sgn F$_\Psi$
(π/2-ξ...2π/3-ξ)	-sgn F$_T$	sgn F$_\Psi$	B
(2π/3-ξ...5π/-ξ 6)	C	sgn F$_\Psi$	sgn F$_T$
(5π/6-ξ... π-ξ)	-sgn F$_\Psi$	D	sgn F$_T$
(π-ξ...7π/6-ξ)	-sgn F$_\Psi$	-sgn F$_T$	A
(7π/6-ξ...4π/3-ξ)	B	-sgn F$_T$	sgn F$_\Psi$
(4π/3-ξ...3π/2-ξ)	sgn F$_T$	C	sgn F$_\Psi$
(3π/3-ξ...5π/3-ξ)	sgn F$_T$	-sgn F$_\Psi$	D
(5π/3-ξ...11π/6-ξ)	A	-sgn F$_\Psi$	-sgn F$_T$
(11π/6-ξ...2π-ξ)	sgn F$_\Psi$	B	-sgn F$_T$

TABLE III.
QUANTITIES A...D FOR TABLE II

sgn F$_\Psi$	1	1	-1	-1
sgn F$_T$	1	-1	1	-1
A	1	0	0	-1
B	0	-1	1	0
C	-1	0	0	1
D	0	1	-1	0

The selected 3L-VSI dc-link voltage value and control table guarantees, that the system will from initial condition selected in (19), (20) reach the sliding manifold (6), (13) in finite time, remain on it and will be robust against the load value, the selected uncertainty of the system parameters.

It must put emphasis on this fact that the IM drive is robust against the uncertainty of the 3L-VSI dc-link voltage and the estimation of the stator-flux angle position, too. The sliding motion would be guaranteed by satisfying the condition (21) that is an inequality. In this case the dc voltage could be variable, only its lowest value is bounded depending on the selected value of the dc-link voltage. As a result the value of the dc-link capacitor and its size could be smaller as by the tradition control.

The same remark can be made to the definition of the angle sectors table. First of all the stator flux position must

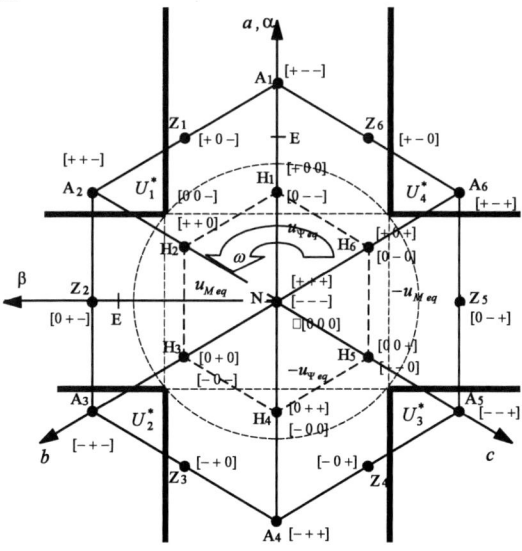

Fig 3. Sliding mode control areas and voltage space vectors of 3L-VSI

estimated within one $(\pi/6)$-sector from 12 sectors. Second, if the dc-link voltage value is higher than the minimum value from (21) the sector border can be estimated not exactly, i.e. the request upon the measuring angle sector table can be lower.

As design result the SMC algorithms (12), (13), (17) – (21) and Tables II, III are obtained that guaranty zero torque- and flux-modulus errors by variation of drive parameters, load and inverter dc-link voltage.

V. SIMULATION RESULTS AND DISCUSSION

In order to verify the operation of the IM drive with the proposed sliding-mode controller, the simulations of a 90 kW IM drive fed by and 3L-VSI taken from [2] was made up. The parameters of the IM are presented in Table IV. The specifications of the 3L-VSI are given as follows: dc-link voltage is $U_d = 422$ V, the control cycle time is 25μs. The sliding-mode controller uses as input signals the error signs (13) and information about the stator-flux position and produces the output signals: the phase switch state, using the Tables II and III.

In this case the sliding-mode controller uses only the full and intermediate voltage space-vectors. The rotating speed if IM was thus held constant at 82% of rated speed, 1200 rpm, which is appropriate, using only these space vectors. All quantities (except time) are given in p. u., with the flux normalized to rated stator-flux amplitude (1.71Wb), torque to rated break-down torque $T_b = 3373$ Nm, rated torque being 35% of break-down torque. Currents are normalized to the rated ideal rotor short-circuit current $I_\infty = \left| \Psi_{sz} \right| / (1 - L_m^2) = 1315$ A and voltages to $2/\pi \cdot U_d = 368$ V.

From the viewpoint of the real 3L-VSI instead of the relay elements the flux and torque hysteresis ones with the bands were set to 7% of rated flux amplitude and break-down torque, respectively are used. In this case the resulting device switching frequency is below 300 Hz.

At $t = 0$, the machine is magnetized to rated flux, at $t = 0.5$ s the torque set value T^* is step up to 35% of break-down torque (Figs. 4 & 5) and at $t = 0.7$ s the torque is step down to 20% of break-down torque (Figs. 6 & 7).

Figs. 4 & 6 display the time functions of torque and torque set value, the α- and β-coordinates of the stator-flux space vector $\Psi_{s\alpha}$, $\Psi_{s\beta}$ and its modulus $\left| \Psi_s \right|$ and the (double of the) α-coordinate of the stator-current space

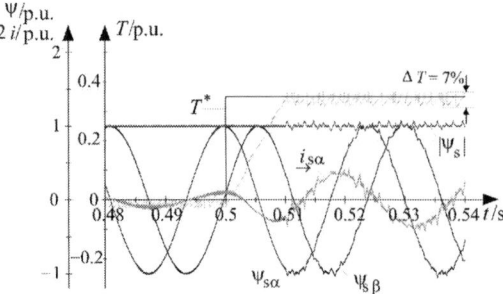

Fig. 3. Torque and torque set value, α/β components of stator-flux space vector and modulus $|\Psi_s|$, stator current (x2); torque set value rising from 0 to 35 % T_b

vector $i_{s\alpha}$ (current in phase a) The torque rise is achieved – due to the reduced voltage margin – only within 10.2 ms, as Dynamic Field Weakening [6] is not yet provided; but the torque reduction shows the dynamics to be expected from direct control, with a sink time approximately 0.12 ms. Torque and flux modulus are guided safely in their bands, the flux and current waveforms are mainly sinusoidal, with small distortions.

Fig. 5 & 7 show the trajectories of the space vectors of stator flux Ψs, the double of the stator current is and the employed full (A) and intermediate (Z) voltage states, belonging to Figs. 4 & 6 accordingly.

The pertaining fundamental current amplitudes (= radii of trajectories) $i_{s0} = \sigma/(1-\sigma)I_\infty$ are:

- no load: $i_{s0} = 0.047 I_\infty$;
- rated load ($T = 0.35 \ T_b$): $i_{s0} = 0.19 I_\infty$
- 57% rated load ($T = 0.2 \ T_b$): $i_{s0} = 0.12 I_\infty$

TABLE IV
PARAMETERS OF IM FOR SIMULATION

Parameter	Symbol	Value
Rated power (kW)	P_r	90
Rated voltage (V)	U_r	330
Rated current (A)	I_r	180
Number of pole pairs	p	2
Rated speed (rpm)	n_r	1480
Rated torque (Nm)	T_r	1160
Rated flux linkage amplitude (Wb)	Ψ	1.71
Stator resistance (mΩ)	R_s	25.9
Rotor resistance (mΩ)	R_r	18
Magnetizing inductance (mH)	L_μ	27.6
Leakage inductance (mH)	L_σ	1.3

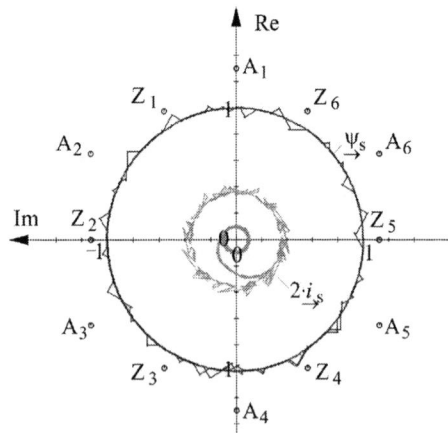

Fig. 5. Space-vector trajectories of stator flux, stator current (x2) and stator voltage, torque set value rising from 0 to 35 % T_b

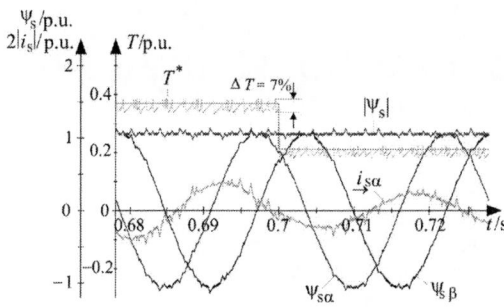

Fig. 6. Torque and torque set value, α/β components of stator-flux space vector and modulus |Ψₛ|, stator current (x2); torque set value reduced from 35 to 20 % T_b

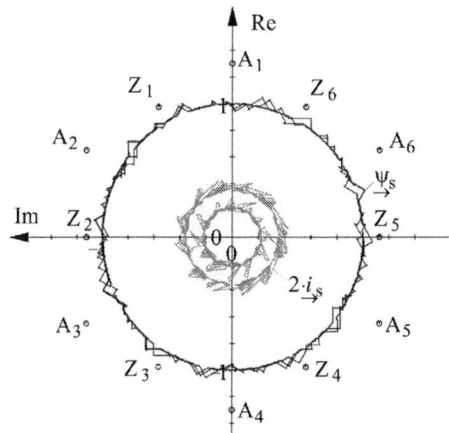

Fig. 7. Space-vector trajectories of stator flux, stator current (x2) and stator voltage, torque set value reduced from 35 to 20 % T_b

V. CONCLUSION

This paper proposed a novel designed by using the sliding-mode technique the control for an IM drive supplied by a 3L-VSI with high-voltage power semiconductor switches. Such control has the following advantages:

1. The control design is simpler and more graphical due the two-step design procedure and the separately using the nonlinearity of the IM and the switching character of the 3L-VSI.

2. The control system has all advantageous characteristics of the control plant with such type of control as high dynamic, low sensitivity to disturbance of both the load and the to plant parameter variations.

3. The requirements to the quality of the input dc-link voltage and as a result to the dc-line filter capacitor are reduced.

4. The measurement precision of the stator flux position is within $(\pi/6)$.

Using the half-voltage and zero-voltage space vectors by proper design of the controller will be the next improvement of the torque control performance. It is very impotent for the lower speed range. It must improve the control quality, but needs the development of the new design technique and receiving the additional information about the position of the equivalent control.

ACKNOWLEDGMENT

The kind sponsorship of the German Academic Exchange Service, Bonn, of the research visit of Prof. Sergey Ryvkin at the Institute for Electrical Power Engineering and Power Electronics of Ruhr-University Bochum, is gratefully acknowledged.

REFERENCES

[1] J. Rodriguez, S. Bernet, B. Wu, J. O. Pontt and S. Kouro, "Multi-level voltage-source-converter topologies for industrial medium-voltage drives," *IEEE Trans. Ind. Electron.*, vol.54, pp.2930–2945, 2007.

[2] E. Krafft, A. Steimel and J. K. Steinke, "Three-level high-power inverters with IGCT and IGBT elements compared on the basis of measurements of the device loses," *Proc. the 8th European Conf. on Power Electronics*, 1999.

[3] W. Leonhard, *Control of electrical drive*, 3rd ed., Springer: Berlin, 2001.

[4] F. Blaschke, "The principle of field orientation as applied to the new Transvektor closed-loop control system for rotating-field machines," *Siemens Review*, vol. 39, pp.217–220, 1972.

[5] M. Depenbrock, "Direct self-control (DSC) of inverter-fed induction machine," *IEEE Trans. Power Electron.*, vol.4, pp. 420–429, 1988.

[6] A. Steimel, "Stator-flux-oriented high performance control in traction," Tutorial H-4, *35th Annu. Meeting IEEE IAS*, Rome 2000

[7] G. S. Buja and M. P. Kazmierkowski, "Direct torque control of PWM inverter-fed AC motor – a survey," *IEEE Trans. Ind. Electron.*, vol. 51, pp. 744–757, 2004.

[8] C. Lascu, I. Boldea and F. Blaabjerg, "Variable-structure direct torque control – a class of fast and robust controllers for induction machine drive," *IEEE Trans. Ind. Electron.*, vol.51, pp. 785–792, 2004.

[9] X. del Toro Garcia, A. Arias, M. G. Jayne and P. A. Witting, "Direct torque control of induction motor utilizing three-level voltage source inverter," *IEEE Trans. Ind. Electron.*, vol. 55, pp. 956–958, 2008.

[10] A. R. Beig, G. Narayanan and V. T. Ranganathan, " Modified SVPWM algorithm for three-level VSI with synchronized and symmetrical waveforms," *IEEE Trans. Ind. Electron.*, vol. 54, pp. 486–494, 2007.

[11] J. Pou, J. Zaragoza, P. Rodriguez, S. Ceballos, V.M. Sala, R.P. Burgos and D. Boroyevich, "Fast-processing modulation strategy for the neutral-point-clamped converter with total elimination of low-frequency voltage oscillations in the neutral point," *IEEE Trans. Ind. Electron.*, vol. 54, pp. 2288–2294, 2007.

[12] V. I. Utkin, J. Güldner and J. Shi, *Sliding mode control in electromechanical system*, Taylor & Francis: London, Philadelphia, 1999.

[13] S. Ryvkin, "Sliding Mode Technique for AC Drive," *Proc. the 10th In. Power Electronics & Motion Control Conf., EPE –PEMC*, 2002.

[14] J. Vittek, M. Stulrajter, P. Makys, I. Skalka and M. Mienkina, "Microprocessor Implementation of Forced Dynamics Control of Permanent Magnet Synchronous Motor Drives," *Proc. 10th Int. Conf. on Optimization of Electrical and Electronic Equipment, OPTIM 2006*, vol. III, pp. 3–8, 2006.

[15] B. S. Heck, "Sliding mode control for singularly perturbed systems," *Int. J. Control*, vol. 53, pp. 985–1001. 1991.

[16] S. Ryvkin, R. Schmidt-Obermoeller and A. Steimel, "Three-Level Inverter Drive with Sliding Mode Control, *Proc. 11th Int. Conf. on Optimization of Electrical and Electronic Equipment, OPTIM 2008*, vol. III, pp. 25–30, 2008.

Analysis and configuration of supercapacitor based energy storage system on-board light rail vehicles

R. Barrero[1], X. Tackoen[2], J. Van Mierlo[3]
[1,3]Vrije Universiteit Brussel, IR-ETEC, Pleinlaan 2, B-1050 Elsene, Belgium
Tel: +32 2 629 2838, +32 2 629 2803. FAX: +32 2 629 3620
rbarrero@vub.ac.be, jvmierlo@vub.ac.be
[2]Université Libre de Bruxelles, CIEM, Avenue F.D. Roosevelt 50, CP 194/7 1050 Bruxelles - Belgique
Tel : +32 2 650 3933. Fax: 02/650.27.83
xtackoen@ulb.ac.be

Abstract— This article will propose different energy storage systems, ranging from 0.91 kWh to 1.56 kWh, suitable for a 30 m long tram. To configure the system regarding energy content, voltage variation, maximum current and power losses, a model of the tram, network and substations power flow has been developed in a Matlab/Simulink environment. Results obtained in energy savings at substation level vary from 24% to 27,6% under the same driving profile and auxiliaries load; while at the end-of-life of supercapacitors, the range varies from 18,1% to 25,1% depending on the super capacitor module used and vehicle load.

The effect of the power converter between the energy storage system and the pantograph will be evaluated in terms of efficiency and rated power.

Keywords— Supercapacitor, Efficiency, Energy Storage, Hybrid electric vehicle (HEV)

I. INTRODUCTION

To reduce emissions, electric powered vehicles are in use in many cities. Although these mass transit vehicles enable large reductions in terms of emissions, their energy efficiency could be significantly improved. This improvement can be reached by the hybridization of their power system, with the inclusion of an energy storage system (ESS) for energy recovery purposes [1,2]. Recent studies have shown that up to 40 % of the energy supplied to electrical rail guided vehicles could be fed back to the grid [3]. This energy can be sent to other vehicles on the line provided that they will consume it simultaneously; but this is infrequent for tram networks where the low traffic density entails a small percentage of braking energy re-use. Furthermore, the nature of city driving, where low speeds, frequent acceleration and sudden braking occur, implies that the public transportation sector is an ideal candidate to benefit from an enhanced ESS. Supercapacitors form an ideal option for this purpose. They can accept high power peaks from regenerative braking, have a long lifetime, need no maintenance and work at a wide temperature range

without suffering from a major negative effect on their lifetime [4].

Hybridizing the tram drive train with supercapacitors can have several purposes, depending on the particular aims, such as: energy savings, peak power shaving, overhead line voltage stabilization [5,6], etc. According to this aim, a specific control strategy and a particular sizing will be needed.

This study will focus on the development of an ESS oriented to energy savings. Therefore, the power flow controller will be optimized for this aim. However there will also be some advantages in terms of voltage stabilization and peak power shaving.

II. METHODOLOGY

Many parameters influence the design of a supercapacitive ESS for a rail vehicle. Features such as tram weight, passenger load, maximum speed, driving cycle, altitude differences and supercapacitor characteristics need to be studied to determine the ESS in terms of energy capacity.

To evaluate the effects of all these parameters, a 'backwards looking' or 'effect-cause' simulation tool [7,8] has been developed in Matlab/Simulink with the objective of determining the power flow at tram level, line voltage and current, and power drawn from substations with and without on-board supercapacitors.

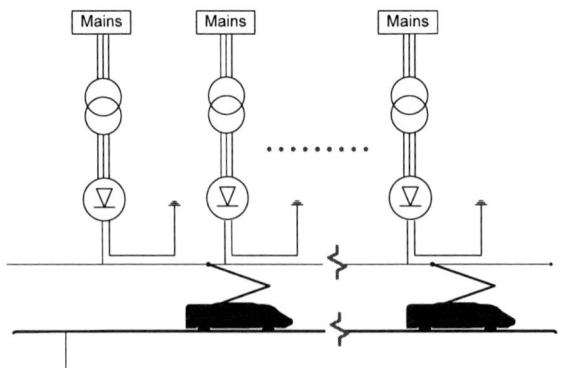

Figure 1. Detail of conventional tram line

The main structure of a tram network is depicted by Figure 1. Several unidirectional substations feed the overhead lines with a separation among them that depends on the line and vehicles characteristics. In the system subject of study the distances are around 1500m.

The model developed treats the vehicles and the network as different blocks that interact with each other as shown by Figure 2.

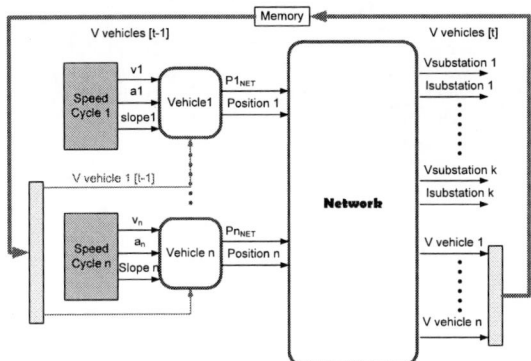

Figure 2. Vehicles and Network model topology

Vehicles are modeled based on the 'effect-cause' method. Starting from a given speed cycle, the power at the wheels is calculated. This power is the needed to overcome the vehicle inertia, slopes, aerodynamic drag and rolling friction. Equations will not be detailed since they are well known and can be found in many references such as [9].

Going upstream the vehicle components and their related efficiencies ($\eta_{component}$), the power requested from the electric grid is determined. The power is calculated according to Eq 1 and Eq 2, depending on the power flow direction.

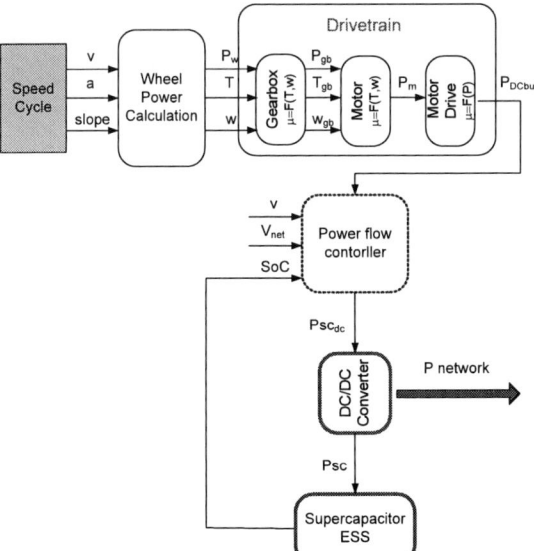

Figure 3. Detail of vehicle hybrid model

$$P_{DCbusTRACTION} = \frac{P_{Wheels}}{\eta_{geearbox} \cdot \eta_{motor} \cdot \eta_{motordrive}} \qquad \text{Eq 1}$$

$$P_{DCbusBRAKING} = P_{Wheels} \cdot \eta_{geearbox} \cdot \eta_{motor} \cdot \eta_{motordrive} \qquad \text{Eq 2}$$

The vehicle can be simulated to be hybrid or conventional.

The traction power requested from the network in conventional vehicles will be equal to that stated in Eq 1, while the braking power will be set to zero. Since the tram line subject of study has a very little energy regeneration rate due to the low traffic density, it has been assumed that the vehicles do not send energy back to the network.

For hybrid vehicles, the power flow controller will decide the amount of power to be provided by the network and by the supercapacitor based ESS.

The network model is based on Figure 4 schematics, with several substations modeled as ideal voltage source with a series resistance *Rsub* and a diode. From the network point of view, the vehicles are current sources whose value is determined at the vehicle block.

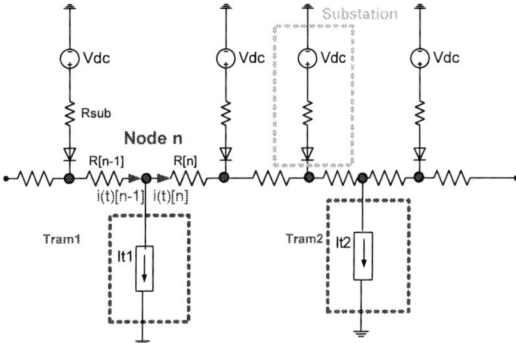

Figure 4. Details of the network model

In this case study, only one vehicle will be running on the line. The power and energy delivered by the substations when a conventional vehicle is running is compared to that of hybrid vehicles in order to assess the energy consumption of each vehicle.

III. POWER FLOW CONTROLLER

The strategy of the power flow controller will have a very important effect on the behaviour of the ESS. The presented solution intends to achieve optimal energy savings trying to have some other added benefits such as voltage drops reductions and peak power shaving. This is done by measuring the vehicle speed and estimating the amount of kinetic energy that the vehicle has at a certain moment.

The proposed controller is based on a proportional control of the state of charge as depicted by Figure 5.

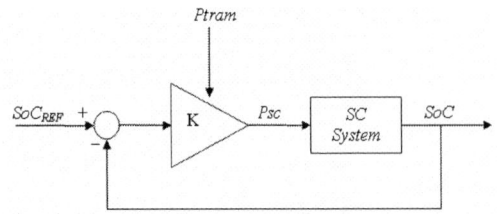

Figure 5. ESS state of charge control

When traction power is needed, the reference State of charge (SoC_{REF}) is determined by the vehicle kinetic energy. Measuring the vehicle speed, the controller estimates the amount of energy that would be generated in a sudden deceleration from the current speed to standstill and determines the desired state of charge (SoC_{REF}) according to Eq 3 and Eq 4.

$$E_{Kinetic\,Re\,coverable} \approx K_1 \cdot E_{Kinetic} \qquad \text{Eq 3}$$

$$SoC_{REF} = 1 - \frac{E_{Kinetic\,Re\,coverable}}{Esc_{MAX}} \qquad \text{Eq 4}$$

The value of K_1 depends on factors such as power auxiliaries, internal losses, etc. and is tuned by examining the simulations. Its value varies normally between 0.5 and 0.6.

When braking power is generated, it will be stored in the supercapacitors as long as the SoC is lower than 1. When supercapacitors are full, the surplus braking power is burnt in the braking resistors.

At standstill and low speeds, supercapacitors will be smoothly recharged with a power inversely proportional to their state of charge.

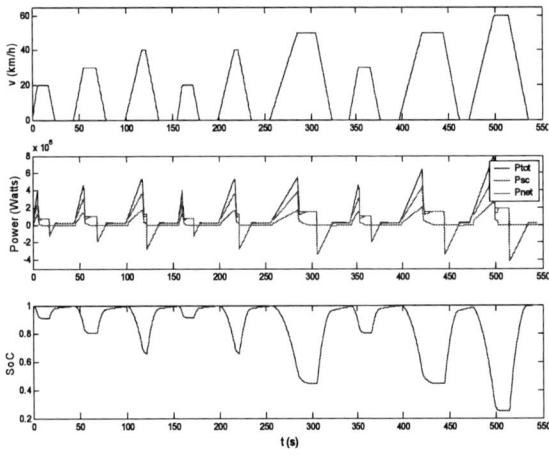

Figure 6. Detail of power flow controller behaviour

Figure 6 shows the power split between the supercapacitors and the network when a hybrid tram is running on the line with an on-board ESS with an energy content of 1.2kWh.

The top graph shows the vehicle speed, the bottom one the SC state of charge and the middle one the power split. In the power graph, the blue line represents the total DC power, the green one is the power provided by the SCs while the red one accounts for the network power. It can be observed that the ESS power increases with the vehicle speed due to the strategy implemented.

It is also noticeable that the SCs are slowly recharged by the network when the vehicle is stopped. Thus they will have energy available to soften the acceleration power peaks requested from the vehicle to the network

IV. ESS SIZING

To assess the influence of the ESS size in the final energy savings calculation, 4 modules will be proposed in Table 1. Each of them will be sized according to the described criteria based on a 31.8 m long tram with an empty weight of 38.6 tons, which is currently in revenue service in Brussels.

Different sizes of energy storage systems have been found in literature for similar tramways with usable energy content varying between 0.97 kWh [10] to 1.5 kWh [11].

Option A	
Criteria:	*The system will have a minimum size in terms of energy equal to that needed to store all the available braking energy of a tram, loaded with 4 persons/m², in a deceleration from 60 to 0 km/h. The minimum energy is 1.13 kWh.*
Cells:	*C=2000F, V_{max}= 2.5V.*
Configuration:	*4 strings x 232 cells in series*
Usable energy:	*1.2 kWh*
Max Voltage:	*580 V*
Cells weight:	*371kg*

Option B	
Criteria:	*same as Option A*
Built-in modules:	*125 V; C=63F.*
Configuration:	*3 strings X 4 modules in series*
Max Voltage:	*500V*
Usable energy:	*1.23 kWh*
Modules weight:	*696 kg (it includes cells, connection inside the modules, Voltage balancing circuits, packaging and cooling)*

Option C	
Criteria:	*The system will have a minimum size in terms of energy equal to that needed to store all the available braking energy of a tram, loaded with 6 persons/m², in a deceleration from 65 to 0 km/h. The minimum energy is 1.52 kWh.*
Cells:	*C=3000F, V_{max}= 2.5V.*
Configuration:	*4 strings x 200 cells in series*
Usable energy:	*1.56 kWh*
Max Voltage:	*500 V*
Cells weight:	*440 kg*

Option D
Criteria: *The system will have a minimum size in terms of energy equal to that needed to store all the available braking energy of a tram, loaded with 4 persons/m², in a deceleration from 50 to 0 km/h. The minimum energy is 0.91 kWh.*
Cells: *C=1500F, V_{max}= 2.5V.* Configuration: *4 strings x 234 cells in series* Usable energy: *0.91 kWh* Max Voltage: *585 V* Cells weight: *300 kg*

Table 1. Energy storage systems proposed

A.   .Design criteria

- The voltage variation of the SC will be kept between 100% and 50% of its maximum voltage. Thus, the available energy of the SC will be 75 % of the total energy stored according to the equation Eq 5.

$$E_{Total} = \frac{1}{2} \cdot C_{Total} \cdot (V_{TOTAL\,max}^2 - V_{TOTAL\,min}^2) \quad \text{Eq 5}$$

- The current of the SC cell will not go over the value $0.12 \cdot I_{ShortCircuit}$ [12].
- Power losses and end-of-life (EoL) of the SC will be taken into account for the energy saving calculation.
- Maximum ESS voltage will be lower than network voltage (700 V at no load) to ease the DC/DC converter design.

V. SIMULATIONS RESULTS

The cycle used for the simulations, is built based on the route of tram line 23 in Brussels. The route includes surface and tunnel sections covering a total distance of 20.4 km. A maximum speed of 60 km/h is assumed for tunnel sections while a maximum speed of 50 and 30 km/h is reached at the surface depending on the distance between stops; the average speed of the cycle is 23 km/h. Stop times of 20 s are implemented. Further measurements in the coming months will allow the simulation of a real driving cycle.

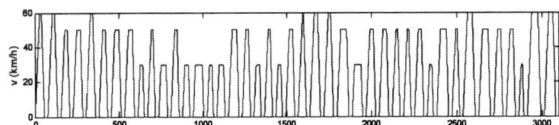

Figure 7. Built-in driving cycle of tram 23 route in Brussels.

A.   Energy savings vs. ESS energy content

It is observed in Figure 8 that on an empty tram, the energy savings are around 23%, almost independently of the energy storage used; while for a tram loaded with 6 persons/m² the energy savings will vary from 23.8% (using option A; 0,91 kWh) to 26% (using option C; 1,52 kWh). The difference between option A and option B is also marked. Although they have almost the same energy

capacity, the results differ significantly. This is due to the fact that option A has a higher efficiency than option B. The current through the cells of option A is lower than in option B, due to the module configuration. This makes option A more efficient than option B and shows that the SC module topology has an important influence on the result. ESS size has a higher impact at the EoL of supercapacitors due to the drop of capacitance as it can be seen at Figure 9.

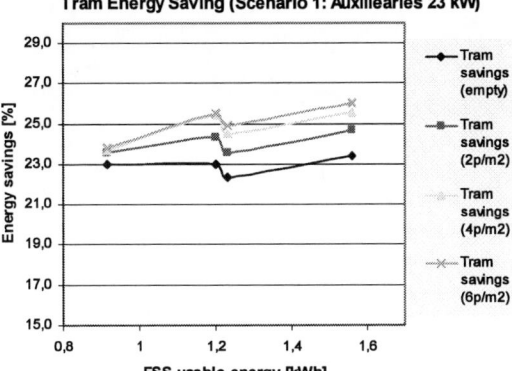

Figure 8. Energy savings as function of ESS size

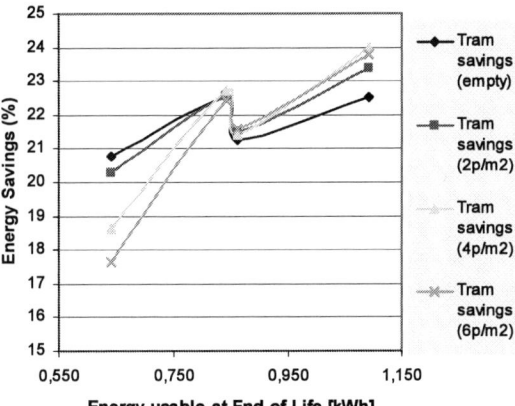

Figure 9. Energy savings as function of ESS size at SC end of life

B.   Energy savings vs. vehicle weight.

Figure 10 and Figure 11 show a comparison between the ESS (option A, C and D) under different load conditions at the beginning and at the end of life. Option C is the most efficient in all conditions but for loads smaller than 4 p/m², there is no big difference between option A and C. At the end of life this differences have increased.

Energy savings shown on Figure 10 and Figure 11 correspond to the savings at tram level. Due to the losses on the line, the energy savings at substation level are between 1 and 2% higher considering only 1 tram running on the line. In the next step the net will be simulated with several trams on it, as it happens daily. In this case, the savings on the line will also be higher due to the higher current flowing through it.

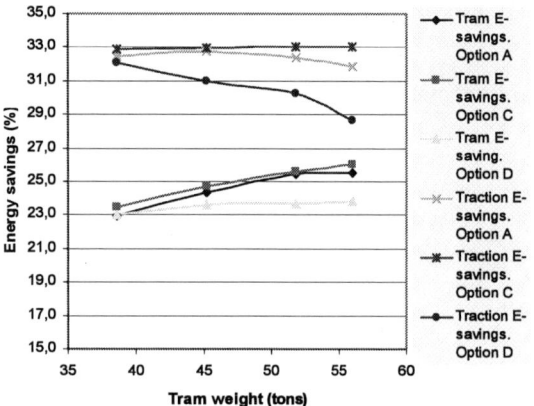

Figure 10. Energy savings vs.vehicle occupancy

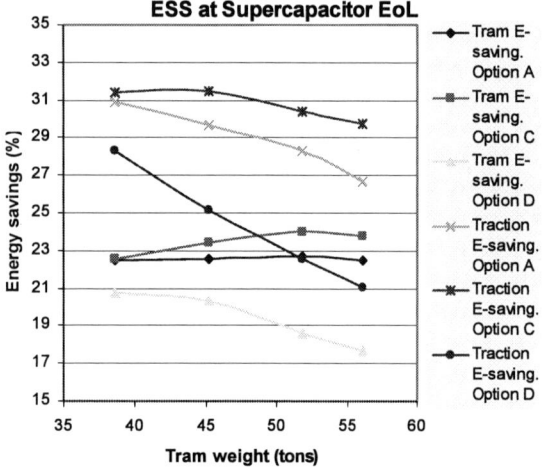

Figure 11. Energy savings vs.vehicle occupancy at SC end of life

Both figures also show that if the power consumption for the auxiliaries' is not considered, then the traction energy savings are much higher, over 30% in some cases.

C. Influence of DC/DC converter

Another important point to consider is the effect of the DC/DC converter on power flow management and on the energy savings [13,14]. For the results gathered up to this point, the efficiency of the converter has been set at 91%. Since the power flows twice through the power converter every cycle, it has a significant influence on the final energy consumption.

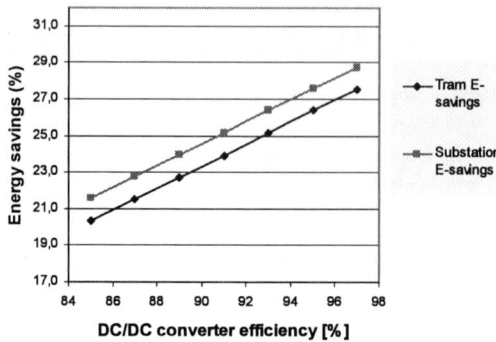

Figure 12. Energy savings vs. DC/DC converter efficiency

Out of the simulations, considering a hybrid tram (with on board SC described by *Option A*) loaded with an average of 2 persons/m2, and consuming 23 kW for auxiliaries (average measured value), the energy savings at tram level will vary from 19.7% (DC/DC efficiency: 85%) up to 27.4% (DC/DC efficiency: 97%) at tram level; and from 21.6% to 28.8% at substation level. This is shown in Figure 12.

Resonant converters seem to be a good option to achieve high efficiencies [15] and values up to 97% have been reported using this topology [16].

Another factor to consider is the rated power of the converter. For this application, observing Figure 13, there is no big benefit in having a converter rated over 400 kW (the nominal power of the tram is 4x105kW with peaks over 600 kW). Smaller converters will be a more economical and less bulky solution, but less efficient since there will be a portion of the braking energy that can not be recovered due to this power limitation.

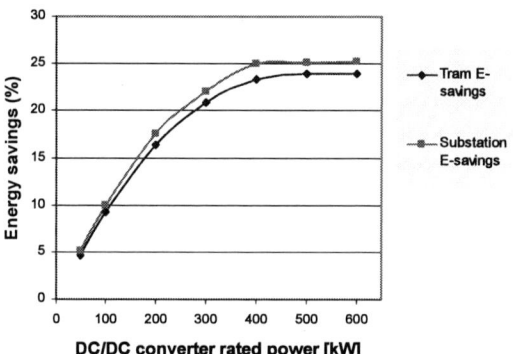

Figure 13. Energy savings vs. converter rated power

VI. CONCLUSIONS

It is not a simple task to choose an ESS which is optimal in all conditions. However assuming that the average number of passengers in the trams is lower than 4p/m^2, Option A (1.2 kWh) is considered a good option since it also behaves well at the end of life.

The main problem of option D (smaller size 0.91 kWh), would be its performance at the end of life, when a drop in capacitance would entail a loss of energy capacity which would reduce the system performance.

Option C (large size, 1.56 kWh) is the best option considering only its performance. But what needs to be assessed is whether the advantageous performance in certain situations (mainly very high, but not usual, loads and end of life scenarios) compensates for the extra cost and weight of this module. This option could also give more flexibility in cases when different strategies (e.g. voltage balancing compensation, peak power, certain autonomy required without overhead line) should be considered.

The behaviour of the DC/DC converter is not only crucial for managing the power from/to the supercapacitors but it also influences strongly the energy savings. High efficient converters are highly encouraged for this application. Moreover, the power rating of the converter has an impact on the amount of braking energy that can be recovered. A compromise between cost/size and performance has to be reached.

ACKNOWLEDGMENT

This research is financed by the Institute for the Encouragement of the Scientific Research and Innovation in Brussels (IWOIB / IRSIB).

REFERENCES

[1] Van Mierlo J, Maggetto G, et al., "How to Define Clean Vehicles? Environmental Impact Rating of Vehicles", *International Journal of Automotive Technology (IJAT)*, *KSAE,SAE*, isbn 1229-9138, Vol 4, Nr 2, Pg 77-86, 2003

[2] Van Mierlo J, Timmermans J.-M. et al., "Environmental Rating of vehicles with different alternative fuels and drive trains: a comparison of two approaches," in Transportation Research Part D, vol. 9, 2004, pp. 387-399, 2004.

[3] Gunselmann W, 'Technologies for Increased Energy Efficiency in Railway Systems', in proceedings of EPE 2005, Dresden. ISBN: 90-75815-08-5.

[4] Burke A, 'Ultracapacitors: why, how, and where is the technology', Journal of power sources. Vol 91 (1), pg 37-50. NOV 2000.

[5] Rufer A, Hotelier D, Barrade P, 'A supercapacitor-based energy storage substation for voltage compensation in weak transportation networks', IEEE Transactions on power delivery, vol. 19 (2), pg 629-636. APR 2004

[6] Rufer A, Barrade P, Hotellier D, Hauser S, 'Sequential supply for electrical transportation vehicles: Properties of the fast energy transfer between supercapacitive tanks', Journal of circuits systems and computers. vol 13 (4), pg 941-955. AUG 2004.

[7] Van Mierlo J, Maggeto G, 'Innovative iteration algorithm for a vehicle simulation program', IEEE transactions on vehicular technology. vol 53 (2). pg 401-412. MAR 2004

[8] Van Mierlo J, Van den Bossche P, Maggetto G, 'Models of energy sources for EV and HEV: fuel cells, batteries, ultracapacitors, flywheels and engine-generators', Journal of power sources. Vol. 128 (1), pg. 76-89, MAR 29 2004

[9] Gazella L., Sciarretta A., 'Vehicle Propulsión Systems. Introduction to Modeling and Optimization', Springer Verlag Berlin Heidelberg 2005. ISBN-10 3-540-25195-2

[10] Destraz B, Barrade P, Rufer A, Klohr, 'Study and Simulation of the energy Balance of an Urban Transportation Network' in proceedings of EPE 2007, Aalborg. ISBN 9789075815108

[11] Steiner M, Scholten J, 'Energy Storage on Board of Railway Vehicles' in Proceedings of EPE 2005 Dresden. ISBN 90-75815-08-5.

[12] Cheng Y, Van Mierlo J, et.al., "Energy sources control and management in hybrid electric vehicles", Proceedings of EPE PEMC 2006, Portoroz, Slovenia, 2006

[13] Ayad M.Y. et al., 'Voltage regulated hybrid DC power source using supercapacitors as energy storage device', Energy Conversion and Management. Vol. 48 (7), pg 2196-2202. JUL 2007

[14] Rufer A, Barrade P, 'A supercapacitor-based energy-storage system for elevators with soft commutated interface', IEEE Transactions on industry applications, vol. 38 (5). pg 1151-1159. SEP-OCT 2002.

[15] Miller J. M, "Propulsion Systems for Hybrid Vehicles", IEE Power and Energy series, vol 45, ISBN 0-86341-336. The Institution of Electrical Engineers 2004

[16] Eno O.A. et.al, "High power resonant topology for dc-dc converter", Proceedings EPE2005, Dresden, 2005. ISBN : 90-75815-08-5

Design of High Power Electronic Building Block based on Parallel of IGBTs for Electric Vehicle

Wen Huiqing[*][†], Liu Jun[*], Zhang Xuhui[*][†], Wen Xuhui[*]

[*]Institute of Electrical Engineering Chinese Academy of Sciences, Beijing, China
[†]Graduate University of Chinese Academy of Sciences, Beijing, China
E-mail: whq@mail.iee.ac.cn

Abstract--**A number of IGBT modules or IGBT converters can be paralleled to enlarge output power level of the whole system, which is attractive in applications such as electrical vehicles, ships and spacecrafts where power density, power quality and cost are very sensitive. Firstly, this paper explains the reason why the NPT IGBT is fit for parallel operation. The principle of IGBT parallel operation and the new evaluation plan are also presented. A power electronic building block (PEBB) based on parallel of IGBT is designed and a Saber-based accurate simulation platform including IGBT model and main parasitic has been developed. The detailed experimental results are present and compared with the simulation results, which show the PEBB module is practical and the new evaluation is valid.**

Keywords--**Parallel Operation; IGBT; Electric Vehicle; System Integration**

I. INTRODUCTION

The power range which can be handled by an inverter circuit applying available IGBT devices is limited. To increase the power handling capacity, two methods are commonly used: inverter circuits in parallel and a number of power devices in parallel [1-3]. The former method is simple and can be easy implemented, but it has the most significant shortcoming of low power density. In high power density requirements applications such as electric vehicle, ships and spacecrafts, this method is obviously not a preferred choice. Recently in the development of an inverter with 250kVA device output power and more than 1000A device output current for electric vehicle, high power density and reliability become extreme difficult problems. Considering power density, device current margin and price, operating a number of power semiconductor devices in parallel is obviously a better choice [4].

II. OPERATING IGBT MODULS IN PARALLEL

Today there are commonly two different types of IGBT devices: the non-punch-through (NPT) IGBTs and the punch-through (PT) IGBTs. NPT IGBTs have positive temperature coefficient of on-stage voltage, and so that the problem of thermal runaway for the parallel NPT IGBT devices is not obvious. Fig.1 shows the influence of junction temperature fluctuation on paralleling characteristics of two NPT IGBT devices. When $IGBT_1$ and $IGBT_2$ are paralleled with the on-stage voltage of V_{CE1}, the corresponding output current at this moment is respectively I_{C1} and I_{C2}. $IGBT_1$ shares the larger part of the load current, which resulting $IGBT_1$ has higher junction temperature than $IGBT_2$. As NPT IGBTs have positive temperature coefficient of on-stage voltage, the working curve of the paralleled module will shift from the working curve marked $IGBT_1$ to the working curve marked $IGBT_1{}'$, where the on-stage voltage is V_{CE2} and the corresponding output current is respectively $I_{C1}{}'$ and $I_{C2}{}'$. The current assumed by $IGBT_1$ is automatically reduced while the current assumed by $IGBT_2$ is automatically increased, so the function of automatically current sharing is achieved. In above analysis the load current is assumed to remain unchanged.

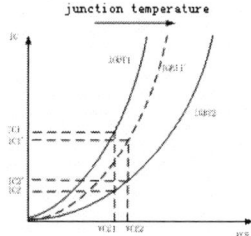

Fig.1. Influence of junction temperature fluctuation on paralleling NPT IGBT devices

Parameters differences among devices such as the propagation delay time from driving modules to gate units and the difference of stray inductances in parallel branches is the main reasons for asymmetric current sharing. Parameter deviations over a wide voltage and temperature range should be considered in the use of IGBT devices. There are several concerns about paralleling IGBT devices:

(1) Choose NPT IGBT power devices with positive temperature coefficient of on-stage voltage, preferably

the IGBT devices are of the same batch.

(2) As delay time and output parameters of the drive circuit have great impact on the dynamic characteristics of paralleled IGBT device, it's recommended to share a common drive circuit. The recommended drive circuit for paralleling IGBT is shown in Fig.2.

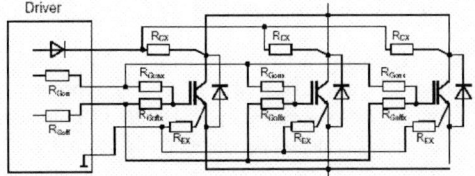

Fig.2. Recommended drive circuit for paralleling IGBT

(3) Small loop inductance and symmetry wiring layout should be maintained in all power circuits and drive circuits. Besides, IGBT devices should keep as close as possible.

(4) It is impossible for all paralleled IGBT devices to achieve a desired static and dynamic balance. Besides, because the internal anti-parallel diode of IGBT device is bipolar device with negative temperature coefficient of on-stage voltage, it is recommended to de-rating parallel IGBTs with factor 15% to 20%.

III. DESIGN OF PARALLELED IGBT MODULE BASED ON PEBB PRINCIPLE

Power Electronic Building Blocks (PEBB) initiated by the office of naval research (ONR) is to achieve such goals: increased power density, "user friendly" design ("plug and play" power modules) and multi-functionality [5-7]. A typical PEBB is shown in Fig.3.

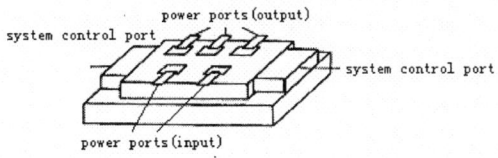

Fig.3. Typical PEBB structure

As a kind of modular product, PEBB has two kinds of interfaces: energy and communication interfaces. PEBB is defined primarily by packaging consideration such as thermal, EMI, interconnections and sensors.

Three main reasons for choosing EconoPACK[TM+] IGBT device produced by EUPEC for paralleling applications are listed as follows: NPT field stop technology, positive temperature coefficient for IGBT and anti-parallel diode in the range of rated current, negligible differences for V_{CEsat}, V_F, V_{Geth} among three chips. Besides, NTC temperature integrated in IGBT devices can provide accurate temperature protection. So three 1700V, 450A IGBT devices can be paralleled to construct a PEBB module with rated value of 1700V, 1350A [8].

Paralleled IGBT devices should have the same drive circuit unit. 2SD300C17S, the latest dual-channel high-power IGBT driving module produced by EUPEC, has function of "soft shutdown" under fault conditions and higher peak output current, which more easily meet the application requirement for paralleled IGBT modules.

RCD circuit is adopted as absorb circuit, which has good integrated characteristic, shortest connected to main circuit and minimize the leakage inductance.

To meet the needs of electric vehicles, an integrated 10W 24V to 15V power supply modules is adopted, taking into account the requirement of 2SD300C17S driving module.

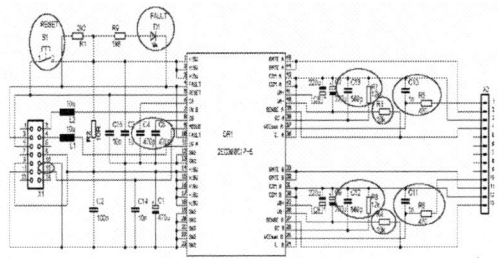

Fig.4. Design of high power IGBT drive circuit

Water-cooled thermal plate with outstanding cooling performance makes the power density of PEBB module increased substantially, which the specific size of water-cooled plate is $15\times15\times10$ cm^3. The output current of the integrated PEBB module can reach 1200A and the power capacity is close to 1.4 MW.

IV. SABER SIMULATION AND PARASITIC

Actually four methods are mainly used to model the power semiconductor devices: the behavioral macro-modeling, the structural macro-modeling, the code modeling and the modeling in Analog Hardware Description Languages (AHDL) [9, 10].

This paper presents a completely behavioral IGBT macro-model by changing the default values of the genetic IGBT component in the template. The equivalent circuit of the IGBT is presented in Fig.5. The parameters and characteristics from datasheets are the starting points for determination the parameters for Saber simulations.

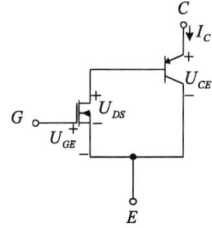

Fig.5. Equivalent circuit of the IGBT

IGBT device FS450R12KE3 is chosen for the verification of the model. The simulated and experimental I-V curves are shown in Fig.6. The tailing of the anode currents simulated by the behavioral model at a constant anode voltage-switching test are also validated.

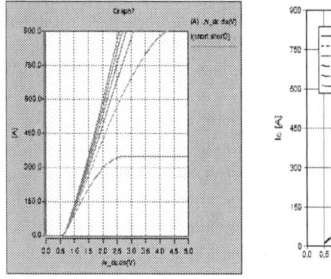

(a)The simulated I-V curves   (b)The experimental I-V curves

Fig.6. Comparison of I-V curves

It is well known that there are parasitic elements in power devices and connecting cables. These parasitic elements should not be omitted in the analysis of the transient switching and paralleling operation of IGBT. LCR meter HP4263B can measure passive components with as little as 0.1% error and the test frequency range is from 100Hz to 100kHz. In the analysis, main parasitic parameters of connecting cables are measured separately. It can be seen in the following experiment analysis that these parasitic elements will have great impact on the current sharing performance of the paralleled IGBTs.

V. EXPERIMENTAL ANALYSIS FOR PARALLELED IGBT MODULE

In order to verify the design of PEBB module, testing on a DC/DC PEBB module shown in Fig.7 is carried out. The equilibrium of layout and the loop leakage inductance should be given more attention. Otherwise the current sharing characteristic will be greatly affected.

A. Main Circuit for Testing

Circuit parameters are shown as following:
IGBT modules: FS450R12KE3
Input inductor: 120uH/350A
Input capacitor: 10000uF/500V
Output capacitor: 3300uF/500V×3
Load: 4.4 ohm resistor with power more than 80kW.

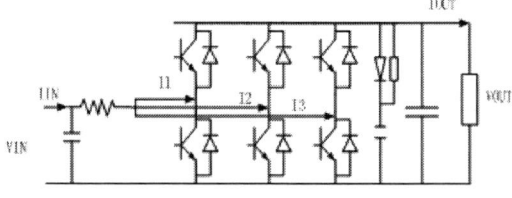

Fig.7. Testing circuit for paralleled IGBT

The bottom three IGBT devices of PEBB module are switching on/off while the anti-diodes of upper devices are conducting during the bottom IGBT devices are of off-stage. The operating mode is similar to a three-phase boost DC/DC converter. Each branch current I_1, I_2 and I_3 are measured and differences among the three branches current are carefully observed. The total input current must limited within 400A considering the limited current capacity of input inductor.

B. Analysis of the Experimental Results

Fig.8 CH1 represents the PWM driving signal issued by DSP control board. CH2, CH3 and CH4 represent driving signals after processed by 2SD300C17S drive modules. The delay time of driving signals from DSP board to 2SD300C17S drive module is about 1us. CH2, CH3 and CH4 signals have nearly the same rise and fall delay time, and the value is all about 0.1us. It can be seen from fig.8 that three driving signals have achieved good consistency.

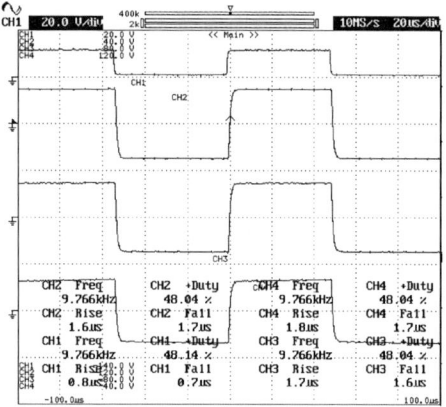

Fig.8. Driving signals for paralleled IGBT modules

Table 1 shows paralleled IGBT analysis below rated current and rated voltage at temperature of 25℃. Parameters are shown as following:

V_{in}: input voltage range of 50V~350V

V_{out}: output voltage with maximum value 450V

I_{in}: input current with maximum value 400A

I_{out}: output current

I_n: input current of IGBT device n (n = 1, 2, 3)

ΔI_{nm}: current difference between IGBT device n and m (n ≠ m, n, m = 1, 2, 3)

P_{in}: input power

P_{out}: output power

η: system efficiency

In order to measure the maximum current error for paralleled IGBT, current maximum mismatching degree (CMMD) can be defined by:

$$I_{mis} = \frac{\max(\Delta I_{12}, \Delta I_{23}, \Delta I_{13})}{I_{total}} = \frac{\max(\Delta I_{12}, \Delta I_{23}, \Delta I_{12})}{I_{in}}$$

(1)

CMMD can also be used to judge the reliability of paralleled IGBT modules.

It can be seen from the experimental results that:

(1) When the input current I_{in} has maximum value of 325.2A, branch current I_1, I_2 and I_3 respectively are 118.9, 109.9A, 93.5A. The biggest difference between branch currents is 25A. The ratio of CMMD value with input current I_{in} can be depicted as Fig.9. It can be seen that CMMD value of paralleled IGBT modules will be within 10% when input current is below 325A. Based on the design requirement for paralleling IGBT, IGBT module normally has 10~15% de-rating factor. Experimental results show that the design of paralleled IGBT modules is feasible. The maximum output current for the PEBB module can be reached 1150A considering 15% de-rating, and the power capacity can be reached 1.38MVA.

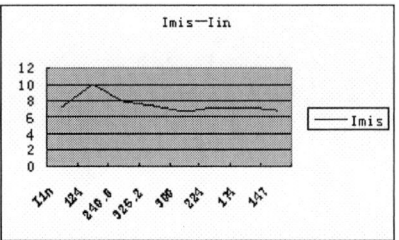

Fig.9. Ratio of CMMD value with I_{in}

Table 1

Paralleled IGBT analysis below rated current and rated voltage at 25℃

V_{in}	I_{in}	I_1	I_2	I_3	P_{in}	ΔI_{12}	ΔI_{13}	ΔI_{23}	I_{mis}	V_{out}	I_{out}	P_{out}	η
(V)	(A)	(A)	(A)	(A)	(Kw)	(A)	(A)	(A)	(%)	(V)	(A)	(Kw)	(%)
51	124	45.3	36.3	41.8	6.32	9.0	3.5	5.5	7.25	165	37.2	6.14	97.1
99	240.8	87.5	69.0	80.2	23.84	18.5	7.3	11.2	9.90	320	71.6	22.91	96.1
139	325.2	118.9	93.5	109.9	45.20	25.4	9.0	16.4	7.81	442	98.3	43.45	96.1
154	300	107	85	101	46.20	22	6.0	16	7.33	429	95.7	41.06	88.9
209	224	80	65	75	46.82	15	5.0	10	6.70	437	96.5	42.17	90.0
257	174	62.4	49.8	58	44.72	12.6	4.4	8.2	7.24	445	98.3	43.74	97.8
300	147	52.5	42	49.6	44.10	10.5	2.9	7.6	7.14	444	98.1	43.56	98.8
345	128.3	44.9	36.2	42.7	44.26	8.7	2.2	6.5	6.78	444	98.2	43.60	98.5

(2) It can be seen from the experimental results that the output current always meet the relationship of $I_1 > I_3 > I_2$. The main reason is that in order to analyze the influence of the circuit layout and connecting ways, three wires with difference thickness and length are used to connect the output ports and inductors. If two wires are exchanged such as wire 1 and wire 2, the relationship of I_1 and I_2 is: $I_2 > I_1$. This shows that the main circuit connecting wires have great impact on the current imbalance.

(3) Static and dynamic tests are all carried out to analyze the influence of switching action. The experimental results are shown in table 2. The static CMMD is about 5.77%, while when IGBTs are switched on/off, the CMMD value is about 7.67%. The current imbalance induced by dynamic error is approximated 1.9%. In can be concluded from the experimental results that the layout of main circuit and driving circuit have more significant impact on CMMD.

Table 2

Static and dynamic tests experimental results for paralleled IGBT

V_{in}	I_{in}	I_1	I_2	I_3	P_{in}	ΔI_{12}	ΔI_{13}	ΔI_{23}	I_{mis}	V_{out}	I_{out}	P_{out}	η
(V)	(A)	(A)	(A)	(A)	(Kw)	(A)	(A)	(A)	(%)	(V)	(A)	(Kw)	(%)
137	29.5	8.6	10.3	10.1	4.04	1.7	1.5	0.2	5.77	137	29.4	4.03	99.7
126	301.4	86.4	109.5	100.3	37.98	23.1	13.9	9.2	7.67	408	88.2	34.96	94.7

C. Simulation Results

A Saber-based accurate simulation platform including IGBT model and main parasitic has been developed. In order to analyze the impact of connecting cables on the current sharing of the paralleled IGBTs, three equivalent circuit of connecting cables based on impedance measurement results of HP4263B are used in simulation. The parasitic inductance is separately 2.13uH, 2.36uH and 4.22uH. The input current is 325A and the output voltage is 442V. The duty cycle of driving signals is calculated to be nearly 70%. Fig.10 shows the simulated results of three branch currents. It can be seen that the current fluctuation in one switching period of three branches is different. The branch with low inductance will have small current fluctuation.

Fig.11 shows driving signal issued by DSP control board and two branches current I_1, I_2 measured by current sensors. CH1 signal is driving signal with duty cycle of 69.87%. It can be seen that the two-channel measurement error is only o.3V, and that is showing that the current error is only 15A, which is very small compared with the total current.

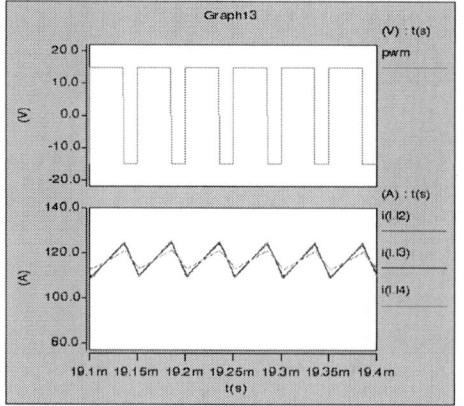

Fig.10. Simulated driving signal and three branch currents

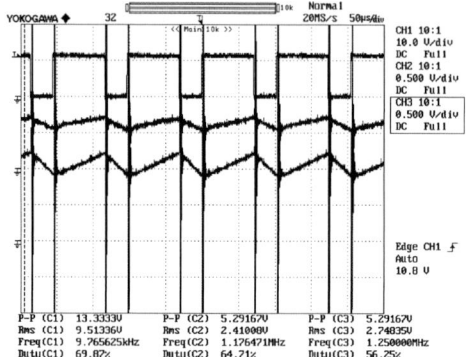

Fig.11. Current waveform for I_1 and I_2

□. CONCLUSIONS

This paper summed up the general principles of paralleled IGBT and explained the reason why NPT IGBTs with positive temperature coefficient of on-stage voltage are fitted for paralleling operation firstly. Then a general indicator CMMD is used in this paper to evaluate IGBT paralleling operation. A high power density PEBB module was designed based on PEBB principle. This PEBB module has rated voltage of 1200V, rated current of 1200A and maximum output capacity of 1.44MW. Drive circuit, protection circuit, sensors and passive devices are all integrated into the single PEBB module.

In order to investigate the practicality and reliability of the module, a 45kW PEBB module paralleled three IGBT devices is constructed. A Saber-based accurate simulation platform including IGBT model and main parasitic has been developed. Experimental and simulation results show that the PEBB module has high practical value in high power density and low cost applications such as electric vehicle, ships and spacecrafts.

ACKNOWLEDGEMENT

This work was supported by the National Nature Science Foundation of China under Contract 50777060.

REFERENCES

[1] N.Y.A Shammas, R.Withanage, D. Chamund, "Review of series and parallel connection of IGBTs Circuit," IEE Proceedings of Circuit, Devices and Systems, pages: 34~39, 2006.

[2] Romeo Letor, "Static and Dynamic Behavior of Paralleled IGBT's," IEEE Transactions on Industry Applications, pages: 395~402, 1992.

[3] J.J.Nelson, G. Venkataramanan, B.C. Beihoff, "Investigation of Parallel Operation of IGBTs," Conference Record of the 37th IAS Annual Meeting, pages:2585~2591, 2002.

[4] C.Keller, Y.Tadros, "Are paralleled IGBT modules or paralleled IGBT inverters the better choice?" Fifth European Conference on Power Electronics and Application, pages: 1~6,1993.

[5] T.Ericsen, A.Tucker, "Power Electronics Building Blocks and potential power modulator applications," Conference Record of the 1998 Twenty-Third International Power Modulator Symposium, pages: 12~15, 1998.

[6] Qian Zhaoming, Zhang Junming, Xie Xiaogao, Gu Yilei, Lu Zhengyu, Wu Xiaobo, "Progress in power electronic system integration," Dian Gong Ji Shu Xue Bao (Transactions of China Electrotechnical Society), pages:1~14, 2006.

[7] Fang Z. Peng, Ju Wang, Fan Zhang, and Zhaoming Qian, "Development of a 1.5 MVA Universal Converter Module for Traction Drive and Utility Applications," Power Electronics Specialists Conference, pages: 2290~2295, 2005.

[8] Power Semiconductors Data-CD 2004 , eupec, 2004.

[9] A.R.Hefner, D.M.Diebolt, "An experimentally verified IGBT model implemented in the Saber circuit simulator," IEEE Transactions on power Electronics, pages: 532~542, 1994.

[10] A.Maxim, G.Maxim, "A novel analog behavioral IGBT spice macromodel," 30th Annual IEEE Power Electronics Specialists Conference, pages: 364~369, 1999.

Stability Analysis on the DC Power Distribution System of More Electric Aircraft

H. Zhang[1,2,3], C. Saudemont[1,2,3], B. Robyns[1,2,3], N. Huttin[1,4], R. Meuret[1,4]

National Center of Technological Research (CNRT FuturElec) of Lille[1]
Laboratory of Electrotechnics and Power Electronics of Lille (L2EP)[2]
Ecole des Hautes Etudes d'Ingénieur (HEI), Lille, France[3]
Hispano-Suiza Group SAFRAN, Moissy Cramayel, France[4]

Abstract—**The local DC Power Distribution System (PDS) of a More Electric Aircraft is one of the cores of the electric power transmission. Generally, the PDS is made up of small power subsystems designed on their own stand-alone operation. But the interaction between the subsystems (sources and loads) induces some problems of stability. This paper compares the different stability behaviours of two load models: resistive load and constant power load. Some tests for the two load models at different operating points are carried out and the results are compared. The small signal theory and the impedance criterion GMPM (Gain Margin Phase Margin) are used to analyse the stability of each system. At conclusion, some statements are expressed to contribute to such electric interconnected systems stability.**

Keywords—**On-board network, Distributed power, DC power supply**

I. INTRODUCTION

The More Electric Aircraft (MEA) is the trend of the future aircraft because of ecologic, economic and sustainable development reasons. The replacement of the mechanical, hydraulic and pneumatic power transfers by electrical power transfer is one of the most important innovative concepts of the MEA [1]. In the European Power Optimized Aircraft project (POA), it has been proven that the use of a local DC Power Distribution System (PDS) makes possible to reach the subsystems optimum in order to get some benefits, like a weight reduction for example. Nevertheless, the use of such a network involves taking precautions in terms of stability. Electric protections, over-voltage and over-current situations due to energy transfers must be considered.

A DC PDS is physically made up of smaller power modules/subsystems. Usually, each subsystem is designed according to its own stand-alone operation, but the interconnection of these subsystems may lead to performance degradation, even system instability.

This paper defines a local DC PDS of MEA. Some test scripts have been selected and simulated on this network. According to these test results, some problems of stability are highlighted. Then a simplified network is defined to focus on the stability problem. We use the theory of small signal to analyze the minor loop gain [2][3] of this network, and to explain the different stability behaviours are led by different types of load model in order to point out the interaction between the source and the loads.

In section II, the definition of the embedded PDS and the method to point out the stability problems are described. Then, the small signal analysis method and the small signal impedance measurement method are treated in section III. Section IV deals with the comparison of the different kinds of load behaviours at different operating points, using the methods described in section III. At last, in section V, some statements explaining instability phenomena in the studied cases are expressed.

II. DEFINITION OF THE EMBEDDED NETWORK AND SELECTION OF THE TEST SCENARIO

A. Definition of the embedded power network

The DC local power distribution network under study is defined in Fig. 1.

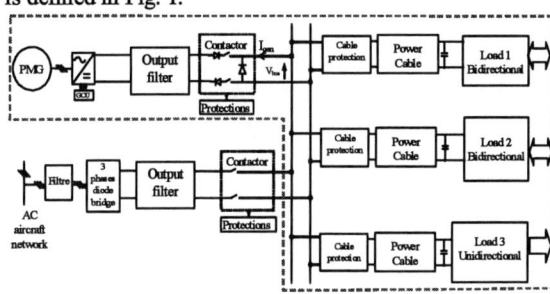

Fig. 1 The definition of DC embedded network.

This network includes two independent sources: the permanent magnet generator with its associated AC/DC converter and the aircraft AC network rectified through a diode bridge. Two bi-directional and one unidirectional loads, in the power range of 2 to 15kW, are used. The loads are connected to the DC bus by power cables and protection elements. The aircraft AC network is only used when the local generator is in fault. Firstly, we only consider that the network is in "Healthy" condition, which means that no fault is present in this network. In this paper, only the broken line framed network part shown in Fig. 1 will be studied. The rated voltage of DC bus is 270V.

B. Finding out the problem

1) Definition of the different load models

Several types of load models are referenced in literature [5][6]. In our study, we will only use two of them: resistive load and constant power load. The two load models are considered as controllable current sources, with different control loop.

The resistive load model is shown in Fig. 2: The value of resistance R_{load} is computed from the rated voltage of the DC bus and the reference power P_{ref}. In this model, the output power varies according to the voltage value V_{load} (since R_{load} is calculated from a constant voltage).

The constant power load model is shown in Fig. 3: The output power follows the evaluation of P_{ref} for any voltage value V_{load}.

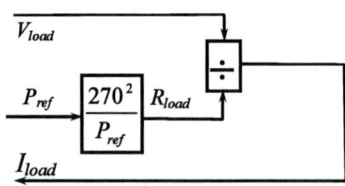

Fig. 2 Model of resistive load

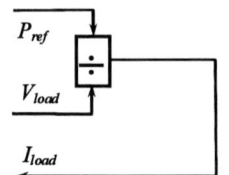

Fig. 3 Model of constant power load

2) Selecting of the most significant test scenario

In the "Healthy" conditions study, we use the known characteristics of the loads to observe the behaviour of the system. It has been estimated that the simulation of all possible healthy states of the network, as well as all different load states will drive to a high number of possibilities (up to 688 in the case of the defined network associated with the three loads).

In order to limit the simulation effort without excluding the essential problem, only the most critical cases that cause malfunction are studied. The different states of the system are classified using their steady state and transient state behaviour. The most critical and the most representative cases are then selected and finally, the loads and network states are combined together, knowing that there is at least one transient state for each combination. Using this method, the simulation cases are reduced to a number of 15. In these 15 cases, the three loads considered are modelled by constant power load in order to better simulate the power demand of the load and limit the influence of bus voltage variation.

3) Highlighted problem

After the simulation of the 15 most significant test scripts, we find that the greater the consumption power (15 kW in our case) is, the more unstable the system becomes. Indeed, in this situation, the transient state of bus voltage is much longer than the others'. Fig. 4 shows the different transient phenomena under different load profiles. We can note that with the same slope and step size of the load profile, when the load reaches at 15kW,

the transient of bus voltage becomes much longer than it reaches at 11kW. This is a sign of instability. The next section will focus on the stability analysis.

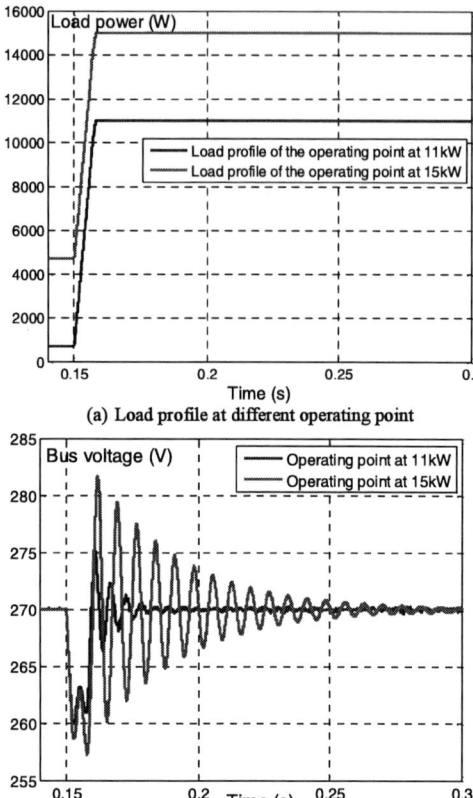

(a) Load profile at different operating point

(b) Bus voltage at different operating point

Fig. 4 Comparison of the bus voltage for 2 different load profiles.

III. ANALYSIS OF INSTABILITY PHENOMENA

A. Influence of the load models on the instability phenomena

In this analysis, we use both above mentioned load models: resistive load and constant power load, for all the three loads mentioned in Fig. 1 and observe their behaviour at different operating points. To visualize the different stability phenomenon, two cases extreme are chosen: 6kW and 17kW loads are considered. To avoid the instable phenomena caused by the hash slope transient state and focus on the unstable problem of steady state, the gentle slope of the load profile is chosen. As shown in Fig. 5 (a), the increasing of load power is slower (17kW in 1.2s) that in Fig. 4 (a)

Fig. 5 shows that, at the 17kW operating point, the system becomes unstable in the case of constant power load whereas it remains stable in the case of resistive load.

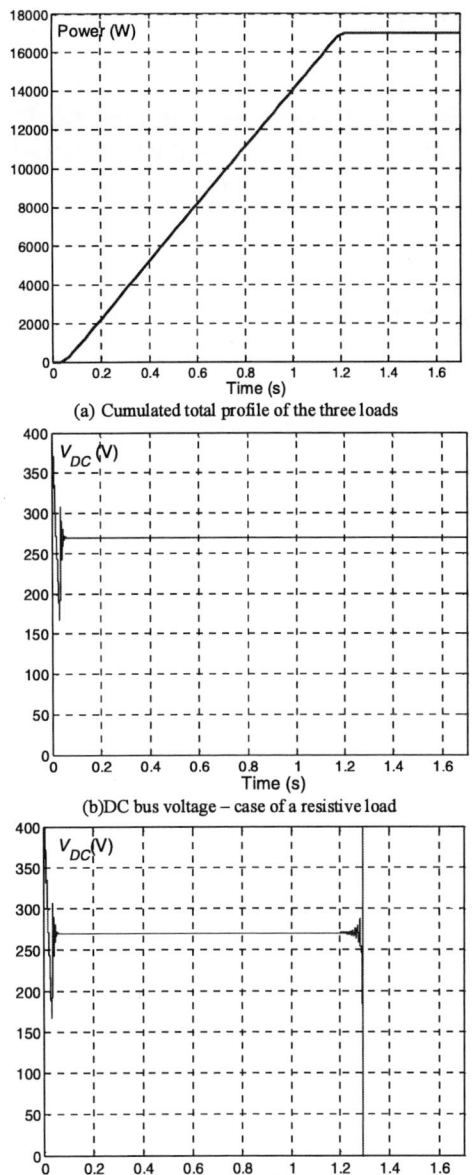

(a) Cumulated total profile of the three loads

(b)DC bus voltage – case of a resistive load

(c) DC bus voltage – case of a constant power load

Fig. 5 The different stability phenomena at the 17kW operating point in the defined network

B. Small signal analysis

The different stability behaviours for different kinds of load at the same operation point are analysed using the small signal analysis theory and the notion of impedance specification.

As the study will concentrate on the influence of the generation system (PMG, AC/DC converter and output filter), the three constant power loads in Fig. 1 are replaced by only one representative load. To avoid the stability influence of the power cable impedance and capacitors, these elements will not be considered in the stability study. So, the network is simplified as illustrated in Fig. 6.

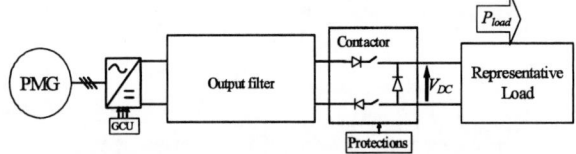

Fig. 6 Considered simplified network for stability study

When the tests of different operating points in Fig. 5 are applied the simplified model, at the operating point 17kW, the different stability phenomena by using different types of load model are shown in Fig. 7. By comparing Fig. 7 (b) and Fig. 5 (c), we can note this simplification does not change the essential of the unstable phenomenon. This validates the simplified model for stability analysis. We can continue the stability analyse based on this simplification.

(a)DC bus voltage – case of a resistive load

(b) DC bus voltage – case of a constant power load

Fig. 7 The different stability phenomena at the 17kW operating point in simplified model

As shown in Fig. 8, we can model the source and load module as two-port networks [3].

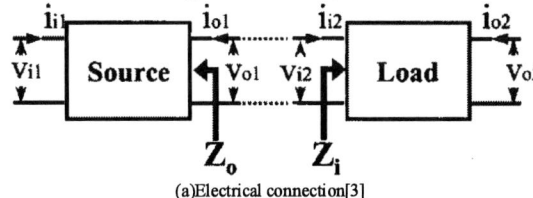

(a)Electrical connection[3]

1525

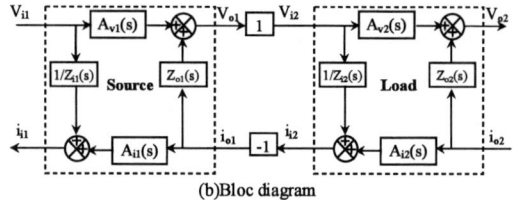

(b)Bloc diagram

Fig. 8 Two-port network model of the system

Because of the existence of non linear elements, the parameters vary according to the different operating points. So the small signal theory is employed, which makes possible to linearize the system at certain operating points and study the stability of this system at this operating point. The linearized voltage loop is shown in Fig. 9 and the transfer function is indicated in (1) where T_{v1} and T_{v2} are the linearized transfer functions of A_{v1}, and A_{v2}. $\hat{z}_o$, $\hat{z}_i$ are respectively the small signal impedances of source and load. $\hat{v}_{o1}$, $\hat{v}_{i1}$, $\hat{v}_{o2}$, $\hat{v}_{i2}$, $\hat{i}_{o1}$, $\hat{i}_{i2}$ are the small signal variables.

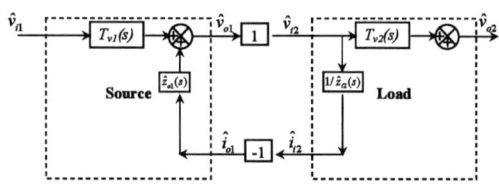

Fig. 9 The voltage loop of the small signal linearized model

$$\frac{\hat{v}_{o2}(s)}{\hat{v}_{i1}(s)} = \frac{T_{v1}(s) \cdot T_{v2}(s)}{1 + T_m(s)} \text{ where } T_m(s) = \frac{\hat{z}_{o1}(s)}{\hat{z}_{i2}(s)} \quad (1)$$

Since the source and the load are individually stable at their operating points, $T_{v1}(s)$, $T_{v2}(s)$ are stable in (1). So the study of the stability problem is focused on the $T_m(s)$.

Many efforts have already been made to develop the impedance criterion to study the stability and the performance of the interconnected system [2][4][5]. In [2], the GMPM criterion has been developed by defining the forbidden region to ensure the system has enough gain margin and phase margin. The definition of forbidden region in polar scheme is shown in Fig. 10 if we define the gain margin should greater than 6dB and the phase margin should be greater than 60°.

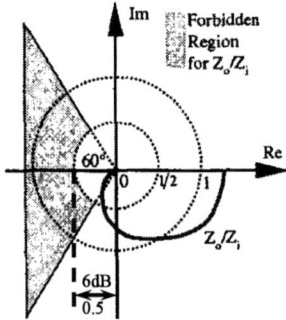

Fig. 10 The description of criterion GMPM

C. Measurement of the impedance

In order to calculate the ratio of impedance, we have to measure the small signal impedance $\hat{z}_o$, $\hat{z}_i$ in certain operating point. Fig. 11 shows the implementation of this measurement. We inject the small signal voltage $\hat{v}_p$ in certain range of frequency into the bus, then measure the current response $\hat{i}$, the voltage response of source $\hat{v}_o$ and load $\hat{v}_i$ to calculate their small signal impedances[3].

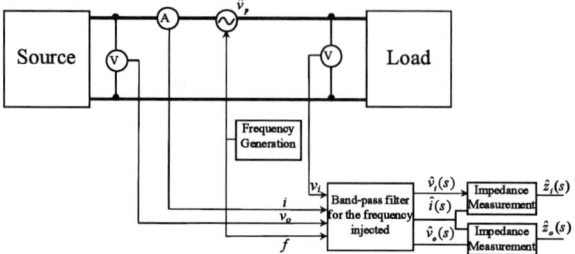

Fig. 11 Implementation of the small signal impedance measurement

$$\hat{z}_o(s) = \frac{\hat{v}_o(s)}{\hat{i}(s)}$$
$$\hat{z}_i(s) = \frac{\hat{v}_i(s)}{\hat{i}(s)} \quad (2)$$

IV. RESULT OF THE ANALYSIS

We compare the result of $\hat{z}_o(s)$ the impedance at source side, $\hat{z}_i(s)$ the impedance at load side and $T_m(s) = \hat{z}_o(s)/\hat{z}_i(s)$ the minor loop gain at different operating points and of different loads in the frequency range10~100kHz.

A. Resistive and constant power load at 6kW

Fig. 12 shows the result of small signal analyses of 6kW resistive load and constant power load. We observe that even the system with 6kW constant power load has not enough phase margin, but it does not enter into the forbidden region, because it has enough gain margin also. In this case, the system can remain stable and reject the disturbance.

B. Resistive and constant power load at 17kW

In the polar scheme of $\hat{z}_o(s)/\hat{z}_i(s)$ of the system with 17kW resistive load which is shown in Fig. 13 (c), the minor loop gain has overstep the reserved gain margin 0.5 (6dB), but the system is still stable as figured in Fig. 7, because it has enough phase margin.

But because of the different load models, as shown in Fig. 13 (d), the minor loop gain of 17kW constant power load not only oversteps the reserved gain margin but also passes the limited phase margin. In consequence, it enters into the forbidden region according to the criterion

1526

GMPM. In this case, the system becomes unstable and Fig. 7 (b) proves this opinion.

C. Comparison of the stability margin between different operation points

Comparing the Fig. 13(c) and Fig. 12(c), even though 17kW resistive load is stable but compared with 6kW resistive load. In this case, the sensibility of phase is increased. The same comparison between the Fig. 13(d) and Fig. 12(d), the 6kW constant power has passed the limit of phase margin, but the gain margin is great enough to maintain the stable, in the 17kW constant power load, the minor loop gain is increasing and overstep in the forbidden region.

We can conclude, the increase of load consumed power decreases the stability margin because of the increasing of minor loop gain.

D. Comparison of the source impedance between 6kW and 17kW

When we compare the impedance of the source at different operating points, we can find out that there are differences. As shown in Fig. 14, the magnitude $\hat{z}_o(s)$ at 17kW is greater than the one at 6kW. Because the increasing of power will cause the decreasing of load impedance (compare the $\hat{z}_i$ in Fig. 12 (b) and Fig. 13(b)), this change of source impedance may push the minor loop gain $\hat{z}_o(s)/\hat{z}_i(s)$ into the forbidden region.

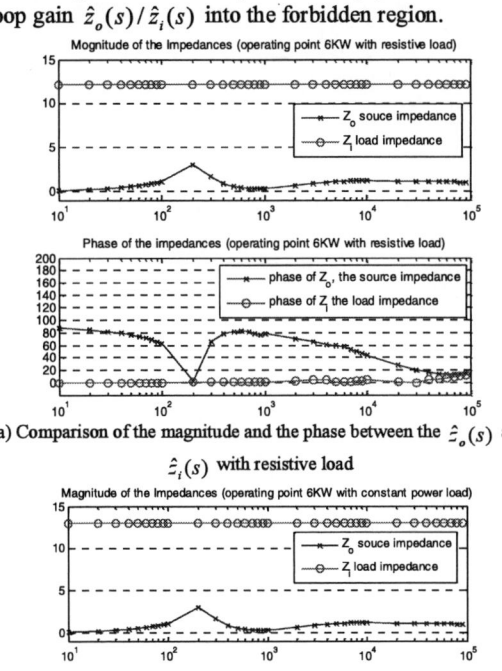

(a) Comparison of the magnitude and the phase between the $\hat{z}_o(s)$ and $\hat{z}_i(s)$ with resistive load

(b) Comparison of the magnitude and the phase between the $\hat{z}_o(s)$ and $\hat{z}_i(s)$ with constant power load

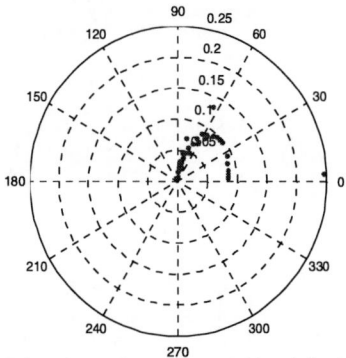

(c) Polar scheme of $\hat{z}_o(s)/\hat{z}_i(s)$ with resistive load

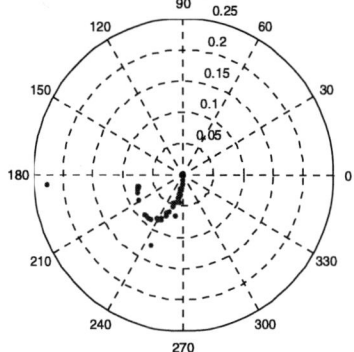

(d) Polar scheme of $\hat{z}_o(s)/\hat{z}_i(s)$ with constant power load

Fig. 12 The comparison of the resistive and constant power load at 6kW

V. CONCLUSION

According to the above analysis, some factors that effect on the stability of DC DPS are pointed out.

Firstly, according to the different stability behaviour of the resistive load and constant power load, we can conclude that different model of load changes the stability phenomena due to their different frequency response.

Secondly, the increase of load consumed power decreases the stability margin.

Thirdly, the non linear component is a factor of instability because the frequency characteristic changes in function of the operating point.

From the above conclusion, we can consider to create a supervision of the network, by measuring the impedance in order to ensure an optimized stability margin value. Some storage system could take the role of impedance adjuster to control this stability margin.

Acknowledgments:

This work was supported by a financing from the regional Council Nord-Pas de Calais, HISPANO-SUIZA Company, and HEI.

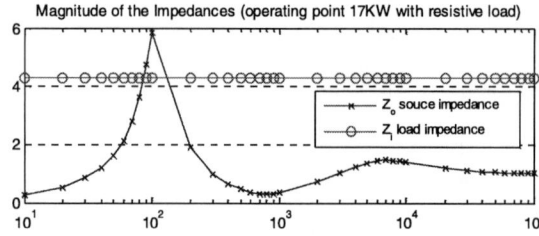

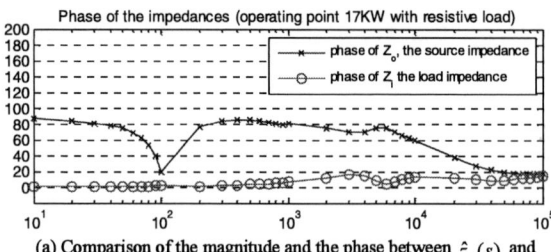

(a) Comparison of the magnitude and the phase between $\hat{z}_o(s)$ and $\hat{z}_i(s)$ with resistive load

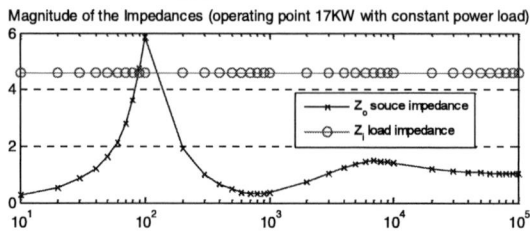

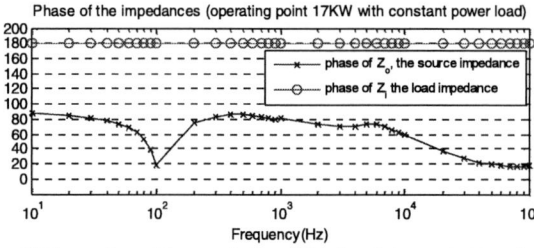

(b) Comparison of the magnitude and the phase between $\hat{z}_o(s)$ and $\hat{z}_i(s)$ with constant power load

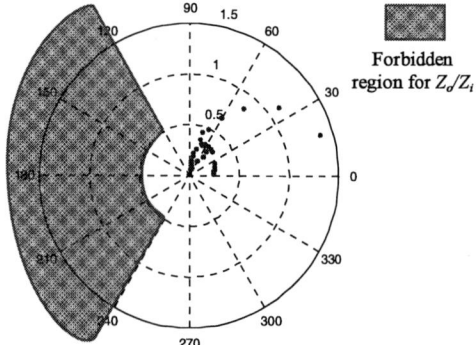

(c) Polar scheme of $\hat{z}_o(s)/\hat{z}_i(s)$ with resistive load

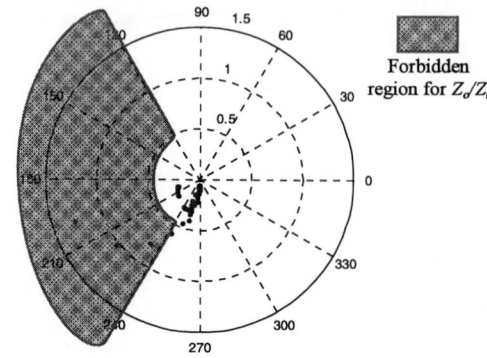

(d) Polar scheme of $\hat{z}_o(s)/\hat{z}_i(s)$ with constant power load

Fig. 13 The comparison of the resistive and constant power load at 17kW

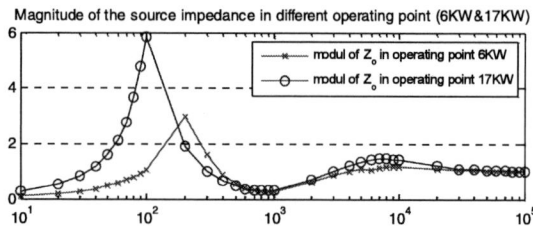

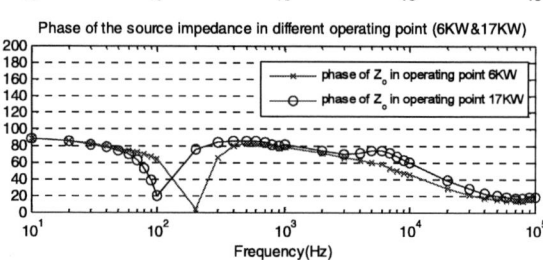

Fig. 14 The comparison for magnitude and phase between the source impedances at different operating points

REFERENCES

[1] D. Van Den Bossche, "more electric control surface actuation" *EPE, 2003 Toulouse*

[2] C.M.Wildrickand F.C.Lee, "A method of defining the load impedance specification for a stable distributed power system," *IEEETrans.Power Electron.*, vol. 10, pp. 280–285, May 1995.

[3] X. Feng and F.C. Lee, "On-line Measurement on Stability Margin 0f DC Distributed Power System" *Applied Power Electronics Conference and Exposition, 2000. APEC 2000. Fifteenth Annual IEEE*, Vol 2, 6-10 Feb 2000, page(s): 1190 - 1196

[4] F. Barruel, A. Caisley, N. Retière, J.L. Schanen, "Stability Approach for Vehicle DC Power Networks: Application to Aircraft On-Board System" *Power Electronics Specialists Conference, 2005. PESC '05. IEEE 36th* Page(s):1163 – 1169

[5] P. Liutanakul, "Stability of Embedded network, Power Interaction – Structure – Command", *Thesis, L'Institut National Polytechnique de Lorraine, 2007*

[6] D. Nilsson, A. Sannino, "Load Modelling for Steady-State and Transient Analysis of Low-Voltage dc Systems" *Industry Applications Conference, 2004. 39th IAS Annual Meeting. Conference Record of the 2004 IEEE* ,Volume 2, 3-7 Oct. 2004 Page(s):774 - 780 vol.2

Design Considerations for Control of Traction Drive with Permanent Magnet Synchronous Machine

Zdeněk Peroutka, Karel Zeman

University of West Bohemia in Pilsen/Dept. of Electromechanics & Power Electronics, Plzeň, Czech Republic,
e-mail: *peroutka@ieee.org, zeman@kev.zcu.cz*

Abstract—This contribution deals with the control of the wheel drive with permanent magnet synchronous machine of the tram car. In order to achieve minimal dimensions of PMSM and on the other hand to be able to provide sufficient torque, the drive is operating in very wide speed range in the field weakening and with the machine load angle of 90°. Thus, the new problem introduced by this drive is requirement for the limitation of the load angle. This paper presents proposed drive control, simulation results of tram drive prototype and experimental evidence of designed laboratory prototype of rated power of 10.7kW.

Keywords—Control of drive, permanent magnet motor, rail vehicle, synchronous motor, traction application, vector control.

I. INTRODUCTION

The motivation for this research has been demand of our industrial partner for the design of the control of the direct wheel drive with permanent magnet synchronous machine (PMSM) for the new generation of the tram cars. The drive employs PMSMs with surface magnets on the rotor. In the literature there are published very interesting papers dealing with PMSM traction drives and their control e.g. [1] – [6]. Especially Alstom and Siemens have introduced very interesting solutions. However, they do not use the direct PMSM based wheel drive.

The explored traction drive with PMSM presents two major problems. The oscillations of the LC filter in the traction converter dc-link are first very difficult problem. However, this problem is common for all traction drives. In order to achieve as small as possible dimensions of PMSM and on the other hand to be able to provide sufficient torque, the drive is operating in very wide speed range in the field weakening and with the load angle of 90° (see Fig. 1). Thus, the new problem introduced by this drive is requirement for the limitation of the load angle.

This paper deals with the design considerations for the control of the above described traction drive and explains the proposed control strategy. Moreover, this contribution also presents the simulation results of the tram drive prototype as well as experimental evidence of designed laboratory prototype of the tram drive of rated power of 10.7kW.

This research has been supported by the Ministry of Industry and Trade of the Czech Republic under the project FT-TA5/084: "Drive systems with permanent magnet synchronous motors".

II. PROPOSED DRIVE CONTROL

Most of the published solutions use for the control of PMSM traction drives the rotor flux oriented vector control with two current controllers – "independent" control of the torque (i_{sq}) and flux component (i_{sd}) of the stator current. The vector control brings advanced possibilities from the dc-link filter oscillations damping point of view (e.g. [7]). Therefore, we have also selected the above described vector control for our application (Fig. 2).

The biggest problem, which we have been dealing with, is the limitation of the load angle in the field weakening region. In order to achieve the minimal dimensions, the drive is operating in a very wide speed range in the field weakening (in our case approx. 65% of the whole speed range). The drive must be able to permanently operate with the load angle of 90° in order to comply with defined torque-speed characteristic. The vector diagram describing the operation of the PMSM with the load angle of 90° is shown in the Fig. 1. If the machine operates with the load angle of 90°, the q-component of the stator voltage vector (u_{sq}) is equal to zero. Therefore, we have at the beginning completed the control system with the u_{sq} controller (R_{Uq}) – Fig. 3, which should via the interventions to the torque component command i_{sqw} provide the limitation of the load angle to 90°. However, this solution is not robust enough to be used in the final application. The reason is that if the machine operates under the load angle of 90°, the position and size of the stator current vector is exactly defined (it is clearly described in the Fig. 1). It means that the condition of the load angle of 90° is accomplished only for one current vector – therefore, i_{sd} and i_{sq} are exactly defined in this mode. The limitation of the load angle is in this case very difficult process, the optimal operating point is found by the incorporation of both current controllers. In addition to the difficulties of this solution, the proper limitation of the load angle also required the adaptive gains of the current controllers. Thus, this solution is not used for the final application due to its poor robustness and complexity.

We have overcome the problem with the limitation of the load angle in the field weakening region using the control described in the Fig. 4. In this case the controller R_{Iq} directly commands the angle between the back EMF vector and the stator voltage vector (if we neglect the stator resistance, the angle between back EMF vector and stator voltage vector is equal to the load angle). This makes us possible very easy limitation of the machine load angle. The stator voltage vector magnitude is calculated using simplified voltage model of the machine – see (1):

$$U_{mw} = \sqrt{\left(p_p \omega_m \Psi_{pm}\right)^2 + \left(p_p \omega_m L_s I_{sqw}\right)^2} \,, \qquad (1)$$

where p_p is number of pole pairs, ω_m is rotor angular speed, Ψ_{pm} is flux excited by permanent magnets on the rotor, L_s is stator inductance. The computed voltage vector magnitude (U_{mw}) is corrected using controller R_{Id}. The proposed solution is very simple and powerful. However, this control can not be used around zero speed. Therefore, we use this control only in the field weakening region. At the standstill and in the low speeds, we use the vector control described in the Fig. 2.

The switching between two employed controls is made based on the rotor speed – the switching is of course realized using the hysteresis block with sufficiently wide hystresis band. It is necessary to properly initiate the control circuits at the moment of switching of the control strategies in order to reduce the current and torque overshoot. Thus, if the control strategy is switching from the circuits depicted in Fig. 2 to control in Fig. 4, the integral part of the controller R_{Iq} must be preset to the value of β_u given by previous control strategy and integral part of the controller R_{Id} must be reset. If we are switching from the control in Fig. 4 to the control in Fig. 2, integral parts of both current controllers (R_{Id} and R_{Iq}) must be reset. In spite of the above described proper initialization of the control circuits in Fig. 4, the simplified calculation of the stator voltage vector magnitude (1) can cause the undesirable current and torque overshoot during the switching of the control strategies under higher trolley wire voltages (it means that control strategies switching occurs out of the field weakening region). This disadvantage can be avoided by improved stator voltage magnitude control, which is presented in the Fig. 5. The recently proposed control uses the modulation depth controller (R_{Urm}), which generates the flux current command i_{sdw} (in the same way like in the Fig. 2). The

stator voltage vector magnitude is calculated using (2):

$$u_{sd} = R_s I_{sdw} - p_p \omega_m L_s I_{sqw}$$
$$u_{sq} = R_s I_{sqw} + p_p \omega_m L_s I_{sdw} + p_p \omega_m \Psi_{pm} \,, \qquad (2)$$
$$U_{mw} = \sqrt{u_{sd}^2 + u_{sq}^2}$$

where R_s is stator resistance.

The control proposed in the Fig. 4 and in Fig. 5, respectively, is simple, robust and makes very easy limitation of the machine load angle. On the other hand, the control of the motor voltage in proposed polar form (control of the voltage magnitude and angle) has generally worse dynamic properties than the control of the voltage vector components (u_{sd} and u_{sq}). However, in the field weakening, where the proposed control (either Fig. 4 or Fig. 5) is employed, the voltage source inverter is operated with maximum modulation depth (the machine stator voltage magnitude is constant). There is only one acting quantity in the field weakening – the motor voltage vector position. Therefore, the dynamic properties of the proposed control are acceptable.

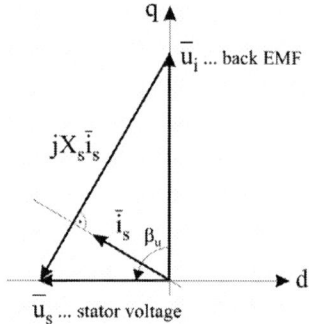

Fig. 1. Vector diagram describing the operation of PMSM with the load angle of 90□ (stator resistance neglected)

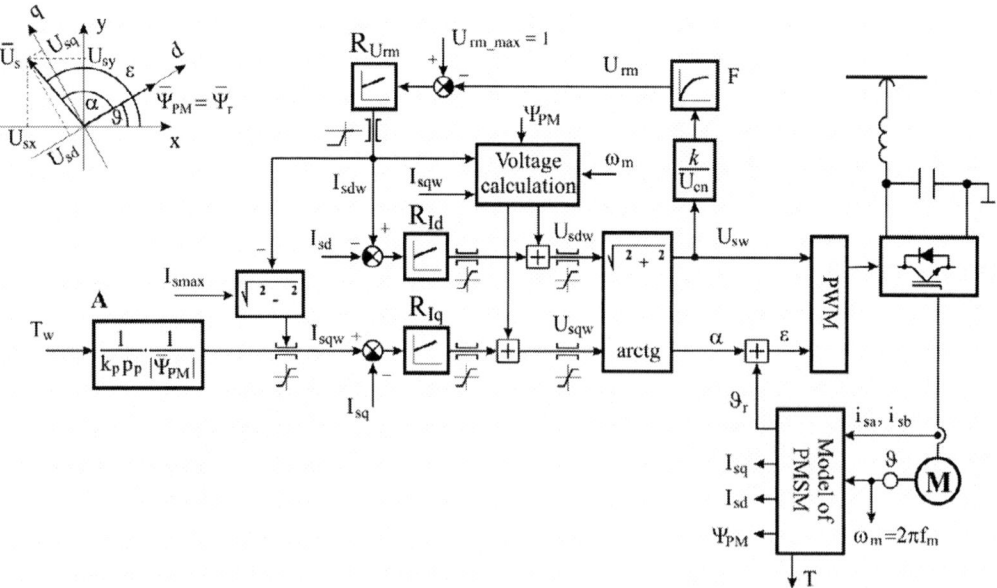

Fig. 2. Rotor flux oriented vector control employed at the standstill and in the low speeds

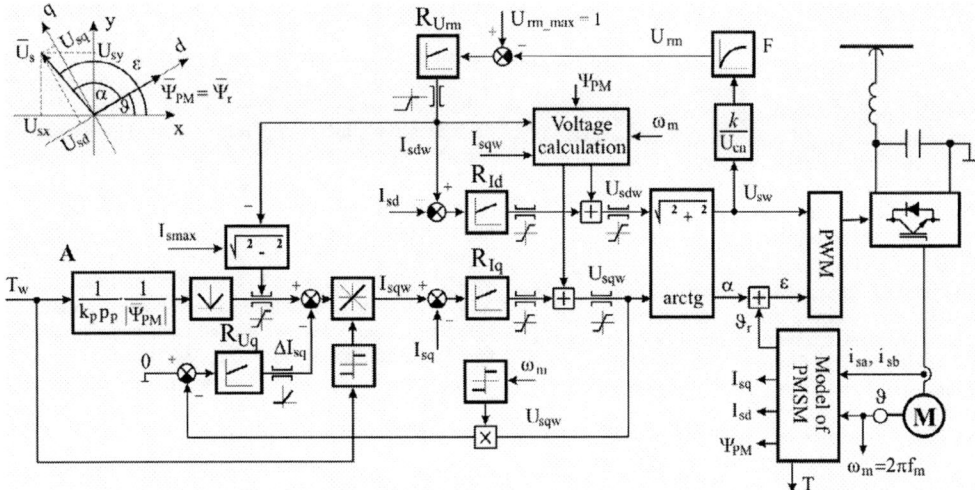

Fig. 3. Rotor flux oriented vector control with load angle limitation using control of q-component of stator voltage vector (u_{sq}) ... controller R_{Uq}

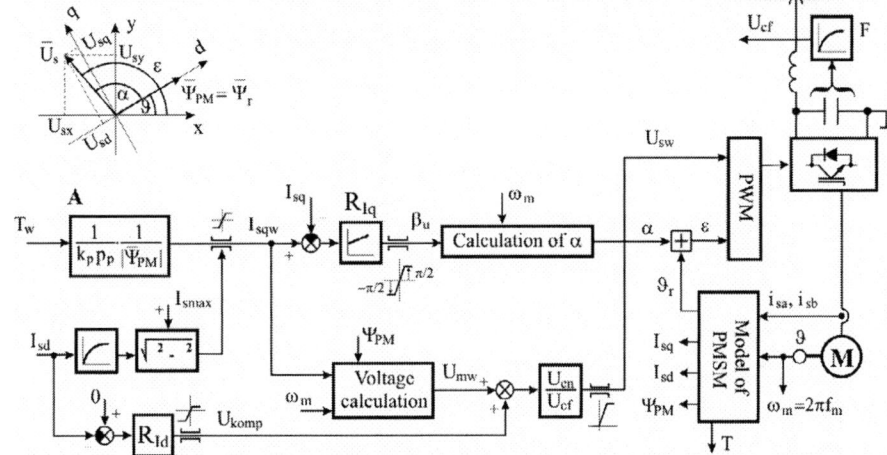

Fig. 4. Proposed drive control in field weakening region – direct control of angle between back EMF vector and stator voltage vector (β_u)

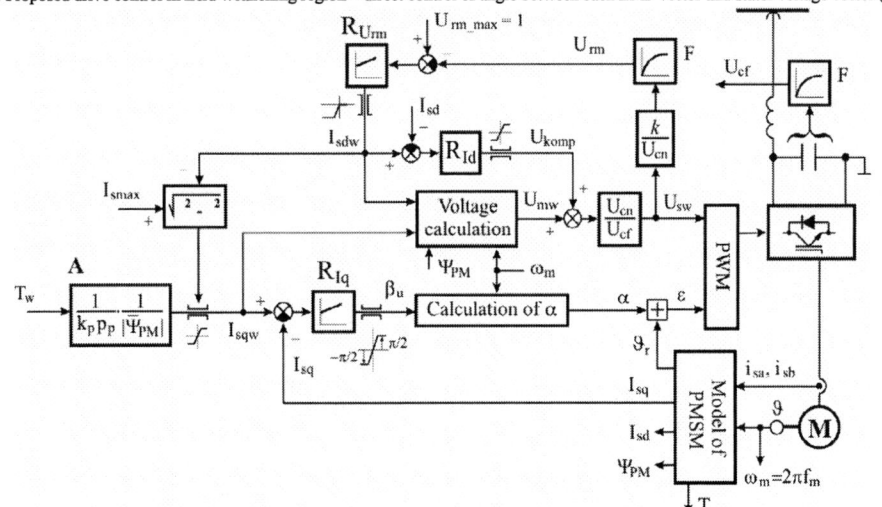

Fig. 5. Proposed drive control in field weakening region:
direct control of angle between back EMF vector and stator voltage vector (β_u) and improved stator voltage magnitude control

III. SIMULATION RESULTS OF TRAM DRIVE PROTOTYPE

The simulations present behavior of the permanent magnet synchronous machine traction drive prototype of rated power of 46.5kW. The explored control employs third-harmonic injected PWM with carrier frequency of 4kHz. The simulation model respects as close as possible properties of the real traction drive (dead-times of 3μs, nonlinear voltage drops on power electronic devices, LC filter in traction converter dc-link, etc.) and microprocessor-based controller.

Fig. 6 shows drive acceleration from standstill to maximum speed of 700rpm with maximum torque. The trolley wire voltage has been during the test considered of $400V_{dc}$, the equivalent moment of inertia of vehicle has been set to $350kgm^2$. Left column displays the motor quantities as a function of time. The right column plots the motor quantities as a function of the rotor speed. The control in Fig. 2 has been employed from standstill to 200rpm. Fig. 6 confirms the proper transition from the

control in Fig. 2 to control in Fig. 5 at the rotor speed of 200rpm. There is neither significant current nor torque overshoot. The simulation shows the reliable saturation of the angle (β_u) between back EMF vector and stator voltage vector to 90°. The machine load angle slightly decreases with increasing rotor speed. This is caused by resistive drops (if we neglect the stator resistance, the angle β_u is equal to the machine load angle). The small difference between the machine load angle and β_u does impact neither dynamic properties nor robustness of the proposed control. We can of course simply change the control strategy in order to exactly control the load angle to 90°. However, it will make the control circuits more complex. In our case, we do not need exact limitation of the load angle to 90°, but we must preserve the controllability of the drive around the load angle of 90°. This function is reliably covered by saturation of the angle β_u as is presented in this paper.

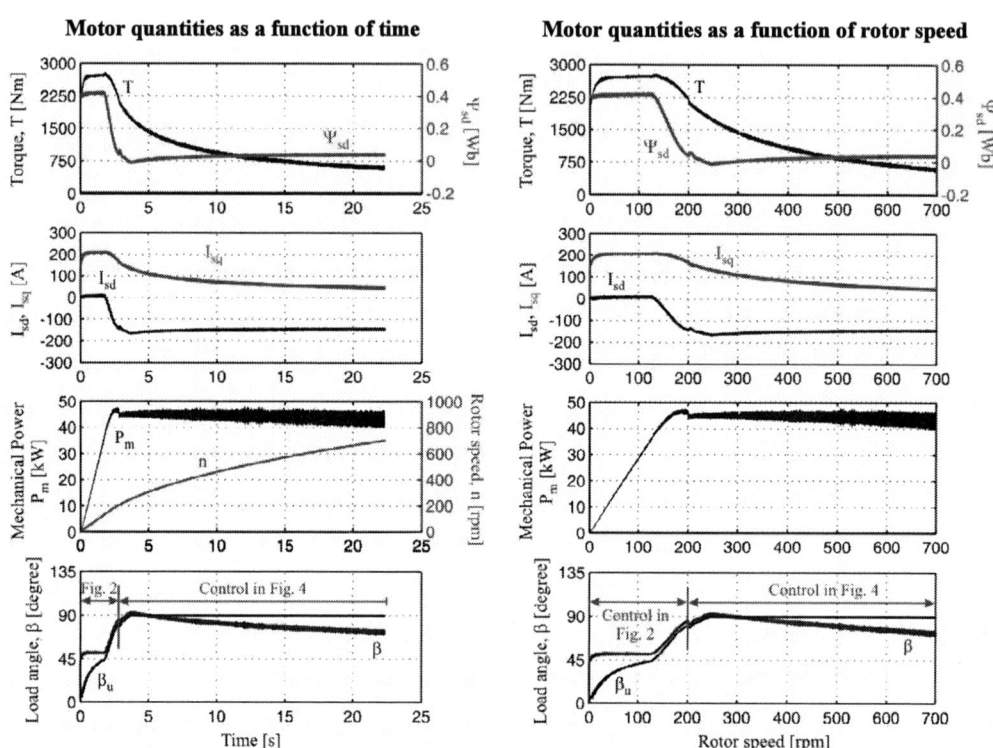

Fig. 6. Tram drive – acceleration from standstill to max. speed of 700rpm with maximum torque: trolley wire voltage of 400V, equivalent moment of inertia of $350kgm^2$

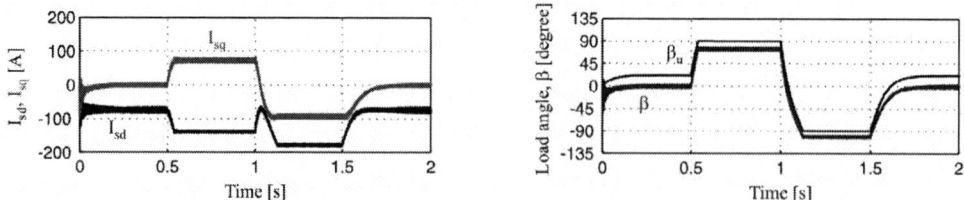

Fig. 7. Tram drive – dynamic behavior of control described in Fig. 5: flying start and step changes of commanded torque (trolley wire voltage of 600V, fixed rotor speed of 700rpm, step changes of $I_{sqw} = \pm I_{smax}$)

Fig. 7 illustrates dynamic behavior of the control strategy described in the Fig. 5. The presented test shows flying start of the drive and extreme step changes of commanded torque. Trolley wire voltage has been set to 600V$_{dc}$ (rated trolley wire voltage), the rotor speed has been fixed to 700rpm. The drive behavior is acceptable – the dynamic properties are satisfactorily and the employed control is able to damp the oscillations of the input trolley wire filter. Nevertheless, the dynamic properties of the control in the field weakening as well as the active damping of the oscillations of the LC filter in the converter dc-link are worse in comparison to control in Fig. 2 as has been discussed in the paragraph II.

The simulation results confirm the proper operation of the proposed tram wheel drive control.

IV. LABORATORY PROTOTYPE OF TRAM DRIVE

The proposed control has been implemented in the fixed-point digital signal processor Texas Instruments TMS320F2812. The control system of the laboratory prototype is the same as the traction prototype to be designed (third-harmonic injected PWM with carrier frequency of 4kHz, dead-times of 3µs, etc.). Selected experimental results of the laboratory prototype of the tram drive with rated power of 10.7kW are shown in the Fig. 8 and Fig. 9.

Fig. 8 analyzes in detail transitions between the control in Fig. 2 and the control in Fig. 5. The tests have been made with the reduced dc-link voltage of 200V and the drive has been operated in the rotor speed control mode. The switching from control in Fig. 2 to control in Fig. 5 occurs at the rotor electrical speed of f_{me} = 90Hz. The switching from control in Fig. 5 to control in Fig. 2 occurs at the rotor electrical speed of f_{me} = 80Hz. Thus, we have employed hysteresis band of 10Hz. Fig. 8a and Fig. 8b analyze the drive startup, acceleration form standstill to electrical rotor speed of 150Hz and return back to standstill. Fig. 8c documents the drive acceleration (speed ramp) from standstill to electrical rotor speed of 120Hz and speed reversal to electrical rotor speed of -120Hz. Fig. 8d presents drive startup and speed reversal – triangular speed profile with electrical rotor speed command of ±120Hz. The dynamic properties of designed drive prototype in the field weakening are illustrated in the Fig. 9, which shows the drive responses to the torque current command changes under different fixed rotor electrical speeds (the drive has been in these tests operated in the torque control mode).

V. CONCLUSIONS

This contribution describes design considerations for the control of the wheel drive with permanent magnet synchronous machine of the tram car and presents the proposed control strategy. Moreover, this contribution also includes the simulation results of the tram drive prototype as well as experimental evidence of designed laboratory prototype of the tram drive of rated power of 10.7kW.

The main attention is in this contribution paid to the problem of the limitation of the machine load angle in the field weakening region. The proposed control system overcomes this problem by direct control of the angle between the back EMF vector and the stator voltage vector (see Fig. 4 and Fig. 5). This approach makes us possible very easy and powerful limitation of the load angle to 90°. Because the proposed control (Fig. 4, Fig. 5) can not be employed in the whole speed range, we use at the standstill and in the low speeds conventional rotor flux oriented vector control (Fig. 2) and the direct load angle control is employed only in the field weakening region. The switching between two employed controls is made based on the rotor speed – the switching is of course realized using the hysteresis block with sufficiently wide hystresis band.

The proposed drive control is going to be tested on the developed tram car during the year 2008. The new tram car with above described drive should be in passenger operation in the year 2009.

ACKNOWLEDGMENT

This project is realized in cooperation with the company Škoda Electric, a.s., Plzeň, Czech Republic (www.skoda.cz).

REFERENCES

[1] Belin, S., Scrooby, M., Masselus, J.E., Jobard, T., Courtine, S., "A PMSM Based Control for Traction Applications," in *EPE 2003*, Toulouse, France, 2003. pp. P.1 – P.15. ISBN: 90-75815-07-7.

[2] Jockel, A., Lowenstein, L., Teichmann, M., Hoffmann, T., v. Wangelin, F., "Syntegra – Innovativer Prototyp einer nächten Triefahrwerk-Generation," *eb - Elektrische Bahnen*, no. 8-9/2006, 2006. pp. 360 – 369.

[3] Frenzke, T., Piepenbreier, B., "Position-Sensorless Control of Direct Drive Permanent Magnet Synchronous Motors for Railway Traction," in *PESC 2004*, Aachen, Germany, pp. 1372 – 1377.

[4] Vas, P., *Sensorless Vector and Direct Torque Control*. Oxford University Press, New York, USA, 1998.

[5] Leonhard, W., *Control of Electrical Drives*, 3rd Edition, Springer, Berlin, Germany, 2001.

[6] Holtz, J., "Sensorless Control of Induction Motors and PM Synchronous Machines," Tutorial in *IEEE International Symposium on Industrial Electronics (ISIE) 2005*. Dubrovnik, Croatia, 2005.

[7] Peroutka, Z., Zeman, K., "New Field Weakening Strategy for AC Machine Drives for Light Traction Vehicles," in *EPE 2007*, Aalborg, Denmark, 2007.

APPENDIX
CONFIGURATION OF THE LABORATORY PROTOTYPE

PMSM	Permanent magnet synchronous machine with surface magnets on the rotor: Ultract II – 1004303F, 10748W/346V/30A, 3000rpm, 38Nm, 8 poles Rotor position sensor: LARM ARC405/12PB
Loading machine	Induction machine: EM Brno 1AY112L-4, 17kW/390V/32A/101.9Hz, 3000rpm Rotor speed sensor: LARM IRC 305/4096PB
Loading converter	Frequency converter ABB ACS 800-11-0016-3 (15kW)

a) Drive startup, acceleration to electrical rotor speed of $f_{mew} = 150Hz$ and return back to standstill (speed ramp)
Ch3: Electrical rotor speed f_{me} (80Hz/V)

b) Drive startup and ramp acceleration from standstill to rotor electrical speed of $f_{mew} = 150Hz$
Ch3: Electrical rotor speed f_{me} (80Hz/V)

c) Acceleration (speed ramp) from standstill to el. rotor speed of $f_{me} = 120Hz$ and speed reversal to $f_{me} = -120Hz$
Ch3: Electrical rotor speed f_{me} (160Hz/V)

d) Drive startup and speed reversal: triangular speed profile ... $f_{mew} = \pm 120Hz$
Ch3: Electrical rotor speed f_{me} (160Hz/V)

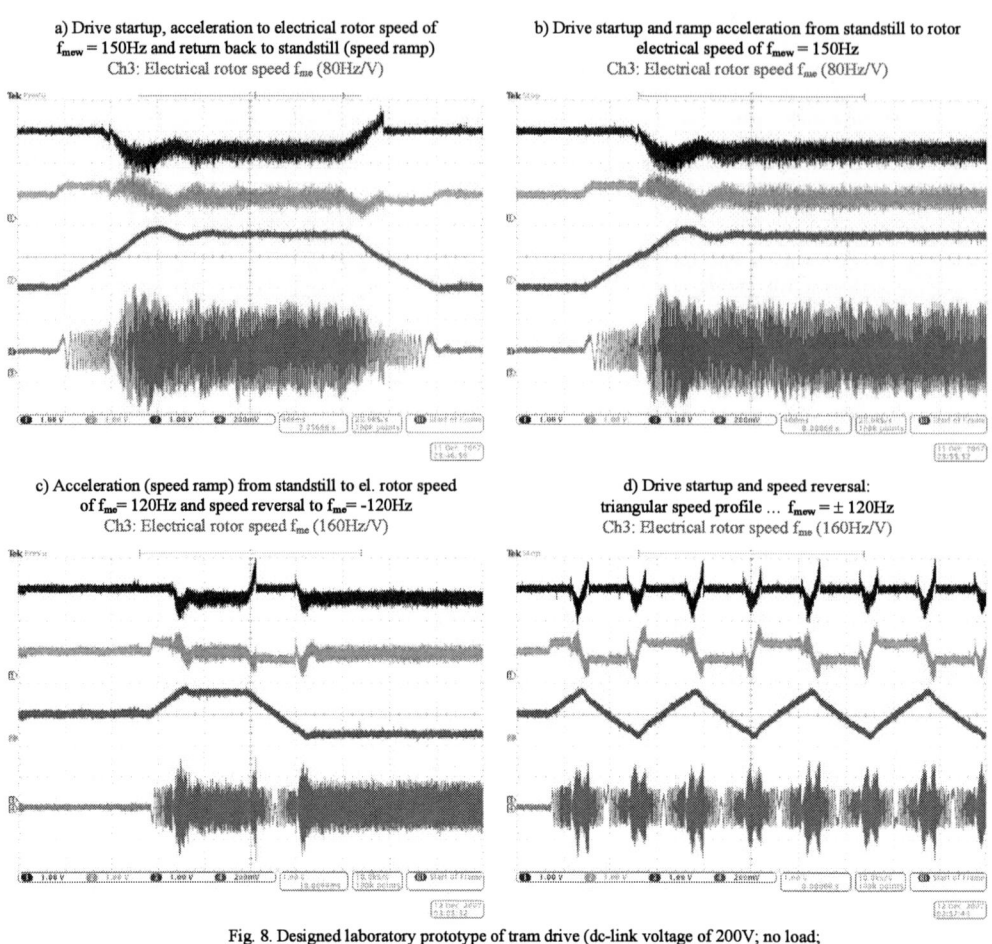

Fig. 8. Designed laboratory prototype of tram drive (dc-link voltage of 200V; no load; transition from control in Fig. 2 to control in Fig. 5 ... $f_{me} = 90Hz$, transition from control in Fig. 5 to control in Fig. 1 ... $f_{me} = 80Hz$)

Ch1: Flux component of stator current I_{sd} (33.94A/V), Ch2: Torque component of stator current I_{sq} (33.94A/V),
Ch4: Stator current I_{sa} (10mV/A)

Fixed electrical rotor speed of $f_{me} = 150Hz$,
ramp change of torque command

Fixed electrical rotor speed of $f_{me} = 120Hz$,
step change of torque command

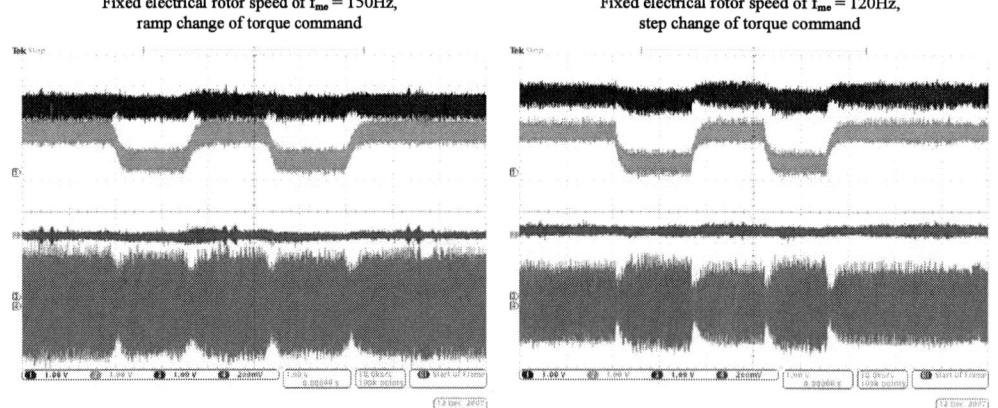

Fig. 9. Laboratory prototype of tram drive – dynamic behavior of drive with control strategy in Fig. 5:
Changes of torque current command ... $I_{sqw} = 20A \rightarrow -20A \rightarrow 20A$ (dc-link voltage of 200V)

Ch1: Flux component of stator current I_{sd} (33.94A/V), Ch2: Torque component of stator current I_{sq} (33.94A/V),
Ch3: Electrical rotor speed f_{me} (160Hz/V), Ch4: Stator current I_{sa} (10mV/A)

Control of Primary Voltage Source Active Rectifiers for Traction Converter with Medium-Frequency Transformer

Vojtěch Blahník, Zdeněk Peroutka, Jan Molnár, Jan Michalík

University of West Bohemia, Department of Electromechanics and Power Electronics, Plzeň, Czech Republic,
e-mail: *lucke@kev.zcu.cz, peroutka@ieee.org, jmolnar@kev.zcu.cz, jmichali@kev.zcu.cz*

Abstract—This paper deals with the new configuration of the traction converter with medium-frequency transformer intended for ac trolley wire fed locomotives. The main attention is paid to the primary serially connected single-phase voltage-source active rectifiers, which are directly connected to the trolley wire. This paper describes the control of the primary active rectifiers, presents simulation results as well as experimental evidence of designed low-voltage laboratory prototype of loco converter of rated power of 12kW.

Keywords—Converter control, high voltage power converters, locomotive, multilevel converters, pulse width modulation (PWM), rail vehicle, traction application.

NOMENCLATURE

C_1, C_2 Capacitance of primary DC-link capacitors [F].

ε Control angle – see Fig. 2 [rad].

I_{load1}, I_{load2} Equivalent current sources representing the load of primary active rectifiers [A].

i_t Trolley wire current [A].

L_{input} Inductance at the converter trolley wire side [H].

MFT   Medium-frequency transformer.

R_L Resistance of the converter input inductor L_{input}[Ω].

ϑ The position of the trolley wire voltage vector in the stationary reference frame [rad].

U_{c1}, U_{c2} Dc-link voltage of primary active rectifiers [V].

U_{cw} Commanded sum of dc-link voltages of primary active rectifiers [V].

U_m Trolley wire voltage magnitude [V]

u_t Trolley wire (supply) voltage [V].

u_v Voltage at the converter ac terminals [V].

I. INTRODUCTION

This project has been made in cooperation with our industrial partner Škoda Plzeň company. The objective of this project has been the research into the prospective configurations of the ac trolley wire fed traction converter for the new generation of the locomotives and especially suburban units. The presented converter configuration with medium-frequency transformer is one of selected promising solutions. The explored traction converter

consists of indirect frequency converters at its input. The input parts of the indirect converters are composed of voltage-source active rectifiers, which are directly connected to the ac trolley wire. Therefore, they are connected in series – see Fig. 1. The medium-frequency transformer is fed by voltage source inverters operated in the six-step mode with the output frequency of 400 Hz. The voltage-source active rectifier (VSAR) at the secondary side of the medium-frequency transformer is fed by square wave voltage with frequency of 400 Hz and the control of this part has been introduced in [1].

In the literature, there are published very interesting papers dealing with the configuration and control of the traction converter for locomotives, which can be directly connected to ac trolley wire – e.g. [2] – [7]. Our considerations are coming out from [2], where is presented the similar configuration of the traction converter for 15kV/16,7Hz fed locomotive.

The main attention is paid in this paper to the primary serially connected single-phase voltage-source active rectifiers, which are directly connected to the trolley wire. This paper describes the proposed control of the primary active rectifiers, presents simulation results as well as experimental evidence of designed low-voltage laboratory prototype of loco converter of rated power of 12kW.

II. CONTROL STRATEGY OF PRIMARY VOLTAGE SOURCE ACTIVE RECTIFIERS

The proposed control strategy of the primary voltage-source active rectifiers is shown in the Fig. 2. We control the total dc-link voltage U_{cw} (it means the sum of dc-link voltages from all rectifiers). The dc-link voltage controller (solved as conventional PS controller) commands the angle ε between the trolley wire voltage (u_t) and voltage at the rectifier ac side terminals (u_v). The magnitude of u_v is calculated from the information about the trolley wire voltage magnitude (U_m) and commanded angle ε. Thus, we employ the conventional model-based control of input rectifiers. The control signal for the modulation (u_v) is calculated using its model-based computed magnitude U_{vm} and the information about the position of the trolley wire voltage vector in the stationary reference frame (ϑ). The position of the trolley wire voltage vector in the stationary reference frame (ϑ) and the trolley wire voltage magnitude (U_m) are evaluated by discrete Fourier transform (DFT). The PWM with shifted carriers is used for the control of primary voltage-source active rectifiers.

978-1-4244-1741-4/08/$25.00 ©2008 IEEE

The major benefit of employed shifted carriers is the significant decrease of the trolley current ripple due to multilevel nature of the voltage at the converter ac terminals. In our particular case with two active rectifiers in series, the voltage at the converter ac side is five-level. The shifted carriers and resulting reduced trolley current ripple have positive impact on the converter input inductance design.

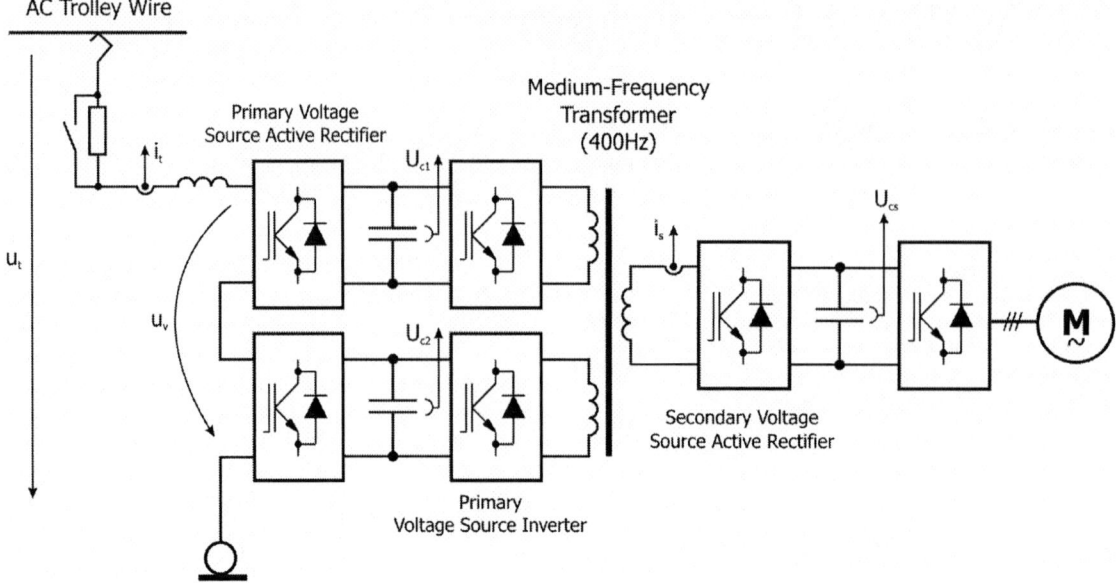

Fig. 1. Configuration of designed low-voltage laboratory prototype of traction converter with medium-frequency transformer

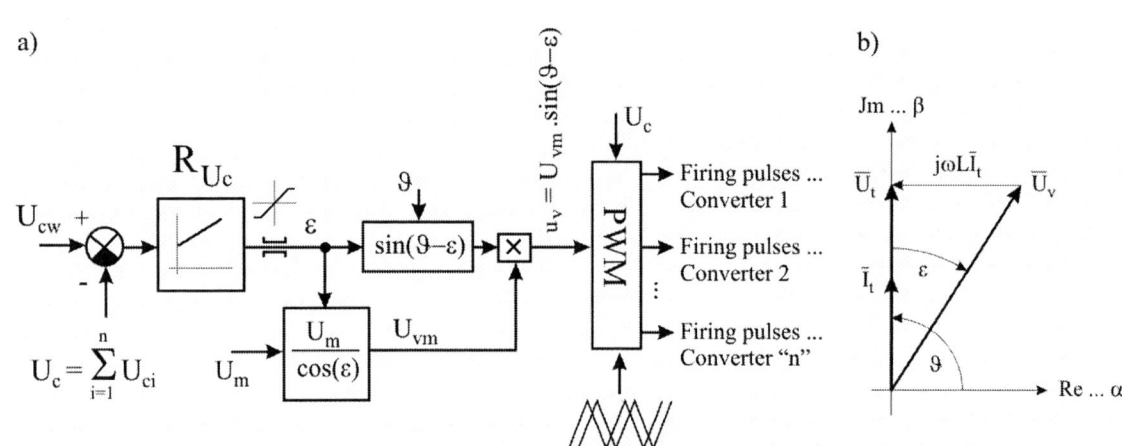

Fig. 2 Proposed control of primary voltage-source active rectifiers:
a) control circuits, b) vector diagram

III. SIMULATION RESULTS

The Fig. 3 presents converter simulation model, which is composed of voltage source (trolley wire voltage), input inductor including its resistance and two primary voltage-source active rectifiers connected in series. The simulation model parameters are the same as the parameters of the laboratory prototype, which are in detail listed in Appendix. The load of active rectifiers has been in simulations replaced by equivalent current sources (I_{load1} and I_{load2}). The simulation model has been designed in language C.

Fig. 4 – Fig. 6 show simulation results of the primary voltage-source active rectifiers operating in rectifier and inverter mode respectively under steady-state conditions. The quantity i_t [A] represents trolley wire current and is harmonic and in phase with trolley wire voltage u_t. The ripple of controlled dc-link voltages (in this case u_{c1}, u_{c2}) has the dominant frequency of double trolley wire frequency. This distortion is coming-out from physical properties of one phase voltage-source active rectifier. Fig. 6 illustrates the multilevel nature of the voltage (u_v) at the converter ac terminals. The multilevel (in this case 5-level) waveform of the voltage u_v is caused by shifted carriers in PWM control of the primary active rectifiers.

1536

The most important transient conditions are presented in Fig. 7 – Fig. 9. Fig. 7 describes the converter startup under no-load conditions. Fig. 8 shows the response of primary active rectifiers to step change of the load from inverter (P = - 7.2 kW) to rectifier mode (P = 8.8 kW). Fig. 9 depicts the response of primary active rectifiers to step change of the load from rectifier (P = 7 kW) to inverter mode (P = -5.4 kW).

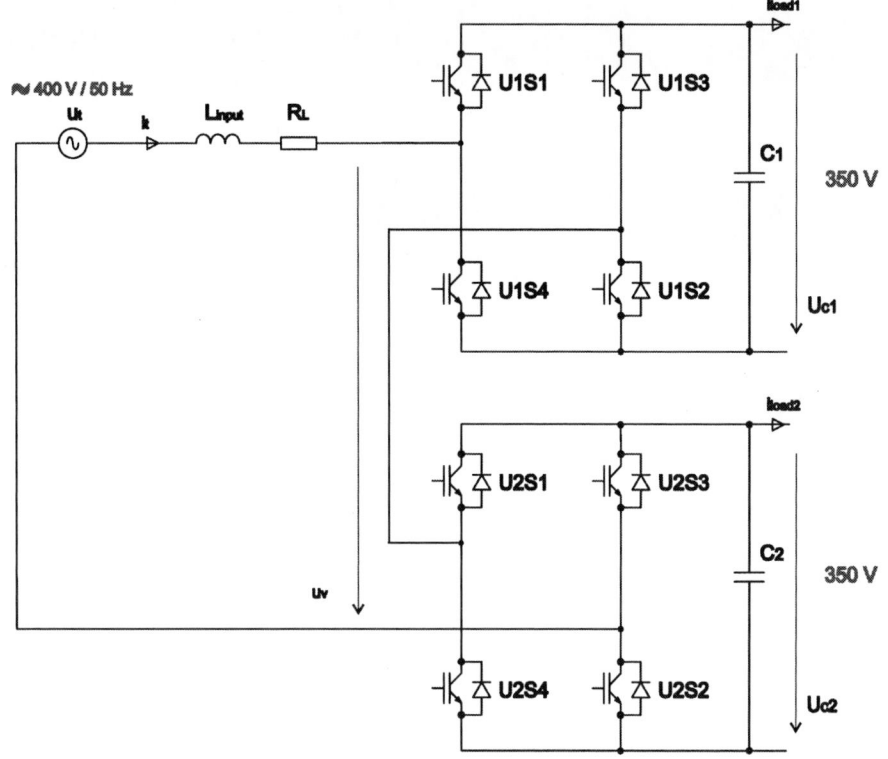

Fig. 3. Equivalent simulation model of primary voltage-source active rectifiers

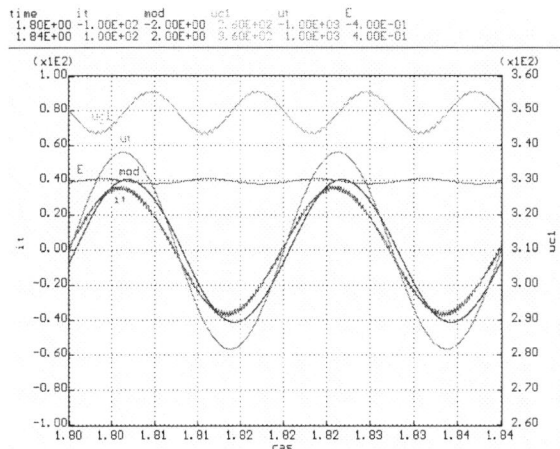

Fig. 4 Behaviour of primary voltage-source active rectifiers under steady-state conditions in rectifier mode
(load P = 10 kW, u_t = 400 V_{rms} / 50 Hz, U_{cw} = 700 V)

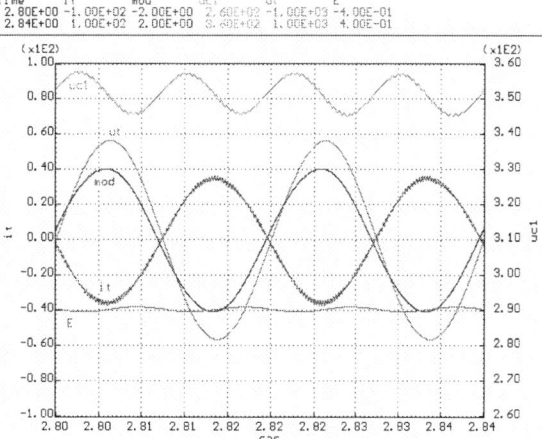

Fig. 5. Behaviour of primary voltage-source active rectifiers under steady-state conditions in inverter mode
(load P = -10 kW, u_t = 400 V_{rms} / 50 Hz, U_{cw} = 700 V)

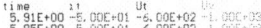
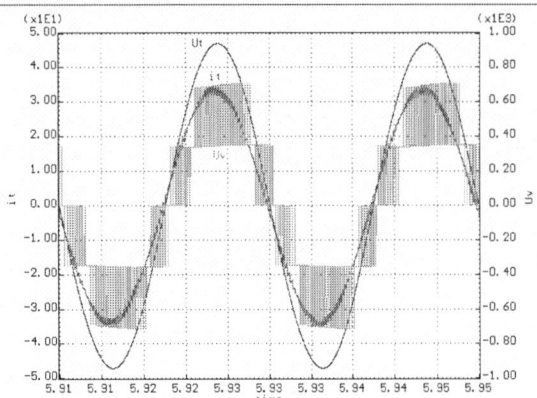
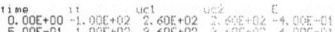
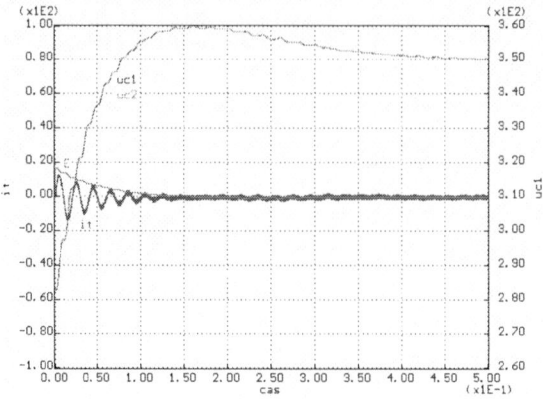

Fig. 6 Behaviour of primary voltage-source active rectifiers under steady-state conditions in rectifier mode
(load P = 9.3 kW, u_t = 400 V_{rms} / 50 Hz, U_{cw} = 700 V)

Fig. 7 Converter startup: charging of dc-link capacitors of primary active rectifiers (load P = 0 kW, u_t = 400 V_{rms} / 50 Hz, U_{cw} = 700 V)

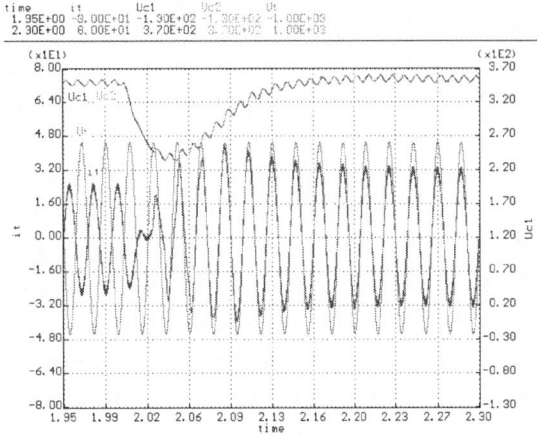
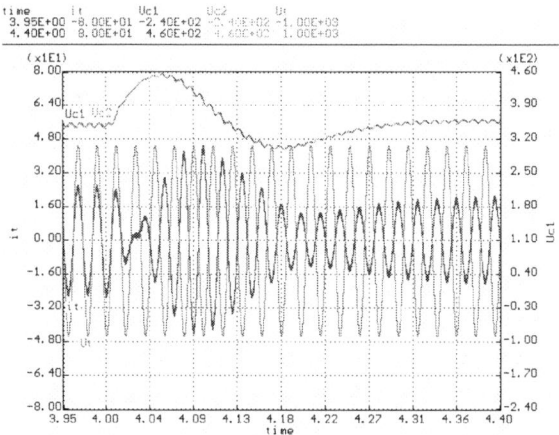

Fig. 8 Transition from inverter to rectifier mode:
Step change of the load ... P = - 7.2 kW → 8.8 kW,
u_t = 400 V_{rms} / 50 Hz, U_{cw} = 700 V

Fig. 9 Transition from rectifier to inverter mode:
Step change of the load ... P = 7 kW → -5.4 kW,
u_t = 400 V_{rms} / 50 Hz, U_{cw} = 700 V

IV. DESIGNED LABORATORY PROTOTYPE OF TRACTION CONVERTER

The proposed converter control has been implemented in the fixed-point digital signal processor Texas Instruments TMS320F2812. The control system and its parameters are the same as for the 100kW-prototype. The synchronization of the control system with the trolley wire voltage (it means evaluation of the magnitude and position of the trolley wire voltage in the stationary reference frame) is realized using DFT. The parameters of designed laboratory prototype of the traction converter are in detail listed in Appendix. The experimental tests were made under the converter rated supply voltage of 400V_{rms}/50Hz.

Fig. 10 – Fig. 12 present selected experimental results of the primary voltage-source active rectifiers operating in rectifier and inverter mode respectively under steady-state conditions. In Fig. 10 and Fig. 11, we can see that the trolley-wire current i_t is harmonic and is in phase

(respectively in anti-phase in inverter mode) with trolley-wire voltage u_t. Fig. 12 shows the multilevel nature of the voltage at the converter ac terminals, which is the result of employed shifted carriers in PWM modulator. In our particular case with two active rectifiers in series, the voltage at the converter ac side is five-level. Fig. 13 describes the startup sequence of primary voltage-source active rectifiers using ramp for required dc-link voltage U_{cw}. The Fig. 14 and Fig. 15 present response of the converter prototype to the step change of the load – these figures exactly document the transition from inverter to rectifier mode (Fig. 14) and the transition from rectifier to inverter mode (Fig. 15). The response of the converter to the demand for transition from rectifier mode to inverter mode and vice-versa are slower than in the simulations. This is caused by reduced controller gains in prototype in order to not overcome the protection limits of the laboratory installation under the fast transients in the whole power range of designed converter.

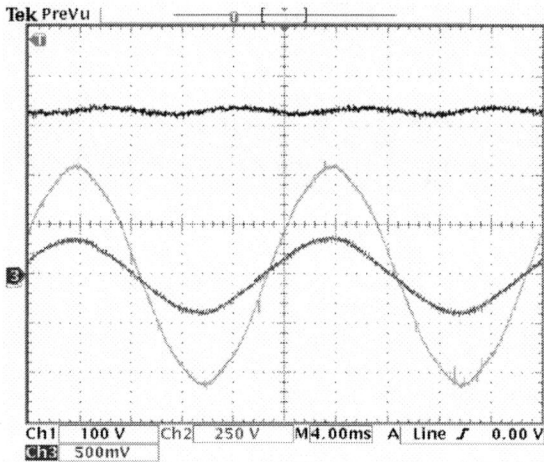

Fig. 10 Steady-state – rectifier mode (load ... P = 6.5kW):
ch1: Voltage U_{c1}, ch2: Trolley voltage U_t,
ch3: Trolley current I_t (10mV/A)

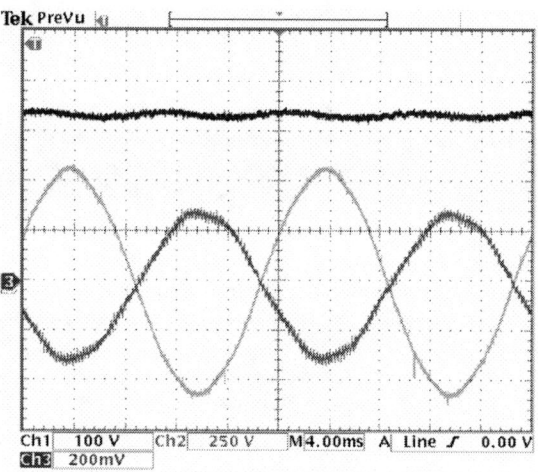

Fig. 11 Steady-state – inverter mode (load ... P = -5.6 kW):
ch1: DC-link voltage U_{c1}, ch2: Trolley voltage U_t,
ch3: Trolley current I_t (10mV/A)

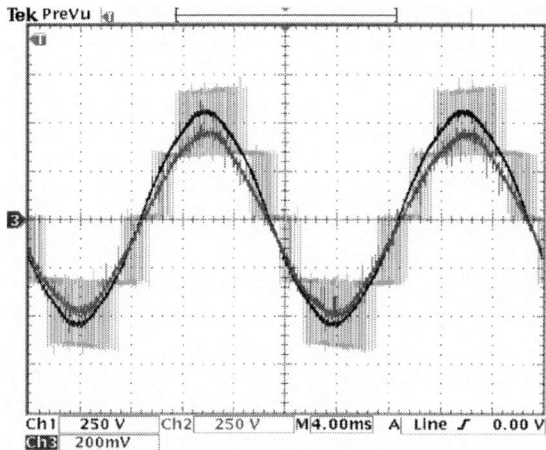

Fig. 12 Steady-state – rectifier mode (load P = 9.3kW):
Ch1: Trolley voltage U_t, Ch2: Sum of voltages at converters
ac terminals u_v, Ch3: Trolley current i_t (10mV/A)

Fig. 13 Charging of dc-link capacitor (load P = 0 kW)
Ch1: Trolley voltage Ut, Ch2: DC-link voltage U_{c1},
Ch3: Trolley current i_t (10mV/A)

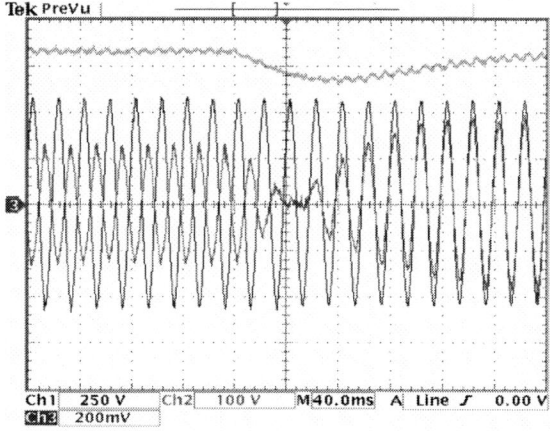

Fig. 14. Transition from inverter to rectifier mode:
Step change of the load ... P = - 7.2 kW □ 8.8 kW, Ch1: Trolley voltage
U_t, Ch2: DC-link voltage U_{c1}, Ch3: Trolley current i_t (10mV/A)

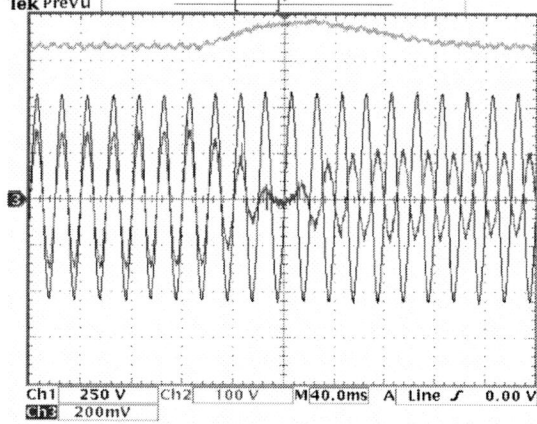

Fig. 15. Transition from rectifier to inverter mode:
Change of the load ... P = 7 kW □ -5.4 kW, Ch1: Trolley voltage U_t,
Ch2: DC-link voltage U_{c1}, Ch3: Trolley current i_t (10mV/A)

V. CONCLUSION

The presented configuration of the traction converter with medium-frequency transformer is one of the promising solutions for the new generation of the locomotives and especially suburban units. This contribution gives the main emphasis on the control of the primary voltage-source active rectifiers, which are directly connected to the ac trolley wire. This paper describes proposed model-based control of primary rectifiers, simulation results under both steady-state and transient conditions and experimental evidence of the designed low-voltage laboratory prototype of the locomotive converter of rated power of 12kW.

The proposed model-based control with PWM is simple, sufficiently robust and ensures very good converter behaviour under steady-state conditions as well as satisfactory dynamic properties. PWM is eligible from the viewpoint of low-frequency EMC disturbances – e.g. problems with railway signaling. The operated shifted carriers provide multi-level voltage on the converter ac terminals (in this case, the converter is five-level – see Fig. 6 and Fig. 12). The multilevel nature of the converter brings positive impact on design of converter input inductor. The disadvantage of presented control is the problem with efficient trolley current limitation, because trolley current is in this case controlled indirectly. The designed laboratory prototype of traction converter has been successfully tested in the whole power range. We have prepared in cooperation with our industrial partner Škoda Electric, a.s. a 100 kW prototype, which is nowadays under the tests.

ACKNOWLEDGMENT

This research has been supported by the Ministry of Industry and Trade of the Czech Republic under the project FT-TA2/035. This project is realized in cooperation with Škoda Electric, a.s., Plze□, Czech Republic (www.skoda.cz). We would like to kindly thanks Prof. Vondrášek, which participated in the converter design, and our colleagues Mr. Tomáš Komrska and Mr. Jan Žák for their assistance during the laboratory testing of designed converter prototype.

REFERENCES

[1] Peroutka, Z., Vondrášek, F., Molnár, J.: *Traction Converter with Medium-Frequency Transformer: Proposed Converter Control.* Research report no. 22160-22-07. University of West Bohemia in Pilsen, Plze□ 2007. (in Czech)

[2] Victor, M.: *Energieumwandlung auf AC – Triebfahrzeugen mit Mittelfrequenztrans-formator.* Fahrzeugtechnik eb 103 (2005) Heft 11, str. 505 - 510. (in German)

[3] Steiner, M., Reinold, H.: *Medium Frequency Topology in Railway Applications.* In: EPE 2007. Aalborg, Denmark.

[4] Glinka, M., Marquardt, R.: *A New AC/AC Multilevel Converter Family.* IEEE Transactions on Industrial Electronics, Vol. 52, No. 3, June 2005, pp. 662 – 669.

[5] Kjellqvist, T.; Norrga, S.; Östlund, S.: *Design considerations for a Medium Frequency Transformer in a Line Side Power Conversion System.* Annual IEEE Power Electronics Specialists Conference 2004. Aachen, pp. 704-710.

[6] Carpita, M.; Pellerin, M.; Herminjard, J.:*Medium Frequency Transformer for Traction Applications making use of Multilevel Converter: Small Scale Prototype Test Results.* SPEEDAM2006, Taormina, pp. S2-17 – S2-22.

[7] Rufer, A.; Schibli, N.; Chabert, C.; Zimmermann, C.: *Configurable front-end converters for multicurrent locomotives operated on 16 2/3 Hz AC and 3 kV DC systems.* In: Power Electronics, IEEE Transactionss on Volume 18, Issue 5, Sept. 2003 Pages: 1186 – 1193, 10.1109/TPEL.2003.816191.

APPENDIX:

PARAMETERS OF LABORATORY PROTOTYPE OF TRACTION CONVERTER

Converter rated power: $P_n = 12$ kW

Rated supply voltage: $U_t = 400$ V_{rms} / 50 Hz

Input inductor: $L_{input} = 8$ mH

$R_L = 0{,}2$ Ω

DC-link capacitors: $C_1 = C_2 = 4$ mF

Rated dc-link voltage of primary active rectifiers:

$$U_{cw1} = 350V$$

$$U_{cw2} = 350V$$

$$U_{cw} = U_{cw1} + U_{cw2} = 700V$$

Switching frequency of IGBTs: 1 kHz

Fig. 16. Illustrative photo of designed laboratory prototype of traction converter with medium-frequency transformer

Energy management strategy for Coupling Supercapacitors and Batteries with DC-DC converters for hybrid vehicle applications

M.B. Camara, F. Gustin, H. Gualous, **A. Berthon**
University of Franche-Comte, FEMTO-ST Enisys
Rue Thierry Mieg F90010 Belfort, FRANCE
alain.berthon@univ-fcomte.fr
Tel. +33 384 58 3604

Abstract—In this paper, the authors propose an approach to the problem of the energy management in ECCE laboratory hybrid vehicle project. ECCE is a series hybrid vehicle, which currently has three sources of energy: two diesel motors each coupled with an alternator and a battery module of rated voltage DC 540V. This contribution is focused on the energy coupling between batteries and two supercapacitors modules. The authors present a strategy for coupling these power sources with the batteries in order to find the best compromise between sizes of the on-board devices, dynamics of the supply and efficiency of the energy storage. The target is to provide 200kW power during 20s from the supercapacitors modules.

Keywords—Hybrid electric vehicle (HEV), Supercapacitors, Energy storage, Energy converters for HEV, Energy system management, Hybrid power integration, Power converters for HEV

I. INTRODUCTION

Since nineties, the cars manufacturers reacted to the public interest regarding the urban pollution by commercializing the electric vehicles. But the obstacle of heavy, expensive battery and of insufficient capacity did not solve the problem. Indeed, the battery must be dimensioned to provide a sufficient energy, to reach the performances of the vehicle and the peak power during transient states. However the possibility of providing at the same time energy and an important power is currently a severe constraint, thus penalizing the batteries. To solve this problem, it is interesting to find other solutions for the implementation of an auxiliary power source to assist batteries during transient states. Among the various solutions of possible hybridization of the sources (batteries, supercapacitors, fly-wheels, fuel cell), the hybrid vehicle equipped with batteries and supercapacitors seems to be the promising solution in the short term. ECCE is a special HEV truck developed to test some new energy solutions. In this paper a short description of this 4 wheel drives truck will be provide. The full power rated of the system is about 160 kW provided mainly by two diesel engines. A 2tons pack of batteries will be connected to the DC common bus. In this paper the purpose is to study the connection between the batteries and a pack of supercapacitors which target will be to provide 200 kW during 20s from a power

peak request. SABER software is used for simulation and a data acquisition system with a microcontroller (PIC18F4431) is setup. The general topology of the hybrid system is presented in Fig1.

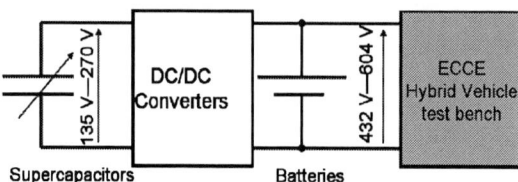

Fig1. General topology of hybrid system

II. BOOST CONVERTERS MODELING

To associate the supercapacitors and batteries in ECCE hybrid vehicle DC-bus the parallel topology of the buck-boost [1], [2], [3] has been proposed and this topology is presented in Fig.2.

This topology ensures the power management between the supercapacitors and battery [4], [5], [6], [7]. The supercapacitors modules are used during the hybrid vehicle transient states. The control strategy of these converters depends of the hybrid vehicle power request. To define the converters models the following assumptions are made:

• The semi conductors are ideal (no losses during switching and conduction states),

• No losses in the connecting devices

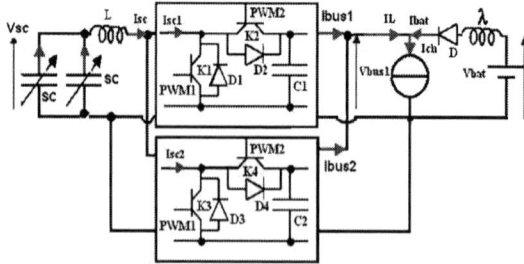

Fig.2: Parallel topology of the buck-boost converters

By convention, the flowing currents I_{sc1}, I_{sc2}, I_L, I_{bus1} and I_{bus2} are positive during the supercapacitors discharge (boost converter mode). In this mode the PWM1 signals are ON and PWM2 signals are OFF. The supercapacitors module provides energy to the DC-bus [8], [9], [10]. The analytical model of the boost converter is given by equation (1); where α_1 variable define the converter average duty cycle.

$$\begin{cases} V_L = L \cdot \dfrac{d}{dt}(i_{sc}) = V_{sc} - (1-\alpha_1) \cdot V_{bus1} \\ V_{sc} = V_{sc1} = V_{sc2} \\ V_\lambda = \lambda \cdot \dfrac{d}{dt}(i_{bat}) = V_{bat} - V_{bus1} \\ I_{ch} = I_L + I_{bat} \end{cases} \quad (1)$$

The voltages drops in the L and λ inductances are respectively given by V_L and V_λ. The converter average model has a nonlinear behavior because of crosses between the control variable (α_1) and the (I_{sc}, V_{bus1}) state variables. The V_{bus1}, V_{sc}, I_{ch} and V_{bat} variables are likely to disturb the control; they must be measured and used in the estimate of the control law to ensure a dynamics of control.

III. CONVERTERS CONTROL STRATEGY

The boost converters control strategy [11], [12] is presented in Fig.3 and the used control law is defined by the boost converters average duty cycle (2).

$$\alpha_1 = 1 - \frac{V_{sc} - V_L}{V_{bat} - V_\lambda} \approx 1 - \frac{I_L}{I_{sc}} \quad (2)$$

This control law is used to generate PWM1 signals. The supercapacitors reference current (3) is obtained from energy management between the supercapacitors modules and DC-bus, where η_{3th} and η_{4th} are the first and second boost converters theoretical efficiencies (95%).

$$I_{scref} = \frac{1}{2} \cdot \left(\frac{\eta_{3th} + \eta_{4th}}{\eta_{3th} \cdot \eta_{4th}} \right) \cdot \frac{V_{bus1}}{V_{sc}} \cdot \left(I_{ch} - I_{batref} \right) \quad (3)$$

The V_{bus1}, V_{sc}, I_{ch}, and V_{bat} variables are likely to disturb the control. They must be measured and used in the estimate of the control law to ensure a dynamics of control.

The control strategy used is to store all measurement errors samples of the supercapacitors and battery currents to estimate the voltage drops in L and λ inductors.

To make these estimations, it is necessary to provide to the system the desired battery current reference (I_{batref}), the PI corrector's coefficients [11], the initial conditions of the V_L and V_λ.

IV. DESIGN AND EXPERIMENTAL SETUP

For reasons of components cost and safety the experimental test bench was carried out at a reduced scale (1/10). This test bench is made of:

- a battery module of 4 cells in series,
- two supercapacitors modules of 10 cells (Maxwell -MC2600) in series for each one,
- an active load who is used to simulate hybrid vehicle power request,
- two buck-boost converters in parallel, which ensure power management between the batteries and supercapacitors.

The supercapacitors modules voltages must be between 27V and 13.5V. The batteries module, which imposes the DC-bus voltage presents, a rated voltage of 48V and the DC-link voltage level must be between 43V and 60V.

The control of the system is ensured by a Microchip's microcontroller (PIC18F4431) who presents 9 ADC input channels. The boost converters control frequency used is 10 kHz. The experimental parameters of the system and converter setup are respectively presented in TABLE I and Fig.4.

TABLE I
EXPERIMENTAL TEST BENCH PARAMETERS

Symbol	Value	Name
L	50μH	I_{sc} smoothing standard inductances
$C_1 = C_2$	1500μF	V_{bus1} smoothing standard capacitances
λ	25μH	I_{bat} smoothing inductance
V_{sc}	13.5 V-27 V	Supercapacitors modules voltages
V_{bus1}	43 V-60 V	DC-link voltage
T_e	100μs	Sampling period

Fig.4: Experimental setup

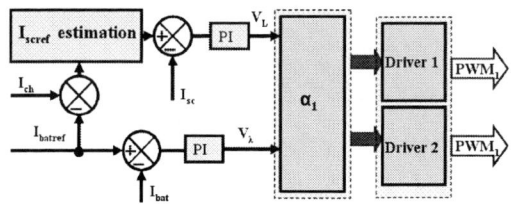

Fig.3 : Boost converters control strategy

V. SIMULATION AND EXPERIMENTAL RESULTS

The boost converters experimental test is carried out in the following conditions: during the supercapacitors discharge, the batteries current reference (I_{batref}) is fixed at 14A so that, the supercapacitors modules provide hybrid vehicle power request during the transient states. For these tests, the hybrid vehicle request (I_{ch}) was fixed at 54A. The experimental and simulations results of the batteries current are compared in Fig.5. This figure shows the batteries current control strategy is satisfactory during 40s after that it becomes insufficient because the supercapacitors discharge. In other words these curves show that the simulation model developed describes correctly the behavior of the experimental system during the first 40 seconds. After the first 40 seconds, the difference that appears between experimental and simulation results can be explained by the fact that one supercapacitors module between the two parallel sets has been characterized. This difference can be eliminated by adjusting the supercapacitors capacitance value or characterizing the two supercapacitors modules.

Fig.6 shows the supercapacitors experimental and simulations results during hybrid vehicle traction states. These results show that simulation model developed with SABER software is satisfactory for hybrid vehicle test bench behavior description.

The DC-bus current (I_L=40A) presented in Fig.7 is ensured by two supercapacitors modules. The first boost converter (K1-D2) ensures 50% and the second (K3-D4) ensures also 50% of the DC-bus current. In other words the two super capacitors modules ensure a (I_L) current of 40A to hybrid vehicle and 14A only is provided by the batteries.

Fig.8 and Fig.10 show I_{sc1} and I_{sc2} currents are not the same because of the supercapacitors modules dispersion. The simulation and experimentation results of the hybrid vehicle request (I_{ch}) current are well fitted and these results are presented in Fig.9.

Experimental and simulations results of the supercapacitors modules current and DC-bus voltage are respectively presented in Fig.11 and Fig.12. The first figure show the control strategy of the supercapacitors global current (I_{sc}=I_{sc1}+I_{sc2}) is satisfactory during hybrid vehicle traction state.

The experimental tests is carried out with a classical battery module not entirely charged this is why appears a difference of 5V approximately between the DC-bus voltage (V_{bus1}) experimental and simulations results (Fig.12). This difference is also caused by the simulations model which is ideal (no losses).

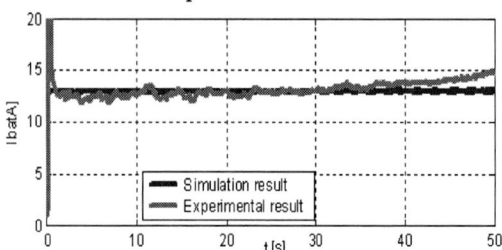

Fig.5: Battery current control results

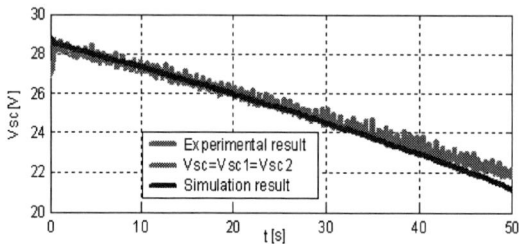

Fig.6: Two supercapacitors modules voltages

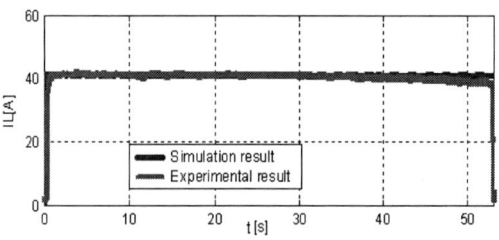

Fig.7 : Hybrid vehicle DC-bus full current

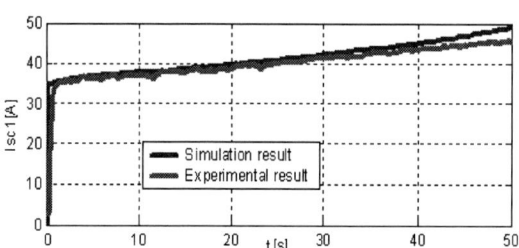

Fig.8: First supercapacitors module current

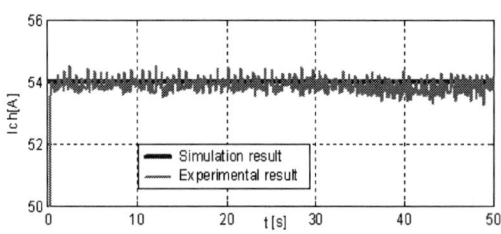

Fig.9: Hybrid vehicle request current

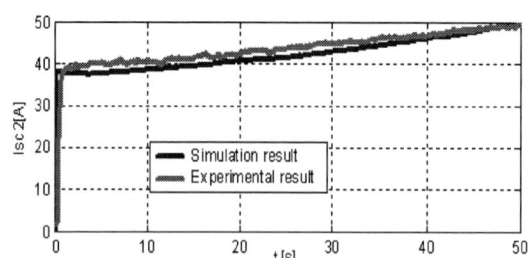

Fig.10: Second supercapacitors module current

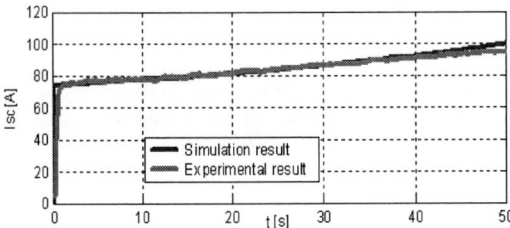

Fig.11: Supercapacitors modules full current

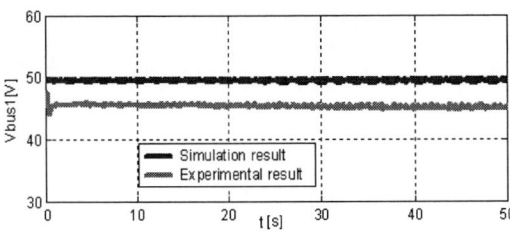

Fig.12: DC-bus voltage

VI. POWER SEMI-CONDUCTORS LOSSES ESTIMATION

The power semi-conductors (IGBT) present two main types of losses [13]. The first losses are due to semi-conductors dynamics resistances (internal) and second are due to IGBT frequency switching [14] which are not studied in this paper. To illustrate and simplify IGBT and diodes conduction losses estimation, the voltage drops in these components have been considered constant. The power semi-conductors (SKM400GB124D) parameters extracts in manufacturer datasheet are summarized in TABLE.II

To estimate the semiconductors losses, it is necessary to know the DC/DC converters (boost) duty cycle average value which is defined by (4).

$$d_1 = d_2 = \frac{V_{bus1} - V_{sc}}{V_{bus1}} \tag{4}$$

TABLE II
POWER SEMI-CONDUCTORS PARAMETERS

Symbol	Value	Name
$V_{CE(T0)}$	1.25 V	IGBT threshold voltage
R_{CE}	5.3mΩ	Dynamic resistance of IGBT
$V_{(T0)}$	1.2V	Diodes threshold voltage
R_T	3.5mΩ	Diodes dynamic resistance
ΔI_{SC1}	10% I_{SC1}	I_{SC1} variation current
ΔI_{SC2}	10% I_{SC2}	I_{SC2} variation current
$\langle I_{sc1} \rangle, \langle I_{sc2} \rangle$	I_{SC1}, I_{SC2}	I_{SC1} and I_{SC2} average values

The conduction losses of the D2 and D4 diodes can be estimated by using equation (5).

$$\begin{cases} P_{D2} = (1-d_1) \cdot \langle I_{sc1} \rangle \cdot V_{(T0)} + R_T \cdot I_{D2eff}^2 \\ P_{D4} = (1-d_2) \cdot \langle I_{sc2} \rangle \cdot V_{(T0)} + R_T \cdot I_{D4eff}^2 \end{cases} \tag{5}$$

$$\begin{cases} I_{D2eff}^2 = (1-d_1) \cdot \left(1 + \frac{1}{12} \cdot \left(\frac{\Delta I_{sc1}}{\langle I_{sc1} \rangle}\right)^2\right) \cdot \langle I_{sc1} \rangle^2 \\ I_{D4eff}^2 = (1-d_2) \cdot \left(1 + \frac{1}{12} \cdot \left(\frac{\Delta I_{sc2}}{\langle I_{sc2} \rangle}\right)^2\right) \cdot \langle I_{sc2} \rangle^2 \end{cases}$$

For K1 and K3 semi-conductors (IGBT) conduction losses, they can be estimated by using (6).

$$\begin{cases} P_{K1} = d_1 \cdot \langle I_{sc1} \rangle \cdot V_{CE(T0)} + R_{CE} \cdot I_{K1eff}^2 \\ P_{K3} = d_2 \cdot \langle I_{sc2} \rangle \cdot V_{CE(T0)} + R_{CE} \cdot I_{K3eff}^2 \end{cases} \tag{6}$$

$$\begin{cases} I_{K1eff}^2 = d_1 \cdot \left(1 + \frac{1}{12} \cdot \left(\frac{\Delta I_{sc1}}{\langle I_{sc1} \rangle}\right)^2\right) \cdot \langle I_{sc1} \rangle^2 \\ I_{K3eff}^2 = d_2 \cdot \left(1 + \frac{1}{12} \cdot \left(\frac{\Delta I_{sc2}}{\langle I_{sc2} \rangle}\right)^2\right) \cdot \langle I_{sc2} \rangle^2 \end{cases}$$

$V_{(T0)}$, $V_{CE(T0)}$, R_{CE} and R_T parameters are respectively the diode threshold voltage , IGBT threshold voltage, IGBT internal resistance and the diode resistance.

The experimental results of K1 and D2 semi-conductors conduction losses for first boost converter are shown in Fig.13 and for second boost converter (K3, D4) in Fig.14. The P_{K1} and P_{K3} values increase when the supercapacitors modules voltages decrease.

The boost converter efficiency [15], [16] is defined by the ratio between the DC-bus power and supercapacitors power (7).

This efficiency depends to supercapacitors current. The progressive discharge of the supercapacitors modules is compensated by supercapacitors current progressive increase; which ensures increase the semi-conductors losses.

$$n_{t\_PI} = \frac{(I_{bus1} + I_{bus1}) \cdot V_{bus1}}{(I_{sc1} + I_{sc2}) \cdot V_{sc}} \tag{7}$$

The boost converters experimental efficiency resulting from this equation is presented in Fig.15. This figure shows that for I_{sc}=76A the converters efficiency is maximum (87.5%). However, it is possible to improve this efficiency by optimising the wiring and using low losses power semi-conductors (IGBT, diode).

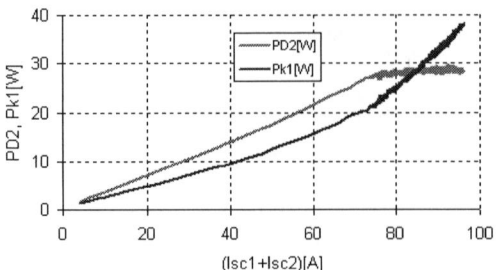

Fig.13: Losses of the first boost converter

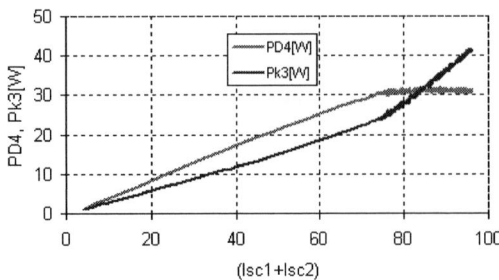

Fig.14: Losses of the second boost converter

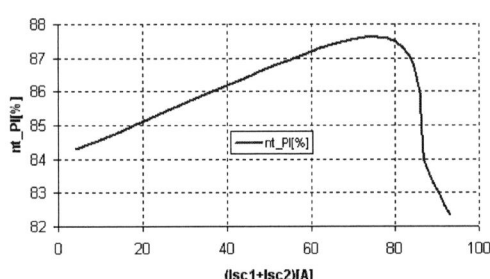

Fig.15. DC/DC converter efficiency

VII. CONCLUSION

This paper presents a possibility of coupling the battery module and the supercapacitors in the ECCE hybrid vehicle DC-bus by using the parallel topology of the buck-boost converters. This topology enables to decrease the supercapacitors current smoothing inductances number.

In this paper, the authors present the boost converter modeling and the original control strategy to share energy between supercapacitors and batteries. To control the whole system, a boost converter law control is established. The supercapacitors reference current is calculated by the strategy of energy management in the hybrid vehicle. The resulting algorithm is very robust and simple to realize with microcontroller or DSP devices.

For cost and safety reasons, the experimental test bench is carried out at reduced scale of (1/10). Simulations and experimental results have been compared. The battery current follows the reference value except when the supercapacitors modules voltages exceed the minimum voltage.

ACKNOWLEDGMENT

This work is under the continuity of the work started in L2ES Laboratory within the ECCE framework program in collaboration with Electronic and Electrical engineering Research Center of BELFORT (CREEBEL).

REFERENCES

[1] L.Solero, L. Lidozzi, J-A. Pomilio, Design of Multiple-Input Power converter for Hybrid Vehicles, IEEE Trans. On power Electronics, vol:20, No5, September 2005

[2] Camara M.B, Gustin F, Gualous H, Berthon A., "Studies and realization of the buck-boost and full bridge converters with multi sources system for the hybrid vehicle applications", Second European Symposium on Supercapacitors and Applications, ESSCAP2006, 2-3 November 2006, Lausanne, Switzerland, Proceedings CD

[3] Camara, M.B.; Gualous, H.; Gustin, F.; Berthon, A.; "Control strategy of Hybrid sources for Transport applications using Supercapacitors and batteries", Power Electronics and Motion Control Conference, 2006. IPEMC'06. CES/IEEE 5th International, Volume 1, Aug. 2006, Pages: 1– 5

[4] Camara M.B, Gustin F, Gualous H, Berthon A., "Supercapacitors and Batteries powers management for Hybrid Vehicles Applications/ Using multi boost and multi full bridge converters", 12th European Conference on Power Electronics and Applications - EPE 2007, ISBN: 9789075815108, 2-5 September 2007, Denmark, Proceedings CD

[5] Camara, M.B.; Gualous, H.; Gustin, F.; Berthon, A.; "Experimental study of Buck-Boost converters with polynomial Control strategy for Hybrid Vehicles Applications", International Review of Electrical Engineering (IREE), ISSN 1827-6660,Vol.2, No.4, Pages:601-611, July-August 2007

[6] B. Michael, B.Burnett, L.J. Borle, A power system combining batteries and supercapacitors in a solar/hydrogen hybrid electric vehicle, IEEE,vol:3, Sept. 2005

[7] M. Marchesoni, C. Vacca, A New DC-DC Converter Structure for Power Flow Management in Fuel-Cell Electric Vehicles with Energy Storage Systems, 2004 35thAnnual IEEE Conf. Power Electronics Specialists , Germany

[8] P.Thounthong,S.Raël,B.Davat, Control strategy of fuel cell/supercapacitors hybrid power sources for electric vehicle, Journal of Power Sources September 2005

[9] P. Mestre, S.Astier, Utilization of Ultracapacitors as auxiliary power source in Electric Vehicle, EPE ,vol:4, pp4670-4673, Sept.1997

[10] A. Di Napoli, F.Crescimbini, L. Solero, F. Caricchi, F.G. Capponi, Multiple-Input DC-DC power for power-flow management in hybrid vehicles, IEEE,vol:3,pp1578-1585,oct 2002

[11] Camara M.B., Gualous H., Gustin F. and Berthon A., "Design and New Control of DC/DC Converters to share energy between Supercapacitors and Batteries in Hybrid Vehicle", IEEE Transactions on Vehicular Technology Vol 57 sept. 2008

[12] Wenxun Xiao, Bo Zhang, Dongyuan Qiu, "Analysis and Design of an Automatic-Current Sharing Control Based on Average-Current Mode for Parallel Boost Converters", IPEMC2006, vol.2,pp1-5

[13] Carlos A. Canesin, and Ivo Barbi, "Comparaison of experimental losses among six different topologies for a 1.6kW Boost converter, Using IGBT'S",IEEE, PESC1995, vol.2,pp1265-1271

[14] Scott Castagno, Randy D. Curry, and Ellis Loree "Analysis and Comparaison of a Fast Turn-On Series IGBT Stack and High-Voltage-Rated Commercial IGBTS", IEEE Trans. On plasma science, vol.34,No.5, pp1692-1696

[15] Hein van der Broeck, Ibrahim Tezcan, "1KW Dual Interleaved Boost Converter for Low Voltage Applications", IPEMC2006, vol.3,pp1-5

[16] Dong Li, and Xinbo Ruan "A High Efficiency Boost Converter With Power Fractor Correction",IEEE Power Electronics Specialists Conference, 2004, vol.2, pp1653-1657

Dr. Mamadou Baïlo Camara was born in Mamou, Guinea. He got his engineer degree from Polytechnic Institute of Conakry (IPC), Guinea in 2003, and the Master degree in 2004 from the University of Franche-Comté, France, where he received his Ph.D degree in 2007. Since 2004, he has been working in power electronics and electric vehicle research projects, involving static converter topologies, supercapacitors, batteries, and electrical power management for Hybrid Vehicle Applications.

Dr. Frederic Gustin received his Ph.D. degree in 2000 from the University of Franche-Comté, Belfort France. His research activities involve power electronics, converters and simulation methods at the Femto-ST laboratory. He is presently associate Professor at the I.U.T (Institute of technology) of Belfort, France.

Dr. Hamid Gualous was born in Morocco on January 1967,. He received his Ph.D in electronic from the University Paris XI Orsay, France, in 1994. Since 1996, he is Associate Professor at the University of Franche-Comté France, L2ES laboratory. His main research activities are concerning supercapacitor and fuel cell dedicated to transport applications, power electronics and energy management.

Pr. Alain Berthon, received his Dipl. engineer in electrical engineering from ENSEM in Nancy in 1972 and his Dr. Ing. degree from the University of Franche-Comté. He is currently Professor at IUT and ENISYS Department of FEMTO-ST laboratory University of Franche-Comte. He is working in the field of power electronics converters, drives and electrical power management for electrical vehicle.

Dual-Source Fed Multiphase Traction System with Standard and Non-Standard Control Regimes Based on Synchronized PWM

Valentin Oleschuk[†] and Marian P. Kazmierkowski[*]

[*] Warsaw University of Technology/Inst. of Control and Ind. Electr., Warsaw, Poland, e-mail: *mpk@isep.pw.edu.pl*
[†] Academy of Sc. of Moldova, Kishinau, Moldova, and Politecnico di Torino, Italy, e-mail: *oleschukv@hotmail.com*

Abstract — This paper presents results of application of algorithms of synchronized pulsewidth modulation (PWM) for control of hybrid electric vehicle drive supplied by dual thee-phase inverter system with two DC sources. Both standard linear *V/F=const* and non-standard non-linear control regimes of hybrid vehicle drive with synchronized PWM have been analyzed and compared. Simulation results are given for the system with basic continuous and discontinuous schemes of synchronized PWM. The spectra of the phase voltages of hybrid electric vehicle with synchronized PWM do not contain even harmonics and sub-harmonics in both standard and non-standard modes, which is especially important for the high power/current applications.

Keywords: Voltage source inverters, converter control, pulse width modulation, power converters for EV, multiphase drive.

I. INTRODUCTION

Dual three-phase induction motor drives are a subject of an increasing interest in the last years due to some advantages compared with conventional three-phase adjustable speed drives [1]-[5]. Dual three-phase induction motor can be supp-lied by from two voltage source inverters fed by different DC sources. This solution can be used for traction drive systems for electrical vehicles powered by fuel cells [5].

Electrical vehicles are ones of the most perspective applications of six-phase drives [5]-[7]. Fig. 1 presents a topology of the electrical vehicle system on the base of the asymmetrical six-phase induction motor supplied by two inverters with two different DC links: 1) Battery DC link with the V_{dc1} voltage, and 2) Fuel Cell DC link with the V_{dc2} voltage [5]. The induction motor has in this case two sets of winding spatially shifted by 30 electrical degrees with isolated neutral points [4],[5].

Dual three-phase induction motor drives have several advantages over their three-phase counterparts, such as: 1) reduction of torque pulsations; 2) reduction of the rotor harmonic losses; 3) the rated current of power switches is halved; 4) improved reliability at system level, and 5) the possibility to supply more than one machine from a single inverter to get a multi-motor, multi-phase drive [1]-[5].

Operation of power electronic converters is based on switchings of semiconductor devices (power transistors and thyristors), and principle of modulation of pulse signals is the basic for their control. Characteristics of power conversion systems are in dependence of the used methods of modulation. Two last decades have been marked by fast application of methods of space-vector modulation, based on space vector representation of the set of output signals of converters, which are ones of the most suitable for adjustable speed induction motor drives [8]-[16].

Almost all versions of classical space-vector modulation are based on the asynchronous principle, which results in sub-harmonics (of the fundamental frequency) in the spectrum of the output voltage of converters, that are very undesirable in most applications [8],[17]. It is known that for high power drive systems it is necessary to synchronize the output voltage waveforms of power converters to eliminate undesirable sub-harmonics of voltage and current [8],[17]. In order to avoid asynchronism of standard space-vector PWM, novel methods of continuous synchronized pulsewidth modulation have been recently proposed for drive converters [18]-[20].

Some results of application of novel methods of synchronized PWM for regulation of vehicle drives with standard control modes are in [6],[7]. So, this paper presents results of investigation of vehicle drive system with synchronized PWM in both standard and some non-standard control regimes.

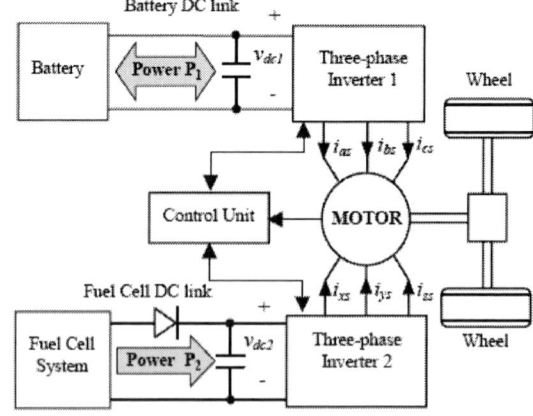

Fig. 1. Dual-source fed six-phase motor drive for electrical vehicles [5].

II. Synchronized Schemes of Space-Vector Modulation

In order to avoid asynchronism of conventional versions of voltage space-vector modulation, a novel method (methodology) of synchronized PWM can be used for control of each inverter in dual three-phase drive system [18]-[20].

Fig. 2 presents typical switching state sequences of standard three-phase voltage source inverter inside the interval 0^0-90^0 for continuous version of space-vector PWM, which are used typically in adjustable speed drive systems. The upper trace in Fig. 2 is switching state sequence (in accordance with conventional designation [18]), then - control signals for the cathode switches of the phases a, b, c and x, y, z of each inverter. The lower trace in Fig. 2 shows the corresponding quarter-wave of the line output voltage of inverters. Signals β_j represent total switch-on durations during switching sub-interval τ, signals γ_k are generated on the borders of the corresponding β. Widths of notches λ_k represent duration of zero sequences.

Special signals λ' (λ_5) (with the neighbouring β'' (β_5)) are formed in the clock-points (0^0,60^0,120^0..) of the output curve. They are reduced simultaneously until zero at the boundary frequencies F_i, providing a continuous adjustment of the output voltage with smooth pulses-ratio changing [18]. Fig. 3 illustrates this process more in details, showing switching state sequences and line-to-line output voltage of the inverter with continuous synchronized PWM in the border part of the voltage half-wave. It shows a step-by-step variation of the β'', λ', γ_1 signals, accompanied by smooth variation of the waveform of the line-to-line output voltage.

Table I shows generalised properties and basic control functions of new method of synchronized PWM for standard three-phase inverter, which are compared here with conventional space-vector PWM [18]. The proposed method can be based or on approximate algebraic or on precise trigonometric control laws for determination of the voltage pulse patterns in the function of the fundamental frequency of drive system.

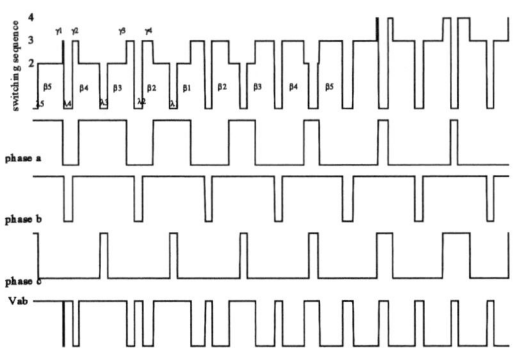

Fig. 2. Control and output signals for quarter-period of three-phase inverter with continuous synchronized PWM.

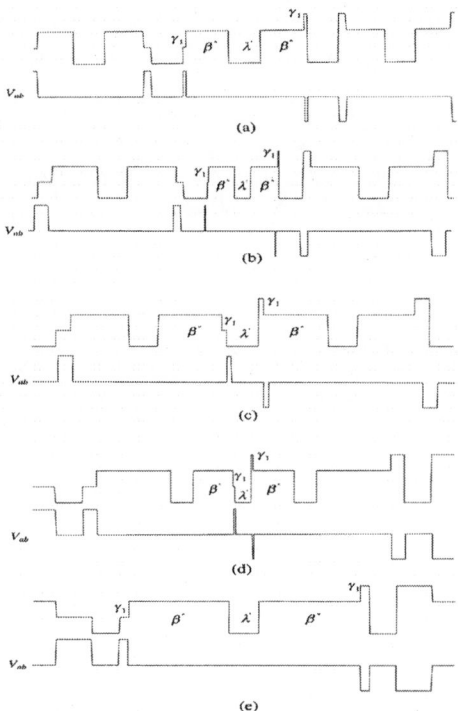

Fig. 3. Switching state sequences and line voltage on the border of the voltage half-wave of the inverter with continuous synchronized PWM.

TABLE I.
BASIC PARAMETERS OF PWM METHODS

Control (modulation) parameter	Conventional schemes of vector PWM	Proposed method of modulation	
Operating and max parameter	Operating & max voltage V and V_m	Operating & maximum fundamental frequency F and F_m	
Modulation index m	V/V_m	F/F_m	
Duration of sub-cycles	T	τ	
Center of the k-signal	α_k (angles/degr.)	$\tau(k-1)$ (sec)	
		Algebraic PWM	Trigonometric PWM
Switch-on durations	$T_{ak} = 1.1mT[\sin(60^0 - \alpha_k) + \sin\alpha_k]$ $t_{ak} = 1.1mT\sin\alpha_k$ $t_{bk} = 1.1mT \times \sin(60^0 - \alpha_k)$	$\beta_k = \beta_1[1 - A \times (k-1)\tau F K_{ov1}]$ $\gamma_k = \beta_{i-k+1}[0.5 - 6(i-k)\tau F]K_{ov2}$ $\beta_k - \gamma_k$	$\beta_k = \beta_1 \times \cos[(k-1)\tau K_{ov1}]$ $\gamma_k = \beta_{i-k+1}[0.5 - 0.9\tau n(i-k)\tau]K_{ov2}$ $\beta_k - \gamma_k$
Switch-off states (zero voltage)	$t_{0k} = T - t_{ak} - t_{bk}$	$\lambda_k = \tau - \beta_k$	
Special parameters providing synchronization of the process of PWM		$\beta'' = \beta_1[1 - A \times (k-1)\tau F K_{ov1}]K_s$ $\lambda' = (\tau - \beta'') \times K_{ov1}K_s$	$\beta'' = \beta_1 \times \cos[(k-1)\tau K_{ov1}]K_s$ $\lambda' = (\tau - \beta'') \times K_{ov1}K_s$

An important parameter of the proposed PWM scheme is the width of the β_1 signal (the central active switching state, see Fig. 2). In particular, for standard scalar $V/F=const$ control mode of the drive system it is reasonable to determine its width as $\beta_1 = 1.1m\tau$, where m – modulation index [18],[19]. In this case it provides simple linear control of the fundamental voltage of adjustable speed drive until the zone of overmodulation.

For control modes, different from the standard ones, this dependence should be correspondingly modified. Table II presents the dependencies for three basic control regimes of the drive system, including standard scalar $V/F=const$ control, and also $V^2/F = constant$ and $V^{3/2}/F = constant$ control modes. The last two functions are typical for control of adjustable speed ac drives with non-linear dependencies of the fundamental voltage from the fundamental frequency, and these control modes can be applied for control of the drive systems with some specific loads [9],[10]. Table II includes also the corresponding values of two threshold frequencies F_{ov1} and F_{ov2} [8],[18], which are the basic for control in the zone of overmodulation for the systems with equal DC voltages $V_{dc1} = V_{dc2}$ (it is represented here as a relative part of the maximum fundamental frequency F_m, at which modulation index $m=1$).

TABLE II.
CONTROL PARAMETERS FOR DIFFERENT CONTROL MODES

Control mode	β_1	F_{ov1}	F_{ov2}
$V/F=const$	$1.1m\tau$	$0.907F_m$	$0.952F_m$
$V^2/F=const$	$1.1\sqrt{m}\,\tau$	$0.823F_m$	$0.907F_m$
$V^{3/2}/F=const$	$1.1\sqrt[3]{m^2}\,\tau$.	$0.866F_m$	$0.931F_m$

Fig. 4 presents the variation of the fundamental voltage versus fundamental frequency (with the maximum fundamental frequency $F_m = 50$ Hz) for three-phase inverter with three control regimes presented in Table II. Different intermediate control regimes can also be provided in this case, with other special β_1 functions. As an example, the dotted line in Fig. 4 shows voltage variation under other non-standard control mode of the system, when $V^{4/3}/F=constant$ (in this case $\beta_1 = 1.1\sqrt[4]{m^3}\,\tau$).

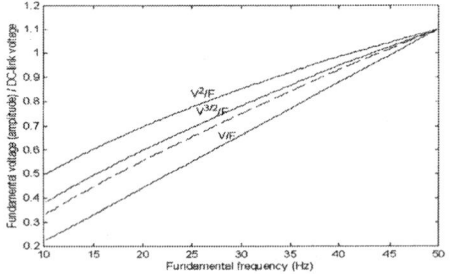

Fig. 4. Fundamental voltage versus frequency for four control modes.

III. OPERATION OF DUAL THREE-PHASE VEHICLE DRIVE WITH SYNCHRONIZED PWM

Control of dual three-phase induction machine drives is based on the 30^0-phase-shifting of control and output signals of two inverters [1],[2],[4]. In accordance with the theory of vector space decomposition, the basic six-dimensional space (*as, bs, cs, xs, ys, zs*) of a dual-three phase induction machine with isolated neutral points can be transformed into two orthogonal two-dimensional subspaces (*sa, sb*) and (*m1, m2*) [1]. Voltage components V_{sa}, V_{sb}, V_{m1} and V_{m2} in these subspaces, and also the phase voltages V_{as} and V_{xs}, are calculated as [4]:

$$V_{sa}=0.333(V_a-0.5V_b-0.5V_c+0.866V_x -0.866V_y) \quad (1)$$

$$V_{sb}=0.333(0.866V_b -0.866V_c +0.5V_x +0.5V_y -V_z) \quad (2)$$

$$V_{m1}=0.333(V_a -0.5V_b -0.5V_c -0.866V_x +0.866V_y) \quad (3)$$

$$V_{m2}=0.333(-0.866V_b+0.866V_c +0.5V_x + 0.5V_y -V_z) \quad (4)$$

$$V_{as}=V_{sa} +V_{m1}=V_a -0.333(V_a + V_b + V_c) \quad (5)$$

$$V_{xs}=V_{sb}+V_{m2}=V_x - 0.333(V_x + V_y + V_z), \quad (6)$$

where V_a, V_b, V_c, V_x, V_y, V_z are the corresponding pole voltages of each inverter.

In this case, the V_{sa} and V_{sb} components, which produce useful rotating MMF k-th order voltage harmonics ($k = 12m\pm1$, $m=1,2,3,..$), are the useful components. But the V_{m1} and V_{m2} components, which generate loss-producing harmonics ($k = 6m\pm1$, $m=1,3,5,..$), are the undesirable voltage components [1].

Fig. 5 and Fig. 6 present basic voltage waveforms of dual three-phase vehicle drive with two DC sources with different voltages ($V_{dc1} = 0.5V_{dc2}$) under non-standard $V^2/F=const$ control mode during a period of the fundamental frequency. In order to provide power balancing between DC sources, the corresponding dependence between modulation indices of two inverters ($m_1V^2_{dc1}=m_2V^2_{dc2}$) should be performed in this case. Fig. 5 shows basic voltages of the drive system with continuous synchronized PWM (CPWM), and Fig. 6 presents voltage waveforms for the system with discontinuous synchronized PWM with the 30^0-non-switching intervals (DPWM). The switching and fundamental frequencies of each inverter are equal to *1 kHz* and *38 Hz*.

Figs. 7 - 12 show spectra of the V_{as}, V_{xs} and V_{sa} voltages of split-phase drive system with both continuous (Figs. 7, 9, 11) and discontinuous (Figs. 8, 10, 12) synchronized pulsewidth modulation, operating under standard $V/F=const$ control mode (Figs. 7 – 8), and under the non-linear $V^{3/2}/F=constant$ control mode (Figs. 9 – 10) and $V^2/F=constant$ control regime (Figs. 11 – 12). Regarding the system with non-standard $V^{3/2}/F=constant$ control mode, special relationship between modulation indices of two inverters ($m_1V^{3/2}_{dc1}=m_2V^{3/2}_{dc2}$) should be provided in order to achieve power balancing capability between DC sources.

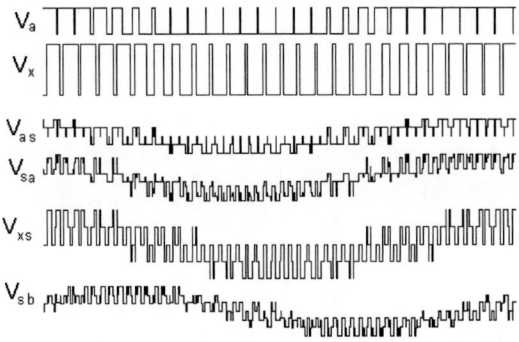

Fig. 5. Pole voltages V_a and V_x, phase voltages V_{as} and V_{xs}, and the V_{sa} and V_{sb} voltages, in the system with continuous synchronized PWM (CPWM, $V_{dc1}=0.5V_{dc2}$).

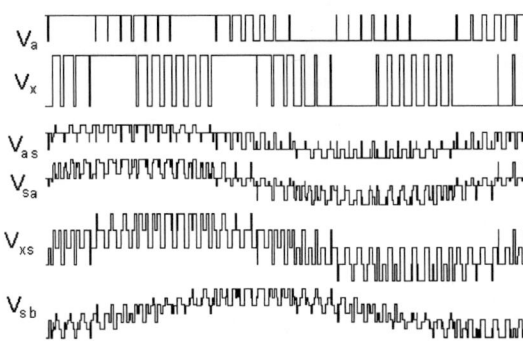

Fig. 6. Pole voltages V_a and V_x, phase voltages V_{as} and V_{xs}, and the V_{sa} and V_{sb} voltages, in the system with discontinuous synchronized PWM (DPWM, $V_{dc1}=0.5V_{dc2}$).

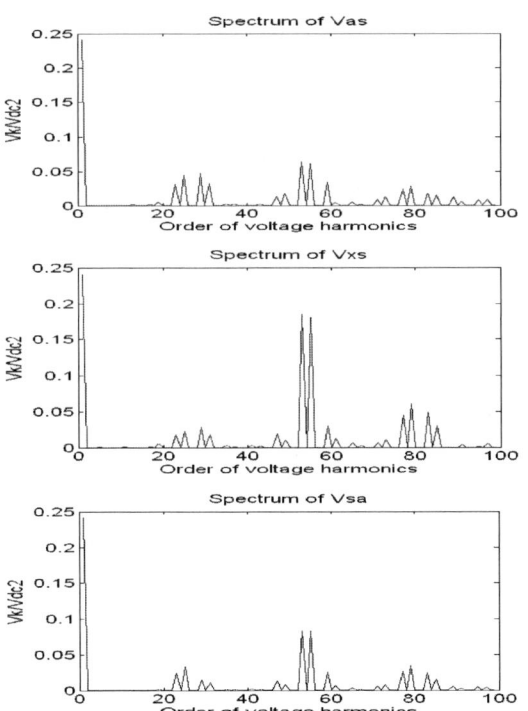

Fig. 7. Voltage spectra in the system with synchronized CPWM under standard $V/F=const$ control.

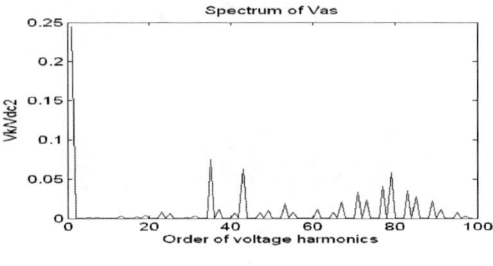

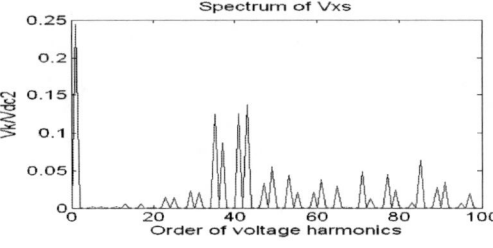

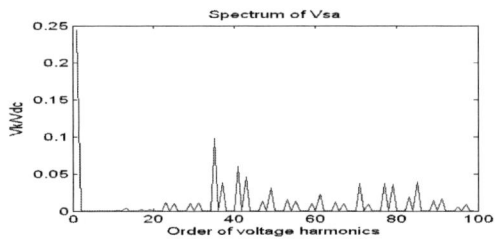

Fig. 8. Voltage spectra in the system with synchronized DPWM under standard $V/F=const$ control.

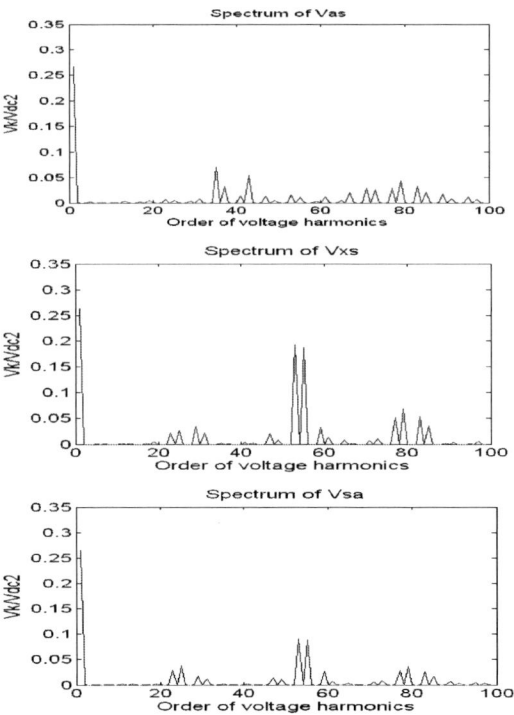

Fig. 9. Voltage spectra in the system with synchronized CPWM under non-linear $V^{3/2}/F=const$ control mode.

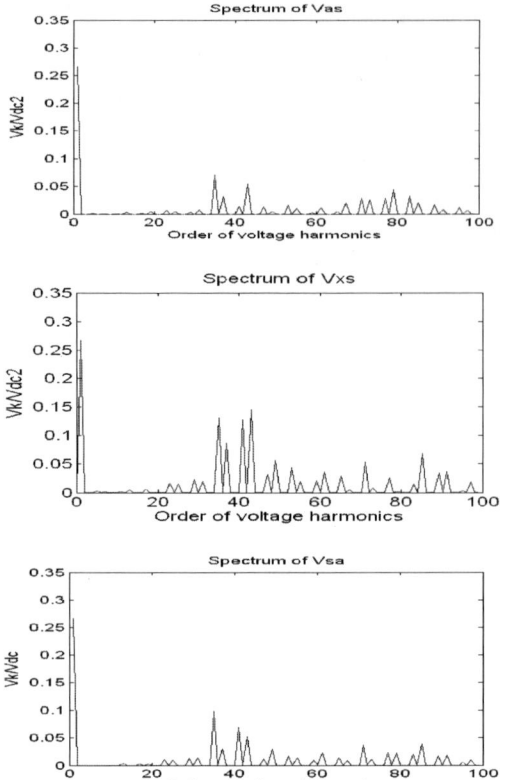

Fig. 10. Voltage spectra in the system with synchronized DPWM under non-linear $V^{3/2}/F=const$ control mode.

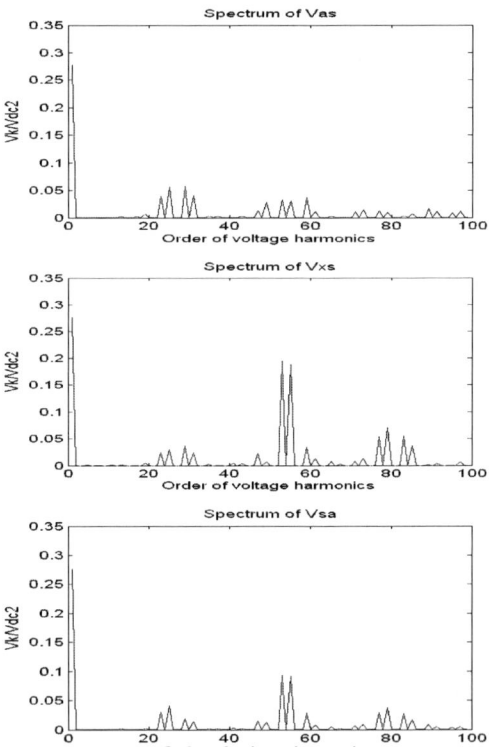

Fig. 11. Voltage spectra in the system with synchronized CPWM under non-linear $V^2/F=const$ control mode.

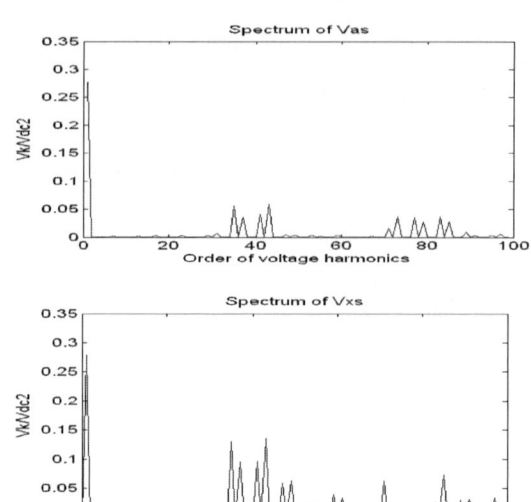

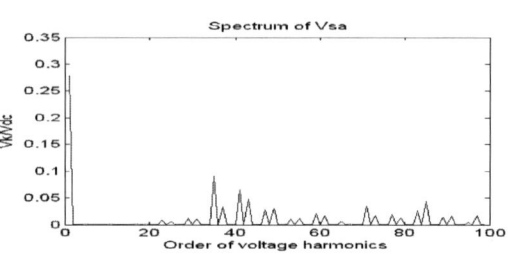

Fig. 12. Voltage spectra in the system with synchronized DPWM under non-linear $V^2/F=const$ control mode.

The motor phase voltages V_{as} and V_{xs}, and also its useful components V_{sa} and V_{sb}, of dual three-phase vehicle drive with both continuous and discontinuous synchronized PWM have symmetry during the whole control range. The corresponding voltage spectra (Figs. 7--12) do not include even harmonics and sub-harmonics for any linear or non-linear control mode and for any (integral or fractional) ratios between the switching and fundamental frequencies. In particular, for all mentioned above PWM variants of dual three-phase drive system this frequency ratio is equal to 26.3 (*1kHz/38Hz*).

The proposed method of continuous synchronisation of voltage waveforms of inverters can also be applied to dual three-phase systems with unbalanced power distribution between two DC sources. In particular, Figs. 13-14 show basic voltage waveforms of dual three-phase drive with two DC sources with different voltages ($V_{dc1}=0.5V_{dc2}$), controlled without power balancing between two sources. Fig. 13 shows basic voltages of the drive system with continuous synchronized PWM, and Fig. 14 - for the system with discontinuous synchronized PWM. The switching and fundamental frequencies of each inverter are equal to *1kHz* and *38Hz* (modulation indices of two inverters are equal to each other ($m_1=m_2=0.76$) in this case). Figs. 15-16 present the spectra of basic voltage waveforms of the system, corresponding to this control regime, which do not contain even harmonics and subharmonics of the fundamental frequency.

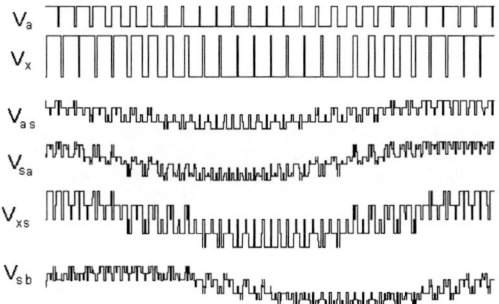

Fig. 13. Pole voltages V_a and V_x, phase voltages V_{as} and V_{xs}, and the V_{sa} and V_{sb} voltages, in the system with continuous synchronized PWM with unbalanced power distribution between DC sources ($m_1=m_2=0.76$).

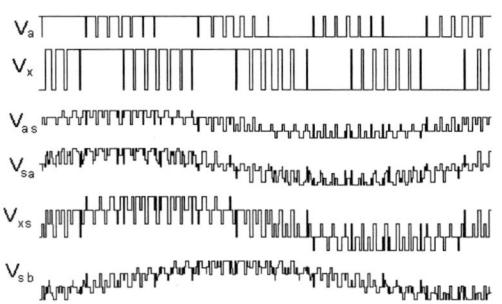

Fig. 14. Pole voltages V_a and V_x, phase voltages V_{as} and V_{xs}, and the V_{sa} and V_{sb} voltages, in the system with discontinuous synchronized PWM with unbalanced power distribution between DC sources ($m_1=m_2=0.76$).

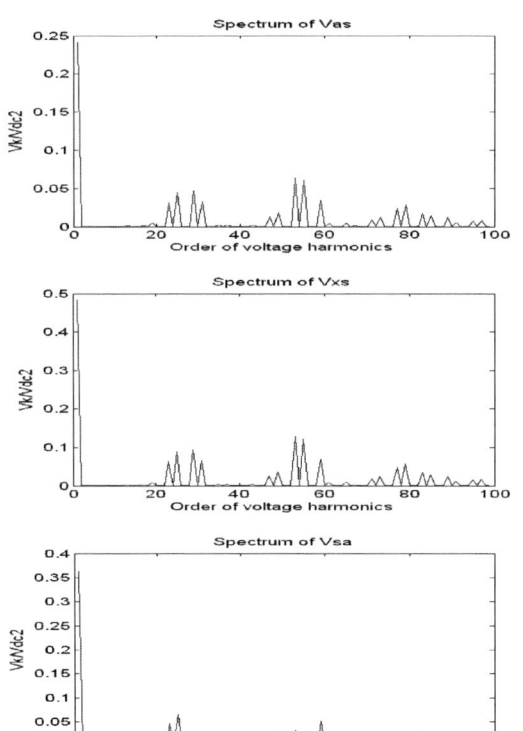

Fig. 15. Voltage spectra in the system with synchronized CPWM with unbalanced power distribution between DC sources ($m_1=m_2=0.76$).

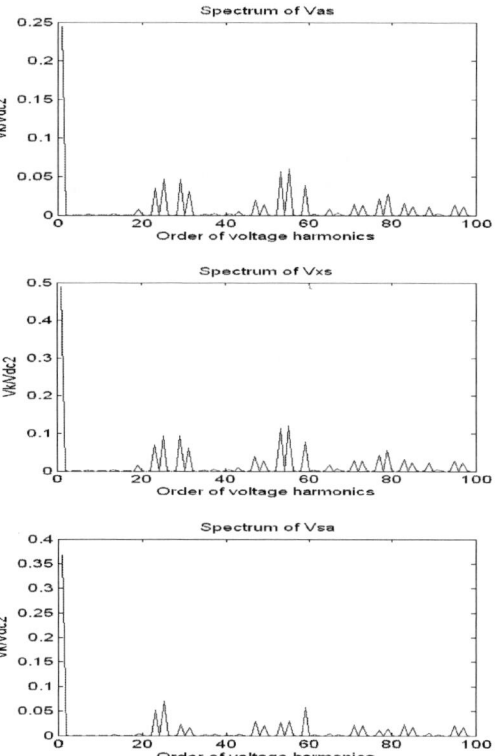

Fig. 16. Voltage spectra in the system with synchronized DPWM with unbalanced power distribution between DC sources ($m_1=m_2=0.76$).

Fig. 17 and Fig. 18 present calculation results of the Weighted Total Harmonic Distortion factor (*WTHD*) of the phase voltage V_{as} and its basic useful component V_{sa} (averaged values of $WTHD = (1/V_1)\sqrt{\sum\limits_{k=2}^{1000} (V_k/k)^2}$) for dual three-phase drive with two separate DC voltage sources ($V_{dc1}=0.5V_{dc2}$), with power balancing capability between DC sources. Fig. 17 corresponds to the system operating in the zone of undermodulation under standard *V/F=const* control mode. Modulation indices for the first and the second inverters are in linear dependence $m_2=0.5m_1$ in this case.

Results presented in Fig. 18 correspond to dual three-phase system operating under non-standard $V^2/F=const$ control regime with strong non-linearity between the fundamental voltage and fundamental frequency. The average switching frequency of each inverter is *1 kHz*. Modulation indices for the first and the second inverters $m_2=0.25m_1$ in this case.

Results of analysis of the *WTHD* factor of the basic voltage waveforms of dual three-phase drive system emphasize the fact, that in the zone of low fundamental frequencies algorithms of continuous PWM provide better voltage spectral composition for the systems with both standard and non-standard control modes. And at medium and higher fundamental frequencies algorithms of discontinuous PWM are more preferable. The phase voltages of dual three-phase systems with synchronized PWM have symmetry during the whole control range.

1553

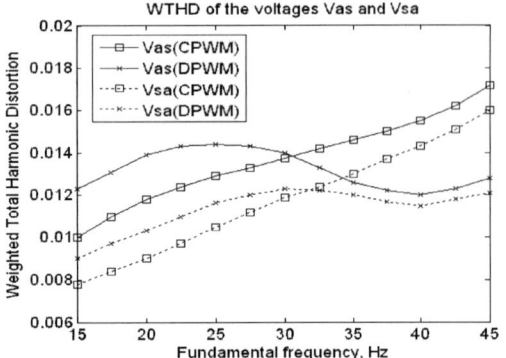

Fig. 17. Averaged *WTHD* of the basic voltages versus fundamental frequency for standard *V/F=const* control of dual three-phase drive system ($V_{dc1}=0.5V_{dc2}$).

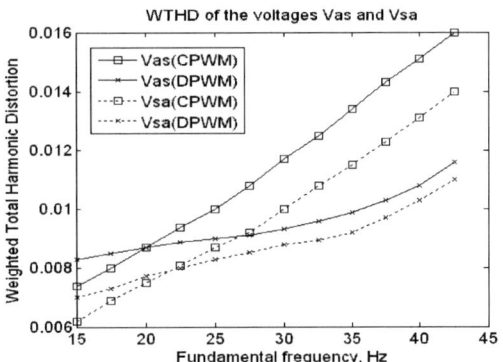

Fig. 18. Averaged *WTHD* of the basic voltages versus fundamental frequency for non-linear $V^2/F=const$ control mode of dual three-phase system ($V_{dc1}=0.5V_{dc2}$).

IV. CONCLUSION

A novel method of synchronized PWM has been applied to an asymmetrical dual three-phase vehicle drive with different control regimes. It has been shown possibility of dissemination of the proposed principles of continuous voltage synchronization to six-phase drive systems with two DC sources with control modes based on any linear or non-linear dependences between the fundamental voltage and fundamental frequency. Also possibility of application of the proposed method of synchronized PWM to the systems with unbalanced power distribution between DC sources has been illustrated. Phase voltages of dual three-phase vehicle drive with synchronized PWM have symmetry during the whole control range, and its spectra do not contain even harmonics, and also sub-harmonics (of the fundamental frequency), which is especially important for the medium-power and high-power applications.

REFERENCES

[1]   Y. Zhao and T.A. Lipo, "Space vector PWM control of dual three-phase induction machine using vector decomposition", *IEEE Tr. on Industry Appl.*, vol.31, no.5, 1995, pp.1100-1109.

[2]   R. Bojoi, A. Tenconi, F. Profumo, G. Griva and D. Martinello, "Complete analysis and comparative study of digital modulation techniques for dual three-phase AC motor drives", *Proc. of the 2002 IEEE Power Electronics Specialists Conf.*, pp.851-857.

[3]   D. Hadiouche, L. Baghli and A. Rezzoug, "Space vector PWM techniques for dual three-phase AC machine: analysis, performance evaluation and DSP implementation", *Proc. of the 2003 IEEE Industrial Application Soc. Conf.*, pp.648-655.

[4]   R. Bojoi, F. Farina, F. Profumo and A. Tenconi, "Dual-three phase induction machine drives control – a survey", *CD-ROM Proc. of the 2005 IEEE Int'l Power Electronics Conf.*, 10 p.

[5]   R. Bojoi, A. Tenconi, F. Farina and F. Profumo, "Dual-source fed multiphase induction motor drive for fuel cell vehicles: topology and control", *Proc. of the 2005 IEEE Power Electronics Specialists Conf.*, pp.2676-2683.

[6]   V. Oleschuk, R. Bojoi, F. Profumo, A. Tenconi and A.M. Stankovic, "Multifunctional six-phase motor drives with algorithms of synchronized PWM", *Proc. of the 2006 IEEE Ind. Electr. Soc. Conf.*, pp. 1852-1859.

[7]   V. Oleschuk, R. Bojoi, V. Ermuratski and F. Profumo, "Synchronous control of six-phase traction drive with two DC links", *Proc. of the 2007 IEEE-IEEJapan Power Conversion Conf.*, pp. 942-948.

[8]   J. Holtz, "Pulsewidth modulation for electronic power conversion", *Proc. of IEEE*, vol. 82, no. 8, pp. 1194-1213, 1994.

[9]   B.K. Bose, *Modern Power Electronics and AC Drives*. Prentice Hall, 2002.

[10] M. P. Kazmierkowski, R. Krishnan and F. Blaabjerg. *Control in Power Electronics*, Academic Press, 2002.

[11] M. P. Kazmierkowski, "Control strategies for PWM rectifier/inverter-fed induction motors", *Proc. of the 2000 IEEE Int'l Symp. Ind. Electr.*, pp. TU15-TU23.

[12] M. P. Kazmierkowski, R. Bracha and M. Malinowski, "Web based teaching of pulse width modulation methods for three-phase two-level converters", *Proc. of the 2006 EPE Power Electr. and Motion Contr. Conf.*, pp. 2134-2139.

[13] I. Nagy, "Control of bidirectional power conversion", *Proc. of the 2003 IEEE Int'l Symp. Ind. Electr.*, pp. 176-182.

[14] Z. Suto and I. Nagy, "Analysis of nonlinear phenomena and design aspect of three-phase space vector modulated converters", *IEEE Trans. on Circuits and System, I. Fundamental Theory and Applications*, vol. 50, no. 8. pp. 1064-1071, 2003.

[15] S. Halasz, A.A.M. Hasan and B.T. Huu, "Optimal control of three-level PWM inverters", *IEEE Trans. Ind. Electr.*, vol. 44, no. 1, pp. 96-106, 1997.

[16] A.M. Hava, R.J. Kerkman and T. Lipo, "Simple analytical and graphical methods for carrier-based PWM-VSI drives", *IEEE Trans. Power Electron.*, vol. 14, no. 1, pp. 49-61, 1999.

[17] N. Mohan, T.M. Undeland and W.P. Robbins, *Power Electronics*, 3rd ed., John Wiley & Sons, 2003.

[18] V. Oleschuk and F. Blaabjerg, "Direct synchronized PWM techniques with linear control functions for adjustable speed drives", *Proc. of the 2002 IEEE Appl. Power Electronics Conf.*, p. 76-82.

[19] V. Oleschuk and F. Blaabjerg, "Synchronised PWM schemes for three-level inverters with zero common-mode voltage", *Int'l Journal of Electrical Engineering*, vol. 2, no. 1, pp. 43-48, 2002.

[20] V. Oleschuk, F. Blaabjerg and B. K. Bose, "Triphase cascaded converters with direct synchronous pulsewidth modulation", *Automatika*, vol. 44, no. 1-2, pp. 27-33, 2003.

Author Index

A

Abbatelli, L. 61
Abbey, Chad 2178
Abdelhamid, Tamer H. 606
Abdellatif, Meriem 938
Abe, Seiya 393
Abourida, Simon 1077
Abroshan, Mohammad 1117
Abroushan, Mohammad 365
Abuishmais, Ibrahim 867
Abu-Rub, Haithem 1084, 1382
Adabi, Jafar 718, 903
Adamidis, Georgios 1840
Adamowicz, Marek 1729
Adzic, Evgenije 1957
Ahmadi, Muhammad 1847
Ahmed, M.M.R. 1866, 2472
Ahn, Jonng-Bo 2524
Ahn, ong-Bo 2492
Ait-Ahmed, Mourad 1740
Akhondi, Hamidreza 2071
Alarcón, E. 2108
Albert, Laurent 2037
Al-Diab, Ahmad 1710
Alexandrov, Alexandar 787
Al-Khayat, Nazar 2150
Allard, Bruno 2457
Al-Othman, A. K. 606
Amelon, Nicolas 1740
Anaya-Lara, O. 1784, 1941
Andersen, Michael A.E. 127
Ando, Kenji 614
Andrzejewski, Andrzej 1090
Areerak, K-N 2049
Arellano-Padilla, J. 1173
Arellano-Padilla, Jesus 769
Armstrong, S. 1688
Aroudi, A. El 2108
Aroudi, Abdelali El 2115, 2120
Arshad, Waqas M. 867
Asher, G. M. 2261
Asher, G.M. 2049
Asher, Greg 2300
Asiltürk, Ilhan 967
Aurel, Campeanu 893
Averberg, Andreas 213

B

Baalbergen, Freek J.F. 2170
Baghaee, H.R. 313, 629, 750
Bahri, I. 1365
Bailey, Chris 76
Bakas, Panagiotis 1840
Balazovic, Peter 1402

Balouktsis, Anastasios 1840
Baluta, Gh. 2043
Ban, Drago 818
Barai, Mukti 674
Barakat, Georges 1834, 810
Baranowski, Jerzy 1432, 1446
Barbosa, Fabián H. 637
Barlik, Roman 84
Barrero, R. 1512
Bartelt, R. 521
Baskys, Algirdas 1140
Bastiani, A. 1293
Baszynski, Marcin 1779
Bauer, Pavel 422
Bauer, Pavol 2170, 2354, 2368, 2371
Beck, Hans-Peter 1243
Bekbudov, Radiy 337
Bekishev, Anatoly 663
Bélanger, Jean 1077, 1475
Belfkira, Rachid 1834
Belkhodja, I. Slama 1149
Bellini, Armando 490
Bellmunt, Oriol 731
Belter, D. 1044
Benadero, Luis 2115
Bendkowski, Lukas 250
Benecke, Marcel 1280
Benkhoris, Mohamed-Fouad 1740
Bennani, A.Ben Abdelghani 1149
Beran, Leos 782
Bergas-Jane, Joan 731
Bergogne, Dominique 2457
Berthon, A. 1542
Bertoluzzo, Manuele 1491
Bertram, Torsten 1215
Betz, R.E. 1293
Bevilacqua, Pascal 2457
Bifaretti, Stefano 1771, 490, 561
Binder, Andreas 1625, 2385
Binkowski, T. 714
Birolleau, Damien 2037
Biswas, Jayanta 674
Bizon, Nicu 621
Blahník, Vojtech 1535
Blanco, M. 2481
Blazic, B. 2510
Böcker, J. 1598
Böcker, Joachim 159
Bodora, A. 326
Bogalecka, Elzbieta 1975, 804
Bojda, Petr 422
Bolgov, Viktor 154
Bolognani, Silverio 1097
Boora, Arash A 468, 723
Bossche, A. Van den 1326
Botan, Corneliu 1111

Author Index

Botsali, FatihM. 949
Bouafia, Abdelouahab 703
Boucherit, M.S. 1987
Bouhalli, Nadia 281
Bozhko, S.V. 2049
Brand1tetter, Pavel 1375
Braslavsky, I.Ya. 1050
Breban, Stefan. 1896
Brown, Neil L. 2150
Bruno, Francois. 2205
Bucher, Alexander 244, 250
Buja, Giuseppe. 1491
Bukatov, Alexander 1872
Bulic, Neven 556
Buonomo, S. 61
Buss, Martin 2312

C

C., Ilioudis Vasilios 1105
Caballero, M. 1555
Cabrita, Carlos M. P. 1646
Calado, Maria R. A. 1646
Camara, M.B. 1542
Cambál, Marek 982
Candusso, Denis 734
Cartes, D.A. 793
Case, Michael James 1798
Castaing, Ambroise. 2464
Catalão, J. P. S. 1682
Cédl, Marek 1593, 372
Ceglia, Gerardo 268
Cepisca, Costin 1963, 908
Cernat, Mihai 1748
Cernohorský, Josef 1009
Cerovský, Zdenk 982
Cha, Gil-Ro. 383
Champenois, Gérard 2015
Chan, Paul K.W. 1688
Chang, Hao-Chi 1652
Chang, Lon-Kou 320
Chang, Yuan-Chih 456
Chante, Jean-Pierre 2457
Charaabi, L. 1365
Chekhet, Eduard 307
Chen, Anyuan 799
Chen, Junling 2000
Chen, Yonggang 1981, 2000, 405, 515
Chen, Zhe. 2325
Chen, Zong-Jie 1704
Cheng, K.W.E. 576
Cherif, M. Ghodbane 1149
Cheung, N. C. 1221
Chien, Sywe-Bin 1652
Chillet, Christian 2037
Chimento, F. 61

Chlodnicki, Zdzislaw 2150
Choi, Heung-Kwan 2524
Choi, Jaeho 2498
Choi, Uk-Don 1421
Chou, Ming-Chang 1652
Chrenko, Daniela 2156
Chrzan, Piotr J. 144
Chudzik, Piotr 1568
Chun, Tae-Won 1421
Clare, J. 1326
Clare, Jon C 207
Clare, Jon C. 229, 561
Clare, Jon 1771, 307
Comnac, Vasile. 1748
Cook, B.J. 1293
Cook, D. 1326
Coquery, Gérard 2192
Correa, Pablo 451, 699
Courtecuisse, Vincent 1896, 2184
Cousineau, Marc 281
Cuk, Vladimir 1426
Cychowski, Marcin 2241
Czapp, Stanislaw 2059

D

Dabroom, A.M. 1337
Dakyo, B. 1911
Dannehl, J. 444
Darie, Eleonora 1963, 908
Darie, Emanuel 1963, 908
Davey, J. 1918
De Bernardinis, Alexandre 2192
De Castro, M.R. 2126
De Gersem, Herbert 2385
de Kock, H.W. 859
De Souza, Kleber C.A. 1951
Deaconu, Sorin 1409
Debowski, Andrzej 1568, 2289
Degeratu, Sonia 893
Delaney, Kieran 2241
Demenko, Andrzej 2412
Denny, Ernest Edward 1798
Depernet, Daniel 734
Derbel, Nabil 2120
Deskur, Jan 1204, 2227
Deuse, Jacques. 2184
Dheilly, Nicolas 2457
Di, Lu 2205
Dianov, Anton 1002
Díaz, Nelson L. 637
Diblík, Martin 1676
Diguet, Marc 1382
Dilevs, Guntis 1811
Dimitrakakis, Georgios S. 1301
Dinkhauser, Vincenz 1819

Author Index

Dobrucky, Branislav 1402
Dockhorn, Matthias 1734
Dodds, Stephen J. 2551, 2559
Dodds, Stephen James 2543
Doebbelin, Reinhard 1280
Doi, Nobuaki 744
Dong, P. 576
Dong, Wei 1716
Dontchev, Dimitar 787
Drábek, Pavel 1593, 372
Draganov, Denis 1610
Dubowski, Marian Roch 1090
Dudak, Juraj 2368
Dudrik, Jaroslav 295
Duerbaum, Thomas 244, 250
Dufour, Christian 1077, 1475
Duke, Richard 528
Dumur, Guillaume 1475
Durovsky, Frantisek 961
Dybkowski, Mateusz 2211, 2306
Dzieniakowski, Maciej A. 2082

E

Eberhard, Andreas 1371
Eckel, Hans-Guenter 48
Edrington, C.S. 793
Egan, Michael G. 1249
Egorov, Mikhail 1257
Ehsan, Mehdi 1847
Eilenberger, Andreas 945
Elmoctar, Mohamed Y. Ould 810
Empringham, Lee 207, 229, 388
Endo, Tsunehiro 924
Eno, Otu A. 114
Erceg, Gorislav 556
Etxeberria-Otadui, I. 1555

F

Fabianowski, Jan 2082
Fabijanski, Pawel 1040, 2055, 2087
Fahrni, C. 256
Fakham, Hicham 2142
Fan, Yue 1771
Farhangi, Sh. 173
Farshad, Siamak 1575
Fedák, Viliam 2354
Fedak, Viliam 961
Fedyczak, Zbigniew 165, 236
Feki, Moez 2120
Fernández, Herman 1947
Fernandez-Mola, Josep-Maria 731
Ferreira, Jan Abraham 187
Ferreira, Luís António Fialho Marcelino 2076
Fetyko, Jan 961
Filchev, T. 1326

Filho, Braz Jesus Cardoso 1345
Filka, Roman 1402
Fisher, R. 1293
Fleisch, Karl 48
Fodor, D. 2096
Foft, Jiří 1593
Foo, Gilbert 2269
Forster, Stefan 2420
Fotouhi, Reza 1575
Francois, Bruno 2142, 2184
Franke, W. Toke 69
Franko, Marek 2538
Friedli, T. 27
Fröhleke, Norbert 159
Fuchs, F.W. 444
Fuchs, Friedrich W. 1390, 1819, 69
Fujita, Y. 275
Fukushima, Kentaro 148
Funabashi, Toshihisa 2478, 2487
Funato, Hirohito 479
Funayama, Koichi 1020
Futami, Motoo 2337

G

Gabriela, Petropol Serb 893
Gan, W. C. 1221
Gao, Fanqiang 515
Gao, Q. 2261
Gao, Qiang 1058
García-Tabarés, L. 2481
Gardecki, Arkadiusz 1193
Gasiewski, Marcin 1562
Gaubert, Jean-Paul 703
Gavranic, Ivica 818
Gaztañaga, H. 1555
Gelezevicius, Vilius Antanas 1144
Gennadevich, Kiselev Michail 428
Gennadevich, Lepanov Michail 428
Gerada, C. 1173
Gerada, Chris 1058, 388, 769, 887
Ghaedi, Azam 1054
Gharehpetian, G.B. 313, 629, 750
Ghosh, Arindam 468, 723, 903
Gímenez, María Isabel 1947
Giménez, María 268
Giral, Roberto 2115
Gizinski, Zygmunt 1562
Glasberger, Tomál 1268
Glavin, M.E. 1688
Glushkin, Evgeny 1872
Gnacinski, Piotr 826
Gobis, Vitoldas 1140
Goeldel, C. 2126
Gomis-Bellmunt, Oriol 1670
González-Hernández, S. 1784

Author Index

Gorbounov, Yassen 787
Goto, Hiroki 1163, 1168
Grabic, Stevan 1957
Grad, M. 714
Grecki, Filip 1440
Grigaitis, Arunas 1144
Grigans, Linards 2066
Grossi, Federica 874
Grzesiak, Lech M. 1071
Grzesik, B. 956
Gualous, H. 1542
Guo, Hai-Jiao 1163, 1168
Gustin, F. 1542
Gustin, Frederic 734
Guy, Owen J. 2464
Guzinski, Jaroslaw 1382, 994
Guzmán, Víctor 1947, 268
Gwózdz, Michal 728

H

Haan, Sjoerd de 187
Habetler, Thomas G. 21
Hadas, Zdenek 1665
Hadjov, Kliment 787
Hájek, Vítezslav 2371
Halasz, S. 682
Halgos, Jan 2368
Hamada, Tomoyuki 1884
Hamar, J. 1755
Hameyer, K. 2393
Hameyer, Kay 2412
Harada, Yosuke 148
Hartansky, Rene 2368
Hartnett, Kevin J. 1249
Hasegawa, Masaru 614
Hashimoto, Seiji 932
Hayashi, Kenta 589
Hayashi, Yusuke 2445
Hayes, John G. 1249
Heising, C. 521
Helmut, Weiss 1722, 1934
Henrotte, F. 2393
Henze, Olaf 2385
Hercog, Darko 2349
Hicham, Fakham 2205
Himmelstoss, Felix. A. 331
Hiraki, Eiji 119, 1877
Hirokawa, Masahiko 393
Hissel, Daniel 2156
HISSEL, Daniel 734
Hmasic, N. 2134
Ho, S.L. 576
Hoffmann, Frank 1215
Hõimoja, H. 2005
Hõimoja, Hardi 1581

Hojo, Masahide 2487
Holtz, Joachim 1084
Holub, Marcin 195
Horen, Yoram 776
Horga, V. 2043
Horga, Vasile 1111
Hrasko, Martin 2538
Hu, Weihao 2325
Hubik, Vladimir 1620
Huiqing, Wen 1518, 417
Hurley, W.G 1688
Huttin, N. 1523

I

I., Margaris Nikolaos 1105
Iannuzzi, Diego 1469
Ibach, Robert 2082
Ibáñez, Fernando 268
Ichinokura, Osamu 1163, 1168, 758
Ichinose, Masaya 2337
Ide, Kazumasa 2337
Igic, P. M. 2464
Iida, Takahiko 595
Ikeda, Yoshiko 498
Ikhouane, Faycal 1670
Iman-Eini, H. 173
Inoue, Yukinori 1859
Ion, Petropol Serb 893
Iov, Florin 1771, 561
Ishikawa, Kazumi 1020
Iskhakov, Albert 663
Ito, Fumio 1309
Itoh, Jun-ichi 581
Itoi, M. 275
Ivanovic, Zoran 1957
Iwaji, Yoshitaka 924
Iwanski, Grzegorz 1440, 2164
Iwase, Yuta 2487
Izadbakhsh, Alireza 2102

J

Jalakas, T. 1263
Jalakas, Tanel 1257
Jan, Mucko 1316
Ján, Vittek 2219
Jang, Gil-Soo 2498
Jang, Su-Jin 1924
Jansen, Uwe 88
Janson, Kuno 154
Járdán, Rafael K. 916
Jardan, Rafael Kalman 2360
Jasim, O. 1173
Jasim, Omar 887
Jasinski, Marek 1904
Javurek, Jiri 1465

Author Index

Jedryczka, Cezary 2406
Jennings, Michael R. 2464
Jeon, Jin-Hong 2492, 2524
Jezernik, Karel 2283, 2349, 432
Ji, Young-Hyok 1929
Jian, Xiao 1722
Jin, Zhao 1128
Johnson, C Mark 76
Joós, Géza 2178
Joost, M. 1064
Judek, Slawomir 1497
Jufer, Marcel 1
Jun, Liu 1518, 417
Jung, Doo-Yong 1929
Jung, Yong-Chae 181, 1924, 1929, 383

K

Kalatchikov, P. 837
Kalisiak, Stanislaw 195
Kallaste, Ants 154
Kallenbach, E. 1598
Kalyoncu, Mete 1132, 949, 974
Kamata, Yuki 498
Kamiski, Bartlomiej 2378
Kamper, M.J. 859
Kampisios, Konstantinos 887
Kaneko, Daigo 924
Kanerva, Sami 867
Kaplon, Andrzej 377
Karaffy, Z. 2096
Karsli, Vedat M. 850
Karwowski, Krzysztof 1497
Kasa, Nobuyuki 595
Kasinski, A. 1044
Kasprowicz, Andrzej 1332
Katic, Vladimir 1957
Kato, Koji 581
Katsura, Seiichiro 1187, 1604, 1614
Kawamura, Atsuo 7, 924
Kayhan, Ince 1934
Kazimierz, Jaracz 912
Kazmierkowski, Marian P. 1548, 1904
Kelemen, Franjo 855
Kennel, R.M. 859
Kennel, Ralph 1239
Khaldi, B.S 1987
Kim, Eel-Hwan 2498
Kim, Heung-Gun 1421
Kim, Jae-Hong 2498
Kim, Jae-Hyung 1924, 1929
Kim, Jong-Yul 2492
Kim, Se-Ho 2498
Kim, Seul-Ki 2492, 2524
Kimura, Kensuke 1168
Kimura, Noriyuki 1884

Kinoshita, Hirotaka 2337
Kireev, V. 1598
Klimczak, Pawel 108
Klug, O. 2096
Klyachko, Leonid 663
Klytta, Marius 165
Knop, André 69
Kobayashi, Yukinori 479
Kobougias, Ioannis C. 1274
Koczara, Wlodzimierz 1440, 2150, 2164, 2254
Koda, Noriaki 1877
Kolar, J. W. 27
Kolesnikov, Artem 1872
Kolomeitsev, L. 1598
Kompa, K. 695
Komura, Akiyoshi 2337
Kondo, Masaki 1614
Koneke, Thies 1458
Kong, S.T. 43
Konstantinovich, Rozanov Yurie 428
Korondi, Peter 2360
Korotyeyev, Igor 236
Koskin, Y. 837
Kosmecki, Michal 1975
Kostylev, A.V. 1050
Kotodziejek, Piotr 804
Kouzou, A. 1987
Kowalski, Czeslaw T. 1359
Kraeftner, Wilhelm 331
Kraynov, D. 1598
Krettek, Johannes 1215
Krim, Fateh 703
Krismer, F. 27
Krykowski, K. 326
Krystkowiak, Michal 728
Krzeminski, Zbigniew 1382, 2294
Kubiak, Andrzej 2452
Kubin, Jiri 1815
Kubota, Sachio 1309
Kuchta, Jozef 2538
Kudarauskas, Sigitas 2200
Kuebrich, Daniel 244
Kuhn, Harald 1458
Kuisma, M. 1233
Kulka, Arkadiusz 657
Kumar, Dinesh 207
Kuperman, Alon 776
Kurokawa, Fujio 2434, 2504
Kürschner, Daniel 1696, 1734
Kuß, H. 695
Kusserow, Wolf 1239
Kütt, Lauri 154
kuwata, M. 275
Kyritsis, A.C. 1287

Author Index

L

Laczynski, Tomasz 569, 649
Lafoz, M. 2481
Lagoda, Ryszard 1040, 2055, 2087
Laloya, Eduardo..845
Lange, E. 2393
Lapointe, Vincent 1077
Lastowiecki, Jozef 1440
Latka, M...714
Latkovskis, Leonards 2066
Laugis, J. 1263
Laugis, Juhan 1017
Laur, R. 1064
Lazar, C. 2043
Lazar, Mihai..2457
Ledwich, Gerard 468, 723, 903
Lee, Joo-Hyuk...1924
Lee, Tzung-Lin 1704
Lehtla, Madis 1581
Lehtla, T. 2011
Leidhold, Roberto 1353
Leszek, Szychta 2091
Leuchter, Jan 422
Levins, Nikolajs 1811
Lewandowski, Daniel 2289, 669
Lewicki, Arka diusz 1382
Lewis, A.W. 1790
Leyva, R...2108
Li, Kaihang 97
Li, Rongyuan 159
Li, Yaohua 1981, 2000, 405, 515
Li, Zixin 1981, 2000, 405, 515
Liaw, Chang-Ming......................... 1652, 456
Lie, Xu 229
Liffran, Florent 409
Lillo, Liliana de 388
Lindemann, Andreas......................... 1280, 2420
Lingemann, M. 2134
Lis, Jacek D. 1359
Lisik, Zbigniew..2452
Lisowski, Grzegorz..669
Liu, Congwei 405
Liu, Li 793
Lladó, Juan..845
Lodzinski, Michal 2464
Lopez-de-Heredia, A. 1555
Lorenz, Robert D. 903
LU, Di...2142
Lu, Hua...76
Lu, Y. 1221
Luft, Miroslaw...463
Luiz, Alex-Sander Amavel 1345
Luniewski, Piotr..88
Lyons, Brendan J. 1249
Lyskawinski, Wieslaw.......................................2406

M

Macek-Kaminska, Krystyna 1193
Madawala, U. K...139
Madawala, Udaya K. 1918
Maga, Dusan...2368
Mahmoudi, M.O. 1987
Mahyob, Amin...810
Mailat, Adrian..1748
Majidi, Behrooz...763
Maksimovic, Dragan 498
MAKYS, Pavol..2538
Malekian, Kaveh.................. 1117, 1123, 2071, 365, 763
Malska, W. 714
Man, T.K. 400, 475
Mandache, Lucian 1585
Mandra, Slawomir 1071
Mandrek, Slawomir 144
Marek, Stulrajter...2219
Margaliot, M. 260
Mariano, Sílvio José Pinto Simões 2076
Marouchos, Christos 1967
Martín-del-Brío, Bonifacio 845
Martínez, Abelardo.............................. 1947, 845
Martinez, Itziar 437
Martins, Denizar C. 1951
Masada, E. 1755
Mascibrodzki, Ireneusz.....................................1562
Mathis, W. 132
Matsui, Keiju 614
Matsui, Nobumasa 2504
Mawby, P.A. 2472
Mawby, Philip A. 2464
McEachern, Alex 1371
Mecke, Rudolf 1734
Melício, R. 1682
Mendes, V. M. F. 1682
Mertens, A. 132
Mertens, Axel 1458, 213, 569, 649
Meuret, R. 1523
Meynard, Thierry..281
Michalík, Jan 1535, 550
Michalke, N. 695
Mierlo, J. Van 1512
Milanovic, Miro..301
Milimonfared, Jafar 1117, 2071, 365, 763
Mimura, Yasuhiro...2434
Mirsalim, M. 313, 629, 750
Mirsalim, Mojtaba 1123
Mirzaeva, G. 1155
Mishima, Tomokazu 119
Mitani, Tetsuya 2428
Mladenovic, I. 2022
Mohd, A. 2134
Mokrovica, Josipa..855
Mõlder, Heigo...154

Author Index

Molinas, Marta 2318
Möller, T. 2005
Mollov, Stefan V. 350
Molnár, Jan 1535, 550
Mondzik, Andrzej 345
Monmasson, E. 1365
Montesinos-Miracle, Daniel 1670, 731
Morel, Herve 2457
Moreno-Font, Vanessa 2115
Moreno-Goytia, E. 1784, 1941
Morimoto, Shigeo 1859
Morino, Kimio 2478
Morizane, Toshimitsu 1884
Morton, D. 2134
Mouni, Emile 2015
Mukhopadhyay, Siddhartha 485
Munk-Nielsen, Stig 108
Murata, Toshiaki 2337
Musallam, Mahera 76
Mustonen, P. 1233
Musumeci, S. 61
Muszynski, Roman 2227
Mutschler, Peter 1353
Müür, M. 2005
Mysinski, Wojciech 1321

N

Nagy, I. 1755
Nagy, Istvan 2360
Nagy, István 916
Naka, Toshiyuki 498
Nakagawa, Akio 498
Nakamura, Kazutoshi 498
Nakamura, Kenji 758
Nakaoka, M. 275
Nakaoka, Mutsuo 119
Nakayama, Hiroaki 1877
Nanakos, Anastasios Ch. 1827
Naouar, M-W. 1365
Narayanan, E.M. Sankara 43
Narjiss, Abdellah 734
Nasser, Mehdi 1896
Navarro, Daniel 437
Nawaz, Muhammed 2472
Nekoui, Mohammad Ali 1054
Ngwendson, L. 43
Ni, Bingchang 2331
Nichita, C. 1911
Nichita, Cristian 1834
Nicolae, Ileana-Diana 1585
Nicolae, Petre Marian 1585
Nicolae, Petre-Marian 1181
Niechaj, Marek 1890
Niemelä, Markku 1763
Nikolic, Aleksandar 1426

Nilssen, Robert 799
Ninomiya, Tamotsu 148, 393
Nishida, Yasuyuki 2530
Nishikata, Shoji 2343
Nishimiya, Ayumu 1163
Nishioka, Kunihiro 1309
Nitta, Mayumi 932
Noda, Shuji 1877
Norigoe, Isami 148
Novák, Jaroslav 982
Novák, Martin 982
Nowak, Lech 2400
Nowak, Mietek 84
Numata, Shigeo 2478
Nuutinen, Pasi 1763
Nyczkowski, Lukasz 740
Nymand, Morten 127
Nysveen, Arne 799

O

O'Sullivan, D.L. 1790
Ogiwara, H. 275
Ohashi, Hiromichi 2428, 2445, 54
Ohishi, Kiyoshi 1187, 1604, 1614
Ohsaki, H. 1755
Ohyama, Kazuhiro 2300
Okamatsu, Masashi 2434
Oleschuk, Valentin 1548
Omari, O. 2134
OMORI, Hideki 2530
Ondrusek, Cestmir 1665
ONEN, Umit 949
OPROESCU, Mihai 621
Orlik, B. 1064, 830
Orlowska-Kowalska, Teresa 2211, 2306
Ortjohann, E. 2134
Oyarbide, Estanislao 845

P

Pacas, Mario 2248
Pajchrowski, Tomasz 1198, 1204
Pakhomin, S. 1598
Palis, Frank 1610
Palis, Stefan 1660
Panoiu, Caius 1409
Panoiu, Manuela 1409
Papanikolaou, N.P. 1287
Papic, I. 2510
Paquin, Jean-Nicolas 1475
Park, JuneHo 2492
Park, Sang-Hoon 181, 383
Park, So-Ri 181
Parkatti, P. 201
Parker-Allotey, N-A. 2472
Patel, N. D. 139

Author Index

Patra, Pradipta.................................485
Patra", Amit..................................485
Pavelka, Jiri.................................221
Pavelka, Jirí.................................988
Pavlitov, Constantin..........................787
Pavlovsky, Martin..............................7
Pavol, Makys.................................2219
Pavoni, Alessandro...........................1491
Peftitsis, Dimosthenis.......................1840
Peltoniemi, Pasi.............................1763
Peplinski, Marcin.............................826
Pera, Marie-Cecile...........................2156
Perez, Francisco.............................845
Perez-Tomas, Amador..........................2464
Peric, Nedjeljko.............................2235
Peroutka, Zdenek.........1268, 1529, 1535, 550
Peter, Bris..................................2219
Peter, Zaucher...............................1722
Petit, Marc..................................2184
Petrella, Roberto............................1097
Petrisor, Anca...............................893
Piatek, Pawel................................1446
Pietrzak-David, Maria.........................938
Piróg, Stanislaw.............................1779
Pittermann, Martin......................1593, 372
Planson, Dominique...........................2457
Poljugan, Alen...............................1058
Pollán, Tomás................................845
Popa, Anca Sorana............................1225
Popa, Mircea.................................1225
Porada, Ryszard..............................740
Pospelov, Vladimir...........................663
Pronin, M....................................837
Pugachevs, Vladislavs........................1811
Pyrhönen, Juha...............................1763

Q
Quiroga, J...................................793

R
Rabkowski, Jacek..............................84
Raciti, A.....................................61
Radomski, Grzegorz...........................504
Raducu, Marian...............................621
Radulescu, Mircea M..........................1896
Rafecas-Sabate, Josep........................731
Rafiei, S.M.R................................2102
Rahman, M.F..................................2269
Rahnamaee, Arash........................1117, 365
Rao, Sachit..................................2312
Rathge, Christian............................1696
Ratoi, Marcel................................1111
Rawicki, Stanislaw...........................1481
Raynaud, Christophe..........................2457
Rednov, F....................................1598

Reghem, Pascal..........................1834, 810
Rerucha, Vladimir............................422
Rezaei, Mohammad Mehdi.......................1123
Reznikov, B..................................260
Ribickis, Leonids............................1811
Richter, F...................................1398
Riipinen, T..................................1233
Risteiu, Mircea..............................1243
Riz, A.......................................2096
Roasto, I....................................2011
Robert, B.G.M................................2126
Robert, Bruno Gerard Michel..................2120
Robinson, Jonathan...........................2178
Robyns, B....................................1523
Robyns, Benoît...............................1896
Robyns, Benoit...............................2184
Rodic, Miran.................................2283
Rodriguez, E.................................2108
Rodriguez, Jose.........................451, 699
Rojas, A.....................................1155
Rojko, Andreja...............................2349
Rolek, Jaroslaw..............................377
Rompelman, Otto..............................2354
Ronkowski, Mieczyslaw........................880
Rosin, A.....................................2005
Rothenhagen, Kai........................1390, 1904
Round, S. D...................................27
Ru1scin, Vladimír............................295
Ruderman, A..................................260
Ruderman, Michael............................1215
Rufer, A.....................................256
Ruger, N. E..................................132
Rusinov, Radoslav............................787
Rylko, Marek S...............................1249
Ryvkin, Sergey...............................1505
Rzasa, Janina................................357

S
Saadi, S.....................................1987
Sabirin, Chip Rinaldi........................1625
Saito, Makoto................................2439
Saito, Tsuyoshi..............................744
Sajkowski, M.................................956
Sakamoto, Kiyoshi............................924
Sakamoto, Yosei..............................288
Salo, M......................................201
Salonen, Pasi................................1763
Samanta, Susovon.............................485
Samuelsen, Dag...............................1416
Sanada, Masayuki.............................1859
Sánchez, Beatriz.............................845
Sánchez, Carlos..............................268
Sang-Joon, Lee...............................1002
Sang-Taek, Lee...............................1002
Sanjari, M. J.........................313, 629, 750

Author Index

San-Sebastian, J. 1555
Santo, António Espírito 1646
Sarraute, Emmanuel 281
Sasaki, Masahiro 2434
Sato, Muneo 1309
Saudemont, C. 1523
Sayed, Mahmoud A. 542
Sayeef, S. 2269
Schallschmidt, Thomas 1610, 1660
Schanen, JL. 173
Schmelter, A. 2134
Schmid, Markus 244, 250
Schmidt, Istvan 1803
Schmidt-Obermoeller, Richard 1505
Schmitt, Günter 1239
Schneider, T. 1598
Schnick, O. 132
Schrödl, Manfred 2275
Schroedl, Manfred 945
Schuffenhauer, U. 695
Scollo, R. 61
Sengupta, Sabyasachi 674
Seppä, L. 1233
Shao, S. 1293
Shapoval, Ivan 307
Sharma, R. 1918
She, X. 710
She, Yun 710
Shieh, Fa-Hwa 1652
Shimaoka, Yoshihiro 1309
Shimizu, Takaaki 498
Shimizu, Toshihisa 2428, 2445, 288, 600
Shimoda, Eisuke 2478
Shiraishi, Keiichi 2504
Shonin, O. 837
Shoyama, Masahito 148, 393
Shyu, Juei Lung 643
Siatkowski, M. 830
Silea, Ioan 1225
Silventoinen, P. 1233
Simetzberger, Christian 2275
Simon, Miklós G. 916
Singule, Vladislav 1620, 1665
Sinsukthavorn, W. 2134
Siostrzonek, Tomasz 1779
Sîrbu, Ioana-Gabriela 1181, 1585
Siroky, Peter 2368
Sitar, Jan 2368
Sivkov, Oleg 221
Skovpen, Sergey 663
Skuta, Ondfej 1375
Slama-Belkhodja, I. 1365
Slama-Belkhodja, Ilhem 938
Smet, Bart 102
Sobczuk, Dariusz 2378
Sobczynski, D. 714

Sochacki, Mariusz 2452
Soltani, Hamid 718
Song, Sang-Hoon 383
Song, Seung-Ho 2498
Soroudi, Alireza 1847
Sosa-Ruiz, J. 1941
Souad, Rafa 1209
Sourkounis, C. 1398
Sourkounis, Constantinos 1633, 1710, 2331
Sozanski, Krzysztof Piotr 1995
Stadler, Paul Andreas 2543
Stala, Robert 1852, 345
Stamann, Mario 1660
Stanescu, Dan-Gabriel 1181
Staudt, V. 521
Staudt, Volker 2371
Stefanutti, Fabio 1097
Steimel, A. 521
Steimel, Andreas 1505, 2371
Stenzel, T. 956
Stepanyuk, D.P. 1050
Stepien, P. 1293
Stocco, Piero 1097
Strac, Leonardo 855
Strzelecki, Ryszard Michal 1332
Strzelecki, Ryszard 1729
Stumpf, P. 1755
Stumpf, Péter 916
Sugai, T. 275
Sugimasa, Junji Tamura Masatoshi 2337
Suissa, Uri 776
Sulkowski, Waldemar 1416
Sumida, Yuichi 2434
Sumina, Damir 556
Sumiyoshi, Shinichiro 2530
Summers, T.J. 1293
Sumner, M. 1173, 2261
Sumner, Mark 1058, 2300, 769
Sun, Z. G. 1221
Susluoglu, Berrin 850
Suul, Jon Are 2318
Sveda, Martin 1620
Sweet, M. 43
Sykulski, Jan K. 2383
Szabat, Krzysztof 2211, 2241
Szamel, Laszlo 1033
Szczeniak, Pawel 165
Szczepankowski, Pawel 1332
Szczesniak, Pawel 236
Szelag, Wojciech 2406
Sziebig, Gabor 2360
Szmidt, Jan 2452
Szubert, Krzysztof 536
Szweda, Mariusz 826
Szychta, Elzbieta 463
Szychta, Leszek 463

Author Index

Szymanski, B. J. 695

T

Tackoen, X. 1512
Tae-Ho, Yoon 1002
Taguchi, Toyoki 498
Takahashi, Nobuo 1877
Takahashi, Rion 2337
Takao, Kazuto 2445, 54
Takeshita, Takaharu 542
Takeuchi, Nobuhito 614
Takeuchi, Toshihiro 924
Tan, Longcheng 1981, 2000, 405, 515
Tanabe, Takayuki 2478
Tanaka, Toshihiko 1877
Taniguchi, Katsunori 1884
Taniguchi, Satoshi 600
Tankari, A.M. 1911
Tao, Zhou 2205
Tapuchi, Saad 776
Tarczewski, Tomasz 1071
Tatakis, E.C. 1287
Tatakis, Emmanuel C. 1274, 1301, 1827
Tatsuta, Fujio 2343
Theodoridis, Michael P. 350
Thomas, D.W.P. 2049
Thomas, David W.P. 1716
Thompson, David S. 114
Tinkir, Mustafa 1132, 949, 974
Tnani, Slim 2015
Tournier, Dominique 2457
Tran, Quang-Vinh 1421
Trentin, Andrew 887
Trujillo, Cesar L. 637
Tsai, Jih-Run 1652
Tseng, K.J. 2516
Tsukakoshi, Kenta 148
Tsuruta, Yukinori 7
Tulbure, Adrian 1243
Turner, Robert W. 528
Tutaj, Andrzej 1432
Tuusa, H. 201

U

Ueda, Yoshinobu 2478, 2487
Ummaneni, Ravindra. B. 799
Undeland, Tore 2318, 657
Ünüvar, Ali 967
Urabe, R. 275
Urbanski, Konrad 1454
Utkin, Vadim 2312, 512

V

Väisänen, V. 1233

Valchev, V. 1326
van Duivenbode, Jeroen 102
Vasak, Mario 2235
Vedrana, Jerkovic 690
Vekic, Marko 1957
Vergnol, Arnaud 1896
Veszpremi, Karoly 1803
Vicuña, Javier 845
Villanueva, Elena 451
Villwock, Sebastian 2248
Vinnikov, D. 1263, 2011
Vinnikov, Dmitri 1257
Viscarret, U. 1555
Vittek, Jan 2551
Vladimír, Vavrus 2219
Vodovozo, Valery 1017
Vorontsov, A. 837
Vrana, Petr 1465

W

Wada, Keiji 2428, 288, 600
Walas, K. 1044
Walter, Julio 268
Walton, Simon 528
Wang, Ping 1981, 2000, 405, 515
Wang, Yi 187
Wang, Yue 2325
Wang, Zhaoan 2325
Weidinger, Thomas 2028
Weiland, Thomas 2385
Weindl, Ch. 2022
Werner, Timur 649
Wheeler, P. 1326
Wheeler, Patrick W. 207
Wheeler, Patrick W. 229
Wheeler, Patrick 388
Wiktor, Hudy 912
Willis, K. 1293
Winternheimer, Stefan 1872
Wisniewski, Janusz 2254
Wlas, Miroslaw 1084
Won, Chung-Yuen 181, 1924, 1929, 383
Wong, L.K. 400, 475
Wu, Dongming 97

X

XiaoyanHuang, ... 388
Xu, Wei 2000, 405
Xuhui, Wen 1518, 417
Xuhui, Zhang 1518, 417

Y

Yaguchi, Hiroyuki 1020
Yamanouchi, Wataru 1187

Author Index

Yang, Lingling 97
Yang, Liu 1128
Yang, Ru-Shiuan 320
Yin, Chunyan 76
Yokokura, Yuki 1187, 1604
Yokoyama, Tomoki 589, 744
Young-Kwan, Kim 1002
Yousefi, Ashkan 1847

Z

Zakrzewski, Zbigniew 1332
Zamma, Toshihiro.......................................1020
Zanasi, Roberto...874
Zanchetta, Pericle 1716, 1771, 561, 887
Zare, Firuz 468, 718, 723, 903
Zaring, Carina..2472
Zarko, Damir 855
Zarko, Damirarko 818
Zaskalicka, Maria 899
Zaskalicky, Pavel 899
Zatocil, Heiko 1024
Zawirski, Krzysztof 1198, 1204, 1454
Zdenek, Jiri 1638
Zdravko, Valter...690
Zeljko, Spoljaric 690
Zeman, Karel 1529
Zeroug, Houcine 1209
Zhang, H. 1523
Zhang, S..2516
Zhao, S. W. 1221
Zhou, Tao..2142
Zhu, Haibin 515
Zielinski, K. 1064
Zigic, Aleksandar..1426
Zinoviev, Genady Stepanovic.....................1332
Zlosnikas, Valerijus 1140
Zouhar, Jan 1665
Zulawnik, Marcin 1562
Zych, Michal..1562
Zymmer, Krzysztof.......................... 1332, 1562

CURRAN ASSOCIATES INC.
proceedings
.com

9781424417414